Properties of ferrous metals.

Material	Density, kg/m^3	Modulus of elasticity, psi × 10^6 (GPa)	Yield strength, ksi (MPa)	Ultimate strength, ksi (MPa)	Ductility, %EL in 2 in.	Poisson's ratio	Thermal conductivity, W/m-°C	Coefficient of thermal expansion, $(°C)^{-1} \times 10^{-6}$
Iron	7 870	30 (207)	19 (130)	38 (260)	45	0.29	80	11.8
Gray cast iron	7 150	Variable	—	18 (125)	—	Variable	46	10.8
Nodular cast iron	7 120	24 (165)	40 (275)	60 (415)	18	0.28	33	11.8
Malleable cast iron	7 200–7 450	25 (172)	32 (220)	50 (345)	10	0.26	51	11.9
Low-carbon steel (AISI 1020)	7 860	30 (207)	43 (295)	57 (395)	37	0.30	52	11.7
Medium-carbon steel (1040)	7 850	30 (207)	51 (350)	75 (520)	30	0.30	52	11.3
High-carbon steel (AISI 1080)	7 840	30 (207)	55 (380)	89 (615)	25	0.30	48	11.0
Stainless steels								
Ferritic, type 446	7 500	29 (200)	50 (345)	80 (552)	20	0.30	21	10.4
Austenitic, type 316	8 000	28 (193)	30 (207)	80 (552)	60	0.30	16	16.0
Martensitic, type 410	7 800	29 (200)	40 (275)	70 (483)	30	0.30	25	9.9

Properties of nonferrous metals.

Material	Density, kg/m^3	Modulus of elasticity, psi ×10^6 (GPa)	Yield strength, ksi (MPa)	Ultimate strength, ksi (MPa)	Ductility, %EL in 2 in.	Poisson's ratio	Thermal conductivity, W/m-°C	Coefficient of thermal expansion, $(°C)^{-1} \times 10^{-6}$
Aluminum (>99.5%)	2 710	10 (69)	2.5 (17)	8 (55)	25	0.33	231	23.6
Aluminum alloy 2014	2 800	10.5 (72)	14 (97)	27 (186)	18	0.33	192	22.5
Copper (99.95%)	8 940	16 (110)	10 (69)	32 (220)	45	0.35	398	16.5
Brass (70Cu–30Zn)	8 530	16 (110)	11 (75)	44 (303)	68	0.35	120	20.0
Bronze (92Cu–8Sn)	8 800	16 (110)	22 (152)	55 (380)	70	0.35	62	18.2
Magnesium (>99%)	1 740	6.5 (45)	6 (41)	24 (165)	14	0.29	122	27.0
Molybdenum (>99%)	10 220	47 (324)	82 (565)	95 (655)	35	—	142	4.9
Nickel (>99%)	8 900	30 (207)	20 (138)	70 (483)	40	0.31	80	13.3
Silver (>99%)	10 490	11 (76)	8 (55)	18 (125)	48	0.37	418	19.0
Titanium (>99%)	4 510	15.5 (107)	35 (240)	48 (330)	30	0.34	17	9.0

SI units and prefixes.

(a) SI units Quantity	Unit	SI symbol	Formula
SI base units			
Length	meter	m	—
Mass	kilogram	kg	—
Time	second	s	—
Temperature	kelvin	K	—
SI supplementary unit			
Plane angle	radian	rad	—
SI derived units			
Energy	joule	J	N-m
Force	newton	N	kg-m/s^2
Power	watt	W	J/s
Pressure	pascal	Pa	N/m^2
Work	joule	J	N-m

(b) SI prefixes

Multiplication factor	Prefix	SI symbol for prefix
1 000 000 000 000 = 10^{12}	tera	T
1 000 000 000 = 10^9	giga	G
1 000 000 = 10^6	mega	M
1 000 = 10^3	kilo	k
100 = 10^2	hecto	h
10 = 10^1	deka	da
0.1 = 10^{-1}	deci	d
0.01 = 10^{-2}	centi	c
0.001 = 10^{-3}	milli	m
0.000 001 = 10^{-6}	micro	μ
0.000 000 001 = 10^{-9}	nano	n
0.000 000 000 001 = 10^{-12}	pico	p

Conversion factors and definitions.

(a) Fundamental conversion factors

	English unit	Exact SI value	Approximate SI value
Length	1 in.	0.0254 m	—
Mass	1 lbm	0.453 592 37 kg	0.4536 kg
Temperature	1 deg R	5/9 K	—

(b) Definitions

Acceleration of gravity	1g = 9.8066 m/s^2 (32.174 ft/s^2)
Energy	Btu (British thermal unit) ≡ amount of energy required to raise 1 lbm of water 1 deg F (1 Btu = 778.2 ft-lbf)
	kilocalorie ≡ amount of energy required to raise 1 kg of water 1 K (1 kcal = 4187 J)
Length	1 mile = 5280 ft; 1 nautical mile = 6076.1 ft
Power	1 horsepower ≡ 550 ft-lbf/s
Pressure	1 bar ≡ 10^5 Pa
Temperature	degree Fahrenheit $t_F = 9/5\, t_C + 32$ (where t_C is degrees Celsius)
	degree Rankine $t_R = t_F + 459.67$
	Kelvin $t_K = t_C + 273.15$ (exact)
Kinematic viscosity	1 poise ≡ 0.1 kg/m-s
	1 stoke ≡ 0.0001 m^2/s
Volume	1 cubic foot = 7.48 gal

(c) Useful conversion factors

1 ft = 0.3048 m
1 lbf = 4.448 N
1 lbf = 386.1 lbm-in./s^2
1 kgf = 9.807 N
1 lbf/in.^{2} = 6895 Pa
1 ksi = 6.895 MPa
1 Btu = 1055 J
1 ft-lbf = 1.356 J
1 hp = 746 W = 2545 Btu/hr
1 kW = 3413 Btu/hr
1 quart = 0.000946 m^3 = 0.946 liter
1 kcal = 3.968 Btu

Case studies are presented in almost every chapter of FUNDAMENTALS OF MACHINE ELEMENTS. They are included to further expose students to a synthesis of concepts in each chapter, including the design process. It is the authors' hope that students will be motivated to learn more as they encounter realistic and practical case studies directly tied to concepts in the chapter. Following is a list of case studies and their location in the text. There are also three system design projects and 19 additional design projects to assign in Chapter 19 of the text.

Fundamentals of Machine Elements

Bernard J. Hamrock
Ohio State University

Bo O. Jacobson
Lund University
Lund, Sweden

Steven R. Schmid
The University of Notre Dame

WCB McGraw-Hill

Boston • Burr Ridge, IL • Dubuque, IA • Madison, WI • New York • San Francisco • St. Louis
Bangkok • Bogotá • Caracas • Lisbon • London • Madrid
Mexico City • Milan • New Delhi • Seoul • Singapore • Sydney • Taipei • Toronto

WCB/McGraw-Hill

A Division of The ***McGraw-Hill*** *Companies*

FUNDAMENTALS OF MACHINE ELEMENTS
International Editions 1999

1 2 3 4 5 6 7 8 9 10 KKP UPE 2 0 9

Library of Congress Cataloging-in-Publication Data

Hamrock, Bernard J.
Fundamentals of machine elements / Bernard J. Hamrock, Bo O. Jacobson, Steven R. Schmid.
p. cm.
ISBN 0-256-19069-0
1. Machine design. 2. Mechanical movements. I. Jacobson, Bo O., 1942- II. Schmid, Steven R. III. Title.
TJ230.H245 1999
621.8'15–dc21 98-34942

www.mhhe.com

When ordering this title, use ISBN 0-07-116374-3

Printed in Singapore

PREFACE

This book is intended to provide the undergraduate student with a clear and thorough understanding of both the theory and application of the fundamentals of machine elements. It is expected that this book will also be used as a reference by practicing engineers. Familiarity with differential and integral calculus is needed to comprehend the material presented. The design of machine elements involves a great deal of geometry as well. Therefore, the ability to sketch the various configurations that arise, as well as to draw a free-body diagram of the loads acting on a component, is also needed. The material covered in this text is appropriate as a third- or fourth-year engineering course for students who have studied basic engineering sciences, including physics, engineering mechanics, and materials and processes.

The book is divided into two parts. Part 1 (Chapters 1 to 8) presents the fundamentals, and Part 2 (Chapters 9 to 19) uses the fundamentals in considering the design of various machine elements. The material in Part 1 is sequential; material presented in early chapters is needed in subsequent chapters. This building-block approach provides the foundation necessary to design the various machine elements considered in Part 2.

LEARNING TOOLS

The following pedagogical devices are used in each chapter to improve understanding and motivate the student:

- **Symbol list**—defines the symbols used within the chapter and gives their units for use in unit checks within equations.
- **Quotation and photograph**—open each chapter as an introduction to the topics in the chapter
- **Introduction**—previews the material covered in the chapter.
- **Key words**—presented in bold when first used and listed along with definitions at the end of the chapter.
- **Worked examples**—presented when a new concept is developed to reinforce student understanding. There are over 200 worked examples, and each one uses a consistent problem-solving format.
- **Consistent problem-solving methodology**—each example and problem is solved according to a consistent methodology. Students are encouraged to follow these four steps in solving examples and problems:
  1. *Sketch*—gives graphical description of problem.
  2. *Given*—presents the information from the problem statement in symbol form.
  3. *Find*—states what needs to be determined.
  4. *Solution*—indicates the method, procedure, and equations used to solve the problem.

- **Case studies**—presented in select chapters, the case studies are design oriented and combine multiple concepts from the chapter. Many of the case studies involve situations encountered by practicing engineers. There are 25 case studies total. In addition there are three system design projects at the end of Chapter 19 in complete solution form, as well as 19 additional design projects to assign.
- **Summary**—recapitulates the information from the chapter.
- **Recommended readings and references**—these lists are located at the end of each chapter as sources of further information and greater detail. The author-date system is used for reference citations.
- **End-of-chapter problems**—over 600 homework problems to solidify understanding of the chapter material and stimulate creativity. The problems range from simple to complex and many provide design-related opportunities for the student. Solutions to the homework problems can be found in the Instructor's Solutions Manual, available to professors who adopt the text.

MACHINE DESIGN CD-ROM

The publisher and authors researched the types of teaching and learning aids that would be useful to users of *Fundamentals of Machine Elements.* Based on that research, a CD-ROM containing various tools was developed and packaged with every copy of this text. The CD-ROM contains the following items:

- PowerPoint files containing all the figures and tables from the text. These PowerPoint files can be used for the creation of print transparencies, or they can be loaded into an instructor's PowerPoint presentation for electronic use in lecture.
- Video clips and brief animations from industry to highlight concepts from the text such as gears, failure analysis, and bearings.
- Design tutorials to serve as interactive case studies. These tutorials will provide real-world design situations, along with exercises to test problem-solving and design skills.
- Full-color animations of select machine elements to help the student better visualize the motion and dynamics of the element being studied.
- Additional rolling element bearing data to allow more flexible design problems to be examined.

WEB SITE

A web site containing other book-related resources can be found at *http://www.mhhe.com/hamrock.* The web site provides reported errata, web links to related sites of interest, password-protected solutions to homework problems for instructors, a bulletin board, and information about ordering books and supplements.

CONTENTS

Chapter 1 introduces machine design and machine elements and covers a number of topics, such as safety factors, statistics, units, unit checks, and significant figures. In designing a machine element it is important to evaluate the kinematics, loads, and stresses at the critical section. Chapter 2 describes the applied loads (normal, torsional, bending, and transverse shear) acting on a machine element with respect to time, the area over which the load is applied, and the location and method of application. The importance of support reaction, application of static force and moment equilibrium, and proper use of free-body diagrams is highlighted. Shear and moment diagrams applied to beams for various types of singularity function are also considered. Chapter 2 then describes stress and strain separately.

Chapter 3 focuses on the properties of solid engineering materials, such as the modulus of elasticity. (Appendix A gives properties of ferrous and nonferrous metals, ceramics, polymers, and natural rubbers. Appendix B explores the stress-strain relationships for uniaxial, biaxial, and triaxial stress states.) Chapter 4 describes the stresses and strains that result from the types of load described in Chapter 2, while making use of the general Hooke's law relationship developed in Appendix B. Chapter 4 also considers straight and curved members under these four types of load.

Certainly, ensuring that the design stress is less than the yield stress for ductile materials and less than the ultimate stress for brittle materials is important for a safe design. However, attention must also be paid to displacement (deformation), since a machine element can fail by excessive elastic deformation. Chapter 5 attempts to quantify the deformation that might occur in a variety of machine elements. Some approaches investigated are the integral method, the singularity function, the method of superposition, and Castigliano's theorem. These methods are applicable for distributed loads.

Stress raisers, stress concentrations, and stress concentration factors are investigated in Chapter 6. An important cause of machine element failure is cracks within the microstructure. Therefore, Chapter 6 covers stress levels, crack-producing flaws, and crack propagation mechanisms and also presents failure prediction theories for both uniaxial and multiaxial stress states. The loading throughout Chapter 6 is assumed to be static (i.e., load is gradually applied and equilibrium is reached in a relatively short time). However, most machine element failures involve loading conditions that fluctuate with time. Fluctuating loads induce fluctuating stresses that often result in failure by means of cumulative damage. These topics, along with impact loading, are considered in Chapter 7.

Chapter 8 covers lubrication, friction, and wear. Not only must the design stress be less than the allowable stress and the deformation not exceed some maximum value, but lubrication, friction, and wear (tribological considerations) also must be properly understood for machine elements to be successfully designed. Stresses and deformations for concentrated loads, such as those that occur in rolling-element bearings and gears, are also determined in Chapter 8. Simple expressions are developed for the deformation at the center of the contact as well as for the maximum stress. Chapter 8 also describes the properties of fluid film lubricants used in a number of machine elements. Viscosity is an important parameter for establishing the load-carrying capacity and performance of fluid-film lubricated machine elements. Fluid viscosity is greatly affected by temperature, pressure, and shear rate. Chapter 8 considers not only lubricant viscosity but also pour point and oxidation stability, greases and gases, and oils.

Part 2 (Chapters 9 to 19) relates the fundamentals to various machine elements. Chapter 9 deals with columns, which receive special consideration because yielding and excessive deformation do not accurately predict the failure of long columns. Because of their shape (length much larger than radius) columns tend to deform laterally upon loading; and if deflection becomes critical, they fail catastrophically. Chapter 9 establishes failure criteria for concentrically and eccentrically loaded columns.

Chapter 10 considers cylinders, which are used in many engineering applications. The chapter covers tolerancing of cylinders; stresses and deformations of thin-walled, thick-walled, internally pressurized, externally pressurized, and rotating cylinders; and press and shrink fits.

Chapter 11 considers shafting and associated parts, such as keys and flywheels. A shaft design procedure is applied to static and cyclic loading; thus, the material presented in Chapters 6 and 7 is directly applied to shafting. Chapter 11 also considers critical speeds of rotating shafts.

Chapter 12 presents the design of hydrodynamic bearings—both thrust and journal configurations—as well as design procedures for the two most commonly used slider bearings. The procedures provide an optimum pad configuration and describe performance parameters, such as normal applied load, coefficient of friction, power loss, and lubricant flow through the bearing. Similar design information is given for plain and nonplain journal bearings. The chapter also considers squeeze film and hydrostatic bearings, which use different pressure-generating mechanisms. The last section of the chapter presents the design of gas-lubricated bearings.

Rolling-element bearings are presented in Chapter 13. Statically loaded radial, thrust, and preloaded bearings are considered, as well as loaded and lubricated rolling-element bearings, fatigue life, and dynamic analysis. The use of the elastohydrodynamic lubrication film thickness is integrated with the rolling-element bearing ideas developed in this chapter.

Chapter 14 covers the design of spur gears and briefly describes helical gears. Stress failures are also considered. The transmitted load is used to establish the design bending stress in a gear tooth, which is then compared with an allowable stress to establish whether failure will occur. Chapter 14 also considers fatigue failures. The Hertzian contact stress with modification factors is used to establish the design stress, which is then compared with an allowable stress to determine whether fatigue failure will occur. If an adequate protective elastohydrodynamic lubrication film exists, gear life is greatly extended.

Chapter 15 covers threaded, riveted, welded, and adhesive joining of members, as well as power screws. Riveted and threaded fasteners in shear are treated alike in design and failure analysis. Four failure modes are presented: bending of member, shear of rivet, tensile failure of member, and compressive bearing failure. Fillet welds are highlighted, since they are the most frequently used type of weld. A brief stress analysis for lap and scarf adhesively bonded joints is also given.

Chapter 16 treats the design of springs, especially helical compression springs. Because spring loading is most often continuously fluctuating, Chapter 16 considers the design allowance that must be made for fatigue and stress concentration. Helical extension springs are also covered in Chapter 16. The chapter ends with a discussion of torsional and leaf springs.

Brakes and clutches are covered in Chapter 17. The brake analysis focuses on the actuating force, the torque transmitted, and the reaction forces in the hinge pin. Two theories relating to clutches are studied: the uniform pressure model and the uniform wear model.

Chapter 18 deals with flexible machine elements. Flat belts and V-belts, ropes, and chains are covered. Methods of effectively transferring power from one shaft to another while using belts, ropes, and chains are also presented. Failure modes of these flexible machine elements are considered.

Chapter 19 presents three system design projects in complete solution form. When possible, some of the design engineer's reasons for pursuing alternatives are discussed in detail. This chapter is the culmination of machine element design. At the end of the chapter 26 additional projects are suggested.

Bernard J. Hamrock
Ohio State University

Bo O. Jacobson
Lund University
Lund, Sweden

Steven R. Schmid
The University of Notre Dame

ACKNOWLEDGMENTS

The idea and outline for this text were worked on for several years before the original contract was signed in 1994 with Richard D. Irwin, Inc. In the third year of the project McGraw-Hill purchased Irwin. In all, this text has been in progress for more than six years.

Many people helped to produce this textbook. The authors would first like to thank Rosemary Hamrock for many hours spent typing most of the text and for providing encouragement and support. The equations were typed by Ho Sheng Hsiao. The authors are greatly indebted to Carol Vidoli for her careful editing. A number of faculty members from the mechanical engineering department at Ohio State University provided comments about the text. In particular the authors would like to thank Professors Anthony Luscher, Si Lee, Necip Berme, and Bharat Bhushan. The authors would like to give recognition to the following people who helped with producing the solution manual: Mohsen Esfanian, Shashi Sharma, Ho-Sheng-Hsiao, Hans Jacobson, Birgitta Tell, and Yun-Hsiao. Special thanks are also due to Triodyne, Inc. personnel, especially Ralph L. Barnett and Brian D. King, whose insights were invaluable. A number of resources from Triodyne were kindly made available in the preparation of the text. The authors would also like to recognize the Jacob Wallenberg Foundation, who provided a grant that was used in producing the text. Jerry Durkin needs to be recognized for his help with the cover of the text. Many of the staff from Irwin and McGraw-Hill also need to be recognized. The authors are particularly grateful for help from Kelley Butcher and Beth Cigler, who have been associated with the project from the beginning. Betsy Jones, editor with Irwin, provided initial encouragement and support. Thanks are due, too, Per Sverdrup, who as an undergraduate student at Ohio State University, provided insightful comments from a student perspective.

Many reviewers have been associated with this project. The authors would like to acknowledge them. We are grateful for their time, expertise, and advice. The list of reviewers are Robert W. Ellis, Lawrence Technological University; William G. Ovens, Rose-Hulman Institute of Technology; John D. Reid, University of Nebraska—Lincoln; Steven Y. Liang, Georgia Institute of Technology; Steven A. Velinsky, University of California—Davis; Thierry Blanchet, Rensselaer Polytechnic Institute; K. V. C. Rao, Michigan Technological University; Alexander G. Liniecki, San Jose State University; Jesa Kreiner, California State University—Fullerton; Clarence Maday, North Carolina State University; J. Darrell Gibson, Rose-Hulman Institute of Technology; Terry F. Lehnhoff, University of Missouri—Rolla; Yu Michael Wang, University of Maryland; B. K. Rao, Idaho State University; W. Brent Hall, University of Illinois—Urbana–Champaign; N. Duke Perreira, Lehigh University; Gordon Pennock, Purdue University; Richard E. Dippery, Kettering University; Anthony Luscher, Ohio State University; Michael Peterson, Colorado State University; Gary McDonald, The University of Tennessee at Chattanooga; Wayne D. Milestone, University of Wisconsin—Madison; and Dean Taylor, Cornell University. The authors are also grateful to focus group participants Yu Michael Wang, University of

Maryland; J. Darrell Gibson, Rose-Hulman Institute of Technology; K. V. C. Rao, Michigan Technological University; Alexander G. Liniecki, San Jose State University; Jesa Kreiner, California State University—Fullerton; Clarence Maday, North Carolina State University; J. Darrell Gibson, Rose-Hulman Institute of Technology; and Steven A. Velinsky, University of California—Davis.

The authors also acknowledge the use of tables and illustrations from the following publishers: American Society of Mechanical Engineers, American Society for Testing and Materials, BHRA Fluid Engineering, Buttersworths, Elsevier Science Publishing Company, Engineering Sciences Data Unit, Ltd., Heinemann (London), Hemisphere Publishing Corporation, Macmillan Publishing Company, Inc., McGraw-Hill Higher Education, Mechanical Technology Incorporated, Non-Ferrous Founders Society, Oxford University Press, Inc., Penton Publishing Inc., Society of Automotive Engineers, Society of Tribologists and Lubrication Engineers, VCH Publishers, John Wiley & Sons, and Wykeham Publications (London), Ltd. The specific sources are identified in the text.

About the Authors

Bernard J. Hamrock

Bernard J. Hamrock joined the staff of Ohio State University as a professor in the Department of Mechanical Engineering in September 1985. Prior to that he spent 18 years as a research consultant in the Tribology Branch of the NASA Lewis Research Center in Cleveland, Ohio. He received his Ph.D. and Doctor of Engineering degrees from the University of Leeds, Leeds, England. Professor Hamrock's research has resulted in a book with Duncan Dowson, *Ball Bearing Lubrication,* published in 1982 by Wiley Interscience, three separate chapters for handbooks, and over 125 archival publications. His second book, *Fundamentals of Fluid Film Lubrication,* was published in 1993 by McGraw-Hill. His awards include the 1976 Melville Medal from the American Society of Mechanical Engineers, the NASA Exceptional Achievement Medal in 1984, and the 1998 Jacob Wallenberg Award given by The Royal Swedish Academy of Engineering Sciences.

Bo O. Jacobson

Bo O. Jacobson received his Ph.D. and D.Sc. degrees from Lund University in Sweden. From 1973 until 1987 he was Professor of Machine Elements at Luleå Technology University in Sweden. In 1987 he joined SKF Engineering & Research Centre in the Netherlands, while retaining a professorship at Chalmers University from 1987 to 1991 and at Luleå Technical University from 1992 to 1997. In 1997 he was appointed Professor of Machine Elements at Lund University, Sweden. Professor Jacobson was a NRC Research Fellow at NASA Lewis Research Center from 1981 to 1982. He has published four compendia used at Swedish universities. His text *Rheology and Elastohydrodynamic Lubrication* was published by Elsevier in 1991. Professor Jacobson has more than 60 archival publications. His awards include the prestigious Gold Medal given by the Institution of Mechanical Engineers, England.

Steven R. Schmid

Steven R. Schmid received his B.S. degree in Mechanical Engineering from the Illinois Institute of Technology in 1986. He then joined Triodyne, Inc., where his duties included investigation of machinery failures and consultation in machine design. He earned his Master's Degree from Northwestern University in 1989 and his Ph.D. in 1993, both in mechanical engineering. In 1993 he joined the faculty at the University of Notre Dame, where he teaches and conducts research in the fields of design and manufacturing. Dr. Schmid holds professional engineering (PE) and certified manufacturing engineer (C Mfg.E) licenses. He has authored more than 30 papers.

Contents in Brief

CONTENTS

Chapter 11

SHAFTING AND ASSOCIATED PARTS 423

Chapter 12

HYDRODYNAMIC AND HYDROSTATIC BEARINGS AND SEALS 467

Chapter 13

ROLLING-ELEMENT BEARINGS 539

Chapter 14

GEARS 613

Chapter 15

FASTENERS AND POWER SCREWS 665

Chapter 16

SPRINGS 735

Chapter 17

BRAKES AND CLUTCHES 781

Chapter 18

FLEXIBLE MACHINE ELEMENTS 826

Chapter 19

DESIGN PROJECTS 867

Appendix A

MATERIAL PROPERTIES 899

Appendix B

STRESS-STRAIN RELATIONSHIPS 904

Index 923

PART

1

FUNDAMENTALS

PART OUTLINE

CHAPTER

1

INTRODUCTION

A modern printing press. Gutenberg described the idea of a printing press as "coming like a ray of light." (AP/Wide World Photos)

Th' invention all admir'd, and each, how he
To be th' inventor miss'd; so easy it seem'd,
Once found, which yet unfound most would have thought
Impossible

John Milton

SYMBOLS

n_s safety factor
n_{sx} safety factor involving quality of materials, control over applied load, and accuracy of stress analysis
n_{sy} safety factor involving danger to personnel and economic impact
σ_{all} allowable normal stress, Pa
σ_d design normal stress, Pa

1.1 WHAT IS DESIGN?

Design means different things to different people. A clothing manufacturer believes that incorporating different materials or colors into a new tuxedo style constitutes design. A potter paints designs onto china to complement its surroundings. An architect designs ornamental facades for residences. An engineer chooses a bearing from a catalog and incorporates it into a speed-reducer assembly. These design activities, although they appear to be fundamentally different, share a common thread: they all require significant creativity, practice, and vision to be done well.

"Engineering," wrote Thomas Tredgold [Florman (1987)], "is the art of directing the great sources of power in nature for the use and convenience of man." It is indeed significant that this definition of engineering is more than 60 years old—few people now use the words "engineering" and "art" in the same sentence, let alone in a definition. However, many products are successful for nontechnical reasons, reasons that cannot be proved mathematically. On the other hand, many problems are mathematically tractable, but usually because they have been inherently overconstrained. Design problems are, almost without exception, open-ended problems combining hard science and creativity. Engineering is indeed an art, even though parts of engineering problems lend themselves well to analysis.

For the purposes of this textbook **design** is the transformation of concepts and ideas into useful machinery. A **machine** is a combination of mechanisms and other components that transforms, transmits, or uses energy, load, or motion for a specific purpose. If Tredgold's definition of engineering is accepted, design of machinery is the fundamental practice in engineering.

A machine comprises several different machine elements properly designed and arranged to work together as a whole. Fundamental decisions regarding loading, kinematics, and the choice of materials must be made during the design of a machine. Other factors, such as strength, reliability, deformation, tribology (friction, wear, and lubrication), cost, and space requirements, also need to be considered. The objective is to produce a machine that not only is sufficiently rugged to function properly for a reasonable time but is also economically feasible. Further, nonengineering decisions regarding marketability, product liability, ethics, politics, etc., must be integrated into the design process early. Since few people have the necessary tools to make all these decisions, machine design in practice is a discipline-blending human endeavor.

This textbook emphasizes one of the disciplines necessary in design, mechanical engineering. It therefore involves calculation and consideration of forces, energies, temperatures, etc.—concepts instilled into an engineer's psyche.

To "direct the great sources of power in nature" in machine design, the engineer must recognize the functions of the various machine elements and the types of load they transmit. A **machine element** may **function** as a normal load transmitter, a torque transmitter, an energy absorber, or a seal. Some normal load transmitters are rolling-element bearings, hydrodynamic bearings, and rubbing bearings. Some torque transmitters are gears, traction drives, chain drives, and belt drives. Brakes and dampers are energy absorbers. All the machine elements in Part II can be grouped into one of these classifications.

Engineers must produce safe, workable, good designs, as stated in the first fundamental canon in the *Code of Ethics for Engineers* [ASME (1997)]:

> Engineers shall hold paramount the safety, health, and welfare of the public in the performance of their professional duties.

Designing reasonably safe products involves many design challenges to ensure that components are large enough, strong enough, or tough enough to survive the loading environment. One subtle concept, but of huge importance, is that the engineer has a duty to protect the welfare of the general public. Welfare includes economic well-being, and it is well known that successful engineering innovations lead to wealth and job production. However, products that are too expensive are certain to fail in a competitive marketplace. Similarly, products that do not perform their function well will fail. Economics and functionality are always pressing concerns, and good design inherently means safe, economical, and functional design.

1.2 Design of Mechanical Systems

A **mechanical system** is a synergistic collection of machine elements. It is synergistic because as a design it represents an idea or concept greater than the sum of the individual parts. For example, a wristwatch, although merely a collection of gears, springs, and cams, also represents the physical realization of a time-measuring device. Mechanical system design requires considerable flexibility and creativity to obtain good solutions. Creativity seems to be aided by familiarity with known successful designs, and mechanical systems are often collections of well-designed components from a finite number of proven classes.

Designing a mechanical system is a different type of problem than selecting a component. Often, the demands of the system make evident the functional requirements of a component. However, designing a large mechanical system, potentially comprising thousands or even millions of machine elements, is a much more open, unconstrained problem.

To design superior mechanical systems, an engineer must have a certain sophistication and experience regarding machine elements. Studying the design and selection of machine elements affords an appreciation for the strengths and limitations of classes of components. They can then be more easily and appropriately incorporated into a system. For example, a mechanical system cannot incorporate a worm gear or a Belleville spring if the designer does not realize that these devices exist.

A toolbox analogy of problem solving can be succinctly stated as, "If your only tool is a hammer, then every problem is a nail." The purpose of studying machine element design is to fill the toolbox so that problem solving and design synthesis activities can be flexible and unconstrained.

1.3 DESIGN AS MULTIDISCIPLINARY ENDEAVOR

The quality revolution that hit the manufacturing sector in the early 1980s has forever changed the approach companies and engineers take toward product development. A typical design process of the recent past [Fig. 1.1(a)] shows that the skills involved in machine element design played an essential role in the process. This approach was commonly used in the United States in the postwar era. However, major problems with this approach became apparent in the 1970s and 1980s—some market driven, others product driven.

The term "over-the-wall engineering" (OTW) has been used to describe this design approach. Basically, someone would apply a particular skill and then send the product "over

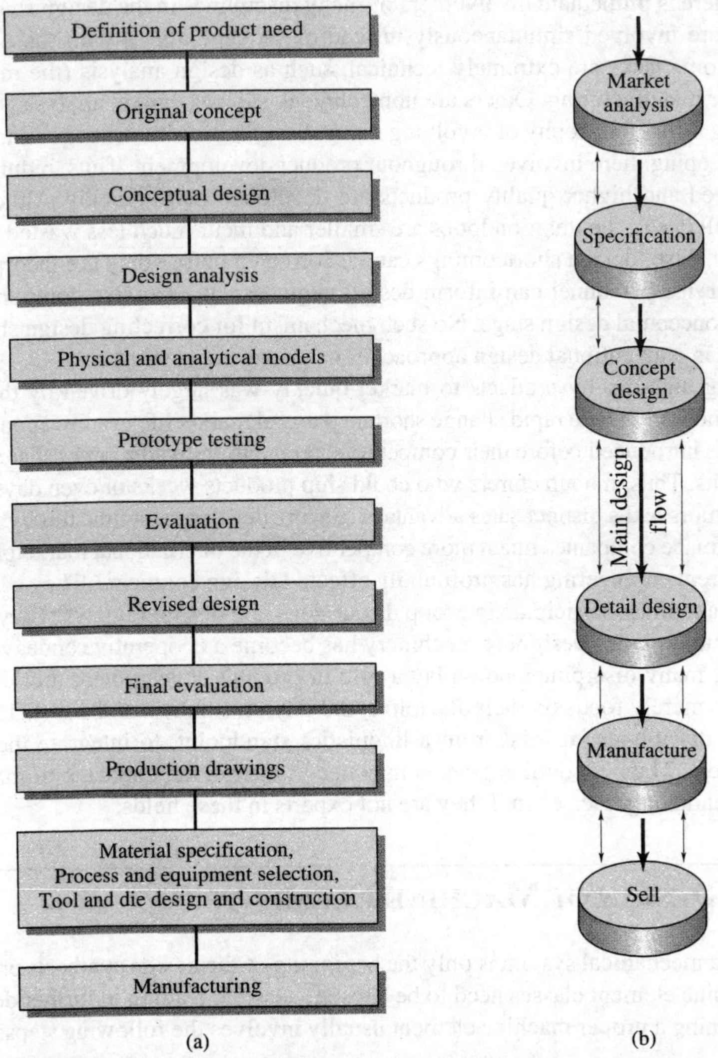

Figure 1.1 Approaches to product development.
(a) Over-the-wall engineering approach [Adapted from Kalpakjian (1997)].
(b) Concurrent engineering approach [Adapted from Pugh (1996)].

the wall" to the next step in development. A product design could sometimes flow smoothly from one step to the next and into the marketplace within weeks or months. This was rarely the case, however, as usually a problem would be discovered. For example, a manufacturing engineer might ask that workpieces be more easily clamped into a milling machine fixture. The design engineer would then alter the design and send the product back downstream. A materials scientist might then point out that the material chosen had drawbacks and suggest a different choice. The design engineer would make the alteration and resubmit the design. This process could continue ad infinitum, with the result that the product would take a long time to develop.

Figure 1.1(b) shows a more modern design approach. Here, there is still the recognized general flow of information from product conception through introduction into the marketplace, but there is immediate involvement of many disciplines in the design stage. Different disciplines are involved simultaneously instead of sequentially as with the over-the-wall approach. Some tasks are extremely technical, such as design analysis (the main focus of this book) or manufacturing. Others are nontechnical, such as market analysis. **Concurrent engineering** is the philosophy of involving many disciplines from the beginning of a design effort and keeping them involved throughout product development. Thus, redundant efforts are minimized and higher quality products are developed more quickly. Although design iterations still occur, the iteration loops are smaller and incur much less wasted time, effort, and expense. Also, design shortcomings can be corrected before they are incorporated. For example, service personnel can inform design engineers of excessive component failures during the conceptual design stage. No such mechanism for correcting design shortcomings ever existed in conventional design approaches or management structures.

Bringing high-quality products to market quickly was largely driven by the consumer electronics industry, where rapid change shortened useful market life to a few months. Further, new products introduced before their competitors' products enjoyed a larger share of the market and profits. Thus, manufacturers who could ship products weeks or even days faster than their competitors had a distinct sales advantage. Saving development time through concurrent engineering made companies much more competitive in the international marketplace.

Concurrent engineering has profoundly affected design engineers. They can no longer work alone and must participate in group discussions and design reviews. They need good communications skills. Designing machinery has become a cooperative endeavor.

Clearly, many disciplines now play a role in product development, but design engineers cannot merely focus on their discipline and rely on experts for the rest. They need to know other disciplines, at least from a linguistics standpoint, to integrate them into the design process. Thus, modern engineers may need to speak the language of materials science, law, marketing, etc., even if they are not experts in these fields.

1.4 Design of Machine Elements

Suggesting a mechanical system is only the beginning of the design synthesis process. Particular machine element classes need to be chosen, possibly leading to further design iterations. Designing a proper machine element usually involves the following steps:

1. Selecting a suitable type of machine element from consideration of its function
2. Estimating the size of the machine element that is likely to be satisfactory

3. Evaluating the machine element's performance against the requirements
4. And then modifying the design and the dimensions until the performance is near to whichever optimum is considered most important

The last two steps in the process can be handled fairly easily by someone who is trained in analytical methods and understands the fundamental principles of the subject. The first two steps, however, require some creative decisions and, for many, represent the most difficult part of design.

After a suitable type of machine element has been selected for the required function, the specific machine element is designed by analyzing kinematics, load, and stress. These analyses, coupled with proper material selection, will enable a stress-strain-strength evaluation in terms of a safety factor (covered in Sec. 1.5.1.1). A primary question in designing any machine element is whether it will fail in service. Most people, including engineers, commonly associate **failure** with the actual breaking of a machine element. Although breaking is one type of failure, a design engineer must have a broader understanding of what really determines whether a part has failed.

A machine element is considered to have failed

1. When it becomes completely inoperable
2. When it is still operable but is unable to perform its intended function satisfactorily
3. When serious deterioration has made it unreliable or unsafe for continued use, necessitating its immediate removal from service for repair or replacement

The role of the design engineer is to predict the circumstances under which failure is likely to occur. These circumstances are stress-strain-strength relationships involving the bulk of the solid members and such surface phenomena as friction, wear, lubrication, and environmental deterioration.

The principles of design are universal. An analysis is equally valid regardless of the size, material, and loading. However, an analysis by itself should not be looked on as an absolute and final truth. An analysis is limited by the assumptions imposed and by its range of applicability. Experimental verification of an analysis is always the preferred approach.

Design analysis attempts to predict the strength or deformation of a machine element so that it can safely carry the imposed loads for as long as required. Assumptions have to be made about the material properties under different loading types (axial, bending, torsion, and transverse shear, as well as various combinations) and classes (static, sustained, impact, or cyclic). These loading constraints may vary throughout the machine as they relate to different machine elements, an important factor for the design engineer to keep in mind.

1.5 FUNDAMENTAL DESIGN CONSIDERATIONS

1.5.1 SAFETY IN MECHANICAL DESIGN

The code of Hammurabi, a Babylonian doctrine over 3000 years old, had this requirement [Petroski (1992*a*)]:

> If a builder build a house for a man and do not make its construction firm, and the house which he has built collapse and cause the death of the owner of the house, that builder shall be put to death.

It could be argued that engineers are getting off a lot easier these days. Modern legal doctrines do not call for the death of manufacturers of unsafe products or of the engineers who designed them. Regardless of the penalty, however, engineers have a moral and legal obligation to produce reasonably safe products. A number of fundamental concepts and tools are available to assist them in meeting this challenge.

1.5.1.1 Safety Factor If 500 tension tests are performed on a specimen of one material, 500 different yield strengths will be obtained if the precision and accuracy of measurement are high enough. With some materials a wide range of strengths can be achieved; in others a reasonable guaranteed minimum strength can be found. However, this strength does not usually represent the stress that engineers apply in design.

Using small-scale tension tests a design engineer prescribes a stress somewhat less than the semiempirical strength of a material. The **safety factor** can be expressed as

$$n_s = \sigma_{\text{all}}/\sigma_d \quad (1.1)$$

where

σ_{all} = allowable normal stress, Pa
σ_d = design normal stress, Pa

If $n_s > 1$, the design is adequate. The larger n_s is, the safer the design is. If $n_s < 1$, the design may be inadequate and redesign may be necessary. In later chapters, especially Chapter 7, more will be said about σ_{all} and σ_d. The rest of this section focuses on the left side of Eq. (1.1).

It is difficult to accurately evaluate the various factors involved in engineering design problems. One factor is the shape of a part: for an irregularly shaped part there may be no design equations available for accurate stress computation. Another factor is the consequences of part failure: life-threatening consequences require more consideration than non-life-threatening consequences.

Engineers use a safety factor to ensure against such uncertain or unknown conditions. The engineering student is often asked, What safety factor was used in the design, and which value should be used? Safety factors are sometimes prescribed by code, but usually they are rooted in design experience. That is, design engineers have established through a product's performance that a safety factor is sufficient or insufficient. Future designs are often based on safety factors found adequate in previous products for similar applications.

Particular design experience for specific applications does not form a basis for the rational discussion of illustrative examples. The Pugsley method for determining the safety factor is used in this text, although the reader should again be warned that safety factor selection is somewhat nebulous in the real world. Pugsley (1966) systematically determined the safety factor from

$$n_s = n_{sx}\, n_{sy} \quad (1.2)$$

where

n_{sx} = safety factor involving characteristics A, B, and C
A = quality of materials, workmanship, maintenance, and inspection

B = control over load applied to part
C = accuracy of stress analysis, experimental data, or experience with similar devices
n_{sy} = safety factor involving characteristics D and E
D = danger to personnel
E = economic impact

Table 1.1 gives n_{sx} values for various A, B, and C conditions. To use this table, estimate each characteristic for a particular application as being very good (vg), good (g), fair (f), or poor (p). Table 1.2 gives n_{sy} values for various D and E conditions. To use this table, estimate each characteristic for a particular application as being either very serious (vs), serious (s), or not serious (ns). Putting the values of n_{sx} and n_{sy} into Eq. (1.2) yields the safety factor.

Although a simple procedure to obtain safety factors, the Pugsley method illustrates the concerns present in safety factor selection. Many parameters, such as material strength and applied loads, may not be well known, and confidence in the engineering analysis may be suspect. For these reasons the safety factor has been called an "ignorance factor," as it compensates for ignorance of the total environment, a situation all design engineers encounter to some extent. Also, the Pugsley method is merely a guideline and is not especially conservative; most engineering safety factors are much higher than those resulting from Eq. (1.2), as illustrated in Example 1.1.

Table 1.1 Safety factor characteristics A, B, and C

Characteristic[a]		B = vg	B = g	B = f	B = p
A = vg	C = vg	1.1	1.3	1.5	1.7
	C = g	1.2	1.45	1.7	1.95
	C = f	1.3	1.6	1.9	2.2
	C = p	1.4	1.75	2.1	2.45
A = g	C = vg	1.3	1.55	1.8	2.05
	C = g	1.45	1.75	2.05	2.35
	C = f	1.6	1.95	2.3	2.65
	C = p	1.75	2.15	2.55	2.95
A = f	C = vg	1.5	1.8	2.1	2.4
	C = g	1.7	2.05	2.4	2.75
	C = f	1.9	2.3	2.7	3.1
	C = p	2.1	2.55	3.0	3.45
A = p	C = vg	1.7	2.15	2.4	2.75
	C = g	1.95	2.35	2.75	3.15
	C = f	2.2	2.65	3.1	3.55
	C = p	2.45	2.95	3.45	3.95

[a]vg = very good, g = good, f = fair, and p = poor.
A = quality of materials, workmanship, maintenance, and inspection.
B = control over load applied to part.
C = accuracy of stress analysis, experimental data, or experience with similar parts.

Table 1.2 Safety factor characteristics D and E

Characteristic[a]	D = ns	D = s	D = vs
E = ns	1.0	1.2	1.4
E = s	1.0	1.3	1.5
E = vs	1.2	1.4	1.6

[a]vs = very serious, s = serious, and ns = not serious
D = danger to personnel.
E = economic impact

Example 1.1

GIVEN A wire rope is used on an elevator transporting people to the 20th floor of a building. The design of the elevator can be 50% overloaded before the safety switch shuts off the motor.

FIND What is the safety factor?

Solution

A = vg, because life threatening

B = f to p, since large overloads are possible

C = vg, due to being highly regulated

D = vs, people could die if the elevator fell from the 20th floor

E = s, possible lawsuits

From Tables 1.1 and 1.2 the safety factor is

$$n_s = n_{sx} n_{sy} = (1.6)(1.5) = 2.4$$

By improving factors over which there is some control n_{sx} can be reduced from 1.6 to 1.0 according to the Pugsley method, thus reducing the safety factor to 1.5.

The Pugsley method could conceivably suggest that a passenger elevator wire rope be designed with a safety factor of 1.5. In actuality, the safety factor is prescribed by an industry standard (see Sec. 1.5.2) and cannot be lower than 7.6 and may be as high as 11.9 [ANSI (1995)].

1.5.1.2 Failure Mode and Effects Analysis and Fault Trees Some common tools available to design engineers are **failure mode and effects analysis (FMEA)** and fault tree analysis. FMEA addresses component failure effects on the entire system. It forces the design engineer to exhaustively consider reasonably foreseeable failure modes for every component and its alternatives.

FMEA is flexible, allowing spreadsheets to be tailored for particular applications. For example, an FMEA can also be performed on the steps taken in assembling components to identify critical needs for training and/or warning.

In **fault tree analysis** statistical data are incorporated into the failure mode analysis to help identify the most likely (as opposed to possible) failure modes. Often, hard data are not available, and the engineer's judgment qualitatively identifies likely failure modes.

Machine designers are legally required to provide reasonably safe products and to consider the product's intended uses as well as foreseeable misuses. FMEA and fault tree analysis help identify unforeseeable misuses as well. For example, an aircraft designer may identify aircraft-meteorite collision as a possible loading of the structure. However, because no aircraft accidents have resulted from meteorite collisions and the probability of such occurrences is extremely low, the design engineer ignores such hypotheses, recognizing them as not reasonably foreseeable. On the other hand, the classic example of a reasonably foreseeable misuse is that of a chair, intended for use as a seat, that someone stands on to change a light bulb. Although the actual use was not the designer's intended use, the misuse is foreseeable and the designer has to produce a chair that can be safely stood and sat upon. Indeed, this is the public expectation as well.

1.5.1.3 LOAD REDISTRIBUTION, REDUNDANCY, FAIL SAFE, AND THE DOCTRINE OF MANIFEST DANGER One potential benefit of failure mode and effects analysis and fault tree analysis is that they force the design engineer to think of minimizing the effects of individual component failures. A common goal is that the failure of a single component not result in a catastrophic accident. The design engineer can ensure this by designing the system so that, upon a component failure, loads are redistributed to other components without exceeding their nominal strengths—a philosophy known as **redundancy** in design. For example, a goose or other large bird sucked into an aircraft engine may cause several components to fail and shut down the engine. This type of accident is not unheard of and is certainly reasonably foreseeable. Thus, modern aircraft are designed with sufficient redundancy to allow a plane to fly and land safely with one or more engines shut down.

Many designs are redundant. Redundant designs can be active (where two or more components are in use but only one is needed) or passive (where one component is inactive until the first component fails). An active redundant design is using two deadbolt locks on a door: both systems serve to keep the door locked. A passive redundant design is adding a chain lock on a door having a deadbolt lock: if the deadbolt lock fails, the chain will keep the door closed.

An often-used philosophy is to design machinery with **fail-safe** features. For example, a brake system can be designed so that a pneumatic cylinder pushes the brake pads or shoes against a disk or drum, respectively. Alternatively, a spring could maintain pressure against the disk or drum and a pneumatic system could work against the spring to release the brake. If the pressurized air supply were interrupted, such a system would force brake actuation and prevent machinery motion. This alternative design assumes that the spring is far more reliable than the pneumatic system.

The doctrine of **manifest danger** [Barnett (1992)] is a powerful tool used by machinery designers to prevent catastrophic losses. If danger becomes manifest, troubleshooting is straightforward and repairs can be quickly made. Thus, if a system can be designed so that imminent failure is detectable or so that single-component failure is detectable before other elements fail in turn, a safer design results. A classic application of the doctrine of manifest danger is in the design of automotive braking systems, where the brake shoe consists of a friction material held onto a metal backing plate by rivets. By making the rivets long enough, an audible and tactile indication is given to the car driver when the brake system needs service (i.e., the friction material has worn to where the rivet contacts the disk or drum, indicating maintenance is required) long before braking performance is compromised.

1.5.1.4 RELIABILITY Section 1.5.1.1 discussed safety factors as a way of compensating for variances in loading and material properties. Another way that can be extremely successful in certain circumstances is the application of reliability methods.

Manufacturing tension test coupons from the same extruded billet of aluminum would result in little difference in material properties from one test specimen to another. Thus, aluminum in general (as well as most metals) is a deterministic material, and deterministic methods can be used in designing aluminum structures if the load is known. For example, in a few hundred tensile tests a guaranteed minimum strength can be defined that is below the minimum strength of any test specimen and that would not vary much from one test population to another. Deterministic methods are the approach used in most solid mechanics and mechanics courses. That is, a material has a single strength and the loading is always well defined.

Most ceramics, however, would have a significant range of any given material property. Thus, ceramics are probabilistic, and an attempt to define a minimum strength for a population of ceramic test coupons would be an exercise in futility. There would not be a guaranteed minimum strength. One can only treat ceramics in terms of a likelihood or probability of strength exceeding a given value.

Some loadings, on the other hand, are well known and never vary much. Examples are the stresses inside intravenous bags during sterilization, the load supported by counterweight springs, and the load on bearings supporting centrifugal fans. Other loads can vary significantly, such as the force exerted on automotive shock absorbers (depends on the size of the pothole and the speed at impact) or on wooden pins holding a chair together (depends on the weight of the seated person or persons) or the impact force on the head of a golf club.

For situations where a reasonable worst-case scenario cannot be defined, reliability methods are sometimes a reasonable design approach. In reliability design methods the goal is to achieve a reasonable likelihood of survival under the loading conditions during the intended design life. This approach has its difficulties as well, including the following:

1. To use statistical methods, a reasonable approximation of an infinite test population must be defined. That is, mean values and standard deviations about the mean, and even the nature of the distribution about the mean, must be known. However, they are not usually known very well after only a few tests. After all, if only a few tests were needed to quantify a distribution, deterministic methods would be a reasonable, proper, and less mathematically intensive approach.
2. Even if strengths and loadings are known well enough to quantify their statistical distributions, defining a desired reliability is as nebulous a problem as defining a desired safety factor. A reliability of 99% might seem acceptable, unless that were the reliability of an elevator you happened to be occupying. A reliability of 100% is not achievable, or else deterministic methods would be used. A reliability of 99.9999...% should be recognized as an extremely expensive affair, as indicative of overdesign as a safety factor of 2000.
3. The mathematical description of the data has an effect on reliability calculations. A quantity may be best described by a Gaussian or normal distribution, a lognormal distribution, a binary distribution, or a Weibull distribution, etc. Often, one cannot know a priori which distribution is best. Some statisticians recommend using a normal distribution until it is proved ineffective.

The implications are obvious: reliability design is a complicated matter and even when applied does not necessarily result in the desired reliability if calculated from insufficient or improperly reduced data.

This textbook will emphasize deterministic methods for the most part. The exceptions are the treatments of rolling-element bearings and reliability in fatigue design. For more information on reliability design refer to the excellent text by Lewis (1987), among others.

1.5.2 Government Codes and Industry Standards

In many cases engineers must rely on government codes and industry-promulgated standards for design criteria. Some of the most common sources for industry standards are

1. ANSI, the American National Standards Institute
2. ASME, the American Society of Mechanical Engineers
3. ASTM, the American Society for Testing and Materials
4. AGMA, the American Gear Manufacturers Association
5. AISI, the American Iron and Steel Institute
6. AISC, the American Institute of Steel Construction
7. ISO, the International Standards Organization
8. NFPA, the National Fire Protection Association

UL, Underwriters Laboratories government codes, are published annually in the Code of Federal Regulations (CFR) and periodically in the Federal Register (FR) at the national level. States and local cities and towns have codes as well, although most relate to building standards and fire prevention.

Code compliance is important for many reasons, some of which have already been stated. However, one essential goal of industry standards is conformability. For example, bolt geometries are defined in ANSI standards so that bolts have fixed thread dimensions and bolt diameters. Therefore, bolts can be mass produced, and manufacturing economy of scale results in inexpensive, high-quality threaded fasteners. Also, maintenance is simplified in that standard bolts can be purchased anywhere, making replacement parts readily available.

1.5.3 MANUFACTURING

Design and manufacturing are difficult to consider apart from one another. The tenet of "form follows function" suggests that shapes are derived only from applied loads in the design environment. As Petroski (1992*b*) eloquently discussed, this is not always the case, and the shapes of products are often natural progressions from arbitrary beginnings.

Design for manufacturability (DFM) is a well-established and needed tool for design engineers. Manufacturability plays a huge role in the success of commercial products. After all, a brilliant concept that cannot be manufactured cannot be a design (per our definition in Sec. 1.1). Also, because most manufacturing costs are determined by decisions made early in the design process, market success depends on minimizing these decisions. Individual components should be designed to be easily fabricated, assembled, and constructed (design for assembly, DFA). Although manufacturing and assembly are outside the scope of this text, Fig. 1.2 shows their effect on design.

Engineers must wear many hats. Some predominant concerns of a design engineer have been discussed, but many more exist, including these:

1. *Environmental design*—This issue addresses whether products can be produced that are less harmful to the environment. Biocompatible or easily recycled materials may need to be selected.
2. *Economics*—Deciding whether a product will lead to corporate profitability is of utmost concern.
3. *Legal considerations*—Violating patents and placing unreasonably dangerous products into the marketplace are not only ethically wrong but legal suicide.

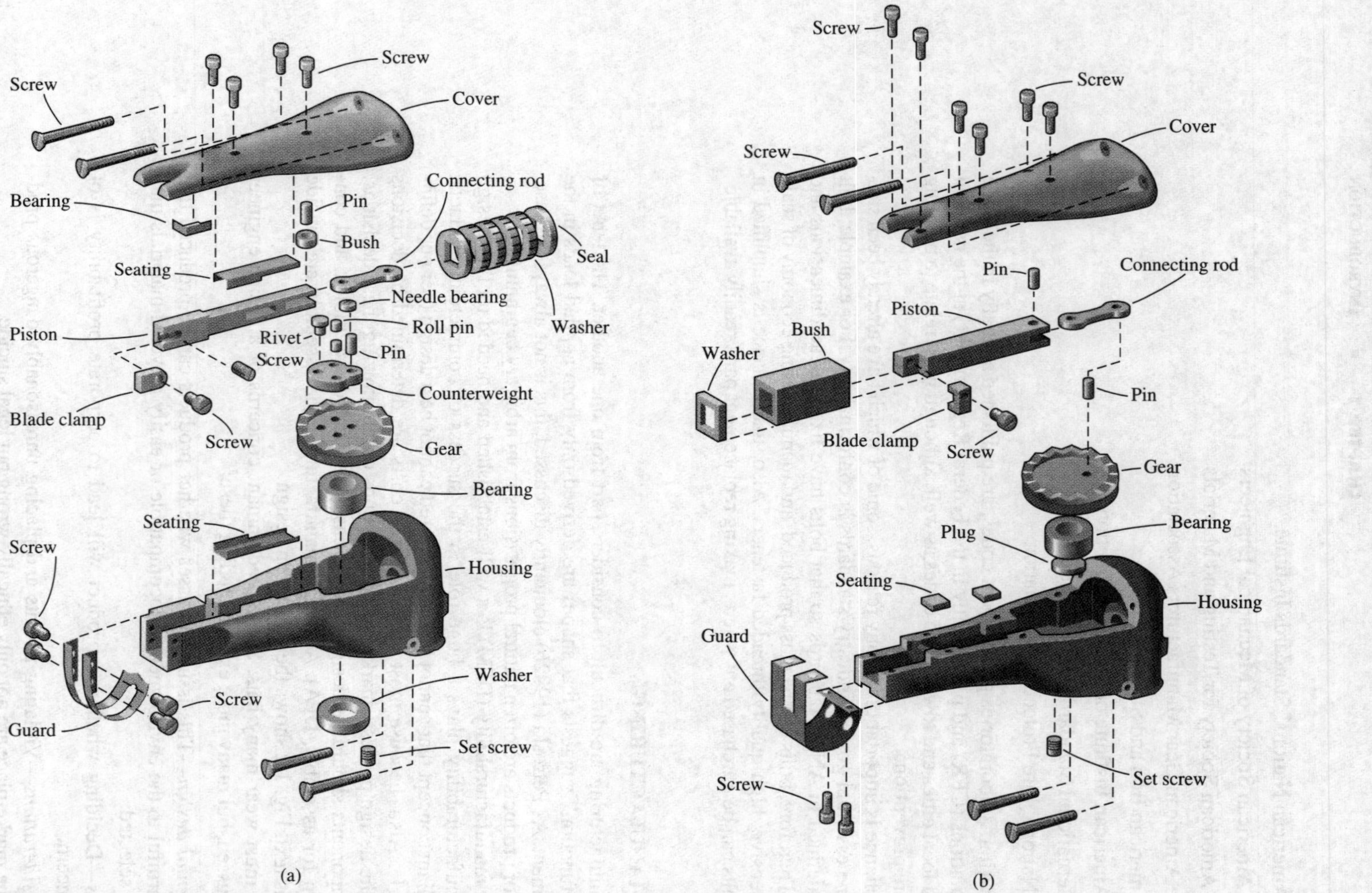

Figure 1.2 Effect of manufacturing and assembly considerations on design of a reciprocating power saw. (a) Original design, with 41 parts and 6.37-min assembly time; (b) modified design, with 29 parts and 2.58-min assembly time. [Adapted from Boothroyd (1992).]

4. *Marketing*—The features of a product that attract consumers and how the product is presented to the marketplace play a significant role in a product's success.
5. *Serviceability*—If a part breaks, can repairs be done in the field, or must customers send the product back to the manufacturer at excessive expense? Unless such concerns are incorporated into design, long-term customer loyalty is compromised.
6. *Quality*—Approaches such as total quality engineering and Taguchi methods have been successfully applied to make certain that no defects are shipped.

These are merely a few of the concerns faced by design engineers.

The design process may appear so elaborate and involved that no one can master it. In actuality, one most important skill makes the design process flow smoothly. Communication between diverse disciplines involved in the product life cycle ensures that all voices are heard. Effective communication skills, written and oral, are today the most important trait of a good engineer. Although this text emphasizes the more analytical and technical sides of design, always remember that total design is not merely an analytical effort but one of human interaction.

1.6 COMPUTERS IN DESIGN

Computer-aided design (CAD) also means different things to different people, but in this text it is the application of computer technology to planning, performing, and implementing the design process. Computers allow us to virtually integrate all phases of the design process, whether technical or managerial activities. With sophisticated hardware and software, manufacturers can now minimize design costs, maximize efficiency, improve quality, reduce development time, and maintain an edge in domestic and international markets.

CAD allows the design engineer to visualize geometries without making costly models, iterations, or prototypes. These systems can now analyze designs from simple brackets to complex structures quickly and easily. Designs can be optimized and modified directly and easily at any time. Information stored by computer can be accessed and retrieved from anywhere within the organization.

Whereas some restrict the term "CAD" to drafting activities, others will generically group all computer-assisted functions arbitrarily as CAD. **Artificial intelligence (AI)** attempts to duplicate how the human mind works and apply it to processes on the computer. Sometimes, AI is used to describe the cases where computers are used as more than mere drafting tools and actually help in the intellectual design tasks. **Expert systems** are rule-based computer programs that solve specialized problems on an expert level and provide problem-solving skills to the design engineer. For example, an expert system could analyze a part drafted on a computer system for ease in manufacturing. If an excessively small tolerance is found, the expert system warns the engineer that manufacturing difficulties will ensue and suggests easing the tolerance. Similarly, an expert system could analyze a design to standardize parts (e.g., to make sure that an assembly uses only one bolt size instead of the optimum size for each location, thereby easing inventory and maintenance difficulties). Artificial intelligence is a more elaborate form of an expert system; AI really refers to computer systems that can learn new information.

The advent of computer numerical control (CNC) has made group technology (GT) and cellular manufacturing (CM) increasingly valuable. Group technology involves classi-

Case Study 1.1 CAD IN BOEING'S 777 AIRCRAFT DESIGN [GOTTSCHALK (1994)]

The Boeing 777 was the first aircraft ever designed totally by computer—no drawings, physical mockups, or prototypes were produced during development. As can be imagined, the millions of individual parts in an aircraft represent an assembly nightmare, and it had been common aerospace industry practice to assemble a prototype and modify designs as errors became known.

A revolutionary task was planned in the design of the 777. It was realized that the design of one component could have a ripple effect on adjacent or affected components throughout the aircraft. Normally, a week's time delay was expected after one component or assembly was modified until the affected elements were detected. To reduce these development time delays, engineers from the United States and Japan were given access to all 3.5 terabytes of CAD data stored in a common database. Design changes could be quickly accommodated.

Other design shortcomings were rooted out without constructing expensive prototypes. Three-dimensional graphics and rendering software assisted in laying out cockpits and in human factors studies. Expert systems analyzed the various systems and checked for interference. When a problem was detected, the design engineers corrected it immediately. Boeing estimated that they had achieved their goal of reducing error and rework by 50% from previous aircraft development efforts.

fying and codifying parts so that those with similar design attributes can be quickly retrieved on a computer system. GT is extremely valuable when a desired design is merely an extension or modification of an existing component. The existing component's drawing can be quickly retrieved because of the logical classification, and the file can be modified with minimal redundant effort. Cellular manufacturing is the organization of machine tools so that a wide variety of designs can be fabricated under CNC with minimal lag time.

Rapid prototyping is another computer-driven technology that produces parts from computer geometry description files in hours or even minutes. Rapid prototyping has been especially helpful in design visualization and rapid detection of design errors. For example, a casting with an excessively thin wall is easily detected when a solid model is held in the hand, a subtlety difficult to discern when viewing a part drawing on a two-dimensional computer screen.

Finite element analysis (FEA) is the most prevalent computational method for solid and fluid mechanics analysis. The finite element computational method solves complex shapes, such as those found in machinery, and replaces the complex shape with a set of simple elements interconnected at a finite set of node points. In FEA a part geometry is sectioned into many subsections or elements. The stiffness of each element is known and is expressed in terms of a stiffness matrix for that element. By combining all the stiffness matrices, applying kinematic and stress boundary conditions, and solving for unknown stresses or displacements, complicated geometries and loading conditions can be easily analyzed.

1.7 CATALOGS AND VENDORS

Manufacturing concerns are latent in all design problems. Clearly, many machine elements are mass produced because economic arguments favor hard automation for large production runs. Hard automation generally results in higher quality, tighter tolerance parts than does soft automation or hand manufacture. In fact, the industry standards mentioned in Sec. 1.5.2 exist to prescribe geometries that can be mass produced in order to advocate mass production. For example, a centerless grinder can produce many high-quality 1/2-in.-diameter bushings, whereas a single 1/2-in. bushing is difficult to manufacture. Therefore, machinery design often involves selecting mass-produced elements from catalogs.

Mechanical designers know the importance of good vendor identification and readily available and up-to-date catalog information. Such catalogs are rarely available for the engineering student and are not needed if a mere familiarity with the design process is the main goal of study. However, for the design practitioner such material is invaluable.

Machine element vendors can be most easily identified through the Thomas Register, a listing of corporations in the United States. Most, and probably all, engineering libraries maintain a copy of the Thomas Register, and this is an excellent resource.

1.8 UNITS

The solutions to engineering problems must be given in specific units that correspond to the specific parameter being evaluated. Two systems of units are mainly used in this text:

1. **Système International d'Unites (SI units):** Force is measured in newtons, length in meters, time in seconds, mass in kilograms, and temperature in degrees Kelvin.
2. **English units:** Force is measured in pounds force, length in inches, time in seconds, mass in pounds mass, and temperature in degrees Rankine.

In Chapter 8 an additional measure, viscosity, is given in centimeter-gram-seconds (cgs).

Example 1.2

GIVEN A supercomputer has a calculation speed of 1 gigaflop = 10^9 floating-point operations per second. The calculation speed is determined only by the length of the wires. The electron speed for coaxial cables is 0.9 times the speed of light.

FIND How long can the connecting wires be?

Solution

If the speed is determined only by the cable length,

$$l = (0.9)(3)(10^8)/(10^9) = 0.27\text{ m} = 27\text{ cm}$$

The mean cable length must be less than 27 cm.

Example 1.3

GIVEN The distance from Earth to α-Centauri is 4 light-years.

FIND How many terameters is it?

Solution

$$1 \text{ year} = (365)(24)(3600) \text{ s} = (3.1536)(10^7) \text{ s}$$

$$\text{Speed of light} = (3)(10^8) \text{ m/s}$$

The distance from Earth to α-Centauri is

$$(4)(3.1536)(10^7) \text{ s } (3)(10^8) \text{ m/s} = 3.784 \times 10^{16} \text{ m}$$

From Table 1.3(b), $1\text{T} = 10^{12}$. Therefore,

$$(3.784)(10^{16}) \text{ m} = 37\,840 \text{ Tm}$$

The data needed to solve problems are not always in the same system of units. It is therefore necessary to convert from one system to another. The SI units, prefixes, and symbols used throughout the text are shown in Table 1.3 as well as inside the front cover. The primary units of the text are SI, but problems are given in English units as well as in SI units to enable the student to handle either system.

Basic SI units, some definitions, and fundamental and other useful conversion factors are given in Table 1.4, which is also inside the front cover.

1.9 UNIT CHECKS

Unit checks should always be performed during engineering calculations to make sure that each term of an equation is in the same system of units. The importance of knowing the units of the various parameters used in an equation cannot be overemphasized. In this text a symbol list giving the units of each parameter is provided at the beginning of each chapter. If no units are given for a particular phenomenon, it is dimensionless. This symbol list can be used as a partial check during algebraic manipulations of an equation.

Example 1.4

GIVEN The centrifugal force P acting on a car going through a curve with a radius r at a velocity v is $m_a v^2/r$, where m_a is the mass of the car. Assume a 1.3-ton car running at 100 km/hr through a 100-m-radius bend.

FIND Calculate the centrifugal force.

Solution

Rewriting all units in metric units gives

$$m_a = 1.3 \text{ ton} = 1300 \text{ kg}$$

$$v = 100 \text{ km/hr} = (100)(1000) \text{ m}/3600 \text{ s} = 27.78 \text{ m/s}$$

Table 1.3 SI units and prefixes

(a) SI units

Quantity	Unit	SI symbol	Formula
SI base units			
Length	meter	m	—
Mass	kilogram	kg	—
Time	second	s	—
Temperature	kelvin	K	—
SI supplementary unit			
Plane angle	radian	rad	—
SI derived units			
Energy	joule	J	N-m
Force	newton	N	kg-m/s^2
Power	watt	W	J/s
Pressure	pascal	Pa	N/m^2
Work	joule	J	N-m

(b) SI prefixes

Multiplication factor	Prefix	SI symbol for prefix
1 000 000 000 000 = 10^{12}	tera	T
1 000 000 000 = 10^9	giga	G
1 000 000 = 10^6	mega	M
1 000 = 10^3	kilo	k
100 = 10^2	hecto	h
10 = 10^1	deka	da
0.1 = 10^{-1}	deci	d
0.01 = 10^{-2}	centi	c
0.001 = 10^{-3}	milli	m
0.000 001 = 10^{-6}	micro	μ
0.000 000 001 = 10^{-9}	nano	n
0.000 000 000 001 = 10^{-12}	pico	p

$$r = 100 \text{ m}$$

The centrifugal force is

$$P = \frac{m_a v^2}{r} = \frac{(1300)[\text{kg}](27.78)^2[\text{m/s}]^2}{100[\text{m}]}$$

$$P = 10\ 030 \text{ kg-m/s}^2 = 10\ 030 \text{ N}$$

The centrifugal force is 10 030 N.

Table 1.4 Conversion factors and definitions

(a) Fundamental conversion factors

	English unit	Exact SI value	Approximate SI value
Length	1 in.	0.0254 m	—
Mass	1 lbm	0.453 592 37 kg	0.4536 kg
Temperature	1 deg R	5/9 K	—

(b) Definitions

Acceleration of gravity	1g = 9.8066 m/s^2 (32.174 ft/s^2)
Energy	Btu (British thermal unit) ≡ amount of energy required to raise 1 lbm of water 1 deg F (1 Btu = 778.2 ft-lbf)
	kilocalorie ≡ amount of energy required to raise 1 kg of water 1 K (1 kcal = 4187 J)
Length	1 mile = 5280 ft; 1 nautical mile = 6076.1 ft
Power	1 horsepower ≡ 550 ft-lbf/s
Pressure	1 bar ≡ 10^5 Pa
Temperature	degree Fahrenheit $t_F = 9/5\, t_C + 32$ (where t_C is degrees Celsius)
	degree Rankine $t_R = t_F + 459.67$
	Kelvin $t_K = t_C + 273.15$ (exact)
Kinematic viscosity	1 poise ≡ 0.1 kg/m-s
	1 stoke ≡ 0.0001 m^2/s
Volume	1 cubic foot = 7.48 gal

(c) Useful conversion factors

1 ft = 0.3048 m
1 lbf = 4.448 N
1 lbf = 386.1 lbm-in./s^2
1 kgf = 9.807 N
1 lbf/in.^{2} = 6895 Pa
1 ksi = 6.895 MPa
1 Btu = 1055 J
1 ft-lbf = 1.356 J
1 hp = 746 W = 2545 Btu/hr
1 kW = 3413 Btu/hr
1 quart = 0.000946 m^3 = 0.946 liter
1 kcal = 3.968 Btu

1.10 SIGNIFICANT FIGURES

The accuracy of a number is specified by how many significant figures it contains. Throughout this text four significant figures will be used. For example, 8201 and 30.51 each has four significant figures. When numbers begin or end with a zero, however, it is difficult to tell how many significant figures there are. To clarify this situation, the number should be reported by using powers of 10. Thus, the number 8200 can be expressed as 8.200×10^3 to represent four significant figures. Also, 0.005012 can be expressed as 5.012 $\times 10^{-3}$ to represent four significant figures.

Case Study 1.2 | WHEELCHAIR DESIGN AS MULTIDISCIPLINARY ENDEAVOR

Figure 1.3 shows a modern manual wheelchair with significant improvements over its predecessors. In many ways the design attributes of the wheelchair have been determined by the unique desires and needs of the users. Many disciplines have contributed to the design, some of which are identified here:

1. *Biomechanics*—Since pressure sores are a common malady for wheelchair users, especially those suffering paralysis in their lower extremities, the seat material is an advanced multilayered foam with a rigid pan to distribute the user's weight more uniformly. (Original wheelchairs used a cloth sling.) This seat leads to lower local pressures and better blood flow through tissue, greatly reducing the likelihood of pressure sores and increasing the user's comfort.
2. *Marketing*—Wheelchair users, just like bicyclists and car drivers, care very much about appearance. Here, vibrant colors and a sleek look have been emphasized. Product information is disseminated to an identified class of wheelchair users—paraplegics with active lifestyles who would most fully utilize the design features. User input is brought to the design stage indirectly through surveys and evaluations of existing designs.
3. *User surveys*—Wheelchair users are directly consulted from the beginning of the product development cycle, and their evaluations and recommendations are incorporated throughout the design process.
4. *Structural design*—Because a wheelchair must be lifted and moved from time to time, the lightweight design is fabricated from composite materials and lightweight metals. The wheel axles incorporate quick-disconnect couplings so that they can be stowed separately from the frame.
5. *Mechanical design*—Quiet, high-performance bearings have been selected, and a four-bar linkage holds the drive wheel secure when actuated. Also, because the intended users are paraplegics with active lifestyles, the drive wheels have been tilted toward the user to make pushing easier, and the wheelbase is narrow for cornering in hallways or maneuvering behind desks. The frame is adjustable to fit the dimensions of the individual user. Anti-tipping wheels are available to hinder tip-over accidents.
6. *Nursing*—Nurses were brought in as consultants during the design stage and emphasized the difficulties in nurses, or the patients themselves, lifting patients over the arm rests to transfer to a bed or toilet. The arm rests are removable on this wheelchair because of this input.

Since it is difficult for one person to bring all these insights into concept design, multidisciplinary efforts benefit the design of products.

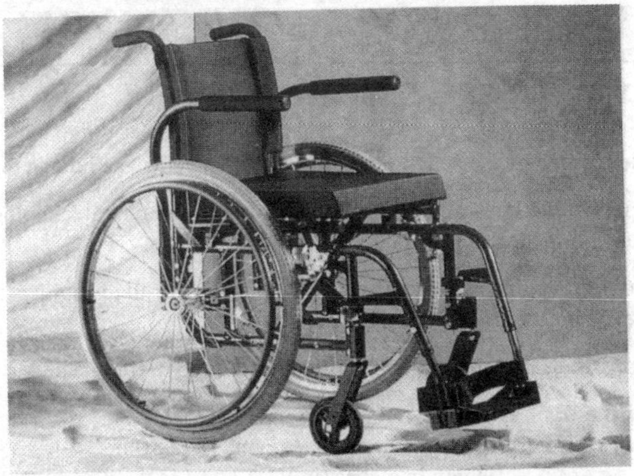

Figure 1.3 Wheelchair. (Courtesy of Sunrise Medical Equipment)

Example 1.5 **GIVEN** A car weighing 1502 kg is accelerated by a force of 14.0 N.

FIND Calculate the acceleration and answer with significant figures.

Solution

Newton's equation gives that acceleration equals the force divided by the mass

$$a = \frac{P}{m_a} = \frac{14.0}{1502} = 0.00932091 \ \text{m/s}^2$$

Since force is accurate to three figures, the acceleration can only be calculated with the accuracy of

$$\pm\frac{0.5}{140} = \pm 0.004 = \pm 0.4\%$$

The acceleration is 0.00932 m/s^2.

1.11 SUMMARY

This chapter introduced the concept of design as it applies to machines and machine elements. The most important goal of the design process is to ensure that the design does not fail. The design engineer must predict the circumstances under which failure is first likely to occur. These circumstances are stress-strain relationships involving the material properties as well as surface phenomena, including friction, wear, lubrication, and environmental deterioration.

The concept of failure was quantified by using a safety factor, which is simply the allowable stress established for the material being used divided by the maximum design stress that will occur. If the safety factor is less than 1, the design is inadequate and redesign is necessary. Besides the simple safety factor, other failure models, such as failure mode and effects analysis and fault tree analysis, were presented. Design was found to be a cooperative endeavor where multidisciplinary approaches are invaluable. Failure prevention and engineering analysis of machine elements are two essential requirements of successful design teams. These topics are the focus of this text.

KEY WORDS

artificial intelligence (AI) attempts to duplicate how the human mind works in computer processes

computer-aided design (CAD) application of computer technology to planning, performing, and implementing the design process

concurrent engineering design approach wherein all disciplines involved with a product are in the development process from beginning to end

design transformation of concepts and ideas into useful machinery

English units system of units where
force is measured in pounds force (lbf)
length in inches (in.)
time in seconds (s)

mass in pounds mass (lbm)
temperature in degrees Rankine (°R)

expert systems computer programs that solve specialized problems on an expert level

fail-safe design approach where no catastrophic loss can occur as a result of a component failure

failure the condition of a machine element when it is completely inoperable, cannot perform its intended function adequately, or is unreliable for continued safe use

failure mode and effects analysis (FMEA) systematic consideration of component failure effects on the entire system

fault tree analysis statistical data used to identify the most likely failure modes

finite element analysis (FEA) computational method used for solving complex shapes, such as those found in machinery; replaces the complex shape with a set of simple elements interconnected at a finite set of node points

machine combination of mechanisms and other components that transform, transmit, or use energy, load, or motion for a specific purpose

machine element function normal load transmitter, torque transmitter, energy absorber, or seal

manifest danger design approach where needed service is made apparent before catastrophic failure

mechanical system synergistic collection of machine elements

rapid prototyping parts produced quickly from computer geometry description files

redundancy additional capacity or incorporation of backup systems so that a component failure does not lead to catastrophic loss

safety factor (n_s) ratio of allowable stress to design stress

SI units system of units where
force is measured in newtons (N)
length in meters (m)
time in seconds (s)
mass in kilograms (kg)
temperature in degrees Kelvin (K)

RECOMMENDED READINGS

GENERAL ENGINEERING

Florman, S.C. (1976)*The Existential Pleasures of Engineering,* St. Martin's Press, New York.
Petroski, H. (1992) *To Engineer Is Human,* Vintage Books, New York.

MANUFACTURING/DESIGN FOR MANUFACTURE

Boothroyd, G., Dewhurst, P., and Knight, W. (1994) *Product Design for Manufacture and Assembly,* Marcel Dekker, New York.
Boothroyd, G. (1992) *Assembly Automation and Product Design,* Marcel Dekker, New York.

DeGarmo, E.P., Black, J.T., and Kohser, R.A. (1997) *Materials and Processes in Manufacturing,* 8th ed., Prentice-Hall, New Jersey.
Dieter, G.E. (1991) *Engineering Design: A Materials and Processing Approach,* McGraw-Hill, New York.
Design for Manufacturability (1992) *Tool and Manufacturing Engineer's Handbook,* vol. 6, Society of Manufacturing Engineers, New York.
Groover, M.K. (1996) *Fundamentals of Modern Manufacturing,* Prentice-Hall, New Jersey.
Kalpakjian, S. (1997) *Manufacturing Processes for Engineering Materials,* 3rd ed., Addison-Wesley, New York.

CONCURRENT ENGINEERING

Clausing, D. (1994) *Total Quality Development,* Marcel Dekker, New York.
Nevins, J.L., and Whitney, D.E. (eds.) (1989) *Concurrent Design of Products and Processes,* McGraw-Hill, New York.
Prasad, B. (1996) *Concurrent Engineering Fundamentals,* Prentice-Hall, New Jersey.
Pugh, S. (1996) *Creating Innovative Products Using Total Design,* Addison-Wesley, New York.
Pugh, S. (1991) *Total Design,* Addison-Wesley, New York.

COMPUTER-AIDED DESIGN

Groover, M., and Zimmer, W. (1984) *CAD/CAM Computer-Aided Design and Manufacture,* Prentice-Hall, New Jersey.
Wilson, C.E. (1997) *Computer Integrated Machine Design,* Prentice-Hall, New Jersey.

REFERENCES

ANSI (1995) A17.1 "Minimum Safety Requirements for Passenger Elevators," American National Standards Institute, New York.
ASME (1997) *Code of Ethics for Engineers,* Board on Professional Practice and Ethics, American Society of Mechanical Engineers, New York.
Barnett, R.L. (1992) *The Doctrine of Manifest Danger,* Triodyne Inc., Niles, IL.
Boothroyd, G. (1992) *Assembly Automation and Product Design,* Marcel Dekker, New York, pp. 239–240.
Burke, J. (1985) *The Day the Universe Changed,* Little, Brown and Company, Boston, pp. 303–305.
Florman, S.C. (1987) *The Civilized Engineer,* St. Martin's Press, New York, p. 44.
Gottschalk, M. (1994) "How Boeing Got to 777th Heaven," *Design News,* vol. 49, Sept. 12, 1994, pp. 50–56.
Kalpakjian, S. (1997) *Manufacturing Processes for Engineering Materials,* Addison-Wesley, New York, 3rd ed., p. 6.
Lewis, E.E.(1987) *Introduction to Reliability Engineering,* Wiley, New York.
Petroski, H. (1992*a*) *To Engineer Is Human,* Vintage Books, New York, p. 3.
Petroski, H. (1992*b*) *The Evolution of Useful Things,* Alfred A. Knopf, New York.
Pugh, S. (1996) *Creating Innovative Products Using Total Design,* Addison-Wesley, New York, p. 11.
Pugsley, A.G. (1966) *The Safety of Structures,* Arnold, New York.

PROBLEMS

Section 1.1

1.1 Design transport containers for milk in 1 gallon and 1 liter sizes.

1.2 Design a simple coat hanger to be used by dry cleaners. It should be able to accommodate both a winter coat and a pair of trousers. It should function well but also be as inexpensive as possible.

Section 1.3

1.3 Journal bearings on train boxcars in the early 19th century used a "stink additive" in their lubricant. If the bearing got too hot, it would attain a noticeable odor, and an oiler would give the bearing a squirt of lubricant at the next train stop. What design philosophy does this illustrate? Explain.

1.4 A safety chain is used in automotive towing applications. Which design philosophy is being incorporated into the towing system design?

Section 1.4

1.5 A hand-held drilling machine has a bearing to take up radial and thrust load from the drill. Depending on the number of hours the drill is expected to be used before it is scrapped, different bearing arrangements will be chosen. A rubbing bushing has a 50-hr life. A small ball bearing has a 300-hr life. A two-bearing combination of a ball bearing and a cylindrical roller bearing has a 10 000-hr life. The cost ratios for the bearing arrangements are 1:5:20. What is the optimum bearing type for a simple drill, a semiprofessional drill, and a professional drill?

1.6 Using the hand-held drill described in Problem 1.5, if the solution with the small ball bearing was chosen for a semiprofessional drill, the bearing life could be estimated to be 300 hr until the first spall forms in the race. The time from first spall to when the whole rolling-contact surface is covered with spalls is 200 hr, and the time from then until a ball cracks is 100 hr. What is the bearing life

a) If high precision is required?
b) If vibrations are irrelevant?
c) If an accident can happen when a ball breaks?

Section 1.5

1.7 The dimensions of skis used for downhill competition need to be determined. The maximum force transmitted from one foot to the ski is 2500 N, but the snow conditions are not known in advance, so the bending moment acting on the skis is not known. Estimate the safety factor needed.

1.8 A crane has a loading hook that is hanging in a steel wire. The allowable normal tensile stress in the wire gives an allowable force of 100 000 N. Find the safety factor that should be used.

a) If the wire material is not controlled, the load can cause impact, and fastening the hook in the wire causes stress concentrations. (If the wire breaks, people can be seriously hurt and expensive equipment can be destroyed.)
b) If the wire material is extremely well controlled, no impact loads are applied, and the hook is fastened in the wire without stress concentrations. (If the wire breaks, no people or expensive equipment can be damaged.)

Denotes open-ended problem. (Problems 1.1, 1.2 and 1.7 are marked with this symbol.)

1.9 Give three examples of fail-safe and three examples of fail-unsafe products.

1.10 An acid container will damage the environment and people around it if it leaks. The cost of the container is proportional to the container wall thickness. The safety can be increased either by making the container wall thicker or by mounting a reserve tray under the container to collect the leaking acid. The reserve tray costs 10% of the thick-walled container cost. Which is less costly, to increase the wall thickness or to mount a reserve tray under the container?

Section 1.8

1.11 The unit for dynamic viscosity in the SI system is newton-seconds per square meter, or pascal-seconds ($N\text{-}s/m^2$ = Pa-s). How can that unit be rewritten by using the basic relationships described by Newton's law for force and acceleration?

1.12 The unit for dynamic viscosity in Problem 1.11 is newton-seconds per square meter ($N\text{-}s/m^2$) and the kinematic viscosity is defined as the dynamic viscosity divided by the fluid density. Find at least one unit for kinematic viscosity.

1.13 A square surface has sides 1 m long. The sides can be split into decimeters, centimeters, or millimeters, where 1 m = 10 dm, 1 dm = 10 cm, and 1 cm = 10 mm. How many millimeters, centimeters, and decimeters equal 1 m? Also, how many square millimeters, square centimeters , and square decimeters equal a square meter?

1.14 A volume is 1 tera (mm^3) large. Calculate how long the sides of a cube must be to contain that volume.

1.15 A ray of light travels at a speed of 300 000 km/s = 3×10^8 m/s. How far will it travel in 1 ps, 1 ns, and 1 μs?

Section 1.9

1.16 Two smooth flat surfaces are separated by a 10-μm-thick lubricant film. The viscosity of the lubricant is 0.100 Pa-s. One surface has an area of 1 dm^2 and slides over the plane surface with a velocity of 1 km/hr. Determine the friction force due to shearing of the lubricant film. Assume the friction force is the viscosity times the surface area times the velocity of the moving surface and divided by the lubricant film thickness.

1.17 A firefighter sprays water on a house. The nozzle diameter is small relative to the hose diameter, so the force on the nozzle from the water is

$$F = v\frac{dm_a}{dt}$$

where v is the water velocity and dm_a/dt is the water mass flow per unit time. Calculate the force the firefighter needs to hold the nozzle if the water mass flow is 3 tons/hr and the water velocity is 100 km/hr.

1.18 The mass of a car is 1346 kg. The four passengers in the car weigh 65.55 kg, 75.23 kg, 88.66 kg, and 91.32 kg. It is raining and the additional weight due to the water on the car is 1.349 kg. Calculate the total weight of the car, including the weight of the passengers and the weight of the water, using four significant figures.

1.19 During an acceleration test of a car the acceleration was measured to be 1.4363 m/s^2. Because slush and mud adhered to the bottom of the car, the weight was estimated to be 1400 ± 100 kg. Calculate the force driving the car and indicate the accuracy.

CHAPTER

2

LOAD, STRESS, AND STRAIN

A dragline hoist of a large load in a mining operation. (Courtesy of Bucyrus-Erie)

When I am working on a problem, I never think about beauty. I think only of how to solve the problem. But when I have finished, if the solution is not beautiful, I know it is wrong.

Richard Buckminster Fuller

SYMBOLS

A	area, m^2
d	diameter, m
g	gravitational acceleration, 9.807 m/s^2
l	length, m
M	moment, N-m
m_a	mass, kg
n	any integer
P	force, N
p	normal pressure, Pa
q	load intensity function, N/m
R	reaction force, N
r	radius of Mohr's circle, m
$\mathbf{S}$	stress tensor
$\mathbf{S}'$	principal stress tensor
s	shear pressure, Pa
$\mathbf{T}$	strain tensor
T	torque, N-m
V	transverse shear force, N
W	normal applied load, N
w_0	constant load per unit length, N/m
x, y, z	Cartesian coordinate system, m
x', y', z'	rotated Cartesian coordinate system, m
γ	shear strain
δ	elongation, m
ε	normal strain
θ	angle representing deviation from initial right angle or angle at which force is applied, radian
μ	coefficient of friction
σ	normal stress, Pa
τ	shear stress, Pa
$\tau_{1/2}$, $\tau_{2/3}$, $\tau_{1/3}$	principal shear stresses in triaxial stress state, Pa
ϕ	angle of oblique plane, deg

Subscripts

a	axial
avg	average
b	biaxial stress
c	center
cr	critical
e	von Mises
l	left
max	maximum
min	minimum
n	normal surface
oct	octahedral planes
r	right
t	triaxial stress; transverse
x, y, z	Cartesian coordinates
x', y', z'	rotated Cartesian coordinates
θ	angle representing deviation from initial right angle
σ	normal stress
τ	shear stress
ϕ	angle of oblique plane
1, 2, 3	principal axes

2.1 INTRODUCTION

The focus of this text is the design and analysis of machines and machine elements. Since machine elements carry **loads,** it follows that an analysis of loads is essential in machine element design. Proper selection of a machine element often is a simple matter of calculating the stresses or deformations expected in service and then choosing a proper size so that critical stresses or deformations are not exceeded. The first step in calculating the stress or deformation of a machine element is to accurately determine the load. Load, stress, and strain in all its forms is the focus of this chapter, and the information developed here is used throughout the text.

2.2 CRITICAL SECTION

To determine when a machine element will fail, the designer evaluates the stress, strain, and strength at the critical section. The **critical section,** or the location in the design where the largest internal load is developed and failure is most likely, is often not intuitively known a priori. To establish the critical section and the critical loading, the designer

1. Considers the external loads applied to a machine (e.g., a gyroscope)
2. Considers the external loads applied to an element within the machine (e.g., a ball bearing)
3. Locates the critical section within the machine element (e.g., the inner race)
4. Determines the loading at the critical section (e.g., contact stresses)

The first and second steps arise from system design. The third step is quite challenging and may require analysis of a number of locations or failure modes before the most critical mode is found. For example, a beam subjected to a distributed load might conceivably exceed the maximum deflection at a number of locations; thus, the deflection would need to be calculated at more than one position in the beam. The fourth step is a topic for future chapters.

In general, the critical section will often occur at locations of geometric nonuniformity, such as where a shaft changes its diameter along a fillet. Also, locations where load is applied or transferred are often critical locations. Finally, areas where the geometry is most critical are candidates for analysis. This topic will be expanded upon in Chapter 6.

Example 2.1

GIVEN A simple crane shown in Fig. 2.1(a) consists of a horizontal beam loaded vertically at one end with load of 10 kN. At the other end the beam is pinned. The force at the pin and roller must not be larger than 30 kN to satisfy other design constraints.

FIND The location of the critical section and also whether the load of 10 kN can be applied without damage to the crane.

Solution

Forces acting on the horizontal beam are shown in Fig. 2.1(b). Summation of moments about the pin gives

$$(1.0)\,P = (0.25)\,W_r$$

$$\therefore W_r = 40 \text{ kN}$$

Summation of forces gives

$$-W_p + W_r - 10 \text{ kN} = 0$$

$$\therefore W_p = 30 \text{ kN}$$

The critical section is at the roller, since $W_r > W_p$. Also, since $W_r > W_{\text{all}}$, failure will occur. To avoid failure, the load at the end of the horizontal beam must be changed from 10 kN to

$$(10)\left(\frac{30}{40}\right)(10^3)\text{ N} = 7.5(10^3)\text{ N} = 7.5\text{ kN}$$

ANSWER The critical section occurs at the roller, where failure will occur unless the load applied to the end of the beam is 7.5 kN or less.

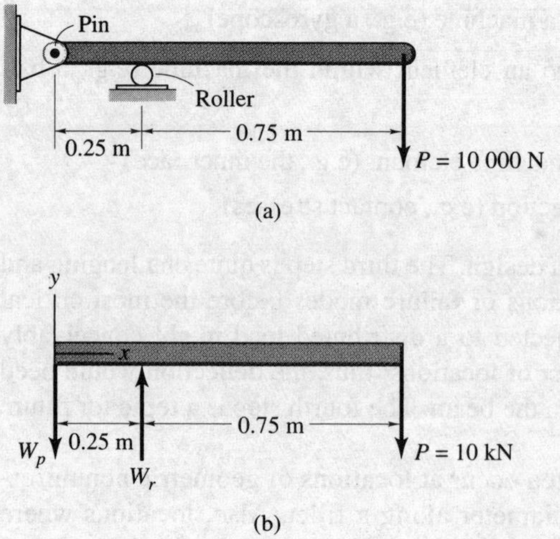

Figure 2.1 A simple crane and forces acting on it. (a) Assembly drawing; (b) free-body diagram of forces acting on the beam.

2.3 LOAD CLASSIFICATION AND SIGN CONVENTION

Any applied load can be classified with respect to time in the following ways:

1. *Static load*—Load is gradually applied and equilibrium is reached in a relatively short time. The structure experiences no dynamic effects.
2. *Sustained load*—Load, such as the weight of a structure, is constant over a long time.
3. *Impact load*—Load is rapidly applied. An impact load is usually attributed to an energy imparted to a system.
4. *Cyclic load*—Load can vary and even reverse itself in sign and has a characteristic period with respect to time.

The load can also be classified with respect to the area over which it is applied:

1. *Concentrated load*—Load is applied to an area much smaller than the loaded member, as presented for nonconformal surfaces in Chapter 8 (Sec. 8.2). An example would be the contact between a caster and a support beam on a mechanical crane, where the contact area is 100 times smaller than the surface area of the caster. For these cases the applied force can be considered to act at a point on the surface.
2. *Distributed load*—Load is spread along the entire area. An example would be the weight of a concrete bridge floor of uniform thickness.

Loads can be further classified with respect to location and method of application. Also, the coordinate direction must be determined before the sign of the loading can be established:

1. *Normal load*—The load passes through the centroid of the resisting section. Normal loads may be tensile [Fig. 2.2(a)] or compressive [Fig. 2.2(b)]. The sign convention has tensile load being positive and compressive load negative.
2. *Shear load*—The force P is assumed to be collinear with the transverse shear force V. The separated bar in Fig. 2.2(c) illustrates the action of positive shearing. A shear force is positive if the force direction and the normal direction are both positive or both negative. The shear force shown on the left surface of Fig. 2.2(c) is in the positive

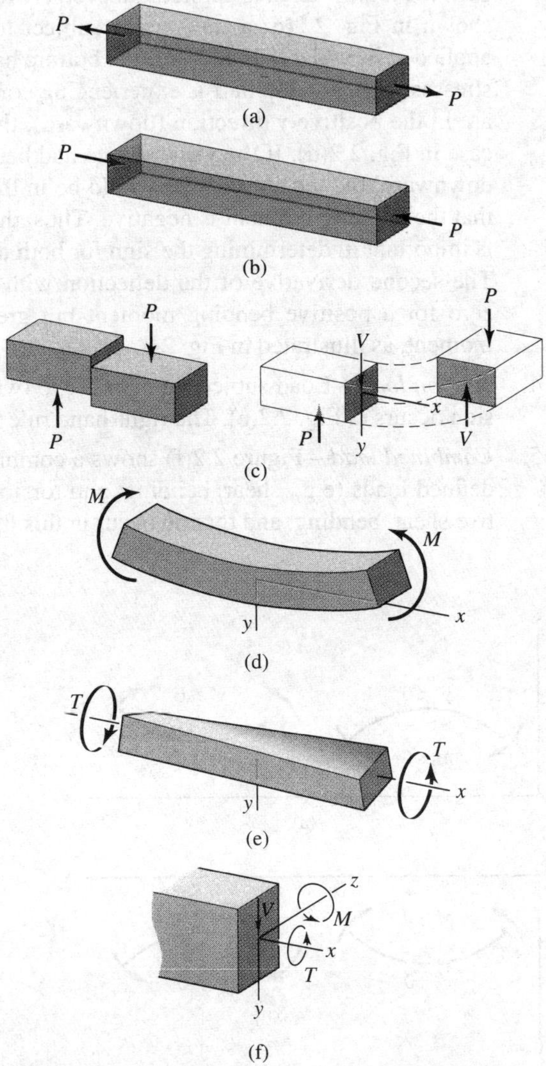

Figure 2.2 Load classified as to location and method of application. (a) Normal, tensile; (b) normal, compressive; (c) shear; (d) bending; (e) torsion; (f) combined.

y direction, which is downward, and the normal to the surface is in the positive x direction. Thus, the shear force is positive. On the right surface of Fig. 2.2(c) the shear force is also positive, since the direction of the shear force and the normal to the surface are both negative. A shear force is negative if the force direction and the normal direction have different signs. If the positive y coordinate had been chosen to be upward (negative) rather than downward (positive) in Fig. 2.2(c), the shear force would be negative rather than positive. Thus, to establish whether a shear force is positive or negative, the positive x and y coordinates must be designated.

3. *Bending load*—Load is applied transversely to the longitudinal axis of the member. As shown in Fig. 2.2(d) a member is subject to equal and opposite moments that are applied at its ends. In Fig. 2.2(d) the bottom half of the member is experiencing tensile stresses while the top half is experiencing compressive stresses. If the tensile stresses are in the positive y direction (downward), the bending moment is positive, as is the case in Fig. 2.2(d). If the y coordinate had been chosen upward (negative) rather than downward, the tensile stresses would be in the negative y direction and would imply that the bending moment is negative. Thus, the designation of the x and y coordinates is important in determining the sign for both the shear force and the bending moment. The second derivative of the deflection with respect to the x coordinate is less than zero for a positive bending moment but greater than zero for a negative bending moment, as illustrated in Fig. 2.3.
4. *Torsion load*—Load subjects a member to twisting motion [Fig. 2.2(e)]. Positive torsion occurs in Fig. 2.2(e). The right-hand rule is applicable here.
5. *Combined load*—Figure 2.2(f) shows a combination of two or more of the previously defined loads (e.g., shear, bending, and torsion acting on a member). Note that positive shear, bending, and torsion occur in this figure.

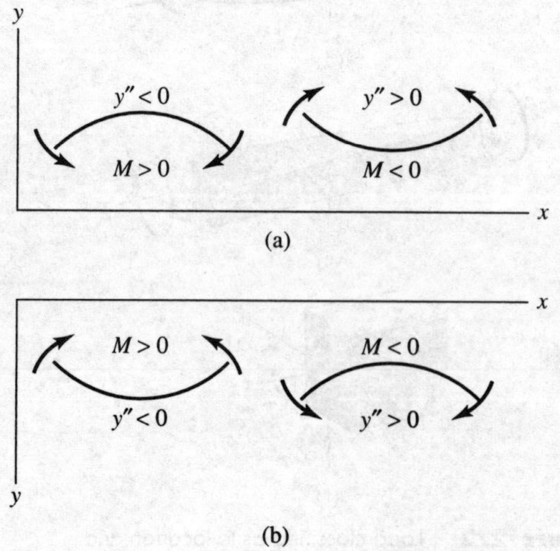

Figure 2.3 Sign convention used in bending. (a) y coordinate upward; (b) y coordinate downward.

Example 2.2

GIVEN The lever assembly shown in Fig 2.4(a).

FIND The normal, shear, bending, and torsional loads acting at section B

Solution

Figure 2.4(b) shows the various loads acting on the lever, all in the positive direction. To the right of the figure expressions are given for the loading at section B of the lever shown in Fig. 2.4(a).

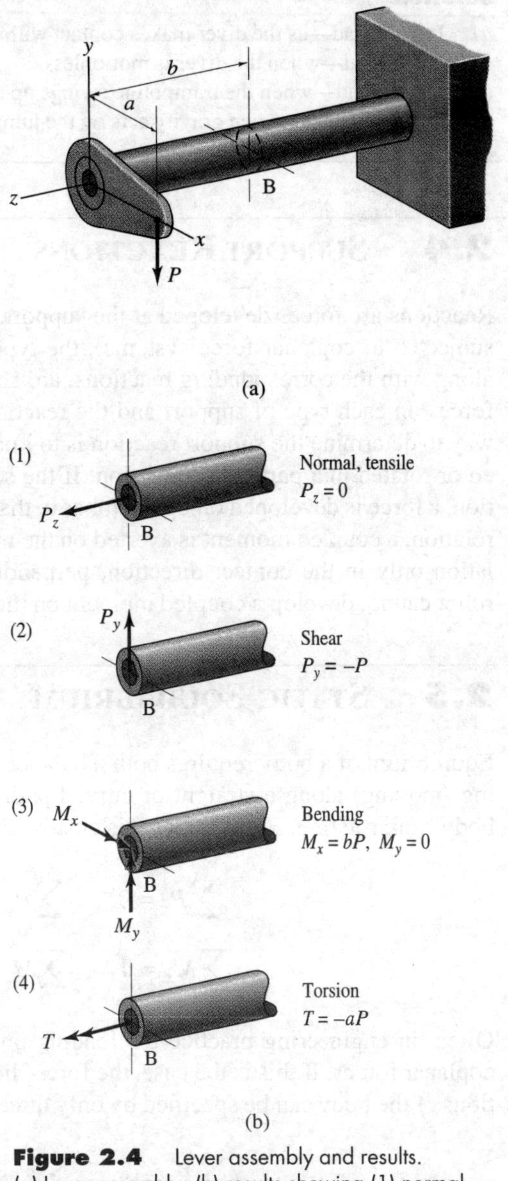

Figure 2.4 Lever assembly and results. (a) Lever assembly; (b) results showing (1) normal, tensile, (2) shear, (3) bending, (4) torsion on section B of lever assembly.

Example 2.3 **GIVEN** A diver jumping on a trampoline.

FIND

a) The load type when the diver lands on the trampoline
b) The load type when the diver stands motionless waiting for the signal to jump
c) The load type on the jumping tower just as the diver jumps
d) The load type on the jumping tower against the ground when no dynamic loads are acting

Solution

a) Impact load—as the diver makes contact with the trampoline
b) Static load—when the diver is motionless
c) Cyclic load—when the trampoline swings up and down just after the dive
d) Sustained load—when gravity acts on the jumping tower, pressing it against the ground

2.4 SUPPORT REACTIONS

Reactions are forces developed at the support. For two-dimensional problems (i.e., bodies subjected to coplanar force systems), the types of support most commonly encountered, along with the corresponding reactions, are shown in Table 2.1. (Note the direction of the forces on each type of support and the reaction they exert on the attached member.) One way to determine the support reaction is to imagine the attached member as being translated or rotated in a particular direction. If the support prevents translation in a given direction, a force is developed on the member in that direction. Likewise, if the support prevents rotation, a coupled moment is averted on the member. For example, a roller prevents translation only in the contact direction, perpendicular (or normal) to the surface; thus, the roller cannot develop a coupled moment on the member at the point of contact.

2.5 STATIC EQUILIBRIUM

Equilibrium of a body requires both a balance of forces, to prevent the body from translating (moving) along a straight or curved path, and a balance of moments, to prevent the body from rotating. From statics it is customary to present these equations as

$$\sum P_x = 0 \qquad \sum P_y = 0 \qquad \sum P_z = 0 \tag{2.1}$$

$$\sum M_x = 0 \qquad \sum M_y = 0 \qquad \sum M_z = 0 \tag{2.2}$$

Often, in engineering practice the loading on a body can be represented as a system of coplanar forces. If this is the case, the forces lie in the x-y plane and the equilibrium conditions of the body can be specified by only three equations:

$$\sum P_x = 0 \qquad \sum P_y = 0 \qquad \sum M_z = 0 \tag{2.3}$$

Table 2.1 Four types of support with their corresponding reactions

Type of support	Reaction
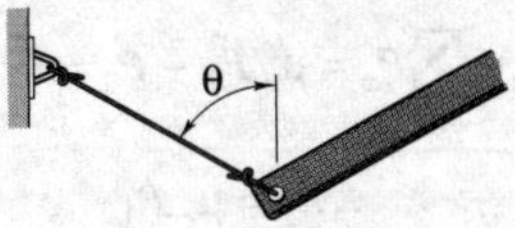 Cable	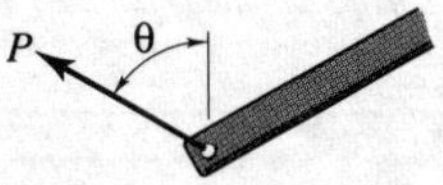
 Roller	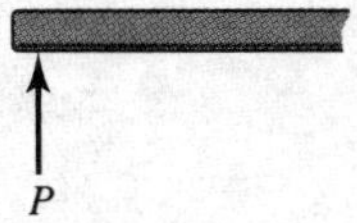
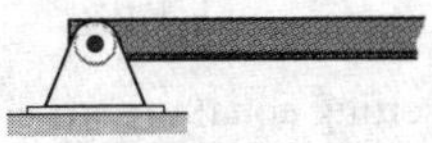 Pin	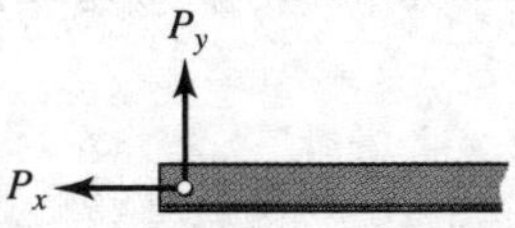
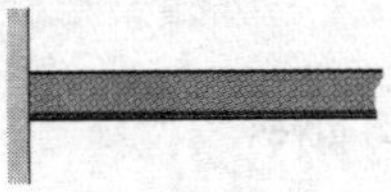 Fixed support	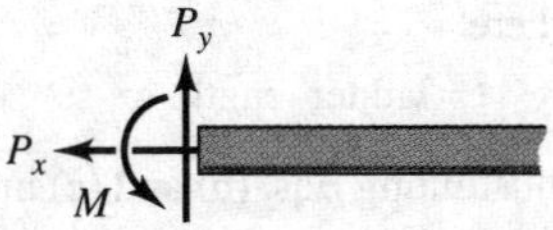

Note that the moment M_z is perpendicular to the plane that contains the forces. Successful application of the equilibrium equations requires complete specification of all the known and unknown forces acting on the body.

Example 2.4

GIVEN A painter stands on a ladder that leans against the wall of a house. Assume the painter is at the midheight of the ladder. The ladder stands on a horizontal surface with a coefficient of friction of 0.3 and leans at an angle of 20° against the house, which also has a coefficient of friction of 0.3.

FIND Whether the painter and ladder are in static equilibrium and what critical coefficient of friction μ_{cr} will not provide static equilibrium.

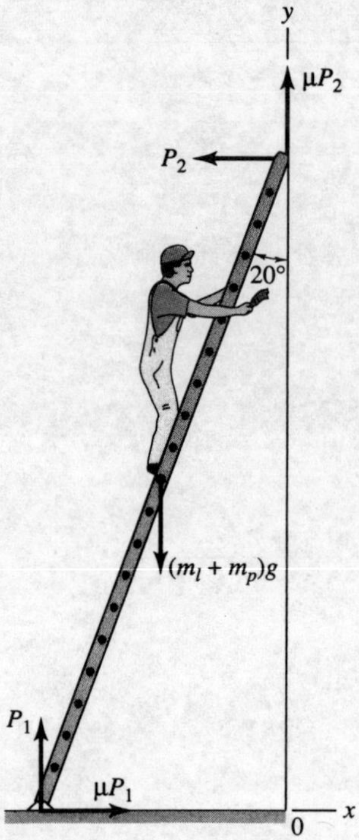

Figure 2.5 Ladder having contact with the house and the ground while having a painter on the ladder.

Solution

Figure 2.5 shows a diagram of the forces acting on the ladder due to the weight of the painter as well as the weight of the ladder. The mass of the ladder is m_l and the mass of the painter is m_p.

If the ladder starts to slide down, the friction force will counteract the motion. Summation of horizontal and vertical forces gives

$$\sum P_x = \mu_{cr} P_1 - P_2 = 0 \qquad \textbf{(a)}$$

$$\therefore P_2 = \mu_{cr} P_1$$

$$\sum P_y = P_1 - (m_l + m_p)g + \mu_{cr} P_2 = 0$$

Making use of Eq. (a) gives

$$P_1(1 + \mu_{cr}^2) = (m_l + m_p)g$$

$$\therefore P_1 \frac{(m_l + m_p)g}{1 + \mu_{cr}^2} \qquad \textbf{(b)}$$

Making use of Eq. (a) again gives

$$P_2 = \frac{\mu_{cr}(m_l + m_p)g}{1 + \mu_{cr}^2} \qquad \textbf{(c)}$$

Taking the moment about point 0 and setting equal to zero gives

$$P_1 l \sin 20° - P_2 l \cos 20° - (m_l + m_p)\, g \frac{l}{2} \sin 20° = 0 \qquad \textbf{(d)}$$

where

l = ladder length, m

Substituting Eqs. (b) and (c) into Eq. (d) gives

$$0 = \frac{(m_l + m_p)gl \sin 20°}{1 + \mu_{cr}^2} - \frac{\mu_{cr}(m_l + m_p)gl \cos 20°}{1 + \mu_{cr}^2} - (m_l + m_p)g \frac{l}{2} \sin 20°$$

$$0 = \frac{1}{1 + \mu_{cr}^2} - \frac{\mu_{cr}}{\tan 20° \,(1 + \mu_{cr}^2)} - \frac{1}{2}$$

$$0.5 = \frac{\tan 20° - \mu_{cr}}{\tan 20° \,(1 + \mu_{cr}^2)}$$

$$0.5 \tan 20° + \mu_{cr}^2\, 0.5 \tan 20° = \tan 20° - \mu_{cr}$$

$$\mu_{cr}^2 + \frac{\mu_{cr}}{0.5 \tan 20°} - 1 = 0$$

$$\mu_{cr} = 0.1763$$

Since μ is 0.3, the ladder will not move because $\mu_{cr} < \mu$, or $0.1763 < 0.3$.

ANSWER The painter and ladder are in static equilibrium. The critical coefficient where the ladder starts to slide is 0.1763.

2.6 FREE-BODY DIAGRAM

An entire machine, any individual machine element, or any part of a machine element can be represented as a free body. Static equilibrium is assumed at each level. The best way to account for the forces and moments in the equilibrium equations is to draw the free-body diagram. For the equilibrium equations to be correct, the effects of all the applied forces and moments must be represented in the free-body diagram.

A **free-body diagram** is a sketch of a machine, a machine element, or part of a machine element that shows all acting forces, such as applied loads and gravity forces, and all reactive forces. The reactive forces are supplied by the ground, walls, pins, rollers, cables, or other means. The sign of the reaction is initially guessed. If after the static equilibrium analysis the sign of the reactive force is positive, the direction initially guessed is correct; if it is negative, the direction is opposite to that initially guessed.

Example 2.5

FIND Draw a free-body diagram of each component of the external rim brake shown in Fig. 2.6(a).

Solution

Figure 2.6(b) shows each brake component as well as the forces acting on it. The static equilibrium of each component must be preserved, and the friction force acts opposite to the direction of motion on the drum and in the direction of motion on both shoes. The $4W$ value in Fig 2.6(b) was obtained from the moment equilibrium of the lever. Details of brakes are considered in Chapter 17, but in this chapter it is important to be able to draw the free-body diagram of each component.

Example 2.6

GIVEN A steel sphere [Fig 2.7(a)] with a mass of 10 kg hangs from two wires. A spring attached to the bottom gives a downward force of 150 N.

FIND The forces acting on the two wires. Also, draw a free-body diagram showing the forces acting on the sphere.

Solution

Figure 2.7(b) shows the free-body diagram of the forces acting on the sphere. Summation of the vertical forces gives

$$2P \cos 60^\circ - m_a g - 150 = 0$$

or

$$P = \frac{(10)(9.807) + 150}{2 \cos 60^\circ} = 248.1 \text{ N}$$

2.7 SUPPORTED BEAMS

A **beam** is a structural member designed to support loading applied perpendicular to its longitudinal axis. In general, beams are long, straight bars having a constant cross-sectional area. Often, they are classified by how they are supported. Three major types of support are shown in Figure 2.8:

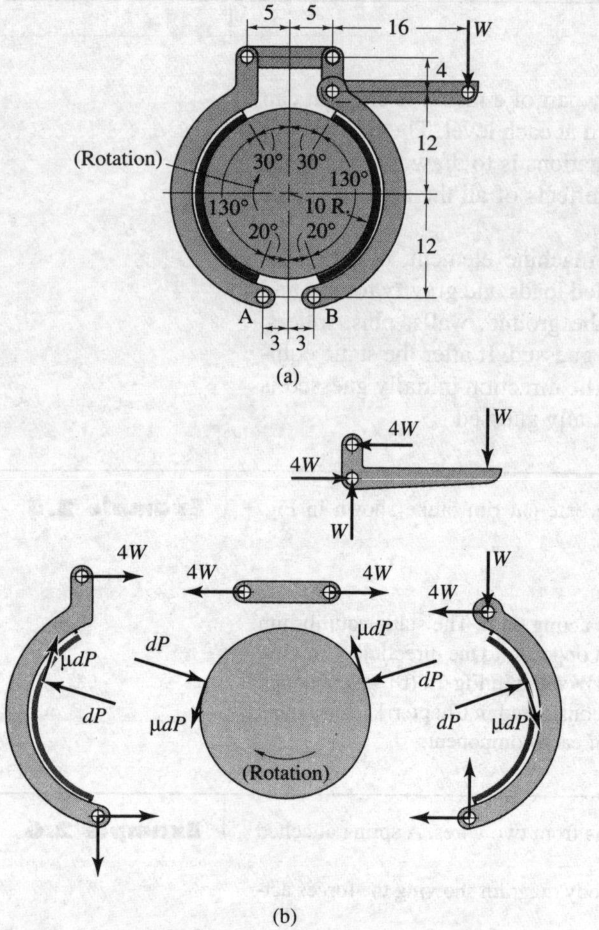

Figure 2.6 External rim brake and forces acting on it. (a) External rim brake; (b) external rim brake with forces acting on each part. (Linear dimensions are in millimeters.)

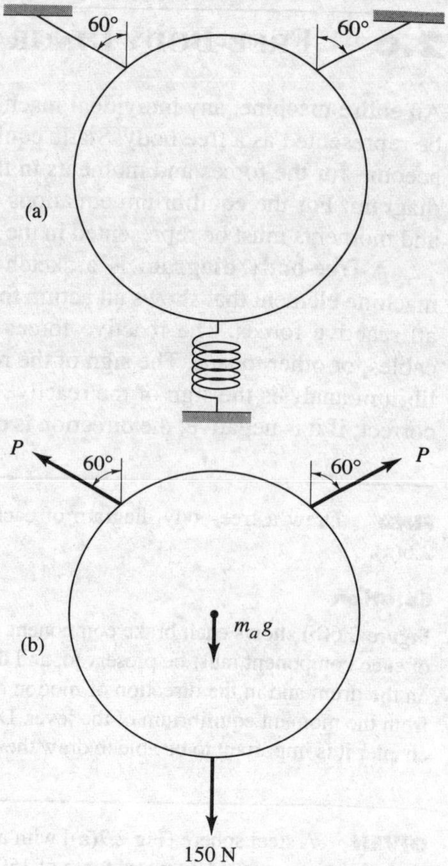

Figure 2.7 Sphere and forces acting on it. (a) Sphere supported with wires from top and spring at bottom; (b) free-body diagram of forces acting on sphere.

1. A **simply supported beam** [Fig. 2.8(a)] is pinned at one end and roller supported at the other.
2. A **cantilevered beam** [Fig. 2.8(b)] is fixed at one end and free at the other.
3. An **overhanging beam** [Fig. 2.8(c)] has one or both of its ends freely extending past the support.

Two major parameters used in evaluating beams are strength and deflection. Shear and bending are the two primary modes of beam loading. However, if the height of the beam is large relative to its width, elastic instability becomes important and the beam can twist under loading [see unstable equilibrium in Chapter 9 (Sec. 9.2.3)].

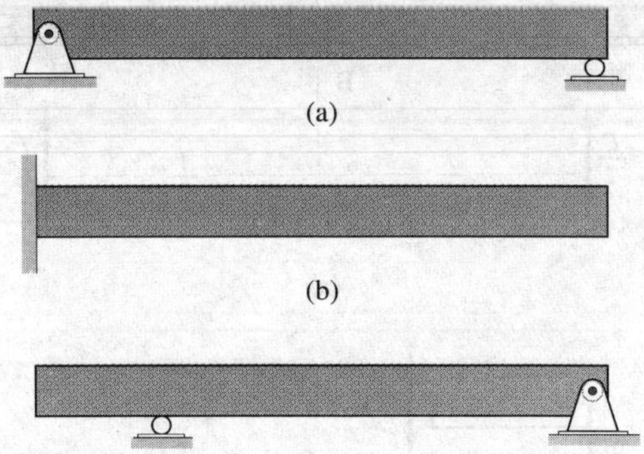

Figure 2.8 Three types of beam support.
(a) Simply supported; (b) cantilevered;
(c) overhanging.

2.8 SHEAR AND MOMENT DIAGRAMS

Designing a beam on the basis of strength requires first finding its maximum shear and moment. One way to do this is to express the transverse shear force V and the moment M as functions of an arbitrary position x along the beam's axis. These shear and moment functions can then be plotted as shear and moment diagrams from which the maximum values of V and M can be obtained.

The procedure for drawing shear and moment diagrams is as follows:

1. Draw a free-body diagram and determine all the support reactions. Resolve the forces into components acting perpendicular and parallel to the beam's axis.
2. Choose a position x between the origin and the length of the beam l, thus dividing the beam into two segments. The origin is chosen at the beam's left end to ensure that any x chosen will be positive.
3. Draw a free-body diagram of the two segments and use the equilibrium equations to determine the transverse shear force V and the moment M.
4. Plot the shear and moment functions versus x. Note the location of the maximum moment. Generally, it is convenient to show the shear and moment diagrams directly below the free-body diagram of the beam.

Example 2.7

GIVEN The bar shown in Fig. 2.9(a).

FIND Draw the shear and moment diagrams.

Solution

For $0 \le x < l/2$ the free-body diagram of the bar section would be as shown in Fig. 2.9(b). The unknowns V and M are positive. Applying the equilibrium equations gives

$$\sum P_y = 0 \rightarrow V = \frac{P}{2} \qquad \textbf{(a)}$$

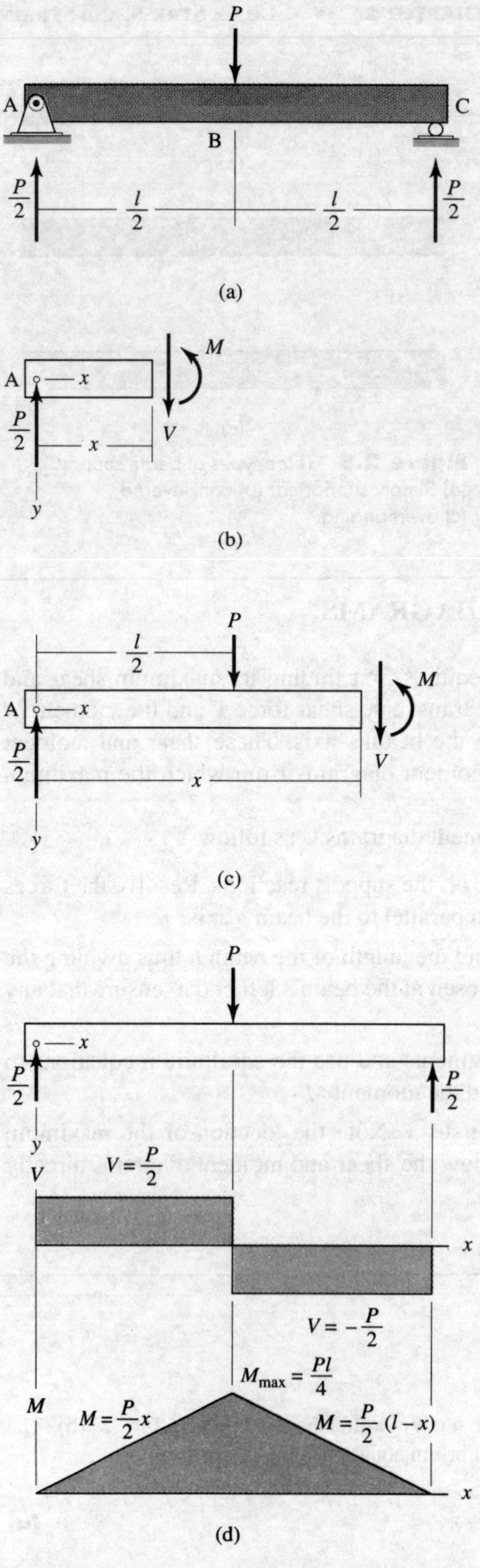

Figure 2.9 Simply supported bar. (a) Midlength load and reactions; (b) free-body diagram for $0 < x < l/2$; (c) free-body diagram for $l/2 \leq x < l$; (d) shear and moment diagrams.

$$\sum M_z = 0 \rightarrow M = \frac{P}{2}x \qquad \textbf{(b)}$$

For $l/2 \le x < l$, the free-body diagram is shown in Fig. 2.9(c). Again, V and M are shown in the positive direction.

$$\sum P_y = 0 \;\rightarrow\; \frac{P}{2} - P - V = 0 \qquad \therefore V = -P/2 \qquad \textbf{(c)}$$

$$\sum M_z = 0 \;\rightarrow\; M + P\left(x - \frac{l}{2}\right) - \frac{P}{2}x = 0$$

$$\therefore M = \frac{P}{2}(l - x) \qquad \textbf{(d)}$$

The shear and moment diagrams [Fig. 2.9(d)] can be obtained directly from Eqs. (b) to (d). As a check, $V = dM/dx$, which is indeed valid for $0 \le x < l$.

Note that if $q(x)$ is the load intensity function in the y direction, the transverse shear force is

$$V(x) = -\int_{-\infty}^{x} q(x)dx \qquad \textbf{(2.4)}$$

and the bending moment is

$$M(x) = \int_{-\infty}^{x} V(x)dx = -\int_{-\infty}^{x}\int_{-\infty}^{x} q(x)dxdx \qquad \textbf{(2.5)}$$

2.9 SINGULARITY FUNCTIONS

If the loading is simple, the method for obtaining shear and moment diagrams described in Sec. 2.8 can be used. Usually, however, this is not the situation. Differentiating or integrating across a discontinuity, such as a concentrated load or moment, presents problems. For more complex loading, methods such as **singularity functions** can be used. A singularity function of x is written as

$$f_n(x) = \langle x - a \rangle^n \qquad \textbf{(2.6)}$$

where

n = any integer (positive or negative) including zero

a = constant distance on x axis equal to value of x at initial boundary, m

Some general rules relating to singularity functions and angular brackets are

1. If $n > 0$ and the expression inside the angular brackets is positive (i.e., $x \ge a$), then $f_n(x) = (x - a)^n$. Note that the angular brackets to the right of the equal sign in Eq. (2.6) are now parentheses.

2. If $n > 0$ and the expression inside the angular brackets is negative (i.e., $x < a$), then $f_n(x) = 0$.
3. If $n < 0$, then $f_n(x) = \begin{cases} 1 \text{ for } x = a \\ 0 \text{ for } x \neq a \end{cases}$
4. If $n = 0$, then $f_n(x) = 1$ when $x \geq a$ and $f_n(x) = 0$ when $x < a$.
5. If $n \geq 0$, the integration rule is $\int_{-\infty}^{x} \langle x - a \rangle^n \, dx = \dfrac{\langle x - a \rangle^{n+1}}{n + 1}$, the same as if there were parentheses instead of angular brackets.
6. If $n < 0$, the integration rule is $\int_{-\infty}^{x} \langle x - a \rangle^n \, dx = \langle x - a \rangle^{n+1}$.
7. When $n \geq 1$, then $(d/dx) \langle x - a \rangle^n = n \langle x - a \rangle^{n-1}$.

The advantage of using a singularity function is that it permits writing an analytical expression directly for the transverse shear and moment over a range of discontinuities.

Table 2.2 shows six singularity and load intensity functions along with corresponding graphs and expressions. Note in particular the inverse ramp example. A unit step is constructed beginning at $x = a$, and the ramp beginning at $x = a$ is subtracted. To have the negative ramp discontinued at $x = a + b$, a ramp beginning at this point is constructed.

The procedure for drawing the shear and moment diagrams by making use of singularity functions is as follows:

1. Draw a free-body diagram with all the singularities acting on the beam, and determine all support reactions. Resolve the forces into components acting perpendicular and parallel to the beam's axis.
2. Write an expression for the load intensity function $q(x)$ that describes all the singularities acting on the beam.
3. Integrate the negative load intensity function over the beam length to get the shear force. Integrate the shear force over the beam length to get the moment.
4. Draw shear and moment diagrams from the expressions developed.

Example 2.8

GIVEN The same conditions as in Example 2.7.

FIND Draw the shear and moment diagrams by using a singularity function for a concentrated force located midway on the beam.

Solution

The load intensity function for a simply supported beam [Fig. 2.10(a)] is

$$-q(x) = \frac{P}{2} \langle x \rangle^{-1} - P\left\langle x - \frac{l}{2} \right\rangle^{-1} + \frac{P}{2} \langle x - l \rangle^{-1}$$

The shear expression is

$$V(x) = \int_{-\infty}^{x} \left[\frac{P}{2} \langle x \rangle^{-1} - P\left\langle x - \frac{l}{2} \right\rangle^{-1} + \frac{P}{2} \langle x - l \rangle^{-1} \right] dx$$

$$V(x) = \frac{P}{2}\langle x\rangle^0 - P\left\langle x - \frac{l}{2}\right\rangle^0 + \frac{P}{2}\langle x - l\rangle^0$$

$$V(x) = V_1 - V_2 + V_3$$

Table 2.2 Singularity and load intensity functions with corresponding graphs and expressions

Singularity	Graph of $q(x)$	Expression for $q(x)$
Concentrated moment	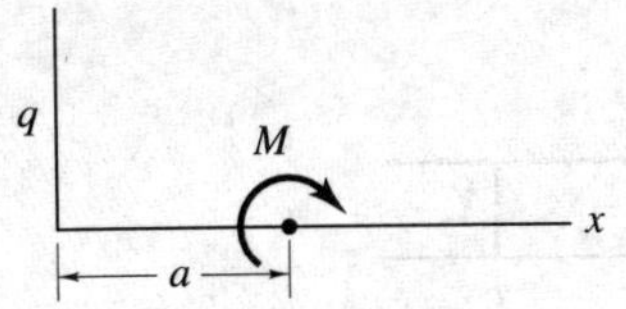	$q(x) = M\langle x - a\rangle^{-2}$
Concentrated force	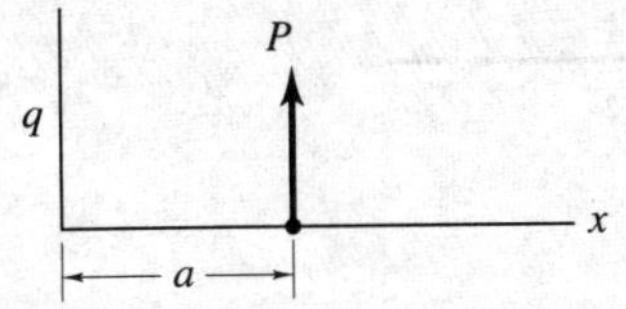	$q(x) = P\langle x - a\rangle^{-1}$
Unit step	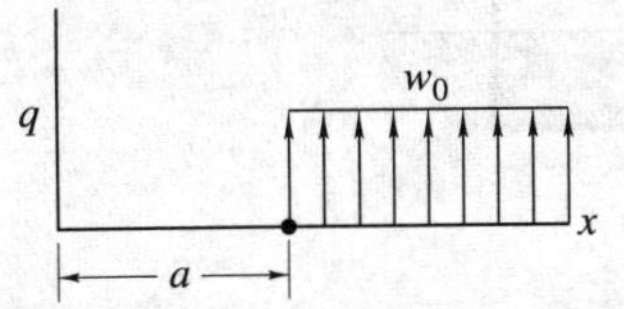	$q(x) = w_0\langle x - a\rangle^0$
Ramp	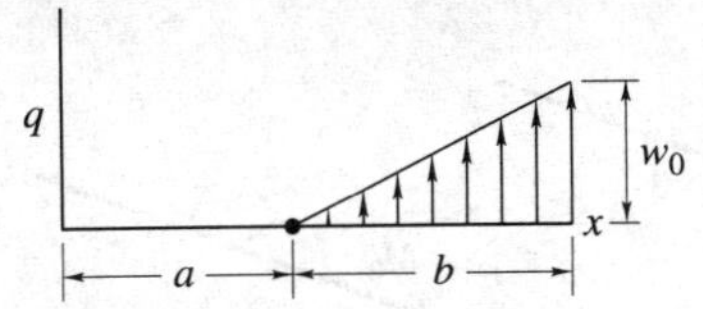	$q(x) = \frac{w_0}{b}\langle x - a\rangle^1$
Inverse ramp	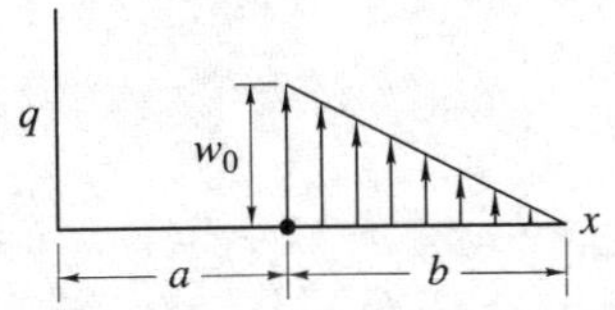	$q(x) = w_0\langle x - a\rangle^0 - \frac{w_0}{b}\langle x - a\rangle^1$
Parabolic shape	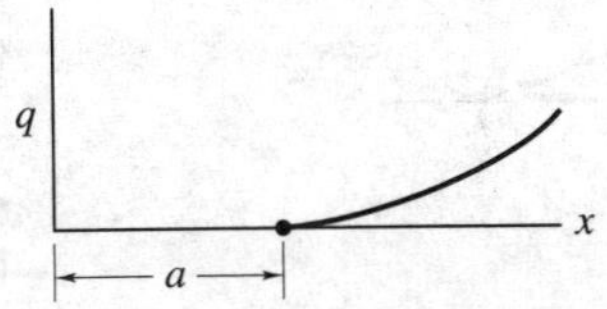	$q(x) = \langle x - a\rangle^2$

Figure 2.10(a) shows the resulting shear diagrams. The diagram at the top shows individual shear, and the diagram below shows the composite of these shear components. The moment expression is

$$M(x) = \int_{-\infty}^{x} \left[\frac{P}{2}\langle x\rangle^{0} - P\left\langle x - \frac{l}{2}\right\rangle^{0} + \frac{P}{2}\langle x - l\rangle^{0} \right] dx$$

$$M(x) = \frac{P}{2}\langle x\rangle^{1} - P\left\langle x - \frac{l}{2}\right\rangle^{1} + \frac{P}{2}\langle x - l\rangle^{1}$$

$$M(x) = M_1 - M_2 + M_3$$

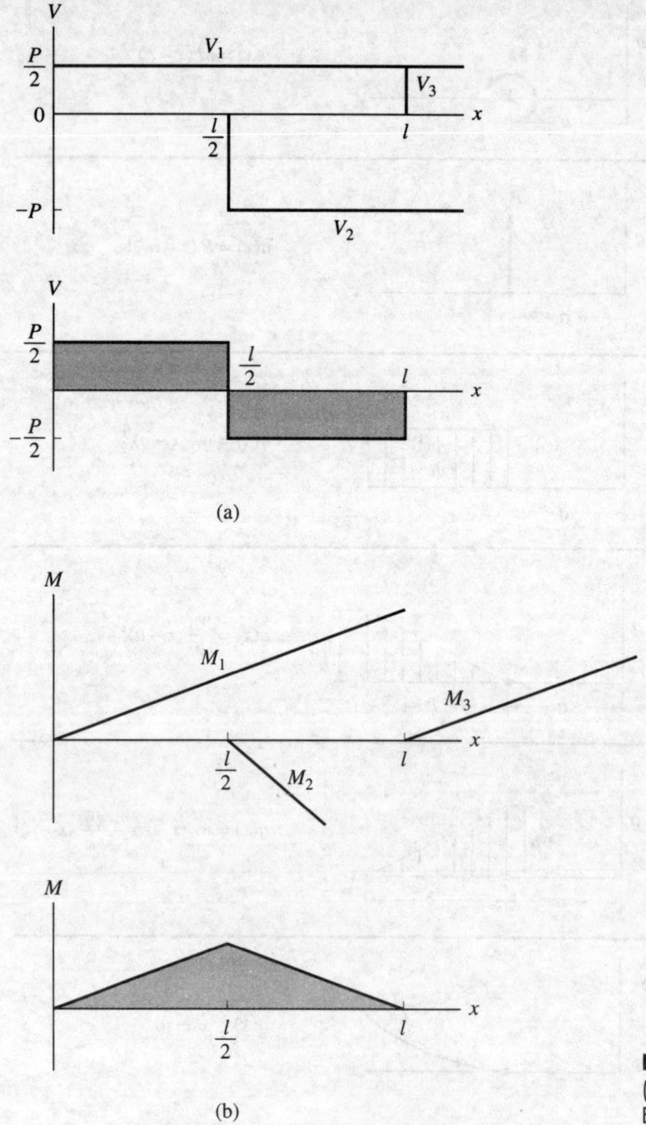

Figure 2.10 (a) Shear and (b) moment diagrams for Example 2.8.

Figure 2.10(b) shows the moment diagrams. The diagram at the top shows individual moments; the diagram at the bottom is the composite moment diagram. The slope of M_2 is twice that of M_1 and M_3, which are equal. The resulting shear and moment diagrams are the same as those found in Example 2.7.

Example 2.9

GIVEN A simply supported beam shown in Fig. 2.11(a) when

$$P_1 = 8 \text{ kN} \qquad P_2 = 5 \text{ kN} \qquad w_0 = 4 \text{ kN/m} \qquad l = 12 \text{ m}$$

FIND The shear and moment expressions as well as their corresponding diagrams while using singularity functions.

Solution

The first task is to solve for the reactions at $x = 0$ and $x = l$. The force representation is shown in Fig. 2.11(b) Note that w_0 is defined as the load per unit length for the central part of the beam. From force equilibrium

$$0 = R_1 + P_1 + P_2 + R_2 - \frac{w_0 l}{2} - \frac{w_0 l}{8} \qquad \textbf{(a)}$$

In Fig. 2.11(b) and Eq. (a) it can be shown that the unit step w_0 over a length of $l/2$ produces a resultant force of $w_0 l/2$ and that the positive ramp over the length of $l/4$ can be represented by a resultant vector of

$$w_0\left(\frac{l}{4}\right)\left(\frac{1}{2}\right) \quad \text{or} \quad \frac{w_0 l}{8}$$

Also, note that the resultant vector acts at

$$x = \left(\frac{2}{3}\right)\left(\frac{l}{4}\right) = \frac{l}{6}$$

$$\therefore\ R_1 + R_2 = -P_1 - P_2 + \frac{5w_0 l}{8} \qquad \textbf{(b)}$$

Making use of moment equilibrium and the moment of a triangular section load gives

$$\frac{(P_1 + 2P_2)l}{4} - \frac{w_0 l^2}{4} - \frac{w_0 l}{8}\left(\frac{l}{6}\right) + R_2 l = 0$$

$$\therefore\ R_2 = \frac{13w_0 l}{48} - \frac{P_1 + 2P_2}{4} \qquad \textbf{(c)}$$

Substituting Eq. (c) into Eq. (b) gives

$$R_1 = -\frac{3P_1}{4} - \frac{P_2}{2} + \frac{17w_0 l}{48} \qquad \textbf{(d)}$$

Substituting the given values for P_1, P_2, w_0, and l gives

$$R_1 = 8.5 \text{ kN} \quad \text{and} \quad R_2 = 8.5 \text{ kN} \qquad \textbf{(e)}$$

The load intensity function can be written as

$$-q(x) = R_1\langle x\rangle^{-1} - \frac{w_0}{l/4}\langle x\rangle^{1} + \frac{w_0}{l/4}\left\langle x - \frac{l}{4}\right\rangle^{1} + P_1\left\langle x - \frac{l}{4}\right\rangle^{-1}$$
$$+P_2\left\langle x - \frac{l}{2}\right\rangle^{-1} + w_0\left\langle x - \frac{3l}{4}\right\rangle^{0} + R_2\langle x - l\rangle^{-1} \qquad \textbf{(f)}$$

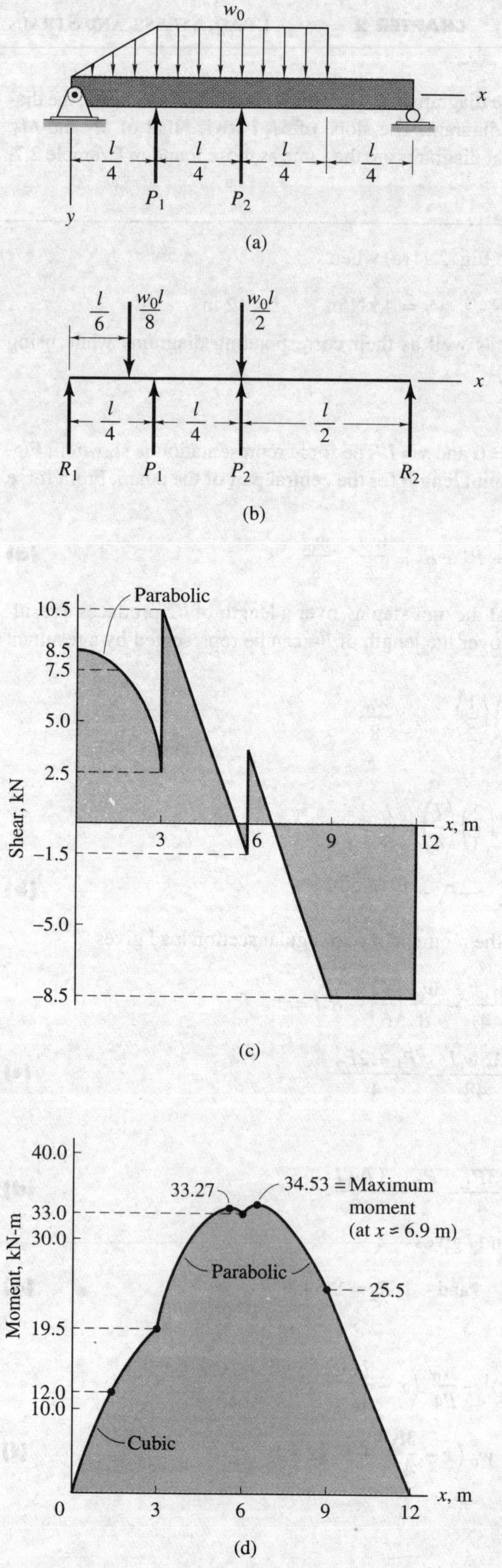

Figure 2.11 Simply supported beam. (a) Forces acting on beam when P_1 = 8 kN, P_2 = 5 kN; w_0 = 4 kN/m; l = 12 m; (b) free-body diagram showing resulting forces; (c) shear and (d) moment diagrams of Example 2.9.

Note that a unit step beginning at $l/4$ is created by initiating a ramp at $x = 0$ acting in the negative direction and summing it with another ramp starting at $x = l/4$ acting in the positive direction while assuming that the slopes of the ramps are the same. The second and third terms on the right side of Eq. (f) produce this effect. The sixth term on the right side of the equation turns off the unit step.

Integrating the load intensity function gives the shear force as

$$V(x) = R_1\langle x\rangle^0 - \frac{2w_0}{l}\langle x\rangle^2 + \frac{2w_0}{l}\left\langle x - \frac{l}{4}\right\rangle^2 + P_1\left\langle x - \frac{l}{4}\right\rangle^0 + P_2\left\langle x - \frac{l}{2}\right\rangle^0 + w_0\left\langle x - \frac{3l}{4}\right\rangle^1 + R_2\langle x - l\rangle^0$$

Integrating the shear force gives the moment, and substituting the values for w_0 and l gives

$$M(x) = 8.5\langle x\rangle^1 - \frac{2}{9}\langle x\rangle^3 + \frac{2}{9}\langle x - 3\rangle^3 + 8\langle x - 3\rangle^1 + 5\langle x - 6\rangle^1 + 2\langle x - 9\rangle^2 + 8.5\langle x - 12\rangle^1$$

The shear and moment diagrams are shown in Fig. 2.11(c) and (d), respectively.

2.10 STRESS

One of the fundamental problems in engineering is determining the effect of a loading environment on a part. This determination is an essential part of the design process; one cannot choose a dimension or a material without first understanding the intensity of force inside the component being analyzed. **Stress** is the term used to define the intensity and direction of the internal forces acting at a given point on a particular plane. Strength, on the other hand, is a property of a material or a part and will be covered in later chapters.

For normal loading on a load-carrying member in which the external load is uniformly distributed over a cross-sectional area of a part, the magnitude of the average normal stress can be calculated from

$$\sigma_{avg} = \frac{\text{Average force}}{\text{Cross-sectional area}} = \frac{P}{A} \quad \textbf{(2.7)}$$

Thus, the unit of stress is force per unit area, or newtons per square meter. Consider a small area ΔA on the cross section, and let ΔP represent the internal forces transmitted by this small area. The average intensity of the internal forces transmitted by the area ΔA is obtained by dividing ΔP by ΔA. If the internal forces transmitted across the section are assumed to be continuously distributed, the area ΔA can be made increasingly smaller and will approach a point on the surface in the limit. The corresponding force ΔP will also become increasingly smaller. The stress at the point on the cross section to which ΔA converges can be defined as

$$\sigma = \lim_{\Delta A \to 0} \frac{\Delta P}{\Delta A} = \frac{dP}{dA} \quad \textbf{(2.8)}$$

The stress at a point acting on a specific plane is a vector and thus has a magnitude and a direction. Its direction is the limiting direction ΔP as area ΔA approaches zero. Similarly, the shear stress can be defined in a specific plane. Thus, a stress must be defined with respect to a direction.

Example 2.10

GIVEN As shown in Fig. 2.12(a) a 3-m-long bar is supported at the left end (B) by a 6-mm-diameter steel wire and at the right end (C) by a 10-mm-diameter steel cylinder. The bar carries a mass m_{a1} = 200 kg and the mass of the bar itself m_{a2} = 50 kg.

FIND Determine the stresses in the wire and in the cylinder.

Solution

$$A_B = \frac{\pi}{4}(6)^2 = 28.27\ \text{mm}^2$$

$$A_C = \frac{\pi}{4}(10)^2 = 78.54\ \text{mm}^2$$

$$m_{a1} = 200\ \text{kg}$$

$$m_{a2} = 50\ \text{kg}$$

Figure 2.12(b) shows a free-body diagram of the forces acting on the bar. From force equilibrium

$$R_B - m_{a1}g - m_{a2}g + R_C = 0$$

$$\therefore R_B + R_C = g(m_{a1} + m_{a2}) = 9.81(200 + 50) = 2452.5\ \text{N}$$

$$\therefore R_B = 2453 - R_C$$

Moment equilibrium about point C gives

$$3R_B = 2(200)(9.81) + 1.5(50)(9.81) = 4660\ \text{N}$$

$$\therefore R_B = 1553\ \text{N}$$

$$R_C = 2453 - 1553 = 900\ \text{N}$$

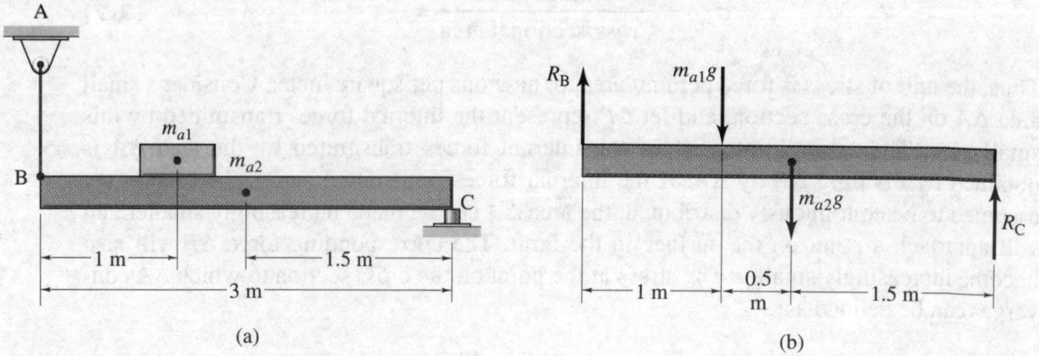

Figure 2.12 Figures used in Example 2.10.
(a) Load assembly drawing; (b) free-body diagram.

The stresses at points B and C are

$$\sigma_B = \frac{R_B}{A_B} = \frac{1553}{28.27} = 54.94\ \text{N/mm}^2 = 54.94\ \text{MPa}$$

$$\sigma_C = -\frac{R_C}{A_C} = -\frac{900}{78.54} = -11.46\ \text{N/mm}^2 = -11.46\ \text{MPa}$$

2.11 STRESS ELEMENT

Figure 2.13 shows a stress element with the origin of stress placed inside the element. Across each of the mutually perpendicular surfaces there are three stresses, yielding a total of nine stress components. Of the three stresses acting on a given surface the normal stress is denoted by σ and the shear stress by τ. A normal stress will receive a subscript indicating the direction in which the stress acts (e.g., σ_x). A shear stress requires two subscripts, the first to indicate the plane of the stress and the second to indicate its direction (e.g., τ_{xy}). The **sign convention for normal stress** distinguishes positive for tension and negative for compression. A positive shear stress points in the positive direction of the coordinate axis denoted by the second subscript if it acts on a surface with an outward normal in the positive direction. The **sign convention for shear stress** is directly associated with the coordinate directions. If both the normal from the surface and the shear are in the positive direction or if both are in the negative direction, the shear stress has a positive sign. Any other combinations of the normal and the direction of shear will produce a negative shear stress. The surface stresses of an element have the following relationships:

1. The normal and shear stress components acting on opposite sides of an element must be equal in magnitude but opposite in direction.
2. Moment equilibrium requires that the shear stresses be symmetric, implying that the subscripts can be reversed in order, or

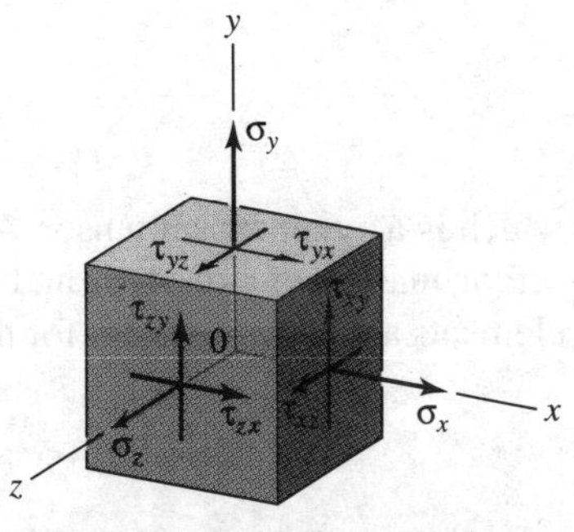

Figure 2.13 Stress element showing general state of three-dimensional stress with origin placed in center of element.

$$\tau_{xy} = \tau_{yx} \qquad \tau_{xz} = \tau_{zx} \qquad \tau_{yz} = \tau_{zy} \tag{2.9}$$

thus reducing the nine different stresses acting on the element to six: three normal stresses σ_x, σ_y, σ_z and three shear stresses τ_{xy}, τ_{yz}, τ_{xz}.

Example 2.11 **GIVEN** The stress element shown in Fig. 2.13 is put into a pressure vessel pressurized to 10 MPa. The shear stress acting on the bottom surface is directed in the positive x direction.

FIND Are the stresses positive or negative?

Solution

The definition of a positive normal stress is a tensile stress, and a positive shear stress is directed in the positive coordinate direction when the normal to the surface is directed in the positive coordinate direction. A shear stress acting on a surface with the normal in the negative coordinate direction is positive when the stress is directed in the negative coordinate direction. The normal stress here is thus

$$\sigma = -10 \text{ MPa}$$

The shear stress acts on a surface with the normal in the negative y direction, but the stress is directed in the positive x direction. Thus, the shear is negative.

The general laws of stress transformation, given in Appendix B (Sec. B.1), enable the determination of stresses acting on any new orthogonal coordinate system.

2.12 Stress Tensor

In engineering courses it is common to encounter scalar quantities, those that have numerical value. Vectors, such as force, have a magnitude as well as a direction. Stress requires six quantities for its definition; thus, stress is a tensor. From the stress element of Fig. 2.13 and Eqs. (2.9) the stress tensor is

$$\mathbf{S} = \begin{pmatrix} \sigma_x & \tau_{xy} & \tau_{xz} \\ \tau_{xy} & \sigma_y & \tau_{yz} \\ \tau_{xz} & \tau_{yz} & \sigma_z \end{pmatrix} \tag{2.10}$$

which is a symmetrical tensor. A property of a **symmetrical tensor** is that there exists an orthogonal set of axes 1, 2, and 3 (called principal axes) with respect to which the tensor elements are all zero except for those in the principal diagonal; thus,

$$\mathbf{S}' = \begin{pmatrix} \sigma_1 & 0 & 0 \\ 0 & \sigma_2 & 0 \\ 0 & 0 & \sigma_3 \end{pmatrix} \tag{2.11}$$

where σ_1, σ_2, and σ_3 = principal stresses ($\sigma_1 \geq \sigma_2 \geq \sigma_3$). Note that no principal shear stresses occur in Eq. (2.11).

The combination of the applied normal and shear stresses that produces the maximum normal stress is called the maximum principal stress σ_1. The combination of the applied stresses that produces the minimum normal stress is called the minimum principal stress σ_3. Knowing the maximum and minimum **principal normal stresses** is important in engineering design. If the design satisfies these limits, it will satisfy any other stress.

2.13 PLANE STRESS

Many cases of stress analysis can be simplified to the case of plane stress, where one surface is comparatively free of stress. Another common practice is to instrument strain gages onto structural members. Since strain gages are easily placed onto free surfaces, they are placed in locations of plane stress. Although other locations in the geometry may have more complicated loadings, such measurements are often used to confirm loading environments or to ensure that stresses or strains are not excessive. Most stress analyses of machine elements involve one surface that is comparatively free of stress. The third direction can thus be neglected, and all stresses on the stress element act on two pairs of faces rather than three, as shown in Fig. 2.14. This two-dimensional stress state is sometimes called **biaxial stress** or **plane stress.**

In comparing the two views of the plane stress element shown in Fig. 2.14, note that all stresses shown in Fig. 2.14(b) act on planes perpendicular to the paper, with the paper being designated as either the *x-y* plane or the *z* plane. The stresses shown in Fig. 2.14 all have positive signs in compliance with the conventions presented in Sec. 2.11.

The magnitude of stress depends greatly on the orientation of the coordinate system. For example, consider the stress element shown in Fig. 2.15(a). When a uniform stress is applied to the sheet, the stress state is clearly $\sigma_x = \sigma_0$, $\sigma_y = 0$, and $\tau_{xy} = 0$. However, if the original orientation of the element were as shown in Fig. 2.15(b), this would no longer be the case, and all stress components in the plane would be nonzero. A profound question can be raised at this point, How does the material know the difference between these stress

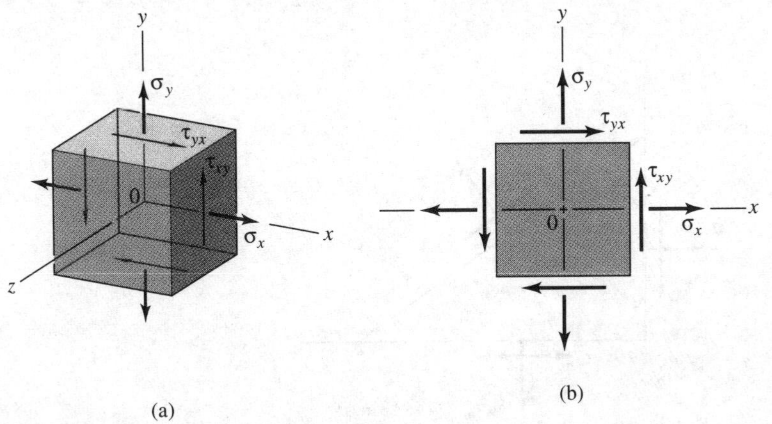

Figure 2.14 Stress element showing two-dimensional state of stress. (a) Three-dimensional view; (b) plane view.

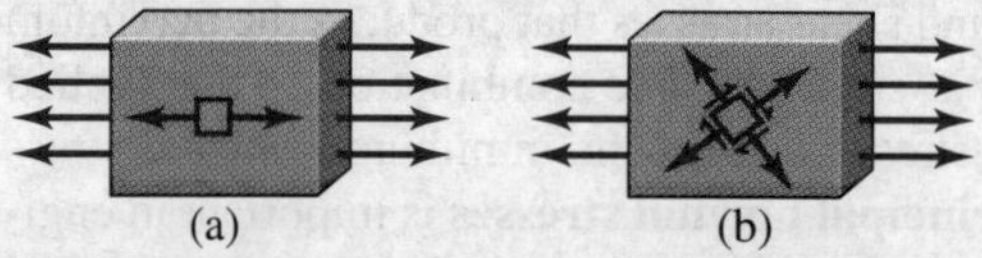

Figure 2.15 Illustration of equivalent stress states; (a) Stress element oriented in the direction of applied stress; (b) stress element oriented in different (arbitrary) direction.

states? The answer is that there is no difference between the stresses of Fig. 2.15, so that these stress states are equivalent. Obviously, it is of great importance to be able to transform stresses from one orientation to another, and the resultant stress transformation equations will be of great use throughout the remainder of the text.

If instead of the stresses acting as shown in Fig. 2.14(b) they act in an oblique plane at angle ϕ as shown in Fig. 2.16, the stresses σ_x, σ_y, and τ_{xy} must be found in terms of the stresses on an inclined surface whose normal stress makes an angle ϕ with the x axis, as shown in Fig. 2.16. Angle ϕ is arbitrarily chosen, and the object of the analysis that follows is to establish what values σ and τ have on the inclined surface.

Note from Fig. 2.16 that if the area of the inclined surface is A (length of the surface times the thickness into the paper), the area of the horizontal side of the triangular element will be $A \sin \phi$, and the area of the vertical side, $A \cos \phi$. From force equilibrium

$$\sigma_\phi A = \tau_{xy} \sin \phi A \cos \phi + \tau_{yx} \cos \phi A \sin \phi + \sigma_x \cos \phi A \cos \phi + \sigma_y \sin \phi A \sin \phi$$

This reduces to

$$\sigma_\phi = 2\tau_{xy} \sin \phi \cos \phi + \sigma_x \cos^2\phi + \sigma_y \sin^2\phi \tag{2.12}$$

By using trigonometric identities for the double angle Eq. (2.12) can be written as

$$\sigma_\phi = \frac{\sigma_x + \sigma_y}{2} + \frac{\sigma_x - \sigma_y}{2} \cos 2\phi + \tau_{xy} \sin 2\phi \tag{2.13}$$

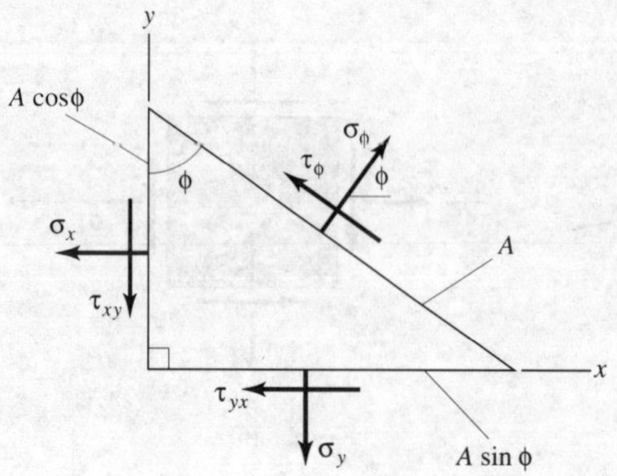

Figure 2.16 Stresses in oblique plane at angle ϕ.

Similarly, from force equilibrium the shear stress in the oblique plane can be expressed as

$$\tau_\phi = \tau_{xy} \cos 2\phi - \frac{\sigma_x - \sigma_y}{2} \sin 2\phi \tag{2.14}$$

Example 2.12

GIVEN A thin, square steel plate has normals to the sides in the x and y directions. A tensile stress σ acts on the four sides. Thus, $\sigma_x = \sigma_y = \sigma$.

FIND The normal and shear stresses acting on the diagonal of the square.

Solution

From Eq. (2.12)

$$\sigma_\phi = \frac{\sigma_x + \sigma_y}{2} + \frac{\sigma_x - \sigma_y}{2} \cos 2\phi + \tau_{xy} \sin 2\phi$$

$$\therefore \sigma_{45^\circ} = \frac{\sigma + \sigma}{2} + \frac{\sigma - \sigma}{2} \cos 90^\circ + \tau_{xy} \sin 90^\circ$$

$$\sigma_{45^\circ} = \sigma$$

Similarly, from Eq. (2.14)

$$\tau_\phi = \tau_{xy} \cos 2\phi - \frac{\sigma_x - \sigma_y}{2} \sin 2\phi$$

$$\therefore \tau_{45^\circ} = \tau_{xy} \cos 90^\circ - \frac{\sigma - \sigma}{2} \sin 90^\circ$$

$$\tau_{45^\circ} = 0$$

Equations (2.13) and (2.14) have maximum and minimum values that are of particular interest in stress analysis. The angle ϕ_σ, which gives the extreme value of σ_ϕ, can be determined by differentiating σ_ϕ with respect to ϕ and setting the result equal to zero, giving

$$\frac{d\sigma_\phi}{d\phi} = -(\sigma_x - \sigma_y) \sin 2\phi_\sigma + 2\tau_{xy} \cos 2\phi_\sigma = 0$$

$$\tan 2\phi_\sigma = \frac{2\tau_{xy}}{\sigma_x - \sigma_y} \tag{2.15}$$

where ϕ_σ = angle ϕ where normal stress is extreme. Equation (2.15) has two roots, since the tangents of angles in diametrically opposite quadrants are the same. The roots are 180° apart, and for the double-angle nature of the left side of Eq. (2.15) this suggests roots of ϕ_σ being 90° apart. One of these roots corresponds to the maximum value of normal stress; the other to the minimum value.

Substituting Eq. (2.15) into Eqs. (2.13) and (2.14) gives the following after some algebraic manipulation:

$$\sigma_1, \sigma_2 = \frac{\sigma_x + \sigma_y}{2} \pm \sqrt{\tau_{xy}^2 + \frac{(\sigma_x - \sigma_y)^2}{4}} \quad \textbf{(2.16)}$$

$$\tau_{\phi_\sigma} = 0 \quad \textbf{(2.17)}$$

At this stress element orientation, where the normal stresses are extreme, the shear stress is zero. The axes that define this orientation are called the principal axes, and the normal stresses from Eq. (2.16) are called the **principal normal stresses**. Principal stresses are given numerical subscripts to differentiate them from stresses at any other orientation. A common convention is to order the principal stresses according to $\sigma_1 \geq \sigma_2 \geq \sigma_3$. In plane stress one of the principal stresses is always zero.

Another orientation of interest is the one where the shear stress takes an extreme value. Differentiating Eq. (2.14) with respect to ϕ and solving for τ gives the orientation ϕ_τ, where the shear reaches an extreme value. The stresses that result at this orientation are

$$\tau_{max}, \tau_{min} = \tau_1, \tau_2 = \pm\sqrt{\left(\frac{\sigma_x - \sigma_y}{2}\right)^2 + \tau_{xy}^2} \quad \textbf{(2.18)}$$

$$\sigma_{\phi_\tau} = \frac{\sigma_x + \sigma_y}{2}$$

The shear stresses from Eq. (2.18) are called **principal shear stresses**. Thus, on the stress element oriented to achieve a maximum shear stress, the normal stresses on the two faces are equal. Also, it can be shown that

$$|\phi_\tau - \phi_\sigma| = \pi/4 \quad \textbf{(2.19)}$$

In summary, for a plane stress situation where σ_x, σ_y, and τ_{xy} are known,

1. The normal and shear stresses σ_ϕ and τ_ϕ can be determined for any oblique plane at angle ϕ from Eqs. (2.13) and (2.14).
2. The principal normal and shear stresses σ_1, σ_2, τ_1, and τ_2 can be determined from Eqs. (2.16) and (2.18).

If the principal normal stresses σ_1 and σ_2 are known, the normal and shear stresses at any oblique plane at angle ϕ can be determined from the following equations:

$$\sigma_\phi = \frac{\sigma_1 + \sigma_2}{2} + \frac{\sigma_1 - \sigma_2}{2}\cos 2\phi \quad \textbf{(2.20)}$$

$$\tau_\phi = \frac{\sigma_1 - \sigma_2}{2}\sin 2\phi \quad \textbf{(2.21)}$$

In Eq. (2.21) a second subscript is not present and is not needed because τ_ϕ represents a shear stress acting on any oblique plane at angle ϕ as shown in Fig. 2.16.

2.14 MOHR'S CIRCLE

Mohr's circle for a triaxial state of stress at a point was first constructed by a German engineer, Otto Mohr [Mohr (1914)], who noted that Eqs. (2.13) and (2.14) define a circle in a σ-τ plane. This circle is used extensively as a convenient method of graphically visualizing the state of stress acting in different planes passing through a given point. The approach used in this text is first to apply Mohr's circle to a two-dimensional stress state and then in Sec. 2.15 to apply it to a three-dimensional stress state. Indeed, Mohr's circle is most useful for stress visualization in plane stress situations.

The main value of Mohr's circle is that it makes visual the realities of the stress state in a specific situation. Figure 2.17 shows a typical Mohr's circle diagram. Note that the circle is symmetrical about the normal stress axis (the abscissa).

A number of points should be made about the Mohr's circle diagram:

1. Normal stresses are plotted along the abscissa (x axis), and shear stresses are plotted as the ordinate (y axis).
2. The circle defines all stress states that are equivalent.

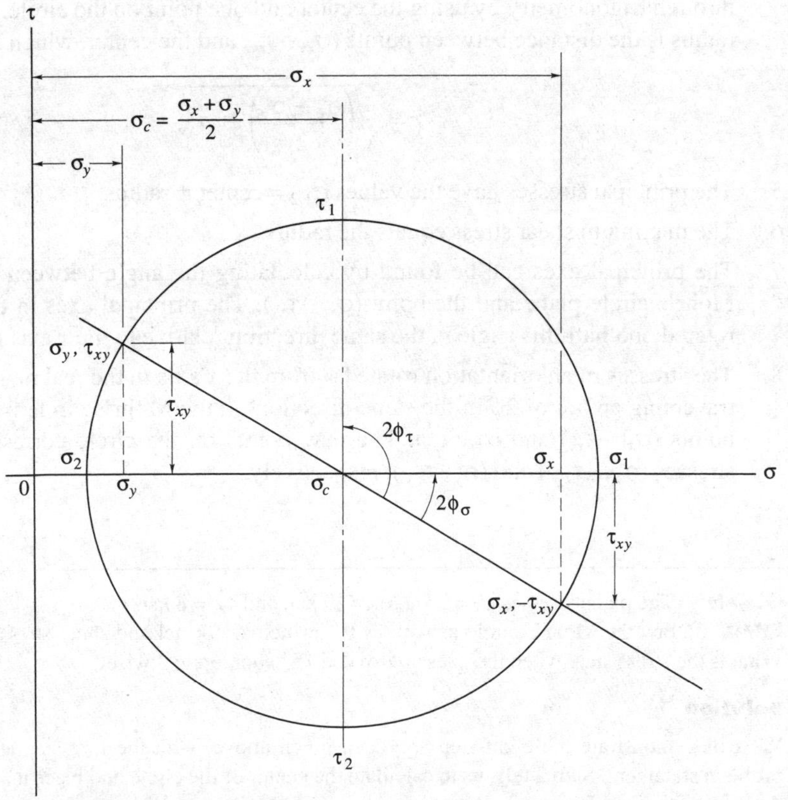

Figure 2.17 Mohr's circle diagram of Eqs. (2.13) and (2.14).

3. The biaxial stress state for any direction can be scaled directly from the circle.
4. The principal normal stresses (i.e., the extreme values of normal stress) are at the locations where the circle intercepts the x axis.
5. The maximum shear stress equals the radius of the circle.
6. A rotation from a reference stress state in the real plane of ϕ corresponds to a rotation of 2ϕ from the reference points in the Mohr's circle plane.

The steps in constructing and using Mohr's circle in two dimensions are as follows:

1. Calculate the plane stress state for any x-y coordinate system so that σ_x, σ_y, and τ_{xy} are known.
2. The center of the Mohr's circle can be placed at

$$\left(\frac{\sigma_x + \sigma_y}{2}, 0\right) \tag{2.22}$$

3. Two points diametrically opposite to each other on the circle correspond to the points $(\sigma_x, -\tau_{xy})$ and (σ_y, τ_{xy}). Using the center and either point allows one to draw the circle.
4. The radius of the circle can be calculated from stress transformation equations or through trigonometry by using the center and one point on the circle. For example, the radius is the distance between points $(\sigma_x, -\tau_{xy})$ and the center, which directly leads to

$$r = \sqrt{\left(\frac{\sigma_x - \sigma_y}{2}\right)^2 + \tau_{xy}^2} \tag{2.23}$$

5. The principal stresses have the values $\sigma_{1,2}$ = center ± radius.
6. The maximum shear stress equals the radius.
7. The principal axes can be found by calculating the angle between the x axis in the Mohr's circle plane and the point $(\sigma_x, -\tau_{xy})$. The principal axes in the real plane are rotated one-half this angle in the same direction relative to the x axis in the real plane.
8. The stresses in an orientation rotated ϕ from the x axis in the real plane can be read by traversing an arc of 2ϕ in the same direction on the Mohr's circle from the reference points $(\sigma_x, -\tau_{xy})$ and (σ_y, τ_{xy}). The new points on the circle correspond to the new stresses $(\sigma_{x'}, -\tau_{x'y'})$ and $(\sigma_{y'}, \tau_{x'y'})$, respectively.

Example 2.13

GIVEN The plane stresses $\sigma_x = 9$ ksi, $\sigma_y = 19$ ksi, and $\tau_{xy} = 8$ ksi.

FIND Draw the Mohr's circle as well as the principal normal and shear stresses on the x-y axis. What is the stress state when the axes are rotated 15° counterclockwise?

Solution

We will demonstrate the eight-step approach given above, with the first step already done in the problem statement. Step 2 tells us to calculate the center of the circle and place it at $(\sigma_c, 0)$, where

$$\sigma_c = \frac{\sigma_x + \sigma_y}{2} = \frac{(9 + 19)10^3}{2} \text{ psi} = 14 \text{ ksi}$$

Step 3 tells us to use either point $(\sigma_x, -\tau_{xy})$ or (σ_y, τ_{xy}) to draw the circle. We use the point $(\sigma_x, -\tau_{xy})$ = (9 ksi, −8 ksi) to draw the circle as shown. From step 4 and from the triangle defined by the x axis and the point $(\sigma_x, -\tau_{xy})$ we can calculate the radius as

$$r = \sqrt{(9-14)^2 + (-8)^2} = 9.43 \text{ ksi}$$

From step 5 the principal stresses have the values $\sigma_{1,2} = 14 \pm 9.43$, or $\sigma_1 = 23.43$ ksi and $\sigma_2 = 4.57$ ksi. Step 6 tells us that the maximum shear stress equals the radius, or $\tau_{max} = 9.43$ ksi. We can calculate the principal stress orientation, if desired, from trigonometry. In the Mohr's circle plane (Fig. 2.18(a)) the point $(\sigma_x, -\tau_{xy})$ makes an angle of $2\phi = \tan^{-1}(8/5) = 58°$ with the x axis. To reach the point on the x axis, we need to sweep an arc of this angle in the clockwise direction on the Mohr's circle. Thus, the principal plane is $\phi = 29°$ clockwise from the x axis. Finally, the stresses at an angle of 15° can be obtained by rotating 30° on the Mohr's circle, leading to

$$\sigma_{x'} = 14 \text{ ksi} - (9.43 \text{ ksi}) \cos 28° = 5.67 \text{ ksi}$$

$$\sigma_{y'} = 14 \text{ ksi} + (9.43 \text{ ksi}) \cos 28° = 22.32 \text{ ksi}$$

$$\tau_{x'y'} = (9.43 \text{ ksi}) \sin 28° = 4.43 \text{ ksi}$$

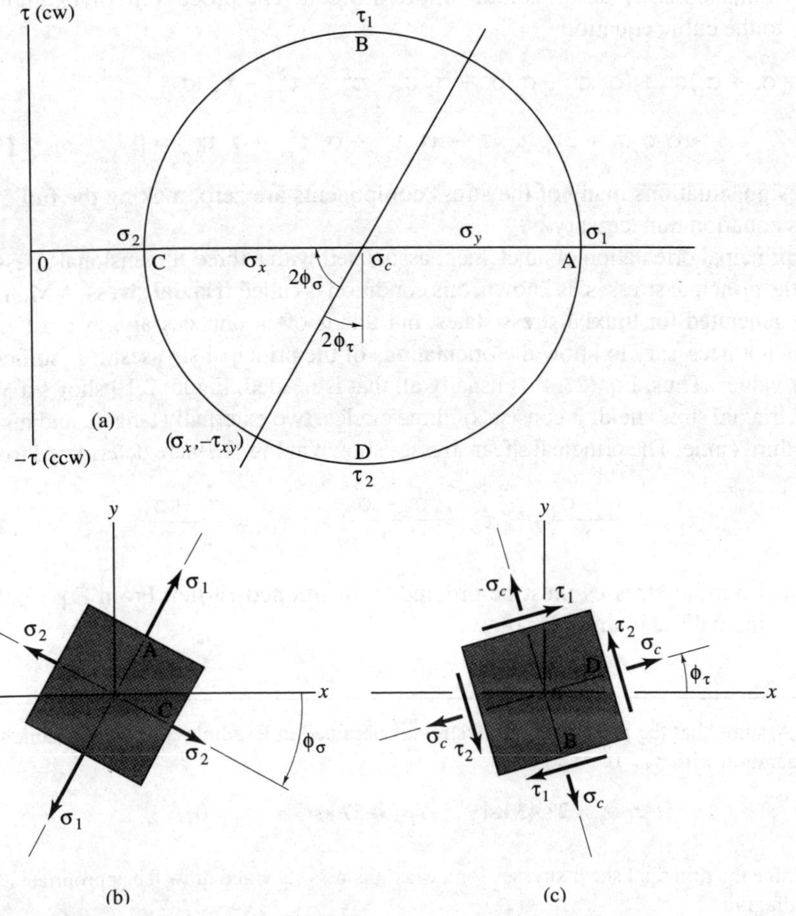

Figure 2.18 Results from Example 2.13.
(a) Mohr's circle diagram; (b) stress element for principal normal stresses shown in x-y coordinates; (c) stress element for principal shear stresses shown in x-y coordinates.

Figure 2.18(b) shows an element of the principal normal stresses as well as the appropriate value of ϕ_σ. Figure 2.18(c) shows an element of the principal shear stresses as well as the appropriate value of ϕ_τ. The stress at the center of the Mohr's circle diagram is also represented in Fig. 2.18(c) along with the principal shear stresses. Also, the stress element is always oriented from the *x* axis to the principal shear stress on the Mohr's circle diagram.

2.15 Three-Dimensional Stresses

Considering the general situation shown in Fig. 2.13, the stress element has six faces, implying that there are three principal directions and three principal stresses σ_1, σ_2, and σ_3. Six stress components (σ_x, σ_y, σ_z, τ_{xy}, τ_{xz}, and τ_{yz}) are required to specify a general state of stress in three dimensions, in contrast to the three stress components (σ_x, σ_y, and τ_{xy}) that were used for two-dimensional (plane or *biaxial*) stress. Determining the principal stresses for a three-dimensional situation is much more difficult. The process involves finding the three roots to the cubic equation

$$\sigma^3 - (\sigma_x + \sigma_y + \sigma_z)\sigma^2 + (\sigma_x\sigma_y + \sigma_x\sigma_z + \sigma_y\sigma_z - \tau_{xy}^2 - \tau_{yz}^2 - \tau_{zx}^2)\sigma$$

$$-(\sigma_x\sigma_y\sigma_z + 2\tau_{xy}\tau_{yz}\tau_{zx} - \sigma_x\tau_{yz}^2 - \sigma_y\tau_{zx}^2 - \sigma_z\tau_{xy}^2) = 0 \qquad \textbf{(2.24)}$$

In most design situations many of the stress components are zero, making the full evaluation of this equation unnecessary.

If the principal orientation of an element associated with a three-dimensional stress state, as well as the principal stresses, is known, this condition is called **triaxial stress**. A Mohr's circle can be generated for triaxial stress states, but this is often unnecessary. In most circumstances it is not necessary to know the orientations of the principal stresses; it is sufficient to know their values. Thus, Eq. (2.24) is usually all that is needed. Figure 2.19 shows a Mohr's circle for a triaxial stress field. It consists of three circles, two externally tangent and inscribed within the third circle. The principal shear stresses shown in Fig. 2.19 are determined from

$$\tau_{1/2} = \frac{\sigma_1 - \sigma_2}{2} \qquad \tau_{2/3} = \frac{\sigma_2 - \sigma_3}{2} \qquad \tau_{1/3} = \frac{\sigma_1 - \sigma_3}{2} \qquad \textbf{(2.25)}$$

The principal normal stresses must be ordered as mentioned earlier. From Eq. (2.25) the maximum principal shear stress is $\tau_{1/3}$.

Example 2.14

GIVEN Assume that the principal normal stresses obtained in Example 2.13 are the same for triaxial consideration with $\sigma_3 = 0$.

$$\therefore \sigma_1 = 23.43 \text{ ksi} \qquad \sigma_2 = 4.57 \text{ ksi} \qquad \sigma_3 = 0$$

FIND

a) Determine the principal shear stresses for a triaxial stress state and draw the appropriate Mohr's circle diagram.

b) If the shear stress τ_{xy} is changed from 8 to 16 ksi, show how the Mohr's circles for the biaxial and triaxial stress states change.

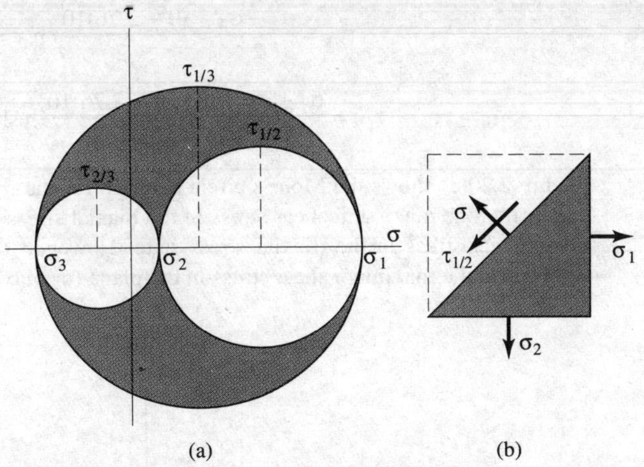

Figure 2.19 Mohr's circle for triaxial stress state. (a) Mohr's circle representation; (b) principal stresses on two planes.

Solution

a) From Eq. (2.25) the principal shear stresses in a triaxial stress state are

$$\tau_{1/2} = \frac{\sigma_1 - \sigma_2}{2} = \frac{(23.43 - 4.57)10^3}{2} \text{ psi} = 9.43 \text{ ksi}$$

$$\tau_{2/3} = \frac{\sigma_2 - \sigma_3}{2} = \frac{(4.57)10^3}{2} \text{ psi} = 2.285 \text{ ksi}$$

$$\tau_{1/3} = \frac{\sigma_1 - \sigma_3}{2} = \frac{(23.43 - 0)10^3}{2} \text{ psi} = 11.765 \text{ ksi}$$

Figure 2.20(a) shows the appropriate Mohr's circle diagram for the triaxial stress state.

b) If the shear stress in Example 2.13 is doubled (τ_{xy} = 16 ksi instead of 8 ksi), Eq. (2.23) gives

$$\tau_1, \tau_2 = \pm\sqrt{\tau_{xy}^2 + \left(\frac{\sigma_x - \sigma_y}{2}\right)^2} = \pm 10^3\sqrt{16^2 + \left(\frac{9 - 19}{2}\right)^2} \text{ psi} = \pm 16.76 \text{ ksi}$$

The principal normal stresses for the biaxial stress state are

$$\sigma_1 = \sigma_c + \tau_1 = 30.76 \text{ ksi}$$

$$\sigma_2 = \sigma_c + \tau_2 = -2.76 \text{ ksi}$$

Figure 2.20(b) shows the resultant Mohr's circle diagram for the biaxial stress state. In a triaxial stress state that is ordered σ_1 = 30.76 ksi, σ_2 = 0, and σ_3 = –2.76 ksi, from Eq. (2.25) the principal shear stresses can be written as

$$\tau_{1/2} = \frac{\sigma_1 - \sigma_2}{2} = \frac{(30.76 - 0)10^3}{2} \text{ psi} = 15.38 \text{ ksi}$$

$$\tau_{2/3} = \frac{\sigma_2 - \sigma_3}{2} = \frac{(0 + 2.76)10^3}{2} \text{ psi} = 1.38 \text{ ksi}$$

$$\tau_{1/3} = \frac{\sigma_1 - \sigma_3}{2} = \frac{(30.76 + 2.76)10^3}{2} \text{ psi} = 16.76 \text{ ksi}$$

Figure 2.20(c) shows the Mohr's circle diagram for the triaxial stress state. From Fig. 2.20(b) and (c) τ_1, the maximum shear stress in the biaxial stress state, is equivalent to $\tau_{1/3}$, the maximum shear stress in the triaxial stress state. However, comparing Figs. 2.18(a) and 2.20(a) shows that the maximum shear stress in the plane (or biaxial) stress state is not equal to that in

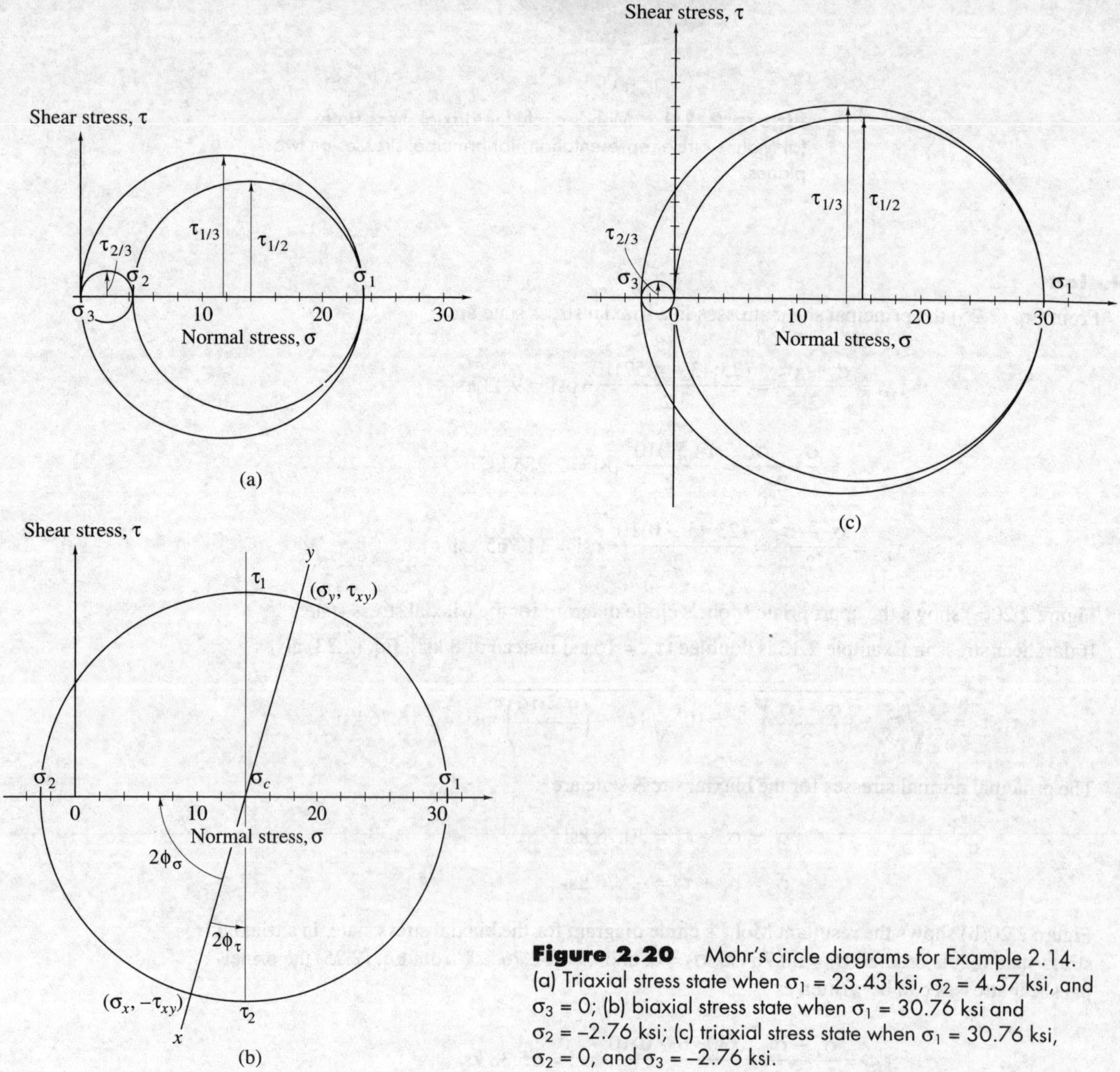

Figure 2.20 Mohr's circle diagrams for Example 2.14. (a) Triaxial stress state when $\sigma_1 = 23.43$ ksi, $\sigma_2 = 4.57$ ksi, and $\sigma_3 = 0$; (b) biaxial stress state when $\sigma_1 = 30.76$ ksi and $\sigma_2 = -2.76$ ksi; (c) triaxial stress state when $\sigma_1 = 30.76$ ksi, $\sigma_2 = 0$, and $\sigma_3 = -2.76$ ksi.

the triaxial stress state. Furthermore, the maximum triaxial stress is larger than the maximum biaxial stress. Thus, if σ_1 and σ_2 have the same sign in the biaxial stress state, the triaxial maximum stress $\tau_{1/3}$ must be used for design considerations. However, if σ_1 and σ_2 have opposite signs in the biaxial stress state, the maximum biaxial and triaxial shear stresses will be the same and either one can be used in the analysis.

2.16 OCTAHEDRAL STRESSES

Sometimes it is advantageous to represent the stresses on an octahedral stress element rather than on a conventional cubic element of principal stresses. Figure 2.21 shows the orientation of one of the eight octahedral planes that are associated with a given stress state. Each **octahedral plane** cuts across a corner of a principal element, so that the eight planes together form an octahedron (Fig. 2.21). The following characteristics of the stresses on a octahedral plane should be noted:

1. Identical normal stresses act on all eight planes. Thus, the normal stresses tend to compress or enlarge the octahedron but not distort it.
2. Identical shear stresses act on all eight planes. Thus, the shear stresses tend to distort the octahedron without changing its volume.

The fact that the normal and shear stresses are the same for the eight planes is a powerful tool in failure analysis.

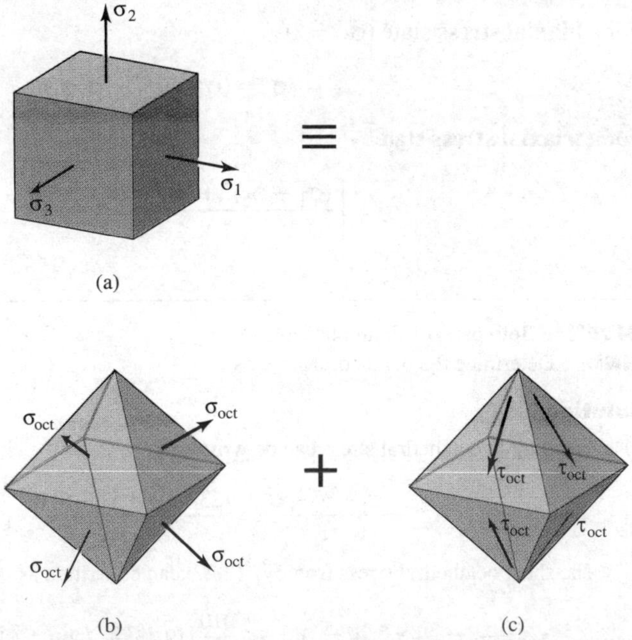

Figure 2.21 Stresses acting on octahedral planes.
(a) General state of stress; (b) normal stress; (c) octahedral shear stress.

The normal octahedral stress can be expressed in terms of the principal normal stresses, or the stresses in the x, y, z coordinates, as

$$\sigma_{oct} = \frac{\sigma_1 + \sigma_2 + \sigma_3}{3} = \frac{\sigma_x + \sigma_y + \sigma_z}{3} \tag{2.26}$$

$$\tau_{oct} = \frac{1}{3}[(\sigma_1 - \sigma_2)^2 + (\sigma_2 - \sigma_3)^2 + (\sigma_3 - \sigma_1)^2]^{1/2}$$

$$+ \frac{2}{3}[\tau_{1/2}^2 + \tau_{2/3}^2 + \tau_{1/3}^2]^{1/2} \tag{2.27}$$

$$\tau_{oct} = \frac{1}{3}[(\sigma_x - \sigma_y)^2 + (\sigma_y - \sigma_z)^2 + (\sigma_z - \sigma_x)^2 + 6(\tau_{xy}^2 + \tau_{yz}^2 + \tau_{xz}^2)]^{1/2} \tag{2.28}$$

Equations (2.26) to (2.28) are derived in Durelli et al. (1958).

Note that the octahedral shear stress can be expressed in terms of the von Mises stress as

$$\tau_{oct} = \frac{\sqrt{2}}{3}\sigma_e \tag{2.29}$$

where

σ_e = von Mises stress, Pa

For a **uniaxial stress** state ($\sigma_2 = \sigma_3 = 0$)

$$\sigma_e = \sigma_1 \tag{2.30}$$

For a **biaxial stress** state ($\sigma_3 = 0$)

$$\sigma_e = (\sigma_1^2 + \sigma_2^2 - \sigma_1\sigma_2)^{\frac{1}{2}} \tag{2.31}$$

For a **triaxial stress** state

$$\sigma_e = \left[\frac{(\sigma_1 - \sigma_2)^2 + (\sigma_1 - \sigma_3)^2 + (\sigma_2 - \sigma_3)^2}{2}\right]^{\frac{1}{2}} \tag{2.32}$$

Example 2.15

GIVEN Both parts of Example 2.14.

FIND Determine the octahedral stresses.

Solution

a) The normal octahedral stress can be written as

$$\sigma_{oct} = \frac{\sigma_1 + \sigma_2 + \sigma_3}{3} = \frac{(23.43 + 4.57 + 0)10^3}{3} \text{ psi} = 9.33 \text{ ksi}$$

The shear octahedral stress from Eq. (2.27) can be written as

$$\tau_{oct} = \frac{2}{3}[\tau_{1/2}^2 + \tau_{2/3}^2 + \tau_{1/3}^2]^{1/2} = \frac{(2)10^3}{3}[9.43^2 + 2.29^2 + 11.77^2]^{1/2} \text{ psi} = 10.17 \text{ ksi}$$

Comparing these results with those from Examples 2.13 and 2.14 gives

$$\sigma_{1,b} = \sigma_{1,t} > \sigma_{oct} \quad \text{and} \quad \tau_{1,b} < \tau_{oct} < \tau_{1,t}$$

where

subscript b = biaxial stress
subscript t = triaxial stress

b) Only the shear stresses change in this part of the example.

$$\tau_{oct} = \frac{2}{3}[\tau_{1/2}^2 + \tau_{2/3}^2 + \tau_{1/3}^2]^{1/2} = \frac{(2)10^3}{3}[15.38^2 + 1.38^2 + 16.76^2]^{1/2}\ \text{psi} = 15.19\ \text{ksi}$$

Comparing this result with those from Examples 2.10 and 2.11 gives

$$\tau_{oct} < \tau_{1,b} = \tau_{1,t}$$

Thus, in comparing the principal triaxial stresses with the octahedral stresses,

$$\sigma_{1,t} > \sigma_{oct} \quad \text{and} \quad \tau_{1,t} = \tau_{1/3} > \tau_{oct}$$

The triaxial maximum principal normal and shear stresses are greater than their octahedral counterparts.

2.17 STRAIN

Strain may be defined as the displacement per length produced in a solid as the result of stress. In designing a machine element, not only do engineers need to ensure that the design is adequate when considering the stress relative to the strength, but they also need to ensure that the displacements, or deformations, are not excessive and are within limits from the standpoint of fracture or yielding. Depending on the application, these deformations may be either highly visible or practically unnoticeable. For the same stress large deformations are clearly visible in components made of natural rubber but much less visible in the same components made of steel.

Just as the direction and intensity of the stress at any given point are important with respect to a specific plane passing through that point, the same is true for strain. Thus, as for stress the state of strain is a tensor. Also, just as there are normal and shear stresses, so too there are normal and shear strains. **Normal strain,** designated by the symbol ε, is used to describe a measure of the elongation or contraction of a linear segment of an element in which stress is applied. The average normal strain is

$$\varepsilon_{avg} = \frac{d_{avg}}{l} = \frac{\text{Average elongation}}{\text{Original length}} \tag{2.33}$$

Note that strain is dimensionless. Furthermore, the strain at a point is

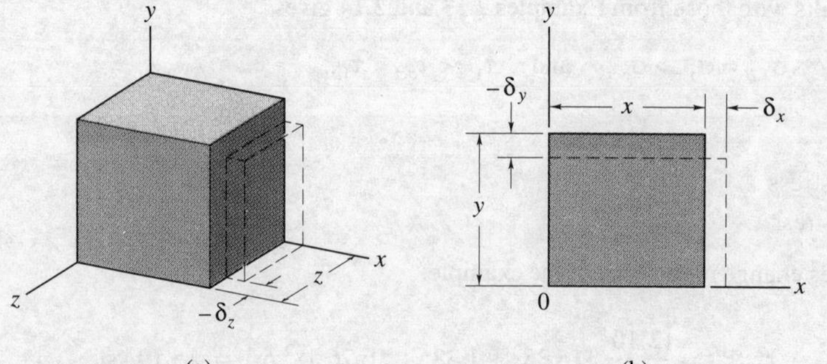

Figure 2.22 Normal strain of cubic element subjected to uniform tension in x direction. (a) Three-dimensional view; (b) two-dimensional (or plane) view.

$$\varepsilon = \lim_{\Delta l \to 0} \frac{\Delta\delta_{avg}}{\Delta l} = \frac{d\delta_{avg}}{dl} \tag{2.34}$$

Figure 2.22 shows the strain on a cubic element subjected to uniform tension in the x direction. The element elongates in the x direction while simultaneously contracting in the y and z directions, a phenomenon known as the Poisson effect. From Eq. (2.34) the normal strain components can be written as

$$\varepsilon_x = \lim_{x\to 0} \frac{\delta_x}{x} \qquad \varepsilon_y = \lim_{y\to 0} \frac{\delta_y}{y} \qquad \text{and} \qquad \varepsilon_z = \lim_{z\to 0} \frac{\delta_z}{z} \tag{2.35}$$

Example 2.16 **GIVEN** A 12-in.-long aluminum bar 2.25 in. in diameter has an axial load of 32 000 lbf applied to it. The axial elongation is 0.00938 in. and the diameter is decreased by 0.000585 in.

FIND The transverse and axial strains in the bar.

Solution

The axial strain is

$$\varepsilon_a = \frac{\delta}{l} = \frac{0.00938}{12} = (0.7817)(10^{-3})$$

The transverse strain is

$$\varepsilon_t = \frac{\delta_t}{d} = \frac{-(0.585)(10^{-3})}{2.25} = -(0.2600)(10^{-3})$$

The sign for the transverse strain is negative because the diameter decreased after the bar was loaded. The axial strain is positive because the axial length increased after loading.

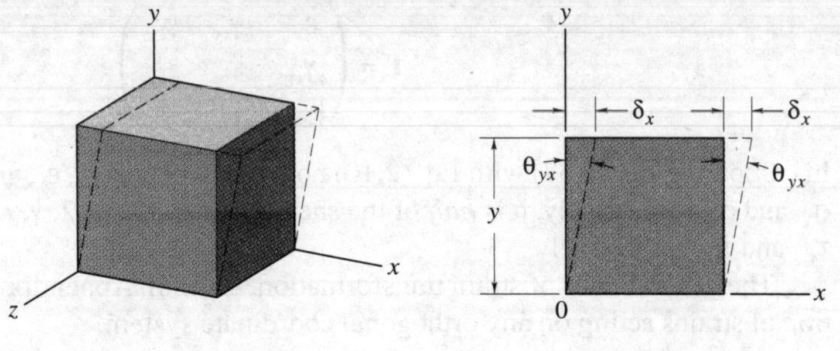

Figure 2.23 Shear strain of cubic element subjected to shear stress. (a) Three-dimensional view; (b) two-dimensional (or plane) view.

Figure 2.23 shows the shear strain of a cubic element due to shear stress in both a three-dimensional view and a two-dimensional (or plane) view. The **shear strain**, designated by γ, is used to measure angular distortion (the change in angle between two lines that are orthogonal in the undeformed state). The shear strain as shown in Fig. 2.23 is defined as

$$\gamma_{yx} = \lim_{y \to 0} \frac{\delta_x}{y} = \tan \theta_{yx} \approx \theta_{yx} \tag{2.36}$$

where

θ_{yx} = angle representing deviation from initial right angle

The subscripts used to define the shear strains are like those used to define the shear stresses in Sec. 2.11. The first subscript designates the coordinate direction perpendicular to the plane in which the strain acts, and the second subscript designates the coordinate direction in which the strain acts. For example, γ_{yx} is the strain resulting from taking adjacent planes perpendicular to the y axis and displacing them relative to each other in the x direction. The sign conventions for strain follow directly from those developed for stress. A positive stress produces a positive strain and a negative stress produces a negative strain. The shear strain shown in Fig. 2.23 and described in Eq. (2.36) is positive. The strain of the cubic element thus contains three normal strains and six shear strains, just as found for stresses. Symmetry reduces the shear strain elements from six to three, as for stress.

2.18 STRAIN TENSOR

For strains within the elastic range the equations relating normal and shear strains with the orientation of the cutting plane are analogous to the corresponding equations for stress given in Eq. (2.10). Thus, the state of strain can be written as a tensor, as

$$\mathbf{T} = \begin{pmatrix} \varepsilon_x & \frac{1}{2}\gamma_{xy} & \frac{1}{2}\gamma_{xz} \\ \frac{1}{2}\gamma_{xy} & \varepsilon_y & \frac{1}{2}\gamma_{yz} \\ \frac{1}{2}\gamma_{xz} & \frac{1}{2}\gamma_{yz} & \varepsilon_z \end{pmatrix} \qquad \textbf{(2.37)}$$

In comparing Eq. (2.37) with Eq. (2.10) note that whereas ε_x, ε_y, and ε_z are analogous to σ_x, σ_y, and σ_z, respectively, it is *half* of the shear strain $\gamma_{xy}/2$, $\gamma_{yz}/2$, $\gamma_{zx}/2$ that is analogous to τ_{xy}, τ_{yz}, and τ_{zx}, respectively.

The general laws of strain transformation, given in Appendix B, enable the determination of strains acting on any orthogonal coordinate system.

2.19 PLANE STRAIN

Instead of the six strains for the complete strain tensor, here the effects of the components ε_x, γ_{xz}, and γ_{yz} are ignored, and thus only two normal strain components ε_z and ε_y and one shear strain component γ_{xy} are considered. Figure 2.24 shows the deformation of an element caused by each of the three strains considered in plane strain. The normal strain components ε_x and ε_y, shown in Fig. 2.24(a) and (b), are produced by changes in element length in the x and y directions, respectively. The shear strain γ_{xy}, shown in Fig. 2.24(c), is produced by the relative rotation of two adjacent sides of the element. Figure 2.24(c) also helps to explain the physical significance of the fact that τ is analogous to $\gamma/2$ rather than to γ. Each side of an element changes in slope by an angle $\gamma/2$ when subjected to pure shear.

The following **sign convention** is to be used for strains:

1. **Normal strains** ε_x and ε_y are positive if they cause elongation along the x and y axes, respectively. In Fig. 2.24(a) and (b), ε_x and ε_y are positive.
2. **Shear strain** γ_{xy} is positive when the interior angle A0B in Fig. 2.24(c) becomes smaller than 90°.

The principal strains, planes, and directions are directly analogous to those found earlier for principal stresses. The principal normal strains in the x-y plane, the maximum

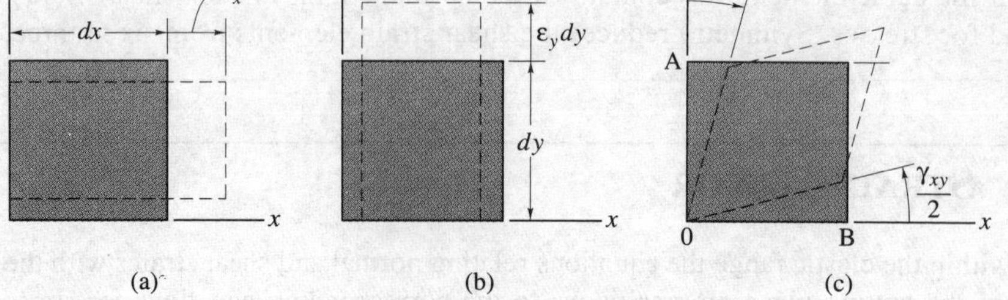

Figure 2.24 Graphical depiction of plane strain element.
(a) Normal strain ε_x; (b) normal strain ε_y; and (c) shear strain γ_{xy}.

shear strain in the *x-y* plane, and the orientation of the principal axes relative to the *x* and *y* axes are

$$\varepsilon_1, \varepsilon_2 = \frac{\varepsilon_x + \varepsilon_y}{2} \pm \sqrt{\left(\frac{1}{2}\gamma_{xy}\right)^2 + \left(\frac{\varepsilon_x - \varepsilon_y}{2}\right)^2} \quad \textbf{(2.38)}$$

$$\gamma_{max} = \pm 2\sqrt{\left(\frac{1}{2}\gamma_{xy}\right)^2 + \left(\frac{\varepsilon_x - \varepsilon_y}{2}\right)^2} \quad \textbf{(2.39)}$$

$$2\phi = \tan^{-1}\left(\frac{\gamma_{xy}}{\varepsilon_x - \varepsilon_y}\right) \quad \textbf{(2.40)}$$

From here there are two important classes of problem. If the principal strains are known and it is desired to find the strains acting at a plane oriented at angle ϕ from the principal plane (numbered 1), the equations are

$$\varepsilon_\phi = \frac{\varepsilon_1 + \varepsilon_2}{2} + \frac{\varepsilon_1 - \varepsilon_2}{2}\cos 2\phi \quad \textbf{(2.41)}$$

$$\gamma_\phi = (\varepsilon_1 - \varepsilon_2)\sin 2\phi \quad \textbf{(2.42)}$$

In Eq. (2.42), γ_ϕ represents a shear strain acting on the ϕ plane and directed 90° from the ϕ axis. Just as for stress the second subscript is omitted for convenience and no ambiguity results. A Mohr's circle diagram can also be used to represent the state of strain.

The second problem of interest is the case where a normal strain component has been measured in three different but specified directions and it is desired to obtain the strains ε_x, ε_y, and γ_{xy} from these readings. In this case the equation

$$\varepsilon_\theta = \varepsilon_x \cos^2\theta + \varepsilon_y \sin^2\theta + \gamma_{xy}\sin\theta\cos\theta \quad \textbf{(2.43)}$$

is of great assistance. Here, ε_θ is the measured strain in the direction rotated θ counterclockwise from the *x* axis, and ε_x, ε_y, and γ_{xy} are the desired strains. Thus, measuring a strain in three different directions gives three equations for the three unknown strains and is sufficient for their quantification. Strain gages are often provided in groups of three, called rosettes, for such purposes.

Example 2.17

GIVEN A 0°-45°-90° rosette as shown in Fig. 2.25 is attached to a structure with the 0° gage placed along a reference (*x*) axis. Upon loading, the strain in the 0° direction reads +50 μm/m, the strain gage in the 45° direction reads -27×10^{-6}, and the gage in the 90° direction reads 0.

FIND The strains ε_x, ε_y, and γ_{xy}.

Solution

Equation (2.43) can be applied three times to obtain three equations:

$$\varepsilon_{0°} = \varepsilon_x \cos^2(0°) + \varepsilon_y \sin^2(0°) + \gamma_{xy}\sin 0°\cos 0° \qquad \varepsilon_x = \varepsilon_0 = 50 \times 10^{-6}$$

$$\varepsilon_{90°} = \varepsilon_x \cos^2(90°) + \varepsilon_y \sin^2(90°) + \gamma_{xy}\sin(90°)\cos(90°) \qquad \varepsilon_y = 0$$

$$\varepsilon_{45°} = \varepsilon_x \cos^2(45°) + \varepsilon_y \sin^2(45°) + \gamma_{xy}\sin(45°)\cos(45°)$$

$$\gamma_{xy} = -54 \times 10^{-6} - 50 \times 10^{-6} - 0 = -104 \times 10^{-6}$$

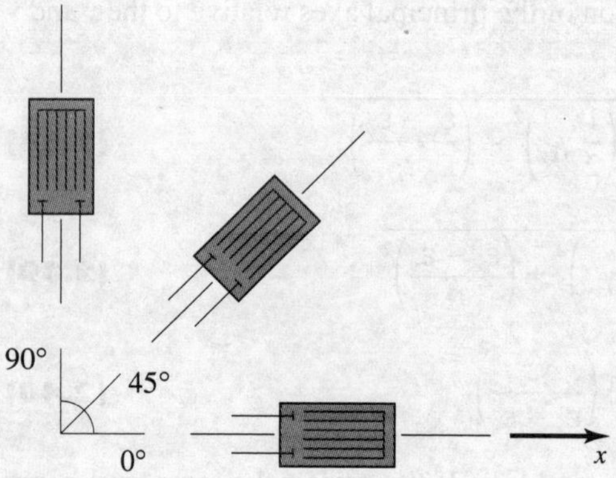

Figure 2.25 Strain gage rosette used in Example 2.17.

Case Study 2.1 GLUE SPREADER

Figure 2.26 shows the expansion process used in making honeycomb materials. The process begins from the left and continues to the right. Sheets are cut from the roll and an adhesive is applied in parallel strips along the width of the sheet. These sheets are stacked to form a block and cured in an oven, enabling strong bonds to develop at the adhesive joints. The block is then cut into slices of desired dimension and stretched to produce a honeycomb structure. These stiff and lightweight construction materials are often used in aerospace applications.

This case study is particularly interested in the adhesive application to the sheet, or the glue spreader. This machine [Fig. 2.27(a)] spreads glue onto a sheet by exposing four polymer rollers to a polymer bath and then bringing them into contact with the sheet. By spacing the rollers properly, the desired glue pattern can be obtained. At the ends of the shaft are ball bearings as shown in Fig. 2.27(a).

Figure 2.27(b) shows a free-body diagram of the forces acting on the shaft along with the spacing between rollers. Here, the roller is represented by a concentrated force rather than a unit step. From moment and force equilibrium $R_A = R_B = 800$ N.

FIND

a) Using singularity functions determine the shear and moment expressions and also give the corresponding diagrams and the maximum moment.

b) If the distance between the bearing and the first roller is *a*, as is also the distance between the last roller and the bearing, and the distance between rollers is *b*, find a general expression for the shear and moment for *n* rollers.

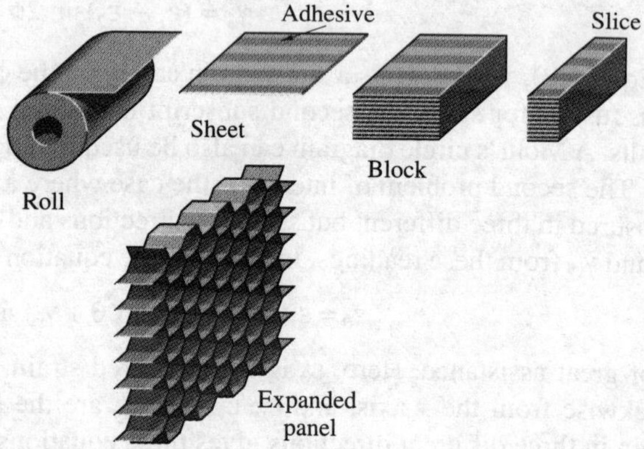

Figure 2.26 Expansion process used in honeycomb materials. [Adapted from Kalpakjian (1991).]

Solution

a) From Table 2.2 for a concentrated force the load intensity function for the glue spreader shown in Fig. 2.27(b) (when *x* is expressed in meters) is

$$-q(x) = -800\langle x\rangle^{-1} + 400\langle x-0.075\rangle^{-1} + 400\langle x-0.145\rangle^{-1} + 400\langle x-0.215\rangle^{-1} + 400\langle x-0.285\rangle^{-1} - 800\langle x-0.360\rangle^{-1}$$

(continued)

Case Study 2.1 (CONCLUDED)

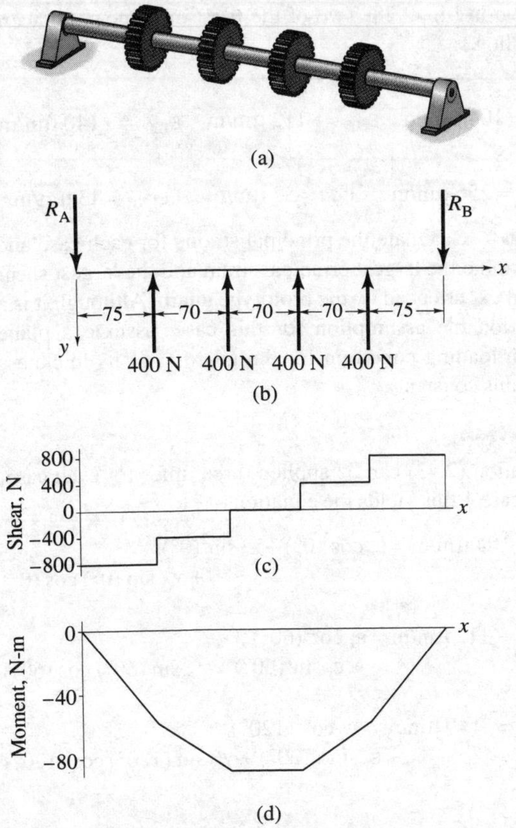

Figure 2.27 Glue spreader case study. (a) Machine; (b) free-body diagram; (c) shear diagram; (d) moment diagram.

The shear force and moment expressions are

$$V(x) = -800[\langle x\rangle^0 + \langle x - 0.360\rangle^0] + 400[\langle x - 0.075\rangle^0 + \langle x - 0.145\rangle^0 + \langle x - 0.215\rangle^0 + \langle x - 0.285\rangle^0]$$

$$M(x) = -800[\langle x\rangle^1 + \langle x - 0.360\rangle^1] + 400[\langle x - 0.075\rangle^1 + \langle x - 0.145\rangle^1 + \langle x - 0.215\rangle^1 + \langle x - 0.285\rangle^1]$$

Figure 2.27(c) and (d) shows the shear and bending moment diagrams. The maximum bending moment M_{max} is 88 N-m.

b) The general expressions for the load intensity, shear force, and moment are

$$-q(x) = -\frac{nP}{2}[\langle x\rangle^{-1} + \langle x - 2a - nb\rangle^{-1}] + P\sum_{i=1}^{n}\langle x - a - b(i-1)\rangle^{-1}$$

$$V(x) = -\frac{nP}{2}[\langle x\rangle^{0} + \langle x - 2a - nb\rangle^{0}] + P\sum_{i=1}^{n}\langle x - a - b(i-1)\rangle^{0}$$

$$M(x) = -\frac{nP}{2}[\langle x\rangle^{1} + \langle x - 2a - nb\rangle^{1}] + P\sum_{i=1}^{n}\langle x - a - b(i-1)\rangle^{1}$$

where

n = number of rollers
a = distance from bearing to roller
b = distance between rollers

Case Study 2.2 | STRAINS ON A SNOWMOBILE GUARD

The drive belt of a snowmobile transmits power from the engine to the shafting driving the tracks. Because of space limitations the drive belt is near the driver, just behind the dashboard. When the drive belt is excessively worn, it may rupture; and if it hits the dashboard with sufficient force, it may fracture it. Such incidents have been known to fragment the dashboard, and projectiles are a serious hazard to the driver and/or passenger of the snowmobile.

A design solution has been proposed and is sketched in Fig. 2.28, wherein a guard will be fabricated to enclose the drive belt. The loading in this case is unknown and difficult to determine, let alone estimate, analytically because the belt can rupture anywhere relative to a pulley and can strike the guard with many different angles of incidence. However, an experimental program was conducted wherein notches were cut into new drive belts so that they would fail in a laboratory snowmobile when the motor was accelerating. A prototype guard was affixed to the belt area. Strain gages placed on the guard measured the strain during belt rupture. By performing many tests a distribution of strains could be obtained, and with sufficiently delicate instrumentation the energy absorbed by the guard could be directly measured as well. These data then allowed application of a failure criterion (to be covered in Chapters 6 and 7) to determine a minimum safety factor for the guard.

Twenty-five ruptures tests were conducted and the strains measured. An equiangular rosette was used as shown in Fig. 2.28. Two of the most extreme cases were as follows:

Case 1:

$\varepsilon_{0^\circ} = 104\ \mu\text{m/m} \quad \varepsilon_{60^\circ} = -112\ \mu\text{m/m} \quad \varepsilon_{120^\circ} = -140\ \mu\text{m/m}$

Case 2:

$\varepsilon_{0^\circ} = 256\ \mu\text{m/m} \quad \varepsilon_{60^\circ} = 57\ \mu\text{m/m} \quad \varepsilon_{120^\circ} = 43\ \mu\text{m/m}$

FIND Calculate the principal strains for each case and determine the largest principal strain and the largest shear strain experienced by the prototype guard. Although it is a questionable assumption for this case, assume a plane strain loading condition for the guard (i.e., its thickness remains constant).

Solution

Equation (2.43) can be applied three times for each case. For case 1 this yields the equations

$$\varepsilon_{0^\circ} = 104\ \mu\text{m/m} = \varepsilon_x \cos^2(0^\circ) + \varepsilon_y \sin^2(0^\circ) + \gamma_{xy} \sin(0^\circ)\cos(0^\circ)$$

$$\varepsilon_{60^\circ} = -112\ \mu\text{m/m} = \varepsilon_x \cos^2(60^\circ) + \varepsilon_y \sin^2(60^\circ) + \gamma_{xy} \sin(60^\circ)\cos(60^\circ)$$

$$\varepsilon_{120^\circ} = -140\ \mu\text{m/m} = \varepsilon_x \cos^2(120^\circ) + \varepsilon_y \sin^2(120^\circ) + \gamma_{xy} \sin(120^\circ)\cos(120^\circ)$$

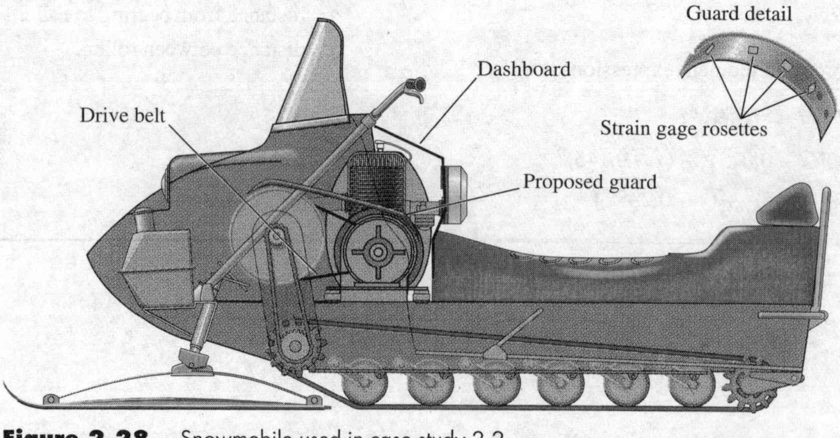

Figure 2.28 Snowmobile used in case study 2.2.

(continued)

Case Study 2.2 (CONCLUDED)

yielding $\varepsilon_x = 104$ μm/m, $\varepsilon_y = -133$ μm/m, and $\gamma_{xy} = -88$ μm/m. For case 2 the resultant strains are $\varepsilon_x = 256$ μm/m, $\varepsilon_y = 16$ μm/m, and $\gamma_{xy} = -44$ μm/m.

A Mohr's circle of strain can be constructed for these combinations of strain. For case 1 the circle will have its center at

$$\frac{\varepsilon_x + \varepsilon_y}{2} = \frac{104 - 133}{2}(10^{-6}) = -14.5\ \mu\text{m/m}$$

One point on the circle, for example, is given by $-\gamma_{xy}/2$ as shown in Fig. 2.24(c). From this point and the center the circle's radius is found to be 126.5 μm/m. Thus, the extreme normal strains in the plane of the strain gage are given by

$$\varepsilon_1 = -14.5\ \mu\text{m/m} + 126.5\ \mu\text{m/m} = 112\ \mu\text{m/m}$$

$$\varepsilon_2 = -14.5\ \mu\text{m/m} - 126.5\ \mu\text{m/m} = -141\ \mu\text{m/m}$$

The largest shear strain in the plane is given by

$$\frac{\gamma_{max}}{2} = r \qquad \gamma_{max} = 2(126.5)(10^{-6}) = 253\ \mu\text{m/m}$$

For case 2 a circle with its center at 136 μm/m and a radius of 122 μm/m is determined. The extreme normal strains in this case are 258 μm/m and 0 μm/m. The smallest normal strain is zero because a plane strain circumstance was assumed. Therefore, the largest normal strain did not take place in the plane of the strain gage, and the experiment is suspect. However, since the smallest stress in the plane is approximately 14 μm/m (~5% error from zero strain). the maximum shear stress is approximately the circle radius, or 122 μm/m.

2.20 SUMMARY

This chapter described how load, stress, and strain affect the design of machine elements. If the proper type of machine element has been selected, after considering its function, the primary cause of failure is the design stress exceeding the strength of the machine element. Therefore, it is important to evaluate the stress, strain, and strength of the machine element at the critical section. To do so requires first a determination of load in all its forms. The applied load on a machine element was described with respect to time, the area over which load is applied, and the location and method of application. Furthermore, the importance of support reactions, application of static force and moment equilibrium, and proper use of free-body diagrams was pointed out.

The chapter then focused on shear and moment diagrams applied to a beam. Singularity functions introduced by concentrated moment, concentrated force, unit step, ramp, inverse ramp, and parabolic shape were considered. Various combinations of these singularity functions can exist within a beam. Integrating the load intensity function for the various beam singularity functions over the beam length establishes the shear force. Integrating the shear force over the beam length determines the moment. From these analytical expressions the shear and moment diagrams can be readily constructed.

Stress defines the intensity and direction of the internal forces at a particular point and acting on a given plane. The stresses acting on an element have normal and shear components. Across each mutually perpendicular surface there are two shear stresses and one normal stress, yielding a total of nine stresses (three normal stresses and six shear stresses). The sign conventions for both the normal and shear stresses were presented. The nine stress components may be regarded as the components of a second-order Cartesian tensor. It was found that the stress tensor is symmetrical, implying that the tensor can be written with zero shear stress and the principal normal stresses.

In many engineering applications the stress analysis assumes that the surface is free of stress or that the stress in one plane is small relative to the stresses in the other two planes. The two-dimensional stress situation is called the biaxial (or plane) stress state and can be expressed in terms of two normal stresses and one shear stress (σ_x, σ_y, and τ_{xy}). That the stresses can be expressed in any oblique plane is important in describing the Mohr's circle diagram for a biaxial stress. It was found impossible to express analytically a general three-dimensional stress state. If the principal normal stresses are known, the triaxial stress state can be represented by a Mohr's circle diagram when the shear stresses are expressed in terms of the principal normal stresses. It was found advantageous to represent stresses by an octahedral stress element, where the normal and shear stresses acting on all eight planes are equal. The triaxial maximum principal normal and shear stresses are greater than their octahedral counterparts.

KEY WORDS

beam structural member designed to support loads perpendicular to its longitudinal axis

bending load load applied transversely to longitudinal axis of member

biaxial or plane stress condition where one surface is comparatively free of stress

cantilevered beam support where one end is fixed and other end is free

combined load combination of two or more previously defined loads

concentrated load load applied to small nonconformal area

critical section section where largest internal stress occurs

cyclic load load varying throughout cycle

distributed load load distributed over entire area

free-body diagram sketch of part showing all acting forces

impact load load rapidly applied

loads force, moment, or torque applied to mechanism or structure

Mohr's circle method used to graphically visualize state of stress acting in different planes passing through given point

normal load load passing through centroid of resisting section

normal strain elongation or contraction of linear segment of element in which stress is applied

octahedral plane plane that cuts across corner of principal element so that eight planes form octahedron

overhanging beam support where one or both ends freely extend past support

principal normal stresses combination of applied normal and shear stresses that produces maximum principal normal stress or minimum principal normal stress, with a third principal stress between or equivalent to the extremes

principal shear stresses combination of applied and shear stresses that produces maximum principal shear stress or minimum principal shear stress

shear load load collinear with transverse shear force

shear strain measure of angular distortion in which shear stress is applied

sign convention for normal strain positive if elongation is in direction of positive axes

sign conversion for normal stress positive for tension and negative for compression

sign convention for shear strain positive if interior angle becomes smaller after shear stress is applied

sign conversion for shear stress positive if both normal from surface and shear are in positive or negative direction; negative for any other combination

simply supported beam support where one end is pinned and other is roller-supported

singularity functions functions used to evaluate shear and moment diagrams, especially when discontinuities, such as concentrated load or moment, exist

static load load gradually applied and equilibrium reached in short time

strain displacement produced in solid as result of stress

stress intensity and direction of internal force acting at given point on particular plane

sustained load load constant over long time

symmetrical tensor condition where principal normal stresses exist while all other tensor elements are zero

torsion load load subjected to twisting motion

triaxial stress stress where all surfaces are considered

uniaxial stress condition where two perpendicular surfaces are comparatively free of stress

RECOMMENDED READINGS

Beer, F.P., and Johnson, E.R. (1981) *Mechanics of Materials,* McGraw-Hill, New York.

Craig, R.R. (1996) *Mechanics of Materials,* Wiley, New York.

Crandall, S.H., and Dahl, H.C. (1954) *An Introduction to the Mechanics of Solids,* McGraw-Hill, New York.

Juvinall, R.C. (1967) *Stress, Strain, and Strength,* McGraw-Hill, New York.

Lardner, T.J., and Archer, R.R. (1994) *Mechanics of Solids: An Introduction,* McGraw-Hill, New York.

Popov, E.P. (1968) *Introduction to Mechanics of Solids,* Prentice-Hall, Englewood Cliffs, NJ.

REFERENCES

Durelli, A.J., Phillips, E.A., and Tsao, C.H. (1958) *Introduction to the Theoretical and Experimental Analysis of Stress and Strain,* McGraw-Hill, New York.

Kalpakjian, S. (1991) *Manufacturing Processes for Engineering Materials*, Addison-Wesley, New York.

Mohr, O. (1914) *Abhandlungen aus dem Gebiete der technischen Mechanik,* 2nd ed., pp. 192–203, Wilhelm Ernst und Sohn, Berlin.

PROBLEMS

Section 2.2

2.1 The stepped shaft A-B-C shown in sketch *a* is loaded with the forces P_1 and/or P_2. Note that P_1 gives a tensile stress σ in B-C and $\sigma/4$ in A-B and that P_2 gives a bending stress σ at B and 1.5σ at A. What is the critical section

a) If only P_1 is applied?
b) If only P_2 is applied?
c) If both P_1 and P_2 are applied?

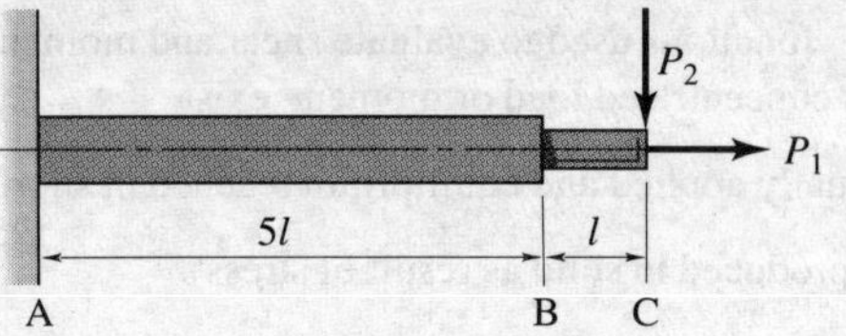

Sketch *a*, used in Problems 2.1 and 2.2

Section 2.3

2.2 The stepped shaft in Problem 2.1 (sketch *a*) has loads P_1 and P_2. Find the load classification if P_1's variation is sinuous and P_2 is the load from a weight

a) If only P_1 is applied
b) If only P_2 is applied
c) If both P_1 and P_2 are applied

Section 2.4

2.3 The sign convention for bending moments can be derived from the differential equation for the elastic line

$$EIy'' = -M$$

where

E = modulus of elasticity, Pa
I = area moment of inertia, m^4
M = bending moment, N-m
y'' = d^2y/dx^2 = second derivative of y with respect to x, m^{-1}
y = deflection, m

This gives a positive (tensile) stress in the fibers on the beam in the positive (y) direction. Find how the slope of the beam changes with increasing x coordinate for positive and negative bending.

Section 2.5

2.4 A bar hangs freely from a frictionless hinge. A horizontal force P is applied at the bottom of the bar until it inclines 30° from the vertical direction. Calculate the horizontal and vertical components of the force on the hinge if the acceleration due to gravity is g, the bar has a constant cross section along its length, and the total mass is m_a.

2.5 Sketch *b* shows the forces acting on a rectangle. Is the rectangle in equilibrium?

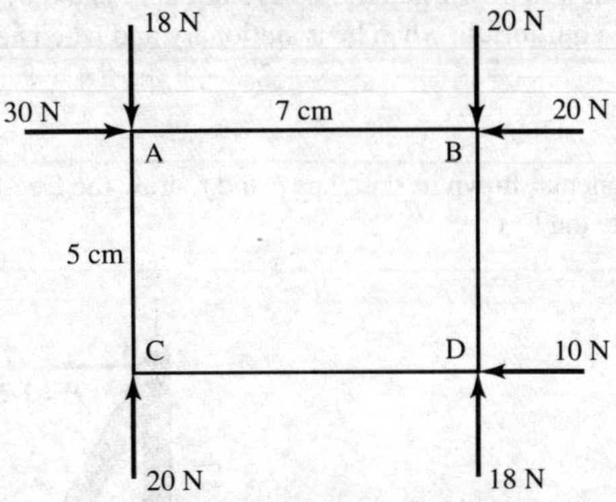

Sketch *b*, used in Problem 2.5

2.6 Sketch *c* shows the forces acting on a triangle. Is the triangle in equilibrium?

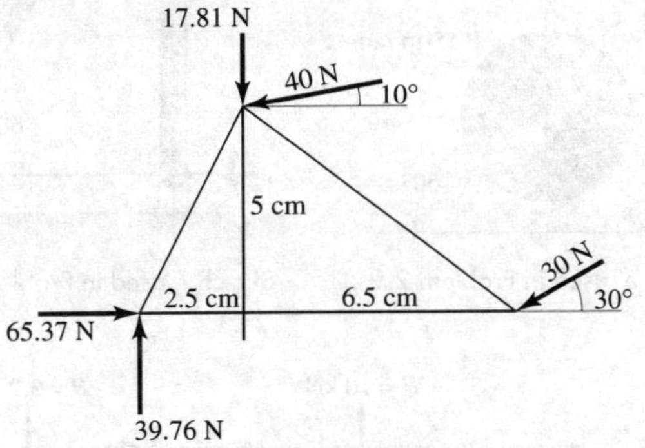

Sketch *c*, used in Problem 2.6

2.7 Sketch *d* shows a cube with side lengths *a* and eight forces acting at the corners. Is the cube in equilibrium?

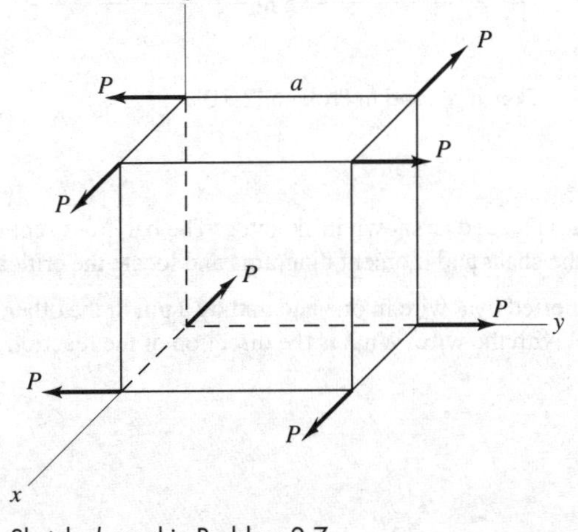

Sketch *d*, used in Problem 2.7

2.8 A downhill skier stands on a slope with a 5° inclination. The coefficient of friction between the skis and the snow is 0.10 when the ski is stationary and 0.07 when the ski starts to slide. Is the skier in static equilibrium when he is stationary and when he slides down the slope?

Section 2.6

2.9 Given the components shown in sketches *e* and *f*, draw the free-body diagram of each component and calculate the forces.

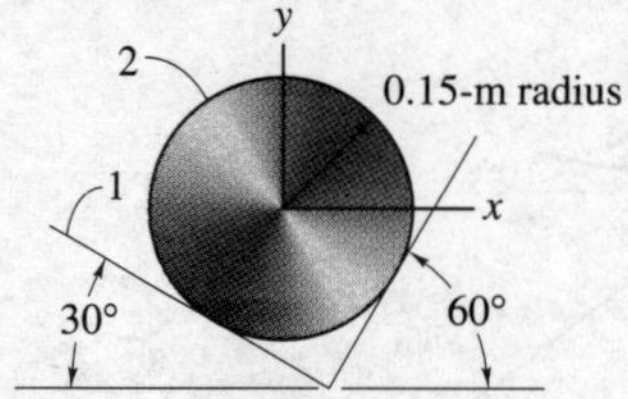

Sketch *e*, used in Problem 2.9

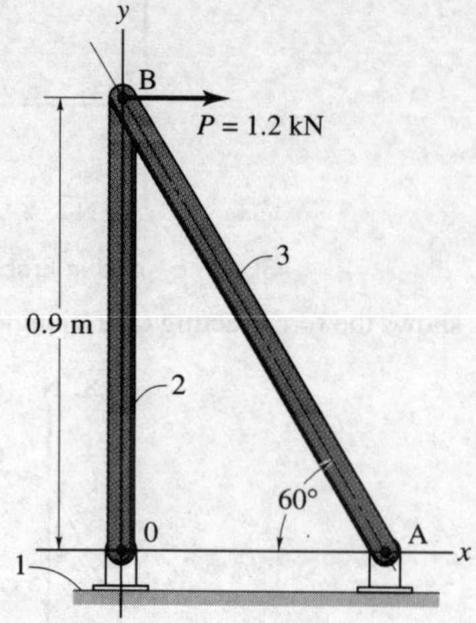

Sketch *f*, used in Problem 2.9

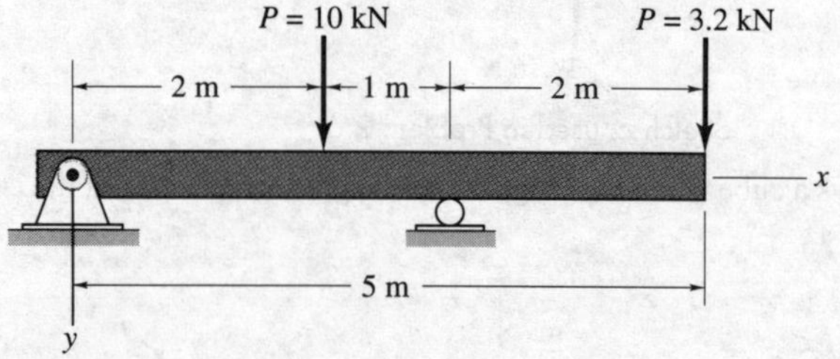

Sketch *g*, used in Problem 2.10

Section 2.8

2.10 A 5-m-long bar is loaded as shown in sketch *g*. The bar cross section is constant along its length. Draw the shear and moment diagrams and locate the critical section.

2.11 A beam is supported by a wire in one end and by a pin in the other end. The beam is loaded by a force parallel with the wire. What is the direction of the reaction force in the pin?

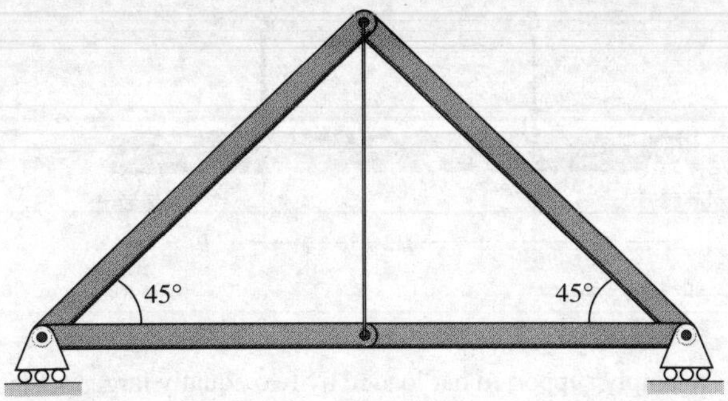

Sketch *h*, used in Problem 2.12

2.12 Sketch *h* shows a simple bridge. The midpoint of the bridge is held up by a wire leading to the top hinge of the two equal beams. When a truck runs over the bridge, the force in the wire is measured to be 500 000 N. Determine the forces in the beams and the horizontal force component in the middle of the bridge.

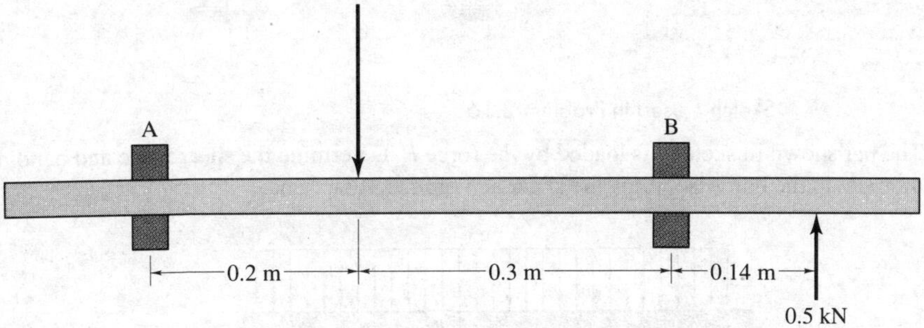

Sketch *i*, used in Problem 2.13

2.13 Sketch *i* shows a 0.05-m-diameter steel shaft supported by self-aligning bearings (which can provide radial but not bending loads on the shaft). A gear causes each force to be applied as shown. All length dimensions are in meters. The shaft can be considered weightless. Determine the forces at A and B and the maximum bending stress. Draw shear and moment diagrams.

2.14 Sketch *j* shows a simply supported bar loaded with a force P at a position one-third of the length from one of the support points. Determine the shear force and bending moment in the bar. Also, draw the shear and moment diagrams.

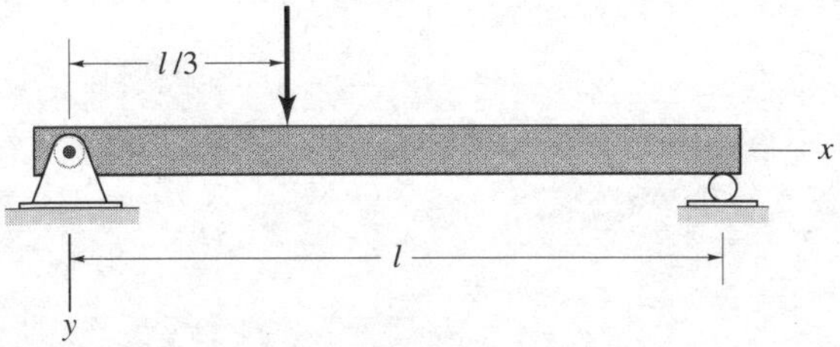

Sketch *j*, used in Problem 2.14

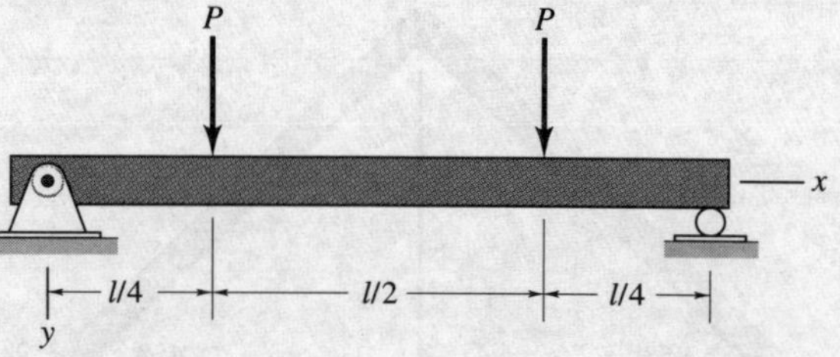

Sketch *k*, used in Problem 2.15

2.15 Sketch *k* shows a simply supported bar loaded by two equally large forces P at a distance $l/4$ from its ends. Determine the shear force and bending moment in the bar, and find the critical section with the largest bending moment. Also, draw the shear and moment diagrams.

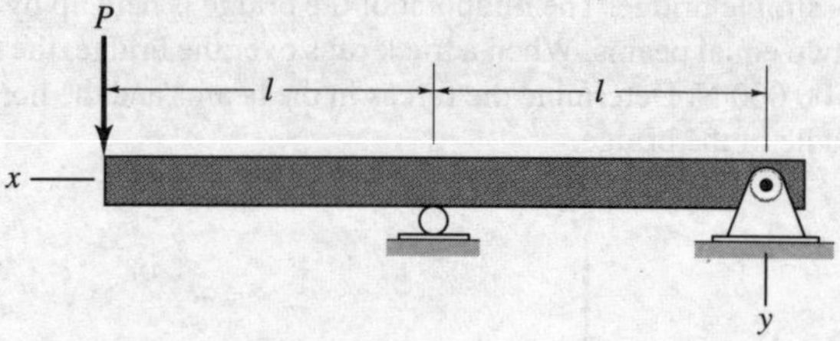

Sketch *l*, used in Problem 2.16

2.16 The bar shown in sketch l is loaded by the force P. Determine the shear force and bending moment in the bar. Also, draw the shear and moment diagrams.

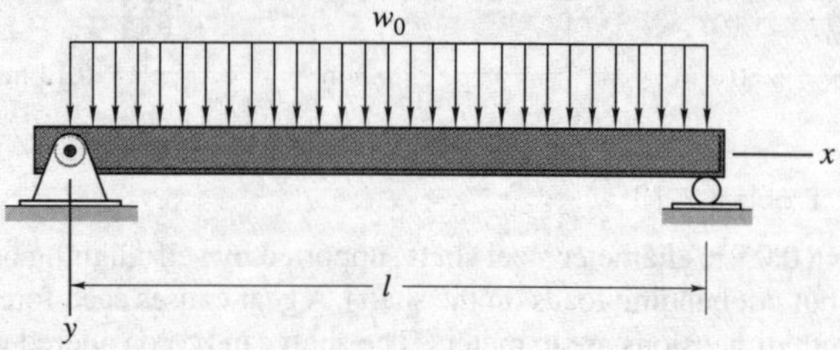

Sketch *m*, used in Problem 2.17

2.17 Sketch *m* shows a simply supported bar with a constant load per unit length w_0 imposed over its entire length. Determine the shear force and bending moment as functions of x. Draw a graph of these functions. Also, find the critical section with the largest bending moment.

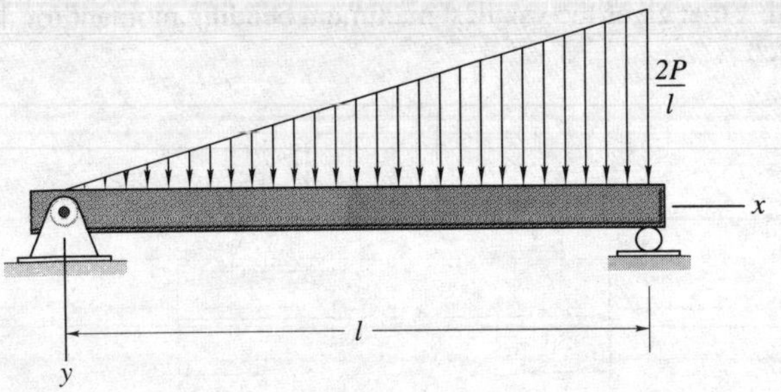

Sketch *n*, used in Problem 2.18

2.18 Sketch *n* shows a simply supported beam loaded with a ramp function over its entire length, the largest value being 2 *P*/*l*. Determine the shear force and the bending moment and the critical section with the largest bending moment. Also, draw the shear and moment diagrams.

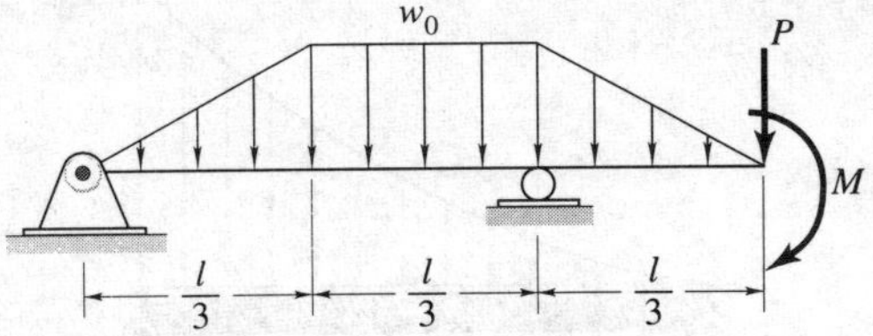

Sketch *o*, used in Problems 2.19 and 2.20

2.19 Draw the shear and moment diagrams and determine the critical section for the loading conditions shown in sketch *o*. Neglect the weight of the bar. Let $w_0 = 100$ N/m, $P = 5$ kN, $M = 2 \times 10^4$ N-mm, and $l = 300$ mm.

2.20 Redo Problem 2.19 while considering the weight of the bar per unit length to be w_0.

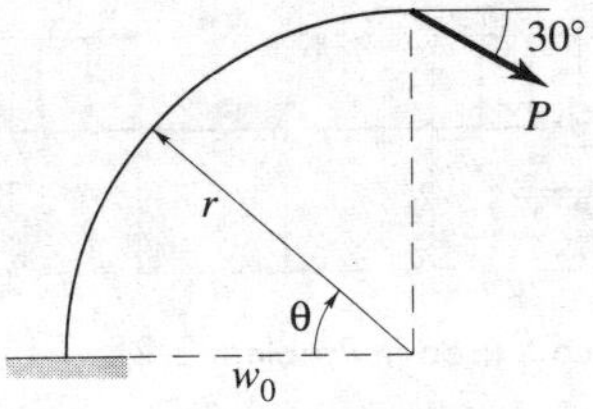

Sketch *p*, used in Problem 2.21

2.21 Find the expressions for the shear force, bending moment, and axial load *P* for the curved member shown in sketch *p*. Neglect the weight of the bar. Let $P = 10$ N and $r = 1$ m.

2.22 Sketch *q* shows a sinusoidal distributed force applied to a beam. Determine the shear force and bending moment for each section of the beam.

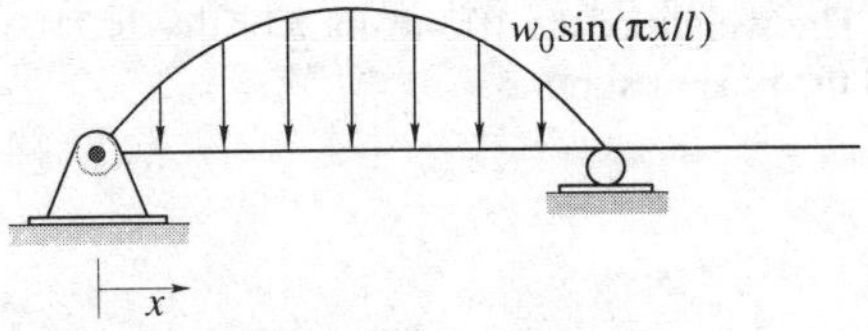

Sketch *q*, used in Problem 2.22

2.23 Find the length x that gives the smallest maximum bending moment for the load distribution shown in sketch r.

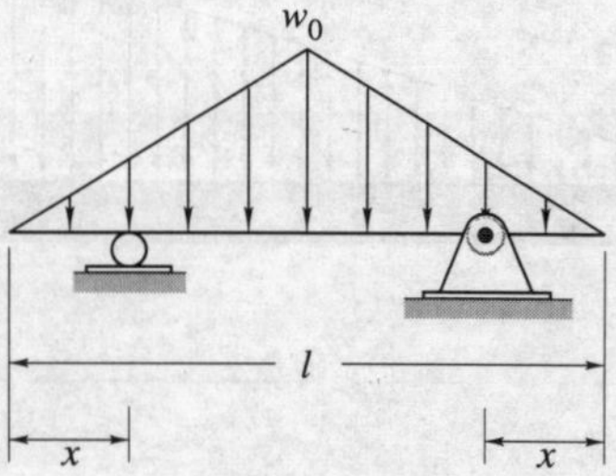

Sketch r, used in Problem 2.23

2.24 Draw the shear and moment diagrams for the load distribution shown in sketch s and the load intensity of $q(x) = \langle x \rangle^n$ when $n = 2$ and 3. Also, calculate the reaction forces.

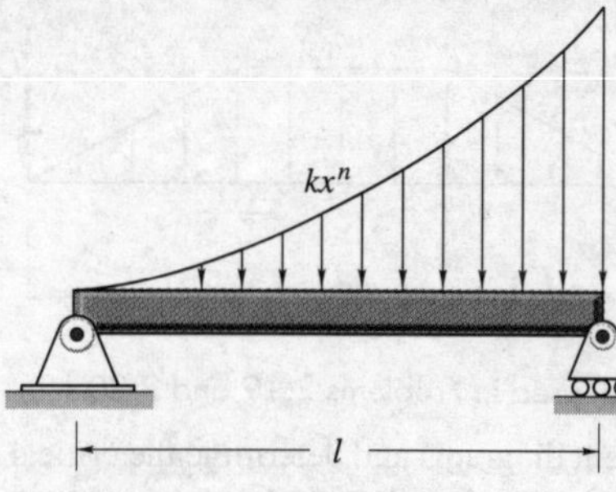

Sketch s, used in Problem 2.24

2.25 Draw the shear and moment diagrams and give the reaction forces for the load distribution shown in sketch t.

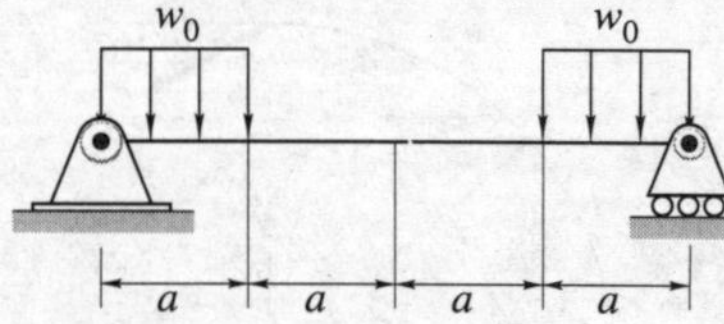

Sketch t, used in Problem 2.25

2.26 The loading acting on a beam consists of a distributed load of 2500 N/m over the entire beam, and three concentrated loads, each of 5000 N, located at $l/4$, $l/2$, and $3l/4$. The beam length is 8 m. Choose the type of beam support and the position. Also, draw the shear and moment diagrams.

2.27 The maximum load for a bridge crossing a river occurs when a fully loaded truck passes over. The maximum load is 550 000 N distributed over 22 m (the farthest distance between the wheels of a vehicle). The weight of the 100-m-long bridge is 210 tons. Choose the position of the bridge pillars and the type of support.

Section 2.9

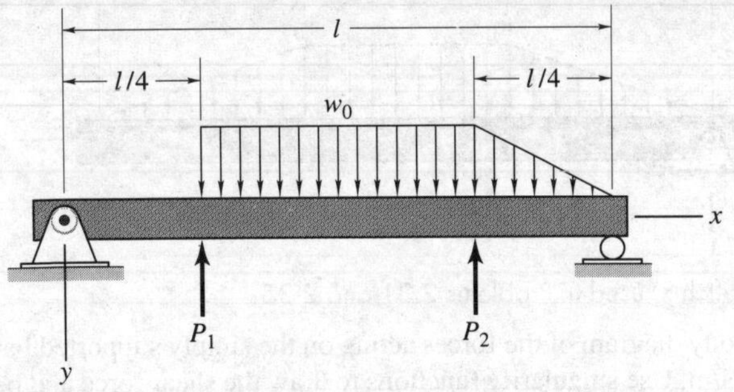

Sketch *u*, used in Problem 2.28

2.28 The simply supported bar shown in sketch *u* has $P_1 = 5$ kN, $P_2 = 8$ kN, $w_0 = 4$ kN/m, and $l = 12$ m. Use singularity functions to determine the shear force and bending moment as functions of *x*. Also, draw your results.

2.29 Use singularity functions for the force system shown in sketch *v* to determine the load intensity, the shear force, and the bending moment. From a force analysis determine the reaction forces R_1 and R_2. Also, draw the shear and moment diagrams.

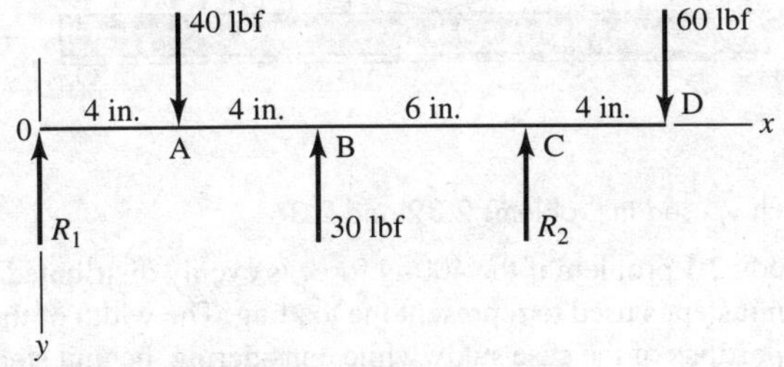

Sketch *v*, used in Problem 2.29

2.30 Use singularity functions for the force system shown in sketch *w* to determine the load intensity, the shear force, and the bending moment. Draw the shear and moment diagrams. Also, from a force analysis determine the reaction forces R_1 and R_2.

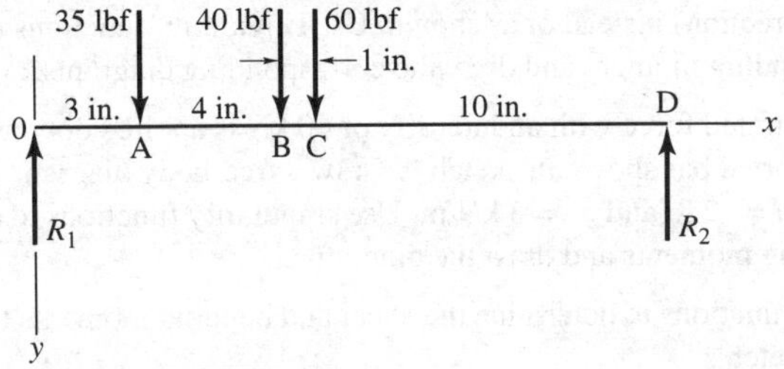

Sketch *w*, used in Problem 2.30

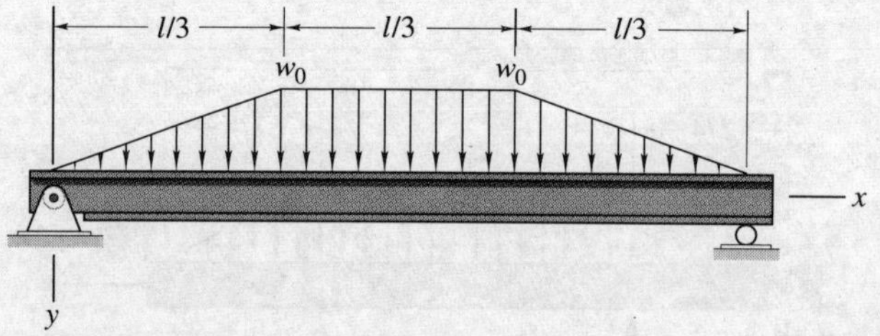

Sketch *x*, used in Problems 2.31 and 2.35

2.31 Draw a free-body diagram of the forces acting on the simply supported bar shown in sketch *x*, with $w_0 = 4$ kN/m. Use singularity functions to draw the shear force and bending moment diagrams.

2.32 Sketch *y* shows a simply supported bar with $w_0 = 5$ kN/m and $l = 12$ m. Draw a free-body diagram of the forces acting along the bar as well as the coordinates used. Use singularity functions to determine the shear force and the bending moment.

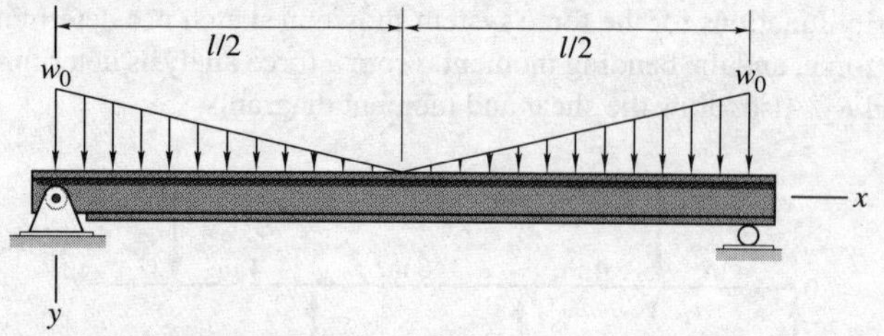

Sketch *y*, used in Problems 2.32 and 2.36

2.33 Redo the case study 2.1 problem if the 400-N force is evenly distributed along the width of the roller and a unit step is used to represent the loading. The width of the roller is 30 mm. Do both (*a*) and (*b*) portions of the case study while considering the unit step representation.

2.34 A simply supported beam has a parabolic load distribution beginning at $x = 0$. Use singularity functions to draw the shear and moment diagrams.

2.35 Draw a free-body diagram of the forces acting on the simply supported beam shown in sketch *x* and used in Problem 2.31. The central section ($l/3 \le x \le 2l/3$) has the load acting upward (in the negative *y* direction) instead of as shown. Use singularity functions to determine the shear force and the bending moment and draw the corresponding diagrams.

2.36 An extra concentrated force with an intensity of 60 kN is applied downward at the center of the simply supported bar shown in sketch *y*. Draw a free-body diagram of the forces acting on the bar. Assume $l = 12$ m and $w_0 = 5$ kN/m. Use singularity functions to determine the shear force and bending moments and draw the diagrams.

2.37 Use singularity functions to determine the shear and bending moments for the loading situation shown in sketch *z*.

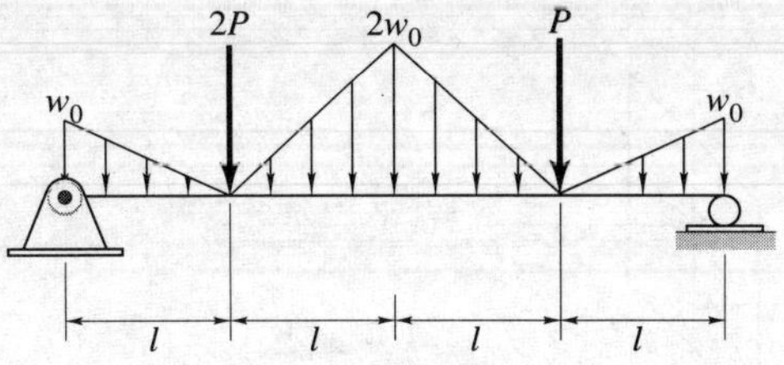

Sketch *z*, used in Problem 2.37

2.38 Obtain the shear force and moment expressions using singularity functions for a pinned beam at both ends with loading conditions described in sketch *aa*. Draw a free-body diagram of the forces acting on the bar. Draw the shear force and moment along the length of the bar and give tabular results. Assume $w_0 = 4$ kN/m, $P = 3$ kN, and $l = 12$ m.

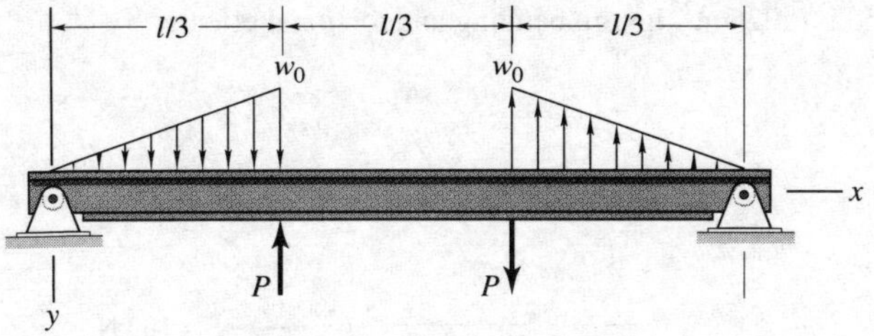

Sketch *aa*, used in Problem 2.38

Section 2.11

2.39 A steel bar is loaded by a tensile force $P = 25$ kN. The cross section of the bar is circular with a radius of 7 mm. What is the normal tensile stress in the bar?

2.40 A stainless-steel bar of square cross section has a tensile force of $P = 15$ kN acting on it. Calculate how large the side l of the cross-sectional area must be to give a tensile stress in the bar of 120 MPa.

2.41 What is the maximum length l_{max} that a copper wire can have if its weight should not give a higher stress than 75 MPa when it is hanging vertically? The density of copper is 8900 kg/m^3, and the density of air is so small relative to that of copper that it may be neglected. The acceleration of gravity is 9.81 m/s^2.

2.42 A machine weighing 5 tons will be lifted by a steel rod with an ultimate tensile strength of 860 MPa. A safety factor of 5 is to be used. Determine the diameter needed for the steel rod.

2.43 The largest measured ocean depth (near the Philippines in the Pacific) is 11 km. What ultimate strength is required for a steel wire to reach the bottom without being torn apart by its own weight? The density of steel is 7860 kg/m^3 and that of water is 1000 kg/m^3.

2.44 In Problem 2.43 a steel wire is used to measure the ocean depth. Find a way to do the measurements without risking breaking the wire due to tensile stresses being too high.

2.45 A string on a guitar is made of nylon and has a cross-sectional diameter of 0.6 mm. It is tightened to a force $P = 15$ N. What is the stress in the string?

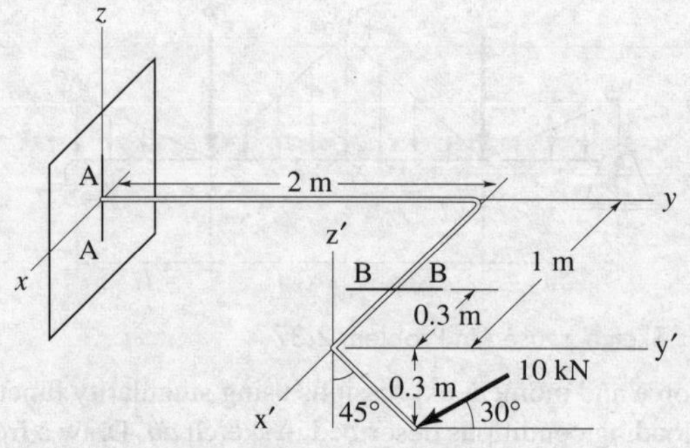

Sketch *bb*, for Problem 2.46

2.46 Determine the normal and shear stresses in sections A and B of sketch *bb*. The cross-sectional area of the rod is 0.025 m^2. Ignore bending and torsional effects.

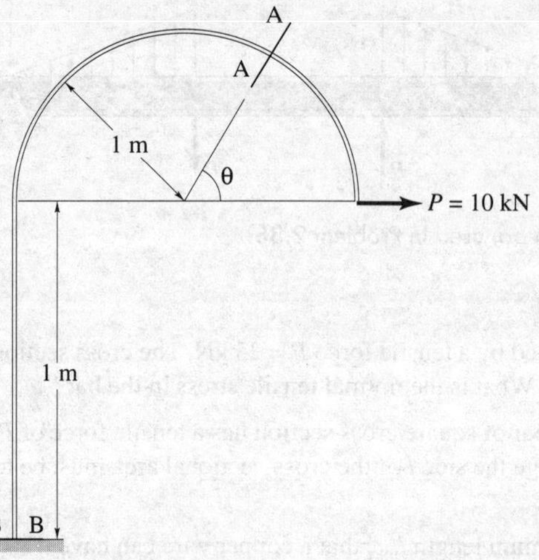

Sketch *cc*, for Problem 2.47

2.47 Determine the normal and shear stresses due to axial and shear forces at sections A and B in sketch *cc*.

2.48 For the stress element given in Fig. 2.13 determine how large the shear stress will become if all three normal stresses are doubled.

2.49 The normal stresses in the *x*, *y*, and *z* directions are all equal to 100 MPa, and the shear stresses are all zero. Find the stresses acting on a plane having the normal in the direction of the space diagonal $\left(\frac{1}{\sqrt{3}}; \frac{1}{\sqrt{3}}; \frac{1}{\sqrt{3}}\right)$.

2.50 If the stresses are such that $\sigma_x = -\sigma_y$ and $\sigma_z = \tau_{xy} = \tau_{xz} = \tau_{yz} = 0$, find the shear stress acting on the plane diagonal between the *x* and *y* axes and parallel to the *z* direction.

Section 2.12

2.51 A stressed element with shear stress τ_{yx} acting in the x direction on the surface has the normal in the y direction. Determine the shear stress acting in the y direction on the surface having the normal in the x direction.

Section 2.13

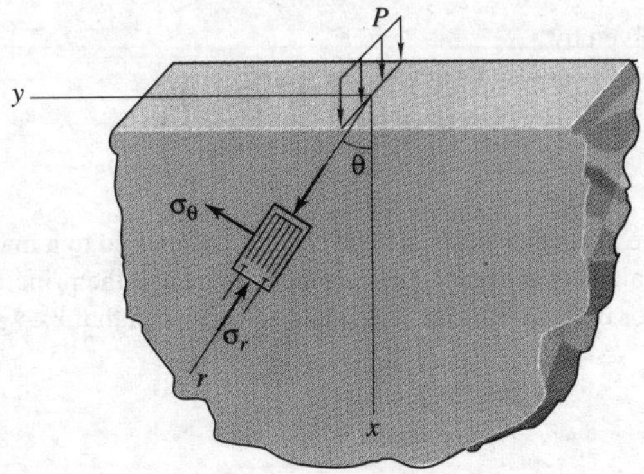

Sketch *dd*, used in Problem 2.52

2.52 Sketch *dd* shows a distributed load on a semi-infinite plane. The stress in polar coordinates based on the plane stress assumption is

$$\sigma_r = -\frac{2w_0 \cos\theta}{\pi r}$$

$$\sigma_\theta = \tau_{r\theta} = \tau_{\theta r} = 0$$

Determine the expressions σ_x, σ_y, and τ_{xy} in terms of r and θ.

2.53 Sketch *ee* shows loading of an extremely thin and infinitely long plane. Determine the angle θ needed so that the stress element will have no shear stress. For plate thickness t_w, modulus of elasticity E, and Poisson's ratio ν, find the reduction in thickness of the plate.

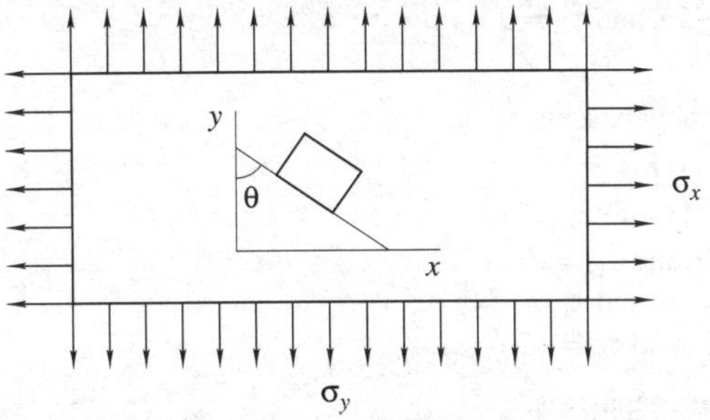

Sketch *ee*, used in Problem 2.53

2.54 The stress state in a machine element is given by

$$\mathbf{S} = \begin{pmatrix} 167 & 40 & -30 \\ 40 & 25 & 12 \\ -30 & 12 & -25 \end{pmatrix}$$

where all the values are in megapascals. Determine the stress tensor if the coordinates are rotated such that $x \to y = x'$; $y \to -x = y'$; and $z \to z = z'$.

2.55 A stress tensor is given by

$$\mathbf{S} = \begin{pmatrix} 178 & -83 & 0 \\ -83 & 12 & 0 \\ 0 & 0 & 0 \end{pmatrix}$$

where all values are in megapascals. The stress state is applied to a machine element made of AISI 1020 steel. Calculate the principal normal stresses and the principal shear stresses. Also, find the stress tensor if the coordinate system is rotated such that $x \to z = x'$; $y \to y = y'$; and $z \to -x = z'$.

Section 2.14

2.56 A thin, square steel plate has normals to the sides in the x and y directions. A tensile stress σ acts in the x direction, and a compressive stress $-\sigma$ acts in the y direction. Determine the normal and shear stresses on the diagonal of the square.

2.57 A thin, rectangular brass plate has normals to the sides in the x and y directions. A tensile stress σ acts on the four sides. Determine the principal normal and shear stresses.

2.58 Given the thin, rectangular brass plate in Problem 2.57, but with the stress in the y direction being $\sigma_y = -\sigma$ instead of $+\sigma$, determine the principal normal and shear stresses and their directions.

2.59 For the following stress states draw the appropriate Mohr's circle, determine the principal stresses and their directions, and show the stress elements:

a) $\sigma_x = 30$, $\sigma_y = -20$, and $\tau_{xy} = 10$
b) $\sigma_x = 30$, $\sigma_y = -30$, and $\tau_{xy} = 10$
c) $\sigma_x = 50$, $\sigma_y = -50$, and $\tau_{xy} = 0$
d) $\sigma_x = \sigma_y = -10$

All stresses are in megapascals.

2.60 Repeat Problem 2.59 for

a) $\sigma_x = 55$, $\sigma_y = -15$, and $\tau_{xy} = 40$
b) $\sigma_x = 0$, $\sigma_y = 30$, and $\tau_{xy} = 20$
c) $\sigma_x = -20$, $\sigma_y = 40$, and $\tau_{xy} = -40$
d) $\sigma_x = 30$, $\sigma_y = 0$, and $\tau_{xy} = -20$

All stresses are in megapascals.

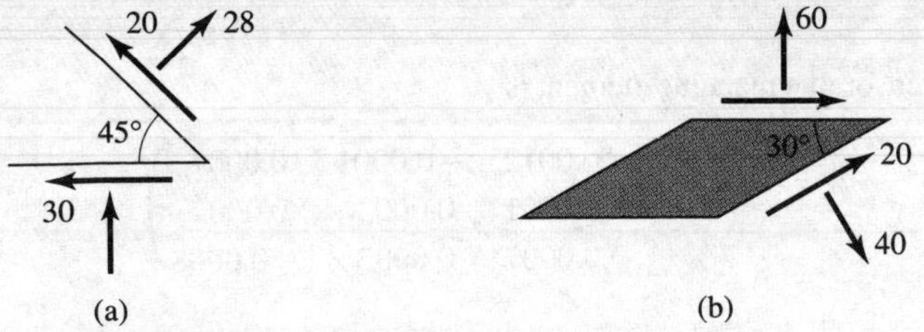

Sketch *ff*, for Problem 2.61

2.61 Given the state of stresses shown in the two parts of sketch *ff* determine the principal stresses and their directions by using a Mohr's circle and the stress equations. Show the stress elements. All stresses in sketch *ff* are in megapascals.

2.62 Given the normal and shear stresses $\sigma_x = 12$ ksi, $\sigma_y = 6$ ksi, and $\tau_{xy} = -4$ ksi, draw the Mohr's circle diagram and the principal normal and shear stresses on the *x-y* axis. Determine the triaxial stresses and give the corresponding Mohr's circle diagram. Also, determine the octahedral stresses.

2.63 Given the normal and shear stresses $\sigma_x = 16$ ksi, $\sigma_y = 9$ ksi, and $\tau_{xy} = 5$ ksi, draw the Mohr's circle diagram and the principal normal and shear stresses on the *x-y* axis. Determine the triaxial stresses and give the corresponding Mohr's circle diagram. Also, determine the octahedral stresses.

2.64 Given the normal and shear stresses $\sigma_x = 10$ ksi, $\sigma_y = 24$ ksi, and $\tau_{xy} = -6$ ksi, draw the Mohr's circle diagram and the principal normal and shear stresses on the *x-y* axis. Determine the triaxial stresses and give the corresponding Mohr's circle diagram. Also, determine the octahedral stresses.

Section 2.15

2.65 In a three-dimensional stress field used to describe the stresses below the surface of a ball in a ball bearing, the principal stresses are given as $\sigma_1 = 300$ MPa, $\sigma_2 = 200$ MPa, and $\sigma_3 = -2800$ MPa. Calculate and draw the Mohr's circle diagram.

Section 2.16

2.66 Given the normal and shear stresses $\sigma_x = -10$ ksi, $\sigma_y = 15$ ksi, and $\tau_{xy} = 5$ ksi, determine or draw the following:

a) Two-dimensional Mohr's circle diagram
b) Normal principal stress element on *x-y* axis
c) Shear principal stress on *x-y* axis
d) Three-dimensional Mohr's circle diagram and corresponding principal normal and shear stresses
e) Octahedral normal and shear stresses

2.67 A thin plate is loaded with stresses $\sigma_x = \sigma$, $\sigma_y = -\sigma$, $\sigma_z = 0$ (perpendicular to the plate), $\tau_{xy} = \tau_{yx} = \sigma/\sqrt{2}$, and $\tau_{xz} = \tau_{zx} = \tau_{yz} = \tau_{zy} = 0$. Calculate the octahedral normal and shear stresses.

Section 2.18

2.68 The strain tensor in a machine element is

$$\mathbf{T} = \begin{pmatrix} 0.0012 & -0.0001 & 0.0007 \\ 0.0001 & 0.0003 & 0.0002 \\ 0.0007 & 0.0002 & -0.0008 \end{pmatrix}$$

Find the strain in the x, y, and z directions, in the direction of the space diagonal $\left(\frac{1}{\sqrt{3}}; \frac{1}{\sqrt{3}}; \frac{1}{\sqrt{3}}\right)$, and in the direction ε_x, ε_y, and ε_z.

2.69 The strain tensor is

$$\mathbf{T} = \begin{pmatrix} 0.0023 & 0.0006 & 0 \\ 0.0006 & 0.005 & 0 \\ 0 & 0 & 0 \end{pmatrix}$$

Calculate the maximum shear strain and the principal strains.

Section 2.19

2.70 For the stresses in a stress element it is possible to find coordinate directions where the shear stresses are zero and the normal stresses are parallel with the new coordinate directions. Is it possible to determine the stress state in a machine element just by using two strain gages?

CHAPTER

3

SOLID MATERIALS

Molten iron is poured from a furnace to form a billet. (Courtesy of Weirton Steel Corporation)

Iron is taken from the earth, and copper is smelted from ore.
Man puts an end to the darkness;
he searches the farthest recesses for ore in the darkness.

The Bible (Job 28:2–3)

SYMBOLS

A	area, m^2
$\bar{a}$	linear thermal expansion coefficient, $(°C)^{-1}$
C	reference line
C_p	specific heat of material, J/(kg-°C)
C_R	relative cost
d	fiber diameter, m
E	modulus of elasticity, Pa
%EL	elongation, percent
G	shear modulus, Pa
K	bulk modulus, Pa
K_A	Archard wear constant, $(Pa)^{-1}$
K_t	thermal conductivity, W/(m-°C)
k	spring rate, N/m
l	length, m
l_{cr}	critical length, m
l_{fr}	length of specimen at fracture, m
l_s	sliding distance, m
l_0	length of specimen without load, m
m_a	mass of body, kg
P	force, N
p	normal pressure, Pa
p_l	limiting pressure, Pa
$\hat{Q}$	quantity of heat, J
r_i	inner radius, m
r_m	mean radius, m
r_o	atom size, m
S	strength, Pa
S_{fr}	fracture strength, Pa
S_u	ultimate strength , Pa
S_y	yield strength , Pa
T	temperature, °C
ΔT	temperature change, °C
t_h	thickness, m
U_r	modulus of resilience, Pa
v	volume fraction
W	weight, N
W_r	wear rate, m^2
x	volume of carbon fiber in plastic
α	rotational angle, deg
γ	shear strain
δ	deformation; deflection, m
ε	strain
ν	Poisson's ratio
ρ	density, kg/m^3
σ	normal stress, Pa
τ	shear stress, Pa
τ_f	fiber-matrix bond strength, Pa

Subscripts

a	axial
all	allowable
avg	average
c	composite; cross sectional
cr	critical
f	fiber
fr	at fracture
g	glass transition
i	inner
m	matrix; magnesia
max	maximum
s	steel
t	transverse
0	without load
1,2,3	principal axes

Superscripts

c	compression
t	tension

3.1 INTRODUCTION

Choosing the solid material is an important step in designing machine elements. Being able to exploit a material's potential and characteristics is essential to ensuring that the best material is used for a particular machine element. Therefore, knowing the properties of solid materials is essential. This chapter will classify, give the properties of, and select solid materials in a general sense. No manufacturing or availability considerations are discussed, even though such information weighs heavily on the minds of design engineers. Chapters 9 to 19, dealing with different machine elements, present more specific information regarding material selection for particular cases. The general knowledge obtained from this chapter will be of great use in later chapters.

3.2 DUCTILE AND BRITTLE MATERIALS

3.2.1 DUCTILE MATERIALS

Ductility measures the degree of plastic deformation sustained at fracture. A ductile material can be subjected to large strains before it ruptures. Designers often use ductile materials because they can absorb shock (or energy) and, if they become overloaded, will usually exhibit large deformations before failing. Also, stress concentrations (covered in Chapter 7) can be partially relieved through the deformations that can be achieved with ductile materials.

One way to specify a material as ductile is by the percent elongation (%EL) or

$$\%EL = \left(\frac{l_{fr} - l_0}{l_0}\right) \times 100\% \tag{3.1}$$

where

l_{fr} = length of specimen at fracture, m

l_0 = length of specimen without load, m

A **ductile material** is one with a large %EL before failure, arbitrarily defined as 5% or higher for the purposes of this text. Table A.1 (Appendix A) shows that the %ELs for low-carbon, medium-carbon, and high-carbon steels are 37%, 30%, and 25%, respectively. (Note that some of the material from tables in Appendix A also appears on the inside of the cover.) Thus, steel is ductile because it far exceeds the 5% elongation described in Eq. (3.1). Also note from Eq. (3.1) that the original length of the specimen l_0 is an important value because a significant portion of the plastic deformation at fracture is confined to the neck region. Thus, the magnitude of %EL will depend on the specimen length. The shorter l_0, the greater the fraction of total elongation from the neck and, consequently, the higher the value of %EL. Therefore, l_0 should be specified when percent elongation values are cited. The length of a specimen without load is commonly defined as 50 mm (Table A.1).

Figure 3.1(a) shows a test specimen of a ductile material in which necking (decreasing cross-sectional area) is occurring. Figure 3.1(b) shows the same specimen just as fracture occurs. Note the considerable amount of plastic deformation at fracture.

Example 3.1

GIVEN A flat plate is rolled into a cylinder with an inner radius of 100 mm and a wall thickness of 60 mm.

FIND Determine which of the three stainless steels in Table A.1 *cannot* be formed cold to the cylinder. Assume that the midplane of the plate does not experience either tension or compression and will thus not experience any elongation. Also, calculate the %EL for the three stainless steels.

Solution

The length without load is at the midplane of the plate, or

$$l_0 = 2\pi\left(r_i + \frac{t_h}{2}\right) = 2\pi\left(100 + \frac{60}{2}\right)(10^{-3}) \text{ m} = 816.8 \text{ mm}$$

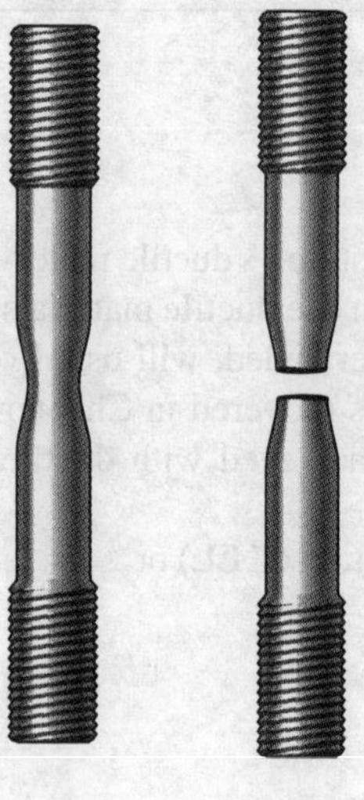

Figure 3.1 Ductile material from a standard tensile test apparatus. (a) Necking; (b) failure.

The length at fracture at the outer diameter of the cylinder is

$$l_{\text{fr}} = 2\pi r_o = 2\pi(r_i + t_h) = 2\pi(100 + 60)(10^{-3})\text{ m} = 1005\text{ mm}$$

Thus, from Eq. (3.1) the percent elongation is

$$\%\text{EL} = \left(\frac{l_{\text{fr}} - l_0}{l_0}\right)(100\%) = \left(\frac{1005 - 816.8}{816.8}\right)100\% = 23.1\%$$

From Table A.1 the ferritic type of AISI 446 steel can experience only 20% EL. Therefore, this material would crack if it was cold formed with a 23.1% deformation.

3.2.2 Brittle Materials

A **brittle material** exhibits little (%EL<5%) or no yielding before failure. Gray cast iron is an example of a brittle material whose %EL is so small that no listing is given in Table A.1. Figure 3.2 shows a brittle test specimen at failure. Little or no necking occurred prior to failure, in contrast to Fig. 3.1.

Example 3.2

GIVEN A malleable cast iron plate has bent due to an uneven cooling rate through the thickness of the plate during casting. The plate is 20 mm thick and is bent to a mean radius of 750 mm. Elastic springback is neglected.

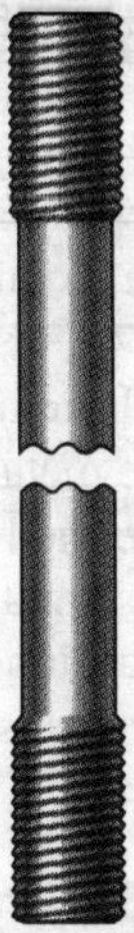

Figure 3.2 Failure of a brittle material from a standard tensile test apparatus.

FIND Is it possible to flatten the plate without cracking it?

Solution

$$\%EL = \left(\frac{t_h/2}{r_m - t_h/2}\right)100\% = \left(\frac{10}{750-10}\right)100\% = 1.33\%$$

From Table A.1 the %EL = 10% for malleable cast iron. Since 1.33% < 10%, the plate can be bent flat without risk of cracking.

3.3 CLASSIFICATION OF SOLID MATERIALS

Engineering materials fall into four major classes: metals, ceramics and glasses, polymers and elastomers, and composites. The members of each class generally have the following common features:

1. Similar properties, such as chemical makeup and atomic structure
2. Similar processing routes
3. Similar applications

3.3.1 METALS

Metals are combinations of metallic elements. They have large numbers of nonlocalized electrons (i.e., electrons not bound to particular atoms). Metals are extremely good conductors of electricity and heat and are not transparent to visible light; a polished metal surface has a lustrous appearance. Furthermore, metals are strong and usually deformable, making them extremely important materials in machine design.

Metals can be made stronger by alloying and by mechanical and heat treatment, and they are usually ductile. High-strength alloys, such as spring steel, can have ductilities as low as 2% or percent elongation as low as 2%, but even this is enough to ensure that the material yields before it fractures. Some cast parts can have very low ductility, however. Because metals are usually ductile, they are often used in circumstances where cyclic loading is applied (more in Chapter 7), so that they often fail from fatigue; and they are generally resistant to corrosion. Ductile materials, such as steel, accommodate stress concentrations by deforming in a way that redistributes the load more evenly; therefore, they can be used under static loads within a small margin of their yield strength.

3.3.2 Ceramics and Glasses

Ceramics are compounds of metallic and nonmetallic elements, most frequently oxides, nitrides, and carbides. For example, aluminum oxide (also known as alumina, carborundum, or in single crystal form, sapphire), is Al_2O_3. **Glasses** are made up of metallic and nonmetallic elements just as are ceramics, but glasses typically have no clear crystal structure. A typical soda-lime glass consists of approximately 70 wt% silicon dioxide (SiO_2), the balance being mainly soda (Na_2O) and lime (CaO). Both ceramics and glasses typically insulate against the passage of electricity and heat and are more resistant to high temperatures and harsh environments than are metals and polymers.

Ceramics and glasses, like metals, have high density. However, instead of being ductile like metals, ceramics and glasses (at room temperature) are brittle. They are typically 15 times stronger in compression than in tension. Ceramics and glasses cannot deform and so are not as easy to design machine elements from as are metals. Despite this, they have attractive features. They are stiff, hard, and abrasion resistant (hence their use for bearings and cutting tools). Thus, they must be considered as an important class of engineering material for use in machine elements.

Example 3.3

GIVEN A piece of ceramic is to be implanted into a stainless steel ring. The ceramic to be used is stabilized zirconia (ZrO_2), and the fit between the stainless steel and the zirconia is a medium press fit at room temperature. When the temperature fluctuates from room temperature to 500°C, the zirconia should not loosen.

FIND The correct type of stainless steel to be used from those given in Appendix A (Sec. A.1).

Solution

From Table A.3 (Appendix A) zirconia has a coefficient of thermal expansion of 1×10^{-5}/°C. Thus, since the zirconia should not loosen from the steel when the temperature is increased, a slightly smaller coefficient of thermal expansion, but very close to the zirconia value, is desired for the stainless steel. From Table A.1 the martensitic type of stainless steel (AISI 410) has a coefficient of thermal expansion of 0.99×10^{-5}/°C and would therefore be the type of stainless steel preferred.

3.3.3 POLYMERS AND ELASTOMERS

Polymers and **elastomers** include plastic and rubber materials. Many polymers are organic compounds chemically based on carbon, hydrogen, and other nonmetallic elements. Furthermore, they have large molecular structures.

Polymers are of two basic types: thermoplastics and thermosets. In general, thermoplastics are more ductile than thermosets, and at elevated temperatures they soften significantly and will melt. Thermosets are more brittle, do not soften as much as thermoplastics, and will usually degrade chemically before melting. **Thermoplastics** are long-chain molecules where the strength arises from interference between chains, similar to a hair ball. **Thermosets** are a networked structure, more like a sponge.

Elastomers have a networked structure, but not as elaborate as that for thermosets, so they will see large deformations at relatively light loads. A common elastomer is a rubber band, which displays the typical characteristics of large elastic deformation but brittle fracture. Further, the elastic properties of rubber bands are highly nonlinear.

Polymers and elastomers can be extremely flexible with large elastic deformations. Polymers are roughly five times less dense than metals but have nearly equivalent strength/weight ratio. Because polymers creep (the time-dependent permanent deformation that occurs under stress) even at room temperature, a polymer machine element under load may, with time, acquire a permanent set. The properties of polymers and elastomers change greatly with variations in temperature. For example, a polymer that is tough and flexible at 20°C may be brittle at the 4°C of a household refrigerator and yet creep rapidly at the 100°C of boiling water.

The mechanical properties of polymers are specified with many of the same parameters used for metals (i.e., modulus of elasticity and tensile, impact, and fatigue strengths). However, polymers vary much more in strength, stiffness, etc., than do metals. The main reasons for this variation are that, even with the same chemical constituents, two polymers may have different chain lengths and a varying number of atoms may be in a crystalline versus amorphous state. In addition, the mechanical characteristics of polymers, for the most part, are highly sensitive to the rate of deformation, the temperature, and the chemical nature of the environment (the presence of water, oxygen, organic solvents, etc.). Therefore, the particular values given for polymer mechanical properties should be used with caution. For most polymeric materials the simple stress-strain test is used to characterize some of these parameters.

Polymers are as strong per unit weight as metals. They are easy to shape: complicated parts performing several functions can be molded from a polymer in a single operation. However, injection molding operations are costly and can only be justified for large production runs. Large elastic deflections allow the design of polymer components that snap together, making assembly fast and cheap. Polymers are corrosion resistant and have low coefficients of friction.

3.3.4 COMPOSITES

Figure 3.3 compares a number of materials from a minimum-weight design standpoint (i.e., a larger strength/density ratio leads to a lighter design). The fibrous materials are far

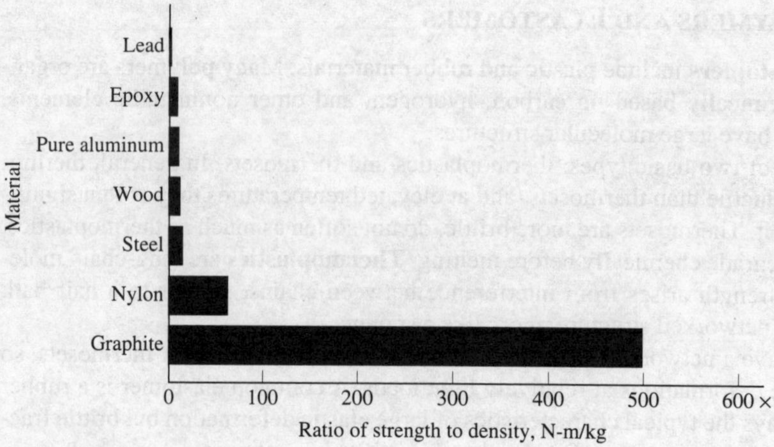

Figure 3.3 Strength/density ratio for various materials.

lighter than conventional extruded bars, molded plastics, and sintered ceramics. However, fibrous materials are notoriously susceptible to corrosion, even in air. For example, graphite fibers will oxidize readily in air and cannot provide their exceptional strength for long in an oxygen environment.

Many modern technologies require machine elements with unusual combinations of properties that cannot be met by conventional metal alloys, ceramics, and polymeric materials. Present-day technologies require solid materials that have low densities, are strong, stiff, and abrasion and impact resistant, and are not easily corroded. This combination of characteristics is rather formidable, considering that strong materials are usually relatively dense and that increasing stiffness generally decreases impact strength. Further, although fibers display some of these characteristics, they are easily corroded.

Composite materials combine the attractive properties of two or more material classes while avoiding drawbacks. A composite is designed to display a combination of the best characteristics of each component material. For example, graphite-reinforced epoxy acquires strength from the graphite fibers while the epoxy protects the graphite from oxidation. The epoxy also helps support shear stresses and provides toughness.

The three main types of composite material are

1. *Particle reinforced*—particles with approximately the same dimensions in all directions in a matrix, such as concrete
2. *Discontinuous fiber reinforced*—fibers of limited length-to-diameter ratio in a matrix, such as Fiberglas
3. *Continuous fiber reinforced*—continuous fibers constructed into a part by layers, such as graphite tennis rackets

Figure 3.4 shows a cross section of a fiber-reinforced composite material where the fibers are assumed to be long relative to their diameter. Most such composites contain

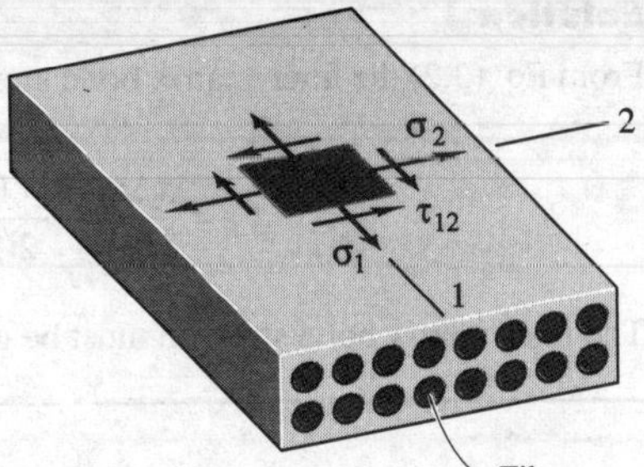

Figure 3.4 Cross section of fiber-reinforced composite material.

glass or carbon fibers and a polymer matrix. These composites cannot be used above 250°C because the polymer softens, but at room temperature their performance can be outstanding. Other disadvantages of using composites in machine elements are that they raise the price of the component considerably and are relatively difficult to form and join.

Composite materials have many characteristics that are different from those of the other three classes of material considered. Whereas metals, polymers, and ceramics are **homogeneous** (properties are not a function of position in the solid), **isotropic** (properties are the same in all directions at a point in the solid), or **anisotropic** (properties are different in all directions at a point in the solid), composites are *nonhomogeneous* and **orthotropic.** An orthotropic material has properties that are different in three mutually perpendicular directions at a point in the solid but has three mutually perpendicular planes of material symmetry. Consideration in this text is limited to simple, unidirectional, fiber-reinforced orthotropic composite materials, such as shown in Fig. 3.4.

An important parameter in fiber-reinforced composites is the fiber length. Some critical fiber length is necessary for effective strengthening and stiffening of the composite material. The critical length l_{cr} of the fiber depends on the fiber diameter d, its ultimate strength S_u, and the fiber-matrix bond strength τ_f, according to

$$l_{cr} = \frac{S_u d}{2\tau_f} \qquad (3.2)$$

The 2 occurs in Eq. (3.2) because the fiber is embedded in the matrix and splits into two parts at failure. For a number of glass- and carbon-fiber-reinforced composites this critical length is about 1 mm, or 20 to 150 times the fiber diameter.

Example 3.4

GIVEN Fiber-reinforced plastic contains carbon fibers having an ultimate strength of 1 GPa and a modulus of elasticity of 150 GPa. The fibers are 3 mm long with a diameter of 30 μm.

FIND How strong the fiber-matrix bond must be to ensure that the ultimate strength is safe.

Solution

From Eq. (3.2) the fiber-matrix bond strength can be expressed as

$$\tau_f = \frac{S_u d}{2l_{cr}} = \frac{10^9(30)(10^{-6})}{2(3)(10^{-3})} = 5\ 000\ 000\ \text{Pa} = 5\ \text{MPa}$$

The fiber-matrix bond strength must be greater than 5 MPa.

3.4 STRESS-STRAIN DIAGRAMS

The stress-strain diagram is important in designing machine elements because it yields data about a material's strength without regard to the material's physical size or shape. Because stress-strain diagrams differ considerably for the different classes of material covered in the previous sections, each class will be treated separately. The exception is that a stress-strain diagram for composites will not be presented because of the diverse natures of these materials.

3.4.1 METALS

Figure 3.5 shows the stress-strain diagram for a ductile metal. Although the stress shown in the figure is tensile, the stress-strain diagrams for metals are essentially the same for compression and tension. This feature is not true when dealing with polymers or ceramics.

Figure 3.6 better clarifies what is happening near the yield stress. A number of points presented in Figs. 3.5 and 3.6 need to be described:

1. **Proportional limit** (point P)—stress at which the stress-strain curve first deviates from a straight line (or the limit of linear-elastic theory, otherwise known as Hooke's law)
2. **Elastic limit** (point E)—highest stress the material can withstand and still return exactly to its original length when unloaded.

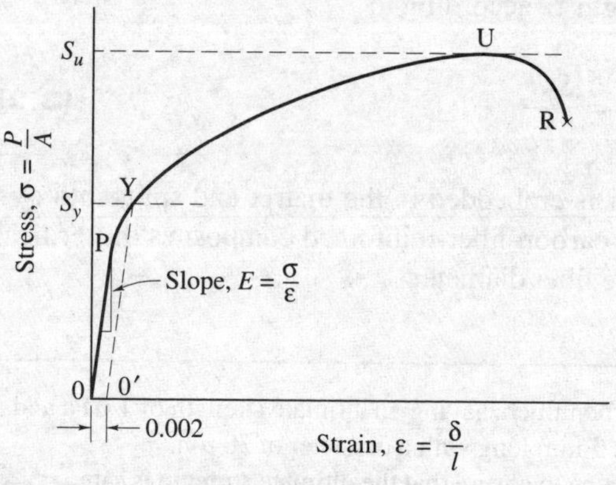

Figure 3.5 Stress-strain diagram for ductile metal.

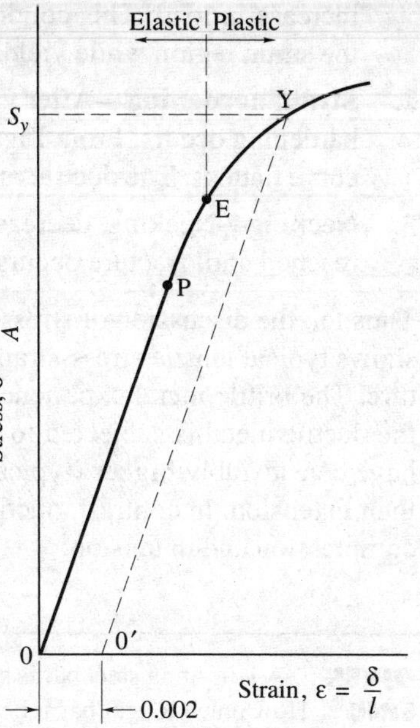

Figure 3.6 Typical stress-strain behavior for ductile metal showing elastic and plastic deformations and yield strength S_y.

3. **Yield strength** (point Y)—stress at which significant deformation first occurs. If the load is removed after yielding to point Y, the specimen exhibits a 0.2% permanent elongation.
4. **Ultimate strength** (point U)—maximum stress reached in the stress-strain diagram
5. **Fracture stress** (point R)—stress at the time of fracture or rupture. For some materials the ultimate stress is greater than the fracture stress.

Note that the elastic limit (point E) is not shown in Fig. 3.5 but is shown in Fig. 3.6 and clarifies what is happening near the elastic-plastic demarcation point. The true yield stress is difficult to determine experimentally. A solution is to use a small strain, usually 0.2%, and draw a straight line with its slope equal to the initial Young's modulus from the experiment. The point where this line intersects the stress-strain curve is known as the 0.2% offset yield point. In both figures the yield strength S_y is determined by 0.002 strain. That is, the distance in Fig. 3.6 between 0 and 0′ is 0.002 and is the permanent set. Loading occurs along 0PEY; unloading occurs along Y0′.

The strain portion of Figs. 3.5 and 3.6 can be divided into elastic and plastic behavior. The demarcation point is point E, the elastic limit. Three different phenomena occur within the plastic behavior:

1. **Yielding**—A slight increase in stress above the elastic limit (point E) will break down the material and cause it to deform permanently (plastic deformation). For stress just slightly larger than the yield stress the specimen continues to elongate with little

increase in load. This condition is not accurately depicted in Fig. 3.5 or 3.6 because the strain region while yielding is relatively small.

2. **Strain hardening**—After yielding and before reaching the ultimate strength S_u, strain hardening occurs. From Fig. 3.5 note that as the metal approaches S_u, the stress-strain curve flattens. The ductile metal becomes harder and stronger as it plastically deforms.
3. **Necking**—Necking, decreasing cross-sectional area, occurs after the ultimate stress is reached until fracture occurs, as shown in Fig. 3.1.

Thus far, the discussion of stress-strain diagrams has focused on ductile metals. Figure 3.7 shows typical tensile stress-strain diagrams for a brittle and a ductile metal loaded to fracture. The brittle metal experiences little or no plastic deformation upon fracture, whereas the ductile metal is subjected to large strains before fracture occurs. Also, brittle materials have considerably higher (typically 10 times or greater) ultimate strength in compression than in tension. In contrast, ductile materials have essentially the same ultimate strength in compression and in tension.

Example 3.5

GIVEN A 1-in.-long steel bar is stressed to its yield point and then released.
FIND How much longer has it become by this loading?

Solution

The definition of yield stress is that it leaves a permanent deformation of 0.2% of the original length.

$$\frac{2}{1000} \times 1 = 0.002 \text{ in.}$$

This is the permanent deformation.

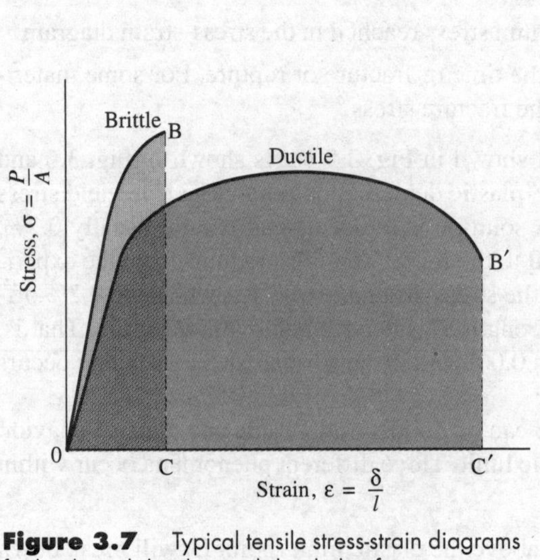

Figure 3.7 Typical tensile stress-strain diagrams for brittle and ductile metals loaded to fracture.

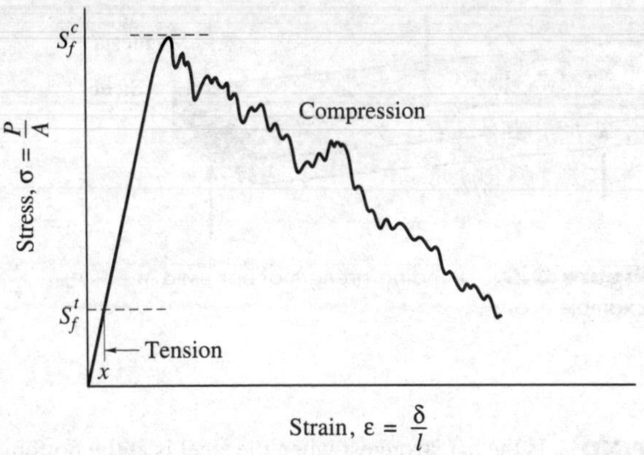

Figure 3.8 Stress-strain diagram for ceramic in tension and in compression.

3.4.2 CERAMICS

The stress-strain behavior of ceramics is not determined by the tensile test used for metals. The reason for this is twofold. First, it is difficult to prepare and test specimens having the required geometry; and second, there is a significant difference in results obtained from tests conducted in compression and in tension. Therefore, a more suitable transverse bending test is most frequently used, in which a specimen rod having either a circular or rectangular cross section is bent until fracture. Stress is computed from the specimen thickness, the bending moment, and the moment of inertia of the cross section. The maximum stress, or the stress at fracture, is known as the **modulus of rupture,** an important parameter used in determining the strength of ceramics.

The elastic stress-strain behavior for ceramic materials obtained in the transverse bending test is similar to the tensile test results for metals. A linear relationship exists for stress and strain. In Fig. 3.8 the stress-strain diagram for a ceramic shows that strength depends strongly on whether the loading is compressive or tensile. This is understandable because ceramics are notoriously brittle and, if loaded in tension, imperfections in the material become fracture initiation and propagation sites. In compression, on the other hand, defects such as microcracks are squeezed shut, so that they do not compromise the material's strength. This kind of behavior can also be seen with cast metals, which have large numbers of voids in their lattices. Strength for ceramics means fracture strength S_{fr}^t in tension and crushing strength S_{fr}^c in compression; typically, $S_{fr}^c = 15S_{fr}^t$. Once the crushing strength is reached, the strain increases significantly but the stress decreases. The x in the figure designates the elastic strain when the fracture strength in tension is reached.

Example 3.6

GIVEN A bar, shown in Fig. 3.9, consists of equal lengths and cross sections of magnesia and steel (AISI 1080). The magnesia and steel sections are glued together so that they bend together as one. The bar is simply supported and bending is tested by applying a force at its center.

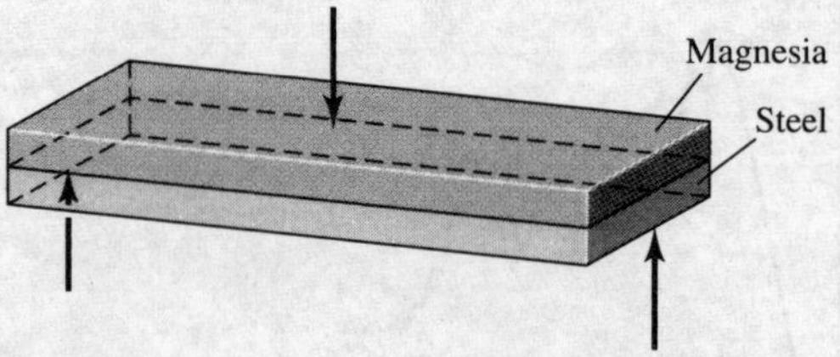

Figure 3.9 Bending strength of bar used in Example 3.6.

FIND Is the bar strongest when the steel is at the bottom (as shown in Fig. 3.9) or at the top?

Solution

From Table A.1 for AISI 1080 steel

$$E_s = 207 \text{ GPa} \qquad S_{ys} = 380 \text{ MPa} \qquad \text{and } S_{us} = 615 \text{ MPa}$$

From Table A.3 for magnesia

$$E_m = 207 \text{ GPa} \qquad \text{and } S_{um} = 105 \text{ MPa}$$

An important feature of ceramics is that they are 15 times stronger in compression than in tension, whereas steels have the same yield stress in compression or tension. In Fig. 3.9 the top member is in compression while the bottom member is in tension. Thus, for the steel at the bottom and magnesia at the top, as shown in Fig. 3.9, the magnesia is in compression with a compressive strength of $(15)(105)(10^6) = 1575$ MPa and is much stronger than the steel, which is in tension with a strength of 380 MPa. If the magnesia were at the bottom instead of the top, it would be in tension with a strength of 105 MPa and thus would be weaker than the steel, which would be in compression with a strength of 380 MPa. Taking the weakest members of the two bar combinations (since that is where it will fail), we find that the bar is about 380/105 = 3.642 times stronger when the steel is at the bottom in Fig. 3.9.

3.4.3 Polymers

The testing techniques and specimen configurations used to develop stress-strain diagrams for metals must be modified for polymers, especially for highly elastic materials, such as rubbers. It also helps to distinguish between thermoplastics and thermosets. Thermosets typically have little plastic deformation, so that their stress-strain diagrams in tension are essentially the same as Fig. 3.8.

Thermoplastics behave very differently. Figure 3.10 is a stress-strain diagram for a thermoplastic polymer below, at, and above its glass transition temperature. The glass transition temperature T_g is the temperature at which, upon cooling, a polymer transforms from a supercooled liquid to a solid. For $T \ll T_g$ the stress-strain diagram shows that the polymer fractures after relatively small strains. For $T \approx T_g$ the plastic can undergo enormous strains.

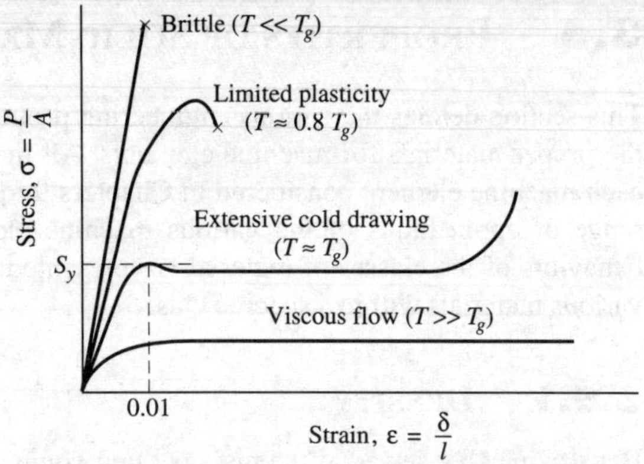

Figure 3.10 Stress-strain diagram for polymer below, at, and above its glass transition temperature T_g.

Elastomers, like thermosetting polymers, are brittle but can survive large strains before fracture (as can be seen with a rubber band). Elastomers may be elastic, but they are highly nonlinear. At the yield strength S_y the stress-strain diagram becomes markedly nonlinear, having typically a strain of 0.01, as shown in Fig. 3.10. This nonlinearity may be caused either by *shear yielding,* the irreversible slipping of molecular chains, or by *crazing,* the formation of low-density, crack-like volumes that scatter light, making the polymer look white. Polymers are slightly stronger (~20%) in compression than in tension. Finally, the deformation displayed for $T<<T_g$ is totally elastic.

These stress-strain diagrams for metals, polymers, and ceramics reveal the following characteristics of the ultimate strength S_u and the yield strength S_y of these materials:

1. For brittle solids (ceramics, glasses, and brittle polymers) the ultimate strength equals the yield strength ($S_u = S_y$).
2. For metals, ductile polymers, and most composites the ultimate strength is larger than the yield strength by a factor of 1.1 to 4. The reason for this is mainly work hardening or, in the case of composites, load transfer to the reinforcement.

These are important observations to keep in mind when selecting solid materials.

Example 3.7

GIVEN A plastic cup is made of polymethylmethacrylate. The room-temperature elongation at breakage is 5% and the glass transition temperature is 90°C. The cup is sterilized by being washed with 100°C superheated steam at a high pressure, stressing the plastic to 30 MPa.

FIND Can the cup be expected to maintain its shape during the washing?

Solution

Since the washing temperature is 10°C above the glass transition temperature, the plastic will deform by more than 5% (see Fig. 3.10). The stress is approximately half the ultimate strength at room temperature. It can therefore be concluded that the cup will deform during washing.

3.5 PROPERTIES OF SOLID MATERIALS

This section defines the various engineering properties of solid materials needed to select the proper materials for machine elements. All these properties may not be important for each machine element considered in Chapters 9 to 19, but they are important for the wide range of applications of the various machine elements. For each property the relative behaviors of the classes of material are presented as well as the relative behaviors of the various materials within a specific class.

3.5.1 DENSITY

Density may be viewed as the mass per unit volume. The SI unit of density is kilograms per cubic meter. In the English system density is in pounds-mass per cubic inch. Typical densities of solid materials lie between 10^3 and 10^4 kg/m^3. Figure 3.11 illustrates the density ranking of

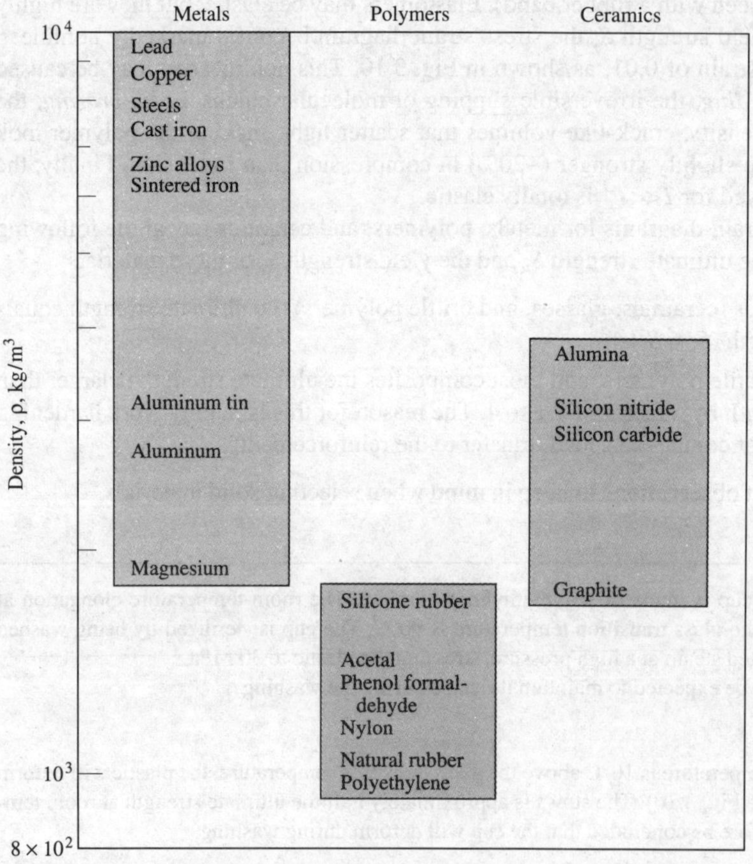

Figure 3.11 Density for various metals, polymers, and ceramics at room temperature (20°C; 68°F). [From ESDU Data Item No. 84041 (1984).]

Table 3.1 Density for various metals, polymers, and ceramics at room temperature (20°C; 68°F). [From ESDU Data Item No. 84041 (1984).]

Material	Density, ρ kg/m^3	lbm/in.^{3}
Metals		
Aluminum and its alloys[a]	2.7×10^3	0.097
Aluminum tin	3.1×10^3	.11
Babbitt, lead-based white metal	10.1×10^3	.36
Babbitt, tin-based white metal	7.4×10^3	.27
Brasses	8.6×10^3	.31
Bronze, aluminum	7.5×10^3	.27
Bronze, leaded	8.9×10^3	.32
Bronze, phosphor (cast)[b]	8.7×10^3	.31
Bronze, porous	6.4×10^3	.23
Copper	8.9×10^3	.32
Copper lead	9.5×10^3	.34
Iron, cast	7.4×10^3	.27
Iron, porous	6.1×10^3	.22
Iron, wrought	7.8×10^3	.28
Magnesium alloys	1.8×10^3	.065
Steels[c]	7.8×10^3	.28
Zinc alloys	6.7×10^3	.24
Polymers		
Acetal (polyformaldehyde)	1.4×10^3	.051
Nylons (polyamides)	1.14×10^3	.041
Polyethylene, high density	$.95 \times 10^3$	.034
Phenol formaldehyde	1.3×10^3	.047
Rubber, natural[d]	1.0×10^3	.036
Rubber, silicone	1.8×10^3	.065
Ceramics		
Alumina (Al_2O_3)	3.9×10^3	.14
Graphite, high strength	1.7×10^3	.061
Silicon carbide (SiC)	2.9×10^3	.10
Silicon nitride (Si_3N_4)	3.2×10^3	.12

[a]Structural alloys.
[b]Bar stock, typically 8.8×10^3 kg/m^3 (0.03 lbm/in.^{3}).
[c]Excluding "refractory" steels.
[d]"Mechanical" rubber.

various metals, polymers, and ceramics. Metals, such as lead, copper, and steel, have the highest mass density. Polymers, such as nylon, natural rubber, and polyethylene, have the lowest mass density. Table 3.1 gives quantitative values of density at room temperature (20°C).

Metals are dense because they are made of heavy atoms in an efficient packing. Ceramics, for the most part, have lower densities than metals because they contain oxygen, nitrogen, and carbon atoms, which are light, but they are also efficiently packed. Polymers have low densities because they are mainly made of carbon (atomic weight, 12) and hydrogen, and they are never totally crystalline. Thus, there is a lot of "wasted space" in a polymer.

Alloying changes the density of metals only slightly. To a first approximation the highest density of an alloy (metallic solid resulting from dissolving two or more molten metals in each other) is given by the **rule of mixtures** (i.e., by a linear interpolation between the densities of the alloy concentrations). The rule of mixtures may also be applied to composites.

3.5.2 MODULUS OF ELASTICITY, POISSON'S RATIO, AND SHEAR MODULUS

The **modulus of elasticity** (sometimes called the elastic constant or **Young's modulus**) is defined as the slope of the linear-elastic part of the stress-strain curve. In Fig. 3.6 the linear portion of the stress-strain curve is between the origin and point P, or stress less than the proportional limit stress. Recall the description of average normal stress in a uniaxial stress state in Eq. (2.6) and average normal strain in Eq. (2.33). The modulus of elasticity can be written as

$$E = \frac{\sigma_{avg}}{\varepsilon_{avg}} \tag{3.3}$$

Recall that the stress acts axially or normally. The SI unit of modulus of elasticity is that of stress, namely, pascals or newtons per square meter. Figure 3.12 illustrates the modulus of elasticity ranking for various metals, polymers, and ceramics at room temperature (20°C). The moduli of elasticity for metals and ceramics are high and quite similar, but those for polymers are considerably lower. Table 3.2 gives quantitative values of the moduli of elasticity, respectively, for various metals, polymers, and ceramics at room temperature. Note that in Table 3.2 the modulus of elasticity is in gigapascals and million pounds per square inch.

The moduli of elasticity of most materials depend on two factors: bond stiffness and bond density per unit area. A bond is like a spring: it has a spring rate k (in newtons per meter). The modulus of elasticity E is roughly

$$E = k/r_o$$

where

r_o = atom size (this can be obtained from the mean atomic volume $4\pi r_o^3/3$, which is generally known)

The wide range of modulus of elasticity in Fig. 3.12 and Table 3.2 is largely caused by the range of k in materials. The covalent bond is stiff (k = 20 to 200 N/m); the metallic and ionic bonds are somewhat less stiff (k = 15 to 100 N/m). Diamond, although not shown in Fig. 3.12 or Table 3.2, has a very high modulus of elasticity because the carbon atom is small (giving a high bond density) and its atoms are linked by extremely strong bonds (200 N/m). Metals have a high modulus of elasticity because close packing gives a high bond

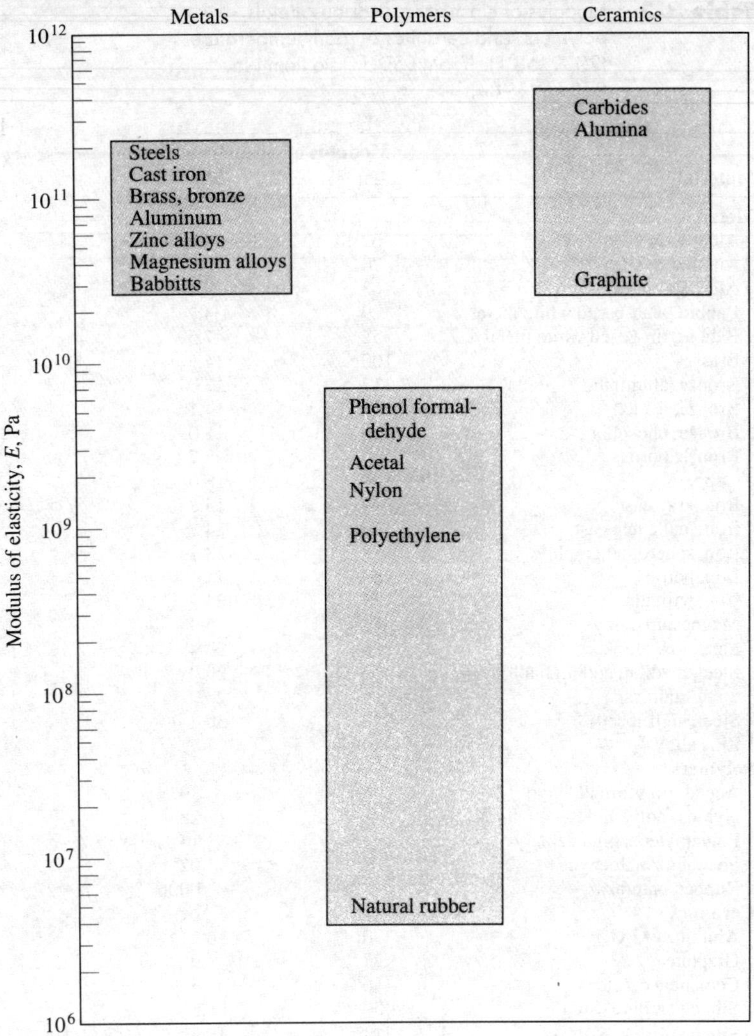

Figure 3.12 Modulus of elasticity for various metals, polymers, and ceramics at room temperature (20°C; 68°F). [From ESDU Data Item No. 84041 (1984).]

density and the bonds are strong, although not as strong as those of diamond. Polymers contain both strong diamond-like covalent bonds and weak hydrogen (or van der Waals) bonds ($k = 0.5$ to 2 N/m); these weak bonds stretch when a polymer is deformed, giving a lower modulus of elasticity. Elastomers have a low modulus of elasticity because the weak secondary bonds have melted, leaving only the extremely weak entropic restoring force, which is associated with tangled, long-chain molecules when the material is loaded.

Table 3.2 Modulus of elasticity for various metals, polymers, and ceramics at room temperature (20°C; 68°F). [From ESDU Data Item No. 84041 (1984).]

Material	Modulus of elasticity, E GPa	Mpsi
Metals		
Aluminum	62	9.0
Aluminum alloys[a]	70	10.2
Aluminum tin	63	9.1
Babbitt, lead-based white metal	29	4.2
Babbitt, tin-based white metal	52	7.5
Brasses	100	14.5
Bronze, aluminum	117	17.0
Bronze, leaded	97	14.1
Bronze, phosphor	110	16.0
Bronze, porous	60	8.7
Copper	124	18.0
Iron, gray cast	109	15.8
Iron, malleable cast	170	24.7
Iron, spheroidal graphite[b]	159	23.1
Iron, porous	80	11.6
Iron, wrought	170	24.7
Magnesium alloys	41	5.9
Steel, low alloys	196	28.4
Steel, medium and high alloys	200	29.0
Steel, stainless[c]	193	28.0
Steel, high speed	212	30.7
Zinc alloys[d]	50	7.3
Polymers		
Acetal (polyformaldehyde)	2.7	.39
Nylons (polyamides)	1.9	.28
Polyethylene, high density	.9	.13
Phenol formaldehyde[e]	7.0	1.02
Rubber, natural[f]	.004	.0006
Ceramics		
Alumina (Al_2O_3)	390	56.6
Graphite	27	3.9
Cemented carbides	450	65.3
Silicon carbide (SiC)	450	65.3
Silicon nitride (Si_3N_4)	314	45.5

[a]Structural alloys.
[b]For bearings.
[c]Precipitation-hardened alloys up to 211 GPa (30 Mpsi).
[d]Some alloys up to 96 GPa (14 Mpsi).
[e]Filled.
[f]25%-carbon-black "mechanical" rubber.

Although no stress acts transversely to the axial direction, there will nevertheless be dimensional changes in the transverse direction, for as a bar experiences an axial loading and extends axially, it contracts transversely. The transverse average strain $\varepsilon_{t,\text{avg}}$ is related to the axial average strain $\varepsilon_{a,\text{avg}}$ by **Poisson's ratio** ν such that

$$\varepsilon_{t,\text{avg}} = -\nu\varepsilon_{a,\text{avg}} \tag{3.4}$$

Table 3.3 Poisson's ratio for various metals, polymers, and ceramics at room temperature (20°C; 68°F). [From ESDU Data Item No. 84041 (1984).]

Material	Poisson's ratio, ν
Metals	
Aluminum and its alloys[a]	0.33
Aluminum tin	—
Babbitt, lead-based white metal	—
Babbitt, tin-based white metal	—
Brasses	.33
Bronze	.33
Bronze, porous	.22
Copper	.33
Copper lead	—
Iron, cast	.26
Iron, porous	.20
Iron, wrought	.30
Magnesium alloys	.33
Steels	.30
Zinc alloys	.27
Polymers	
Acetal (polyformaldehyde)	—
Nylons (polyamides)	.40
Polyethylene, high density	.35
Phenol formaldehyde	—
Rubber	.50
Ceramics	
Alumina (Al_2O_3)	.28
Graphite, high strength	—
Cemented carbides	.19
Silicon carbide (SiC)	.19
Silicon nitride (Si_3N_4)	.26

[a]Structural alloys.

The negative sign simply means that the transverse deformation will be in the opposite sense to the axial deformation. Poisson's ratio is dimensionless. Table 3.3 gives quantitative values of Poisson's ratio for various metals, polymers, and ceramics at room temperature. The highest Poisson's ratio is 0.5 for rubber, and the lowest is 0.19 for silicon carbide and cemented carbides. In fact, Poisson's ratio cannot be less than zero (or else the second law of thermodynamics would be violated), nor can it exceed 0.5 (or else a material's volume would increase when compressed).

Shear stress and strain are proportional to each other through the expression

$$\tau_{avg} = G\gamma_{avg} \tag{3.5}$$

where

G = shear modulus or modulus of rigidity, N/m^2

This relation is only true for the linear-elastic region of the shear stress-strain curve (from 0 to P in Fig. 3.6). The unit of G is the same as that for the modulus of elasticity E because γ is measured in radians, a dimensionless quantity.

The three material properties E, G, and ν are related by the following equation:

$$G = \frac{E}{2(1+\nu)} \tag{3.6}$$

Thus, when two parameters are known, the third can easily be determined from Eq. (3.6). That is, if for a particular material the modulus of elasticity is obtained from Table 3.1 and Poisson's ratio from Table 3.3, the shear modulus can be obtained from Eq. (3.6). The material presented thus far in this section (Sec. 3.5.2) is valid for metals, polymers, or ceramics. To establish the modulus of elasticity for a unidirectional fiber-reinforced composite in the direction of the fibers, it is assumed that the fiber-matrix interfacial bond is good, so that deformation of both matrix and fibers is the same. Under these conditions the total load sustained by the composite P_c is equal to the loads carried by the matrix P_m and the fiber P_f, as

$$P_c = P_m + P_f \tag{3.7}$$

where subscripts c, m, and f refer to composite, matrix, and fiber, respectively.

Substituting Eq. (3.7) into the definition of stress [Eq. (2.7)] gives

$$\sigma_{c,\text{avg}} = \sigma_{m,\text{avg}} \frac{A_m}{A_c} + \sigma_{f,\text{avg}} \frac{A_f}{A_c} \tag{3.8}$$

If the composite, matrix, and fiber lengths are equal, Eq. (3.8) becomes

$$\sigma_{c,\text{avg}} = \sigma_{m,\text{avg}} v_m + \sigma_{f,\text{avg}} v_f \tag{3.9}$$

where

v_m = volume fraction of matrix

v_f = volume fraction of fiber

Because the same deformation of matrix and fibers was assumed and the composite consists of only matrix and fibers (i.e., $v_m + v_f = 1$), Eq. (3.9) can be rewritten in terms of the modulus of elasticity, instead of the stress, as

$$E_c = E_m v_m + E_f v_f \tag{3.10}$$

or

$$E_c = E_m(1 - v_f) + E_f v_f \tag{3.11}$$

Thus, Eq. (3.11) enables the composite's modulus of elasticity to be determined when the moduli of elasticity of the matrix and the fiber and the volume fractions of each are known. It can also be shown that the ratio of the load carried by the fibers to that carried by the matrix is

$$\frac{P_f}{P_m} = \frac{E_f v_f}{E_m v_m} \tag{3.12}$$

Recall that Eqs. (3.7) to (3.12) are only applicable for unidirectional, fiber-reinforced composites.

Example 3.8

GIVEN A fiber-reinforced plastic contains 10 vol% glass fibers ($E = 70$ GPa, $S_u = 0.7$ GPa).
FIND Calculate how this fiber percentage has to be changed to give the same elastic properties if the glass fibers are changed to carbon fibers ($E = 150$ GPa, $S_u = 1$ GPa). For the matrix material $E_m = 2$ GPa.

Solution

According to Eq. (3.10) the modulus of elasticity for a fiber composite is

$$E_c = E_m v_m + E_f v_f$$

where

E_m = modulus of elasticity for matrix material, which has volume fraction v_m

E_f = modulus of elasticity for fiber material, which has volume fraction v_f

$$E_c = 2\times10^9 \times 0.9 + 70\times10^9 \times 0.1 = 8.8\times10^9 \text{ Pa} = 8.8 \text{ GPa}$$

This composite modulus of elasticity should also be maintained for the carbon-reinforced plastic.

$$8.8\times10^9 = 2\times10^9 \times x + 150\times10^9(1-x)$$

$$148x = 141.2$$

$$x = 0.954 \quad \text{or} \quad 1-x = 0.046$$

The plastic should thus contain 4.6 vol% carbon fibers and 95.4 vol% matrix material to get the same elastic properties as the glass-fiber-reinforced plastic.

3.5.3 STRENGTH

The strength of a machine element depends on the class, treatment, and geometry of the specimen and also on the type of loading that the machine element will experience. This section focuses on the strength of the various classes of material and then deals with the type of loading.

The various types of loading that a material experiences are important. Design deals with allowable stresses, or reduced value of strength. The allowable normal stress σ_{all} and the allowable shear stress τ_{all} for ferrous and nonferrous metals for various types of loading may be represented by

Tension: $$0.45S_y \le \sigma_{all} \le 0.60S_y \tag{3.13}$$

Shear: $$\tau_{all} = 0.40S_y \tag{3.14}$$

Bending: $$0.60S_y \le \sigma_{all} \le 0.75S_y \tag{3.15}$$

Bearing: $$\sigma_{all} = 0.9S_y \tag{3.16}$$

These relationships can also be applied to polymers and ceramics if ultimate strength at the break and fracture strength, respectively, are substituted for yield strength in Eqs, (3.13) to (3.16). It should be recalled from Sec.1.5.1.1 that other factors besides type of loading are incorporated into the safety factor.

3.5.3.1 METALS Metals can be divided into ferrous and nonferrous alloys. Ferrous alloys are those in which iron is the primary component, but carbon as well as other alloying elements may be present. Nonferrous alloys are all those alloys that *are not* iron based.

The strength of metals will be directly associated with the yield strength of the material. The yield strength S_y is determined by 0.002 of the strain after unloading (see Fig. 3.6). The strength of metals is essentially equal in compression or in tension. Tables A.1 and A.2 show the yield strengths for ferrous and nonferrous metals, respectively.

3.5.3.2 POLYMERS The strength of polymers is determined by a total strain of 0.010, as opposed to a nonrecoverable plastic strain of 0.002 for metals. Also, when dealing with polymers the strength of interest is the ultimate strength at the break rather than the yield strength as for metals. The tensile strengths at the break for selected thermoplastic and thermosetting polymers are given in Table A.4. The other unique characteristic of polymers is that they are stronger (~20%) in compression than in tension.

3.5.3.3 CERAMICS The strength of interest for ceramics is the fracture strength. Ceramics, being brittle materials, are much stronger in compression (typically, 15 times) than in tension. Table A.3 gives fracture strength in tension for selected ceramic materials.

Example 3.9

GIVEN For the contact pressure distribution between a sphere and a flat surface, H. Hertz calculated in 1881 that the maximum contact pressure in the middle of the contact area was approximately 10 times larger than the maximum tensile stress at the edge of the contact area. When a steel ball is dropped on a thick glass plate, a circular crack is formed, the size of which approximately equals the maximum Hertzian contact size.

FIND How much stronger is the glass in compression than in tension?

Solution

Since the glass is not crushed by the high contact pressure but by the tensile stress, it is concluded that the compressive strength of the glass is at least 10 times the tensile strength.

3.5.4 RESILIENCE AND TOUGHNESS

3.5.4.1 RESILIENCE **Resilience** is the capacity of a material to absorb energy when it is deformed elastically and then, upon unloading, to release this energy. The modulus of resilience U_r is the strain energy per unit volume required to stress a material from an unloaded state to the point of yielding. Mathematically, this is expressed as

$$U_r = \int_0^{\varepsilon_y} \sigma d\varepsilon \qquad \textbf{(3.17)}$$

where

ε_y = strain corresponding to yield strength S_y

For the linear-elastic region the area below the stress-strain diagram is the modulus of resilience or

$$U_r = S_y \varepsilon_y / 2 \tag{3.18}$$

The SI unit of modulus of resilience is pascals (Pa).

Making use of Eq. (3.3) while dropping the "average" subscript gives Eq. (3.18) as

$$U_r = S_y^2 / 2E \tag{3.19}$$

Thus, resilient materials are those having high yield strengths and low moduli of elasticity. An example of a material with high modulus of resilience is high-carbon steel. Table A.1 shows that the yield strength for high-carbon steel is highest when the modulus of elasticity is high, but the difference between the highest and lowest values given is small. Thus, high-carbon steels have a high modulus of resilience. This property is extremely useful in selecting a material for springs (see Chapter 16), making high-carbon steel alloys primary candidate materials for springs.

3.5.4.2 TOUGHNESS **Toughness** is a material's ability to absorb energy up to fracture. Specimen geometry and manner of load application are important in determining the toughness of a material. Fracture toughness indicates a material's resistance to fracture when a crack is present.

For the static (low strain rate) situation, toughness can be obtained from the stress-strain curve (e.g., see Fig. 3.5) up to the point of fracture or rupture. Resilience is the strain energy per unit volume up to the yield strength of the material (point Y in Fig. 3.5); whereas toughness is energy per unit volume to rupture (point R in Fig. 3.5). The unit of toughness is the same as that for resilience (pascals).

For a material to be tough it must display both strength and ductility; and often, ductile materials are tougher than brittle materials, as demonstrated in Fig. 3.7. Even though the brittle material has higher yield and ultimate strengths, by virtue of lack of ductility, it has lower toughness than the ductile material. That is, the area of a ductile material (given as 0B′C′) is considerably larger than the area of a brittle material (given as 0BC).

Example 3.10

GIVEN In a mining operation the iron ore is dumped into a funnel for further transport by train. The inside of the funnel wears rapidly because the impact of the ore makes plastic deformations in the funnel surface. A change of surface material is considered.

FIND Which is the most efficient choice of funnel surface material, hard steel or rubber?

Solution

A key parameter to be used in this evaluation is the resilience of the two materials. For high-carbon steel (AISI 1080) from Table A.1

$$S_y = 380 \text{ MPa} \quad \text{and} \quad E = 207 \text{ GPa}$$

From Eq. (3.19) the modulus of resilience for steel is

$$(U_r)_{\text{steel}} = \frac{S_y^2}{2E} = \frac{(380)^2(10^{12})}{2(207)(10^9)} = 348\ 800 \text{ Pa} = 0.3488 \text{ MPa}$$

For natural rubber from Tables A.5 and Eq. (3.4)

$$S_u = 30 \text{ MPa} \quad \text{and} \quad E = 0.004 \text{ GPa}$$

$$\therefore (U_r)_{\text{rubber}} = \frac{(30)^2(10^{12})}{2(4)(10^6)} = 112.5(10^6) \text{ Pa} = 112.5 \text{ MPa}$$

Rubber is over two orders of magnitude more resilient than steel. The rubber is able to elastically deform on impact, whereas the steel plastically deforms and thus has a higher wear rate. The inside of the funnel should have a rubber lining.

3.5.5 THERMAL CONDUCTIVITY

The rate at which heat is conducted through a solid at steady state (meaning that temperature does not vary with time) is a measure of the **thermal conductivity** K_t. When two bodies at different temperatures are brought together, the faster-moving molecules of the warmer body collide with the slower-moving molecules of the cooler body and transfer some of their motion to the latter. The warmer body loses energy (drops in temperature) while the cooler one gains energy (rises in temperature). The transfer process stops when the two bodies reach the same temperature. This transfer of molecular motion through a material is called heat conduction. Materials differ in how fast they let this transfer go on. The SI unit of thermal conductivity K_t is watts per meter-Celsius; and the English unit is British thermal units per foot-hour-Fahrenheit.

Figure 3.13 illustrates the thermal conductivity ordering for various metals, polymers, and ceramics. The metals and ceramics in general are good conductors (high K_t) and the polymers are good insulators (low K_t). Table 3.4 quantifies the thermal conductivity results given in Fig. 3.13. In Fig. 3.13 and Table 3.4, unless otherwise stated, the temperature is assumed to be room temperature (20°C; 68°F).

3.5.6 LINEAR THERMAL EXPANSION COEFFICIENT

Different materials expand at different rates when heated. A solid object elongates by a certain fraction for each degree rise in temperature. This rate is constant over a fairly large temperature range and, once measured, can be used to determine how much an object will expand for a given change in temperature. This rate is given for each material by a number called linear expansivity or linear **thermal expansion coefficient** $\bar{a}$. The SI unit of $\bar{a}$ is $(°\text{C})^{-1}$; the English unit is $(°\text{F})^{-1}$. If the material is isotropic, the volume expansion per degree is $3\bar{a}$. Figure 3.14 illustrates the linear thermal expansion coefficient ordering for various metals, polymers, and ceramics applied over the temperature range 20 to 200°C (68 to 400°F). The polymers have the highest $\bar{a}$, followed by the metals and then the ceramics. Table 3.5 gives quantitative values of $\bar{a}$ for various metals, polymers, and ceramics from 20 to 200°C (68 to 392°F).

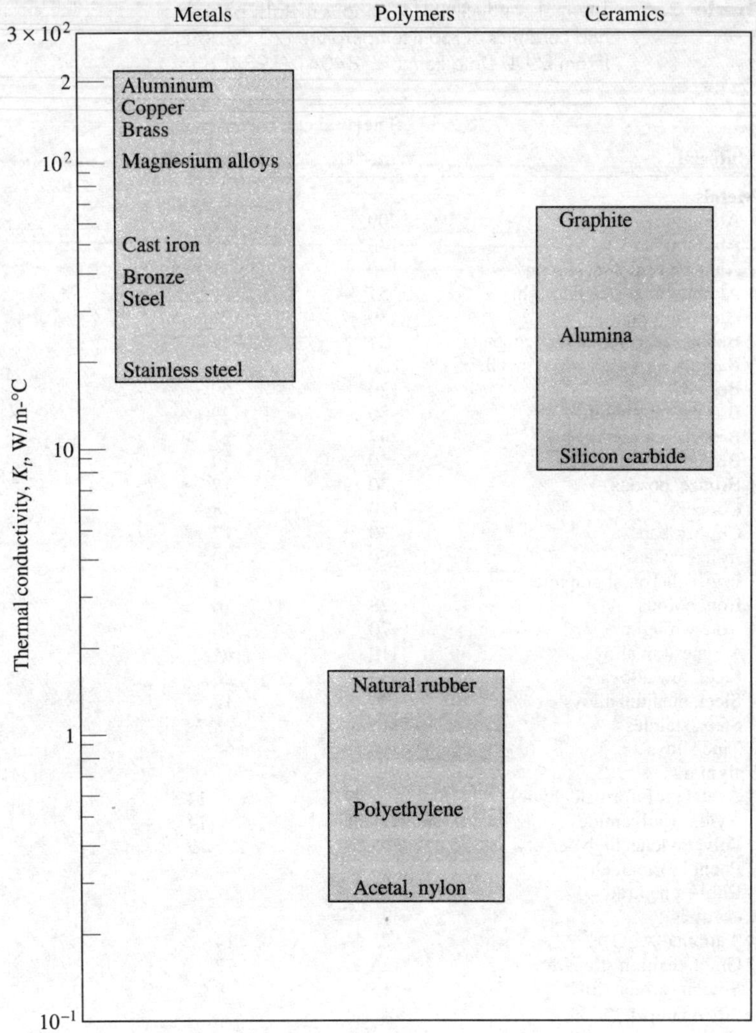

Figure 3.13 Thermal conductivity for various metals, polymers, and ceramics at room temperature (20°C; 68°F). [From ESDU Data Item No. 84041 (1984).]

3.5.7 SPECIFIC HEAT CAPACITY

The nature of a material determines the amount of heat transferred to or from a body when its temperature changes by a given amount. Imagine an experiment in which a cast iron ball and a babbitt (lead-based white metal) ball of the same size are heated to the temperature of boiling water and then laid on a block of wax. The cast iron ball would melt a considerable amount of wax, but the babbitt ball, in spite of its greater mass, would melt hardly any. It therefore would seem that different materials, in cooling through the same temperature range, give up different amounts of heat.

Table 3.4 Thermal conductivity for various metals, polymers, and ceramics at room temperature (20°C; 68°F). [From ESDU Data Item No. 84041 (1984).]

Material	Thermal conductivity, K_t W/m-°C	Thermal conductivity, K_t Btu/ft-hr-°F
Metals		
Aluminum	209	120
Aluminum alloys, cast[a]	146	84
Aluminum alloys, silicon[b]	170	98
Aluminum alloys, wrought[c]	151	87
Aluminum tin	180	100
Babbitt, lead-based white metal	24	14
Babbitt, tin-based white metal	56	32
Brasses[a]	120	69
Bronze, aluminum[a]	50	29
Bronze, leaded	47	27
Bronze, phosphor (cast)[d]	50	29
Bronze, porous	30	17
Copper[a]	170	98
Copper lead	30	17
Iron, gray cast	50	29
Iron, spheroidal graphite	30	17
Iron, porous	28	16
Iron, wrought	70	40
Magnesium alloys	110	64
Steel, low alloys[e]	35	20
Steel, medium alloys	30	17
Steel, stainless[f]	15	8.7
Zinc alloys	110	64
Polymers		
Acetal (polyformaldehyde)	.24	.14
Nylons (polyamides)	.25	.14
Polyethylene, high density	.5	.29
Phenol formaldehyde	—	—
Rubber, natural	1.6	.92
Ceramics		
Alumina (Al_2O_3)[g]	25	14
Graphite, high strength	125	72
Silicon carbide (SiC)	15	8.6
Silicon nitride (Si_3N_4)	—	—

[a] At 100°C.
[b] At 100°C (~150 W/m-°C at 25°C).
[c] 20 to 100°C.
[d] Bar stock, typically 69 W/m-°C.
[e] 20 to 200°C.
[f] Typically 22 W/m-°C at 200°C.
[g] Typically 12 W/m-°C at 400°C.

The quantity of heat energy given up or taken on when a body changes its temperature is proportional to the mass of the object, the amount that its temperature changes, and a characteristic number called the **specific heat capacity** of the material the body is made from:

$$\hat{Q} = C_p m_a (\Delta T) \quad \textbf{(3.20)}$$

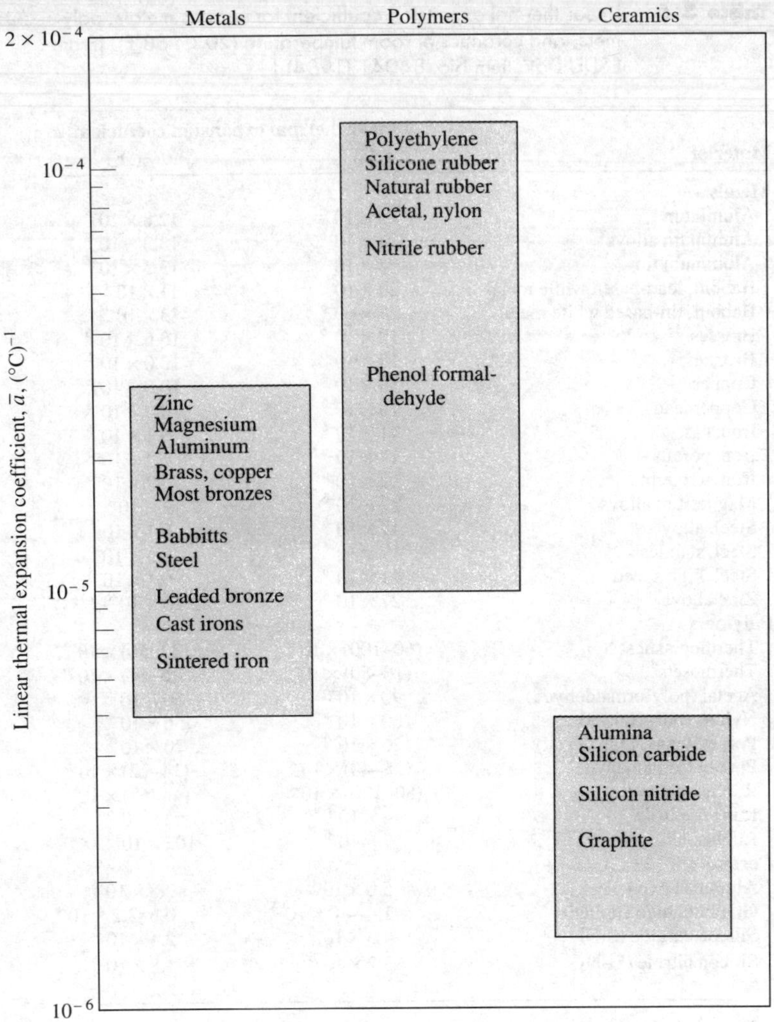

Figure 3.14 Linear thermal expansion coefficient for various metals, polymers, and ceramics applied over temperature range 20 to 200°C (68 to 392°F). [From ESDU Data Item No. 84041 (1984).]

where

$\hat{Q}$ = quantity of heat, J

C_p = specific heat of material, J /(kg-°C)

m_a = mass of body, kg

ΔT = temperature change, °C

Figure 3.15 illustrates the specific heat capacity ordering of various metals, polymers, and ceramics at room temperature (20°C; 68°F). Polymers have considerably higher specific heat than metals or ceramics. Table 3.6 quantifies the information presented in Fig. 3.15.

Table 3.5 Linear thermal expansion coefficient for various metals, polymers, and ceramics at room temperature (20°C; 68°F). [From ESDU Data Item No. 84041 (1984).]

Material	Linear thermal expansion coefficient, a $(°C)^{-1}$	$(°F)^{-1}$
Metals		
Aluminum	23×10^{-6}	12.8×10^{-6}
Aluminum alloys[a]	24×10^{-6}	13.3×10^{-6}
Aluminum tin	24×10^{-6}	13.3×10^{-6}
Babbitt, lead-based white metal	20×10^{-6}	11×10^{-6}
Babbitt, tin-based white metal	23×10^{-6}	13×10^{-6}
Brasses	19×10^{-6}	10.6×10^{-6}
Bronzes	18×10^{-6}	10.0×10^{-6}
Copper	18×10^{-6}	10.0×10^{-6}
Copper lead	18×10^{-6}	10.0×10^{-6}
Iron, cast	11×10^{-6}	6.1×10^{-6}
Iron, porous	12×10^{-6}	6.7×10^{-6}
Iron, wrought	12×10^{-6}	6.7×10^{-6}
Magnesium alloys	27×10^{-6}	15×10^{-6}
Steel, alloy[b]	11×10^{-6}	6.1×10^{-6}
Steel, stainless	17×10^{-6}	9.5×10^{-6}
Steel, high speed	11×10^{-6}	6.1×10^{-6}
Zinc alloys	27×10^{-6}	15×10^{-6}
Polymers		
Thermoplastics[c]	$(60–100) \times 10^{-6}$	$(33–56) \times 10^{-6}$
Thermosets[d]	$(10–80) \times 10^{-6}$	$(6–44) \times 10^{-6}$
Acetal (polyformaldehyde)	90×10^{-6}	50×10^{-6}
Nylons (polyamides)	100×10^{-6}	56×10^{-6}
Polyethylene, high density	126×10^{-6}	70×10^{-6}
Phenol formaldehyde[e]	$(25–40) \times 10^{-6}$	$(14–22) \times 10^{-6}$
Rubber, natural[f]	$(80–120) \times 10^{-6}$	$(44–67) \times 10^{-6}$
Rubber, nitrile[g]	34×10^{-6}	62×10^{-6}
Rubber, silicone	57×10^{-6}	103×10^{-6}
Ceramics		
Alumina (Al_2O_3)[h]	5.0×10^{-6}	2.8×10^{-6}
Graphite, high strength	$1.4–4.0 \times 10^{-6}$	$0.8–2.2 \times 10^{-6}$
Silicon carbide (SiC)	4.3×10^{-6}	2.4×10^{-6}
Silicon nitride (Si_3N_4)	3.2×10^{-6}	1.8×10^{-6}

[a]Structural alloys.
[b]Cast alloys can be up to $15 \times 10^{-6} (°C)^{-1}$.
[c]Typical bearing materials.
[d]$25 \times 10^{-6} (°C)^{-1}$ to $80 \times 10^{-6} (°C)^{-1}$ when reinforced.
[e]Mineral filled.
[f]Fillers can reduce coefficients.
[g]Varies with composition.
[h]0 to 200°C.

From the simple experiment given at the beginning of this section and the values in Table 3.6, the specific heat capacity C_p of cast iron is 2.8 times that of lead babbitt. Table 3.1 shows lead babbitt to have 1.4 times the mass density of cast iron. Because cast iron's specific heat ratio is considerably higher relative to its mass than the babbitt's, it is clear why the cast iron ball melts a considerable amount of wax and the babbitt ball melts hardly any.

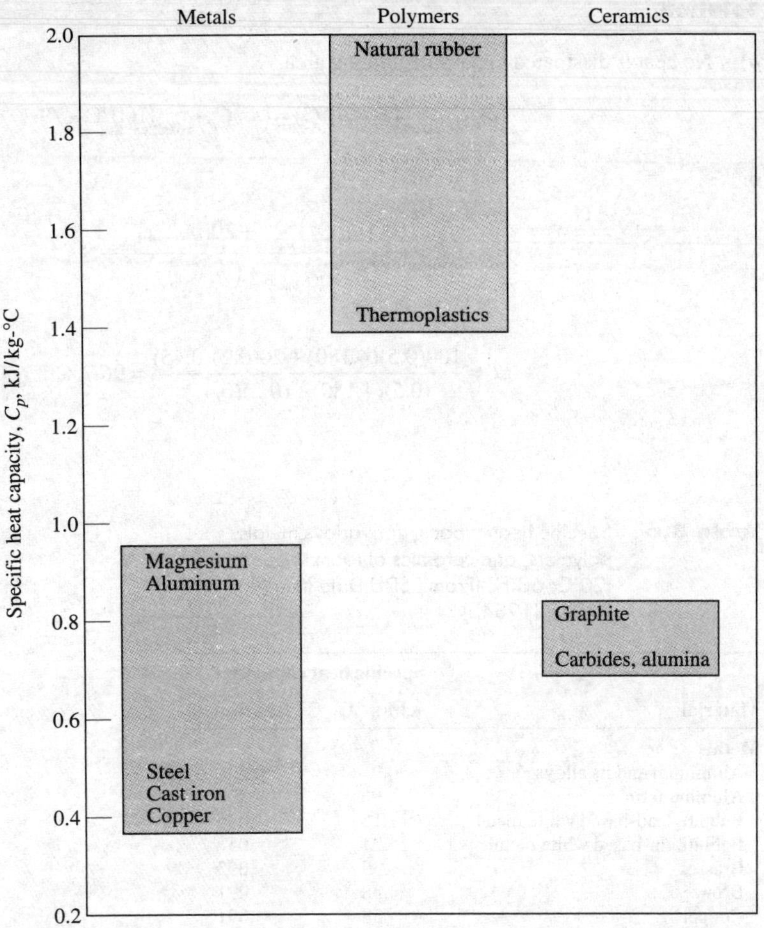

Figure 3.15 Specific heat capacity for various metals, polymers, and ceramics at room temperature (20°C; 68°F). [From ESDU Data Item No. 84041 (1984).]

Example 3.11

GIVEN A thermos made of steel where the inner bottle weighs 200 g is filled with 500 g of boiling water. The initial temperature of the thermos is 20°C. The specific heat capacity for water C_p = 4180 J/kg-°C.

FIND

a) The maximum temperature of the water in the thermos when the heat has spread to the thermos walls

b) The maximum temperature if the first hot water is emptied from the bottle and replaced by 500 g of new boiling water

Solution

a) No heat is dissipated to the surrounding area.

$$(m_a C_p)_{\text{steel}}(T - 20°\text{C}) = (m_a C_p)_{\text{water}}(100°\text{C} - T)$$

or

$$T = \frac{100(m_a C_p)_{\text{water}} + 20(m_a C_p)_{\text{steel}}}{(m_a C_p)_{\text{water}} + (m_a C_p)_{\text{steel}}}$$

$$T = \frac{100(0.5)(4.180) + 20(0.2)(0.45)}{(0.5)(4.180) + (0.2)(0.45)} = 96.7°\text{C}$$

Table 3.6 Specific heat capacity for various metals, polymers, and ceramics at room temperature (20°C; 68°F). [From ESDU Data Item No. 84041 (1984).]

	Specific heat capacity, C_p	
Material	**kJ/kg-°C**	**Btu/lbm-°F**
Metals		
Aluminum and its alloys	0.9	0.22
Aluminum tin	.96	.23
Babbitt, lead-based white metal	.15	.036
Babbitt, tin-based white metal	.21	.05
Brasses	.39	.093
Bronzes	.38	.091
Copper[a]	.38	.091
Copper lead	.32	.076
Iron, cast	.42	.10
Iron, porous	.46	.11
Iron, wrought	.46	.11
Magnesium alloys	1.0	.24
Steels[b]	.45	.11
Zinc alloys	.4	.096
Polymers		
Thermoplastics	1.4	.33
Thermosets	—	—
Rubber, natural	2.0	.48
Ceramics		
Alumina (Al_2O_3)	—	—
Graphite	.8	.2
Cemented carbides	.7	.17
Silicon carbide (SiC)	—	—
Silicon nitride (Si_3N_4)	—	—

[a]Aluminum bronze up to 0.48 kJ/kg-°C (0.12 Btu/lbm-°F).
[b]Rising to 0.55 kJ/kg-°C (0.13 Btu/lbm-°F) at 200°C (392°F).

b) The only change in the above is that 20°C is replaced with 96.7°C.

$$T = \frac{(100)(0.5)(4.180) + (96.7)(0.2)(0.45)}{(0.5)(4.180) + (0.2)(0.45)} = 99.9°\text{C}$$

There is no need to empty the bottle and replace with new boiling water.

3.5.8 ARCHARD WEAR CONSTANT

Wear is harder than the other seven properties of solid materials to quantify, partly because it is a surface phenomenon and not a bulk phenomenon and partly because wear involves interactions between two materials, not just the property of one material. When solids slide, the volume of material lost from one surface, per unit distance slid, is called the wear rate W_r. The wear resistance of the surface is characterized by the **Archard wear constant** K_A (SI unit of square meters per newton, or inverse pascals, and English unit of square inches per pound-force) defined by the following equation:

$$\frac{W_r}{A} = K_A p \qquad \textbf{(3.21)}$$

where

A = area of surface, m^2

p = normal pressure pressing surfaces together, Pa

The equations of wear are covered in greater detail in Chapter 8.

3.6 STRESS-STRAIN RELATIONSHIPS

As discovered earlier in this chapter the stress-strain diagrams for most engineering materials exhibit a linear relationship between stress and strain within the elastic limit. Therefore, an increase in stress causes a proportionate increase in strain. First discovered by Robert Hooke in 1678, this linear relationship between stress and strain in the elastic range is known as Hooke's law. Thomas Young, in 1807, suggested that the ratio of stress to strain is a measure of a material's stiffness. This ratio, called Young's modulus or the modulus of elasticity, is the slope of the straight-line portion of the stress-strain diagram and is given in Eq. (3.3). Thus, for a uniaxial stress state Hooke's law can be expressed as

$$\sigma = E\varepsilon \qquad \textbf{(3.22)}$$

This equation establishes the proportionality between stress and strain in simple tension or compression discussed in Sec. 3.5.2.

Chapter 2 discussed stress and strain separately, and material in this chapter discusses material properties such as the modulus of elasticity. Appendix B explores the stress-strain relationships for biaxial and triaxial states. Here, the discussion is limited to solids loaded in the elastic range. Furthermore, only isotropic materials are considered (i.e., materials having identical physical properties, in particular, elastic properties, in any direction). Most engineering materials are isotropic, with the notable exceptions of wood and reinforced concrete.

Example 3.12

GIVEN A 1-m-long rigid beam shown in Fig. 3.16 is pinned at its left end, carries a 3-kN vertical load at its right end, and is kept horizontal by a vertical pillar located 0.3 m from the left end. The pillar is a 0.5-m-long steel tube with an outer diameter of 0.1 m and a wall thickness of 5 mm. The modulus of elasticity of the steel used throughout is 205 GPa.

FIND How much does the right end of the beam sink due to the force?

Solution

Moment equilibrium about the hinge pin gives

$$P_1 l_1 - P_2 l_2 = 0$$

$$\therefore P_2 = \frac{P_1 l_1}{l_2} = \frac{(3000)(1)}{0.2} = 15\ 000\ \text{N}$$

Because the beam can rotate only about the hinge pin, the deflection δ_1 at P_1 can be described by the rotational angle α, and this angle is the same when describing the deflection δ_2 at P_2.

$$\therefore \delta_1 = l_1 \alpha \quad \text{and} \quad \delta_2 = l_2 \alpha$$

This implies that

$$\delta_1 = \frac{l_1 \delta_2}{l_2} \tag{a}$$

The compression of the vertical pillar δ_2 can be obtained by using Hooke's law or

$$\delta_2 = \frac{l_3 P_2}{EA} = \frac{l_3 P_2}{\pi E\left(r_o^2 - r_i^2\right)} \tag{b}$$

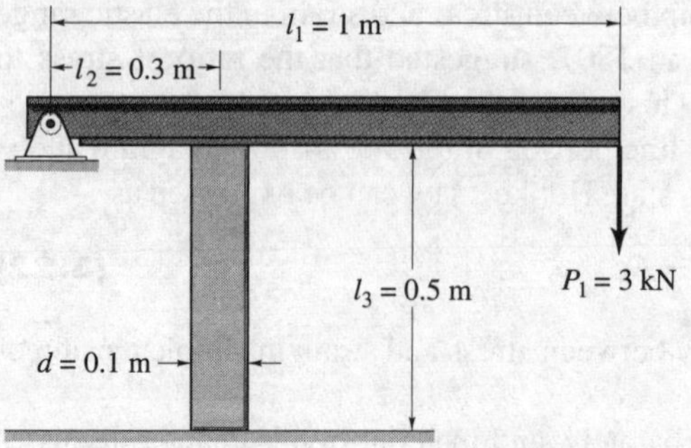

Figure 3.16 Rigid beam assembly used in Example 3.12.

Substituting Eq. (b) into Eq. (a) gives

$$\delta_1 = \frac{l_1 l_3 P_2}{\pi l_2 E\left(r_o^2 - r_i^2\right)} = \frac{1(0.5)(15\ 000)}{\pi(0.3)(205)\left(10^9\right)\left(0.05^2 - 0.045^2\right)} = 8.17 \times 10^{-5}\ \text{m} = 81.7\ \mu\text{m}$$

3.7 TWO-PARAMETER MATERIALS CHARTS

Material properties limit the performance and life of machine elements. Performance and life seldom depend on just one property of the solid materials. Thus, the information presented in Sec. 3.5 on the individual properties of solid materials is not adequate for selecting a material for a particular application. Instead, one or several combinations of properties are needed. Some important property combinations are

1. Stiffness versus density (E versus ρ)
2. Strength versus density (S versus ρ)
3. Stiffness versus strength (E versus S)
4. Wear rate versus limiting pressure (K_A versus p_l)

A number of other combinations might be useful in material selection, but these are the primary considerations in designing machine elements. The material presented in this section comes from Ashby (1992).

3.7.1 STIFFNESS VERSUS DENSITY

Modulus of elasticity and density (corresponding to stiffness and weight) are familiar properties in selecting a solid material. Lead is heavy; steel is stiff; rubber is compliant: these are the effects of modulus of elasticity. Figure 3.17 shows the full range of modulus of elasticity E and density ρ for engineering materials. Data for members of a particular class of material cluster together and are enclosed by a heavy line. The same class cluster appears on all the diagrams. Table 3.7 shows the various classes and their members with the short name for each member.

Figure 3.17 shows that the moduli of elasticity for engineering materials span five decades from 0.01 to 1000 GPa (1.47 to 147×10^3 ksi); the density spans a factor of 200, from less than 100 to 20 000 kg/m^3. The shear modulus $G \approx 3E/8$, and the bulk modulus $K \approx E$ for all materials except elastomers, for which they can be approximated as $G = E/3$ and $K >> E$.

The chart helps in common problems of material selection for applications in which weight must be minimized. For example, consider a simple tension member where the weight is to be minimized under a constraint that the strain cannot exceed a given value ε_{cr}. If the material is loaded up to its yield point, the stress is given by $\sigma = P/A$. Also, from Hooke's law

$$\sigma = E\varepsilon_{cr}$$

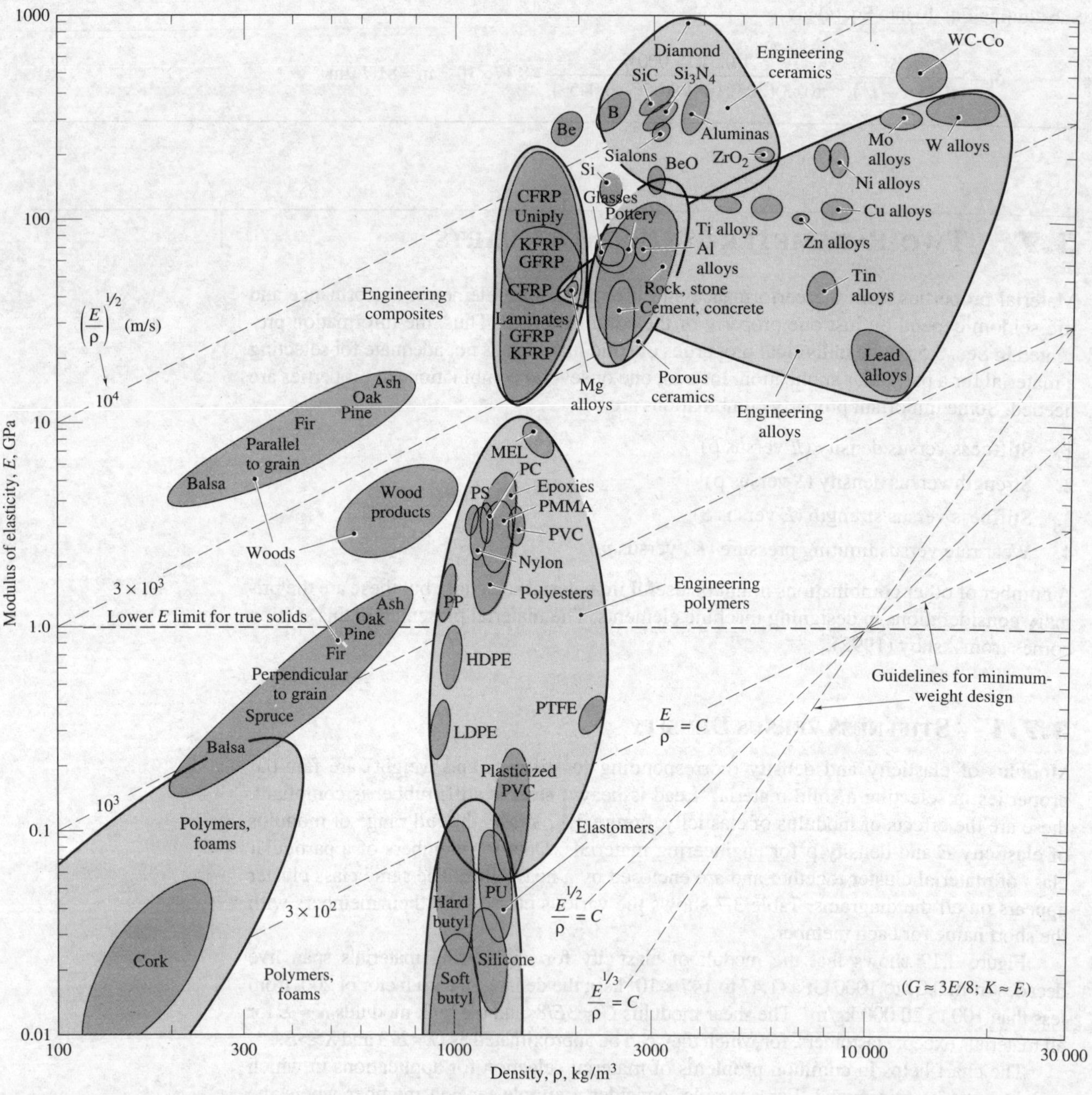

Figure 3.17 Modulus of elasticity plotted against density. The heavy envelopes enclose data for a given class of material. The diagonal contours show the longitudinal wave velocity. The guidelines of constant E/ρ, $E^{1/2}/\rho$, and $E^{1/3}/\rho$ allow selection of materials for minimum-weight, deflection-limited design. [M.F. Ashby, *Materials Selection in Mechanical Design*, Pergamon Press, 1992.]

Table 3.7 Material classes and members and short names of each member. [M.F. Ashby, *Materials Selection in Mechanical Design*, Pergamon Press, 1992.]

Class	Members	Short name
Engineering alloys (the metals and alloys of engineering)	Aluminum alloys	Al alloys
	Copper alloys	Cu alloys
	Lead alloys	Lead alloys
	Magnesium alloys	Mg alloys
	Molybdenum alloys	Mo alloys
	Nickel alloys	Ni alloys
	Steels	Steels
	Tin alloys	Tin alloys
	Titanium alloys	Ti alloys
	Tungsten alloys	W alloys
	Zinc alloys	Zn alloys
Engineering polymers (the thermoplastics and thermosets of engineering)	Epoxies	EP
	Melamines	MEL
	Polycarbonate	PC
	Polyesters	PEST
	Polyethylene, high density	HDPE
	Polyethylene, low density	LDPE
	Polyformaldehyde	PF
	Polymethylmethacrylate	PMMA
	Polypropylene	PP
	Polytetrafluoroethylene	PTFE
	Polyvinyl chloride	PVC
Engineering ceramics (fine ceramics capable of load-bearing application)	Alumina	Al_2O_3
	Diamond	C
	Sialons	Sialons
	Silicon carbide	SiC
	Silicon nitride	Si_3N_4
	Zirconia	ZrO_2
Engineering composites (the composites of engineering practice) A distinction is drawn between the properties of a ply (uniply) and of a laminate (laminates).	Carbon-fiber-reinforced polymer	CFRP
	Glass-fiber-reinforced polymer	GFRP
	Kevlar-fiber-reinforced polymer	KFRP
Porous ceramics (traditional ceramics, cements, rocks, and minerals)	Brick	Brick
	Cement	Cement
	Common rocks	Rocks
	Concrete	Concrete
	Porcelain	Pcln
	Pottery	Pot
Glasses (ordinary silicate glass)	Borosilicate glass	B-glass
	Soda glass	Na-glass
	Silica	SiO_2
Woods Separate clusters describe properties parallel to the grain and normal to it and wood products.	Ash	Ash
	Balsa	Balsa
	Fir	Fir
	Oak	Oak
	Pine	Pine
	Wood products (ply, etc.)	Wood products
Elastomers (natural and artificial rubbers)	Natural rubber	Rubber
	Hard butyl rubber	Hard butyl
	Polyurethanes	PU
	Silicone rubber	Silicone
	Soft butyl rubber	Soft butyl
Polymer foams (foamed polymers of engineering)	Cork	Cork
	Polyester	PEST
	Polystyrene	PS
	Polyurethane	PU

Equating the stresses gives

$$A = P/E\varepsilon_{cr} \tag{3.23}$$

The weight of the member is

$$W = Al\rho$$

Substituting Eq. (3.23) into the above equation gives

$$W = \left(\frac{Pl}{\varepsilon_{cr}}\right)\left(\frac{1}{E/\rho}\right) \tag{3.24}$$

Note that the first fraction contains design constraints and the second fraction contains material properties. Thus, the optimum material for minimizing Eq. (3.24) is one that maximizes the quantity E/ρ. In Fig. 3.17 one would draw lines parallel to the $E/\rho = C$ reference. Those materials with the greatest value in the direction normal to these lines are the optimum materials (i.e., those farthest toward the top and left of the charts). In Fig. 3.17 the reference lines refer to the minimum-weight design subjected to strain requirements under the following conditions:

$E/\rho = C$	Minimum-weight design of stiff tension members
$E^{1/2}/\rho = C$	Minimum-weight design of stiff beams and columns
$E^{1/3}/\rho = C$	Minimum-weight design of stiff plates
$(E/\rho)^{1/2} = C$	Wave speed in material

Example 3.13 **GIVEN** A fishing rod is to be made of a material that gives low weight and high stiffness.

FIND From Fig. 3.17 determine which is better, a rod made of plastic (without fiber reinforcement) or a split-cane rod (bamboo fibers glued together).

Solution

Figure 3.17 shows that only very special polymers have moduli of elasticity as high as the best wooden fibers. The polymers are also two to three times more dense than wood. A split-cane rod will therefore give a lower weight for a given stiffness than any plastic.

3.7.2 Strength Versus Density

Weight is represented by density. Strength, on the other hand, means different things for different classes of solid material. For metals and polymers it is the *yield strength,* which is the same in tension and compression. For brittle ceramics it is the *crushing strength in compression,* not that in tension, which is about 15 times smaller. For elastomers strength means the *tear strength.* For composites it is the *tensile failure strength* (compressive strength can be lower because of fiber buckling).

Figure 3.18 shows these strengths, for which the symbol S is used (despite the different failure mechanisms involved), plotted against density ρ. The class clusters for brittle

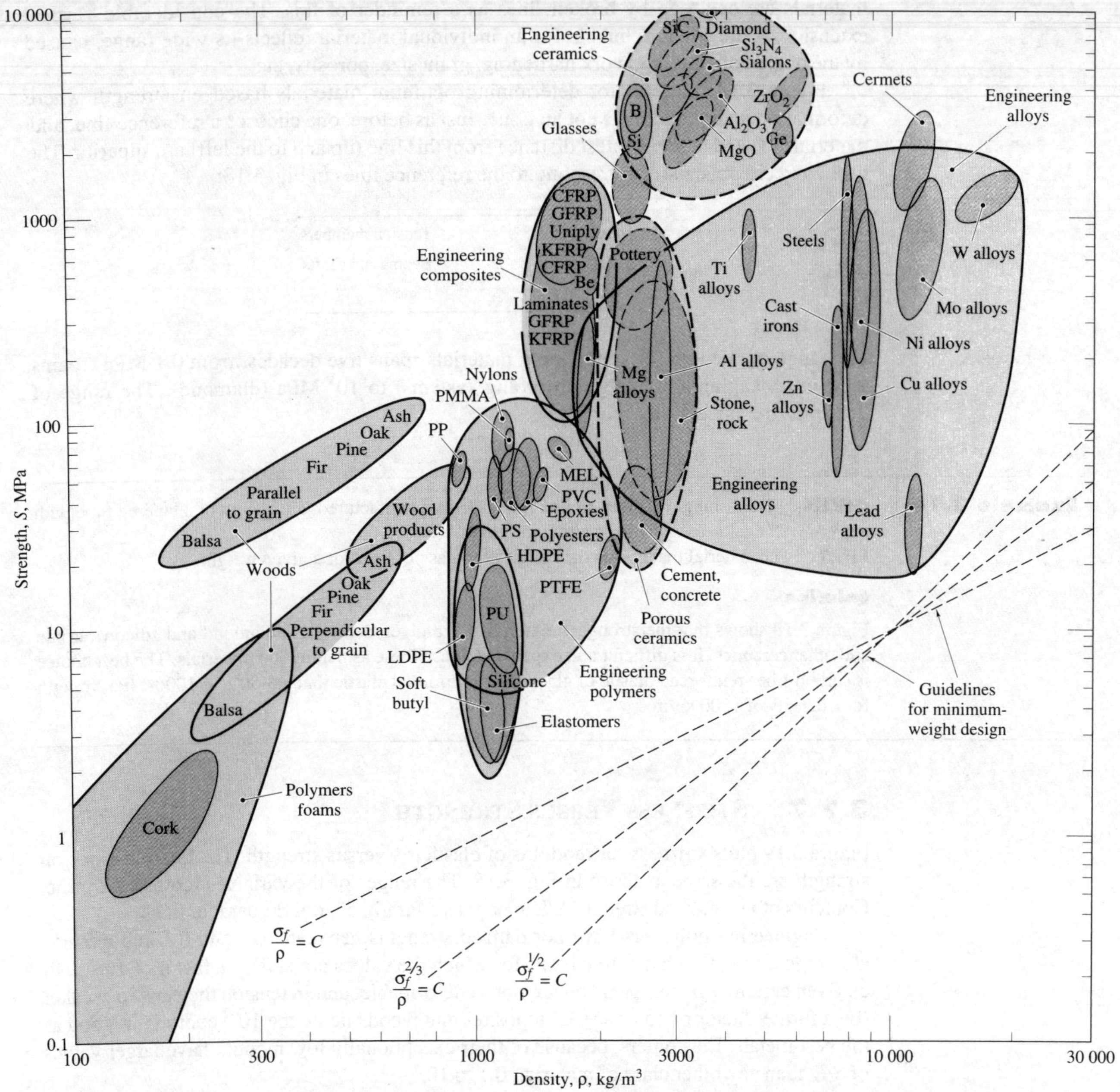

Figure 3.18 Strength plotted against density (yield strength for metals and polymers, compressive strength for ceramics, tear strength for elastomers, and tensile strength for composites). The guidelines of constant S/ρ, $S^{2/3}/\rho$, and $S^{1/2}/\rho$ allow selection of materials for minimum-weight, yield-limited design. [M.F. Ashby, *Materials Selection in Mechanical Design*, Pergamon Press, 1992.]

materials are enclosed by broken lines as a reminder of this. The considerable vertical extension of the strength bubble for an individual material reflects its wide range, caused by the degree of alloying, work hardening, grain size, porosity, etc.

Figure 3.18 is useful for determining optimum materials based on strength where deformation under loading is not an issue. Just as before, one chooses a reference line, and materials located at the greatest distance from this line (up and to the left) are superior. The following circumstances correspond to the reference lines in Fig. 3.18:

$\sigma/\rho = C$	Tension members
$\sigma^{2/3}/\rho = C$	Beams and shafts
$\sigma^{1/2}/\rho = C$	Plates

The range of strength for engineering materials spans five decades, from 0.1 MPa (foams used in packaging and energy-absorbing systems) to 10^4 MPa (diamond). The range of density is the same as in Fig. 3.17.

Example 3.14

GIVEN The fishing rod given in Example 3.13 is manufactured in the form of a tapered tube with a given wall thickness distributed along its length.

FIND The material that makes the rod as strong as possible for a given weight.

Solution

Figure 3.18 shows that the strongest materials for a given density are diamond and silicon carbide and other ceramics. It is difficult and expensive to use these as fishing rod materials. The best choice is carbon-fiber-reinforced plastic or glass-fiber-reinforced plastic that has 800- to 1000-MPa strength for a density of 1500 kg/m^3.

3.7.3 STIFFNESS VERSUS STRENGTH

Figure 3.19 plots stiffness, or modulus of elasticity, versus strength. The qualifications on strength are the same as those in Fig. 3.18. The ranges of the variables too are the same. Contours of normalized strength *S/E* appear as a family of straight parallel lines.

Engineering polymers have normalized strengths between 10^{-2} and 0.1 and remarkably high elasticity relative to metals, for which the values are at least a factor of 10 smaller. Even ceramics in compression are not as deformable, and in tension they are far weaker (by a further factor of 15 or so). Composites and woods lie on the 10^{-2} contour, as good as the best metals. Elastomers, because of their exceptionally low moduli, have larger values of *S/E* than any other class of material, 0.1 to 10.

The reference lines in Fig. 3.19 are useful for the following circumstances:

$S/E = C$	Design of seals and hinges
$S^{3/2}/E = C$	Elastic components such as knife-edges and diaphragms
$S^2/E = C$	Elastic energy storage per volume (for compact energy adsorption)

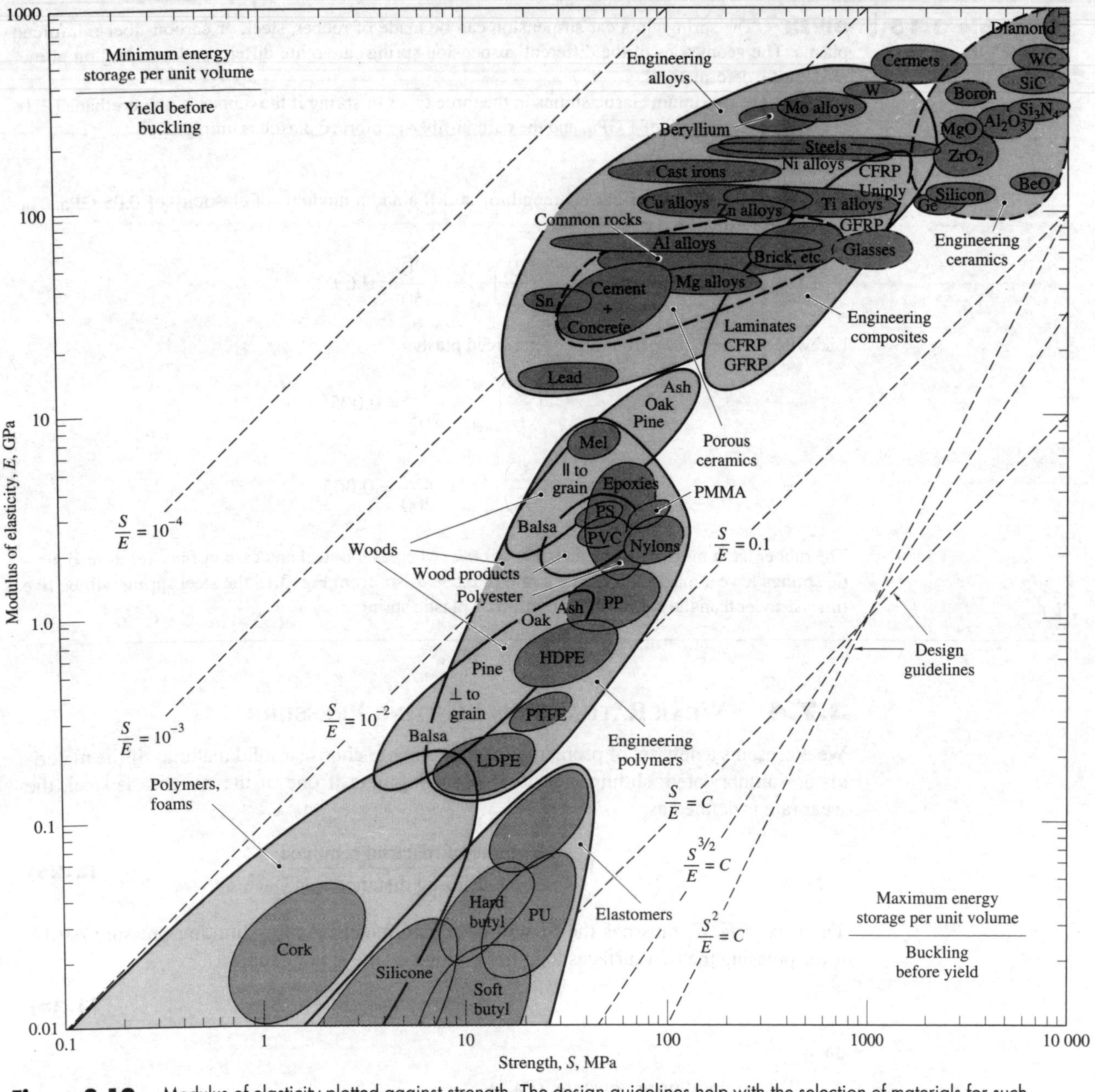

Figure 3.19 Modulus of elasticity plotted against strength. The design guidelines help with the selection of materials for such machine elements as springs, knife-edges, diaphragms, and hinges. [M.F. Ashby, *Materials Selection in Mechanical Design*, Pergamon Press, 1992.]

Example 3.15 **GIVEN** The springs in a car suspension can be made of rubber, steel, or carbon-fiber-reinforced plastic. The geometries of the different suspension springs are quite different, depending on allowable elastic deformations.

FIND The maximum elastic strains in the three types of spring if the rubber is polyurethane (PU), the steel has a strength of 1 GPa, and the carbon-fiber-reinforced plastic is uniply.

Solution

From Fig. 3.19 PU rubber has a strength of 30 MPa and a modulus of elasticity of 0.05 GPa. The maximum elastic strain is

$$\left(\frac{S}{E}\right)_{\text{rubber}} = \frac{30}{50} = 0.60$$

Likewise, for steel and carbon-fiber-reinforced plastic

$$\left(\frac{S}{E}\right)_{\text{steel}} = \frac{1}{205} = 0.005$$

$$\left(\frac{S}{E}\right)_{\text{plastic}} = \frac{1}{200} = 0.005$$

The rubber has a maximum elastic strain of 60%, whereas the steel and carbon-fiber-reinforced plastic springs have a maximum elastic strain of 0.5%. Also, from Fig. 3.18 the steel spring will be five times heavier than the carbon-fiber-reinforced plastic spring.

3.7.4 Wear Rate versus Limiting Pressure

Wear presents a new set of problems in attempting to choose a solid material. If the materials are unlubricated, sliding motion is occurring; and if one of the surfaces is steel, the wear rate is defined as

$$W_r = \frac{\text{Volume of material removed}}{\text{Sliding distance}} \tag{3.25}$$

The wear rate W_r thus has the SI unit of square meters. At low limiting pressure p_l (the force pressing the two surfaces together divided by the area of contact)

$$W_r = K_A A p_l \tag{3.26}$$

where

K_A = Archard wear constant, (Pa^{-1})
A = area of contact, m^2
p_l = limiting pressure, Pa

Figure 3.20 shows the constant K_A plotted against limiting pressure p_l. Each class cluster shows the constant value of K_A at low p_l and the steep rise as p_l is approached. Materials cannot be used above p_l.

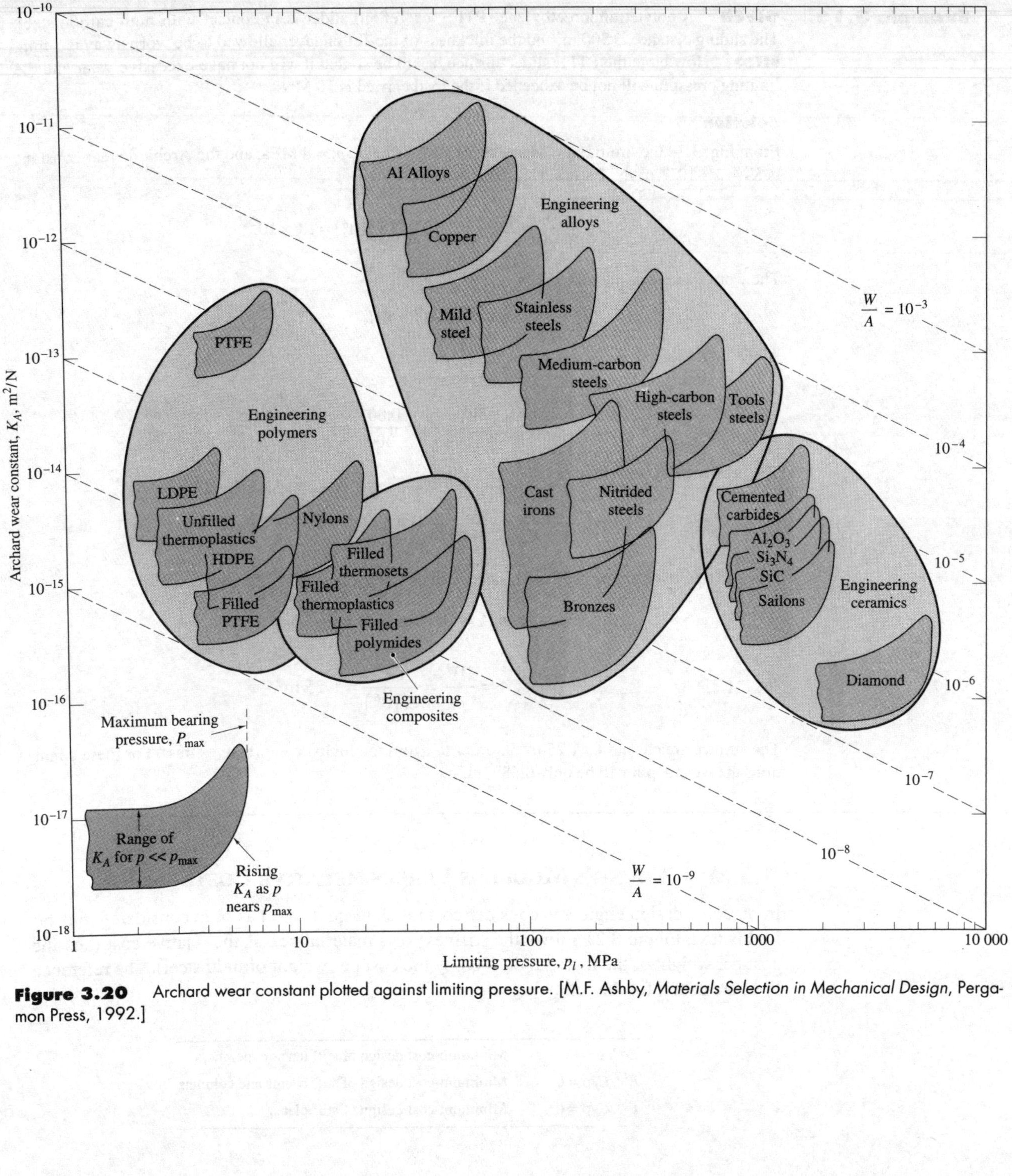

Figure 3.20 Archard wear constant plotted against limiting pressure. [M.F. Ashby, *Materials Selection in Mechanical Design*, Pergamon Press, 1992.]

Example 3.16 **GIVEN** A polytetrafluoroethylene (PTFE, or Teflon) slider is in contact with high-carbon steel. The sliding distance is 300 m, and the thickness of the Teflon layer allowed to be worn away is 3 mm.

FIND How large the PTFE slider surface has to be so that it will not have excessive wear and the limiting pressure will not be exceeded if the load carried is 10 MN.

Solution

From Fig. 3.20 the limiting pressure for PTFE on steel is $p_l = 8$ MPa, and the Archard wear constant is $K_A = 2 \times 10^{-13}$ m²/N. From Eq. (3.26)

$$\frac{W_r}{A} = K_A p_l = 2\left(10^{-13}\right)(8)\left(10^6\right) = 1.6 \times 10^{-6}$$

The worn volume of the material is

$$At_h = W_r l_s$$

where

l_s = sliding distance, m

$$\therefore \frac{W_r}{A} = \frac{t_h}{l_s} = \frac{0.003}{300} = 10^{-5}$$

The pressure can be written as

$$p = \left(\frac{W_r}{A}\right)\frac{1}{K_A} = \frac{10^{-5}}{2 \times 10^{-13}} = 0.5 \times 10^8 \text{ Pa} = 50 \text{ MPa}$$

Since $p >> p_l$, the limiting pressure is needed to determine the size of the slider.

$$\therefore p_l A = (10)\left(10^6\right) \text{ N} = 10^7 \text{ N}$$

$$\therefore A = \frac{10^7}{p_l} = \frac{10^7}{8\left(10^6\right)} = 1.25 \text{ m}^2$$

The surface area has to be 1.25 m² in order to avoid too high a compressive stress. For these conditions the wear depth will be only 0.48 mm.

3.7.5 Young's Modulus Versus Relative Cost

In practice, design engineers consider cost much more than it has been considered thus far in this text. Figure 3.21 shows the stiffness of a material versus the relative cost (i.e., the cost per weight of the material divided by the cost per weight of mild steel). The reference lines are useful for the following:

$E/C_R\rho = C$	Minimum-cost design of stiff tension members
$E^{1/2}/C_R\rho = C$	Minimum-cost design of stiff beams and columns
$E^{1/3}/C_R\rho = C$	Minimum-cost design of stiff plates

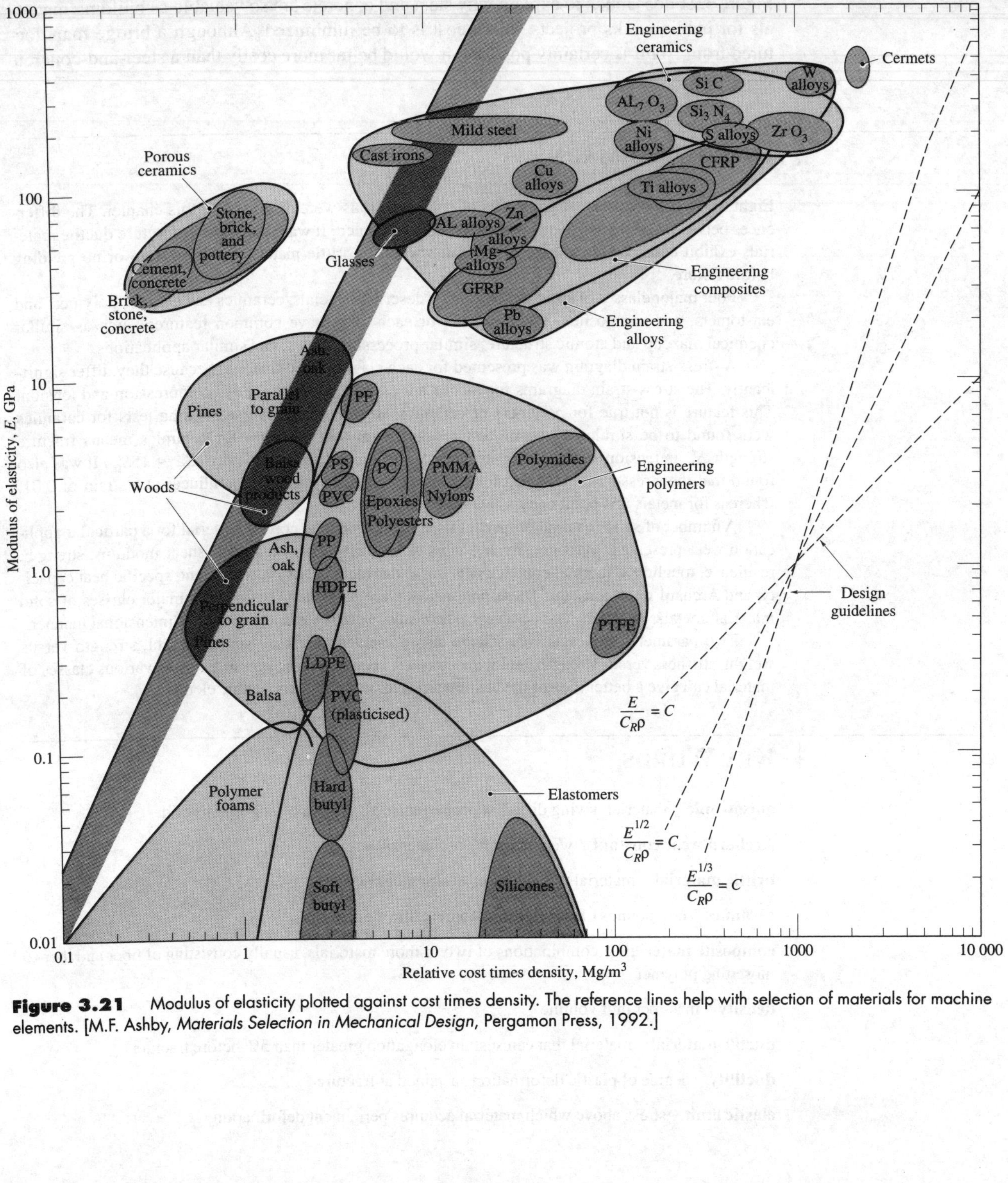

Figure 3.21 Modulus of elasticity plotted against cost times density. The reference lines help with selection of materials for machine elements. [M.F. Ashby, *Materials Selection in Mechanical Design*, Pergamon Press, 1992.]

Figure 3.21 does much to explain why steel and concrete are so valuable as building materials for public works projects where cost is to be minimized. Although a bridge manufactured from PTFE is certainly possible, it would be far more costly than a steel-and-concrete bridge.

3.8 SUMMARY

Eight important mechanical properties of solid materials were discussed in this chapter. The differences between ductile and brittle materials were presented. It was found that at fracture ductile materials exhibit considerable plastic deformation whereas brittle materials exhibit little or no yielding before failure.

Four major classes of solid material were described: metals, ceramics and glasses, polymers and elastomers, and composites. The members of each class have common features, such as similar chemical makeup and atomic structure, similar processing routes, and similar applications.

A stress-strain diagram was presented for each class of solid material because they differ significantly. The stress-strain diagrams for metals are essentially the same for compression and tension. This feature is not true for polymers or ceramics. Results of transverse bending tests for ceramics were found to be similar to tensile test results for metals. Strength for ceramics means fracture strength S_{fr}^{t} in tension and crushing strength S_{fr}^{c} in compression; typically, $S_{\text{fr}}^{c} \approx 15 S_{\text{fr}}^{t}$. It was also found that the stress-strain diagram for a polymer becomes markedly nonlinear at a strain of 0.01, whereas for metals this point occurs at 0.002.

A number of solid material properties used in choosing the correct material for a particular application were presented: mass density, modulus of elasticity, Poisson's ratio, shear modulus, strength, resilience, toughness, thermal conductivity, linear thermal expansion coefficient, specific heat capacity, and Archard wear constant. These parameters were presented for the three major classes of solid material: metals, polymers, and ceramics. The results were represented in a one-dimensional manner.

Two-parameter materials charts were also presented. Stiffness versus weight, strength versus weight, stiffness versus strength, and wear constant versus limiting pressure for the various classes of material can give a better idea of the best material for a particular machine element.

KEY WORDS

anisotropic material having different properties in all directions at point in solid

Archard wear constant wear property of material

brittle material material that fractures at strain below 5%

ceramics compounds of metallic and nonmetallic elements

composite materials combinations of two or more materials, usually consisting of fiber and thermosetting polymer

density mass per unit volume

ductile material material that can sustain elongation greater than 5% before fracture

ductility degree of plastic deformation sustained at fracture

elastic limit stress above which material acquires permanent deformation

elastomers polymers with intermediate amount of cross-linking

fracture stress stress at time of fracture or rupture

glasses compounds of metallic and nonmetallic elements with no crystal structure

homogeneous material having properties not function of position in solid

isotropic material having same properties in all directions at point in solid

metals combinations of metallic elements

modulus of elasticity proportionality constant between stress and strain

modulus of rupture stress at rupture from bending test, used to determine strength of ceramics

necking decreasing cross-sectional area that occurs after ultimate stress is reached and before fracture

orthotropic material having different properties in three mutually perpendicular directions at point in solid and having three mutually perpendicular planes of material symmetry

Poisson's ratio absolute value of ratio of transverse to axial strain

polymers compounds of carbon and other elements forming long-chain molecules

proportional limit stress above which stress is no longer linearly proportional to strain

resilience capacity of material to release absorbed energy

rule of mixtures linear interpolation between densities of alloy concentration

specific heat capacity ratio of heat stored per mass to change in temperature of material

strain hardening increase in hardness and strength of ductile material as it is plastically deformed

thermal conductivity ability of material to transmit heat

thermal expansion coefficient ratio of elongation in material to temperature rise

thermoplastics polymers without cross-links

thermosets polymers with highly cross-linked structure

toughness ability to absorb energy up to fracture

ultimate strength maximum stress achieved in stress-strain diagram

yield strength stress level defined by intersection of reference line (with slope equal to initial material elastic modulus and x intercept of 0.2%) and material stress-strain curve

yielding onset of plastic deformation

Young's modulus (see modulus of elasticity)

RECOMMENDED READINGS

Ashby, M.J. (1992) *Materials Selection in Mechanical Design,* Pergamon Press, Oxford.
ASM Metals Handbook, 8th ed. (1973) American Society for Metals, Metals Park, OH.

Budinski, K. (1979) *Engineering Materials, Properties and Selection,* Prentice-Hall, Englewood Cliffs, NJ.

Cottrell, A.H. (1964) *Mechanical Properties of Matter,* Wiley , New York.

Crane, F.A.A., and Charles, J.A. (1984) *Selection and Use of Engineering Materials,* Butterworths, London,

Dieter, G. E. (1976) *Mechanical Metallurgy,* McGraw-Hill, New York.

Farag, M.M. (1990) *Selection of Materials and Manufacturing Processes for Engineering Materials,* Prentice-Hall, Englewood Cliffs, NJ.

Flinn, R. A., and Trojan, P. K. (1986) *Engineering Materials and Their Applications,* Houghton Mifflin, Boston.

Lewis, G. (1990) *Selection of Engineering Materials,* Prentice-Hall, Englewood Cliffs, NJ.

REFERENCES

Ashby, M. F. (1992) *Materials Selection in Mechanical Design,* Pergamon Press, Oxford.

Properties of common engineering materials. Data Item 84041, ESDU International, London, 1984.

PROBLEMS

Section 3.2

3.1 A 2-m-long polycarbonate tensile rod has a cross-sectional diameter of 13 mm. It is used to lift a tank weighing 65 tons from a 1.8-m-deep ditch onto a road. The vertical motion of the crane's arc is limited to 4.2 m. Will it be possible to lift the tank onto the road?

Section 3.3

3.2 Materials are normally classified according to their properties, processing routes, and applications. Give examples of common metal alloys that do not show some of the typical metal features in their applications.

3.3 Equation (B.56) gives the relationship between stresses and strains in isotropic materials. For a polyurethane rubber the elastic modulus at 100% elongation is 7 MPa. When the rubber is exposed to a hydrostatic pressure of 10 MPa, the volume shrinks 0.5%. Calculate Poisson's ratio for the rubber.

3.4 A fiber-reinforced plastic has fiber-matrix bond strength $\tau_f = 10$ MPa and fiber ultimate strength $S_u = 1$ GPa. The fiber length is constant for all fibers at $l = 3$ mm. The fiber diameter $d = 30$ μm. Find whether the fiber strength or the fiber-matrix bond will determine the strength of the composite.

3.5 Having the same material as in Problem 3.4 but with fiber length $l = 1$ mm, calculate if it is possible to increase the fiber stress to $S_u = 1$ GPa by making the fiber rectangular instead of circular, maintaining the same cross-sectional area for each fiber.

Section 3.4

3.6 A copper bar is stressed to its ultimate strength $S_u = 250$ MPa. The cross-sectional area of the bar before stressing is 100 mm^2, and the area at the deformed cross section where the bar starts to break at the ultimate strength is 60 mm^2. How large a force is needed to reach the ultimate strength?

3.7 AISI 440C stainless steel has ultimate strength $S_u = 807$ MPa and fracture strength $S_{fr} = 750$ MPa. At the ultimate strength the cross-sectional area of a tension bar made of AISI 440C is 80% of its undeformed value. At the fracture point the minimum cross-sectional area has shrunk to 70%. Calculate the real stresses at the point of ultimate strength and at fracture.

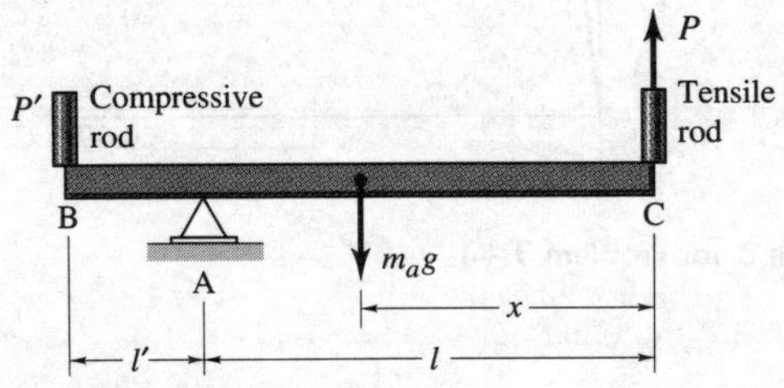

Sketch *a*, for Problem 3.8.

3.8 According to sketch *a* a beam is supported at point A and at either B or C. At C the silicon nitride tensile rod is lifting the beam end with force $P = S_{fr}^{t}$, where A_c is the cross-sectional area of the rod. Find the distance A-B such that the silicon nitride rod would be crushed if it took up a compressive force at B instead of a tensile force at C. Note that $S_{fr}^{c} = 15S_{fr}^{t}$ for silicon nitride. Also, find the reaction forces at A for the two load cases.

3.9 Polymers have different properties depending on the relationship between the local temperature and the polymer's glass transition temperature T_g. The rubber in a bicycle tire has $T_g = -12$°C. Could this rubber be used in tires for an Antarctic expedition at temperatures down to −70°C?

Section 3.5

3.10 Given an aluminum bronze with 20 wt% aluminum and 80 wt% copper, find the density of the aluminum bronze.

3.11 The glass-fiber-reinforced plastic in Example 3.8 (Sec. 3.5.2) is used in an application where the bending deformations, caused by the applied static load, will crack the plastic by overstressing the fibers. Will a carbon-fiber-reinforced plastic also crack if it has the same elastic properties as the glass-fiber-reinforced plastic?

3.12 In Problem 3.11 carbon fibers were used to reinforce a polymer matrix. The concentration of fibers was decreased in Example 3.8 (Sec. 3.5.2) to give the same elastic properties for the carbon-fiber-reinforced polymer as for the glass-fiber-reinforced polymer. If instead the fiber concentration were kept constant at 10% when the glass fibers were changed to carbon fibers, how much smaller would the deformation be for the same load, and would the fibers be overstressed or not? The material properties are the same as in Example 3.8.

3.13 A bent beam, shown in sketch *b*, is loaded with force $P = 128\,000$ N. The beam has a square cross section a^2. The length of a side $a = 30$ mm. The length $l_1 = 50$ mm and $l_2 = 100$ mm. The yield strength $S_y = 350$ MPa (medium-carbon steel). Find whether the stresses in tension and shear are below the allowable stresses.

Sketch *b*, for Problem 3.13.

3.14 A tough material, such as soft stainless steel (AISI 316), has yield strength $S_y = 207$ MPa, ultimate strength $S_u = 552$ MPa, and 60% elongation. Find the ratio of the material toughness to the resilience assuming that the stress-strain curve consists of two straight lines according to sketch *c*.

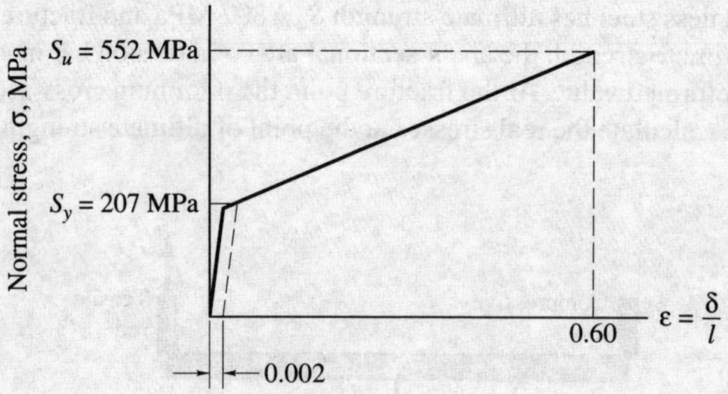

Sketch c, for Problem 3.14.

Section 3.6

3.15 A steel cube has side length $l = 0.1$ m, modulus of elasticity $E = 206$ GPa, and Poisson's ratio $\nu = 0.3$. Find the compressive stresses needed on four of the cube faces to give the same elongation perpendicular thereto as a stress σ in that perpendicular direction.

3.16 For the stressed steel cube in Problem 3.15 calculate the volume ratio ($v_{\text{tension}}/v_{\text{compression}}$) when $\sigma = 500$ MPa.

3.17 Hooke's law describes the relationship between uniaxial stress and uniaxial strain. How large is the ratio of deformation for a given load within the group of materials considered in Chapter 3?

Section 3.7

3.18 According to Archard's wear equation the wear depth is proportional to the sliding distance and the contact pressure. How will the contact pressure be distributed radially for a disk brake if the wear rate is the same for all radii?

3.19 Given a brake pad for a disk brake on a car, and using Archard's wear constant, determine how the wear is distributed over the brake pad if the brake pressure is constant over the pad.

CHAPTER

4

NORMAL, TORSIONAL, BENDING, AND TRANSVERSE SHEAR STRESSES AND STRAINS

The failed Hyatt Regency Hotel walkway (Kansas City, 1981) that was directly attributable to changes in design that overstressed structural members of the walkway. (AP/Wide World Photos)

I am never content until I have constructed a mechanical model of the subject I am studying. If I succeed in making one, I understand; otherwise I do not.

William Thomson (Lord Kelvin)

SYMBOLS

A	cross-sectional area, m^2
A'	portion of cross-sectional area, m^2
a	width of cross-sectional area, m
b	height of cross-sectional area, m
c	distance from neutral axis to outer fiber of solid, m
d_x, d_y	distance between two parallel axes one of which contains centroid of area, m
E	modulus of elasticity, Pa
e	eccentricity, distance separating centroidal and neutral radii of curved member, m
G	shear modulus of elasticity, Pa
h	height of triangular cross-sectional area, m
h_p	power, W
I	area moment of inertia, m^4
I_m	mass moment of inertia, kg-m^2
J	polar area moment of inertia, m^4
$\bar{J}$	polar area moment of inertia about centroidal coordinates, m^4
k	spring rate, N/m
k_a	angular spring rate, N-m
l	length, m
M	bending moment, N-m
m_a	mass, kg
N_a	rotational speed, rpm
P	force, N
Q	first moment about neutral axis, m^3
R	reaction force, N
r	radius, m
$\bar{r}$	centroidal radius, m
r_c	radius of cross section, m
r_g	radius of gyration, m
r_n	radius of neutral axis, m
T	torque, N-m
u	velocity, m/s
u_f	velocity, ft/min
V	transverse shear force, N
w_t	width, m
x, y, z	Cartesian coordinate system, m
$\bar{x}, \bar{y}, \bar{z}$	centroidal coordinate system, m
x', y'	coordinates parallel to x and y axes, respectively, m
Z_m	section modulus I/c, m^3
α	angle of sector, rad
γ	shear strain
ε	normal strain
θ	angle of twist, rad
σ	normal stress, Pa
τ	shear stress, Pa
ω	rotational speed, rad/s

Subscripts

avg	average
i	inner
max	maximum
o	outer
x, y, z	Cartesian coordinates
$\bar{x}, \bar{y}, \bar{z}$	centroidal coordinates
x', y'	coordinates parallel to x and y axes

4.1 INTRODUCTION

In Chapter 2 (Sec. 2.3) of this book the different types of loading were described: normal, torsional, bending, and transverse shear. This chapter describes the stresses and strains resulting from these types of loading while making use of the general Hooke's law relation developed in Chapter 3. The general theory developed in this chapter is applicable to any machine element. For the purposes of this chapter, however, the member is assumed to be straight, to have a symmetrical cross section, and to be made of a homogeneous, linear-elastic material. Later in the chapter a curved member is also considered for bending. More complicated geometries will be considered in Chapter 6.

First, six concepts important to this chapter as well as to the overall text need to be defined.

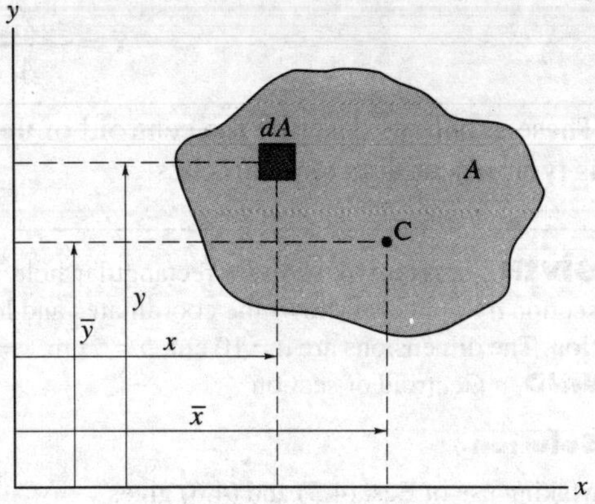

Figure 4.1 Centroid of area.

4.2 DEFINITIONS

The centroid of an area, the moment of inertia of an area, the parallel-axis theorem, the radius of gyration, the section modulus, and mass moment of inertia are important concepts that need to be defined before the main topics of this chapter can be studied. These concepts will be used throughout the text.

4.2.1 CENTROID OF AREA

The **centroid of an area** (Fig. 4.1), or the center of gravity of an area, refers to the point that defines the geometric center of the area. Mathematically, it is that point at which the sum of the first moments of an area about the axis through it is zero, or

$$\int_A (y-\bar{y})dA = 0 \tag{4.1}$$

$$\int_A (x-\bar{x})dA = 0 \tag{4.2}$$

Solving for $\bar{x}$ and $\bar{y}$ gives

$$\bar{y} = \frac{\int_A y dA}{\int_A dA} = \frac{\int_A y dA}{A} \tag{4.3}$$

$$\bar{x} = \frac{\int_A x dA}{\int_A dA} = \frac{\int_A x dA}{A} \tag{4.4}$$

A complicated area can usually be divided into simple subareas, and Eqs. (4.3) and (4.4) can be applied by making the numerator equal to the sum of the first moment integrals of the separate pairs. The denominator is the total area.

$$\therefore \bar{y} = \frac{A_1\bar{y}_1 + A_2\bar{y}_2 + \ldots}{A_1 + A_2 + \ldots} \tag{4.5}$$

$$\bar{x} = \frac{A_1\bar{x}_1 + A_2\bar{x}_2 + \dots}{A_1 + A_2 + \dots} \tag{4.6}$$

These equations describe the centroid of the composite area. The unit of centroid distance is typically millimeters or inches.

Example 4.1

GIVEN Figure 4.2 shows a rectangular hole having dimensions $c \times d$ within a rectangular cross section $a \times b$. It also shows the coordinates and location of this hole within the rectangular cross section. The dimensions are $a = 10$ cm, $b = 5$ cm, $c = 3$ cm, $d = 1$ cm, $e = 9$ cm, and $f = 3$ cm.

FIND Centroid of section

Solution

Making use of Eqs. (4.5) and (4.6) gives

$$\bar{y} = \frac{ab(b/2) - cd(f + d/2)}{ab - cd} = \frac{(10)(5)(5/2) - (3)(1)(3+1/2)}{(10)(5) - (3)(1)} = 2.436 \text{ cm}$$

$$\bar{x} = \frac{ab(a/2) - cd(e - c/2)}{ab - cd} = \frac{(10)(5)(10/2) - (3)(1)(9-3/2)}{(10)(5) - (3)(1)} = 4.840 \text{ cm}$$

4.2.2 AREA MOMENT OF INERTIA

The terms **"area moment of inertia"** and "second moment of an area" are used interchangeably. In the last section the first moment of an area $\int_A y dA$ was associated with the centroid of an area, and in this section the second moment of an area $\int_A y^2 \, dA$ is associated with the moment of inertia of an area.

Figure 4.3 shows the coordinates that describe the area moments of inertia, which are designated by the symbol I. Thus, the moments of inertia with respect to the x and y axes, respectively, can be expressed as

$$I_x = \int_A y^2 dA \qquad \text{and} \qquad I_y = \int_A x^2 dA \tag{4.7}$$

When the reference axis is normal to the plane of the area, through 0 in Fig. 4.3, the integral is called the polar moment of inertia J and can be written as

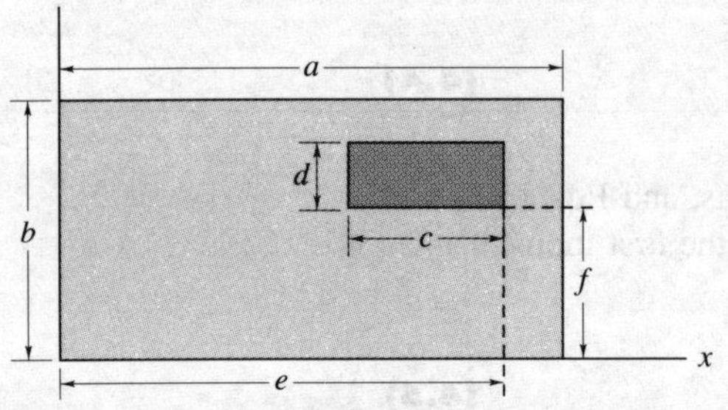

Figure 4.2 Rectangular hole within a rectangular section used in Example 4.1.

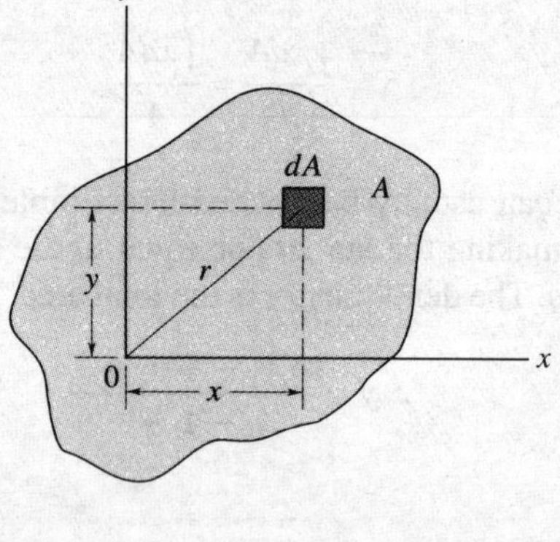

Figure 4.3 Area with coordinates used in describing area moment of inertia.

$$J = \int_A r^2 dA = \int_A (x^2 + y^2) dA = \int_A x^2 dA + \int_A y^2 dA \tag{4.8}$$

$$\therefore \quad J = I_x + I_y \tag{4.9}$$

Note that the x and y axes can be any two mutually perpendicular axes intersecting at zero. The unit of moment of inertia and polar moment of inertia is length (commonly in meters or inches) raised to the fourth power.

Example 4.2

GIVEN Figure 4.4 shows a circular cross section with radius r and x-y coordinates.

FIND The area moment of inertia of the circular area about the x and y axes and the polar moment of inertia about the centroid.

Solution

From Eq. (4.7) the area moment of inertia of the circular area about the x axis is

$$I_x = \int y^2 dA = \int y^2 2r\cos\phi \, dy$$

But because $y = r \sin \phi$ and $dy = r \cos \phi \, d\phi$,

$$I_x = 2r^4 \int_{-\pi/2}^{\pi/2} \sin^2\phi \cos^2\phi \, d\phi = \frac{r^4}{2} \int_{-\pi/2}^{\pi/2} \frac{1 - \cos 4\phi}{2} d\phi = \frac{\pi r^4}{4}$$

Because of symmetry the moment of inertia with respect to the y axis is

$$I_y = \frac{\pi r^4}{4}$$

Thus, the area moments of inertia about the x and y axes are identical and equivalent to $\pi r^4/4$. From Eq. (4.9) the polar moment of inertia about the centroid is

$$J_z = I_x + I_y = 2I_x = \pi r^4/2$$

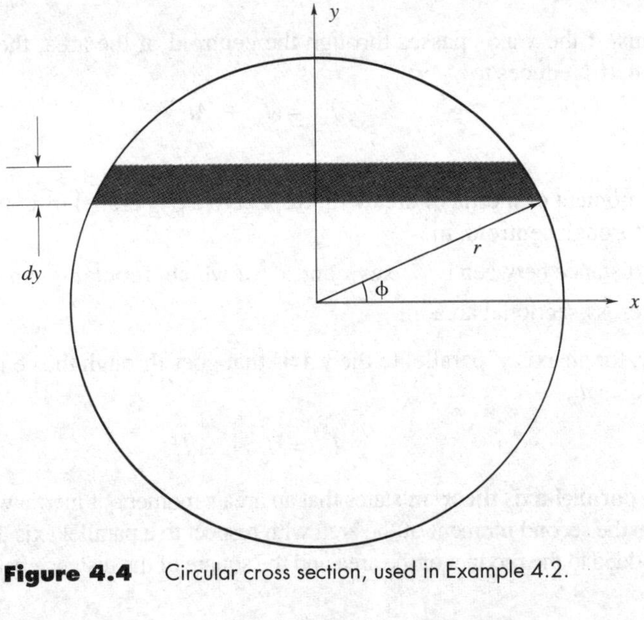

Figure 4.4 Circular cross section, used in Example 4.2.

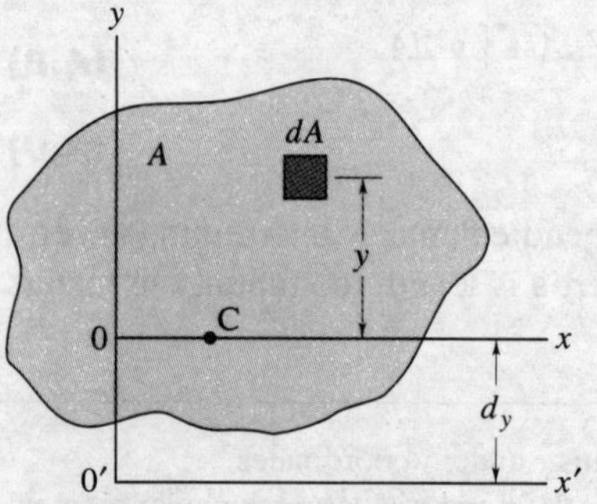

Figure 4.5 Coordinates and distance used in describing parallel-axis theorem.

4.2.3 PARALLEL-AXIS THEOREM

When an area's moment of inertia has been determined with respect to a given axis, the moment of inertia with respect to a parallel axis can be obtained by means of the parallel-axis theorem, provided that one of the axes passes through the centroid of the area. Figure 4.5 shows the coordinates and distances to be used in deriving the parallel-axis theorem.

The moment of inertia of the area A about the axis x' is

$$I_{x'} = \int_A (y + d_y)^2\, dA = \int_A y^2 dA + 2d_y \int_A y dA + d_y^2 \int_A dA$$

$$I_{x'} = I_x + 2d_y \int_A y dA + A d_y^2 \qquad \textbf{(4.10)}$$

But the integral $\int_A y dA$ is the first moment of the area with respect to the x axis, or the centroid. Thus, if the x axis passes through the centroid of the area, the first moment is zero and Eq. (4.10) reduces to

$$I_{x'} = I_x + A d_y^2 \qquad \textbf{(4.11)}$$

where

I_x = moment of inertia of area with respect to axis parallel to x' axis and passing through centroid, m^4

d_y = distance between two axes x and x', of which x contains centroid of area, m

A = cross-sectional area, m^2

Similarly, for an axis y' parallel to the y axis that goes through the centroid and is separated by a distance d_x,

$$I_{y'} = I_y + A d_x^2 \qquad \textbf{(4.12)}$$

Thus, the **parallel-axis theorem** states that an area's moment of inertia with respect to any axis is equal to the second moment of the area with respect to a parallel axis through the centroid of the area added to the product of the area and the square of the distance between the two axes.

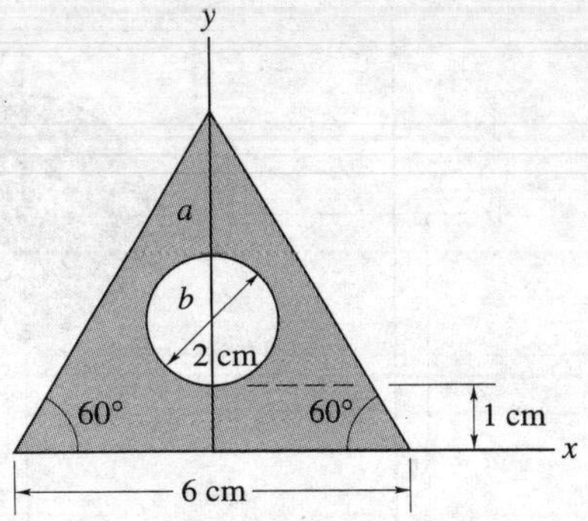

Figure 4.6 Triangular cross section with circular hole within it, used in Example 4.3.

Example 4.3

GIVEN Figure 4.6 shows a triangular cross section having a hole within it.

FIND The area moment of inertia and the centroid.

Solution

Assume that the y axis starts at the midwidth of the base and is positive in the upward direction. The height of the triangle is

$$h = \frac{6}{2}\tan 60° = 5.196 \text{ cm}$$

The triangle is defined as a and the circle as b. The centroids and areas of the triangle and the circle can be expressed separately as

$$\bar{y}_a = \frac{h}{3} = \frac{5.196}{3} = 1.732 \text{ cm}$$

$$A_a = \frac{1}{2}bh = \frac{1}{2}(6)(5.196) = 15.59 \text{ cm}^2$$

$$\bar{y}_b = 2 \text{ cm}$$

$$A_b = \frac{\pi d^2}{4} = \frac{\pi(2)^2}{4} = \pi = 3.142 \text{ cm}^2$$

The centroid of the composite figure is

$$\bar{y} = \frac{\bar{y}_a A_a - \bar{y}_b A_b}{A_a - A_b} = \frac{(1.732)(15.59) - (2)(3.142)}{15.59 - 3.142} = 1.664 \text{ cm}$$

The moments of inertia of the areas of the triangle and circle are

$$I_a = \frac{bh^3}{36} = \frac{6(5.196)^3}{36} = 23.38 \text{ cm}^4$$

$$I_b = \frac{\pi d^4}{64} = \frac{\pi(2)^4}{64} = 0.7854 \text{ cm}^4$$

From the parallel-axis theorem the moment of inertia of the composite area about the centroidal axis is

$$I_x = I_a + (\bar{y} - \bar{y}_a)^2 A_a - I_b - (\bar{y} - \bar{y}_b)^2 A_b$$

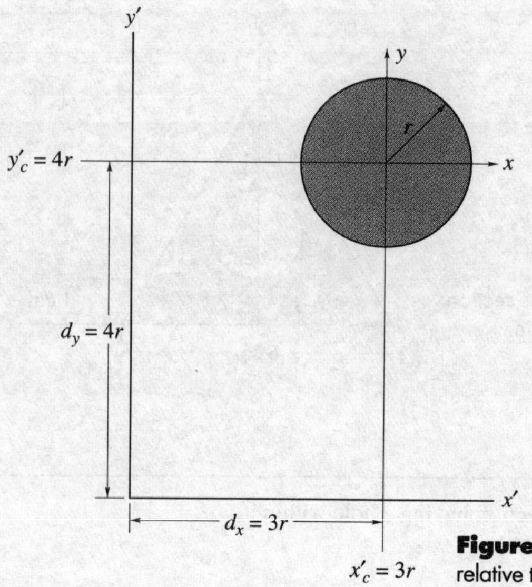

Figure 4.7 Circular cross-sectional area relative to x′-y′ coordinates, used in Example 4.4.

$$I_x = 23.38 + (1.664 - 1.732)^2(15.59) - 0.7854 - (1.664 - 2)^2(3.142)$$

$$I_x = 22.31\ cm^4 = 2.231 \times 10^{-7}\ m^4$$

Example 4.4

GIVEN Figure 4.7 shows a circular cross section of radius r relative to x'-y' coordinates.

FIND The area moments of inertia $I_{x'}$, $I_{y'}$ and the polar moment of inertia $J_{z'}$, relative to the x', y', and z' axes.

Solution

Using Eqs. (4.11) and (4.12) and the results from Example 4.2 gives

$$I_{x'} = I_x + Ad_y^2 = \frac{\pi r^4}{4} + \pi r^2(4r)^2 = 16.25\pi r^4$$

$$I_{y'} = I_y + Ad_x^2 = \frac{\pi r^4}{4} + \pi r^2(3r)^2 = 9.25\pi r^4$$

$$J_{z'} = I_{x'} + I_{y'} = 25.5\pi r^4$$

4.2.4 Radius of Gyration

An area's radius of gyration with respect to a specific axis is the length that, when squared and multiplied by the area, will give the area's moment of inertia with respect to the specific axis:

$$I = r_g^2 A \tag{4.13}$$

This then is another way of expressing the area's moment of inertia. The radius of gyration can be written as

$$r_g = \sqrt{\frac{I}{A}} \tag{4.14}$$

The **radius of gyration** *is not* the distance from the reference axis to a fixed point in the area (such as the centroid), but it is a useful property of the area and the specified axis. The radius of gyration has the typical unit of millimeters or inches.

Example 4.5

GIVEN The same situation as in Example 4.4.

FIND The radius of gyration with respect to the x', y', and z' axes.

Solution

Using Eq. (4.14) and the results from Example 4.4 gives

$$r_{gx'} = \sqrt{\frac{I_{x'}}{A}} = \sqrt{\frac{16.25\pi r^4}{\pi r^2}} = 4.03r$$

$$r_{gy'} = \sqrt{\frac{I_{y'}}{A}} = \sqrt{\frac{9.25\pi r^4}{\pi r^2}} = 3.04r$$

$$r_{gz'} = \sqrt{\frac{J_{z'}}{A}} = \sqrt{\frac{25.25\pi r^4}{\pi r^2}} = 5.02r$$

When the circle is small relative to the distances d_x and d_y, the radii of gyration are just a little larger than the centroidal distances $3r$, $4r$, and $5r$.

4.2.5 Section Modulus

The **section modulus** is simply equivalent to an area's moment of inertia divided by the farthest distance from the centroidal axis to the outer fiber of the solid c, or

$$Z_m = \frac{I}{c} \tag{4.15}$$

The typical unit of section modulus is cubic millimeters or cubic inches.

The area moment of inertia I in Eqs. (4.13) to (4.15) can be applicable to I_x and I_y with appropriate changes in r_g and Z_m. Thus, r_{gx} and Z_{mx} would correspond to the use of I_x, and r_{gy} and Z_{my} would correspond to the use of I_y.

Table 4.1 gives the centroid, area, and moment of inertia for seven different cross sections. Using the subscripts $\bar{x}$ and $\bar{y}$ for the area moment of inertia means that the area moment of inertia was taken with respect to the centroidal axis so designated. Also, $\bar{J}$ implies that the area polar moment of inertia was taken with the centroidal coordinates; J without an overbar indicates that the area polar moment of inertia was taken with respect to coordinates x and y. The formulas expressed in this table are dependent on the coordinate location relevant to the specific cross section.

Table 4.1 Centroid, area moment of inertia, and area for seven cross sections.

Cross section	Centroid	Area moment of inertia	Area
Circular area 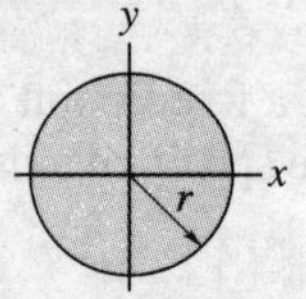	$\bar{x}=0$ $\bar{y}=0$	$I_x = I_{\bar{x}} = \frac{\pi}{4}r^4$ $I_y = I_{\bar{y}} = \frac{\pi}{4}r^4$ $J = \frac{\pi}{2}r^4$	$A=\pi r^2$
Hollow circular area 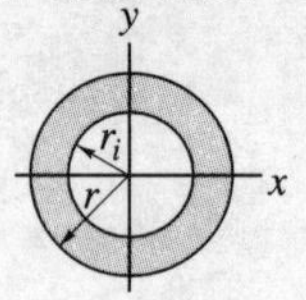	$\bar{x}=0$ $\bar{y}=0$	$I_x = I_{\bar{x}} = \frac{\pi}{4}\left(r^4 - r_i^4\right)$ $I_y = I_{\bar{y}} = \frac{\pi}{4}\left(r^4 - r_i^4\right)$ $J = \frac{\pi}{2}\left(r^4 - r_i^4\right)$	$A = \pi\left(r^2 - r_i^2\right)$
Triangular area 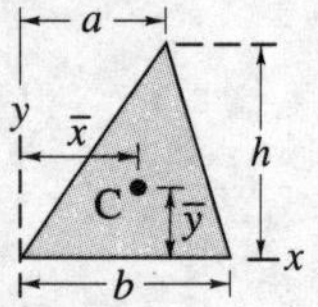	$\bar{x} = \frac{a+b}{3}$ $\bar{y} = \frac{h}{3}$	$I_x = \frac{bh^3}{12},\ I_{\bar{x}} = \frac{bh^3}{36}$ $I_y = \frac{bh\left(b^2+ab+a^2\right)}{12}$ $I_{\bar{y}} = \frac{bh\left(b^2-ab+a^2\right)}{36}$ $\bar{J} = \frac{bh}{36}\left(b^2+h^2\right)$	$A = \frac{bh}{2}$
Rectangular area 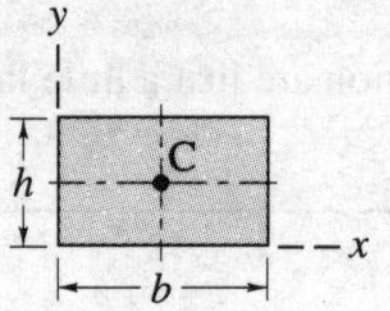	$\bar{x} = \frac{b}{2}$ $\bar{y} = \frac{h}{2}$	$I_x = \frac{bh^3}{3},\ I_{\bar{x}} = \frac{bh^3}{12}$ $I_y = \frac{hb^3}{3},\ I_{\bar{y}} = \frac{hb^3}{12}$ $\bar{J} = \frac{bh}{12}\left(b^2+h^2\right)$	$A = bh$
Area of circular sector 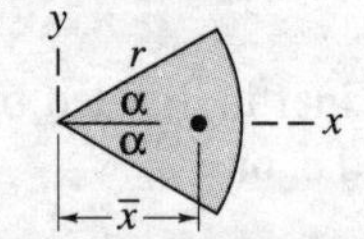	$\bar{x} = \frac{2}{3}\frac{r\sin\alpha}{\alpha}$	$I_x = \frac{r^4}{4}\left(\alpha - \frac{1}{2}\sin 2\alpha\right)$ $I_y = \frac{r^4}{4}\left(\alpha + \frac{1}{2}\sin 2\alpha\right)$ $J = \frac{1}{2}r^4\alpha$	$A = r^2\alpha$
Quarter-circular area 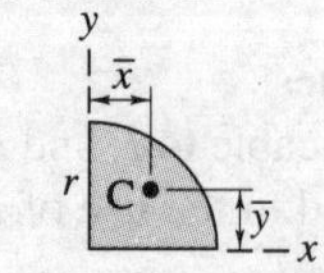	$\bar{x} = \bar{y} = \frac{4r}{3\pi}$	$I_x = I_y = \frac{\pi r^4}{16}$ $I_{\bar{x}} = I_{\bar{y}} = \left(\frac{\pi}{16} - \frac{4}{9\pi}\right)r^4$ $J = \frac{\pi r^4}{8}$	$A = \frac{\pi r^2}{4}$
Area of elliptical quadrant 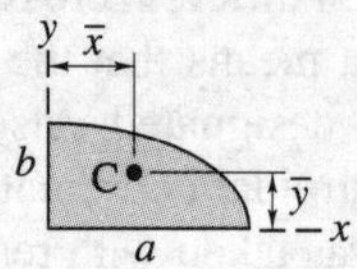	$\bar{x} = \frac{4a}{3\pi}$ $\bar{y} = \frac{4b}{3\pi}$	$I_x = \frac{\pi ab^3}{16},\ I_{\bar{x}} = \left(\frac{\pi}{16} - \frac{4}{9\pi}\right)ab^3$ $I_y = \frac{\pi a^3 b}{16},\ I_{\bar{y}} = \left(\frac{\pi}{16} - \frac{4}{9\pi}\right)a^3 b$ $J = \frac{\pi ab}{16}\left(a^2+b^2\right)$	$A = \frac{\pi ab}{4}$

4.2.6 MASS MOMENT OF INERTIA

The mass moment of inertia of an element of mass is the product of the element's mass and the square of the element's distance from the axis. Figure 4.8 shows a mass element in three-dimensional coordinates and the distance from the three axes. From this figure the **mass moments of inertia** can be expressed with respect to the x, y, and z axes as

$$I_{mx} = \int (y^2 + z^2)dm_a \tag{4.16}$$

$$I_{my} = \int (x^2 + z^2)dm_a \tag{4.17}$$

$$I_{mz} = \int (x^2 + y^2)dm_a \tag{4.18}$$

The SI unit of mass moment of inertia is kilogram–meter squared and the English unit is pound mass–inch squared. If instead of the three-dimensional coordinates given in Fig. 4.8 we have a thin plate (see Fig. 4.9) so that only the x-y plane needs to be considered, the mass moments of inertia become

$$I_{mx} = \int y^2 dm_a \tag{4.19}$$

$$I_{my} = \int x^2 dm_a \tag{4.20}$$

$$J_0 = \int r^2 dm_a = \int (x^2 + y^2) dm_a = I_{mx} + I_{my} \tag{4.21}$$

The polar mass moment of inertia is given in Eq. (4.21) as J_0. Table 4.2 gives the mass and mass moment of inertia of six commonly used shapes. Note in this table that the origin and coordinates are important for the equations relative to a specific shape.

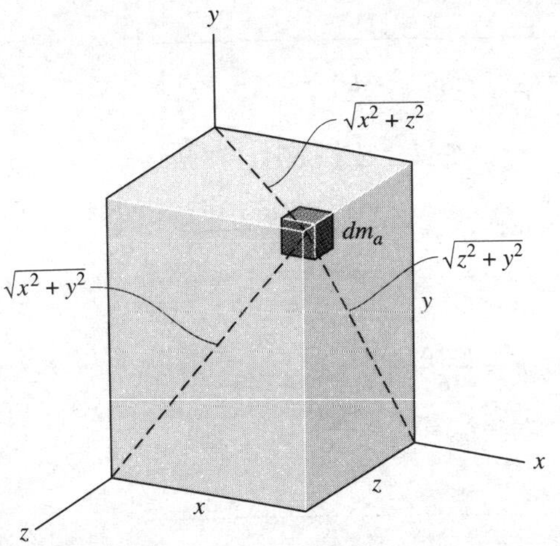

Figure 4.8 Mass element in three-dimensional coordinates and distance from the three axes.

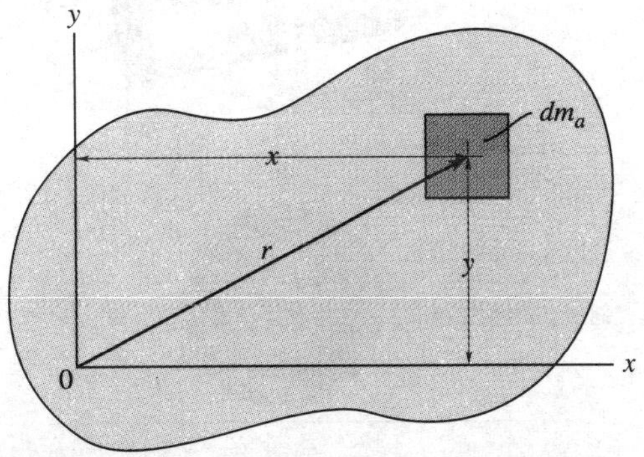

Figure 4.9 Mass element in two-dimensional coordinates and distance from the two axes.

Table 4.2 Mass and mass moment of inertia of six solids.

Shape	Equations
Rod 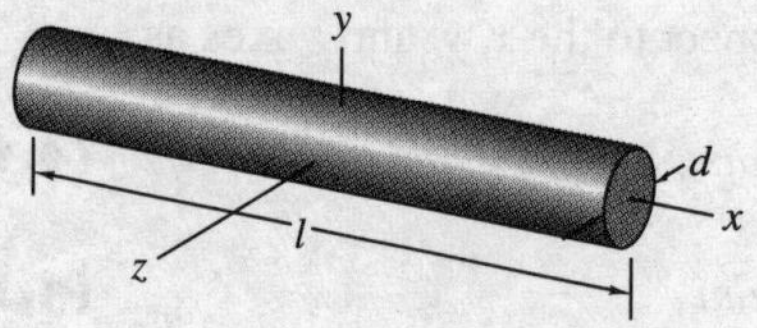	$m_a = \dfrac{\pi d^2 l \rho}{4}$ $I_{my} = I_{mz} = \dfrac{m_a l^2}{12}$
Disk 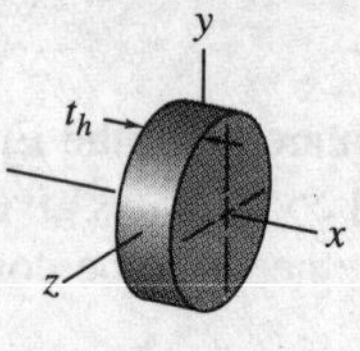	$m_a = \dfrac{\pi d^2 t_m \rho}{4}$ $I_{mx} = \dfrac{m_a d^2}{8}$ $I_{my} = I_{mz} = \dfrac{m_a d^2}{16}$
Rectangular prism 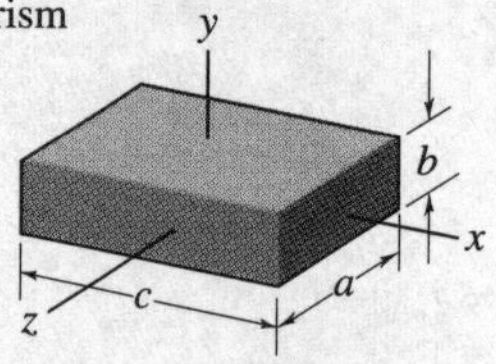	$m_a = abc\rho$ $I_{mx} = \dfrac{m_a(a^2 + b^2)}{12}$ $I_{my} = \dfrac{m_a(a^2 + c^2)}{12}$ $I_{mz} = \dfrac{m_a(b^2 + c^2)}{12}$
Cylinder 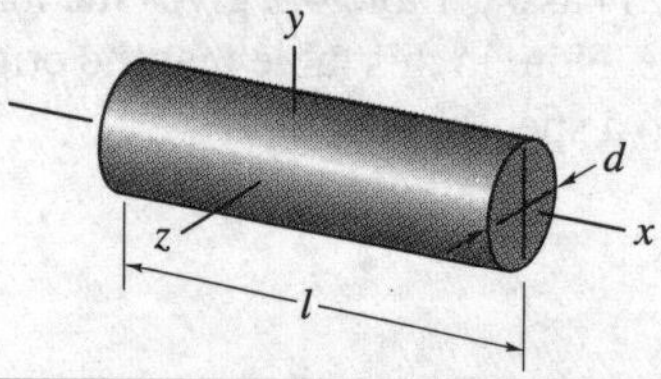	$m_a = \dfrac{\pi d^2 l \rho}{4}$ $I_{mx} = \dfrac{m_a d^2}{8}$ $I_{my} = I_{mz} = \dfrac{m_a(3d^2 + 4l^2)}{48}$
Hollow cylinder 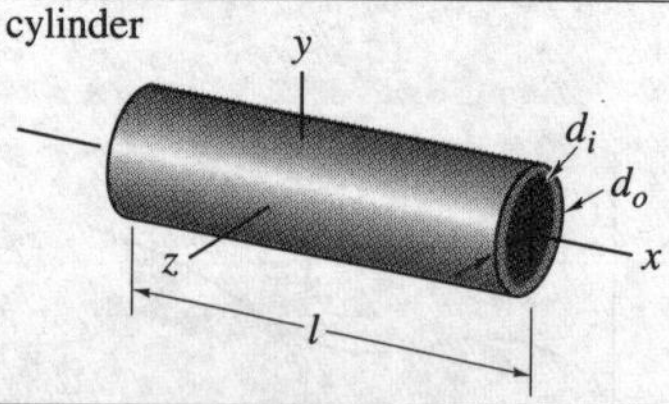	$m_a = \dfrac{\pi l \rho(d_o^2 - d_i^2)}{4}$ $I_{mx} = \dfrac{m_a(d_o^2 - d_i^2)}{8}$ $I_{my} = I_{mz} = \dfrac{m_a(3d_o^2 + 3d_i^2 + 4l^2)}{48}$
Sphere 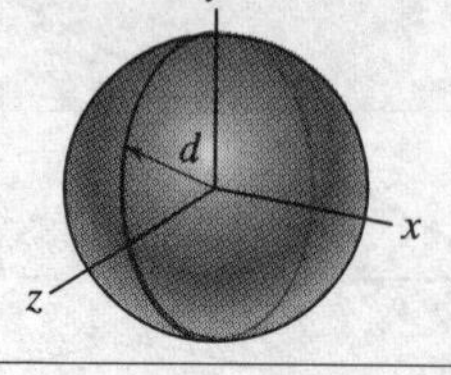	$m_a = \dfrac{\pi d^3 \rho}{6}$ $I_{mx} = I_{my} = I_{mz} = \dfrac{m_a d^2}{10}$

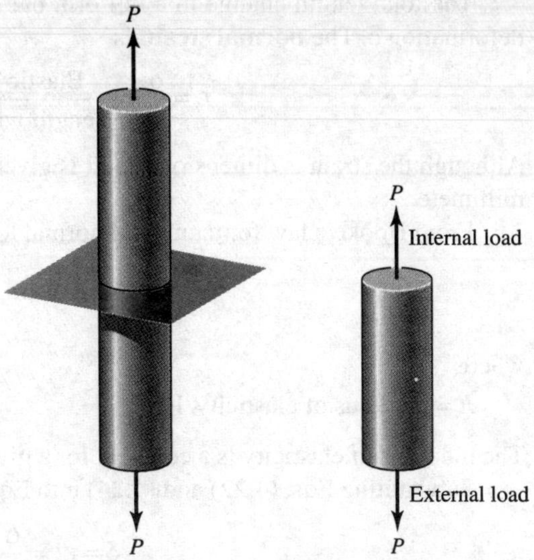

Figure 4.10 Circular bar with tensile load applied.

4.3 NORMAL STRESS AND STRAIN

The bar shown in Fig. 4.10 supports a force P and is in tension. Such a force increases the length of the bar. For a section away from the ends an average intensity of the normal force, which is also known as the **average normal stress** σ_{avg} on the cross section, can be written as

$$\sigma_{avg} = \frac{P}{A} \tag{4.22}$$

where

A = cross-sectional area, m^2

A subtlety is associated with the real application of loads. Had the section been cut near the ends, where the shape is no longer a prism, the situation would be more complicated and the stress system would no longer be simple tension uniformly distributed over the cross section. Fortunately, the stresses farther from the load application point are fairly uniform [i.e., stress concentrations (covered in detail in Chapter 6) are relieved as the distance from them increases]. This tendency was first noted by St.-Venant, so it is known as St.-Venant's principle. Its importance should not be underestimated; it makes the application of this chapter's equations possible. This chapter deals with circumstances far from the stress concentrations where the stresses are uniform.[1] The stress concentration itself will be dealt with in Chapter 6.

Consistent with the sign convention presented in Chapter 2, a positive sign is used to designate a tensile normal stress; and a negative sign, a compressive normal stress. A compressive stress decreases the length of the bar in the direction of the bar.

[1]The exact distance that one must travel from a stress concentration before stresses can be considered uniform varies with material and can range from one characteristic length (diameter, grip length, etc.) to over ten characteristic lengths for some composite materials.

The total length change in a uniform bar caused by an axial load is called the elastic deformation δ. The **normal strain** is

$$\varepsilon = \frac{\delta}{l} = \frac{\text{Elastic deformation}}{\text{Length without any load}} \tag{4.23}$$

Although the strain is dimensionless, it is given in terms of inch per inch or millimeter per millimeter.

From Hooke's law for a uniaxial normal loading,

$$\sigma = \varepsilon E \quad \text{or} \quad \varepsilon = \frac{\sigma}{E} \tag{4.24}$$

where

E = modulus of elasticity, Pa

The modulus of elasticity is a constant for a given material.

Substituting Eqs. (4.22) and (4.24) into Eq. (4.23) gives

$$\delta = \varepsilon l = \frac{\sigma}{E} l = \frac{Pl}{AE} \tag{4.25}$$

Equations (4.24) and (4.25) are valid for either tension or compression. Tensile stress, corresponding to a length increase, is considered positive; compressive stress, corresponding to length decrease, is considered negative.

The **spring rate** is the normal load divided by the elastic deflection, or for normal axial loading

$$k = \frac{P}{\delta} = \frac{AE}{l} \tag{4.26}$$

The unit of spring rate for axial loading is newtons per meter or pounds per inch. Equations (4.22) to (4.26) are valid for any cross-sectional area as long as the area remains constant over the length l.

Example 4.6

GIVEN A hollow carbon steel shaft 50 mm long must carry a normal force of 5000 N at a normal stress of 100 MPa. The inside diameter is 0.65 of the outside diameter.

FIND The outside diameter, the axial deformation, and the spring rate.

Solution

The cross-sectional area can be expressed as

$$A = \frac{\pi}{4}\left(d_o^2 - d_i^2\right) = \frac{\pi d_o^2}{4}\left[1 - \left(\frac{d_i}{d_o}\right)^2\right] = 0.4536 d_o^2$$

From Eq. (4.22)

$$A = \frac{P}{\sigma} = \frac{5000}{10^8} = 5 \times 10^{-5}\,\text{m}^2$$

Therefore,

$$d_o^2 = 1.102 \times 10^{-4}\,\text{m}^2$$

$$d_o = 1.05 \times 10^{-2}\,\text{m} = 0.0105\,\text{m} = 10.5\,\text{mm}$$

For carbon steel the modulus of elasticity is 2.07×10^{11} Pa. From Eq. (4.25) the elastic deformation is

$$\delta = \frac{Pl}{AE} = \frac{(5000)(0.05)}{(5)(10^{-5})(2.07)(10^{11})} = 24.15 \times 10^{-6}\,\text{m} = 24.15\,\mu\text{m} = 0.02415\,\text{mm}$$

From Eq. (4.26) the spring rate is

$$k = \frac{P}{\delta} = \frac{5000}{0.02415} = 2.07 \times 10^5\,\text{N/mm} = 2.07 \times 10^8\,\text{N/m}$$

Example 4.7

GIVEN A fisherman catches a salmon with a lure fastened to his 0.45-mm-diameter nylon line. When the fish bites, the line is 46 m long from the fish to the reel. The modulus of elasticity of the line is 4 GPa, and its ultimate strength is 70 N. The salmon pulls with a force of 50 N.

FIND The elastic elongation of the line, the spring rate, and the tensile stress in the line.

Solution

The cross-sectional area of the line is

$$A = \frac{\pi d^2}{4} = \frac{\pi(0.45)^2}{4} = 0.1590\,\text{mm}^2$$

The stress being exerted on the line by the salmon is

$$\sigma = \frac{P}{A} = \frac{(50)(10^6)}{0.159} = 0.3144\,\text{GPa}$$

The elongation of the line is

$$\delta = \varepsilon l = \frac{\sigma l}{E} = \frac{0.3144(46)}{4} = 3.615\,\text{m}$$

The spring rate of the line is

$$k = \frac{P}{\delta} = \frac{50}{3.615} = 13.83\,\text{N/m}$$

4.4 TORSION

A shaft is a slender element that is mainly loaded by an axial moment or twist (torque), which causes torsional deformation and shear stresses. Thus, **torsion** is loading resulting in twisting of the shaft. Shafting in general will be dealt with in Chapter 11; the focus here is on the twisting, or torsion, that shafts are subjected to and the stresses which result. A major use of the shaft is to transfer, or transmit, mechanical power from one point to another. Engineers are interested primarily in the twisting moment that can be transmitted by the shaft without damaging the material or exceeding maximum deformations. Hence, they wish to know the stresses in the shaft and the angle of twist. Solid circular members are of primary concern because most torque-transmitting shafts are of this shape. For cross sections that are not circular, or for hollow cross sections, the interested reader is referred to the classic text by Timoshenko and Goodier (1970).

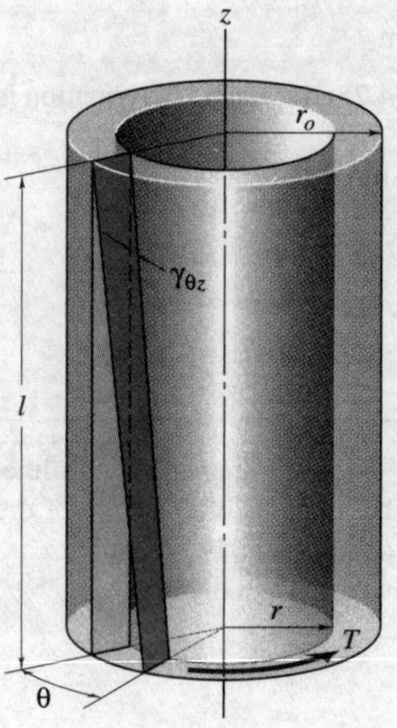

Figure 4.11 Twisting of member due to applied torque.

4.4.1 Stress and Strain

Torque is a moment that tends to twist a member about its longitudinal axis. Figure 4.11 shows the twisting of a member due to an applied torque. The circular shaft deforms such that each plane cross section originally normal to the axis remains plane and normal and does not distort within its own plane. The shaft is fixed at the top and a torque is applied to the bottom end. The angle of twist is defined as θ. The extension strains are assumed to be zero. Thus, the five strains are zero ($\varepsilon_r = \varepsilon_\theta = \varepsilon_z = \gamma_{r\theta} = \gamma_{rz} = 0$), and the only nonzero strain is

$$\gamma_{\theta z} = r\frac{d\theta}{dz} \approx \frac{r\theta}{l} \tag{4.27}$$

Equation (4.27) can also be obtained from Fig. 4.11 by observing the common surface as it relates $l\gamma_{\theta z}$ and $r\theta$.

From Hooke's law the stress is related to the strain by

$$\tau_{\theta z} = G\gamma_{\theta z} = Gr\frac{d\theta}{dz} \approx \frac{Gr\theta}{l} \tag{4.28}$$

where

G = shear modulus of elasticity, Pa

In Eq. (4.28) the shear strain $\gamma_{\theta z}$ and the shear stress $\tau_{\theta z}$ vary linearly with respect to the rate of twist $d\theta/dz \approx \theta/l$. Also, the shear stress does not change in the θ direction (because of symmetry) or in the z direction (because the deformation and the stress pattern are uniform along the shaft's length).

The rate of twist $d\theta/dz$, or θ/l in Eq. (4.28), is still unknown. To solve for it, the stresses must meet the conditions of equilibrium. The applied twisting moment, or torque, is

$$T = \int_A r(\tau_{\theta z}\, dA) = \frac{G\theta}{l}\int_A r^2\, dA \tag{4.29}$$

But the area polar moment of inertia is

$$J = \int_A r^2 dA \tag{4.30}$$

Substituting Eq. (4.30) into Eq. (4.29) gives the torque, or the angle of twist, as

$$T = \frac{G\theta J}{l} \qquad \text{or} \qquad \theta = \frac{Tl}{GJ} \tag{4.31}$$

From Table 4.1 for a solid circular shaft of outer radius r_o and diameter d,

$$J = \frac{\pi r_o^4}{2} = \frac{\pi d^4}{32} \tag{4.32}$$

Also, substituting Eqs. (4.29) and (4.30) into Eq. (4.28) gives

$$\tau_{\theta z} = \frac{rT}{J} \tag{4.33}$$

The maximum stress is

$$\tau_{max} = \frac{cT}{J} \tag{4.34}$$

where

c = distance from neutral axis to outer fiber, m

The angular spring rate can be expressed as

$$k_a = \frac{T}{\theta} = \frac{JG}{l} \tag{4.35}$$

Example 4.8

GIVEN A 50-mm-long, circular, hollow shaft made of carbon steel must carry a torque of 5000 N-m at a maximum shear stress of 70 MPa. The inside diameter is 0.5 of the outside diameter.

FIND The outside diameter, the angle of twist, and the angular spring rate.

Solution

From Table 4.1 for a hollow cylinder

$$J = \frac{\pi}{2}\left(r^4 - r_i^4\right) = \frac{\pi r^4}{2}\left[1 - \left(\frac{r_i}{r}\right)^4\right] \tag{a}$$

From Eq. (4.15) the section modulus is

$$Z_m = \frac{J}{c} = \frac{T}{\tau_{max}} = \frac{5000}{70(10^6)} = 71.43 \times 10^{-6}\,\text{m}^3 \tag{b}$$

Note that

$$\frac{r_i}{r} = \frac{r_i}{r_o} = 0.5 \qquad \text{and} \qquad c = r = \frac{d_o}{2} \tag{c}$$

Therefore, by making use of Eqs. (a) and (c), Eq. (b) becomes

$$\frac{\pi r_o^3}{2}\left[1 - (0.5)^4\right] = 71.43 \times 10^{-6}\,\text{m}^3$$

$$r_o^3 = 48.51 \times 10^{-6}\, \text{m}^3$$

$$r_o = 0.03647\ \text{m} = 36.47\ \text{mm}$$

$$d_o = 72.94\ \text{mm}$$

The shear modulus of rigidity for carbon steel is 80 GPa. Making use of Eq. (4.31) gives the angular twist due to torsion as

$$\theta = \frac{Tl}{GJ} = \frac{(5000)(2)(0.050)}{8(10^{10})(71.43)(10^{-6})(0.07294)} = 0.0012\ \text{rad}$$

Recall that $1° = \pi/180$ rad and 1 rad = 57.296°.

$$\therefore \theta = 0.0687°$$

From Eq. (4.35) the angular spring rate is

$$k_a = \frac{T}{\theta} = \frac{5000}{0.0012} = 4.167 \times 10^6\ \text{N-m/rad}$$

4.4.2 Power Transfer

It is convenient to present power transfer directly after considering torsional stresses and deformation. One of the most common uses of a circular shaft is in power transfer; therefore, no discussion of torsion would be adequate without including this topic.

Power is the rate of doing work or

$$h_p = \text{Force} \times \text{Velocity} = Pu \tag{4.36}$$

4.4.2.1 English Units The English unit of power is horsepower, which is defined as 33 000 ft-lbf/min.

$$h_p = \frac{Pu_f}{33\ 000} \tag{4.37}$$

where

P = force, lbf

u_f = velocity, ft/min

Equation (4.37) is applicable only for the unit specified.

For rotating systems the force can be replaced by torque if

$$P = \frac{T}{r} \tag{4.38}$$

Also, the linear velocity u_f at radius r can be expressed as

$$u_f = \frac{2\pi r N_a}{12} \quad \text{ft/min} \tag{4.39}$$

where

N_a = rotational speed, rpm

r = radius, in.

Substituting Eqs. (4.38) and (4.39) into Eq. (4.37) gives

$$h_p = \frac{\left(\frac{T}{r}\right)\left(\frac{2\pi r N_a}{12}\right)}{33\ 000} = \frac{TN_a}{63\ 025} \quad \text{horsepower (or hp)} \tag{4.40}$$

or

$$T = \frac{63\ 025 h_p}{N_a} \quad \text{lbf-in.}$$

4.4.2.2 SI UNITS The SI unit of work is a joule, or a newton-meter. The SI unit of power is a watt, or a joule per second.

$$\therefore h_p = Pu = T\omega \tag{4.41}$$

where

P = force, N

u = velocity, m/s

T = torque, N-m

ω = rotational speed, rad/s

or

$$T = \frac{h_p}{\omega} \tag{4.42}$$

The torque obtained from Eq. (4.40) is dependent on units of power and rotational speed, but the torque given in Eq. (4.42) is in newton-meters, which is the SI unit of torque.

Example 4.9

GIVEN A shaft carries a torque of 10 000 in.-lbf and turns at 900 rpm.
FIND The power transmitted in both horsepower and kilowatts.

Solution

From Eq. (4.40)

$$h_p = \frac{TN_a}{63\ 025} = \frac{(10\ 000)(900)}{63\ 025} = 142.8 \text{ hp} \tag{a}$$

The torque can be expressed in newton-meters if

$$T = 10\ 000 \text{ in.-lbf}\left(\frac{4.448 \text{ N}}{\text{lbf}}\right)\left(\frac{0.0254 \text{ m}}{\text{in.}}\right) = 1130 \text{ N-m}$$

From Eq. (4.42)

$$h_p = T\omega = 1130\frac{900(2\pi)}{60}$$
$$h_p = 1.065\times 10^5 \text{ W} = 106.5 \text{ kW} \tag{b}$$

A conversion from English to SI units is provided by

$$\text{Horsepower} = \frac{\text{Kilowatts}}{0.7457} \tag{c}$$

For example, 10 hp is equal to 7.457 kW. Thus, a check on the solution provided in Eqs. (a) and (b) is

$$142.8 \text{ hp} = \frac{106.5 \text{ kW}}{0.7457} = 142.8 \text{ hp}$$

Therefore, a check is made.

4.5 BENDING STRESS AND STRAIN

The bending of a long, thin member, commonly called a beam, is often a concern to engineers. Bending occurs in horizontal members of buildings (joists) subjected to vertical floor loading, in leaf springs on a truck, and in wings of an airplane in supporting the weight of the fuselage. In each of these applications the stress and deformation are important design considerations.

Throughout this section the following assumptions are made:

1. The cross section is symmetrical in the plane of loading (about the y axis).
2. The solid material that the beam is made of is homogeneous and linear elastic.

A straight member is considered initially and then a curved member.

4.5.1 STRAIGHT MEMBER

Figure 4.12 shows bending occurring in a highly deformable material, such as rubber, which is well suited for demonstration purposes. Figure 4.12(a) shows an undeformed bar with square cross sections marked by longitudinal and transverse grid lines. In Fig. 4.12(b) a moment is applied. The longitudinal lines become curved while the transverse lines remain straight and yet undergo a rotation. The longitudinal lines have a radius when the

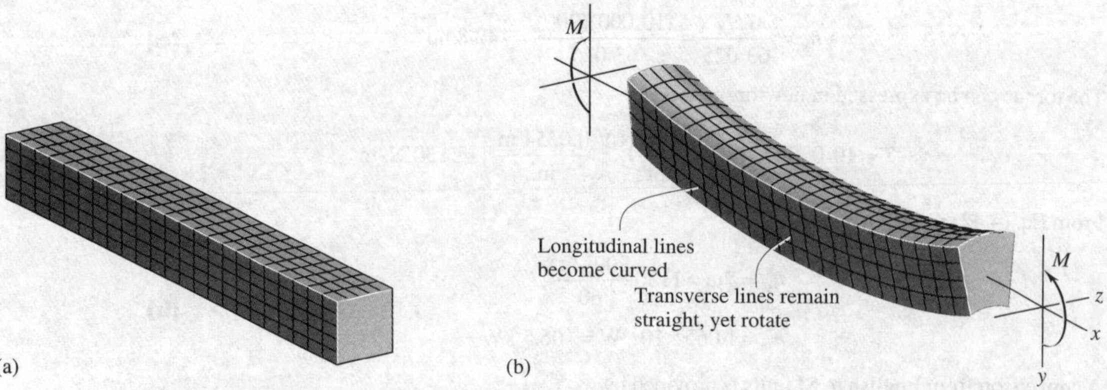

Figure 4.12 Bar made of elastomeric material to illustrate effect of bending. (a) Undeformed bar; (b) deformed bar.

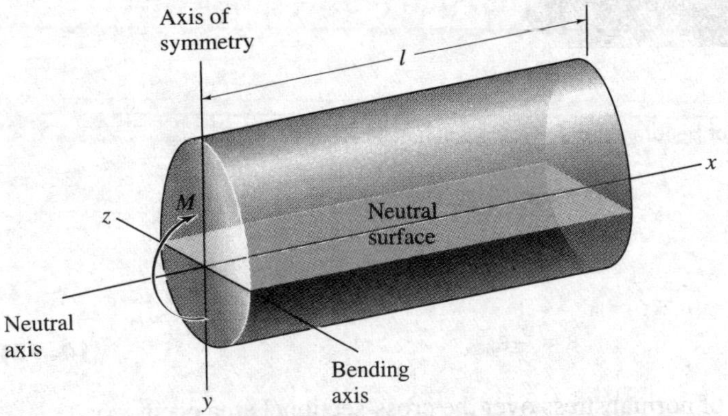

Figure 4.13 Bending occurring in cantilevered bar, showing neutral surface.

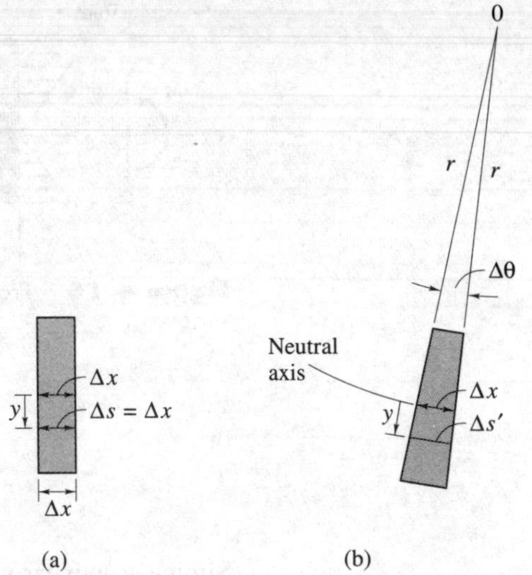

Figure 4.14 Undeformed and deformed elements in bending.

bar is deformed, even though initially they were straight. Figure 4.13 further demonstrates the effect of bending. The bending moment causes the material in the bottom portion of the member to stretch, or be in tension, and that in the top to be in compression. Consequently, between these two regions there must be a surface, called the neutral surface, in which the longitudinal fibers of the material will not undergo a change in length. On this neutral surface no bending stress is occurring, and it is neither in tension nor in compression.

Figure 4.14 shows undeformed and deformed elements when bending occurs. The normal strain along line segment Δs is

$$\varepsilon = \lim_{\Delta s \to 0} \frac{\Delta s' - \Delta s}{\Delta s} \tag{4.43}$$

From Fig. 4.11, where r is the radius of curvature of the element's longitudinal axis,

$$\Delta x = \Delta s = r\Delta\theta$$
$$\Delta s' = (r + y)\Delta\theta$$

Substituting these equations into Eq. (4.43) gives

$$\varepsilon = \lim_{\Delta s \to 0} \frac{(r + y)\Delta\theta - r\Delta\theta}{r\Delta\theta} = \frac{y}{r} \tag{4.44}$$

The longitudinal normal strain will vary linearly with y from the neutral axis. The farthest distance from the neutral axis to the outer fiber is defined as c. The maximum strain occurs at the outermost fiber, located at c from the neutral axis.

$$\therefore \frac{\varepsilon}{\varepsilon_{\max}} = \frac{y/r}{c/r}$$

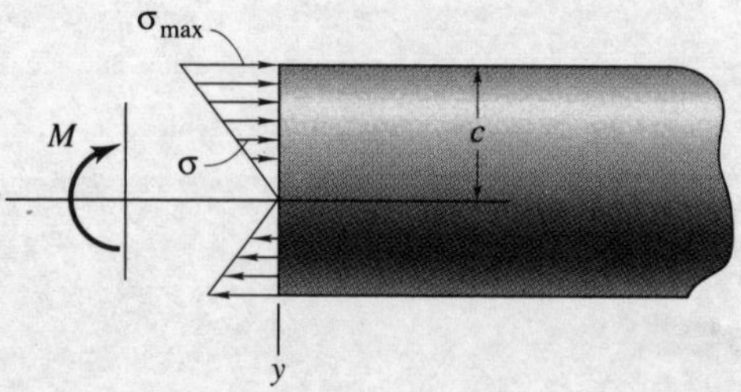

Figure 4.15 Profile view of bending stress variation.

or

$$\varepsilon = \frac{y}{c}\varepsilon_{max} \tag{4.45}$$

Similarly, a linear variation of normal stress over the cross-sectional area occurs, or

$$\sigma = \frac{y}{c}\sigma_{max} \tag{4.46}$$

Figure 4.15 shows a profile view of the normal stress. For positive y the normal stress is tensile and for negative y the normal stress is compressive. The normal stress is zero at the neutral axis.

From force equilibrium

$$0 = \int_A dP = \int_A \sigma dA = \int_A \frac{y}{c}\sigma_{max} dA = \frac{\sigma_{max}}{c}\int_A ydA$$

Because σ_{max}/c is not equal to zero,

$$\int_A ydA = 0 \tag{4.47}$$

This equation implies that the first moment of the member's cross-sectional area about the neutral axis must be zero.

The moment may be expressed as

$$M = \int_A ydP = \int_A y\sigma dA = \frac{\sigma_{max}}{c}\int_A y^2 dA$$

But the area moment of inertia is

$$I = \int_A y^2 dA$$

$$\therefore \sigma_{max} = \frac{Mc}{I} \tag{4.48}$$

The stress at any intermediate distance y is

$$\sigma = \frac{My}{I} \tag{4.49}$$

Making use of Eqs. (3.22), (4.44), and (4.49) gives

$$\frac{1}{r} = \frac{M}{EI} \tag{4.50}$$

From Eq. (4.50) when the bending moment is positive, the curvature is positive [i.e., convex in the y direction (see Fig. 2.2)].

If the distribution of the normal stress in a given cross section is not affected by the deformation caused by the shear stresses, the moment can be rewritten as Vxy. Thus, Eq. (4.49) becomes

$$\sigma = \frac{Vxy}{I} \tag{4.51}$$

where

V = transverse shear force, N

The normal stress is thus proportional to the distance x from the load to the section considered. Thus, the maximum compressive stress in the bar occurs at $x = l$ at the bottom of the bar and at the outer fiber. At $x = l$ the bar is mounted to a wall, thus producing a cantilever arrangement.

Example 4.10

GIVEN An aluminum alloy beam with the cross section shown in Fig. 4.16 experiences positive bending by an applied moment M. The allowable stress is 150 MPa.

FIND

a) The maximum moment that can be applied to the beam

b) Stresses at points A, B, and C when the maximum moment is applied

Solution

a) The cross-sectional area, if subscript 1 refers to a horizontal bar and subscript 2 refers to a vertical bar, is

$$A = A_1 + 2A_2 = 8(80 - 16 + 240) = 2432 \text{ mm}^2$$

The centroids of the cross section are

$$\bar{y} = \frac{\bar{y}_1 A_1 + 2\bar{y}_2 A_2}{A} = \frac{4(64)(8) + 2(60)(8)(120)}{2432} = 48.21 \text{ mm}$$

$$\bar{x} = 0$$

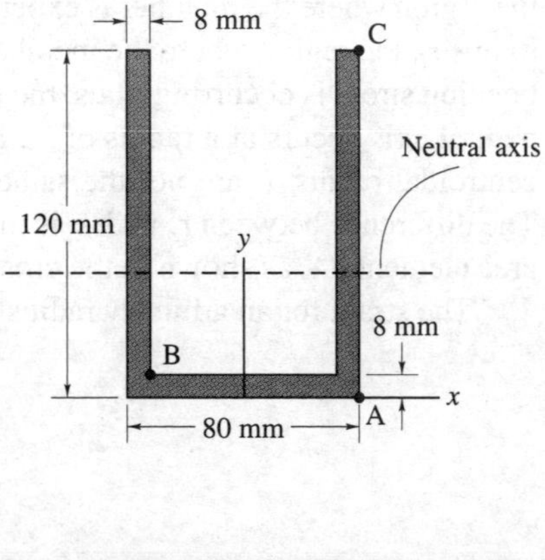

Figure 4.16 U-shaped cross section experiencing bending moment, used in Example 4.10.

The distances from the neutral axis to the centroids of the horizontal and vertical bars are

$$d_{n1} = 48.21 - 4 = 44.21 \text{ mm}$$
$$d_{n2} = 60 - 48.21 = 11.79 \text{ mm}$$

The area moment of inertia of the composite structure is

$$I = I_1 + A_1 d_{n1}^2 + 2\left(I_2 + A_2 d_{n2}^2\right)$$

$$I = \frac{64(8)^3}{12} + (64)(8)(44.2)^2 + 2\left[\frac{8(120)^3}{12} + 8(120)(11.79)^2\right] = 3\ 574\ 335 \text{ mm}^4 = 357.4 \text{ cm}^4$$

The distances from the neutral axis to the points where the stress is to be evaluated are

$$d_{nA} = 48.21 \text{ mm} \qquad d_{nB} = 48.21 - 8 = 40.21 \text{ mm} \qquad d_{nC} = 120 - 48.21 = 71.79 \text{ mm}$$

Point C is the farthest from the neutral axis and is thus the location where the stress is the largest. Furthermore, from Chapter 2 (Sec. 2.3) positive bending implies that the portion of the composite structure above the neutral axis is in tension while the portion below it is in compression. From Eq. (4.48) the maximum moment is

$$M_{max} = \frac{\sigma_{all} I}{d_{nC}} = \frac{(150)(10^6)(357.4)(10^{-8})}{(71.79)(10^{-3})} = 7468 \text{ N-m}$$

b) The stresses at the various points of interest are

$$\sigma_A = -\frac{M_{max} d_{nA}}{I} = -\frac{(7468)(48.21)(10^{-3})}{(357.4)(10^{-8})} = -100.7 \text{ MPa}$$

$$\sigma_B = -\frac{M_{max} d_{nB}}{I} = -\frac{(7468)(40.21)(10^{-3})}{(357.4)(10^{-8})} = -84.0 \text{ MPa}$$

$$\sigma_C = \sigma_{all} = 150 \text{ MPa}$$

4.5.2 Curved Member

In a straight member the normal stress and strain vary linearly from the neutral axis, as shown in Eqs. (4.45) and (4.46), respectively. In a curved member the stress and strain are not linearly related. Examples of curved members are hooks and chain links, which are not slender but have a high degree of geometric curvature.

Figure 4.17 shows a curved member in bending. Positive bending causes surface dc to rotate through $d\phi$ to $d'c'$. Note that y is positive from the neutral axis to the outer radius r_o, the region where the member is experiencing tension, and from the neutral axis to the inner radius r_i, the region where the member is experiencing compression. At the neutral axis no bending stress is occurring; thus, the member is neither in compression nor in tension. The neutral axis occurs at a radius of r_n. From Fig. 4.17 note that the neutral radius r_n and the centroidal radius $\bar{r}$ are not the same, although they are the same for a straight member. The difference between r_n and $\bar{r}$ is the eccentricity e. Also, the radius r locates an arbitrary area element dA, as shown in the cross-sectional view of Fig. 4.17.

The strain for an arbitrary radius r can be expressed as

$$\varepsilon = \frac{(r - r_n)d\phi}{r\phi} \qquad \textbf{(4.52)}$$

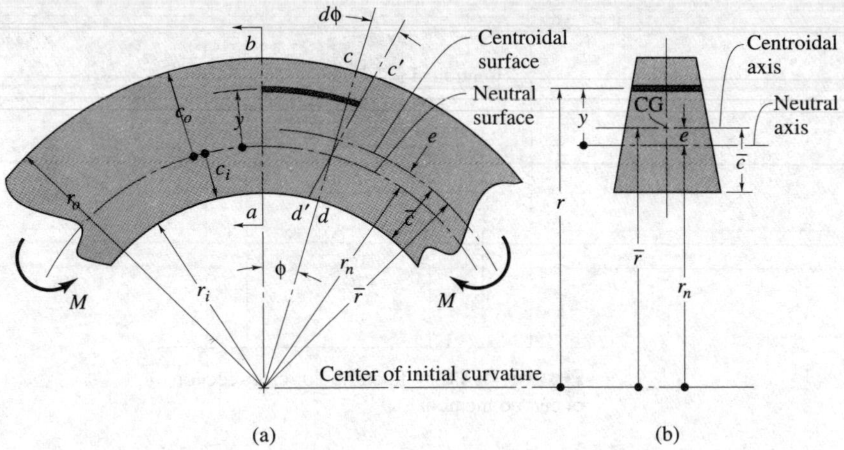

Figure 4.17 Curved member in bending. (a) Circumferential view; (b) cross-sectional view.

The strain is zero when r is equivalent to the neutral radius r_n and is largest at the outer fiber, or $r = r_o$. The normal stress can be simply written as

$$\sigma = \varepsilon E = \frac{E(r - r_n)d\phi}{r\phi} \tag{4.53}$$

For r less than r_n the stress is compressive, and for r greater than r_n the stress is tensile.

Solving for the strain and stress in Eqs. (4.52) and (4.53) requires the location of the neutral axis. This location is obtained by taking the sum of the normal stresses acting on the section and setting it to zero. Thus, making use of Eq. (4.53) gives

$$\int_A \sigma dA = \frac{Ed\phi}{\phi}\int_A \frac{r - r_n}{r} dA = 0 \tag{4.54}$$

This equation reduces to

$$A - r_n \int_A \frac{dA}{r} = 0$$

or

$$r_n = \frac{A}{\int_A \frac{dA}{r}} \tag{4.55}$$

Equation (4.55) clearly indicates that the neutral radius is a function of the sectional area.

4.5.2.1 Rectangular Cross Sections Figure 4.18 shows a rectangular cross section of a curved member, with its centroidal and neutral axes. From Eq. (4.55) the neutral radius for a rectangular cross section is

$$r_n = \frac{b(r_o - r_i)}{\int_{r_i}^{r_o} \frac{bdr}{r}} = \frac{r_o - r_i}{\ln\left(\frac{r_o}{r_i}\right)} \tag{4.56}$$

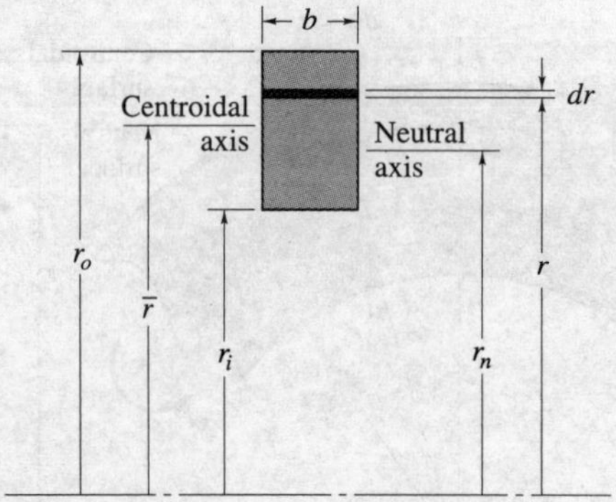

Figure 4.18 Rectangular cross section of curved member.

The centroidal radius is

$$\bar{r} = \frac{r_i + r_o}{2} \tag{4.57}$$

The eccentricity is

$$e = \bar{r} - r_n = \frac{r_i + r_o}{2} - \frac{r_o - r_i}{\ln\left(\frac{r_o}{r_i}\right)} \tag{4.58}$$

4.5.2.2 Circular Cross Sections The neutral radius for a circular cross section is

$$r_n = 2\pi\left[\bar{r} - \sqrt{(\bar{r})^2 - r_c^2}\right] \tag{4.59}$$

where

r_c = radius of cross section, m
$\bar{r}$ = centroid of cross section or center of circular cross section, m

Having established the location of the neutral radius for two different cross sections of a curved member, we are ready to return to the stress equation given in Eq. (4.53). The bending moment is just the moment arm $(r - r_n)$ multiplied by the force σdA while integrating over the cross-sectional area, or

$$M = \int (r - r_n)(\sigma dA)$$

Making use of Eq. (4.53) gives

$$M = \frac{Ed\phi}{\phi}\int \frac{(r - r_n)^2\, dA}{r} = \frac{Ed\phi}{\phi}\int \frac{(r^2 - 2rr_n + r_n^2)dA}{r}$$

$$M = \frac{Ed\phi}{\phi}\left(\int r dA - r_n A - r_n A + r_n^2 \int \frac{dA}{r}\right)$$

From Eq. (4.55) this equation reduces to

$$M = \frac{Ed\phi}{\phi}\left(\int rdA - r_n A\right) \tag{4.60}$$

From the definition of a centroid

$$\bar{r} = \frac{1}{A}\int rdA$$

Equation (4.60) then becomes

$$M = E\frac{d\phi}{\phi}Ae \tag{4.61}$$

where

$$e = \bar{r} - r_n \tag{4.62}$$

From Eq. (4.53), Eq. (4.61) becomes

$$M = \frac{r\sigma Ae}{r - r_n}$$

or

$$\sigma = \frac{M(r - r_n)}{Aer} = \frac{My}{Ae(r_n + y)} \tag{4.63}$$

where

$$y = r - r_n \tag{4.64}$$

The stress distribution is hyperbolic in shape. The maximum stress occurs at either the inner or outer surface:

$$\sigma_i = -\frac{Mc_i}{Aer_i} \tag{4.65}$$

$$\sigma_o = \frac{Mc_o}{Aer_o} \tag{4.66}$$

Example 4.11

GIVEN A rectangular cross section of a curved member, as shown in Fig. 4.18, has the dimensions $b = 1$ in. and $h = r_o - r_i = 3$ in. and is subjected to a pure bending moment of 20 000 in.-lbf. No other type of loading is acting on the member. Positive bending occurs (compression of the inner fiber).

FIND The maximum stress for the following geometries:

a) A straight member

b) A member whose centroidal axis has a radius of 15 in.

c) A member whose centroidal axis has a radius of 3 in.

Solution

a) For a straight member

$$I = \frac{bh^3}{12} \qquad c = \frac{h}{2} \qquad \therefore \frac{I}{c} = \frac{bh^2}{6}$$

$$\sigma = \frac{Mc}{I} = \frac{(20\ 000)6}{(1)(3)^2} = 13\ 333 \text{ psi}$$

$$\therefore \sigma_i = -13\ 333 \text{ psi}$$

$$\sigma_o = 13\ 333 \text{ psi}$$

b) The outer and inner radii relative to the centroidal radius of 15 in. are

$$r_o = \bar{r} + \frac{h}{2} = 15 + \frac{3}{2} = 16.5 \text{ in.}$$

$$r_i = \bar{r} - \frac{h}{2} = 15 - \frac{3}{2} = 13.5 \text{ in.}$$

From Eq. (4.56) the neutral radius is

$$r_n = \frac{r_o - r_i}{\ln\left(\frac{r_o}{r_i}\right)} = \frac{3}{\ln\left(\frac{16.5}{13.5}\right)} = 14.95 \text{ in.}$$

$$e = \bar{r} - r_n = 15 - 14.95 = 0.05 \text{ in.}$$

The distances from the neutral axis to the inner and outer fibers are

$$c_o = \frac{h}{2} + e = 1.5 + 0.05 = 1.55 \text{ in.}$$

$$c_i = \frac{h}{2} - e = 1.5 - 0.05 = 1.45 \text{ in.}$$

The corresponding normal stresses are

$$\sigma_i = -\frac{Mc_i}{Aer_i} = -\frac{(20\ 000)(1.45)}{3(1)(0.05)(13.5)} = -14\ 321 \text{ psi}$$

$$\sigma_o = \frac{Mc_o}{Aer_o} = \frac{(20\ 000)(1.55)}{3(1)(0.05)(16.5)} = 12\ 525 \text{ psi}$$

Because the radius of curvature is so large, there is little difference from the stresses found for a straight beam in part a.

c) The outer and inner radii relative to the centroidal radius of 3 in. are

$$r_o = \bar{r} + \frac{h}{2} = 3 + \frac{3}{2} = 4.5 \text{ in.}$$

$$r_i = \bar{r} - \frac{h}{2} = 3 - \frac{3}{2} = 1.5 \text{ in.}$$

$$r_n = \frac{r_o - r_i}{\ln\left(\frac{r_o}{r_i}\right)} = \frac{4.5 - 1.5}{\ln\left(\frac{4.5}{1.5}\right)} = 2.73 \text{ in.}$$

$$e = \bar{r} - r_n = 3 - 2.73 = 0.27 \text{ in.}$$

$$\therefore c_o = \frac{h}{2} + e = 1.5 + 0.27 = 1.77 \text{ in.}$$

$$c_i = \frac{h}{2} - e = 1.5 - 0.27 = 1.23 \text{ in.}$$

$$\sigma_i = -\frac{Mc_i}{Aer_i} = -\frac{(20\ 000)(1.23)}{(1)(3)(0.27)(1.5)} = -20\ 247 \text{ psi}$$

$$\sigma_o = \frac{Mc_o}{Aer_o} = \frac{(20\ 000)(1.77)}{(1)(3)(0.27)(4.5)} = 9712 \text{ psi}$$

Thus, the more curved the surface is, the more different the stresses are from the straight-beam results.

4.6 TRANSVERSE SHEAR STRESS AND STRAIN

In addition to the bending stresses considered in Sec. 4.5, the loads on a member can also cause shear stresses within the member. Figure 4.19 attempts to illustrate how transverse shear is developed. Figure 4.19(a) shows three boards that are not bonded together. The application of a force P will cause the boards to slide relative to one another, and the beam will deflect as shown, with the ends no longer flush as they were when no load was applied. On the other hand, if the boards are bonded together, the longitudinal shear stress between the boards will prevent them from sliding, and consequently the beam will act as a single unit, as shown in Fig. 4.19(b). In a solid beam the elements do not slide on each other, but the shear stress tending to make them do so is present. As a result of the internal shear stress distribution, shear strains will be developed and these will tend to distort the cross section in a rather complex manner. An undeformed bar, shown in Fig. 4.20(a), made of a

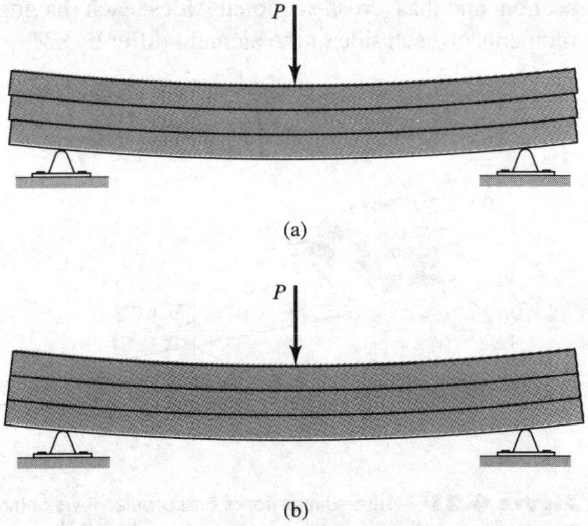

Figure 4.19 How transverse shear is developed. (a) Boards not bonded together; (b) boards bonded together.

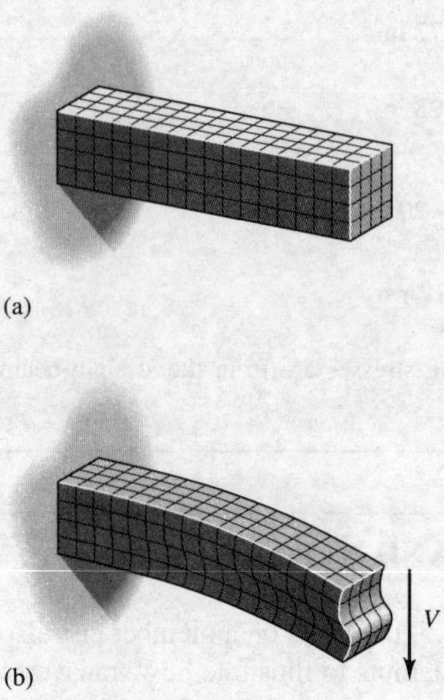

Figure 4.20 Cantilevered bar made of highly deformable material and marked with horizontal and vertical grid lines to show deformation due to transverse shear. (a) Undeformed; (b) deformed.

highly deformable material and marked with horizontal and vertical grid lines tends to deform when a shear force is applied, changing these lines into the pattern shown in Fig. 4.20(b). The squares near the top and bottom of the bar retain their undeformed shapes. The strain on the center square of the bar will cause it to have the greatest deformation. The transverse shear stress causes the cross section to warp, or to not remain plane. Figure 4.20(b) shows that the deformation caused by transverse shear is much more complex than that caused by another type of loading (axial, torsion, or bending).

The transverse shear formulas also apply to Fig. 4.21. The shaded top segment of the element has been sectioned at y' from the neutral axis. This section has a width w_t at the section and has cross-sectional sides each having an area A'. Because the resultant moments on each side of the element differ by dM,

$$\sum P_x = 0$$

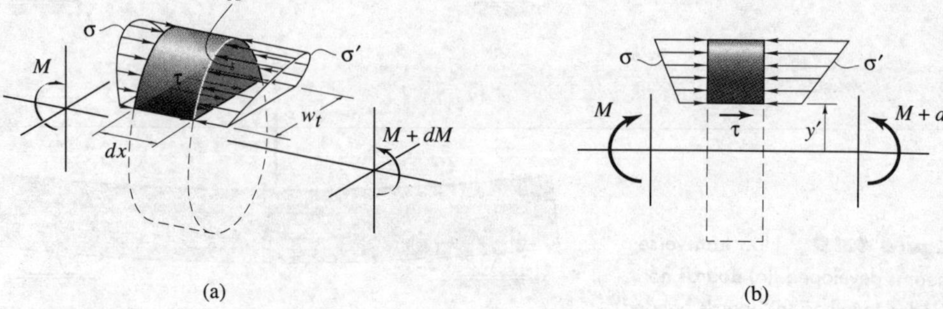

Figure 4.21 Three-dimensional and profile views of moments and stresses associated with shaded top segment of element that has been sectioned at y' about neutral axis. (a) Three-dimensional view; (b) profile view.

will only be satisfied from Fig. 4.21 if a longitudinal shear stress τ acts over the bottom face of the segment. This longitudinal shear stress is also responsible for the results shown in Fig. 4.19(a). The expression for the shear stress in the member at the point located a distance y' from the neutral axis is

$$\tau = \frac{VQ}{Iw_t} \tag{4.67}$$

where

V = transverse shear force, N

I = moment of inertia of entire cross section computed about neutral axis, m^4

w_t = width at point where τ is determined, m

Q = first moment about neutral axis of shaded portion of Fig. 4.21 = $\int_{A'} y\,dA = \bar{y}'A'$, m^3

A' = cross-sectional area of top portion shown in Fig. 4.21, m^2

$\bar{y}'$ = distance to centroid of A', measured from neutral axis, m

Applying the shear formula given in Eq. (4.67) for a rectangular cross section gives

$$\tau = \frac{6V}{bh^3}\left(\frac{h^2}{4} - y^2\right) \tag{4.68}$$

where

b = base of rectangular section, m

h = height of rectangular section, m

From Eq. (4.68) the shear stress intensity for a rectangular cross section varies from zero at the top and bottom ($y = \pm h/2$) to a maximum at the neutral axis ($y = 0$). At other intermediate values it is parabolic in shape. For $A = bh$ and $y = 0$, Eq. (4.68) reduces to

$$\tau_{max} = 1.5\frac{V}{A} \tag{4.69}$$

The maximum shear stress depends on the shape of the cross section. Table 4.3 summarizes maximum values for common cross-sectional shapes. In all cases, shear stress is zero on extreme fibers, is maximum on the neutral axis, and obtains a parabolic distribution through the thickness.

Example 4.12

GIVEN A cantilevered beam having a square cross section is loaded by a shear force perpendicular to the beam centerline at the free end. The sides of the square cross section are 50 mm, and the shear force is 10 000 N.

FIND Calculate the bending shear stress at the beam centerline

a) If the shear force is parallel with two sides of the square cross section

b) If the shear force is parallel with one of the diagonals of the square cross section

Solution

The area moment of inertia is

$$I = \frac{bh^3}{12} = \frac{(50)(50)^3}{12} = 520\ 833\ \text{mm}^4$$

Table 4.3 Maximum shear stress for different beam cross sections

Cross section	Maximum shear stress
 Rectangular	$\tau_{max} = \dfrac{3V}{2A}$
 Circular	$\tau_{max} = \dfrac{4V}{3A}$
 Round tube	$\tau_{max} = \dfrac{2V}{A}$
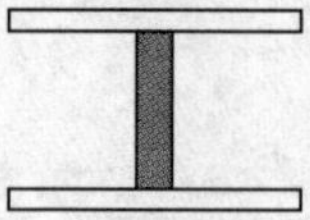 I-beam	$\tau_{max} = \dfrac{V}{A_{web}}$

This moment is valid in all directions due to symmetry. Thus, the shear force and the area moment of inertia are the same for parts a and b.

a) The width of the point where τ is applied is 50 mm. The evaluation of Q is

$$Q = \int_{A'} y\,dA = \int_0^{25} y50\,dy = 50(25)^2/2 = 15\,625 \text{ mm}^3$$

From Eq. (4.67) the bending shear stress is

$$\tau = \frac{VQ}{Iw_t} = \frac{(10\,000)(15\,625)}{(520\,833)(50)} = 6.0 \text{ N/mm}^2 = 6 \text{ MPa}$$

b) The width at the point where τ is applied is

$$w_t = \sqrt{2(50)^2} = 50\sqrt{2} = 70.71 \text{ mm}$$

$$Q = \int_0^{25\sqrt{2}} 2y\left(25\sqrt{2} - y\right)dy = \left[25\sqrt{2}y^2 - 2\left(25\sqrt{2}\right)^3/3\right]_{y=0}^{y=25\sqrt{2}} = 14\,731 \text{ mm}^3$$

From Eq. (4.67) the bending shear stress is

$$\tau = \frac{VQ}{Iw_t} = \frac{(10\,000)(14\,731)}{(520\,833)(70.71)} = 4.0 \text{ N/mm}^2 = 4 \text{ MPa}$$

Case Study | DESIGN OF SHAFT FOR COIL SLITTER

GIVEN Flat rolled sheets are produced in wide rolling mills, but many products are manufactured from strip stock. Figure 4.22(a) depicts a coil slitting line, where large sheets are cut into ribbons or strips. Figure 4.22(b) shows a shaft supporting the cutting blades. The rubber rollers ensure that the sheet does not wrinkle. For such slitting lines the shafts that support the slitting knives are a highly stressed and critical component. Figure 4.22(c) is a free-body diagram of a shaft for a short slitting line where a single blade is placed in the center of the shaft and a motor drives the shaft through a pulley at the far right end.

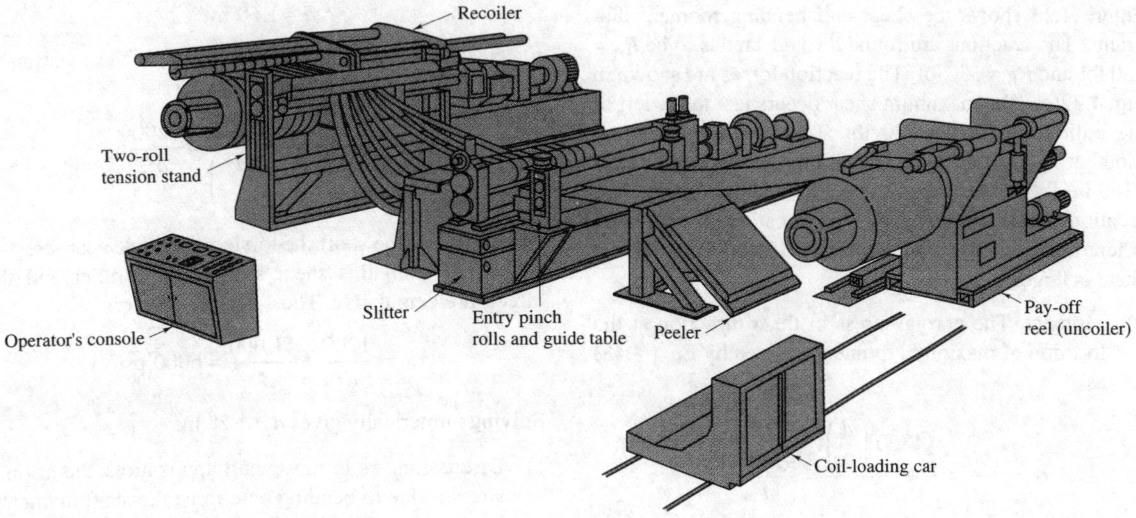

(a)

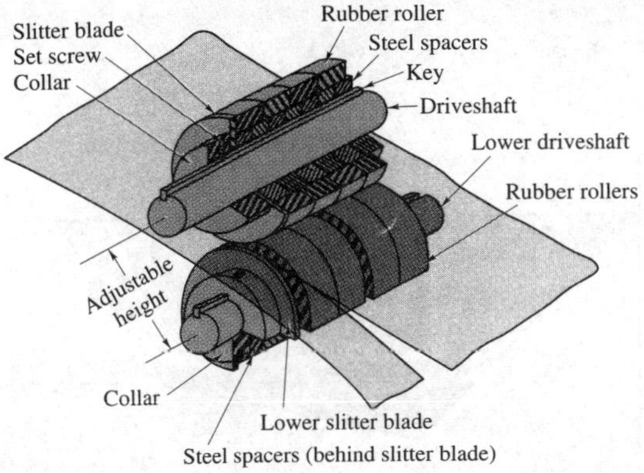

(b)

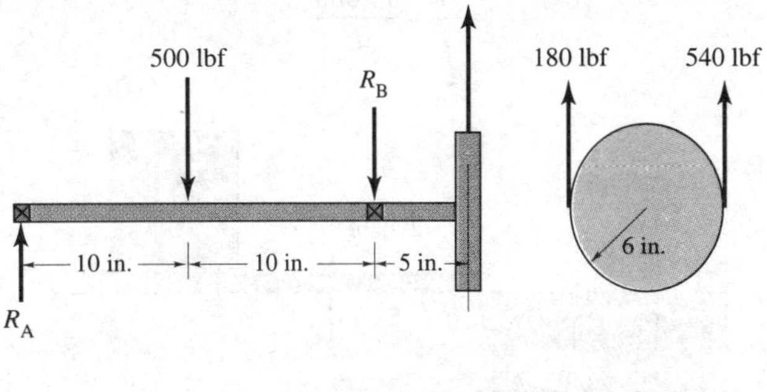

(c)

Figure 4.22 Design of shaft for coil slitting line. (a) Illustration of coil slitting line; (b) knife and shaft detail; (c) free-body diagram of simplified shaft for case study. Illustrations (a) and (b) are adapted from *Tool and Manufacturing Engineers Handbook, Fourth Edition, Volume 2 Forming,* (1984). Reprinted with permission of the Society of Manufacturing Engineers, copyright 1984.

(continued)

Case Study (CONTINUED)

FIND If the maximum shear stress is 6000 psi and the largest gage sheet causes a blade force of 500 lbf, what shaft diameter is needed? HINT: Use a Mohr's circle to determine the maximum shear stress as a function of diameter and equate this to the shear stress limit.

Solution

Figure 4.23 shows the shear and bending moment diagrams. The reactions are found through statics to be R_A = 430 lbf and R_B = 650 lbf. The reaction forces are shown in Fig. 4.22(c). The maximum shear occurs just to the left of the pulley and equals 720 lbf. The maximum bending moment is 4300 in.-lbf. In addition, there is a torque of 2160 in.-lbf between the pulley and the knife blade. Two locations must be analyzed: the location in the shaft where the moment is largest, and the location where the shear is largest.

a) Moment. The normal stress in the x direction at the location of maximum moment is given by Eq. (4.48) as

$$\sigma_x = \frac{Mc}{I} = \frac{(4300)\left(\frac{d}{2}\right)}{\frac{\pi d^4}{64}} = \frac{43\ 800}{d^3}$$

The shear stress due to the torque exerted on the pulley is, from Eq. (4.34),

$$\tau_{xy} = \frac{Tc}{J} = \frac{11\ 000}{d^3}$$

A Mohr's circle can be constructed as discussed in Chapter 2. The Mohr's circle for this case is shown in Fig. 4.24 and has a radius of 24 500 in.-lbf/d^3. Setting this equal to the maximum allowable shear stress of 6000 psi yields

$$\frac{24\ 500}{d^3} = 6000$$

$$\therefore d = 1.60 \text{ in.}$$

b) Shear. The maximum shear stress at the location of maximum shear is, from Table 4.1,

$$\tau_{max} = \frac{4V}{3A} = \frac{4(720)}{3\left(\frac{\pi d^2}{4}\right)} = \frac{1200}{d^2}$$

At one end of the shaft the torsion-induced shear stress is subtracted from this shear stress; at the other end the effects are cumulative. Thus, the total shear is

$$\tau_{tot} = \frac{1200}{d^2} + \frac{11\ 000}{d^3} = 6000 \text{ psi}$$

Solving numerically gives d = 1.28 in.

c) Discussion. As in most shaft applications the normal stresses due to bending determine the shaft diameter, so that a shaft with a diameter not less than 1.60 in. should be used. A number of points should be made regarding this analysis:

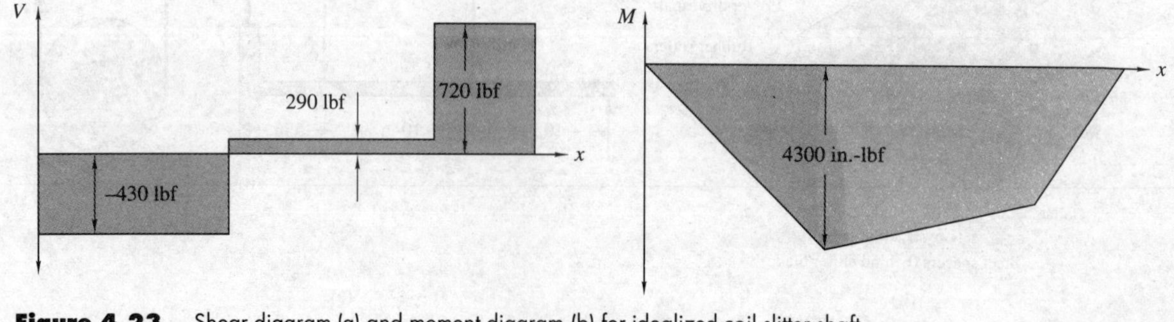

Figure 4.23 Shear diagram (a) and moment diagram (b) for idealized coil slitter shaft.

(continued)

Case Study | (CONCLUDED)

- When the normal stress due to bending was calculated, the shear stress due to shear was neglected, even though there was shear in the shaft at that location. However, the distribution of normal stress is such that it is extreme at the top and the bottom, where the shear stress is zero.
- When the maximum shear stress due to vertical shear was calculated, the effects of bending were ignored. The bending stress is zero at the neutral axis, the location of the maximum shear stress.
- There are two general shaft applications. Some shafts are extremely long, as in this problem, whereas others are made much shorter to obtain compact designs. A coil slitter can have shafting more than 20 ft long, but more supporting bearings would be needed for stiffness. This shaft was used only as an illustrative example; in actuality the supporting bearings would be placed much closer to the load application and more than two bearing packs would probably be appropriate.

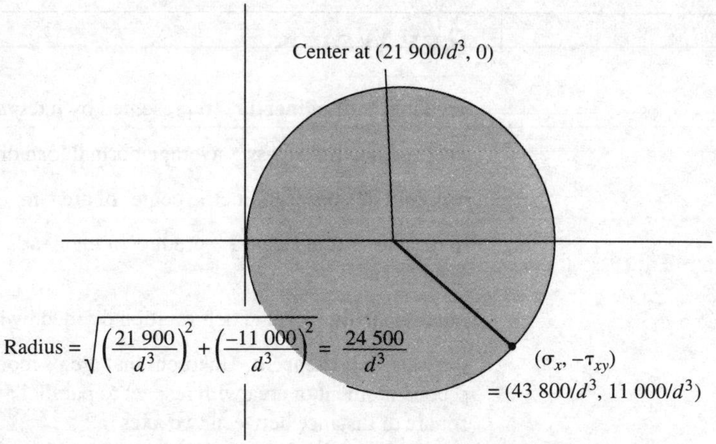

Figure 4.24 Mohr's circle for location of maximum bending stress.

- The 10-in. clearance between the leftmost bearing and the slitting knives is totally unnecessary and would lead to larger shaft deflections, as will be discussed in Chapter 5.

4.7 SUMMARY

The first four chapters have provided the essentials needed to describe the stress and strain for the four types of loading (normal, torsional, bending, and transverse shear) that might occur. These stresses and strains will be used in designing machine elements later in the text. Each chapter attempted to build off the knowledge learned from the previous chapters. It was assumed throughout the first four chapters that the member experiencing one of the four types of loading had a symmetrical cross section and was made of a homogeneous linear-elastic material. Chapter 4 started by defining centroid, moment of inertia, parallel-axis theorem, radius of gyration, and section modulus. These concepts needed to be understood before stresses and strains resulting from normal, torsional, bending, or transverse shear could be explored. These definitions, which will be used throughout the text, were followed by the stresses and strains found in the four types of loading. While evaluating the stress and strain due to normal loading, it was found convenient to include the axial displacement and the spring rate. For torsional loading the stresses and strains as well as the angular twist and the angular spring rate were presented. Also, while discussing torsion the relevant equations associated with power transfer were given in both English and SI units. The stresses and strains associated with bending were explained, as well as the importance of the neutral axis. Both a straight and curved member were analyzed. Deformations associated with bending and combined transverse loading were not covered in this chapter but will be considered more fully in Chapter 5. The last section of Chapter 4 defined the stress associated with transverse shear.

KEY WORDS

area moment of inertia represented by integral $\int r^2 dA$, m^4

average normal stress average normal load divided by cross-sectional area, Pa

centroid of area geometric center of area, m

mass moment of inertia product of element's mass and square of element's distance from the axis, $kg\text{-}m^2$

normal strain elastic deformation divided by length without any load

parallel-axis theorem theorem that area's moment of inertia with respect to any axis is equal to second moment of area with respect to parallel axis through centroid added to product of area and square of distance between two axes

power rate of doing work, or force times velocity, N-m/s

radius of gyration radius that when squared and multiplied by area gives area moment of inertia, m

section modulus area's moment of inertia divided by farthest distance from centroidal axis to outer fiber of solid, m^3

spring rate normal load divided by elastic deformation, N/m

torsion loading resulting in twisting of shaft

RECOMMENDED READINGS

Beer, F.P., and Johnson, E.R. (1981) *Mechanics of Materials,* McGraw-Hill, New York..
Craig, R.R. (1996) *Mechanics of Materials,* Wiley, New York.
Juvinall, R.C. (1967) *Stress, Strain, and Strength,* McGraw-Hill, New York.
Lardner, T.J., and Archer, R.R. (1994) *Mechanics of Solids: An Introduction,* McGraw-Hill, New York.
Popov, E. (1968) *Introduction to Mechanics of Solids,* Prentice-Hall, Englewood Cliffs, NJ.
Timoshenko and Goodier (1970) *Theory of Elasticity,* 3rd ed., McGraw-Hill, New York.
Ugural, A.C., and Fenster, S.K. (1995) *Advanced Strength and Applied Elasticity,* 3rd ed., Prentice-Hall, Englewood Cliffs, NJ.

REFERENCES

Marshall, R.D., et al. (1982) *Investigation of the Kansas City Hyatt Regency Walkways Collapse,* U.S. Department of Commerce, National Bureau of Standards, Report NBSIR 82–2465.
Timoshenko and Goodier (1970) *Theory of Elasticity,* 3rd ed., McGraw-Hill, New York.
Tool and Manufacturing Engineers Handbook (1984), Society of Manufacturing Engineers, Dearborn, MI.

PROBLEMS

Section 4.2

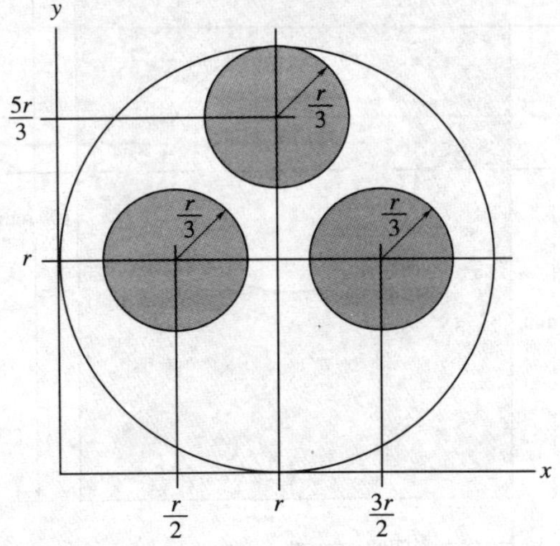

Sketch *a*, for Problems 4.1 and 4.2

4.1 An area in the *x-y* coordinate system, as shown in sketch *a*, consists of a large circle having radius *r* out of which are cut three smaller circles having radii *r*/3. Find the *x* and *y* coordinates for the centroid. The radius *r* is 10 cm.

4.2 The circular surface in sketch *a* has circular cutouts glued onto it diagonally below the top cutout such that the centroids of the three cutous are at $\bar{x} = r$ and $\bar{y} = r/3$. Find the *x* and *y* coordinates for the centroid ($r = 10$ cm).

4.3 The rectangular area shown in sketch *b* is situated symmetrically in the *x-y* coordinate system. The lengths of the rectangle's sides are *a* in the *x* direction and *b* in the *y* direction. Find the moments of inertia I_x and I_y and the polar moment of inertia *J* for the rectangular surface.

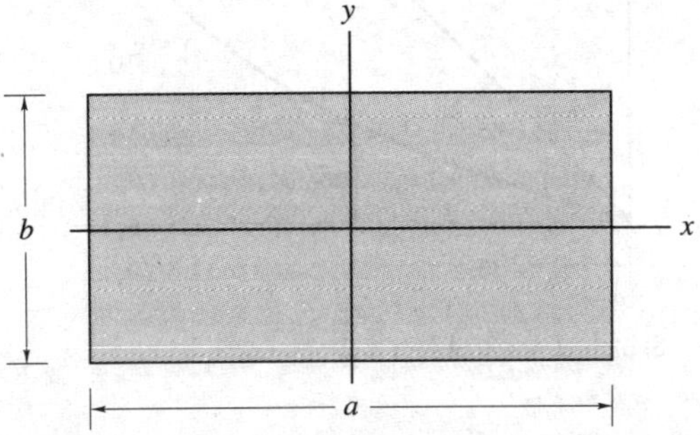

Sketch *b*, for Problem 4.3

4.4 Derive the area moment of inertia for the hollow circular area shown in Table 4.1.

4.5 Derive the area moment of inertia for the elliptical quadrant shown in Table 4.1.

4.6 Derive the area moment of inertia for the triangular section shown in Table 4.1.

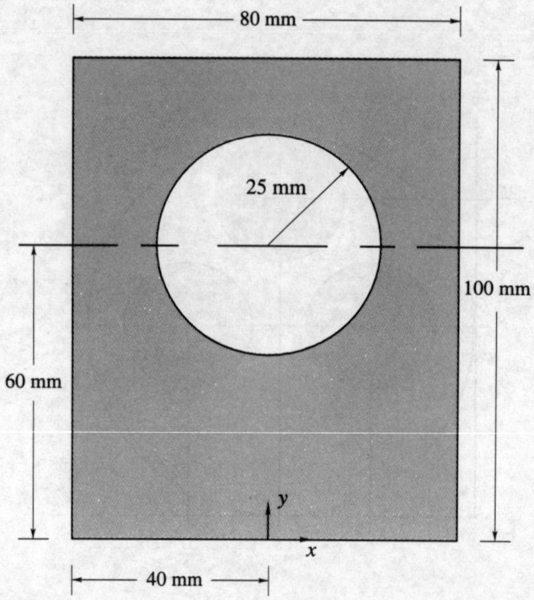

Sketch *c*, for Problem 4.7

4.7 Derive the area moment of inertia for a rectangular section with a cutout as shown in sketch *c*.

4.8 Derive the area moment of inertia and the polar moment of inertia for the two relationships shown in sketch *d*.

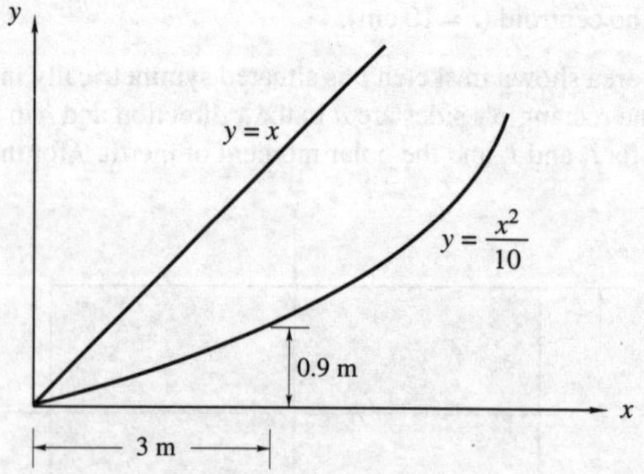

Sketch *d*, for Problem 4.8

Section 4.3

4.9 An elevator is hung by a steel rope. The rope has a cross-sectional area of 200 mm^2 and a modulus of elasticity of 70 GPa. The upward acceleration when the elevator starts is 3 m/s^2. The rope is 100 m long and the elevator weighs 1000 kg. Determine the stress in the rope, the elongation of the rope due to the elevator's weight, and the extra elongation due to the acceleration.

4.10 The 700-m-long cables in a suspension bridge are stressed to 200-MPa tensile stress. The total force in each cable is 10 MN. Calculate the cross-sectional area, the total elongation of each cable, and the spring rate when the modulus of elasticity is 70 GPa.

4.11 A 1.28-m-long steel piston in a hydraulic cylinder exerts a force of 4000 kgf. The piston is made of AISI 1080 steel and has a diameter of 50 mm. Calculate the stress in the piston, the elongation, and the spring rate.

4.12 A steel pillar supporting a highway bridge is 14 m high and made of AISI 1040 steel tubing having an outer diameter of 1.5 m and a wall thickness of 30 mm. The weight carried by the pillar is 12 MN. Calculate the deformation of the pillar, the spring rate, and the stress in the pillar.

4.13 The foundation of a bronze statue is made of a 3-m-high conical tube of constant wall thickness (8 mm). The tube's outer diameter is 200 mm at the top (just under the statue) and 400 mm at the ground. The tube material is AISI 316 stainless steel. The statue weighs 1630 kg. Calculate the deformation of the tube, the spring rate, and the maximum and minimum compressive stresses in the tube.

4.14 Calculate the deflection at point A of the hanging cone shown in sketch *e*. The cone has a density of 2000 kg/m^3.

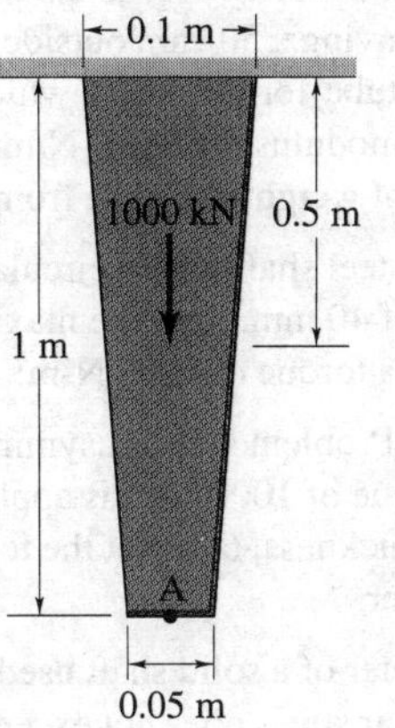

Sketch *e*, for Problem 4.14

4.15 An aluminum core having a diameter d_i of 30 mm is placed within a tubular steel shaft having a diameter d_o of 50 mm. See sketch *f*. A flange is welded to the end of the shaft, and a pulley force of 100 kN is applied. The shaft is 100 mm long. Find the deflection at the end of the shaft and the stresses induced in the aluminum and steel sections of the shaft. Assume that the moduli of elasticity are 2×10^{11} Pa for steel and 0.7×10^{11} Pa for aluminum.

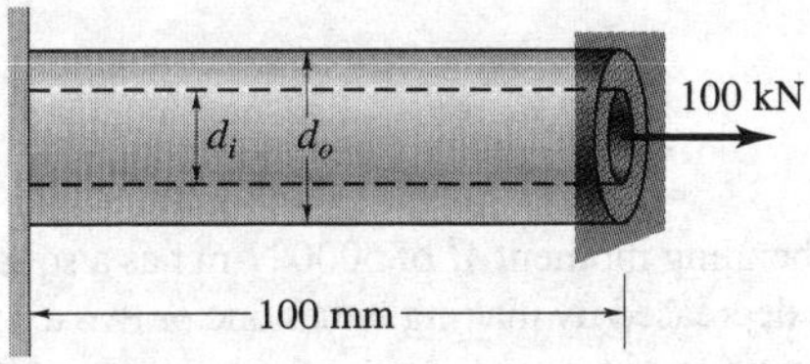

Sketch *f*, for Problem 4.15

4.16 A bar of weight W is supported horizontally by three weightless rods as shown in sketch g. Assume that the cross-sectional areas, the moduli of elasticity, and the yield stresses are the same for the three rods. What is the maximum weight that can be supported?

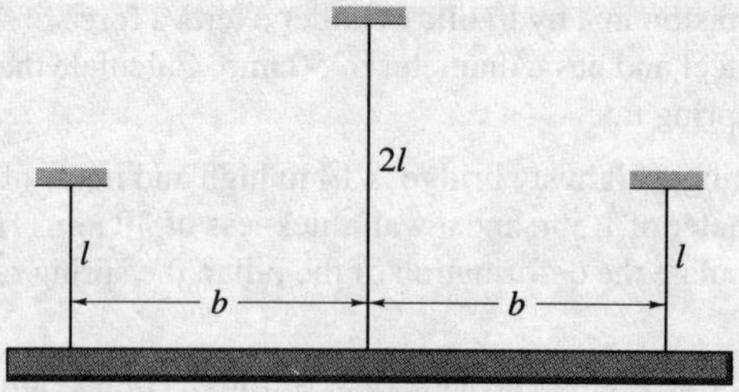

Sketch g, for Problem 4.16

Section 4.4

4.17 An electric motor transmits 100 kW to a gearbox through a 50-mm-diameter solid steel shaft that rotates at 1000 rpm. Find the torque transmitted through the shaft and the angular torsion of the 1-m-long shaft.

4.18 The torque-transmitting shaft in Problem 4.17 is too heavy for the application, so it is exchanged for a circular tube having a 50-mm outside diameter and a 40-mm inside diameter. Find the angular torsion of the tube-formed shaft, which is 1 m long, when 100 kW is transmitted at 1000 rpm. The shear modulus is 80 000 N/mm^2. Also find the maximum shear stress in the tube and the percentage of weight decrease from the solid shaft.

4.19 A torque-transmitting, hollow steel shaft with a circular cross section has an outer diameter of 50 mm and an inner diameter of 40 mm. Find the maximum length possible for the shaft if the torsion should be below 10° at a torque of 2000 N-m.

4.20 The bronze statue described in Problem 4.13 is asymmetrical, so that when a gale force wind blows against it, a twisting torque of 1000 N-m is applied to the tube. Calculate how much the statue twists. The tube's wall thickness, 6 mm at the top and 12 mm at the ground, is assumed to be proportional to its diameter.

4.21 Determine the minimum diameter of a solid shaft used to transmit 500 kW of power from a 2000-rpm motor so that the shear stress does not exceed 50 MPa.

4.22 A steel coupling is used to transmit a torque of 30 000 N-m. The coupling is connected to the shaft by a number of 0.05 m-diameter bolts placed equidistant on a pitch circle of 0.3-m diameter. The inner diameter of the coupling is 0.1 m. The allowable shear stress on the bolts is 500 MPa. Find the minimum number of bolts needed.

4.23 A shaft and a coupling are to transmit 50 kW of power at an angular speed of 1000 rpm. The coupling is connected to the shaft by 10 bolts, 20 mm in diameter, placed on a pitch circle of 200 mm. For an allowable stress on the bolts of 100 MPa, are the bolts able to transfer this power?

Section 4.5

4.24 A beam transmitting a bending moment M of 5000 N-m has a square cross section 100 by 100 mm. The weight is decreased by making either one or two axial circular holes along the beam. Determine whether one or two holes gives the lowest weight for the beam at a given bending stress. Neglect the transverse shear stress.

4.25 A straight beam is loaded at the ends by moments M: The area moment of inertia for the beam is $I = a^3b/12$. Find the bending stress distribution in the beam when $M = 1000$ N-m, $a = 3$ cm, and $b = 6$ cm. Also find the radius of curvature to which the beam is bent. The beam's modulus of elasticity is 2.05×10^{11} Pa.

4.26 The beam in Problem 4.25 is bent in a perpendicular direction so that $I = ab^3/12$. Find the bending stress distribution and the radius of curvature to which the beam is bent.

4.27 A curved bar has a rectangular cross section with height $h = r_o - r_i = 50$ mm and width $b =$ 100 mm. Its inner radius is 100 mm. Find the distance between the neutral axis and the centroid.

4.28 The curved bar in Problem 4.27 is loaded with a bending moment of 3000 N-m. Find the stress at the innermost and outermost radii.

4.29 Two beams with rectangular cross sections $a \times b$ are placed on top of each other to form a beam having height $2a$ and width b. Find the area moments of inertia I_x and I_y for the two beams

a) When they are welded together along the length
b) When they are not welded together

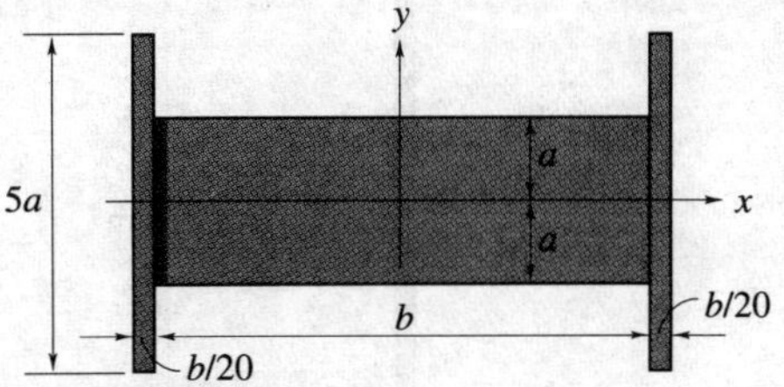

Sketch *h*, for Problem 4.30

4.30 Two thin steel plates having width $b/20$ and height $5a$ are placed one at each side of the two beams in Problem 4.29. See sketch h. Find the moments of inertia around the x and y axes

a) When the plates are not welded together
b) When they are welded together as in sketch h
c) When they are welded together to form a closed tube $5a \times 1.1b$.

Section 4.6

4.31 A cantilevered beam with a rectangular cross section is loaded by a force perpendicular to the beam centerline at the free end. The cross section is 75 mm high and 25 mm wide. The vertical load at the beam end is 20 000 N. Calculate how long the beam should be to give tensile and compressive stresses 10 times higher than the maximum shear stress. Also, calculate these stresses.

4.32 A cantilevered beam with a circular-tube cross section has an outer diameter of 100 mm and a wall thickness of 10 mm. The load perpendicular to the beam is 15 000 N, and the beam is 1.2 m long from the point of force to the wall where the beam is fastened. Calculate the maximum bending and shear stresses.

4.33 A fishing rod is made of glass-fiber-reinforced plastic in the form of a tube having an outer diameter of 10 mm and a wall thickness of 1.5 mm. The glass fibers are parallel to the tube axis, so that the bending shear stress is carried by the plastic and the bending stresses are carried by the fibers. The fishing rod is 2 m long. Determine whether the rod fails from tensile stresses in the fibers or from shear overstressing in the plastic. The bending strength of the fiber-reinforced plastic is 800 MPa, and its shear strength is 3.2 MPa.

4.34 Two aluminum beams, like the one shown in Fig. 4.16, are welded together to form a closed cross section with dimensions of 240 by 80 mm. The weld was badly done, so that the wall is only 2 mm thick instead of 8 mm. The allowable shear stress in the weld is 50 MPa, whereas the maximum allowable bending stress is 150 MPa. Find whether the 2-m-long beam fails first at the outermost fibers or at the welds.

4.35 To decrease as much as possible the weight of beams subjected to bending, the center of gravity of the cross section is placed as far away from the beam center of gravity as possible. A beam supporting a floor has a cross section as shown in sketch *i*. The bending moment acting on the beam is 3000 N-m and the shear force is 1000 N. Calculate the maximum bending stress and maximum shear stress.

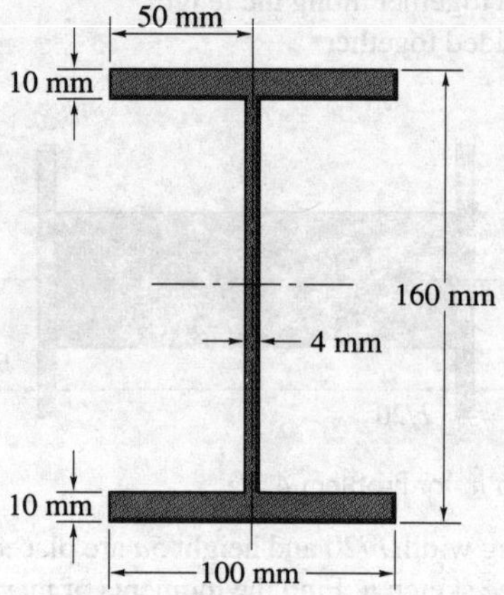

Sketch *i*, for Problem 4.35

CHAPTER

5

DEFORMATION

High-speed photographs of a golf ball just before and after impact. (Courtesy of Cordin Camera Company)

Let me tell you the secret that has led me to my goal. My strength lies in my tenacity.

Louis Pasteur

SYMBOLS

A	area, m^2
a	length dimension, m
b	length dimension, m
C_1, C_2	constants
c	distance from neutral axis to outer fiber of solid, m
E	modulus of elasticity, Pa
G	shear modulus of elasticity, Pa
h	height, m
I	area moment of inertia, m^4
J	polar area moment of inertia, m^4
l	length, m
M	moment, N-m
P	force, N
Q	load applied at point of deformation, N
q	load intensity, N/m
R	reaction force, N
T	torque, N-m
U	strain energy, N-m
V	shear force, N
v	volume, m^3
w_0	unit step load distribution, N/m
x,y,z	Cartesian coordinates, m
γ	shear strain rate
δ	deformation (deflection), m
δ_{max}	maximum deformation (deflection), m
δ_P	deflection at location of applied load, m
ε	strain ratio
θ	slope
ν	Poisson's ratio
σ	normal stress, Pa
τ	shear stress, Pa

Subscripts

A, B, C	points at which reactions and moments occur
a, b	solids a and b
H	horizontal
max	maximum
V	vertical
x,y,z	coordinates

5.1 INTRODUCTION

The focus of Chapters 2 to 4 has been on describing load, stress, and strain for various conditions that may occur in machine elements. Knowing the design stress and making sure it is less than the yield stress for ductile materials and less than the ultimate stress for brittle materials are important for a safe design. However, attention also needs to be paid to strain limitation and displacement, since a machine element can fail if a part elastically deforms excessively. For example, in high-speed machinery with close tolerances, excessive deflections can cause interference between moving parts. This chapter attempts to quantify the deformation that may occur in a great variety of machine elements. It should be noted that in this paragraph displacement, deflection, and deformation are used to mean essentially the same thing.

Chapter 4 described the deformation for normal stresses [Eq. (4.25)] and defined the angle of twist for torsional stress [Eq. (4.31)] as well as the spring rate and the angular spring rate for normal and torsional stresses [Eqs. (4.26) and (4.35), respectively]. These derivations were described in Chapter 4 because they are simple and could easily be included with the stress development. The derivations are not repeated in this chapter.

Chapter 5 focuses on describing the deformation for distributed loading such as occurs in a beam. Some approaches investigated are the integral method, the singularity function, the method of superposition, and Castigliano's theorem.

5.2 M…

Figure 4.14(b) s… …nding. The **radius of curvature** *r* ca…

$$\frac{\ldots}{r} = \frac{\ldots\; dx^2}{\left[1+\left(\frac{dy}{dx}\right)^2\right]^{\frac{3}{2}}} \quad \textbf{(5.1)}$$

But dy/dx is much less than 1.

$$\therefore \frac{1}{r} = -\frac{d^2y}{dx^2} \quad \textbf{(5.2)}$$

Substituting Eq. (5.2) into Eq. (4.50) gives

$$\frac{d^2y}{dx^2} = -\frac{M}{EI} \quad \textbf{(5.3)}$$

This equation relates the transverse displacement to a bending moment. Even though an approximation to the curvature was used in reducing Eq. (5.1) to Eq. (5.2) that is valid only for small bending angles, this is a reasonable approximation for most beams in machinery element applications. Equation (5.3) is the **moment-curvature relation** and is sometimes referred to as "the equation of the elastic line."

It is convenient to display the load intensity, shear force, moment, slope, and deformation in the following group of ordered derivatives:

$$\frac{q}{EI} = \frac{d^4y}{dx^4} \quad \textbf{(5.4)}$$

$$-\frac{V}{EI} = \frac{d^3y}{dx^3} \quad \textbf{(5.5)}$$

$$-\frac{M}{EI} = \frac{d^2y}{dx^2} \quad \textbf{(5.3)}$$

$$\theta = \frac{dy}{dx} \quad \textbf{(5.6)}$$

$$y = f(x) \quad \textbf{(5.7)}$$

If SI units are used in these equations, the appropriate units are newtons per meter for load intensity, newtons for shear force, newton-meters for moment, and meters for deformation. Slope is dimensionless, meaning that it is given in radians.

Integrating Eq. (5.3) gives the dimensionless slope at any point x as

$$EI\frac{dy}{dx} = EI\theta = -Mx + C_1 \tag{5.8}$$

Integrating Eq. (5.8) again gives the deflection at any point x as

$$EIy(x) = -M\frac{x^2}{2} + C_1 x + C_2 \tag{5.9}$$

Note that integrating the load intensity function will produce the negative of the shear force distribution, or

$$-V(x) = \int_0^x q(x)dx \tag{5.10}$$

Furthermore, integrating the shear force gives the moment

$$M(x) = \int_0^x V(x)dx \tag{5.11}$$

Equations (5.3) to (5.11) will be used in directly obtaining the deflection due to any type of loading.

Example 5.1 **GIVEN** A perpendicular force P acts at the end of a cantilevered beam with length l, as shown in Fig. 5.1. Assume that the cross section is constant along the beam and that the material is the same throughout, implying that the area moment of inertia I and the modulus of elasticity E are constant.

FIND The deformation for any x, the slope of the deformation, and the location and value of the maximum slope.

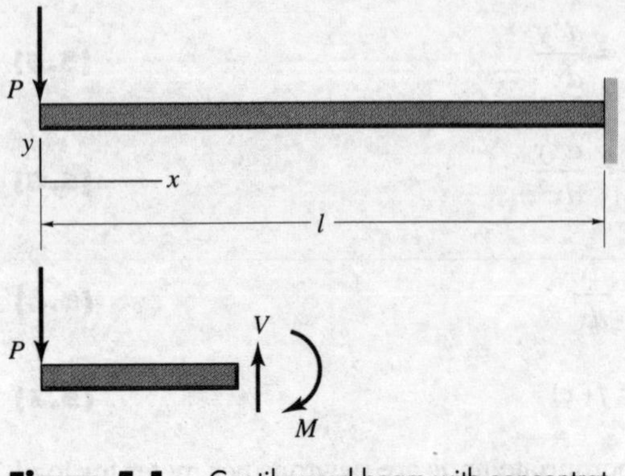

Figure 5.1 Cantilevered beam with concentrated force applied at free end. Used in Example 5.1.

Solution

The moment is $M = Px$. From Eq. (5.3)

$$\frac{d^2y}{dx^2} = -\frac{M}{EI} = -\frac{Px}{EI}$$

Integrating once gives

$$\frac{dy}{dx} = -\frac{Px^2}{2EI} + C_1$$

Integrating again gives

$$y = -\frac{Px^3}{6EI} + C_1x + C_2$$

The boundary conditions are

1. $x = l \qquad y = 0$
2. $x = l \qquad dy/dx = 0$

From boundary condition 2

$$C_1 = \frac{Pl^2}{2EI}$$

$$\therefore \quad y = -\frac{Px^3}{6EI} + \frac{Pl^2x}{2EI} + C_2$$

Making use of boundary condition 1 gives

$$C_2 = \frac{Pl^3}{6EI} - \frac{Pl^3}{2EI} = -\frac{Pl^3}{3EI}$$

$$\therefore \quad y = \frac{P}{6EI}\left(-x^3 + 3l^2x - 2l^3\right)$$

and

$$\frac{dy}{dx} = \frac{P\left(l^2 - x^2\right)}{2EI}$$

The maximum slope occurs at $x = 0$ and is

$$\left(\frac{dy}{dx}\right)_{x=0} = \frac{Pl^2}{2EI}$$

5.3 SINGULARITY FUNCTIONS

A **singularity function** permits expressing in one equation what would normally be expressed in several separate equations with boundary conditions. The deflection can be obtained while making use of singularity functions by the following procedure:

1. Draw a free-body diagram showing the forces acting on the system.
2. Use force and moment equilibrium to establish reaction forces acting on the system.
3. Write an expression for the load intensity function for all the loads acting on the system while making use of Table 2.2.
4. Integrate the negative load intensity function to give the shear force and then integrate the shear force to give the moment.
5. Make use of Eq. (5.9) to describe the deflection at any value.
6. Plot the following as a function of x:

   a) Shear
   b) Moment
   c) Slope
   d) Deflection

The following three examples show how to determine the deflection by using singularity functions. These three examples are for general situations that will be used throughout the text.

Example 5.2

GIVEN A bar with a unit weight between the simply supported ends.

FIND The deflection for any x by using singularity functions.

Solution

Figure 5.2 shows a free-body diagram of the simply supported ends for a complete bar and a portion of the bar. From Table 2.2 for concentrated forces the load intensity equation for the forces shown in Fig. 5.2(b) can be written as

$$q(x) = \frac{Pb}{l}\langle x\rangle^{-1} - P\langle x-a\rangle^{-1} \quad \textbf{(a)}$$

Integrating twice gives the moment as

$$-M(x) = \frac{Pb}{l}\langle x\rangle^{1} - P\langle x-a\rangle^{1} \quad \textbf{(b)}$$

Making use of Eq. (5.3) gives

$$EI\frac{d^2y}{dx^2} = -M(x) = \frac{Pb}{l}\langle x\rangle^{1} - P\langle x-a\rangle^{1} \quad \textbf{(c)}$$

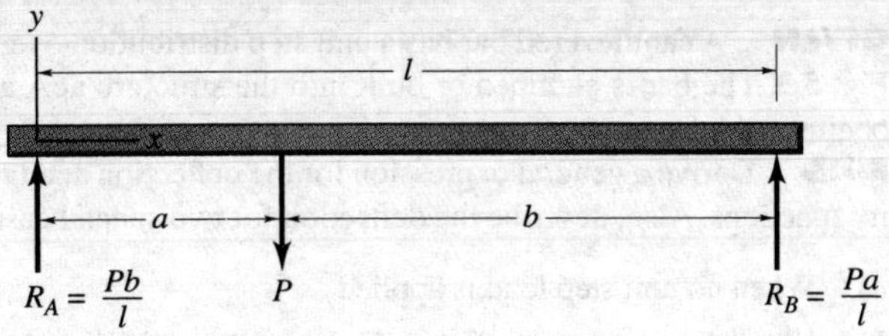

(a)

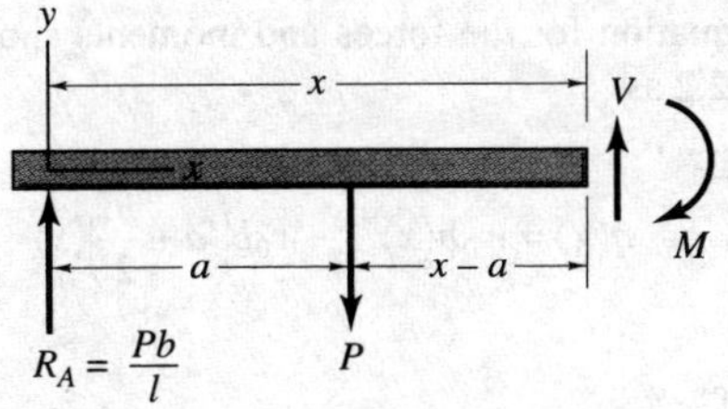

(b)

Figure 5.2 Free-body diagram of force anywhere between simply supported ends. (a) Complete bar; (b) portion of bar.

Because EI is constant along the bar, integrating Eq. (c) gives

$$EI\frac{dy}{dx} = \frac{Pb}{2l}\langle x\rangle^2 - \frac{P}{2}\langle x-a\rangle^2 + C_1 \tag{d}$$

Integrating again gives

$$EIy = \frac{Pb}{6l}\langle x\rangle^3 - \frac{P}{6}\langle x-a\rangle^3 + C_1x + C_2 \tag{e}$$

The boundary conditions are

1. $y = 0$ at $x = 0 \rightarrow C_2 = 0$
2. $y = 0$ at $x = l$ gives

$$C_1 = -\frac{Pbl}{6} + \frac{Pb^3}{6l} = \frac{Pb}{6l}(b^2 - l^2) = -\frac{Pb}{6l}(l^2 - b^2) \tag{f}$$

$$\therefore y(x) = -\frac{P}{6EI}\left[\frac{xb}{l}(l^2 - x^2 - b^2) + \langle x-a\rangle^3\right] \tag{g}$$

Note from Eq. (g) that when $x \le a$ the last term on the right side of the equal sign is zero and when $x > a$ the angular brackets become round brackets. For more information about angular brackets see Chapter 2 (Sec. 2.9). Also note that from Fig. 5.2 positive y is upward but that in Eq. (g) y is negative, meaning it is downward.

Example 5.3 **GIVEN** A cantilevered bar has a unit step distribution over a part of the bar as shown in Fig. 5.3. The bar is clamped or built into the structure at A and free at C. The unit step distribution begins at the free end.

FIND Derive a general expression for the deflection at any x and at the free end by using singularity functions. Also, describe the deflection for two special cases:

a) When no unit step load is applied

b) When the unit step load is applied over the entire length

Solution

The load intensity equation for the forces and moments shown in Fig. 5.3(c) can be expressed by making use of Table 2.2 as

$$q(x) = w_0 b\langle x\rangle^{-1} - w_0 b\left(a+\frac{b}{2}\right)\langle x\rangle^{-2} - w_0\langle x-a\rangle^{0} \quad \textbf{(a)}$$

Integrating twice gives

$$-M(x) = w_0 b\langle x\rangle^{1} - w_0 b\left(a+\frac{b}{2}\right)\langle x\rangle^{0} - \frac{w_0}{2}\langle x-a\rangle^{2} \quad \textbf{(b)}$$

Making use of Eq. (5.3) gives

$$EI\frac{d^2y}{dx^2} = -M(x) = w_0 b\langle x\rangle^{1} - w_0 b\left(a+\frac{b}{2}\right)\langle x\rangle^{0} - \frac{w_0}{2}\langle x-a\rangle^{2}$$

Integrating once gives

$$EI\frac{dy}{dx} = \frac{w_0 b}{2}\langle x\rangle^{2} - w_0 b\left(a+\frac{b}{2}\right)\langle x\rangle^{1} - \frac{w_0}{6}\langle x-a\rangle^{3} + C_1 \quad \textbf{(c)}$$

Integrating again gives

$$EIy(x) = \frac{w_0 b}{6}\langle x\rangle^{3} - \frac{w_0 b}{2}\left(a+\frac{b}{2}\right)\langle x\rangle^{2} - \frac{w_0}{24}\langle x-a\rangle^{4} + C_1 x + C_2 \quad \textbf{(d)}$$

The boundary conditions are

1. $\frac{dy}{dx} = 0$ at $x = 0 \rightarrow C_1 = 0$

2. $y = 0$ at $x = 0 \rightarrow C_2 = 0$

$$y = \frac{w_0}{EI}\left[\frac{bx^3}{6} - \frac{bx^2}{2}\left(a+\frac{b}{2}\right) - \frac{1}{24}\langle x-a\rangle^{4}\right] \quad \textbf{(e)}$$

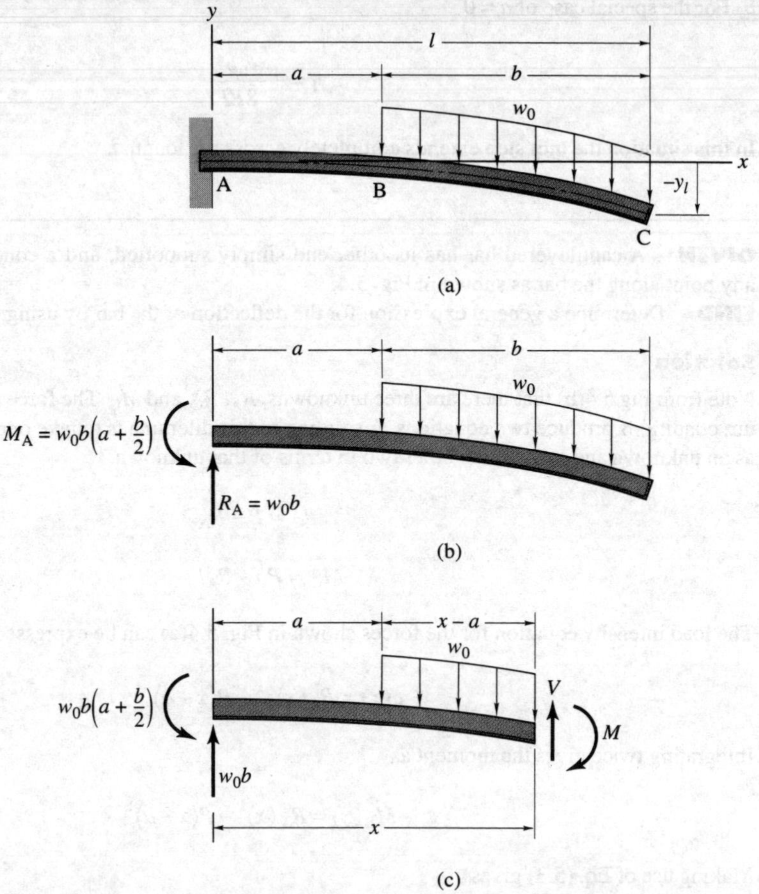

Figure 5.3 Cantilevered bar with unit step distribution over part of bar. (a) Loads and deflection acting on cantilevered bar; (b) free-body diagram of forces and moments acting on entire bar; (c) free-body diagram of forces and moments acting on portion of bar.

The deflection at the free end, or where $x = a + b$, is

$$y_l = -\frac{w_0}{EI}\left[\frac{b^4}{24} + \frac{b(a+b)^2}{2}\left(a + \frac{b}{2}\right) - \frac{b(a+b)^3}{6}\right]$$

$$= -\frac{w_0 b}{EI}\left[\frac{a^3}{3} + \frac{3}{4}a^2 b + \frac{ab^2}{2} + \frac{b^3}{8}\right] \qquad \textbf{(f)}$$

a) For the special case of $b = 0$, $y_{max} = 0$ (no deflection occurs).

b) For the special case of $a = 0$,

$$y_l = -\frac{w_0 b^4}{8EI} \tag{g}$$

In this situation the unit step extends completely across the length l.

Example 5.4

GIVEN A cantilevered bar has its other end simply supported, and a concentrated force acts at any point along the bar as shown in Fig. 5.4.

FIND Determine a general expression for the deflection of the bar by using singularity functions.

Solution

Note from Fig 5.4(b) that there are three unknowns, R_A, R_C, and M_C. The force and moment equilibrium conditions produce two equations. A solution to this dilemma is to take one of the reaction forces as an unknown and express the other two in terms of that unknown.

$$\therefore R_C = P - R_A \tag{a}$$

$$M_C = Pb - R_A l \tag{b}$$

The load intensity equation for the forces shown in Fig. 5.4(a) can be expressed as

$$q(x) = R_A \langle x \rangle^{-1} - P\langle x - a \rangle^{-1} \tag{c}$$

Integrating twice gives the moment as

$$-M_B(x) = R_A \langle x \rangle^{1} - P\langle x - a \rangle^{1} \tag{d}$$

Making use of Eq. (5.3) gives

$$EI\frac{d^2y}{dx^2} = -M(x) = R_A x - P\langle x - a \rangle^{1} \tag{e}$$

Integrating once gives

$$EI\frac{dy}{dx} = R_A\frac{x^2}{2} - \frac{P}{2}\langle x - a \rangle^{2} + C_1 \tag{f}$$

Integrating again gives the deflection as

$$EIy = R_A\frac{x^3}{6} - \frac{P}{6}\langle x - a \rangle^{3} + C_1 x + C_2 \tag{g}$$

The boundary conditions are

1. $y = 0$ at $x = 0 \rightarrow C_2 = 0$
2. $\frac{dy}{dx} = 0$ at $x = l$ gives

$$C_1 = -R_A\frac{l^2}{2} + \frac{P}{2}(l - a)^2 \tag{h}$$

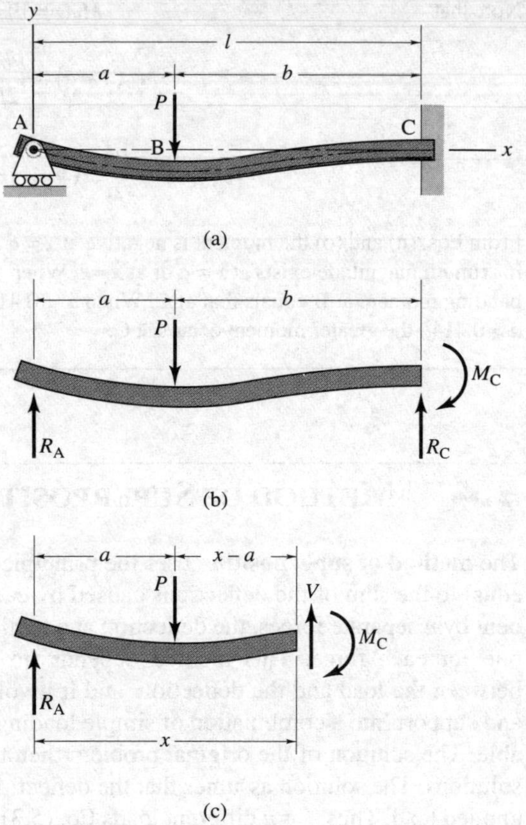

Figure 5.4 Cantilevered bar with other end simply supported and with concentrated force acting anywhere along bar. (a) Sketch of assembly; (b) free-body diagram of entire bar; (c) free-body diagram of part of bar.

$$\therefore EIy = R_A \frac{x^3}{6} - \frac{P}{6}\langle x-a\rangle^3 - \frac{R_A l^2 x}{2} + \frac{Px}{2}(l-a)^2 \quad \textbf{(i)}$$

3. $y = 0$ at $x = l$

$$R_A = \frac{Pb^2}{2l^3}(3l-b) \quad \textbf{(j)}$$

The general expression for the deformation is

$$y = \frac{P}{6EI}\left[\frac{xb^2}{2l^3}\left(-3l^3 + 3lx^2 + 3bl^2 - bx^2\right) - \langle x-a\rangle^3\right] \quad \textbf{(k)}$$

Substituting Eq. (j) into Eq. (d) gives

$$-M_B(x) = \frac{Pxb^2}{2l^3}(3l-b) - P\langle x-a\rangle^1 \quad \textbf{(l)}$$

Note that

$$M_B(x=0)=0 \quad \textbf{(m)}$$

$$-M_B(x=a)=\frac{Pb^2a}{2l^3}(3l-b) \quad \textbf{(n)}$$

$$-M_B(x=l)=\frac{Pb^2l}{2l^2}(3l-b)-Pb=-\frac{Pab}{2l^2}(2l-b) \quad \textbf{(o)}$$

From Eqs. (n) and (o) the moment is negative at $x=a$ and positive a t $x=l$. It is not clear whether the maximum magnitude exists at $x=a$ or at $x=l$. When $a=(\sqrt{2}-1)l=0.414l$, the magnitude of the bending moment at B equals that at C. When $a<0.414l$, the greater moment occurs at B; and when $a>0.414l$, the greater moment occurs at C.

5.4 METHOD OF SUPERPOSITION

The **method of superposition** uses the principle that the deflection *at any point* in a bar is equal to the sum of the deflections caused by each load acting separately. Thus, if a bar is bent by *n* separate forces, the deflection at a particular point is the sum of the *n* deflections, one for each force. This method depends on the linearity of the governing relations between the load and the deflection, and it involves reducing complex conditions of load and support into a combination of simple loading conditions for which solutions are available. The solution of the original problem then takes the form of a superposition of these solutions. The solution assumes that the deflection of the bar is linearly proportional to the applied load. Thus, for *n* different loads Eq. (5.3) can be written as

$$-EI\frac{d^2y}{dx^2}=-EI\frac{d^2}{dx^2}(y_1+y_2+\cdots+y_n)=M_1+M_2+\cdots+M_n \quad \textbf{(5.12)}$$

Table 5.1 gives solutions for some simple bar deflection situations that may be combined to produce the deflection for a more complex situation.

Example 5.5

GIVEN Figure 5.5 shows a cantilevered bar fixed at one end and free at the other end. A moment is applied at the free end, and a concentrated force is applied at any distance from the fixed end.

FIND Use the method of superposition to determine the deflection at the free end.

Solution

Figure 5.5(b) depicts the deflection with one end fixed and one end free for a concentrated force at any point within the length of the bar. Figure 5.5(c) shows a moment applied to the free end and the deformation. From Table 5.1 the individual deflections can be obtained directly as

$$y_{l,1}=-\frac{Pa^2}{6EI}(3l-a) \quad \textbf{(a)}$$

Table 5.1 Deflection for three different situations when one end is fixed and one end is free and two different situations of simply supported ends

Type of loading	Deflection for any x
Concentrated load at any x 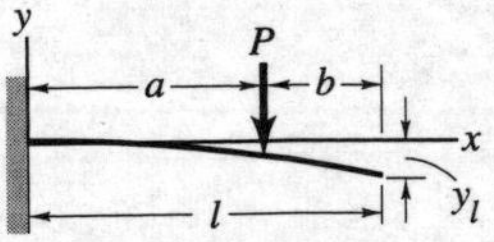	$y = -\frac{P}{6EI}(\langle x-a\rangle^3 - x^3 + 3x^2a)$
Unit step distribution over part or all of bar 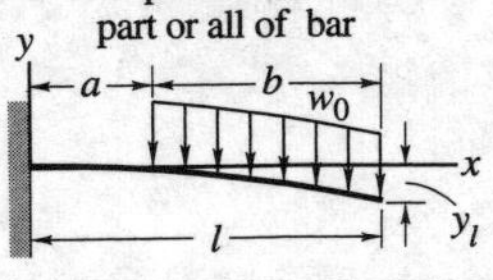	$y = \frac{w_0}{EI}\left[\frac{bx^3}{6} - \frac{bx^2}{2}\left(a + \frac{b}{2}\right) - \frac{1}{24}\langle x-a\rangle^4\right]$
Moment applied to free end 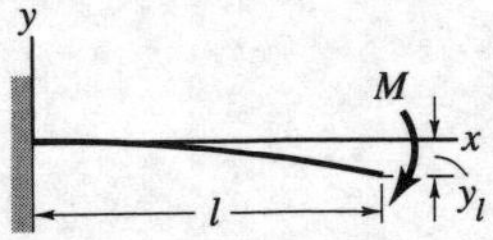	$y = -\frac{Mx^2}{2EI}$

(a)

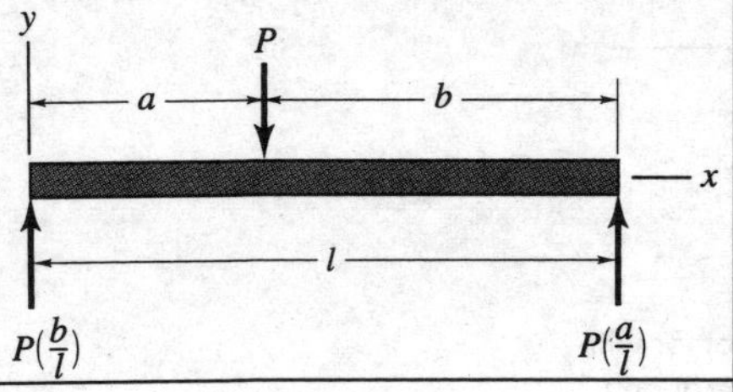	$y = \frac{P}{6EI}\left[\frac{b}{l}\langle x\rangle^3 - \langle x-a\rangle^3 + 3a^2x - 2alx - \frac{a^3x}{l}\right]$
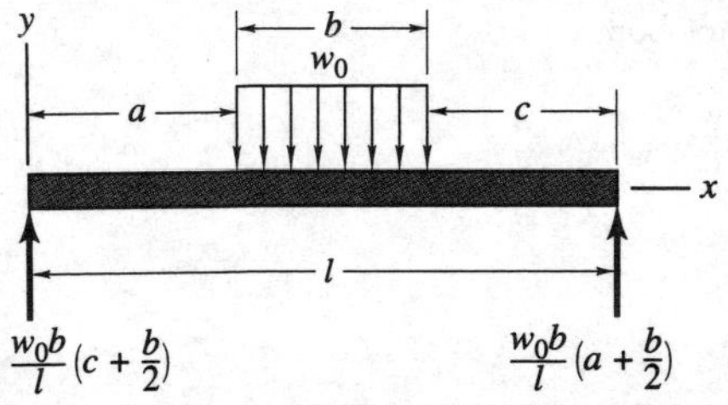	$y = -\frac{w_0b}{24lEI}\left\{4\left(c+\frac{b}{2}\right)\langle x\rangle^3 - \frac{l}{b}\left[\langle x-a\rangle^4 - \langle x-a-b\rangle^4\right] + x\left[b^3 + 6bc^2 + 4b^2c + 4c^3 - 4l^2\left(c + \frac{b}{2}\right)\right]\right\}$

(b)

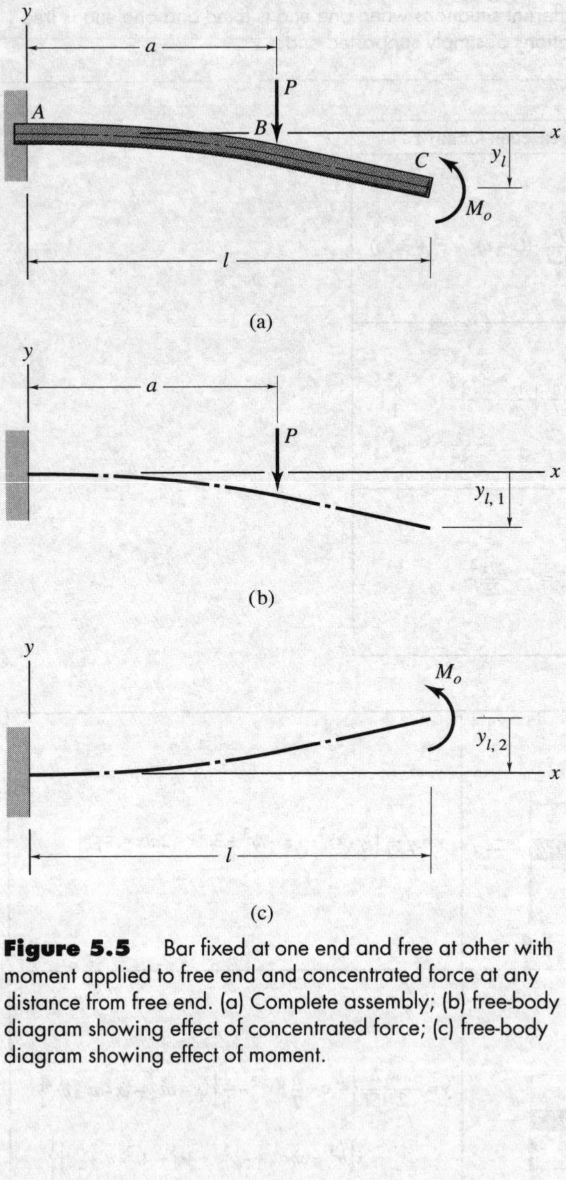

Figure 5.5 Bar fixed at one end and free at other with moment applied to free end and concentrated force at any distance from free end. (a) Complete assembly; (b) free-body diagram showing effect of concentrated force; (c) free-body diagram showing effect of moment.

and

$$y_{l,2} = \frac{Ml^2}{2El} \tag{b}$$

The resultant deflection from applying the method of superposition is

$$y_l = y_{l,1} + y_{l,2} = \frac{-Pa^2(3l - a) + 3Ml^2}{6EI} \tag{c}$$

The deflection at any point on the bar is

$$y = -\frac{P}{6EI}\left[\langle x-a\rangle^3 - x^3 + 3x^2 a\right] + \frac{Mx^2}{2EI} \qquad \textbf{(d)}$$

5.5 STRAIN ENERGY

Statically indeterminate beams and beams of varying material properties or cross sections cannot be successfully analyzed by using the methods discussed so far. Also, when a loading is energy related, such as an object striking a beam with a given initial velocity, the exact forces in the loadings are not known. For this reason energy methods are often extremely useful.

When loads are applied to a machine element, the material of the machine element will deform. In the process the external work done by the loads will be converted by the action of either normal or shear stress into internal work called **strain energy,** provided that no energy is lost in the form of heat. This strain energy is stored in the body. The unit of strain energy is newton-meters in SI units and pounds force–inch in English units. Strain energy is always positive even if the stress is compressive because stress and strain are always in the same direction. The symbol U is used to designate strain energy.

5.5.1 NORMAL STRESS

When a tension test specimen is subjected to an axial load, a volume material element (shown in Fig. 5.6) is subjected to an axial stress, and the stress develops a force

$$dP = \sigma_z dA = \sigma_z dx dy \qquad \textbf{(5.13)}$$

on the top and bottom faces of the element after the element undergoes a vertical elongation $\varepsilon_z dz$.

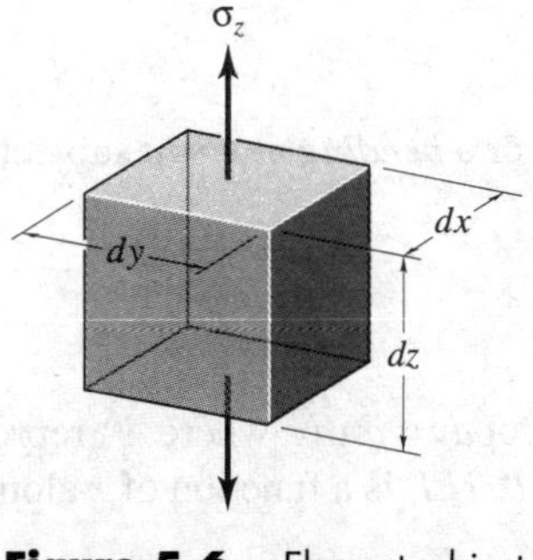

Figure 5.6 Element subjected to normal stress.

Work is determined by the product of the force and the displacement in the direction of the force. Because the force ΔP is increased uniformly from zero to its final magnitude dP when the displacement $\varepsilon_z dz$ is attained, the average force magnitude $dP/2$ times the displacement $\varepsilon_z dz$ is the strain energy or

$$dU = \left(\frac{1}{2}dP\right)\varepsilon_z dz$$

Making use of Eq. (5.13) and the fact that $dv = dxdydz$ gives

$$dU = \frac{1}{2}\sigma_z \varepsilon_z dv \tag{5.14}$$

In general then, if the member is subjected only to a uniaxial normal stress, the strain energy is

$$U = \int_v \frac{\sigma\varepsilon}{2} dv \tag{5.15}$$

Also, if the material acts in a linear-elastic manner, Hooke's law applies and Eq. (3.3) can be substituted into Eq. (5.15) to give

$$U = \int_v \frac{\sigma^2}{2E} dv \tag{5.16}$$

For *axial loading* of a bar of length l, making use of Eq. (2.7) gives

$$U = \int_v \frac{P^2}{2EA^2} dv \tag{5.17}$$

But $dv = dAdx$

$$\therefore U = \int_0^l \frac{P^2}{2AE} dx = \frac{P^2 l}{2AE} \tag{5.18}$$

For a *bending moment* subjected to a given loading, making use of Eq. (4.49) gives

$$U = \int_v \frac{\sigma^2}{2E} dv = \int \frac{M^2 y^2}{2EI^2} dv$$

But $dv = dAdx$, where dA represents an element of the cross-sectional area. Also, recall that $M^2/2EI^2$ is a function of x alone; then,

$$U = \int_0^l \frac{M^2}{2EI^2}\left(\int y^2 dA\right) dx$$

Making use of Eq. (5.7) gives

$$U = \int_0^l \frac{M^2}{2EI} dx \tag{5.19}$$

5.5.2 SHEAR STRESS

Consider the volume element shown in Fig. 5.7. The shear stress causes the element to deform such that only the shear force $dV = \tau dxdy$, acting on the top of the element, is displaced γdz relative to the bottom surface. Only the vertical surfaces rotate, and therefore the shear forces on these surfaces do not contribute to the strain energy. Thus, the strain energy stored in the element due to a shear stress is

$$dU = \frac{1}{2}(\tau dxdy)\gamma dz$$

or

$$dU = \frac{1}{2}\tau\gamma dv \tag{5.20}$$

Integrating over the entire volume gives the strain energy stored in the member due to shear stress as

$$U = \int_v \frac{\tau\gamma}{2} dv \tag{5.21}$$

Making use of Hooke's law for shear stress, expressed in Eq. (4.28), gives

$$U = \int_v \frac{\tau^2}{2G} dv \tag{5.22}$$

For strain energy in *torsion* of circular shafts, making use of Eqs. (4.27) to (4.33) gives

$$U = \int_v \frac{T^2 r^2}{2GJ^2} dv$$

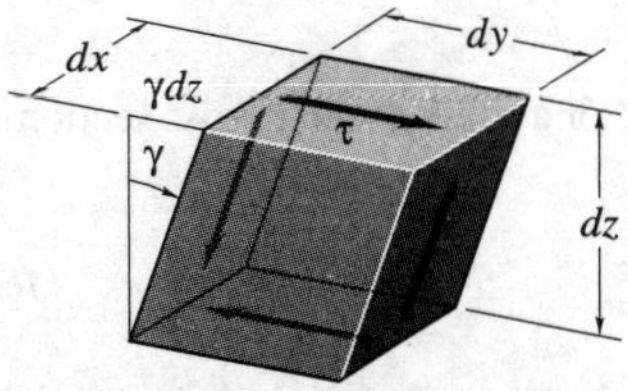

Figure 5.7 Element subjected to shear stress.

But $dv = dAdx$, and $T^2/2GJ^2$ is a function alone of x:

$$\therefore U = \int_0^l \frac{T^2}{2GJ^2}\left(\int r^2 dA\right)dx$$

Making use of Eq. (4.30) gives

$$U = \int_0^l \frac{T^2}{2GJ}dx \tag{5.23}$$

If the shaft has a uniform cross section,

$$U = \frac{T^2 l}{2GJ} \tag{5.24}$$

Recall that Table 4.1 contains values of J for a number of different cross sections, but Eq. (5.24) can be used only for cross sections with circular symmetry.

Example 5.6

GIVEN A 1-m-long solid shaft with circular cross section has a diameter of 40 mm over 0.5 m of the length and 30 mm over the rest of the length. The shaft is loaded with a torque of 1100 N-m. The shaft material is AISI 1080 high-carbon steel.

FIND Calculate the strain energy in the shaft. Also calculate the ratio of the strain energies stored in the thinner and thicker parts of the shaft.

Solution

From Eq. (5.23)

$$U = \int_0^l \frac{T^2}{2GJ}dx = \frac{T^2}{2G}\left(\frac{1}{J_1}\int_0^{0.5} dx + \frac{1}{J_2}\int_{0.5}^{1.0} dx\right)$$

From Table A.1 for AISI 1080 high-carbon steel

$$E = 207 \text{ GPa} \qquad \text{and} \qquad \nu = 0.3$$

From Eq. (3.6) the shear modulus is

$$G = \frac{E}{2(1+\nu)} = \frac{207(10^9)}{2(1+0.3)} \text{ Pa} = 79.6 \text{ GPa}$$

For a solid circular cross section

$$J_1 = \frac{\pi r^4}{2} = \frac{\pi}{2}(0.02)^4 = 25.13(10^{-8}) \text{ m}^4$$

$$J_2 = \frac{\pi}{2}(0.015)^4 = 7.952(10^{-8}) \text{ m}^4$$

The strain energy becomes

$$U = \frac{1100^2}{2(79.6)(10^9)}\left[\frac{10^8}{25.13} + \frac{10^8}{7.952}\right]\left(\frac{1}{2}\right) = 62.91 \text{ N-m}$$

The ratio of the energy stored in the two parts can be expressed in terms of the two terms in brackets as

$$\frac{U_1}{U_2} = \frac{J_2}{J_1} = \frac{7.952(10^{-8})}{25.13(10^{-8})} = 0.3164$$

5.5.3 TRANSVERSE SHEAR STRESS

The strain energy due to the shear stress can be obtained from Eq. (4.68). For a rectangular cross section with width b and height h,

$$\tau = \frac{3V}{2A}\left(1 - \frac{y^2}{c^2}\right) = \frac{3V}{2bh}\left(1 - \frac{y^2}{c^2}\right) \tag{5.25}$$

Substituting this equation into Eq. (5.22) and integrating give

$$U = \frac{1}{2G}\left(\frac{3V}{2bh}\right)^2 \int \left(1 - \frac{y^2}{c^2}\right)^2 dv$$

Setting $dv = bdxdy$ results in

$$U = \frac{9V^2}{8Gbh^2} \int_{-c}^{c} \left(1 - \frac{2y^2}{c^2} + \frac{y^4}{c^4}\right) dy \int_0^l dx$$

Integrating yields

$$U = \frac{9V^2 l}{8Gbh^2}\left(y - \frac{2y^3}{3c^2} + \frac{y^5}{5c^4}\right)\Bigg|_{y=-c}^{y=c}$$

Putting in the limits gives

$$U = \frac{6V^2 lc}{5Gbh^2}$$

Recall that $c = h/2$.

$$\therefore U = \frac{3V^2 l}{5Gbh} \tag{5.26}$$

Equation (5.26) gives the strain energy due to transverse shear stress for a rectangular cross section.

Table 5.2 Strain energy for four types of loading

Loading type	Factors involved	Strain energy for special case where all three factors are constant with x	General expression for strain energy
Axial	P, E, A	$U=\dfrac{P^2 l}{2EA}$	$U=\int_0^l \dfrac{P^2}{2EA}dx$
Bending	M, E, I	$U=\dfrac{M^2 l}{2EI}$	$U=\int_0^l \dfrac{M^2}{2EI}dx$
Torsion	T, G, J	$U=\dfrac{T^2 l}{2GJ}$	$U=\int_0^l \dfrac{T^2}{2GJ}dx$
Transverse shear (rectangular section)	V, G, A	$U=\dfrac{3V^2 l}{5GA}$	$U=\int_0^l \dfrac{3V^2}{5GA}dx$

Table 5.2 summarizes the strain energy for the four types of loading. Recall that the transverse shear is valid only for a rectangular cross section. For torsion Table 4.1 should be used for J, the polar area moment of inertia for a circular cross section. For bending the I corresponds to I_x in Table 4.1.

5.5.4 General State of Stress

The total strain energy due to a general state of stress can be expressed as

$$U=\int_v \left(\frac{\sigma_x\varepsilon_x}{2}+\frac{\sigma_y\varepsilon_y}{2}+\frac{\sigma_z\varepsilon_z}{2}+\frac{\tau_{xy}\gamma_{xy}}{2}+\frac{\tau_{yz}\gamma_{yz}}{2}+\frac{\tau_{xz}\gamma_{xz}}{2}\right)dv \tag{5.27}$$

Making use of Eq. (B.44) gives

$$U=\int_v \left[\frac{1}{2E}\left(\sigma_x^2+\sigma_y^2+\sigma_z^2\right)-\frac{\nu}{E}\left(\sigma_x\sigma_y+\sigma_y\sigma_z+\sigma_z\sigma_x\right)+\frac{1}{G}\left(\tau_{xy}^2+\tau_{yz}^2+\tau_{zx}^2\right)\right]dv \tag{5.28}$$

If only the principal stresses σ_1, σ_2, and σ_3 act on the elements, Eq. (5.28) reduces to

$$U=\int_v \left[\frac{1}{2E}\left(\sigma_1^2+\sigma_2^2+\sigma_3^2\right)-\frac{\nu}{E}\left(\sigma_1\sigma_2+\sigma_2\sigma_3+\sigma_3\sigma_1\right)\right]dv \tag{5.29}$$

5.6 Castigliano's Theorem

Often, it is necessary to calculate the elastic deformation of distributed loads that are not as simple as those already presented. Castigliano's theorem can handle a wide range of deflection problems. Extensive use is made here of the strain energy material presented in Sec. 5.5.

Castigliano's theorem states that when a body is elastically deformed by a system of loads, the deflection at any point p in any direction a is equal to the partial derivative of the strain energy (with the system of loads acting) with respect to a load at p in the direction a, or

$$y_i = \frac{\partial U}{\partial Q_i} \tag{5.30}$$

The load Q_i is applied to a particular point of deformation and therefore is not a function of x. Thus, it is permissible to take the derivative with respect to Q_i before integrating for the general expressions for the strain energy. Also, the load may be any of the loads presented first in Chapter 2 (Sec. 2.3) and throughout the text: normal, shear, bending, and transverse shear. Table 5.2 shows the strain energy for the various types of loading.

The following procedure is to be employed in using Castigliano's theorem:

1. Obtain an expression for the *total strain energy* including
   a) Loads (P, M, T, V) acting on the device (use Table 5.2)
   b) A fictitious force Q acting at the point and in the direction of the desired deflection
2. Obtain deflection from $y = \partial U/\partial Q$.
3. If Q is fictitious, set $Q = 0$ and solve the resulting equation.

The best way to understand how to apply Castigliano's theorem is to observe how it is used in a number of different examples. The four examples below demonstrate various features of Castigliano's approach.

Example 5.7

GIVEN The simply supported bar shown in Fig. 5.2 with the force P applied at $x = l/2$.

FIND Determine the deflection at the location of the applied force δ_P by using Castigliano's theorem. Consider both bending and transverse shear.

Solution

Because of symmetry the deflection at the point of applied force can be obtained by doubling the solution from zero to $l/2$. The two types of loading being applied to the bar are

a) Bending with

$$M = \frac{Px}{2} \quad \text{and} \quad \frac{dM}{dP} = \frac{x}{2} \tag{a}$$

b) Transverse shear with

$$V = \frac{P}{2} \quad \text{and} \quad \frac{dV}{dP} = \frac{1}{2} \tag{b}$$

By making use of Table 5.2 the total strain energy can be expressed as

$$U = 2\int_0^{l/2} \frac{M^2}{2EI}dx + 2\int_0^{l/2} \frac{3V^2}{5GA}dx \tag{c}$$

By making use of Eqs. (a) and (b) the above equation becomes

$$U = \int_0^{l/2} \frac{P^2 x^2}{4EI} dx + \int_0^{l/2} \frac{3P^2}{10GA} dx$$

From Castigliano's theorem

$$-\delta_P = \frac{\partial U}{\partial P} = +\int_0^{l/2} \frac{Px^2}{2EI} dx + \int_0^{l/2} \frac{3P}{5GA} dx$$

Because *P, E, I, G,* and *A* are not functions of *x*, this equation becomes

$$\delta_P = -\frac{Pl^3}{48EI} - \frac{3Pl}{10GA} \tag{d}$$

The first term on the right side is due to bending and the second is due to transverse shear.

To get a better understanding of the bending and transverse shear contributions to the total deflection at the applied force, assume that the bar has a rectangular cross section. From Table 4.1 for a rectangular shape $I = bh^3/12$ and $A = bh$. Substituting these expressions into Eq. (d) gives

$$\delta_P = -\frac{Pl^3}{4Ebh^3} - \frac{3Pl}{10Gbh} \tag{e}$$

which can be rewritten as

$$\delta_P = -\frac{3Pl}{10Gbh}\left[\left(\frac{5}{6}\right)\left(\frac{G}{E}\right)\left(\frac{l}{h}\right)^2 + 1\right] \tag{f}$$

Recall that the first term within the brackets is due to bending and the second is due to transverse shear.

For carbon steel $G/E = 0.383$. The length-to-height ratio l/h of a bar is typically at least 10. Assume the smallest value, $l/h = 10$. After putting the values for G/E and l/h into Eq. (f), the first term within the brackets is 32 times the second term. Thus, in most applications the transverse shear term will be considerably smaller than the bending moment term (typically, less than 3%).

Example 5.8

GIVEN A cantilevered bar with a concentrated force acting at a distance *b* from the free end as shown in Fig. 5.8.

FIND Determine the deflection at the free end by using Castigliano's theorem. Transverse shear can be neglected.

Solution

Note from Fig. 5.8 that, since no force was acting at the free end, a fictitious force was created. The moment for any *x* can be expressed as

$$M = Qx + P\langle x - b\rangle$$

The only force contributing to the total strain rate is the bending moment, and from Table 5.2

$$U = \int_0^l \frac{M^2 dx}{2EI} = \frac{1}{2EI}\int_0^l \left[Q^2x^2 + 2QxP\langle x-b\rangle + \left(P\langle x-b\rangle\right)^2\right]dx$$

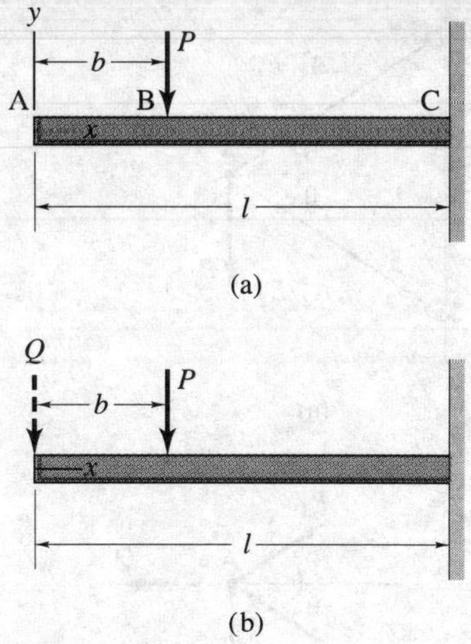

Figure 5.8 Cantilevered bar with concentrated force acting at distance b from free end. (a) Coordinate system and important points shown; (b) fictitious force shown along with concentrated force.

From Castigliano's theorem

$$\delta_A = \frac{\partial U}{\partial Q} = \frac{1}{2EI}\left[\frac{2Ql^3}{3} + 2P\int_0^l x\langle x-b\rangle dx\right] \qquad \textbf{(a)}$$

Setting $Q = 0$ and integrating gives

$$\delta_A = \frac{P}{6EI}\left[l^2(2l-3b)+b^3\right] \qquad \textbf{(b)}$$

Example 5.9

GIVEN The cable assembly shown in Fig. 5.9(a) can be made of different materials and have different cross-sectional areas in its two equal parts, denoted by the subscripts 1 and 2.

FIND Calculate the horizontal displacement at the point of vertical force application by using Castigliano's theorem.

Solution

Note from the free-body diagram in Fig. 5.9(b) that, since no force was acting horizontally, a fictitious force Q had to be created. From force equilibrium of the vertical and horizontal forces

$$\sum P_V = 0 \rightarrow -P + P_1 \sin\theta + P_2 \sin\theta = 0$$

$$\sum P_H = 0 \rightarrow Q - P_1 \cos\theta + P_2 \cos\theta = 0$$

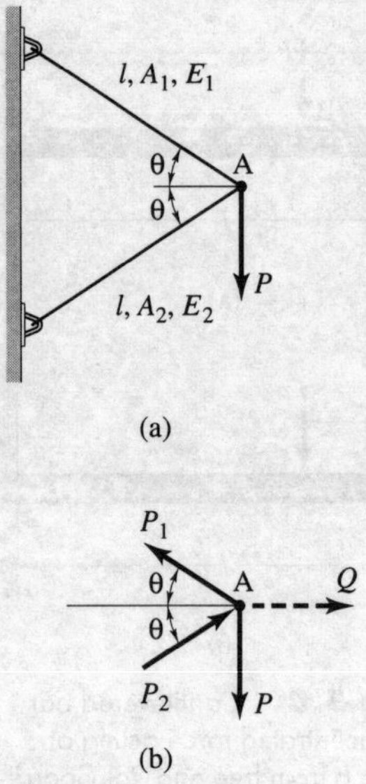

Figure 5.9 Cable system arrangement. (a) Entire assembly; (b) free-body diagram of forces acting at point A.

Solving for P_1 and P_2 gives

$$P_1 = \frac{1}{2}\left(\frac{P}{\sin\theta} + \frac{Q}{\cos\theta}\right) \quad \text{and} \quad \frac{\partial P_1}{\partial Q} = \frac{1}{2\cos\theta} \tag{a}$$

$$P_2 = \frac{1}{2}\left(\frac{P}{\sin\theta} - \frac{Q}{\cos\theta}\right) \quad \text{and} \quad \frac{\partial P_2}{\partial Q} = -\frac{1}{2\cos\theta} \tag{b}$$

From Table 5.2 the total strain energy for axial loading can be written as

$$U = \frac{P_1^2 l}{2A_1E_1} + \frac{P_2^2 l}{2A_2E_2}$$

From Castigliano's theorem the horizontal displacement at point A is

$$\delta_{A,H} = \frac{\partial U}{\partial Q} = \frac{l}{2}\left(\frac{2P_1\dfrac{\partial P_1}{\partial Q}}{A_1E_1} + \frac{2P_2\dfrac{\partial P_2}{\partial Q}}{A_2E_2}\right)$$

Substituting Eqs. (a) and (b) into the above equation gives

$$\delta_{A,H} = \frac{l}{4A_1E_1}\left(\frac{P}{\sin\theta} + \frac{Q}{\cos\theta}\right)\frac{1}{\cos\theta} - \frac{l}{4A_2E_2}\left(\frac{P}{\sin\theta} - \frac{Q}{\cos\theta}\right)\frac{1}{\cos\theta}$$

Setting $Q = 0$ gives

$$\delta_{A,H} = \frac{lP}{4\sin\theta\cos\theta}\left(\frac{1}{A_1E_1} - \frac{1}{A_2E_2}\right) \quad \textbf{(c)}$$

Note that if the cable sections 1 and 2 have the same cross-sectional area and are made of the same material (implying that the modulus of elasticity is the same), Eq. (c) gives $\delta_{A,H} = 0$. In that case there would not be any horizontal displacement.

Example 5.10

GIVEN Figure 5.10 shows a cantilevered bar with a 90° bend acted on by a horizontal force P at the free end.

FIND Calculate the vertical deflection at the free end if transverse shear is neglected. Use Castigliano's theorem.

Solution

Note from Fig. 5.10 that, since no vertical force exists at the free end, a fictitious force was created. Thus, the four components used to define the total strain energy are

a) Bending in AB, where $M_{AB} = Py$
b) Bending in BC, where $M_{BC} = Qx + Ph$
c) Axial load in AB of magnitude Q
d) Axial load in BC of magnitude P

From Table 5.2 the total strain energy can be written as

$$U = \int_0^h \frac{P^2y^2}{2EI}dy + \int_0^l \frac{(Qx + Ph)^2}{2EI}dx + \int_0^h \frac{Q^2dy}{2EA} + \int_0^l \frac{P^2dx}{2EA} \quad \textbf{(a)}$$

The bar's material and cross sections are the same in sections AB and BC. From Castigliano's theorem

$$\delta_{A,V} = \frac{\partial U}{\partial Q} = \int_0^l \frac{(Qx + Ph)xdx}{EI} + \int_0^h \frac{Qdy}{EA} = \frac{Ql^3}{3EI} + \frac{Phl^2}{2EI} + \frac{Qh}{EA}$$

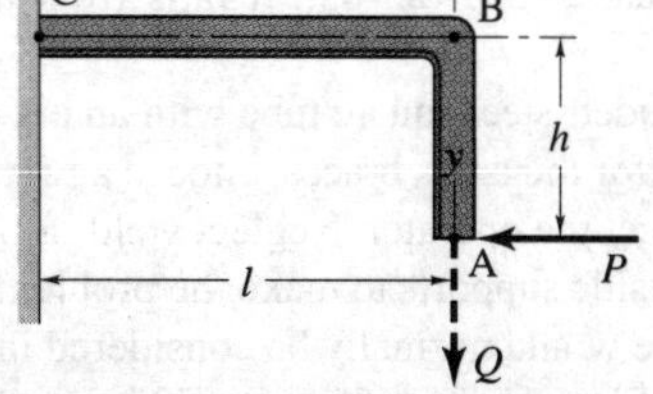

Figure 5.10 Cantilevered bar with 90° bend acted upon by horizontal force at free end.

Setting $Q = 0$ gives

$$\delta_{A,V} = \frac{Phl^2}{2EI} \tag{b}$$

Thus, the four examples just presented demonstrate the wide range of deflection problems to which Castigliano's theorem can be applied.

Case Study

DESIGN OF FALLING-OBJECT PROTECTIVE STRUCTURE FOR FORKLIFT TRUCK

Figure 5.11 shows a typical industrial forklift truck, commonly used for materials handling in factories and warehouses. A necessary design feature of a forklift truck is that the mast consists of telescoping cylinders, so that the forklift can deposit and retrieve objects stored in racks far above ground level, thereby maximizing the use of floor space in storage facilities. A serious safety problem exists, however, when the forklift is used at large mast extensions. If the operator has slightly misaligned the forks with the racks or storage bins, they may strike the rack, jarring the load on the forks. Boxes or objects on the forks can topple because of this jarring and fall onto the operator, causing severe injury. To prevent such occurrences, a falling-object protective structure (FOPS) is used to keep the operator safe within the cab of the forklift. This case study investigates part of the design process in selecting FOPS dimensions.

GIVEN The FOPS under consideration will consist of two supports to the right and left of the operator with cross braces between them. The design goal is a maximum deflection less than 3 in. if an object falls from a height of 20 ft.

FIND Using an extruded-steel square tube with an outside dimension of 2 in. for the cross braces, choose a tube thickness that will protect the operator. Neglect weld failure and buckling of the side supports to make the problem tractable, although these would normally be considered in an industrial design. In Fig 5.11, $a = 36$ in., $b = 50$ in., and $c = 84$ in.

Solution

One important approach used by designers is a worst-case scenario. A designer will consider all reasonable loading scenarios, material strengths, product abuses, etc., and will choose all of these variables so as to *minimize* the safety of the system. If the design *still* survives under the worst conditions, the designer is reasonably assured that it will survive in applications that are unlikely to approach this level of harshness.

In this case no force has been specified, so the designer must choose one. Note that, although a forklift can lift a very large weight from a pallet, heavy loads will not usually be stacked higher than the backing of the forks. Thus, heavy loads will not be candidates for falling onto the operator. Therefore, it is reasonable to choose a load of 200 lbf (chosen because it is a heavy weight but one that an operator can still move and stack onto the forks). Thus, the input energy is 200 lbf times the distance the force falls before striking the FOPS, or (200 lbf)(25 ft − 7 ft)(12 in./1 ft) = 43 200 in.-lbf.

A typical box will usually strike more than one cross brace and will usually strike closer to one side support or the other. However, the worst-case scenario occurs when the box strikes the center of one cross brace.

The vertical supports are much stiffer than the cross braces, which are loaded in tension, and are therefore assumed to be rigid. Although this may appear to be contrary to the worst-case scenario, in practice, damage to a forklift truck FOPS is either in the welds connecting the cross bracing to the side supports or in bending deformation of the cross braces. The latter case is being considered here.

Case Study (CONTINUED)

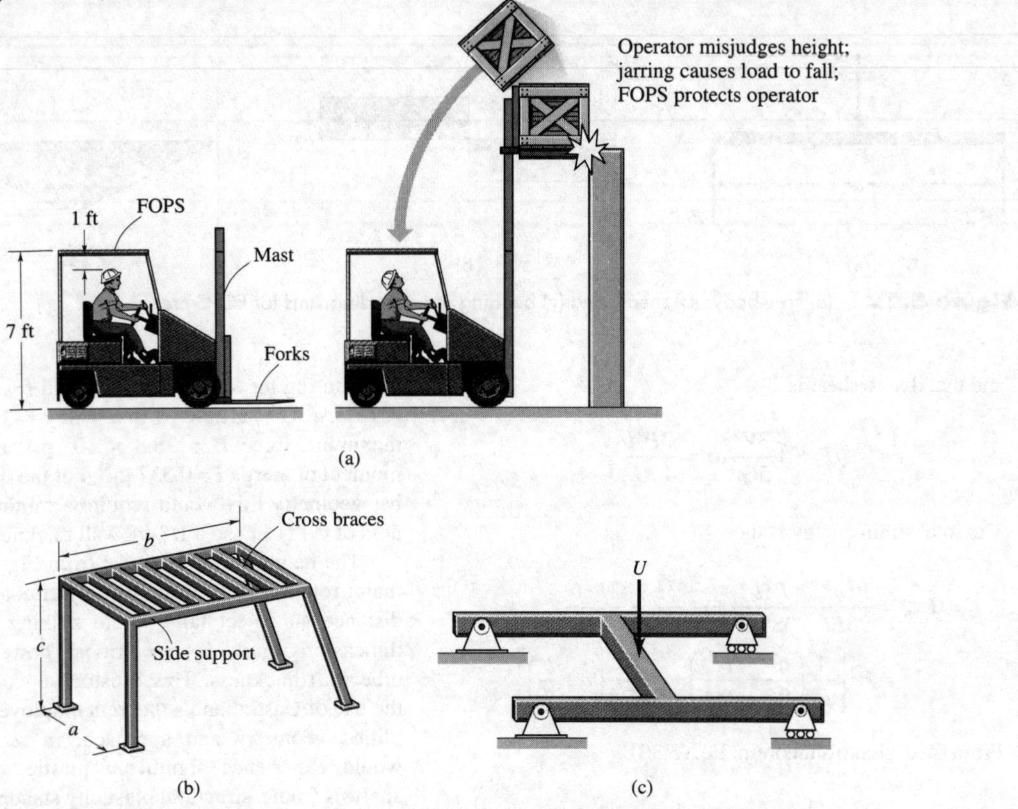

Figure 5.11 Figures used in case study. (a) Forklift truck; (b) FOPS details (not to scale); (c) idealized FOPS for analysis.

A strain energy U is applied at the center of a single cross brace, leading to a maximum dynamic force P (unknown) in accordance with the worst-case scenario described above. To make the deformations larger than they actually would be in practice, the beam is assumed to be simply supported. Actually, the welds at the cross brace/side support interface would add to the strain energy, but by neglecting this amount the predicted deformations will be on the high side. Figure 5.12 shows a free-body diagram and shear and bending moment diagrams of the central cross brace. The strain energy due to bending can be calculated from Eq. (5.19) as

$$U = 2\left[\int_0^{a/2} \frac{M^2}{2EI}\,dx\right] = \frac{1}{EI}\int_0^{a/2} \frac{P^2}{4}x^2\,dx = \frac{P^2a^3}{96EI}$$

The shear strain energy is difficult to obtain for a hollow tube. As a worst-case analysis the tube will be taken as a solid square with the same cross-sectional area as the square hollow tube. Then, from Eq. (5.26) the strain energy for transverse shear in the cross brace is

$$U = \int_0^a \frac{3V^2}{5GA}\,dx = \frac{3P^2a}{20GA}$$

For the side supports the strain energy due to bending is

$$U = 4\left[\int_0^{b/2} \frac{M^2}{2EI}\,dx\right] = \frac{2}{EI}\int_0^{b/2} \frac{P^2}{16}x^2\,dx = \frac{P^2b^3}{384EI}$$

(continued)

Case Study (CONCLUDED)

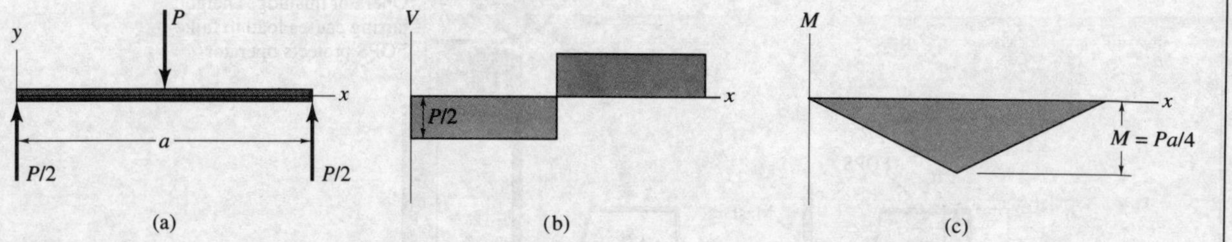

Figure 5.12 (a) Free-body; (b) shear; and (c) bending moment diagrams for FOPS cross brace.

and that due to shear is

$$U = 2\int_0^b \frac{3V^2}{5GA}dx = \frac{3P^2b}{40GA}$$

The total strain energy is then

$$U_{\text{tot}} = \frac{P^2a^3}{96EI} + \frac{P^2b^3}{384EI} + \frac{3P^2a}{20GA} + \frac{3P^2b}{40GA}$$

$$= P^2\left[\frac{1}{EI}\left(\frac{a^3}{96} + \frac{b^3}{384}\right) + \frac{3}{20GA}\left(a + \frac{b}{2}\right)\right]$$

From Castigliano's theorem, Eq. (5.30),

$$y = \frac{\partial U}{\partial P} = \frac{\partial}{\partial P}\left\{P^2\left[\frac{1}{EI}\left(\frac{a^3}{96} + \frac{b^3}{384}\right) + \frac{3}{20GA}\left(a + \frac{b}{2}\right)\right]\right\}$$

$$= P\left\{\frac{1}{EI}\left(\frac{a^3}{48} + \frac{b^3}{192}\right) + \frac{3}{10GA}\left(a + \frac{b}{2}\right)\right\}$$

Substituting the values for U, a, and b and using $E = 30 \times 10^6$ psi, $G = 11.5 \times 10^6$ psi, and $y = 3$ in. result in the maximum force $P = 28.8 \times 10^3$ psi and the required moment of inertia $I = 0.537$ in.^{4}. For the defined extruded-bar geometry this would require a minimum wall thickness of 0.121 in., so a 1/8-in. wall thickness is chosen.

The height of the FOPS, c from Fig. 5.11, played no major role in the analysis. Its only effect is to change the distance an object falls before striking the FOPS. The dimensions a and b were critical, however, as was the tube wall thickness. This illustrative example regarding the use of Castigliano's theorem employed a greatly simplified geometry and approach. In reality, the FOPS would experience significant plastic strain, requiring analysis from a structural plasticity standpoint.

5.7 SUMMARY

The three main ways a machine element will fail are from being overstressed, from the lack of a tribological film, and from excessive elastic deformations. This chapter attempted to describe the deformations that machine elements may experience. Deformations due to distributed and concentrated loads were both considered. For a distributed load four major approaches to describing the deformations were presented: the moment-curvature relation, singularity functions, the method of superposition, and Castigliano's theorem. Each has its particular strengths. The type of load being applied (normal, bending, shear, or transverse shear) determines the approach. Castigliano's theorem is the most versatile of the four approaches considered, since it can be applied to a wide range of deflection problems.

KEY WORDS

Castigliano's theorem theorem that when a body is elastically deformed by a system of loads, deflection at any point p in any direction a is equal to partial derivative of strain energy (with system of loads acting) with respect to load at p in direction a.

method of superposition principle that deflection at any point in bar is equal to sum of deflections caused by each load acting separately

moment-curvature relation relationship between beam curvature and bending moment, given by (for small bend angles) $\frac{d^2y}{dx^2} = -\frac{M}{EI}$

radius of curvature geometrically represented in Cartesian coordinates as $\frac{1}{r} = \frac{-\frac{d^2y}{dx^2}}{\left[1 + \left(\frac{dy}{dx}\right)^2\right]^{\frac{3}{2}}}$

singularity function function that permits expressing in one equation what would normally be expressed in several separate equations with boundary conditions

strain energy internal work that was converted from external work done by applying load

RECOMMENDED READINGS

Beer, F.P., and Johnson, E.R. (1981) *Mechanics of Materials,* McGraw-Hill, New York.

Craig, R.R. (1996) *Mechanics of Materials* , Wiley, New York.

Crandall, S.H., and Dahl, H.C. (1954) *An Introduction to the Mechanics of Solids,* McGraw-Hill, New York.

Fung, Y.C. (1965) *Foundations of Solid Mechanics,* Prentice-Hall, Englewood Cliffs, NJ.

Juvinall, R.C. (1967) *Stress, Strain, and Strengths,* McGraw-Hill, New York.

Lardner, T.J., and Archer, R.R. (1994) *Mechanics of Solids: An Introduction,* McGraw-Hill, New York.

Norton, R.L. (1996) *Machine Design—An Integrated Approach,* Prentice-Hall, Englewood Cliffs, NJ.

Popov, E.P. (1968) *Introduction to Mechanics of Solids,* Prentice-Hall, Englewood Cliffs, NJ.

Shigley, J., and Mischke, C. (1989) *Mechanical Engineering Design,* McGraw-Hill, New York.

Timoshenko, S., and Goodier, J. (1951) *Theory of Elasticity,* McGraw-Hill, New York.

PROBLEMS

Section 5.2

5.1 A beam is loaded by a concentrated bending moment M at the free end. The size of the moment equals the maximum moment, or Pl. Find the vertical and angular deformations along the beam by using the equation of the elastic line, Eq. (5.3).

5.2 A simply supported beam of length l carries a force P. Find the ratio between the bending stresses in the beam when P is concentrated in the middle of the beam and evenly distributed along it. Use the moment-curvature relation given in Eq. (5.3). Also, calculate the ratio of the deformations at the middle of the beam.

5.3 A simply supported beam with length l is centrally loaded with a force P. How large a moment needs to be applied at the ends of the beam

a) To maintain the slope angle of zero at the supports?

b) To maintain the midpoint of the beam without deformation when the load is applied?

Use the equation of the elastic line, Eq. (5.3).

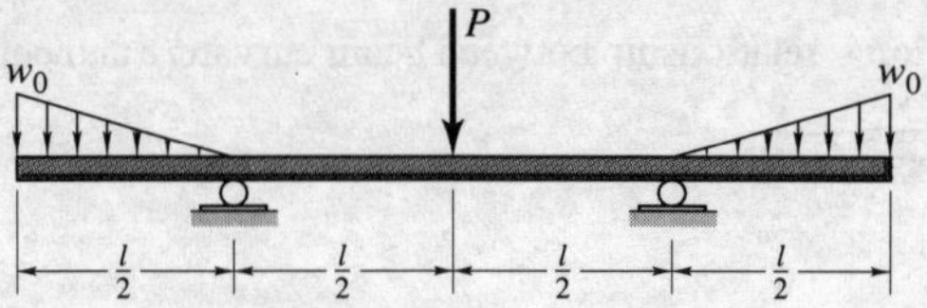

Sketch *a*, for Problem 5.4

5.4 Find the relation between P and w_0 so that the slope of the deflected beam is zero at the supports for the loading conditions shown in sketch *a*. Assume that E and A are constant.

Section 5.3

5.5 Given a simply supported beam with two concentrated forces acting on it as shown in sketch *b*, determine the expression for the elastic deformation of the beam for any x by using singularity functions. Assume that E and I are constant. Also determine the location of maximum deflection and derive an expression for it.

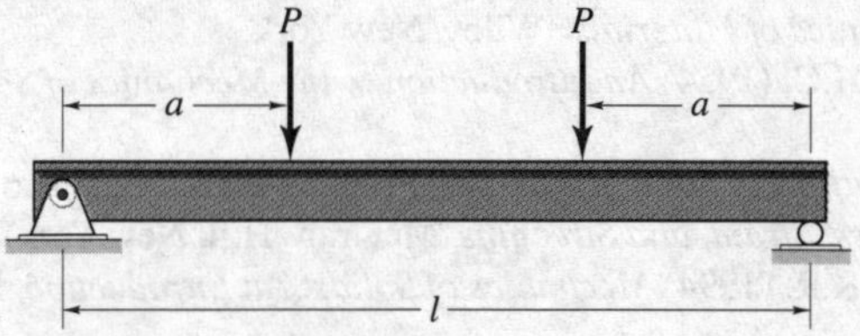

Sketch *b*, for Problem 5.5

5.6 For the loading condition described in sketch *c* obtain the internal shear force $V(x)$ and the internal moment $M(x)$ by using singularity functions. Draw $V(x)$, $M(x)$, $\Theta(x)$, and $y(x)$ as a function of x. Assume that $w_0 = 9$ kN/m and $l = 3$ m.

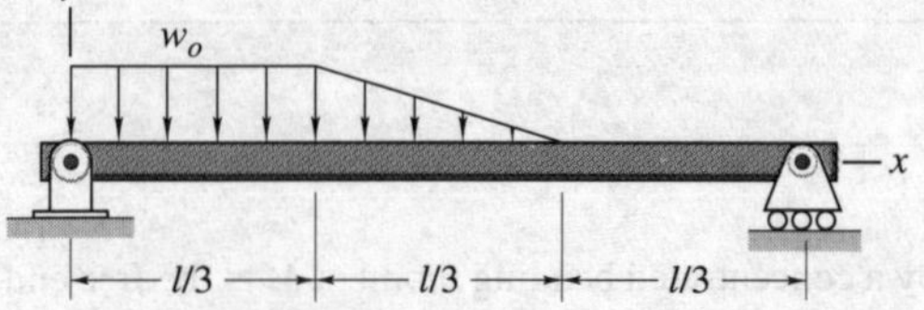

Sketch *c*, for Problem 5.6

5.7 A simply supported bar is shown in sketch *d* with $w_0 = 4$ kN/m and $l = 12$ m.

a) Draw the free-body diagram of the bar.

b) Use singularity functions to determine shear force, bending moment, slope, and deflection.

Provide both tabular results and a diagram of the shear force, the bending moment, the slope, and the deflection.

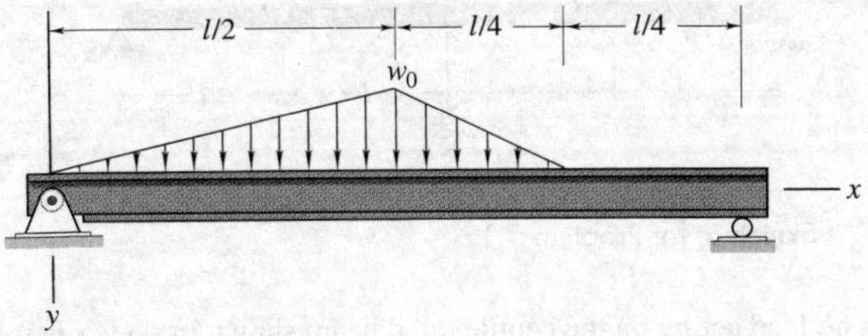

Sketch *d*, for Problem 5.7

5.8 The simply supported beam in Problem 5.5 is altered so that instead of a concentrated force *P* a concentrated moment *M* is applied at the same location. The moments are positive and act parallel with each other. Determine the deformation of the beam for any position *x* along it by using singularity functions. Assume that *E* and *I* are constant. Also determine the location of the maximum deflection.

5.9 The simply supported beam considered in Problem 5.8 has moments applied in opposite directions so that the moment at $x = a$ is M_0 and that at $x = l - a$ is $-M_0$. Find the elastic deformation of the beam by using singularity functions. Also, determine the location and size of the maximum deflection.

5.10 Given the loading condition shown in sketch *e*, find the deflection at the center and ends of the beam. Assume that $EI = 750$ kNm^2.

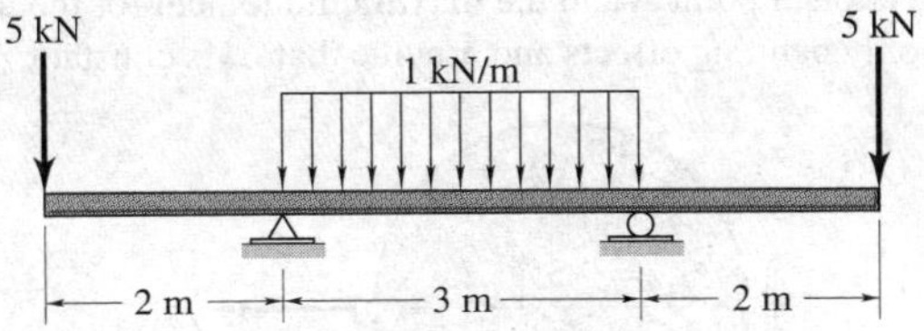

Sketch *e*, for Problem 5.10

5.11 Given the loading condition shown in sketch *f* obtain an expression for the deflection at any location on the beam. Assume that *EI* is constant.

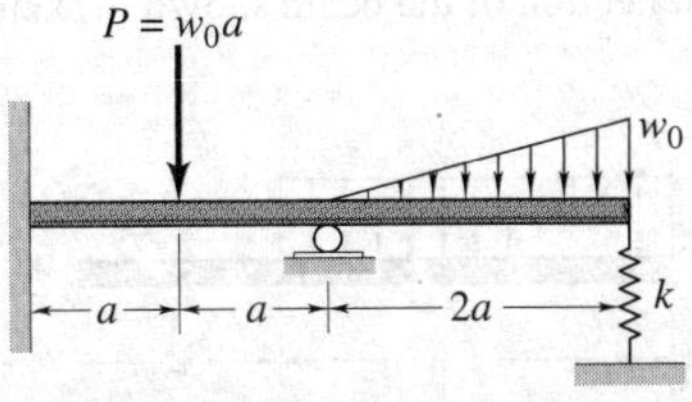

Sketch *f*, for Problem 5.11

5.12 Given the loading condition and spring shown in sketch *g* determine the stiffness of the spring so that the bending moment at point B is zero. Assume that *EI* is constant.

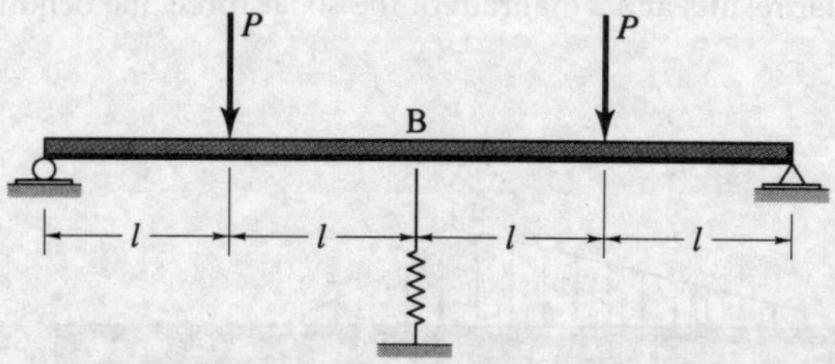

Sketch *g*, for Problem 5.12

5.13 When there is no load acting on the cantilevered beam shown in sketch *h*, the spring has zero deflection. When there is no spring and a force of 20 kN is applied at point C, a deflection of 50 mm occurs. When there is a spring, if a 50-kN load is applied at the location shown in sketch *h*, what will be the deflection of the bar? Assume that the stiffness of the spring is 450 kN/m.

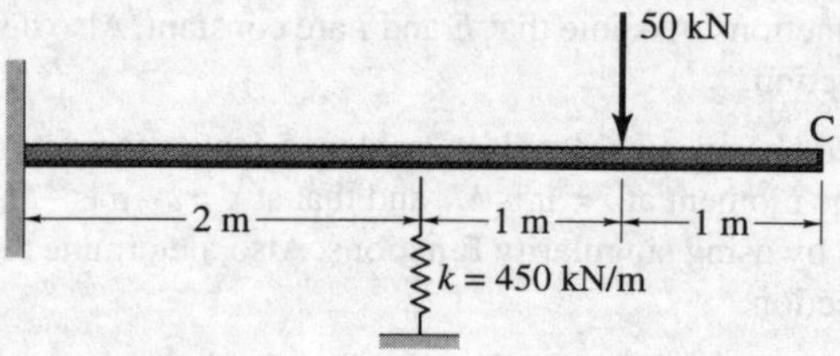

Sketch *h*, for Problem 5.13

5.14 Determine the deflection at point A and the maximum moment for the loading shown in sketch *i*. Consider only bending effects and assume that *EI* is constant.

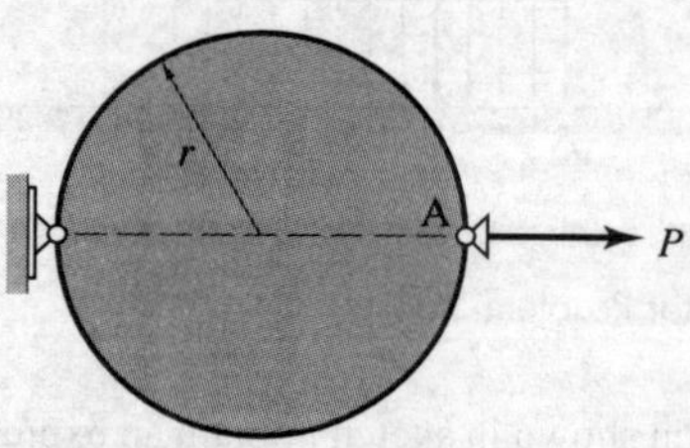

Sketch *i*, for Problem 5.14

5.15 Determine the maximum deflection of the beam shown in sketch *j*.

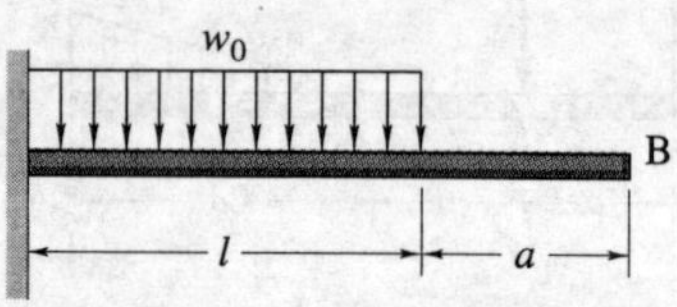

Sketch *j*, for Problem 5.15

5.16 Determine the deflection at any point in the beam shown in sketch k. Use singularity functions.

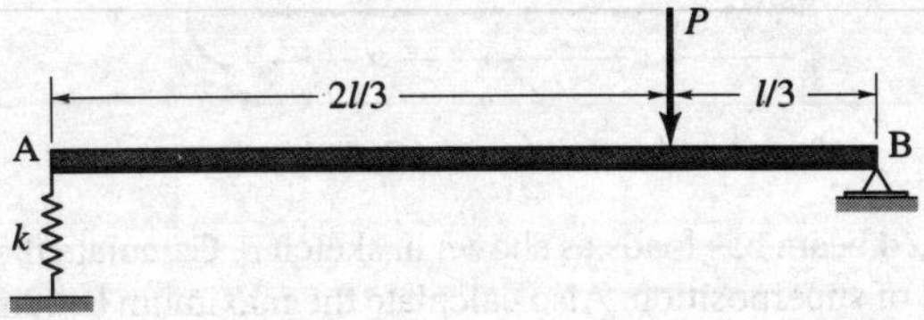

Sketch k, for Problem 5.16

Section 5.4

5.17 Determine the deformation of a cantilevered beam with loading shown in sketch l as a function of x. Also determine the maximum bending stress in the beam and the maximum deflection. Assume that $E = 207$ GPa, $I = 250\ \text{cm}^4$, $P = 1000$ N, $w_0 = 3000$ N/m, $a = 0.5$ m, $b = 0.15$ m, $c = 0.45$ m. The distance from the neutral axis to the outermost fiber of the beam is 0.040 m.

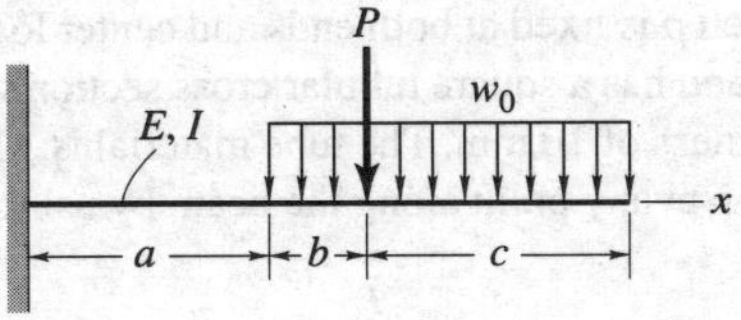

Sketch l, for Problem 5.17

5.18 Given the loading shown in sketch m let $a = 0.6$ m, $b = 0.7$ m, $M = 6500$ N-m, and $w_0 =$ 20 000 N/m. The beam has a square cross section with sides of 75 mm, and the beam material has a modulus of elasticity of 207 GPa. Determine the beam deformation by using the method of superposition. Also, calculate the maximum bending stress and maximum beam deformation.

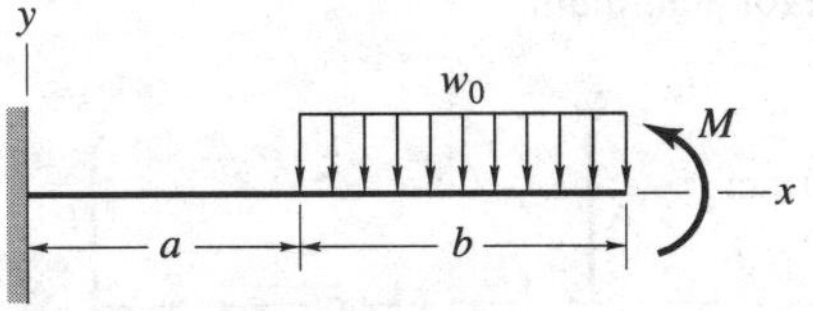

Sketch m, for Problem 5.18

5.19 The cantilevered beam shown in sketch n has both a concentrated force and a moment acting on it. Let $a = 1$ m, $b = 0.7$ m, $P = 8700$ N, and $M = 4000$ N-m. The beam cross section is rectangular with a height of 80 mm and a width of 35 mm. Also, $E = 207$ GPa. Calculate the beam deformation by using the method of superposition. Find how large M has to be to give zero deformation at $x = a$.

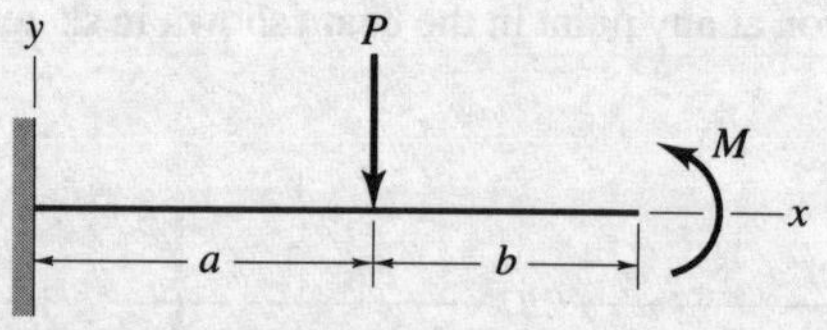

Sketch *n*, for Problem 5.19

5.20 A simply supported beam has loads as shown in sketch *o*. Calculate the beam deformation by using the method of superposition. Also calculate the maximum bending stress and the maximum beam deformation. Assume that $E = 207$ GPa and that the beam has a rectangular cross section with a height of 30 mm and a width of 100 mm. Also, $P = 1200$ N, $w_0 = 10\,000$ N/m, $a = 0.2$ m, $b = 0.1$ m, $c = 0.4$ m, and $d = 0.2$ m.

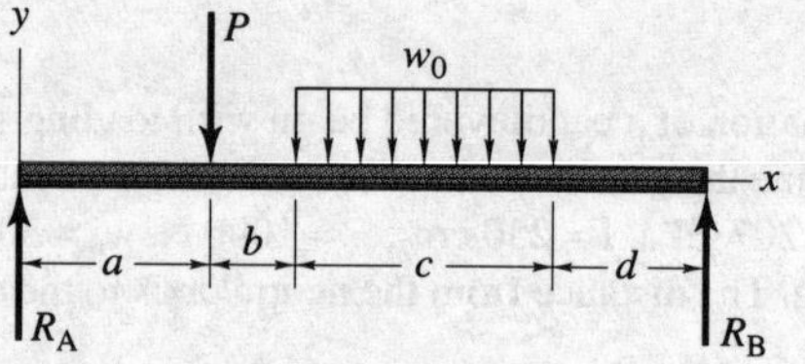

Sketch *o*, for Problem 5.20

5.21 The beam shown in sketch *p* is fixed at both ends and center loaded with a force of 2300 N. The beam is 3.2 m long and has a square tubular cross section with an outside width of 130 mm and a wall thickness of 10 mm. The tube material is AISI 1080 high-carbon steel. Calculate the deformation at any point along the beam by using the method of superposition.

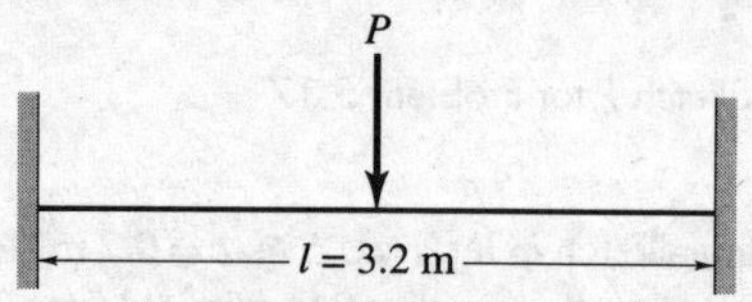

Sketch *p*, for Problem 5.21

5.22 Bar A shown in sketch *q* is a 13-mm-diameter aluminum bar; bar B is an 8-mm-diameter steel bar. The lower member is of uniform cross section and is assumed to be rigid. Find the distance *x* if the lower member is to remain horizontal. Assume that the modulus of elasticity for steel is three times that for aluminum.

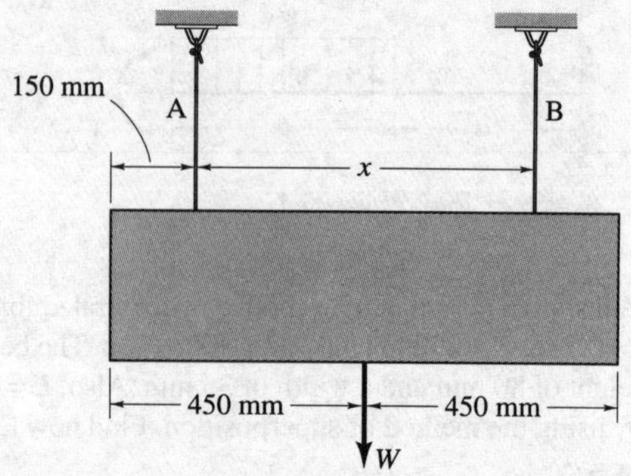

Sketch *q*, for Problem 5.22

5.23 An aluminum rod 3/4 in. in diameter and 48 in. long and a nickel steel rod 1/2 in. in diameter and 32 in. long are spaced 60 in. apart and fastened to a horizontal beam that carries a 2000-lbf load, as shown in sketch *r*. The beam is to remain horizontal after load is applied. Assume that the beam is weightless and absolutely rigid. Find the location *x* of the load and determine the stresses in each rod.

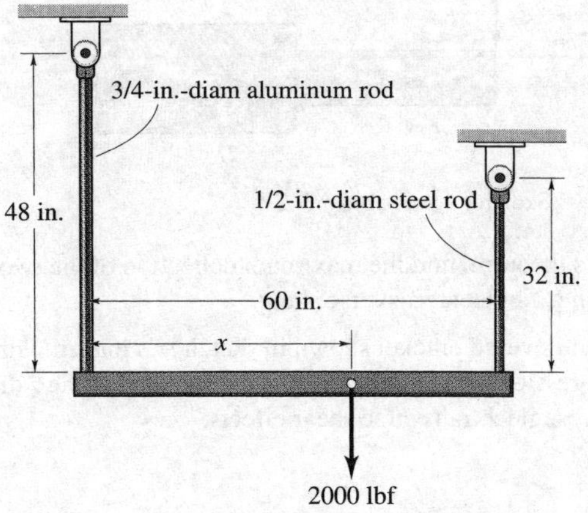

Sketch *r*, for Problem 5.23

5.24 Find the force on each of the vertical bars shown in sketch *s*. The weight is assumed to be rigid and horizontal, implying that the three vertical bars are connected at the weight in a straight line. Also, assume that the support at the top of the bars is rigid. The bar materials and its circular cross-sectional area are given in the sketch.

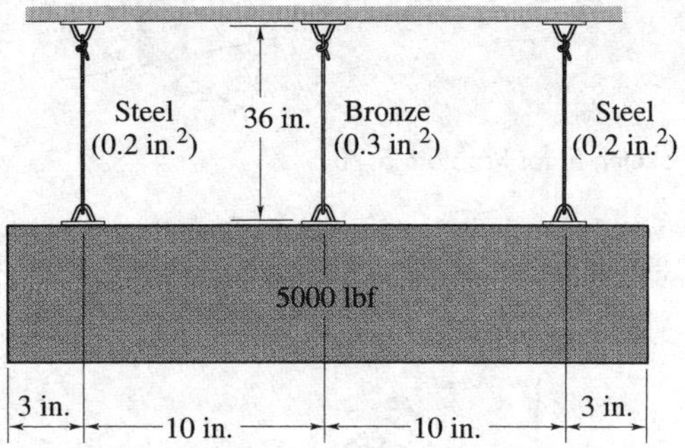

Sketch *s*, for Problem 5.24

5.25 Two solid spheres, one made of aluminum alloy 2014 and the other made of AISI 1040 medium-carbon steel, are lowered to the bottom of the sea at a depth of 8000 m. Both spheres have a diameter of 0.3 m. Calculate the elastic energy stored in the two spheres when they are at the bottom of the sea if the density of water is 1000 kg/m^3 and the acceleration of gravity is 9.807 m/s^2. Also calculate how large the steel sphere has to be to have the same elastic energy as the 0.3-m-diameter aluminum sphere.

Section 5.6

5.26 Use Castigliano's approach instead of singularity functions to solve Problem 5.5. Assume that transverse shear is negligible.

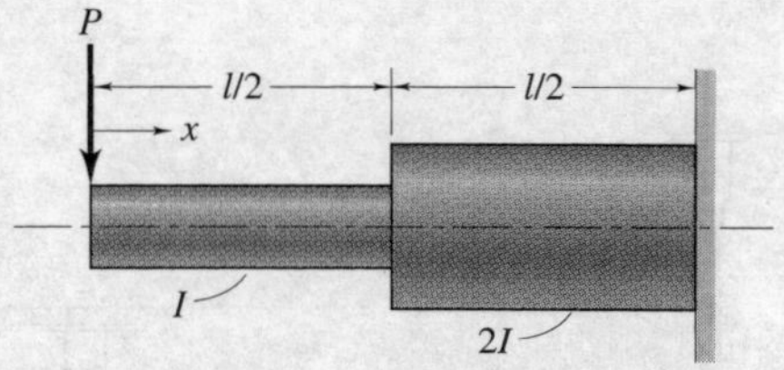

Sketch *t*, for Problem 5.27

5.27 Using Castigliano's theorem, find the maximum deflection of the two-diameter cantilevered bar shown in sketch *t*. Neglect transverse shear.

5.28 The right-angle-cantilevered bracket shown in sketch *u* is loaded with force P in the z direction. Derive an expression for the deflection of the free end in the z direction by using Castigliano's theorem. Neglect transverse shear effects.

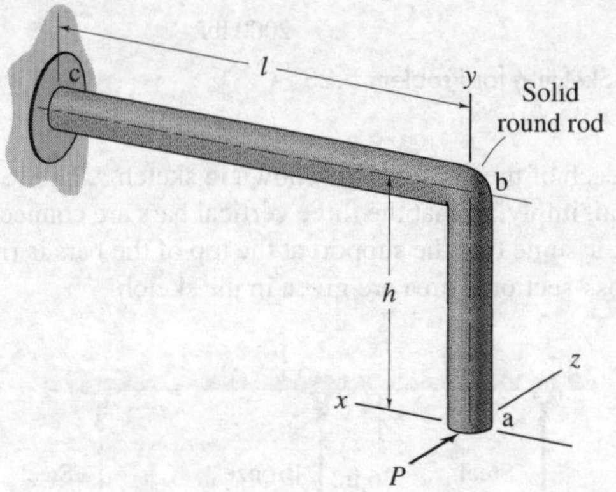

Sketch *u*, for Problem 5.28

5.29 A triangular cantilevered plate is shown in sketch *v*. Use Castigliano's theorem to derive an expression for the deflection at the free end, assuming that transverse shear is neglected.

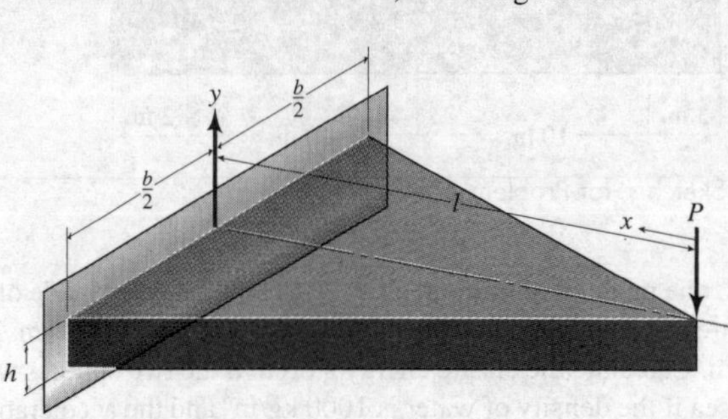

Sketch *v*, for Problem 5.29

5.30 A right-angle-cantilevered bracket with concentrated load and torsional loading at the free end is shown in sketch *w*. Using Castigliano's theorem, find the deflection at the free end in the *z* direction. Neglect transverse shear effects.

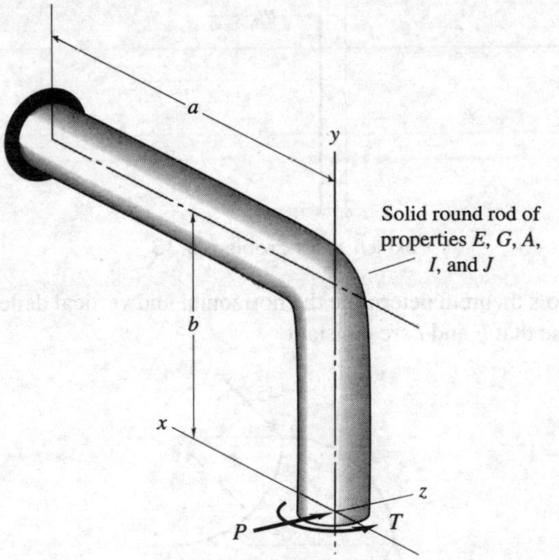

Sketch *w*, for Problem 5.30

5.31 A cantilevered I-beam has a concentrated load applied to the free end as shown in sketch *x*. What upward force at point S is needed to reduce the deflection at S to zero? Use Castigliano's theorem. Transverse shear can be neglected.

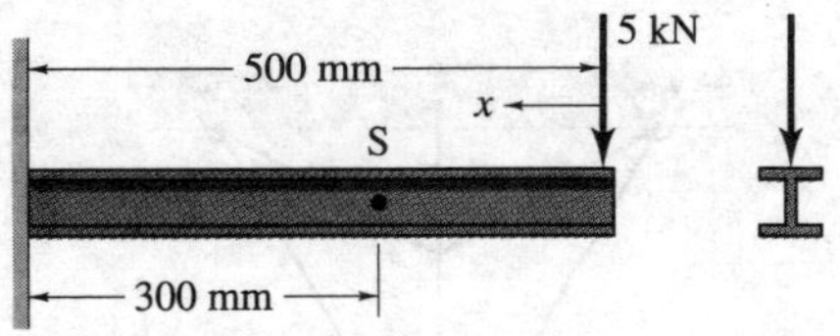

Sketch *x*, for Problem 5.31

5.32 Using Castigliano's theorem calculate the horizontal and vertical deflections at point A shown in sketch *y*. Assume that *E* and *A* are constant.

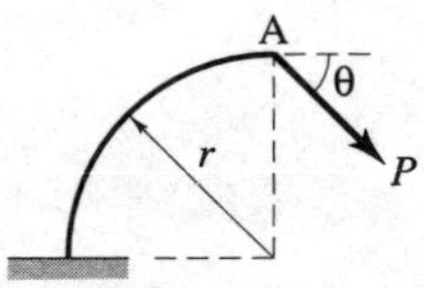

Sketch *y*, for Problem 5.32

5.33 Calculate the deflection at the point of load application and in the load direction for a load applied as shown in sketch *z*. Assume that *E* and *I* are constant.

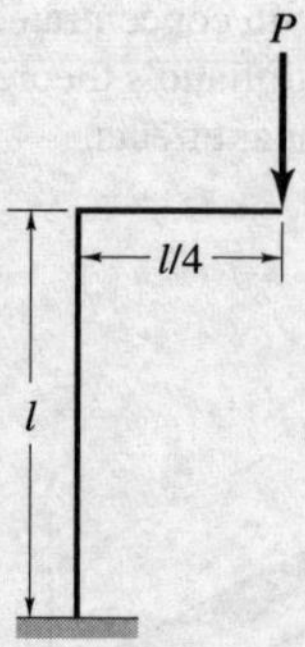

Sketch *z*, for Problem 5.33

5.34 Using Castigliano's theorem determine the horizontal and vertical deflections at point A of sketch *aa.* Assume that *E* and *I* are constant.

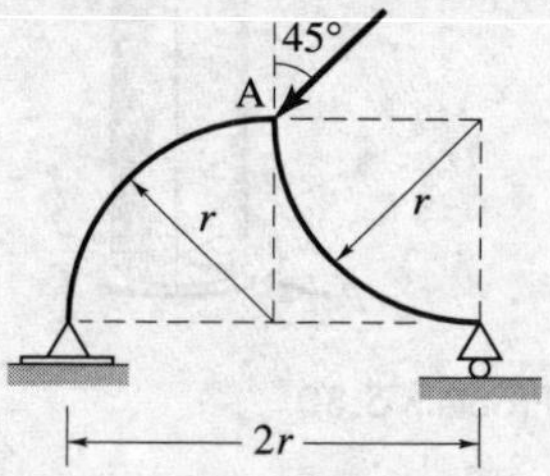

Sketch *aa*, for Problem 5.34

5.35 For the structure shown in sketch *bb* find the force in each member and determine the deflection at point A. Assume that *E* and *A* are the same in each member.

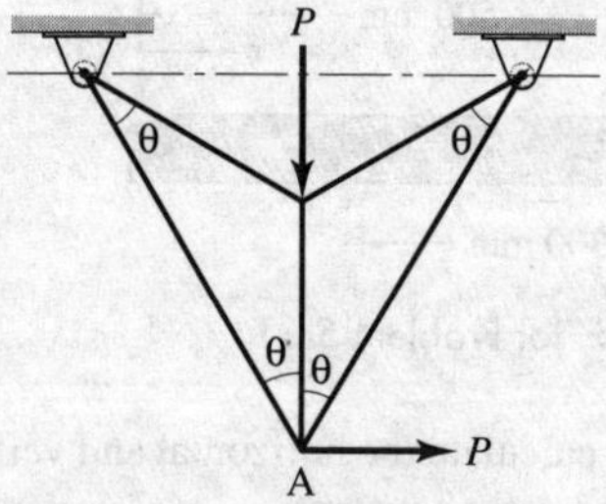

Sketch *bb*, for Problem 5.35

CHAPTER

6

FAILURE PREDICTION FOR STATIC LOADING

The liberty bell, a classic case of brittle fracture. (AP/Wide World Photos)

The concept of failure is central to the design process, and it is by thinking in terms of obviating failure that successful designs are achieved.

Henry Petroski, *Design Paradigms*

SYMBOLS

A	area, m^2
a	half of crack length, m
b	plate width, m
c	distance from neutral axis to outer fiber, m
D	major diameter, m
d	diameter, m
H	major height, m
h	minor height, m
I	area moment of inertia, m^4
J	polar area moment of inertia, m^4
K_c	stress concentration factor
K_{ci}	fracture toughness, MPa $\sqrt{m}$
K_i	stress intensity factor, MPa $\sqrt{m}$
l	length, m
M	bending moment, N-m
n_s	safety factor
P	force, N
q	volume flow, m^3/s
r	radius, m
S_{uc}	ultimate stress in compression, Pa
S_{ut}	ultimate stress in tension, Pa
S_y	yield stress, Pa
S_{yt}	yield stress in tension, Pa
T	torque, N-m
V	shear force, N
u	velocity, m/s
Y	dimensionless correction factor that accounts for geometry of part containing a crack
τ	shear stress, Pa
σ	normal stress, Pa
σ_e	von Mises stress, Pa
σ_1, σ_2, σ_3	principal normal stresses, Pa, where $\sigma_1 \geq \sigma_2 \geq \sigma_3$

Subscripts

all	allowable
avg	average
d	design
max	maximum
nom	nominal
oct	octahedral
x, *y*, *z*	Cartesian coordinates
1,2,3	principal axes

6.1 INTRODUCTION

A machine element may fail at sites of local stress concentration caused by geometrical or microstructural discontinuities. Stress concentration, stress raisers, and stress concentration factors are investigated in this chapter. The presence of cracks within a microstructure is also an important feature in understanding the failure of machine elements. Fracture mechanics is a technique of fracture analysis used to determine the stress level at which preexisting cracks of known size will propagate, leading to fracture. Materials, stress levels, crack-producing flaws, and crack propagation mechanisms are considered while studying fracture toughness and critical crack length. The chapter ends with failure prediction theories for both uniaxial and multiaxial stress states. Various theories are presented for which there are experimental data. Each theory has its own strengths and shortcomings and is best suited for a particular class of material. The loading throughout the chapter is assumed to be static, thus implying that the load is gradually applied and equilibrium is reached in a relatively short time. Thus, load is not a function of time.

6.2 STRESS CONCENTRATION

Stresses at or near a discontinuity, such as a hole in a plate, are higher than if the discontinuity did not exist. Figure 6.1 shows a rectangular bar with a hole under an axial load. The stress is largest near the hole; therefore, failure will first occur at the hole. The same can be deduced for any other discontinuity, such as a fillet (a narrowing in the width of a plate), a notch (a sharp groove or cut especially intended to initiate failure), an inclusion (such as a discontinuous fiber in a polymer matrix), or an area of load application.

A **stress raiser** is any discontinuity in a part that alters the stress distribution near the discontinuity so that the elementary stress equation no longer describes the state of stress in the part. **Stress concentration** is the region in which stress raisers are present. **Stress concentration factor** K_c is the factor used to relate the actual maximum stress at the discontinuity to the average stress without the discontinuity:

$$K_c = \frac{\text{Actual maximum stress}}{\text{Average stress}} \tag{6.1}$$

The stress concentration factor assumes that the stress distribution shown in Fig. 6.1 can be represented by an average stress and that the change to the stress-strain equation can be obtained by using the stress concentration factor. Static loading conditions are assumed. The maximum stress occurs at the smallest cross-sectional area. The value of K_c is difficult to calculate and is usually determined by some experimental technique, such as photoelastic analysis of a plastic model of a part or by numerical simulation of the stress field.

6.2.1 CHARTS

The stress concentration factor is a function of the type of discontinuity (hole, fillet, or groove), the geometry of the discontinuity, and the type of loading being experienced.

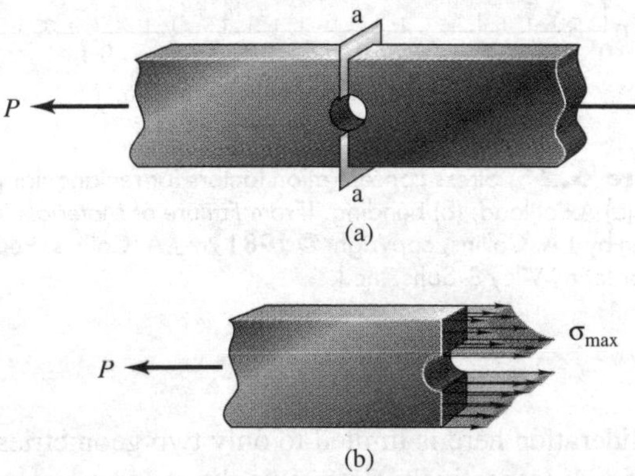

Figure 6.1 Rectangular plate with hole subjected to axial load. (a) Plate with cross-sectional plane. (b) Half of plate with stress distribution.

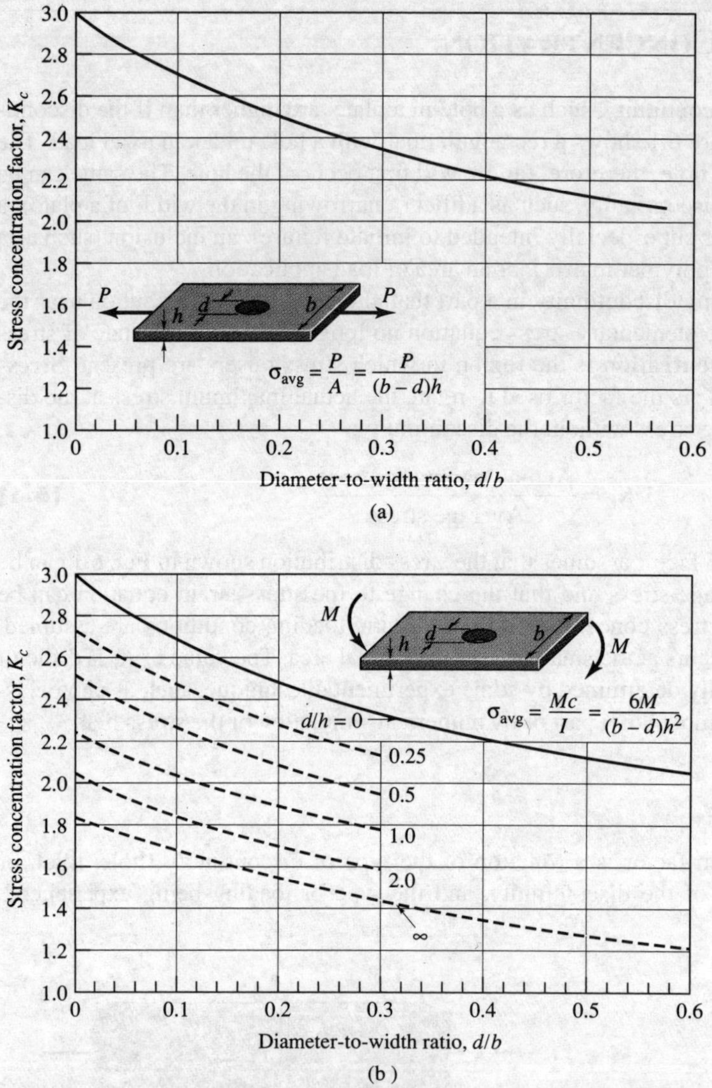

Figure 6.2 Stress concentration factors for rectangular plate with central hole. (a) Axial load; (b) bending. [From *Failure of Materials in Mechanical Design* by J.A. Collins, copyright © 1981 by J.A. Collins. Reprinted by permission of John Wiley & Sons, Inc.]

Consideration here is limited to only two geometries, a flat plate and a round bar. Figures 6.2 to 6.4, respectively, determine the stress concentration factor due to bending for a flat plate with a hole, a fillet, or a groove, each axially loaded. Figures 6.5 and 6.6, respectively, determine the stress concentration factor due to bending and due to torsion for a round bar with a fillet and a groove axially loaded. These are by no means all the possible geo-

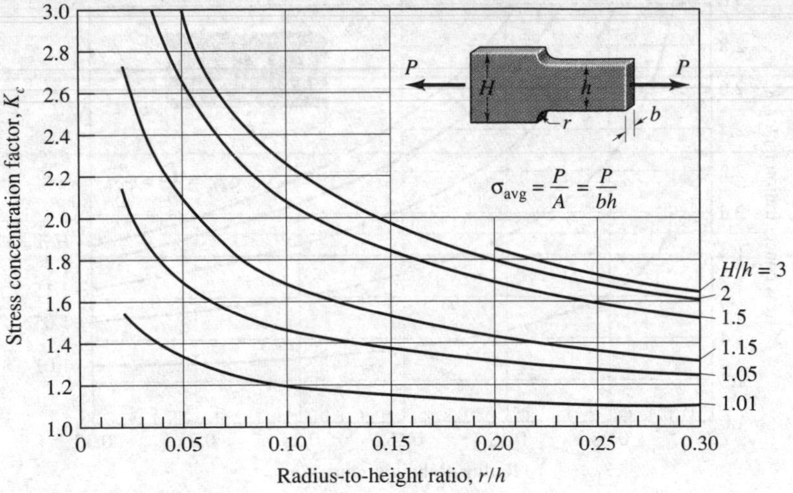

(a)

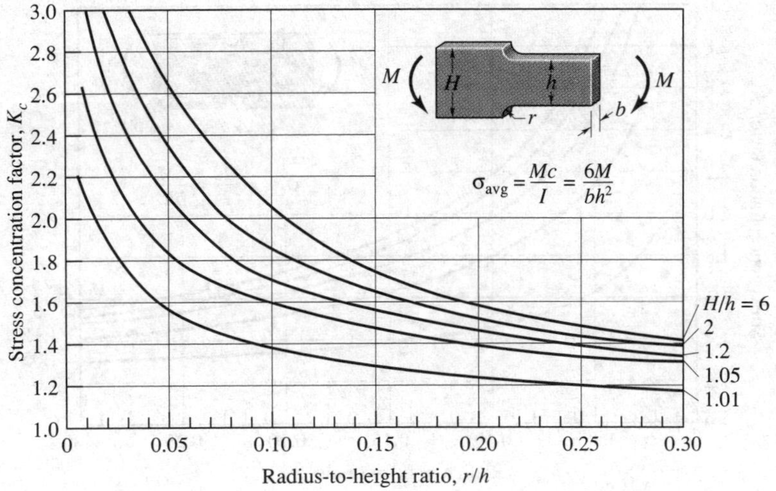

(b)

Figure 6.3 Stress concentration factors for rectangular plate with fillet. (a) Axial load; (b) bending. [From *Failure of Materials in Mechanical Design* by J.A. Collins, copyright © 1981 by J.A. Collins. Reprinted by permission of John Wiley & Sons, Inc.]

metries but are those most often used in practice. For other geometries refer to Peterson (1974) or Young (1989).

From these figures a number of observations can be made about the stress concentration factor:

1. The stress concentration factor K_c is *independent* of the part's material properties.
2. It is significantly affected by the geometry. Note from Fig. 6.5 that as the radius r of the discontinuity is decreased, the stress concentration is increased.

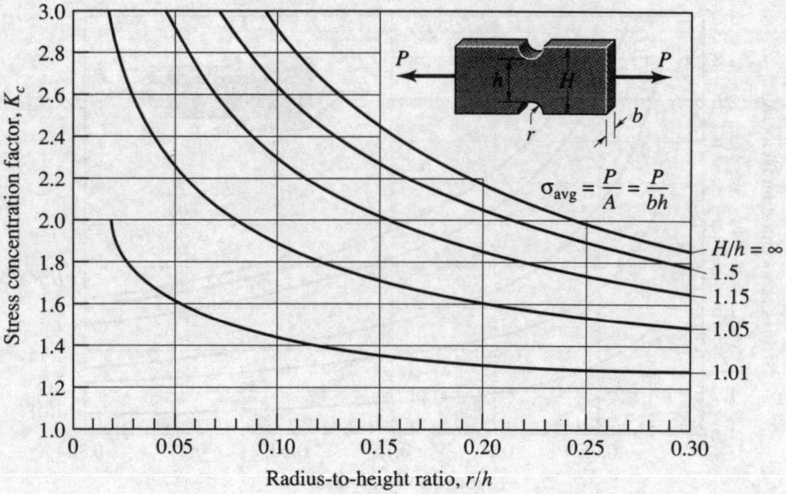

(a)

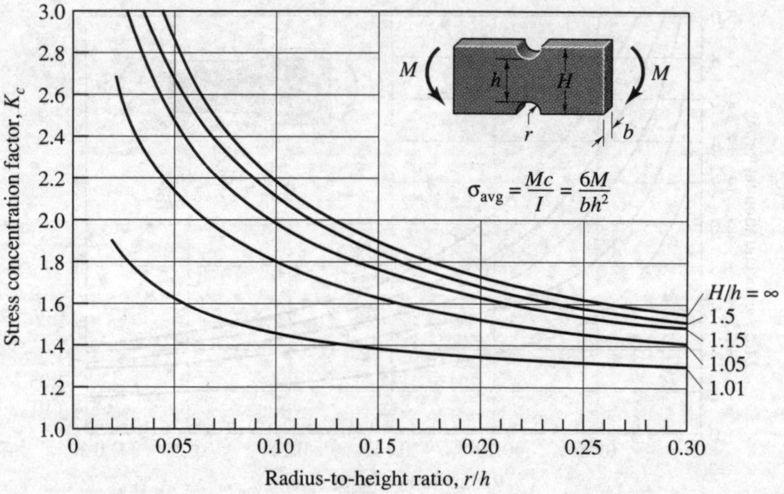

(b)

Figure 6.4 Stress concentration factors for rectangular plate with groove. (a) Axial load; (b) bending. [From *Failure of Materials in Mechanical Design* by J.A. Collins, copyright © 1981 by J.A. Collins. Reprinted by permission of John Wiley & Sons, Inc.]

3. It is also affected by the type of discontinuity; the stress concentration factor is considerably less for a fillet (Fig. 6.3) than for a hole (Fig. 6.2).

These observations relate to reducing stresses in a part.

The stress concentration factors given in Figs. 6.2 to 6.6 were determined on the basis of static loading, with the additional assumption that the stress in the material does not exceed the proportional limit. If the material is brittle, the proportional limit is the rupture stress, so failure for this material will begin at the point of stress concentration when the

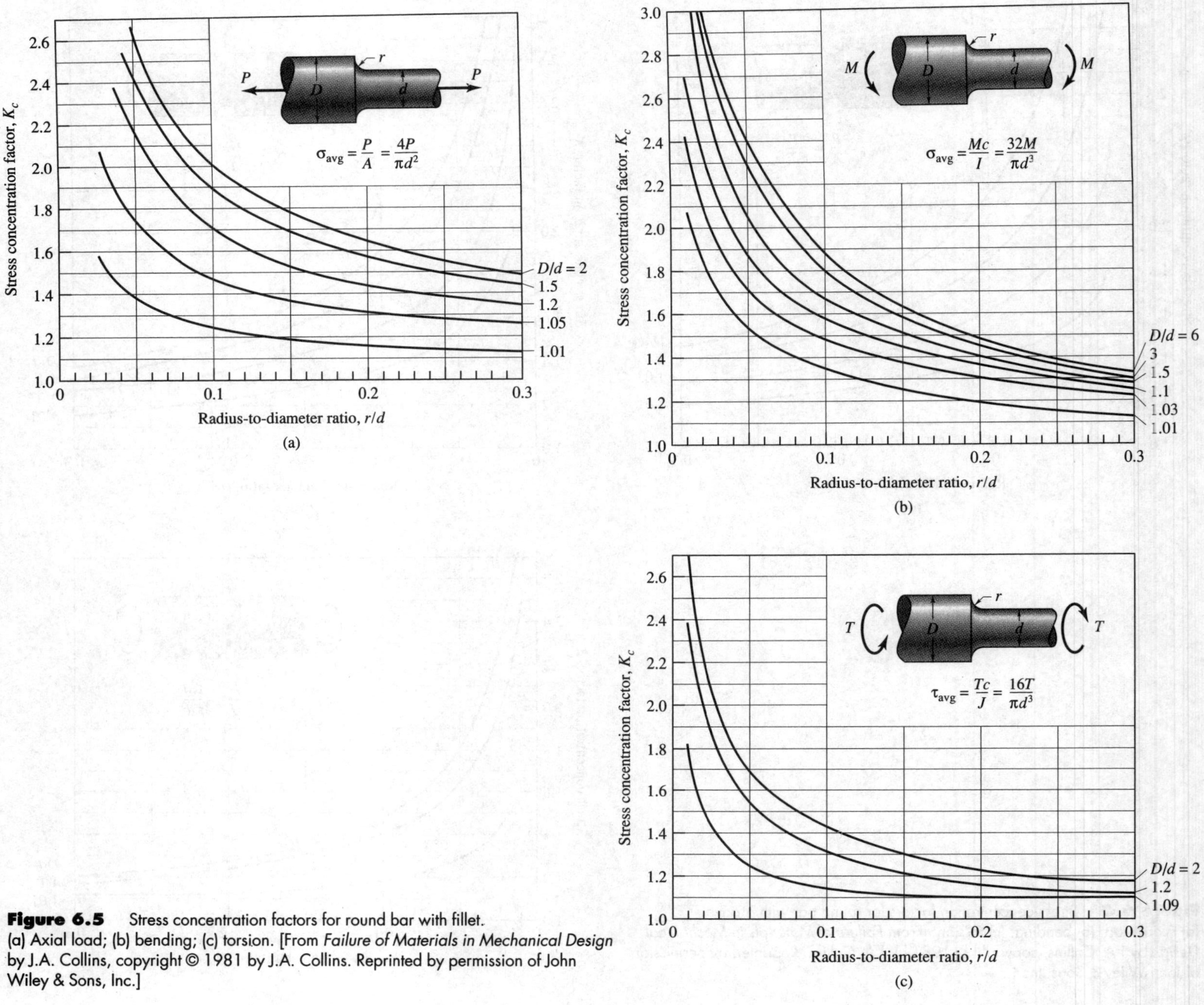

Figure 6.5 Stress concentration factors for round bar with fillet.
(a) Axial load; (b) bending; (c) torsion. [From *Failure of Materials in Mechanical Design* by J.A. Collins, copyright © 1981 by J.A. Collins. Reprinted by permission of John Wiley & Sons, Inc.]

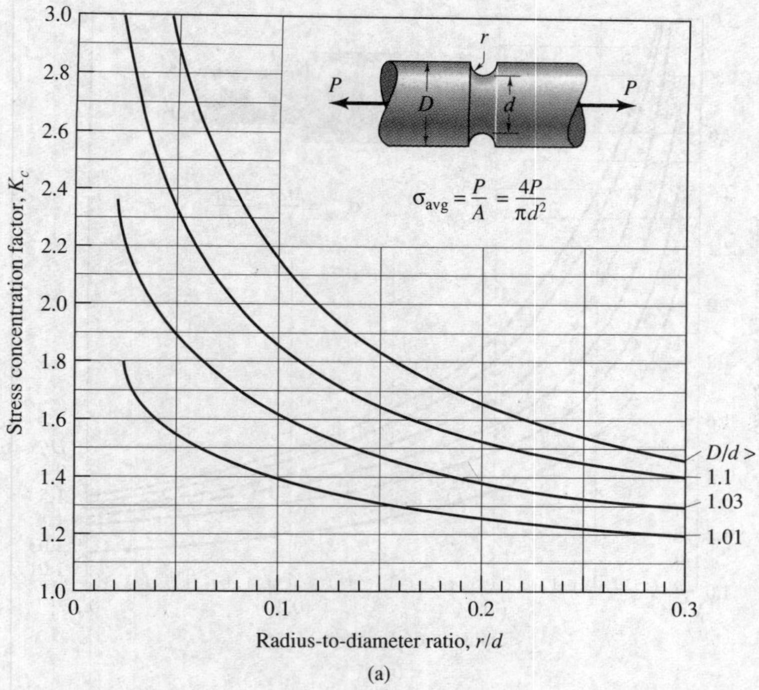

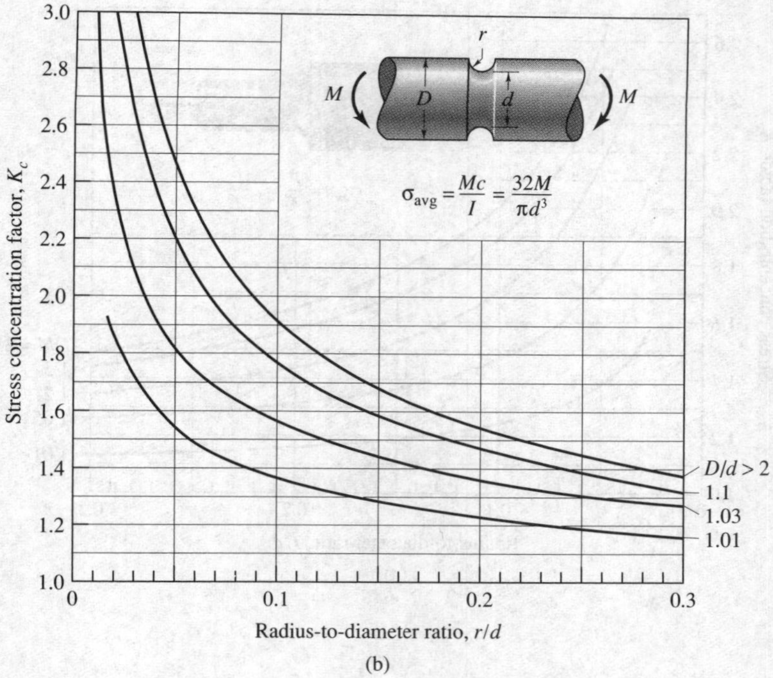

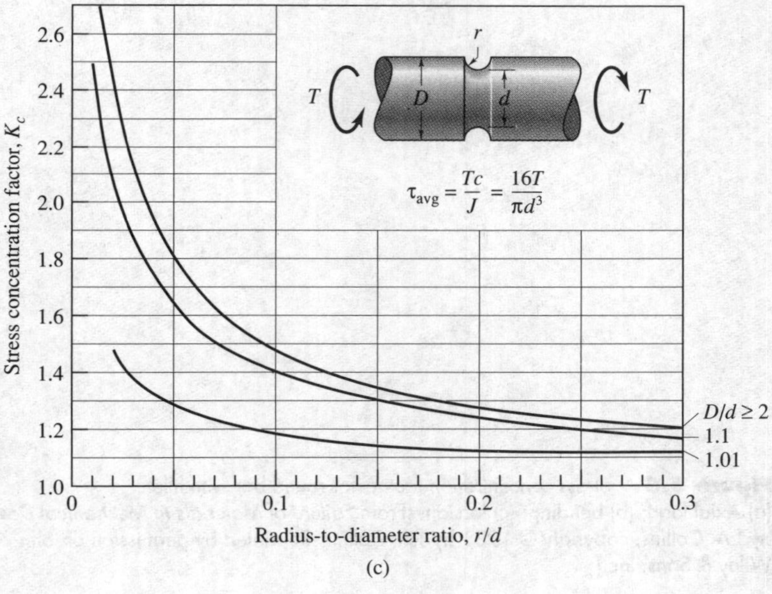

Figure 6.6 Stress concentration factors for round bar with groove. (a) Axial load; (b) bending; (c) torsion. [From *Failure of Materials in Mechanical Design* by J.A. Collins, copyright © 1981 by J.A. Collins. Reprinted by permission of John Wiley & Sons, Inc.]

proportional limit is reached. It is thus important to apply stress concentration factors when using brittle materials. On the other hand, if the material is ductile and subjected to a static load, designers often neglect stress concentration factors, since a stress that exceeds the proportional limit will not result in a crack. Instead, the ductile material will have reserve strength due to yielding and strain hardening. Further, as a material yields near a stress concentration, deformation results in blunting of notches, so that the stress concentration is relieved. In applications where stiff designs and tight tolerances are essential, stress concentration will be considered regardless of material ductility.

Example 6.1

GIVEN A flat plate made of a brittle material and with a major height H of 4.5 in., a minor height h of 2.5 in., and a fillet radius r of 0.5 in.

FIND The stress concentration factor and the maximum stress for the following conditions:

a) Axial loading
b) Pure bending
c) Axial loading but with fillet radius changed to 0.25 in.

Solution

a) Axial loading

$$\frac{H}{h} = \frac{4.5}{2.5} = 1.80 \qquad \frac{r}{h} = \frac{0.5}{2.5} = 0.2$$

From Fig. 6.3(a), $K_c = 1.8$. From Eq. (6.1) the maximum stress is

$$\sigma_{\text{max}} = 1.8\left(\frac{P}{A}\right) = \frac{1.8P}{bh}$$

where

b = width of flat plate

b) Pure bending. From Fig. 6.3(b), $K_c = 1.5$. The maximum stress is

$$\sigma_{\text{max}} = 1.5\frac{6M}{bh^2} = \frac{9M}{bh^2}$$

c) Axial loading but with fillet radius changed to 0.25 in.

$$\frac{r}{h} = \frac{0.25}{2.5} = 0.1$$

From Fig. 6.3(a), $K_c = 2.2$. The maximum stress is

$$\sigma_{\text{max}} = \frac{2.2P}{bh}$$

Thus, reducing the fillet radius by one-half increased the maximum stress by 22%.

Example 6.2

GIVEN A 50-mm-wide, 5-mm-high rectangular plate has a 5-mm-diameter central hole. The allowable stress due to applying a tensile force is 700 MPa.

FIND

a) The maximum tensile force that can be applied
b) The maximum bending moment that can be applied to reach the maximum stress
c) The maximum tensile force and the maximum bending moment if the hole is not put in the plate. Express the results as a ratio when compared to parts a and b.

Solution

a) The diameter-to-width ratio is $d/b = 5/50 = 0.1$. The cross-sectional area with the hole is

$$A = (b-d)h = (50-5)5 = 225\ \text{mm}^2 = 0.225\left(10^{-3}\right)\ \text{m}^2$$

From Fig. 6.2(a) for $d/b = 0.1$ the stress concentration factor $K_c = 2.70$ for axial loading. The maximum tensile force is

$$P_{\max} = \frac{\sigma_{\text{all}} A}{K_c} = \frac{700\left(10^6\right)(0.225)\left(10^{-3}\right)}{2.70} = 58\ 330\ \text{N} = 58.33\ \text{kN}$$

b) From Fig. 6.2(b) for bending when $d/b = 0.1$ and $d/h = 5/5 = 1$ the stress concentration factor $K_c = 2.04$. The maximum bending moment is

$$M_{\max} = \frac{Ah\sigma_{\text{all}}}{6K_c} = \frac{0.225\left(10^{-3}\right)5\left(10^{-3}\right)700\left(10^6\right)}{6(2.04)} = 64.34\ \text{N-m}$$

c) The cross-sectional area without the hole is

$$A = bh = (50)5 = 250\ \text{mm}^2 = 0.250\left(10^{-3}\right)\ \text{m}^2$$

The force ratio is

$$\frac{P_{\max}}{\left(P_{\max}\right)_{\text{hole}}} = \frac{175}{58.33} = 3.00$$

For bending

$$M_{\max} = \frac{\sigma_{\text{all}} bh^2}{6} = \frac{\sigma_{\text{all}} Ah}{6} = \frac{700\left(10^6\right)0.25\left(10^{-3}\right)5\left(10^{-3}\right)}{6} = 145.8\ \text{N-m}$$

The bending moment ratio is

$$\frac{M_{\max}}{\left(M_{\max}\right)_{\text{hole}}} = \frac{145.8}{64.34} = 2.266$$

6.2.2 FLOW ANALOGY

Good design practice drives the mechanical engineer to reduce the stress concentration as much as possible. Recommending ways of reducing the stress concentration requires better understanding of what occurs at the discontinuity to increase the stress. One way of achieving this understanding is to observe similarity between the flow velocity of fluid in a channel and the stress distribution of an axially loaded plate when the channel dimensions are comparable to the size of the plate. The analogy is accurate, since the equations of flow potential in fluid mechanics and stress potential in solid mechanics are of the same form.

If the channel has constant dimensions throughout, the velocities are uniform and the streamlines are equally spaced. For a bar of constant dimensions under axial load the stresses are uniform and equally spaced. At any point within the channel the flow must be constant, where the volume flow is

$$q = \int u dA \tag{6.2}$$

From solid mechanics the force must be constant at any location in the plate,

$$P = \int \sigma dA \tag{6.3}$$

If the channel section changes sharply, the flow velocity increases near the shape change and, to maintain equal flow, the streamlines must narrow and crowd together. In a stressed member of the same cross section the increase in stress is analogous to the increase in velocity, or inversely to the change in space between the streamlines. Figure 6.7(a) shows the stress distribution around the sharp corners of an axially loaded flat plate. The situation in Fig. 6.7(a) produces a stress concentration factor greater than 3. Recall from Eq. (6.1) that this implies that the maximum stress is more than three times greater than the average stress. However, this stress concentration factor can be reduced, typically from 3 to 1.5, by rounding the corners as shown in Fig. 6.7(b). A still further reduction in stress concentration

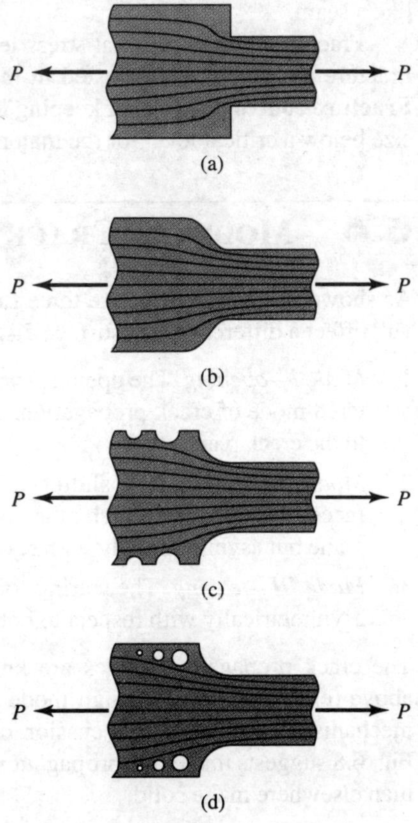

Figure 6.7 Bar with fillet axially loaded showing stress contours through a flat plate for (a) square corners, (b) rounded corners, (c) small grooves, and (d) small holes.

factor can be achieved by introducing small grooves or holes, as shown in Fig. 6.7(c) and (d), respectively. In Fig. 6.7(b) to (d) the design helps to reduce the rigidity of the material at the corners, so that the stress and strain are more evenly spread throughout the flat plate. The improvements in these designs can be gleaned from the flow analogy.

6.3 Fracture Mechanics

Structural studies that consider crack extension as a function of applied load are performed in fracture mechanics. A **crack** is a microscopic flaw that always exists under normal conditions on the surface and within the body of the material. These cracks (or dislocations) on and within the surface are like a dropped stitch in knitting. Under applied stress the crack moves easily through the material, causing a small slip in the plane in which it moves. Materials come undone more easily at these locations. No materials or manufacturing processes yield defect-free crystal structures, so such microscopic imperfections are always present.

Less stress is required to propagate a crack than to initiate it. Propagating a crack is like tearing a cloth. Once you start a little tear, it will propagate rather easily right across the cloth. However, the tear stops at a seam or other interruption of the regular weave of the fabric. So too, crack propagation can be prevented by introducing discontinuities to act like a seam.

Fracture failures occur at stress levels well below the yield stress of a solid material. Fracture mechanics is interested in the critical crack length that will make the part fail. **Fracture control** consists of keeping the combination of nominal stress and existing crack size below a critical level for the material being used in a given machine element.

6.4 Modes of Crack Displacement

As shown in Fig. 6.8 there are three fundamental **modes of crack propagation**, and each will effect a different crack surface displacement:

1. *Mode I—opening*. The opening (or tensile) mode [Fig. 6.8(a)] is the most often encountered mode of crack propagation. The crack faces separate symmetrically with respect to the crack plane.
2. *Mode II—sliding*. The sliding (or in-plane shearing) mode occurs when the crack faces slide relative to each other symmetrically with respect to the normal to the crack plane but asymmetrically with respect to the crack plane [Fig. 6.8(b)].
3. *Mode III—tearing*. The tearing (or antiplane) mode occurs when the crack faces slide asymmetrically with respect to both the crack plane and its normal [Fig. 6.8(c)].

The crack propagation modes are known by their Roman numeral designations given above (e.g., mode I). Although mode I is the easiest to visualize as a crack-propagating mechanism, applying the discussion of stress raisers in Chapter 5 to geometries such as Fig. 6.8 suggests that crack propagation will occur when stresses are higher at the crack tip than elsewhere in the solid.

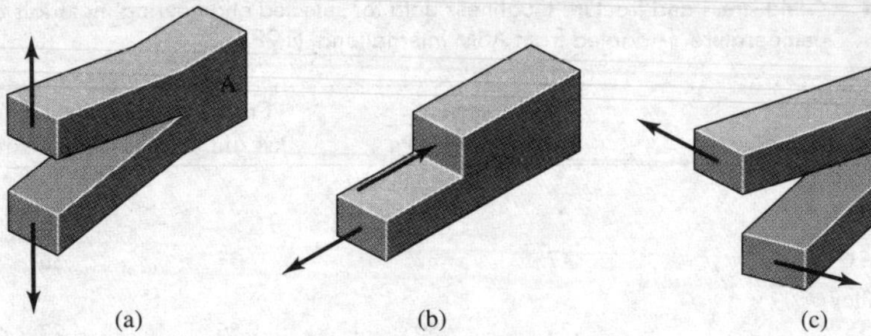

Figure 6.8 Three modes of crack displacement.
(a) Mode I, opening; (b) mode II, sliding; (c) mode III, tearing.

6.5 FRACTURE TOUGHNESS

Here, considerations of fracture toughness are restricted to mode I crack displacement. But first we need to determine what is meant by **stress intensity factor**. The stress intensity factor K_i specifies the stress intensity at the crack tip [see point *A* in Fig. 6.8(a)]. The SI unit of K_i is megapascals times meters$^{1/2}$ (MPa $\sqrt{\text{m}}$). **Fracture toughness**, on the other hand, is the critical value of stress intensity at which crack extension occurs. Fracture toughness is used as a design criterion in fracture prevention for brittle materials, just as yield strength is used as a design criterion in yielding prevention for ductile materials under static loading. Because the stresses near a crack tip [point A in Fig. 6.8(a)] can be defined in terms of the stress intensity factor, a critical value of fracture toughness K_{ci} exists that can be used to determine the condition for brittle fracture. In general, the equation for fracture toughness is

$$K_{ci} = Y\sigma_{\text{nom}}\sqrt{\pi a} \tag{6.4}$$

where

Y = dimensionless correction factor that accounts for geometry of part containing crack
σ_{nom} = nominal stress at fracture, MPa
a = half of crack length, m

Some assumptions imposed in deriving Eq. (6.4) are that the load is applied far from the crack and that the crack length $2a$ is small relative to the plate width.

The differences between stress intensity factor K_i and fracture toughness K_{ci} need to be made clear. The stress intensity factor K_i represents the stress level at the crack tip [point A in Fig. 6.8(a)] in a part containing a crack. Fracture toughness K_{ci} is the highest stress intensity that the part can withstand without fracturing. Thus, stress intensity K_i has many values, whereas fracture toughness K_{ci} is a particular value. That the units of stress intensity factor and fracture toughness are the same may seem somewhat strange, but they should be viewed as a combination of units of stress and the square root of crack length.

Table 6.1 shows room-temperature yield stress and fracture toughness (mode I) data for selected engineering materials. Note that the fracture toughness K_{ci} depends on many

Table 6.1 Yield stress and fracture toughness data for selected engineering materials at room temperature. [Adapted from ASM International (1989).]

Material	Yield stress, S_y		Fracture toughness, K_{ci}	
	ksi	MPa	ksi $\sqrt{\text{in.}}$	MPa $\sqrt{\text{m}}$
Metals				
Aluminum alloy 2024–T351	47	325	33	36
Aluminum alloy 7075–T651	73	505	26	29
Alloy steel 4340 tempered at 260°C	238	1640	45.8	50.0
Alloy steel 4340 tempered at 425°C	206	1420	80.0	87.4
Titanium alloy Ti-6Al-4V	130	910	40–60	44–66
Ceramics				
Aluminum oxide	—	—	2.7–4.8	3.0–5.3
Soda-lime glass	—	—	0.64–0.73	0.7–0.8
Concrete	—	—	0.18–1.27	0.2–1.4
Polymers				
Polymethyl methacrylate	—	—	0.9	1.0
Polystyrene	—	—	0.73–1.0	0.8–1.1

factors, with the most important being temperature, strain rate, and microstructure. The magnitude of K_{ci} diminishes with increasing strain rate and decreasing temperature. Furthermore, enhancing yield strength by a material process, such as strain hardening, produces a corresponding decrease in K_{ci}. Recall that mode I, the opening mode, is assumed.

Example 6.3

GIVEN The following two materials:

a) AISI 4340 steel tempered at 260°C (500°F)

b) Aluminum alloy 7075–T651

Use the assumptions imposed in deriving Eq. (6.4). Also, assume that the fracture stress is 0.8 times the yield stress and that the dimensionless correction factor is unity.

FIND The critical crack length at room temperature.

Solution

a) From Table 6.1 for AISI 4340

$$S_y = 238 \text{ ksi} \quad \rightarrow \quad \sigma_{\text{nom}} = 0.8S_y = 190.4 \text{ ksi} \quad \text{and} \quad K_{ci} = 45.8 \text{ ksi} \sqrt{\text{in.}}$$

From Eq. (6.4)

$$a = \frac{1}{\pi}\left(\frac{K_{ci}}{Y\sigma_{\text{nom}}}\right)^2 = \frac{1}{\pi}\left[\frac{(45.8)(10^3)}{(1)(190.4)(10^3)}\right]^2 = 0.01842 \text{ in.}$$

b) From Table 6.1 for aluminum alloy 7075–T651.

$$S_y = 73 \text{ ksi} \rightarrow \sigma_{\text{nom}} = 0.8S_y = 58.4 \text{ ksi} \quad \text{and} \quad K_{ci} = 26 \text{ ksi}\sqrt{\text{in.}}$$

From Eq. (6.4)

$$a = \frac{1}{\pi}\left(\frac{K_{ci}}{Y\sigma_{\text{nom}}}\right)^2 = \frac{1}{\pi}\left[\frac{26(10^3)}{(1)(58.4)(10^3)}\right]^2 = 0.06309 \text{ in.}$$

The stronger material (the steel) will fail first, since it has a smaller critical crack length. Thus, the weaker material (the aluminum) is a tougher material as far as crack propagation is concerned.

Example 6.4

GIVEN A container used for compressed air is made of aluminum alloy 2024–T351. The required safety factor against yielding is 1.6, and the largest crack allowed through the thickness of the material is 6 mm. The form of the cracks gives the dimensionless correction factor $Y = 1$.

FIND

a) The stress intensity factor and the safety factor guarding against brittle fracture

b) Whether a higher safety factor will be achieved if the material is changed to a stronger aluminum alloy 7075–T651. Assume the same crack.

Solution

a) From Table 6.1 for aluminum alloy 2024–T351

$$S_y = 325 \text{ MPa} \quad \text{and} \quad K_{ci} = 36 \text{ MPa}\sqrt{\text{m}}$$

The nominal stress is

$$\sigma_{\text{nom}} = \frac{S_y}{n_s} = \frac{325}{1.6} = 203.1 \text{ MPa}$$

The crack half-length is 3 mm. The stress intensity factor from Eq. (6.4) is

$$K_i = Y\sigma_{\text{nom}}\sqrt{\pi a} = 1(203.1)(10^6)\sqrt{\pi 3(10^{-3})} = 19.72 \text{ MPa}\sqrt{\text{m}}$$

The safety factor for brittle fracture is

$$n_s = \frac{K_{ci}}{K_i} = \frac{36}{19.72} = 1.83$$

b) From Table 6.1 for stronger aluminum alloy 7075–T651

$$S_y = 505 \text{ MPa} \quad \text{and} \quad K_{ci} = 29 \text{ MPa}\sqrt{\text{m}}$$

The safety factor guarding against yielding is

$$1.6\left(\frac{505}{325}\right) = 2.49$$

Thus, the increased strength of 7075–T651 results in a higher safety factor guarding against yielding. The safety factor guarding against crack propagation is

$$\frac{K_{ci}}{K_i} = \frac{29}{19.72} = 1.47$$

Thus, the stronger material will fail more easily from crack propagation.

6.6 FAILURE PREDICTION FOR UNIAXIAL STRESS STATE

Normal experimental data exist for the axial loading in a uniaxial stress state. Failure is predicted if the design stress σ_d is greater than the allowable stress σ_{all}. Thus, $\sigma_d > \sigma_{\text{all}}$. Recall from Chapter 3 [Eqs. (3.13) to (3.16)] that the allowable stress depends on the type of load being imposed as well as on whether the material is ductile or brittle. Thus, when either the yield strength or the ultimate strength is known, the allowable stress can be determined for the uniaxial stress state.

The safety factor from Eq. (1.1) in a uniaxial stress state can be expressed as

$$n_s = \frac{\sigma_{\text{all}}}{\sigma_d} \tag{6.5}$$

The type of loading and the type of material need to be incorporated into Eq. (6.5). Therefore, for a uniaxial stress state the allowable stress is divided by the design stress; if it is greater than 1, the design is adequate. Of course, the larger n_s is, the safer is the design, and values of $n_s < 1$ mean that a redesign is necessary. For a redesign, depending on the machine element, it is important to recognize what would reduce the design stress.

Example 6.5

GIVEN The leaf springs of a truck's rear wheels are loaded in pure bending. The 8-ton axle load is taken up by the two springs, giving a bending moment of 9800 N-m in each spring at the load application point. The steel used for the spring is AISI 4340 tempered at 260°C. The dimensions of the leaf spring are such that the width is 10 times the thickness. Assume a safety factor of 5.

FIND The cross section of the leaf spring

Solution

Table 6.1 for AISI 4340 steel tempered at 260°C gives $S_y = 1640$ MPa. Using the lower limit of Eq. (3.15) (to be conservative) yields

$$\sigma_{\text{all}} = 0.6S_y = 0.6(1640) = 984 \text{ MPa}$$

The design stress from Eq. (1.1) is

$$\sigma_d = \frac{\sigma_{\text{all}}}{n_s} = \frac{984}{5} = 196.8 \text{ MPa} \tag{a}$$

From Table 4.1 for a rectangular section

$$I = \frac{bh^3}{12} \quad \text{and} \quad c = h/2$$

We are given that $b = 10h$. Substituting the above into Eq. (4.48) gives the bending design stress as

$$\sigma_d = \frac{Mc}{I}$$

Making use of Eq. (a) gives

$$196.8\left(10^6\right) = \frac{5880}{h^3}$$

or

$$h^3 = \frac{5880}{196.8\left(10^6\right)} = 29.88\left(10^{-6}\right)$$

$$h = 0.031 \text{ m} = 31.0 \text{ mm}$$

The cross section of the leaf spring is 31 by 310 mm.

6.7 FAILURE PREDICTION FOR MULTIAXIAL STRESS STATE

A multiaxial stress state can be either a biaxial or triaxial stress state. In practice, it is difficult to devise experiments to cover every possible combination of critical stresses because each test is expensive and a large number are required. Therefore, a theory is needed that compares the normal and shear stresses σ_x, σ_y, σ_z, τ_{xy}, τ_{yz}, and τ_{xz} with the uniaxial stress, for which experimental data are relatively easy to obtain. Several failure prediction theories are presented for a multiaxial stress state while assuming static loading.

6.7.1 DUCTILE MATERIALS

Ductile materials include most metals and polymers. In general, metal castings are not as ductile as wrought or cold-worked metal parts and are treated in Sec. 6.7.2.2. Ductile materials typically have the same tensile strength as compressive strength and are not as susceptible to stress raisers as are brittle materials. For the purposes of this text a ductile material is considered to have failed when it yields. Although in some applications a certain amount of plastic deformation may be acceptable, this is rarely the case in machinery elements. Two popular theories of yield criteria are presented: the maximum-shear-stress theory and the distortion-energy theory.

6.7.1.1 MAXIMUM-SHEAR-STRESS THEORY The **maximum-shear-stress theory** (MSST) was first proposed by Coulomb (1773) but was independently discovered by Tresca (1868) and is therefore often called the **Tresca yield criterion.** Tresca noted that titanium exhibited bright shear bands under small strains, indicating that metals deformed under shear in all circumstances and that the shear was localized on well-defined planes. His observations led to the MSST, which states that a part subjected to any combination of loads will fail (by yielding or fracturing) whenever the maximum shear stress exceeds a critical value. The critical value can be determined from standard uniaxial tension tests.

Experimental evidence verifies that the MSST is a good theory for predicting the yielding of ductile materials, and it is an extremely popular approach in design. If the nomenclature $\sigma_1 \geq \sigma_2 \geq \sigma_3$ is used for the principal stresses, the maximum shear stress says that yielding will occur when

$$\sigma_1 - \sigma_3 = \frac{S_y}{n_s} \tag{6.6}$$

where

S_y = yield stress of material
n_s = safety factor

For a three-dimensional stress state the maximum-shear-stress theory provides an envelope describing the stress combinations that cause yielding, as illustrated in Fig. 6.9. The curve defined by the yield criterion is known as the yield locus. Any stress state in the interior of the yield locus results in the material being deformed elastically. Points outside the yield locus are not possible because any such stress state would cause yielding in the solid before these stresses could be attained. Therefore, any calculations that predict a stress state outside the yield locus predict failure. If a situation arises that would increase the material strength (such as strain rate effects or work hardening), the yield locus expands, so that fracture may not necessarily be the result.

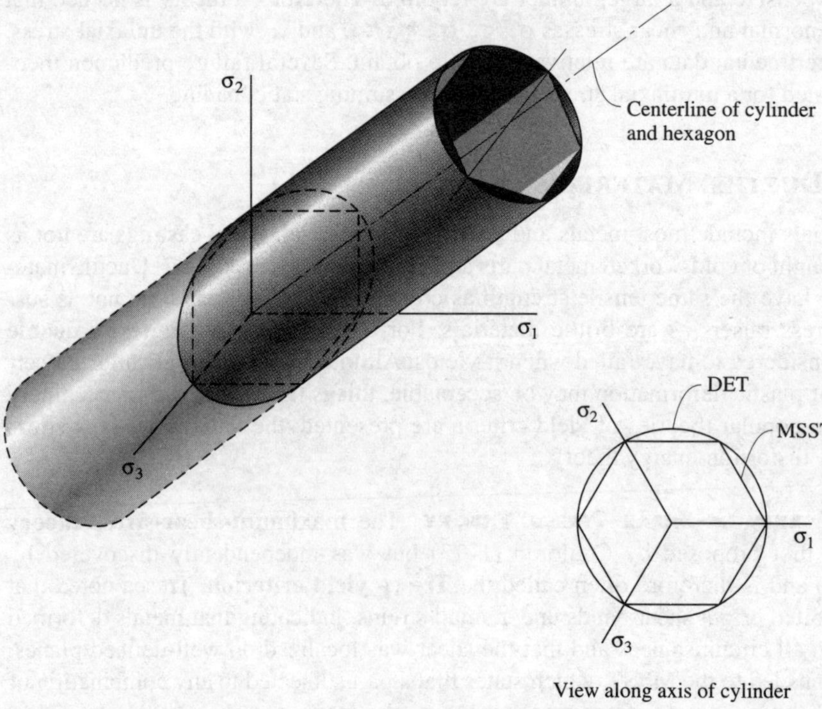

Figure 6.9 Three-dimensional yield locus for MSST and DET. [Adapted from Popov (1968).]

It is helpful to present the yield criterion in a plane stress circumstance, for which there will be two principal stresses in the plane as well as a principal stress equal to zero perpendicular to the plane. Figure 6.10 graphically depicts failure prediction in the plane stress state by the maximum-shear-stress theory. The principal stresses used in the figure are labeled as σ_1 and σ_2, indicating that the normal ordering of stresses ($\sigma_1 \geq \sigma_2 \geq \sigma_3$) is not being enforced. Remember that the principal stress outside the page is zero. In the first quadrant, where σ_1 and σ_2 are guaranteed to be positive, this means that the value of σ_3 in Eq. (6.6) would be zero and that yielding would occur whenever σ_1 or σ_2 reached the uniaxial yield strength S_y. In the second quadrant, where σ_1 is guaranteed to be negative and σ_2 is positive, Eq. (6.6) would result in a line as shown in Fig. 6.10. The third and fourth quadrants of the curve follow the same reasoning in their development.

The MSST is extremely popular because it is simple and, more importantly, because it works well.

6.7.1.2 Distortion-Energy Theory

The **distortion-energy theory** (DET), also known as the **von Mises criterion,** postulates that failure is caused by the elastic energy associated with shear deformation. This theory is valid for ductile materials and predicts yielding under combined loading with greater accuracy than any other recognized theory (although the differences between the DET and the MSST are small).

The DET can be derived mathematically in a number of ways, but one of the more straightforward is to use the concept of octahedral stresses from Chapter 2. From Eq. (2.28) the octahedral shear stress produced by uniaxial ($\sigma_2 = \sigma_3 = 0$) tension is

$$\tau_{oct} = \frac{\sqrt{2}}{3}\sigma_1 \qquad (6.7)$$

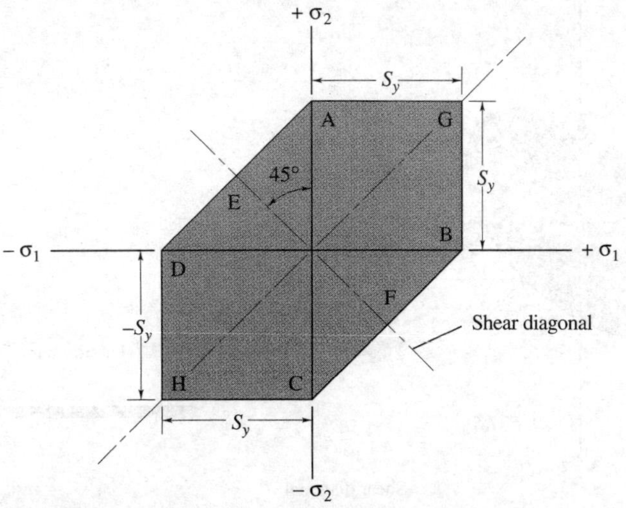

Figure 6.10 Graphical representation of maximum-shear-stress theory (MSST) for biaxial stress state ($\sigma_z = 0$).

The maximum octahedral stress occurs at

$$(\tau_{oct})_{limit} = \frac{\sqrt{2}}{3}\sigma_e \quad \textbf{(6.8)}$$

where

σ_e = **von Mises stress**, Pa

For a triaxial stress state

$$\sigma_e = \frac{1}{\sqrt{2}}\left[(\sigma_2-\sigma_1)^2+(\sigma_3-\sigma_1)^2+(\sigma_3-\sigma_2)^2\right]^{1/2} \quad \textbf{(6.9)}$$

For a biaxial stress state assuming $\sigma_3 = 0$

$$\sigma_e = \left(\sigma_1^2+\sigma_2^2-\sigma_1\sigma_2\right)^{1/2} \quad \textbf{(6.10)}$$

Thus, the DET predicts failure if

$$\sigma_e \geq \frac{S_y}{n_s} \quad \textbf{(6.11)}$$

The DET yield locus is shown in Fig. 6.9 for a three-dimensional stress state and in Fig. 6.11 for plane stress loading (biaxial stress state). Relative to the MSST the DET has the advantage that the yield criterion is continuous in its first derivative, an important consideration for application in plasticity.

Example 6.6

GIVEN In the rear wheel suspension of the Volkswagen "Beetle" the spring motion was provided by a torsion bar fastened to an arm on which the wheel was mounted. See Fig. 6.12 for more details. The torque in the torsion bar was created by the 2500-N force acting on the wheel from the ground

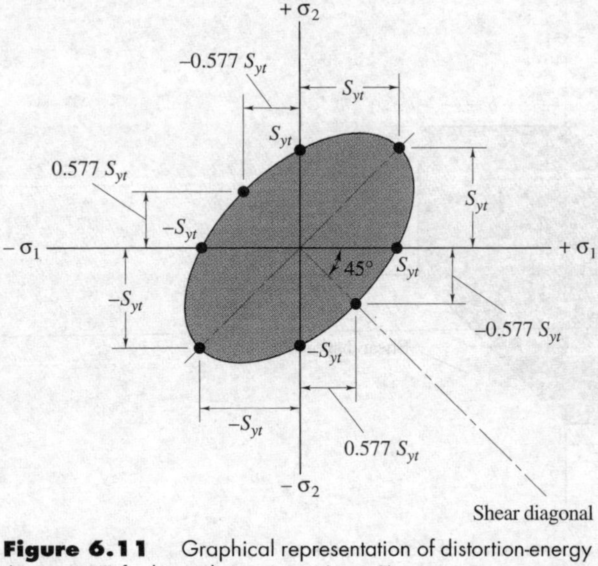

Figure 6.11 Graphical representation of distortion-energy theory (DET) for biaxial stress state (σ_z = 0).

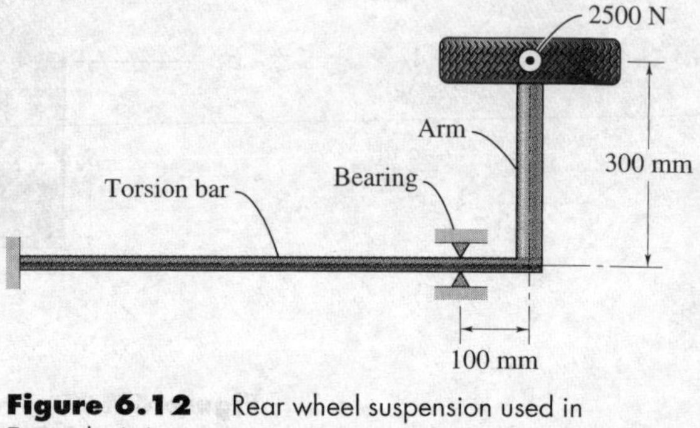

Figure 6.12 Rear wheel suspension used in Example 6.6.

through a 300-mm lever arm. Because of space limitations the bearing holding the torsion bar was situated 100 mm from the wheel shaft. The diameter of the torsion bar was 28 mm.

FIND The stresses in the torsion bar at the bearing by using the distortion-energy theory

Solution

The stresses acting on the torsion bar are a shear stress from the torsion and a perpendicular tensile/compressive stress from the bending. Using Eq. (4.34) gives the shear stress from torsion as

$$\tau = \frac{Tc}{J} = \frac{2500(0.3)(0.014)32}{\pi(0.028)^4} \text{ Pa} = 174.0 \text{ MPa}$$

Using Eq. (4.48) gives the tensile stress from bending as

$$\sigma = \frac{Mc}{I} = \frac{2500(0.1)(0.014)64}{\pi(0.028)^4} \text{ Pa} = 116.0 \text{ MPa}$$

From Eq. (2.16) the principal normal stresses are

$$\sigma_1, \sigma_2 = \frac{\sigma_x + \sigma_y}{2} \pm \sqrt{\tau_{xy}^2 + \left(\frac{\sigma_x - \sigma_y}{2}\right)^2} = \frac{116.0}{2} \pm \sqrt{(174.0)^2 + \left(\frac{116.0}{2}\right)^2}$$

$$\sigma_1 = 241.4 \text{ MPa} \quad \text{and} \quad \sigma_2 = -125.4 \text{ MPa}$$

From Eq. (6.10) the von Mises stress is

$$\sigma_e = \left(\sigma_1^2 + \sigma_2^2 - \sigma_1\sigma_2\right)^{0.5} = \left[(241.4)^2 + (-125.4)^2 - 241.4(-125.4)\right]^{0.5} = 322.9 \text{ MPa}$$

This is the stress produced as given by the distortion-energy theory.

Example 6.7

GIVEN A round, cantilevered bar made of a ductile material experiences torsion applied to the free end.

FIND Determine when yielding will occur by using a) the MSST and b) the DET.

Solution

Figure 6.13 shows the cantilevered bar, the stresses acting on an element, and a Mohr's circle representation of the stress state. The principal stresses are

$$\sigma_1 = \tau_1 \quad \text{and} \quad \sigma_3 = -\tau_1$$

a) Using Eq. (6.6) the MSST predicts failure if

$$|\sigma_1 - \sigma_3| = 2\tau_{max} \geq S_y / n_s$$

$$\therefore \tau_{max} \geq S_y / 2n_s \qquad \textbf{(a)}$$

b) Using Eq. (6.10) yields

$$\sigma_e = \left[\sigma_1^2 - \sigma_1\sigma_2 + \sigma_2^2\right]^{1/2} = \left[\tau_1^2 - \tau_1(-\tau_1) + (-\tau_1)^2\right]^{1/2} = \sqrt{3}\tau_1 = \sqrt{3}\tau_{max} \qquad \textbf{(b)}$$

Therefore, using Eq. (6.11) the DET predicts failure if

$$\sigma_e > S_y / n_s \qquad \textbf{(c)}$$

$$\therefore \tau_{max} \geq \frac{S_y}{\sqrt{3} n_s} = 0.5774 \frac{S_y}{n_s} \qquad \textbf{(d)}$$

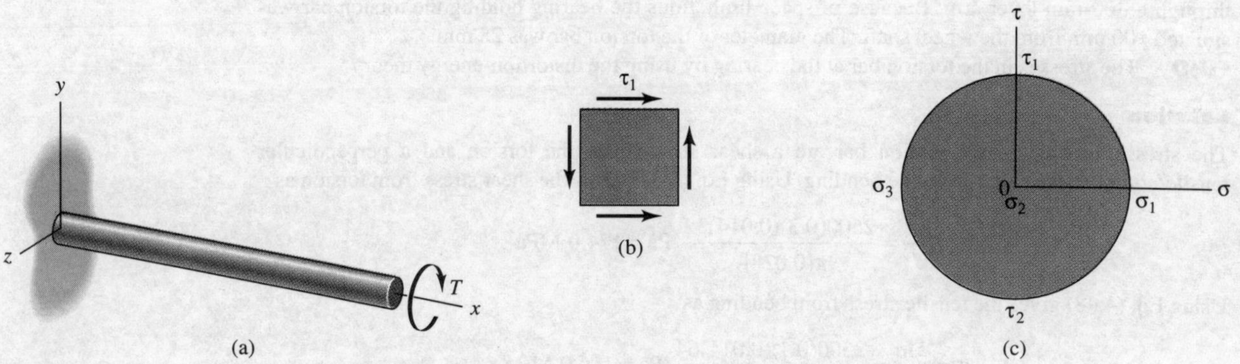

Figure 6.13 Cantilevered, round bar with torsion applied to free end (used in Example 6.7). (a) Bar with coordinates and load; (b) stresses acting on element; (c) Mohr's circle representation of stresses.

Equations (a) to (d) show that the MSST and the DET are in good agreement. This circumstance, a loading of pure shear, is actually the greatest difference between the MSST and the DET, suggesting that both theories will give close to the same results.

Example 6.8

GIVEN A round, cantilevered bar, similar to that considered in Example 6.7, is subjected not only to torsion but also to a transverse load at the free end, as shown in Fig. 6.14(a). The bar is made of a ductile material having a yield stress of 50 000 psi. The transverse force is 500 lbf and the torque is 1000 lbf-in. applied to the free end. The bar is 5 in. long and a safety factor of 2 is assumed. Transverse shear can be neglected.

FIND Determine what the minimum diameter should be to avoid yielding by using both a) the MSST and b) the DET.

Solution

Figure 6.14(b) shows the stresses acting on an element on the top of the bar at the wall. Note that in this example $\sigma_z = 0$. The critical section occurs at the wall. By using Eqs. (4.48) and (4.34) the normal and shear stresses can be written as

$$\sigma_x = \frac{Mc}{I} = \frac{Pl(d/2)}{\pi d^4/64} = \frac{32Pl}{\pi d^3}$$

$$\tau_{xz} = \frac{Tc}{J} = \frac{T(d/2)}{\pi d^4/32} = \frac{16T}{\pi d^3}$$

From Eq. (2.16) the principal normal stresses in a biaxial stress field can be written as

$$\sigma_1, \sigma_2 = \frac{\sigma_x}{2} \pm \sqrt{\left(\frac{\sigma_x}{2}\right)^2 + \tau_{xz}^2} = \frac{16Pl}{\pi d^3} \pm \sqrt{\left(\frac{16Pl}{\pi d^3}\right)^2 + \left(\frac{16T}{\pi d^3}\right)^2}$$

$$\therefore \sigma_1, \sigma_2 = \frac{16}{\pi d^3}\left(Pl \pm \sqrt{(Pl)^2 + T^2}\right)$$

Putting all the information given into the above equation yields

$$\sigma_1, \sigma_2 = \frac{16}{\pi d^3}\left[(500)(5) \pm \sqrt{(500)^2(5)^2 + (1000)^2}\right]$$

$$\therefore \sigma_1 = \frac{26\,450}{d^3} \quad \text{and} \quad \sigma_2 = -\frac{980.8}{d^3}$$

We immediately see that the stresses are in the wrong order; to ensure $\sigma_1 \geq \sigma_2 \geq \sigma_3$, we rearrange them so that $\sigma_1 = 26\,450/d^3$, $\sigma_2 = 0$, and $\sigma_3 = -980.8/d^3$.

From Eq. (2.18) the maximum and principal shear stresses can be written as

$$\tau_1, \tau_2 = \pm\sqrt{\tau_{xz}^2 + \frac{(\sigma_x - \sigma_z)^2}{4}}$$

$$\tau_{max} = \tau_1 = \frac{16}{\pi d^3}\sqrt{(Pl)^2 + T^2} = \frac{16}{\pi d^3}\sqrt{(500)^2(5)^2 + (1000)^2}$$

$$\tau_{max} = \tau_1 = \frac{13\,710}{d^3}$$

a) Using Eq. (6.6) the MSST predicts that failure will be avoided if

$$|\sigma_1 - \sigma_3| = 2\tau_1 = 2\tau_{max} < \frac{S_y}{n_s}$$

$$\therefore \left|\frac{26\,450}{d^3} + \frac{980.8}{d^3}\right| < \frac{50\,000}{2}$$

or

$$d^3 > \frac{27\,430}{25\,000} = 1.097 \text{ in.}^3$$

$$d > 1.031 \text{ in.}$$

b) Using Eq. (6.10) yields

$$\sigma_e = \left[\sigma_1^2 - \sigma_1\sigma_2 + \sigma_2^2\right]^{1/2} = \left[\left(\frac{26\,450}{d^3}\right)^2 + \left(\frac{26\,450}{d^3}\right)\left(\frac{980.8}{d^3}\right) + \left(\frac{980.8}{d^3}\right)^2\right]^{1/2} = \frac{26\,950}{d^3}$$

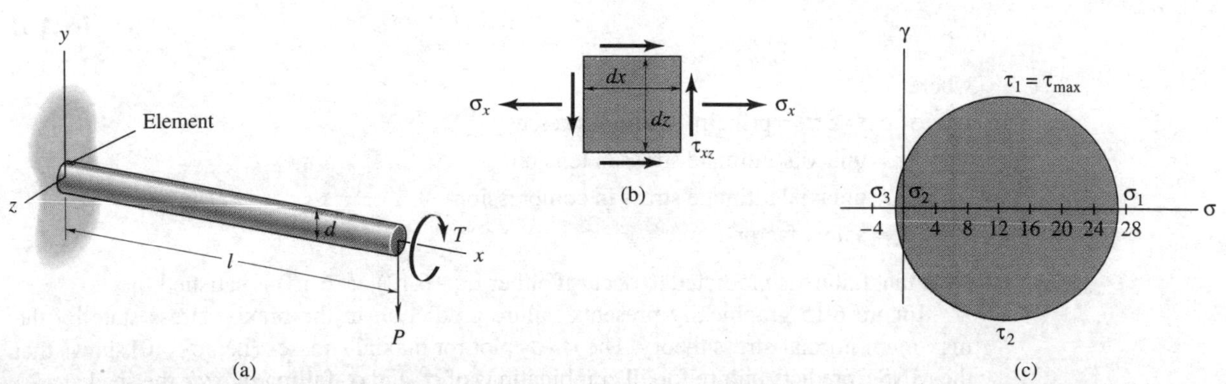

Figure 6.14 Cantilevered, round bar with torsion and transfer force applied to free end (used in Example 6.8). (a) Bar with coordinates and loads; (b) stresses acting on element at top of bar and at wall; (c) Mohr's circle representation of stresses.

Thus, using Eq. (6.11) the DET predicts that failure will be avoided if

$$\sigma_e < S_y/n_s$$

$$\frac{26\,950}{d^3} < \frac{50\,000}{2}$$

$$\therefore d^3 > \frac{26\,950}{25\,000} = 1.078 \text{ in.}^3$$

$$d > 1.025 \text{ in.}$$

Here, both theories give approximately the same solution.

6.7.2 Brittle Materials

As discussed in Chapter 3, brittle materials do not yield; they fracture. Thus, a failure criterion applied to brittle materials really addresses the circumstances under which the material will literally break. One important consideration with brittle materials is that their strengths in compression are usually much greater than their strengths in tension. Therefore, the failure criterion will show a difference in tensile and compressive behavior. Three failure criteria are presented: the maximum-normal-stress theory, the internal friction theory, and the modified Mohr theory.

6.7.2.1 Maximum-Normal-Stress Theory The **maximum-normal-stress theory** (MNST) states that a part subjected to any combination of loads will fail whenever the greatest positive principal stress exceeds the tensile yield strength or whenever the greatest negative principal stress exceeds the compressive yield strength. This theory works best for fibrous brittle materials and some glasses, but it works reasonably well for brittle materials in general and is therefore popular. Failure will occur using the MNST theory if

$$\sigma_1 \geq S_{ut}/n_s \tag{6.12}$$

$$\sigma_3 \leq S_{uc}/n_s \tag{6.13}$$

where

$\sigma_1 \geq \sigma_2 \geq \sigma_3$ = principal normal stresses
S_{ut} = uniaxial ultimate stress in tension
S_{uc} = uniaxial ultimate stress in compression
n_s = safety factor

Note that failure is predicted to occur if either Eq. (6.12) or (6.13) is satisfied.

Figure 6.15 graphically presents failure prediction in the biaxial stress state by the maximum-normal-stress theory. The σ_1-σ_2 plot for biaxial stresses (i.e., $\sigma_3 = 0$) shows that the MNST predicts failure for all combinations of σ_1 and σ_2 falling *outside* the shaded area of Fig. 6.15.

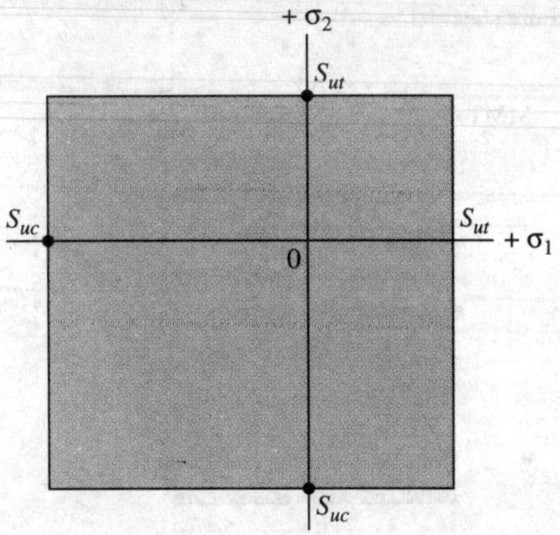

Figure 6.15 Graphical representation of maximum-normal-stress theory (MNST) for biaxial stress state ($\sigma_z = 0$).

6.7.2.2 Internal Friction Theory The MNST is most useful for fibrous brittle materials and glasses where the microstructure would orient in the direction of maximum normal stress before fracture could occur. Many brittle materials, such as ceramics and cast metals, do not have this capability, and a different fracture theory is needed. Further, the maximum-shear-stress-theory is difficult to apply to brittle materials, since the strength in compression is so much larger than the tensile strength of most brittle materials, such as ceramics and metal castings. A logical extension to the MSST is to separate compressive and tensile strengths, or in mathematical terms,

If $\sigma_1 > 0$ and $\sigma_3 < 0$,
$$\frac{\sigma_1}{S_{ut}} + \frac{\sigma_3}{S_{uc}} = \frac{1}{n_s} \tag{6.14}$$

If $\sigma_3 > 0$,
$$\sigma_1 = \frac{S_{ut}}{n_s} \tag{6.15}$$

If $\sigma_1 < 0$,
$$\sigma_3 = \frac{S_{uc}}{n_s} \tag{6.16}$$

where

$\sigma_1 \geq \sigma_2 \geq \sigma_3$ = are the ordered principal stresses
S_{ut} = fracture strength in tension, Pa
S_{uc} = fracture strength in compression, Pa
n_s = safety factor

Although this would appear to be an arbitrary extension of the MSST, Eqs. (6.14) to (6.16) can be derived analytically if internal friction is considered [Marin (1962)]. For this reason, this fracture criterion is known as the **internal friction theory (IFT)** and is also known as the **Coulomb-Mohr theory**. The IFT has the advantage of being more accurate for brittle materials that have a pronounced difference in tensile and compressive strengths

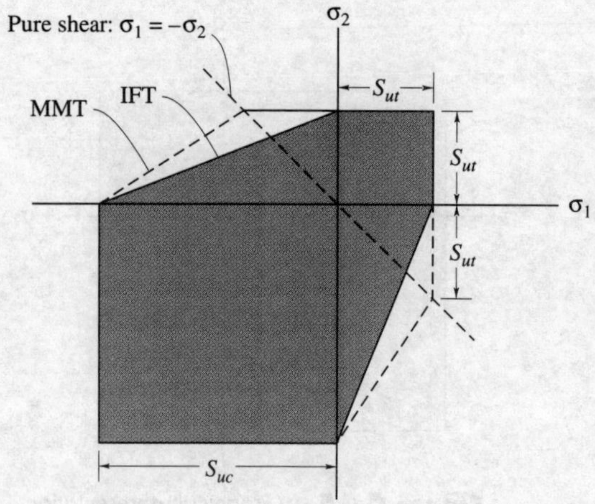

Figure 6.16 Internal friction theory and modified Mohr theory for failure prediction of brittle materials.

than either the MNST or MSST. Figure 6.16 depicts the IFT for a two-dimensional stress state ($\sigma_3 = 0$).

6.7.2.3 Modified Mohr Theory The IFT has an analytical basis, but the **modified Mohr theory** arose through efforts at fitting test data. The modified Mohr theory, or MMT, predicts brittle material behavior best, especially in the fourth quadrant. The MMT can be expressed by

If $\sigma_1 > 0$ and $\sigma_3 < -S_{ut}$,
$$\sigma_1 - \frac{S_{ut}\sigma_3}{S_{uc} - S_{ut}} = \frac{S_{uc}S_{ut}}{n_s S_{uc} - S_{ut}} \tag{6.17}$$

If $\sigma_3 > -S_{ut}$,
$$\sigma_1 = \frac{S_{ut}}{n_s} \tag{6.18}$$

If $\sigma_1 < 0$,
$$\sigma_3 = \frac{S_{uc}}{n_s} \tag{6.19}$$

Figure 6.16 depicts the MMT, along with the IFT, to demonstrate that there is only a slight difference between the two criteria.

Example 6.9

GIVEN Repeat Example 6.7 but with the cantilever constructed from a brittle material.

FIND Determine when fracture will occur by using a) the MNST, b) the IFT, and c) the MMT. Assume that the compressive strength is twice the tensile strength.

Solution

Just as in Example 6.7 the stress state is $\sigma_1 = \tau_1$, $\sigma_2 = 0$, and $\sigma_3 = -\tau_1$.

a) Using Eq. (6.12) the MNST predicts failure if

$$\sigma_1 > \frac{S_{ut}}{n_s} \quad \text{or} \quad \tau_1 = \tau_{max} \geq \frac{S_{ut}}{n_s} \qquad \textbf{(a)}$$

Tensile fracture stress was used in Eq. (a) because the material will fail in tension before it fails in compression.

b) When using the IFT, σ_1 is positive and σ_3 is negative. Thus, Eq. (6.13) yields

$$\frac{\sigma_1}{S_{ut}} - \frac{\sigma_3}{S_{uc}} = \frac{1}{n_s} \qquad \frac{\tau_1}{S_{ut}} - \frac{(-\tau_1)}{2S_{ut}} = \frac{1}{n_s} \qquad \tau_1 = \frac{2}{3}\frac{S_{ut}}{n_s} \qquad \textbf{(b)}$$

c) Figure 6.16 shows that for pure shear there is no difference between the MMT and the IFT, so that the prediction using the MMT is the same as in part b.

6.7.3 Selecting Failure Criterion

Selecting a failure criterion to use in design is somewhat of an art. Figure 6.17 shows some available test data on biaxial failures or yielding for several materials. The DET fits ductile materials *slightly* better than does the MSST, but most of the data fall between the two curves. Although the DET works well for ductile solids and results in a firm mathematical foundation when investigating problems in plasticity, the MSST is often applied. The MSST is mathematically simple and conservative; for a given circumstance the MSST will predict yielding at lower loads than does the DET. Thus, because there is an added safety factor in using the MSST, it is often used in design. On the other hand, many commercial stress analyses and finite element codes use a von Mises stress to obtain stress profiles, and there is a natural tendency to use a DET in such circumstances.

Ductile materials are produced with a high repeatability in terms of strength, so designers can use either failure criterion with a fair amount of confidence. Brittle materials are somewhat more difficult to analyze because their material properties vary much more than those of ductile materials. Owing to the statistical nature of brittle material properties,

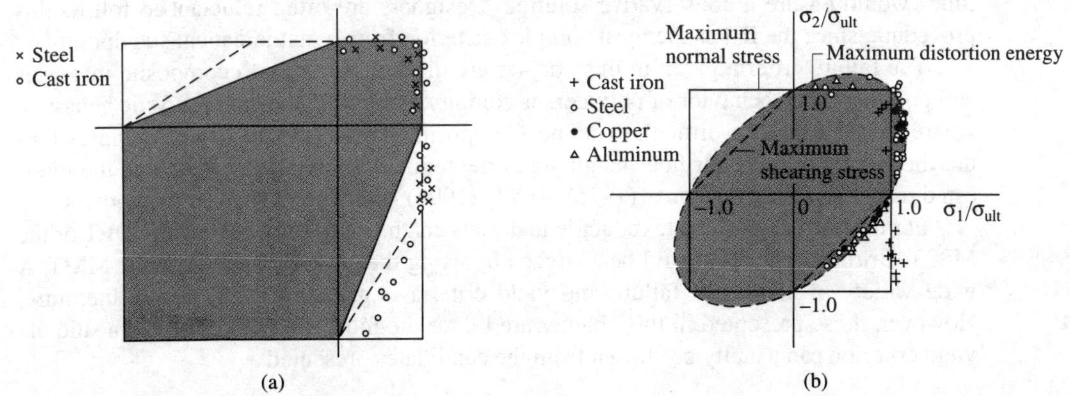

Figure 6.17 Experimental verification of yield and fracture criteria for several materials. (a) Brittle fracture. (b) Ductile yielding. [Data from Grassi and Cornet (1949), Dowling (1993), and Murphy (1964).]

experimental strength verification as in Fig. 6.17 is usually preferred over theoretical predictions regardless of test data.

When a ductile material is tested, as in a tension test, measured yield stresses deviate little among specimens, especially if the specimens have been obtained from the same batch in a manufacturing process. Thus, only a few experiments are needed to specify the strength of a ductile material with good certainty. For most brittle materials, however, many tests must be performed to accurately assess the strength distribution. Test specimens *do not* fail at the same stress, even if they were manufactured in the same batch with the same processes. It is difficult to give a strength value for a brittle material with any great certainty. This distinction leads to an entirely different field of study, that of probabilistic design. In deterministic design, the approach followed in this text, it is assumed that an assured minimum strength can be defined (i.e., that all samples of the material in question will have at least this strength). This is certainly possible with metals (even internal flaws are limited in size or else the manufacturing process will be unsuccessful, such as when a bar breaks during extrusion because of a large void). With brittle materials, such as ceramics and castings, a guaranteed minimum strength is difficult, and at times impossible, to define. Thus, the engineer needs to apply concepts of probability in design. Since the failure criteria defined here are deterministic but brittle materials are inherently probabilistic, it could be argued that these failure criteria should never be used with brittle materials. At the very least they should be applied with great caution. The greatest difficulty in using these deterministic criteria is in determining the strengths (especially the tensile strength) to be used in the equations.

A student of engineering design often wonders why the IFT theory is presented at all if the MMT fits brittle material data better, as is suggested by Fig. 6.17. Given the variation in strengths between brittle test specimens, the difference in the three failure criteria presented here is insignificant. To the question, "Which theory is better or best?" no absolute answer can be given, and a designer or organization should use the one with which it has the most experience and history.

The IFT and the MSST are identical for metals because the yield strengths of metals in compression and tension are approximately equal. Thus, applying the IFT theory at all times would ensure a conservative solution. Designers are often reluctant to follow this procedure, since the IFT is the most complex criterion from a mathematical standpoint.[1]

The failure criteria given in this chapter are difficult to apply to composite materials and polymers. The behavior of polymers is complex, including their viscoplastic behavior, where "a" yield point is difficult to define. Composite materials require more complex failure theories to account for fiber length and orientation with respect to load, and the interested reader is referred to Jones (1975), Reddy (1996), and Kaw (1997), among others.

Tutorial 1

For the purposes of this text ductile materials can be analyzed by using the DET or the MSST. Brittle materials should be analyzed by using the MNST, the IFT, or the MMT. A wide variety of additional failure and yield criteria is given in the technical literature. However, those presented in this chapter are by far the most often applied, and a suitable yield criterion can usually be chosen from the candidates presented.

[1]As Galileo Galilei said: *Entia non multiplicantur sunt prater necessitatum* (rough translation: Don't make things more difficult than necessary). Modern designers refer to this dictum as KISS (Keep It Simple, Stupid).

Table 6.2 Safety factors from using different criteria for three different materials used in Example 6.10.

Part	Criterion	Equation used	Safety factor
a)	MSST	(6.6)	1.5
	DET	(6.11)	1.73
b)	MSST	(6.6)	1.71
	DET	(6.11)	1.75
c)	MNST	(6.13)	3
	IFT	(6.14)	1.6
	MMT	(6.17)	1.634

Example 6.10

GIVEN The following materials and loadings:

a) Pure aluminum: $S_y = 30$ MPa, $\sigma_x = 10$ MPa, $\sigma_y = -10$ MPa, $\tau_{xy} = 0$

b) 0.2% carbon steel: $S_y = 65$ ksi, $\sigma_x = -5$ ksi, $\sigma_y = -35$ ksi, $\tau_{xy} = 10$ ksi

c) Gray cast iron: $S_{ut} = 30$ ksi, $S_{uc} = 120$ ksi, $\sigma_x = -35$ ksi, $\sigma_y = 10$ ksi, $\tau_{xy} = 0$

FIND The safety factors for the ductile and brittle materials.

Solution

The results are summarized in Table 6.2.

Case Study — SAFETY FACTOR FOR TOTAL HIP REPLACEMENT

GIVEN Figure 6.18 shows a total hip replacement inserted into a human femur and hip. Such devices are commonly used to treat painful arthritic conditions that result in loss of mobility. The insertion operation consists of sawing off portions of the femur, reaming the femoral cavity to allow for implant insertion, and hammering the implant into the femur. The hip portion of the implant is similarly installed, often with screws as shown. The femur portion of the implant includes a stem onto which the highly polished ball is attached. All components are available in a number of sizes so that the operating physician can optimize the components for a particular patient. Figure 6.19 is a drawing of a femoral implant, complete with typical dimensions. The commonly used implant materials are cast cobalt chromium, forged stainless steel, and Ti-6Al-4V (titanium alloy). The allowable stresses in these materials are as follows:

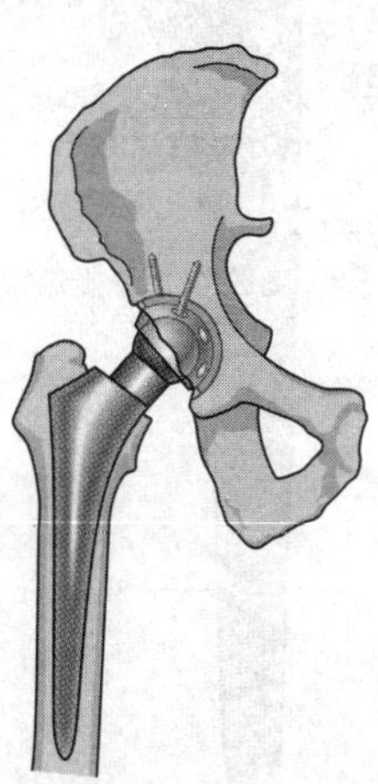

Figure 6.18 Inserted total hip replacement.

(continued)

Case Study (CONTINUED)

Material	Allowable stress, ksi
Cast cobalt chromium (CC)	80 (170 in compression)
Forged stainless steel (SS)	120
Ti-6Al-4V titanium alloy (Ti)	160

FIND Ascertain whether or not the implant is safe for application from a static load standpoint.

Solution

The most difficult part of the problem was obtaining the stresses; applying a failure criterion was a relatively short part of the analysis. The hip was analyzed at the sections indicated in Fig. 6.20. These sections were selected because they have geometrical features that act as stress raisers and because their locations maximize the stresses associated with the applied loads. The following dimen-

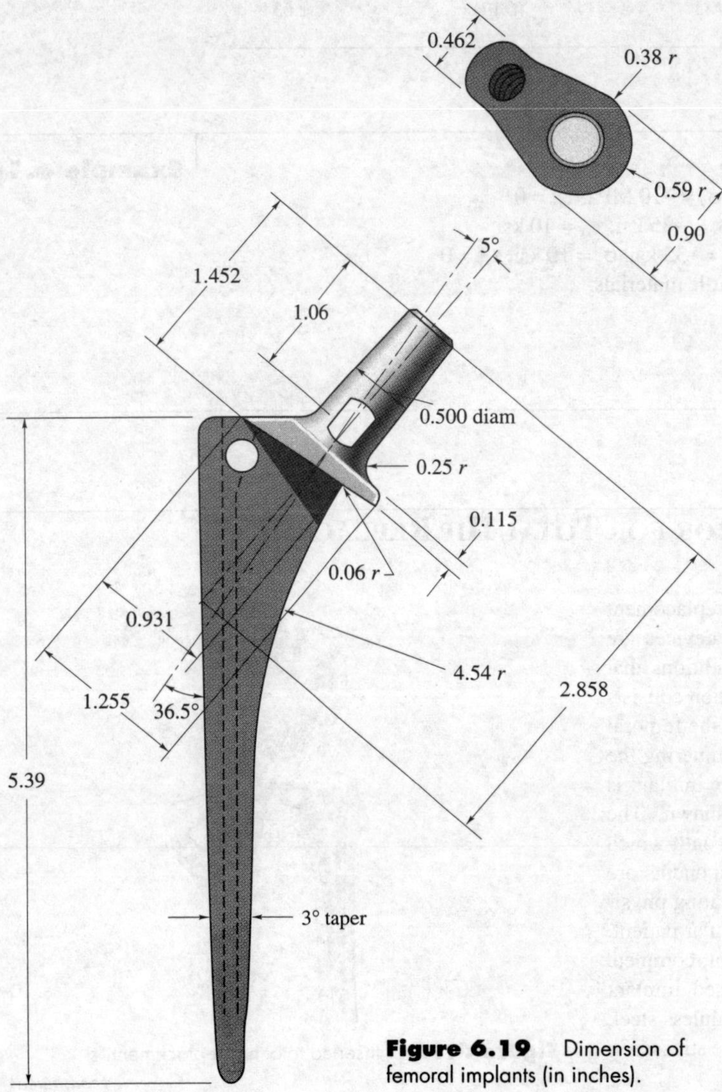

Figure 6.19 Dimension of femoral implants (in inches).

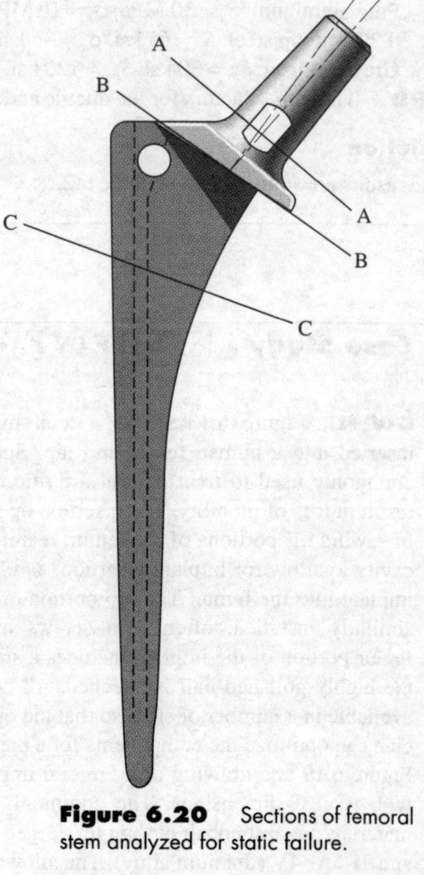

Figure 6.20 Sections of femoral stem analyzed for static failure.

(continued)

Case Study (CONTINUED)

sions were obtained for the three sections analyzed, and the stress concentrations were obtained from Fig. 6.5.

Section	d	D	r	K_c (axial)	K_c (bending)
A–A	0.50	1.255	0.25	1.8	1.2
B–B	**1.0**	1.255	0.06	2.0	1.8
C–C	**0.75**	**0.8**	4.54	1	1

Numbers given in boldface were interpolated from the drawing or extrapolated from the charts. To obtain the stress concentration factor, it was assumed that the geometry is circular in cross section and that it can be taken as a straight rod with fillets at each section. Although this assumption is not strictly true, the resultant stress concentration factors should be close to those actually seen in the hip. Also, St.-Venant's principle was liberally applied. That is, the stresses at the cross sections were assumed to be free of end effects; and the stress concentrations not to overlap. A more detailed analysis could be done by methods such as finite element analysis, but this approach will get a fairly accurate result.

The actual load applied to a hip joint is extremely complicated and varies from person to person. Gait, step length, etc., all play a role in the biomechanics of walking. Recognizing that the loading can be complex, the load was taken as four times the user's body weight, a peak force measured during a walking step. Running was not considered for this implant, although it can be a factor for some recipients (note that most implant recipients are elderly or have a less active lifestyle). Further, although the direction of the applied force can vary by as much as 30°, it was assumed that the load direction is vertical and centered on the stem (this is also a worst-case assumption—any inclination of the load reduces the bending moment at any section). Load direction could be varied to obtain a better understanding of the effect of stresses on the hip implant. A 200-lbm user was considered, or an 800-lbf load on the implant. This load too is high; most implant recipients weigh less, and obese patients are required to lose weight before receiving an implant.

The weight leads to a normal force, a shear force, and a bending moment at each section. These values were directly obtained through statics and by using the measured dimensions from Fig. 6.19. The resultant loads on each cross section were as follows:

Section	Normal force, lbf	Shear force, lbf	Moment, in.-lbf
A–A	800 cos 36.5° = 640	800 sin 36.5° = 475	800 (1.06/2) sin 36.5° = 252
B–B	640	475	345
C–C	800	0	680

Note that section C–C was actually loaded far less severely than is suggested here; part of the load is supported by bone, and the implant is never exposed to these stresses. Regardless, the resultant maximum stresses were calculated from

$$\sigma_{\max} = -K_{\text{axial}}\frac{P}{A} \pm K_{\text{bending}}\frac{Mc}{I}$$

$$\tau = \frac{4V}{3A}$$

yielding

Section	Maximum normal stress, psi	Maximum shear stress, psi
A–A	−30 500 or +18 700	3220
B–B	−7955 or +4690	1100
C–C	−18 200 or +14 608	0

Clearly, section A–A has the most severe loading, so it was analyzed with respect to failure.

At the extreme ends of the cross section there is no shear stress, so the stresses given are principal stresses. For uniaxial stress states the safety factor is merely the ratio of allowable stress to applied stress. For the location of maximum tensile stress the safety factor is given by $n_s = S_{ut}/\sigma_1$ and equals 4.2 for CC, 6.4 for SS, and 8.5 for Ti. In compression the safety factor is $n_s = S_{uc}/\sigma_3$ and equals 5.6 for CC, 3.9 for SS, and 5.2 for Ti. At the center of the cross section there is a normal stress component P/A in addition to the maximum shear. Thus, at section A–A the principal stresses (from a Mohr's circle analysis) are $\sigma_1 = 9700$ psi, $\sigma_2 = 0$, and $\sigma_3 = -3800$ psi. For cobalt chromium a brittle

(continued)

Case Study (CONCLUDED)

fracture theory must be chosen, so that the IFT is arbitrarily selected. From Eq. (6.14)

$$\frac{\sigma_1}{S_{ut}} + \frac{\sigma_3}{S_{uc}} = \frac{1}{n_s} \qquad \therefore\ n_s = 7.0$$

For stainless steel and titanium either the MSST or the DET can be used. Using the MSST gives

$$\sigma_1 - \sigma_3 = \frac{S_y}{n_s}$$

so that $n_s = 8.8$ for SS and 11.8 for Ti.

These safety factors may seem high, but they are quite reasonable for such critical applications, where failure requires immediate surgery and should be avoided whenever possible. Further, this confirms the field experience that most implanted hips do not fail statically, rather they need to be replaced because of fatigue, biological, or tribological issues.

6.8 SUMMARY

The main emphasis of this chapter was on failures due to static loading. It was discovered that stress concentrations, such as holes, grooves, and fillets, reduce the safety factor for the part. Using the stress concentration factor enables more accurate depiction of stresses for components with holes, grooves, and fillets. A flow analogy was also presented that suggested some design changes which might reduce the stress near the discontinuity.

Fracture mechanics was discussed in considering crack extension as a function of applied load. Cracks were viewed as microscopic flaws that always exist on the surface and within the body of a solid material. Fracture failures due to crack propagation occur at stress levels well below the yield stress of the material. Interest was in establishing the critical crack length that will make the part fracture. To do so required understanding fracture toughness. Fracture toughness is used as a design criterion in preventing fracture, just as yield strength is used as a design criterion in preventing yielding in a ductile material.

Failure predictions for uniaxial and multiaxial stress states were presented. Failure theories for multiaxial stress states were considered for ductile and brittle materials. The maximum-shear-stress theory states that a part subjected to any combination of loads will yield whenever the maximum shear stress exceeds the shear strength of the material. The distortion-energy theory postulates that yielding of ductile materials is caused by the elastic energy associated with shear deformation. The maximum-normal-stress theory states that a brittle material subjected to any combination of loads will fail whenever the greatest positive (or negative) principal stress exceeds the tensile (or compressive) yield strength of the material. Another failure criterion for brittle materials was the internal friction theory, which is similar to the maximum-shear-stress theory but takes into account the higher strength of brittle materials in compression versus tension. The modified Mohr theory deviates from the internal friction theory slightly to better account for the behavior of some brittle materials. In each of these theories a criterion was proposed such that a critical stress was established that was then compared with a critical uniaxial stress for which experimental data are known. This approach may be incompatible with some materials, suggesting that probabilistic methods, or at the very least generous safety factors, may be called for.

KEY WORDS

Coulomb-Mohr theory theory identical to internal friction theory

crack microscopic flaw, always present, that can compromise material strength

distortion-energy theory postulate that failure is caused by elastic energy associated with shear deformation

fracture control maintenance of nominal stress and crack size below critical level

fracture toughness critical value of stress intensity at which crack extension occurs

internal friction theory failure criterion accounting for difference between compressive and tensile strengths of brittle materials

maximum-normal-stress theory theory that yielding will occur whenever greatest positive principal stress exceeds tensile yield strength or whenever greatest negative principal stress exceeds compressive yield strength

maximum-shear-stress theory theory that yielding will occur when largest shear stress exceeds critical value

modes of crack propagation principal mechanisms for cracks to enlarge: mode I, opening through tension; mode II, sliding or in-plane shearing; mode III, tearing

modified Mohr theory failure postulate similar to Coulomb-Mohr theory, except that curve is altered in quadrants II and IV of plane stress plot of principal stresses

stress concentration region where stress raiser is present

stress concentration factor factor used to relate actual maximum stress at discontinuity to average stress without discontinuity

stress intensity factor stress intensity at crack tip

stress raiser discontinuity that alters stress distribution so as to increase maximum stress

Tresca yield criterion criterion identical to maximum-shear-stress theory

von Mises criterion criterion identical to distortion-energy theory

von Mises stress effective stress based on von Mises criterion, given by

$$\sigma_e = \frac{\sqrt{2}}{2}\left[(\sigma_2-\sigma_1)^2+(\sigma_3-\sigma_1)^2+(\sigma_3-\sigma_2)^2\right]^{1/2}$$

RECOMMENDED READINGS

Caddell, R.M. (1980) *Deformation and Fracture of Solids*, Prentice-Hall, Englewood Cliffs, NJ.
Calladine, C.R. (1969) *Engineering Plasticity*, Pergamon Press, Elmsford, NY.
Dowling, N.E. (1993) *Mechanical Behavior of Materials*, Prentice-Hall, Englewood Cliffs, NJ.
Norton, R.L. (1996) *Machine Design—An Integrated Approach*, Prentice-Hall, Englewood Cliffs, NJ.

Popov, E.P. (1968) *Introduction to Mechanics of Solids*, Prentice-Hall, Englewood Cliffs, NJ.
Shigley, J., and Mischke, C. (1989) *Mechanical Engineering Design*, McGraw-Hill, New York.

REFERENCES

ASM International (1989) *Guide to Selecting Engineering Materials*, American Society for Metals, Materials Park, OH.

Collins, J.A. (1981) *Failure of Materials in Mechanical Design*, Wiley, New York.

Coulomb, C.A. (1773) *Sur une Application des Regles de maximmis et minimus a quelques problemes de statique relatifs a l'architecture.*

Dowling, N.E. (1993) *Mechanical Behavior of Materials*, Prentice-Hall, Englewood Cliffs, NJ.

Grassi, R.C., and Cornet, I. (1949) "Fracture of Gray Cast Iron Tubes Under Biaxial Stresses," *Journal of Applied Mechanics*, vol. 16, pp. 178–183.

Jones, R.M. (1975) *Mechanics of Composite Materials*, Hemisphere Publishing Co., New York.

Kaw, A.K. (1997) *Mechanics of Composite Materials*, CRC Press, Boca Raton, FL.

Marin, J. (1962) *Mechanical Behavior of Engineering Materials*, Prentice-Hall, Englewood Cliffs, NJ.

Murphy, G. (1964) *Advanced Mechanics of Materials*, McGraw-Hill, New York.

Peterson, R.E. (1974) *Stress Concentration Factors*, Wiley, New York.

Popov, E.P. (1968) *Introduction to Mechanics of Solids*, Prentice-Hall, Englewood Cliffs, NJ.

Reddy, J.N. (1996) *Mechanics of Laminated Composite Plates*, CRC Press, Boca Raton, FL.

Tresca, H. (1868) "Mem. prenetes par divers savants," vol. 59, p. 754, *Comptes Rendus Acad. Sci.*, Paris.

Young, W.C. (1989) *Roark's Formulas for Stress and Strain*, 6th ed., McGraw-Hill, New York.

PROBLEMS

Section 6.2

6.1 Given that the stress concentration factor is 3.12 for a machine element made of steel with a modulus of elasticity of 207 GPa, find the stress concentration factor for an identical machine element made of aluminum instead of steel. The modulus of elasticity for aluminum is 69 GPa.

6.2 A flat part with constant thickness b is loaded in tension as shown in Fig. 6.3(a). The height changes from 50 to 100 mm with a radius $r = 10$ mm. Find how much higher a load can be transmitted through the bar if the height increases from 50 to 70 mm and the radius decreases from 10 to 3 mm.

6.3 A flat steel plate axially loaded as shown in sketch *a* has two holes for electric cables. The holes are situated beside each other and each has a diameter d. To make it possible to draw more cables, the two holes are replaced with one hole having twice the diameter $2d$, as shown in sketch b. Assume that the ratio of diameter to width is $d/b = 0.2$ for the two-hole plate. Which plate will fail first?

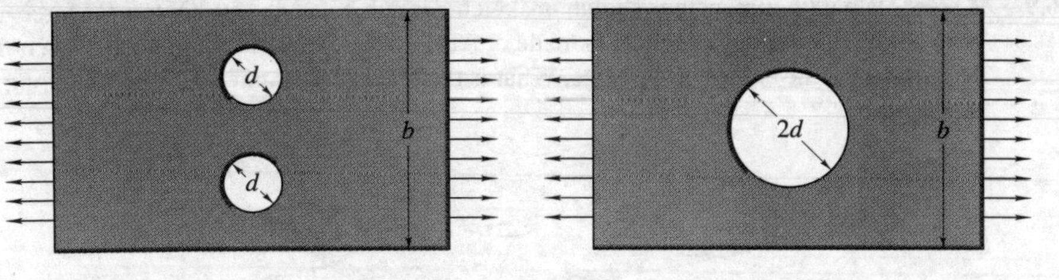

(a) (b)

Sketches *a* and *b*, for Problem 6.3

6.4 A load-carrying beam is loaded with a bending moment M. The minor beam height is h and the width in the perpendicular direction is b. This beam, which carries a balcony on the wall of a house, is welded together with a beam in the house structure. That beam has a major height H. Because of manufacturing problems the radius r of the connections cannot be made larger than r_{max} but can be made smaller, down to zero. Find how the load-carrying beam should be placed on the house beam to give maximum strength.

6.5 A round bar has a fillet with $r/d = 0.15$ and $D/d = 1.5$. The bar transmits both bending moment and torque. A new construction is considered to make the shaft stiffer and stronger by making it equally thick on each side of the fillet or groove. Determine whether that is a good idea.

6.6 A rectangular plate has a central hole. The width of the plate is $b = 50$ mm and the diameter of the hole is $d = 10$ mm. The plate is axially loaded with a force $P = 1500$ N, and the minor height of the plate is $h = 5$ mm. Is it possible to find the stress concentration factor for the plate without using Fig. 6.2?

6.7 A machine has three circular shafts, each with fillets giving stress concentrations. The ratio of fillet radius to shaft diameter is 0.1 for all three shafts. One of the shafts transmits a tensile force, one transmits a bending torque, and one transmits torsion. Because they are stressed exactly to the stress limit ($n_s = 1$), a design change is proposed doubling the notch radii to get a safety factor greater than 1. How large will the safety factors be for the three shafts if the diameter ratio is 2 ($D/d = 2$)?

6.8 The shaft shown in sketch *c* is subjected to tensile, torsional, and bending loads. Determine the principal stresses at the location of stress concentration.

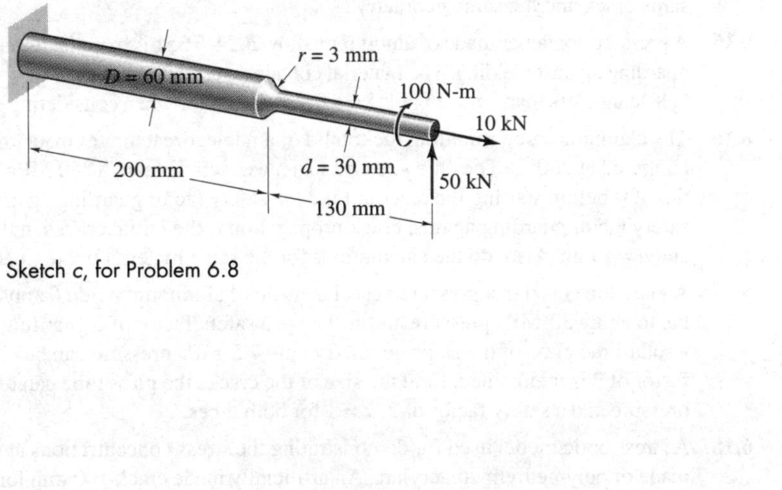

Sketch *c*, for Problem 6.8

6.9 A steel plate with dimensions shown in sketch *d* is subjected to 150-kN tensile force and 300-N-m bending moment. The plate is made of AISI 1080 steel and is kept at 20°C. A hole is to be punched in the center of the plate. What is the maximum diameter of the hole for a safety factor of 1.5?

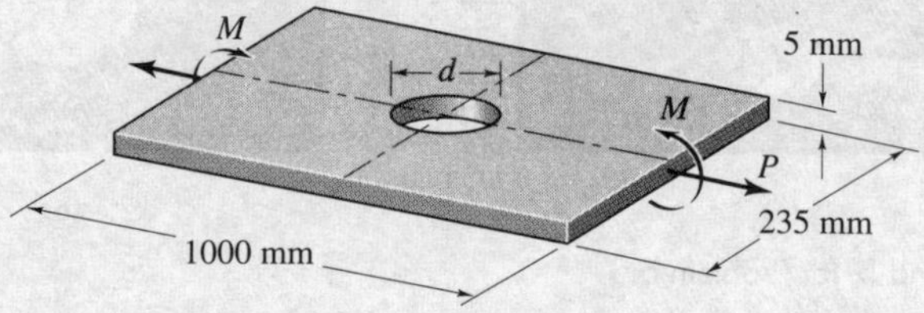

Sketch *d*, for Problem 6.9

Section 6.3

6.10 A Plexiglas plate with dimensions 1 m × 1 m × 1 cm is loaded by a nominal tensile stress of 55 MPa in one direction. The plate contains a small crack perpendicular to the load direction. At this stress level a safety factor of 1.3 against crack propagation is obtained. Find how much larger the crack can get before it grows catastrophically.

6.11 A pressure container is made of AISI 4340 steel. The wall thickness is such that the tensile stress in the material is 1100 MPa. The dimensionless geometry correction factor $Y = 1$ for the given geometry. Find how big the largest crack can be without failure if the steel is tempered

a) At 260°C

b) At 425°C

6.12 Two tensile test rods are made of AISI 4340 steel tempered at 260°C and aluminum alloy 2024–T351. The dimensionless geometry correction factor $Y = 1$. Find how high a stress each rod can sustain if there is a 2-mm crack in each of them.

6.13 A plate made of titanium alloy Ti-6Al-4V has the dimensionless correction factor $Y = 1$. How large can the largest crack in the material be if it still should be possible to plastically deform the plate in tension?

6.14 A Plexiglas model of a gear has a 1-mm crack formed in its fillet curve (where the tensile stress is maximum). The model is loaded until the crack starts to propagate. $Y = 1.5$. How much higher a load can a gear made of AISI 4340 steel tempered to 425°C carry with the same crack and the same geometry?

6.15 A pressure container made of aluminum alloy 2024–T351 is manufactured for a safety factor of 2 guarding against yielding. The material contains cracks through the wall thickness with a crack half-length less than 3 mm. $Y = 1$. Find the safety factor when considering crack propagation.

6.16 The clamping screws holding the top lid of a nuclear reactor are made of AISI 4340 steel tempered at 260°C. They are stressed to a maximum level of 1250 MPa during a pressurization test before starting the reactor. Find the safety factor guarding against yielding and the safety factor guarding against crack propagation if the initial cracks in the material have $Y = 1$ and $a = 1$ mm. Also, do the calculations for the same material but tempered to 425°C.

6.17 A glass tube used in a pressure vessel is made of aluminum oxide (sapphire) to make it possible to apply 30-MPa pressure and still have a safety factor of 2 guarding against fracture. For a soda-lime glass of the same geometry only 7.5-MPa pressure can be allowed if a safety factor of 2 is maintained. Find the size of the cracks the glass tube can tolerate at 7.5-MPa pressure and a safety factor of 2. $Y = 1$ for both tubes.

6.18 A stress optic model used for demonstrating the stress concentrations at the ends of a crack is made of polymethylmethacrylate. An artificially made crack 100 mm long is perpendicular to the loading direction. $Y = 1$. Calculate the highest tensile stress that can be applied to the model without propagating the crack.

6.19 A passengerless airplane requires wings that are lightweight and the prevention of cracks more than 2 mm long. The dimensionless geometry correction factor Y is usually 1.5 for a safety factor of 2. What is the appropriate alloy for this application? If Y is increased to 4.5, what kind of alloy from Table 6.1 should be used?

Section 6.6

6.20 The anchoring cables carrying a suspension bridge are made of cylindrical AISI 1080 steel bars 210 mm in diameter. The force transmitted from the cable to the steel bar is 3.5 MN. Calculate the safety factor range guarding against yielding.

6.21 The arm of a crane has two steel plates connected with a rivet that transfers the force in pure shear. The rivet is made of AISI 1040 steel and has a circular cross section with a diameter of 25 mm. The load on the rivet is 10 kN. Calculate the safety factor.

6.22 A telescope stands on four feet, each carrying a load of 200 kN. The feet are made of AISI 1020 steel, and the floor is a thick plate of AISI 1080 steel on a concrete base. Calculate how large the diameter of the feet needs to be if they are circular and flat. The safety factor is 2.5.

Section 6.7

6.23 A machine element is loaded so that the principal normal stresses at the critical location for a biaxial stress state are $\sigma_1 = 20$ ksi and $\sigma_2 = -15$ ksi. The material is ductile with a yield strength of 60 ksi. Find the safety factor according to

a) The maximum-shear-stress theory (MSST)

b) The distortion-energy theory (DET)

Which theory would be expected to agree most closely with actual test data?

6.24 A glass bottle is filled with a liquid having a compressibility similar to (~1% lower than) the compressibility of the glass. To what depth within an ocean can the bottle be lowered without cracking?

6.25 A bolt is tightened, subjecting its shank to a tensile stress of 80 ksi and a torsional shear stress of 50 ksi at a critical point. All the other stresses are zero. Find the safety factor at the critical point by using the DET and the MSST. The material is high-carbon steel (AISI 1080). Will the bolt fail because of the static loading?

6.26 A torque is applied to a piece of chalk used in a classroom until the chalk cracks. Using the maximum-normal-stress theory (MNST) and assuming the tensile strength of the chalk to be small relative to its compressive strength, determine the angle of the cross section at which the chalk cracks.

6.27 A cantilevered bar 500 mm long with square cross section has 25-mm sides. Three perpendicular forces are applied to its free end, a 100-N force is applied in the y direction, and an equally large force of 100 N is applied in the z direction. Calculate the equivalent stress at the clamped end of the bar by using the DET when the sides of the square cross section are parallel with the y and z directions.

6.28 A shaft transmitting torque from the gearbox to the rear axle of a truck is unbalanced, so that a centrifugal load of 500 N acts at the middle of the 3-m-long shaft. The AISI 1040 tubular steel shaft has an outer diameter of 70 mm and an inner diameter of 58 mm. Simultaneously, the shaft transmits a torque of 6000 N-m. Use the DET to determine the safety factor guarding against yielding.

6.29 The right-angle-cantilevered bracket used in Problem 5.30, sketch *w*, has a concentrated force of 1000 N and a torque of 300 N-m. Calculate the safety factor. Use the DET and neglect transverse shear. Assume that the bracket is made of AISI 1040 steel and use the following values: $a = 0.5$ m, $b = 0.3$ m, $d = 0.035$ m, $E = 205$ GPa, and $\nu = 0.3$.

6.30 A 10-cm-diameter shaft is subjected to a 10-kN-m steady bending moment, an 8-N-m steady torque, and a 150-kN axial force. The yield strength of the shaft material is 600 MPa. Use the MSST and the DET to determine the safety factors for the various types of loading.

6.31 Use the MSST and the DET to determine the safety factor for 2024 aluminum alloys for each of the following stress states:

a) $\sigma_x = 10$ MPa, $\sigma_y = -60$ MPa
b) $\sigma_x = \sigma_y = \tau_{xy} = -30$ MPa
c) $\sigma_x = -\sigma_y = 20$ MPa, and $\tau_{xy} = 10$ MPa
d) $\sigma_x = 2\sigma_y = -70$ MPa, and $\tau_{xy} = 40$ MPa

6.32 Four different stress elements, each made of the same material, are loaded as shown in sketches *e*, *f*, *g*, and *h*. Use the MSST and the DET to determine which element is the most critical.

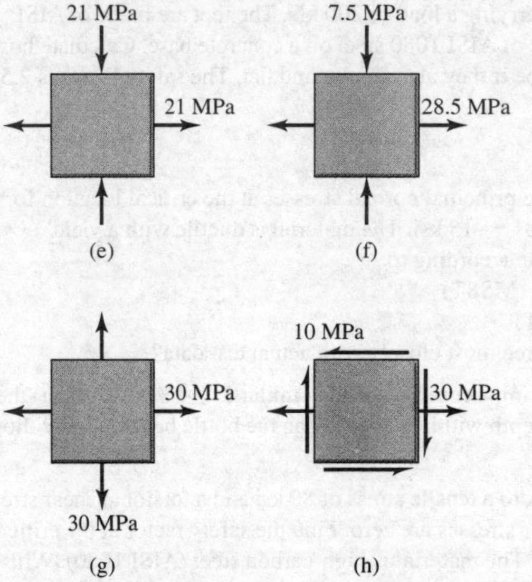

Sketches *e*, *f*, *g*, and *h*, for Problem 6.32

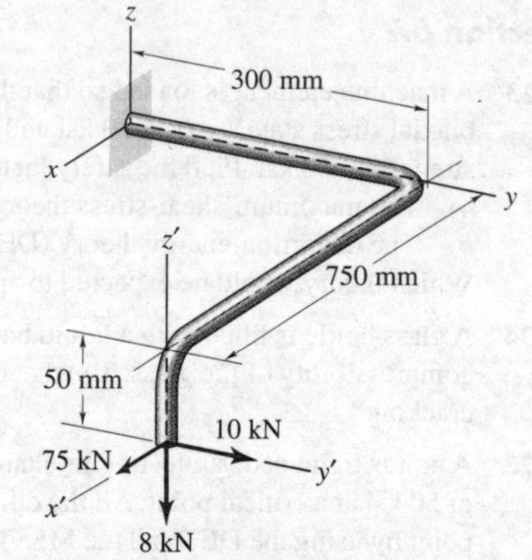

Sketch *i*, for Problem 6.33

6.33 The rod shown in sketch *i* is made of AISI 1040 steel and has two 90° bends. Use the MSST and the DET to determine the minimum rod diameter for a safety factor of 2 at the most critical section.

6.34 The shaft shown in sketch *j* is made of AISI 1020 steel. Determine the most critical section by using the MSST and the DET. Dimensions of the various diameters shown in sketch *j* are $d = 30$ mm, $D = 45$ mm, and $d_2 = 40$ mm.

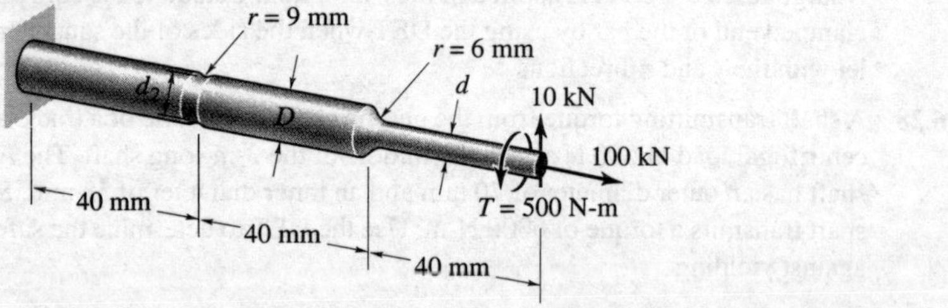

Sketch *j*, for Problem 6.34

CHAPTER

7

FAILURE PREDICTION FOR CYCLIC AND IMPACT LOADING

Aloha Airlines flight 243, a Boeing 737–200, taken April 28, 1988. The midflight fuselage failure was caused by corrosion-assisted fatigue. (Steven Minkowski/Gamma Liaison)

All machine and structural designs are problems in fatigue because the forces of Nature are always at work and each object must respond in some fashion

Carl Osgood, *Fatigue Design*

SYMBOLS

A	area, m^2
A_a	amplitude ratio, σ_a/σ_m
a	fatigue strength exponent
b	width, m
b_s	slope
C	constant used in Eq. (7.35)
C_1, C_2	integration constants
$\overline{C}$	intercept

c	distance from neutral axis to outer fiber of solid, m
d	diameter, m
E	modulus of elasticity, Pa
e, f	factors used in Eq. (7.21)
f	exponent
g	gravitational acceleration, 9.807 m/s^2 (386.1 in./s^2)
H	height including two notch radii, m
h	height without two notch radii, m
I	area moment of inertia, m^4
I_m	impact factor
K	stress intensity factor
K_c	stress concentration factor
K_f	fatigue stress concentration factor
k	spring rate, N/m
ΔK	stress intensity range
k_f	surface finish factor
k_m	miscellaneous factor
k_r	reliability factor
k_s	size factor
k_t	temperature factor
k_0	stress concentration factor
l	length, m
l_c	crack length, m
M	bending moment, N-m
m	constant used in Eq. (7.35)
N	fatigue life in cycles
N_c	number of stress cycles
N'	number of cycles to failure at a specific stress
N_t'	total number of cycles to failure
n'	number of cycles at a specific stress when $n' < N'$
n_s	safety factor
P	force, N
$P_{\max}$	maximum force, N
q_n	notch sensitivity factor
R_a	arithmetic average, m
R_s	stress ratio, $\sigma_{\min}/\sigma_{\max}$
r	notch radius, m
S_e	modified endurance limit, Pa
S_e'	endurance limit, Pa
S_f	modified fatigue strength, Pa
S_f'	fatigue strength, Pa
S_l'	strength at 10^3 cycles for ductile material, Pa
S_l'	fatigue strength at which high-cycle fatigue begins, Pa
S_u	ultimate strength, Pa
S_{uc}	ultimate strength in compression, Pa
S_{ut}	ultimate strength in tension, Pa
S_y	yield strength, Pa
S_{yt}	yield strength in tension, Pa
u	sliding velocity, m/s
V	shear force, N
W	weight, N
Y	plate size correction factor
x,y	Cartesian coordinates, m
α	fatigue ductility exponent
α_i	cyclic ratio n_i'/N_i'
δ	deflection, m
$\delta_{\max}$	maximum impact deflection, m
δ_{st}	static deflection, m
ε_f'	fatigue ductility coefficient
$\Delta\varepsilon$	total strain
σ	stress, Pa
σ_f'	stress at fracture, Pa
σ_r	stress range, Pa

Subscripts

a	alternating
f	final
i	integer
m	mean
max	maximum
min	minimum
ref	reference

7.1 INTRODUCTION

Most machine element failures involve load conditions that fluctuate with time. However, the static load conditions covered in the last chapter are important, since they provide the foundation for understanding this chapter. This chapter focuses on understanding and predicting component failure under cyclic and impact loading. Rather than considering a general variation of loading with time (fluctuating load), we have considered only cyclic loads, or a sequence of loading that repeats itself.

Fluctuating loads induce fluctuating (cyclic) stresses that often result in failure by means of cumulative damage. To better understand failures due to fluctuating stresses, consider the yielding caused by the back and forth bending of a paper clip. Bending results in compressive and tensile stresses on opposite sides of the paper clip, and these stresses reverse when the bending direction is changed. Thus, the stress at any point around the wire of the paper clip will vary as a function of time. Bending the paper clip will eventually exhaust the ductility of the material, resulting in failure.

Stresses and deflections in impact loading are much greater than those found in static loading. Thus, dynamic loading effects are important. A material's physical properties are a function of loading speed. Fortunately, the more rapid the load, the higher the yield and ultimate strengths. Some examples in which impact loading would need to be considered are colliding automobiles, driving a nail with a hammer, and an air hammer breaking up concrete.

7.2 FATIGUE

In the last century the industrial revolution was in full swing. Coal was mined, converted into coke, and used to smelt iron from ore. The iron was used to manufacture bridges, railroads, and trains that brought more coal to the cities, allowing more coke to be produced, etc., in a spiral of ever-increasing production. But then bridges, the monuments to engineering, began failing. Even worse, they would fail in extremely confusing ways. At 10 o'clock a 50-ton locomotive would pass over a bridge without incident; at 11 o'clock a farmer driving a horse-drawn wagon full of hay would cause the bridge to collapse. Fear gripped the populace, as people believed death awaited on the bridges (Fig. 7.1).

Today, we no longer fear our "aging infrastructure." And the road construction and improvements that halt traffic from time to time make collapses of civil engineering works, such as bridges, truly rare. But fatigue failures, as we now know them to be, are not at all rare. Fatigue is the single largest cause of failure in metals, estimated to be 90% of all metallic failures. Fatigue failures, especially in structures, are catastrophic and insidious, occurring suddenly and often without warning. For that reason engineers must apply fatigue considerations in design.

Figure 7.1 "On the Bridge," an illustration from *Punch* magazine in 1891 warning the populace that death was waiting for them on the next bridge. Note cracks in iron bridge. (Puck, 1891)

Fatigue is applicable on both microscopic and macroscopic scales. That is, although a rolling-element bearing failing by surface spalling and a ship breaking in two are quite different events, both the bearing and the ship have failed due to fatigue. The failure stresses were considerably lower than the yield strengths of the materials. The rolling-element bearing suffered surface fatigue failure; the ship, structural fatigue failure. Thus, fatigue failure occurs at relatively low stress levels to a component or structure subjected to fluctuating or cyclic stresses. Fatigue failure looks brittle even in ductile metals; little if any gross plastic deformation is associated with it.

Fatigue is a complex phenomenon. Essentially, it is crack propagation, initially on a microscale and then extremely rapid as the fatigue crack reaches a critical length. Fatigue is a concern whenever cyclic stresses are present. Experimentalists have found that fatigue cracks generally begin at a surface and propagate through the bulk, unless large subsurface flaws or stress raisers exist in the substrate. Fatigue cracks begin at several sites simultaneously and propagate when one flaw becomes dominant and grows more rapidly than the others. The total life of a component or structure is the time it takes a crack to begin plus the time it needs to propagate through the cross section. Design life can then be maximized in the following ways:

1. *By minimizing initial flaws, especially surface flaws.* Great care is taken to produce fatigue-susceptible surfaces through processes, such as grinding or polishing, that leave exceptionally smooth surfaces. These surfaces are then carefully protected before being placed into service.
2. *By maximizing initiation time.* Surface residual stresses are imparted (or at least tensile residual stresses are relieved) through manufacturing processes, such as shot peening or burnishing, or by a number of surface treatments.
3. *By maximizing propagation time.* Substrate properties, especially those that retard crack growth, are also important. For example, fatigue cracks propagate more quickly along grain boundaries than through grains (because grains have much more efficient atomic packing). Thus, using a material that does not present elongated grains in the direction of fatigue crack growth can extend fatigue life (e.g., by using cold-worked components instead of castings).
4. *By maximizing the critical crack length.* Fracture toughness is an essential ingredient. (The material properties that allow for larger internal flaws are discussed in Chapter 6.)

Given a finite number of resources, which way should the designer emphasize? Ever smoother surfaces can be manufactured at ever-increasing costs but at ever-decreasing payoff as the surface valleys become smaller. Maximizing initiation time allows parts to function longer with little or no loss in performance. Maximizing propagation time can allow cracks to be detected before they become catastrophic, the approach previously called the "doctrine of manifest danger." Maximizing the critical crack length can lead to long life with a greater likelihood of recognizing an imminent failure. The proper emphasis, just as with the safety factor, is product specific and is best decided through experience.

The designer's job is to select a material and processing route that will lead to successful products. However, fatigue is an extremely complicated subject. This chapter introduces approaches to design for cyclic loading. In practice extreme caution should be used in applying any of these approaches without generous safety factors.

7.3 CYCLIC STRESSES

Cyclic stress is a function of time, but the variation is such that the stress sequence repeats itself. The stresses may be axial (tensile or compressive), flexural (bending), or torsional (twisting). Figure 7.2 shows the cyclic variation of nonzero mean stress with time. Also shown are several parameters used to characterize fluctuating cyclic stress. A stress cycle ($N_c = 1$) constitutes a single application and removal of a load and then another application and removal of the load in the opposite direction. Thus, $N_c = 1/2$ means that the load is applied once and then released. The stress amplitude alternates about a **mean stress** σ_m, defined as the average of the maximum and minimum stresses in the cycle, or

$$\sigma_m = \frac{\sigma_{\max} + \sigma_{\min}}{2} \tag{7.1}$$

The **stress range** σ_r is the difference between $\sigma_{\max}$ and $\sigma_{\min}$, namely,

$$\sigma_r = \sigma_{\max} - \sigma_{\min} \tag{7.2}$$

The **stress amplitude** σ_a is one-half of the stress range, or

$$\sigma_a = \frac{\sigma_r}{2} = \frac{\sigma_{\max} - \sigma_{\min}}{2} \tag{7.3}$$

The **stress ratio** is the ratio of minimum and maximum stress amplitudes, or

$$R_s = \frac{\sigma_{\min}}{\sigma_{\max}} \tag{7.4}$$

Finally, the **amplitude ratio** is the ratio of the stress amplitude to the mean stress, or

$$A_a = \frac{\sigma_a}{\sigma_m} = \frac{\sigma_{\max} - \sigma_{\min}}{\sigma_{\max} + \sigma_{\min}} = \frac{1 - R_s}{1 + R_s} \tag{7.5}$$

The four most frequently used patterns of constant-amplitude cyclic stress are

1. Completely reversed ($\sigma_m = 0, R_s = -1, A_a = \infty$)
2. Nonzero mean (as shown in Fig. 7.2)
3. Released tension ($\sigma_{\min} = 0, R_s = 0, A_a = 1, \sigma_m = \sigma_{\max}/2$)
4. Released compression ($\sigma_{\max} = 0, R_s = \infty, A_a = -1, \sigma_m = \sigma_{\min}/2$)

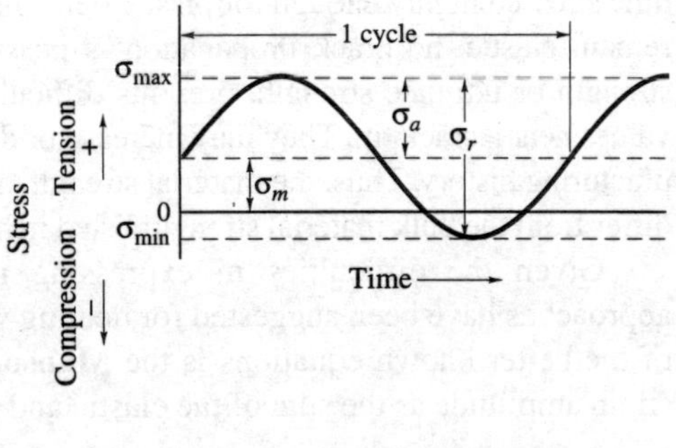

Figure 7.2 Variation in nonzero cyclic mean stress.

Example 7.1

GIVEN Before the thread-formed additions at the top third of high chimneys were invented, chimneys swayed perpendicular to the motion of the wind. This swaying motion appeared when the wind speed and direction were constant for some time. In a certain chimney the bending stress was ± 100 MPa.

FIND The mean stress, the range of stress, the stress amplitude, the stress ratio, and the amplitude ratio for that swaying steel chimney.

Solution

The drag force acting on the chimney is in the wind direction, and it gives a maximum tensile stress on the front side and a maximum compressive stress on the back side of the chimney. The stress perpendicular to the wind is zero. The swinging motion, which gives the highest stress perpendicular to the wind direction, will thus be symmetrical relative to the equlibrium position.

The mean stress is

$$\sigma_m = \frac{\sigma_{max} + \sigma_{min}}{2} = \frac{100 - 100}{2} = 0$$

The range of stress is

$$\sigma_r = \sigma_{max} - \sigma_{min} = 100 - (-100) = 200 \text{ MPa}$$

The alternating stress is

$$\sigma_a = \frac{\sigma_r}{2} = \frac{200}{2} = 100 \text{ MPa}$$

The stress ratio is

$$R_s = \frac{\sigma_{min}}{\sigma_{max}} = \frac{-100}{100} = -1$$

The amplitude ratio is

$$A_a = \frac{\sigma_a}{\sigma_m} = \frac{\sigma_{max} - \sigma_{min}}{\sigma_{max} + \sigma_{min}} = \frac{1 - R_s}{1 + R_s} = \frac{1-(-1)}{1+(-1)} = \frac{2}{0} = \infty$$

7.4 Strain Life Theory of Fatigue

Fatigue is a damage accumulation process that manifests itself through crack propagation, but no crack propagation is possible without plastic deformation at the crack tip. Although the volume stressed highly enough for plastic deformation can be extremely small, if the stress fields remain elastic, no crack propagation is possible. Using material properties, such as yield strength or ultimate strength, presents difficulties because cyclic loadings can change these values near a crack tip. They may increase or decrease depending on the material and its manufacturing history. Thus, the material strength at the location where cracks are propagating can differ from the bulk material strength listed in handbooks or obtained from tension tests.

Given the difficulties in expressing material strengths near crack tips, several approaches have been suggested for dealing with the strain encountered at a crack tip. One of the better known equations is the **Manson-Coffin relationship,** which gives the total strain amplitude as the sum of the elastic and plastic strain amplitudes:

$$\frac{\Delta\varepsilon}{2} = \frac{\sigma'_f}{E}(2N')^a + \varepsilon'_f(2N')^\alpha \quad (7.6)$$

where

$\Delta\varepsilon$ = total strain

σ'_f = stress at fracture in one stress cycle, Pa

E = elastic modulus of material, Pa

N' = number of cycles that will occur before failure

ε'_f = fatigue ductility coefficient (true strain corresponding to fracture in one stress cycle)

a = fatigue strength exponent

α = fatigue ductility exponent

Table 7.1 gives some typical values of the material properties a and α.

Landgraf (1968) and Mitchell (1978) have surveyed the industrial applications of the strain life approach to fatigue, but as pointed out by Shigley and Mitchell (1983), the Manson-Coffin relationship is difficult to use because the total strain $\Delta\varepsilon$ is difficult to determine.

Table 7.1 Cyclic properties of some metals [J.E. Shigley and L.D. Mitchell, *Mechanical Engineering Design*, 4E, © 1983 by McGraw-Hill, reproduced with permission of The McGraw-Hill Companies; and S. Suresh, *Fatigue of Materials*, Cambridge University Press, 1991.]

Material	Condition	Yield strength, S_y, MPa	Fatigue strength σ'_f, MPa	Fatigue ductility coefficient, ε'_f	Fatigue strength exponent, a	Fatigue ductility exponent, α
			Steel			
1015	Normalized	228	827	0.95	−0.110	−0.64
4340	Tempered	1172	1655	.73	−.076	−.62
1045	Q&T[a] 80°F	—	2140	—	−.065	−1.00
1045	Q&T 360°F	1720	2720	.07	−.055	−.60
1045	Q&T 500°F	1275	2275	.25	−.080	−.68
1045	Q&T 600°F	965	1790	.35	−.070	−.69
4142	Q&T 80°F	2070	2585	—	−.075	−1.00
4142	Q&T 400°F	1720	2650	.07	−.076	−.76
4142	Q&T 600°F	1340	2170	.09	−.081	−.66
4142	Q&T 700°F	1070	2000	.40	−.080	−.73
4142	Q&T 840°F	900	1550	.45	−.080	−.75
			Aluminum			
1100	Annealed	97	193	1.80	−.106	−0.69
2014	T6	462	848	0.42	−.106	−.65
2024	T351	379	1103	0.22	−.124	−.59
5456	H311	234	724	0.46	−.110	−.67
7075	T6	469	1317	0.19	−.126	−.52

[a]Quenched and tempered.

Stress concentration factors, such as those presented in Chapter 6, are readily available in the technical literature, but strain concentration factors, especially in the plastic range, are nowhere to be found. The advantage of the Manson-Coffin equation is that it gives insight into important properties in fatigue strength determination. It shows the importance of strength as well as ductility, and it leads to the conclusion that as long as there is a cyclic plastic strain, no matter how small, eventually there will be failure.

7.5 FATIGUE STRENGTH

7.5.1 ROTATING-BEAM EXPERIMENTS

Fatigue is inherently probabilistic (i.e., there is a great range of performance within samples prepared from the same materials). In previous problems and case studies a valuable approach called the worst-case scenario was described. To apply this approach to fatigue, a designer would select surface finishes, notch sizes, initial flaw sizes, etc., that minimize the fatigue strength of the candidate specimen. However, this process would necessarily result in fatigue specimens with zero strength, a situation that does nothing to aid designers. Thus, data on fatigue often reflect the best-case scenario, and the designer is strongly cautioned that great care must be taken in applying fatigue design theories to critical applications.

Because fatigue is basically a damage accumulation phenomenon, initial flaws have a great effect on performance. No manufacturing process produces flaw-free parts; indeed, no parts are possible without thousands, even millions, of flaws per cubic inch. Analytical approaches that derive fatigue strengths from first principles are thus very difficult, and most knowledge on material fatigue is experimentally based.

Experimental approaches to fatigue use either exemplars or idealized, standard specimens. The former are more reliable and best for critical specimens. The latter are often used when a direct simulation of the loading environment is cost prohibitive.

To establish the fatigue strength of an exemplar, a series of tests is performed. The test apparatus duplicates as nearly as possible the stress conditions (stress level, time frequency, stress pattern, etc.) in practice. The exemplar duplicates as nearly as possible any manufacturing and treatment processes. Such experiments give the most direct indication of a component's survivability in the actual loading environment.

To test idealized, standard specimens, a rotating-beam fatigue testing machine is used, most commonly the R.R. Moore rotating-beam machine. The specimen is subjected to pure bending, and no transverse shear is imposed. The specimen has specific dimensions (Fig. 7.3) and a highly polished surface, with a final polishing in the axial direction to avoid circumferential scratches. If the specimen breaks into two equal pieces, the test is

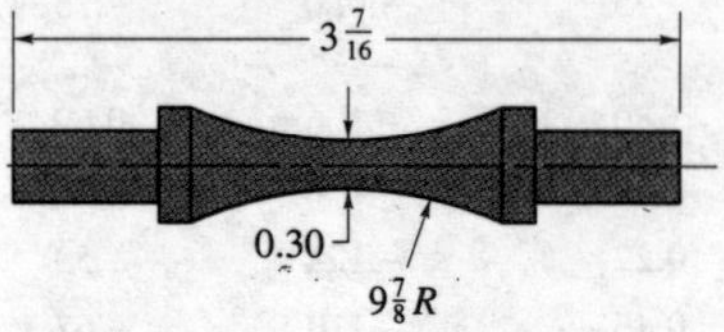

Figure 7.3 R.R. Moore machine fatigue test specimen. Dimensions in inches.

indicative of the material's fatigue strength. If the pieces are unequal, a material or surface flaw has skewed the results. The test specimen is subjected to completely reversed ($\sigma_m = 0$) stress cycling at a relatively large maximum stress amplitude, usually two-thirds of the static ultimate strength, and the cycles to failure are counted. Thus, for each specimen at a specific stress level the test is conducted until failure occurs. The procedure is repeated on other identical specimens, progressively decreasing the maximum stress amplitude.

7.5.2 S–N DIAGRAMS

Data from reversed bending experiments are plotted as the fatigue strength versus the logarithm of the total number of cycles to failure N_t' for each specimen. These plots are called **S–N diagrams** or **Wöhler diagrams** after August Wöhler, a German engineer who published his fatigue research in 1870. They are a standard method of presenting fatigue data and are extremely common and informative. Two general patterns for two classes of material, those with and those without endurance limits, emerge when plotting the fatigue strength versus the logarithm of the number of cycles to failure. Figure 7.4 shows typical results for several materials. Figure 7.4(a) presents test data for wrought steel. Note the large amount of scatter in the data, even with the great care in preparing test specimens. Thus, life predictions based on curves such as those in Fig. 7.4 are all subject to a great deal of error. Figure 7.4(a) also shows a common result. For some materials with **endurance limits**, such as ferrous (iron based) and titanium alloys, a horizontal straight line occurs at low stress levels, implying that an endurance limit S_e' is reached below which failure will not occur. This endurance limit S_e' represents the largest fluctuating stress that will *not* cause failure for an infinite number of cycles. For many steels the endurance limit ranges between 35 and 60% of the material's ultimate strength.

Most nonferrous alloys (e.g., aluminum, copper, and magnesium) *do not* have a significant endurance limit. Their fatigue strength continues to decrease with increasing cycles. Thus, fatigue will occur regardless of the stress amplitude. The fatigue strength for these materials is the stress level at which failure will occur for some specified number of cycles (e.g., 10^7 cycles).

Determining the endurance limit experimentally is quite lengthy and expensive. The Manson-Coffin relationship [Eq. (7.6)] demonstrated that the fatigue life will depend on the material's fracture strength during a single load cycle, suggesting a possible relationship between static material strength and strength in fatigue. Such a relationship has been noted by several researchers (e.g., Fig. 7.5). The stress endurance limits of steel for three types of loading can be approximated as

$$
\begin{aligned}
S_e' &= 0.5S_u \qquad \text{bending} \\
S_e' &= 0.45S_u \qquad \text{axial} \\
S_e' &= 0.29S_u \qquad \text{torsion}
\end{aligned}
\tag{7.7}
$$

Students may use these equations to approximate the endurance limits for other ferrous alloys but must recognize that the limits can differ significantly from experimentally determined endurance limits. Since the ultimate stress and the type of loading are known for various materials, their endurance limits can be approximated.

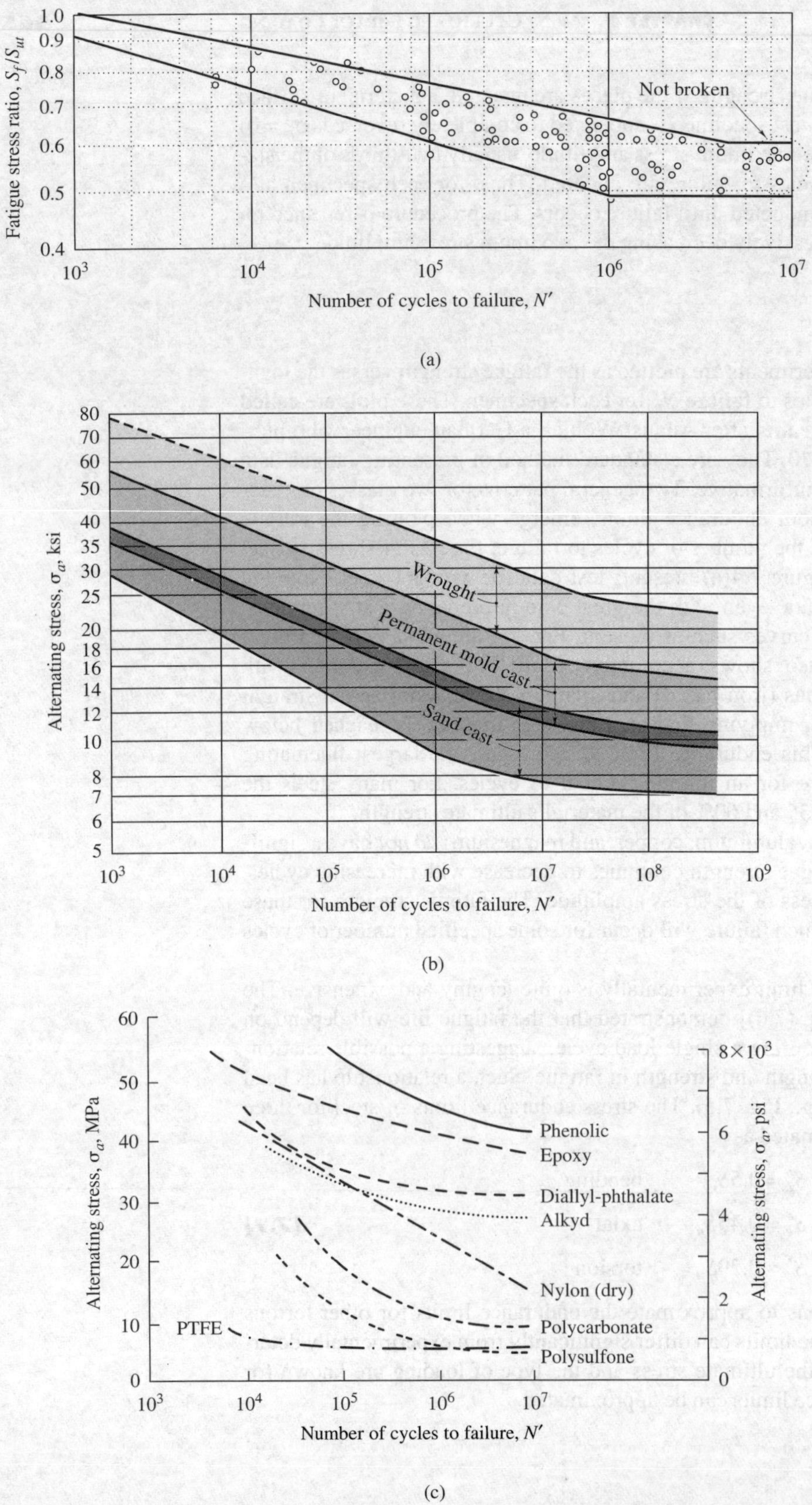

Figure 7.4 Fatigue strengths as a function of number of loading cycles. (a) Ferrous alloys, showing clear endurance limit [Adapted from Lipson and Juvinall (1963).]; (b) aluminum alloys, with less pronounced knee and no endurance limit [Adapted from Juvinall and Marshek (1991).]; (c) selected properties of assorted polymer classes [Adapted from Norton (1996).].

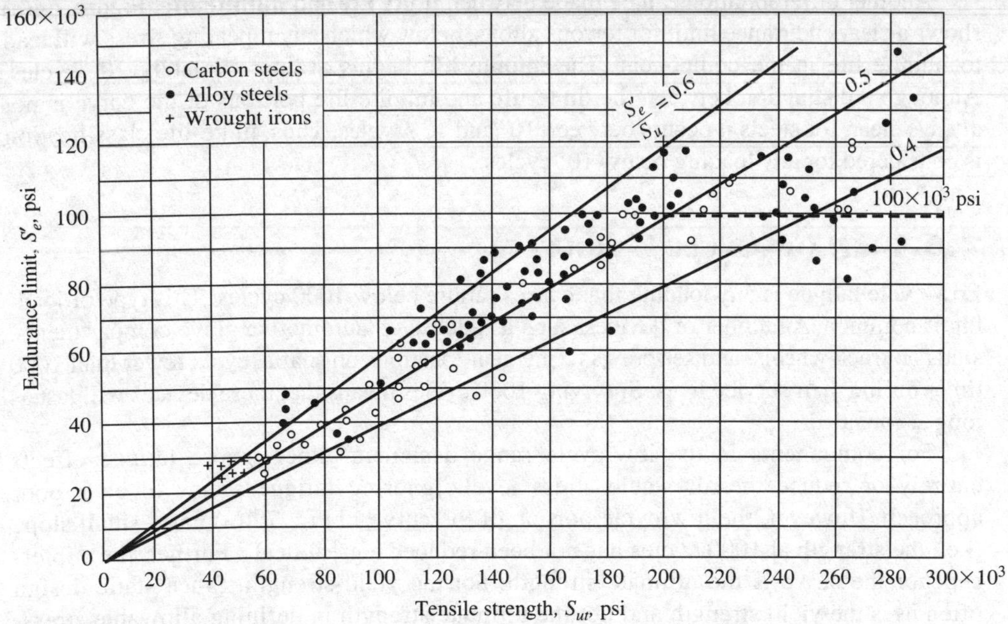

Figure 7.5 Endurance limit as function of ultimate strength for wrought steels. [Adapted from Shigley and Mitchell (1983).]

Table 7.2 Approximate endurance limit for various materials. [Adapted from Juvinall and Marshek (1991).]

Material	Number of cycles	Relation
Magnesium alloys	10^8	$S'_e = 0.35S_u$
Copper alloys	10^8	$0.25S_u < S'_e < 0.5S_u$
Nickel alloys	10^8	$0.35S_u < S'_e < 0.5S_u$
Titanium	10^7	$0.45S_u < S'_e < 0.65S_u$
Aluminum alloys	5×10^8	$S'_e = 0.45S_u$ ($S_u < 48$ ksi) $S'_e = 19$ ksi ($S_u \geq 48$ ksi)

Other materials, for which there is much less experience, are finding increasing uses in fatigue applications. Table 7.2 gives the approximate strengths in fatigue for various material classes. Figure 7.4(c) gives some stress-life curves for common polymers. Because polymers have a much greater variation in properties than metals do, Fig. 7.4(c) should be seen as illustrative of fatigue properties and not used for quantitative data.

7.6 FATIGUE REGIMES

The S–N diagram [Fig. 7.4(a)] shows different types of behavior as the number of cycles to failure increases. Two basic regimes are **low-cycle fatigue** (generally below 10^3 stress cycles) and **high-cycle fatigue** (more than 10^3 but less than 10^6 stress cycles). The slope of the line is much less in low-cycle fatigue than in high-cycle fatigue.

Another differentiation can be made between **finite life** and **infinite life.** Figure 7.4(a) shows a clear endurance limit for ferrous alloys, below which any repeating stress will lead to infinite life in the component. The infinite life begins at approximately 10^6 cycles. Although a distinction between the finite-life and infinite-life portions of the curve is not always clear, for steels it occurs between 10^6 and 10^7 cycles. Thus, finite-life classification is considered for any loading below 10^7 cycles.

7.6.1 Low-Cycle Fatigue

Low-cycle fatigue is any loading that causes failure below 1000 cycles. This type of loading is common. A number of devices, such as latches on automotive glove compartments, studs on truck wheels, and setscrews fixing gear locations on shafts, cycle fewer than 1000 times during their service lives. Surviving 1000 cycles means that these devices will last as long as intended.

For components in the low-cycle range designers either ignore fatigue effects entirely or reduce the allowable stress level. Ignoring fatigue seems to be a poor approach. However, the low-cycle portion of the curve in Fig. 7.4(a) has a small slope (i.e., the strength at 1000 cycles has not been reduced a great deal). Further, the *y* intercept for the curve is the ultimate strength, not the yield strength. Since static design often uses the yield strength and not the ultimate strength in defining allowable stresses, static approaches are acceptable for designing low-cycle components. In fact, the safety factor compensates for the uncertainty in material strength due to cyclic loading.

Taking low-cycle effects into account allows modifying the material strength based on experimental data. The fatigue strength for steel at which high-cycle fatigue begins can be approximated as

$$\begin{aligned} S_l' &= 0.9S_u && \text{bending} \\ S_l' &= 0.75S_u && \text{axial} \\ S_l' &= 0.72S_u && \text{torsional} \end{aligned} \qquad \textbf{(7.8)}$$

Again, the fatigue strengths of other ferrous alloys can be approximated by these equations but they may differ from what is experienced in practice.

7.6.2 High-Cycle, Finite-Life Fatigue

In many applications the number of stress cycles placed on a component during its service life is between 10^3 and 10^7. Examples include car door hinges, aircraft body panels, and aluminum softball bats. Because the strength drops quickly in this range (Fig. 7.4), an approach not accounting for this drop is inherently flawed.

The fatigue strength at any location between S_l' and S_e' can generally be expressed as

$$\log S_f' = b_s \log N_t' + \overline{C} \qquad \textbf{(7.9)}$$

where

b_s = slope
$\overline{C}$ = intercept

At the end points Eq. (7.9) becomes

$$\log S_l' = b_s \log(10^3) + \overline{C} = 3b_s + \overline{C} \tag{7.10}$$

$$\log S_e' = b_s \log(10^6) + \overline{C} = 6b_s + \overline{C} \tag{7.11}$$

Subtracting Eq. (7.11) from Eq. (7.10) gives

$$b_s = -\frac{1}{3}\log\left(\frac{S_l'}{S_e'}\right) \tag{7.12}$$

Substituting Eq. (7.12) into Eq. (7.11) gives

$$\overline{C} = 2\log\left(\frac{S_l'}{S_e'}\right) + \log S_e' = \log\left[\frac{(S_l')^2}{S_e'}\right] \tag{7.13}$$

Thus, by using Eqs. (7.7) and (7.8) the slope b_s and the intercept $\overline{C}$ can be determined for a specific type of loading. Knowing the slope and the intercept from Eq. (7.9) yields the fatigue strength as

$$S_f' = 10^{\overline{C}}\left(N_t'\right)^{b_s} \qquad \text{for} \qquad 10^3 \le N_t' \le 10^6 \tag{7.14}$$

If the fatigue strength is given and the number of cycles until failure is desired,

$$N_t' = \left(S_f' 10^{-\overline{C}}\right)^{1/b_s} \qquad \text{for} \qquad 10^3 \le N_t' \le 10^6 \tag{7.15}$$

Thus, by using Eqs. (7.7), (7.8), (7.12), and (7.13) the fatigue strength can be obtained from Eq. (7.14) or the number of cycles to failure can be determined from Eq. (7.15).

Example 7.2

GIVEN The pressure vessel lids of nuclear power plants are bolted down to seal the high pressure exerted by the water vapor (in a boiler reactor) or the pressurized water (in a pressurized water reactor). The bolts are so heavily stressed that they are replaced after the reactors are opened 25 times. A 20% decrease in stress would give 10 000 cycles of life. The ultimate strength is 1080 MPa.

FIND How low the stress needs to be to give a life of 10 000 cycles.

Solution

Equations (7.7) and (7.8) for axial loading give $S_e' = 0.45S_u$ and $S_l' = 0.75S_u$. Note that S_l' is for 1000 cycles and S_e' is for a life of 10^6 cycles. Equation (7.12) gives the slope as

$$b_s = -\frac{1}{3}\log\left(\frac{S_l'}{S_e'}\right) = -\frac{1}{3}\log\left(\frac{0.75S_u}{0.45S_u}\right) = -0.07395$$

From Eq. (7.13) the intercept is

$$\overline{C} = \log\left[\frac{(S_l')^2}{S_e'}\right] = \log\left[\frac{(0.75)^2(1080)^2}{(0.45)(1080)}\right] = 3.130$$

Knowing the slope and intercept Eq. (7.14) gives the fatigue strength as

$$S'_f = 10^{\bar{C}}\left(N'_t\right)^{b_s} = 10^{3.13}(10\ 000)^{-0.07395} = 682.7 \text{ MPa}$$

The stress needs to be decreased to 682.7 MPa to give 10 000 cycles.

7.6.3 High-Cycle, Infinite-Life Fatigue

A number of applications call for infinite life, defined for steels as the number of cycles above which an endurance limit can be defined, usually taken as 10^6 cycles. If a material does not have an endurance limit, it cannot be designed for infinite life. Thus, aluminum alloys, for example, will always be designed for finite life (using the approach in Sec. 7.6.2).

For ferrous and titanium alloys, however, an infinite-life design approach can be followed. Basically, the designer determines an endurance limit and uses this strength as the allowable stress. Then, sizing and selection of components can proceed just as in static design. This approach, which is fairly complex, is addressed in the next section.

7.7 Endurance Limit Modification Factors

Fatigue experiments assume that the best circumstances exist for promoting long fatigue lives. However, this situation cannot be guaranteed for design applications, so the component's endurance limit must be modified or reduced from the material's best-case endurance limit. The endurance limit modification factors covered in this text are for completely reversed loading ($\sigma_m = 0$). The **modified endurance limit** can be expressed as

$$S_e = k_f k_s k_r k_t k_m S'_e \tag{7.16}$$

where

S'_e = endurance limit from experimental apparatus under idealized conditions, Pa
k_f = surface finish factor
k_s = size factor
k_r = reliability factor
k_t = temperature factor
k_m = miscellaneous factor

The type of loading has been incorporated into S'_e as presented in Eq. (7.7). Many effects can influence the fatigue strength of a part. Some of those most frequently encountered are described in this section.

7.7.1 Stress Concentration Effects

The Manson-Coffin equation [Eq. (7.6)] showed that the life of a component has a direct correlation with the strain to which it is subjected. Because locations of stress concentration are also locations of strain concentration, these locations can be seen as prime candidates for the promotion of fatigue crack initiation and growth. However, the stress concentration factor developed in Chapter 6 cannot be directly applied to fatigue applications, since many materials will relieve stresses near a crack tip through plastic flow. That is, because some materials flow plastically near crack tips, fracture is avoided and the crack's growth is retarded.

For *static* loading the stress concentration factor K_c is used, and for *fatigue* loading the fatigue stress concentration factor K_f is used, where

$$K_f = \frac{\text{Endurance limit for notch-free specimen}}{\text{Endurance limit for notched specimen}} \quad \textbf{(7.17)}$$

A notch or stress concentration may be a hole, fillet, or groove. Recall from Sec. 6.3 that the stress concentration factor is just a function of geometry. In this section the fatigue stress concentration factor is not only a function of geometry but also a function of material and type of loading. The consideration of the material is often dealt with by using a **notch sensitivity** factor q_n defined as

$$q_n = \frac{K_f - 1}{K_c - 1} \quad \textbf{(7.18)}$$

or

$$K_f = 1 + (K_c - 1)q_n \quad \textbf{(7.19)}$$

From Eq. (7.18) note that the range of q_n is between zero (when $K_f = 1$) and unity (when $K_f = K_c$). From Eq. (7.19) observe that obtaining the fatigue stress concentration factor requires knowing the material's notch sensitivity and the type of loading.

Figure 7.6 is a plot of notch sensitivity versus notch radius for some commonly used materials and for various types of loading. For all the materials considered, the notch sensitivity approaches zero as the notch radius approaches zero. Also, the harder and stronger steels tend to be more notch sensitive (have a large value of q_n). This is not too surprising, since notch sensitivity is a measure of material ductility and the hardest steels have limited ductility. Figure 7.6 also shows that a given steel is slightly more notch sensitive for torsional loading than for bending and axial loading.

The endurance limit modification factor taking stress concentrations into account is

$$k_0 = \frac{1}{K_f} \quad \textbf{(7.20)}$$

where K_f is obtained from Eq. (7.19).

To apply the effects of stress concentrations, designers can either reduce the endurance limit by k_0 or increase the stress by K_f. Either approach is equally valid. In this text neither factor is included in the modified endurance limit equation [Eq. (7.16)]. It is more convenient to deal with these factors separately, since in later chapters, such as that on shafting (Chapter 11), these effects are handled differently for ductile and brittle materials.

Example 7.3

GIVEN The drive shaft for a formula-one racing car has a diameter of 30 mm and a half-circular notch with a 1-mm radius. The shaft was dimensioned to give a coefficient of friction between the tires and the ground of 1.5 for equal shear and bending stresses. By mounting spoilers and a wing on the car, the load on the tires can be doubled at high speed without increasing the car's mass. The shaft material has an ultimate strength of 965 MPa. Assume from the distortion-energy theory that the equivalent stress is proportional to the square root of the bending stress squared plus three times the shear stress squared $\left(\sigma_e = \sqrt{\sigma^2 + 3\tau^2}\right)$.

FIND Determine the fatigue stress concentration factors for bending and torsion of the drive shaft. Also, find if increased acceleration or increased curve handling will give the higher risk of drive shaft failure.

Solution

From Fig. 7.6 for a notch radius of 1 mm and ultimate strength of 965 MPa the notch sensitivity is 0.825 for bending and 0.853 for torsion. From Fig. 6.6(b) when $r/d = 1/28 = 0.0357$ and $D/d = 30/28 = 1.0714$ the stress concentration factor is 2.2 for bending, and from Fig. 6.6(c) the stress concentration factor is 1.8 for torsion.

From Eq. (7.19) the fatigue stress concentration factor due to *bending* is

$$K_f = 1 + \left(K_c - 1\right)q_n = 1 + (2.2 - 1)(0.825) = 1.99$$

The fatigue stress concentration factor due to *torsion* is

$$K_f = 1 + (1.8 - 1)(0.853) = 1.68$$

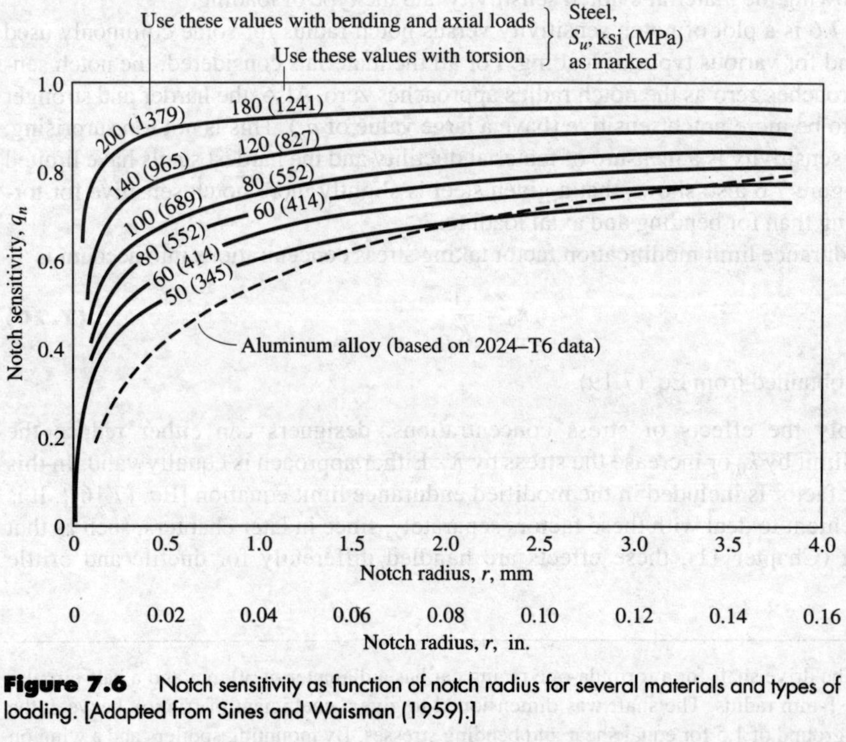

Figure 7.6 Notch sensitivity as function of notch radius for several materials and types of loading. [Adapted from Sines and Waisman (1959).]

Let σ_{e1} be the equivalent stress for increased curve handling and σ_{e2} be the equivalent stress for increased acceleration. Doubling the load and using the distortion-energy theory results in

$$\frac{\sigma_{e1}}{\sigma_{e2}} = \frac{\sqrt{(2\sigma)^2 + 3\tau^2}}{\sqrt{\sigma^2 + 3(2\tau)^2}} = \frac{2\sqrt{1+0.75(\tau/\sigma)^2}}{\sqrt{1+12(\tau/\sigma)^2}}$$

Recall that the shaft was dimensioned such that the shear and bending stresses are equal ($\tau = \sigma$).

$$\therefore \frac{\sigma_{e1}}{\sigma_{e2}} = \frac{2\sqrt{1+0.75}}{\sqrt{1+12}} = 0.7338$$

$$\sigma_{e2} = \frac{\sigma_{e1}}{0.7338} = 1.363\sigma_{e1}$$

Therefore, increased acceleration gives a higher risk of drive shaft failure than increased curve handling.

7.7.2 SURFACE FINISH FACTOR

The specimen shown in Fig. 7.3 has a highly polished surface finish with final polishing in the axial direction to smooth any circumferential scratches. Machine elements do not usually have such a high-quality finish. The modification factor to incorporate the finish effect depends on the process used to generate the surface and on the ultimate strength. Once that process is known, Fig. 7.7(a) enables the surface finish factor to be obtained when the

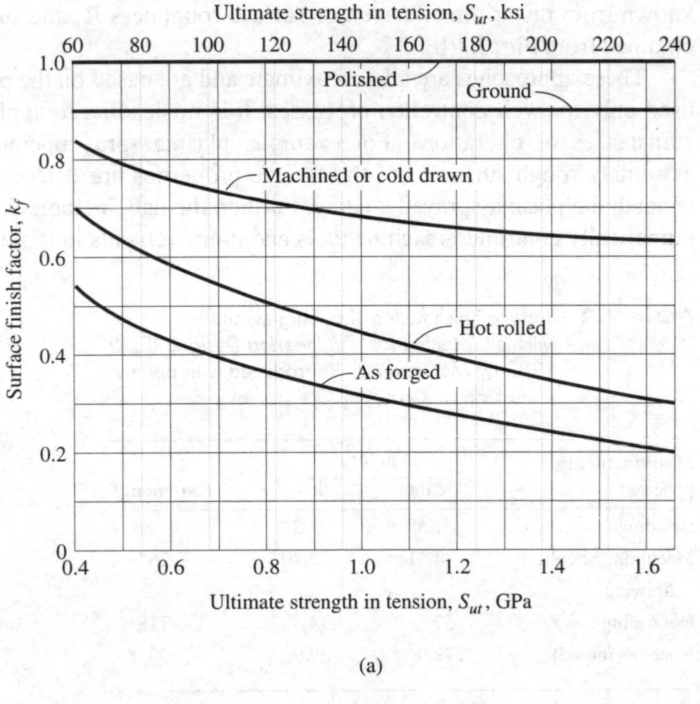

Figure 7.7 Surface finish factors for steel. (a) As function of ultimate strength in tension for different machine processes. [Adapted from Shigley and Mitchell (1983).]

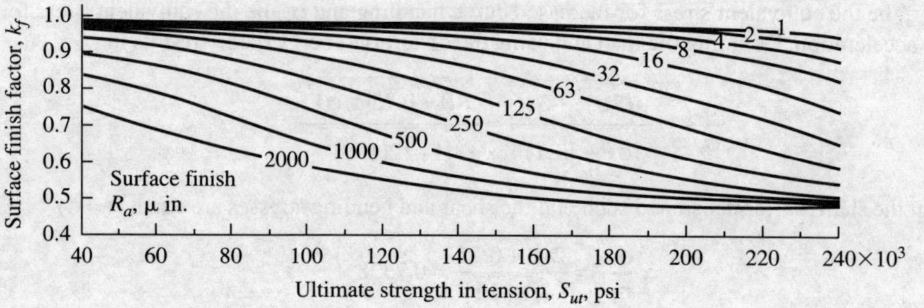

(b)

Figure 7.7 (b) As function of ultimate strength and surface roughness as measured with a stylus profilometer. [Adapted from Johnson (1967).]

ultimate strength in tension is known, or else the coefficients from Table 7.3 can be used with the equation

$$k_f = eS_{ut}^f \tag{7.21}$$

where

k_f = surface finish factor
S_{ut} = ultimate tensile strength of material
e and f = coefficients defined in Table 7.3

If the process used to obtain the surface finish is not known but the quality of the surface is known from the arithmetic average surface roughness R_a, the surface finish factor can be obtained from Fig. 7.7(b).

These approaches are all approximate and are based on the processes in the charts and used only for well-controlled processes. It is misleading to apply Table 7.3 for other circumstances or operations. For example, plasma spray operations tend to provide an extremely rough surface, but the fatigue properties are determined by the surface layer beneath the plasma-sprayed coating. Further, the data in Table 7.3 are dated. With modern numerically controlled machine tools and improvements in tooling materials, superior fin-

Table 7.3 Surface finish factor. [J.E. Shigley and L.D. Mitchell, *Mechanical Engineering Design*, 4E, © 1983 by McGraw-Hill. Reproduced with permission of the McGraw-Hill Companies.]

Manufacturing process	Factor e MPa	Factor e ksi	Exponent f
Grinding	1.58	1.34	−0.085
Machining or cold drawing	4.51	2.70	−.265
Hot rolling	57.7	14.4	−.718
None (as forged)	272.0	39.9	−.995

ishes are routinely produced today that will give slightly better performance from a fatigue standpoint.

7.7.3 SIZE FACTOR

The high-cycle fatigue apparatus used to obtain the endurance limit S_e' was for a specific diameter, namely, 0.30 in., and uses extruded or drawn bar stock for metals. For metals such extrusions have pronounced grain elongation in the direction opposite to the fatigue crack growth. Also, the degree of cold work is high and the likelihood of large flaws is low. Similar effects are seen for ceramics and castings but for different reasons (smaller shrinkage pores, etc). Regardless of the manufacturing process larger parts are more likely to contain flaws and will not demonstrate the strength of the 0.3-in.-diameter shaft. The size factor for a round bar is affected by the method of loading. For bending or torsion the size factor is

$$k_s = \begin{cases} 0.869d^{-0.112} & 0.3 \text{ in.} < d < 10 \text{ in.} \\ 1 & d < 0.3 \text{ in. or } d \leq 8 \text{ mm} \\ 1.189d^{-0.112} & 8 \text{ mm} < d \leq 250 \text{ mm} \end{cases} \qquad \textbf{(7.22)}$$

For axial loading $k_s = 1$.

For components that are not circular in cross section the size factor is difficult to determine. However, because the reasons for strength reductions hold for such components, it is reasonable to define an effective diameter based on an equivalent circular cross section. The effective diameter can be defined by using equivalent cross sections defined by stress distributions as proposed by Shigley and Mischke (1989). A simpler approach is merely to define the diameter from the effective circular cross section with an area equal to that of the cross section in question. Both approaches are too simple. They do not differentiate between material and manufacturing processes and could cause a mistaken belief that the correction factor due to size effects can be calculated with accuracy.

7.7.4 RELIABILITY FACTOR

Table 7.4 shows the reliability factor for various percentages of the probability of survival, the probability of surviving to the life indicated at a particular stress. This table is based on the fatigue limit's having a standard deviation of 8%, generally the upper limit for steels.

Table 7.4 Reliability factors for six probabilities of survival.

Probability of survival, percent	Reliability factor, k_r
50	1.00
90	.90
95	.87
99	.82
99.9	.75
99.99	.70

The reliability factor as obtained from Table 7.4 can be considered only as a guide because the actual law of distribution is more complicated than the simple approach used here. We will also assume that the table can be applied to materials other than steel, for which it was originally determined.

7.7.5 Temperature Factor

Many high-cycle fatigue applications take place under extremely high temperatures, such as in aircraft engines, where the material may be much weaker than at room temperature. Conversely, in some applications, such as automobile axles in Alaska during January, the metal may be less ductile than at room temperature.

In either case the Manson-Coffin relationship [Eq. (7.6)] would suggest that the major factor affecting fatigue life is the strength in one loading cycle σ_f'. It is therefore reasonable to follow one of two approaches. The designer can either (1) modify the ultimate strength of the material based on its properties at the temperature of interest before determining a material endurance limit in Eq. (7.16) or (2) use a temperature factor:

$$k_t = \frac{S_{ut}}{S_{ut,\text{ref}}} \tag{7.23}$$

where

S_{ut} = ultimate tensile strength of material at desired temperature

$S_{ut,\text{ref}}$ = ultimate tensile strength at reference temperature, usually room temperature

7.7.6 Miscellaneous Effects

Many more phenomena can affect a component's strength. Whereas the preceding sections outlined methods for numerically approximating some effects, other considerations defy quantification. Among these are the following:

1. *Manufacturing history.* Manufacturing processes play a huge role in determining the fatigue life characteristics of engineering materials. This role is manifested in the size factor discussed in Sec. 7.7.3, but there are other effects as well. Because fatigue crack growth is quicker along grain boundaries than through grains, any manufacturing process affecting grain size and orientation can affect fatigue. Because some operations, such as rolling, extrusion, and drawing, lead to elongated grains, the material's fatigue strength will vary in different directions. With extrusion and drawing this effect is usually beneficial, since the preferred direction of crack propagation becomes the axial direction and crack propagation through the thickness is made more difficult by grain orientation and elongation in metals. Annealing a metal component relieves residual stresses, causes grains to become equiaxed, and may cause grain growth. Relieving tensile residual stresses at a surface is generally beneficial, but equiaxed or larger grains are detrimental from a fatigue standpoint.
2. *Residual stresses* can result from manufacturing processes. A **residual stress** is caused by elastic recovery after nonuniform plastic deformation across a component's thickness.

Compressive residual stress in a surface retards crack growth; tensile residual stress can encourage crack growth. Compressive residual stresses can be obtained through shot peening and roller burnishing and may be obtained in forging, extrusion, or rolling. Tensile residual stresses can result from forging, extrusion, rolling, cutting, or grinding.

3. *Coatings* can greatly affect fatigue. Some operations, such as carburizing, lead to a high carbon content in steel surface layers (and thus a high fracture strength) and impart a compressive residual stress on the surface. Electroplated surfaces can be extremely porous and promote crack growth, reducing fatigue strengths by as much as 50%. Zinc plating is the main exception where the fatigue strength is not seriously affected. Anodized oxide coatings are also usually porous, reducing fatigue strength. Coatings applied at high temperatures, such as in chemical vapor deposition processes or hot dipping, may thermally induce tensile residual stresses at the surface.
4. *Corrosion.* It is not surprising that materials operating in corrosive environments have lowered fatigue strengths. The main culprits in corroding metals are hydrogen and oxygen. Hydrogen diffuses into a material near a crack tip, aided by large tensile stresses at the tip, embrittling the material and aiding crack propagation. Oxygen causes coatings to form that are brittle or porous, aiding crack initiation and growth. High temperatures in corrosive environments speed diffusion-based processes.

Tutorial 2

Example 7.4

GIVEN Figure 7.8 shows a tensile-loaded bar. In Fig. 7.8(a) the bar is not notched and in Fig. 7.8(b) the bar is notched, but the smallest height is the same in both figures. The bar is machine made of low-carbon steel (AISI 1020).

FIND The modified endurance limits for the notched and unnotched bars as well as the fatigue strengths at 10^4 cycles

Solution

From Table A.1 for low-carbon steel (AISI 1020) the ultimate strength is

$$S_u = 57 \text{ ksi}$$

The loading is axial, so from Eqs. (7.7) and (7.8)

$$S'_e = 0.45S_u = (0.45)(57)\left(10^3\right) \text{ psi} = 25.65 \text{ ksi}$$

$$S'_l = 0.75S_u = (0.75)(57)\left(10^3\right) \text{ psi} = 42.75 \text{ ksi}$$

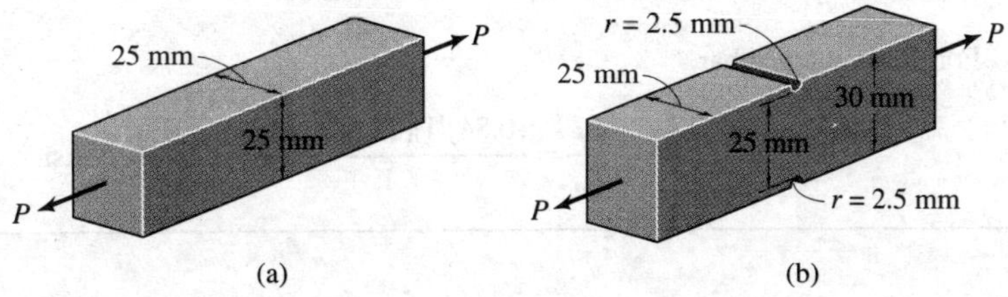

Figure 7.8 Tensile-loaded bar. (a) Unnotched; (b) notched.

From Eqs. (7.12) and (7.13) the slope and intercept are

$$b_s = -\frac{1}{3}\log\left(\frac{S'_l}{S'_e}\right) = -\frac{1}{3}\log\left(\frac{42.75}{25.65}\right) = -0.07400$$

$$\overline{C} = \log\left[\frac{(S'_l)^2}{S'_e}\right] = \log\left[(1.667)(42.75)(10^3)\right] = 4.853$$

From Fig. 7.7(a) for machined surfaces the surface finish factor $k_f = 0.84$. For axial loading the size factor $k_s = 1$. For the unnotched bar the fatigue stress concentration factor $K_f = 1$. For the notched bar with a groove Fig. 6.4 is to be used. From Fig. 7.8

$$H = 30 \text{ mm} \qquad r = 2.5 \text{ mm} \qquad b = 25 \text{ mm} \qquad \text{and} \qquad h = 25 \text{ mm}$$

$$\therefore \frac{H}{h} = \frac{30}{25} = 1.2 \qquad \text{and} \qquad \frac{r}{h} = \frac{2.5}{25} = 0.10$$

From Fig. 6.4(a) for $H/h = 1.2$ and $r/h = 0.1$ the stress concentration factor K_c is 2.35. From Fig. 7.6 for $S_u = 57$ ksi and $r = 2.5$ mm, the notch sensitivity is 0.73. From Eq. (7.19) the fatigue stress concentration can be expressed as

$$K_f = 1 + (K_c - 1)q_n = 1 + (2.35 - 1)(0.73) = 1.986$$

From Eq. (7.16) the modified endurance limit for the notched bar is

$$\frac{S_e}{K_f} = \frac{k_f k_s S'_e}{K_f} = \frac{(0.84)(1)(25.65)(10^3)}{1.986} \text{ psi} = 10.85 \text{ ksi}$$

For the unnotched bar the modified endurance limit is

$$\frac{S_e}{K_f} = \frac{(0.84)(1)(25.65)(10^3)}{1} \text{ psi} = 21.55 \text{ ksi}$$

Thus, even though the notched and unnotched bars have the same height, width, and length, the presence of the notch decreases the modified endurance limit by one-half.

The fatigue strength at 10^4 cycles from Eq. (7.14) is

$$S'_f = 10^{\overline{C}}(N'_t)^{b_s} = 10^{4.853}(10^4)^{-0.074} = 10^{4.557} \text{ psi} = 36.06 \text{ ksi}$$

Thus, for the notched bar

$$\frac{S_f}{K_f} = \frac{k_f k_s S'_f}{K_f} = \frac{(0.84)(1)(36.06)(10^3)}{1.986} \text{ psi} = 15.25 \text{ ksi}$$

For the unnotched bar

$$\frac{S_f}{K_f} = \frac{(0.84)(1)(36.06)(10^3)}{1} \text{ psi} = 30.29 \text{ ksi}$$

7.8 CUMULATIVE DAMAGE

In constructing the solid curve in Fig. 7.4(c) it was assumed that the cyclic variation was completely reversed ($\sigma_m = 0$). Furthermore, for any stress level between the strengths S_l' and S_e', say S_1', the maximum stress level in the completely reversed variation was kept constant until failure occurred at N_1' cycles. Operating at stress amplitude S_1' for a number of cycles $n_1' < N_1'$ produced a smaller damage fraction. Because cyclic variations are not constant in practice, engineers must deal with several different levels of completely reversed stress cycles. Operating over stress levels between S_l' and S_e', say S_i', at a number of cycles $n_i' < N_i'$ resulted in the damage fraction n_i'/N_i'. When the damage fraction due to different levels of stress exceeds unity, failure is predicted. Thus, *failure is predicted if*

$$\frac{n_1'}{N_1'} + \frac{n_2'}{N_2'} + \ldots \geq 1 \qquad \textbf{(7.24)}$$

This formulation is frequently called the **linear damage rule,** since it states that the damage at any stress level is directly proportional to the number of cycles (when each cycle does the same amount of damage). The rule also assumes that the stress sequence does not matter and that the rate of damage accumulation at a stress level is independent of the stress history. Despite these shortcomings the linear damage rule (sometimes called **Miner's rule**) remains popular, largely because it is so simple.

If N_t' is the total number of cycles to failure when there are different cyclic patterns (all of which are completely reversed), the ratio of the number of cycles at a specific stress level to the total number of cycles to failure is

$$\alpha_i = \frac{n_i'}{N_t'} \quad \text{or} \quad n_i' = \alpha_i N_t' \qquad \textbf{(7.25)}$$

Substituting Eq. (7.25) into Eq. (7.24) gives that failure will occur if

$$\sum \frac{\alpha_i}{N_i'} \geq \frac{1}{N_t'} \qquad \textbf{(7.26)}$$

Example 7.5

GIVEN For the unnotched bar in Example 7.4, the fatigue stress will be 25 ksi for 20% of the time, 30 ksi for 30%, 35 ksi for 40%, and 40 ksi for 10%.

FIND The number of cycles until cumulative failure.

Solution

Note that $S_1' = 25$ ksi is less than S_e', which was found to be 25.65 ksi. Therefore, $N_1' = \infty$, implying that at this strength level failure will not occur. From Eq. (7.14) for the other three fatigue strength levels

$$N_2' = \left(S_f' 10^{-\bar{C}}\right)^{1/b_s} = \left[(30\ 000)(10)^{-4.853}\right]^{-1/0.074} = 1.201 \times 10^5 \text{ cycles}$$

$$N_3' = \left[(35\ 000)(10)^{-4.853}\right]^{-1/0.074} = 1.495 \times 10^4 \text{ cycles}$$

$$N_4' = \left[(40\ 000)(10)^{-4.853}\right]^{-1/0.074} = 2.461 \times 10^3 \text{ cycles}$$

Making use of Eq. (7.26) gives

$$\frac{\alpha_1}{N_1'}+\frac{\alpha_2}{N_2'}+\frac{\alpha_3}{N_3'}+\frac{\alpha_4}{N_4'}=\frac{1}{N_t'}$$

$$\frac{0.2}{\infty}+\frac{0.3}{1.201\times10^5}+\frac{0.4}{1.495\times10^4}+\frac{0.1}{2.461\times10^3}=\frac{1}{N_t'}$$

$$\frac{1}{N_t'}=0+2.498\times10^{-6}+2.676\times10^{-5}+4.063\times10^{-5}$$

$$N_t'=14\ 310\ \text{cycles}$$

The cumulative number of cycles to failure is 14 310.

7.9 Influence of Nonzero Mean Stress

Other than in classifying cyclic behavior, completely reversed ($\sigma_m = 0$) stress cycles have been assumed. Many machine elements involve fluctuating stresses about a nonzero mean. The experimental apparatus used to generate the results shown in Fig. 7.4 could not apply mean and alternating stresses. Because data are not available, the influence of nonzero mean stress must be estimated by using one of several empirical relationships that determine failure at a given life when both alternating and mean stresses are nonzero.

7.9.1 Ductile Materials

Figure 7.9 shows how four empirical relationships estimate the influence of nonzero mean stress on fatigue life for ductile materials tensilely loaded. On the ordinate the yield strength S_{yt} is plotted, reminding us that yielding rather than fatigue might be the failure criterion. On the abscissa the yield strength in tension S_{yt} and the ultimate strength in tension S_{ut} are plotted. Failure is predicted for values of σ_a and σ_m above the different curves.

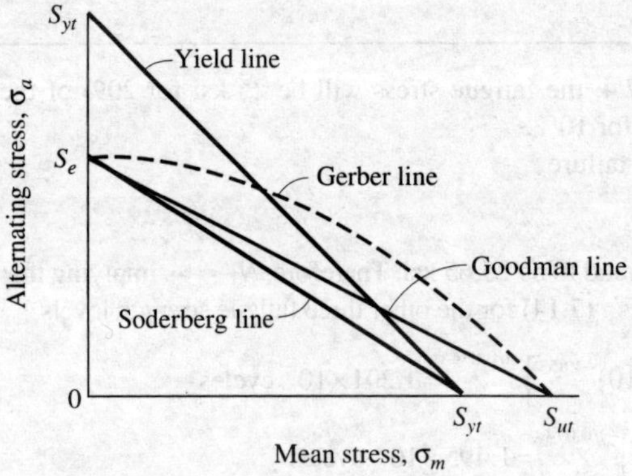

Figure 7.9 Influence of nonzero mean stress on fatigue life for tensile loading as estimated by four empirical relationships.

7.9.1.1 GERBER LINE The **Gerber line** is sometimes called the Gerber parabolic relationship because the equation is

$$\frac{K_f n_s \sigma_a}{S_e} + \left(\frac{n_s \sigma_m}{S_{ut}} \right)^2 = 1 \qquad \textbf{(7.27)}$$

where

S_e = modified endurance limit, Pa
S_{ut} = ultimate strength in tension, Pa
n_s = safety factor
σ_a = alternating stress, Pa
σ_m = mean stress, Pa
K_f = fatigue stress concentration factor

This line passes through the central portion of the experimental failure points and hence should be the best predictor of failure. The parabolic nature of Eq. (7.27) poses problems for the practical implementation of this equation.

7.9.1.2 GOODMAN LINE The **Goodman line** proposes connecting the modified endurance limit on the alternating stress axis with the ultimate strength in tension on the mean stress axis in Fig. 7.9 by a straight line, or

$$\frac{K_f \sigma_a}{S_e} + \frac{\sigma_m}{S_{ut}} = \frac{1}{n_s} \qquad \textbf{(7.28)}$$

Note the linearization of Eq. (7.28) relative to Eq. (7.27). Equation (7.28) fits experimental data reasonably well and is simpler to use than Eq. (7.27). The starting and ending points for the Goodman and Gerber lines are the same in Fig. 7.9, but between these points the Goodman line is linear and the Gerber line is parabolic.

Example 7.6

GIVEN A straight, circular rotating beam with a 30-mm diameter and a 1-m length has an axial load of 30 000 N applied at the end and a stationary radial load of 400 N. The beam surface is cold drawn and the material is AISI 1040 steel. Assume that $k_s = k_r = k_t = k_m = 1$.

FIND The safety factor for infinite life by using the Goodman line.

Solution

From Table A.1 for AISI 1040 steel S_u = 520 MPa. From Fig. 7.7(a) for the cold-drawn process and S_u = 520 MPa the surface finish factor is 0.78. From Eq. (7.16) the modified endurance limit while making use of Eq. (7.7) is

$$S_e = k_f k_s k_r k_t k_m S_e' = (0.78)(1)(1)(1)(1)(0.45)(520) = 182.5 \text{ MPa}$$

The bending stress from Eq. (4.48) gives the alternating stress that the beam experiences as

$$\sigma_a = \frac{Mc}{I} = \frac{(64)(400)(1)(0.03/2)}{\pi(0.03)^4} \text{ Pa} = 150.9 \text{ MPa}$$

The mean stress due to the axial load is

$$\sigma_m = \frac{P_a}{A} = \frac{(30\ 000)(4)}{\pi(0.03)^2}\ \text{Pa} = 42.44\ \text{MPa}$$

For an unnotched beam $K_f = 1$. From Eq. (7.28)

$$\frac{K_f\sigma_a}{S_e} + \frac{\sigma_m}{520} = \frac{1}{n_s}$$

$$\therefore \frac{(1)(150.9)}{182.5} + \frac{42.44}{520} = \frac{1}{n_s}$$

$$n_s = \frac{1}{0.9084} = 1.101$$

Using the Goodman line the safety factor for infinite life is 1.101.

7.9.1.3 SODERBERG LINE The **Soderberg line** is conservative and is given as

$$\frac{K_f\sigma_a}{S_e} + \frac{\sigma_m}{S_{yt}} = \frac{1}{n_s} \tag{7.29}$$

Note from Fig. 7.9 and Eqs. (7.28) and (7.29) that the ultimate strength in the Goodman relationship has been replaced with the yield strength in the Soderberg relationship. The Soderberg relationship is usually not recommended.

7.9.1.4 YIELD LINE To complete the possibilities, the **yield line** is given. It is used to define yielding on the first cycle, or

$$\frac{\sigma_a}{S_{yt}} + \frac{\sigma_m}{S_{yt}} = \frac{1}{n_s} \tag{7.30}$$

This completes the description of the theories presented in Fig. 7.9.

7.9.1.5 MODIFIED GOODMAN DIAGRAM The Goodman relationship given in Eq. (7.28) is modified by combining fatigue failure with failure by yield. The complete **modified Goodman diagram** is shown in Fig. 7.10. Thus, all points inside a modified Goodman diagram ABCDEFGH correspond to fluctuating stresses that should cause neither fatigue failure nor yielding. The word "complete" is used to indicate that the diagram is valid for tension and compression. The word "modified" designates that the Goodman line shown in Fig. 7.9 has been modified in Fig. 7.10. That is, in Fig. 7.9 the Goodman line extends from the endurance limit on the alternating stress ordinate to the ultimate strength on the mean stress abscissa. In the modified Goodman diagram in Fig. 7.10 the Goodman line AB is modified such that for stresses larger than the yield strength the yield line BC is used. Thus, the modified Goodman diagram combines fatigue criteria as represented by the Goodman line and yield criteria as represented by the yield line. Note in Fig. 7.10 that lines AB, DE, EF, and HA are Goodman lines and that lines BC, CD, FG, and GH are yield lines. The static load is represented by line CG. Table 7.5 gives the equations and range of applicability for the construction of the Goodman and yield lines of the complete modified Goodman diagram. Note that when the ultimate and yield strengths are known for a

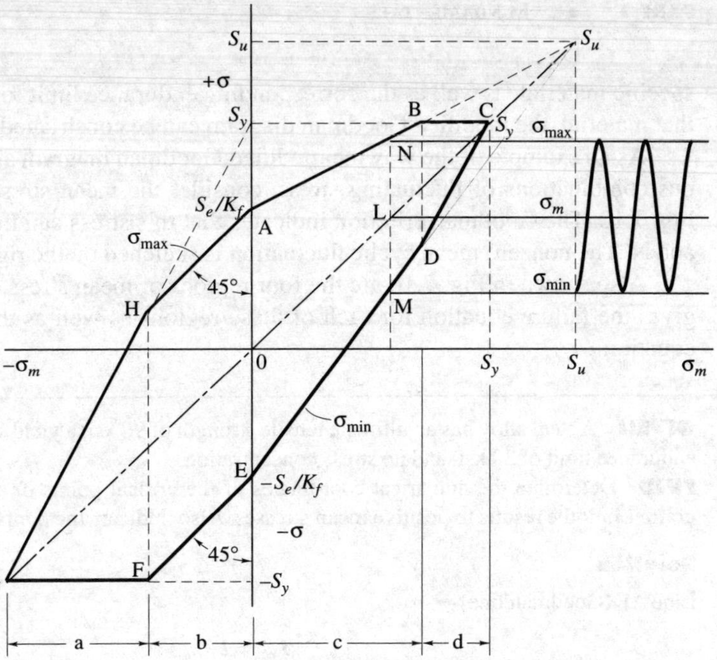

Figure 7.10 Complete modified Goodman diagram, plotting stress as ordinate and mean stress as abscissa.

Table 7.5 Equations and range of applicability for construction of complete modified Goodman diagram.

Line	Equation	Range
AB	$\sigma_{max} = \dfrac{S_e}{K_f} + \sigma_m\left(1 - \dfrac{S_e}{S_u K_f}\right)$	$0 \le \sigma_m \le \dfrac{S_y - S_e/K_f}{1 - \dfrac{S_e}{K_f S_u}}$
BC	$\sigma_{max} = S_y$	$\dfrac{S_y - \dfrac{S_e}{K_f}}{1 - \dfrac{S_e}{K_f S_u}} \le \sigma_m \le S_y$
CD	$\sigma_{min} = 2\sigma_m - S_y$	$\dfrac{S_y - \dfrac{S_e}{K_f}}{1 - \dfrac{S_e}{K_f S_u}} \le \sigma_m \le S_y$
DE	$\sigma_{min} = \left(1 + \dfrac{S_e}{K_f S_u}\right)\sigma_m - S_e/K_f$	$0 \le \sigma_m \le \dfrac{S_y - \dfrac{S_e}{K_f}}{1 - \dfrac{S_e}{K_f S_u}}$
EF	$\sigma_{min} = \sigma_m - \dfrac{S_e}{K_f}$	$\dfrac{S_e}{K_f} - S_y \le \sigma_m \le 0$
FG	$\sigma_{min} = -S_y$	$-S_y \le \sigma_m \le \dfrac{S_e}{K_f} - S_y$
GH	$\sigma_{max} = 2\sigma_m + S_y$	$-S_y \le \sigma_m \le \dfrac{S_e}{K_f} - S_y$
HA	$\sigma_{max} = \sigma_m + \dfrac{S_e}{K_f}$	$\dfrac{S_e}{K_f} - S_y \le \sigma_m \le 0$

specific material, as well as the corresponding endurance limit for a particular part made of that material, the modified Goodman diagram can be constructed.

As an example of the way the modified Goodman diagram aids in visualizing the various combinations of fluctuating stress, consider the mean stress indicated by point L of Fig. 7.10. The Goodman criterion indicates that this stress can fluctuate between points M and N. The nonzero mean cyclic fluctuation is sketched on the right of the figure.

Also shown in Fig. 7.10 are the four regions of mean stress on the abscissa. Table 7.6 gives the failure equation for each of these regions as well as the validity limits for each equation.

Tutorial 3

Example 7.7

GIVEN A steel alloy has an ultimate tensile strength of 90 ksi, a yield strength of 60 ksi, a modified endurance limit of 30 ksi, and no stress concentration.

FIND Determine the numerical coordinates of the critical points on the modified Goodman diagram. Limit the results to positive mean stresses. Also, indicate the nonfailure region of the diagram.

Solution

Line AF (Goodman line)

$$\sigma_{\max} = b_s \sigma_m + \overline{C} \qquad \textbf{(a)}$$

where

b_s = slope

$\overline{C}$ = intercept

But

$$\sigma_m = 0 \qquad \sigma_{\max} = 30 \text{ ksi} \qquad \therefore \overline{C} = 30 \text{ ksi}$$

$$\therefore \sigma_{\max} = b_s \sigma_m + 30 \text{ ksi} \qquad \textbf{(b)}$$

But

$$\sigma_{\max} = 90 \text{ ksi when } \sigma_m = 90 \text{ ksi}$$

Substituting this into Eq. (b) gives $b_s = 2/3$. Therefore, line AB is given as

$$\sigma_{\max} = (2/3)\sigma_m + 30 \text{ ksi}$$

$$\therefore \text{Point A} = (0, 30 \text{ ksi}) \qquad \textbf{(c)}$$

$$\text{Point B, } \sigma_{\max} = 60 \text{ ksi}$$

From Eq. (c)

$$60 \text{ ksi} = (2/3)\sigma_m + 30 \text{ ksi}$$

$$\therefore \sigma_m = 45 \text{ ksi}$$

$$\text{Point B} = (45 \text{ ksi}, 60 \text{ ksi})$$

$$\text{Point C} = (60 \text{ ksi}, 60 \text{ ksi})$$

Line EF (Goodman line)

$$\sigma_{\min} = b_s \sigma_m + \overline{C} \qquad \textbf{(d)}$$

Table 7.6 Failure equations and validity limits of equations for four regions of complete modified Goodman relationship

Region in Fig. 7.10	Failure equation	Validity limits of equation
a	$\sigma_{max} - 2\sigma_m = S_y / n_s$	$-S_y \le \sigma_m \le \frac{S_e}{K_f} - S_y$
b	$\sigma_{max} - \sigma_m = \frac{S_e}{n_s K_f}$	$\frac{S_e}{K_f} - S_y \le \sigma_m \le 0$
c	$\sigma_{max} + \sigma_m\left(\frac{S_e}{K_f S_u} - 1\right) = \frac{S_e}{n_s K_f}$	$0 \le \sigma_m \le \frac{S_y - \frac{S_e}{K_f}}{1 - \frac{S_e}{K_f S_u}}$
d	$\sigma_{max} = \frac{S_y}{n_s}$	$\frac{S_y - \frac{S_e}{K_f}}{1 - \frac{S_e}{K_f S_u}} \le \sigma_m \le S_y$

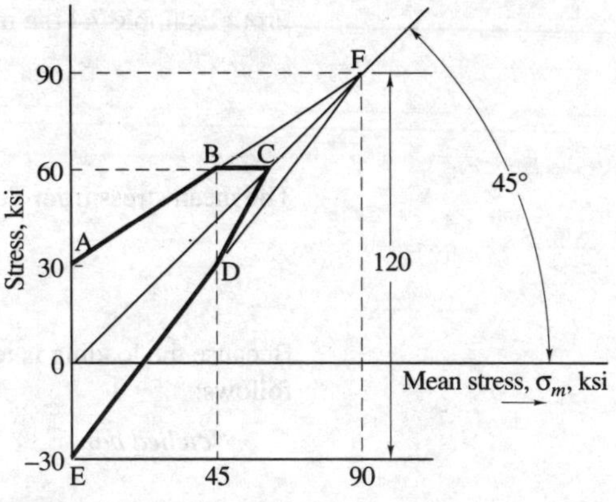

Figure 7.11 Modified Goodman diagram for Example 7.7.

Given that

$$\sigma_m = 0 \qquad \sigma_{min} = -30 \text{ ksi}$$

$$\therefore \overline{C} = -30 \text{ ksi} \tag{e}$$

and

$$\sigma_{min} = b_s \sigma_m - 30 \text{ ksi}$$

Given that

$$\sigma_m = 90 \text{ ksi} \qquad \sigma_{min} = 90 \text{ ksi}$$

$$\therefore 90 \text{ ksi} = b_s 90 \text{ ksi} - 30 \text{ ksi}$$

$$\therefore b_s = 4/3 \tag{f}$$

Substituting Eqs. (e) and (f) into Eq. (d) gives

$$\sigma_{min} = (4/3)\sigma_m - 30 \text{ ksi}$$

$$\therefore \text{ Point E} = (0, -30 \text{ ksi})$$

$$\text{Point D} = (45 \text{ ksi}, 30 \text{ ksi})$$

Figure 7.11 shows a plot of these points as well as the shaded nonfailure region.

Example 7.8

GIVEN For the same two bars given in Example 7.4 the load is cyclically varying between 2 and 10 ksi.

FIND Using the modified Goodman relationship, determine the safety factor guarding against fatigue failure.

Solution

From Table A.1 for low-carbon steel (AISI 1020)

$$S_u = 57 \text{ ksi} \qquad \text{and} \qquad S_y = 43 \text{ ksi} \tag{a}$$

From Example 7.4 the modified endurance limit is

$$S_e/K_f = 10.85 \text{ ksi} \qquad \text{for notched bar} \tag{b}$$

$$S_e/K_f = 21.55 \text{ ksi} \qquad \text{for unnotched bar} \tag{c}$$

The mean stress from Eq. (7.1) is

$$\sigma_m = \frac{\sigma_{\max} + \sigma_{\min}}{2} = 6 \text{ ksi} \tag{d}$$

Because the loading is tensile, the region in Fig. 7.10 is either *c* or *d*. To establish which, evaluate as follows:

Notched bar

$$\frac{S_y - S_e/K_f}{1 - S_e/S_u K_f} = \frac{(43 - 10.85)10^3}{1 - 10.85/57} \text{ psi} = 39.71 \text{ ksi}$$

$$\therefore 0 \le \sigma_m \le \frac{S_y - S_e/K_f}{1 - S_e/S_u K_f} \qquad \text{or} \qquad 0 \le 6 \text{ ksi} \le 39.71 \text{ ksi}$$

$$\therefore \text{ Region c}$$

Unnotched bar

$$\frac{S_y - S_e/K_f}{1 - S_e/S_u K_f} = \frac{(43 - 21.55)10^3}{1 - 21.55/57} \text{ psi} = 34.491 \text{ ksi}$$

$$\therefore 0 \le \sigma_m \le \frac{S_y - S_e/K_f}{1 - S_e/S_u K_f} \qquad \text{or} \qquad 0 \le 6 \text{ ksi} \le 34.49 \text{ ksi}$$

$$\therefore \text{ Region c}$$

From Table 7.5 for region *c* failure will occur if

$$\sigma_{\max} - \sigma_m = \frac{S_e}{n_s K_f}\left(1 - \frac{\sigma_m}{S_u}\right)$$

Notched bar

$$10 - 6 = \frac{10.85}{n_s}\left(1 - \frac{6}{57}\right)$$

$$\therefore n_s = 2.427$$

Unnotched bar

$$10 - 6 = \frac{21.55}{n_s}\left(1 - \frac{6}{57}\right)$$

$$\therefore n_s = 4.820$$

The notched bar will fail before the unnotched bar.

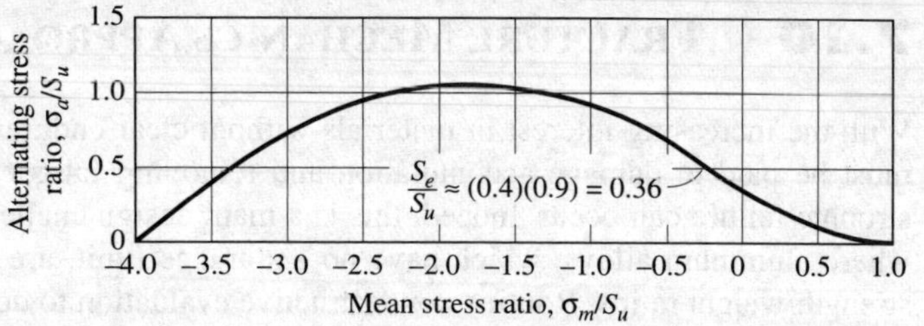

Figure 7.12 Alternating stress ratio as function of mean stress ratio for axially loaded cast iron.

7.9.2 BRITTLE MATERIALS

Figure 7.12 shows the alternating stress ratio as a function of the mean stress ratio for axially loaded cast iron. The figure is skewed, since the compressive strength is typically several times greater than the tensile strength. Until recently, the use of brittle materials in a fatigue environment has been limited to gray cast iron in compression. Now, however, carbon fibers and ceramics have had significant usage in fatigue environments.

In Fig. 7.12 the dimensionalization of the alternating and mean stresses is with respect to the ultimate strength rather than to the yield strength as for ductile materials. Also, the compressive mean stress permits large increases in alternating stress.

For *brittle* materials a stress raiser increases the likelihood of failure under either steady or alternating stresses, and it is customary to apply a stress concentration factor to both. Thus, designers apply the fatigue stress concentration factor K_f to the alternating component of stress for ductile materials but apply the stress concentration factor K_c to both the alternating and mean components of stress for brittle materials.

For a single normal stress in brittle materials the equation for the safety factor with a steady stress σ_m and with the ultimate tensile strength S_{ut} as the basis of failure is

$$n_s = \frac{S_{ut}}{K_c \sigma_m} \tag{7.31}$$

With an alternating stress σ_a and a modified endurance limit S_e

$$n_s = \frac{S_e}{K_c \sigma_a} \tag{7.32}$$

For a single shear stress on a brittle component and with a *steady* shear stress τ_m, the safety factor is

$$n_s = \frac{S_{ut}}{K_{cs} \tau_m \left[1 + \dfrac{S_{ut}}{S_{uc}}\right]} \tag{7.33}$$

With an alternating shear stress τ_a and a modified endurance limit S_e

$$n_s = \frac{S_e}{K_{cs} \tau_a \left[1 + \dfrac{S_{ut}}{S_{uc}}\right]} \tag{7.34}$$

7.10 FRACTURE MECHANICS APPROACH TO FATIGUE

With the increasing interest in materials without clear endurance limits, special attention must be paid to damage accumulation and removing fatigued components before catastrophic failure can occur. Indeed, this is a main design challenge in the aircraft industry, where aluminum alloys, which have no endurance limit, are used because of their high strength/weight ratios. Routine nondestructive evaluation to determine the size of flaws in stress-bearing members is conducted to identify and remove suspect components.

Paris et al. (1961) hypothesized that crack growth in a cyclic loading should follow the rule

$$\frac{dl_c}{dN} = C(\Delta K)^m \tag{7.35}$$

where dl_c/dN is the change in crack length per load cycle (l_c is the crack length and N is the number of stress cycles), C and m are empirical constants, and ΔK is the stress intensity range defined as

$$\Delta K = K_{\max} - K_{\min} \tag{7.36}$$

where $K_{\max}$ and $K_{\min}$ are the maximum and minimum stress intensity factors, respectively, around a crack during a loading cycle. For a center-cracked plate with crack length $2l_c$

$$K_{\max} = Y\sigma_{\max}\sqrt{\pi l_c} \tag{7.37}$$

where $\sigma_{\max}$ is the maximum far-field stress and Y is a correction factor to account for finite plate sizes. Figure 7.13 gives values of Y. Equation (7.35) is known as the **Paris power law** and is the most widely used equation in fracture mechanics approaches to fatigue problems. Suresh (1991) derives the life of a component based on the Paris power law as

$$N = \frac{2}{(m-2)CY^m(\Delta\sigma)^m \pi^{m/2}} \left\{ \frac{1}{(l_{c0})^{(m-2)/2}} - \frac{1}{(l_{cf})^{(m-2)/2}} \right\} \tag{7.38}$$

unless $m = 2$, when the fatigue life is

$$N = \frac{1}{CY^2(\Delta\sigma)^2 \pi} \ln \frac{l_{cf}}{l_{c0}} \tag{7.39}$$

where

N = fatigue life in cycles
C and m = material constants
Y = correction factor to account for finite plate sizes
$\Delta\sigma$ = range of far-field stresses to which component is subjected
l_{c0} = initial crack size
l_{cf} = critical crack size based on fracture mechanics

<table>
<tr><th>Geometry</th><th>Plate size correction factor, Y</th></tr>
<tr><td>Single-edge-cracked tension specimen
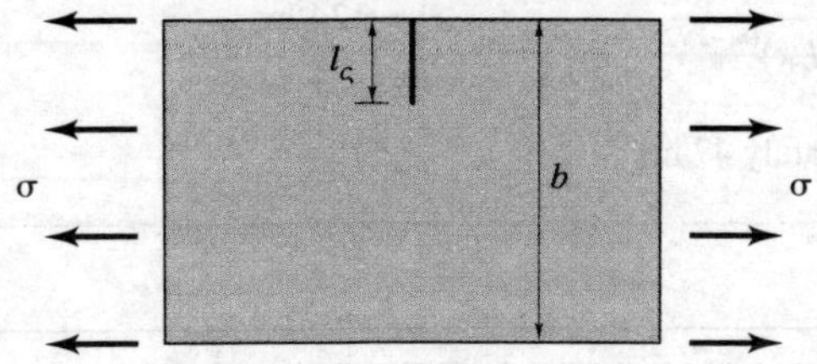
</td><td>$1.99 - 0.41\left(\frac{l_c}{b}\right) + 18.7\left(\frac{l_c}{b}\right)^2 - 38.48\left(\frac{l_c}{b}\right)^3 + 53.85\left(\frac{l_c}{b}\right)^4$</td></tr>
<tr><td>Center-cracked tension specimen
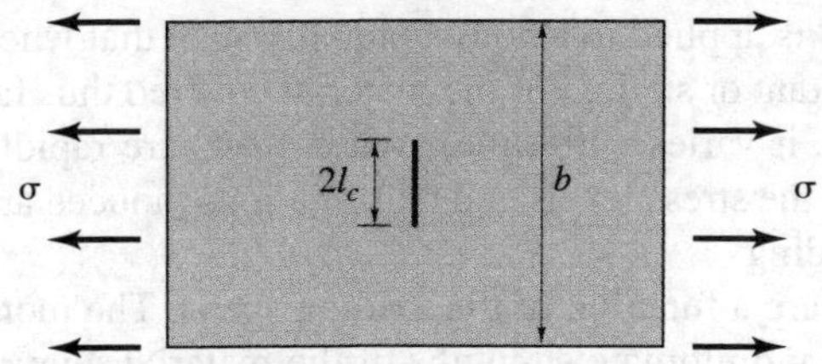
</td><td>$\left[\sec\left(\frac{\pi l_c}{b}\right)\right]^{1/2}$</td></tr>
<tr><td>Double-edge-cracked tension specimen
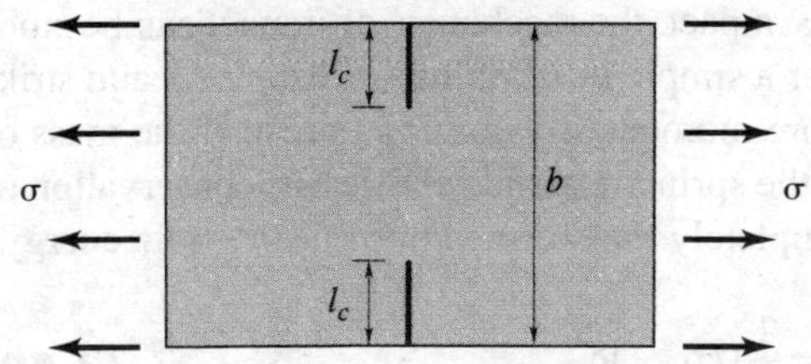
</td><td>$1.99 + 0.76\left(\frac{l_c}{b}\right) - 8.48\left(\frac{l_c}{b}\right)^2 + 27.36\left(\frac{l_c}{b}\right)^3$</td></tr>
</table>

Figure 7.13 Correction factor Y to compensate for plate width in fracture mechanics approach to fatigue crack propagation. [S. Suresh, *Fatigue in Materials*, Cambridge University Press, 1991.]

Example 7.9

GIVEN An aluminum alloy aircraft component in the form of a 100-mm-wide plate sees a 100-MPa stress during pressurization of the aircraft cabin. Superimposed on this stress is a fluctuation arising from vibration, which has an amplitude of 10 MPa and a frequency of 45 Hz. Nondestructive crack detection techniques do not detect any flaws, but the smallest detectable flaw is 0.2 mm.

FIND Determine the minimum expected life of the component. If upon reinspection a centered 1.1-mm-long crack is found, what is the expected life from the time of inspection? Use a fracture toughness of 29 MPa $\sqrt{\text{m}}$ and a fracture stress of 260 MPa and assume $m = 2.5$ and $C = 6.9 \times 10^{-12}$ m/cycle for $\Delta\sigma$ in megapascals.

Solution

The critical crack length l_{cf} is given by Eq. (6.4) and is found to be 1.60 mm. A worst-case scenario would occur if the largest undetectable flaw resulting in the largest value of Y were located in the geometry. This flaw occurs for the double-edge-cracked tension specimen, where Y equals 2.0. Therefore, the life is found from Eq. (7.38) as

$$N = \frac{2}{(m-2)CY^m(\Delta\sigma)^m \pi^{m/2}} \left\{ \frac{1}{(l_{c0})^{(m-2)/2}} - \frac{1}{(l_{cf})^{(m-2)/2}} \right\} = 324 \text{ hr}$$

If $l_{c0} = 1.1$ mm, the life until fracture is approximately 47 hr.

7.11 LINEAR IMPACT STRESSES AND DEFORMATIONS

This chapter has thus far focused on cyclic load variation. This section focuses on **impact loading.** Throughout most of this text a load is applied to a body gradually such that when the load reaches a maximum it remains constant or static. For the material covered thus far in this chapter the loading is **dynamic** (i.e., it varies with time). When loads are rapidly applied to a body as in impact loading, often the stress levels and deformations induced are much larger than those in static or cyclic loading.

The physical properties of the material are a function of the loading speed. The more rapid the loading, the higher both the yield and ultimate strengths of the material. Figure 7.14 shows the variation of the mechanical properties with loading speed for a typical mild steel. For average strain rates from 10^{-1} to 10^3 s^{-1} the yield strength increases significantly.

If no energy is assumed to be lost during impact, the mechanics of impact can be studied by using conservation of energy. Consider a simple block falling a distance h and striking a spring compressed a distance δ_{max} before momentarily coming to rest. If the mass of the spring is neglected and it is assumed that the spring responds elastically, conservation of energy requires that the kinetic energy be completely transformed into elastic strain energy:

$$W(h + \delta_{max}) = \frac{1}{2}(k\delta_{max})\delta_{max} \qquad \textbf{(7.40)}$$

where

k = spring constant, N/m

W = weight of block, N

Equation (7.40) can be expressed as a quadratic equation:

$$\delta_{max}^2 - \frac{2W}{k}\delta_{max} - 2\left(\frac{W}{k}\right)h = 0 \qquad \textbf{(7.41)}$$

Solving for δ_{max} gives the maximum root as

$$\delta_{max} = \frac{W}{k} + \sqrt{\left(\frac{W}{k}\right)^2 + 2\left(\frac{W}{k}\right)h} \qquad \textbf{(7.42)}$$

If the weight is applied statically (or gradually) to the spring, the static displacement is

$$\delta_{st} = \frac{W}{k} \qquad \textbf{(7.43)}$$

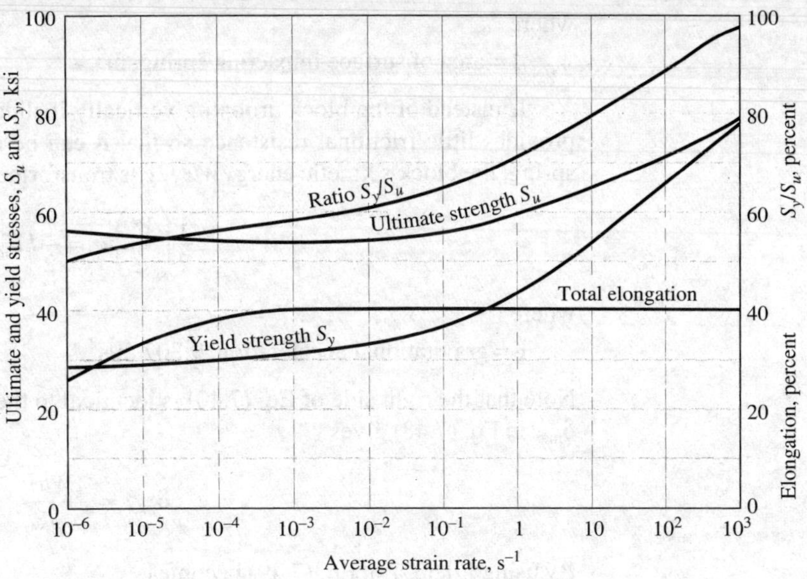

Figure 7.14 Mechanical properties of mild steel at room temperature as function of average strain rate. [M.J. Manjoine, "Influence of Rate of Strain and Temperature on Yield Stresses of Mild Steel," *Journal of Applied Mechanics*, Vol. 66, pp. A-22–A-218. Reprinted with permission by The American Society of Mechanical Engineers International.]

Substituting Eq. (7.43) into Eq. (7.42) gives

$$\delta_{max} = \delta_{st} + \sqrt{(\delta_{st})^2 + 2\delta_{st}h}$$

or

$$\delta_{max} = \delta_{st}\left(1 + \sqrt{1 + \frac{2h}{\delta_{st}}}\right) \qquad \textbf{(7.44)}$$

From the maximum displacement in Eq. (7.44) the maximum force is

$$P_{max} = k\delta_{max} \qquad \textbf{(7.45)}$$

Recall that in dropping the block from some distance h the maximum force P_{max} on impact is instantaneous. The block will continue to vibrate until the motion dampens and the block assumes the static position. All this assumes that when the block first makes contact with the spring, the block does not rebound from the spring (does not separate from the spring). Making use of the spring constant k in Eqs. (7.43) and (7.45) relates the static and dynamic effects.

The impact factor can be expressed as

$$I_m = \frac{\delta_{max}}{\delta_{st}} = \frac{P_{max}}{W} = 1 + \sqrt{1 + \frac{2h}{\delta_{st}}} \qquad \textbf{(7.46)}$$

Once the impact factor is known, the impact load, stresses, and deflections can be calculated. The impact stress is

$$\sigma = \frac{P_{max}}{A} \qquad \textbf{(7.47)}$$

where

A = area of surface impacting spring, m^2

If instead of the block dropping vertically it slides with a velocity u on a surface that provides little frictional resistance so that it can be neglected and the block impacts the spring, the block's kinetic energy $Wu^2/2g$ is transformed into stored energy in the spring, or

$$\frac{1}{2}\left(\frac{W}{g}\right)u^2 = \frac{1}{2}k\delta_{max}^2 \qquad \textbf{(7.48)}$$

where

g = gravitational acceleration, 9.807 m/s^2

Note that the right side of Eq. (7.40) is identical to the right side of Eq. (7.48). Solving for δ_{max} in Eq. (7.48) gives

$$\delta_{max} = \sqrt{\frac{Wu^2}{gk}} \qquad \textbf{(7.49)}$$

By using Eq. (7.43), Eq. (7.49) becomes

$$\delta_{max} = \sqrt{\frac{\delta_{st}u^2}{g}} \qquad \textbf{(7.50)}$$

In both situations the moving body (the block) is assumed to be rigid, and the stationary body (the spring) is assumed to be deformable. The material is assumed to behave in a linear-elastic manner. Thus, whether a block falls a distance h and impacts on a spring or a block moves at a velocity u and impacts on a spring, the formulation for the deformation, the impact force, or the impact stress can be determined.

Example 7.10

GIVEN A diver jumps up 2 ft on the free end of a diving board before diving into the water. Figure 7.15 shows a sketch of the diver and the dimensions of the diving board. The supported end of the diving board is fixed. The modulus of elasticity is 10.3×10^6 psi and the yield strength is 30 ksi. The weight of the diver is 200 lbf.

FIND The safety factor based on yielding for impact loading.

Solution

Equations (7.42) and (7.43) should be used to determine the maximum deflection at impact at the end of the diving board. The spring rate is not given and will need to be determined. The forces acting are shown in Fig. 7.15(c). From Eq. (6.3)

$$\frac{d^2y}{dx^2} = -\frac{M}{EI} = -\frac{Px}{EI}$$

Integrating this equation gives

$$EI\frac{dy}{dx} = -\frac{Px^2}{2} + C_1$$

where

C_1 = integration constant

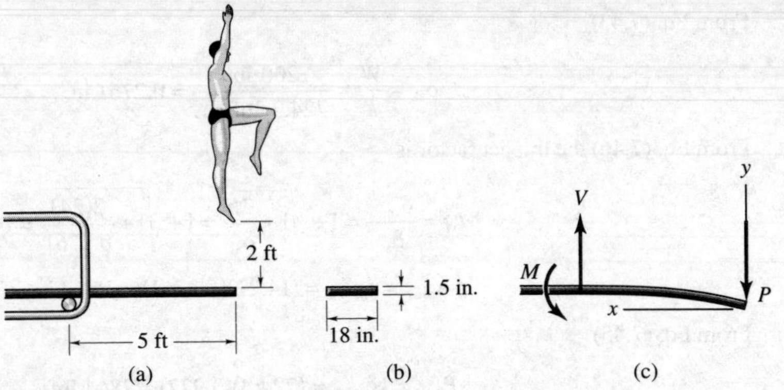

Figure 7.15 Diver impacting diving board, used in Example 7.5.
(a) Side view; (b) front view; (c) side view showing forces and coordinates.

Integrating again gives

$$EIy = -\frac{Px^3}{6} + C_1 x + C_2$$

where

C_2 = integration constant

The boundary conditions are

1. $x = l,\ y' = 0 \quad \rightarrow C_1 = \frac{Pl^2}{2}$

2. $x = l,\ y = 0 \quad \rightarrow C_2 = -\frac{Pl^3}{3}$

The deflection at the end of the board is of interest when $x = 0$.

$$\therefore EIy = -\frac{Px^3}{6} + \frac{Pl^2 x}{2} - \frac{Pl^3}{3}$$

or

$$\frac{6EIy}{P} = -x^3 + 3l^2 x - 2l^3$$

$$\therefore \delta = y\big|_{x=0} = -\frac{Pl^3}{3EI}$$

The spring constant is

$$k = -P/\delta = 3EI/l^3$$

where

$$I = \text{Area moment of inertia} = \frac{bh^3}{12} = \frac{18(1.5)^3}{12} = 5.063 \text{ in.}^4$$

$$l = \text{Length} = 5 \text{ ft} = 60 \text{ in.}$$

$$E = \text{Modulus of elasticity} = 10.3 \times 10^6 \text{ psi}$$

$$k = \frac{3(10.3)(10^6)(5.063)}{(60)^3} = 724.3 \text{ lbf/in.}$$

From Eq. (7.43)

$$\delta_{st} = \frac{W}{k} = \frac{200 \text{ lbf}}{724.3 \text{ lbf/in.}} = 0.2761 \text{ in.}$$

From Eq. (7.46) the impact factor is

$$I_m = \frac{\delta_{max}}{\delta_{st}} = 1 + \sqrt{1 + \frac{2h}{\delta_{st}}} = 1 + \sqrt{1 + \frac{2(24)}{0.2761}} = 14.22$$

$$\therefore \delta_{max} = I_m \delta_{st} = (14.22)(0.2761) = 3.927 \text{ in.}$$

From Eq. (7.45)

$$P_{max} = k\delta_{max} = (724.3)(3.927) = 2844 \text{ lbf}$$

The maximum bending stress from Eq. (4.48) is

$$\sigma_{max} = \frac{Mc}{I} = \frac{(2844)(5)(12)(0.75)}{5.063} \text{psi} = 25.28 \text{ ksi}$$

The yield stress was given as 30 ksi. The safety factor for yielding is

$$n_s = \frac{S_y}{\sigma_{max}} = \frac{(30)(10^3)}{(25.20)(10^3)} = 1.187$$

Thus, $n_s > 1$ (so failure should not occur), but because the safety factor is just above 1, the margin of safety is a minimum.

Case Study 7.1 ONE-HOSS SHAY

The One-Hoss Shay
or
The Deacon's Masterpiece
A Logical Story
by Oliver Wendell Holmes

Have you heard of the wonderful one-hoss shay,
That was built in such a logical way
It ran a hundred years to a day,
And then, of a sudden, it—ah, but stay,
I'll tell you what happened without delay,
Scaring the parson into fits,
Frightening people out of their wits,—
Have you ever heard of that, I say?

Seventeen hundred and fifty-five,
Georgius Secundus was then alive,—
Snuffy old drone from the German hive.
That was the year when Lisbon-town
Saw the earth open and gulp her down.
And Braddock's army was done so brown
Left without a scalp to its crown.
It was on the terrible Earthquake-day
That the Deacon finished the one-hoss shay.

Now, in building of chaises, I tell you what,
There is always somewhere a weakest spot,—
In hub, tire, felloe, in spring or thill,
In panel, or crossbar, or floor or sill,
In screw, bolt, thoroughbrace,—lurking still,
Find it somewhere you must and will,—
Above or below, or within or without,—
And that's the reason, beyond a doubt,
A chaise *breaks down*, but doesn't *wear out*.

But the Deacon swore (as Deacons do),
With an "I dew vum" or an "I tell yeou,"
He would build one shay to beat the taown
'N' the keounty 'n' all the kentry raoun';

(continued)

Case Study 7.1 | (CONCLUDED)

It should be so built that it *couldn'* break daown:
—"Fur," said the Deacon, "t's mighty plain
That the weakes' place mus' stan' the strain;
'N' the way t' fix it, uz I maintain,
Is only jest
T' make that place uz strong uz the rest."

So the Deacon inquired of the village folk
Where he could find the strongest oak,
That couldn't be split nor bent nor broke,—
That was for spokes and floor and sills;
He sent for lancewood to make the thills;
The crossbars were ash, from the straightest trees,
The panels of white-wood, that cuts like cheese,
But lasts like iron for things like these;
The hubs of logs from the "Settler's ellum,"—
Last of its timber,—they couldn't sell 'em,
Never an axe had seen their chips,
And the wedges flew from between their lips,
Their blunt ends frizzled like celery tips;
Step and prop-iron, bolt and screw,
Spring, tire, axle, and linchpin, too,
Steel of the finest, bright and blue;
Thoroughbrace bison-skin, thick and wide;
Boot, top, dasher, from tough old hide
Found in the pit when the tanner died.
That was the way he "put her through."—
"There!" said the Deacon, "naow she'll dew!"

Do! I tell you, I rather guess
She was a wonder, and nothing less!
Colts grew horses, beards turned gray,
Deacon and Deaconess dropped away,
Children and grandchildren—where were they?
But there stood the stout old-one-hoss shay
As fresh as on Lisbon-earthquake day!

EIGHTEEN HUNDRED;—it came and found
The Deacon's masterpiece strong and sound.
Eighteen hundred increased by ten;—
"Hahnsum kerridge" they called it then.
Eighteen hundred and twenty came;—
Running as usual; much the same.
Thirty and forty at last arrive,
and then come fifty and FIFTY-FIVE.

Little of all we value here
Wakes on the morn of its hundreth year
Without both feeling and looking queer.
In fact, there's nothing that keeps its youth,
So far as I know, but a tree and truth.
(This is a moral that runs at large;
Take it,—You're welcome.—No extra charge,)

FIRST OF NOVEMBER, the-Earthquake-day,—
There are traces of age in the one-hoss-shay,
A general flavor of mild decay,
But nothing local, as one may say,
There couldn't be,—for the Deacon's art
Had made it so like in every part
That there wasn't a chance for one to start.
For the wheels were just as strong as the thills
And the floor was just as strong as the sills,
And the panels just as strong as the floor,
And the whipple-tree neither less nor more,
And the back-crossbar as strong as the fore
And spring and axle and hub *encore*.
And yet, as a *whole*, it is past a doubt,
In another hour, it will be *worn out*!

First of November, 'Fifty-five!
This morning the parson takes a drive.
Now, small boys, get out of the way!
Here comes the wonderful one-hoss shay,
Drawn by a rat-tailed, ewe-necked bay.
"Huddup," said the parson. Off went they.
The parson was working his Sunday text,—
Had got to *fifthly* and stopped perplexed
At what the—Moses—was coming next.
All at once the horse stood still,
Close by the meet'n'-house on the hill.
—First a shiver, and then a thrill,
Then something decidedly like a spill,—
And the parson was sitting upon a rock,
At half-past nine by the meet'n'-house clock,—
Just the hour of the Earthquake shock!
—What do you think the Parson found
When he got up and stared around?
The poor old chaise in a heap or mound,
As if it had been to the mill and ground!

You see, of course, if you're not a dunce,
How it went to pieces all at once,—
All at once and nothing first,—
Just as bubbles do when they burst.
End of the wonderful one-hoss shay.
Logic is logic. That's all I say.

Case Study 7.2 POWER PRESS BRAKE STUD DESIGN REVIEW

GIVEN An engineer working for a manufacturer of mechanical power presses is assisting in the development of a new size of machine. When starting to design a new brake stud, the engineer finds the information in Fig. 7.16 from an existing computer-assisted drawing (CAD) of a machine with the closest capacity to the new project. The brake stud supports the brake on the cam shaft of the mechanical power press. The brake actuates with every cycle of the press, stopping the ram at the top-dead-center position so that an operator can remove a workpiece from the dies and insert another workpiece for the next cycle. If the stud fails, the press could continue to coast and could result in a serious injury.

Focusing on section A–A the engineer recognizes that no fillet radius had been specified for the brake stud for machines sold previously. Conversations with machinists who routinely worked on the part led to the conclusion that the common practice was to undercut the fillet to make sure assembly was complete; the fillet radius in effect was the radius of the machine tool insert, a value as low as one-eighth of an inch. The immediate concern is whether a product recall is in order, since a brake stud failure could result in a machine operator's hands being in a die while the machine fails to stop after a cycle. The supervisor shares the engineer's fears but assigns one immediate task.

FIND Is the design safe as manufactured? What concerns exist for this machine in service? Is a recall necessary? The carbon steel used for the stud has a minimum ultimate strength of 74.5 ksi, and no yield strength was prescribed in the drawing.

Solution

a) Applied stresses. Figure 7.16 shows that the power press frame and the brake stud share the loading if the stud-retaining nut is tight. If the nut is loose, the stud can conceivably carry the entire loading. Existing brake design calculations reveal that the brake acts over one-quarter of the camshaft revolution and has a relatively constant peak load of almost 1000 lbf. Figure 7.17 shows the shear and bending moment diagrams for the stud and the loading with respect to time. The critical loading is bending (shear must also be considered, but in this case, as with most shafts, bending stresses are the critical con-

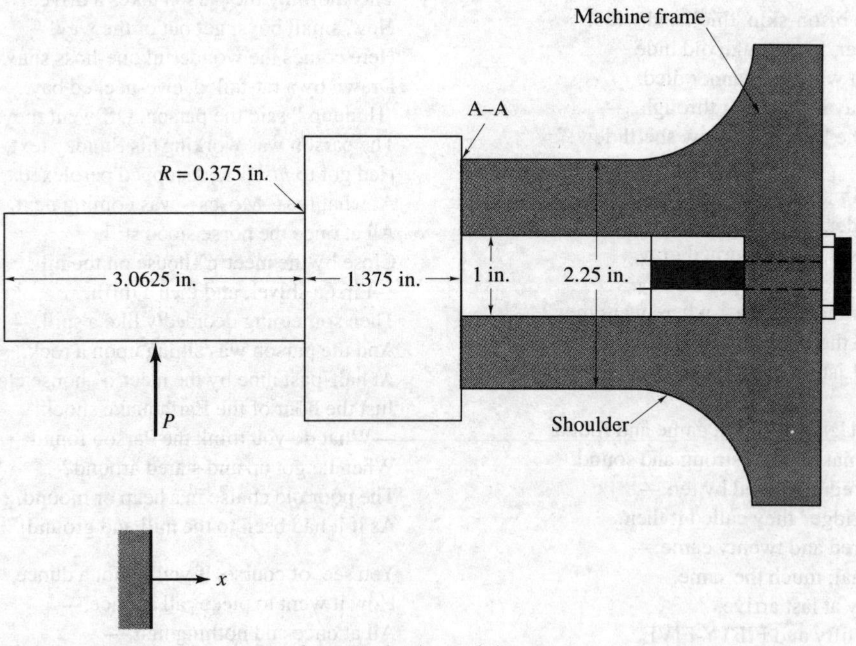

Figure 7.16 Dimensions of existing brake stud design. Note that no radius has been specified at point O.

(continued)

Case Study 7.2 (CONTINUED)

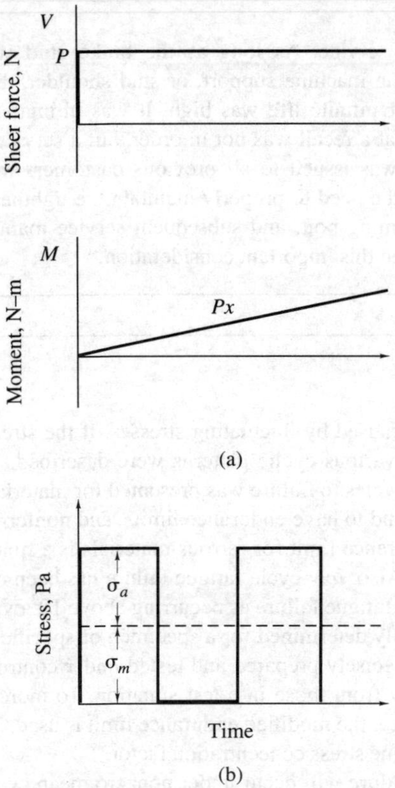

Figure 7.17 Press brake.
(a) Shear and bending-moment diagrams for applied load; (b) stress cycle.

sideration). The maximum stress in the absence of stress raisers is given by

$$\sigma_{max} = \frac{Mc}{I} = \frac{(1000)\left(\frac{3.0625}{2} + 1.375\right)}{\frac{\pi}{4}(0.5)^4} \text{ psi} = 29.6 \text{ ksi}$$

The loading is not completely reversing; the mean stress is 14.8 ksi and the stress amplitude is 14.8 ksi (Fig. 7.17).

The stress raiser cannot be ignored in this case. From Fig. 6.5 the theoretical stress concentration factor is approximately 1.7. The notch sensitivity is obtained from Fig. 7.6 as 0.85. Thus, from Eq. (7.19)

$$K_f = 1 + (1.7 - 1)0.85 = 1.6$$

Recall that for ductile materials a stress concentration factor is not used for static loads, with the rationale that plastic flow (even at microscopic scales) will relieve stress concentrations.

The brake actuates every time the press cycles. Some mechanical presses operate at more than 100 cycles per minute, but the design speed for the subject machine is 40 cycles per minute. Still, even when used for only one 8-hr shift per day, the number of loading cycles on the stud would be 10^7 after 20 years. Many mechanical presses have expected lives of twice this. Clearly, the stud must be designed for infinite life.

b) Fatigue strength calculation. The steel has an ultimate strength of 74.5 ksi, but no yield strength was given. Thus, the Goodman line will be used for fatigue analysis. Although this may seem conservative, if a plastic deformation in the stud were possible, it would manifest itself in trials conducted on the machines before shipping. Therefore, the design reviewer feels reasonably confident that the main failure mode is fatigue.

From Eq. (7.7) the approximate value of an unmodified endurance limit is $S_e' = 0.5S_{ut} = 37.25$ ksi. The following correction factors are obtained:

- Surface finish: a machined finish so that from Eq. (7.21)

$$k_f = 2.70(74.5)^{-0.265} = 0.86$$

- Size factor: from Eq. (7.22)

$$k_s = 0.869(1)^{-0.112} = 0.869$$

- Reliability factor: $k_r = 0.82$ (chosen from Table 7.3 for a 99% reliability)

The endurance limit for the steel brake stud is then given by Eq. (7.16) as

$$S_e = (37.25)(0.86)(0.869)(0.82) = 22.8 \text{ ksi}$$

c) Fatigue criterion. Using the Goodman line [Eq. (7.28)] gives the safety factor as

$$\frac{K_f\sigma_a}{S_e} + \frac{\sigma_m}{S_{ut}} = \frac{1.6(14.8)}{22.8} + \frac{14.8}{74.5} = \frac{1}{n_s}$$

$$\therefore\ n_s = 0.8$$

Thus, the design is inadequate. However, if there were no stress concentration at all, the safety factor would increase to 1.17, certainly a low safety factor regardless.

(continued)

Case Study 7.2 (CONCLUDED)

Discussion

The machine manufacturer had not experienced failures in the brake stud, even though the existing design had been on the market for over three decades. Through laboratory testing it was determined that if the brake-stud-retaining bolt was loose, the brake would make a noticeably loud squeal and the stud would clank. Such problems were undoubtedly promptly identified by maintenance personnel, and the stud bolt was retightened whenever it became loose in service. As long as the brake stud was snug against the machine support, or stud shoulder, the safety factor for infinite life was high. It was ultimately determined that a recall was not in order, but a service memorandum was issued to all previous customers informing them of the need to properly maintain the tightness of the stud-retaining bolt, and subsequent service manuals also mentioned this important consideration.

7.12 SUMMARY

Failures in components or structures are often caused by fluctuating stresses. If the stress variation sequence repeats itself, it is called cyclic. The various cyclic patterns were described. The fatigue strength versus the logarithm of the number of cycles to failure was presented for materials with and without an endurance limit. Ferrous materials tend to have endurance limits, and nonferrous materials tend not to have endurance limits. The endurance limit for ferrous materials is a function of the type of loading the component is subjected to. Also, *low*-cycle fatigue failure has been classified as occurring at less than 10^3 cycles, and *high*-cycle fatigue failure as occurring above 10^3 cycles.

The endurance limit has been experimentally determined for a specimen of specific size with a mirror-like surface finish. The specimen was precisely prepared and tested under controlled conditions. In practice, conditions differ significantly from those in a test situation. To more accurately characterize the conditions that prevail in practice, the modified endurance limit is used. The surface finish and size factors are used along with a fatigue stress concentration factor.

Various approaches for estimating when failure will occur under nonzero mean cyclic stresses were also considered. The Soderberg, Gerber, Goodman, and modified Goodman theories were presented. These theories allow the variation of mean and alternating stresses. The construction of a complete modified Goodman diagram was shown, and fatigue and yield criteria were used. Failure equations were given for specific regions of the modified Goodman diagram.

Most of the results presented were for ductile materials, but the behavior of brittle materials was briefly described. The major difference between brittle and ductile materials is that the compressive and tensile strengths are nearly identical for ductile materials whereas for brittle materials the compressive strength is several times greater than the tensile strength.

Impact loading has to be considered when loads are rapidly applied. The stress levels and deformations induced are much larger than with static or cyclic loading. The more rapid the loading, the higher the yield and ultimate strengths of a material. By equating the kinetic and elastic strain energies, two types of impact loading were considered: a block vertically falling a given height onto a spring, and a block sliding into a spring. An impact factor, maximum and static deformations, and maximum and static loads were obtained.

KEY WORDS

amplitude ratio ratio of stress amplitude to mean stress

cyclic stress stress sequence that repeats over time

dynamic adjective indicating variation with time

endurance limit stress level below which infinite life can be realized

fatigue failure, at relatively low stress levels, of structures that are subject to fluctuating and cyclic stresses

finite life stress level that causes failure due to fatigue

Gerber line parabolic relationship taking mean and alternating stresses into account

Goodman line theory connecting modified endurance limit and ultimate strength on plot of alternating stress versus mean stress

high-cycle fatigue fatigue failure that occurs above 10^3 cycles but below 10^6 cycles

impact loading load rapidly applied to body

infinite life stress levels that do not cause fatigue failure

linear damage rule theory that damage at any stress level is proportional to number of cycles

low-cycle fatigue fatigue failure that occurs below 10^3 cycles

Manson-Coffin relationship theoretical approach to fatigue based on strain

mean stress average of minimum and maximum stresses in cycle

Miner's rule same as linear damage rule

modified endurance limit corrections for endurance limit based on surface finish, material, specimen size, loading type, temperature, etc.

modified Goodman diagram diagram that defines all stress states not resulting in fatigue failure or yielding

notch sensitivity material property that reflects ability of ductile materials to be less susceptible to stress raisers in fatigue

Paris power law postulate that crack growth in cyclic loading follows power law

residual stress internal stress usually caused by manufacturing process

Soderberg line theory connecting modified endurance limit and yield strength on plot of alternating stress versus mean stress

S–N diagram plot of stress level versus number of cycles before failure

stress amplitude one-half of stress range

stress range difference between maximum and minimum stresses in cycle

stress ratio ratio of minimum and maximum stresses

Wöhler diagram same as S–N diagram

yield line failure criterion that postulates yielding on first cycle of cyclic loading with nonzero mean

Recommended Readings

Juvinall, R.C. (1967) *Engineering Considerations of Stress, Strain, and Strengths,* McGraw-Hill, New York.

Madayag, A.F. (1969) *Metal Fatigue: Theory and Design,* Wiley, New York.

Rice, R.C., ed. (1988) *Fatigue Design Handbook,* Society of Automotive Engineers, Warrendale, PA.

Suresh, S. (1991) *Fatigue of Materials,* Cambridge University Press, Cambridge, UK.

References

Dowling, N.E. (1993) *Mechanical Behavior of Materials,* Prentice-Hall, Englewood Cliffs, NJ.

Johnson, R.C. (1967) *Predicting Part Failures Part I* "Machine Design," vol. 45, p. 108.

Juvinall, R.C. (1967) *Engineering Considerations of Stress, Strain, and Strength,* McGraw-Hill, New York.

Juvinall, R.C., and Marshek, K.M. (1991) *Fundamentals of Machine Component Design,* Wiley, New York.

Landgraf, R.W. (1968) Cyclic Deformation and Fatigue Behavior of Hardened Steels, Report no. 330, Department of Theoretical and Applied Mechanics, University of Illinois, Urbana.

Lipson, C., and Juvinall, R.C. (1963) *Handbook of Stress and Strength*, Macmillan, New York.

Manjoine, M.J. (1944) "Influence of Rate of Strain and Temperature on Yield Stresses of Mild Steel," *Journal of Applied Mechanics,* vol. 66, pp. A–211 to A–218.

Mitchell, M. (1978) "Fundamentals of Modern Fatigue Analysis for Design," *Fatigue and Microstructure,* M. Meschii, ed., American Society for Metals, Metals Park, OH, pp. 385–437.

Norton, R.L. (1996) *Machine Design*, Prentice-Hall, New Jersey.

Parris, P.C., Gomez, M.P., and Anderson, W.P. (1961) *A Rational Analytic Theory of Fatigue,* Trend Engineering, University of Washington, vol. 13, no. 1, pp. 9–14.

Petroski, H. (1992) *To Engineer Is Human,* Cambridge University Press, Cambridge, UK.

Shigley, J.E., and Mischke, C.R. (1989) *Mechanical Engineering Design,* McGraw-Hill, New York.

Shigley, J.E., and Mitchell, L.D. (1983) *Mechanical Engineering Design,* McGraw-Hill, New York.

Sines G., and Waisman, J.L. (1959) *Metal Fatigue,* McGraw-Hill, New York.

Suresh, S. (1991) *Fatigue of Materials,* Cambridge University Press, Cambridge, UK.

Problems

Section 7.3

7.1 A tuning fork is hit with a pencil and starts to vibrate with a frequency of 440 Hz. The maximum bending stress in the tuning fork is 4 MPa at the end positions. Calculate the mean stress, the range of stress, the stress amplitude, the stress ratio, and the amplitude ratio. Also calculate how much stress the tuning fork can sustain without being plastically deformed if it is made of AISI 1080 steel.

Section 7.5

7.2 The jack for a Volvo consists of a mechanism in which the lift screw extends horizontally through two corners of the mechanism while the other two corners apply a force between the ground and the car to be lifted. The maximum compressive stress in the jack is 190 MPa when the Volvo is jacked up so high that both wheels on one side of the car are in the air and the load on the jack is

8000 N. How many times can the jack be used for a small truck that weighs 6 tons and loads the jack to 17 000 N before the jack fails from fatigue? The jack material is AISI 1080 steel.

Section 7.6

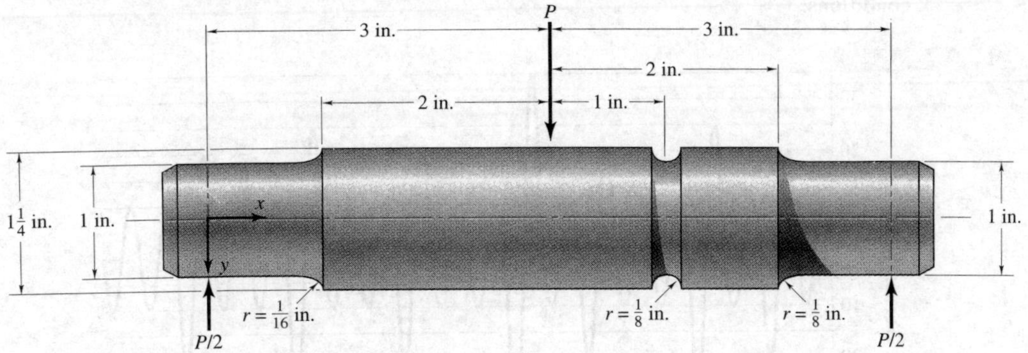

Sketch *a*, for Problem 7.3

7.3 The shaft shown in sketch *a* rotates at high speed while the imposed loads remain static. The shaft is machined from ground, high-carbon steel (AISI 1080). If the loading is sufficiently large to produce a fatigue failure after 1 million cycles, where would the failure most likely occur? Show all necessary computations and reasoning.

Section 7.7

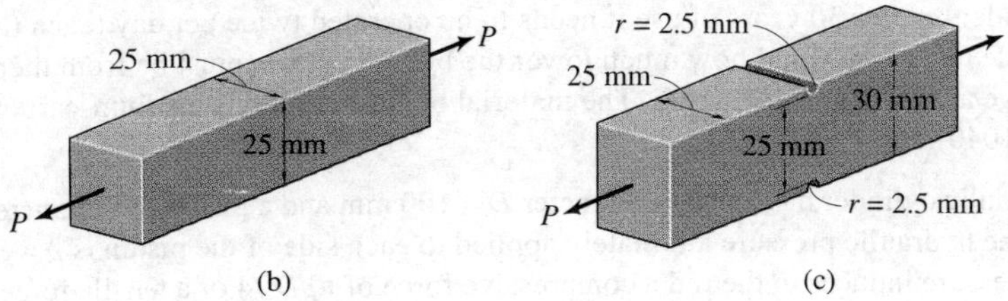

Sketches *b* and *c*, for Problem 7.4

7.4 For each of the bars shown in sketches *b* and *c* determine
a) The static tensile load causing fracture
b) The alternating (completely reversing) axial load $\pm P$ that would be just on the verge of producing eventual fatigue failure

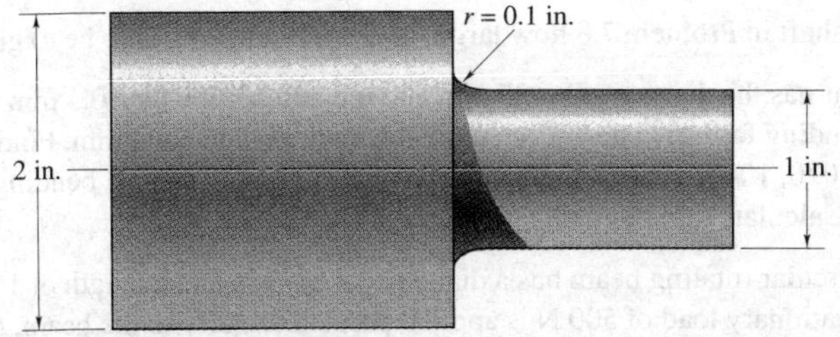

Sketch *d*, for Problem 7.5

7.5 A stepped shaft, as shown in sketch *d*, has a major diameter of 2 in., a minor diameter of 1 in., and a notch radius of 0.1 in. It was machined from high-carbon steel (AISI 1080). The loading is one of completely reversed torsion. During a typical 30 s of operation under overload conditions the nominal stress in the 1-in.-diameter section was measured to be as shown in sketch *e*. Estimate the life of the shaft when it is operating continually under these conditions.

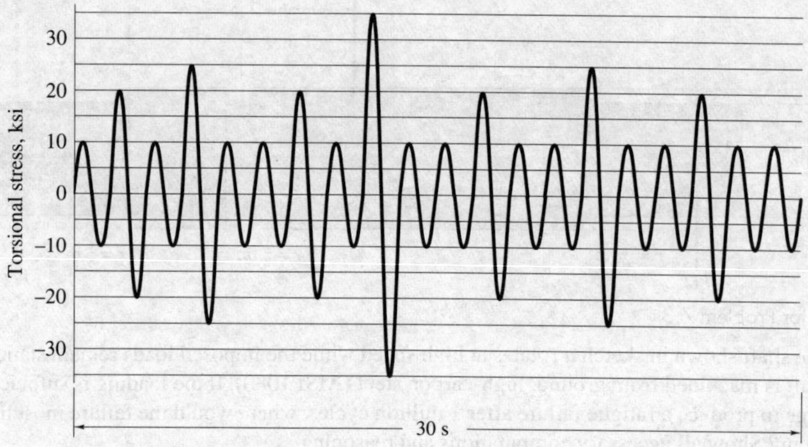

Sketch *e*, for Problem 7.5

7.6 A flood-protection dam gate is supposed to operate only once per week for 100 years, but suddenly after 30 years of use it needs to be operated twice per day (each time the high tide comes in). Find how much lower the bending stress must be from then on to still give a total life of 100 years. The material being fatigued is medium-carbon steel (AISI 1040).

7.7 A hydraulic cylinder has a piston diameter $D = 100$ mm and a piston rod diameter $d = 32$ mm. The hydraulic pressure alternately applied to each side of the piston is $p = 300$ bars. This pressure induces in the rod a compressive force of $\pi p D^2/4$ or a tensile force of $\pi p(D^2 - d^2)/4$, depending on which side of the piston the pressure is applied. The piston rod material is martensitic stainless steel (AISI 410). Find the endurance life of the piston rod.

7.8 A 20-mm-diameter shaft transmits a variable torque of 500 ± 400 N-m. The frequency of the torque variation is 0.1 s^{-1}. The shaft is made of high-carbon steel (AISI 1080). Find the endurance life of the shaft.

7.9 Find for the shaft in Problem 7.8 how large the shaft diameter has to be to give infinite life.

7.10 A notched bar has the diameter $D = 25$ mm and the notch radius $r = 0.5$ mm. It experiences rotational bending fatigue. The bottom diameter of the notch is 24 mm. Find which of the steels AISI 1020, 1040, 1080, or 316 will give the highest allowable bending moment for infinite life. Calculate that moment.

7.11 A straight, circular rotating beam has a diameter of 30 mm and a length of 1 m. At the end of the beam a stationary load of 500 N is applied perpendicularly to the beam, giving a bending deformation. Find which surface machining method can be used to give infinite life to the rotating beam if it is made of

a) AISI 1020 steel

b) AISI 1080 steel

Note that $k_s = k_r = k_m = 1$.

7.12 The pedals on a bicycle are screwed into the crank with opposite threads, one left-hand thread and one right-hand thread, to ensure that the pedals do not accidentally unscrew. Just outside the thread is a 0.75-mm-radius fillet connecting the 12.5-mm-diameter threaded part with an 11-mm-diameter central shaft in the pedal. The shaft is made of AISI 4340 alloy steel tempered at a low temperature to give an ultimate stress of 2 GPa. Find the fatigue stress concentration factor for the fillet, and calculate the maximum allowable pedal force for infinite life if the force from the foot is applied 70 mm from the fillet.

7.13 A saddle pole for a mountain bike has a fillet of 5-mm radius connecting the top part of the pole (which has a diameter of 35 mm) with the bottom part of the pole (which has a diameter of 25 mm). The saddle pole is solid and made of 2024 aluminum. Find the fatigue stress concentration factors for axial load and bending.

7.14 During the development of a new submarine a 1-to-20 model was used. The model was tested to find if there was any risk of fatigue failure in the full-size submarine. To be on the safe side, the stresses in the model were kept 15% higher than the stresses in the full-size submarine. Is it possible to conclude that the full-size submarine will be safe if the model was safe at a 15% higher stress level?

7.15 During the development of a new car it was found that the power from the motor was too high for the gearbox to handle. When maximum torque was transmitted through the gearbox 50% of the gearboxes failed within the required lifetime of 800 hr. How much stronger should the gearbox be made to ensure that only one gearbox out of 1000 would fail within 800 hr?

Section 7.8

7.16 Truck gearboxes are dimensioned for infinite life at maximum torque regarding contact stresses and bending stresses for all gear steps. Car gearboxes are dimensioned for finite life according to typical running conditions. Maximum torque for the first gear can typically be maintained only 3 to 6 s for each acceleration before the maximum speed is reached. If a driver accelerates at full power 20 times per day for 20 years on the first gear, the required life is only 60 to 120 hr. The normal load spectrum gives a life of 200 000 km for 99% of the gearboxes. A driver uses the car twice a year to move his 10-ton boat 50 km, the distance between his home and the harbor, during the 10-year life of the car. Calculate how much of the 200 000-km nominal life of the gearbox is consumed by the boat moving if the life is inversely proportional to the load raised to the 3.5 power. During the move the gearbox load is four times higher than normal.

7.17 The hand brakes of a bicycle are often the same type for the front and rear wheels. The high center of gravity (compared with the distance between the wheels) will increase the contact force between the front wheel and the ground and will decrease the contact force between the rear wheel and the ground when the brakes are applied. If equal force is applied to each of the two brakes, the rear wheel will start sliding while the front wheel is still rolling. The manufacturer of the brake wires did not know about this difference between the front and rear wheels, so the wires were dimensioned as if the wheels had equal contact force and thus needed equal force in the two brake wires. The wires were originally dimensioned to withstand 20 years' use with brake applications at a force necessary to lock the two (equally loaded) wheels. How long will the life of the wires be if the friction between the front wheel and the ground is just enough to lift the real wheel from the ground? Assume the endurance life is proportional to the stress raised to the −10 power.

7.18 Ball bearings often run at varying load and speed. Life is inversely proportional to the contact stress raised to the 10/3 power for a number of bearing types. For ball bearings the contact stresses are approximately proportional to the third root of the load. A ball bearing in a gearbox is dimensioned to have a life of 1752 million revolutions, half of which are at half of the maximum motor torque, one-quarter at full motor torque, and one-quarter at 75% of full motor torque. Calculate the bearing life if the motor torque is kept constant at the maximum level.

7.19 A cold-drawn steel bar with a solid cross section is subjected to a cyclic force that ranges from a maximum of 80 000 lbf in tension to a minimum of 30 000 lbf in compression. The ultimate strength is 158 000 psi and the yield strength is 133 000 psi. The critical safety factor is 2. Determine the following:

a) The modified endurance limit

b) The cross-sectional area that will produce fatigue failure and the corresponding diameter

c) The region (a, b, c, or d), assuming modified Goodman criteria, and why

7.20 Both bars used in Problem 7.4 and shown in sketches *b* and *c* are made from cold-drawn, medium-carbon steel (AISI 1040). Determine the safety factor for each bar. The load varies from 10 to 40 kN. Use the modified Goodman relationship.

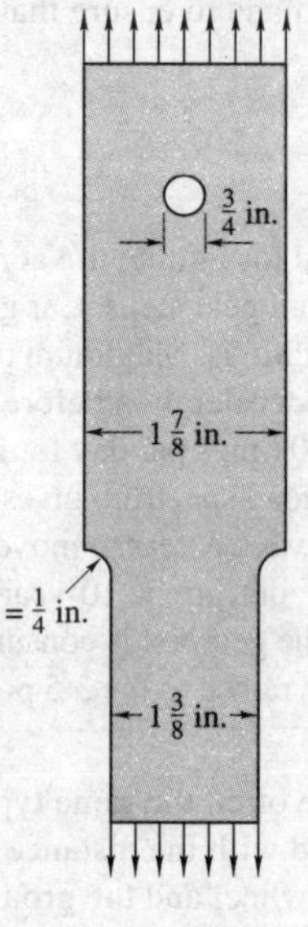

Sketch *f*, for Problem 7.21

7.21 The 7/16th-in.-thick component in sketch *f* is designed with a fillet and a hole. The load varies from 12 000 to 2000 lbf. The following strengths are given: $S_u = 56$ ksi, $S_y = 41$ ksi. Using the modified Goodman failure theory, determine the safety factor for the hole as well as for the fillet. At which location will failure occur first? Will the component fail?

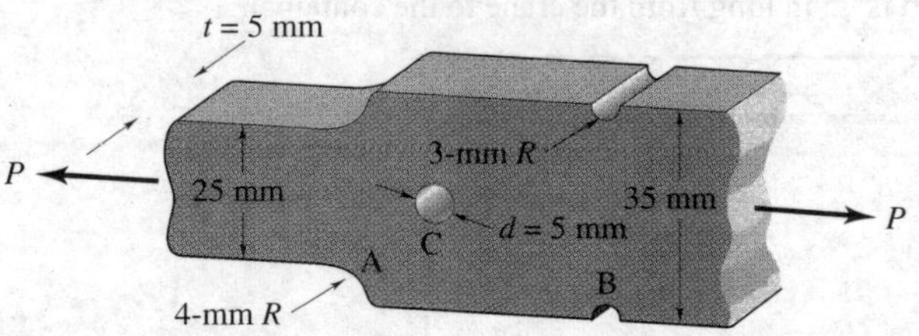

Sketch *g*, for Problem 7.22

7.22 The flat bar shown in sketch *g* is made of cold-drawn, high-carbon steel (AISI 1080). The cyclic, nonzero, mean axial load variation is from a minimum of 2 kN to a maximum of 10 kN. Using the modified Goodman failure theory, determine the safety factors for the hole, the fillet, and the groove. Also, indicate where the flat bar will first fail.

7.23 A straight, circular rotating beam with a diameter of 30 mm and a length of 1 m has an axial load of 30 000 N applied at the end and a stationary radial load of 400 N. The beam surface is cold drawn and the material is titanium. Note that $k_s = k_r = k_m = 1$. Find the safety factor n_s for infinite life by using the Goodman line.

7.24 A 10 000-lbm elevator is supported by a stranded steel cable having a 2.5-in.^{2} cross section and an effective modulus of elasticity of 12×10^6 psi. As the elevator is descending at a constant velocity of 400 ft/min, an accident causes the cable, 60 ft above the elevator, to suddenly stop. Determine the static elongation of the cable, the impact factor, the maximum elongation, and the maximum tensile stress developed in the cable.

7.25 A person is planning a bungee jump from a 40-m-high bridge. Under the bridge is a river with crocodiles, so the person does not want to be submerged into the water. The rubber rope fastened to the ankles has a spring constant of 3600 N/m. The distance from the ankles to the top of the head is 175 cm, and the person weighs 80 kg. Calculate how long the rope should be.

7.26 A toy with a bouncing 50-mm-diameter steel ball has a compression spring with a spring constant of 100 000 N/m. The ball falls from a 3-m height down onto the spring, which can be assumed to be weightless, and bounces away and lands in a hole. Calculate the maximum force on the spring and the maximum deflection during the impact. The steel ball density is 7840 kg/m^3.

7.27 At a building site a 1-ton container hangs from a crane in a wire and is then placed on the floor so that the wire becomes unloaded. The container is pushed to the elevator shaft where it is to be lowered as shown in sketch *h*. By mistake there is a 1-m slack in the wire from the crane when the container falls into the elevator shaft. Calculate the maximum force in the wire if it has a cross-sectional steel surface of 500 mm^2 and an effective modulus of elasticity of 70 GPa and is 25 m long from the crane to the container.

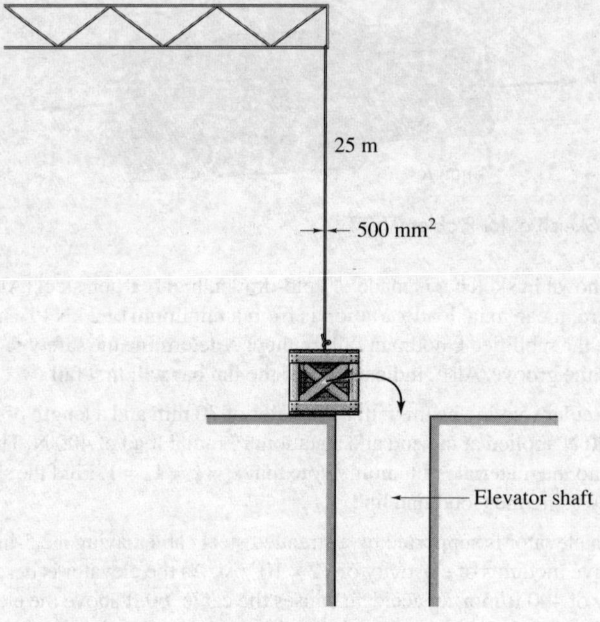

Sketch *h*, for Problem 7.27

7.28 Modern kitchen drawers have small rubber springs mounted onto the sides of the inside of the front plate to take up the force and stop the drawer when it is being closed. The spring constant for each of the two rubber springs is 400 kN/m. The drawer is full of cutlery, which weighs 5 kg, and is closed with a speed of 0.5 m/s. Calculate the maximum force in each rubber spring if the drawer itself weighs 1 kg and

a) The cutlery is in a container that is fixed to the drawer so that it moves with the drawer

b) The cutlery is in a plastic container that can slide 3 cm with a coefficient of friction of 0.25 inside the drawer

7.29 Car doors are easy to slam shut but difficult to press shut by hand force. The door lock has two latches, the first easily engaged and the second requiring the rubber seal around the door to be quite compressed before it can engage. The rubber seal gives a spring constant of 50 000 N/m at the lock position for the door, and the mass moment of inertia for the door around its hinges is 2.5 kg-m^2. The distance between the lock and the hinges is 1 m. Calculate what force is needed to press the car door shut at the lock if a speed of 0.8 m/s at the lock slams it shut.

CHAPTER

8

LUBRICATION, FRICTION, AND WEAR

The space shuttle takes off. Complicated systems such as this require considerable tribological sophistication. (AP/Wide World Photos)

...among all those who have written on the subject of moving forces, probably not a single one has given sufficient attention to the effect of friction in machines...

Guillaume Amontons (1699)

SYMBOLS

A	area, m^2
A_o	sum of projected areas in sliding, m^2
A_r	sum of real areas in contact, m^2
b^*	contact semiwidth, m
D_x	diameter of contact ellipse along x axis, m
D_y	diameter of contact ellipse along y axis, m
d	diameter, m
E	modulus of elasticity, Pa
E'	effective elastic modulus, Pa
$\mathcal{E}$	complete elliptic integral of second kind
F	friction force, N
$\mathcal{F}$	complete elliptic integral of first kind
H	hardness of softer material, Pa
h	film thickness, m
h_{min}	minimum film thickness, m
k_e	ellipticity parameter, D_y/D_x
k_1	adhesive wear constant
k_2	abrasive wear constant
L	sliding distance, m
l	length, m
m_g	gravitational mass, kg
m_p	mass of person, kg
N	number of measurements of z_i
P_1, P_2	forces, N
p	pressure, Pa
p_a	ambient pressure, Pa
p_H	maximum Hertzian contact pressure, Pa
p_{max}	maximum pressure, Pa
p_s	supply pressure, Pa
R	radius, m; curvature sum, m
R_a	arithmetic average surface roughness, m
R_q	root-mean-square surface roughness, m
R_d	curvature difference
r	radius, m
S_y	yield stress in tension, Pa
s	shear strain rate, s^{-1}
t_m	temperature, °C
u	velocity, m/s
u_b	velocity of upper surface, m/s
v	wear volume, m^3
v_{abr}	abrasive wear volume, m^3
W	normal load, N
W'	dimensionless load for rectangular contact
w'	load per unit width, N/m
w_a	squeeze velocity, m/s
x, y, z	Cartesian coordinate system, m
z	coordinate in direction of film, m
z_i	height from reference line, m
α_r	radius ratio, R_y/R_x
δ_{max}	maximum deformation, m
η	absolute viscosity, Pa-s
η_k	kinematic viscosity, m^2/s
η_0	absolute viscosity at $p = 0$ and at constant temperature, Pa-s
θ	cone angle, rad
Λ	dimensionless film parameter
μ	coefficient of sliding friction
μ_d	coefficient of dynamic friction
μ_r	coefficient of rolling friction
ν	Poisson's ratio
ξ	pressure-viscosity coefficient, m^2/N
ρ	density, kg/m^3
σ	normal stress, Pa
τ	shear stress, Pa

Subscripts

a	solid a
b	solid b
x,y,z	Cartesian coordinate system

8.1 INTRODUCTION

Tribology may be defined as the study of the lubrication, friction, and wear of moving or stationary parts. Lubrication, friction, and wear play a significant role in machine element life. Not only must the design stress be less than the allowable stress and the deformation not exceed some maximum value, but also lubrication, friction, and wear (tribological considerations) must be properly understood for machine elements to be successfully designed. The interaction of surfaces in relative motion considered in tribology should not be regarded as a special subject. Like the strength of materials, tribology is basic to every engineering design of machine elements. There are few machine elements that do not depend on tribological considerations.

Fluid film lubrication occurs when opposing surfaces are completely separated by a lubricant film and no asperities are in contact. The applied load is carried by pressure generated within the fluid, and frictional resistance to motion arises entirely from the shearing of the viscous fluid.

This chapter presents the lubrication, friction, and wear considerations that are important in the successful design of machine elements. It includes the concept of conformal and nonconformal surfaces, or geometrical conformity, which is used throughout the text. This chapter also considers the various lubrication regimes, surface and film parameters, lubricant viscosity, deformation due to concentrated loading, friction, and wear. All these topics are important to machine elements.

8.2 CONFORMAL AND NONCONFORMAL SURFACES

Conformal surfaces fit snugly into each other with a high degree of geometrical conformity so that the load is carried over a relatively large area. For example, the lubrication area of a journal bearing would be 2π times the radius times the length. The load-carrying surface area remains essentially constant while the load is increased. Fluid film journal bearings (Fig. 8.1) and slider bearings have conformal surfaces. In journal bearings the radial clearance between the journal and the sleeve is typically one-thousandth of the journal diameter; in slider bearings the inclination of the bearing surface to the runner is typically one part in a thousand. Another example of conformal surfaces is a human hip joint.

Many fluid-film-lubricated machine elements have surfaces that do not conform to each other well. The full burden of the load must then be carried by a small lubrication area. The lubrication area of a nonconformal conjunction is typically three orders of magnitude less than that of a conformal conjunction. In general, the lubrication area between **nonconformal surfaces** enlarges considerably with increasing load, but it is still smaller than the lubrication area between conformal surfaces. Some examples of nonconformal surfaces are mating gear teeth, cams and followers, and rolling-element bearings (Fig. 8.2).

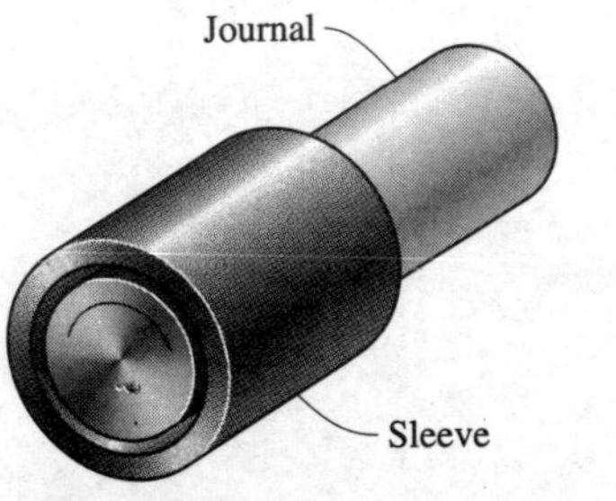

Figure 8.1 Conformal surfaces.
[B.J. Hamrock and W.J. Anderson, *Rolling-Element Bearings,* NASA Reference Publication 1105, 1983.]

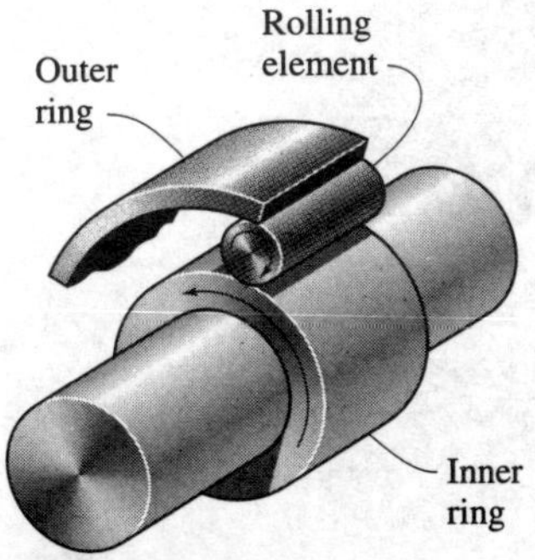

Figure 8.2 Nonconformal surfaces.
[B.J. Hamrock and W.J. Anderson, *Rolling-Element Bearings,* NASA Reference Publication 1105, 1983.]

8.3 LUBRICATION

Features that distinguish the four lubrication regimes present in machine elements a sidered here.

8.3.1 HYDRODYNAMIC LUBRICATION

Hydrodynamic lubrication (HL) is generally characterized by conformal surfac fluid film lubrication. A positive pressure develops in a hydrodynamically lubricate nal or thrust bearing because the bearing surfaces converge and their relative moti the viscosity of the fluid separate the surfaces. The existence of this positive p implies that a normal applied load may be supported. The magnitude of the pressure oped (usually less than 5 MPa) is not generally large enough to cause significant deformation of the surfaces. The minimum film thickness in a hydrodynamically lu ed bearing is a function of normal applied load W, velocity u_b, lubricant absolute vi η_0, and geometry (R_x and R_y). Figure 8.3 shows some of these characteristics of hy namic lubrication. Minimum film thickness h_{min} as a function of u_b and W for motion is given as

$$(h_{min})_{HL} \propto \left(\frac{u_b}{W}\right)^{\frac{1}{2}}$$

The minimum film thickness normally exceeds 1 μm.

In hydrodynamic lubrication the films are generally thick, so that opposing sol faces are prevented from coming into contact. This condition is often called the idea of lubrication because it provides low friction and high resistance to wear. The lubri of the solid surfaces is governed by the bulk physical properties of the lubricant, n the viscosity; and the frictional characteristics arise purely from the shearing of the v lubricant.

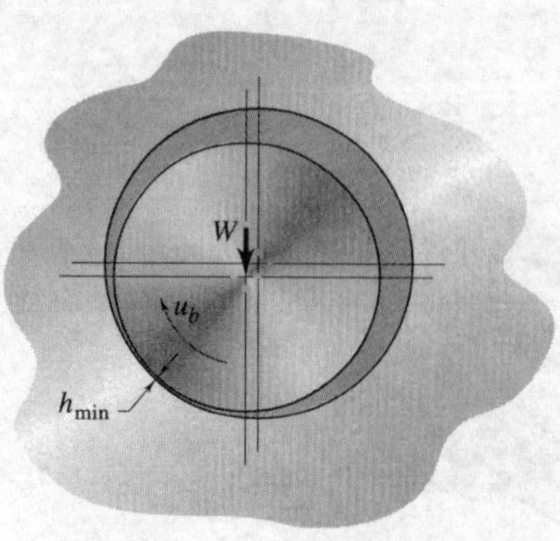

Conformal surfaces
$p_{max} \approx 5$ MPa
$h_{min} = f(W, u_b, \eta_0, R_x, R_y) > 1$ μm
No elastic effect

Figure 8.3 Characteristics of hydrodynamic lubrication. [B.J. Hamrock, *Fundamentals of Fluid Film Lubrication*, NASA Reference Publication 1255, 1991.]

For a normal load to be supported by a bearing, positive pressure profiles must be developed over the bearing length. Figure 8.4 illustrates three ways of developing positive pressure in hydrodynamically lubricated bearings. For positive pressure to be developed in a slider bearing [Fig. 8.4(a)] the lubricant film thickness must be decreasing in the sliding direction. In a squeeze film bearing [Fig. 8.4(b)] the bearing surfaces approach each other with velocity w_a, providing a valuable cushioning effect; positive pressures will be generated only when the film thickness is diminishing. In an externally pressurized bearing, sometimes called a hydrostatic bearing [Fig. 8.4(c)], the pressure drop across the bearing supports the load. The load-carrying capacity is independent of bearing motion and lubricant viscosity. There is no surface contact wear at starting and stopping as there is with the slider bearing.

8.3.2 Elastohydrodynamic Lubrication

Elastohydrodynamic lubrication (EHL) is a form of hydrodynamic lubrication where elastic deformation of the lubricated surfaces becomes significant. The features important in a hydrodynamically lubricated slider bearing [Fig. 8.4(a)]—converging film thickness, sliding motion, and a viscous fluid between the surfaces—are also important here. Elastohydrodynamic lubrication is normally associated with nonconformal surfaces and fluid film lubrication. There are two distinct forms of EHL.

8.3.2.1 Hard EHL Hard EHL relates to materials of high elastic modulus, such as metals. In this form of lubrication the elastic deformation and the pressure-viscosity effects are equally important. Figure 8.5 gives the characteristics of hard elastohydrodynamically lubricated conjunctions. The maximum pressure is typically between 0.5 and 4 GPa; the minimum film thickness normally exceeds 0.1 μm. These conditions are dramatically different from those found in a hydrodynamically lubricated conjunction (Fig. 8.3). At loads normally experienced in nonconformal machine elements the elastic deformations are two orders of magnitude larger than the minimum film thickness. Furthermore, the lubricant viscosity can vary by as much as 20 orders of magnitude within the lubricated conjunction. The minimum film thickness is a function of the same parameters as for hydrodynamic lubrication (Fig. 8.3) but with the additions of the effective elastic modulus E' and the pressure-viscosity coefficient ξ of the lubricant. The effective modulus of elasticity is

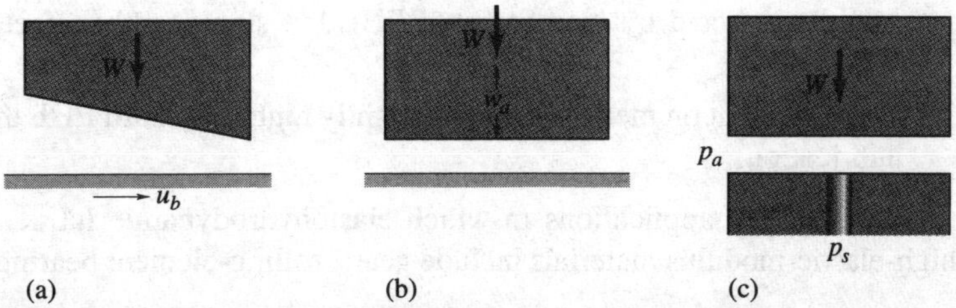

Figure 8.4 Mechanism of pressure development for hydrodynamic lubrication. (a) Slider bearing; (b) squeeze film bearing; (c) externally pressurized bearing. [B.J. Hamrock, *Fundamentals of Fluid Film Lubrication*, NASA Reference Publication 1255, 1991.]

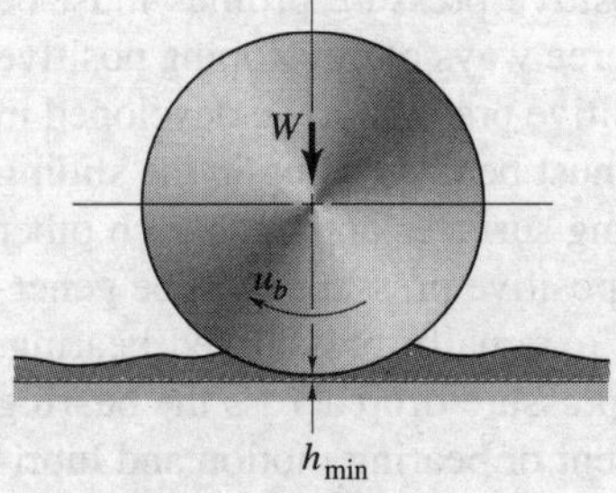

Nonconformal surfaces
High-elastic-modulus material (e.g., steel)
$p_{max} \approx 0.5$ to 4 GPa
$h_{min} = f(W, u_b, \eta_0, R_x, R_y, E', \xi) > 0.1\ \mu m$
Elastic and viscous effects both important

Figure 8.5 Characteristics of hard elastohydrodynamic lubrication. [B.J. Hamrock, *Fundamentals of Fluid Film Lubrication*, NASA Reference Publication 1255, 1991.]

$$E' = \frac{2}{\dfrac{1-\nu_a^2}{E_a} + \dfrac{1-\nu_b^2}{E_b}} \tag{8.2}$$

where

ν = Poisson's ratio
E = modulus of elasticity, Pa

The relationships between the minimum film thickness h_{min} and the normal applied load W and upper surface velocity u_b for hard EHL as obtained from Hamrock and Dowson (1977) are

$$(h_{min})_{HEHL} \propto W^{-0.073} \tag{8.3}$$

$$(h_{min})_{HEHL} \propto u_b^{0.68} \tag{8.4}$$

Comparing the results for hard EHL [Eqs. (8.3) and (8.4)] with those for hydrodynamic lubrication [Eq. (8.1)] yielded the following conclusions:

1. The exponent on the normal applied load is nearly seven times larger for hydrodynamic lubrication than for hard EHL. This difference implies that the film thickness is only slightly affected by load for hard EHL but significantly affected for hydrodynamic lubrication.
2. The exponent on mean velocity is slightly higher for hard EHL than for hydrodynamic lubrication.

Engineering applications in which elastohydrodynamic lubrication is important for high-elastic-modulus materials include gears, rolling-element bearings, and cams.

8.3.2.2 Soft EHL Soft EHL relates to materials of low elastic modulus, such as rubber. Figure 8.6 shows the characteristics of soft-EHL materials. In soft EHL the elastic distortions are large, even with light loads. The maximum pressure for soft EHL is 0.5 to

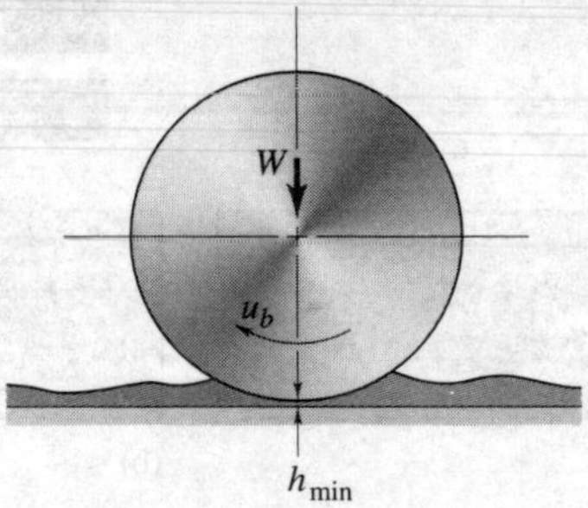

Nonconformal surfaces (e.g., nitrite rubber)
$p_{max} \approx 0.5$ to 4 MPa
$h_{min} = f(W, u_b, \eta_0, R_x, R_y, E') \approx 1$ μm
Elastic effects predominate

Figure 8.6 Characteristics of soft elastohydrodynamic lubrication. [B.J. Hamrock, *Fundamentals of Fluid Film Lubrication*, NASA Reference Publication 1255, 1991.]

4 MPa (typically 1 MPa), in contrast to 0.5 to 4 GPa (typically 1 GPa) for hard EHL (Fig. 8.5). This low pressure has a negligible effect on the viscosity variation throughout the conjunction. The minimum film thickness is a function of the same parameters as in hydrodynamic lubrication with the addition of the effective elastic modulus. The minimum film thickness for soft EHL is typically 1 μm. Engineering applications in which elastohydrodynamic lubrication is important for low-elastic-modulus materials include seals, human joints, tires, and a number of lubricated machine elements that use rubber as a material. The common features of hard and soft EHL are that the local elastic deformation of the solids provides coherent fluid films and that asperity interaction is largely prevented. Lack of asperity interaction implies that the frictional resistance to motion is due to lubricant shearing.

8.3.3 BOUNDARY LUBRICATION

Because in **boundary lubrication** the solids are not separated by the lubricant, fluid film effects are negligible and there is considerable asperity contact. The contact lubrication mechanism is governed by the physical and chemical properties of thin surface films of molecular proportions. The properties of the bulk lubricant are of minor importance, and the coefficient of friction is essentially independent of fluid viscosity. The frictional characteristics are determined by the properties of the solids and the lubricant film at the common interfaces. The surface films vary in thickness from 1 to 10 nm—depending on the molecular size.

Figure 8.7 illustrates the film conditions existing in fluid film and boundary lubrication. The surface slopes and film thicknesses in this figure are greatly distorted for purposes of illustration. To scale, real surfaces would appear as gently rolling hills rather than sharp peaks. The surface asperities are not in contact for fluid film lubrication but are in contact for boundary lubrication.

Figure 8.8 shows the behavior of the coefficient of friction [see Eq. (8.33)] in the different lubrication regimes. In boundary lubrication, although the friction is much higher than in the hydrodynamic regime, it is still much lower than for unlubricated surfaces. The

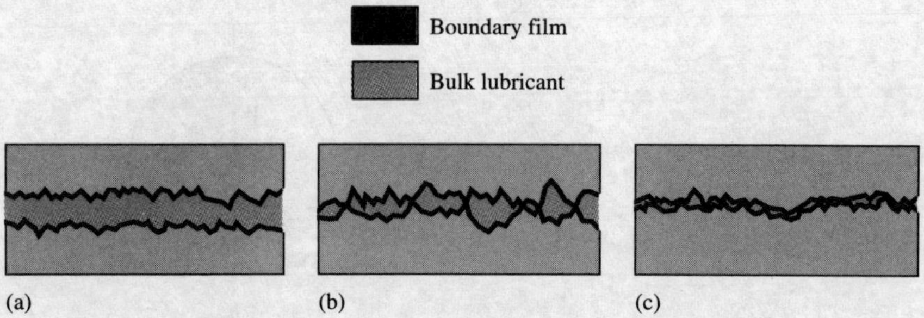

Figure 8.7 Film conditions required for lubrication.
(a) Fluid film lubrication—surfaces separated by bulk lubricant film; (b) partial lubrication—both bulk lubricant and boundary film play a role; (c) boundary lubrication—performance depends essentially on boundary film. [B.J. Hamrock, *Fundamentals of Fluid Film Lubrication*, NASA Reference Publication 1255, 1991.]

mean coefficient of friction increases a total of three orders of magnitude in going from the hydrodynamic to the elastohydrodynamic to the boundary to the unlubricated regime.

Figure 8.9 shows the wear rate in the various lubrication regimes as determined by the operating load. In the hydrodynamic and elastohydrodynamic regimes there is little or no wear because there is no asperity contact. In the boundary lubrication regime the degree of asperity interaction and the wear rate increase as the load increases. The transition from boundary lubrication to an unlubricated condition is marked by a drastic change in wear

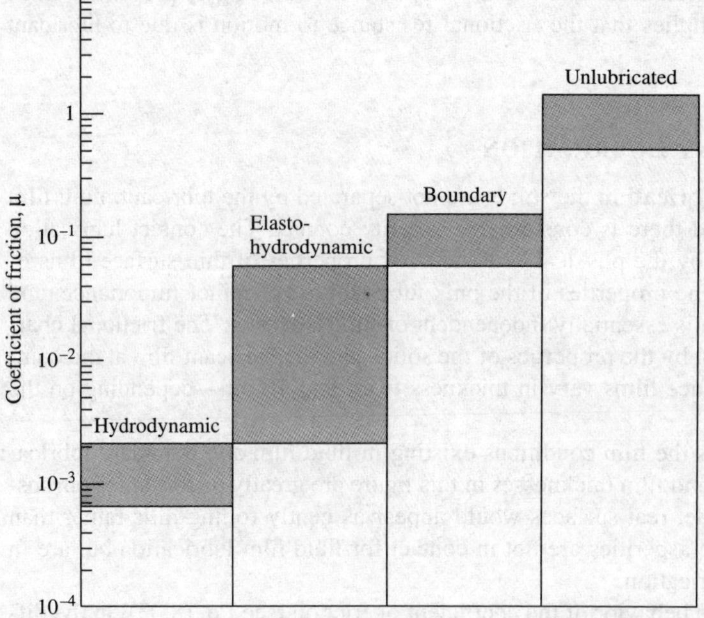

Figure 8.8 Bar diagram showing coefficient of friction for various lubrication conditions. [B.J. Hamrock, *Fundamentals of Fluid Film Lubrication*, NASA Reference Publication 1255, 1991.]

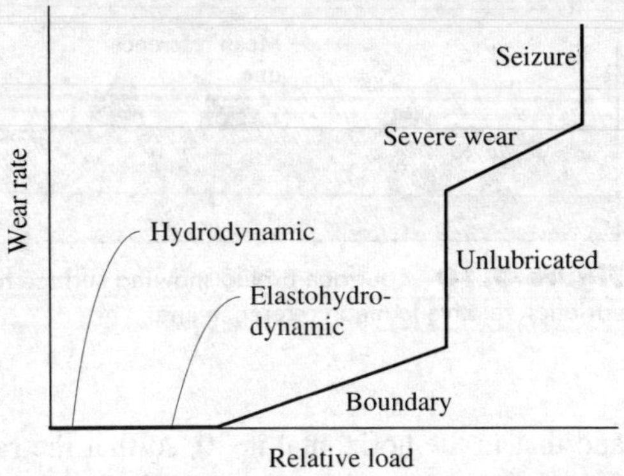

Figure 8.9 Wear rate for various lubrication regimes. [A. Beerbower, "Boundary Lubrication," GRU.IGBEN.72, Report on Scientific and Technical Application Forecasts, (Available NTIS AD-747336), 1972. Reprinted by permission of Exxon Chemical Company.]

rate. As the relative load is increased in the unlubricated regime, the wear rate increases until scoring or seizure occurs and the machine element can no longer operate successfully. Most machine elements cannot operate long with unlubricated surfaces. Together Figs. 8.8 and 8.9 show that the friction and wear of unlubricated machine element surfaces can be greatly decreased by providing boundary lubrication.

Boundary lubrication is used for machine elements with heavy loads and low running speeds, where fluid film lubrication is difficult to attain. Mechanisms such as door hinges operate under conditions of boundary lubrication. Other applications where low cost is of primary importance use boundary lubrication in rubbing sleeve bearings.

8.3.4 PARTIAL LUBRICATION

If the pressures in elastohydrodynamically lubricated machine elements are too high or the running speeds are too low, the lubricant film will be penetrated. Some contact will take place between the asperities, and partial lubrication (sometimes called **mixed lubrication**) will occur. The behavior of the conjunction in a partial lubrication regime is governed by a combination of boundary and fluid film effects. Interaction takes place between one or more molecular layers of boundary lubricating films. A partial fluid film lubrication action develops in the bulk of the space between the solids. The average film thickness in a **partial lubrication** conjunction is less than 1 μm and greater than 0.01 μm.

It is important to recognize that the transition from elastohydrodynamic to partial lubrication does not take place instantaneously as the severity of loading is increased, but rather a decreasing proportion of the load is carried by pressures within the fluid that fills the space between the opposing solids. As the load increases, a larger part of the load is supported by the contact pressure between the asperities of the solids. Furthermore, for conformal surfaces the regime of lubrication goes directly from hydrodynamic to partial lubrication.

8.4 SURFACE PARAMETERS

Designing many machine elements is ultimately a problem of two surfaces that are either in contact or separated by a thin fluid film. In either case the surface texture is important in ensuring long component life. In Fig. 8.10 the magnification in the vertical direction is 1000

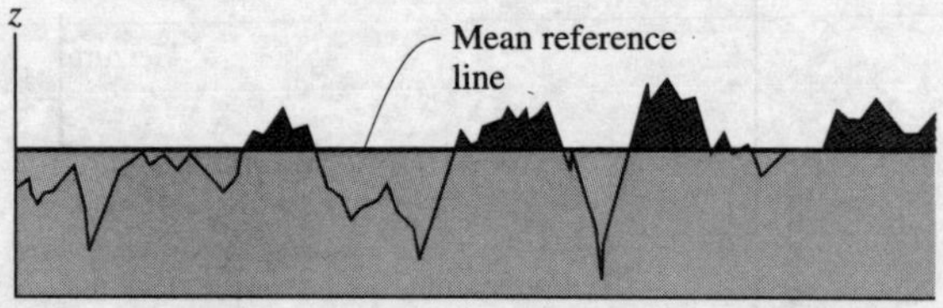

Figure 8.10 Surface profile showing surface height variation relative to mean reference line.

and that in the horizontal is 20, so that the ratio of vertical to horizontal magnification is 50:1. It is therefore important to take into account this difference in magnification. This is often not done, and thus a false impression of the nature of surfaces has become prevalent.

Figure 8.10 is a graph of a surface profile generated by a stylus profilometer. The surface profile shows the surface height variation relative to a mean reference line. The mean, or M system, reference line is based on selecting the mean line as the centroid of the profile. Thus, the areas above and below this line are equal, so that the average of z_i is zero. The white area below the reference line equals the shaded area above the reference line.

Two different surface parameters may be computed:

1. Centerline average or arithmetic average surface roughness, denoted by $\boldsymbol{R_a}$,

$$R_a = \frac{1}{N}\sum_{i=1}^{N}|z_i| \tag{8.5}$$

where

z_i = height from reference line, m

N = number of height measurements taken

2. Root-mean-square (rms) surface roughness, denoted by $\boldsymbol{R_q}$,

$$R_q = \left(\frac{1}{N}\sum_{i=1}^{N} z_i^2\right)^{\frac{1}{2}} \tag{8.6}$$

If a Gaussian height distribution is assumed, R_q has the advantage of being the standard deviation of the profile.

For a simple sinusoidal distribution the ratio of R_q to R_a is

$$\frac{R_q}{R_a} = \frac{\pi}{2\sqrt{2}} = 1.11 \tag{8.7}$$

Table 8.1 gives typical values of the arithmetic average for various processes and components. Note from this table that as higher precision processes are applied, the R_a values decrease significantly. The ultimate process is lapping, which produces an R_a between 0.005 and 0.1 μm. This table contains R_a values for a number of machine elements considered in Chapters 9 to 19.

Table 8.1 Typical arithmetic average surface roughness for various processes and machine components. [B.J. Hamrock, *Fundamentals of Fluid Film Lubrication*, NASA Reference Publication 1255, 1991.]

	Arithmetic average surface roughness, R_a	
	μm	μin.
Processes		
Sand casting; hot rolling	12.5–25	500–1000
Sawing	3.2–25	128–1000
Planing and shaping	.8–25	32–1000
Forging	3.2–12.5	128–500
Drilling	1.6–6.3	64–250
Milling	.8–6.3	32–250
Boring; turning	.4–6.3	16–250
Broaching; reaming; cold rolling; drawing	.8–3.2	32–128
Die casting	.8–1.6	32–64
Grinding, coarse	.4–1.6	16–64
Grinding, fine	.1–.4	4–16
Honing	.03–.4	1.2–16
Polishing	.02–.2	.8–8
Lapping	.005–.1	.2–4
Components		
Gears	.25–10	10–400
Plain bearings—journal (runner)	.12–.5	5–20
Plain bearings—bearing (pad)	.25–1.2	10–50
Rolling bearings—rolling elements	.015–.12	.6–5
Rolling bearings—tracks	.1–.3	4–12

8.5 FILM PARAMETER

When such machine elements as rolling-element bearings, gears, hydrodynamic journal and thrust bearings, and seals (all of which are considered later in this text) are adequately designed and fluid film lubricated, the lubricated surfaces are completely separated by a lubricant film. Endurance testing of ball bearings, for example, as reported by Tallian et al. (1967) has demonstrated that when the lubricant film is thick enough to separate the contacting bodies, bearing fatigue life is greatly extended. Conversely, when the film is not thick enough to provide full separation between the asperities in the contact zone, bearing life is adversely affected by the high shear stresses resulting from direct metal-to-metal contact.

The four lubrication regimes were defined in Sec. 8.3, and calculation methods for determining the rms surface finish were introduced in Sec. 8.4. This section introduces a **film parameter** and describes its range of values for the four lubrication regimes. The relationship between the dimensionless film parameter Λ and the minimum film thickness h_{min} is

$$\Lambda = \frac{h_{\min}}{\left(R_{qa}^2 + R_{qb}^2\right)^{\frac{1}{2}}} \tag{8.8}$$

where

$R_{q\,a}$ = rms surface roughness of surface a

$R_{q\,b}$ = rms surface roughness of surface b

The film parameter is used to define the four important lubrication regimes. The range for these four regimes is

1. *Hydrodynamic lubrication,* $5 \le \Lambda < 100$
2. *Elastohydrodynamic lubrication,* $3 \le \Lambda < 10$
3. *Partial lubrication,* $1 \le \Lambda < 5$
4. *Boundary lubrication,* $\Lambda < 1$

These values are rough estimates. The great differences in geometric conformity in going from hydrodynamically lubricated conjunctions to elastohydrodynamically lubricated conjunctions make it difficult for clear distinctions to be made.

Running-in is a process that affects the film parameter. This process allows wear to occur so that the mating surfaces can adjust to each other to provide for smooth running. This type of wear may be viewed as beneficial. The film parameter will increase with running-in, since the composite surface roughness will decrease. Running-in also has a significant effect on the shape of the asperities that is not captured by the composite surface roughness. With running-in the tips of the asperities in contact become flattened.

Example 8.1

GIVEN Gears for an excavator are manufactured by sand casting. The surface roughness is measured to have a centerline average of 18 μm. This high surface roughness makes the gear wear rapidly. The film thickness for the grease-lubricated gears is calculated to be 1.6 μm.

FIND How should the sand-cast gears be machined to give a film parameter of 1?

Solution

Using Eq. (8.8) while assuming that the roughnesses are equal on the two surfaces gives

$$\Lambda = \frac{h_{\min}}{R_q\sqrt{2}} \quad \text{or} \quad R_q = \frac{h_{\min}}{\Lambda\sqrt{2}} = \frac{1.6}{1\sqrt{2}} = 1.131\ \mu\text{m}$$

Table 8.1 shows that for a surface roughness of 1 μm grinding is the fastest and cheapest method of achieving these surface finishes. Smoother surfaces can be manufactured by honing, polishing, and lapping, but these processes are considerably more expensive.

8.6 LUBRICANT VISCOSITY

A **lubricant** is any substance that reduces friction and wear and provides smooth running and a satisfactory life for machine elements. Most lubricants are liquids (such as mineral oils, synthetic esters, silicone fluids, and water), but they may be solids (such as polytetrafluoroethylene, or PTFE) for use in dry bearings, greases for use in rolling-element bearings, or gases (such as air) for use in gas bearings. The physical and chemical interactions between the lubricant and the lubricating surfaces must be understood in order to provide the machine elements with satisfactory life.

A lubricant is a solid, liquid, or gas that is interposed between solid surfaces to facilitate their relative sliding or rolling, thus controlling friction and wear. Separation of the surfaces is, however, not the only function of a lubricant. Liquid lubricants have desirable secondary properties and characteristics:

1. They can be drawn between moving parts by hydrodynamic action.
2. They have relatively high heat-sink capacity to cool the contacting parts.
3. They are easily mixed with chemicals to give a variety of properties, such as corrosion resistance, detergency, or surface-active layers.
4. They can remove wear particles.

Liquid lubricants can be divided into those of petroleum origin, known as mineral oils, and those of animal or vegetable origin, known as fatty oils. Synthetic oils are often grouped with the latter. For a lubricant to be effective it must be viscous enough to maintain a lubricant film under operating conditions but should be as fluid as possible to remove heat and to avoid power loss due to viscous drag. The most important lubricant property, the viscosity, is considered more fully in the next four subsections.

8.6.1 ABSOLUTE VISCOSITY

Absolute or dynamic **viscosity** can be defined in terms of the simple model shown in Fig. 8.11, which depicts two parallel flat plates separated by a distance h with the upper plate moving with velocity u_b and the lower plate stationary. Oil molecules are visualized as small spheres that roll along in layers between the flat plates. To move the upper plate of area A at a constant velocity u_b across the surface of the oil and cause adjacent layers to flow past each other, a tangential force must be applied. Since the oil will "wet" and cling to the two surfaces, the bottommost layer will not move at all, the topmost layer will move

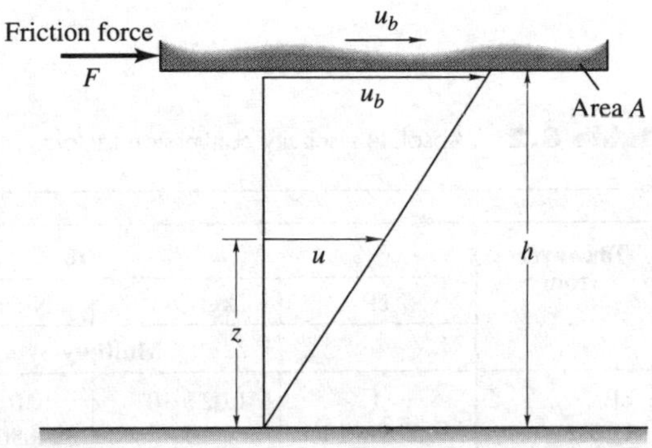

Figure 8.11 Slider bearing illustrating absolute viscosity.

with a velocity equal to the velocity of the upper plate, and the layers in between will move with velocities directly proportional to their distances z from the stationary plate. This type of orderly movement in parallel layers is known as streamline, laminar, or viscous flow.

The shear stress on the oil causing relative movement of the layers is equal to F/A. The shear strain rate s of a particular layer is defined as the ratio of its velocity to its perpendicular distance from the stationary surface z and is constant for each layer.

$$\therefore s = \frac{u}{z} = \frac{u_b}{h} \tag{8.9}$$

The shear strain rate has the unit of reciprocal seconds.

Newton correctly deduced that the force required to maintain a constant velocity u_b of the upper surface is proportional to the area and the shear strain rate or

$$F = \eta A u_b / h \tag{8.10}$$

where

η = absolute viscosity, Pa-s

By rearranging Eq. (8.10) the absolute viscosity can be expressed as

$$\eta = \frac{F/A}{u_b/h} = \frac{\text{Shear stress}}{\text{Shear strain rate}} \tag{8.11}$$

It follows from Eq. (8.11) that the unit of viscosity must be the unit of shear stress divided by the unit of shear strain rate. The units of viscosity η for three different systems are

1. SI units: newton-second per square meter or, since a newton per square meter is also called a pascal, pascal-second
2. cgs units: dyne-second per square centimeter or centipoise, where 1 cP = 10^{-2} P
3. English units: pound force–second per square inch, called a reyn in honor of Osborne Reynolds

Conversion of absolute viscosity from one system to another can be facilitated by Table 8.2. To convert from a unit in the column on the left side of the table to a unit at the top of the table, multiply by the corresponding value given in the table.

Table 8.2 Absolute viscosity conversion factors.

To convert from	To: cP	kgf-s/m^2	N-s/m^2	lbf-s/in.^{2}
	Multiply by			
cP	1	1.02×10^{-4}	10^{-3}	1.45×10^{-7}
kgf-s/m^2	9.807×10^{3}	1	9.807	1.422×10^{-3}
N-s/m^2	10^{3}	1.02×10^{-1}	1	1.45×10^{-4}
lbf-s/in.^{2}	6.9×10^{6}	7.03×10^{2}	6.9×10^{3}	1

Example 8.2

GIVEN An absolute viscosity of 0.04 N-s/m^2.
FIND The absolute viscosity in reyn, centipoise, and poise.

Solution

Using Table 8.2 while recognizing that a reyn = lbf-s/in.^{2} gives

$$\eta = 0.04\ \text{N-s/m}^2 = (0.04)(1.45)(10^{-4})\ \text{lbf-s/in.}^2 = 5.8\times10^{-6}\ \text{lbf-s/in.}^2$$

Note also that

$$\eta = 0.04\ \text{N-s/m}^2 = 0.04\ \text{Pa-s} = 5.8\times10^{-6}\ \text{reyn}$$

and

$$\eta = 0.04\ \text{N-s/m}^2 = (0.04)(10^3)\ \text{cP} = 40\ \text{cP} = 0.4\ \text{P}$$

Example 8.3

GIVEN An absolute viscosity of 1 reyn.
FIND The absolute viscosity in centipoise.

Solution

$$1\ \text{reyn} = 1\ \text{lbf-s/in.}^2 = 7.03\times10^2\ \text{kgf-s/in.}^2 = 6.9\times10^6\ \text{cP}$$

$$\therefore 1\ \mu\text{reyn} = 1\times10^{-6}\ \text{reyn} = 6.9\ \text{cP}$$

8.6.2 KINEMATIC VISCOSITY

In many situations it is convenient to use **kinematic viscosity** rather than absolute viscosity. The kinematic viscosity η_k is defined as

$$\eta_k = \frac{\text{Absolute viscosity}}{\text{Density}} = \frac{\eta}{\rho} = \frac{\text{N-s/m}^2}{\text{N-s}^2/\text{m}^4} = \text{m}^2/\text{s} \qquad \textbf{(8.12)}$$

The ratio given in Eq. (8.12) is literally kinematic; all trace of force or mass cancels out. The units of kinematic viscosity are

1. SI units: square meters per second
2. cgs units: square centimeters per second, called a stoke
3. English units: square inches per second

Example 8.4

GIVEN Both mercury and water have an absolute viscosity of 1.5 cP at 5°C, but mercury has 13.6 times higher density than water. Assume that no change in density occurs when the temperature is increased.
FIND The kinematic viscosities of mercury and water at 5°C and at 90°C.

Solution

Figure 8.12 gives the absolute viscosities of various fluids as a function of temperature. At 90°C the absolute viscosity of water is 0.32 cP, and for mercury it is 1.2 cP. The kinematic viscosity of *mercury* at 5°C is

$$\eta_k = \frac{\eta}{\rho} = \frac{1.5(10^{-3})}{13\ 600} = 0.110(10^{-6})\ \text{m}^2/\text{s}$$

At 90°C it is

$$\eta_k = \frac{\eta}{\rho} = \frac{1.2(10^{-3})}{13\ 600} = 0.0882(10^{-6})\ \text{m}^2/\text{s}$$

The kinematic viscosity of *water* at 5°C is

$$\eta_k = \frac{\eta}{\rho} = \frac{1.5(10^{-3})}{1000} = 1.50(10^{-6})\ \text{m}^2/\text{s}$$

At 90°C it is

$$\eta_k = \frac{\eta}{\rho} = \frac{0.32(10^{-3})}{1000} = 0.32(10^{-6})\ \text{m}^2/\text{s}$$

Although mercury has the same absolute viscosity as water at 5°C and 3.75 times higher absolute viscosity at 90°C, the kinematic viscosities for mercury are much lower than those for water because of mercury's high density.

8.6.3 Viscosity-Pressure Effects

As long ago as 1893, Barus proposed the following formula for the isothermal viscosity-pressure dependence of liquids:

$$\ln \frac{\eta}{\eta_0} = \xi p \tag{8.13}$$

where

$\ln$ = natural, or Napierian, logarithm, $\log_e$

η_0 = absolute viscosity at $p = 0$ and at a constant temperature, cP

ξ = pressure-viscosity coefficient of lubricant dependent on temperature, m²/N

p = pressure, Pa

Table 8.3 lists the kinematic viscosities in square meters per second and the absolute viscosities in centipoise of 11 fluids at zero pressure and 3 temperatures. These values of the absolute viscosity correspond to η_0 in Eq. (8.13) for the particular fluid and temperature used. The manufacturer and the manufacturer's designation for these 11 fluids are given in Table 8.4; and the pressure-viscosity coefficients ξ, expressed in square meters per newton, in Table 8.5. The values correspond to ξ in Eq. (8.13).

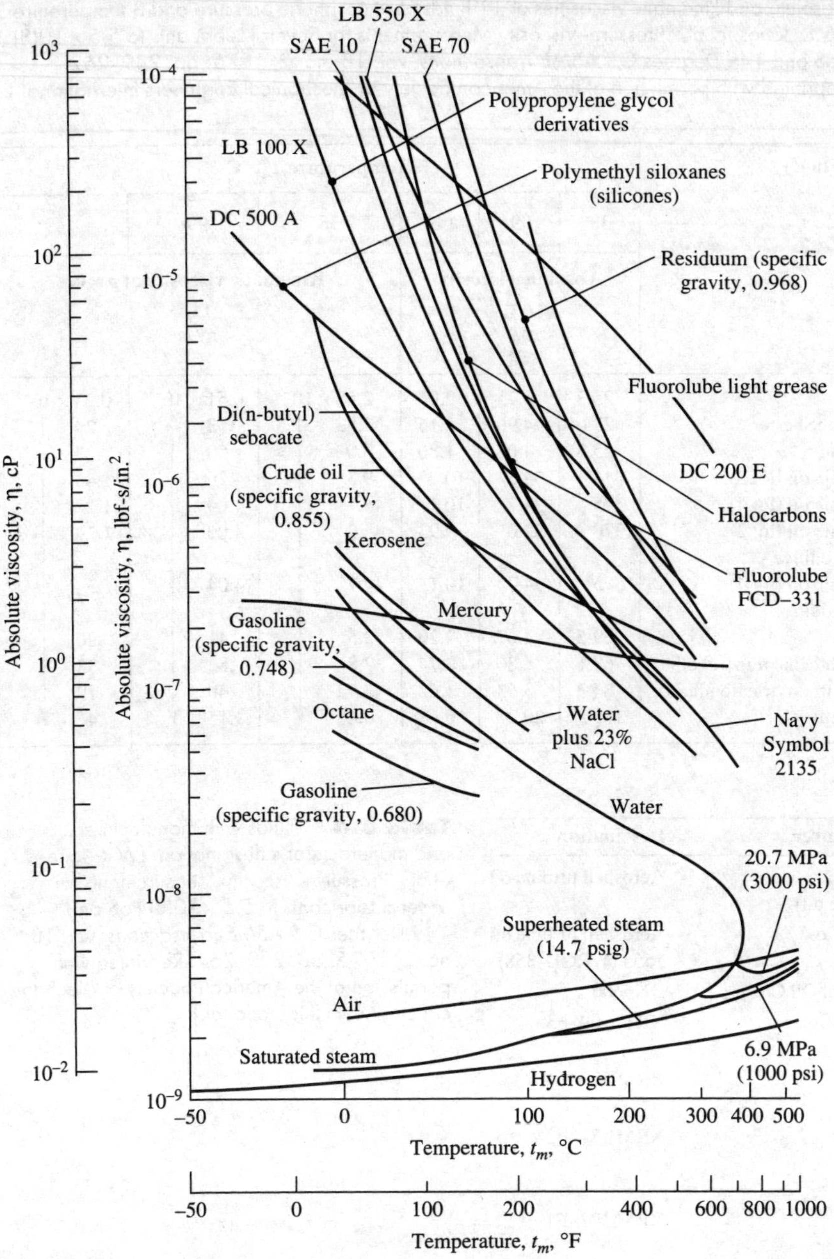

Figure 8.12 Absolute viscosities of a number of fluids for wide range of temperatures.

Table 8.3 Absolute and kinematic viscosities of 11 fluids at atmospheric pressure and 3 temperatures. [W.R. Jones, et al, "Pressure–Viscosity Measurements for Several Lubricants to 5.5 × 10(8) PaS and 149 Degrees C," *ASME Transactions*, vol. 18, no. 2, 1975, pp. 249–262. Reprinted with permission of the American Society of Mechanical Engineers International.]

Fluid	Temperature, t_m, °C					
	38	99	149	38	99	149
	Absolute viscosity at $p = 0$, η_0, cP			Kinematic viscosity at $p = 0$, η_k, m^2/s		
Advanced ester	25.3	4.75	2.06	2.58×10^{-5}	0.51×10^{-5}	0.23×10^{-5}
Formulated advanced ester	27.6	4.96	2.15	2.82	.53	.24
Polyalkyl aromatic	25.5	4.08	1.80	3.0	.50	.23
Synthetic paraffinic oil (lot 3)	414	34.3	10.9	49.3	4.26	1.4
Synthetic paraffinic oil (lot 4)	375	34.7	10.1	44.7	4.04	1.3
Synthetic paraffinic oil (lot 2) plus antiwear additive	370	32.0	9.93	44.2	4.00	1.29
Synthetic paraffinic oil (lot 4) plus antiwear additive	375	34.7	10.1	44.7	4.04	1.3
C-ether	29.5	4.67	2.20	2.5	.41	.20
Superrefined naphthenic mineral oil	68.1	6.86	2.74	7.8	.82	.33
Synthetic hydrocarbon (traction fluid)	34.3	3.53	1.62	3.72	.40	.19
Fluorinated polyether	181	20.2	6.68	9.66	1.15	.4

Table 8.4 Fluids with manufacturer and manufacturer's designation. [W.R. Jones, et al, "Pressure–Viscosity Measurements for Several Lubricants to 5.5 × 10(8) PaS and 149 Degrees C," *ASME Transactions*, vol. 18, no. 2, 1975, pp. 249–262. Reprinted with permission of the American Society of Mechanical Engineers International.]

Fluid	Manufacturer	Designation
Advanced ester	Shell Oil Co. 555 (base oil)	Aeroshell turbine oil
Formulated advanced ester	Shell Oil Co.	Aeroshell turbine oil 555 (WRGL–358)
Polyalkyl aromatic	Continental Oil Co.	DN–600
Synthetic paraffinic oil (lot 3)	Mobil Oil Co.	XRM 109–F3
Synthetic paraffinic oil (lot 4)		XRM 109–F4
Synthetic paraffinic oil (lot 2) plus antiwear additive		XRM 177–F2
Synthetic paraffinic oil (lot 4) plus antiwear additive		XRM 177–F4
C-ether	Monsanto Co.	MCS–418
Superrefined naphthenic mineral oil	Humble Oil and Refining Co.	FN 2961
Synthetic hydrocarbon (traction fluid)	Monsanto Co.	MCS–460
Fluorinated polyether	DuPont Co.	PR 143 AB (lot 10)

Table 8.5 Pressure-viscosity coefficients of 11 fluids at 3 temperatures. [W.R. Jones, et al, "Pressure–Viscosity Measurements for Several Lubricants to 5.5 × 10(8) PaS and 149 Degrees C," *ASME Transactions*, vol. 18, no. 2, 1975, pp. 249–262. Reprinted with permission of the American Society of Mechanical Engineers International.]

Fluid	Temperature, t_m, °C		
	38	99	149
	Pressure-viscosity coefficient, ξ, m^2/N		
Advanced ester	1.28×10^{-8}	0.987×10^{-8}	0.851×10^{-8}
Formulated advanced ester	1.37	1.00	.874
Polyalkyl aromatic	1.58	1.25	1.01
Synthetic paraffinic oil (lot 3)	1.77	1.51	1.09
Synthetic paraffinic oil (lot 4)	1.99	1.51	1.29
Synthetic paraffinic oil (lot 2) plus antiwear additive	1.81	1.37	1.13
Synthetic paraffinic oil (lot 4) plus antiwear additive	1.96	1.55	1.25
C-ether	1.80	.980	.795
Superrefined naphthenic mineral oil	2.51	1.54	1.27
Synthetic hydrocarbon (traction fluid)	3.12	1.71	.939
Fluorinated polyether	4.17	3.24	3.02

Example 8.5

GIVEN A synthetic paraffin has an absolute viscosity of 32 cP at 99°C, and a fluorinated polyether has an absolute viscosity of 20.2 cP at 99°C.

FIND

a) The pressure at which the two oils have the same absolute viscosity

b) The pressure at which the paraffin is 100 times less viscous than the polyether

Solution

a) From Table 8.5 the pressure-viscosity coefficient at 99°C for the paraffin is 1.37×10^{-8} m^2/s and for the polyether it is 3.24×10^{-8} m^2/s. Using the Barus equation [(Eq. 8.13)] while equating the viscosity gives

$$32e^{1.37\times10\ ^{8}p} = 20.2e^{3.24\times10^{-8}p}$$

$$1.58e^{1.37\times10^{-8}p} = e^{3.24\times10^{-8}p}$$

But

$$1.58 = e^{0.46}$$

$$\therefore\ e^{0.46+1.37\times10^{-8}p} = e^{3.24\times10^{-8}p}$$

$$p = 24.6\times10^{6}\ \text{Pa} = 24.6\ \text{MPa}$$

b) Using the Barus equation [Eq. 8.13)] while letting the paraffin be 100 times less viscous than the polyether gives

$$1.58(100)e^{1.37\times10^{-8}p} = e^{3.24\times10^{-8}p}$$

But

$$158 = e^{5.065}$$

$$\therefore \quad e^{5.065+1.37\times10^{-8}p} = e^{3.24\times10^{-8}p}$$

$$p = 2.71\times10^{8} \text{ Pa} = 0.271 \text{ GPa}$$

8.6.4 Viscosity-Temperature Effects

The viscosity of mineral and synthetic oils decreases with increasing temperature. Therefore, the temperature at which the viscosity was measured must be quoted with every viscosity reported. Figures 8.12 and 8.13 show how absolute viscosity varies with temperature. Figure 8.12 presents the absolute viscosity of several fluids for a wide temperature range. The interesting point of this figure is how drastically the slope and level of viscosity change for different fluids. The viscosity varies by five orders of magnitude, with the slope being highly negative for the Society of Automotive Engineers (SAE) oils and positive for gases. Figure 8.13 gives the viscosity-temperature effect for the SAE oils. Although in Fig. 8.13 the absolute viscosity is given in newton-second per meter squared, by using Table 8.2 the absolute viscosity in other units can be obtained. According to Table 1.4 (also given on the inside right cover) the conversion to degrees Fahrenheit is $t_F = (9/5)\, t_C + 32$ (where t_c is degrees Celsius).

Example 8.6

GIVEN An SAE 10 oil is used in a car motor where the working temperature is 110°C.

FIND How many times more viscous would the oil be if the motor were in arctic conditions where the working temperature is –30°C?

Solution

Figure 8.13 gives the viscosity at 110°C to be 3.3 cP and by extrapolating to be 10^4 cP at –30°C. The viscosity ratio is $10^4/3.3 = 3030$. The oil is 3030 times more viscous at –30° C than at 110° C.

8.7 Concentrated Loading—Deformations and Stresses

The entire text up to this point has been focused on distributed loading with the exception of Sec. 2.3, which pointed out the differences between distributed and concentrated loading. This section describes not only the deformation due to concentrated loading but also the surface and subsurface stresses. The focus is mostly on elliptical contacts, but at the end of this section rectangular contact situations are also presented.

8.7.1 ELLIPTICAL CONTACTS

The undeformed geometry of nonconformal contacting solids (see Sec. 8.2) can be represented in general terms by two ellipsoids, as shown in Fig. 8.14. Two solids with different radii of curvature in a pair of principal planes (x and y) passing through the conjunction between the solids make contact at a single point under the condition of zero applied load. Such a condition is called point contact and is shown in Fig. 8.14, where the radii of curvature are denoted by r's. It is assumed throughout this book that convex surfaces like those in Fig. 8.14 exhibit positive curvature and concave surfaces, negative curvature. Therefore, if the center of curvature lies within the solid, the radius of curvature is positive; if the center of curvature lies outside the solid, the radius of curvature is negative. Figure 8.15 shows the elements and bearing races. The importance of the sign of the radius of curvature presents itself later in the chapter.

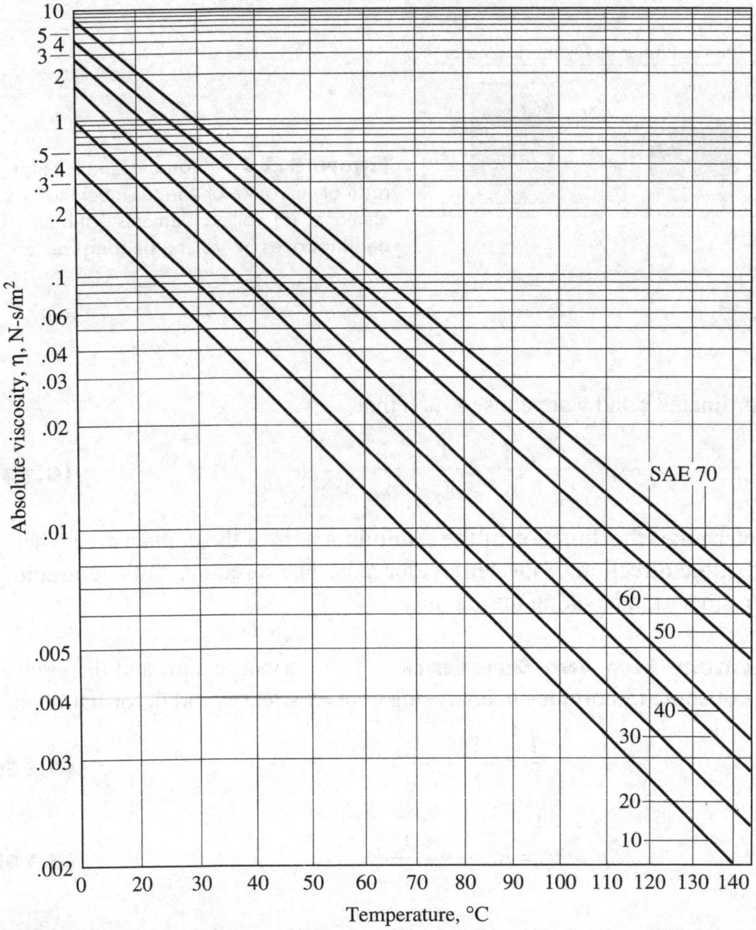

Figure 8.13 Absolute viscosities of SAE lubricating oils at atmospheric pressure.

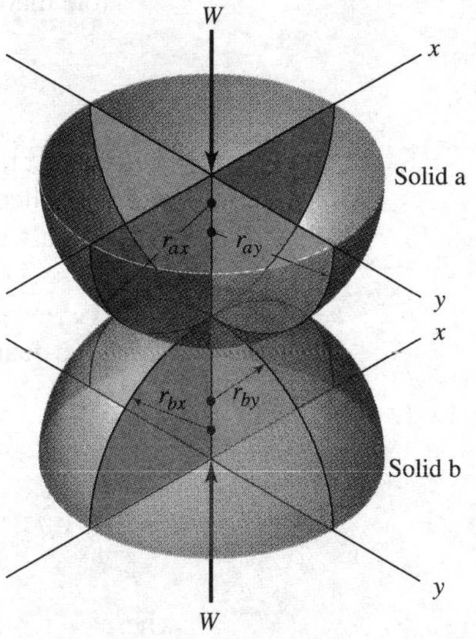

Figure 8.14 Geometry of contacting elastic solids.

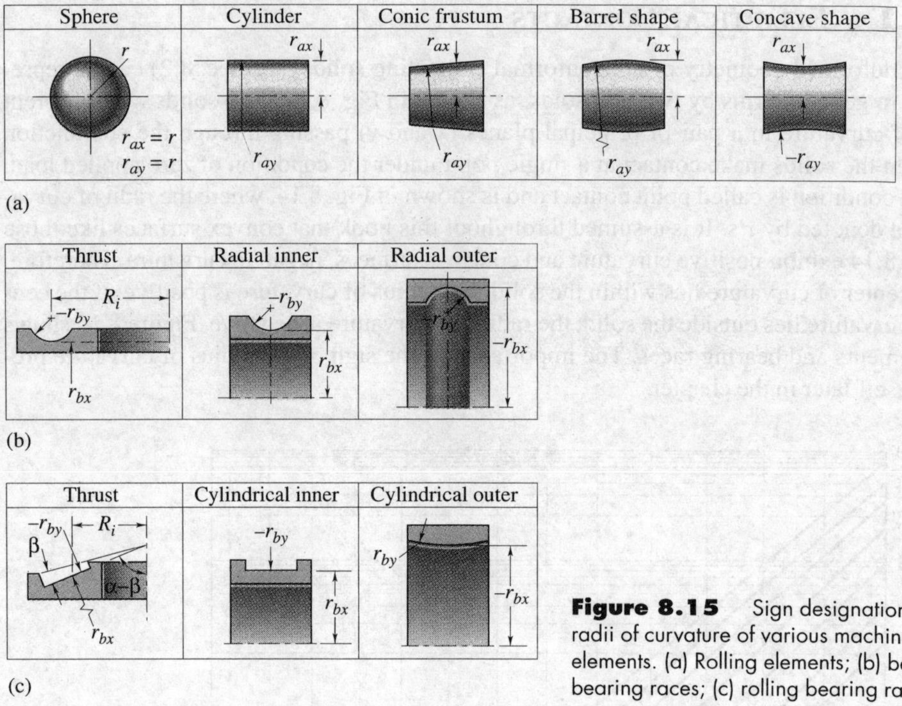

Figure 8.15 Sign designations for radii of curvature of various machine elements. (a) Rolling elements; (b) ball bearing races; (c) rolling bearing races.

Note that if coordinates *x* and *y* are chosen such that

$$\frac{1}{r_{ax}} + \frac{1}{r_{bx}} \geq \frac{1}{r_{ay}} + \frac{1}{r_{by}} \tag{8.14}$$

coordinate *x* then determines the direction of the semiminor axis of the contact area when a load is applied; and *y*, the direction of the semimajor axis. The direction of the entraining motion is always considered to be along the *x* axis.

8.7.1.1 Curvature Sum and Difference The curvature sum and difference, which are quantities of some importance in analyzing contact stresses and deformation, are

$$\frac{1}{R} = \frac{1}{R_x} + \frac{1}{R_y} \tag{8.15}$$

$$R_d = R\left(\frac{1}{R_x} - \frac{1}{R_y}\right) \tag{8.16}$$

where

$$\frac{1}{R_x} = \frac{1}{r_{ax}} + \frac{1}{r_{bx}} \tag{8.17}$$

$$\frac{1}{R_y} = \frac{1}{r_{ay}} + \frac{1}{r_{by}} \tag{8.18}$$

Equations (8.17) and (8.18) effectively redefine the problem of two ellipsoidal solids approaching one another in terms of an equivalent solid of radii R_x and R_y approaching a plane. Note that the curvature difference expressed in Eq. (8.16) is dimensionless.

The radius ratio α_r is

$$\alpha_r = \frac{R_y}{R_x} \tag{8.19}$$

Thus, if Eq. (8.14) is satisfied, $\alpha_r > 1$; and if it is not satisfied, $\alpha_r < 1$.

Machine elements (considered later in the text) with nonconformal surfaces (concentrated loading) have a complete range of radius ratios from 0.03 to 100. That is, for a traction drive simulated by a disk rolling on a plane, $\alpha_r = 0.03$; for a ball-on-plane contact, $\alpha_r = 1.0$, and for a contact approaching a nominal line contact, $\alpha_r \to 100$, such as in a barrel-shaped roller bearing against a plane. Some further examples of machine elements with radius ratios less than 1 are Novikov gear contacts, locomotive wheel-rail contacts, and roller-flange contacts in a radial loaded roller bearing. Some further examples of machine elements with radius ratios greater than 1 are rolling-element bearings and most gears. These topics are considered later in the text, but the general information on contact geometry of nonconformal surfaces is an important feature to consider here.

8.7.1.2 ELLIPTICITY PARAMETER The **ellipticity parameter** k_e is defined as the elliptical contact diameter in the y direction (transverse direction) divided by the elliptical contact diameter in the x direction (direction of entraining motion), or

$$k_e = \frac{D_y}{D_x} \tag{8.20}$$

If $\alpha_r \geq 1$, the contact ellipse will be oriented with its major diameter transverse to the direction of motion, and consequently $k_e \geq 1$. Otherwise, the major diameter would lie along the direction of motion with both $\alpha_r < 1$ and $k_e < 1$. To avoid confusion, the commonly used solutions to the surface deformation and stresses are presented only for $\alpha_r > 1$.

Note that the ellipticity parameter is a function only of the solids' radii of curvature:

$$k_e = f\left(r_{ax}, r_{bx}, r_{ay}, r_{by}\right) \tag{8.21}$$

That is, as the load increases, the semiaxes in the x and y directions of the contact ellipse increase proportionately to each other so that the ellipticity parameter remains constant.

8.7.1.3 CONTACT PRESSURE When an elastic solid is subjected to a load, stresses are produced that increase as the load is increased. These stresses are associated with deformations, which are defined by strains. Unique relationships exist between stresses and their corresponding strains, as shown in Appendix B. For elastic solids the stresses are linearly related to the strains, with the modulus of elasticity (or the proportionality constant) being an elastic constant that adopts different values for different materials, as covered in Sec. 3.6. The modulus of elasticity E and Poisson's ratio ν are two important parameters defined in Chapter 3 that are used in this chapter to describe contacting solids.

As the stresses increase within the material, elastic behavior is replaced by plastic flow in which the material is permanently deformed. The stress state at which the transition from elastic to plastic behavior occurs, known as the yield stress, has a definite value for a given material at a given temperature. In this book only elastic behavior is considered.

When two elastic nonconformal solids are brought together under a load, a contact area develops whose shape and size depend on the applied load, the elastic properties of the materials, and the curvature of the surfaces. When the two solids shown in Fig. 8.14 have a normal load applied to them, the contact area is elliptical. It has been common to refer to elliptical contacts as point contacts, but since under load these contacts become elliptical, they are referred to herein as such. For the special case where $r_{ax} = r_{ay}$ and $r_{bx} = r_{by}$ the resulting contact is a circle rather than an ellipse. When r_{ay} and r_{by} are both infinite, the initial line contact develops into a rectangle when load is applied.

Hertz (1881) considered the stresses and deformations in two perfectly smooth, ellipsoidal, contacting solids much like those shown in Fig. 8.14. His application of the classical elasticity theory to this problem forms the basis of stress calculations for machine elements, such as ball and roller bearings, gears, and cams and followers. Hertz made the following assumptions:

1. The materials are homogeneous and the yield stress is not exceeded.
2. No tangential forces are induced between the solids.
3. Contact is limited to a small portion of the surface such that the dimensions of the contact region are small relative to the radii of the ellipsoids.
4. The solids are at rest and in equilibrium.

Making use of these assumptions, Hertz (1881) was able to obtain the following expression for the pressure distribution within an ellipsoidal contact (shown in Fig. 8.16):

$$p_H = p_{\max}\left[1-\left(\frac{2x}{D_x}\right)^2-\left(\frac{2y}{D_y}\right)^2\right]^{1/2} \tag{8.22}$$

where

D_x = diameter of contact ellipse in x direction, m

D_y = diameter of contact ellipse in y direction, m

If the pressure is integrated over the contact area, the maximum pressure is

$$p_{\max} = \frac{6W}{\pi D_x D_y} \tag{8.23}$$

where W is the normal applied load. Equation (8.22) determines the distribution of pressure or compressive stress on the common interface, which is clearly a maximum at the contact center and decreases to zero at the periphery.

8.7.1.4 SIMPLIFIED SOLUTIONS The classical Hertzian solution requires the calculation of the ellipticity parameter k_e as well as the complete elliptic integrals of the first and second kinds $\mathcal{F}$ and $\mathcal{E}$. This calculation entails finding a solution to a transcendental equation relating k_e, $\mathcal{F}$, and $\mathcal{E}$ to the geometry of the contacting solids and is usually

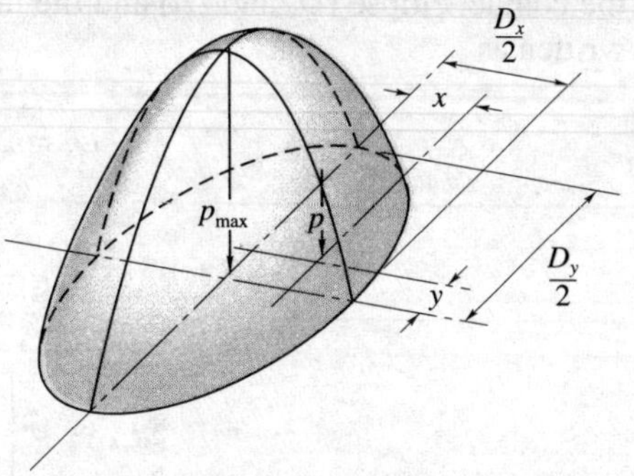

Figure 8.16 Pressure distribution in ellipsoidal contact.

accomplished by some iterative numerical procedure. Hamrock and Brewe (1983) used a linear regression by the method of least squares to power fit the pairs of data and arrived at the simplified equations shown in Table 8.6. From this table the ellipticity parameter for the complete radius ratio range can be expressed as

$$k_e = \alpha_r^{2/\pi} \tag{8.24}$$

Figure 8.17 shows the ellipticity parameter and the elliptic integrals of the first and second kinds for a radius ratio ($\alpha_r = R_y/R_x$) range usually encountered in nonconformal conjunctions. Note from Fig. 8.17 that $\mathcal{F} = \mathcal{E} = \pi/2$ when $\alpha_r = 1$. Also, both $\mathcal{F}$ and $\mathcal{E}$ are discontinuous at $\alpha_r = 1$. Thus, the reason for the two columns presented in Table 8.6.

When the ellipticity parameter k_e, the normal applied load W, Poisson's ratio ν, and the modulus of elasticity E of the contacting solids are known, the major and minor axes of

Table 8.6 Simplified equations. [B.J. Hamrock and D.E. Brewe, "Simplified Solutions to Stresses and Deformations," *Journal of Lubrication Technology,* vol. 105, no. 2, 1983, pp. 171–77. Reprinted by permission of the American Society of Mechanical Engineers International.]

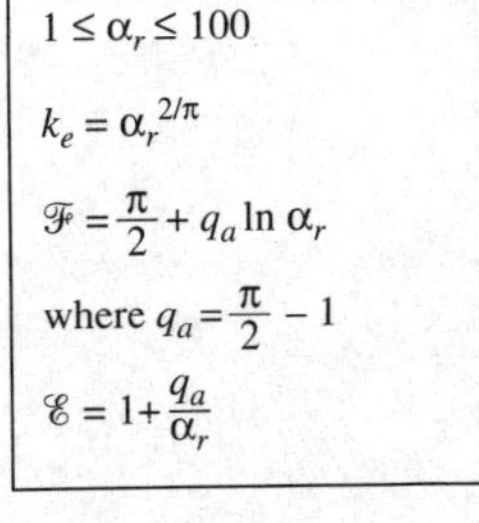	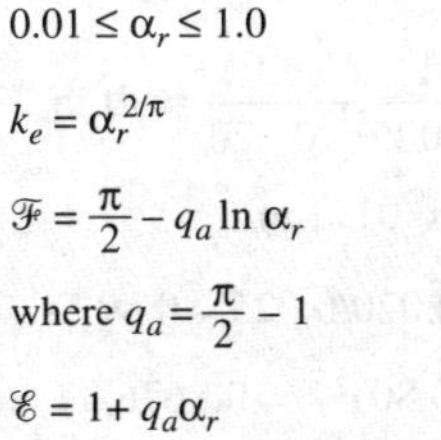
$1 \le \alpha_r \le 100$	$0.01 \le \alpha_r \le 1.0$
$k_e = \alpha_r^{2/\pi}$	$k_e = \alpha_r^{2/\pi}$
$\mathcal{F} = \frac{\pi}{2} + q_a \ln \alpha_r$	$\mathcal{F} = \frac{\pi}{2} - q_a \ln \alpha_r$
where $q_a = \frac{\pi}{2} - 1$	where $q_a = \frac{\pi}{2} - 1$
$\mathcal{E} = 1 + \frac{q_a}{\alpha_r}$	$\mathcal{E} = 1 + q_a \alpha_r$

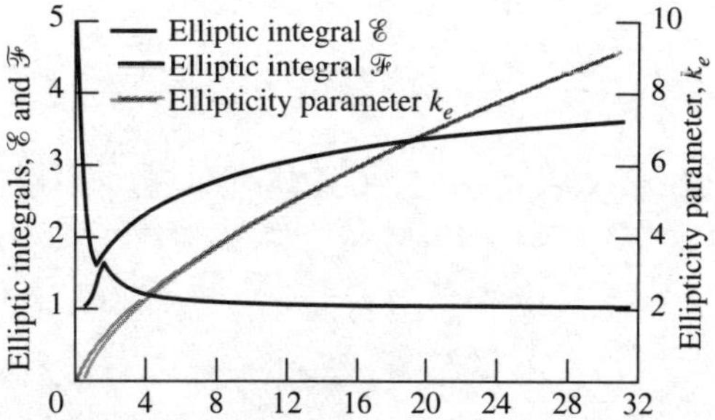

Figure 8.17 Variation of ellipticity parameter and elliptic integrals of first and second kinds as function of radius ratio.

the contact ellipse (D_y and D_x) and the maximum deformation at the contact center can be written as

$$D_y = 2\left(\frac{6k_e^{\,2}\mathscr{E}WR}{\pi E'}\right)^{1/3} \tag{8.25}$$

$$D_x = 2\left(\frac{6\mathscr{E}WR}{\pi k_e E'}\right)^{1/3} \tag{8.26}$$

$$\delta_{\max} = \mathscr{F}\left[\frac{9}{2\mathscr{E}R}\left(\frac{W}{\pi k_e E'}\right)^2\right]^{1/3} \tag{8.27}$$

$$E' = \frac{2}{\dfrac{\left(1-\nu_a^2\right)}{E_a} + \dfrac{\left(1-\nu_b^2\right)}{E_b}} \tag{8.28}$$

In these equations D_y and D_x are proportional to $W^{1/3}$, and $\delta_{\max}$ is proportional to $W^{2/3}$.

Example 8.7

GIVEN A solid sphere of 2-cm radius ($r_{ax} = r_{by} = 2$ cm) rolls on the inside of an outer race with a 10-cm internal radius ($r_{bx} = -10$ cm) and a large width in the axial direction. The sphere is made of silicon nitride and the outer race is made of stainless steel. The normal applied load is 1000 N.

FIND

a) Curvature sum R
b) Ellipticity parameter k_e
c) Elliptic integrals of the first and second kinds $\mathscr{F}$ and $\mathscr{E}$
d) Effective modulus of elasticity E'
e) Dimensions of elliptical contact D_x and D_y
f) Maximum Hertzian pressure or stress on surface $p_{\max}$
g) Maximum deformation $\delta_{\max}$

Solution

a) The curvature sum is

$$\frac{1}{R_x} = \frac{1}{r_{ax}} + \frac{1}{r_{bx}} = \frac{1}{0.02} - \frac{1}{0.10} = 40 \text{ m}^{-1}$$

$$\therefore R_x = 0.025 \text{ m}$$

$$\frac{1}{R_y} = \frac{1}{r_{ay}} + \frac{1}{r_{by}} = \frac{1}{0.02} + \frac{1}{\infty} = \frac{1}{0.02} \text{ m}^{-1}$$

$$\therefore R_y = 0.02 \text{ m}$$

$$\frac{1}{R} = \frac{1}{R_x} + \frac{1}{R_y} = \frac{1}{0.025} + \frac{1}{0.020} = 90 \text{ m}^{-1}$$

$$\therefore R = 0.0111 \text{ m}$$

b) The ellipticity parameter, where $\alpha_r = R_y/R_x = 0.020/0.025 = 0.80$, is

$$k_e = \alpha_r^{2/\pi} = (0.80)^{2/\pi} = 0.8676$$

c) From Table 8.6 the elliptical integrals of the first and second kinds are

$$\mathcal{F} = \frac{\pi}{2} - \left(\frac{\pi}{2} - 1\right)\ln(0.8) = 1.698$$

$$\mathcal{E} = 1 + \left(\frac{\pi}{2} - 1\right)(0.8) = 1.457$$

d) From Table 3.2 $E_a = 314$ GPa and $E_b = 193$ GPa. From Table 3.3 $\nu_a = 0.26$ and $\nu_b = 0.30$. The effective modulus of elasticity is

$$E' = \frac{2}{\dfrac{(1-\nu_a^2)}{E_a} + \dfrac{(1-\nu_b^2)}{E_b}} = \frac{2}{\dfrac{(1-0.26^2)}{3.14\times 10^{11}} + \dfrac{(1-0.32^2)}{1.93\times 10^{11}}} = 2.60\times 10^{11}\ \text{N/m}^2$$

e) The dimensions of the elliptical contact, from Eqs. (8.25) and (8.26), are

$$D_y = 2\left(\frac{6k_e^2\mathcal{E}WR}{\pi E'}\right)^{1/3} = 2\left[\frac{6(0.8676)^2(1.457)(1000)(0.0111)}{\pi(2.60)(10)^{11}}\right]^{1/3} = 0.8943\times 10^{-3}\ \text{m}$$

$$D_x = 2\left(\frac{6\mathcal{E}WR}{\pi k_e E'}\right)^{1/3} = 2\left[\frac{6(1.457)(1000)(0.0111)}{\pi(0.8676)(2.6)(10)^{11}}\right]^{1/3} = 1.0308\times 10^{-3}\ \text{m}$$

f) The maximum pressure, from Eq. (8.23), is

$$p_{\max} = \frac{6W}{\pi D_x D_y} = \frac{6(1000)}{\pi(0.8943)(1.0308)(10)^{-6}}\ \text{Pa} = 2.07\ \text{GPa}$$

g) From Eq. (8.27) the maximum deformation at the center of the conjunction is

$$\delta_{\max} = \mathcal{F}\left[\frac{9}{2\mathcal{E}R}\left(\frac{W}{\pi k_e E'}\right)^2\right]^{1/3} = 1.6981\left\{\frac{4.5}{(1.4566)(0.0111)}\left[\frac{1000}{\pi(0.8676)(2.6)(10)^{11}}\right]^2\right\}^{1/3}$$

$$= 13.95\times 10^{-6}\ \text{m}$$

Example 8.8

GIVEN The balls in a deep-groove ball bearing have a 17-mm diameter. The groove in the inner race has an 8.84-mm radius; the radius from the center of the race to the bottom of the groove is 27.5 mm. The load being applied is 20 000 N. The ball and race material is steel.

FIND

a) The dimensions of the contact ellipse
b) The maximum deformation

Solution

a) From Eqs. (8.17) and (8.18) the curvature sums in the *x* and *y* directions are

$$\frac{1}{R_x} = \frac{1}{r_{ax}} + \frac{1}{r_{bx}} = \frac{1}{8.5\times 10^{-3}} + \frac{1}{27.5\times 10^{-3}} = 154.0\ \text{m}^{-1}$$

$$\frac{1}{R_y} = \frac{1}{r_{ay}} + \frac{1}{r_{by}} = \frac{1}{8.5\times10^{-3}} - \frac{1}{8.84\times10^{-3}} = 4.525 \text{ m}^{-1}$$

From Eqs. (8.15) and (8.16) the curvature sum and difference are

$$\frac{1}{R} = \frac{1}{R_x} + \frac{1}{R_y} = 158.5 \text{ m}^{-1}$$

$$R_d = R\left(\frac{1}{R_x} - \frac{1}{R_y}\right) = \frac{1}{158.5}(154.0 - 4.525) = 0.9432$$

Using Eq. (8.19) gives the radius ratio as

$$\alpha_r = \frac{R_y}{R_x} = \frac{154.0}{4.525} = 34.03$$

From Table 8.6 the elliptical parameter and the elliptic integrals of the first and second kinds are

$$k_e = \alpha_r^{2/\pi} = 34.03^{2/\pi} = 9.446$$

$$\mathcal{F} = \frac{\pi}{2} + \left(\frac{\pi}{2} - 1\right)\ln \alpha_r = \frac{\pi}{2} + \left(\frac{\pi}{2} - 1\right)\ln 34.03 = 3.584$$

$$\mathcal{E} = 1 + \frac{\frac{\pi}{2} - 1}{34.03} = 1.017$$

The effective modulus of elasticity is

$$E' = \frac{E}{1-\nu^2} = \frac{207\times10^9}{1-0.3^2} \text{ Pa} = 227.5 \text{ GPa}$$

From Eqs. (8.25) and (8.26) the dimensions of the contact ellipse are

$$D_y = 2\left(\frac{6k_e^2\mathcal{E}WR}{\pi E'}\right)^{1/3} = 2\left[\frac{6(9.446)^2(1.017)(20\ 000)}{158.5\pi(227.5)(10^9)}\right]^{1/3} \text{ m} = 9.162 \text{ mm}$$

$$D_x = \frac{D_y}{k_e} = \frac{9.162}{9.446} = 0.9699 \text{ mm}$$

b) From Eq. (8.27) the maximum deformation in the center of contact is

$$\delta_{max} = \mathcal{F}\left[\frac{9}{2\mathcal{E}R}\left(\frac{W}{\pi k_e E'}\right)^2\right]^{1/3} = 3.584\left[\frac{9(158.5)}{2(1.017)}\left(\frac{20\ 000}{\pi(9.446)(227.5)(10^9)}\right)^2\right]^{1/3}$$

$$= 65.69\times10^{-6} \text{ m} = 65.69\ \mu\text{m}$$

8.7.2 RECTANGULAR CONTACTS

For rectangular conjunctions the contact ellipse discussed throughout this chapter is of infinite width in the transverse direction ($D_y \to \infty$). This type of contact is exemplified by a cylinder loaded against a plane, a groove, or another parallel cylinder or by a roller loaded against an inner or outer race. In these situations the contact semiwidth is given by

$$b^* = R_x\left(\frac{8W'}{\pi}\right)^{1/2} \tag{8.29}$$

where the dimensionless load is

$$W' = \frac{w'}{E'R_x} \tag{8.30}$$

and w' is the load per unit width along the contact. The maximum deformation for a rectangular conjunction can be written as

$$\delta_{max} = \frac{2W'R_x}{\pi}\left[\ln\left(\frac{2\pi}{W'}\right)-1\right] \tag{8.31}$$

The maximum Hertzian contact pressure in a rectangular conjunction can be written as

$$p_{max} = E'\left(\frac{W'}{2\pi}\right)^{1/2} \tag{8.32}$$

Example 8.9

GIVEN A solid cylinder of 2-cm radius ($r_{ax} = 2$ cm and $r_{ay} = \infty$) rolls around the inside of an outer race with a 10-cm internal radius ($r_{bx} = -10$ cm) and a large width in the axial (y) direction ($r_{by} = \infty$). The cylinder is made of silicon nitride and the outer race is made of stainless steel. The normal applied load per unit width is 1000 N/m.

FIND

a) Curvature sum R
b) Semiwidth of contact b^*
c) Maximum Hertzian contact pressure p_{max}
d) Maximum deformation δ_{max}

Also, compare the results with those of Example 8.8.

Solution

a) From Example 8.8 $E'=2.60\text{x}10^{11}$ N/m^2. The curvature sum is

$$\frac{1}{R_x} = \frac{1}{r_{ax}} + \frac{1}{r_{bx}} = \frac{1}{0.02} - \frac{1}{0.10} = 40 \text{ m}^{-1} \qquad \therefore R_x = 0.025 \text{ m}$$

The dimensionless load is

$$W' = \frac{w'}{E'R_x} = \frac{1000}{(2.6)(10^{11})(2.5)(10^{-2})} = 1.538\times10^{-7}$$

b) The semiwidth of the contact is

$$b^* = R_x\left(\frac{8W'}{\pi}\right)^{1/2} = (0.025)\left[\frac{8(1.538)(10^{-7})}{\pi}\right]^{1/2} = 15.6\times10^{-6}\text{ m} = 15.6\ \mu\text{m}$$

c) The maximum contact pressure is

$$p_{max} = E'\left(\frac{W'}{2\pi}\right)^{1/2} = 2.6(10^{11})\left[\frac{(1.538)(10^{-7})}{2\pi}\right]^{1/2} = 40.68\times10^{6}\text{ Pa}$$

$$= 40.68 \text{ MPa} = 0.0407 \text{ GPa}$$

d) The maximum deformation at the center of the contact is

$$\delta_{max} = \frac{2W'R}{\pi}\left[\ln\left(\frac{2\pi}{W'}\right)-1\right] = \frac{2(1.538)(10^{-7})(2.5)(10^{-2})}{\pi}\left[\ln\left(\frac{2\pi}{0.1538\times10^{-6}}\right)-1\right]\text{ m}$$

$$= 0.0405\ \mu\text{m}$$

8.8 FRICTION

Friction may be viewed as the force resisting relative movement between surfaces in contact. The two main classes of friction, sliding and rolling, are shown in Fig. 8.18. Rolling friction involves nonconformal surfaces, whereas sliding surfaces are conformal. However, most rolling contacts do experience some sliding.

In both rolling and sliding contacts a tangential force F in the direction of motion is needed to move the upper body over the stationary lower body. The ratio between the tangential force and the normal applied load W is known as the coefficient of friction and is

$$\mu = \frac{F}{W} \qquad \textbf{(8.33)}$$

Throughout this text μ designates sliding friction and μ_r designates rolling friction. Also, the term "coefficient of friction" refers to the coefficient of sliding friction; whenever the coefficient of rolling friction is used, it will be so designated. Rolling friction is much smaller (by at least one order of magnitude) than sliding friction. The complete range of rolling and sliding friction goes from 0.001 in lightly loaded rolling-element bearings to greater than 10 for clean metals sliding against themselves in vacuum. For most common materials sliding in air the value of μ lies in a narrow range from approximately 0.1 to 2.0. Table 8.7 gives typical coefficients of sliding friction for unlubricated metals in air. Note that the range is 0.2 to 2.5, one order of magnitude.

8.8.1 LOW FRICTION

In a number of situations low friction is desirable, for example:

1. Turbines and generators in electric or hydroelectric power stations, which use oil-lubricated hydrodynamic bearings
2. Gyroscopes, spinning at very high speeds, which use gas-lubricated hydrodynamic bearings

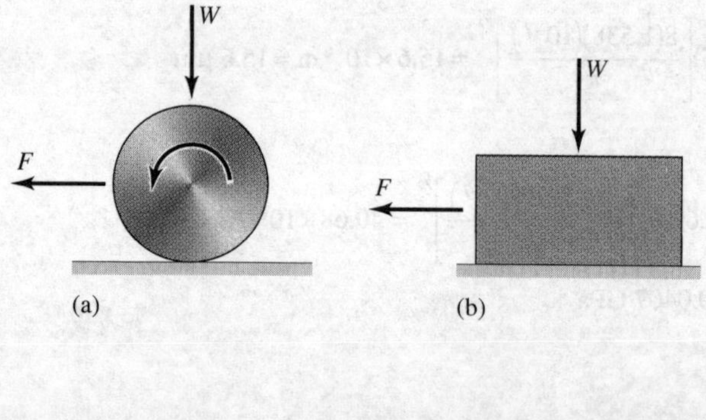

Figure 8.18 Friction force in (a) rolling and (b) sliding.

Table 8.7 Typical coefficients of friction for combinations of unlubricated metals in air. [I.M. Hutchings, *Tribology—Friction and Wear of Engineering Materials,* Edward Arnold, 1992.]

	Coefficient of friction, μ
Self-mated metals in air	
Gold	2.5
Silver	0.8–1
Tin	1
Aluminum	0.8–1.2
Copper	0.7–1.4
Indium	2
Magnesium	0.5
Lead	1.5
Cadmium	0.5
Chromium	0.4
Pure metals and alloys sliding on steel (0.13% carbon) in air	
Silver	0.5
Aluminum	0.5
Cadmium	0.4
Copper	0.8
Chromium	0.5
Indium	2
Lead	1.2
Copper–20% lead	0.2
Whitemetal (tin based)	0.8
Whitemetal (lead based)	0.5
α-brass (copper–30% zinc)	0.5
Leaded α/β brass (copper–40% zinc)	0.2
Gray cast iron	0.4
Mild steel (0.13% carbon)	0.8

Friction can be reduced by using special low-friction materials, by lubricating the surface if it is not already being done, or by using clever designs that convert sliding into rolling. Reducing the wear and heat produced by friction is probably the most important factor in extending machine element life.

8.8.2 HIGH FRICTION

In some situations high controlled friction is desirable, for example:

1. Brakes
2. Interaction between shoe and floor
3. Interaction between tire and road

4. Interaction between a nail and the wood into which it is hammered
5. Grip between a nut and a bolt

It must be emphasized that in these high-friction applications the friction must be controlled.

8.8.3 LAWS OF DRY FRICTION

The three **laws of friction** may be simply stated as follows:

1. The friction force is proportional to the normal load.
2. The friction force is not dependent on the apparent area of the contacting solids; that is, it is independent of the size of the solid bodies.
3. The friction force is independent of the sliding velocity.

These laws are applicable for most sliding conditions in the absence of a lubricant. The first two laws are normally called Amonton's laws. He rediscovered them in 1699, with Leonardo da Vinci normally given the credit for discovering them some 200 years earlier.

The first two laws are found today to be generally satisfied for metals but are violated when polymers are the solid materials in contact. The third law of friction is less well founded than the first two laws. The friction force needed to begin sliding is usually greater than that necessary to maintain it. Thus, the coefficient of sliding friction μ is greater than the coefficient of dynamic friction μ_d. However, once sliding is established, μ_d is often nearly independent of sliding velocity.

The phenomenon important to machine element design is stick-slip. When stick-slip occurs, the friction force fluctuates in a regular manner between two well-defined extreme positions. All stick-slip processes are caused by the fact that the friction force does not remain constant as a function of some other variable, such as distance, time, or velocity.

8.8.4 SLIDING FRICTION OF METALS

Bowden and Tabor's (1973) approach is based on the recognition that surfaces in contact touch only at points of asperity interaction and that very high stresses induced in such small areas would lead readily to local plastic deformation. The penetration of an asperity into the opposing surface can be likened to a miniature hardness test, and the mean normal stress σ over the real areas of asperity contact A_r can be represented to all intents by the hardness H of the softer material. Likewise, if τ represents the shear stress of the asperity junctions, the normal applied load W, the friction force F, and the coefficient of friction μ can be expressed as

$$\mu = \frac{F}{W} = \frac{A_r \tau}{A_r H} = \frac{\tau}{H} \tag{8.34}$$

This expression is an important step in understanding friction, although it is incomplete. It neglects the more complex nature of asperity interactions and deformations and accounts only for the adhesive element of friction. The limitations are apparent when it is recognized that for metals $\tau \approx 0.5S_y$ and $H \approx 3S_y$, where S_y is the yield stress in tension. Thus, from Eq. (8.34) all clean metals should have a universal coefficient of friction of 1/6, which is qualitatively consistent with the statement of Leonardo da Vinci but unrepresentative of more sensitive experimental findings.

Another approach to friction is that a force would be required to move hard asperities through or even over another surface and that microcutting motion represents the friction process. The idea is that if the sum of the projected areas of the indenting asperities perpendicular to the sliding direction is A_o and if the mean stress resisting plastic deformation of the softer material that is being cut is equal to the hardness, the friction force F can be expressed as

$$F = A_o H \tag{8.35}$$

The normal applied load W carried on a number of asperities can be expressed as

$$W = AH \tag{8.36}$$

where

A = contact area, m^2

The coefficient of friction becomes

$$\mu = \frac{F}{W} = \frac{A_o}{A} \tag{8.37}$$

Figure 8.19 shows a conical asperity having a mean angle θ. The coefficient of friction for these considerations is

$$\mu = \frac{A_o}{A} = \frac{2}{\pi}\tan\theta \tag{8.38}$$

Only the front end of the conical asperity shown in Fig. 8.19 is in contact.

Both molecular (adhesive) and deformation (ploughing) actions are effective, and the coefficient of friction is a function of both, or

$$\mu = \frac{\tau}{H} + \frac{2}{\pi}\tan\theta \tag{8.39}$$

Experiments have demonstrated that the adhesion term [first term on the right side of Eq. (8.39)] plays a major role in determining the friction between metals.

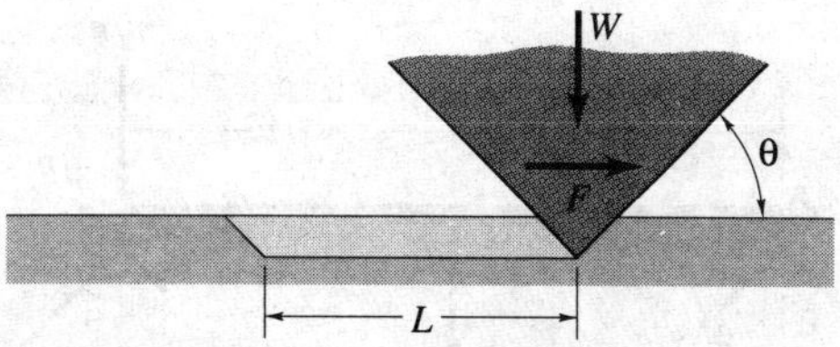

Figure 8.19 Conical asperity having mean angle θ and its effect on normal and frictional loads. Also simulates abrasive wear.

Example 8.10

GIVEN Two roofs having equal slope θ will be provided with a gangway making it possible to walk from one roof to the other as shown in Fig. 8.20(a). The weight of the gangway is evenly distributed along its length. The resultant force of the gangway and the person is shown in Fig. 8.20(b).

FIND Determine the maximum angle θ for the gangway when a person is walking along it without risking that it will slide down. Assume a coefficient of friction of 0.1.

Solution

The most unstable position for the person is at one of the ends of the gangway. Figure 8.20(c) shows a free-body diagram of the forces acting on the gangway. Vertical force equilibrium requires that

$$P_1 \cos\theta - m_g g - m_p g + P_2 \cos\theta - \mu P_1 \sin\theta + \mu P_2 \sin\theta = 0 \qquad \textbf{(a)}$$

Horizontal force equilibrium requires that

$$P_1 \sin\theta + \mu P_1 \cos\theta - P_2 \sin\theta + \mu P_2 \cos\theta = 0 \qquad \textbf{(b)}$$

Collecting like terms in Eqs. (a) and (b) gives

$$P_1(\cos\theta - \mu \sin\theta) + P_2(\cos\theta + \mu \sin\theta) = g(m_p + m_g) \qquad \textbf{(c)}$$

$$P_1(\sin\theta + \mu \cos\theta) = P_2(\sin\theta - \mu \cos\theta) \qquad \textbf{(d)}$$

Solving for P_1 in Eq. (d) and substituting the result into Eq. (c) give

$$\frac{g(m_p + m_g)}{P_2} = \frac{\sin 2\theta(1+\mu^2)}{\sin\theta + \mu \cos\theta} \qquad \textbf{(e)}$$

(a) (b)

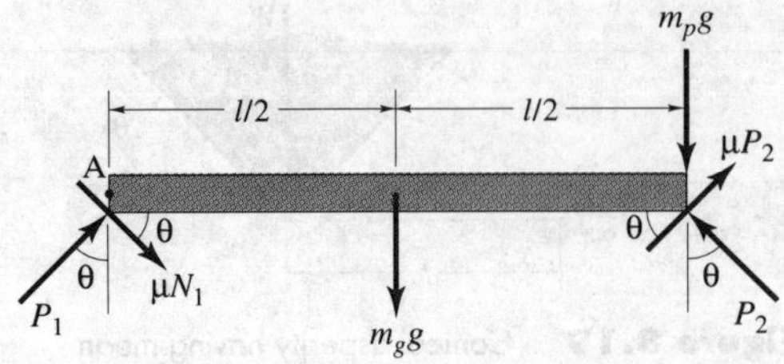

(c)

Figure 8.20 Illustration used in Example 8.10. (a) Two roofs having slope θ with gangway. (b) Resultant force of gangway and person. (c) Free-body diagram of all forces acting on gangway.

Moment equilibrium about point A gives

$$m_g g l/2 + m_p g l - l\mu P_2 \sin\theta - l P_2 \cos\theta = 0$$

Solving for P_2 yields

$$P_2 = \frac{g(m_p + m_g/2)}{\mu \sin\theta + \cos\theta} \qquad \textbf{(f)}$$

Substituting Eq. (f) into Eq. (e) results in

$$m_p = \frac{2\mu m_g}{(1+\mu^2)\sin 2\theta - 2\mu} \qquad \textbf{(g)}$$

Thus, the stability of the gangway determines the maximum allowable weight for the person. From Eq. (g) note that the gangway is stable if

$$2\mu > (1+\mu^2)\sin 2\theta$$

$$\sin 2\theta < \frac{2\mu}{1+\mu^2}$$

If the coefficient of friction is assumed to be 0.1,

$$\sin 2\theta < \frac{2(0.1)}{1+0.01} = \frac{0.2}{1.01}$$

$$\therefore\ \theta < 5.711°$$

This gives a stable gangway.

8.8.5 SLIDING FRICTION OF POLYMERS AND PLASTICS

The coefficient of friction is lower for machine elements made of polymers, such as polyethylene, Plexiglas (Perspex), polystyrene, and nylon, than for those made of metals. For clean copper on steel in air the coefficient of friction is 1.0, but for most plastics it is typically 0.4. Adhesion at the interface is so strong that shear actually occurs in the polymer itself and the strength properties of the polymer determine the friction.

One major difference between the frictional behavior of polymers and what was found for metals in Sec. 8.8.4 is the effect of load and geometry. Geometry refers to the shape of the surfaces, whether they are flat or curved and, if they are curved, how sharp the curvature is. With metals the area of true contact is determined only by the load and the yield pressure and not by the shape of the surfaces. With polymers it is otherwise. These materials deform viscoelastically, implying that deformation depends on the normal applied load W, the geometry, and the time of loading. For a fixed time of loading and fixed geometry—a sphere on a flat surface—the area of true contact is *not* proportional to the load as it is with metals, it is proportional to W^n, where n is less than 1 and usually near 3/4.

$$F \propto W^{3/4} \quad \text{for polymers} \qquad \textbf{(8.40)}$$

$$F \propto W \quad \text{for metals} \qquad \textbf{(8.41)}$$

Thus, the difference is quite significant.

8.8.6 SLIDING FRICTION OF RUBBER

Rubber is an extreme example of a material that deforms elastically. Regardless of the amount that rubber gets distorted, it will return to its original shape when the deforming force is removed, as long as the rubber is not torn or cut. In contrast, polymers deform elastically and also flow viscously, and metals deform elastically over a limited range and then flow plastically. Thus, the friction of rubber often deviates appreciably from the laws of friction. The friction depends markedly on the load and the geometry of the rubber surfaces.

For rubber the relationship between friction force and the normal applied load is

$$F \propto W^{2/3} \tag{8.42}$$

Rubber is truly an elastic solid and this relationship is valid over a large load range.

8.9 WEAR

Wear has long been recognized as a most important and usually detrimental process in machine elements, yet scientific studies of the wear phenomenon began only relatively recently. The historical order of tribological studies has been friction, lubrication, and then wear. Indeed, most research on wear has been packed into the last 50 years—a mere drop in the sea of time.

This chronology is all the more remarkable in view of the importance of wear to machine element life. In a more general sense practically everything made by man wears out, yet the fundamental actions that govern the process remain elusive. Initially, in the tribological chronology attention was focused on friction, since it governs the ability of machine elements to function at all. The role of lubricants in controlling friction and wear for some machine elements was then explored, with little attention to the detailed mechanism of wear. In fact, it remains true that today's engineer is better equipped to design a machine element to withstand known loads than to specify a given life of the machine element—mainly due to the lack of understanding of wear.

Wear may be viewed as the progressive loss of substance from the operating surface of a body occurring as a result of loading and relative motion at the surface. Wear can be classified by the physical nature of the underlying process, such as abrasion, adhesion, and fatigue. These three main types of wear are considered in this section. Other types of wear, such as erosion and corrosion, are beyond the scope of interest of this text.

8.9.1 ADHESIVE WEAR

Figure 8.21 shows the mechanism of **adhesive wear,** where material is transferred from one surface to another by solid-phase welding. Adhesive wear is the most common type of wear and the least preventable. Observe from Fig. 8.20 that as the moving asperities pass, the microscopic weld often breaks and material is removed from the surface having the lower yield strength.

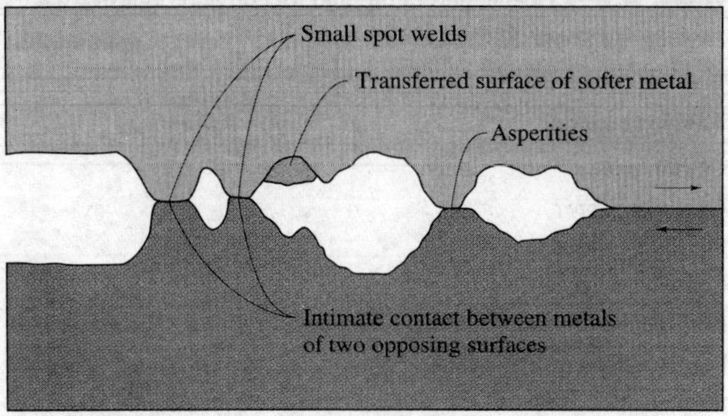

Figure 8.21 Adhesive wear simulation.

The volume of material removed by wear v is directly proportional to the sliding distance L and the normal applied load W and is inversely proportional to the hardness H of the softer of the two materials, or

$$v \propto \frac{WL}{H} \tag{8.43}$$

However, there are many exceptions to this formula. After a dimensionless adhesive wear coefficient k_1 is introduced, Eq. (8.43) becomes

$$v = \frac{k_1 WL}{3H} \tag{8.44}$$

The dimensionless adhesive wear coefficient is a measure of the probability that any interaction between asperities of the two surfaces in contact will produce a wear particle due to any of the wear mechanisms. Thus, if k_1 is 1, every junction involving surface contact will produce a wear fragment; if $k_1 = 0.1$, one-tenth of the contacts between surfaces will produce a wear fragment. Studies have shown that for unlubricated surfaces the lowest value of k_1 is obtained for polyethylene sliding on steel. For this situation $k_1 = 1 \times 10^{-7}$, which means that 1 in 10 million contacts will produce a wear fragment.

Table 8.8 shows some typical values of k_1 for several materials in contact. As presented in Eq. (8.44) the wear volume is a function not only of k_1 but also of the hardness of the softer of the two materials in contact. For example, although the k_1 value of polyethylene on steel is 10 times smaller than that for tungsten carbide sliding on itself, the wear volume is 10 times larger because tungsten carbide wears 10 times less than the polymer. Therefore, the volume of the wear fragments is considerably smaller because tungsten is several orders of magnitude harder than polyethylene.

Table 8.8 also gives the coefficient of friction as well as the adhesive wear constant. The wear constant varies by eight orders of magnitude for the various rubbing materials, but the coefficient of friction varies by only one order of magnitude.

Table 8.8 Coefficients of rubbing friction and adhesive wear constant for nine rubbing materials.

Rubbing materials	Coefficient of friction, μ	Adhesive wear constant, k_1
Gold on gold	2.5	0.1–1
Copper on copper	1.2	0.01–0.1
Mild steel on mild steel	.6	10^{-2}
Brass on hard steel	.3	10^{-3}
Lead on steel	.2	2×10^{-5}
Polytetrafluoroethylene on steel	.2	2×10^{-5}
Stainless steel on hard steel	.5	2×10^{-5}
Tungsten carbide on tungsten carbide	.35	10^{-6}
Polyethylene on steel	.5	10^{-8}–10^{-7}

Example 8.11

GIVEN A rubbing journal bearing in a dam gate moves slowly and operates under a high load. The entire bearing is made of AISI 1040 steel. To increase the bearing wear life, a material change is considered for one of the surfaces.

FIND Which of the following materials will give the longest wear life: brass, lead, polytetrafluoroethylene (PTFE), or polyethylene? The hardness for lead is 30 MPa, for PTFE it is 50 MPa, and for polyethylene it is 70 MPa.

Solution

From Eq. (8.44) the wear volume is

$$v = \frac{k_1 W L}{3H}$$

The wear volume is a minimum when H/k_1 is a maximum. Note that $H \approx 3S_y$. Making use of Table 8.8 for the materials being considered gives

Brass

$$\frac{H}{k_1} = \frac{3S_y}{k_1} = \frac{3(75)(10^6)}{10^{-3}} = 225(10^9)\text{ Pa} = 225\text{ GPa}$$

Lead

$$\frac{H}{k_1} = \frac{30(10^6)}{2(10^{-5})} = 1500(10^9)\text{ Pa} = 1500\text{ GPa}$$

PTFE

$$\frac{H}{k_1} = \frac{50(10^6)}{2(10^{-5})} = 2500(10^9)\text{ Pa} = 2500\text{ GPa}$$

Polyethylene

$$\frac{H}{k_1} = \frac{70(10^6)}{10^{-7}} = 700\ 000(10^9)\ \text{Pa} = 700\ 000\ \text{GPa}$$

Polyethylene will give much longer life than any of the other materials.

8.9.2 ABRASIVE WEAR

Abrasive wear arises when two interacting surfaces are in direct physical contact and one is significantly harder than the other. Under a normal load the asperities of the harder surface penetrate the softer surface, producing plastic deformations and implying that permanent deformation is occurring. Figure 8.19 shows the essence of the abrasive wear model. A hard conical asperity with slope θ under a normal load W causes transfer of a softer surface. The amount of material transferred by abrasive wear is

$$v_{abr} \approx \frac{2k_1}{\pi}\frac{\tan\theta}{H}WL = \frac{k_1 k_2 WL}{H} \qquad \textbf{(8.45)}$$

where

$$k_2 = \frac{2\tan\theta}{\pi} = \textbf{abrasive wear constant} \qquad \textbf{(8.46)}$$

k_1 = adhesive wear constant, given in Table 8.8

For most asperities θ is small, typically between 5° and 10°.

A similarity of form is evident between Eqs. (8.44) and (8.45), and it is clear that the general **laws of wear** can be stated as follows:

1. Wear increases with sliding distance.
2. Wear increases with normal applied load.
3. Wear decreases as the hardness of the sliding surface increases.

These laws, along with Eqs. (8.44) and (8.45), reveal the role of the principal variables, but there are sufficient exceptions to the predicted behavior to justify a measure of caution in their use. Furthermore, wear rates, or the life of rubbing components, cannot be determined without some knowledge of the wear coefficients, which are experimentally determined.

8.9.3 FATIGUE WEAR

For nonconformal contacts in such machine elements as rolling-element bearings, gears, friction drives, cams, and tappets, a prevalent form of failure is fatigue wear. In these situations the removal of material results from a cyclic load variation. Figure 8.22 illustrates the stresses developed on and below the surface that deform and weaken the metal. **Fatigue wear** is caused by the propagation of subsurface damage to the surface due to cyclic loadings.

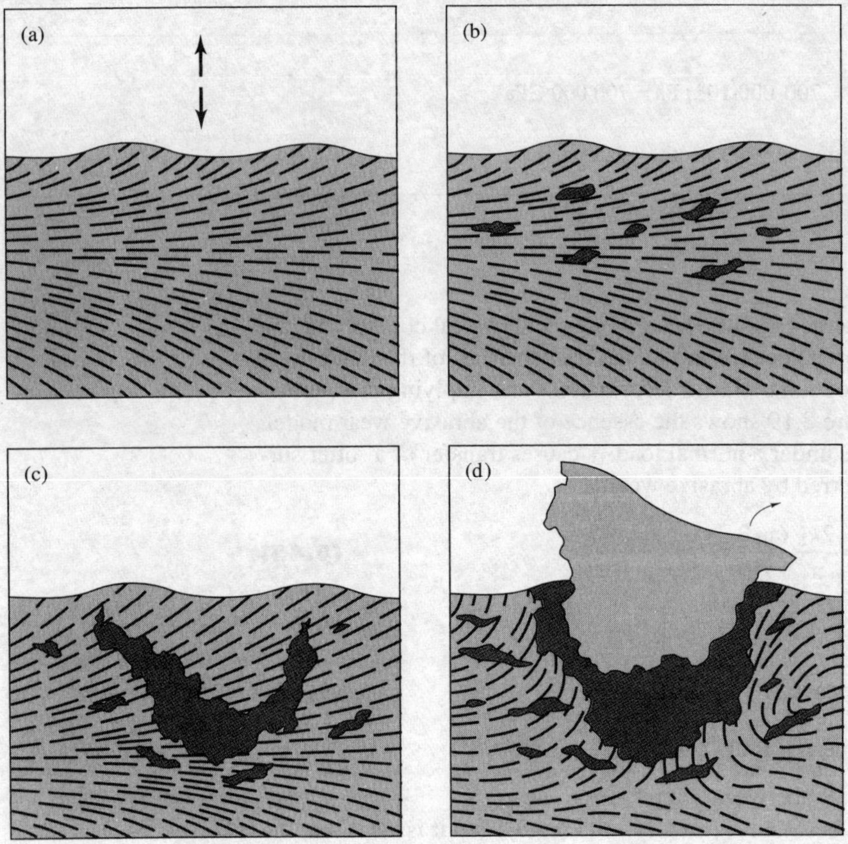

Figure 8.22 Fatigue wear simulation.

As cyclic loading continues, faults or cracks form below the surface. Eventually, the faults coalesce at and near the surface. Material at the surface is then easily broken away, degrading the component surface and releasing work-hardened particulate contaminants.

Fatigue wear occurs in nonconformal machine elements, such as rolling-element bearings, gears, and cam and followers, even in well-lubricated situations. Cyclic loading of nonconformal surfaces can cause extremely high stresses to the solid material, although the surfaces may never make direct contact, due to the loading and their nonconformal nature. After a few million such encounters the asperities may get "tired"; they will fatigue and a piece of material may fall out of the surface.

A group of apparently identical nonconformal machine elements subjected to identical loads, speeds, lubrication, and environmental conditions may exhibit wide variation in failure times. For this reason the fatigue wear process must be treated statistically. That is, the fatigue life of a bearing is normally defined in terms of its statistical ability to survive a certain period of time or by the allowable wear before the component ceases to function adequately. The only requirement for fatigue failure is that the surface material should be loaded. The other wear mechanisms require not only loading but also physical contact between the surfaces. If the surfaces are separated by a lubricant film, adhesive and abrasive wear are virtually eliminated, but fatigue wear can still occur.

Case Study — PIN-ON-DISK TESTS

Many times, engineers need to determine the tribological performance of mating materials. In theory, this can be done by manufacturing prototypes of candidate materials and evaluating them in service. However, this approach is costly, time consuming, and laborious, especially for a large number of materials. The alternative is to perform a simulation in a simplified environment and infer the performance. This notoriously difficult practice usually requires significant experience to make inferences with confidence.

GIVEN A common test procedure is the pin-on-disk method illustrated in Fig. 8.23. The pin is constructed from one material in the tribo-pair and the disk is constructed from the other material. By driving the disk and pushing the pin against the disk, a sliding situation can be simulated. The wear on the pin can be measured with a toolmaker's microscope by measuring the size of the wear scar, and is given by applying the test situation to Eq. (8.44) or

$$v = \frac{k_1}{3H} WL = \frac{\pi}{64}\frac{d^4}{R} \qquad \textbf{(8.47)}$$

where

d = diameter of wear flat [see Fig. (8.23(b)], m

R = radius of pin, m

The test data from the pin-on-disk apparatus are shown in Table 8.9.

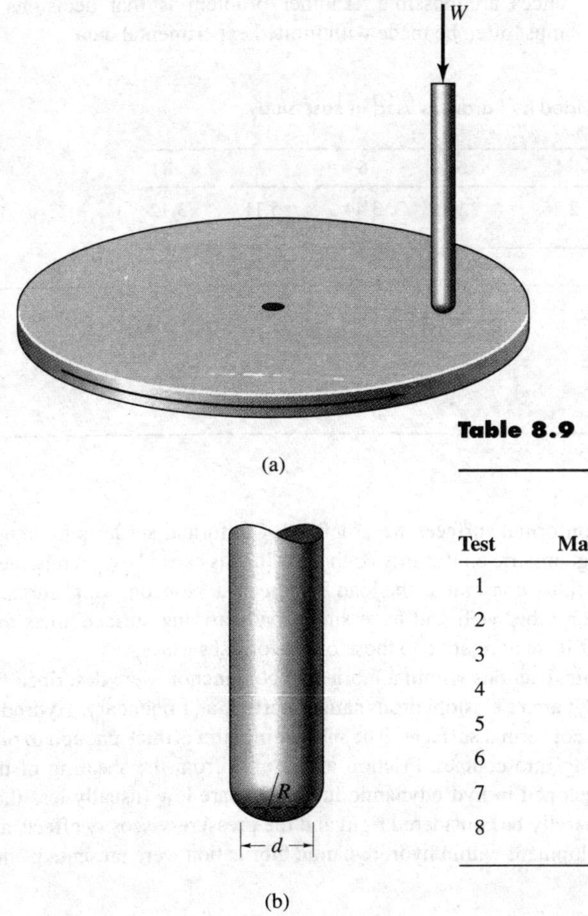

Figure 8.23 (a) Illustration of a pin-on-disk wear test. (b) Detail of pin, with dashed line showing the worn surface.

Table 8.9 Test data for aircraft braking system used in case study.

Test	Material	Sliding speed of disk, rpm	Contact force, N	Test duration, min	Wear scar, mm
1	A	30	100	350	9.70
2	A	30	200	100	8.81
3	A	60	200	60	16.01
4	B	30	100	480	12.63
5	B	60	100	480	15.27
6	C	60	100	480	15.89
7	D	60	200	480	20.83
8	E	60	100	240	13.02

(continued)

Case Study (CONCLUDED)

FIND A material recommendation regarding a low-wear component for an aircraft braking system. Use a distance of 150 mm from the pin to the center of disk rotation and an initial pin radius of 100 mm. Because the pin diameter is approximately 30 mm, all tests are valid.

Solution

Let $W = 100$ N, $R = 100$ mm, $d = 9.7$ mm, and $L = \pi(300)$ mm. Note that L is per revolution. Therefore, substituting the above into Eq. (8.47) gives

$$k_1 = 1.317 \times 10^{-8}\, H,\ \text{Pa}^{-1}$$

Performing the calculation for all trials gives the results shown in Table 8.10.

From these results the best performer was material A in the first test. However, the extremely large wear coefficient for the same material in test 3 is clearly a concern. A dramatic increase in wear rate such as this is common for materials with a protective oxide surface. When the speed or stress becomes large enough, the oxide is removed and does not have time to become reestablished. If the application can be reasonably assumed to avoid this condition, material A may be the best. Otherwise, material B is superior.

This case study demonstrates typical problems encountered by tribologists. Wear is a complicated phenomenon; and to apply test data, one must be certain that the test properly simulates the actual loading environment. The consequences cannot be overstated: If the test does not simulate the actual loading environment, it has little use because there can be no certainty that the wear mode will be the same as in the test. With experience such assurances are possible. Another problem is that decisions must often be made with limited experimental data.

Table 8.10 Test results of wear coefficient divided by hardness used in case study.

Test	1	2	3	4	5	6	7	8
$k_1/H \times 10^{-8}$	1.317	1.57	14.26	2.76	2.95	3.44	5.11	3.12

8.10 SUMMARY

In this chapter conformal and nonconformal surfaces were defined. Conformal surfaces fit snugly into each other with a high degree of geometric conformity so that the load is carried over a relatively large area and the area remains essentially constant as the load is increased. Nonconformal surfaces do not geometrically conform to each other well and have small load-carrying surface areas that enlarge with increasing load but are still small relative to those of conformal surfaces.

A lubricant's physical and chemical actions within a lubricated conjunction were described for the four lubrication regimes: hydrodynamic, elastohydrodynamic, partial, and boundary. Hydrodynamic lubrication is characterized by conformal surfaces. The lubricating film is thick enough to prevent the opposing solids from coming into contact. Friction arises only from the shearing of the viscous lubricant. The pressures developed in hydrodynamic lubrication are low (usually less than 5 MPa), so that the surfaces may generally be considered rigid and the pressure-viscosity effects are small. Three modes of pressure development within hydrodynamic lubrication were presented: slider, squeeze, and external pressurization. For hydrodynamic lubrication with sliding motion the minimum film thickness is quite sensitive to load, being inversely proportional to the square root of the normal applied load.

Elastohydrodynamic lubrication (EHL) is characterized by nonconformal surfaces, and again there is no asperity contact of the solid surfaces. Two modes of elastohydrodynamic lubrication exist: hard and soft. Hard EHL is characterized by metallic surfaces; and soft EHL, by surfaces made of elastomeric materials. The pressures developed in hard EHL are high (typically between 0.5 and 4 GPa), so that elastic deformation of the solid surfaces becomes important as do the pressure-viscosity effects of the lubricant. As with hydrodynamic lubrication, friction is due to the shearing of the viscous lubricant. The minimum film thickness for hard EHL is relatively insensitive to load because the contact area increases with increasing load, thereby providing a larger lubricated area to support the load. For soft EHL the elastic distortions are large, even for light loads, and the viscosity within the conjunction varies little with pressure because the pressures are relatively low and the elastic effect predominates. Both hydrodynamic and elastohydrodynamic lubrication are fluid film lubrication phenomena in that the film is thick enough to prevent opposing solid surfaces from coming into contact.

In boundary lubrication considerable asperity contact occurs, and the lubrication mechanism is governed by the physical and chemical properties of thin surface films that are of molecular proportion (from 1 to 10 nm). The frictional characteristics are determined by the properties of the solids and the lubricant film at the common interfaces. Partial lubrication (sometimes called mixed lubrication) is governed by a mixture of boundary and fluid film effects. Most of the scientific unknowns lie in this lubrication regime.

Because in many lubricated machine elements the solids are separated by thin films, the physical nature of the solids' surface topographies must be understood. In computing the parameters that define surface texture the height measurements were made from a defined reference line. The mean, or M system, reference line was used; it is based on selecting the mean line as the centroid of the profile. The area of peaks above this line equals the area of valleys below this line.

Two ways of computing surface parameters were also discussed:

1. Centerline average or arithmetic average roughness, denoted by R_a:

$$R_a = \frac{1}{N}\sum_{i=1}^{N}|z_i| \tag{8.5}$$

2. Root-mean-square (rms) roughness, denoted by R_q:

$$R_q = \left(\frac{1}{N}\sum_{i=1}^{N} z_i^2\right)^{\frac{1}{2}} \tag{8.6}$$

The film parameter Λ was defined as the ratio of the minimum film thickness to the composite surface roughness. The film parameter was used to define the four important lubrication regimes.

For concentrated loads both elliptical and rectangular contacts were discussed. It was emphasized that a number of machine elements considered later in the text, such as rolling-element bearings and gears, experience concentrated loading. The major features of concentrated loading (first presented in Chapter 2) are that load is applied to an extremely small area and that the load per unit area is much larger than normally found in distributed loading situations. Equations for curvature sum and maximum contact pressure were presented, as well as simplified expressions that enable the dimensions of the contact, the maximum pressure, and the maximum deformation to be determined.

Features of rolling and sliding friction were presented, as well as the laws of friction. The coefficient of friction was defined as the ratio of the friction force to the normal applied load. Sliding friction of metals, polymers, plastics, and rubber was discussed. For these various classes of material the relationship between the friction force and the normal applied load varies significantly.

The useful life of engineering machine components is limited by breakage, obsolescence, and *wear.* Practically everything made will wear out. Wear is the progressive loss of substance from the operating surface of a body occurring as a result of loading and relative motion at the surface. The most common forms of wear were discussed:

1. In *adhesive wear* material transfers from one surface to another due to solid-phase welding.
2. In *abrasive wear* material is displaced by hard particles.
3. In *fatigue wear* material is removed by cyclic stress variations even though the surfaces may not be in contact.

KEY WORDS

abrasive wear wear caused by physical damage from penetration of hard surface into softer one

absolute viscosity shear stress divided by shear strain rate having SI units of pascal-second

adhesive wear wear caused by fracture of solid-state weld junctions of two surfaces

boundary lubrication lubrication condition where considerable asperity interaction occurs between solids and lubrication mechanism is governed by properties of thin surface films that are of molecular proportion

conformal surfaces surfaces that fit snugly into each other with high degree of conformity as in journal bearings

elastohydrodynamic lubrication lubrication condition where *nonconformal* surfaces are completely separated by lubricant film and no asperities are in contact

ellipticity parameter diameter of contact ellipse in y direction divided by diameter of contact ellipse in x direction

fatigue wear wear caused by propagation of subsurface damage to surface due to cyclic loadings

film parameter minimum film thickness divided by composite surface roughness

fluid film lubrication lubrication condition where lubricated surfaces are completely separated by a lubricant film and no asperities are in contact

friction force resisting relative movement between surfaces in contact

hydrodynamic lubrication fluid film lubrication of *conformal* surfaces as in journal bearings

kinematic viscosity absolute viscosity divided by density with SI units of meter squared per second

laws of friction

1. Friction force is proportional to normal load.
2. Friction force is not dependent on apparent area of contact solids; that is, it is independent of size of solid bodies.
3. Friction force is independent of sliding velocity.

laws of wear

1. Wear increases with sliding distance.
2. Wear increases with normal applied load.
3. Wear decreases as hardness of sliding surface increases.

lubricant any substance that reduces friction and wear and provides smooth running and satisfactory life for machine elements

mixed lubrication same as partial lubrication

nonconformal surfaces surfaces that do not conform to each other very well as in rolling-element bearings

partial lubrication lubrication condition where load between two surfaces in contact is transmitted partially through lubricant film and partially through asperity contact

R_a centerline average or arithmetic average surface roughness

R_q root-mean-square surface roughness

running-in process through which beneficial wear causes surfaces to adjust to each other and improve performance

tribology study of lubrication, friction, and wear of moving or stationary parts

wear progressive loss of substance from operating surface of body occurring as result of loading and relative motion of surface

RECOMMENDED READINGS

Arnell, R.D., Davies, P.B., Halling, J., and Wholmes, T.L. (1991) *Tribology: Principles and Design Applications,* Springer-Verlag, New York.

Dowson, D. (1979) *History of Tribology,* Longman, London.

Hamrock, B.J. (1994) *Fluid Film Lubrication,* McGraw-Hill, New York.

Johnson, K.L. (1985) *Contact Mechanics,* Cambridge University Press, Cambridge, UK.

Ludema, K. (1996) *Friction, Wear, Lubrication: A Textbook in Tribology,* CRC Press, Boca Raton, FL.

Rabinowicz, E. (1995) *Friction and Wear of Materials,* 2nd ed., Wiley, New York.

REFERENCES

Bowden, F.P., and Tabor, D. (1973) *Friction—An Introduction to Tribology,* Anchor Press/Doubleday, New York.

Hamrock, B.J. (1993) *Fundamentals of Fluid Film Lubrication,* McGraw-Hill, New York.

Hamrock, B.J., and Brewe, D.E. (1983) "Simplified Solution for Stresses and Deformations," *J. Lub. Technol.,* Vol. 105, No. 2, pp.171–177.

Hamrock, B.J., and Dowson, D. (1977) "Isothermal Elastohydrodynamic Lubrication of Point Contacts, Part III—Fully Flooded Results," *J. Lub. Technol.,* Vol. 99, No. 2, pp. 264–276.

Hertz, H. (1881) "The Contact of Elastic Solids," *J. Reine Angew. Math.,* Vol. 92, pp. 156–171.

Hutchings, I.M. (1992) *Tribology—Friction and Wear of Engineering Materials,* Edward Arnold, London.

Jones, W.R., et al. (1975) "Pressure-Viscosity Measurements for Several Lubricants to 5.5×10^8 Pa-s and 149°C," *ASLE Trans.,* Vol. 18, No. 4, pp. 249–262.

Tallian, T.E., et al. (1967) "On Computing Failure Modes in Rolling Contacts," *ASLE Trans.,* Vol. 10, No. 4, pp. 418–435.

PROBLEMS

Section 8.2

8.1 Describe the difference between conformal and nonconformal surfaces.

8.2 Determine which of the following contact geometries is conformal and which is nonconformal:
a) Meshing gear teeth
b) Ball and inner race of a ball bearing
c) Journal bearing
d) Railway wheel and rail contact
e) Car making contact with the road
f) Egg and egg cup
g) Eye and socket

Section 8.3

8.3 Describe three applications for each of the four lubrication regimes: hydrodynamic lubrication, elastohydrodynamic lubrication, boundary lubrication, and partial lubrication.

8.4 A hydrodynamic journal bearing is loaded with a normal load W and rotating with a surface velocity u_b. According to Eq. (8.1) find how much higher rotational speed the bearing needs to maintain the same minimum film thickness if the load W is doubled.

8.5 A ball bearing lubricated with a mineral oil runs at 3000 rpm. The bearing is made of AISI 52100 steel and is loaded to a medium-high level. Find how much thinner the oil film will be if the load is increased 10 times. How much must the speed be increased to compensate for the higher load while keeping the oil film thickness constant?

Section 8.4

8.6 Two bearing types, one hydrodynamic journal bearing and one ball bearing, operate with a film parameter of 8.3. Calculate how much lower the speed for each bearing has to be to decrease the film parameter to 5.2.

8.7 A machined steel surface has roughness peaks and valleys. The roughness wavelength is such that 400 peaks and 400 valleys are found during one roughness measurement. Find how much the arithmetic average surface roughness $R_a = 0.1$ μm will change if one of the surface roughness peaks and one of the valleys increases from 0.25 to 2.5 μm.

8.8 A surface with a triangular sawtooth roughness pattern has a peak-to-valley height of 5 μm. Find the R_a value and the R_q value.

8.9 A precision ball bearing has a race root-mean-square surface roughness of 0.07 μm and a ball root-mean-square surface roughness of 0.02 μm. Changing the roughness of the components may give a higher Λ value. Determine which components to smooth if it costs equally as much to halve the roughness of the race as it costs to halve the roughness of the balls. Note that $h_{min} = 0.2$ μm.

8.10 The minimum film thickness in a particular application is 10 μm. Assume that the surface roughnesses of the two surfaces being lubricated are identical. What lubrication regime would you expect the application to be operating in? Also, what is the maximum surface roughness allowed?

Section 8.5

8.11 Two equally rough surfaces with rms surface roughness R_q are lubricated and loaded together. The oil film thickness is such that $\Lambda = 2$. Find the Λ value if one of the surfaces is polished so that $R_q \rightarrow 0$ (i.e., the surface becomes absolutely smooth).

8.12 Two lubricated surfaces have rms surface roughnesses R_q of 0.23 and 0.04 μm, respectively. By using a new honing machinery either surface can be made twice as smooth. For good lubrication, which surface roughness is it most important to decrease?

Section 8.6

8.13 The absolute viscosity of a fluid at atmospheric conditions is 6×10^{-3} kgf-s/m^2. Give this absolute viscosity in

a) Reyns
b) Poise
c) Pound force–seconds per square inch
d) Newton-seconds per square meter

8.14 Given a fluid with an absolute viscosity η between two 1-m^2 surfaces spaced 1 mm apart, find how fast the surfaces will move relative to each other if a 10-N force is applied in the direction of the surfaces when η is

a) 0.001 N-s/m^2 (water)
b) 0.100 N-s/m^2 (a thin oil at room temperature)
c) 10.0 N-s/m^2 (syrup; cold oil)
d) 10^8 N-s/m^2 (asphalt)

Section 8.7

8.15 For a single ball rolling in a groove determine the following:
a) Contact zone dimensions
b) Maximum elastic deformation
c) Maximum pressure

The steel ball has a 10-mm diameter and a 4-N normal force acting on it. The ball and the groove have a 200-GPa modulus of elasticity and a Poisson's ratio of 0.3. Assume a 6.08-mm-radius groove in a semi-infinite steel block.

8.16 A solid cylinder rolls with a load against the inside of an outer cylindrical race. The solid cylinder radius is 20 mm and the race internal radius is 150 mm. The race and the roller have the same axial length. What is the radius of a geometrically equivalent cylinder near a plane? The cylinder is made of silicon nitride ($E = 314$ GPa, $\nu = 0.26$) and the race is made of stainless steel ($E = 193$ GPa, $\nu = 0.30$). If the normal applied load per unit width is 12 000 N/m, determine
a) Contact semiwidth
b) Maximum surface stress
c) Maximum elastic deflection

Also, indicate what these values would be if the silicon nitride cylinder were replaced with a stainless steel cylinder.

8.17 A 10-cm-diameter shaft has a 2-cm-diameter ball rolling around the outside. Find the curvature sum, the maximum contact stress, the maximum deflection, and the contact dimensions if the ball load is 1000 N. The ball is made of silicon nitride ($E = 314$ GPa, $\nu = 0.26$), and the shaft is made of steel ($E = 206$ GPa, $\nu = 0.30$). Also, determine the values if both ball and shaft are made of steel.

8.18 The ball–outer-race contact of a ball bearing has a 17-mm ball diameter, an 8.84-mm outer-race groove radius, and a 44.52-mm radius from the bearing axis to the bottom of the groove. The load on the ball is 25 000 N. Calculate the dimensions of the contact ellipse and the maximum deformation at the center of the contact. The race and the ball are made of steel.

Section 8.8

8.19 A rectangular box stands on an inclined plane with the slope angle α. The center of gravity is in the geometric center of the box, which has height h and length l. Find the α at which the box starts to slide down.

8.20 Find the ratio h/l when the box in Problem 8.19 falls over rather than slides down, and determine how high the coefficient of friction then has to be.

8.21 A polymer box stands on a slope. The box weighs 10 kg and the slope angle is 25°. To pull the box down the slope, a friction force $F = 5$ N is needed. The friction force is proportional to the load raised to 0.75. Find the extra weight needed in the box to make it slide down the slope.

8.22 A cross-country skier who waxes the central part of the skis (under the foot) with a high-friction wax can run up a 5.7° slope without slipping backward. The front and the back parts of the skis are waxed with a low-friction wax having a 0.01 coefficient of friction. Determine the downhill slope needed to continue sliding at constant speed

a) If the skier is so light that only the low-friction parts of the skis contact the snow
b) If the skier is so heavy that half the weight is carried by the central part of the skis

8.23 A moose suddenly jumps out onto a dry asphalt road 80 m in front of a car running at 108 km/hr. The maximum coefficient of friction between the rubber tires and the road is 1.0 when they just start to slide and 0.8 when the locked wheels slide along the road. It takes 1 s for the driver to apply the brakes after the moose jumps onto the road. Find whether the car can stop before it hits the moose

a) With locked wheels sliding along the road
b) With an automatic braking system (ABS) that keeps the sliding speed so low that maximum friction is maintained

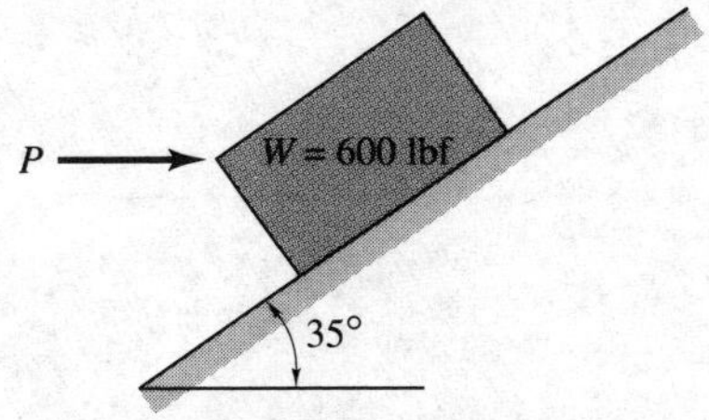

Sketch *a*, for Problem 8.24

8.24 Given a block on an incline as shown in sketch *a*, find the force required

a) To prevent motion downward
b) To cause motion upward

The coefficient of friction is 0.30. Draw a free-body diagram for both situations showing the forces involved.

Section 8.9

8.25 Three equal blocks of plastic are carrying the weight of a steel slider. At first each block carries one-third of the weight, but the central block is made of polytetrafluoroethylene (PTFE) and the two outer blocks are made of polyethylene, so they have different adhesive wear constants (PTFE, $k_1 = 2 \times 10^{-5}$; polyethylene, $k_1 = 2 \times 10^{-8}$). Determine how the load redistributes between the blocks if the hardnesses of the plastics are assumed to be the same.

8.26 Given the plastic blocks in Problem 8.25 and sharing the load so that the wear is equal on each block, find the coefficient of friction.

8.27 A 50-kg copper piece is placed on a flat copper surface as shown in sketch *b*. Assume that copper has a hardness of 275 MPa. Calculate the shear stress, the abrasive wear volume, and the adhesive wear volume.

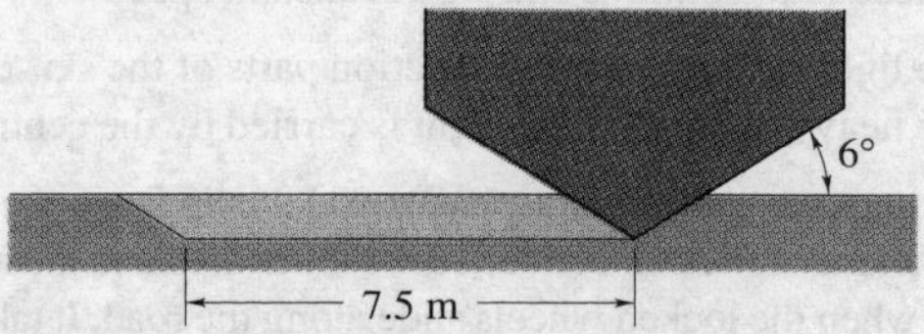

Sketch *b*, for Problem 8.27

PART

2

MACHINE ELEMENTS

PART OUTLINE

CHAPTER

9

COLUMNS

Columns from the Alhambra in Granada, Spain. (Art Resource)

And as imagination bodies forth the forms of things unknown,
The poet's pen turns them to shapes
And gives to airy nothingness a local habitation and a name

William Shakespeare, "A Midsummer Night's Dream"

SYMBOLS

A	cross-sectional area, m^2
a	side of square, m
a_o	outer dimension of hollow square, m
a_i	inner dimension of hollow square, m
b	base dimension of rectangular section, m
C_c	slenderness ratio
C_1, C_2	integration constants
c	distance from neutral axis to outer fiber of column, m
d	diameter, m
E	modulus of elasticity, Pa
E_t	tangent modulus, Pa
e	eccentricity, m
g	gravitational acceleration, 9.807 m/s^2
h	height dimension of rectangular section, m
I	area moment of inertia, m^4
l	length of column, m
l_e	effective length of column, m
M	moment, N-m
M'	statically equivalent moment, N-m
m_a	mass, kg
n	integer, 1, 2, . . .
n_σ	stress reduction factor in AISC criteria
P	force, N
R_e	resultant force, N
r	radius of column, m
r_g	radius of gyration, m
S_y	yield strength, Pa
s	arc length, m
t	time, s
t_h	thickness, m
V	shear force, N
W_e	weight, N
x	length dimension of column, m
y	transverse dimension or deflection of column, m
θ	position angle, deg
ρ	density, kg/m^3
σ	normal stress, Pa
σ_{all}	allowable normal stress, Pa

Subscripts

cr	critical
E	Euler
i	inner
J	Johnson
max	maximum
o	outer
T	tangency point

9.1 INTRODUCTION

The basic understanding of tribological considerations, loads, stresses, and deformations obtained in the preceding chapters is related here to columns. A **column** is a straight and long (relative to its radius) bar that is subjected to compressive, axial loads. The reason for this special consideration of columns is that failures due to yielding, determined from Eq. (4.24), and due to deformation, determined from Eq. (4.25), are not correct in predicting failures of long columns. Specific failure theories for columns need to be developed. Because of their shape (length much larger than radius), columns tend to deform laterally upon loading; and if the deflection becomes larger, they fail catastrophically. This situation is known as **buckling**, which can be defined as a sudden large deformation of a structure due to a slight increase of the existing load, under which the structure had exhibited little, if any, deformation before the increase.

This chapter first describes the meaning of elastic stability and end conditions. It then establishes the failure criteria for concentrically loaded columns and describes the nature of the instability that can occur, thus predicting when buckling will take place. It also establishes the failure criteria for eccentrically loaded columns, so that proper design of concentric and eccentric columns is ensured.

9.2 EQUILIBRIUM REGIMES

To understand why columns buckle, it is first necessary to understand the equilibrium regimes. An important question to be answered is, When an equilibrium position is slightly disturbed, does the component tend to return to that equilibrium position or does it tend to depart even farther? To visualize what is happening, consider Fig. 9.1, which shows the three equilibrium regimes: stable, neutral, and unstable.

9.2.1 STABLE EQUILIBRIUM

Figure 9.1(a) attempts to illustrate what occurs in stable equilibrium. Assume that the surfaces are frictionless and that the sphere has a light weight. The forces on the sphere (gravity and normal surface reaction) are in balance whenever the surface is horizontal. These balanced positions are indicated by a zero. Figure 9.1(a) shows the sphere displaced slightly from its equilibrium position. The forces on it no longer balance, but the resultant imbalance is a restoring force (i.e., gravity is accelerating the sphere back toward the equilibrium position). Such a situation is called *stable equilibrium*.

9.2.2 NEUTRAL EQUILIBRIUM

Figure 9.1(b) considers the sphere in neutral equilibrium. After the sphere has been slightly displaced from the equilibrium position, it is again in equilibrium at the displaced position, and there is no tendency either to return to the previous position or to go to some other position. Equilibrium is always satisfied. Because the surface is flat, the sphere does not move after being placed in another position. This is *neutral equilibrium*.

9.2.3 UNSTABLE EQUILIBRIUM

Figure 9.1(c) shows the case of unstable equilibrium, which is the opposite situation from that presented in Fig. 9.1(a). When the sphere is displaced from its equilibrium position, the resultant imbalance is a disturbing force (i.e., it accelerates the sphere away from its equilibrium position). Such a situation is called *unstable equilibrium*. Gravity and convex surfaces cause the sphere to move farther from the balanced position.

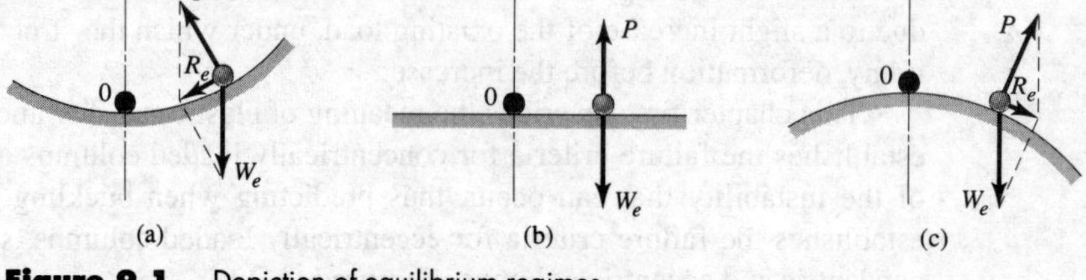

Figure 9.1 Depiction of equilibrium regimes.
(a) Stable; (b) neutral; (c) unstable.

Generalizing from this example, unstable equilibrium occurs if for small displacements from the equilibrium position the disturbing forces tend to accelerate the part away from the equilibrium position. Columns in unstable equilibrium are unreliable and hazardous. A small displacement can cause a catastrophic change in the configuration of a column. Thus, as the load on a column increases, a critical load is reached where unstable equilibrium occurs and the column will not return to its straight configuration. The load cannot be increased beyond this value unless the column is laterally restrained. Thus, for long, slender columns a critical buckling load occurs. This critical buckling load (when defined by the cross-sectional area) gives a critical buckling stress. For columns this critical buckling stress is much less than the yield strength of the material. Thus, columns will generally fail from buckling much before they fail from yielding.

The shape as well as the load establishes if buckling is to occur. A column may be viewed as a straight bar with a large slenderness ratio l/r (typically 100) subjected to axial compression. Buckling occurs in a column with a large slenderness ratio when the column is loaded to a critical load and marked changes in deformation occur that do not result from material yielding.

Example 9.1

GIVEN A simple pendulum (Fig. 9.2) in which a ball is hung by a thin wire and is acted on by gravitational acceleration.

FIND Is the equilibrium neutral, stable, or unstable?

Solution

The force restoring the ball to the center position is

$$P = m_a g \sin\theta \qquad \text{(a)}$$

From simple harmonic motion

$$P = -m_a l \frac{d^2\theta}{dt^2} \qquad \text{(b)}$$

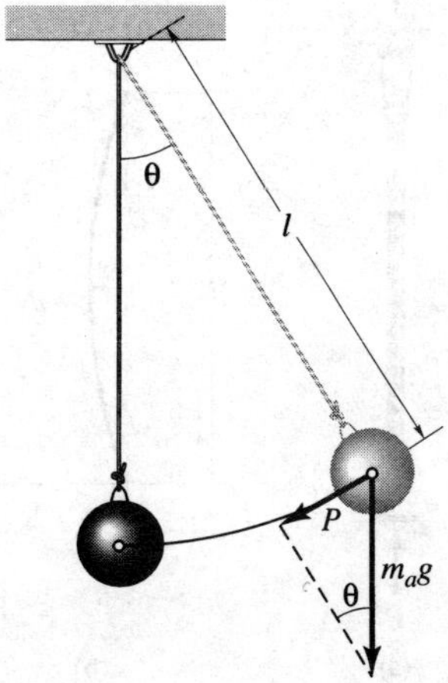

Figure 9.2 Pendulum used in Example 9.1

Combining Eqs. (a) and (b) gives

$$\frac{d^2\theta}{dt^2} = -\frac{g}{l}\sin\theta \tag{c}$$

For any angle θ the angular acceleration $d^2\theta/dt^2$ has the opposite sign from θ for $-\pi \le \theta \le \pi$. Thus, the ball will *always* return to the center position ($\theta = 0$), implying that stable equilibrium prevails.

9.3 CONCENTRICALLY LOADED COLUMNS

9.3.1 LINEAR-ELASTIC MATERIAL

Figure 9.3(a) shows a concentrically loaded column with pinned ends. For such a column the ends are kept in position but are free to rotate. This is one of the simpler situations to analyze. The results for other situations are presented later in this chapter. Assume that the column is straight initially and that the load is concentric as depicted in Fig. 9.3(b). Figure 9.3(c) shows a free-body diagram of the loads acting on the column.

Equation (5.3) describes the moment as

$$M = -EI\frac{d^2y}{dx^2} \tag{5.3}$$

Equilibrium of the section cut from the bar requires that $M = Py$. Substituting this into Eq. (5.3) gives

$$\frac{d^2y}{dx^2} + \frac{P}{EI}y = 0 \tag{9.1}$$

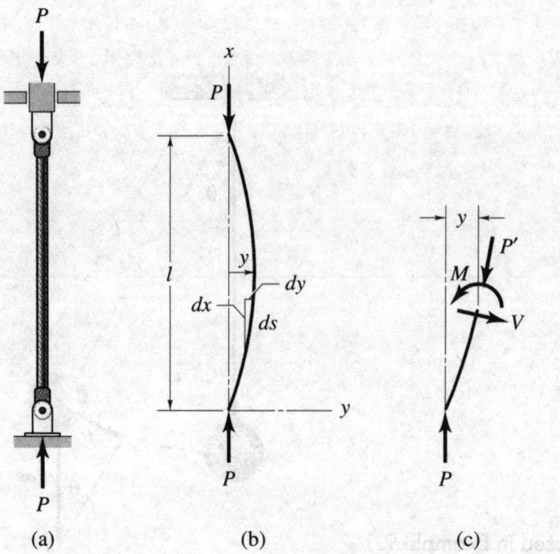

Figure 9.3 Column with pinned ends. (a) Assembly; (b) deformation shape; (c) load acting.

This is a homogeneous, second-order, linear differential equation with constant coefficients. The general solution of this equation is

$$y = C_1 \sin\left(x\sqrt{\frac{P}{EI}}\right) + C_2 \cos\left(x\sqrt{\frac{P}{EI}}\right) \tag{9.2}$$

where

C_1 and C_2 = integration constants

The boundary conditions are

1. $y = 0$ at $x = 0$ $\rightarrow C_2 = 0$
2. $y = 0$ at $x = l$ $\rightarrow C_1 \sin\left(l\sqrt{\frac{P}{EI}}\right) = 0$ **(9.3)**

However, C_1 cannot be equal to zero; otherwise, a trivial solution of $y = 0$ exists and the column will always remain straight, which is contrary to experience. The other possibility of satisfying Eq. (9.3) is for

$$\sin\left(l\sqrt{\frac{P}{EI}}\right) = 0 \tag{9.4}$$

which is satisfied when

$$l\sqrt{\frac{P}{EI}} = n\pi \tag{9.5}$$

$$P = \frac{n^2\pi^2 EI}{l^2} \tag{9.6}$$

where

$n = 1, 2, \ldots$

The smallest value of P is obtained for $n = 1$. Thus, the critical load for a column with pinned ends is

$$P_{cr} = \frac{\pi^2 EI}{l^2} \tag{9.7}$$

This load is sometimes called the **Euler load**, named after the Swiss mathematician Leonhard Euler, who originally solved the problem in 1757.

An interesting aspect of Eq. (9.7) is that the critical load is independent of the material's strength; rather, it depends only on the column's dimensions (expressed in I and l) and the material's stiffness or modulus of elasticity E. For this reason, as far as elastic buckling is concerned, columns made, for example, of high-strength steel offer no advantage over those made of lower strength steel, since both steels have approximately the same modulus of elasticity.

Another interesting aspect of Eq. (9.7) is that the critical load capacity of a column will increase as the moment of inertia of the cross-sectional area increases. Thus, columns are designed so that most of their cross-sectional area is located as far as possible from the section's principal centroidal axes. This implies that a hollow tube is preferred over a solid section.

It is also important to realize that a column will buckle about the principal axis of the cross section having the *least moment of inertia* (the weakest axis). For example, a column having a rectangular cross section, such as a meter stick, as shown in Fig. 9.4, will buckle about the x axis, not the y axis. As a result, engineers prefer square or circular cross sections for columns.

Substituting Eq. (9.5) and $C_2 = 0$ into the deflection equation [Eq. (9.2)] gives the buckling shape as

$$y = C_1 \sin\left(\frac{\pi x}{l}\right) \tag{9.8}$$

When $x = l/2$, $y = y_{max}$ and $C_1 = y_{max}$.

$$\therefore y = y_{max} \sin\left(\frac{\pi x}{l}\right) \tag{9.9}$$

Thus, the buckling shape varies sinusoidally, with zero at the ends and a maximum at half-length.

Engineers are often interested in defining the critical stress of a column. The radius of gyration r_g is given in Eq. (4.14). Substituting this into Eq. (9.7) gives the critical stress for the Euler equation as

$$(\sigma_{cr})_E = \frac{P_{cr}}{A} = \frac{\pi^2 E}{(l/r_g)^2} \tag{9.10}$$

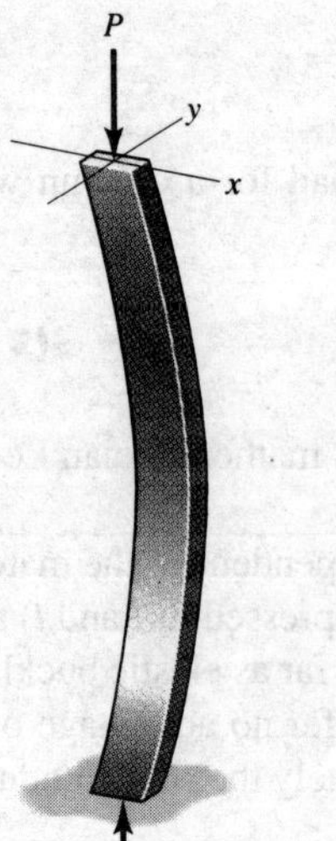

Figure 9.4 Buckling of rectangular section.

The smallest radius of gyration for a section should be used. Also, the critical stress σ_{cr} is an average stress in the column just before the column buckles. This stress is an elastic stress and is therefore less than or equal to the material's yield strength.

9.3.2 INELASTIC BUCKLING

The previous section determined the stress at which a column will buckle if the material in the column is linear elastic. As presented in Chapter 3 this is the case for most materials below the proportional limit. Above the proportional limit and up to the yield point the material may still be elastic, but it will not behave linearly. The effect is that the elastic modulus at the buckling stress in Eq. (9.10) may be significantly lower than one would expect from published values of elastic modulus. Equation (9.10) is often modified as

$$\sigma_{cr} = \frac{\pi^2 E_t}{\left(l / r_g\right)^2} \tag{9.11}$$

where E_t is the **tangent modulus** (i.e., the elastic modulus at the stress level in the column).

Equation (9.11) is called the **tangent-modulus** or **Essenger equation**. It is extremely difficult to apply in design, since the tangent modulus is never well quantified. However, it is mentioned here because of the importance of the tangent-modulus effect.

9.4 END CONDITIONS

Table 9.1 shows four end conditions. The end condition affects the effective length of the column. The effective length is the overall column length minus the portion that takes into account the end conditions. The critical load and stress given in Eqs. (9.7) and (9.10), respectively, are altered simply by replacing l with the effective length l_e for the corresponding end condition. Substituting l_e for l in Eqs. (9.7) and (9.10) gives

$$\left(P_{cr}\right)_E = \frac{\pi^2 EI}{l_e^2} \tag{9.12}$$

$$\left(\sigma_{cr}\right)_E = \frac{\left(P_{cr}\right)_E}{A} = \frac{\pi^2 E}{\left(l_e / r_g\right)^2} \tag{9.13}$$

These equations for the critical load and stress for the Euler criteria are valid for any end conditions. The minimum American Institute of Steel Construction (AISC) recommendations given in Table 9.1 apply for end constructions where "ideal conditions are approximated." The recommended conditions given in Table 9.1 should be used for column calculations.

Example 9.2

GIVEN A column has a square tubular cross section. The wall thickness is 10 mm and the column length is 12 m. The column is axially loaded in compression and has pinned ends. The weight of the tubular column must be no larger than 213.3 kg.

Table 9.1 Effective length for four end conditions.

End condition description	Both ends pinned	One end pinned and one end fixed	Both ends fixed	One end fixed and one end free
Illustration of end condition	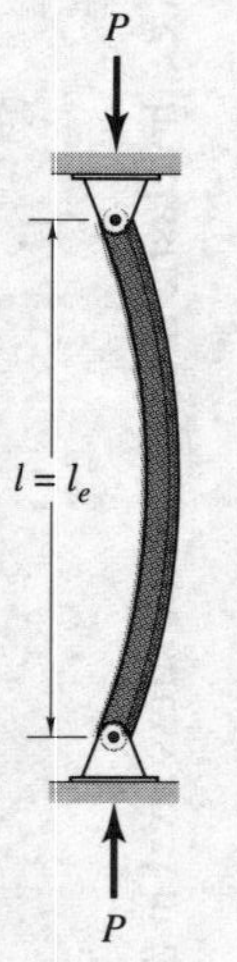	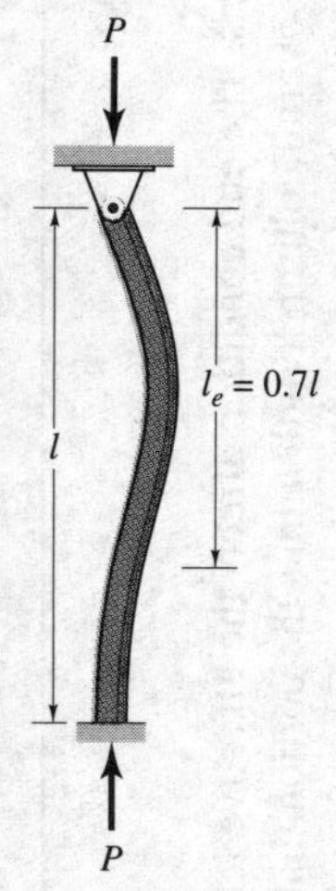	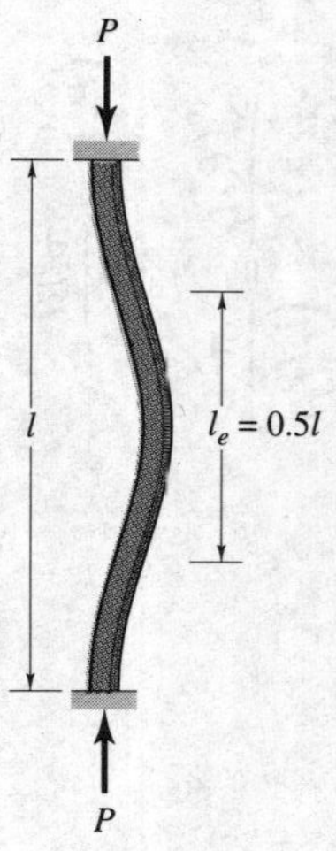	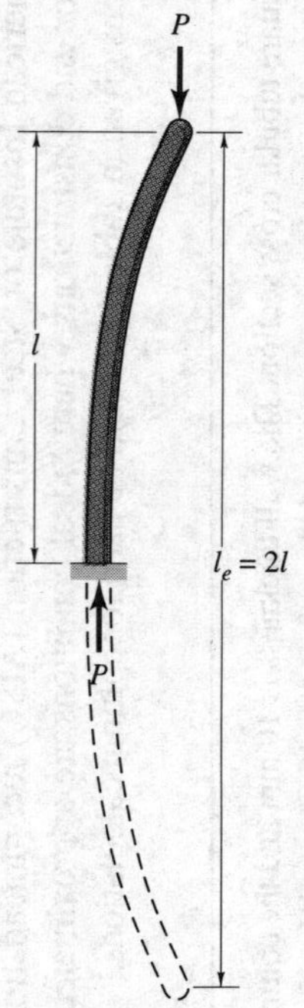
Theoretical effective column length	$l_e = l$	$l_e = 0.7l$	$l_e = 0.5l$	$l_e = 2l$
AISC (1989)– recommended effective column length	$l_e = l$	$l_e = 0.8l$	$l_e = 0.65l$	$l_e = 2.1l$

FIND Determine which metal in Tables A.1 and A.2 gives the highest buckling load. Also, calculate the critical load.

Solution

The area of a square tubular cross section with outside dimension h is

$$A = h^2 - (h-0.02)^2 \quad \textbf{(a)}$$

The weight of the column is

$$m_a = \rho A l = \rho l\left[h^2 - (h-0.02)^2\right] \quad \textbf{(b)}$$

The area moment of inertia for a square tube is

$$I = \frac{\left[h^4 - (h-0.02)^4\right]}{12} = \frac{\left[h^2 - (h-0.02)^2\right]\left[h^2 + (h-0.02)^2\right]}{12} = \frac{m_a\left[h^2 + (h-0.02)^2\right]}{12\rho l} \quad \textbf{(c)}$$

Substituting Eq. (c) into Eq. (9.7) gives the critical load as

$$P_{\text{cr}} = \frac{\pi^2 E m_a}{12\rho l^3}\left[h^2 + (h-0.02)^2\right] \quad \textbf{(d)}$$

The largest P_{cr} would result from the largest value of E/ρ. From Tables A.1 and A.2 observe that magnesium would give the largest E/ρ and that $E = 45$ GPa and $\rho = 1740$ kg/m^3. In Eq. (b), given that $m_a = 213.3$ kg, $l = 12$ m, and $\rho = 1740$ kg/m^3,

$$m_a/\rho l = 0.04(h-0.01)$$

$$\frac{213.3}{(1740)(12)(0.04)} + 0.01 = h$$

$$h = 0.2654 \text{ m} = 265.4 \text{ mm}$$

From Eq. (d) the critical load is

$$P_{\text{cr}} = \frac{\pi^2(45)(10^9)(213.3)}{12(1740)(12)^3}\left[(0.2654)^2 + (0.2454)^2\right] = 3.431\times10^5 \text{ N}$$

9.5 EULER'S BUCKLING CRITERIA

One approach to determining the buckling criteria for columns is simply to say that the critical stress is equivalent to the Euler critical stress as long as it is less than the allowable stress. The American Institute of Steel Construction (1989) assumes that the proportional limit of a material exists at one-half the yield strength, or that the allowable stress for elastic buckling is

$$\sigma_{\text{all}} = 0.5 S_y \quad \textbf{(9.14)}$$

where

S_y = yield strength of material in Table A.1 for several ferrous alloys

Substituting Eq. (9.14) into Eq. (9.13) gives the **slenderness ratio** C_c from Euler's formula as

$$C_c = \left(\frac{l_e}{r_g}\right)_E = \sqrt{\frac{2E\pi^2}{S_y}} \tag{9.15}$$

9.6 JOHNSON'S BUCKLING CRITERIA

Figure 9.5 shows curves for normal stress as a function of slenderness ratio plotted from the Euler and Johnson equations. The abrupt change in the Euler curve [Eq. (9.7)] as it approaches the yield strength (see point A in Fig. 9.5) results in empirical modification at this location. Perhaps the most widely used modification is the parabola proposed by J.B. Johnson around the turn of the century. The **Johnson equation** is

$$(\sigma_{cr})_J = \frac{(P_{cr})_J}{A} = S_y - \frac{S_y^2}{4\pi^2 E}\left(\frac{l_e}{r_g}\right)^2 \tag{9.16}$$

Also shown in Fig. 9.5 is the tangency point T, the point at which the Euler and Johnson equations are equal. This tangency point distinguishes between intermediate columns (Johnson range) and long columns (Euler range). To determine the value of $(l_e/r_g)_T$, equate the Euler and Johnson equations [Eqs. (9.13) and (9.16)] or

$$\left(\frac{l_e}{r_g}\right)_T^4 - \frac{4\pi^2 E}{S_y}\left(\frac{l_e}{r_g}\right)_T^2 + \frac{4\pi^4 E^2}{S_y^2} = 0 \tag{9.17}$$

Solving for $(l_e/r_g)_T$ gives

$$\left(\frac{l_e}{r_g}\right)_T = \sqrt{\frac{2\pi^2 E}{S_y}} = \sqrt{\frac{\pi^2 EA}{P_{cr}}} \tag{9.18}$$

Note from Eqs. (9.15) and (9.18) that $(l_e/r_g)_E = (l_e/r_g)_T$. If $l_e/r_g \le (l_e/r_g)_T$, the Johnson stress equation given in Eq. (9.16) should be used; if $l_e/r_g \ge (l_e/r_g)_T$, the Euler equation given in

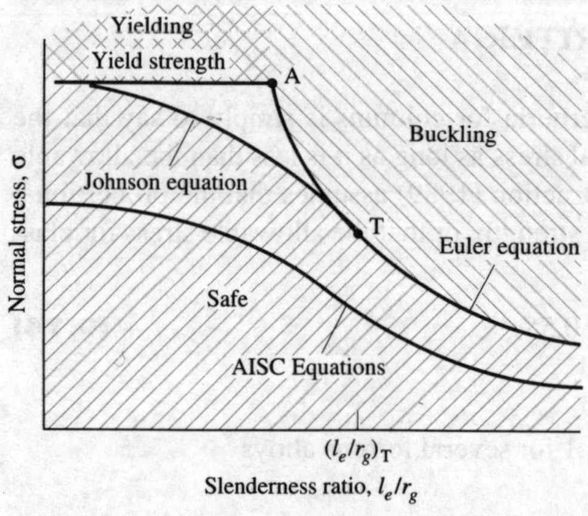

Figure 9.5 Normal stress as function of slenderness ratio obtained by Euler, Johnson, and AISC equations, as well as yield strength.

Eq. (9.13) should be used. Also shown in Fig. 9.5 are three regions: safe, buckling, and yielding. Thus, by knowing the normal stress and the slenderness ratio, one can quickly determine which region a particular design is in.

Example 9.3

GIVEN A column with one end fixed and the other free is to be made of aluminum alloy 2014. The column's cross-sectional area is fixed at 600 mm^2, and its length is fixed at 2.5 m.

FIND The critical column buckling load for the following shapes:

a) A solid round bar
b) A cylindrical tube with a 50-mm outer diameter
c) A square tube with a 50-mm outer dimension
d) A square bar

Solution

a) The cross-sectional area of a solid round bar is

$$A = \frac{\pi d^2}{4} \quad \text{or} \quad d = \sqrt{\frac{4A}{\pi}} = \sqrt{\frac{4(600)}{\pi}} = 27.64 \text{ mm}$$

The area moment of inertia for a circular section is

$$I = \frac{\pi d^4}{64} = \frac{\pi(27.64)^4}{64} = 28\ 650 \text{ mm}^4$$

The radius of gyration is

$$r_g = \sqrt{\frac{I}{A}} = \sqrt{\frac{28\ 650}{600}} = 6.910 \text{ mm}$$

From Table A.2 for aluminum alloy 2014 $E = 72$ GPa and $S_y = 97$ MPa. From Eq. (9.18) $(l_e/r_g)_T = 121.0$. From Table 9.1 for one end fixed and one end free, for the recommended consideration,

$$l_e = 2.1l = 2.1(2.5) = 5.25 \text{ m} = 5250 \text{ mm}$$

$$\frac{l_e}{r_g} = \frac{5250}{6.910} = 759.8$$

Since $l_e/r_g > (l_e/r_g)_T$, the Euler formula in Eq. (9.12) gives the critical load at the onset of buckling as

$$P_{cr} = \frac{\pi^2 EI}{l_e^2} = \frac{\pi^2(72)(10^9)(28\ 650)(10^{-12})}{(5.25)^2} = 738.6 \text{ N}$$

b) The cross-sectional area of a cylindrical tube is $A = \frac{\pi}{4}(d_o^2 - d_i^2)$. Given that

$$d_o = 50 \text{ mm} \quad \text{and} \quad A = 600 \text{ mm}^2,$$

$$\therefore d_i = \sqrt{d_o^2 - \frac{4A}{\pi}} = \sqrt{50^2 - \frac{4(600)}{\pi}} = 41.67 \text{ mm}$$

The area moment of inertia for a hollow circular section is

$$I = \frac{\pi}{64}(d_o^4 - d_i^4) = \frac{\pi}{64}\left[(50)^4 - (41.67)^4\right] = 1.588 \times 10^5 \text{ mm}^4 = 1.588 \times 10^{-7} \text{ m}^4$$

The radius of gyration is

$$r_g = \sqrt{\frac{I}{A}} = \sqrt{\frac{(1.588)(10^5)}{600}} = 16.27 \text{ mm} = 0.01627 \text{ m}$$

$$\therefore l_e / r_g = 5.25 / 0.01627 = 322.7$$

Since $l_e/r_g > (l_e/r_g)_T$, the Euler equation should be used.

$$P_{cr} = \frac{\pi^2 EI}{l_e^2} = \frac{\pi^2 (72)(10^9)(1.588)(10^{-7})}{(5.25)^2} = 4094 \text{ N}$$

In parts a to d the same amount of material is used for the construction of the column. Also, for the cylindrical tube the area moment of inertia and the critical buckling load are 5.5 times larger than found in part a for a solid round bar.

c) The cross-sectional area of a square tube is $A = a_1^2 - a_2^2$. The inner dimension is

$$a_2 = \sqrt{a_1^2 - A} = \sqrt{50^2 - 600} = 43.59 \text{ mm}$$

The area moment of inertia is

$$I = \frac{a_1^4 - a_2^4}{12} = \frac{(50)^4 - (43.59)^4}{12} = 220\ 000 \text{ mm}^4 = 22 \text{ cm}^4$$

The radius of gyration is

$$r_g = \sqrt{\frac{I}{A}} = 10^2 \sqrt{\frac{22}{600}} = 19.15 \text{ mm} = 0.01915 \text{ m}$$

$$\therefore \frac{l_e}{r_g} = \frac{5.25}{0.01915} = 274.2$$

Since $l_e/r_g > (l_e/r_g)_T$, the Euler formula should be used to calculate the critical load:

$$P_{cr} = \frac{\pi^2 EI}{l_e^2} = \frac{\pi^2 (72)(10^9)(0.22)(10^{-6})}{(5.25)^2} = 5672 \text{ N}$$

Comparing results from parts b and c shows that the hollow square had a critical load that was 39% more than found in part b for a hollow circular section.

d) The cross-sectional area of a square bar is $A = a^2$, or $a = \sqrt{A} = \sqrt{600} = 24.49$ mm. The area moment of inertia is

$$I = \frac{a^4}{12} = \frac{(24.5)^4}{12} = 30\ 000 \text{ mm}^4 = 3 \times 10^{-8} \text{ m}^4$$

The radius of gyration is

$$r_g = \sqrt{\frac{I}{A}} = \sqrt{\frac{30\ 000}{600}} = 7.071 \text{ mm}$$

$$\frac{l_e}{r_g} = \frac{5.25}{(7.071)(10^{-3})} = 742.5$$

Since $l_e/r_g > (l_e/r_g)_T$, the Euler formula should be used:

$$P_{cr} = \frac{\pi^2 EI}{l_e^2} = \frac{\pi^2 (72)(10^9)(3)(10^{-8})}{(5.25)^2} = 773.5 \text{ N}$$

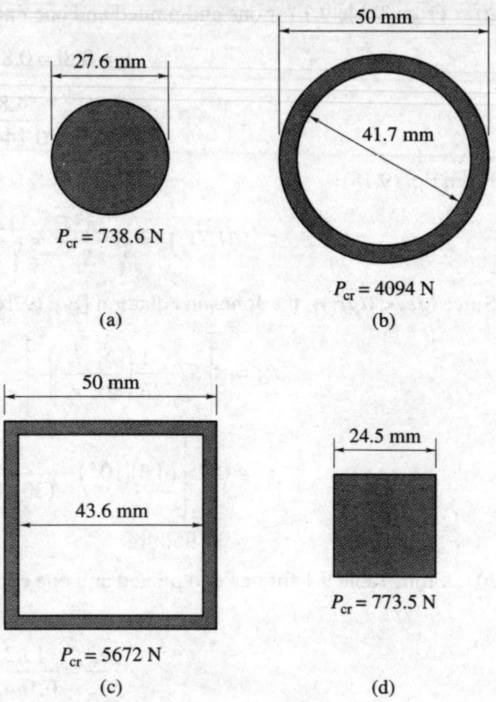

Figure 9.6 Cross-sectional areas drawn to scale from results of Example 9.3, as well as critical buckling load for each cross-sectional area.

Figure 9.6 shows the cross sections of the four shapes considered as well as the critical buckling loads. This figure dramatically illustrates why in designing columns the column's cross-sectional area should be located as far as possible from the section's principal centroidal axes.

Example 9.4

GIVEN A column with one end fixed and one end pinned is made of a low-carbon steel. The column's cross section is rectangular with $h = 0.5$ in. and $b = 1.5$ in.

FIND The buckling load for the following three lengths:

a) 0.5 ft
b) 2 ft
c) 4 ft

Solution

From Table A.1 for low-carbon steel $S_y = 43$ ksi and $E = 30 \times 10^6$ psi. The cross-sectional area is

$$A = bh = (1.5)(0.5) = 0.75 \text{ in.}^2$$

For a rectangular cross section the area moment of inertia is

$$I = \frac{bh^3}{12} = \frac{(1.5)(0.5)^3}{12} = 0.0156 \text{ in.}^4$$

The radius of gyration is

$$r_g = \sqrt{\frac{I}{A}} = \sqrt{\frac{0.0156}{0.75}} = 0.1442 \text{ in.}$$

a) From Table 9.1 for one end pinned and one end fixed and $l = 0.5$ ft,

$$l_e = 0.8l = 0.8(6) = 4.8 \text{ in.}$$

$$\therefore \frac{l_e}{r_g} = \frac{4.8}{0.1442} = 33.28$$

From Eq. (9.18)

$$(l_e/r_g)_T = \sqrt{\frac{2\pi^2 E}{S_y}} = \sqrt{\frac{2\pi^2(30)(10^6)}{43\ 000}} = 117.4$$

Since $l_e/r_g < (l_e/r_g)_T$, the Johnson equation [Eq. (9.16)] should be used:

$$P_{cr} = A\left[S_y - \frac{1}{E}\left(\frac{S_y}{2\pi}\frac{l_e}{r_g}\right)^2\right]$$

$$= 0.75\left[(4.3)(10^4) - \frac{1}{(30)(10^6)}\left(\frac{(4.3)(10^4)(33.28)}{2\pi}\right)^2\right]$$

$$= 30\ 950 \text{ lbf}$$

b) From Table 9.1 for one end pinned and one end fixed and $l = 2$ ft,

$$l_e = 0.80l = 0.8(24) = 19.2 \text{ in.}$$

$$\frac{l_e}{r_g} = \frac{19.2}{0.1442} = 133.1$$

Since $l_e/r_g > (l_e/r_g)_T$, the Euler equation should be used. From Eq. (9.12)

$$P_{cr} = \frac{\pi^2 EI}{l_e^2} = \frac{\pi^2(30)(10^6)(0.0156)}{(19.2)^2} = 12\ 530 \text{ lbf}$$

c) The Euler equation is definitely valid for $l = 4$ ft:

$$l_e = 0.8(4)(12) = 38.4 \text{ in.}$$

The critical load is

$$P_{cr} = \frac{\pi^2 EI}{l_e^2} = \frac{\pi^2(30)(10^6)(0.0156)}{(38.4)^2} = 3132 \text{ lbf}$$

Observe that in part a the column was short, so that yield strength predominated and the Johnson equation was used, but in parts b and c the columns were long enough that the Euler equation was used. Also observe that as the column becomes long, the critical load decreases significantly.

9.7 AISC CRITERIA

The American Institute of Steel Construction [AISC (1989)] has produced design guidelines for elastic stability conditions. A number of subtleties are associated with these recommendations.

As shown by the tangent-modulus equation the elastic modulus based on the linear-elastic portion of a stress-strain curve will lead to erroneous results when applied to buckling problems. However, tangent-modulus data are not readily available in the technical literature and are extremely difficult to obtain experimentally. Thus, the **AISC equations** correct for reductions in elastic modulus as the column stress exceeds the proportional limit.

Further, long columns are much more difficult to design because they are extremely susceptible to defects in straightness or to eccentricity in loading. Therefore, a weighted reduction in the allowable stress is prescribed. The allowable normal stress for elastic buckling is given by

$$\sigma_{\text{all}} = \frac{12\pi^2 E}{23(l_e/r_g)^2} \tag{9.19}$$

For inelastic buckling

$$\sigma_{\text{all}} = \frac{\left\{1 - \left[\dfrac{(l_e/r_g)^2}{2C_c^2}\right]\right\} S_y}{n_\sigma} \tag{9.20}$$

where

C_c = slenderness ratio for Euler buckling defined by Eq. (9.15)
n_σ = reduction in allowable stress given by

$$n_\sigma = \frac{5}{3} + \frac{3(l_e/r_g)}{8C_c} - \frac{(l_e/r_g)^3}{8C_c^3} \tag{9.21}$$

Figure 9.5 shows the AISC results as well as the Euler and Johnson results. The American Association of State Highway and Transportation Officials (AASHTO) uses equations identical to Eqs. (9.19) and (9.20) but requires a constant stress reduction $n_\sigma = 2.12$ for both elastic and inelastic buckling.

Note that n_σ is not a safety factor but a mandatory reduction in a material's allowable stress. Buckling load can be predicted by using these equations, but designs where safety is paramount must use an additional safety factor.

Tutorial 4

9.8 ECCENTRICALLY LOADED COLUMNS

Columns used in applications rarely have the applied load aligned coincidentally with the centroidal axis of the cross section. The distance between the two axes is called the *eccentricity* and is designated by e. It is assumed that the load is always parallel with the centroid of the columns. Just as for concentrically loaded columns (in Sec. 9.3) the analysis of eccentricity is restricted to columns with pinned ends. Figure 9.7(a) shows a pinned-end column subjected to forces acting at a distance e from the centerline of the undeformed column. It is assumed that the load is applied to the column at a short eccentric distance from the centroid of the cross section. This loading on the column is statically equivalent to the axial load and bending moment $M' = Pe$ shown in Fig. 9.7(b). As when considering concentrically loaded columns small deflections and linear-elastic material behavior are assumed. The x-y plane is a plane of symmetry for the cross-sectional area.

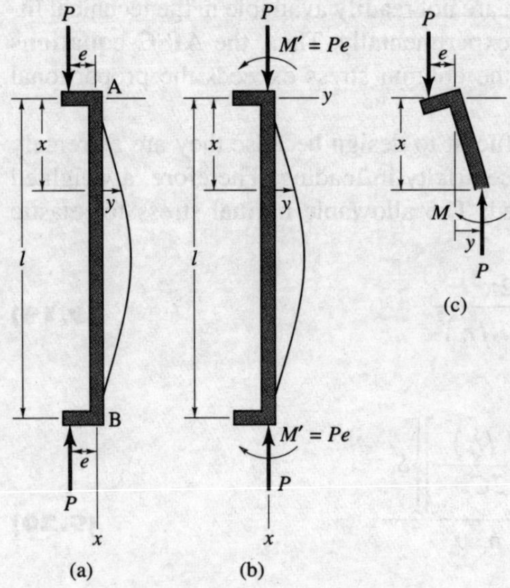

Figure 9.7 Eccentrically loaded column. (a) Eccentricity; (b) statically equivalent bending moment; (c) free-body diagram through arbitrary section.

From a free-body diagram of an arbitrary section shown in Fig. 9.7(c) the internal moment in the column is

$$M = P(e + y) \tag{9.22}$$

The differential equation for the deflection curve is obtained from Eqs. (5.3) and (9.22), or

$$\frac{d^2y}{dx^2} + \frac{Py}{EI} = -\frac{Pe}{EI} = \text{constant} \tag{9.23}$$

If the eccentricity is zero, Eqs. (9.23) and (9.1) are identical.

The general solution to Eq. (9.23) is

$$y = C_1 \sin\left(x\sqrt{\frac{P}{EI}}\right) + C_2 \cos\left(x\sqrt{\frac{P}{EI}}\right) - e \tag{9.24}$$

The boundary conditions are

1. $x = 0, \quad y = 0 \quad \rightarrow C_2 = e$
2. $x = l/2, \quad \partial y/\partial x = 0$

Taking the derivative of Eq. (9.24) gives

$$\frac{dy}{dx} = C_1\sqrt{\frac{P}{EI}} \cos\left(x\sqrt{\frac{P}{EI}}\right) - e\sqrt{\frac{P}{EI}} \sin\left(x\sqrt{\frac{P}{EI}}\right) \tag{9.25}$$

Using Eq. (9.25) and boundary condition 2 gives

$$C_1 = e \tan\left(\frac{l}{2}\sqrt{\frac{P}{EI}}\right) \tag{9.26}$$

Substituting $C_2 = e$ and Eq. (9.26) into Eq. (9.24) gives

$$y = e\left[\tan\left(\frac{l}{2}\sqrt{\frac{P}{EI}}\right)\sin\left(x\sqrt{\frac{P}{EI}}\right) + \cos\left(x\sqrt{\frac{P}{EI}}\right) - 1\right]$$

The maximum deflection occurs at $x = l/2$.

$$\therefore y_{\max} = e\left[\frac{\sin^2\left(\frac{l}{2}\sqrt{\frac{P}{EI}}\right)}{\cos\left(\frac{l}{2}\sqrt{\frac{P}{EI}}\right)} + \cos\left(\frac{l}{2}\sqrt{\frac{P}{EI}}\right) - 1\right] \tag{9.27}$$

$$y_{\max} = e\left[\sec\left(\frac{l}{2}\sqrt{\frac{P}{EI}}\right) - 1\right] \tag{9.28}$$

The maximum stress on the column is caused by the axial load and the moment. The maximum moment occurs at the column's midheight and has a magnitude of

$$M_{\max} = |P(e + y_{\max})|$$

$$M_{\max} = Pe\sec\left(\frac{l}{2}\sqrt{\frac{P}{EI}}\right) \tag{9.29}$$

The maximum stress in the column is compressive and is

$$\sigma_{\max} = \frac{P}{A} + \frac{M_{\max}c}{I}$$

Making use of Eq. (9.29) gives

$$\sigma_{\max} = \frac{P}{A} + \frac{Pec}{I}\sec\left(\frac{l}{2}\sqrt{\frac{P}{EI}}\right)$$

Since the radius of gyration is $r_g^2 = I/A$, the preceding equation becomes

$$\sigma_{\max} = \frac{P}{A}\left[1 + \frac{ec}{r_g^2}\sec\left(\frac{l}{2r_g}\sqrt{\frac{P}{EA}}\right)\right] \tag{9.30}$$

where

P = critical load where buckling will occur in eccentrically loaded column, N

A = cross-sectional area of column, m^2

e = eccentricity of load measured from neutral axis of column's cross-sectional area to load's line of action, m

c = distance from neutral axis to outer fiber of column, m

r_g = radius of gyration, m

l = length before load is applied, m

E = modulus of elasticity of column material, Pa

For end conditions other than pinned, the length is replaced with the effective length (using Table 9.1), and Eqs. (9.28) and (9.30) become

$$y_{max} = e\left[\sec\left(\frac{l_e}{2}\sqrt{\frac{P}{EI}}\right) - 1\right] \quad \textbf{(9.31)}$$

$$\sigma_{max} = \frac{P}{A}\left[1 + \frac{ec}{r_g^2}\sec\left(\frac{l_e}{2r_g}\sqrt{\frac{P}{EA}}\right)\right] \quad \textbf{(9.32)}$$

The parameter ec/r_g^2 is called the **eccentricity ratio**. Equation (9.31) is known as the **secant equation**. Note from Eq. (9.32) that it is not convenient to calculate the load explicitly.

Figure 9.8 shows the effect of slenderness ratio l_e/r_g on normal stress σ for an eccentrically loaded column. These results are for structural grade steel where the modulus of elasticity is 29×10^6 psi and the yield strength is 36 000 psi. Note that as $e \to 0$, $\sigma \to P/A$, where P is the critical column load as established by the Euler formula [Eq. (9.12)].

The curves in Fig. 9.8 indicate that differences in eccentricity ratio have a significant effect on column stress when the slenderness ratio is small. On the other hand, columns with large slenderness ratios tend to fail at the Euler (or critical) load regardless of the eccentricity ratio.

If $\sigma_{max} = \sigma_{all} = S_y/2$, Eq. (9.32) becomes

$$P = \frac{S_y A/2}{1 + \dfrac{ec}{r_g^2}\sec\left(\dfrac{l_e}{2r_g}\sqrt{\dfrac{P}{EA}}\right)} = f(P) \quad \textbf{(9.33)}$$

An initial guess for P is the value obtained from the concentric loading condition or

$$P = P_{cr} \quad \textbf{(9.34)}$$

Using Eq. (9.34) in Eq. (9.33) gives a new value of P. This process is continued until

$$\left|\frac{P - f(P)}{f(P)}\right| \le 1 \times 10^{-4} \quad \textbf{(9.35)}$$

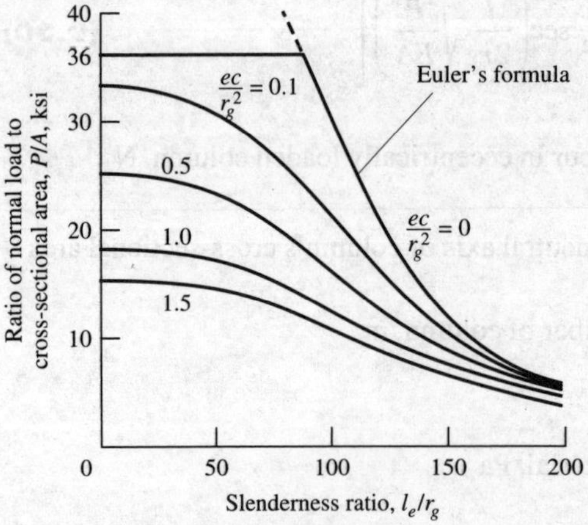

Figure 9.8 Stress variation with slenderness ratio for five eccentricity ratios. Structural steel is assumed with $E = 29 \times 10^6$ psi and $S_y = 36 \times 10^3$ psi.

Example 9.5

GIVEN A 36-in.-long hollow aluminum alloy (2014) tube with a 3.0-in. outside diameter, a 0.03-in. wall thickness, and both ends pinned.

FIND The critical buckling load for

a) Concentric loading

b) Eccentric loading with an eccentricity of 0.15 in.

Solution

a) The inside diameter of the column is

$$d_i = d_o - 2t_h = 3.0 - 2(0.03) = 2.94 \text{ in.}$$

The cross-sectional area is

$$A = \frac{\pi}{4}\left(d_o^2 - d_i^2\right) = \frac{\pi}{4}\left[(3)^2 - (2.94)^2\right] = 0.280 \text{ in.}^2$$

The area moment of inertia is

$$I = \frac{\pi}{64}\left(d_o^4 - d_i^4\right) = \frac{\pi}{64}\left[(3)^4 - (2.94)^4\right] = 0.309 \text{ in.}^4$$

The radius of gyration is

$$r_g = \sqrt{\frac{I}{A}} = \sqrt{\frac{0.309}{0.280}} = 1.050 \text{ in.}$$

From Table 9.1 for a column with both ends pinned the effective length is equal to the actual length ($l = l_e$). From Table A.2 for aluminum alloy 2014

$$E = 10.5 \times 10^6 \text{ psi} \quad \text{and} \quad S_y = 14 \times 10^3 \text{ psi}$$

From Eq. (9.18)

$$\left(\frac{l_e}{r_g}\right)_{\mathrm{T}} = \sqrt{\frac{2\pi^2 E}{S_y}} = \sqrt{\frac{2\pi^2(10.5)(10^6)}{14(10^3)}} = 121.7$$

and

$$\frac{l_e}{r_g} = \frac{36}{1.05} = 34.29$$

Since $l_e/r_g < (l_e/r_g)_{\mathrm{T}}$, the Johnson formula [Eq. (9.16)] should be used:

$$P_{\mathrm{cr}} = AS_y\left[1 - \frac{S_y}{4\pi^2 E}\left(\frac{l_e}{r_g}\right)^2\right] = (0.28)(14)(10^3)\left[1 - \frac{(14)(10^3)(34.29)^2}{4\pi^2(10.5)(10^6)}\right] = 3764 \text{ lbf}$$

b) The distance from the neutral axis to the outer fiber is $c = d_o/2 = 1.5$ in. The eccentricity ratio is

$$\frac{ce}{r_g^2} = \frac{(1.5)(0.15)}{(1.05)^2} = 0.2041$$

$$P = P_{\mathrm{cr}} = 3764 \text{ lbf}$$

From Eq. (9.33)

$$P_1 = \frac{1960}{1 + 0.2041 \sec\left[\dfrac{34.29}{2}\sqrt{\dfrac{3764}{2.940(10^6)}}\right]} = 1569 \text{ lbf}$$

$$P_2 = \frac{1960}{1+0.2041\sec\left[\frac{34.29}{2}\sqrt{\frac{1569}{2.940(10^6)}}\right]} = 1605\text{ lbf}$$

$$P_3 = 1604\text{ lbf}$$

The critical load for the eccentric loading is 1604 lbf and that for the concentric loading is 3764 lbf.

Case Study — DESIGN OF TELESCOPING BOOM FOR HYDRAULIC CRANE

A mobile hydraulic boom crane (Fig. 9.9) is used in various materials handling activities. The boom is a set of telescoping cylinders driven by hydraulic pressure. The load is applied eccentrically from the boom axis.

At low boom angles the crane's capacity is defined by its tipping stability, but at high boom angles boom buckling is a concern. In the manufacture of the telescoping boom great care is taken to ensure that the boom sections are extremely straight; the deviation from straightness is no more than 0.02 in. per 10-ft section. This tolerance is ensured by clamping the boom section in two rigid clamps 10 ft apart and tracing a dial indicator along the entire exposed length of the boom. Any section found to deviate from straightness is carefully deformed plastically with a hydraulic jack until the 0.02-in. criterion is achieved.

GIVEN The inner diameter of the tube must be at least 5 in. based on the hydraulic requirements, and the thickness must be a convenient dimension so that standard extrusions can be used (i.e., an integer multiple of eighths of an inch). The bucket must telescope at least 60 ft and must be constructed from three sections to ensure a reasonable contracted length. The eccentricity is a constant value of 3 ft and cannot be modified, as this dimension is determined by rigging requirements. Use Tri-ten steel ($E = 29 \times 10^6$ psi; $S_y = 30 \times 10^3$ psi) as the boom material.

FIND A tubular cross section that will deform less than 2 ft under a 10 000-lbm load for a fully extended vertical boom.

Figure 9.9 Mobile hydraulic boom crane used in case study. (AP/Wide World Photos)

Solution

From Table 9.1 the effective boom length for the relevant buckling mode is 60 ft times 2, or 120 ft. The boom consists of three sections, each more slender nearer to the load end. As a worst-case analysis the entire boom length will be assumed to have a constant cross section equal to the most slender section of the boom. Also, the deviation from straightness of each section will be assumed to be continuous and cumulative. Thus, if the ultimate design is still acceptable based on deformation and stress requirements, the final boom will be even stronger.

The initial deformation can be taken as that caused by a fictitious load P_i for the load eccentricity. Since the cumulative deviation from straightness is 0.120 in., the fictitious load is determined by solving the secant equation [Eq. (9.31)] for the load P_i, or

$$0.120 = 36\left[\sec\left(\frac{120\times 12}{2}\sqrt{\frac{P_i}{EI}}\right)-1\right]$$

Solving for P_i gives

$$\sqrt{\frac{P_i}{EI}} = \frac{\cos^{-1}\left(\frac{36}{36.12}\right)}{720}$$

$$\therefore P_i = EI(1.28\times 10^{-8})$$

Case Study (CONCLUDED)

The load that causes an additional 24-in. deflection (for 24.12 in. total) is P_i + 10 000 lbf. Again applying the secant equation yields

$$24.120 = 36\left[\sec\left(\frac{120\times 12}{2}\sqrt{\frac{P_i+10}{EI}}\right)-1\right]$$

Numerically solving for I yields $I = 208.8$ in.[4]. Since

$$I=\frac{\pi}{64}\left[(5+2t_h)^4-(5)^4\right]$$

the minimum thickness is 1.68 in. The next largest multiple of 1/8 in. is 1.75 in.

Before this tube wall thickness can be accepted, we must be assured that the allowable boom stress is not exceeded, since the tangent modulus can be much less than the elastic modulus reported for the steel. The iterative approach described in Eqs. (9.33) to (9.35) is used. The cross-sectional area of the tube is $A = \pi[(2.5 + 1.75)^2 - (2.5)^2] = 37.1$ in.^{2}. The tube's radius of gyration is

$$r_g=\sqrt{\frac{I}{A}}=\sqrt{\frac{208.8}{37.1}}=2.37 \text{ in.}$$

From Eq. (9.34)

$$P = P_{cr} = 7200 \text{ lbf}$$

Substituting into Eq. (9.33) yields

$$P=\frac{S_yA/2}{1+\dfrac{ec}{r_g^2}\sec\left(\dfrac{l_e}{2r_g}\sqrt{\dfrac{P}{EA}}\right)}=14\ 000 \text{ lbf}$$

Using this value in Eq. (9.33) yields 9400 lbf, which then yields 12 700 lbf, etc., until eventually a solution of 11 300 lbf is reached. Thus, this value also ensures that the beam remains linearly elastic under the design load.

9.9 SUMMARY

In an attempt to better understand column buckling three equilibrium regimes were studied. Stable, neutral, and unstable equilibrium were explained by observing what happens to a sphere on concave, flat, and convex surfaces. It was concluded that when columns are in unstable equilibrium, they are unreliable and hazardous. A small displacement can cause a catastrophic change in the configuration of the column. Thus, as the load on a column is increased, a critical load is reached where unstable equilibrium will occur and the column will not return to its straight configuration. Thus, for long, slender columns a critical buckling load occurs. The column's critical buckling stress was also determined to be much less than the yield strength of the material. End conditions were found to affect the critical buckling load. The effective length was used to handle four types of end condition. The Euler and Johnson buckling criteria were developed for concentrically loaded columns. Eccentrically loaded columns were also studied and the secant formulas were derived.

KEY WORDS

AISC equations estimations of allowable stress for prevention of buckling in structures (corrections for reductions in elastic modulus as column stress exceeds proportional limit)

buckling sudden large deformation of structure due to slight increase of existing load

column straight and long (relative to a cross-sectional dimension) member subjected to compressive axial loads

eccentricity ratio measure of how far off center load is applied, given by ec/r_g^2

Essenger equation critical buckling load in nonlinear elastic buckling, given by $\sigma_{cr} = \dfrac{\pi^2 E_t}{(l/r_g)^2}$

Euler load critical load of elastic column, given by $P_{cr} = \dfrac{\pi^2 EI}{l^2}$

Johnson equation critical load for inelastic buckling, given by $(\sigma_{cr}) = \dfrac{(P_{cr})_J}{A} = S_y - \dfrac{S_y^2}{4\pi^2 E}\left(\dfrac{l_e}{r_g}\right)^2$

secant equation deflection due to eccentric loading, given by $y_{max} = e\left[\sec\left(\dfrac{l_e}{2}\sqrt{\dfrac{P}{EI}}\right) - 1\right]$

slenderness ratio measure of column slenderness, given by $C_c = \left(\dfrac{l_e}{r_g}\right)_E = \sqrt{\dfrac{2E\pi^2}{S_y}}$

tangent modulus elastic modulus at stress level in column

tangent-modulus equation same as Essenger equation

RECOMMENDED READINGS

Juvinall, R.C., and Marshek, K.M. (1991) *Fundamentals of Machine Component Design*, Wiley, New York.

Ketter, R.L., Lee, G.C., and Prawel, S.P. (1979) *Structural Analysis and Design*, McGraw-Hill, New York.

Manual of Steel Construction, (1980) 8th ed., American Institute of Steel Construction, Chicago.

Norton, R.L. (1996) *Machine Design*, Prentice-Hall, Englewood Cliffs, NJ.

Popov, I. (1968) *Introduction to Mechanics of Solids*, Prentice-Hall, Englewood Cliffs, NJ.

Shigley, J.E., and Mischke, C.R. (1989) *Mechanical Engineering Design*, McGraw-Hill, New York.

Timoshenko, S.P., and Goodier, J.N. (1970) *Theory of Elasticity*, McGraw-Hill, New York.

Uguaral, A.C., and Fenster, S.K. (1995) *Advanced Strength and Applied Elasticity*, 3rd ed., Prentice-Hall, Englewood Cliffs, NJ.

Willems, N., Easley, J., and Rolfe, S. (1981) *Strength of Materials*, McGraw-Hill, New York.

REFERENCE

American Institute of Steel Construction (AISC) (1989) *Manual of Steel Construction*, 9th ed., Chicago.

PROBLEMS

Section 9.2

9.1 A yo-yo begins to move at the origin of the x-y axis (at point A) and is at its farthest point from the origin at point B, as shown in sketch *a*. The string is 0.75 m long, and its weight is small relative to the weight of the yo-yo. Find the equilibrium position for the yo-yo.

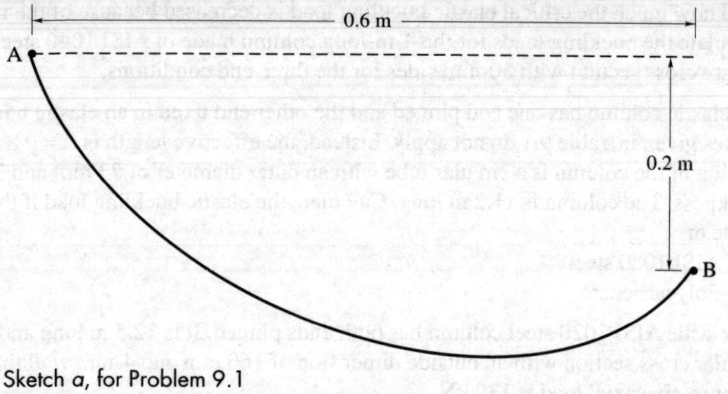

Sketch *a*, for Problem 9.1

9.2 For the yo-yo in Problem 9.1 the central pin has a diameter of 5 mm. The outside diameter of the yo-yo is 50 mm, and the diameter of the string is 1 mm. Will the yo-yo roll from point A to point B if tightened slowly?

9.3 A person rides a bike on a horizontally flat road. Is this a neutral, stable, or unstable equilibrium position?

9.4 A golf ball is placed
a) On top of a small hill
b) On a horizontal flat plane
c) In a shallow groove

Find the type of equilibrium for each condition.

Section 9.3

9.5 A column has pinned ends and is axially loaded in compression. The length is 6 m and the weight is 200 kg, but the form of the cross section can be changed. The column is made of steel (AISI 1040). Find the Euler buckling load for
a) A solid circular section
b) A solid square section
c) A circular tube with outer diameter of 100 mm
d) A square tube with outside dimension of 100 mm

9.6 A column with both ends pinned is 3 m long and has a section with an outer diameter of 30 mm and wall thickness of 5 mm. Find which material given in Tables A.1 through A.4 gives the highest buckling load. Also, give the buckling loads for AISI 1080 steel, aluminum alloy 2014, and molybdenum.

9.7 Determine the critical stresses for the four columns considered in Problem 9.5. Let the column be 8 m long and the cross-sectional area be 50 cm^2. The column material is AISI 1080 steel.

Section 9.4

9.8 A column is axially loaded in compression. The ends were specified to be fixed, but because of a manufacturing mistake they had
a) One end fixed and the other pinned
b) Both ends pinned

Find how much the critical elastic buckling load is decreased because of the mistakes. Also, calculate the buckling loads for the 4-m-long column made of AISI 1040 steel having a solid, square cross section with 30-mm sides for the three end conditions.

9.9 An elastic column has one end pinned and the other end fixed in an elastic bushing so that the values given in Table 9.1 do not apply. Instead, the effective length is $l_e = 0.83l$. The cross section of the column is a circular tube with an outer diameter of 73 mm and 3.2-mm wall thickness. The column is 11.2 m long. Calculate the elastic buckling load if the column is made of

a) AISI 1080 steel
b) Polycarbonate

9.10 An elastic AISI 1020 steel column has both ends pinned. It is 12.5 m long and has a square tubular cross section with an outside dimension of 160 mm and 4-mm wall thickness. Its compressive axial load is 130 kN.

a) Determine the safety factor guarding against buckling and yielding.
b) If the ends are changed to fixed and the material is changed to aluminum alloy 2014, calculate the safety factors guarding against buckling and yielding.

9.11 A beam has both ends mounted in stiff rubber bushings giving bending moments in the beam ends proportional to the angular displacements at the beam ends. Calculate the effective beam length if the angular spring constant at the ends is $\partial M/\partial\theta = 10^5$ N-m/radian in all directions. The beam is 3 m long and has a solid circular cross section of AISI 1080 steel with a 24-mm diameter.

Section 9.5

9.12 Two circular columns are made of different materials, steel and aluminum. The cross-sectional areas are the same and the moduli of elasticity are 207 GPa for the steel and 72 GPa for the aluminum. Find the ratio of the critical buckling lengths for the steel and aluminum columns while assuming that the same buckling load is applied to both columns.

9.13 A television mast shown in sketch *b* consists of a circular tube with an outer diameter d_o and a wall thickness t_h. Calculate how long the distance l between the anchoring points for the cables can be if the television mast should deform plastically rather than buckle.

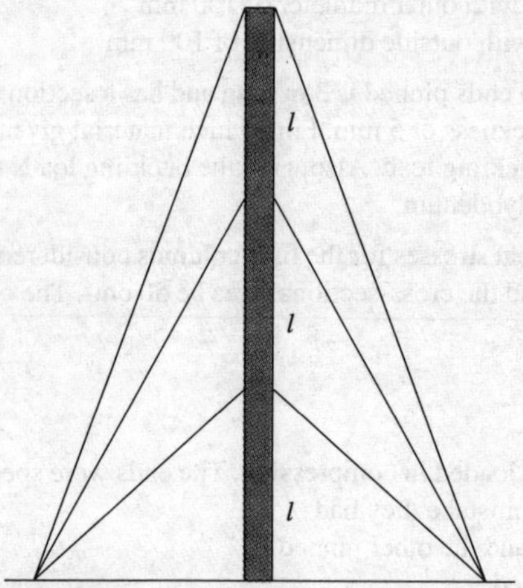

Sketch *b*, for Problem 9.13

Section 9.6

9.14 A 3-m-long column of square, solid cross section with both ends fixed must sustain a critical load of 3×10^5 N. The material is steel with a modulus of elasticity of 207 GPa and a yield stress of 700 MPa. Determine the following:

a) What minimum dimensions of a square, solid cross section are permitted without having failure? Should the Euler or Johnson equation be used?

b) If the critical load is increased by two orders of magnitude ($P_{cr} = 3 \times 10^7$ N), what minimum dimensions of a square, solid cross section are permitted without having failure? Also, indicate whether the Euler or Johnson equation should be used.

9.15 A solid, round column with a length of 2 m and a diameter of 50 mm is fixed at one end and free at the other end. The material's yield strength is 300 MPa and its modulus of elasticity is 207 GPa. Assume concentric loading of the column. Determine the following:

a) The critical load on this column

b) The critical load on the column if the free end is also fixed (i.e., both ends are fixed)

9.16 A circular-cross-section bar with a diameter of 2 in. and a length of 40 in. is axially loaded. The bar is made of medium-carbon steel. Both ends are pinned. Determine

a) Whether the Johnson or Euler formula should be used

b) The critical load

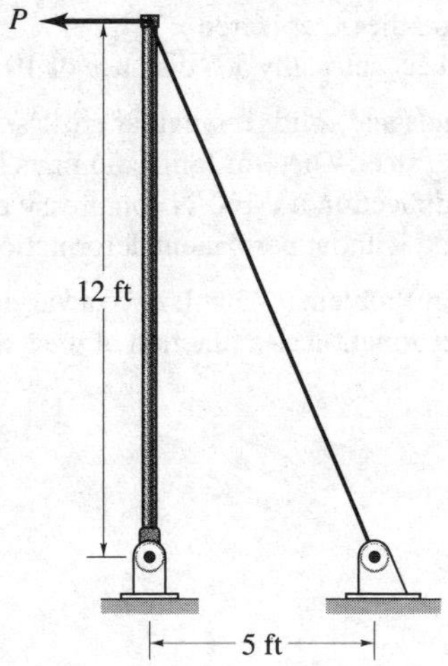

Sketch *c*, for Problem 9.17

9.17 A low-carbon-steel pipe, as shown in sketch *c*, has an outer diameter of 2 in. and a thickness of 0.5 in. If it is held in place by a guy wire, determine the largest horizontal force *P* that can be applied without causing the pipe to buckle. Assume that the ends of the pipe are pin connected.

9.18 A 20-mm-diameter, medium-carbon steel (AISI 1040) rod is loaded as a column with rounded ends. The critical load is 110 kN. How long can the rod be and still carry the following percentages of the critical load: a) 90%, b) 50%, and c) 2%? Also, indicate whether the Johnson or Euler equation should be used.

9.19 Most types of steel have similar moduli of elasticity but can have widely different yield strength properties depending on the alloying and the heat treatment. For a column having a solid, circular cross section with a diameter of 20 mm, the yield strengths for three steels are 300, 600, and 1000 MPa. Find for each steel the critical column length where buckling would not be a problem.

9.20 For a fixed cross-sectional area of a steel column find the geometry that would give the highest buckling load. The minimum column wall thickness is 5 mm.

9.21 A rectangular-cross-section tube is axially loaded in compression. The tube outside dimensions are 70 and 90 mm. The tube is 2.5 m long and both ends are pinned. The wall thickness is 9 mm. Find which of the steels in the table in Appendix A.1 should be used so that plastic buckling is not a problem.

9.22 The column considered in Problem 9.21 is made of AISI 1020 steel. Find the critical buckling load.

9.23 A 2.5-m-long column with one end fixed and one end free is made of aluminum alloy 2014 and has a solid, round cross section. Determine the diameter of the column for the following loads

a) $P = 500$ kN
b) $P = 800$ kN

Section 9.8

9.24 A 1-m-long column with a 60- by 100-mm rectangular cross section and made of aluminum alloy 2014 is subjected to a compressive axial load. Determine the critical buckling load

a) If both ends are fixed
b) If one end is fixed and the other is free
c) If the load is applied eccentrically at a distance of 10 mm for case b

9.25 A column with pinned ends and solid rectangular cross section with dimensions of 35×60 mm is made of AISI 1040 steel. The 4-m-long column is loaded with an eccentricity of 15 mm. Find the elastic deflection if a 100-N compressive load is applied. Also, calculate how large a load can be applied without permanent deformation occurring.

9.26 The column considered in Problem 9.25 will have a varying load in the range 0 to 24 000 N. Calculate and plot the deformation as a function of the axial load.

CHAPTER

10

STRESSES AND DEFORMATIONS IN CYLINDERS

Common beverage cans. Along with food containers, these are the most common pressure vessels. (AP/Wide World Photos)

In all things, success depends on previous preparation. And without such preparation there is sure to be failure.

Confucius, *Analects*

SYMBOLS

A	cross-sectional area, m^2
$\bar{a}$	coefficient of linear thermal expansion, $(°\,C)^{-1}$
C_1, C_2	integration constants
c	radial clearance, m
d	diameter, m
E	modulus of elasticity, Pa
l	length, m
n_s	safety factor
P	force, N
p	pressure, Pa
r	radius, m
S_y	yield strength, Pa

T	torque, N-m	c	circumferential
t_h	thickness, m	e	von Mises
t_l	tolerance, m	f	fit
Δt_m	temperature change, ° C	h	hub
x,y	Cartesian coordinates, m	i	inner or internal
z	axial direction in cylindrical polar coordinates, m	max	maximum
α	cone angle, deg	min	minimum
β	body force per volume, N/m^3	o	outer or external
δ	interference or displacement, m	r	radial
ε	strain	s	shaft
θ	circumferential direction in cylindrical polar coordinates, rad	t	tangential
		t_m	temperature change
μ	coefficient of friction	x,y	Cartesian coordinates
ν	Poisson's ratio	z	axial direction in cylindrical polar coordinates
ρ	density, kg/m^3	θ	circumferential direction in cylindrical polar coordinates
σ	normal stress, Pa	σ	normal stress
ω	angular velocity, rad/s	1,2,3	principal axes

Subscripts

a	axial
avg	average

10.1 INTRODUCTION

Just as the last chapter dealt with a specific shape [a straight and long (relative to its radius) bar] and the unique stresses and strains acting on columns, this chapter also deals with a specific shape in the form of cylinders. Many important engineering applications rely on this shape. This chapter begins with the consideration of tolerancing in the design of cylindrical members, which will prove to be important for a number of machine elements considered later in the text.

Stresses and deformations of thin-walled, thick-walled, internally pressurized, externally pressurized, and rotating cylinders are considered, as well as press and shrink fits. The material developed in this chapter is important to shafting (Chapter 11) and to a number of machine elements, such as rolling-element bearings (Chapter 13) and hydrodynamic bearings (Chapter 12).

10.2 TOLERANCES AND FITS

A number of definitions about tolerancing need to be presented:

1. **Tolerance** t_l is the maximum variation in the size of a part. Two types of tolerance are
   a) **Bilateral tolerance**—A part is permitted to vary both above and below the nominal size, such as 1.5±0.003.
   b) **Unilateral tolerance**—A part is permitted to vary either above or below the nominal size, but not both, such as $1.5^{+0.000}_{-0.003}$.
2. **Nominal diameter** d is the approximate size chosen by the engineer and to which allowances and tolerances are applied.

3. **Allowance** a is the difference between the nominal diameters of mating parts.
4. **Interference** δ is the actual difference in the size of mating parts.

Note that the tolerances of the shaft t_{ls} and the hub t_{lh} may be different.

Because it is impossible to make parts to an exact size, tolerances are used to control the variation between mating parts. Two examples illustrate the importance of tolerancing:

1. In hydrodynamic bearings (considered in Chapter 12) a critical part of the design is the specification of the radial clearance between the journal and the bearing. The typical value is one-thousandth of an inch. However, variations in the journal's outside diameter and the bearing's inside diameter cause additional or smaller clearances. Such variations must be accounted for in analyzing bearing performance. Too small a clearance could cause failure; too large a clearance would reduce the precision of the machine and adversely affect the lubrication. Thus, tolerancing and accuracy of the dimensions can have a significant effect on the performance of hydrodynamic bearings.
2. Rolling-element bearings (considered in Chapter 13) are generally designed to be installed on a shaft (considered more fully in Chapter 11) with an interference fit. The inside diameter of the bearing is smaller than the outside diameter of the shaft where the bearing is to be seated. A significant force is required to press the bearing onto the shaft, thus imposing significant stresses on both the shaft and the bearing. What should be the shaft diameter relative to the bearing bore size to ensure that failure due to overstressing the members does not occur? Proper tolerancing of the members will greatly contribute to successful design. Table 10.1 describes the eight classes of fit, the

Table 10.1 Classes of fit

Class	Description	Type	Applications
1	Loose	Clearance	Where accuracy is not essential, such as in road-building and mining equipment
2	Free	↓	In rotating journals with speeds of 600 rpm or greater, such as in engines and some automotive parts
3	Medium	↓	In rotating journals with speeds under 600 rpm, such as in accurate machine tools and precise automotive parts
4	Snug	↓	Where small clearance is permissible and where moving parts are not intended to move freely under load
5	Wringing	Interference	Where light tapping with a hammer is necessary to assemble the parts
6	Tight	↓	In semipermanent assemblies suitable for drive or shrink fits on light sections
7	Medium	↓	Where considerable pressure is needed to assemble and for shrink fits of medium sections; suitable for press fits on generator and motor armatures and for car wheels
8	Heavy force or shrink	↓	Where considerable bonding between surfaces is required, such as locomotive wheels and heavy crankshaft disks of large engines

type of fit, and their applications. Note that for classes 1 to 4 the members are interchangeable but for classes 5 to 8 the members are not interchangeable.

Having established classes of fit, the next task will be to indicate the tolerance applicable for a specific class. Tables 10.2 and 10.3 give the tolerances for various classes of fit in inches and millimeters, respectively. In these tables the nominal diameter is designated by d.

Having established the allowance, interference, and tolerance, the next step is to establish the maximum and minimum diameters of the hub and shaft. Table 10.4 gives these dimensions for the two types of fit (clearance and interference). This table, along with Tables 10.2 and 10.3, thus establishes the upper and lower limits of the shaft and hub diameters for the eight classes of fit.

Table 10.2 Recommended tolerances in *inches* for classes of fit.

Class	Allowance, a	Interference, δ	Hub tolerance, t_{lh}	Shaft tolerance, t_{ls}
1	$0.0025d^{2/3}$	—	$0.0025d^{1/3}$	$0.0025d^{1/3}$
2	$0.0014d^{2/3}$	—	$0.0013d^{1/3}$	$0.0013d^{1/3}$
3	$0.0009d^{2/3}$	—	$0.0008d^{1/3}$	$0.0008d^{1/3}$
4	0.000	—	$0.0006d^{1/3}$	$0.0004d^{1/3}$
5	—	0.000	$0.0006d^{1/3}$	$0.0004d^{1/3}$
6	—	$0.00025d$	$0.0006d^{1/3}$	$0.0006d^{1/3}$
7	—	$0.0005d$	$0.0006d^{1/3}$	$0.0006d^{1/3}$
8	—	$0.0010d$	$0.0006d^{1/3}$	$0.0006d^{1/3}$

Table 10.3 Recommended tolerances in *millimeters* for classes of fit.

Class	Allowance, a	Interference, δ	Tolerance: Hub, t_{lh}	Tolerance: Shaft, t_{ls}
1	$0.0073d^{2/3}$	—	$0.0216d^{1/3}$	$0.0216d^{1/3}$
2	$0.0041d^{2/3}$	—	$0.0112d^{1/3}$	$0.0112d^{1/3}$
3	$0.0026d^{2/3}$	—	$0.0069d^{1/3}$	$0.0069d^{1/3}$
4	0.000	—	$0.0052d^{1/3}$	$0.0035d^{1/3}$
5	—	0.000	$0.0052d^{1/3}$	$0.0035d^{1/3}$
6	—	$0.00025d$	$0.0052d^{1/3}$	$0.0052d^{1/3}$
7	—	$0.0005d$	$0.0052d^{1/3}$	$0.0052d^{1/3}$
8	—	$0.0010d$	$0.0052d^{1/3}$	$0.0052d^{1/3}$

Table 10.4 Maximum and minimum diameters of shaft and hub for two types of fit.

Type of fit	Hub diameter Maximum, $d_{h,max}$	Minimum $d_{h,\,min}$	Shaft diameter Maximum $d_{s,\,max}$	Minimum $d_{s,\,min}$
Clearance	$d+t_{lh}$	d	$d-a$	$d-a-t_{ls}$
Interference	$d+t_{lh}$	d	$d+\delta+t_{ls}$	$d+\delta$

Example 10.1

GIVEN A medium-force fit is applied to a shaft with a nominal diameter of 3 in.

FIND Determine the interference, the hole and shaft tolerances, and the hub and shaft diameters.

Solution

From Table 10.1 the class of fit is 7 and the type of fit is interference. From Table 10.2 for a class 7 fit

$$\delta = 0.0005d = 0.0005(3) = 0.0015 \text{ in.}$$

$$t_{lh} = t_{ls} = 0.0006d^{1/3} = 0.0006(3)^{1/3} = 0.865\times10^{-3} \text{ in.} \approx 0.0009 \text{ in.}$$

From Table 10.4 for interference fit the hub diameter is

$$d_{h,\max} = d + t_{lh} = 3 + 0.0009 = 3.0009 \text{ in.}$$

$$d_{h,\min} = d = 3.000 \text{ in.}$$

and the shaft diameter is

$$d_{s,\max} = d + \delta + t_{ls} = 3 + 0.0015 + 0.0009 = 3.0024 \text{ in.}$$

$$d_{s,\min} = d + \delta = 3 + 0.0015 = 3.0015 \text{ in.}$$

10.3 Pressurization Effects

This section presents thin-walled and thick-walled cylinders with internal and external pressurization. The material in this section is used for a wide range of applications, some of which are

1. Pressure vessels loaded internally and/or externally
2. Storage tanks
3. Gun barrels
4. Hydraulic and pneumatic cylinders
5. Transmission pipelines
6. Machine elements, such as rolling-element bearings or gears pressed onto shafts

The analysis that follows makes extensive use of material developed in earlier chapters.

A distinction must be made between **thin-walled cylinders** and **thick-walled cylinders.** When a cylinder's inner diameter d_i is 40 times larger than its thickness t_h, thin-wall analysis can be safely used. For smaller ratios of cylinder inner diameter to thickness, thick-wall analysis should be used. Mathematically expressing the above gives

$$\frac{d_i}{t_h} > 40 \quad \rightarrow \quad \text{Thin-walled cylinders} \tag{10.1}$$

$$\frac{d_i}{t_h} < 40 \quad \rightarrow \quad \text{Thick-walled cylinders} \tag{10.2}$$

Thin-wall analysis is more accurate as d_i/t_h increases. Equation (10.1) must be adhered to, but Eq. (10.2) may be valid for a larger range than stated.

10.3.1 THIN-WALLED CYLINDERS

Figure 10.1(a) shows a thin-walled cylinder subjected to internal pressure p_i. The cylinder internal diameter d_i and thickness t_h are also shown in the figure. It is assumed that the stress is uniform throughout the wall thickness. The radial stress is small relative to the circumferential stress because $t_h/d_i << 1$. Thus, a small element can be considered in plane stress with the principal stresses shown in Fig. 10.1(b).

Figure 10.2, the front view of the cylinder shown in Fig. 10.1, shows the forces acting on a small element due to internal pressurization. This element also has a length dl coming out of the paper. Summing forces in the radial direction gives

$$p_i r_i d\theta dl = 2\sigma_{\theta,\text{avg}} \sin\left(\frac{d\theta}{2}\right) t_h dl$$

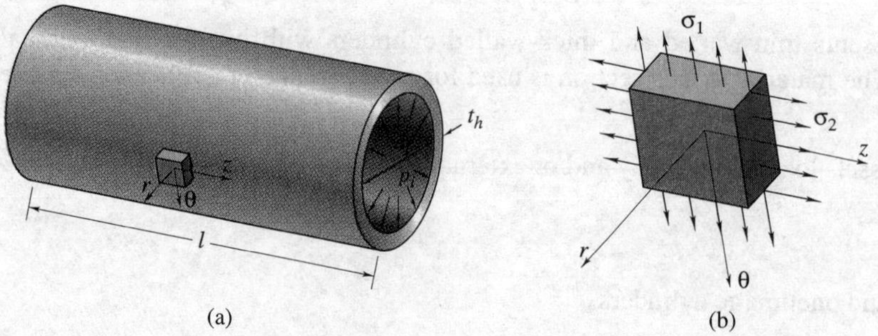

Figure 10.1 Internally pressurized thin-walled cylinder. (a) Stresses acting on cylinder; (b) stresses acting on element.

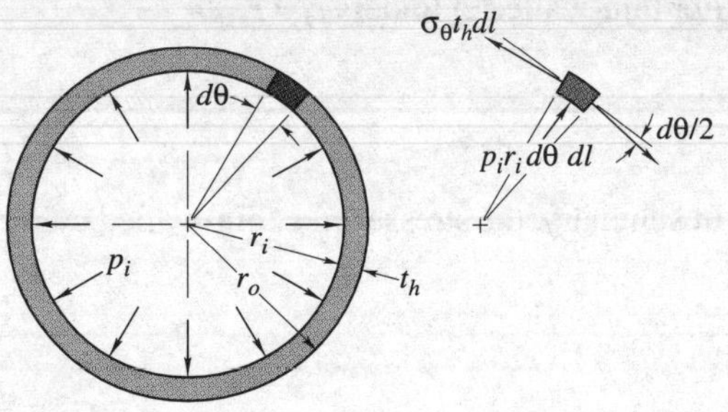

Figure 10.2 Front view of internally pressured, thin-walled cylinder.

Since dθ/2 is small, sin (dθ/2) = dθ/2 and

$$\sigma_{\theta,\text{avg}} = \frac{p_i r_i}{t_h} \tag{10.3}$$

The θ component of stress is sometimes called the tangential stress or the **hoop stress.** The term originates from the manufacture of wooden barrels by shrinking a hot iron ring, or hoop, around closely fitting planks to achieve a seal. The maximum hoop stress can be obtained by modifying Eq. (10.3) such that r_i is replaced with r_{avg}:

$$r_{\text{avg}} = r_i + \frac{t_h}{2} = \frac{d_i + t_h}{2} \tag{10.4}$$

$$\therefore \sigma_{\theta,\max} = \frac{p_i r_{\text{avg}}}{t_h} = \frac{p_i(d_i + t_h)}{2t_h} \tag{10.5}$$

Thus, from Eqs. (10.3) and (10.5) for thin-walled cylinders, where $d_i/t_h > 40$,

$$\sigma_\theta = \sigma_{\theta,\text{avg}} = \sigma_{\theta,\max} = \frac{p_i r}{t_h} \tag{10.6}$$

$$\sigma_r = 0 \tag{10.7}$$

The area subjected to axial stress is

$$A = \pi\left(r_o^2 - r_i^2\right) = 2\pi r_{\text{avg}} t_h \tag{10.8}$$

The average axial tensile stress is

$$\sigma_{z,\text{avg}} = \frac{p_i r_i^2}{r_o^2 - r_i^2} = \frac{p_i r_i^2}{2 r_{\text{avg}} t_h}$$

For thin-walled cylinders $r_i \approx r_{avg} \approx r$

$$\therefore \sigma_z = \frac{p_i r}{2t_h} \tag{10.9}$$

In summary, the stresses in a thin-walled cylinder are

$$\sigma_r = 0 \tag{10.7}$$

$$\sigma_\theta = \frac{p_i r}{t_h} \tag{10.6}$$

$$\sigma_z = \frac{p_i r}{2t_h} \tag{10.9}$$

Thus, the circumferential (hoop) stress is twice the axial stress.

The principal stresses for thin-walled cylinders are

$$\sigma_1 = \sigma_\theta = \frac{p_i r}{t_h} \tag{10.10}$$

Tutorial 5

$$\sigma_2 = \sigma_z = \frac{p_i r}{2t_h} \tag{10.11}$$

Example 10.2

GIVEN A 100-in.-inner-diameter cylinder made of material having a yield strength of 60 ksi is subjected to an inner pressure of 300 psi. Use a safety factor of 3. Assume that thin-wall analysis is adequate.

FIND The wall thickness required, based on

a) Maximum-shear-stress theory (MSST)

b) Distortion-energy theory (DET)

Solution

From Eqs. (10.10) and (10.11)

$$\sigma_1 = \sigma_\theta = \frac{p_i r}{t_h} = \frac{(300)(50)}{t_h}$$

$$\sigma_2 = \sigma_z = \frac{p_i r}{2t_h} = \frac{(300)(50)}{2t_h}$$

$$\sigma_3 = \sigma_r = 0$$

Because $\sigma_1 > \sigma_2 > \sigma_3$, the principal stresses are ordered.

a) Using MSST, from Eq. (6.6) failure will not occur if

$$\sigma_1 - \sigma_3 < \frac{S_y}{n_s} \rightarrow \frac{p_i r}{t_h} < \frac{S_y}{n_s}$$

$$\frac{n_s p_i r}{S_y} < t_h$$

To avoid failure,

$$t_h > \frac{p_i r n_s}{S_y} = \frac{(300)(50)(3)}{(60)(10^3)} = 0.75 \text{ in.}$$

b) Using DET, from Eq. (6.10) the von Mises stress for a biaxial stress state is

$$\sigma_e = (\sigma_1^2 + \sigma_2^2 - \sigma_1\sigma_2)^{1/2} = \left(\frac{3\sigma_1^2}{4}\right)^{1/2} = \sqrt{\frac{3}{4}}\sigma_1$$

$$\therefore \sigma_e = (0.8660)\frac{(300)(50)}{t_h} = \frac{(1.299)(10^4)}{t_h}$$

From Eq. (6.11) failure will not occur if

$$\sigma_e < \frac{S_y}{n_s} \rightarrow \frac{(1.299 \times 10^4)n_s}{S_y} < t_h$$

Thus, to avoid failure,

$$t_h > \frac{(1.299)(10^4)n_s}{S_y} = \frac{(1.299)(10^4)(3)}{(6)(10^4)} = 0.6495 \text{ in.}$$

10.3.2 THICK-WALLED CYLINDERS

It was discovered in Sec. 10.3.1 that if the cylinder walls are thin, the circumferential or hoop stress σ_θ can be assumed to be uniform throughout the wall thickness. This assumption cannot be made for thick-walled cylinders.

Figure 10.3 shows a radially loaded, thick-walled cylinder subjected to internal pressure p_i and external pressure p_o. Because the body and the loading are symmetrical about the axis, shear stresses in the circumferential and radial directions are not present, and only normal stresses σ_θ and σ_r act on the element. The loading is two-dimensional; therefore, only plane stresses will be involved. If an axial loading is superimposed [Eq. (10.9) developed in Sec. 10.3.1], the third principal stress is merely changed from zero to σ_z. Recall that Eq. (10.9) is valid regardless of cylinder thickness.

Figure 10.4 shows a cylindrical polar element before and after deformation. The radial and circumferential (hoop) displacements are given by δ_r and δ_θ, respectively. The radial

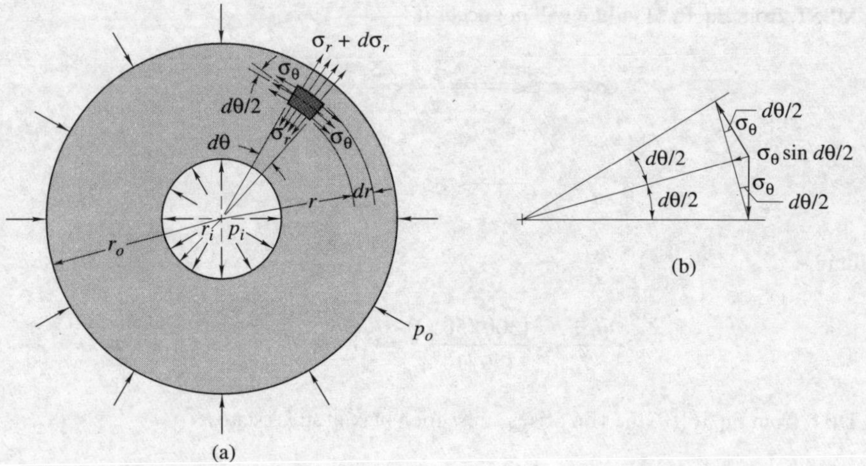

Figure 10.3 Complete front view of thick-walled cylinder internally and externally pressurized. (a) With stresses acting on cylinder; (b) with stresses acting on element.

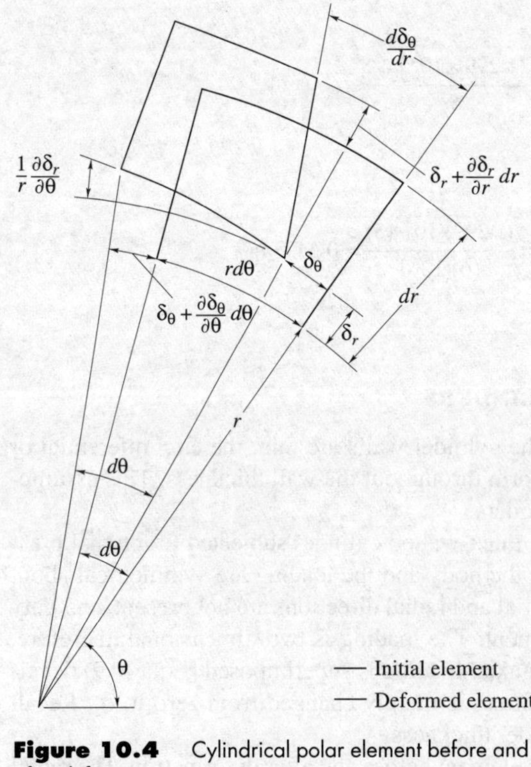

Figure 10.4 Cylindrical polar element before and after deformation.

strain is the change of element thickness in the radial direction divided by the element thickness:

$$\varepsilon_r = \frac{\delta_r + \frac{\partial \delta_r}{\partial r}dr - \delta_r}{dr} = \frac{\partial \delta_r}{\partial r} \tag{10.12}$$

The circumferential strain is affected by the displacement to the new radius $r + \delta_r$, or

$$\varepsilon_\theta = \frac{(r+\delta_r)d\theta - r d\theta}{r d\theta} = \frac{\delta_r}{r} \tag{10.13}$$

Circumferential variations shown in Fig. 10.4 are not present in the press-fit equations because of circular symmetry. From Hooke's law the stress-strain relationship for the biaxial stress state [see Eq. (B.49)], while making use of Eqs. (10.12) and (10.13), gives

$$\varepsilon_r = \frac{\partial \delta_r}{\partial r} = \frac{1}{E}(\sigma_r - \nu\sigma_\theta) \tag{10.14}$$

$$\varepsilon_\theta = \frac{\delta_r}{r} = \frac{1}{E}(\sigma_\theta - \nu\sigma_r) \tag{10.15}$$

The expressions for the radial and circumferential stresses will be obtained from force equilibrium of the element. Figure 10.3(b) gives details of the stresses acting on the element. Summing the forces in the radial direction gives

$$(\sigma_r + d\sigma_r)(r + dr)d\theta dz - \sigma_r r d\theta dz - 2\sigma_\theta \sin\left(\frac{d\theta}{2}\right)dr dz = 0 \tag{10.16a}$$

For small $d\theta$, $\sin(d\theta/2) = d\theta/2$ and Eq. (10.16a) reduces to

$$\sigma_\theta = r\frac{d\sigma_r}{dr} + \sigma_r \tag{10.16b}$$

Equations (10.14), (10.15), and (10.16b) are three equations with three unknowns δ_r, σ_r, and σ_θ. Substituting Eq. (10.16b) into Eqs. (10.14) and (10.15) and then differentiating Eq. (10.15) with respect to r while equating Eqs. (10.14) and (10.15) give

$$0 = 3\frac{d\sigma_r}{dr} + r\frac{d^2\sigma_r}{dr^2} \tag{10.17}$$

Equation (10.17) can be rewritten as

$$0 = 2\frac{d\sigma_r}{dr} + \frac{d}{dr}\left(r\frac{d\sigma_r}{dr}\right) \tag{10.18}$$

Integrating gives

$$0 = 2\sigma_r + r\frac{d\sigma_r}{dr} + C_1$$

This equation can be rewritten as

$$0 = \frac{d}{dr}\left(r^2\sigma_r\right) + C_1 r$$

Integrating gives

$$\sigma_r = -\frac{C_1}{2} - \frac{C_2}{r^2} \tag{10.19}$$

The boundary conditions for thick-walled cylinders pressurized both internally and externally are

1. $\sigma_r = -p_i$ at $r = r_i$
2. $\sigma_r = -p_o$ at $r = r_o$

Because the pressurization is compressive, the negative signs appear in the boundary conditions. Making use of the boundary conditions gives

$$\sigma_r = \frac{p_i r_i^2 - p_o r_o^2 + \left(p_o - p_i\right)\left(\frac{r_o r_i}{r}\right)^2}{r_o^2 - r_i^2} \tag{10.20}$$

$$\frac{d\sigma_r}{dr} = -\frac{2\left(p_o - p_i\right)\left(r_o r_i\right)^2}{r^3\left(r_o^2 - r_i^2\right)} \tag{10.21}$$

Substituting Eqs. (10.20) and (10.21) into Eq. (10.16b) gives

$$\sigma_\theta = \frac{p_i r_i^2 - p_o r_o^2 - \left(\frac{r_i r_o}{r}\right)^2\left(p_o - p_i\right)}{r_o^2 - r_i^2} \tag{10.22}$$

10.3.2.1 INTERNALLY PRESSURIZED If, as for many applications, the outer pressure p_O is zero, Eqs. (10.20) and (10.22) reduce to

$$\sigma_r = \frac{p_i r_i^2\left(1 - \frac{r_o^2}{r^2}\right)}{r_o^2 - r_i^2} \tag{10.23}$$

$$\sigma_\theta = \frac{p_i r_i^2\left(1 + \frac{r_o^2}{r^2}\right)}{r_o^2 - r_i^2} \tag{10.24}$$

Figure 10.5 shows the radial and circumferential (hoop) stresses in an internally pressurized cylinder. The circumferential stress is tensile and the radial stress is compressive. Furthermore, the maximum stresses occur at $r = r_i$ and are

$$\sigma_{r,\max} = -p_i \tag{10.25}$$

$$\sigma_{\theta,\max} = p_i\left(\frac{r_o^2 + r_i^2}{r_o^2 - r_i^2}\right) \tag{10.26}$$

From Eq. (10.15) the circumferential strains for internal pressurization evaluated at the location of maximum stress are

$$\varepsilon_\theta = \frac{\delta_r}{r_i} = \frac{p_i}{E}\left(\frac{r_o^2 + r_i^2}{r_o^2 - r_i^2} + \nu\right) \tag{10.27}$$

where

ν = Poisson's ratio of solid material

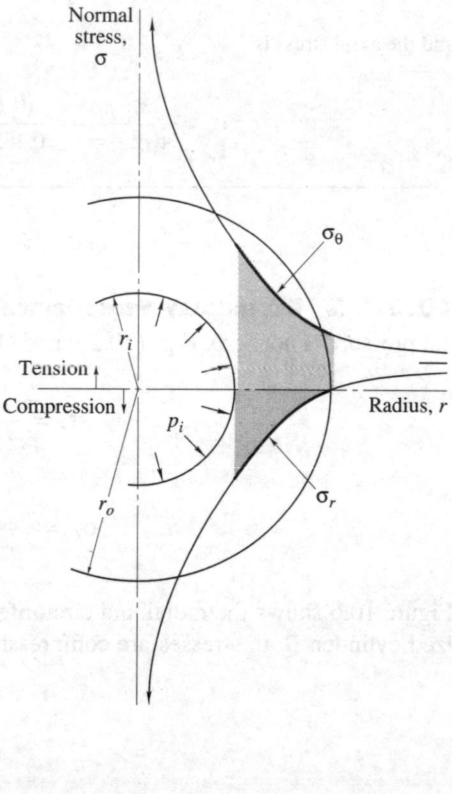

Figure 10.5 Internally pressured, thick-walled cylinder showing circumferential (hoop) and radial stress for various radii. [R. Juvinall, *Stress, Strain, and Strength*, ©1967 by McGraw-Hill. Reproduced with permission of The McGraw-Hill Companies.]

The radial displacement is outward:

$$\delta_r = \frac{p_i r_i}{E}\left(\frac{r_o^2 + r_i^2}{r_o^2 - r_i^2} + \nu\right) \tag{10.28}$$

Example 10.3 **GIVEN** A hydraulic cylinder connected to a high-pressure hydraulic circuit can be pressurized to 100 MPa. The inner radius of the cylinder is 100 mm and the wall thickness is 35 mm.
FIND The tangential and axial stresses in the cylinder wall at the inner diameter for both thin- and thick-wall analysis.

Solution

From the *thin*-wall theory given in Eqs. (10.10) and (10.11) the following can be written:

$$\sigma_1 = \frac{p_i r}{t_h} = \frac{(100)(10^6)(0.1)}{0.035}\ \text{Pa} = 285.7\ \text{MPa}$$

$$\sigma_2 = \frac{p_i r}{2t_h} = 142.9\ \text{MPa}$$

From *thick*-wall theory with internal pressurization Eq. (10.26) gives

$$\sigma_{\theta,\max} = p_i\left(\frac{r_o^2 + r_i^2}{r_o^2 - r_i^2}\right) = (100)(10^6)\left(\frac{0.135^2 + 0.10^2}{0.135^2 - 0.10^2}\right)\ \text{Pa} = 343.2\ \text{MPa}$$

and the axial stress is

$$\sigma_z = \frac{\pi r_i^2 p_i}{\pi\left(r_o^2 - r_i^2\right)} = \frac{(0.10)^2(10^8)}{0.135^2 - 0.10^2}\ \text{Pa} = 121.6\ \text{MPa}$$

10.3.2.2 EXTERNALLY PRESSURIZED If the internal pressure is zero and the external pressure is not zero, Eqs. (10.20) and (10.22) reduce to

$$\sigma_r = \frac{p_o r_o^2}{r_o^2 - r_i^2}\left(\frac{r_i^2}{r^2} - 1\right) \tag{10.29}$$

$$\sigma_\theta = -\frac{p_o r_o^2}{r_o^2 - r_i^2}\left(\frac{r_i^2}{r^2} + 1\right) \tag{10.30}$$

Figure 10.6 shows the radial and circumferential (hoop) stresses in an externally pressurized cylinder. Both stresses are compressive. Furthermore, the maximum circumferential

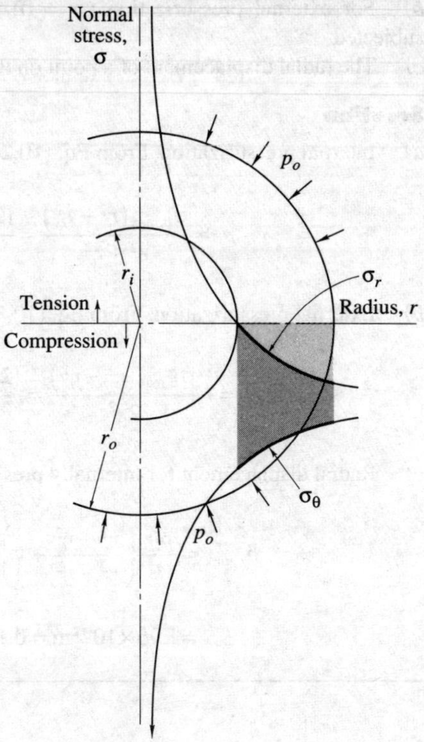

Figure 10.6 Externally pressurized, thick-walled cylinder showing circumferential (hoop) and radial stresses for various radii. [R. Juvinall, *Stress, Strain, and Strength*, ©1967 by McGraw-Hill. Reproduced with permission of The McGraw-Hill Companies.]

(hoop) stress occurs at $r = r_i$ and the maximum radial stress occurs at $r = r_o$. These expressions are

$$\sigma_{r,\max} = -p_o \tag{10.31}$$

$$\sigma_{\theta,\max} = -\frac{2r_o^2 p_o}{r_o^2 - r_i^2} \tag{10.32}$$

Example 10.4

GIVEN A thick-walled cylinder with 0.3-m internal diameter and 0.4-m external diameter has a maximum circumferential (hoop) stress of 250 MPa. The material has a Poisson's ratio of 0.3 and a modulus of elasticity of 207 GPa.

FIND Determine the following:

a) For internal pressurization (p_o = 0) the maximum pressure to which the cylinder may be subjected *b)* For external pressurization ($p_i = 0$) the maximum pressure to which the cylinder may be subjected

c) The radial displacement of a point on the inner surface for the situation presented in part a.

Solution

a) Internal pressurization. From Eq. (10.26)

$$p_i = \frac{\sigma_{\theta,\max}\left(r_o^2 - r_i^2\right)}{r_o^2 + r_i^2} = \frac{(250)\left(10^6\right)\left(0.2^2 - 0.15^2\right)}{0.2^2 + 0.15^2}\text{ Pa} = 70\text{ MPa}$$

b) External pressurization. From Eq. (10.32)

$$p_o = -\frac{\sigma_{\theta,\max}\left(r_o^2 - r_i^2\right)}{2r_o^2} = \frac{(250)\left(10^6\right)\left(0.2^2 - 0.15^2\right)}{2(0.2)^2}\text{ Pa} = 54.69\text{ MPa}$$

c) Radial displacement for internally pressurized cylinders. From Eq. (10.28)

$$\delta_{r,\max} = \frac{p_i r_i}{E}\left(\frac{r_o^2 + r_i^2}{r_o^2 - r_i^2} + \nu\right) = \frac{(70)\left(10^6\right)(0.15)}{(2.07)\left(10^{11}\right)}\left(\frac{0.2^2 + 0.15^2}{0.2^2 - 0.15^2} + 0.3\right)$$

$$= 1.96\times 10^{-4}\text{ m} = 0.196\text{ mm}$$

10.4 ROTATIONAL EFFECTS

Rotating cylinders are encountered in a number of machine elements, such as flywheels, gears, pulleys, and sprockets. This section considers rotation of the cylinder while assuming no pressurization ($p_i = p_o = 0$). Section 10.3 considered pressurization but did not consider rotation.

If a body force had been included when considering the stresses acting on the elements shown in Fig. 10.3, Eq. (10.16b) would have become

$$\frac{\sigma_\theta - \sigma_r}{r} - \frac{d\sigma_r}{dr} - \beta = 0 \qquad \textbf{(10.33)}$$

where

β = body force per volume, N/m^3

Here, the body force is the rotating inertia force that acts radially and is given as

$$\beta = r\omega^2\rho \qquad \textbf{(10.34)}$$

where

ω = rotational speed, s^{-1}

ρ = density, kg/m^3

r = radius of cylinder, m

Two special cases are considered: a cylinder with a central hole and a solid cylinder. Section 10.3 presented a complete derivation of the equations; however, in this section, since the procedures are similar, only the resulting equations are presented.

10.4.1 CYLINDER WITH CENTRAL HOLE

The circumferential (hoop) and radial stresses while considering rotation but neglecting pressurization in a cylinder with a central hole are

$$\sigma_\theta = \frac{3+\nu}{8}\rho\omega^2\left[r_i^2 + r_o^2 + \frac{r_i^2 r_o^2}{r^2} - \frac{1+3\nu}{3+\nu}r^2\right] \tag{10.35}$$

$$\sigma_r = \frac{3+\nu}{8}\rho\omega^2\left(r_i^2 + r_o^2 - \frac{r_i^2 r_o^2}{r^2} - r^2\right) \tag{10.36}$$

where

ν = Poisson's ratio

Figure 10.7 shows how these stress components vary with radius. Both stresses are tensile. The radial stress is zero at the inner and outer radii, with a maximum occurring between r_i

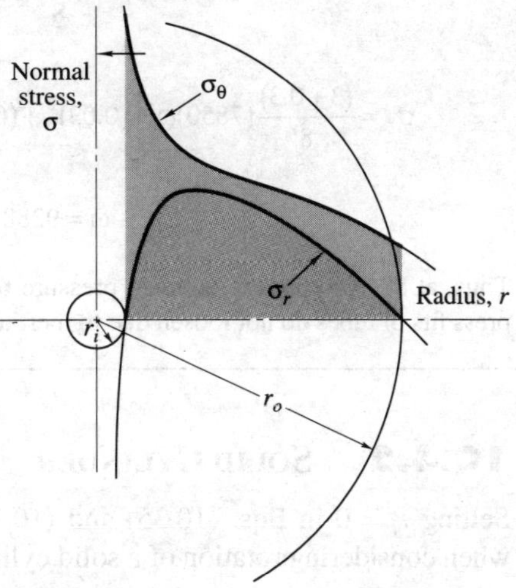

Figure 10.7 Stresses in rotating cylinder with central hole and no pressurization. [R. Juvinall, *Stress, Strain, and Strength*, ©1967 by McGraw-Hill. Reproduced with permission of The McGraw-Hill Companies.]

and r_o. The circumferential stress is a maximum at $r = r_i$ and exceeds the radial stress for any value of radius r. The maximum circumferential stress is

$$\sigma_{\theta,\max} = \frac{3+\nu}{4}\rho\omega^2\left[r_o^2 + \frac{r_i^2(1-\nu)}{3+\nu}\right] \tag{10.37}$$

The maximum radial stress can be obtained by differentiating Eq. (10.36) with respect to r and setting the result equal to zero. Doing this gives

$$\sigma_{r,\max} = \frac{3+\nu}{8}\rho\omega^2(r_i - r_o)^2 \tag{10.38}$$

$$r = \sqrt{r_i r_o} \tag{10.39}$$

Example 10.5

GIVEN Two AISI 1040 steel tubes are press fitted together at 110 MPa. The nominal sizes of the steel tubes are 100-mm outer diameter and 80-mm inner diameter and 80-mm outer diameter and 60-mm inner diameter, respectively.

FIND How fast does the combined tube have to rotate to decrease the press-fit pressure to zero?

Solution

By assuming linear elasticity the press-fit pressure of 110 MPa has to be compensated for by an equally large radial stress at the location of the fit caused by the rotation. Equation (10.36) thus gives

$$\sigma_r = \frac{3+\nu}{8}\rho\omega^2\left(r_i^2 + r_o^2 - \frac{r_i^2 r_o^2}{r^2} - r^2\right)$$

$$\sigma_r = \frac{(3+0.3)}{8}(7850)\omega^2\left[(0.03)^2 + (0.05)^2 - \frac{(0.03)^2(0.05)^2}{(0.04)^2} - (0.04)^2\right] = (110)(10^6)$$

$$\omega = 9288 \text{ rad/s} = 88\ 700 \text{ rpm}$$

Thus, at 88 700 rpm the press-fit pressure becomes zero. This very large speed clearly shows that press fits of tubes do not loosen due to inertial effects in machine elements.

10.4.2 SOLID CYLINDER

Setting $r_i = 0$ in Eqs. (10.35) and (10.36) yields the circumferential and radial stresses when considering rotation of a solid cylinder but neglecting pressurization effects.

$$\sigma_\theta = \frac{3+\nu}{8}\rho\omega^2\left[r_o^2 - \frac{r^2(1+3\nu)}{3+\nu}\right] \tag{10.40}$$

$$\sigma_r = \frac{3+\nu}{8}\rho\omega^2(r_o^2 - r^2) \tag{10.41}$$

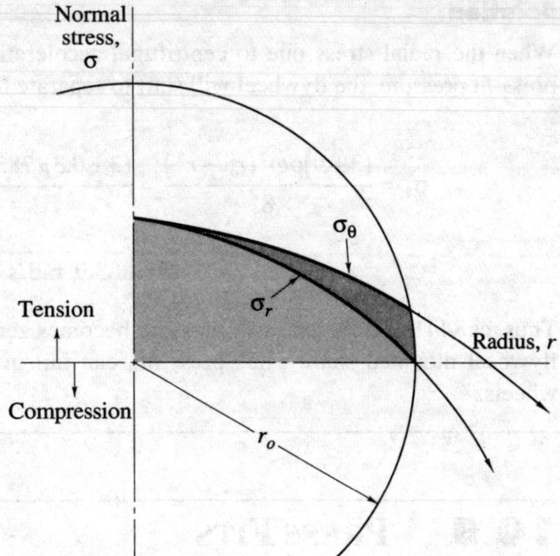

Figure 10.8 Stresses in rotating solid cylinder with no pressurization. [R. Juvinall, *Stress, Strain, and Strength*, ©1967 by McGraw-Hill. Reproduced with permission of The McGraw-Hill Companies.]

Figure 10.8 shows the stress distribution for both stress components as a function of radius for a solid cylinder. Both stresses are tensile, with the maximum occurring at $r = 0$ and both having the same value at that location. The maximum stress is

$$\sigma_{\theta,\max} = \sigma_{r,\max} = \frac{3+\nu}{8}\rho(r_o\omega)^2 \qquad \textbf{(10.42)}$$

Some interesting observations occur when comparing Eqs. (10.37) and (10.38) with Eq. (10.42). As r_i becomes very small, Eqs. (10.38) and (10.42) are in complete agreement. However, as $r_i \to 0$, Eqs. (10.37) and (10.42) differ by a factor of 2 in circumferential stress. This difference is due to the radial stress in the inner portion of the solid cylinder, which decreases the circumferential stress by one-half.

Example 10.6

GIVEN A cylindrical flywheel is press fitted on a solid shaft. Both are made of AISI 1080 steel. The press-fit pressure is 185 MPa, the shaft diameter is 100 mm, and the outside diameter of the flywheel is 550 mm.

FIND The shaft speed when the press-fit pressure disappears due to centrifugal effects.

Solution

When the radial stress due to centrifugal acceleration at the shaft surface is equal to the original press-fit pressure, the flywheel will start to separate from the shaft. From Table A.1 and Eq. (10.41)

$$\sigma_r = \frac{(3+\nu)\rho\omega^2\left(r_o^2 - r^2\right)}{8} = \frac{(3+0.3)(7850)\omega^2\left(0.275^2 - 0.050^2\right)}{8} = 185\times10^6$$

$$\omega = 883.9 \text{ rad/s} = 8441 \text{ rpm}$$

Thus, at 8441 rpm the press-fit pressure becomes zero. This speed is large, but not impossible, for a flywheel mounted shaft. Thus press fits can fail in this manner for overspeeding shafts with flywheels.

10.5 PRESS FITS

In a **press fit** the pressure p_f is caused by the interference between the shaft and the hub. This pressure increases the radius of the hole and decreases the radius of the shaft. Section 10.2 described shaft and hub dimensions in terms of tolerance, which results in specific fits. The present section focuses on the stress and strain found in press fits and uses material developed in Sec. 10.3.2 for thick-walled cylinders.

Figure 10.9 shows a side view of interference in a press fit. There is a radial displacement of the hub δ_{rh} and a radial displacement of the shaft δ_{rs}. Figure 10.10 shows the front view of an interference fit. In Fig. 10.10(a) the cylinders are assembled with an interference fit; in Fig. 10.10(b) the hub and shaft are disassembled and the dimensions of each are clearly shown. Figure 10.10(b) also shows the interference pressure being internal for the hub and external for the shaft. The shaft is shown as hollow in order to present the most general case.

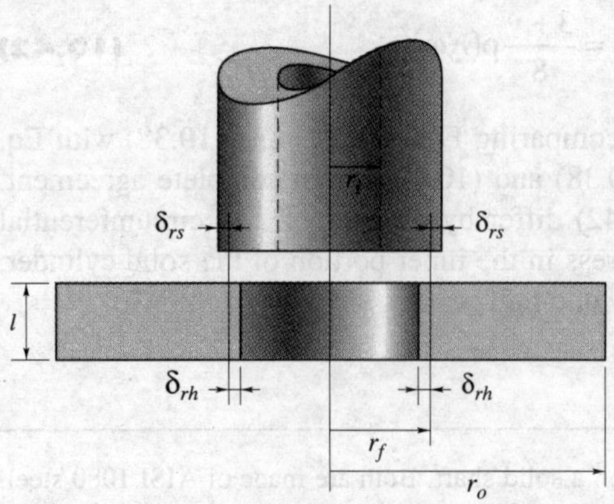

Figure 10.9 Side view showing interference in press fit of hollow shaft to hub.

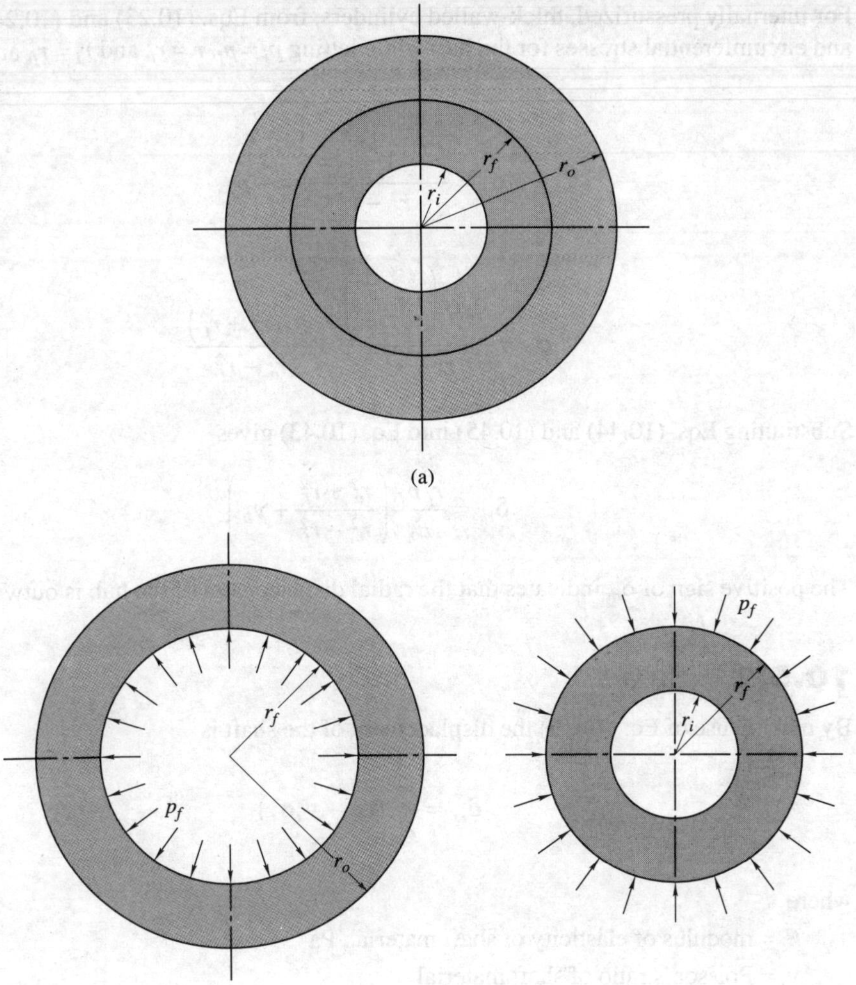

Figure 10.10 Front view showing (a) cylinder assembled with an interference fit and (b) hub and hollow shaft disassembled (also showing interference pressure).

10.5.1 Hub

By making use of Eq. (10.15) the hub displacement is

$$\delta_{rh} = \frac{r_f}{E_h}(\sigma_\theta - \nu_h \sigma_r) \tag{10.43}$$

where

E_h = modulus of elasticity of hub material, Pa

ν_h = Poisson's ratio of hub material

For internally pressurized, thick-walled cylinders, from Eqs. (10.23) and (10.24) the radial and circumferential stresses for the hub while letting $p_i = p_f$, $r = r_f$, and $r_i = r_f$, are

$$\sigma_r = \frac{p_f r_f^2 \left(1 - \dfrac{r_o^2}{r_f^2}\right)}{r_o^2 - r_f^2} = -p_f \tag{10.44}$$

$$\sigma_\theta = \frac{p_f r_f^2 \left(1 + \dfrac{r_o^2}{r_f^2}\right)}{r_o^2 - r_f^2} = \frac{p_f\left(r_o^2 + r_f^2\right)}{r_o^2 - r_f^2} \tag{10.45}$$

Substituting Eqs. (10.44) and (10.45) into Eq. (10.43) gives

$$\delta_{rh} = \frac{r_f p_f}{E_h}\left(\frac{r_o^2 + r_f^2}{r_o^2 - r_f^2} + \nu_h\right) \tag{10.46}$$

The positive sign of δ_{rh} indicates that the radial displacement of the hub is outward.

10.5.2 SHAFT

By making use of Eq. (10.15) the displacement of the shaft is

$$\delta_{rs} = \frac{r_f}{E_s}\left(\sigma_\theta - \nu_s \sigma_r\right) \tag{10.47}$$

where

E_s = modulus of elasticity of shaft material, Pa
ν_s = Poisson's ratio of shaft material

The circumferential and radial stresses for externally pressurized, thick-walled cylinders can be obtained from Eqs. (10.29) and (10.30) while letting $p_o = p_f$, $r_o = r_f$, and $r = r_f$ to give

$$\sigma_r = \frac{p_f r_f^2}{r_f^2 - r_i^2}\left(\frac{r_i^2}{r_f^2} - 1\right) = -p_f \tag{10.48}$$

$$\sigma_\theta = -\frac{p_f r_f^2}{r_f^2 - r_i^2}\left(\frac{r_i^2}{r_f^2} + 1\right) = -\frac{p_f\left(r_f^2 + r_i^2\right)}{r_f^2 - r_i^2} \tag{10.49}$$

Substituting Eqs. (10.48) and (10.49) into Eq. (10.47) gives

$$\delta_{rs} = -\frac{r_f p_f}{E_s}\left(\frac{r_f^2 + r_i^2}{r_f^2 - r_i^2} - \nu_s\right) \tag{10.50}$$

Because the first term in parentheses in Eq. (10.50) is greater than 1 and the Poisson's ratio of any material is less than or equal to 0.5, δ_{rs} is negative. Therefore, shaft displacement is directed inward toward the center of the shaft.

10.5.3 INTERFERENCE FIT

The total radial displacement is shown in Fig. 10.9. Recall that outward deflection (expanding the inside diameter of the hub) is positive in sign and inward deflection (reducing the outside diameter of the shaft) is negative. Thus, the total radial interference is

$$\delta_r = \delta_{rh} - \delta_{rs} = r_f p_f \left[\frac{r_o^2 + r_f^2}{E_h\left(r_o^2 - r_f^2\right)} + \frac{\nu_h}{E_h} + \frac{r_f^2 + r_i^2}{E_s\left(r_f^2 - r_i^2\right)} - \frac{\nu_s}{E_s} \right] \tag{10.51}$$

If the shaft and the hub are made of the same material, $E = E_s = E_h$ and $\nu = \nu_s = \nu_h$ and Eq. (10.51) reduces to

$$\delta_r = \frac{2r_f^3 p_f\left(r_o^2 - r_i^2\right)}{E\left(r_o^2 - r_f^2\right)\left(r_f^2 - r_i^2\right)} \tag{10.52}$$

Furthermore, if the shaft is solid rather than hollow, $r_i = 0$ and Eq. (10.52) further reduces to

$$\delta_r = \frac{2r_f p_f r_o^2}{E\left(r_o^2 - r_f^2\right)} \tag{10.53}$$

In these equations, if displacement is known, the interference pressure may be an unknown and these equations can readily be used.

10.5.4 FORCE AND TORQUE

The maximum force $P_{\max}$ to assemble a press fit varies directly as the thickness of the outer member, the length of the outer member, the difference in diameters of the mating shaft and hub, and the coefficient of friction μ. The maximum stress is

$$\tau_{\max} = p_f \mu = \frac{P_{\max}}{A} = \frac{P_{\max}}{2\pi r_f l} \tag{10.54}$$

The torque is

$$T = P_{\max} r_f = 2\pi\mu r_f^2 l p_f \tag{10.55}$$

The axial and circumferential stresses are related to the maximum stress by

$$\tau_a^2 + \tau_c^2 = \tau_{\max}^2$$

where

$$\tau_a = \frac{P_a}{2\pi r_f l} = \text{axial stress}$$

$$\tau_c = \frac{P_c}{2\pi r_f l} = \text{circumferential stress}$$

Example 10.7

GIVEN A 6-in.-diameter steel shaft is to have a press fit with a 12-in.-outside-diameter cast iron hub. Both the hub and the shaft are 10 in. long. The maximum circumferential stress is to be 5000 psi. The moduli of elasticity are 30 × 10⁶ psi for steel and 15 × 10⁶ psi for cast iron. The Poisson's ratio for both steel and cast iron is 0.3 and the coefficient of friction for the two materials is 0.12. That is,

$$r_f = 3 \text{ in.} \qquad r_i = 0 \qquad r_o = 6 \text{ in.}$$

$$E_s = 30\times10^6 \text{ psi} \qquad E_h = 15\times10^6 \text{ psi} \qquad \nu_s = \nu_h = 0.3$$

$$\mu = 0.12 \qquad l = 10 \text{ in.} \qquad \sigma_{\theta,\max} = 5000 \text{ psi}$$

FIND Determine

a) The interference

b) The axial force required to press the hub on the shaft

c) What torque this press fit can transmit

Solution

a) From Eq. (10.45) the interference pressure is

$$p_f = \frac{\sigma_{\theta,\max}\left(r_o^2 - r_f^2\right)}{r_o^2 + r_f^2} = \frac{5000\left(6^2 - 3^2\right)}{6^2 + 3^2} = 3000 \text{ psi}$$

From Eq. (10.51) the maximum permissible radial interference is

$$\delta_r = r_f p_f \left[\frac{r_o^2 + r_f^2}{E_h\left(r_o^2 - r_f^2\right)} + \frac{\nu_h}{E_h} + \frac{r_f^2 + r_i^2}{E_s\left(r_f^2 - r_i^2\right)} - \frac{\nu_s}{E_s}\right]$$

$$= \frac{3(3000)}{15(10^6)}\left[\frac{6^2 + 3^2}{6^2 - 3^2} + 0.3 + \frac{1}{2} - \frac{0.3}{2}\right] = 1.390\times10^{-3} \text{ in.}$$

b) From Eq. (10.54) the force required for the press fit is

$$P_{\max} = 2\pi\mu r_f l p_f = (2)(\pi)(0.12)(3)(10)(3000) = 67\ 860 \text{ lbf}$$

c) From Eq. (10.55) the torque is

$$T = P_{\max} r_f = (67\ 860)(3) = 203\ 600 \text{ lbf-in.}$$

Example 10.8

GIVEN A wheel hub is press fit on a 105-mm-diameter solid shaft. The coefficient of friction is 0.11 and the hub and shaft material is AISI 1080 steel. The hub's outer diameter is 160 mm and its width is 120 mm. The radial interference between the shaft and the hub is 65 μm (the shaft diameter is 130 μm larger than the inside diameter of the hub).

FIND The axial force necessary to dismount the hub.

Solution

Equation (10.53) gives the relationship between radial displacement and pressure.

$$\delta_r = \frac{2 r_f p_f r_o^2}{E\left(r_o^2 - r_f^2\right)} = (65)\left(10^{-6}\right) = \frac{(2)(0.0525) p_f (0.080)^2}{(2.07)\left(10^{11}\right)\left[(0.080)^2 - (0.0525)^2\right]}$$

$$\therefore p_f = 72.96 \text{ MPa}$$

The axial force necessary to dismount the hub is

$$P = \mu p_f A = (0.11)(72.96)\left(10^6\right)\pi(0.105)(0.120) = 317\ 700 \text{ N} = 317.7 \text{ kN}$$

10.6 Shrink Fits

In producing a **shrink fit** it is common to heat the outer component (hub) in order to expand it beyond the interference and then slip it over the inner component (shaft). Cooling will then contract the outer component. Temperature change produces a strain, called thermal strain, even in the absence of stress. Although thermal strain is not exactly linear with temperature change, for temperature changes of 100 or 200°F the actual variation can be closely described by a linear approximation. According to this linear relationship the temperature difference through which the outer component (hub) must be heated to obtain the required expansion over the undeformed solid shaft is

$$\Delta t_m = \frac{\delta_r}{\bar{a} r_f} \quad \textbf{(10.56)}$$

where

$\bar{a}$ = coefficient of linear thermal expansion (see Table 3.5 and Fig. 3.14)

Equation (10.56) can be expressed in terms of radial strain as

$$\varepsilon_r = \frac{\delta_r}{r_f} = \bar{a}\, \Delta t_m \quad \textbf{(10.57)}$$

The deformation is

$$\delta_r = \varepsilon_r r_f = \bar{a}\ \Delta t_m r_f \tag{10.58}$$

These equations are valid not only for shrink fits of shaft and hub but for a wide range of thermal problems, as illustrated by the following example.

The strain due to a temperature change may be added algebraically to a local strain by using the *principle of superposition.* The principle states that stresses and strains (at a point on a given plane) due to different loads may be computed separately and added algebraically provided that the sum does not exceed the proportionality limit of the material and that the structure remains stable. The method of superposition for different types of loading was covered in Sec. 5.4. Thus, the normal strain due to normal load and temperature effects is

$$\varepsilon = \varepsilon_\sigma + \varepsilon_{t_m} \tag{10.59}$$

where

ε_σ = strain due to normal stress

ε_{t_m} = strain due to temperature change

Thus, the general (triaxial stress state) stress-strain relationship developed in Appendix B, Eq. (B.44), may be expressed while considering thermal strain as

$$\varepsilon_x = \frac{1}{E}\left[\sigma_x - \nu\left(\sigma_y + \sigma_z\right)\right] + \bar{a}\ \Delta t_m$$

$$\varepsilon_y = \frac{1}{E}\left[\sigma_y - \nu\left(\sigma_z + \sigma_x\right)\right] + \bar{a}\ \Delta t_m$$

$$\varepsilon_z = \frac{1}{E}\left[\sigma_z - \nu\left(\sigma_x + \sigma_y\right)\right] + \bar{a}\ \Delta t_m \tag{10.60}$$

Example 10.9

GIVEN A 10-in.-long steel tube has a cross-sectional area of 1 in.^{2} that expands by 0.008 in. from a stress-free condition at 80° F when the tube is heated to 480°F.

FIND The load and stress acting on the steel tube.

Solution

From Table 3.5 for steel alloy $\bar{a} = 6.1 \times 10^{-6}$ °F. From Eq. (10.57)

$$\varepsilon = \bar{a}\ \Delta t_m = (6.1)(10^{-6})(400) = 2.44 \times 10^{-3}$$

$$\Delta l = l\varepsilon = (10)(2.44)(10^{-3}) = 0.0244 \text{ in.}$$

Because the measured expansion was only 0.008 in., the constraint due to compressive normal loading must apply a force sufficient to deflect the tube axially by the following amount:

$$\delta = \frac{Pl}{AE} \rightarrow P = \frac{AE\delta}{l}$$

From Table A.1 the modulus of elasticity is 30 × 10^6 psi. Therefore,

$$P = \frac{(1)(30)(10^6)(0.0164)}{10} = 4.92 \times 10^4 \text{ lbf} = 49\ 200 \text{ lbf}$$

This then is the compressive, normal, axial load being exerted on the steel tube.

Example 10.10

GIVEN A block of aluminum alloy is placed between two rigid jaws of a clamp, and the jaws are brought up snug. The temperature of the entire assembly is raised 250° C in an oven. The cross-sectional areas are 65 mm^2 for the block and 160 mm^2 for the stainless steel screws.

FIND How much stress is induced in the screws and the block?

Solution

Figure 10.11 shows the block-and-screw assembly as well as the forces on these components. From force equilibrium

$$P_a = 2P_s \qquad \textbf{(a)}$$

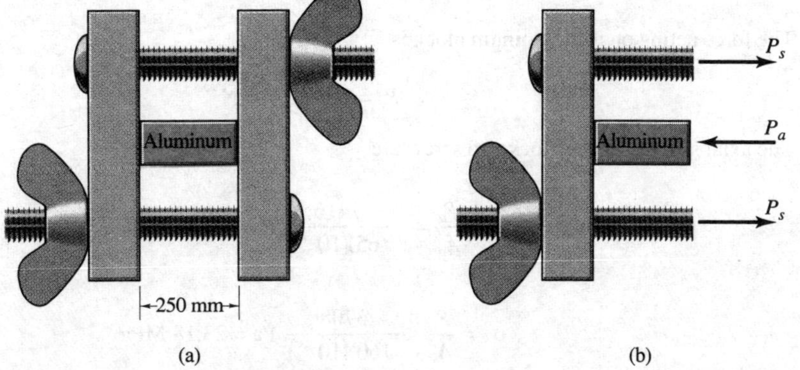

Figure 10.11 Block placed between two rigid jaws of clamp (a) and forces acting (b).

Here subscript a refers to the aluminum block and subscript s refers to the stainless steel screws. Compatibility requires that the length changes of the block and the screws be the same, or

$$\delta_a = \delta_s \tag{b}$$

Thermal expansion will induce an axial force as shown in Fig. 10.11. The displacements of the block and screws are

$$\delta_a = \bar{a}_a l\, \Delta t_m - \frac{P_a l}{E_a A_a} \tag{c}$$

$$\delta_s = \bar{a}_s l\, \Delta t_m + \frac{P_s l}{E_s A_s} \tag{d}$$

Substituting Eqs. (a), (c), and (d) into Eq. (b) gives

$$P_s = \frac{\Delta t_m(\bar{a}_a - \bar{a}_s)}{\dfrac{1}{E_s A_s} + \dfrac{2}{E_a A_a}} \tag{e}$$

From Tables 3.2 and 3.5

$$E_a = 70 \text{ GPa} \qquad E_s = 193 \text{ GPa}$$

$$\bar{a}_a = 24 \times 10^{-6}\ ^\circ\text{C} \qquad \bar{a}_s = 17 \times 10^{-6}\ ^\circ\text{C}$$

We are given that A_a = 65 mm^2 and A_s = 160 mm^2. Substituting these values and the above into Eq. (e) gives the force acting on each screw as

$$P_s = \frac{(250)(24-17)(10^{-6})}{\dfrac{1}{(193)(10^9)(160)(10^{-6})} + \dfrac{2}{(70)(10^9)(65)(10^{-6})}} = 3708 \text{ N}$$

The force acting on the aluminum block is

$$P_a = 2P_s = 7416 \text{ N}$$

The axial stresses of the block and screw are

$$\sigma_a = \frac{P_a}{A_a} = -\frac{7416}{(65)(10^{-6})} \text{ Pa} = -114.1 \text{ MPa}$$

$$\sigma_s = \frac{P_s}{A_s} = \frac{3708}{(160)(10^{-6})} \text{ Pa} = 23.18 \text{ MPa}$$

The stress acting on the aluminum block is compressive, and that acting on the screws is tensile.

Case Study | DESIGN OF HYDRAULIC CYLINDER FOR EXTRUSION PRESS

One of the most demanding manufacturing processes is the extrusion of structural shapes, as illustrated in Fig. 10.12. The extrusion presses needed to complete such operations must provide extremely high forces (up to 16 000 tons in rare circumstances) over long distances (some extrusions are over 30 ft long). This combination makes hydraulic power the natural choice for such machines.

Extrusion presses use a pressurized fluid supply that bears on a piston, as shown in Fig. 10.12. The piston pressure results in a force applied to the workpiece. Once the cylinder is extended, it can be retracted by opening a valve to relieve pressure in the cylinder and then pushing the piston back with either a spring or fluid pressure (through further valving).

One of the important features of such system design is that the expensive subsystems, especially the cylinder, should have a much higher safety factor than the less expensive subsystems, such as the hydraulic lines. Therefore, a failure would not result in the need to replace the most expensive component. Further, many systems will have a specific component that is fail safe (i.e., a component failure results in loss of operation without compromising safety). Ideally, the component intended to fail first should also be relatively easy to replace, such as a keyway (covered in Chapter 11), a brake pad (Chapter 17), or as in this case a hydraulic line. Certainly, overpressurization should not result in the cylinder rupturing.

GIVEN Because hydraulic components are readily available in 3000-psi capacities, this will be the design pressure. A safety factor of 6.0 is specified to ensure that the cylinder will not rupture if the relief valves fail during overpressurization.

FIND Design a cylinder (inner diameter, wall thickness) that can transmit a 1000-ton force over a distance of 8 ft.

Solution

If the pressure in the cylinder is 3000 psi, to generate a 1000-ton (2×10^6-lbf) force,

$$P = pA = p\left(\frac{\pi d^2}{4}\right)$$

$$\therefore\ 2\times 10^6 = (3000)\frac{\pi}{4}d^2 \qquad \textbf{(a)}$$

$$d = 29.1 \text{ in.} \approx 30 \text{ in.}$$

This diameter is referred to as a "30-in. bore," since the inner cylinder will need to be extremely straight with tight tolerances and is therefore produced through a boring operation. Although no material has been specified, such cylinders are almost exclusively manufactured from high-strength steel. For AISI 1040 cold-drawn steel with a yield strength of 100 ksi and the prescribed safety factor of 6.0, the allowable stress is 16.7 ksi.

At this point it is not known whether the cylinder will be a thin-walled or thick-walled pressure vessel. Using the thin-wall equations [Eqs. (10.10) and (10.11)] gives the principal stresses in the wall as

$$\sigma_1 = \frac{p_i r}{t_h} = \frac{p_i d}{2t_h} = \frac{(3000)(30)}{2t_h} = \frac{45\ 000}{t_h} \qquad \textbf{(b)}$$

$$\sigma_2 = \frac{p_i r}{2t_h} = \frac{p_i d}{4t_h} = \frac{22\ 500}{t_h} \qquad \textbf{(c)}$$

Also, a principal approximation in thin-wall analysis is that $\sigma_3 = 0$. Using the DET yield criterion [Eq. (6.10)] gives

$$(\sigma_1-\sigma_2)^2+(\sigma_2-\sigma_3)^2+(\sigma_3-\sigma_1)^2 = 2\sigma_e^2$$

$$(10^6)\left(\frac{45.0}{t_h}-\frac{22.5}{t_h}\right)^2+(10^6)\left(\frac{22.5}{t_h}\right)^2+(10^6)\left(\frac{45.0}{t_h}\right)^2 = 2(16\ 700)^2 \qquad \textbf{(d)}$$

$$t_h = 2.33 \text{ in.}$$

Since the ratio of diameter to wall thickness is approximately 13, we can see from Eq. (10.1) that thin-wall

(continued)

Case Study (CONTINUED)

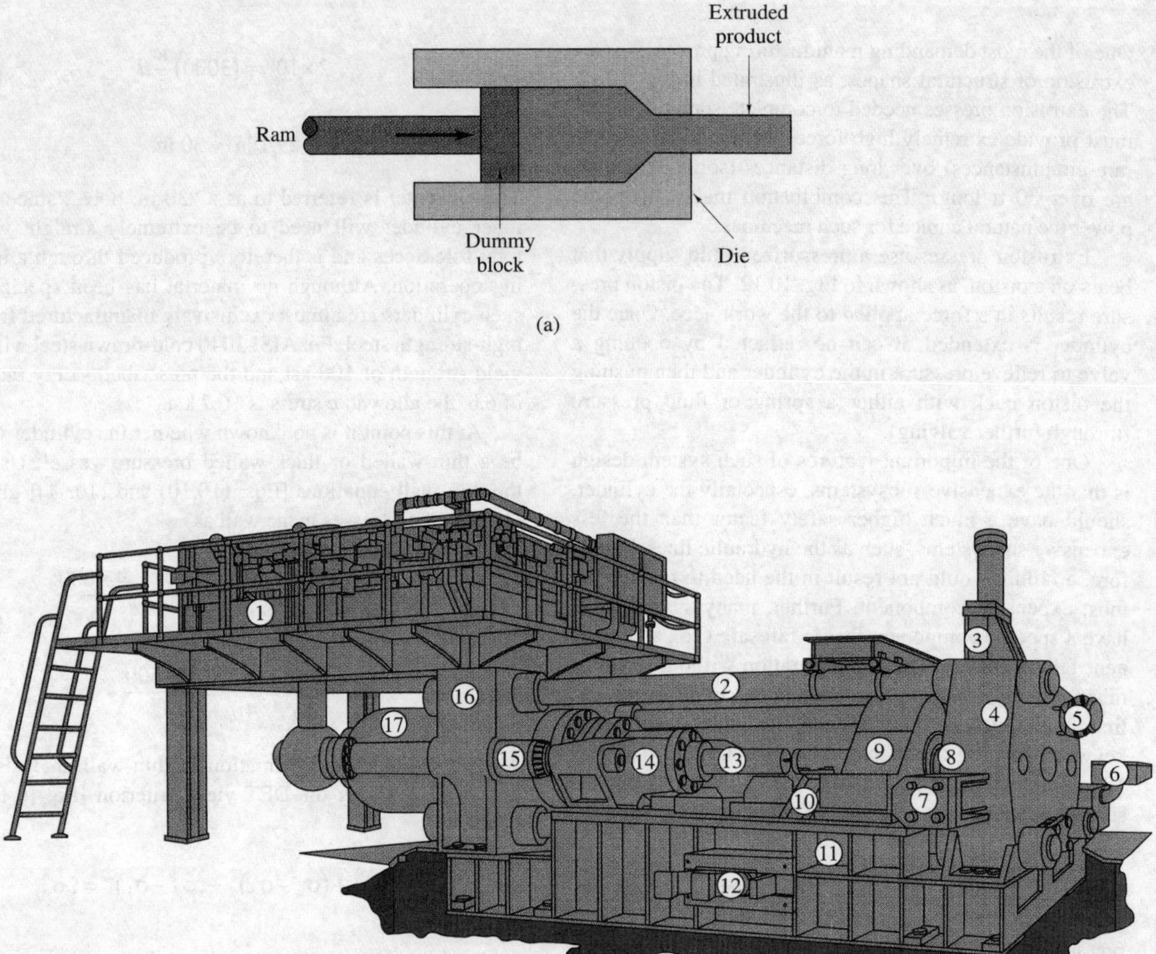

1. Hydraulic power unit
2. Tie rods
3. Butt shear
4. Extrusion platen
5. Container shifting cylinders
6. Swiveling operator's console
7. Die slide
8. Container
9. Container housing
10. Billet loader
11. Press base
12. Billet loader cylinders
13. Pressing stem
14. Crosshead
15. Side cylinder
16. Cylinder platen
17. Main cylinder

Figure 10.12
(a) Schematic of extrusion process. (b) Typical extrusion press.

(continued)

Case Study (CONCLUDED)

analysis does not apply, and we proceed to design a thick-walled cylinder.

For a thick-walled cylinder with internal pressurization, Eqs. (10.23), (10.24), and (10.9) give the principal stresses. Because, from Fig. 10.5, the largest stresses exist at the smallest radius in the cylinder, the internal radius will be the critical location. Thus, the principal stresses are

$$\sigma_r = \frac{p_i r_i^2\left(1-\frac{r_o^2}{r^2}\right)}{r_o^2 - r_i^2} = \frac{(3000)(15)^2\left(1-\frac{r_o^2}{(15)^2}\right)}{r_o^2-(15)^2} \quad \textbf{(e)}$$

$$\sigma_\theta = \frac{p_i r_i^2\left(1+\frac{r_o^2}{r^2}\right)}{r_o^2 - r_i^2} = \frac{(3000)(15)^2\left(1+\frac{r_o^2}{(15)^2}\right)}{r_o^2-(15)^2} \quad \textbf{(f)}$$

$$\sigma_z = \frac{p_i r}{2t_h} = \frac{p_i\left(\frac{r_o+r_i}{2}\right)}{2(r_o-r_i)} = \frac{(3000)\left(\frac{r_o+15}{2}\right)}{2(r_o-15)} \quad \textbf{(g)}$$

Substituting the above equations into Eq. (6.10) gives a complicated equation for r_o. Solving this equation numerically yields r_o = 18.08 in., so that the thickness is t_h = 18.08 – 15 = 3.08 in. To allow for standard stock, the next largest size that is an even multiple of 0.25 in. is selected, so that a 3.25-in. wall thickness is prescribed.

10.7 SUMMARY

In this chapter it was discovered that fits must be specified to ensure the proper mating assembly of a shaft and a hub because it is impossible to manufacture these parts with exactly the desired dimensions. A system was devised to tolerate small dimensional variations of the mating shaft and hub without sacrificing their proper functioning.

Pressurization effects on cylinders were considered. Thin-wall and thick-wall analyses were described for both internal and external pressurization. These effects are important in a great number of applications ranging from pressure vessels to gun barrels. A major assumption made in the thin-wall analysis was that the circumferential (hoop) stress is uniform throughout the wall thickness. This assumption is not valid for thick-wall analysis. Ranges of diameter to thickness were discussed for thin- and thick-wall analyses. For thick-wall situations radial and circumferential variations with radius for both internal and external pressurization were shown. For internal pressurization the maximum stress occurred at the inner radius for both radial and circumferential components, with the radial stress being compressive and the circumferential stress being tensile. For external pressurization the maximum radial stress occurred at the outer radius and the maximum circumferential stress occurred at the inner radius, with both stress components being compressive.

Rotational effects while assuming no pressurization of cylinders were also considered in this chapter. Rotational effects are important in such machine elements as flywheels, gears, and pulleys. The rotating inertial force was considered in establishing the radial and circumferential stresses for both a cylinder with a central hole and a solid cylinder. The chapter ended with consideration of press and shrink fits of a shaft and a hub. The interference, axial force, and torque were obtained for these situations.

KEY WORDS

allowance difference between nominal diameters of mating parts

bilateral tolerance variation above and below nominal size

hoop stress tangential stress in pressure vessel

interference actual difference in size of mating parts

nominal diameter approximate size of element

press fit connections where interfacial pressure is due to interference between mating parts and assembly is accomplished by elastic deformation due to large forces

shrink fit connections where interfacial pressure is due to interference between mating parts and assembly is accomplished by heating outer component

thick-walled cylinder cylinder whose ratio of diameter to thickness is less than 40

thin-walled cylinder cylinder where radial stress is negligible, approximately true for diameter-to-thickness ratios greater than 40

tolerance maximum variation in size of part

unilateral tolerance variation above or below nominal size, but not both

RECOMMENDED READINGS

Beer, F.P., and Johnson, E.R. (1981) *Mechanics of Materials,* McGraw-Hill, New York.

Chuse, R., and Carson, B.E. (1993) *Pressure Vessels,* 7th ed., McGraw-Hill, New York.

Craig, R.R. (1966) *Mechanics of Materials,* Wiley, New York.

Crandall, S.H., and Dahl, H.C. (1954) *An Introduction to the Mechanics of Solids,* McGraw-Hill, New York.

Fung, Y.C. (1965) *Foundations of Solid Mechanics,* Prentice-Hall, Englewood Cliffs, NJ.

Harvey, J.F. (1991) *Theory and Design of Pressure Vessels,* 2nd ed., Van Nostrand Reinhold, New York.

Juvinall, R.C. (1967) *Stress, Strain, and Strength,* McGraw-Hill, New York.

Juvinall, R.C., and Marshek, K.M. (1991) *Fundamentals of Machine Component Design,* Wiley, New York.

Lardner, T.J., and Archer, R.R. (1994) *Mechanics of Solids: An Introduction,* McGraw-Hill, New York.

Megyesy, E.F. (1986) *Pressure Vessel Handbook,* 7th ed., Pressure Vessel Handbook Publishing, Inc., Tulsa, OK.

Mott, R. L. (1985) *Machine Elements in Mechanical Design,* Merrill Publishing Co., Columbus, OH.

Popov, E.P. (1968) *Introduction to Mechanics of Solids,* Prentice-Hall, Englewood Cliffs, NJ.

Shigley, J.E., and Mischke, C.R. (1989) *Mechanical Engineering Design,* McGraw-Hill, New York.

Timoshenko, S., and Goodier, J. (1970) *Theory of Elasticity,* McGraw-Hill, New York.

REFERENCE

Juvinall, R.C. (1967) *Stress, Strain, and Strength,* McGraw-Hill, New York.

PROBLEMS

Section 10.2

10.1 A journal bearing is to be manufactured with optimum geometry for minimum power loss for a given load and speed. The relative clearance $c/r = 0.001$. The journal diameter is 100 mm. Find the accuracy to which the bearing parts have to be manufactured so that there is not more than ±10% error in the relative clearance.

10.2 A press fit between a solid steel shaft and a steel housing is dimensioned to be of class 7. By mistake the shaft is ground at 22° C higher temperature than originally anticipated, so that when the shaft cools, the diameter is slightly too small. The shaft material is AISI 1040 steel. What will be the class of fit between the shaft and the housing because of this mistake? Also, if the grinding temperature were 50° C higher rather than 22° C higher, what would the class of fit be?

Section 10.3

10.3 A cylinder with a 0.3-m inner diameter and a 0.40-m outer diameter is internally pressurized to 140 MPa. Determine the maximum shear stress at the outer surface of the cylinder.

10.4 A rubber balloon has the shape of a cylinder with spherical ends. At low pressure the cylindrical part is 25 cm long with a diameter of 2 cm. The rubber material has constant thickness and is linearly elastic. How long will the cylindrical part of the balloon be when it is inflated to a diameter of 10 cm? Assume that the rubber's modulus of elasticity is constant and that the Poisson's ratio is 0.5.

10.5 A thin-walled cylinder of pressurized gas is fixed by its two ends between rigid walls. Obtain an expression for the wall reactions in terms of cylinder length l, thickness t_h, radius r, and internal pressure p_i.

10.6 A pressurized cylinder has an internal pressure of 1 MPa, a thickness of 8 mm, a length of 4 m, and a diameter of 108 cm. The cylinder is made of AISI 1080 steel. What is the volume increase of the cylinder due to the internal pressure?

10.7 A 1-m-diameter cylindrical container with two hemispherical ends is used to transport gas. The internal pressure is 5 MPa and the safety factor is 2.0. Using the MSST determine the container thickness. Assume that the tangential strains of the cylinder and the sphere are equal, since they are welded joints. The material is such that $E = 200$ GPa, $S_y = 430$ MPa, and $\nu = 0.3$.

Section 10.4

10.8 A solid cylindrical shaft made of AISI 1020 steel rotates at a speed that produces a safety factor of 3 against the stress causing yielding. To instrument the shaft, a small hole is drilled in its center for electric wires. At the same time the material is changed to AISI 1080 steel. Find the safety factor against yielding of the new shaft.

10.9 A flywheel is mounted on a tubular shaft. The shaft's inner diameter is 30 mm and its outer diameter is 50 mm. The flywheel is a cylindrical disk with an inner diameter of 50 mm, an outer diameter of 300 mm, and a thickness of 35 mm. Both the shaft and the flywheel are made of AISI 1080 steel. The flywheel is used to store energy, so that the rotational acceleration is proportional to the angular speed ω. The angular acceleration $\partial\omega/\partial t$ is to be held to within $\pm 0.2\omega$. Calculate the shaft speed ω at which the flywheel starts to slide on the shaft if the press-fit pressure at $\omega = 0$ is 127 MPa and the coefficient of friction between the shaft and the flywheel is 0.13.

10.10 The flywheel and the shaft given in Problem 10.9 are made of aluminum alloy 2014 instead of steel. Find at what speed the flywheel will start to slide if the coefficient of friction is 0.14 and the press-fit pressure is 30 MPa.

10.11 The flywheel and the shaft given in Problem 10.9 are axially loaded by a force of 50 000 N. Calculate the shaft speed at which the flywheel starts to slide on the shaft.

Section 10.5

10.12 A 6-in.-diameter solid steel shaft is to have a press fit with a 12-in.-outer-diameter by 10-in.-long hub made of cast iron. The maximum allowable hoop stress is 5000 psi. The moduli of elasticity are 30×10^6 psi for steel and 20×10^6 psi for cast iron. Poisson's ratio for steel and cast iron is 0.3. The coefficient of friction for both steel and cast iron is 0.11. Determine the following:

a) Total radial interference
b) Axial force required to press the hub on the shaft
c) Torque transmitted with this fit

10.13 A flat, 0.5-m-outer-diameter, 0.1-m-inner-diameter, 0.08-m-thick steel disk shown in sketch *a* is shrink fit onto a shaft. If the assembly can transmit a torque of 100 kN-m, determine the fit pressure and the total radial interference. The coefficient of friction is 0.25.

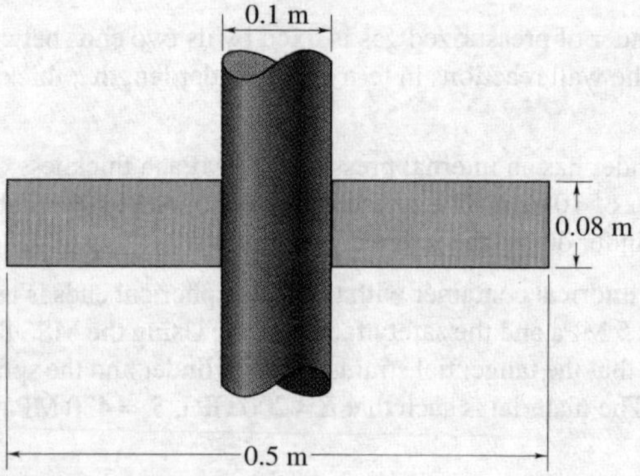

Sketch *a*, for Problem 10.13

10.14 A flat, 0.1-m-outer-diameter, 0.05-m-inner-diameter, 0.12-m-thick steel disk is shrink fit onto a shaft with a 0.02-m inner diameter. If the assembly can transmit a torque of 12 kN-m, determine the fit pressure and the total radial interference. The coefficient of friction is 0.25.

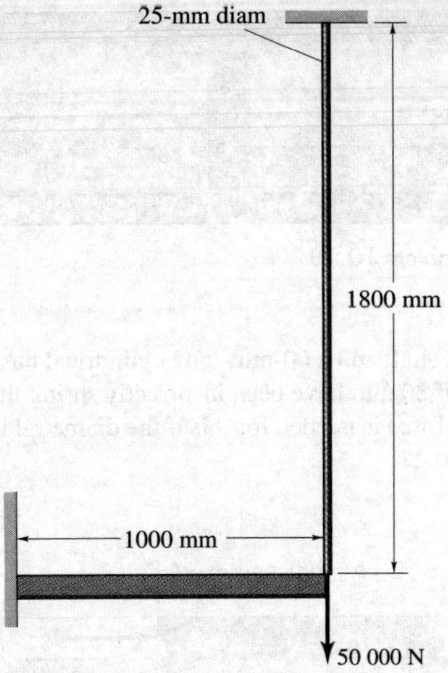

Sketch *b*, for Problem 10.15

10.15 A supported cantilevered beam is loaded as shown in sketch *b*. If at the time of assembly the load was zero and the beam was horizontal, determine the stress in the round rod after the load is applied and the temperature is lowered by 70° C. Beam moment of inertia is $I_{beam} = 9.8 \times 10^7$ (mm)^4. The beam and rod material is high-carbon steel.

10.16 A 10-in.-long tube (with properties $E = 30 \times 10^6$ psi and $\bar{a} = 7 \times 10^{-6}$/° F) having a cross-sectional area of 1 in.^{2} is installed with fixed ends so that it is stress free at 80° F. In operation the tube is heated throughout to a uniform 480° F. Careful measurements indicate that the fixed ends separate by 0.008 in. What loads are exerted on the ends of the tube, and what are the resultant stresses?

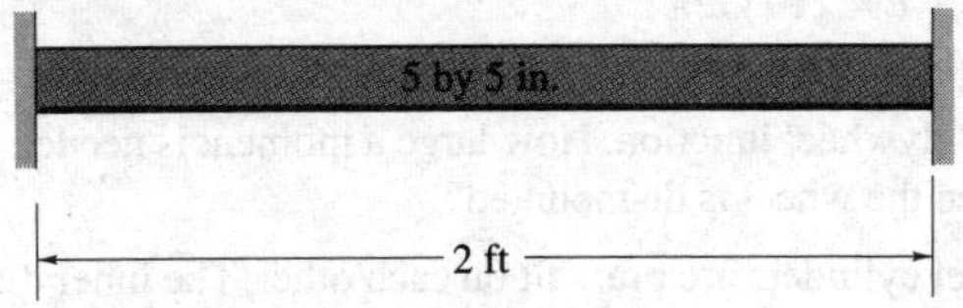

Sketch *c*, for Problem 10.17

10.17 A solid square bar is constrained between two fixed supports as shown in sketch *c*. The square bar has a cross section of 5 by 5 in. and is 2 ft long. The bar just fits between fixed supports at the initial temperature of 60° F. If the temperature is raised to a final temperature of 120° F, determine the average thermal stress developed in the bar. Assume that $E = 30 \times 10^6$ psi and $\bar{a} = 6.5 \times 10^{-6}$/° F.

10.18 Two stiff shafts are connected by a thin-walled elastic tube to a press-fit connection according to sketch *d*. The contact pressure between the shafts and the tube is *p*. The coefficient of friction is μ. Calculate the maximum torque T_{max} that can be transmitted through the press fit. Describe and calculate what happens if the torque decreases from T_{max} to $\theta\, T_{max}$ where $0 < \theta < 1$.

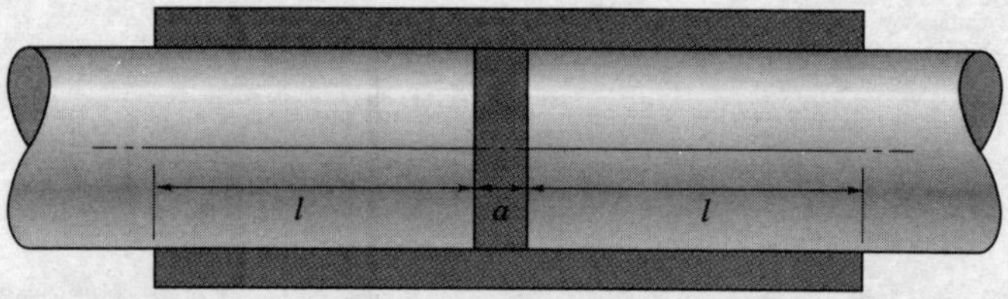

Sketch *d,* for Problem 10.18

10.19 A 50-mm-diameter steel shaft and a 60-mm-long cylindrical bushing in the same material with an outer diameter of 80 mm have been incorrectly shrink fit together and have to be dismounted. What axial force is needed for this if the diametral interference is 50 μm and the coefficient of friction is 0.2?

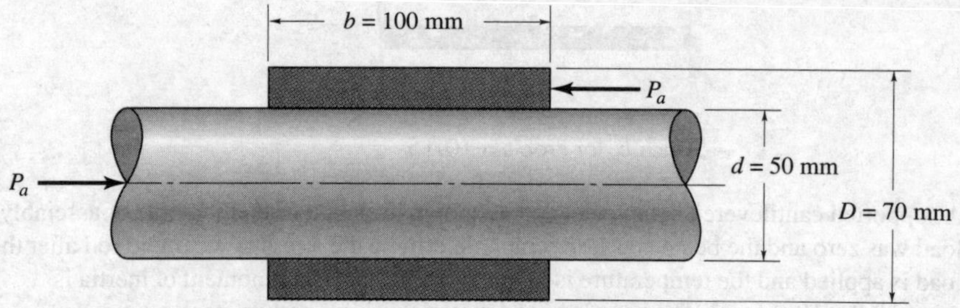

Sketch *e,* for Problem 10.20

10.20 A bushing press fit on a shaft shown in sketch *e* is going to be dismounted. What axial force P_a is needed to dismount the bushing if at the same time the bushing transmits a torque $T =$ 500 N-m? The diametral interference $\delta = 30$ μm, the coefficient of friction $\mu = 0.1$, and the modulus of elasticity $E = 210$ GPa.

10.21 To help in dismounting the wheel in Example 10.8, both an axial force and a moment are applied at the shaft-flywheel junction. How large a moment is needed to decrease the axial force to 50 kN when the wheel is dismounted?

10.22 Two AISI 1040 steel cylinders are press fit on each other. The inner diameter is 200 mm, the common diameter is 300 mm, and the outer diameter is 400 mm. The radial interference of the two cylinders is 0.1 mm.

a) Draw the radial and tangential stress distributions due to the press fit.
b) If the internal pressure is 207 MPa and the external pressure is 50 MPa, determine the radial and tangential stress distributions due only to these pressures.
c) Superimpose the stress distributions from a) and b) to obtain the total radial and tangential stress distribution.

10.23 A thick-walled cylinder is placed freely inside another thick-walled cylinder. What pressure is induced on the surfaces between the two cylinders by internally pressurizing the inner cylinder? *Hint:* Equate the radial displacement of the outer wall of the inner cylinder with that of the inner wall of the outer cylinder.

Section 10.6

10.24 A flat, 0.5-m-outer-diameter, 0.1-m-inner-diameter, 0.1-m-thick disk is shrink fit onto a solid shaft. Both the shaft and the disk are made of high-carbon steel. The assembly transmits 10 MW of power at 1000 rpm. Calculate the minimum temperature to which the disk must be heated for this shrink fit. The coefficient of friction is 0.25.

10.25 Two shafts are connected by a shrink-fit bushing with an outer diameter of 120 mm. The diameter of each shaft is 80 mm and each shaft is 2 m long. Before the shrink-fit bushing was mounted, the diametral interference was $\delta = 80\ \mu m$. The bushing and the shaft are steel with a modulus of elasticity $E = 210$ GPa. Find the temperature increase that the shafts can have without slip in the press fit if a power of 250 kW is transmitted and the far ends of the shafts cannot move axially. The coefficient of thermal expansion is $11.5 \times 10^{-6}/°$ C, and the coefficient of friction is 0.1. The axial length of the bushing press fit on each shaft is 80 mm.

10.26 A 10-mm-thick, 100-mm-wide ring is shrunk onto a 100-mm-diameter shaft. The diametral interference is 75 μm. Find the surface pressure in the shrink fit and the maximum torque that can be transmitted if the coefficient of friction $\mu = 0.10$. The modulus of elasticity $E = 210$ GPa.

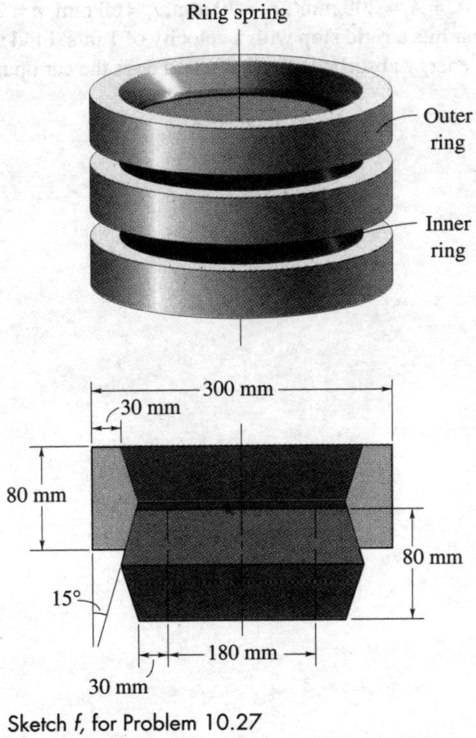

Sketch *f*, for Problem 10.27

10.27 A railway car buffer has a spring consisting of 11 outer rings and 11 inner rings where one of each ring type is a half-ring at the end of the spring. On one occasion a railway car with a total mass of 10 000 kg rolls until it hits a rigid stop. The force stopping the car is equal in the outer and the inner rings. The speed of the car just before the stop is 18 km/hr. Find the compression of the outer and inner rings and the stresses in the rings if the coefficient of friction $\mu = 0.10$ and the modulus of elasticity $E = 210$ GPa. The ring dimensions are shown in sketch *f*. When a spring of this type is compressed, the spring rate k (i.e., force divided by deformation) is

$$k = \frac{P}{\delta} = \frac{\pi E \tan\alpha \tan(\alpha + \gamma)}{n\left(\frac{r_y}{A_y} + \frac{r_i}{A_i}\right)}$$

where

α = cone angle for spring

γ = friction angle

$\tan\gamma = \mu$ = coefficient of friction

r_y, r_i = radii to surface center of gravity

A_y, A_i = cross-sectional areas of rings

10.28 A railway car weighing 20 tons has in each end two buffers of the type described in Problem 10.27. The data are $A_y = A_i = 200$ mm², $r_y = 70$ mm, $r_i = 60$ mm, $n = 22$, $\alpha = 14°$, $\gamma = 7°$, and $E = 206$ GPa. The car hits a rigid stop with a velocity of 1 m/s. Find the maximum force in each buffer and the energy absorbed. At what speed does the car bounce back?

$$\frac{P}{P_1} = \frac{\tan(\alpha + \gamma)}{\tan(\alpha - \gamma)}$$

CHAPTER

11

SHAFTING AND ASSOCIATED PARTS

A crankshaft from an automotive diesel engine. (Courtesy of FAG Kugelfischer Georg Schafer Kga. Schweinfurt, W. Germany)

When a man has a vision, he cannot get the power from the vision until he has performed it on the Earth for the people to see.

Black Elk, Oglala Sioux visionary, as told to John Niedhart

SYMBOLS

A	area, m^2
$\widetilde{A}$	constant defined in Eq. (11.27)
A_1	integration constant
a	length from $x = 0$ to force, m
$\widetilde{B}$	constant defined in Eq. (11.28)
b	length from force to end of beam, m
C_f	coefficient of fluctuation, Eq. (11.70)
c	distance from neutral axis to outer fiber, m
d	diameter, m
E	modulus of elasticity, Pa
g	gravitational acceleration, 9.807 m/s^2
h	height of key, m
h_p	power, W
I	area moment of inertia, m^4
I_m	mass moment of inertia, kg-m^2
J	polar area moment of inertia, m^4
K_c	stress concentration factor
K_e	kinetic energy, N-m
K_f	fatigue stress concentration factor
k	spring rate, N/m
k_f	surface finish factor
k_r	reliability factor
k_s	size factor
l	length, m
M	moment, N-m
M_f	performance index, J/kg
m_a	mass, kg
n_s	safety factor

P	normal force, N
p	pressure, Pa
p_f	interference pressure, Pa
q_n	notch sensitivity factor
R	reaction force, N
r	radius, m
S_e	modified endurance limit, Pa
S_e'	endurance limit, Pa
S_{se}	shear modified endurance limit, Pa
S_{sy}	shear yield stress, Pa
S_u	ultimate stress, Pa
S_{ut}	ultimate tensile stress, Pa
S_y	yield stress, Pa
T	torque, N-m
T_l	load torque, N-m
T_m	mean torque, N-m
t	time, s
t_h	thickness, m
U	potential energy, N-m
u	velocity, m/s
W	load, N
w	width of key, m
x, y, z	Cartesian coordinates, m
δ	deflection, m
ε	normal strain
θ	cylindrical polar coordinate
$\theta_{\omega_{max}}$	location within a cycle where speed is maximum
$\theta_{\omega_{min}}$	location within a cycle where speed is minimum
ν	Poisson's ratio
ρ	density, kg/m^3
σ	normal stress, Pa
σ_e	critical stress using distortion-energy theory, Pa
σ_ϕ	normal stress acting on oblique plane, Pa
τ	shear stress, Pa
τ_ϕ	shear stress acting on oblique plane, Pa
ϕ	oblique angle, deg
ω	angular speed, rad/s
ω_f	fluctuation speed, rad/s

Subscripts

a	alternating
all	allowable
avg	average
c	compression
cr	critical
e	endurance limit
f	flywheel
i	inner
m	mean
max	maximum
min	minimum
o	outer
p	pressure
r	radial
s	shear
x,y,z	Cartesian coordinates
θ	circumferential
ω	speed
1,2,3	principal axes

11.1 INTRODUCTION

A **shaft** is a rotating or stationary member usually having a circular cross section much smaller in diameter than the shaft length and having mounted on it such power-transmitting elements as gears, pulleys, belts, chains, cams, flywheels, cranks, sprockets, and rolling-element bearings. The loading on the shaft can be various combinations of bending (almost always fluctuating); torsion (may or may not be fluctuating); shock; or axial, normal, or transverse shear. All these types of loading are separately considered in Chapter 4. The geometry of the shaft is such that the diameter will generally be the variable used to satisfy the design. In practical applications the shaft may often be stepped instead of having a constant diameter. Some of the main considerations in designing a shaft are strength, using yield or fatigue (or both) as a criterion; deflection; or the dynamics established by the critical speeds. This chapter on shafting is put before the chapters on the various machine elements placed on a shaft mainly because these elements impose a force that can be considered in general and is equally applicable for whatever machine element it represents. This chapter makes extensive use of the failure prediction techniques developed in Chapters

6 and 7. Here combinations of loading are presented, whereas in Chapter 4 each type of loading was considered by itself. The dynamics and the first critical speed are important, since the rotating shaft becomes dynamically unstable and large vibrations are likely to develop. The chapter ends by considering the design of flywheels.

11.2 SHAFT DESIGN PROCEDURE

In the process of transmitting power at a given rotational speed, a shaft may be subjected to a torsional movement, or torque. Thus, a torsional shear stress is developed in the shaft. Also, some machine elements when mounted on a shaft exert forces on it in the transverse direction (perpendicular to the shaft axis). Thus, bending moments are developed in the shaft. A shaft carrying one or more of the various machine elements must be supported by bearings (considered in Chapters 12 and 13). If two bearings can provide radial support to limit shaft bending and deflection to acceptable values, this is highly desirable and simplifies the design. If, on the other hand, three or more bearings must be used to provide adequate support and rigidity, precise alignment of the bearings must be maintained.

The general procedure for shaft design is as follows:

1. Develop a free-body diagram by replacing the various machine elements mounted on the shaft by their statically equivalent load or torque components. To illustrate this, Fig. 11.1(a) shows two gears exerting forces on a shaft, and Fig. 11.1(b) then shows a free-body diagram of the gears acting on the shaft.

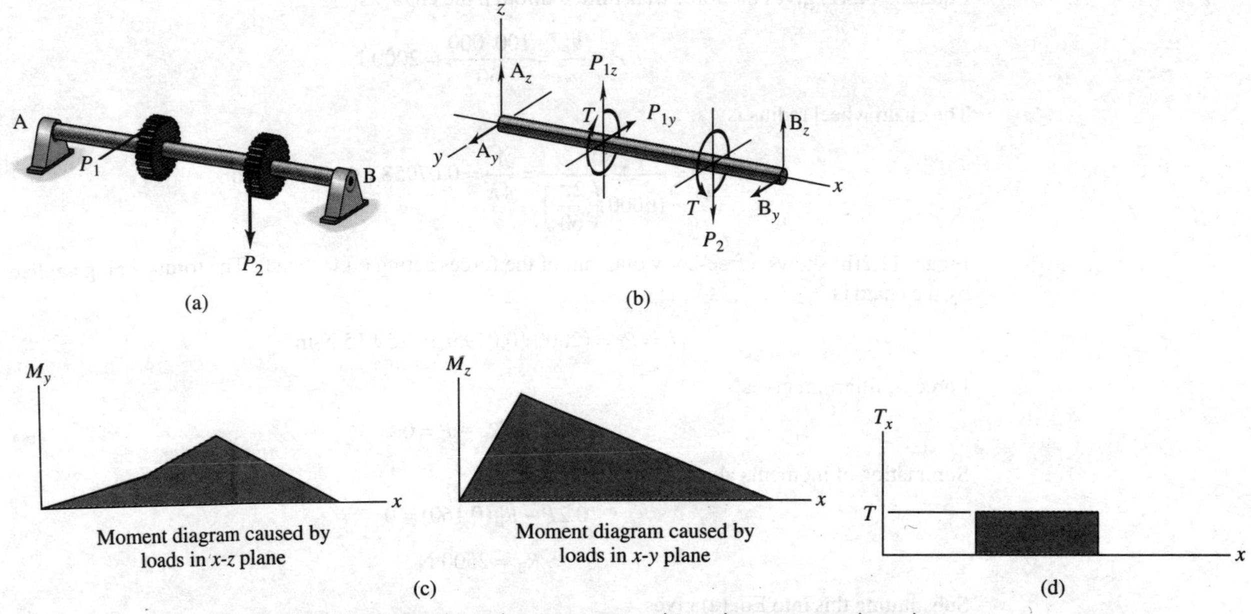

Figure 11.1 Shaft assembly. (a) Shaft with two bearings at A and B and two gears with resulting forces P_1 and P_2; (b) free-body diagram of torque and forces resulting from assembly drawing; (c) moment diagram in *x-z* and *x-y* planes; (d) torque diagram.

2. Draw a bending moment diagram in the *x-y* and *x-z* planes as shown in Fig. 11.1(c). The resultant internal moment at any section along the shaft may be expressed as

$$M_x = \sqrt{M_{xy}^2 + M_{xz}^2} \tag{11.1}$$

3. Develop a torque diagram as shown in Fig. 11.1(d). Torque developed from one power-transmitting element must balance the torque from other power-transmitting elements.
4. Establish the location of the critical cross section, or the *x* location where the torque and moment are the largest.
5. For ductile materials use the maximum-shear-stress theory (MSST), covered in Sec. 6.7.1.1; or the distortion-energy theory (DET), covered in Sec. 6.7.1.2.
6. For brittle materials use the maximum-normal-stress theory (MNST), covered in Sec. 6.7.2.1; the internal friction theory (IFT), covered in Sec. 6.7.2.2; or the modified Mohr theory (MMT), covered in Sec. 6.7.2.3.

Example 11.1

GIVEN The power from the motor of a front-wheel-drive car is transmitted to the gearbox by a chain drive [see Fig. 11.2(a)]. The two chain wheels are the same size. The chain is not prestressed, so the loose chain exerts no force. The safety factor is 4. The shaft is to be made of AISI 1080 steel. The chain transmits 100 kW of power at the chain speed of 50 m/s when the motor speed is 6000 rpm.

FIND The appropriate shaft diameter by using the DET.

Solution

Equation (4.41) gives the force transmitted through the chain as

$$P = \frac{h_p}{u} = \frac{100\ 000}{50} = 2000 \text{ N}$$

The chain wheel radius is

$$r = \frac{u}{\omega} = \frac{50}{(6000)\left(\frac{2\pi}{60}\right)} = \frac{1}{4\pi} = 0.07958 \text{ m} = 79.58 \text{ mm}$$

Figure 11.2(b) shows a free-body diagram of the forces acting on the shaft. The torque being applied by the chain is

$$T = Pr = (2000)(0.07958) = 159.15 \text{ N-m}$$

Force equilibrium gives

$$-R_A + R_B - P = 0 \tag{a}$$

Summation of moments about point A gives

$$0.2P - R_B(0.160) = 0$$

$$\therefore R_B = 2500 \text{ N}$$

Substituting this into Eq. (a) gives

$$R_A = R_B - P = 500 \text{ N}$$

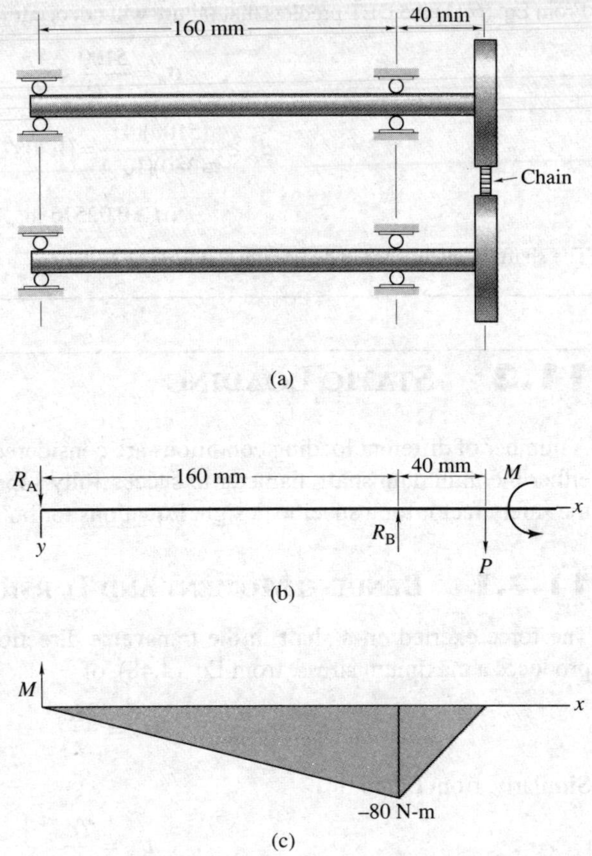

Figure 11.2 Illustration used in Example 11.1.(a) Chain drive assembly; (b) free-body diagram of forces; (c) bending moment diagram.

Figure 11.2(c) shows the bending moment diagram. The maximum bending moment occurs at $x = 0.160$ m and has a value of 80 N-m. The torque is constant along the shaft, so that the critical section occurs at $x = 0.160$ m. From Eqs. (4.34) and (4.48)

$$\tau_{xy} = \frac{cT}{J} = \frac{(d/2)(159.15)}{\dfrac{\pi d^4}{32}} = \frac{2546.4}{\pi d^3}$$

$$\sigma_x = \frac{cM}{I} = \frac{(d/2)(80)}{\dfrac{\pi d^4}{64}} = \frac{2560}{\pi d^3}$$

From Eq. (2.16) the principal stresses when $\sigma_y = 0$ are

$$\sigma_1, \sigma_2 = \frac{\sigma_x}{2} \pm \sqrt{\tau_{xy}^2 + \frac{\sigma_x^2}{4}} = \frac{1280}{\pi d^3} \pm \frac{1}{\pi d^3}\sqrt{(2546.4)^2 + \frac{(2560)^2}{4}}$$

$$\sigma_1 = 4130/\pi d^3 \qquad \text{and} \qquad \sigma_2 = -1570/\pi d^3$$

From Eq. (6.10) the von Mises stress for a biaxial stress state is

$$\sigma_e = \frac{1}{\pi d^3}\left[(4130)^2 + (1570)^2 + (1570)(4130)\right]^{1/2} = \frac{5100}{\pi d^3}$$

From Eq. (6.11) the DET predicts that failure will not occur if

$$\sigma_e = \frac{5100}{\pi d^3} < \frac{S_y}{n_s}$$

$$d^3 \geq \frac{(5100)(4)}{\pi(380)(10^6)} = (1.7087)(10^{-5})$$

$$\therefore d \geq 0.02576 \text{ m}$$

The shaft diameter should be 26 mm.

11.3 STATIC LOADING

A number of different loading conditions are considered. The shaft designer must establish either the minimum shaft diameter to successfully support the loads acting on the shaft or the safety factor for a specific design. Equations for both are presented.

11.3.1 BENDING MOMENT AND TORSION

The force exerted on a shaft in the transverse direction (perpendicular to the shaft axis) produces a maximum stress, from Eq. (4.48), of

$$\sigma_x = \frac{Mc}{I} \tag{11.2}$$

Similarly, from Eq. (4.34)

$$\tau_{xy} = \frac{Tc}{J} \tag{11.3}$$

where

$$c = \frac{d}{2} \qquad I = \frac{\pi d^4}{64} \qquad \text{and} \qquad J = \frac{\pi d^4}{32} \tag{11.4}$$

for a circular cross section. Substituting Eqs. (11.4) into Eqs. (11.2) and (11.3) gives

$$\sigma_x = \frac{64Md}{2\pi d^4} = \frac{32M}{\pi d^3} \tag{11.5}$$

$$\tau_{xy} = \frac{Td/2}{\pi d^4/32} = \frac{16T}{\pi d^3} \tag{11.6}$$

For the plane stress state the principal normal stresses, from Eq. (2.16) when $\sigma_y = 0$, are

$$\sigma_1, \sigma_2 = \frac{\sigma_x}{2} \pm \sqrt{\left(\frac{\sigma_x}{2}\right)^2 + \tau_{xy}^2} \tag{11.7}$$

Substituting Eqs. (11.5) and (11.6) into Eq. (11.7) gives

$$\sigma_1, \sigma_2 = \frac{16M}{\pi d^3} \pm \sqrt{\left(\frac{16M}{\pi d^3}\right)^2 + \left(\frac{16T}{\pi d^3}\right)^2} = \frac{16}{\pi d^3}\left[M \pm \sqrt{M^2 + T^2}\right] \tag{11.8}$$

From Eq. (2.18) the principal shear stresses are

$$\tau_1, \tau_2 = \pm\sqrt{\tau_{xy}^2 + \left(\frac{\sigma_x}{2}\right)^2} \tag{11.9}$$

Substituting Eqs. (11.5) and (11.6) into Eq. (11.9) gives

$$\tau_1, \tau_2 = \pm\frac{16}{\pi d^3}\sqrt{M^2 + T^2} \tag{11.10}$$

11.3.1.1 DISTORTION-ENERGY THEORY As shown by Sec. 6.7.1.2 and Eqs. (6.10) and (6.11) the DET predicts failure if

$$(\sigma_1^2 + \sigma_2^2 - \sigma_1\sigma_2)^{1/2} \geq \frac{S_y}{n_s} \tag{11.11}$$

where

S_y = yield strength of shaft material, Pa

n_s = safety factor

From Eq. (11.8) the DET predicts failure if

$$\frac{16}{\pi d^3}(4M^2 + 3T^2)^{1/2} \geq \frac{S_y}{n_s} \tag{11.12}$$

Thus, the DET predicts the smallest diameter where failure will first start to occur as

$$d = \left(\frac{32 n_s}{\pi S_y}\sqrt{M^2 + \tfrac{3}{4}T^2}\right)^{1/3} \tag{11.13}$$

Often the torque is not explicitly given in Eq. (11.13) and the power transfer equations [Eqs. (4.40) and (4.42)] developed in Chapter 4 need to be used.

If the shaft diameter is known and the safety factor is an unknown, Eq. (11.13) becomes

$$n_s = \frac{\pi d^3 S_y}{32\sqrt{M^2 + \tfrac{3}{4}T^2}} \tag{11.14}$$

11.3.1.2 MAXIMUM-SHEAR-STRESS THEORY As shown by Sec. 6.7.1.1 and Eq. (6.6) the MSST predicts failure for a plane or biaxial stress state ($\sigma_3 = 0$) if

$$|\sigma_1 - \sigma_2| \geq \frac{S_y}{n_s} \tag{11.15}$$

Equation (11.8) gives

$$\frac{32\sqrt{M^2 + T^2}}{\pi d^3} \geq \frac{S_y}{n_s} \tag{11.16}$$

Thus, the MSST predicts the smallest diameter where failure will first start to occur as

$$d = \left(\frac{32 n_s}{\pi S_y}\sqrt{M^2 + T^2}\right)^{1/3} \tag{11.17}$$

If the shaft diameter is known and the safety factor is an unknown, Eq. (11.17) becomes

$$n_s = \frac{\pi d^3 S_y}{32\left(M^2 + T^2\right)^{1/2}} \qquad \textbf{(11.18)}$$

Example 11.2

GIVEN An assembly of belts has tensile forces applied as shown in Fig. 11.3(a) and frictionless journal bearings at locations A and B. The yield strength of the shaft material is 500 MPa and the safety factor is 2.

FIND Determine the smallest safe shaft diameter by using both the DET and the MSST. Also, provide a free-body diagram as well as moment and torque diagrams.

Solution

A free-body diagram is shown in Fig. 11.3(b); a moment diagram in the *x*-*y* plane, in Fig. 11.3(c); and a moment diagram in the *x*-*z* plane, in Fig. 11.3(d). From the moment diagrams the maximum moment is

$$M_{\max} = \sqrt{(118.75)^2 + (37.5)^2} = 124.5 \text{ N-m}$$

Figure 11.3(e) gives the torque diagram.

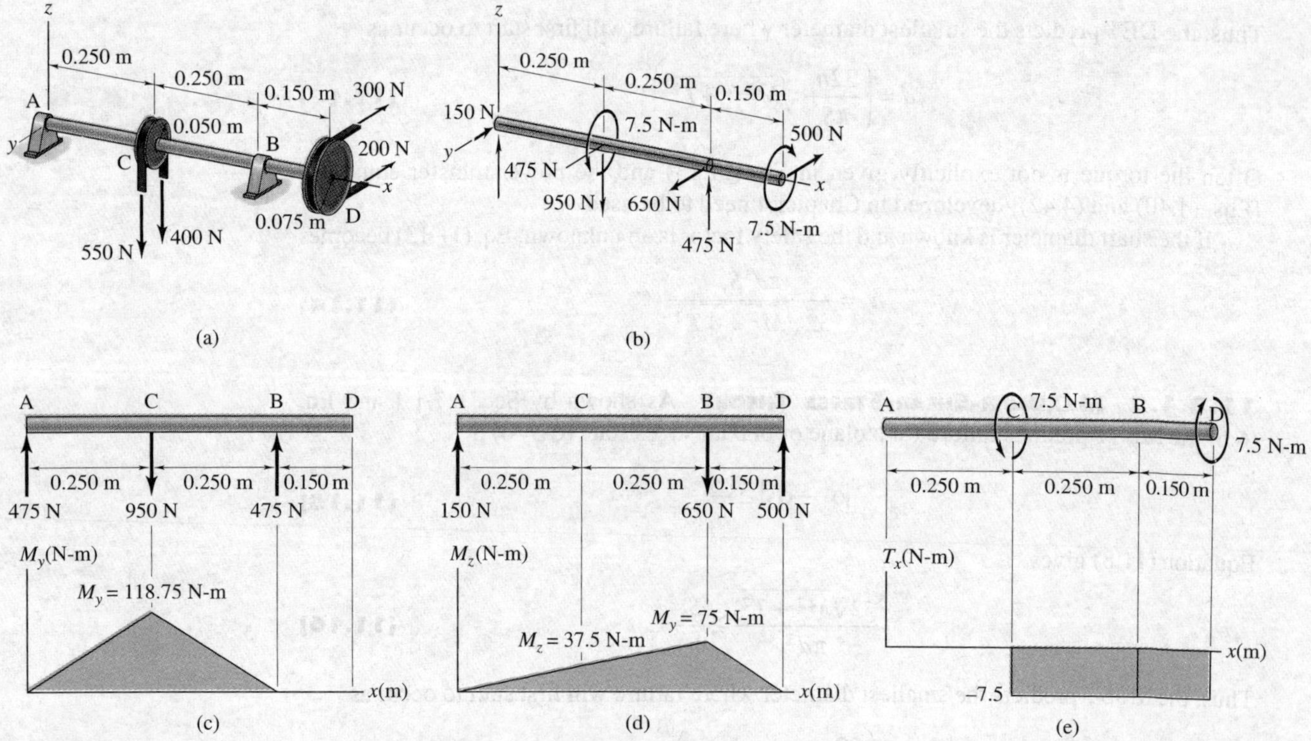

Figure 11.3 Figures used in Example 11.2.
(a) Assembly drawing; (b) free-body diagram; (c) moment diagram in *x*-*y* plane; (d) moment diagram in *x*-*z* plane; (e) torque diagram.

Using the DET [Eq. (11.13)] the smallest safe diameter is

$$d = \left(\frac{32n_s}{\pi S_y} \sqrt{M^2 + 3T^2/4} \right)^{1/3} = \left\{ \frac{32(2)}{\pi(500)(10^6)} \left[124.5^2 + 3(7.5)^2/4 \right]^{1/2} \right\}^{1/3}$$

$$= 0.0172 \text{ m} = 17.2 \text{ mm}$$

Using the MSST [Eq. (11.17)] gives

$$d = \left(\frac{32n_s}{\pi S_y} \sqrt{M^2 + T^2} \right)^{1/3} = \left[\frac{32(2)}{\pi(500)(10^6)} (124.5^2 + 7.5^2)^{1/2} \right]^{1/3}$$

$$= 0.0172 \text{ m} = 17.2 \text{ mm}$$

Since the torque is small relative to the moment, little difference exists between the DET and MSST predictions.

11.3.2 Bending Moment, Torsion, and Axial Loading

If axial loading is added to what was considered in Sec. 11.3.1 [see Eq. (11.5)], the normal stress is

$$\sigma_x = \frac{32M}{\pi d^3} + \frac{4P}{\pi d^2} \tag{11.19}$$

The shear stress is expressed in Eq. (11.6); and the principal normal stresses, in Eq. (11.7). Substituting Eqs. (11.19) and (11.6) into Eq. (11.7) gives

$$\sigma_1, \sigma_2 = \frac{16M}{\pi d^3} + \frac{2P}{\pi d^2} \pm \sqrt{\left(\frac{16M}{\pi d^3} + \frac{2P}{\pi d^2} \right)^2 + \left(\frac{16T}{\pi d^3} \right)^2}$$
$$= \frac{2}{\pi d^3} \left[8M + Pd \pm \sqrt{(8M + Pd)^2 + (8T)^2} \right] \tag{11.20}$$

Substituting Eqs. (11.19) and (11.6) into Eq. (11.9) gives the principal shear stresses as

$$\tau_1, \tau_2 = \pm \frac{2}{\pi d^3} \sqrt{(8M + Pd)^2 + (8T)^2} \tag{11.21}$$

11.3.2.1 Distortion-Energy Theory Substituting Eq. (11.20) into Eq. (11.11) shows that the DET predicts failure if

$$\frac{4}{\pi d^3} \sqrt{(8M + Pd)^2 + 48T^2} \geq \frac{S_y}{n_s} \tag{11.22}$$

Because axial loading was added, an explicit expression for the diameter d cannot be obtained.

11.3.2.2 Maximum-Shear-Stress Theory Substituting Eq. (11.20) into Eq. (11.15) shows that the MSST predicts failure if

$$\frac{4}{\pi d^3} \sqrt{(8M + Pd)^2 + 64T^2} \geq \frac{S_y}{n_s} \tag{11.23}$$

Again, because axial loading was added, an explicit expression for the diameter d cannot be obtained.

11.4 CYCLIC LOADING

In cyclic loading the loads vary throughout a cycle rather than remaining constant as in static loading. Here, a general analysis is derived for the fluctuating normal and shear stresses of ductile materials, and appropriate equations are given for brittle materials.

11.4.1 DUCTILE MATERIALS

Figure 11.4 shows the normal and shear stresses acting on a shaft. In Fig. 11.4(a) the stresses act on a rectangular element, and in Fig. 11.4(b) the stresses act on an oblique plane at angle ϕ. The normal stresses are denoted by σ and the shear stresses by τ. Subscript a designates alternating and subscript m designates mean or steady stress. Also, K_f designates the fatigue stress concentration factor due to normal loading, and K_{fs} designates the fatigue concentration factor due to shear loading. On a rectangular element [Fig. 11.4(a)] the normal stress is $\sigma_m \pm K_f \sigma_a$ and the shear stress is $\tau_m \pm K_{fs} \tau_a$. The largest stress occurs when σ_a and τ_a are in phase, or when the frequency of one is an integral multiple of the frequency of the other. Summing the forces tangent to the diagonal gives

$$-\tau_\phi A + \left(\tau_m + K_{fs}\tau_a\right) A \cos\phi \cos\phi - \left(\tau_m + K_{fs}\tau_a\right) A \sin\phi \sin\phi + \left(\sigma_m + K_f \sigma_a\right) A \cos\phi \sin\phi = 0$$

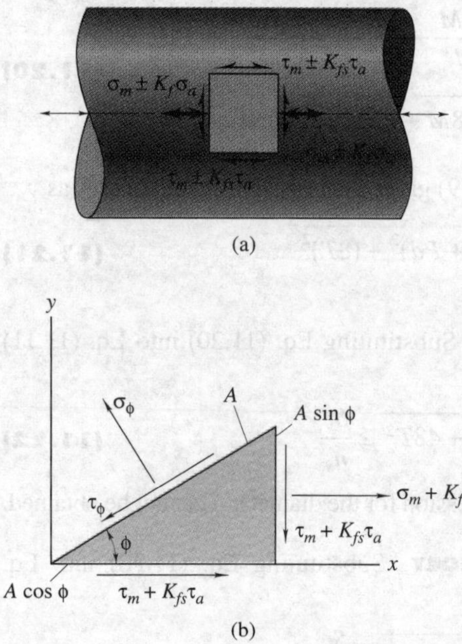

Figure 11.4 Fluctuating normal and shear stresses acting on shaft.
(a) Stresses acting on rectangular element;
(b) stresses acting on oblique plane at angle ϕ.

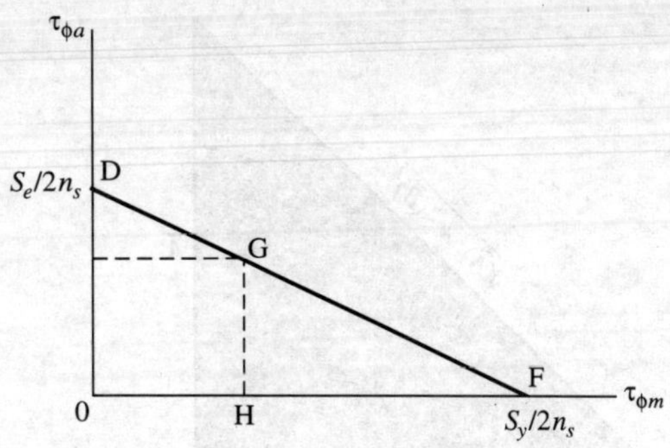

Figure 11.5 Soderberg line for shear stress.

Making use of double angles reduces the above equation to

$$\tau_\phi = \left(\tau_m + K_{fs}\tau_a\right)\cos 2\phi + \frac{1}{2}\left(\sigma_m + K_f\sigma_a\right)\sin 2\phi$$

Separating the mean and alternating components of stress gives the stress acting on the oblique plane as

$$\tau_\phi = \tau_{\phi m} + \tau_{\phi a} = \left(\frac{\sigma_m}{2}\sin 2\phi + \tau_m \cos 2\phi\right) + \left(\frac{K_f\sigma_a}{2}\sin 2\phi + K_{fs}\tau_a\cos 2\phi\right) \quad \textbf{(11.24)}$$

Recall the Soderberg line in Fig. 7.9 for tensile loading. For shear loading the end points of the Soderberg line are $S_{se} = S_e/2n_s$ and $S_{sy} = S_y/2n_s$. Figure 11.5 shows the Soderberg line for shear stress. From the proportional triangles GHF and D0F of Fig. 11.5,

$$\frac{\text{HF}}{\text{0F}} = \frac{\text{HG}}{\text{0D}} \quad \text{or} \quad \frac{S_y/2n_s - \tau_{\phi m}}{S_y/2n_s} = \frac{\tau_{\phi a}}{S_e/2n_s}$$

$$\frac{1}{n_s} = \frac{\tau_{\phi a}}{S_e/2} + \frac{\tau_{\phi m}}{S_y/2} \quad \textbf{(11.25)}$$

Note the similarities between Eq. (7.29) and Eq. (11.25). Substituting into Eq. (11.25) the expressions for $\tau_{\phi a}$ and $\tau_{\phi m}$ gives

$$\frac{1}{n_s} = \frac{\dfrac{K_f\sigma_a}{2}\sin 2\phi + K_{fs}\tau_a\cos 2\phi}{S_e/2} + \frac{\left(\dfrac{\sigma_m}{2}\sin 2\phi + \tau_m\cos 2\phi\right)}{S_y/2} \quad \textbf{(11.26)}$$

$$= \tilde{A}\sin 2\phi + 2\tilde{B}\cos 2\phi$$

where

$$\tilde{A} = \frac{\sigma_m}{S_y} + \frac{K_f\sigma_a}{S_e} \quad \textbf{(11.27)}$$

$$\tilde{B} = \frac{\tau_m}{S_y} + \frac{K_{fs}\tau_a}{S_e} \quad \textbf{(11.28)}$$

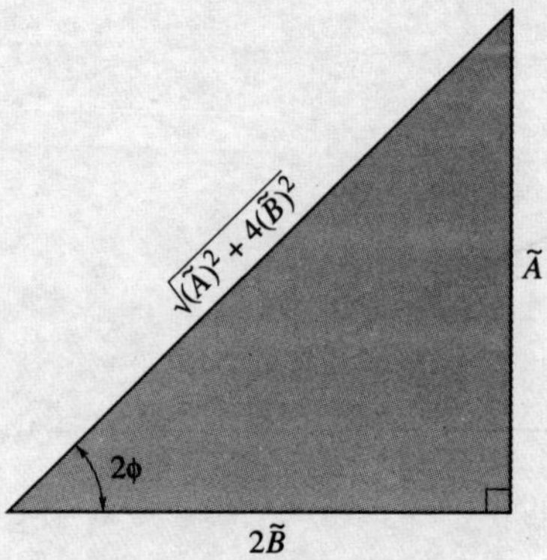

Figure 11.6 Illustration of relationship given in Eq. (11.29).

We are interested in the stress combination that produces the smallest n_s, since this corresponds to a maximum-stress situation. The minimum value of n_s corresponds to a maximum value of $1/n_s$. Differentiating $1/n_s$ in Eq. (11.26) and equating the result to zero give

$$\frac{d}{d\phi}\left(\frac{1}{n_s}\right) = 2\tilde{A}\cos 2\phi - 4\tilde{B}\sin 2\phi = 0$$

$$\therefore \quad \frac{\sin 2\phi}{\cos 2\phi} = \tan 2\phi = \frac{\tilde{A}}{2\tilde{B}} \tag{11.29}$$

This relationship is illustrated in Fig. 11.6, which shows that

$$\sin 2\phi = \frac{\tilde{A}}{\sqrt{\left(\tilde{A}\right)^2 + 4\left(\tilde{B}\right)^2}} \quad \text{and} \quad \cos 2\phi = \frac{2\tilde{B}}{\sqrt{\left(\tilde{A}\right)^2 + 4\left(\tilde{B}\right)^2}} \tag{11.30}$$

Substituting this into Eq. (11.26) gives

$$\frac{1}{n_s} = \frac{\left(\tilde{A}\right)^2}{\sqrt{\left(\tilde{A}\right)^2 + 4\left(\tilde{B}\right)^2}} + \frac{4\left(\tilde{B}\right)^2}{\sqrt{\left(\tilde{A}\right)^2 + 4\left(\tilde{B}\right)^2}} = \sqrt{\left(\tilde{A}\right)^2 + 4\left(\tilde{B}\right)^2}$$

Substituting Eqs. (11.27) and (11.28) into the above equation gives

$$\frac{1}{n_s} = \sqrt{\left(\frac{\sigma_m}{S_y} + \frac{K_f\sigma_a}{S_e}\right)^2 + 4\left(\frac{\tau_m}{S_y} + \frac{K_{fs}\tau_a}{S_e}\right)^2}$$

$$\frac{S_y}{n_s} = \sqrt{\left(\sigma_m + \frac{S_y}{S_e}K_f\sigma_a\right)^2 + 4\left(\tau_m + \frac{S_y}{S_e}K_{fs}\tau_a\right)^2} \tag{11.31}$$

Setting $\sigma_y = 0$, $\sigma_x = \sigma$, and $\tau_{xy} = \tau$ in Eq. (2.18) for biaxial stresses, the maximum shear stress is

$$\tau_{max} = \sqrt{\left(\frac{\sigma}{2}\right)^2 + \tau^2} \tag{11.32}$$

and the safety factor is

$$n_s = \frac{S_y/2}{\tau_{max}} = \frac{S_y/2}{\sqrt{(\sigma/2)^2 + \tau^2}} = \frac{S_y}{\sqrt{\sigma^2 + 4\tau^2}}$$

$$\frac{S_y}{n_s} = \sqrt{\sigma^2 + 4\tau^2} \tag{11.33}$$

Equations (11.33) and (11.31) have the same form and

$$\sigma = \sigma_m + \frac{S_y}{S_e} K_f \sigma_a \qquad \text{and} \qquad \tau = \tau_m + \frac{S_y}{S_e} K_{fs} \tau_a$$

The normal and shear stresses each contain a steady and an alternating component, the latter weighted for the effect of fatigue and stress concentration.

By making use of Eqs. (11.5) and (11.6), Eq. (11.31) becomes

$$n_s = \frac{\pi d^3 S_y}{32\sqrt{\left(M_m + \frac{S_y}{S_e} K_f M_a\right)^2 + \left(T_m + \frac{S_y}{S_e} K_{fs} T_a\right)^2}} \tag{11.34}$$

If instead of the safety factor, we want to know the smallest safe diameter for a specified safety factor, Eq. (11.34) can be rewritten as

$$d = \left[\frac{32 n_s}{\pi S_y}\sqrt{\left(M_m + \frac{S_y}{S_e} K_f M_a\right)^2 + \left(T_m + \frac{S_y}{S_e} K_{fs} T_a\right)^2}\right]^{1/3} \tag{11.35}$$

The shear fatigue stress concentration factor K_{fs} can be estimated by using torsion instead of axial loading in the stress concentration factor.

Equations (11.34) and (11.35) represent the general form of the MSST. Note in Eq. (11.35) that S_y, S_y/S_e, K_f, and K_{fs} depend on the shaft diameter d. The approach to be used is to estimate the range of the shaft diameter and iterate the results if the final design is outside the estimated range.

Peterson (1974) modified Eq. (11.31) by changing the coefficient of the shear stress term from 4 to 3, such that the DET is satisfied and gives

$$\frac{S_y}{n_s} = \sqrt{\left(\sigma_m + \frac{S_y}{S_e} K_f \sigma_a\right)^2 + 3\left(\tau_m + \frac{S_y}{S_e} K_{fs} \tau_a\right)^2} \tag{11.36}$$

By making use of Eqs. (11.5) and (11.6), Eq. (11.36) becomes

$$n_s = \frac{\pi d^3 S_y}{32\sqrt{\left(M_m + \frac{S_y}{S_e} K_f M_a\right)^2 + \frac{3}{4}\left(T_m + \frac{S_y}{S_e} K_{fs} T_a\right)^2}} \tag{11.37}$$

The smallest safe diameter corresponding to a specific safety factor can then be expressed as

$$d = \left[\frac{32 n_s}{\pi S_y}\sqrt{\left(M_m + \frac{S_y}{S_e} K_f M_a\right)^2 + \frac{3}{4}\left(T_m + \frac{S_y}{S_e} K_{fs} T_a\right)^2}\right]^{1/3} \tag{11.38}$$

Tutorial 6

The distinction between Eqs. (11.34) and (11.35) and Eqs. (11.37) and (11.38) needs to be recognized. Equations (11.34) and (11.35) assume that the MSST is valid; Eqs. (11.37) and (11.38) assume that the DET is valid. All four equations are general equations applicable for ductile materials.

Example 11.3

GIVEN When a rear-wheel-drive car accelerates through a bend at high speeds, the drive shafts are subjected to both bending and torsion. The acceleration torque T is reasonably constant at 400 N-m while the bending moment is varying due to cornering and is expressed in newton-meters as

$$M = 250 + 800 \sin \omega t$$

Thus, the mean and alternating moments are $M_m = 250$ N-m and $M_a = 800$ N-m. Assume there is no notch giving a stress concentration. The reliability should be 99% and the safety factor is 4.5. The shaft is forged from AISI 1080 steel.

FIND The shaft diameter while using the MSST.

Solution

From Eq. (7.7) and Table A.1 the bending endurance limit for AISI 1080 steel is

$$S_e' = 0.5 S_u = 0.5(615) = 307.5 \text{ MPa}$$

From Fig. 7.7(a) the surface finish factor for an as-forged surface at $S_{ut} = 615$ MPa is $k_f = 0.42$. In evaluating the size factor the shaft diameter needs to be chosen. Assume $d = 30$ mm. From Eq. (7.22)

$$k_s = 1.189 d^{-0.112} = 1.189(30)^{-0.112} = 0.8124$$

From Table 7.4 for 99% probability of survival the reliability factor is 0.82. Substituting the above into Eq. (7.16) gives the endurance limit as

$$S_e = k_f k_s k_r S_e' = (0.42)(0.8124)(0.82)(307.5)(10^6) = 86.03 \text{ MPa}$$

Equation (11.35) with $T_a = 0$ gives

$$d = \left\{\frac{32 n_s}{\pi S_y}\sqrt{\left(M_m + \frac{S_y}{S_e} K_f M_a\right)^2 + T_m^2}\right\}^{\frac{1}{3}}$$

$$d = \left\{ \frac{32(4.5)}{\pi(380)(10^6)} \sqrt{\left[250 + \frac{(380)(10^6)}{(86.03)(10^6)}(1)(800) \right]^2 + (400)^2} \right\}^{\frac{1}{3}}$$

$$d = 0.07714 \text{ m} = 77.14 \text{ mm}$$

A new calculation needs to be made of the size factor, since initially the shaft diameter was chosen as 30 mm. From Eq. (7.22)

$$k_s = 1.189d^{-0.112} = 1.189(77.14)^{-0.112} = 0.7308$$

$$\therefore S_e = (0.42)(0.7308)(0.82)(307.5)(10^6) = 77.39 \text{ MPa}$$

From Eq. (11.35)

$$d = \left\{ \frac{32(4.5)}{\pi(380)(10^6)} \sqrt{\left[250 + \frac{(380)(10^6)}{(77.39)(10^6)}(1)(800) \right]^2 + (400)^2} \right\}^{\frac{1}{3}} = 0.0797 \text{ m}$$

The shaft diameter should be 80 mm.

11.4.2 Brittle Materials

Although shafts are usually cold-worked metals machined to desired dimensions, there are applications where castings (brittle materials) are used as shafts. As discussed in Chapter 6 this requires a slightly different analysis approach than for ductile materials.

For brittle materials the forces in Fig. 11.4(b) are assumed to be *normal* rather than tangent to the diagonal. Also, the design line as dictated by the maximum-normal-stress theory (MNST) extends from S_e/n_s to S_u/n_s instead of from $S_e/2n_s$ to $S_y/2n_s$ as was true for the MSST. Following procedures similar to those used in obtaining Eq. (11.31) gives

$$\frac{2S_u}{n_s} = K_c\left(\sigma_m + \frac{S_u}{S_e}\sigma_a\right) + \sqrt{K_c^2\left(\sigma_m + \frac{S_u}{S_e}\sigma_a\right)^2 + 4K_{cs}^2\left(\tau_m + \frac{S_u}{S_e}\tau_a\right)^2} \quad \textbf{(11.39)}$$

where

K_c = stress concentration factor

By making use of Eqs. (11.5) and (11.6), Eq. (11.39) can be written as

$$n_s = \frac{\pi d^3 S_u/16}{K_c\left(M_m + \frac{S_u}{S_e}M_a\right) + \sqrt{K_c^2\left(M_m + \frac{S_u}{S_e}M_a\right)^2 + K_{cs}^2\left(T_m + \frac{S_u}{S_e}T_a\right)^2}} \quad \textbf{(11.40)}$$

If we want the minimum safe diameter of the shaft for a specific safety factor,

$$d = \left\{ \frac{16n_s}{\pi S_u}\left[K_c\left(M_m + \frac{S_u}{S_e}M_a\right) + \sqrt{K_c^2\left(M_m + \frac{S_u}{S_e}M_a\right)^2 + K_{cs}^2\left(T_m + \frac{S_u}{S_e}T_a\right)^2} \right] \right\}^{1/3} \quad \textbf{(11.41)}$$

The important difference in the equations developed for the safety factor and the smallest safe diameter is that Eqs. (11.34) and (11.35) are applicable for ductile materials while assuming the MSST, Eqs. (11.37) and (11.38) are also applicable for ductile materials but while assuming the DET, and Eqs. (11.40) and (11.41) are applicable for brittle materials while assuming the MNST. Note the major differences between the equations developed for brittle and ductile materials. For brittle materials [Eqs. (11.40) and (11.41)] the stress concentration factor K_c and the ultimate stress S_u are used, whereas for ductile materials [Eqs. (11.34), (11.35), (11.37), and (11.38)] the fatigue stress concentration factor K_f and the yield stress S_y are used. Having these general expressions, we will now consider specific fluctuating stresses.

A specific cyclic stress variation that occurs in practical applications is reversed bending and steady torsion. From Sec. 7.3 note that reversed bending implies that either $\sigma_m = 0$ or $M_m = 0$. Also, steady torsion implies that either $\tau_a = 0$ or $T_a = 0$. Thus, to apply the condition of reversed bending and steady torsion ($M_m = 0$ and $T_a = 0$), and knowing whether the material is ductile or brittle and which failure theory should be applied, the reduced forms of Eqs. (11.34) and (11.35), (11.37) and (11.38), or (11.40) and (11.41) should be used.

Example 11.4

GIVEN The shaft shown in Fig. 11.7 is subjected to reversed bending and steady torsion. A standard ball bearing is to be placed on diameter d_2 and this surface will therefore be ground to form a good seat for the bearing. The groove between the sections is there to make sure that the large diameter section is not damaged by the grinding operation, and is called a "grinding relief." Assume that standard ball bearing bore sizes are in 5-mm increments in the range 15 to 50 mm. Design so that the relative sizes are approximately $d_2 = 0.75d_3$ and $d_1 = 0.65d_3$. The completely reversed bending moment is 70 N-m, and the steady torsion is 45 N-m. Assume a safety factor of 2.5 and size the shaft for infinite life.

FIND Determine the diameter d_2 by both the MSST and the DET.

Solution

From Table A.1 for high-carbon steel $S_u = 615$ MPa and $S_y = 380$ MPa. From Eq. (7.7) for bending

$$S_e' = 0.5S_u = 307.5 \text{ MPa}$$

From Fig. 7.7(a) the surface finish factor k_f for machined surfaces is 0.75. Assume that the diameter $d_1 = 20$ mm. From Eq. (7.22) the size factor is

$$k_s = 1.189d^{-0.112} = 1.189(20)^{-0.112} = 0.8501$$

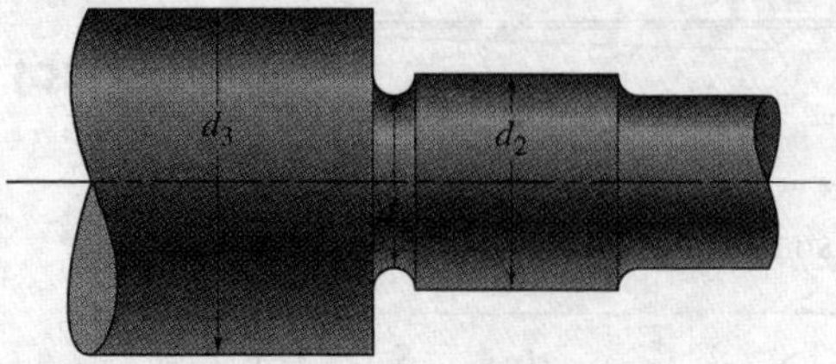

Figure 11.7 Section of shaft used in Example 11.4.

Given that

$$d_3 = \frac{d_2}{0.75} = \frac{d_1}{0.65} \quad \rightarrow \quad \frac{d_2}{d_1} = \frac{0.75}{0.65} = 1.154$$

From Fig. 6.6(b) for

$$\frac{r}{d} = \frac{d_2 - d_1}{2d_1} = \frac{1}{2}\left(\frac{d_2}{d_1} - 1\right) = 0.0769$$

the stress concentration factor $K_c = 1.9$. Recall that

$$r = \frac{d_2 - d_1}{2} = \frac{d_2}{2}\left(1 - \frac{1}{1.154}\right) = 0.0667 d_2$$

$$\therefore r = 1.335 \text{ mm}$$

From Fig. 7.6 for $S_u = 615$ MPa, bending, and $r = 1.335$ mm, the notch sensitivity factor is $q_n = 0.8$. From Eq. (7.19)

$$K_f = 1 + (K_c - 1)q_n = 1 + (1.9 - 1)0.8 = 1.72$$

From Eq. (7.16) the modified endurance limit is

$$S_e = k_f k_s S'_e = (0.75)(0.85)307.5 = 196.0 \text{ MPa}$$

Recall that for completely reversed bending and steady torsion $M_m = 0$ and $T_a = 0$. Using Eq. (11.35) and the MSST gives

$$d_1 = \left[\frac{32(2.5)}{\pi(380)(10^6)}\sqrt{\left(\frac{380}{196}\right)^2 (1.72)^2(70)^2 + (45)^2}\right]^{1/3} = 0.02516 \text{ m} = 25.16 \text{ mm}$$

$$\therefore d_1 = 26 \text{ mm}$$

$$d_3 = \frac{d_1}{0.65} = \frac{26}{0.65} = 40.0 \text{ mm}$$

$$d_2 = 0.75 d_3 = 0.75(40.0) = 30.0 \text{ mm}$$

Having established that $d_1 = 26$ mm instead of the 20 mm initially guessed, we need to recalculate the size factor:

$$k_s = 1.189 d^{-0.112} = (1.189)(26)^{-0.112} = 0.8255$$

$$\frac{r}{d} = \frac{d_2 - d_1}{2d_1} = \frac{30 - 26}{2(26)} = \frac{1}{13} = 0.0769$$

Figure 6.6(b) for $r/d = 0.077$ and $d_2/d_1 = 30/26 = 1.154$ gives a stress concentration of 1.9. Also,

$$r = \frac{d_2 - d_1}{2} = \frac{30 - 26}{2} = 2.0 \text{ mm}$$

From Fig. 7.7 for $S_u = 615$ MPa, bending, and $r = 2.0$ mm, the notch sensitivity $q_n = 0.76$. The fatigue stress concentration factor can be expressed as

$$K_f = 1 + (K_c - 1)q_n = 1 + (1.9 - 1)(0.76) = 1.684$$

From Eq. (7.16) the modified endurance limit is

$$S_e = k_f k_s S'_e = (0.75)(0.8255)307.5(10^6) \text{ Pa} = 190.4 \text{ MPa}$$

Using Eq. (11.35) and the MSST gives

$$d_1 = \left[\frac{32(2.5)}{\pi(380)(10^6)} \sqrt{\left(\frac{380}{190.4}\right)^2 (1.684)^2 (70)^2 + (45)^2} \right]^{1/3} = 0.02523 \text{ m} = 25.23 \text{ mm}$$

$$\therefore d_1 = 26 \text{ mm}, \ d_3 = 40 \text{ mm}, \text{ and } d_2 = 30 \text{ mm}$$

Using Eq. (11.38) and the DET gives

$$d_1 = \left[\frac{32(2.5)}{\pi(380)(10^6)} \sqrt{\left(\frac{380}{190.4}\right)^2 (1.684)^2 (70)^2 + 0.75(45)^2} \right]^{1/3} = 0.02519 \text{ m} = 25.19 \text{ mm}$$

$\therefore\ d_1 = 26$ mm, $d_3 = 40$ mm,and $d_2 = 30$ mm. Thus, the MSST and the DET give the same results.

11.5 CRITICAL SPEED OF ROTATING SHAFTS

All rotating shafts deflect during rotation. The magnitude of the deflection depends on the stiffness of the shaft and its supports, the total mass of the shaft and its attached parts, and the amount of system damping. The **critical speed** of a rotating shaft, sometimes called the **natural frequency**, is the speed at which the rotating shaft becomes dynamically unstable and large vibrations are likely to develop. For any shaft there are an infinite number of critical speeds, but only the lowest (first) and occasionally the second are generally of interest to designers. The others are usually so high as to be well out of the operating range of shaft speed. This text considers only the first critical speed of the shaft. In this section two approximate methods of finding the first critical speed (or lowest natural frequency) of a system are given, one attributed to Rayleigh and the other to Dunkerley.

11.5.1 SINGLE-MASS SYSTEM

The **first critical speed** (or **lowest natural frequency**) can be obtained by observing the rate of interchange between the kinetic (energy of motion) and potential (energy of position) energies of the system during its cyclic motion. A single mass on a shaft can be represented by the simple spring mass shown in Fig. 11.8. The dashed line indicates the static equilibrium position. The potential energy of the system is

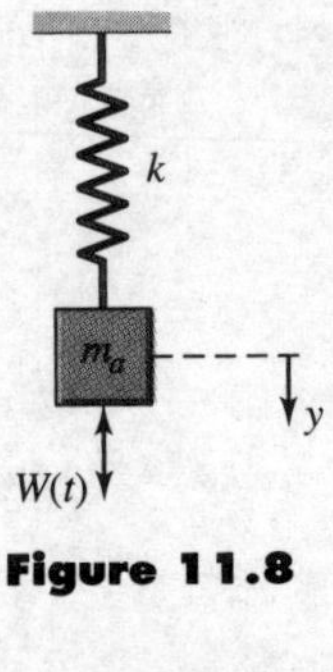

Figure 11.8

$$U = \int_0^{\delta} (m_a g + k\delta)d\delta - m_a g\delta$$

where

m_a = mass, kg
g = gravitational acceleration, 9.807 m/s^2
k = spring rate, N/m
δ = deflection, m

Integrating gives

$$U = \frac{1}{2} k\delta^2 \tag{11.42}$$

The kinetic energy of the system with the mass moving with a velocity of $\dot{\delta}$ is

$$K_e = \frac{1}{2} m_a \left(\dot{\delta}\right)^2 \tag{11.43}$$

Observe the following about Eqs. (11.42) and (11.43):

1. As the mass passes through the static equilibrium position, the potential energy U is zero and the kinetic energy K_e is at a maximum and equal to the total mechanical energy of the system.
2. When the mass is at the position of maximum displacement and is on the verge of changing direction, its velocity is zero. At this point the potential energy U is at a maximum and is equal to the total mechanical energy of the system.

Thus, the total mechanical energy is the sum of the potential and kinetic energies and is constant at any time.

$$\therefore \frac{d}{dt}(U + K_e) = 0 \tag{11.44}$$

Substituting Eqs. (11.42) and (11.43) into Eq. (11.44) gives

$$\frac{d}{dt}\left[\frac{1}{2} k\delta^2 + \frac{1}{2} m_a \left(\dot{\delta}\right)^2\right] = k\delta\dot{\delta} + m_a \dot{\delta}\ddot{\delta} = 0$$

$$\therefore \dot{\delta}\left(m_a \ddot{\delta} + k\delta\right) = 0 \tag{11.45}$$

$$\ddot{\delta} + \omega^2 \delta = 0$$

where

$$\omega = \sqrt{k/m_a}\text{, rad/s} \tag{11.46}$$

The general solution to this differential equation is

$$\delta = A_1 \sin(\omega t + \phi) \tag{11.47}$$

where

A_1 = integration constant

The first critical speed (or lowest natural frequency) is ω.

Substituting Eq. (11.47) into Eqs. (11.42) and (11.43) gives

$$U = \frac{k}{2} A_1^2 \sin^2(\omega t + \phi) \tag{11.48}$$

$$K_e = \frac{m_a}{2} A_1^2 \omega^2 \cos^2(\omega t + \phi) \tag{11.49}$$

Note from Eq. (11.47) that for static deflection if $k = W/\delta$ and $m_a = W/g$, then

$$\omega = \sqrt{\frac{k}{m_a}} = \sqrt{\frac{W/y}{W/g}} = \sqrt{\frac{g}{\delta}} \tag{11.50}$$

11.5.2 MULTIPLE-MASS SYSTEM

From Eq. (11.43) the kinetic energy for n masses is

$$K_e = \frac{1}{2} m_{a1} \left(\dot{\delta}_1\right)^2 + \frac{1}{2} m_{a2} \left(\dot{\delta}_2\right)^2 + \ldots + \frac{1}{2} m_{an} \left(\dot{\delta}_n\right)^2 \tag{11.51}$$

If the deflection is represented by Eq. (11.47), $y_{\max} = A_1$. Also, $\dot{y}_{\max} = A_1\omega = y_{\max}\omega$. Therefore, the above maximum kinetic energy is

$$K_{e,\max} = \frac{\omega^2}{2}\left[m_{a1}\left(\delta_{1,\max}\right)^2 + m_{a2}\left(\delta_{2,\max}\right)^2 + \ldots + m_{an}\left(\delta_{n,\max}\right)^2\right] \tag{11.52}$$

From Eq. (11.42) the potential energy for n masses is

$$U = \frac{1}{2} k_1 \delta_1^2 + \frac{1}{2} k_2 \delta_2^2 + \ldots + \frac{1}{2} k_n \delta_n^2 \tag{11.53}$$

The maximum potential energy is

$$U_{\max} = \frac{1}{2} k_1 \left(\delta_{1,\max}\right)^2 + \frac{1}{2} k_2 \left(\delta_{2,\max}\right)^2 + \ldots + \frac{1}{2} k_n \left(\delta_{n,\max}\right)^2 \tag{11.54}$$

The Rayleigh method assumes that $K_{e,\max} = U_{\max}$ or

$$\frac{1}{2}\omega^2 \sum_{i=1,\ldots,n} m_{ai}\left(\delta_{i,\max}\right)^2 = \frac{1}{2} \sum_{i=1,\ldots,n} k_i \left(\delta_{i,\max}\right)^2$$

$$\omega^2 = \frac{\sum_{i=1,\ldots,n} k_i \left(\delta_{i,\max}\right)^2}{\sum_{i=1,\ldots,n} m_{ai}\left(\delta_{i,\max}\right)^2} \tag{11.55}$$

but

$$k_i = \frac{W_i}{\delta_{i,\max}} \quad \text{and} \quad m_{ai} = \frac{W_i}{g} \tag{11.56}$$

where

W_i = ith weight placed on shaft, N

g = gravitation acceleration, 9.807 m/s^2

Substituting Eq. (11.56) into Eq. (11.55) gives

$$\omega_{\text{cr}} = \sqrt{\frac{g \sum_{i=1,\ldots,n} W_i \delta_{i,\max}}{\sum_{i=1,\ldots,n} W_i \delta_{i,\max}^2}} \tag{11.57}$$

This then is the first critical speed (or first natural frequency) of a multiple-mass system when using the Rayleigh method. Equation (11.57), known as the **Rayleigh equation**, *overestimates* the first critical speed. Because the actual displacements are larger than the static displacements used in Eq. (11.57), the energies in both the denominator and the numerator will be underestimated by the Rayleigh formulation. However, the error in the underestimate will be larger in the denominator, since it involves the square of the approximated displacements. Thus, Eq. (11.57) overestimates (provides an upper bound on) the first critical speed.

The **Dunkerley equation** is another approximation to the first critical speed of a multiple-mass system. It is given as

$$\frac{1}{\omega_{cr}^2} = \frac{1}{\omega_1^2} + \frac{1}{\omega_2^2} + \ldots + \frac{1}{\omega_n^2} \qquad \textbf{(11.58)}$$

where

ω_1 = critical speed if only mass 1 exists
ω_2 = critical speed if only mass 2 exists
ω_n = critical speed if only *n*th mass exists

Recall from Eq. (11.50) that $\omega_i = \sqrt{g/\delta_i}$.

The Dunkerley equation *underestimates* (provides a lower bound on) the first critical speed. The major difference between the Rayleigh and Dunkerley equations is in the deflections. In the Rayleigh equation the deflection at a specific mass location takes into account the deflections due to all the masses acting on the system; in the Dunkerley equation the deflection is due only to the individual mass being evaluated.

Example 11.5

GIVEN Figure 11.9 shows a simply supported shaft arrangement. A solid shaft of 2-in. diameter made of low-carbon steel is used. The following are given: x_1 = 30 in., x_2 = 40 in., x_3 = 20 in., P_A = 80 lbf, and P_B = 120 lbf.

FIND Determine the first critical speed by using

a) The Rayleigh method
b) The Dunkerley method

Solution

From Table 5.1(b) for simply supported ends
For $0 \le x \le a$

$$\delta_x = \frac{Pbx}{6lEI}\left(l^2 - x^2 - b^2\right) \qquad \textbf{(a)}$$

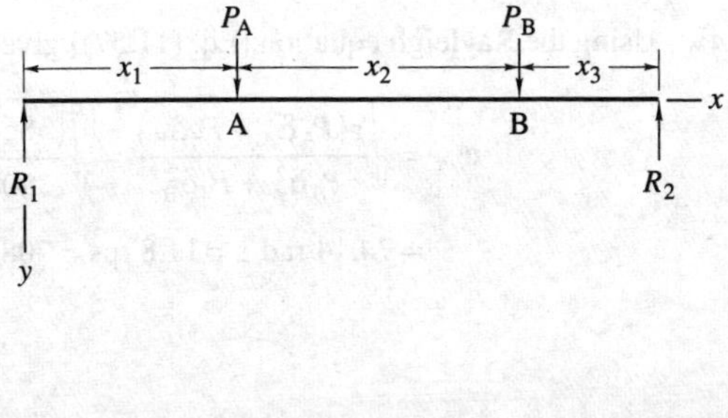

Figure 11.9 Simply supported shaft arrangement used in Example 11.5.

For $a \le x \le l$

$$\delta_x = \frac{Pa(l-x)}{6lEI}\left(2lx - a^2 - x^2\right) \qquad \textbf{(b)}$$

From Table A.1 the modulus of elasticity for carbon steel is 3×10^7 psi. For a solid, round shaft the area moment of inertia is

$$I = \frac{\pi d^4}{64} = \frac{\pi(2)^4}{64} = \frac{\pi}{4} = 0.7854 \text{ in.}^4$$

The deflection at location A due to load P_A from Eq. (a) and $a = x_1 = x = 30$ in. and $b = x_2 + x_3 = 60$ in. is

$$\delta_{AA} = \frac{P_A bx}{6lEI}\left(l^2 - x^2 - b^2\right) = \frac{80(60)(30)\left(90^2 - 30^2 - 60^2\right)}{6(90)(3)\left(10^7\right)(0.7854)} = 0.04074 \text{ in.}$$

Note that in Fig. 11.9 the y direction is downward; thus, δ_{AA} is positive. Also, the first subscript in δ_{AA} designates the location where the deflection occurs, and the second subscript designates the loading that contributes to the deflection.

The deflection at location A due to load P_B from Eq. (a) and $a = x_1 + x_2$, $x = x_1 = 30$ in., and $b = 20$ in. is

$$\delta_{AB} = \frac{P_B bx}{6lEI}\left(l^2 - x^2 - b^2\right) = \frac{120(20)(30)\left(90^2 - 30^2 - 20^2\right)}{6(90)(3)\left(10^7\right)(0.7854)} = 0.03848 \text{ in.}$$

The total deflection at location A is

$$\delta_A = \delta_{AA} + \delta_{AB} = 0.04074 + 0.03848 = 0.07922 \text{ in.}$$

The deflection at location B due to load P_B from Eq. (a) and $x = a = x_1 + x_2 = 70$ in. and $b = 20$ in. is

$$\delta_{BB} = \frac{P_B bx}{6lEI}\left(l^2 - x^2 - b^2\right) = \frac{120(20)(70)\left(90^2 - 70^2 - 20^2\right)}{6(90)(3)\left(10^7\right)(0.7854)} = 0.03697 \text{ in.}$$

The deflection at location B due to load P_A from Eq. (b) and $a = x_1 = 30$ in., $b = x_2 + x_3 = 60$ in., and $x = x_1 + x_2 = 70$ in. is

$$\delta_{BA} = \frac{P_A a(l-x)}{6lEI}\left(2lx - a^2 - x^2\right)$$

$$= \frac{(80)(30)(90-70)\left[2(90)(70) - 30^2 - 70^2\right]}{6(90)(3)\left(10^7\right)(0.7854)} = 0.02565 \text{ in.}$$

The total deflection at location B is

$$\delta_B = \delta_{BA} + \delta_{BB} = 0.02565 + 0.03697 = 0.06262 \text{ in.}$$

a) Using the Rayleigh equation [Eq. (11.57)] gives the first critical speed as

$$\omega_{cr} = \sqrt{\frac{g\left(P_A\delta_A + P_B\delta_B\right)}{P_A\delta_A^2 + P_B\delta_B^2}} = \sqrt{\frac{(386)\left[80(0.07912) + 120(0.06262)\right]}{80(0.07912)^2 + 120(0.06262)^2}}$$

$$= 74.14 \text{ rad/s} = 11.8 \text{ rps} = 708 \text{ rpm}$$

b) Using the Dunkerley equation [Eq. (11.58)] gives

$$\frac{1}{\omega_{cr}^2} = \frac{1}{\omega_{cr,A}^2} + \frac{1}{\omega_{cr,B}^2}$$

where

$$\omega_{cr,A} = \sqrt{\frac{g}{\delta_{AA}}} = \sqrt{\frac{386}{0.04074}} = 97.34 \text{ rad s} = 15.49 \text{ rps} = 929.5 \text{ rpm}$$

$$\omega_{cr,B} = \sqrt{\frac{g}{\delta_{BB}}} = \sqrt{\frac{386}{0.03692}} = 102.2 \text{ rad s} = 16.27 \text{ rps} = 976.4 \text{ rpm}$$

$$\therefore \omega_{cr} = \frac{\omega_{cr,A}\,\omega_{cr,B}}{\sqrt{\omega_{cr,A}^2 + \omega_{cr,B}^2}} = \frac{(929.5)(975.8)}{\sqrt{(929.5)^2 + (975.8)^2}} = 11.22 \text{ rps} = 673 \text{ rpm}$$

The Rayleigh equation gives $\omega_{cr} = 708$ rpm, which overestimates the first critical speed; the Dunkerley equation gives $\omega_{cr} = 673$ rpm, which underestimates the first critical speed. Therefore, the actual first critical speed is between 673 and 708 rpm. The shaft design should avoid this range of operation.

11.6 KEYS

A variety of power-transmitting elements, such as gears, pulleys, and cams, are mounted on rotating shafts. The portion of the mounted member in contact with the shaft is called the **hub**. This section explores one way of attaching the hub to the shaft.

There are a great variety of **keys**, but here the simplest type, a flat key (Fig. 11.10), is analyzed. The main purpose of the key is to prevent motion between the shaft and the connected machine element through which torque is being transmitted. The purpose of using the key is to transmit the *full* torque.

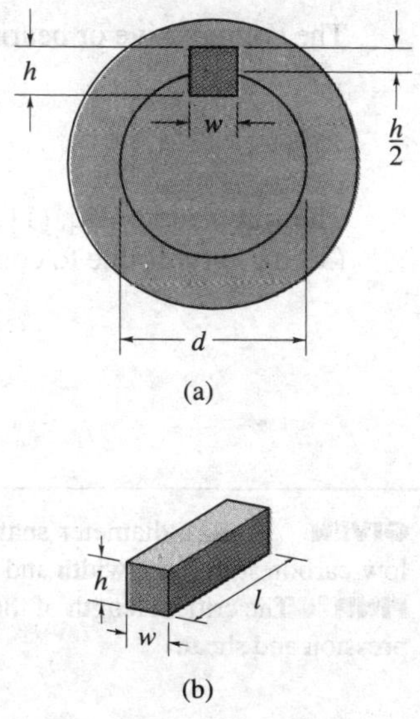

Figure 11.10 Flat key assembly. (a) Cross-sectional view of assembly of shaft, hub, and key; (b) three-dimensional view of flat key.

A key is also intended as a safety system. Most machines have an operating speed and torque that define the required size of the key. However, in the event of a drastic increase in the load conditions, the key will shear before the shaft or machine element (gear, cam, pulley, etc.) fails. Since keys are inexpensive and can be quickly replaced, designers use keys to protect more expensive machinery components.

Note that if $h = w$, the key is square, a special case of a flat key. Keys are usually made of low-carbon steel (AISI 1020) and have a cold-drawn finish. Keys should fit tightly so that key rotation is not possible. In some applications set screws may be required to restrict motion.

Keys fail mainly from shear or compression, both of which are considered here:

1. Failure due to key shear:

$$P = \frac{T}{d/2} = \frac{2T}{d} \quad \textbf{(11.59)}$$

The shear area of the key is $A_s = wl$. The design shear stress is

$$\tau_{\text{design}} = \frac{P}{A_s} = \frac{2T}{dwl} \quad \textbf{(11.60)}$$

The right side of Eq. (11.60) is independent of the height h. To avoid failure due to shear,

$$\tau_{\text{design}} \le \frac{S_{sy}}{n_s} = \frac{\text{Yield stress in shear}}{\text{Safety factor}} \quad \textbf{(11.61)}$$

Recall from Eq. (3.14) that $\tau_{\text{all}} = S_{sy} = 0.40S_y$.

2. Failure due to key compressive or bearing stress: The compression or bearing area is

$$A_c = lh/2 \quad \textbf{(11.62)}$$

The compressive or bearing design stress is

$$\sigma_{\text{design}} = \frac{P}{A_c} = \frac{2T}{dlh/2} = \frac{4T}{dlh} \quad \textbf{(11.63)}$$

The right side of Eq. (11.63) is independent of the width w. While making use of Eq. (3.16), failures due to compressive or bearing stress can be avoided if

$$\sigma_{\text{design}} \le \frac{0.90S_y}{n_s} \quad \textbf{(11.64)}$$

Example 11.6

GIVEN A 4-in.-diameter shaft with a hub is made of high-carbon steel. A square key made of low-carbon steel has a width and height of 1 in. Assume a torque value at a 2-in. radius.

FIND The critical length of the key while assuming a safety factor of 2 and considering both compression and shear.

Solution

From Table A.1 the yield strength for the shaft and hub is $S_y = 55$ ksi. From Eqs. (3.14) and (3.16)

$$\tau_{\text{all}} = S_{ys} = 0.4S_y = 22 \text{ ksi}$$

$$\sigma_{\text{all}} = S_{yc} = 0.9S_y = 49.5 \text{ ksi}$$

Thus, the design stresses are

$$\tau_{\text{design}} = \frac{S_{ys}}{n_s} = \frac{22(10^3)}{2} = 11\ 000 \text{ psi} = 11 \text{ ksi}$$

$$\sigma_{\text{design}} = \frac{S_{yc}}{n_s} = \frac{(49.5)(10^3)}{2} = 24\ 750 \text{ psi} = 24.75 \text{ ksi}$$

For a shaft with a circular cross section

$$J = \frac{\pi d^4}{32} = \frac{\pi(4)^4}{32} = 25.13 \text{ in.}^4$$

The maximum torque acting on the shaft at a radius of 2 in. is

$$T_{\max} = \frac{\tau_{\text{design}} J}{r} = \frac{11(10^3)(23.13)}{2} = 138\ 200 \text{ lbf-in.}$$

The maximum force is

$$P_{\max} = \frac{T_{\max}}{r} = 69\ 100 \text{ lbf}$$

From Table A.1 the yield strength for the key is $S_y = 43$ ksi. From Eqs. (3.14) and (3.16)

$$\tau_{\text{all}} = S_{ys} = 0.4S_y = 17.20 \text{ ksi}$$

$$\sigma_{\text{all}} = S_{yc} = 0.9S_y = 38.70 \text{ ksi}$$

The design stresses are

$$\tau_{\text{design}} = \frac{S_{ys}}{n_s} = \frac{17\ 200}{2} = 8600 \text{ psi}$$

$$\sigma_{\text{design}} = \frac{S_{yc}}{n_s} = \frac{38\ 700}{2} = 19\ 350 \text{ psi}$$

From Eq. (11.60)

$$l_{\text{cr}} = \frac{2T_{\max}}{dw\tau_{\text{design}}} = \frac{2(138\ 200)}{4(1)(8600)} = 8.035 \text{ in.}$$

Thus, to avoid key shear failure, the key should be at least 8.035 in. long. From Eq. (11.63)

$$l_{\text{cr}} = \frac{4T_{\max}}{\sigma_{\text{design}} dh} = \frac{4(138\ 200)}{(19\ 350)(4)(1)} = 7.142 \text{ in.}$$

To avoid key compressive or bearing failure, the key should be at least 7.142 in. long. Therefore, failure will first occur from key shearing, and the key must be at least 8.035 in. long.

11.7 FLYWHEELS

Large variations in acceleration within a mechanism can cause large oscillations in the torque. The peak torque can be so high as to require an overly large motor. However, the average torque over the cycle, due mainly to losses and external work done, may often be much lower than the peak torque. To smooth out the velocity changes and stabilize the back-and-forth flow in energy of rotating equipment, a **flywheel** is often attached to a shaft. The use of a flywheel will therefore allow the following to occur:

1. Reduced amplitude of speed fluctuation
2. Reduced maximum torque required
3. Energy stored and released when needed during cycle

A flywheel with driving (mean) torque T_m and load torque T_l is shown in Fig. 11.11. The flywheel in this figure is a flat circular disk. A motor supplies a torque T_m, which for design purposes should be as constant as possible. The load torque T_l, for such applications as crushing machinery, punch presses, and machine tools, varies considerably.

11.7.1 DYNAMICS

The kinetic energy in a rotating system like that shown in Fig. 11.11 is

$$K_e = I_m \omega^2 / 2 \tag{11.65}$$

where

I_m = mass moment of inertia, kg-m^2
ω = angular speed, rad/s

The mass moments of inertia for six solids are given in Table 4.2, and mass moment of inertia is discussed in Sec 4.2.6. From Newton's second law of motion as applied to the flywheel shown in Fig. 11.11,

$$T_l - T_m = I_m \frac{d\omega}{dt} \tag{11.66}$$

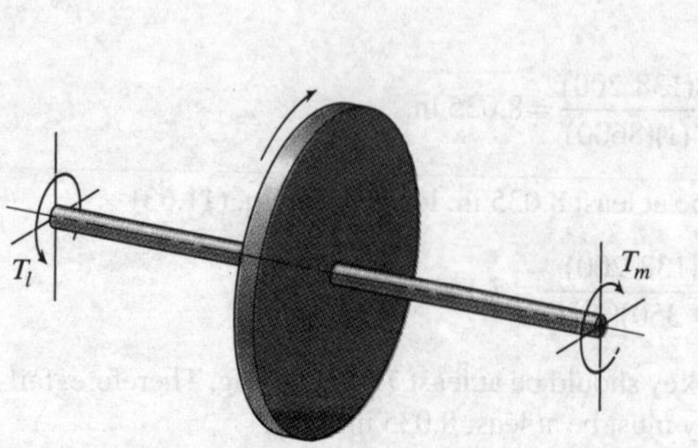

Figure 11.11 Flywheel with driving (mean) torque T_m and load torque T_l.

The design motor torque should be equivalent to the average torque, or $T_m = T_{avg}$. But

$$\frac{d\omega}{dt} = \frac{d\omega}{dt}\frac{d\theta}{d\theta} = \frac{d\theta}{dt}\frac{d\omega}{d\theta} = \omega\frac{d\omega}{d\theta} \tag{11.67}$$

Substituting Eq. (11.67) into Eq. (11.66) gives

$$T_l - T_{avg} = I_m\omega\frac{d\omega}{d\theta}$$
$$\left(T_l - T_{avg}\right)d\theta = I_m\omega d\omega \tag{11.68}$$

This equation can be written in terms of a definite integral as

$$\int_{\theta_{\omega\min}}^{\theta_{\omega\max}}\left(T_l - T_{avg}\right)d\theta = \int_{\omega_{\min}}^{\omega_{\max}} I_m\omega d\omega = \frac{I_m}{2}\left(\omega_{\max}^2 - \omega_{\min}^2\right) \tag{11.69}$$

The *left side* of Eq. (11.69) represents the change in kinetic energy between the maximum and minimum shaft speeds and is equal to the area under the torque-angle diagram between those extreme values of speed. The *far right side* of Eq. (11.69) describes the change in energy stored in the flywheel. The only way to extract energy from the flywheel is to slow it down. Adding energy will speed it up. The best that can be done is to minimize the speed variation $(\omega_{\max} - \omega_{\min})$ by providing a flywheel with a sufficiently large mass moment of inertia I_m.

The location of minimum angular speed $\theta_{\omega_{\min}}$ within a cycle of operation occurs after the maximum positive energy has been delivered from the rotor to the load. The location of maximum angular speed $\theta_{\omega_{\max}}$ within a cycle of operation occurs after the maximum negative energy has been returned to the load, the point where the ratio of the energy summation to the area in the torque pulse is the largest negative value.

11.7.2 FLYWHEEL SIZING

The size of a flywheel required to absorb the energy with an acceptable change in speed needs to be determined. The change in shaft speed during a cycle of operation is called the fluctuation speed ω_f and is expressed as

$$\omega_f = \omega_{\max} - \omega_{\min}$$

The **coefficient of fluctuation** is

$$C_f = \frac{\omega_{\max} - \omega_{\min}}{\omega_{avg}} = \frac{2(\omega_{\max} - \omega_{\min})}{\omega_{\max} + \omega_{\min}} \tag{11.70}$$

Table 11.1 gives coefficients of fluctuation C_f for various types of equipment. The smaller C_f is, the larger the diameter of the flywheel used will have to be. A larger flywheel will add more cost and weight to the system, factors that have to be weighed against the smoothness of the operation desired.

The far right side of Eq. (11.69) describes the kinetic energy of the flywheel and can be rewritten as

$$K_e = \frac{I_m}{2}(\omega_{\max} + \omega_{\min})(\omega_{\max} - \omega_{\min})$$

Table 11.1 Coefficient of fluctuation for various types of equipment. [From *Kent's Mechanical Engineers Handbook: Design & Production*, 12th ed. Copyright © 1969 by Kent. Reprinted by permission of John Wiley & Sons, Inc.]

Type of equipment	Coefficient of fluctuation, C_f
Crushing machinery	0.200
Electrical machinery	0.003
Electrical machinery, direct driven	0.002
Engines with belt transmission	0.030
Flour milling machinery	0.020
Gear wheel transmission	0.020
Hammering machinery	0.200
Machine tools	0.030
Paper-making machinery	0.025
Pumping machinery	0.030–0.050
Shearing machinery	0.030–0.050
Spinning machinery	0.010–0.020
Textile machinery	0.025

Making use of Eq. (11.70) in the above equation gives

$$K_e = I_m \omega_{\text{avg}}^2 C_f \tag{11.71}$$

$$I_m = \frac{K_e}{C_f \omega_{\text{avg}}^2} \tag{11.72}$$

By knowing the desired coefficient of fluctuation for a specific application, obtaining the kinetic energy K_e from the integration of the torque curve, and knowing the average angular velocity, the mass moment of inertia I_m required can be determined. By knowing I_m, the dimension (the diameter) of the flywheel can be determined.

The most efficient flywheel design is obtained by maximizing I_m for the minimum of flywheel material used. Ideally, this design would take the shape of a ring supported by a thin disk, but thin cross sections are extremely difficult to manufacture. Efficient flywheels have their masses concentrated at the largest radius (or in the rim) and their hubs supported on spokes, like bicycle wheels. This configuration puts the greatest part of the mass at the largest radius possible and minimizes the weight for a given mass moment of inertia while making manufacturing problems tractable.

The design procedure for sizing a flywheel is as follows:

1. Plot the load torque T_l versus θ for one cycle.
2. Determine $T_{l,\text{avg}}$ over one cycle.
3. Find the locations $\theta_{\omega_{\max}}$ and $\theta_{\omega_{\min}}$.
4. Determine kinetic energy by integrating the torque curve.
5. Determine ω_{avg}.

6. Determine I_m from Eq. (11.72).
7. Find the dimensions of the flywheel.

Example 11.7

GIVEN The output, or load torque, of a flywheel used in a punch press for each revolution of the shaft is 12 N-m from zero to π and from 3π/2 to 2π and 144 N-m from π to 3π/2. The coefficient of fluctuation is 0.05 about an average speed of 600 rpm. Assume that the flywheel's solid disk is made of low-carbon steel of constant 1-in. thickness.

FIND Determine the following:

a) The average load or output torque
b) The locations $\theta_{\omega_{max}}$ and $\theta_{\omega_{min}}$
c) The energy required
d) The outside diameter of the flywheel

Solution

a) By using the load or output torque variation for one cycle from Fig. 11.12, the average load torque is

$$2\pi T_{l,\text{avg}} = \pi(12) + \frac{\pi}{2}(144) + \frac{\pi}{2}(12)$$

$$T_{l,\text{avg}} = 6 + 36 + 3 = 45 \text{ N-m}$$

b) From Fig. 11.12

$$\theta_{\omega_{max}} = \pi \qquad \text{and} \qquad \theta_{\omega_{min}} = \frac{3\pi}{2}$$

c) From the left side of Eq. (11.69) the kinetic energy for one cycle is

$$K_e = \int_{\theta_{\omega_{min}}}^{\theta_{\omega_{max}}} \left(T_l - T_{\text{avg}}\right) d\theta = (144 - 45)\frac{\pi}{2} = 155.5 \text{ N-m}$$

d) The average angular speed can be expressed in radians per second as

$$\omega_{\text{avg}} = 600 \frac{\text{rev}}{\text{min}} \left(\frac{1 \text{ min}}{60 \text{ s}}\right)\left(\frac{2\pi}{1 \text{ rev}}\right) = 62.83 \text{ rad/s}$$

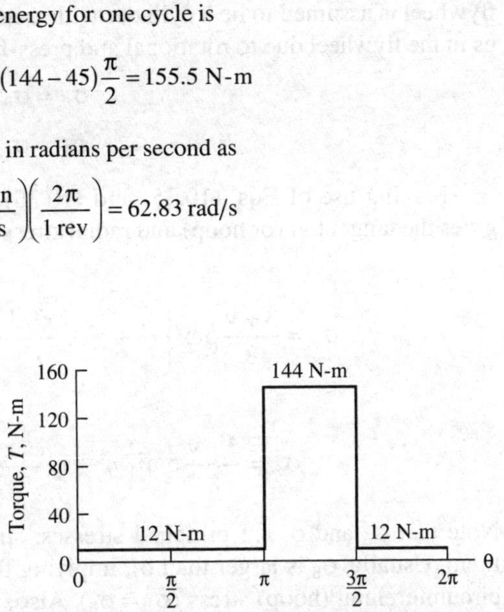

Figure 11.12 Load or output torque variation for one cycle used in Example 11.7.

From Eq. (11.72) the mass moment of inertia is

$$I_m = \frac{K_e}{C_f \omega_{\text{avg}}^2} = \frac{155.5}{(0.05)(62.83)^2} = 0.7879 \text{ kg-m}^2 \tag{a}$$

From Table 4.2 (also given on the inside back cover) the mass moment of inertia for a solid round disk is

$$I_m = \frac{m_a d^2}{8} = \frac{\pi d^2 t_h \rho}{4}\left(\frac{d^2}{8}\right) = \frac{\pi}{32}\rho d^4 t_h$$

For low-carbon steel from Table A.1, $\rho = 7860 \text{ kg/m}^3$.

$$\therefore I_m = \frac{\pi}{32}(7860)(0.0254)d^4 = 19.60 d^4 \tag{b}$$

Equating Eqs. (a) and (b) gives

$$d^4 = \frac{0.7879}{19.60} = 0.0402 \text{ m}^4$$

$$d = 0.4478 \text{ m} = 447.8 \text{ mm} = 17.63 \text{ in.}$$

11.7.3 STRESSES

The total stresses in the flywheel are the sum of the rotational effects and the stresses associated with internal pressurization. Material covered in Chapter 10 will be used, in particular Sec. 10.4 for rotational effects and Sec. 10.3.2.1 for internal pressurization. The flywheel is assumed to be a cylinder with a central hole and constant thickness. The stresses in the flywheel due to rotational and press-fit effects can be written as

$$\sigma_\theta = \sigma_{\theta\omega} + \sigma_{\theta p} \tag{11.73}$$

$$\sigma_r = \sigma_{r\omega} + \sigma_{rp} \tag{11.74}$$

Making use of Eqs. (10.35) and (10.36) and Eqs. (10.23) and (10.24), respectively, gives the tangential (or hoop) and radial stresses as

$$\sigma_\theta = \frac{3+\nu}{8}\rho\omega^2\left[r_i^2 + r_o^2 + \frac{r_i^2 r_o^2}{r^2} - \frac{(1+3\nu)}{(3+\nu)}r^2\right] + \frac{p_i r_i^2\left(1+\frac{r_o^2}{r^2}\right)}{r_o^2 - r_i^2} \tag{11.75}$$

$$\sigma_r = \frac{3+\nu}{8}\rho\omega^2\left(r_i^2 + r_o^2 - \frac{r_i^2 r_o^2}{r^2} - r^2\right) + \frac{p_i r_i^2\left(1-\frac{r_o^2}{r^2}\right)}{r_o^2 - r_i^2} \tag{11.76}$$

Note that σ_θ and σ_r are principal stresses, since there is no shear stress associated with them. Usually, σ_θ is larger than σ_r, implying that the maximum principal stress equals the circumferential (hoop) stress ($\sigma_1 = \sigma_\theta$). Also, the circumferential stress is largest at $r = r_i$ for the flywheel. Thus, for *brittle* materials the design of the flywheel should be such that

$$\sigma_1 < S_u / n_s \tag{11.77}$$

For *ductile* materials a multiaxial failure theory, such as the DET, should be used. If $\sigma_1 = \sigma_\theta$, $\sigma_2 = \sigma_r$, and $\sigma_z = 0$, from Eq. (6.10)

$$\sigma_e = \sqrt{\sigma_\theta^2 + \sigma_r^2 - \sigma_\theta \sigma_r} \qquad \textbf{(6.10)}$$

From Eq. (6.11) the DET predicts failure if

$$\sigma_e \geq \frac{S_y}{n_s} \qquad \textbf{(6.11)}$$

Example 11.8

GIVEN A flywheel made of low-carbon steel has an outside radius of 6 in. and an inside radius of 1 in. A shaft made of low-carbon steel has a radius of 1.003 in. The flywheel is to be assembled onto the shaft and will operate at a speed of 5000 rpm.

FIND Determine the following:

a) Circumferential stresses on the flywheel

b) Radial stresses on the flywheel

c) Speed at which the flywheel will break loose

Solution

From Table A.1 for low-carbon steel

$$\rho = 7860\,\text{kg/m}^3 = 0.2840\,\text{lbm/in.}^2 = 7.358\times10^{-4}\ \text{lbf-s}^2/\text{in.}^4$$

$$E = 3.0\times10^7\ \text{lbf/in.}^2,\ \text{and}\ \nu = 0.30$$

The angular speed is

$$\omega = \frac{2\pi N_a}{60} = \frac{2\pi(5000)}{60} = 523.6\ \text{rad/s}$$

a) Stresses and deformation of the flywheel at the shaft outside radius: For $r = r_o$ and $r_i = 0$ from Eqs. (11.76) and (11.77)

$$\sigma_r = 0$$

$$\sigma_\theta = \frac{\rho\omega^2 r_o^2(1-\nu)}{4} = \frac{(7.358)(10^{-4})(523.6)^2(1.003)^2(1-0.3)}{4} = 35.51\ \text{psi}$$

When the stress is known, the strain can be expressed from Eq. (B.45) as

$$\varepsilon_\theta = \frac{\sigma_\theta}{E} - \frac{\nu\sigma_r}{E} = \frac{\sigma_\theta}{E} = \frac{35.51}{(30)(10^6)} = 1.184\times10^{-6}$$

The deflection can be expressed as

$$\delta = \varepsilon_\theta r = (1.184)(10^{-6})(1.003) = 1.193\times10^{-6}\ \text{in.}$$

It can be concluded that rotational effects do not increase the shaft radius, since δ is so small.

b) Stresses and deformation of the flywheel at the shaft inside radius: The stresses on the inside radius of the flywheel due to rotational effects can be obtained from the first term on the right-hand side of Eqs. (11.75) and (11.76).

$$\sigma_{r\omega} = 0$$

$$\sigma_{\theta\omega} = \frac{(3+\nu)\rho\omega^2}{8}\left[r_i^2 + r_o^2 + \frac{r_i^2 r_o^2}{r^2} - \frac{1+3\nu}{3+\nu}r^2\right]$$

But $r = r_i = 1$ in. and $r_o = 6$ in. Therefore,

$$\sigma_{\theta\omega} = \frac{(3+0.3)(7.358)(10^{-4})(523.6)^2}{8}\left[1+36+36-\frac{(1+0.9)}{(3+0.3)}(1)\right] = 6027 \text{ psi}$$

The expansion of the inside radius of the flywheel due to rotational effects is

$$\delta_\omega = \frac{\sigma_{\theta\omega}}{Er_i} = \frac{6027}{(3)(10^7)(1)} = 2.009\times10^{-4} \text{ in.}$$

Rotational effects expand the inside radius of the flywheel to

$$r_i = 1.000 + 2.009\times10^{-4} = 1.0002 \text{ in.}$$

The interference between the flywheel inside radius and the shaft outside radius is 0.00280 in. (δ_i = 0.0028 in.). From Eq. (10.53) the interference pressure of placing a flywheel on a solid shaft is

$$p_f = \frac{E\delta_r(r_o^2 - r_i^2)}{2r_i r_o^2} = \frac{3(10^7)(2.8)(10^{-3})(36-1)}{2(1)(36)} = 4.083\times10^4 \text{ lbf/in.}^2$$

Substituting the above into Eqs. (11.75) and (11.76) for the dimensions, materials, and flywheel speed given results in the following equations:

$$\sigma_\theta = 4216 + \frac{44\,966}{r^2} - 46.96r^2$$

$$\sigma_r = 4216 - \frac{44\,966}{r^2} - 82.4r^2$$

The circumferential stress is tensile and largest at the inside radius, and the radial stress is compressive and largest at the inside radius.

c) At what speed does the flywheel separate from the shaft? Separation occurs when there is no interference ($\delta = 0$). Recall that the initial interference was $\delta_o = 0.003$ in. Therefore, for the conditions given, the following equations must be satisfied:

$$\delta_f - \delta_s = \delta_o$$

$$\left(\frac{\sigma_\theta}{E}\right)_{\omega f} - \left(\frac{\sigma_\theta}{E}\right)_{\omega s} = \delta_o$$

Making use of Eq. (11.75) (only the ω term) and Eq. (10.35) while recognizing that for the flywheel $r_o = 6$ in. and $r_i = r = 1$ in. and that for the shaft $r_i = 0$ and $r_o = r = 1$ in. gives

$$\omega^2 = \frac{8(0.003)(3)(10^7)}{(72)(7.358)(10^{-4})(3+0.3)}$$

$$\omega = 2118 \text{ rad/s} = 20\,223 \text{ rpm}$$

11.7.4 MATERIALS

The main purpose of a flywheel is to store energy and then use it efficiently. Small flywheels, the type found in children's toys, are often made of lead, but steel and iron are also popular for health reasons. Old steam engines have cast iron flywheels. More recently, flywheels have been proposed for vehicle power storage and regenerative braking systems. High-strength steels and composites will be used for these vehicular flywheels. Thus, a great diversity of materials is being used for flywheels, ranging from lead and cast iron to steel and composites. The obvious question is, What is the best material for use in flywheels?

Recall from Eqs. (11.75) and (11.76) that the speed effect is more significant than the pressurization effect. Thus, the maximum stress is

$$\sigma_{max} = \left(\frac{3+\nu}{8}\right)\rho\omega^2 f(r) \tag{11.78}$$

From Eq. (11.65) the kinetic energy is

$$K_e = \frac{1}{2} I_m \omega^2 \tag{11.79}$$

The mass moment of inertia I_m from Table 4.2 can be represented as

$$I_m = m_a g(r) \tag{11.80}$$

$$\therefore K_e = \frac{1}{2} m_a g(r)\omega^2 \tag{11.81}$$

Assume that $\omega^2 f(r) \approx g(r)\omega^2$. From Eqs. (11.81) and (11.78)

$$\frac{K_e}{m_a} \propto \frac{\sigma_{max}}{\rho}$$

Table 11.2 Materials for flywheels. [M.F. Ashby, *Materials Selection in Mechanical Design*, Pergamon Press, 1992.]

Material		Performance index, M_f kJ/kg	Comment
Ceramics		200–2000 (compression only)	Brittle and weak in tension. Eliminate.
Composites:	CFRP[a]	200–500	The best performance; a good choice.
	GFRP[b]	100–400	Almost as good as CFRP and cheaper; an excellent choice.
Beryllium		300	Good but expensive, difficult to work, and toxic.
High-strength steel		100–200	All about equal in performance; steel and Al-alloys cheaper than Mg and Ti alloys.
High-strength aluminum (Al) alloys		100–200	
High-strength magnesium (Mg) alloys Titanium alloys		100–200	
Lead alloys		3	High density makes these a good (and traditional) selection when performance is velocity limited, not strength limited.
Cast iron		8–10	

[a]Ceramic-fiber-reinforced polymer.

[b]Graphite-fiber-reinforced polymer.

Thus, the best materials for a flywheel are those with high values of the performance index

$$M_f = \frac{\sigma_{max}}{\rho} \quad \text{J/kg} \tag{11.82}$$

Recall that a joule (J) is 1 newton-meter.

Table 11.2 gives performance indices for various flywheel materials. From this table observe that the best materials are ceramics, beryllium, and composites. Lead and cast iron are the worst flywheel materials. The obvious question then is, Why are flywheels for children's toys made of lead? The contrast between these materials is marked, with ceramics, beryllium, and composites being strong and light whereas lead is soft and heavy. Lead is used in children's toys

1. Because flywheels in toys cannot reach burst velocity, since the angular velocity is limited by the drive mechanism, which is generally a pull string
2. Because cost is an important factor, and lead is inexpensive
3. Because the soft feel of lead makes it appealing to children

The important conclusion from these material selection criteria is that a number of factors affect engineering decisions on flywheel materials.

Case Study — DESIGN OF A PUNCH PRESS CAMSHAFT

Mechanical power presses, such as the one shown in Fig. 11.13, are among the oldest and most widely used pieces of equipment in manufacturing. The rating of a mechanical power press is determined by the maximum force that can be developed in a stroke. Thus, the press shown, with a capacity of 80 tons, can develop this force only at the bottom of its stroke, and is referred to as an 80-ton press.

GIVEN It is desired to obtain an 80-ton press with the torque output shown in Fig. 11.14, typical for a sheet-metal-forming operation. The torque curve (Fig. 11.14) is known because the cam has already been designed. The press must be able to cycle 40 times each minute and should not use more than 10% of its energy with each stroke in order to protect against jamming (which leads to catastrophic press failure). The motor drives an 8-in.-wide pinion gear. The flywheel will be a thin disk serving to support a wider cylinder. The cylinder actually has a tooth profile machined into its outer dimension, but this can be safely neglected. The largest bending moment in the camshaft is 100 000 lbf-in.

FIND Determine the following:

a) Whether the coefficient of fluctuation is reasonable

b) A proper flywheel diameter

c) A proper shaft diameter (use a safety factor of 5.0 for the shaft)

Solution

a) Since the machine frame is cast iron, it makes sense to construct the flywheel from cast iron. From Table A.1 the density of cast iron is 7870 kg/m^3, or 0.00885 slug/in.^{3}. The average velocity must be ω_{avg} = 40 rev/min = 4.19 rad/min. The kinetic energy of the rotating flywheel is given by Eq. 11.65) as $I_m\omega^2/2$. It is given that the energy used in each stroke is 10% of the flywheel energy. Therefore,

$$\frac{1}{2}I_m\omega_{min}^2 = 0.9\left(\frac{1}{2}I_m\omega_{max}^2\right) \qquad \omega_{min} = \omega_{max}\sqrt{0.9} \tag{a}$$

Since $\omega_{avg} = (\omega_{min} + \omega_{max})/2 = 4.19$ rad/s, this yields ω_{max} = 4.30 rad/s, ω_{min} = 4.08 rad/s. The coefficient of fluctuation is obtained from Eq. (11.70) as

$$C_f = 2\frac{\omega_{max} - \omega_{min}}{\omega_{max} + \omega_{min}} = 0.0527 \tag{b}$$

From Table 11.1 this coefficient of fluctuation seems high, but it is fairly typical of a stamping press (slightly more

(continued)

Case Study (CONTINUED)

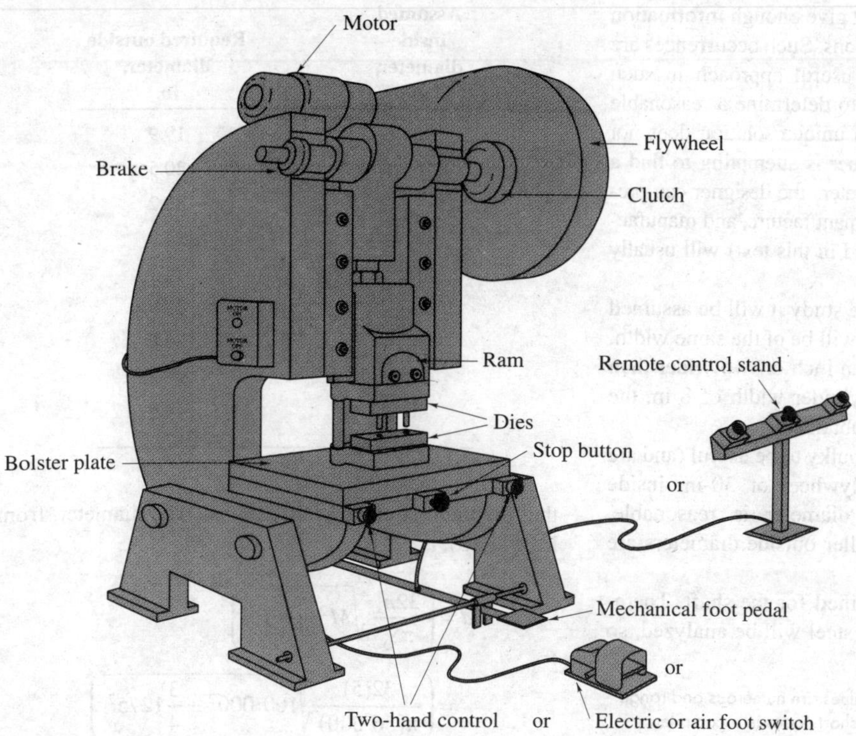

Figure 11.13 Punch press. (Courtesy of Triodyne, Inc.)

demanding than shearing machinery). Thus, the coefficient of fluctuation is not a design concern.

b) It is given that the flywheel should be a cylinder mounted on a disk; in practice, it is quite common to use fairly thin disks, so that the disk's contribution to flywheel inertia is negligible. Thus, the moment of inertia can be taken as (see Table 4.2):

$$I_m = \left(\frac{\pi l \rho\left(d_o^2 - d_i^2\right)}{4}\right)\left(\frac{d_o^2 - d_i^2}{8}\right) \qquad \textbf{(c)}$$

The additional inertia presented by the disk will only serve to give the press an even larger capacity, so that this approximation is consistent with a worst-case scenario approach.

From the torque curve the energy used per stroke of the press is

$$K_e = \int T d\theta = \frac{1}{2}\left(\frac{\pi}{2}\right)(T_{\max}) = 1000 \text{ in.-lbf} \qquad \textbf{(d)}$$

Also,

$$K_e = I_m \omega_{\text{avg}}^2 C_f \qquad \textbf{(e)}$$

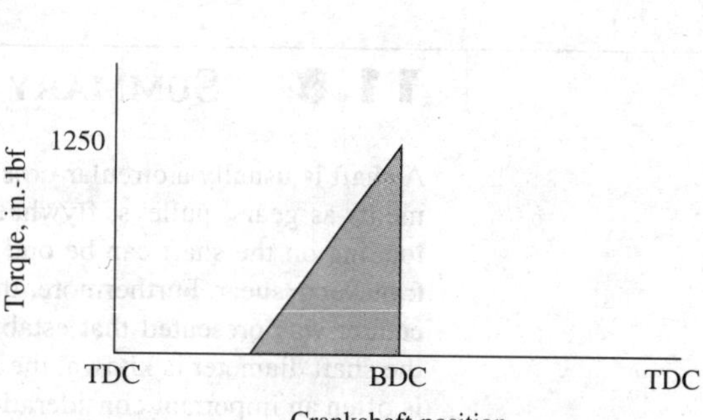

Figure 11.14 Torque requirements of mechanical press. (TDC = top dead center, BDC = bottom dead center.)

(continued)

Case Study (CONCLUDED)

Equations (c), (d), and (e) do not give enough information to solve for the flywheel dimensions. Such occurrences are routine in design problems. A useful approach in such cases is to vary the parameters to determine a reasonable combination of the unknowns. A unique solution does not exist, and at this point the designer is attempting to find a good solution. By varying parameters the designer can present a number of alternatives for manufacture, and manufacturing considerations (not covered in this text) will usually help determine a solution.

For the purposes of this case study it will be assumed that the flywheel and the pinion will be of the same width, although the flywheel is often an inch or two wider than the pinion. Regardless, for a cylinder width of 8 in. the data in the table oppsite can be obtained:[†]

Assumed inside diameter, in.	Required outside diameter, in.
0	19.9
5	20.5
10	22.2
15	24.9
20	28.2
25	31.9
30	36.0
35	40.2
40	44.7

Larger flywheels tend to be too bulky to be useful (and are difficult to manufacture). A flywheel of 30-in. inside diameter and 36-in. outside diameter is reasonable, although slightly larger or smaller outside diameters are reasonable as well.

c) No material has been specified for the shaft. Therefore, an AISI 1015 cold-drawn steel will be analyzed, so that the yield strength is 56 000 psi. The diameter, from Eq. (11.13), is

$$d = \left(\frac{32 n_s}{\pi S_y} \sqrt{M^2 + \frac{3}{4} T^2} \right)^{1/3}$$

$$= \left(\frac{32(5)}{\pi(56\ 000)} \sqrt{100\ 000^2 + \frac{3}{4} 1275^2} \right)$$

$$= 1.96 \text{ in.}$$

So a 2-in. shaft is used.

[†]The approaches to obtaining the values are numerous and range from calculating by hand to writing short codes in a compiler language to using commercial mathematics packages. The proliferation of computers and software has made solving such problems routine but defining the best approach to a solution impossible.

11.8 SUMMARY

A shaft is usually a circular-cross-section rotating member that has such power-transmitting elements as gears, pulleys, flywheels, sprockets, and rolling-element bearings mounted on it. The loading on the shaft can be one or various combinations of bending, torsion, shock, or axial or transverse shear. Furthermore, these types of loading can be either static or cyclic. A design procedure was presented that established the appropriate shaft diameter for specific conditions. If the shaft diameter is known, the safety factor (or the smallest diameter where failure first occurs) is often an important consideration. Three failure prediction theories considering important combinations of loading were presented. These failure theories are by no means all-inclusive but should provide the essential understanding from which any other considerations can be easily obtained.

Shaft dynamics and in particular the first critical speed are important to design, since the rotating shaft becomes dynamically unstable and large vibrations are likely to develop. Both the Rayleigh and the Dunkerley equations for determining the first critical speed of a multiple-mass system were presented. The Rayleigh overestimates and the Dunkerley underestimates the exact solution, thus providing a range of operating speeds that the design should avoid.

The last section was on flywheels. In some applications large variations in acceleration occur that can cause large oscillations in torque. The peak torque can be so high as to require an overly large motor. A flywheel is often used to smooth out the velocity changes and stabilize the back-and-forth energy flow of rotating equipment. A procedure for designing a flywheel was presented as well as flywheel dynamics, sizing, stresses, and material selection.

KEY WORDS

coefficient of fluctuation dimensionless speed, $\dfrac{\omega_{max} - \omega_{min}}{\omega_{avg}}$

critical speed speed at which rotating shaft becomes dynamically unstable

Dunkerley equation relation for first critical speed that underestimates frequency

first critical speed lowest frequency at which dynamic instability occurs

flywheel element that stores energy through rotational inertia

hub portion of member mounted onto shaft that directly contacts shaft

key element that transmits power from shaft to hub

lowest natural frequency same as first critical speed

natural frequency same as critical speed

Rayleigh equation relation for first critical speed that overestimates frequency

shaft rotating or stationary member, usually of circular cross section with small diameter relative to length and used to transmit power through such elements as gears, sprockets, pulleys, and cams

RECOMMENDED READINGS

Juvinall, R.C., and Marshek, K.M. (1991) *Fundamentals of Machine Component Design*, Wiley, New York.

Mott, R. L. (1985) *Machine Elements in Mechanical Design*, Merrill Publishing Co., Columbus, OH.

Norton, R.L. (1996) *Machine Design*, Prentice-Hall, Englewood Cliffs, NJ.

Oberg, E., and Jones, F.D. (1966) *Machinery's Handbook*, 17th ed., Industrial Press, New York.

Piotrowski, J. (1986) *Shaft Alignment Handbook*, Marcel Dekker, New York.

Shigley, J.E., and Mischke, C.R. (1989), *Mechanical Engineering Design*, McGraw-Hill, New York.

REFERENCES

Ashby, M.F. (1992) *Materials Selection in Mechanical Design*, Pergamon Press, Oxford, England.

Kent's Mechanical Engineers Handbook (1969), 12th ed., Design and Production, p. 740.

Peterson, R.E. (1974) *Stress Concentration Factors*, Wiley, New York.

PROBLEMS

Section 11.2

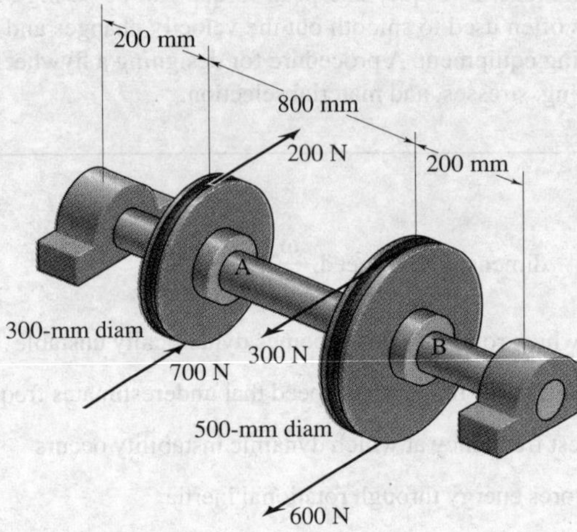

Sketch *a*, for Problem 11.1

11.1 A shaft assembly shown in sketch *a* is driven by a flat belt at location A and drives a flat belt at location B. The drive belt pulley diameter is 300 mm; the driven belt pulley diameter is 500 mm. The belts are long relative to the sheave diameters. The distance between sheaves is 800 mm, and the distance from each sheave to the nearest bearing is 200 mm. The belts are horizontal and load the shaft in opposite directions. Determine the size of the shaft and the types of steel that should be used. Assume a safety factor of 5.

Section 11.3

11.2 The gears in the shaft assembly shown in Fig. 11.1(a) transmit 100 kW of power and rotate at 3600 rpm. Gear wheel 1 is loaded against another gear such that the force P_1 acts in an upward direction at a radius of 80 mm from the shaft center. The force P_2 acts vertically downward at a radius of 110 mm from the shaft center. The distance from bearing A to gear 1 is 100 mm, that from gear 1 to gear 2 is 85 mm, and that from gear 2 to bearing B is 50 mm. Do the following:

a) Draw a free-body diagram with forces acting on the shaft when bearings A and B transmit only radial forces.

b) Give values of force components as well as the resultant force at locations A and B.

c) Give the transmitted torque.

d) Draw a bending moment diagram in the *x*-*y* and *x*-*z* planes along with a torque diagram. Also, indicate what the maximum bending moment and the maximum torque should be.

e) Find the safety factor according to the distortion-energy theory (DET) and the maximum-shear-stress theory (MSST) if the shaft has a diameter of 35 mm and is made of high-carbon steel (AISI 1080).

11.3 The shaft assembly given in Problem 11.2 has an extra loading from thermal expansion of the shaft. The bearings are assumed to be rigid in the axial direction, so that when the shaft heats up it cannot elongate but instead compressive stress builds up. Determine the thermal stress and find the safety factor by using the DET if the shaft heats up

a) 5°C
b) 15°C

11.4 Given the shaft assembly in Problem 11.2 but calculating as if the AISI 1080 steel were brittle [using the maximum-normal-stress theory (MNST)], find the safety factor n_s for fatigue by using the information in Problem 11.2.

11.5 A bridge used for sightseeing is built in the form of a T, where the bottom of the T is fastened into the rock face. The top of the T is situated 7 m horizontally from the fastening point, and the length of the cross bar is 8 m, half on each side of the fastening point. The bridge was built to carry 40 persons with an average weight of 75 kg/person. Assume that the weight is evenly distributed along the crossbar and a safety factor of 8. How large will the safety factor be if all the people gather on one end of the cross bar? The load-carrying structure in the bridge is a circular tube with outer diameter d_o and inner diameter d_i. Use the DET and neglect the weight of the bridge.

11.6 Find the change in safety factor if the load-carrying structure in the bridge in Problem 11.5 is a beam with solid rectangular section of height h and width b instead of the hollow cylinder. Assume that $h/b = 6$ and the shear stress is

$$\tau_{max} = \frac{T(b/2)}{J^*} = \frac{T(b/2)}{0.298hb^3} = \frac{T}{0.596hb^2}$$

11.7 A ladder to a pleasure boat has two vertical aluminum sides to carry the load from the steps bolted to them. By mistake the bolts on one side of one step were changed to black steel and therefore corroded away. Calculate how much the safety factor decreases if people walk at the middle of the steps and all steps except the bad one are fixed to both vertical sides. The vertical sides are much less elastic than the steps. Use MSST. The length of a step (width of the ladder) is 0.600 m, and its cross section is 15 mm thick and 150 mm wide.

Section 11.4

11.8 Gears 3 and 4 act on the shaft shown in sketch *b*. The resultant gear force, $P_A = 600$ lbf, acts at an angle of 20° from the y axis. The yield stress for the shaft, which is made of cold-drawn steel, is 71 000 psi and the ultimate stress is 85 000 psi. The shaft is solid and of constant diameter. The safety factor is 2.6. Assume the DET throughout. Also, for fatigue loading

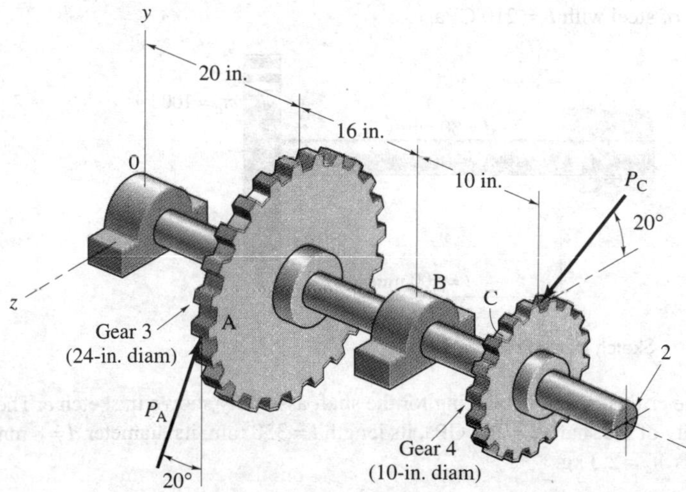

Sketch *b*, for Problem 11.8

conditions assume completely reversed bending with a bending moment amplitude equal to that used for static conditions. The mean torque is zero and the Goodman relationship is assumed. Determine the safe shaft diameter due to static and fatigue loading. Show shear and moment diagrams in the various planes.

11.9 Derive Eq. (11.39) of the text. Start by showing the stresses acting on an oblique plane at angle ϕ. The MNST should be used, implying that the critical stress line extends from S_e/n_s to S_u/n_s.

11.10 The shaft shown in sketch *c* rotates at 7200 rpm and transfers 6 kW of power from input gear A to output gears B and C. The spur gears A and C have pressure angles of 20°. The helical gear has a pressure angle of 20° and a helix angle of 30° and transfers 70% of the input power. All important surfaces are ground. All dimensions in millimeters.

a) Draw a free-body diagram as well as the shear and moment diagrams of the shaft.
b) Which bearing should support the thrust load?
c) Determine the minimum shaft diameter for a safety factor of 2 and 99% reliability.

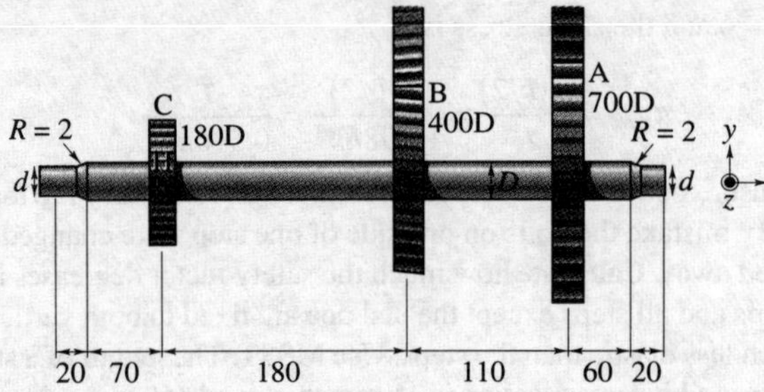

Sketch *c*, for Problem 11.10

11.11 In Problem 11.10 if the shaft diameter is 60 mm and the shaft is made of AISI 1080 cold-drawn steel, what is the safety factor while assuming 90% reliability?

Section 11.5

11.12 The rotor shown in sketch *d* has a stiff bearing on the left. Find the critical speed when the shaft is made of steel with $E = 210$ GPa.

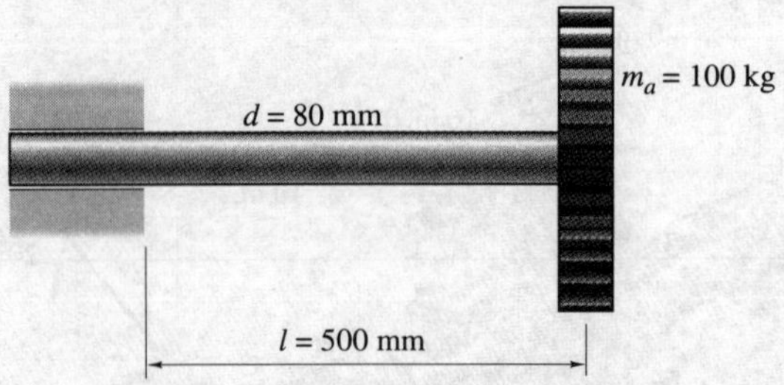

Sketch *d*, for Problem 11.12

11.13 Determine the critical speed in bending for the shaft assembly shown in sketch *e*. The modulus of elasticity of the shaft $E = 207$ GPa, its length $l = 350$ mm, its diameter $d = 8$ mm, and the rotor mass $m_a = 2.3$ kg.

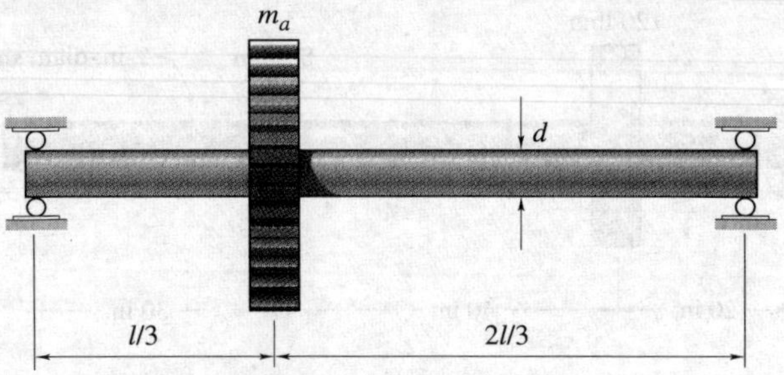

Sketch e, for Problem 11.13

11.14 Calculate the diameter of the shaft in the assembly shown in sketch *f* so that the first critical speed is 9000 rpm. The shaft is made of steel with $E = 207$ GPa. The distance $a = 300$ mm and the mass $m_a = 100$ kg. The mass of the shaft is neglected. Use both the Rayleigh and Dunkerley methods.

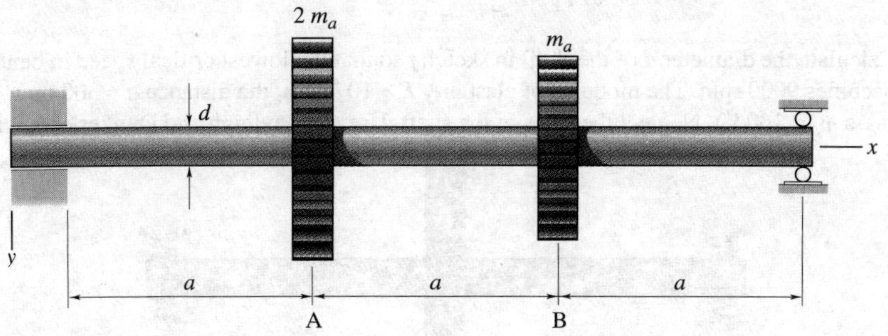

Sketch f, for Problem 11.14

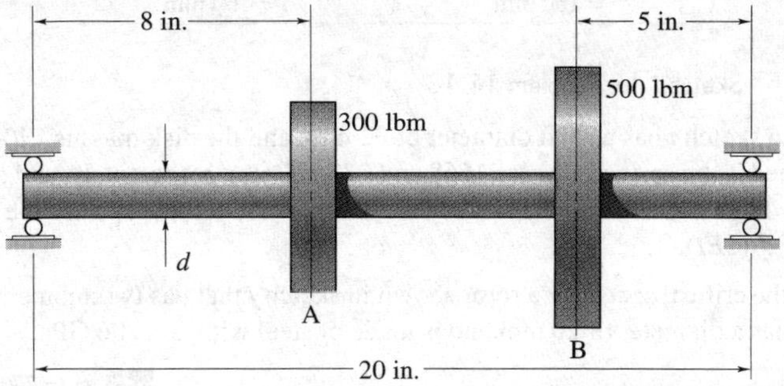

Sketch g, for Problem 11.15

11.15 The simple support shaft shown in sketch *g* has two weights on it. Neglect the shaft weight. Using the Rayleigh method for the stainless steel ($E = 26 \times 10^6$ psi) shaft, determine the safe diameter to ensure that the first critical speed is no less than 3600 rpm.

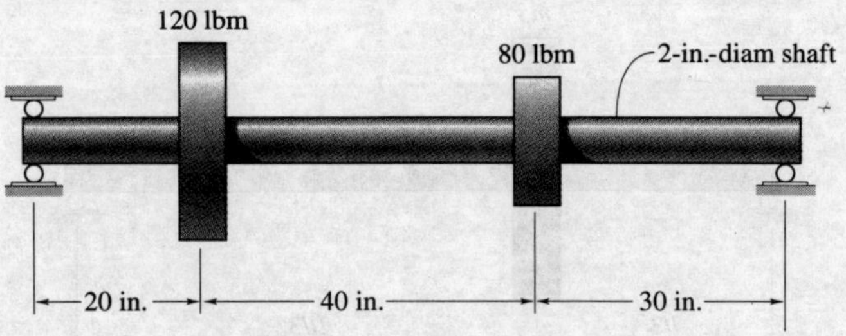

Sketch *h*, for Problem 11.16

11.16 Determine the first critical speed by the Dunkerley and Rayleigh methods for the steel shaft shown in sketch *h*. Neglect the shaft mass. The area moment of inertia is $I = \pi r^4/4$, where $r =$ the shaft radius. The method of superposition may be used with the following given:

$$\delta = \frac{P}{6EI}\left[\frac{bx^3}{l}\right] - \langle x-a\rangle^3 - \frac{xb\left(l^2-b^2\right)}{l}$$

11.17 Calculate the diameter d of the shaft in sketch *f* so that the lowest critical speed in bending becomes 9000 rpm. The modulus of elasticity $E = 107$ GPa, the distance $a = 300$ mm, and the mass $m_a = 100$ kg. Neglect the mass of the shaft. Use the Rayleigh and Dunkerley methods.

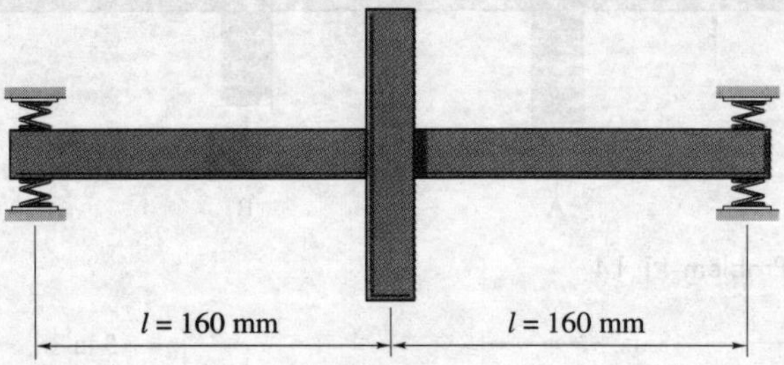

Sketch *i*, for Problem 11.18

11.18 The rotor in sketch *i* has a shaft diameter of 32 mm, and the disk mass is 170 kg. Calculate the critical speed if the spring rates are 1668 and 3335 N/mm, respectively, and are the same in all directions. The elastic modulus $E = 206$ GPa. For a shaft without springs the influence coefficient $\alpha_{11} = l^3/(6EI)$.

11.19 Calculate the critical speed for a rotor shown in sketch *j* that has two moment-free bearings. The shaft has a diameter of 20 mm and is made of steel with $E = 206$ GPa.

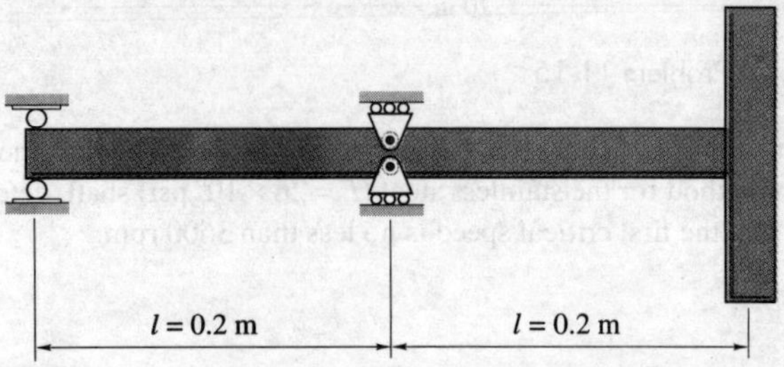

Sketch *j*, for Problem 11.19

Section 11.6

11.20 A flywheel has a hub made of aluminum alloy 2014. The hub is connected to the AISI 1040 steel shaft with a flat, 15-mm-long key made of AISI 1020 steel. Dimension the key so that it can transmit 20 N-m of torque with a safety factor of 3.

11.21 A jaw crusher is used to crush iron ore down to the particle size needed for iron ore pellet production. The pellets are later used in a blast furnace to make steel. If the iron ore pieces coming into the crusher are too large, the crusher is protected by a torque-limiting key made of copper. The torque on the 400-mm-diameter shaft connecting the flywheel with the jaw mechanism should never go above 700 kN-m. Dimension the copper key so that the shaft is not damaged when an oversized ore particle comes into the crusher.

Section 11.7

11.22 The output torque of a flywheel for each revolution of a shaft is 10 N-m from 0 to π and 120 N-m from π to 2π. The coefficient of fluctuation is 0.04 at an average speed of 2000 rpm. Assume that the flywheel disk is an AISI 1040 steel plate of 30-mm constant thickness. Determine the following:

a) Average load torque
b) Locations $\theta_{\omega_{max}}$ and $\theta_{\omega_{min}}$
c) Energy required
d) Outside diameter of flywheel

11.23 The output torque of a flywheel for each revolution of the shaft is 200 in.-lbf from 0 to $\pi/3$, 1600 in.-lbf from $\pi/3$ to $2\pi/3$, 400 in.-lbf from $2\pi/3$ to π, 900 in.-lbf from π to $5\pi/3$, and 200 in.-lbf from $5\pi/3$ to 2π. Input torque is assumed to be constant. The average speed is 860 rpm. The coefficient of fluctuation is 0.10. Find the diameter of the flywheel if it is cut from a 1-in.-thick steel plate.

11.24 A 20-mm-thick flywheel is made of aluminum alloy 2014 and runs at 9000 rpm in a racing car motor. What is the safety factor if the aluminum alloy 2014 is stressed to a quarter of its yield limit at 9000 rpm? To decrease the flywheel outer diameter, higher density exotic materials are investigated. Find the best material from Table A.2 that can be substituted for aluminum alloy 2014 but with the same safety factor. The flywheel is machined from a solid piece of aluminum alloy 2014 with no central hole, and the thickness cannot be larger than 20 mm.

11.25 The aluminum flywheel considered in Problem 11.24 has a mass moment of inertia of 0.0025 kg-m^2. By mistake the motor is accelerated to 7000 rpm in neutral gear, and the throttle sticks in the fully open position. The only way to stop the motor is to disconnect the electric lead from the spark plug. It takes 6 s before the motor is motionless. Neglecting the inertia of all movable parts in the motor except the flywheel, calculate the internal friction moment in the motor and the (mechanical part of the) friction losses in the motor at 5000 rpm. The friction moment is assumed to be constant at all speeds.

11.26 A one-cylinder ignition bulb motor to an old fishing boat has a flywheel that gives the motor a coefficient of fluctuation of 25% when it idles at 180 rpm. The mass moment of inertia for the flywheel is 1.9 kg-m^2. Determine the coefficient of fluctuation at 500 rpm if the compression stroke consumes equally large energy at all speeds. Also, calculate the mass moment of inertia needed to get 20% coefficient of fluctuation at 500 rpm.

11.27 A flywheel for a city bus drive should store as much energy as possible for a given flywheel weight. The diameter must be less than 1.5 m, and the mass must be smaller than 250 kg. Find which material from Tables A.1 and A.2 gives the highest possible stored energy for a safety factor of 4 if the flywheel has constant thickness.

11.28 A flywheel on an AISI 1080 steel shaft is oscillating due to disturbances from a combustion engine. The engine is a four-stroke engine with six cylinders, so that the torque disturbances come three times per revolution. The flywheel shaft has a diameter of 20 mm and is 1 m long, and the flywheel moment of inertia is 0.5 kg-m^2. Find the engine speed at which the large torsional vibrations will appear. The shaft material is AISI 1080 steel.

CHAPTER

12

HYDRODYNAMIC AND HYDROSTATIC BEARINGS AND SEALS

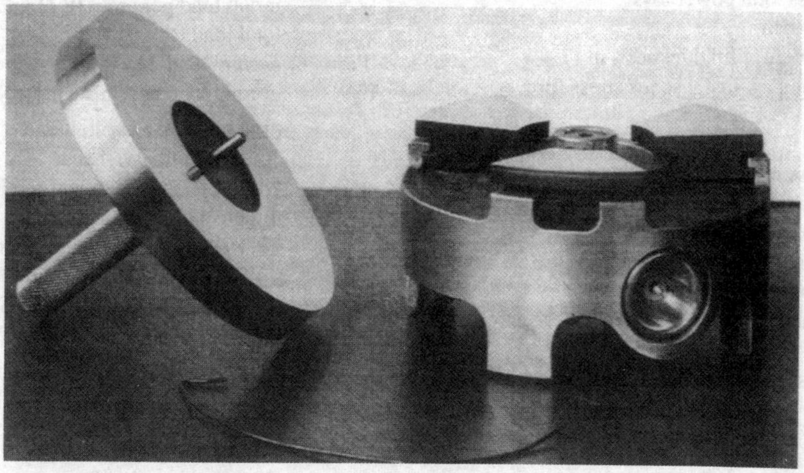

A Kingsbury Bearing. (Albert Kingsbury's Company, 1926)

A cup of tea, standing in a dry saucer, is apt to slip about in an awkward manner, for which a remedy is found in introduction of a few drops of water, or tea, wetting the parts in contact.

Lord Rayleigh (1918)

SYMBOLS

A	area, m^2
$\bar{A}$	integration constant
$\tilde{A}$	integration constant
B	bearing damping constant
B_j	bearing number for journal slider bearing, $\left(\frac{\eta_0 \omega_b r_b w_t}{\pi W_r}\right)\left(\frac{r_b}{c}\right)^2$
B_t	bearing number for thrust slider bearing, $\left(\frac{\eta_0 u_b w_t}{W_z}\right)\left(\frac{l}{s_h}\right)^2$
$\bar{B}$	integration constant
$\tilde{B}$	integration constant
$\tilde{C}$	integration constant
C_p	specific heat of lubricant at constant pressure, J/(kg-°C)
C_s	volumetric specific heat, J/(m^3-°C)

$C_1, \ldots, C_4$ integration constants
c radial clearance in journal bearing, m
c_b bearing clearance at pad minimum film thickness, m
c_s stiffness coefficient
$\tilde{D}$ integration constant
d diameter, m
e eccentricity of journal bearing, m
F friction force, N
g acceleration due to gravity, 9.807 m/s^2 (386.1 in./s^2)
H dimensionless film thickness, h/s_h
H_o outlet film thickness ratio, h_o/s_h
H_p dimensionless total power loss
h film thickness, m
h_o outlet (minimum) film thickness, m
h_p rate of working against viscous stress (power loss), W
k slope of film thickness
L_p dimensionless pivot location, l_p/l
l length in sliding direction, m
l_p length to pivot from inlet, m
m_p preload factor
N_a rotational speed, rps
N_p number of pads
n_s step location
P dimensionless pressure, $\frac{ps_h^2}{\eta_0 u_b l}$
$P_{\max}$ dimensionless maximum film pressure, $\frac{W_r}{2r_b w_t p_{max}}$
p pressure, Pa
p_l lift pressure, Pa
$p_{\max}$ maximum film pressure, Pa
p_r recess pressure, Pa
p_s supply pressure, Pa
Q dimensionless volumetric flow rate, $2\pi q/r_b c w_t \omega_b$
q circumferential volumetric flow rate, m^3/s
q_s side-leakage volumetric flow rate, m^3/s
q_x volumetric flow rate in sliding direction, m^3/s
q_r' radial flow rate per unit width, m^2/s
q_x' volumetric flow rate per unit width in sliding direction, m^2/s
q_y' volumetric flow rate per unit width in transverse direction, m^2/s
R_a centerline average surface roughness
$\bar{R}$ gas constant, m^2/s^2–°C
r radius, m
r_b radius of journal bearing, m
s_h shoulder height, m
T friction torque, N-m
t time, s
Δt time change, s
t_m absolute temperature, °C
$\tilde{t}_m$ mean temperature, °C
Δt_m temperature change, °C
t_w thickness, m
u fluid velocity in sliding direction, m/s
$\tilde{u}$ average velocity in sliding direction, m/s
V volume, m^3
v fluid velocity in transverse direction, m/s
W_r radial load, N
W_r^* radial load per area, $W_r/2r_b w_t$, Pa
W_t total load, N
W_z normal load, N
W_z' normal load per unit width, N/m
$\overline{W}_z$ dimensionless normal load
w squeeze velocity, m/s
w_t width (in side-leakage direction) of bearing, m
X dimensionless coordinate
x Cartesian coordinate in direction of sliding, m
y Cartesian coordinate in side-leakage direction, m
z Cartesian coordinate across film, m
α inclination angle of fixed-incline slider bearing, deg
α_a offset factor
Δ step height, m
ε eccentricity ratio, e/c
η absolute viscosity, Pa-s
η_0 absolute viscosity at $p = 0$ and constant temperature, Pa-s
λ length-to-width ratio, l/w_t
λ_j diameter-to-width ratio, $2r/w_t$
μ coefficient of sliding friction
ρ density, kg/m^3
τ shear stress, Pa
Φ attitude angle, deg
ϕ cylindrical polar coordinate
$\phi_{\max}$ location of maximum pressure, deg
ϕ_0 location of terminating pressure, deg
ω angular velocity, $2\pi N_a$, rad/s

Subscripts

a solid a (upper surface)
all allow
atm atmospheric
b solid b (lower surface)
d disk

i	inlet; inner	o	outlet; outer
m	mean	p	punch
max	maximum	x, y, z	Cartesian coordinates
min	minimum	1,2	regions in thrust slider bearing

12.1 INTRODUCTION

The characterization of **hydrodynamic lubrication** was presented in Sec. 8.3.1 and will not be repeated here. For the bearings considered in this chapter it will be assumed that **fluid film lubrication** occurs, or that the lubricated surfaces are separated by a fluid film. Also, the bearings considered in this chapter are assumed to have conformal surfaces, which were considered in Sec. 8.2. These three topics (characterization of hydrodynamic lubrication, fluid film lubrication, and conformal surfaces), which were covered in Chapter 8, should be understood before proceeding with this chapter.

The history of hydrodynamic lubrication begins with the classical experiments of Tower (1883), who detected the existence of a film from measurements of pressure within the lubricant, and of Petrov (1883), who reached the same conclusion from friction measurements. This work was closely followed by Reynolds' (1886) celebrated analytical paper in which he used a reduced form of the Navier-Stokes equations in association with the continuity equation to generate a second-order differential equation for the pressure in the narrow, converging gap (or conjunction) between bearing surfaces. This pressure enables a load to be transmitted between the surfaces with extremely low friction, since the surfaces are completely separated by a fluid film. In such a situation the physical properties of the lubricant, notably the dynamic viscosity, dictate the behavior in the conjunction.

Designing hydrodynamic bearings is an excellent opportunity to use the basic knowledge obtained from fluid mechanics. Making the connection between bearing design and fluid mechanics is extremely important in the development of an engineer. This chapter hopes to provide such an experience.

This chapter presents the most important aspects of hydrodynamic and hydrostatic bearings. The Reynolds equation is needed to understand these topics. The derivation of the Reynolds equation is presented, and the physical significance of its various terms is explained. The important aspects of self-lubricated thrust and journal bearings and seals are also presented. Useful design charts are given and how these design charts were obtained is explained. The chapter ends with brief considerations of normal squeeze film, hydrostatic, and gas-lubricated bearings.

12.2 REYNOLDS EQUATION

The full Navier-Stokes equation covered in most fluid mechanics texts [e.g., see Currie (1993)] contains inertia, body, pressure, and viscous terms. These equations are sufficiently complicated to prohibit analytical solutions to most practical problems. There is, however, a class of flow conditions known as slow viscous motion in which the pressure and viscous terms predominate. Fluid film lubrication problems are in this class.

The differential equation governing the pressure distribution in fluid film lubrication is known as the **Reynolds equation.** This equation was first derived in a remarkable paper by Osborne Reynolds in 1886. Reynolds' classic paper not only contained the basic differential equation of fluid film lubrication but also directly compared his theoretical predictions with the experimental results obtained by Tower (1883).

12.2.1 DERIVATION OF REYNOLDS EQUATION

From an order-of-magnitude analysis the general Navier-Stokes equations [see any fluid mechanics text, e.g., Currie (1993)] can be considerably reduced. If $h_{\min}$ is the minimum film thickness and is in the z direction, if l is the length and is in the x direction (the direction of motion), and if w_t is the width and is in the y direction, then in hydrodynamic and hydrostatic bearing situations l and w_t are several (usually three) orders of magnitude larger than the minimum film thickness. Thus, if terms of order $(h/l)^2$, $(h/w_t)^2$, h/l, and h/w_t and smaller are neglected and only first-order terms are considered, the reduced Navier-Stokes equations are

$$\frac{\partial p}{\partial x} = \frac{\partial}{\partial z}\left(\eta \frac{\partial u}{\partial z}\right) \tag{12.1}$$

$$\frac{\partial p}{\partial y} = \frac{\partial}{\partial z}\left(\eta \frac{\partial v}{\partial z}\right) \tag{12.2}$$

$$\frac{\partial p}{\partial z} = 0 \quad \rightarrow \quad p = f(x, y, t) \tag{12.3}$$

where

η = absolute viscosity, Pa-s

u = fluid velocity in x direction, m/s

v = fluid velocity in y direction, m/s

p = pressure, Pa

Equations (12.1) and (12.2) can be directly integrated to give the general expressions for the velocity gradients

$$\frac{\partial u}{\partial z} = \frac{z}{\eta}\frac{\partial p}{\partial x} + \frac{\tilde{A}}{\eta} \tag{12.4}$$

$$\frac{\partial v}{\partial z} = \frac{z}{\eta}\frac{\partial p}{\partial y} + \frac{\tilde{C}}{\eta} \tag{12.5}$$

where

$\tilde{A}$ and $\tilde{C}$ = integration constants

The lubricant viscosity might change considerably across the thin film (z direction) as a result of temperature variations that arise in some bearing problems. In this case, progress toward a simple Reynolds equation is considerably complicated. An approach that is satisfactory in most fluid film applications is to treat η as the average viscosity across the film. This approach, which *does not* restrict variation in the x and y directions, is pursued here.

With η representing the average viscosity across the film, integrating Eqs. (12.4) and (12.5) gives the velocity components as

$$u = \frac{z^2}{2\eta}\frac{\partial p}{\partial x} + \tilde{A}\frac{z}{\eta} + \tilde{B} \qquad \textbf{(12.6)}$$

$$v = \frac{z^2}{2\eta}\frac{\partial p}{\partial y} + \tilde{C}\frac{z}{\eta} + \tilde{D} \qquad \textbf{(12.7)}$$

If zero slip at the fluid-solid interface is assumed, the boundary values for the velocity are

1. $z = 0 \quad u = u_b \quad v = v_b$
2. $z = h \quad u = u_a \quad v = v_a$

The subscripts a and b refer to conditions on the upper and lower surfaces, respectively. Therefore, u_a, v_a, and w_a refer to the velocity components of the upper surface in the x, y, and z directions, respectively, and u_b, v_b, and w_b refer to the velocity components of the lower surface in the same directions.

With the boundary conditions applied to Eqs. (12.6) and (12.7) the velocity components are

$$u = -z\left(\frac{h-z}{2\eta}\right)\frac{\partial p}{\partial x} + \frac{u_b(h-z)}{h} + \frac{u_a z}{h} \qquad \textbf{(12.8)}$$

$$v = -z\left(\frac{h-z}{2\eta}\right)\frac{\partial p}{\partial y} + \frac{v_b(h-z)}{h} + \frac{v_a z}{h} \qquad \textbf{(12.9)}$$

The volume flow rates per unit width in the x and y directions are defined as

$$q'_x = \int_0^h u\,dz \qquad \textbf{(12.10)}$$

$$q'_y = \int_0^h v\,dz \qquad \textbf{(12.11)}$$

Substituting Eqs. (12.8) and (12.9) into these equations gives

$$q'_x = -\frac{h^3}{12\eta}\frac{\partial p}{\partial x} + \frac{h(u_a + u_b)}{2} \qquad \textbf{(12.12)}$$

$$q'_y = -\frac{h^3}{12\eta}\frac{\partial p}{\partial y} + \frac{h(v_a + v_b)}{2} \tag{12.13}$$

The first term on the right side of Eqs. (12.12) and (12.13) represents the well-known **Poiseuille** (or pressure) **flow,** and the second term represents the **Couette** (or velocity) **flow.**

From a standard fluid mechanics text [see Currie (1993)] the continuity equation can be expressed as

$$\frac{\partial \rho}{\partial t} + \frac{\partial}{\partial x}(\rho u) + \frac{\partial}{\partial y}(\rho v) + \frac{\partial}{\partial z}(\rho w) = 0 \tag{12.14}$$

where

ρ = density, kg/m^3
w = squeeze velocity, m/s

The density ρ expressed in Eq. (12.14) and the absolute viscosity η expressed in Eqs. (12.8) and (12.9) are functions of x and y but not of z. It is convenient to express the continuity equation [(Eq. 12.14)] in integral form as

$$\int_0^h \left[\frac{\partial \rho}{\partial t} + \frac{\partial}{\partial x}(\rho u) + \frac{\partial}{\partial y}(\rho v) + \frac{\partial}{\partial z}(\rho w)\right] dz = 0 \tag{12.15}$$

Now, a general rule of integration is that

$$\int_0^h \frac{\partial}{\partial x}\left[f(x,y,z)\right] dz = -f(x,y,h)\frac{\partial h}{\partial x} + \frac{\partial}{\partial x}\left[\int_0^h f(x,y,z)\,dz\right] \tag{12.16}$$

Thus, the u component of Eq. (12.15) becomes

$$\int_0^h \frac{\partial}{\partial x}(\rho u) dz = -(\rho u)_{z=h}\frac{\partial h}{\partial x} + \frac{\partial}{\partial x}\left(\int_0^h \rho u dz\right) = -\rho u_a \frac{\partial h}{\partial x} + \frac{\partial}{\partial x}\left(\rho \int_0^h u dz\right) \tag{12.17}$$

Similarly, for the v component

$$\int_0^h \frac{\partial}{\partial y}(\rho v) dz = -\rho v_a \frac{\partial h}{\partial y} + \frac{\partial}{\partial y}\left(\rho \int_0^h v dz\right) \tag{12.18}$$

The w component term can be integrated directly to give

$$\int_0^h \frac{\partial}{\partial z}(\rho w) dz = \rho(w_a - w_b) \tag{12.19}$$

Therefore, after substituting Eqs. (12.17) to (12.19) into Eq. (12.15) the integrated continuity equation becomes

$$h\frac{\partial \rho}{\partial t} - \rho u_a \frac{\partial h}{\partial x} + \frac{\partial}{\partial x}\left(\rho \int_0^h u\,dz\right) - \rho v_a \frac{\partial h}{\partial y} + \frac{\partial}{\partial y}\left(\rho \int_0^h v\,dz\right) + \rho(w_a - w_b) = 0 \quad \textbf{(12.20)}$$

The integrals in this equation represent the volume flow per unit width (q_x' and q_y') given in Eqs. (12.12) and (12.13). Introducing these flow rate expressions into the integrated continuity equation yields the general Reynolds equation

$$0 = \frac{\partial}{\partial x}\left(-\frac{\rho h^3}{12\eta}\frac{\partial p}{\partial x}\right) + \frac{\partial}{\partial y}\left(-\frac{\rho h^3}{12\eta}\frac{\partial p}{\partial y}\right) + \frac{\partial}{\partial x}\left[\frac{\rho h(u_a + u_b)}{2}\right] + \frac{\partial}{\partial y}\left[\frac{\rho h(v_a + v_b)}{2}\right]$$

$$+ \rho(w_a - w_b) - \rho u_a \frac{\partial h}{\partial x} - \rho v_a \frac{\partial h}{\partial y} + h\frac{\partial \rho}{\partial t} \quad \textbf{(12.21)}$$

Example 12.1

GIVEN A piston pump without rings pumps an oil with a viscosity of 0.143 Pa-s at a pressure differential of 20 MPa. The piston moves concentrically in the cylinder so that the radial clearance is constant around the piston. The piston is 100 mm in diameter and 80 mm long and the radial clearance is 50 μm. The piston moves with a speed of 0.2 m/s.

FIND Determine the leakage past the piston when

a) The piston pumps the oil (like a pump)

b) The oil drives the piston (like a hydraulic cylinder)

Solution

The pressure gradient is

$$\frac{\partial p}{\partial x} = \frac{\Delta p}{\Delta x} = \frac{(20)(10^6)}{0.080} = 2.5 \times 10^8 \ \text{N/m}^3 \quad \textbf{(a)}$$

a) When the piston pumps the oil, the cylinder wall passes the piston from high pressure to low pressure. From Eq. (12.12) the pressure flow and the motion flow together are

$$q = \pi d q_x' = \pi d\left[\frac{h^3}{12\eta}\frac{\partial p}{\partial x} + \frac{h(u_a + u_b)}{2}\right]$$

$$= \pi(0.100)\left[\frac{(50)^3(10^{-18})(2.5)(10^8)}{(12)(0.143)} + \frac{(50)(10^{-6})(0.2)}{2}\right] = 7.292 \times 10^{-6} \ \text{m}^3/\text{s}$$

b) When the oil drives the piston, the pressure flow is opposite to the flow from the motion

$$q = \pi d\left[-\frac{h^3}{12\eta}\frac{dp}{dx} + \frac{h(u_a + u_b)}{2}\right]$$

$$= \pi(0.100)\left[-\frac{(50)^3(10^{-18})(2.5)(10^8)}{12(0.143)} + \frac{(50)(10^{-6})(0.2)}{2}\right] = -4.150 \times 10^{-6}\ \text{m}^3/\text{s}$$

Thus, when the piston pumps the oil, the leakage is 7.292 cm^3/s; and when the oil drives the piston, the leakage is –4.150 cm^3/s. Therefore, the cylinder wall motion pulls oil into the high-pressure region.

12.2.2 Physical Significance of Terms in Reynolds Equation

The first two terms in Eq. (12.21) are the Poiseuille terms and describe the net flow rates due to pressure gradients within the lubricated area; the third and fourth terms are the Couette terms and describe the net entraining flow rates due to surface velocities. The fifth to seventh terms describe the net flow rates due to a squeezing motion, and the last term describes the net flow rate due to local expansion. The flows or "actions" can be considered, without any loss of generality, by eliminating the side-leakage terms ($\partial/\partial y$) in Eq. (12.21).

$$\underbrace{\frac{\partial}{\partial x}\left(\frac{\rho h^3}{12\eta}\frac{\partial p}{\partial x}\right)}_{\text{Poiseuille}} = \underbrace{\frac{\partial}{\partial x}\left[\frac{\rho h(u_a + u_b)}{2}\right]}_{\text{Couette}} + \underbrace{\rho\left(w_a - w_b - u_a\frac{\partial h}{\partial x}\right)}_{\text{Squeeze}} + \underbrace{h\frac{\partial \rho}{\partial t}}_{\text{Local expansion}} \tag{12.22}$$

where the Couette term expands as

$$\underbrace{+\frac{h(u_a + u_b)}{2}\frac{\partial \rho}{\partial x}}_{\text{Density wedge}} \quad \underbrace{+\frac{\rho h}{2}\frac{\partial}{\partial x}(u_a + u_b)}_{\text{Stretch}} \quad \underbrace{+\frac{\rho(u_a + u_b)}{2}\frac{\partial h}{\partial x}}_{\text{Physical wedge}}$$

Evidently, the Couette terms leads to three distinct actions. The physical significance of each term within the Reynolds equation is now discussed in detail.

12.2.2.1 Density Wedge Term $[(u_a + u_b)h/2](\partial\rho/\partial x)$ **Density wedge** is concerned with the rate at which lubricant density changes in the sliding direction as shown in Fig. 12.1. If the lubricant density decreases in the sliding direction, the Couette mass flows differ for each location at the three velocity profiles of Fig. 12.1. For continuity of mass flow this discrepancy must be eliminated by generating a balancing Poiseuille flow.

Note from Fig. 12.1 that the density must decrease in the sliding direction if positive pressures are to be generated. This effect could be introduced by raising the temperature of

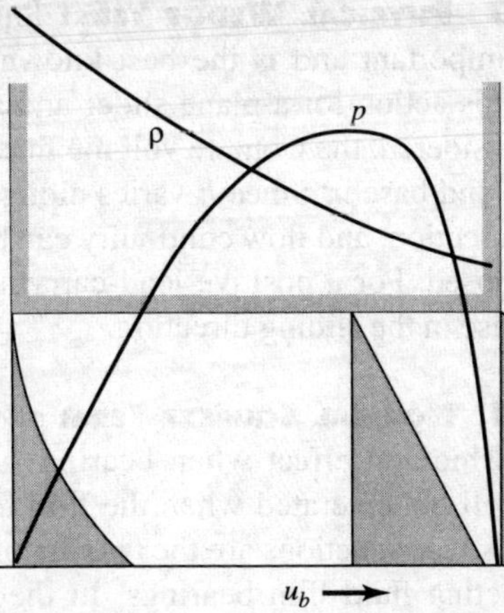

Figure 12.1 Density wedge mechanism. [B. J. Hamrock, *Fundamentals of Fluid Film Lubrication*, NASA Reference Publication 1255, 1991.]

the lubricant as it passes through the bearing. The density wedge (sometimes called the thermal wedge) mechanism is not important in most bearings. It has been suggested that it may play a significant role in the performance of parallel-surface thrust bearings, where the major pressure-generating actions are absent.

12.2.2.2 STRETCH TERM $(\rho h/2)[\partial(u_a + u_b)/\partial x]$ **Stretch** action concerns the rate at which the surface velocity changes in the sliding direction. This effect is produced if the bounding solids are elastic and the extent to which the surfaces are stretched varies through the bearing. For positive pressures to be developed, the surface velocities have to decrease in the sliding direction, as shown in Fig. 12.2. Stretch action is not encountered in conventional bearings.

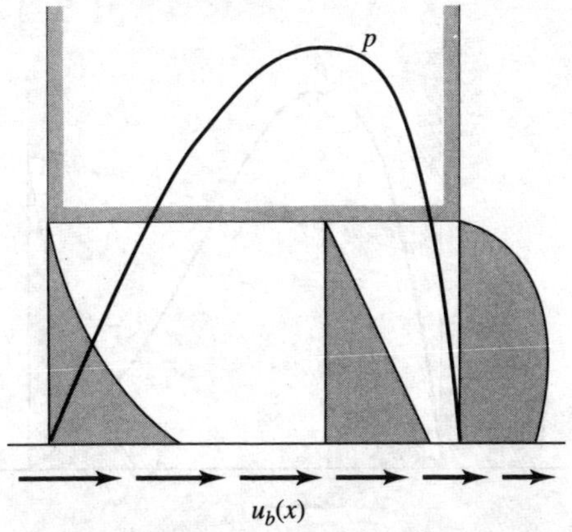

Figure 12.2 Stretch mechanism. [B. J. Hamrock, *Fundamentals of Fluid Film Lubrication*, NASA Reference Publication 1255, 1991.]

12.2.2.3 PHYSICAL WEDGE TERM $[\rho(u_a + u_b)/2](\partial h/\partial x)$ **Physical wedge** action is extremely important and is the best known device for generating pressure. Figure 12.3 illustrates this action for a plane slider and a stationary bearing pad. At each of the three sections considered, the Couette volume flow rate is proportional to the area of the triangle of height h and base u. Since h varies along the bearing, there is a different Couette flow rate at each section, and flow continuity can be achieved only if a balancing Poiseuille flow is superimposed. For a positive load-carrying capacity the thickness of the lubricant film must decrease in the sliding direction.

12.2.2.4 NORMAL SQUEEZE TERM $\rho(w_a - w_b)$ **Normal squeeze** action provides a valuable cushioning effect when bearing surfaces tend to be pressed together. Positive pressures will be generated when the film thickness is diminishing. The physical wedge and normal squeeze actions are the two major pressure-generating devices in hydrodynamic or self-acting fluid film bearings. In the absence of sliding the squeeze effect arises directly from the difference in normal velocities $(w_a - w_b)$ as illustrated in Fig. 12.4. Positive pressures will clearly be achieved if the film thickness is decreasing $(w_b > w_a)$.

12.2.2.5 TRANSLATION SQUEEZE TERM $-\rho u_a(\partial h/\partial x)$ **Translation squeeze** action results from the translation of inclined surfaces. The local film thickness may be squeezed by the sliding of the inclined bearing surface, as shown in Fig. 12.5. The rate at which the film thickness is decreasing is shown in the figure. In this case the pressure profile is moving over the space covered by the fixed coordinate system, the pressure at any fixed point being a function of time.

12.2.2.6 LOCAL EXPANSION TERM $h(\partial\rho/\partial t)$ The local time rate of density change governs the **local expansion** term. The pressure-generating mechanism can be visualized by considering the thermal expansion of the lubricant contained between stationary bearing surfaces, as shown in Fig. 12.6. If heat is supplied to the lubricant, it will expand and the excess volume will have to be expelled from the space between the bearing surfaces. In

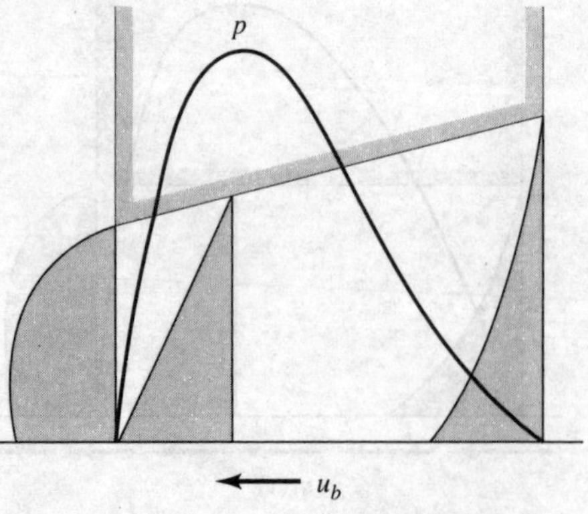

Figure 12.3 Physical wedge mechanism. [B. J. Hamrock, *Fundamentals of Fluid Film Lubrication*, NASA Reference Publication 1255, 1991.]

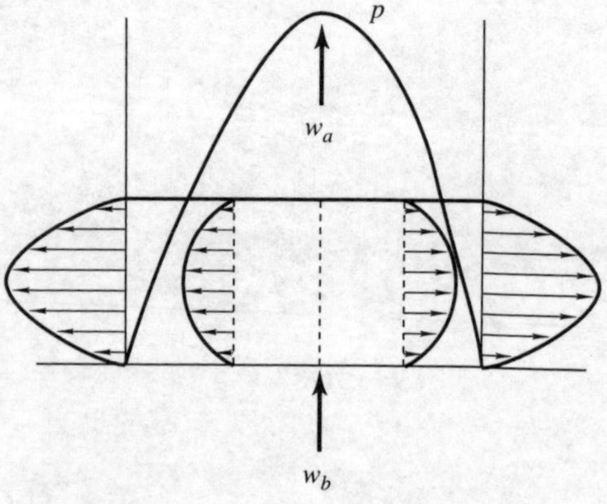

Figure 12.4 Normal squeeze mechanism. [B. J. Hamrock, *Fundamentals of Fluid Film Lubrication*, NASA Reference Publication 1255, 1991.]

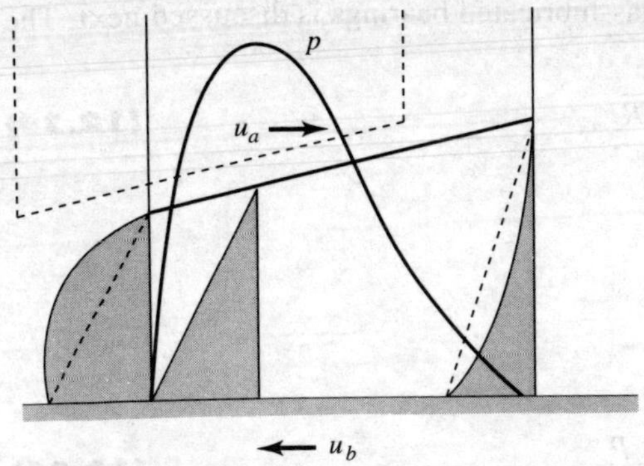

Figure 12.5 Translation squeeze mechanism. [B. J. Hamrock, *Fundamentals of Fluid Film Lubrication*, NASA Reference Publication 1255, 1991.]

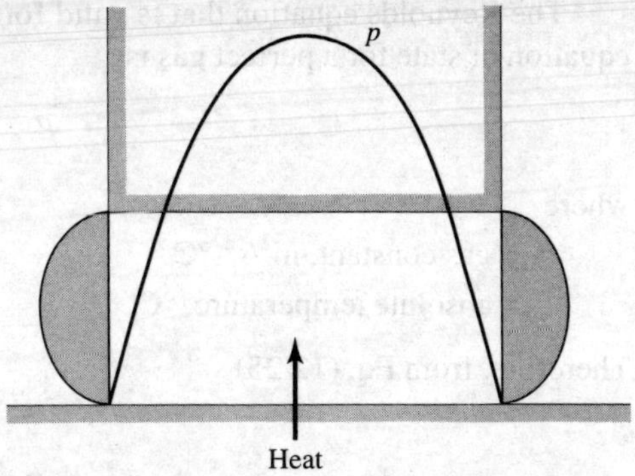

Figure 12.6 Local expansion mechanism. [B. J. Hamrock, *Fundamentals of Fluid Film Lubrication*, NASA Reference Publication 1255, 1991.]

the absence of surface velocities the excess lubricant volume must be expelled by a pressure (Poiseuille) flow action. Pressures are thus generated in the lubricant, and for a positive load-carrying capacity $\partial\rho/\partial t$ must be negative (i.e., the volume of a given mass of lubricant must increase). Local expansion, which is a transient mechanism of pressure generation, is generally of no significance in bearing analysis.

12.2.3 Standard Reduced Forms

For only tangential motion, where $w_a = u_a \partial h/\partial x$ and $w_b = 0$, the Reynolds equation [Eq. (12.21)] becomes

$$\frac{\partial}{\partial x}\left(\frac{\rho h^3}{\eta}\frac{\partial p}{\partial x}\right)+\frac{\partial}{\partial y}\left(\frac{\rho h^3}{\eta}\frac{\partial p}{\partial y}\right)=12\tilde{u}\frac{\partial(\rho h)}{\partial x} \tag{12.23}$$

where

$$\tilde{u} = \frac{u_a + u_b}{2} = \text{constant}$$

This equation is applicable for elastohydrodynamic lubrication. For hydrodynamic lubrication the fluid properties do not vary significantly throughout the bearing and may be considered to be a constant. Thus, the corresponding Reynolds equation is

$$\frac{\partial}{\partial x}\left(h^3\frac{\partial p}{\partial x}\right)+\frac{\partial}{\partial y}\left(h^3\frac{\partial p}{\partial y}\right)=12\tilde{u}\eta_0\frac{\partial h}{\partial x} \tag{12.24}$$

Equation (12.23) not only allows the fluid properties to vary in the x and y directions but also permits the bearing surfaces to be of finite length in the y direction. Side leakage, or flow in the y direction, is associated with the second term in Eqs. (12.23) and (12.24). If the pressure in the lubricant film has to be considered as a function of x and y, Eq. (12.23) can rarely be solved analytically.

The Reynolds equation that is valid for gas-lubricated bearings is discussed next. The equation of state for a perfect gas is

$$p = \rho \bar{R} t_m \tag{12.25}$$

where

$\bar{R}$ = gas constant, m²/s²-°C

t_m = absolute temperature, °C

Therefore, from Eq. (12.25)

$$\rho = \frac{p}{\bar{R} t_m} \tag{12.26}$$

Substituting this equation into Eq. (12.23) yields the Reynolds equation normally used for gas-lubricated bearings for tangential motion only. Because the viscosity of a gas does not vary much, it can be considered to be constant.

$$\frac{\partial}{\partial x}\left(ph^3 \frac{\partial p}{\partial x}\right) + \frac{\partial}{\partial y}\left(ph^3 \frac{\partial p}{\partial y}\right) = 12\tilde{u}\eta_0 \frac{\partial (ph)}{\partial x} \tag{12.27}$$

12.3 THRUST SLIDER BEARINGS

The surfaces of **thrust bearings** are perpendicular to the axis of rotation. A hydrodynamically lubricated slider bearing develops load-carrying capacity by virtue of the relative motion of the two surfaces separated by a fluid film. A slider bearing may be viewed as a bearing that develops a positive pressure due to the physical wedge mechanism presented in Sec. 12.2.2.3. Thrust slider bearings are considered in this section and journal slider bearings in the next main section (Sec. 12.4). The processes occurring in a bearing with fluid film lubrication can be better understood by considering qualitatively the development of oil pressure in a thrust slider bearing.

12.3.1 MECHANISM OF PRESSURE DEVELOPMENT

An understanding of the development of load-supporting pressures in hydrodynamic bearings can be gleaned by considering, from a purely physical point of view, the conditions of geometry and motion required to develop pressure. An understanding of the physical situation can make the mathematics of hydrodynamic lubrication much more meaningful. By considering only what must happen to maintain continuity of flow, much of what the mathematical equations tell us later in this chapter can be deduced.

Figure 12.7 shows velocity profiles for two plane surfaces separated by a constant lubricating film thickness. The plates are extremely wide so that side-leakage flow (into and out of the paper) can be neglected. The upper plate is moving with a velocity u_a, and the bottom plate is held stationary. No slip occurs at the surfaces. The velocity varies uniformly from zero at surface AB to u_a at surface A′B′, thus implying that the rate of shear du/dz throughout the oil

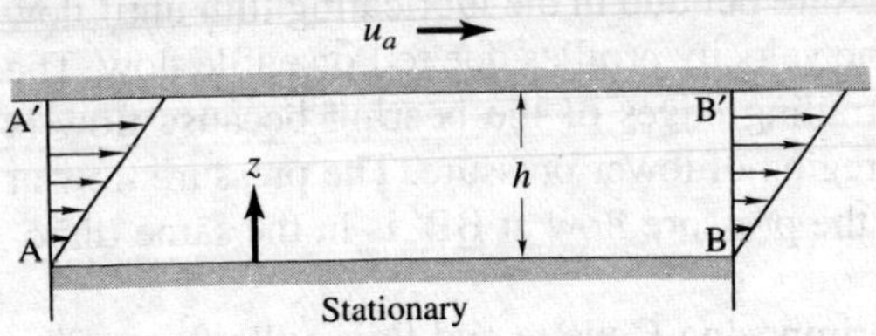

Figure 12.7 Velocity profiles in a parallel-surface slider bearing. [B. J. Hamrock, *Fundamentals of Fluid Film Lubrication*, NASA Reference Publication 1255, 1991.]

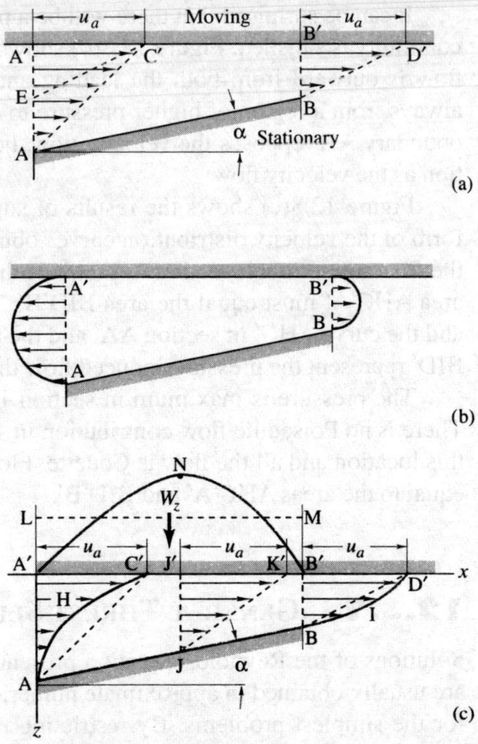

Figure 12.8 Flow within a fixed-incline slider bearing.
(a) Couette flow; (b) Poiseuille flow; (c) resulting velocity profile. [B. J. Hamrock, *Fundamentals of Fluid Film Lubrication*, NASA Reference Publication 1255, 1991.]

film is constant. The volume of fluid flowing across section AA′ in unit time is equal to that flowing across section BB′. The flow crossing the two boundaries results only from velocity gradients, and since the gradients are equal, the flow continuity requirement is satisfied without any pressure buildup within the film. Because the ability of a lubricating film to support a load depends on the pressure buildup in the film, a slider bearing with parallel surfaces is not able to support a load by a fluid film. If any load is applied to the surface AB, the lubricant will be squeezed out and the bearing will operate under conditions of boundary lubrication.

Consider now the case of two nonparallel plates as shown in Fig. 12.8(a). Again, the plates are wide in the direction perpendicular to the motion, so that lubricant flow in this direction is negligibly small. The volume of lubricant that the surface A′B′ tends to carry into the space between the surfaces AB and A′B′ through section AA′ during unit time is AC′A′. The volume of lubricant that the surface tends to discharge from the space through section BB′ during the same time is BD′B′. Because the distance AA′ is greater than the distance BB′, the volume AC′A′ is greater than the volume BD′B′ by the volume AEC′. From flow continuity the actual volume of oil carried into the space must equal the volume discharged from this space.

It can be surmised that there will be a pressure buildup in the lubricating film until flow continuity is satisfied. Figure 12.8(b) shows the velocity profiles due to Poiseuille flow. The flow is outward from both the leading and trailing edges of the bearing because flow is always from a region of higher pressure to a region of lower pressure. The pressure flow at boundary AA′ opposes the velocity flow, but the pressure flow at BB′ is in the same direction as the velocity flow.

Figure 12.8(c) shows the results of superimposing Couette and Poiseuille flows. The form of the velocity distribution curves obtained in this way must satisfy the condition that the flow rate through section AA′ equals the flow rate through section BB′. Therefore, the area AHC′A′ must equal the area BID′B′. The area between the straight, dashed line AC′ and the curve AHC′ in section AA′ and the area between the dashed line BD′ and the curve BID′ represent the pressure-induced flow through these areas.

The pressure is maximum in section JJ′, somewhere between sections AA′ and BB′. There is no Poiseuille flow contribution in section JJ′, since the pressure gradient is zero at this location and all the flow is Couette. Flow continuity is satisfied in that triangle JK′J′ is equal to the areas AHC′A′ and BID′B′.

12.3.2 General Thrust Slider Bearing Theory

Solutions of the Reynolds equation presented in Sec. 12.2 for real bearing configurations are usually obtained in approximate numerical form. Analytical solutions are possible only for the simplest problems. By restricting the flow to two dimensions, say the *x-z* plane, analytical solutions for many common bearings become available. The quantitative value of these solutions is limited, since flow in the third dimension *y,* which is known as *side leakage,* plays an important part in fluid film bearing performance. The two-dimensional solutions have a definite value because they provide a good deal of information about the general characteristics of bearings, information that leads to a clear physical picture of the performance of the lubricating films. The focus of this text is on design and thus this material will not be presented, but it can be found in Hamrock (1993). This text describes the relevant equations used for finite-width bearings, numerically evaluates them, and then obtains useful design charts.

Although side-leakage effects are considered in the results shown in this section, a simplification is to neglect the pressure and temperature effects on the lubricant properties, namely, viscosity and density. Lubricant viscosity is particularly sensitive to temperature (as was demonstrated in Table 8.3 and Fig. 8.12), and since the heat generated in hydrodynamic bearings is often considerable, the limitation imposed by this simplification is at once apparent. The temperature rise within the film can be calculated if it is assumed that all the heat produced by the viscous action is carried away by the lubricant (the adiabatic assumption).

For finite-width bearings having hydrodynamic pressure generated within the lubricant film due to the physical wedge mechanism (see Sec. 12.2.2.3), some flow will take place in the *y* direction. Using the adiabatic assumption (implying that the heat produced in the bearing is carried away by the lubricant), the viscosity and density can be considered constant within the *x-y* plane. The appropriate Reynolds equation from Eq. (12.24) when the velocity of the top surface is zero ($u_a = 0$) is

$$\frac{\partial}{\partial x}\left(h^3\frac{\partial p}{\partial x}\right)+\frac{\partial}{\partial y}\left(h^3\frac{\partial p}{\partial y}\right)=6\eta_0 u_b\frac{\partial h}{\partial x} \qquad \textbf{(12.28)}$$

Thus, for a specific film shape $h(x, y)$ the pressure throughout the bearing can be obtained numerically. Knowing the pressure from the Reynolds equation and integrating over the x-y plane produces the load-carrying capacity.

12.3.3 Hydrodynamic Thrust Bearings—Neglecting Side Leakage

Many loads carried by rotary machinery have components that act in the direction of the shaft's axis of rotation. These thrust loads are frequently carried by self-acting or hydrodynamic bearings of the form shown in Fig. 12.9. A thrust plate attached to, or forming part of, the rotating shaft is separated from the sector-shaped bearing pads by a lubricant film. The load-carrying capacity of the bearing arises entirely from the pressures generated by the geometry of the thrust plate over the bearing pads. The x direction, or direction of motion, is along a mean arc AA′ as shown in Fig. 12.9. The transverse direction, or side-leakage direction, would be perpendicular to arc AA′. All pads are identical and separated by deep lubrication grooves, implying that atmospheric pressure exists completely around the pad. Thus, the total thrust load-carrying capacity is simply the number of pads multiplied by the load-carrying capacity of an individual pad. This section will consider three different slider bearings while neglecting side leakage: parallel slider, fixed-incline slider, and parallel-step slider. These three slider bearings will have three different film shapes across arc AA′ in Fig 12.9.

If side leakage is neglected Eq. (12.28) reduces to

$$\frac{\partial}{\partial x}\left(h^3\frac{\partial p}{\partial x}\right)=6\eta_0 u_b\frac{\partial h}{\partial x} \qquad \textbf{(12.29)}$$

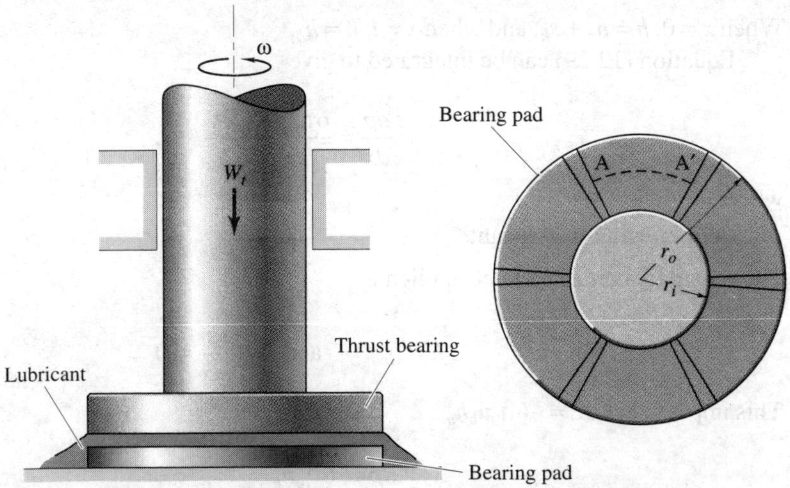

Figure 12.9 Thrust slider bearing geometry. [B. J. Hamrock, *Fundamentals of Fluid Film Lubrication*, NASA Reference Publication 1255, 1991.]

This, then, is the Reynolds equation used for the three types of slider bearing being considered.

12.3.3.1 PARALLEL-SURFACE SLIDER BEARING The film thickness for the parallel-surface slider bearing is constant across the arc length AA′ in Fig. 12.9. Eq. (12.29) becomes

$$\frac{d}{dx}\left(\frac{dp}{dx}\right)=0$$

Integrating twice gives

$$p=\bar{A}x+\bar{B}$$

where
$\bar{A}$ and $\bar{B}$ = integration constants

The boundary conditions are

1. $p=0$ at $x=0 \rightarrow \bar{B}=0$
2. $p=0$ at $x=l \rightarrow \bar{A}=0$

Therefore, the pressure is zero throughout the bearing. The normal load-carrying capacity of the bearing is the integral of the pressure across the length. Since the pressure is zero across the length, the normal applied load would also be zero.

12.3.3.2 FIXED-INCLINE SLIDER BEARING The film shape equation for the fixed-incline slider bearing is

$$h=h_o+s_h\left(1-\frac{x}{l}\right) \tag{12.30}$$

When $x=0$, $h=h_o+s_h$; and when $x=l$, $h=h_o$.

Equation (12.29) can be integrated to give

$$\frac{dp}{dx}=\frac{6\eta_0 u_b}{h^2}+\frac{\bar{A}}{h^3}$$

where
$\bar{A}$ = integration constant

The boundary condition to be applied is

$$h=h_m \qquad \text{and} \qquad \frac{dp}{dx}=0 \tag{12.31}$$

This implies that $\bar{A}=-6\eta_0 u_b h_m$.

$$\therefore \frac{dp}{dx}=\frac{6\eta_0 u_b}{h^3}(h-h_m) \tag{12.32}$$

This equation can be written in dimensionless form if

$$P = \frac{ps_h^2}{\eta_0 u_b l} \qquad H = \frac{h}{s_h} \qquad H_m = \frac{h_m}{s_h} \qquad H_o = \frac{h_o}{s_h} \qquad X = \frac{x}{l}$$

as

$$\frac{dP}{dX} = 6\left(\frac{H - H_m}{H^3}\right) \tag{12.33}$$

where

$$H = \frac{h}{s_h} = H_o + 1 - X \tag{12.34}$$

$$\frac{dH}{dX} = -1 \rightarrow dH = -dX \tag{12.35}$$

Equations (12.34) and (12.35) are valid only for a fixed-incline bearing.

Integrating Eq. (12.33) gives

$$P = 6\int\left(\frac{1}{H^2} - \frac{H_m}{H^3}\right)dX = -6\int\left(\frac{1}{H^2} - \frac{H_m}{H^3}\right)dH$$

$$\therefore P = 6\left(\frac{1}{H} - \frac{H_m}{2H^2}\right) + \overline{B} \tag{12.36}$$

The boundary conditions are

1. $P = 0$ when $X = 0 \rightarrow H = H_o + 1$
2. $P = 0$ when $X = 1 \rightarrow H = H_o$

Making use of boundary conditions 1 and 2 gives

$$H_m = \frac{2H_o(1 + H_o)}{1 + 2H_o} \tag{12.37}$$

$$\overline{B} = -\frac{6}{1 + 2H_o} \tag{12.38}$$

Substituting Eqs. (12.37) and (12.38) into Eq. (12.36) gives

$$P = \frac{6X(1 - X)}{(H_o + 1 - X)^2(1 + 2H_o)} \tag{12.39}$$

The normal applied load per unit width is

$$W_z' = \frac{W_z}{w_t} = \int_0^l p\,dx = \frac{\eta_0 u_b l^2}{s_h^2}\int_0^1 P\,dX$$

The dimensionless normal load is

$$\therefore \overline{W_z} = \frac{W_z s_h^2}{\eta_0 u_b l^2 w_t} = \int_0^1 P\,dX \tag{12.40}$$

Recall that $dH = -dX$ and from Eq. (12.34) note that if

$$X = 0 \rightarrow H = H_o + 1$$

$$X = 1 \rightarrow H = H_o$$

Thus, Eq. (12.40) becomes

$$\overline{W}_z = \frac{W_z s_h^2}{\eta_0 u_b l^2 w_t} = \int_{H_o+1}^{H_o} -P\,dH \qquad \textbf{(12.41)}$$

Substituting Eqs. (12.36) and (12.38) into Eq. (12.41) gives

$$\overline{W}_z = \frac{W_z s_h^2}{\eta_0 u_b l^2 w_t} = 6\ln\left(\frac{H_o+1}{H_o}\right) - \frac{12}{1+2H_o} \qquad \textbf{(12.42)}$$

12.3.3.3 PARALLEL-STEP SLIDER BEARING For the parallel-step slider bearing the film thickness in each of the two regions is constant, implying that the appropriate Reynolds equation is

$$\frac{d^2 p}{dx^2} = 0$$

Integrating gives

$$\frac{dp}{dx} = \text{constant}$$

However, since the film thicknesses between regions differ, their pressure gradients also differ. Thus, there will be no discontinuity at the step.

$$p_{\max} = n_s l\left(\frac{dp}{dx}\right)_i = -(l - n_s l)\left(\frac{dp}{dx}\right)_o \qquad \textbf{(12.43)}$$

The inlet and outlet flow rates at the step must be the same.

$$\therefore q'_{xo} = q'_{xi}$$

Making use of Eq. (12.12) gives

$$-\frac{h_o^3}{12\eta_0}\left(\frac{dp}{dx}\right)_o + \frac{u_b h_o}{2} = -\frac{(h_o + s_h)^3}{12\eta_0}\left(\frac{dp}{dx}\right)_i + \frac{u_b(h_o + s_h)}{2} \qquad \textbf{(12.44)}$$

Equations (12.43) and (12.44) represent a pair of simultaneous equations with unknowns $(dp/dx)_o$ and $(dp/dx)_i$. The solutions are

$$\left(\frac{dp}{dx}\right)_i = \frac{6\eta_0 u_b (1-n_s) s_h}{(1-n_s)(h_o+s_h)^3 + n_s h_o^3} \qquad \textbf{(12.45)}$$

$$\left(\frac{dp}{dx}\right)_o = \frac{-6\eta_0 u_b n_s s_h}{(1-n_s)(h_o+s_h)^3 + n_s h_o^3} \qquad \textbf{(12.46)}$$

The maximum pressure at the step can be found by substituting Eqs. (12.45) and (12.46) into Eq. (12.43) to give

$$p_{\max} = \frac{6\eta_0 u_b l n_s (1-n_s) s_h}{(1-n_s)(h_o + s_h)^3 + n_s h_o^3} \quad \textbf{(12.47)}$$

$$P_{\max} = \frac{p_{\max} s_h^2}{\eta_0 u_b l} = \frac{6n_s(1-n_s)}{(1-n_s)(H_o+1)^3 + n_s H_o^3} \quad \textbf{(12.48)}$$

The pressures in the inlet and outlet regions are

$$P_i = \frac{X P_{\max}}{n_s} = \frac{6X(1-n_s)}{(1-n_s)(H_o+1)^3 + n_s H_o^3} \qquad 0 \le X \le n_s \quad \textbf{(12.49)}$$

$$P_o = \frac{(1-X)P_{\max}}{1-n_s} = \frac{6(1-X)n_s}{(1-n_s)(H_o+1)^3 + n_s H_o^3} \qquad n_s \le X \le 1 \quad \textbf{(12.50)}$$

The normal applied load per unit width is

$$\frac{W_z}{w_t} = \frac{p_{\max} l}{2} = \frac{3\eta_0 u_b l^2 n_s (1-n_s) s_h}{(1-n_s)(h_o+s_h)^3 + n_s h_o^3} \quad \textbf{(12.51)}$$

The dimensionless normal applied load is

$$\overline{W}_z = \frac{W_z}{\eta_0 u_b w_t}\left(\frac{s_h}{l}\right)^2 = \frac{3n_s(1-n_s)}{(1-n_s)(H_o+1)^3 + n_s H_o^3} = \frac{P_{\max}}{2} \quad \textbf{(12.52)}$$

Example 12.2

GIVEN A flat strip of metal emerges from an oil bath having viscosity η_0 and inlet pressure p_i with velocity u_b on passing through a slot of the form shown in Fig. 12.10. In the initial convergent region (region 1) of the slot the film thickness decreases linearly from $h_o + s_h$ to h_o over the length $n_s l$ on each side of the strip. In the final section of the slot the film on each side has a constant thickness h_o over the length $l(1 - n_s)$. The width of the flat strip is large relative to its length, so that it can be considered to be infinitely wide.

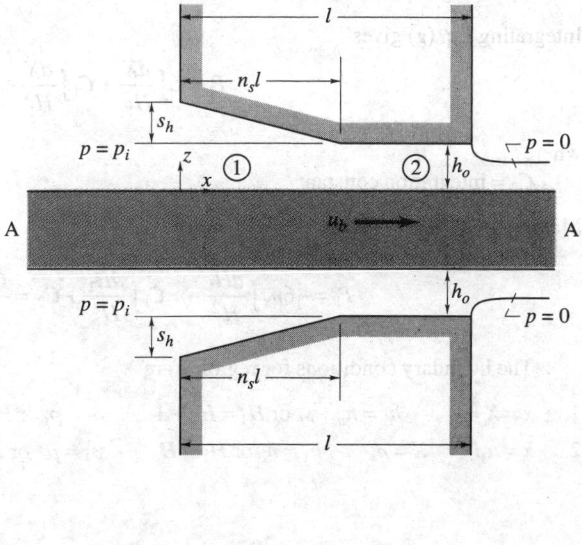

Figure 12.10 Thrust slider bearing used in Example 12.2.

FIND The pressure distribution and volume flow rate in the x direction.

Solution

Let

$$p = \frac{\eta_0 u_b l}{s_h^2} P \qquad h = s_h H \qquad \text{and} \qquad x = lX \tag{a}$$

Eq. (12.29) becomes

$$\frac{\partial}{\partial X}\left(H^3 \frac{\partial P}{\partial X} \right) = 6\frac{\partial H}{\partial X} \tag{b}$$

Equation (b) is the dimensionless Reynolds equation used in region 1 of Fig. 12.10. The appropriate dimensionless Reynolds equation for region 2 when $H = H_o$ = constant is

$$\frac{d^2 P}{dX^2} = 0 \tag{c}$$

Region 1: $0 \le x \le n_s l$ or $0 \le X \le n_s$. The film shape in region 1 is

$$h_1 = h_o + s_h - \frac{x s_h}{n_s l} \tag{d}$$

When $x = 0$, $h_1 = h_o + s_h$; and when $x = n_s l$, $h = h_o$. Equation (d) can be expressed in dimensionless form while making use of Eq. (a) as

$$H_1 = \frac{h_1}{s_h} = H_o + 1 - \frac{X}{n_s} \tag{d}$$

$$\frac{dH_1}{dX} = -\frac{1}{n_s} \tag{f}$$

Integrating Eq. (b) gives

$$\frac{dP_1}{dX} = \frac{6}{H_1^2} + \frac{C_1}{H_1^3} \tag{g}$$

where

C_1 = integration constant

Integrating Eq. (g) gives

$$P_1 = 6\int \frac{dX}{H_1^2} + C_1 \int \frac{dX}{H_1^3} + C_2 \tag{h}$$

where

C_2 = integration constant

Making use of Eq. (f) gives

$$P_1 = -6n_s \int \frac{dH_1}{H_1^2} - n_s C_1 \int \frac{dH_1}{H_1^3} + C_2 = \frac{6n_s}{H_1} + \frac{C_1 n_s}{2H_1^2} + C_2 \tag{i}$$

The boundary conditions for region 1 are

1. $x = X = 0$ $\quad h_1 = h_o + s_h$ or $H_1 = H_o + 1$ $\quad p_1 = p_i$ or $P_1 = P_i$
2. $x = n_s l$ $\quad X = n_s$ $\quad h_1 = h_o$ or $H_1 = H_o$ $\quad p_1 = p^*$ or $P_1 = P^*$

Making use of boundary condition 1 gives

$$P_i = \frac{6n_s}{H_o+1} + \frac{C_1 n_s}{2(H_o+1)^2} + C_2$$

$$\therefore C_2 = P_i - \frac{6n_s}{H_o+1} - \frac{C_1 n_s}{2(H_o+1)^2} \tag{j}$$

Substituting Eq. (j) into Eq. (i) gives

$$P_1 - P_i = 6n_s\left[\frac{1}{H_1} - \frac{1}{H_o+1}\right] + \frac{C_1 n_s}{2}\left[\frac{1}{H_1^2} - \frac{1}{(H_o+1)^2}\right] \tag{k}$$

Making use of boundary condition 2 gives

$$P^* - P_i = \frac{6n_s}{H_o(H_o+1)} + \frac{C_1 n_s(2H_o+1)}{2H_o^2(H_o+1)^2}$$

$$\frac{C_1 n_s}{2} = \frac{H_o^2(H_o+1)^2}{(2H_o+1)}(P^* - P_i) - \frac{6n_s H_o(H_o+1)}{(2H_o+1)} \tag{l}$$

Substituting Eq. (l) into Eq. (k) gives

$$P_1 - P_i = 6n_s\left[\frac{1}{H_1} - \frac{1}{H_o+1}\right] + \left[\frac{H_o^2(H_o+1)^2(P^* - P_i)}{2H_o+1} - \frac{6n_s H_o(H_o+1)}{2H_o+1}\right]\left[\frac{1}{H_1^2} - \frac{1}{(H_o+1)^2}\right] \tag{m}$$

$$\frac{dP_1}{dH_1} = -\frac{6n_s}{H_1^2} - \frac{2}{H_1^3}\left[\frac{H_o^2(H_o+1)^2(P^* - P_i)}{2H_o+1} - \frac{6n_s H_o(H_o+1)}{2H_o+1}\right] \tag{n}$$

$$\left(\frac{dP_1}{dH_1}\right)_{H_1=H_o} = \frac{2}{H_o(2H_o+1)}\left[\frac{3n_s}{H_o} - (H_o+1)^2(P^* - P_i)\right] \tag{o}$$

Region 2: $n_s l \le x \le l$ or $n_s \le X \le 1$. Integrating Eq. (c) twice gives

$$P_2 = C_3 X + C_4 \tag{p}$$

The boundary conditions for region 2 are

1. $x = l$ or $X = 1$ $\quad p_2 = P_2 = 0$
2. $x = n_s l$ or $X = n_s$ $\quad p_2 = p^*$ or $P_2 = P^*$

Making use of boundary condition 1 gives $C_4 = -C_3$, where C_3 and C_4 are integration constants. Substituting this into Eq. (p) gives

$$P_2 = C_3(X-1) \tag{q}$$

Making use of boundary condition 2 gives

$$C_3 = \frac{P^*}{n_s - 1}$$

$$\therefore P_2 = P^*(1-X)/(1-n_s) \tag{r}$$

$$\frac{dP_2}{dX} = -P^*/(1-n_s) \tag{s}$$

In Eqs. (m) and (n) P^*, the dimensionless pressure at the common boundary between regions 1 and 2, is not known. Thus, one more boundary condition is needed. This condition is obtained by equating the volume flow rates at the common boundary, or

$$(q_1')_{x=n_s l} = (q_2')_{x=n_s l} \tag{t}$$

Recall that at $x = n_s l$, $H_1 = H_2 = H_o = h_o/s_h$ and $h_1 = h_2 = h_o$. Making use of Eq. (12.21) and the above changes Eq. (t) to

$$-\frac{h_o^3}{12\eta_0}\left(\frac{dp_1}{dx}\right)_{x=n_s l} + \frac{u_b h_o}{2} = -\frac{h_o^3}{12\eta_0}\left(\frac{dp_2}{dx}\right)_{x=n_s l} + \frac{u_b h_o}{2}$$

This equation implies that

$$\left(\frac{dp_1}{dx}\right)_{x=n_s l} = \left(\frac{dp_2}{dx}\right)_{x=n_s l}$$

$$\left(\frac{dP_1}{dX}\right)_{X=n_s} = \left(\frac{dP_2}{dX}\right)_{X=n_s}$$

$$\left(\frac{dP_1}{dH_1}\right)_{H_1=H_o}\left(\frac{dH_1}{dX}\right) = \left(\frac{dP_2}{dX}\right)_{X=n_s} \tag{u}$$

Substituting Eqs. (f), (o), and (s) into Eq. (u) gives

$$-\frac{6}{H_o^2(2H_o+1)} + \frac{2(H_o+1)^2(P^* - P_i)}{n_s H_o(2H_o+1)} + \frac{P^*}{1-n_s} = 0$$

Solving for P^* gives

$$P^* = \frac{2(1-n_s)}{H_o}\left[\frac{3n_s + H_o(H_o+1)^2 P_i}{2(1-n_s)(H_o+1)^2 + n_s H_o(2H_o+1)}\right] \tag{v}$$

Using Eqs. (m), (r), and (v) enables the pressure in regions 1 and 2 to be obtained. The pressure is a function of n_s, H_o, P_i, and X. Figure 12.11 gives the pressure distributions for $n_s = 0.5$ and $H_o = 1.0$ and two inlet pressures.

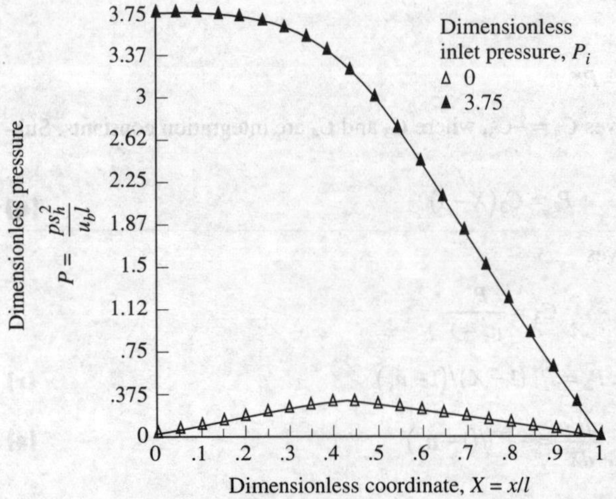

Figure 12.11 Pressure distribution for $n_s = 0.5$ and $H_o = 1.0$ for two inlet pressures obtained from Example 12.2.

The volume flow per unit width is the same at any x location. Thus, making use of Eqs. (12.12), (s), and (v) gives

$$q_2' = \frac{u_b h_o}{2}\left[1+\frac{P^* H_o^2}{6(1-n_s)}\right] = \frac{u_b h_o}{2}\left\{1+\frac{H_o}{3}\left[\frac{3n_s + H_o(H_o+1)^2 P_i}{2(1-n_s)(H_o+1)^2 + n_s H_o(2H_o+1)}\right]\right\}$$

The solution from this example can be reduced to the situation presented earlier for parallel-surface (Sec. 12.3.3.1) and fixed-incline (Sec. 12.3.3.2) slider bearings. If $n_s = 0$ and $P_i = 0$, the parallel-surface slider bearing is simulated and $P = 0$ and $q_2' = u_b h_o/2$. If $n_s = 1$ and $P_i = 0$, the fixed-incline bearing is simulated and

$$q_2' = \frac{u_b h_o}{2}\left[1+\frac{1}{2H_o+1}\right] = \frac{u_b h_o}{2}\left[\frac{2H_o+1+1}{2H_o+1}\right] = \frac{u_b h_o(H_o+1)}{2H_o+1}$$

12.3.4 OPERATING AND PERFORMANCE PARAMETERS

The last section neglected side-leakage effects in the design of slider bearings; the present section will consider these effects. The following general definitions, related to the bearing geometry shown in Fig. 12.12, and relationships are employed: The forces acting on the bearing surfaces of an individual pad can be considered in two groups. The loads, which act in the direction normal to the surface, yield normal loads that can be resolved into components W_x and W_z. The viscous surface stresses, which act in the direction tangent to the surface, yield shear forces on the solids that have components F in the x direction. The shear forces in the z direction are negligible. Once the pressure is obtained from the Reynolds equation [Eq. (12.28)], the force components acting on the bearing surfaces are

$$W_{za} = W_{zb} = \int_0^{w_t}\int_0^{l} p\,dx\,dy \tag{12.53}$$

$$W_{xb} = 0 \tag{12.54}$$

$$W_{xa} = -\int_0^{w_t}\int_{h_o+s_h}^{h_o} p\,dh\,dy = -\int_0^{w_t}\int_0^{l} p\frac{dh}{dx}\,dx\,dy$$

$$= -\int_0^{w_t}(ph)_0^l\,dy + \int_0^{w_t}\int_0^{l} h\frac{dp}{dx}\,dx\,dy = \int_0^{w_t}\int_0^{l} h\frac{dp}{dx}\,dx\,dy \tag{12.55}$$

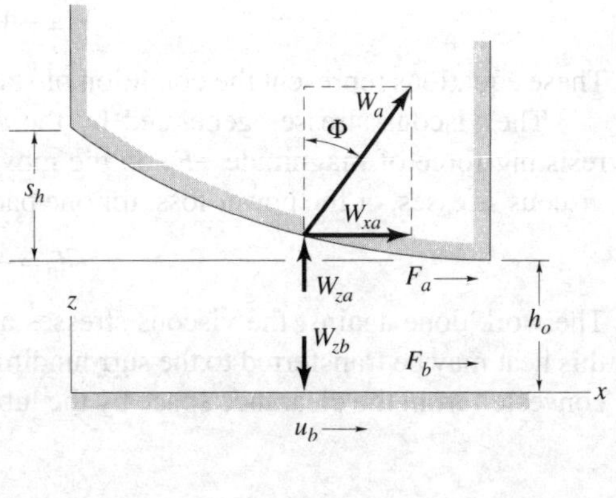

Figure 12.12 Force components and oil film geometry in a hydrodynamically lubricated thrust slider bearing. [B. J. Hamrock, *Fundamentals of Fluid Film Lubrication*, NASA Reference Publication 1255, 1991.]

$$W_b = \sqrt{W_{zb}^2 + W_{xb}^2} = W_{zb} \tag{12.56}$$

$$W_a = \sqrt{W_{za}^2 + W_{xa}^2} \tag{12.57}$$

$$\Phi = \tan^{-1}\left(\frac{W_{xa}}{W_{za}}\right) \tag{12.58}$$

The shear force acting on solid b is

$$F_b = \int_0^{w_t}\int_0^{l} (\tau_{zx})_{z=0}\, dxdy \tag{12.59}$$

But

$$(\tau_{zx})_{z=0} = \left(\eta \frac{\partial u}{\partial z}\right)_{z=0} \tag{12.60}$$

Substituting Eq. (12.8) into Eq. (12.60) gives

$$(\tau_{zx})_{z=0} = -\frac{h}{2}\frac{dp}{dx} - \frac{\eta u_b}{h} \tag{12.61}$$

Substituting Eq. (12.61) into Eq. (12.59) gives

$$F_b = \int_0^{w_t}\int_0^{l}\left(-\frac{h}{2}\frac{dp}{dx} - \frac{\eta u_b}{h}\right)dxdy$$

Making use of Eq. (12.55) gives

$$F_b = -\frac{W_{xa}}{2} - \int_0^{w_t}\int_0^{l} \frac{\eta u_b}{h} dxdy \tag{12.62}$$

Similarly, the shear force acting on solid a is

$$F_a = \int_0^{w_t}\int_0^{l} (\tau_{zx})_{z=h}\, dxdy = -\frac{W_{xa}}{2} + \int_0^{w_t}\int_0^{l} \frac{\eta u_b}{h} dxdy \tag{12.63}$$

Note from Fig. 12.12 that

$$F_b + F_a + W_{xa} = 0 \tag{12.64}$$

$$W_{zb} - W_{za} = 0 \tag{12.65}$$

These equations represent the condition of static equilibrium.

The viscous stresses generated by the shearing of the lubricant film give rise to a resisting force of magnitude $-F_b$ on the moving surface. The rate of working against the viscous stresses, or the power loss, for one pad is

$$h_p = -F_b u_b \tag{12.66}$$

The work done against the viscous stresses appears as heat within the lubricant. Some of this heat may be transferred to the surroundings by radiation or by conduction, or it may be convected from the clearance space by the lubricant flow.

The bulk temperature rise of the lubricant for the case in which all the heat is carried away by convection is known as the **adiabatic temperature rise.** This bulk temperature increase can be calculated by equating the rate of heat generated within the lubricant to the rate of heat transferred by convection

$$h_p = \rho q_x C_p (\Delta t_m)$$

or the adiabatic temperature rise in degrees Celsius may be expressed as

$$\Delta t_m = \frac{h_p}{\rho q_x C_p} \qquad \textbf{(12.67)}$$

where

ρ = density, kg/m^3
q_x = volume flow rate in x direction (direction of motion), m^3/s
C_p = specific heat of lubricant at constant pressure, J/kg-°C

Equations (12.53) to (12.67) are some of the relevant equations used in thrust bearing analysis. Recall that the pressure-generating mechanism used in this section and the next (Secs. 12.3 and 12.4) is the physical wedge, as discussed in Sec. 12.2.2.3. The general thrust analysis will be used to design two types of thrust bearing, the fixed-incline and pivoted-pad bearings. A detailed numerical evaluation will not be made of the Reynolds equation and its load and friction force components, since the emphasis of the present text is on design. If the reader is interested, two references are recommended. The original source is Raimondi and Boyd (1955); a more recent text is Hamrock (1993).

The operating parameters affecting the pressure generation, film thickness, load, and friction components are

1. **Bearing number** for a thrust slider bearing $B_t = \left(\frac{\eta_0 u_b w_t}{W_z}\right)\left(\frac{l}{s_h}\right)^2$ **(12.68)**
2. Length-to-width ratio $\lambda = l/w_t$ **(12.69)**
3. Location of pivot from inlet $X_p = x_p/l$ **(12.70)**

For a fixed-incline bearing only the first two operating parameters affect the pressure generation, film thickness, load, and friction force components. The operating parameters are viewed as the parameters given in the design of a hydrodynamic thrust bearing. The performance parameters that will result from choosing a given set of operating parameters are

1. Outlet (minimum) film thickness h_o
2. Temperature rise due to lubricant shearing Δt_m
3. Power loss h_p
4. Coefficient of sliding friction μ
5. Circumferential and side-leakage volumetric flow rates q and q_s

These performance parameters are needed to evaluate the design of a hydrodynamic thrust slider bearing.

12.3.5 FIXED-INCLINE SLIDER BEARING

The simplest form of fixed-incline bearing provides only straight-line motion and consists of a flat surface sliding over a fixed pad or land having a profile similar to that shown in Fig. 12.13. The fixed-incline bearing depends for its operation on the lubricant being drawn into a wedge-shaped space and thus producing pressure that counteracts the load and prevents contact between the sliding parts. Since the wedge action takes place only when the sliding surface moves in the direction in which the lubricant film converges, the fixed-incline bearing (Fig. 12.13) can carry load only in this direction. If reversibility is desired, a combination of two or more pads with their surfaces sloped in opposite directions is required. Fixed-incline pads are used in multiples as in the thrust bearing shown in Fig. 12.14.

The following procedure assists in designing a fixed-incline thrust bearing:

1. Choose a pad length-to-width ratio. A square pad ($\lambda = 1$) is generally thought to give good performance. If it is known whether maximum load or minimum power loss is more important in the particular application, the outlet film thickness ratio H_o can be determined from Fig. 12.15.
2. Once λ and H_o are known, Fig. 12.16 can be used to obtain the bearing number B_t.
3. From Fig. 12.17 determine the temperature rise due to shear heating for a given λ and B_t. The volumetric specific heat $C_s = \rho C_p$, which is the dimensionless temperature rise parameter, is relatively constant for mineral oils and is equivalent to 1.36×10^6 N/(m^2-°C).
4. Determine lubricant temperature. Mean temperature can be expressed as

$$\tilde{t}_m = t_{mi} + \frac{\Delta t_m}{2} \qquad \textbf{(12.71)}$$

where

t_{mi} = inlet temperature, °C

The inlet temperature is usually known beforehand. Once the mean temperature $\tilde{t}_m$ is known, it can be used in Fig. 8.13 to determine the viscosity of SAE oils, or Fig. 8.12

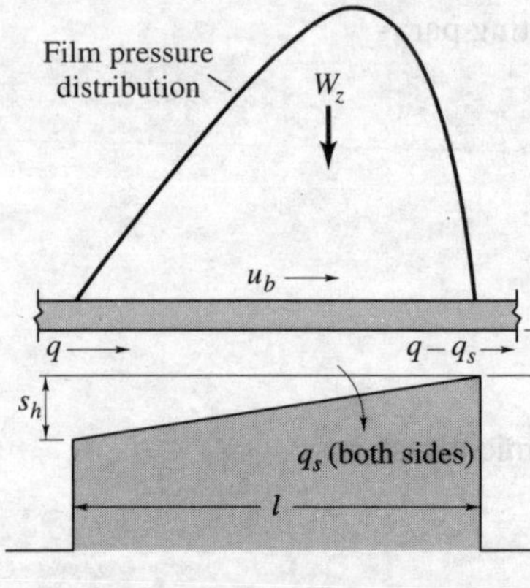

Figure 12.13 Side view of fixed-incline slider bearing. [A. A. Raimondi and J. Boyd, "Applying Bearing Theory to the Analysis and Design of Pad-Type Bearings," *ASME Transactions*, vol. 77, no. 3, 1955, pp. 287–309. Reprinted by permission of the American Society of Mechanical Engineers International.]

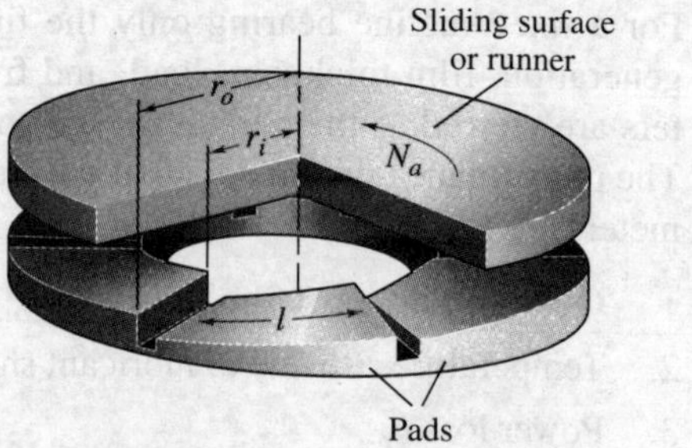

Figure 12.14 Configuration of multiple fixed-incline thrust slider bearing. [A. A. Raimondi and J. Boyd, "Applying Bearing Theory to the Analysis and Design of Pad-Type Bearings," *ASME Transactions*, vol. 77, no. 3, 1955, pp. 287–309. Reprinted by permission of the American Society of Mechanical Engineers International.]

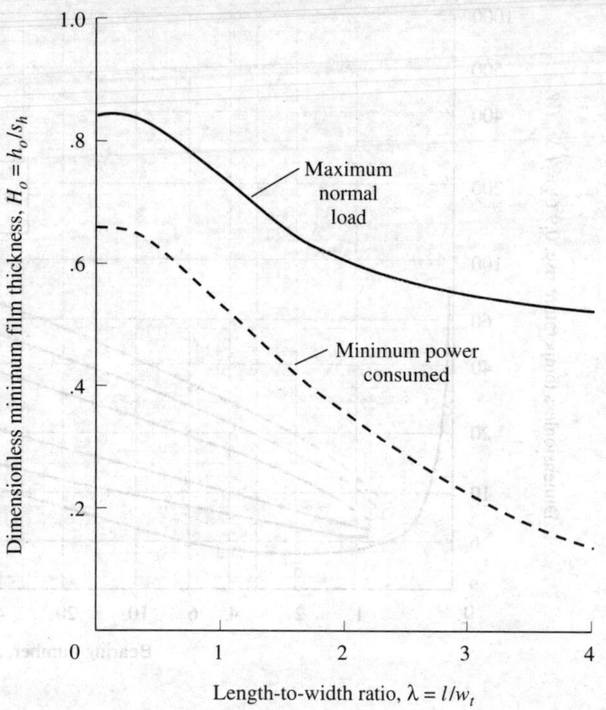

Figure 12.15 Chart for determining minimum film thickness corresponding to maximum load or minimum power loss for various pad proportions in fixed-incline bearings. [A. A. Raimondi and J. Boyd, "Applying Bearing Theory to the Analysis and Design of Pad-Type Bearings," *ASME Transactions*, vol. 77, no. 3, 1955, pp. 287–309. Reprinted by permission of the American Society of Mechanical Engineers International.]

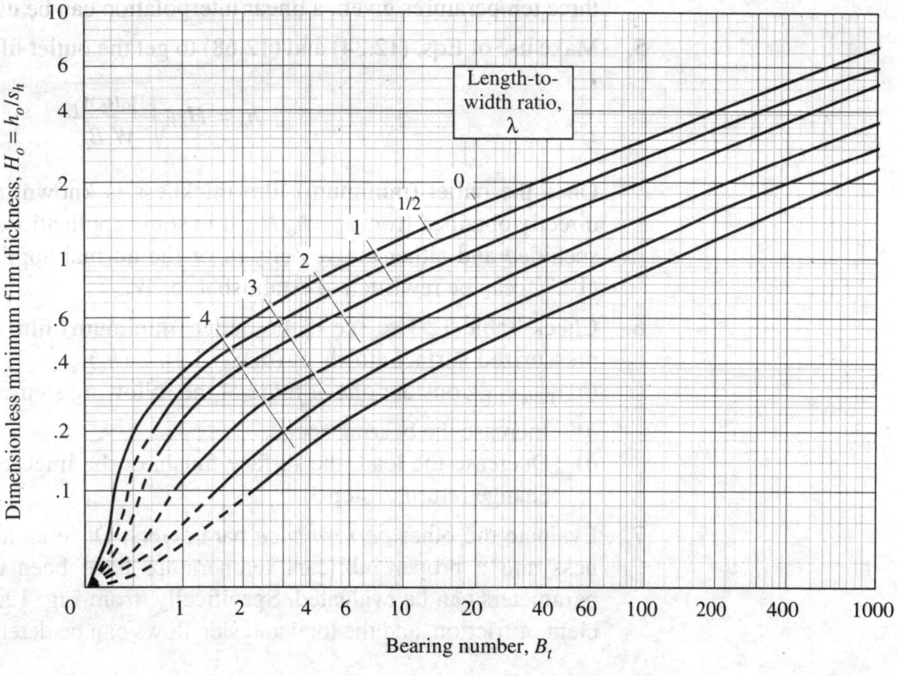

Figure 12.16 Chart for determining minimum film thickness for fixed-incline thrust bearings. [A. A. Raimondi and J. Boyd, "Applying Bearing Theory to the Analysis and Design of Pad-Type Bearings," *ASME Transactions*, vol. 77, no. 3, 1955, pp. 287–309. Reprinted by permission of the American Society of Mechanical Engineers International.]

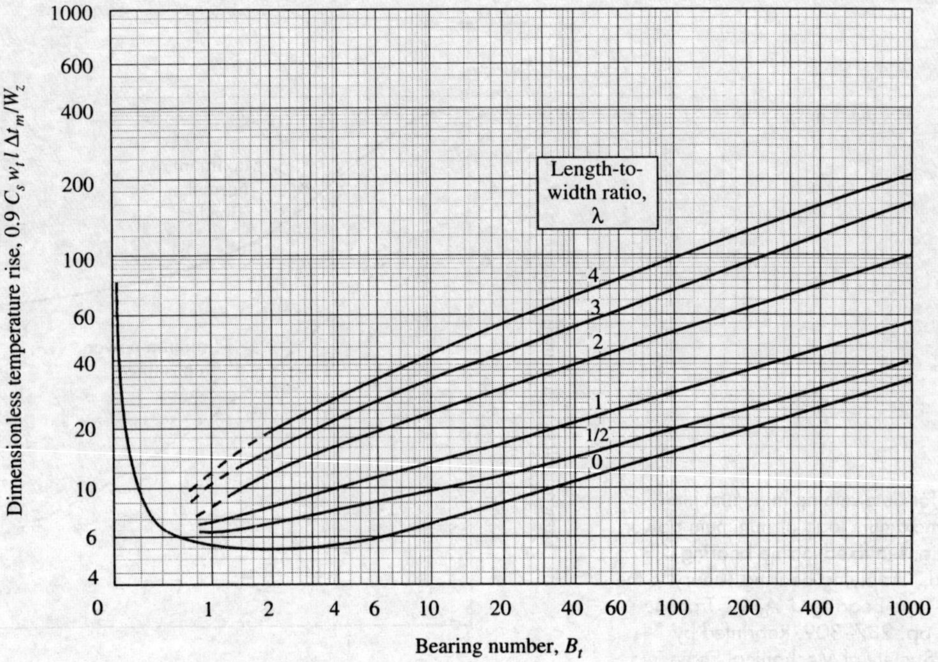

Figure 12.17 Chart for determining dimensionless temperature rise due to viscous shear heating of lubricant in fixed-incline thrust bearings. [A. A. Raimondi and J. Boyd, "Applying Bearing Theory to the Analysis and Design of Pad-Type Bearings," *ASME Transactions*, vol. 77, no. 3, 1955, pp. 287–309. Reprinted by permission of the American Society of Mechanical Engineers International.]

or Table 8.3 can be used. In using Table 8.3 if your temperature is different from the three temperatures given, a linear interpolation can be used.

5. Make use of Eqs. (12.34) and (12.68) to get the outlet film thickness as

$$h_o = H_o l \sqrt{\frac{\eta_0 u_b w_t}{W_z B_t}} \tag{12.72}$$

Once the outlet (minimum) film thickness is known, the shoulder height s_h can be directly obtained from $s_h = h_o/H_o$. If in some applications the outlet film thickness h_o is specified and either the velocity u_b or the normal applied load W_z is not known, Eq. (12.72) can be rewritten to establish u_b or W_z.

6. Check Table 12.1 to see if the outlet (minimum) film thickness is sufficient for the pressurized surface finish. If $(h_o)_{\text{Eq. (12.72)}} \geq (h_o)_{\text{Table 12.1}}$, go to step 7. If $(h_o)_{\text{Eq. (12.72)}} < (h_o)_{\text{Table 12.1}}$, consider one or more of the following steps:
   a) Increase the bearing speed.
   b) Decrease the load, the surface finish, or the inlet temperature. Upon making this change return to step 3.

7. Evaluate the other *performance* parameters. Once an adequate minimum film thickness and a proper lubricant temperature have been determined, the performance parameters can be evaluated. Specifically, from Fig. 12.18 the power loss, the coefficient of friction, and the total and side flows can be determined.

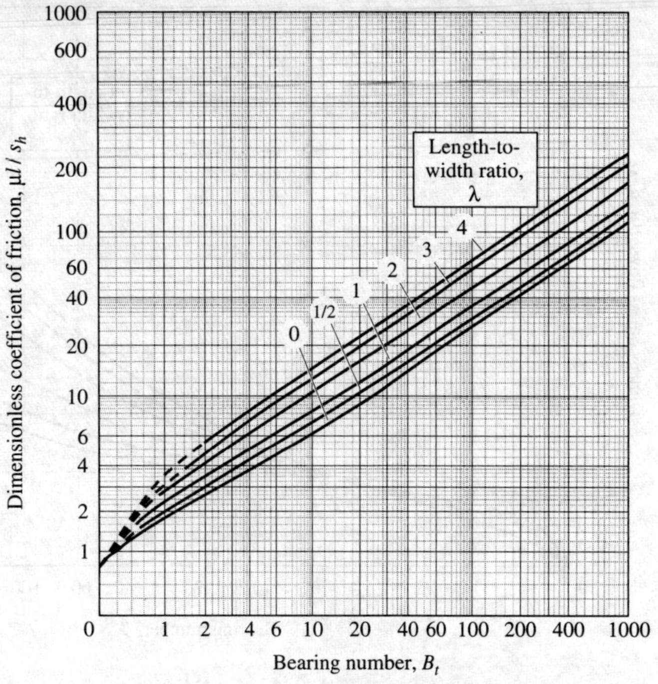

(a)

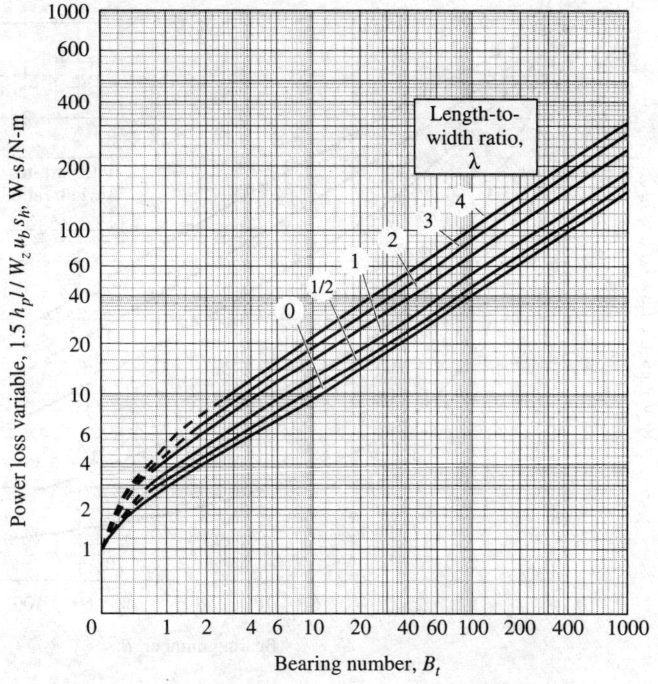

(b)

Figure 12.18 Chart for determining performance parameters of fixed-incline thrust bearings. (a) Friction coefficient; (b) power loss; *(continued)*

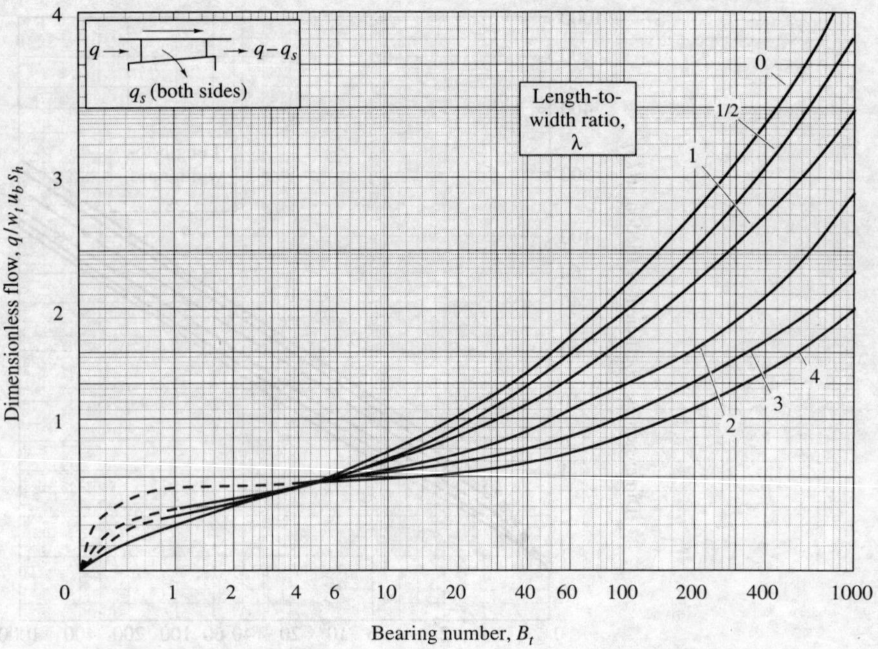

(c)

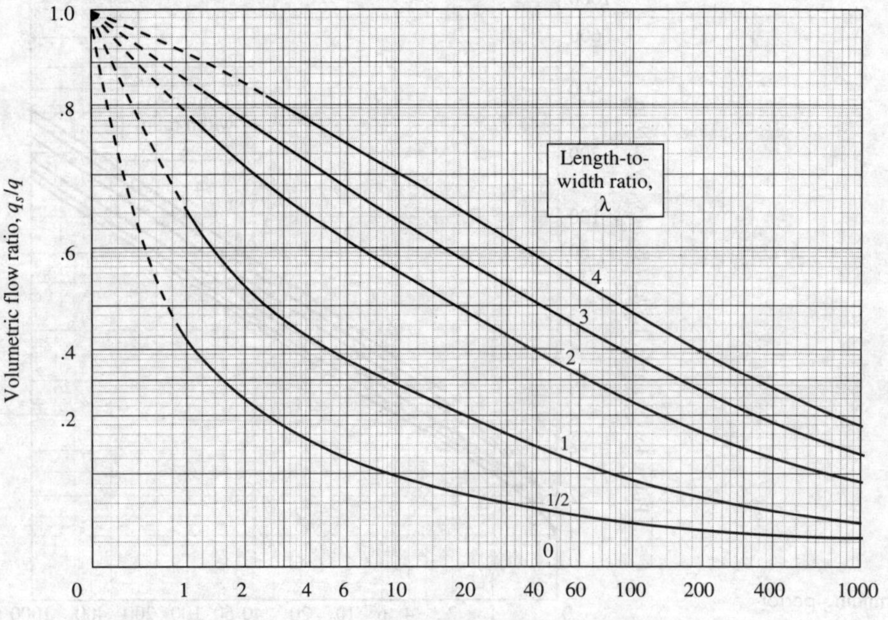

(d)

Figure 12.18 *(concluded)* (c) lubricant flow; (d) lubricant side flow. [A. A. Raimondi and J. Boyd, "Applying Bearing Theory to the Analysis and Design of Pad-Type Bearings," *ASME Transactions*, vol. 77, no. 3, 1955, pp. 287–309. Reprinted by permission of the American Society of Mechanical Engineers International.]

Table 12.1 Allowable outlet (minimum) film thickness for a given surface finish [Properties of common engineering materials. Data Item 84041, ESDU International, London, 1984.]

Surface finish (centerline average), R_a		Description of surface	Examples of manufacturing methods	Approximate relative costs	Allowable outlet (minimum) film thickness[a], h_o	
μm	μin.				μm	μin.
0.1–0.2	4–8	Mirror-like surface without toolmarks; close tolerances	Grind, lap, and super-finish	17–20	2.5	100
.2–.4	8–16	Smooth surface without scratches; close tolerances	Grind and lap	17–20	6.2	250
.4–.8	16–32	Smooth surface; close tolerances	Grind, file, and lap	10	12.5	500
.8–1.6	32–63	Accurate bearing surface without toolmarks	Grind, precision mill, and file	7	25	1000
1.6–3.2	63–125	Smooth surface without objectionable toolmarks; moderate tolerances	Shape, mill, grind, and turn	5	50	2000

[a] The values of film thickness are given only for guidance. They indicate the film thickness required to avoid metal-to-metal contact under clean oil conditions with no misalignment. It may be necessary to take a larger film thickness than that indicated (e.g., to obtain an acceptable temperature rise). It has been assumed that the average surface finish of the pads is the same as that of the runner.

Example 12.3

GIVEN A fixed-incline slider thrust bearing has the following operating parameters: W_z = 3600 lbf, u_b = 1200 in./s, l = 3 in., w_t = 3 in., SAE 10 oil, and t_{mi} = 40°C.

FIND For the maximum load condition determine the following performance parameters: s_h, h_o, Δt_m, μ, h_p, q, and q_s.

Solution

From Fig 12.15 for $\lambda = l/w_t = 3/3 = 1$ and maximum normal load the outlet film thickness ratio H_o = 0.73. Therefore, from Fig. 12.16 the bearing number B_t = 17.0, and from Fig. 12.17

$$\frac{0.9\rho C_p l w_t \Delta t_m}{W_z} = 16$$

Inserting the input parameters and the fact that

$$C_s = \rho C_p = 1.36\times 10^6 \text{ N/(m}^2\text{-°C)} = 197.2 \text{ lbf/(in.}^2\text{-°C)}$$

$$\Delta t_m = \frac{16W_z}{0.9\rho C_p l w_t} = \frac{16(3.6)(10^3)}{(0.9)(197.2)(3)(3)} = 36.06\text{°C}$$

From Eq. (12.71)

$$t_m = t_{mi} + \Delta t_m/2 = 40 + 36.06/2 = 58.03\text{°C}$$

From Fig. 8.13 for SAE 10 oil at 58°C the absolute viscosity is 5×10^{-6} lbf-s/in.^{2}. Making use of Eq. (12.72) gives

$$h_o = H_o l\sqrt{\frac{\eta_0 u_b w_t}{W_z B_t}} = 0.73(3)\sqrt{\frac{(5)(10^{-6})(1.2)(10^3)(3)}{(3.6)(10^3)(17)}} = 1.188\times 10^{-3} \text{ in.}$$

$$s_h = h_o / H_o = (1.188)(10^{-3})/0.73 = 1.627 \times 10^{-3} \text{ in.}$$

All but the last row of surface finish will ensure that $h_o > (h_o)_{\text{all}}$. Thus, as long as the surface finish is less than 1×10^{-3} in., the design is adequate.

The performance parameters for $\lambda = 1$ and $B_t = 17.0$ are as follows:

a) From Fig. 12.18(a)

$$\frac{\mu l}{s_h} = 12$$

$$\therefore \mu = \frac{(12)(1.627)(10^{-3})}{3} = 0.0065$$

b) From Fig 12.18(b)

$$\frac{1.5 h_p l}{W_z u_b s_h} = 18$$

$$\therefore h_p = \frac{18 W_z u_b s_h}{1.5 l} = \frac{18(3.6)(10^3)(1.2)(10^3)(1.627)(10^{-3})}{1.5(3)} = 84.34 \times 10^3 \text{ W} = 84.34 \text{ kW}$$

c) From Fig. 12.18(c)

$$\frac{q}{w_t u_b s_h} = 0.88$$

$$\therefore q = 0.88 w_t u_b s_h = 0.88(3)(1.2)(10^3)(1.627)(10^{-3}) = 5.154 \text{ in.}^3/\text{s}$$

d) $q_s = (0.29)q = 1.484$ in.^{3}/s

12.3.6 PIVOTED-PAD SLIDER BEARING

The simplest form of pivoted-pad bearing provides only for straight-line motion and consists of a flat surface sliding over a pivoted pad (or shoe) as shown in Fig. 12.19. If the pad is assumed to be in equilibrium under a given set of operating conditions, any change in these conditions, such as a change in load, speed, or viscosity, will alter the pressure distribution and thus momentarily shift the center of pressure, creating a moment that causes the pad to change its inclination and shoulder height s_h. A pivoted-pad slider bearing is thus supported at a single point so that the angle of inclination becomes a variable. It has much better stability than a fixed-incline slider bearing under varying operating conditions. The location of the shoe's pivot point can be found from the equilibrium of moments acting on the shoe about the point. For all practical purposes, only two significant forces may be considered in the moment equation: the resultant force due to film pressure and the reaction force normal to the shoe surface. The force due to friction in the pivot is ignored.

Pivoted pads are used in multiples as pivoted-pad thrust bearings (Fig. 12.20). Calculations are carried through for a single pad, and the properties for the complete bearing are found by combining these calculations in the proper manner, as discussed in Sec. 12.3.4.

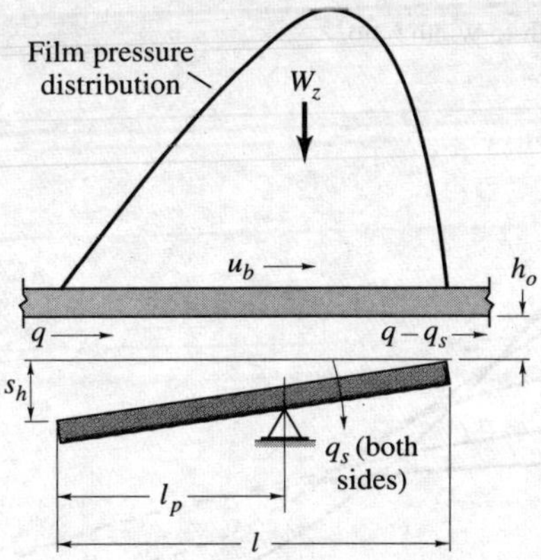

Figure 12.19 Side view of pivoted-pad thrust bearing. [A. A. Raimondi and J. Boyd, "Applying Bearing Theory to the Analysis and Design of Pad-Type Bearings," *ASME Transactions*, vol. 77, no. 3, 1955, pp. 287–309. Reprinted by permission of the American Society of Mechanical Engineers International.]

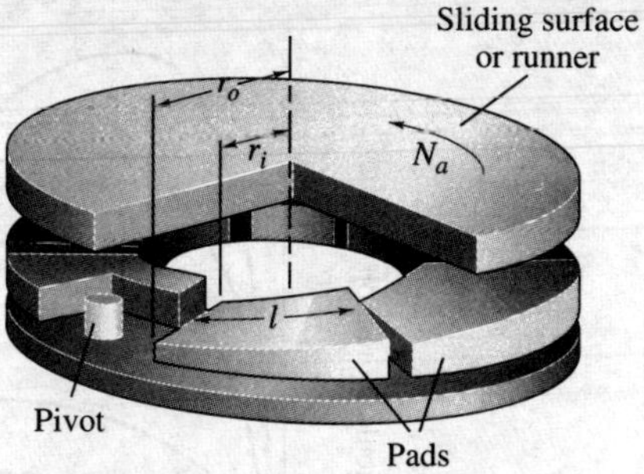

Figure 12.20 Configuration of multiple-pivoted-pad thrust bearing. [A. A. Raimondi and J. Boyd, "Applying Bearing Theory to the Analysis and Design of Pad-Type Bearings," *ASME Transactions*, vol. 77, no. 3, 1955, pp. 287–309. Reprinted by permission of the American Society of Mechanical Engineers International.]

Normally, a pivoted pad will carry load if the pivot is placed somewhere between the center of the pad and the outlet edge ($0.5 \le L_p \le 1.0$). (Note that $L_p = l_p/l$.) For bidirectional operation the pivot is located at the center of the pad, $L_p = 0.5$.

The following procedure helps in designing pivoted-pad thrust bearings:

1. Having established whether minimum power or maximum load is more critical in a particular case and having chosen a pad length-to-width ratio λ, establish the pivot location from Fig. 12.21.
2. For a fixed λ and a dimensionless location of the pivot from the inlet L_p obtained from Fig. 12.21, obtain the dimensionless function $H_o/\sqrt{B_t}$ from Fig. 12.22.

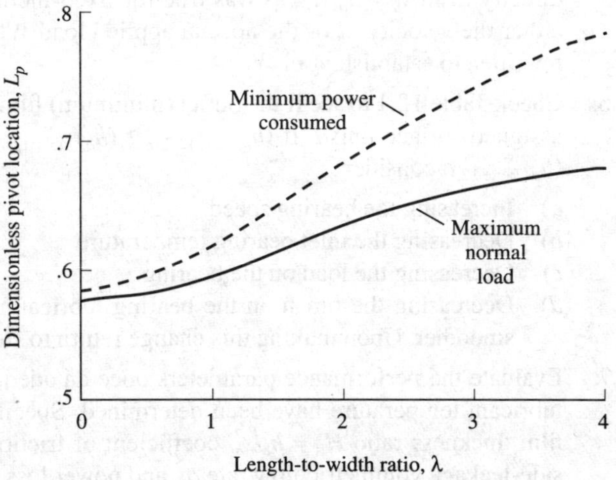

Figure 12.21 Chart for determining pivot location corresponding to maximum load or minimum power loss for various pad proportions of pivoted-pad bearings. [A. A. Raimondi and J. Boyd, "Applying Bearing Theory to the Analysis and Design of Pad-Type Bearings," *ASME Transactions*, vol. 77, no. 3, 1955, pp. 287–309. Reprinted by permission of the American Society of Mechanical Engineers International.]

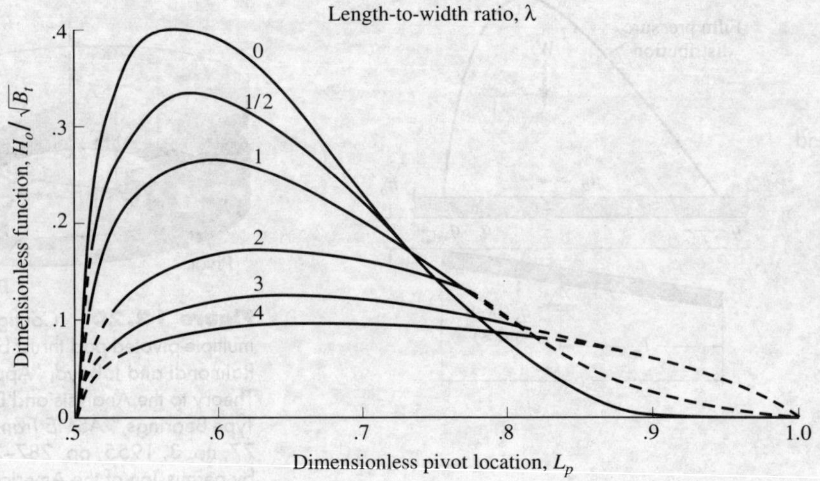

Figure 12.22 Chart for determining outlet film thickness for pivoted-pad thrust bearings. [A. A. Raimondi and J. Boyd, "Applying Bearing Theory to the Analysis and Design of Pad-Type Bearings," *ASME Transactions*, vol. 77, no. 3, 1955, pp. 287–309. Reprinted by permission of the American Society of Mechanical Engineers International.]

3. From Fig. 12.23 determine the temperature rise due to shear heating for a given λ and $H_o/\sqrt{B_t}$.
4. Just as was done for a fixed-incline thrust bearing, establish the lubricant temperature by using Eq. (12.71). Once the temperature is known, obtain the viscosity as discussed for a fixed-incline bearing.
5. Obtain the outlet (minimum) film thickness from

$$h_o = H_o l \sqrt{\frac{\eta_0 u_b w_t}{W_z B_t}} \tag{12.73}$$

Once h_o is known, in some applications the shoulder height s_h can be determined directly from $s_h = h_o/H_o$, as was true for fixed-incline bearings. If h_o is specified and either the velocity u_b or the normal applied load W_z is not known, Eq. (12.73) can be rewritten to establish u_b or W_z.

6. Check Table 12.1 to see if the outlet (minimum) film thickness is sufficient for the preassigned surface finish. If $(h_o)_{\text{Eq. (12.73)}} \geq (h_o)_{\text{Table 12.1}}$, go to step 7. If If $(h_o)_{\text{Eq. (12.73)}} < (h_o)_{\text{Table 12.1}}$, consider
   a) Increasing the bearing speed
   b) Decreasing the inlet bearing temperature
   c) Decreasing the load on the bearing
   d) Decreasing the finish on the bearing lubricating surfaces, thereby making them smoother. Upon making this change return to step 2.
7. Evaluate the performance parameters once an adequate minimum film thickness and lubricant temperature have been determined. Specifically, from Fig. 12.24 the outlet film thickness ratio $H_o = h_o/s_h$, coefficient of friction μ, total volumetric flow rate q, side-leakage volumetric flow rate q_s, and power loss h_p can be determined.

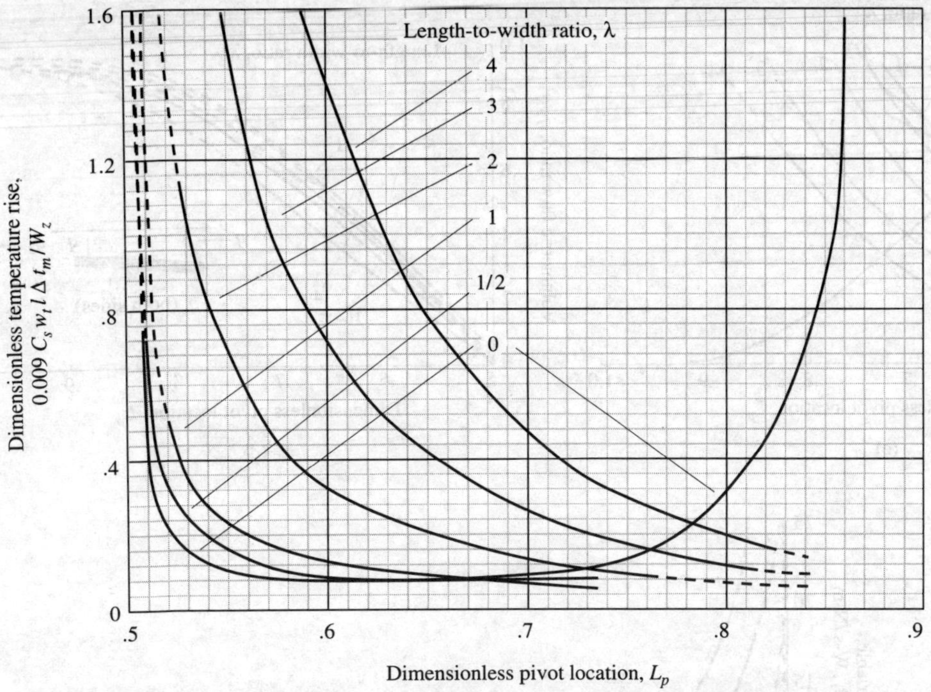

Figure 12.23 Chart for determining dimensionless temperature rise due to viscous shear heating of lubricant for pivoted-pad thrust bearings. [A. A. Raimondi and J. Boyd, "Applying Bearing Theory to the Analysis and Design of Pad-Type Bearings," *ASME Transactions*, vol. 77, no. 3, 1955, pp. 287–309. Reprinted by permission of the American Society of Mechanical Engineers International.]

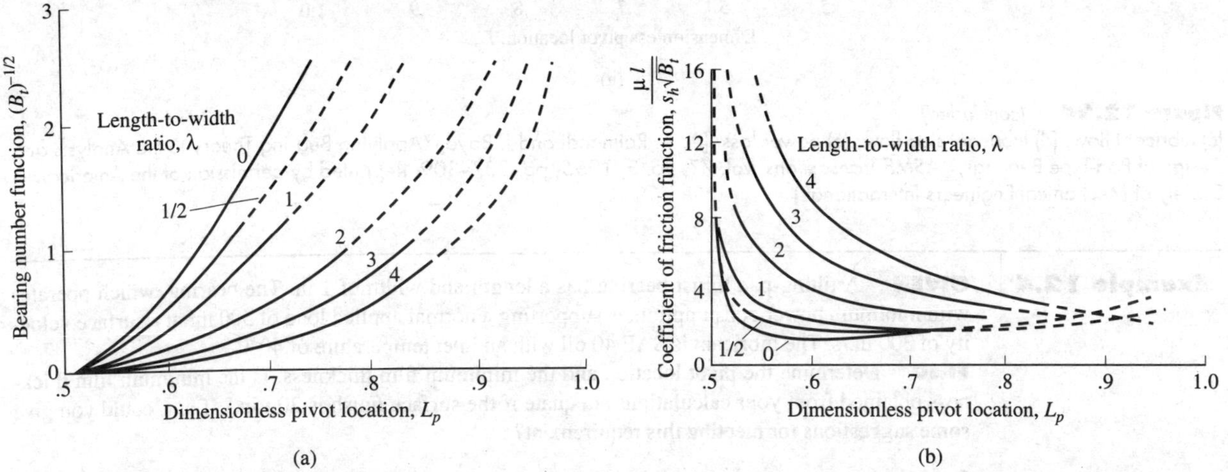

Figure 12.24 Chart for determining performance parameters of pivoted-pad thrust bearings. (a) Film thickness ratio; (b) coefficient of friction; *(continued)*

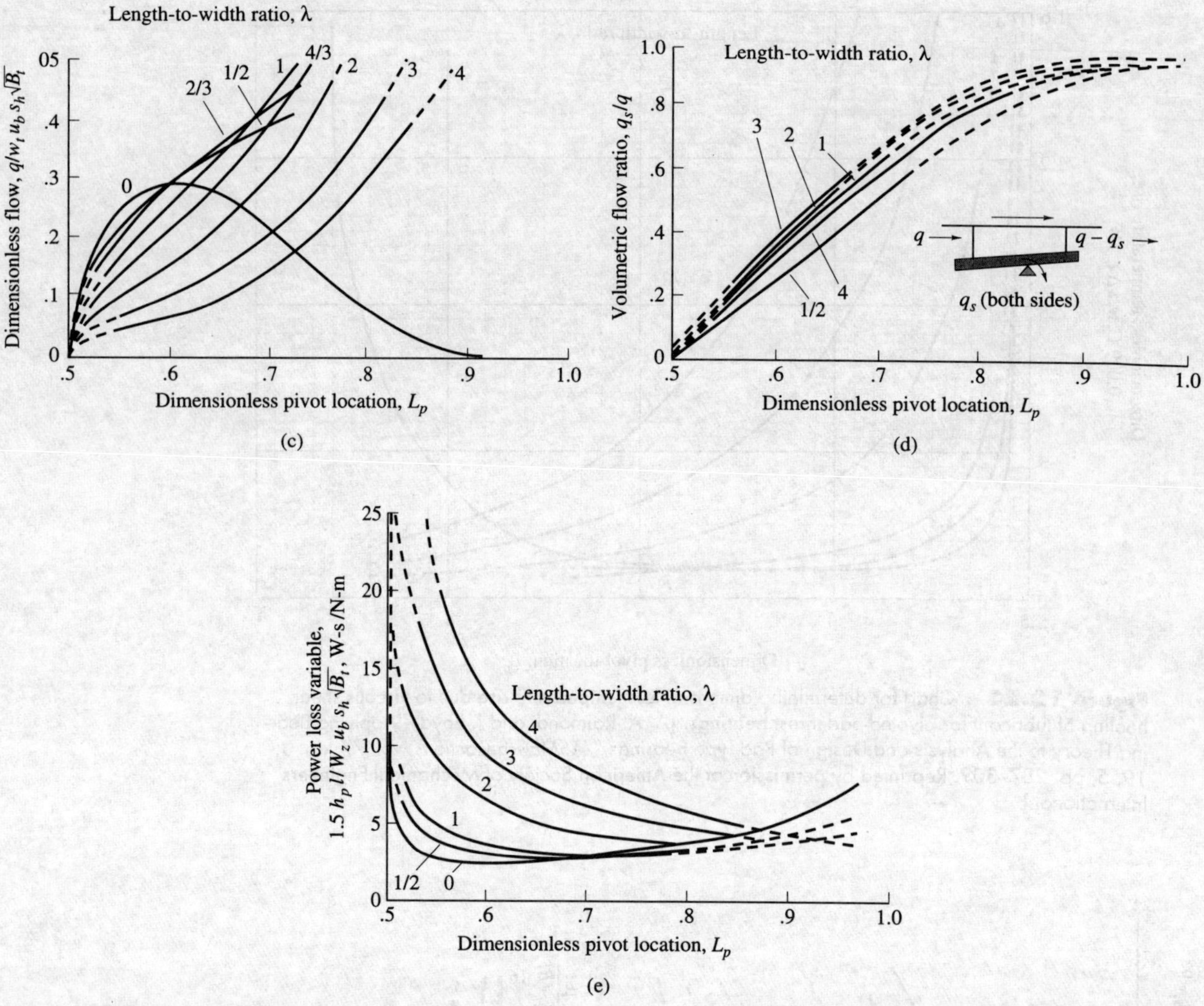

Figure 12.24 *(concluded)*
(c) lubricant flow; (d) lubricant side flow; (e) power loss. [A. A. Raimondi and J. Boyd, "Applying Bearing Theory to the Analysis and Design of Pad-Type Bearings," *ASME Transactions*, vol. 77, no. 3, 1955, pp. 287–309. Reprinted by permission of the American Society of Mechanical Engineers International.]

Example 12.4

GIVEN A tilting-pad thrust bearing has a length and width of 1 in. The bearing, which operates with minimum power consumption, is supporting a normal applied load of 500 lbf at a surface velocity of 300 in./s. The lubricant is SAE 40 oil with an inlet temperature of 40°C.

FIND Determine the pivot location and the minimum film thickness. Is the minimum film thickness obtained from your calculations adequate if the surface finish is 20 μin.? If not, could you give some suggestions for meeting this requirement?

Solution

From Fig. 12.21 for $\lambda = 1$ and minimum power consumption, $L_p = 0.62$. From Fig. 12.22 for $L_p = 0.62$ and $\lambda = 1$

$$\frac{H_o}{\sqrt{B_t}} = 0.26$$

From Fig. 12.23 for $L_p = 0.62$ and $\lambda = 1$

$$\frac{0.009 C_s w_t l \Delta t_m}{W_z} = 0.11$$

$$\therefore \Delta t_m = \frac{(0.11)(500.0)}{(0.009)(197.2)(1)(1)} = 30.99°\text{C}$$

From Eq. (12.71)

$$\tilde{t}_m = t_{mi} + \frac{\Delta t_m}{2} = 40 + \frac{30.99}{2} = 55.49°\text{C}$$

From Fig. 8.13 for SAE 40 oil at 55.5°C the absolute viscosity is 8×10^{-6} lbf-s/in.^{2}.

Making use of Eqs. (12.68) and (12.73) while solving for the outlet (minimum) film thickness gives

$$h_o = 0.26 l \sqrt{\frac{\eta_0 u_b w_t}{W_z}}$$

$$\therefore h_o = 0.26(1)\sqrt{\frac{(8)(10^{-6})(300)(1)}{500}} = 5.70 \times 10^{-4} \text{ in.} = 570\ \mu\text{in.}$$

From Fig. 12.24(a) for $L_p = 0.62$ and $\lambda = 1$

$$\frac{1}{\sqrt{B_t}} = 0.45$$

$$\therefore H_o = 0.26\sqrt{B_t} = \frac{0.26}{0.45} = 0.578$$

$$s_h = \frac{h_o}{H_o} = \frac{(5.70)(10^{-4})}{0.578} = 9.864 \times 10^{-4} \text{ in.} = 986 \times 10^{-6} \text{ in.}$$

From Table 12.1 for a surface finish of 20 μin. the allowable minimum film thickness is 500 μin. Thus,

$$(h_o)_{\text{all}} = 5 \times 10^{-4} \text{ in.}$$

Since the calculated minimum film thickness is greater than the allowable minimum film thickness $(h_o)_{\min}$ obtained from Table 12.1, the pad design is adequate.

12.3.7 Thrust Slider Bearing Geometry

This chapter has dealt with the performance of an individual pad of a thrust bearing. Normally, a number of identical pads are assembled in a thrust bearing as shown, for example, in Figs. 12.14 and 12.20. The length, width, speed, and load of an individual pad can be related to the geometry of a thrust bearing by the following formulas:

$$w_t = r_o - r_i \tag{12.74}$$

$$l = \frac{r_o + r_i}{2}\left(\frac{2\pi}{N_p} - \frac{\pi}{36}\right) \tag{12.75}$$

$$u_b = \frac{(r_o + r_i)\omega}{2} \tag{12.76}$$

$$W_t = N_p\, W_z \tag{12.77}$$

where N_p is the number of identical pads placed in the thrust bearing (usually between 3 and 20). The $\pi/36$ portion of Eq. (12.75) accounts for feed grooves between the pads. These are deep grooves that ensure that ambient pressure is maintained between the pads. Also, in Eq. (12.77) W_t is the total thrust load on the bearing.

12.4 JOURNAL SLIDER BEARINGS

The last section dealt with slider bearing pads as used in thrust bearings. The surfaces of *thrust bearings* are *perpendicular* to the axis of rotation, as shown in Fig. 12.9. This section deals with **journal bearings,** where the bearing surfaces are *parallel* to the axis of rotation, as shown in Fig. 12.25. Journal bearings are used to support shafts and to carry radial loads with minimum power loss and minimum wear. The journal bearing can be represented by a plain cylindrical sleeve (bushing) wrapped around the journal (shaft) but can adopt a variety of forms. The lubricant is supplied at some convenient point in the bearing through a hole or a groove. If the bearing extends around the full 360° of the journal, it is described as a *full* journal bearing. If the wrap angle is less than 360°, the term "*partial* journal bearing" is used.

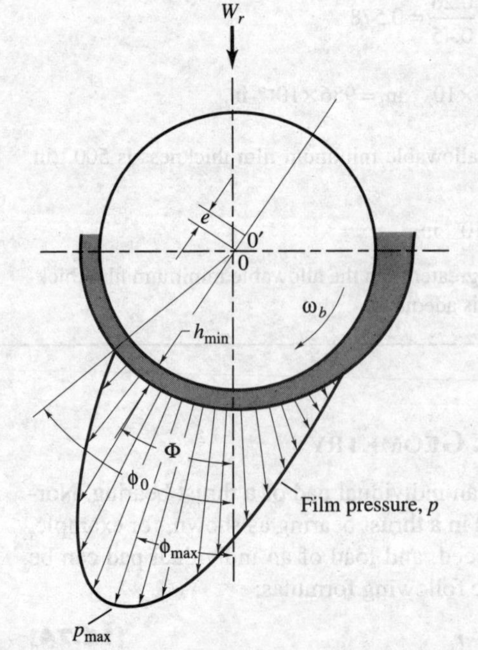

Figure 12.25 Pressure distribution around a journal bearing.

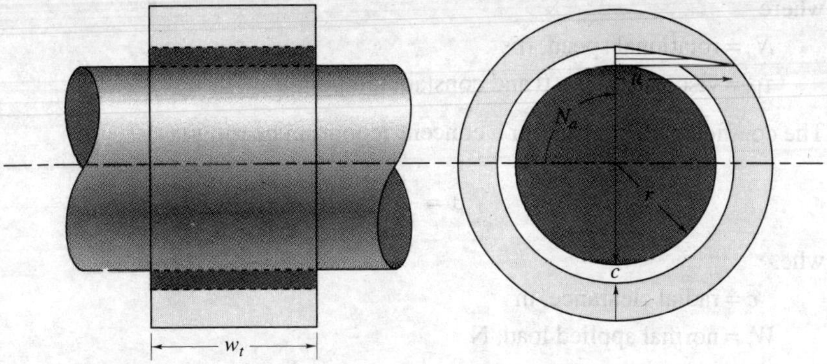

Figure 12.26 Concentric journal bearing.

12.4.1 PETROV'S EQUATION

Petrov's equation applies Newton's postulate (Sec. 8.6.1) to a full journal running concentrically with the bearing (Fig. 12.26). As shown later (Sec.12.4.3) the journal will run concentrically with the bearing only when one of the following conditions prevails: (1) the radial load acting on the bearing is zero, (2) the viscosity of the lubricant is infinite, or (3) the speed of the journal is infinite. None of these conditions are practically possible. However, if the load is light enough, if the viscosity is sufficiently high, and if the journal has a sufficiently high speed, the eccentricity of the journal relative to the bearing may be so small that the oil film around the journal can be considered practically to be of uniform thickness.

The oil film in a journal bearing is always thin relative to the bearing radius. Therefore, the curvature of the bearing surface may be ignored, and the film may be considered as an unwrapped body having a thickness equal to the radial clearance, a length equal to $2\pi r$, and width w_t. Assume that the viscosity throughout the oil film is constant. In Fig. 12.27 the bottom surface is stationary and the top surface is moving with constant velocity u. Petrov (1883) assumed no slip at the interface between the lubricant and the solids.

Making use of Newton's postulate as expressed in Eq. (8.10) gives the friction force in a concentric journal bearing as

$$F = \eta_0 A \frac{u}{c} = \eta_0 2\pi r w_t \frac{2\pi r N_a}{c} = \frac{4\pi^2 \eta_0 r^2 w_t N_a}{c} \qquad \textbf{(12.78)}$$

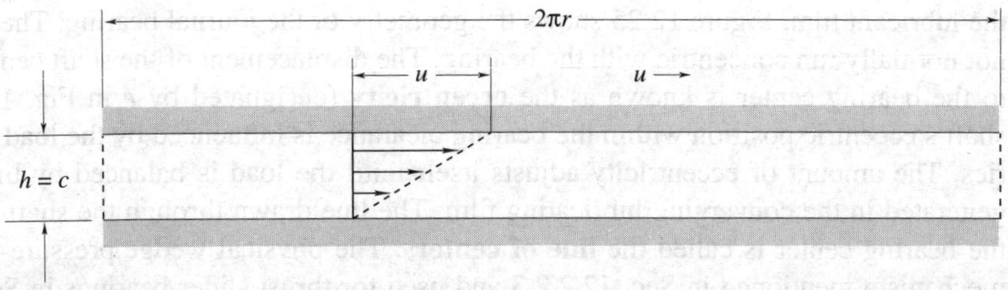

Figure 12.27 Developed journal and bearing surfaces for a concentric journal bearing.

where

N_a = rotational speed, rps

η_0 = viscosity at $p = 0$ and constant temperature

The coefficient of friction for a concentric journal bearing is

$$\mu = \frac{F}{W_r} = \frac{4\pi^2 \eta_0 r^2 w_t N_a}{c W_r} \tag{12.79}$$

whe.˒

c = radial clearance, m

W_r = normal applied load, N

The friction torque for a concentric journal bearing can be written as

$$T = Fr = \frac{4\pi^2 \eta_0 r^3 w_t N_a}{c} = \frac{2\pi \eta_0 r^3 w_t \omega}{c} \tag{12.80}$$

where

$\omega = 2\pi N_a$ = angular velocity, rad/s

Equation (12.80) is generally called **Petrov's equation** (after N. Petrov, who suggested a similar equation for torque in his work published in 1883).

The power loss is the torque times the angular velocity. The power loss for a concentric (lightly loaded) journal bearing can be expressed in horsepower as

$$h_p = \frac{8\pi^3}{(12)(550)} \frac{\eta_0 r^3 w_t N_a^2}{c} = (0.03758) \frac{\eta_0 r^3 w_t N_a^2}{c} \tag{12.81}$$

where

η_0 = viscosity at $p = 0$ and constant temperature, lbf-s/in.^{2}

r = radius of journal, in.

w_t = width of journal, in.

N_a = rotational speed, rps

c = radial clearance, in.

Equation (12.81) is valid only for these units.

12.4.2 JOURNAL SLIDER BEARING OPERATION

Journal bearings rely on shaft sliding motion to generate the load-supporting pressures in the lubricant film. Figure 12.25 shows the geometry of the journal bearing. The shaft does not normally run concentric with the bearing. The displacement of the shaft center relative to the bearing center is known as the **eccentricity** (designated by e in Fig. 12.25). The shaft's eccentric position within the bearing clearance is influenced by the load that it carries. The amount of eccentricity adjusts itself until the load is balanced by the pressure generated in the converging lubricating film. The line drawn through the shaft center and the bearing center is called the **line of centers.** The physical wedge pressure-generating mechanism mentioned in Sec. 12.2.2.3 and used for thrust slider bearings in Secs. 12.3.5 and 12.3.6 is also valid in this section.

The pressure generated and therefore the load-carrying capacity of the bearing depend on the shaft eccentricity, the angular velocity, the effective viscosity of the lubricant, and the bearing dimensions and clearance:

$$W_r = f(e, \omega, \eta_0, r, w_t, c)$$

The load and the angular velocity are usually specified and the minimum shaft radius is often predetermined. To complete the design, it will be necessary to calculate the bearing width and clearance and to choose a suitable lubricant if one is not already specified.

12.4.3 OPERATING AND PERFORMANCE PARAMETERS

The Reynolds equation used for thrust slider bearings [Eq. (12.28)] is altered for a journal slider bearing with $x = r_b\phi$ and $\tilde{u} = u_b/2 = r_b\omega_b/2$.

$$\therefore \quad \frac{\partial}{\partial \phi}\left(h^3 \frac{\partial p}{\partial \phi}\right) + r_b^2 \frac{\partial}{\partial y}\left(h^3 \frac{\partial p}{\partial y}\right) = 6\eta_0 \omega_b r_b^2 \frac{\partial h}{\partial \phi} \qquad \textbf{(12.82)}$$

The film thickness around the journal is expressed as

$$h = c(1 + \varepsilon \cos\phi) \qquad \textbf{(12.83)}$$

where

$\varepsilon = e/c =$ eccentricity ratio

Substituting Eq. (12.83) into Eq. (12.82) gives

$$\frac{\partial}{\partial \phi}\left(h^3 \frac{\partial p}{\partial \phi}\right) + r_b^2 h^3 \frac{\partial^2 p}{\partial y^2} = -6\eta_0 \omega_b r_b^2 e \sin\phi \qquad \textbf{(12.84)}$$

Analytical solutions to Eq. (12.84) are not normally available and numerical methods are needed. Equation (12.84) is often solved by using a relaxation method in which the first step is to replace the derivatives in Eq. (12.84) by finite difference approximations. The lubrication area is covered by a mesh, and the numerical method relies on the fact that a function can be represented with sufficient accuracy over a small range by a quadratic expression. The Reynolds boundary condition that requires that $p = 0$ and $dp/dx = 0$ at $\phi = \phi^*$ (the outlet boundary) is used. Only the results obtained by using this numerical method are presented in this section.

The *operating* parameters for hydrodynamic journal bearings are

1. Bearing number for journal bearings (also called the **Sommerfeld number**)

$$B_j = \left(\frac{\eta_0 \omega_b r_b w_t}{\pi W_r}\right)\left(\frac{r_b}{c}\right)^2 \qquad \textbf{(12.85)}$$

2. Diameter-to-width ratio

$$\lambda_j = 2r_b/w_t \qquad \textbf{(12.86)}$$

3. Angular extent of journal (full or partial)

The *performance* parameters for hydrodynamic journal bearings are

1. Eccentricity e
2. Location of minimum film thickness, sometimes called attitude angle Φ
3. Coefficient of sliding friction μ
4. Total and side-leakage volumetric flow rates q and q_s
5. Angle of maximum pressure ϕ_{max}
6. Location of terminating pressure ϕ_0
7. Temperature rise due to lubricant shearing Δt_m

The parameters Φ, ϕ_{max}, and ϕ_0 are described in Fig. 12.25, which gives the pressure distribution around a journal bearing. Note from this figure that if the bearing is concentric ($e = 0$), the film shape around the journal is constant and equal to c and no fluid film pressure is developed. At heavy loads, the other extreme, the journal is forced downward and the limiting position is reached when $h_{min} = 0$ and $e = c$ (i.e., the journal is touching the bearing).

Temperature rise due to lubricant shearing will be considered here as was done in Sec. 12.3.3 for thrust bearings. In Eq. (12.84) the lubricant viscosity corresponds to the viscosity when $p = 0$ but can vary as a function of temperature. Since work is done on the lubricant as the fluid is being sheared, the temperature of the lubricant is higher when it leaves the conjunction than on entry. In Chapter 8 (Figs. 8.12 and 8.13 as well as Table 8.3) it was shown that oil viscosity drops off significantly with rising temperature. This decrease is compensated for by using a mean of the inlet and outlet temperatures

$$\tilde{t}_m = t_{mi} + \frac{\Delta t_m}{2} \tag{12.87}$$

where

t_{mi} = inlet temperature
Δt_m = temperature rise of lubricant from inlet to outlet

The viscosity used in the bearing number B_j and other performance parameters is at the mean temperature $\tilde{t}_m$. The temperature rise of the lubricant from inlet to outlet Δt_m can be determined from the performance charts provided in this section.

12.4.4 DESIGN PROCEDURE

Now that the operating and performance parameters have been defined, the design procedure for a hydrodynamic journal bearing can be presented. The results are for a *full* journal bearing. Results for a *partial* journal bearing can be obtained from Raimondi and Boyd (1958).

Figure 12.28 shows the effect of the bearing number B_j on the minimum film thickness for four diameter-to-width ratios. The following relationship should be observed:

$$h_{min} = c - e \tag{12.88}$$

In dimensionless form

$$H_{\min} = \frac{h_{\min}}{c} = 1 - \varepsilon \tag{12.89}$$

where

$$\varepsilon = e/c = \text{eccentricity ratio} \tag{12.90}$$

The bearing number for journal bearings is expressed in Eq. (12.85). In a given design the bearing number is affected by

1. Absolute lubricant viscosity η_0
2. Angular shaft speed ω_b
3. Radial load W_r
4. Radial clearance c
5. Journal dimensions r_b and w_t

All these parameters affect the bearing number and thus the design of the journal bearing.

In Fig. 12.28 a recommended operating eccentricity ratio, or minimum film thickness, is indicated as well as a preferred operating area. The left boundary of the shaded zone defines the optimum eccentricity ratio for a minimum coefficient of friction; and the right boundary, the optimum eccentricity ratio for maximum load. The recommended operating eccentricity is midway between these two boundaries.

Figure 12.29 shows the effect of the bearing number on the attitude angle Φ [angle between the load direction and a line drawn through the centers of the bearing and journal

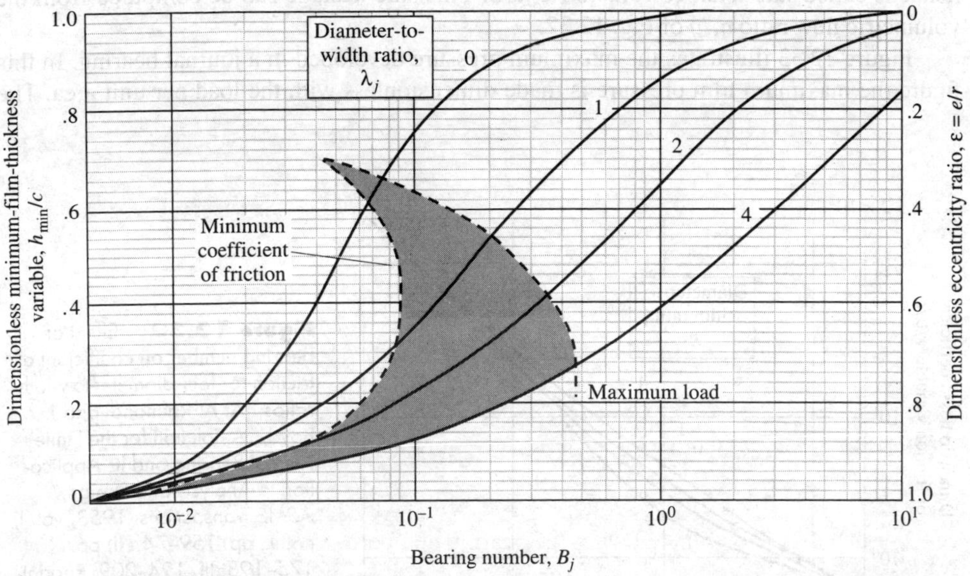

Figure 12.28 Effect of bearing number on minimum film thickness for four diameter-to-width ratios. [A. A. Raimondi and J. Boyd, "A Solution for the Finite Journal Bearing and Its Application to Analysis and Design," *ASME Transactions*, 1958, vol. I, no. 1, pp. 159–74.: II, pp. 175–193: III, 194–209. Reprinted with permission of the American Association of Mechanical Engineers International.]

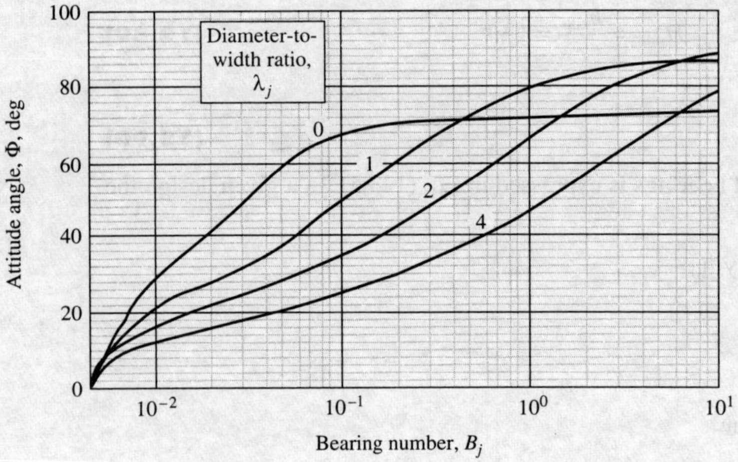

Figure 12.29 Effect of bearing number on attitude angle for four diameter-to-width ratios. [A. A. Raimondi and J. Boyd, "A Solution for the Finite Journal Bearing and Its Application to Analysis and Design," *ASME Transactions,* 1958, vol. I, no. 1, pp. 159–74.: II, pp. 175–193: III, 194–209. Reprinted with permission of the American Association of Mechanical Engineers International.]

(see Fig. 12.25)] for four values of λ_j. This angle establishes where the minimum and maximum film thicknesses are located within the bearing.

Figure 12.30 shows the effect of the bearing number on the coefficient of friction for four values of λ_j. The effect is small for a complete range of dimensionless load parameters.

Figure 12.31 shows the effect of bearing number on the dimensionless volumetric flow rate $Q = 2\pi q / r_b c w_t \omega_b$ for four values of λ_j. The dimensionless volumetric flow rate Q that is pumped into the converging space by the rotating journal can be obtained from this figure. Of the volume of oil q pumped by the rotating journal, an amount q_s flows out the ends and hence is called *side-leakage volumetric flow.* This side leakage can be computed from the volumetric flow ratio q_s/q of Fig. 12.32.

Figure 12.33 illustrates the maximum pressure developed in a journal bearing. In this figure the maximum film pressure is made dimensionless with the load per unit area. The

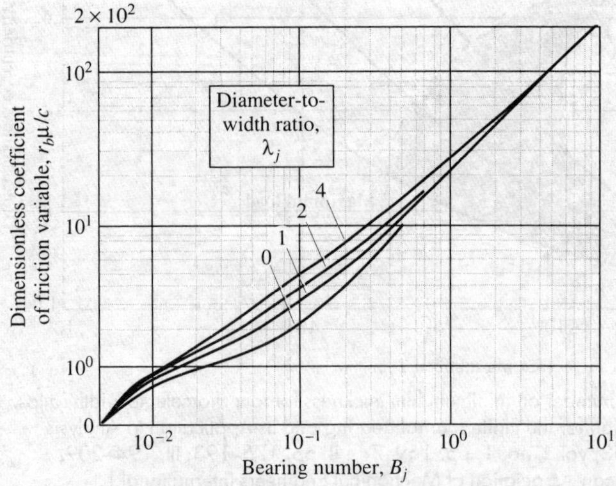

Figure 12.30 Effect of bearing number on coefficient of friction for four diameter-to-width ratios. [A. A. Raimondi and J. Boyd, "A Solution for the Finite Journal Bearing and Its Application to Analysis and Design," *ASME Transactions,* 1958, vol. I, no. 1, pp. 159–74.: II, pp. 175–193: III, 194–209. Reprinted with permission of the American Association of Mechanical Engineers International.]

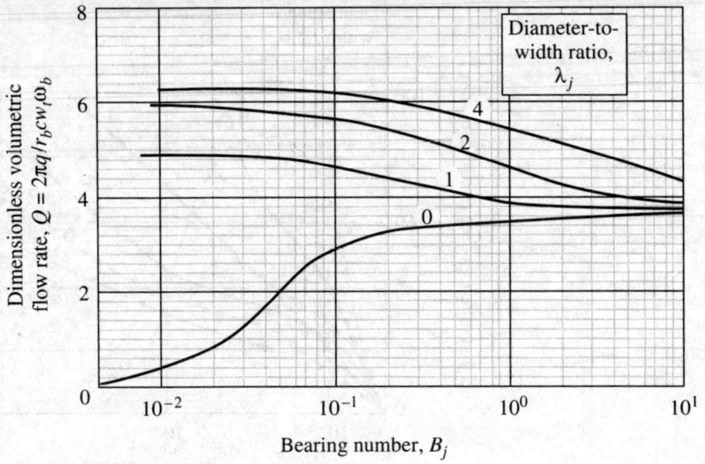

Figure 12.31 Effect of bearing number on dimensionless volumetric flow rate for four diameter-to-width ratios. [A. A. Raimondi and J. Boyd, "A Solution for the Finite Journal Bearing and Its Application to Analysis and Design," *ASME Transactions,* 1958, vol. I, no. 1, pp. 159–74.: II, pp. 175–193: III, 194–209. Reprinted with permission of the American Association of Mechanical Engineers International.]

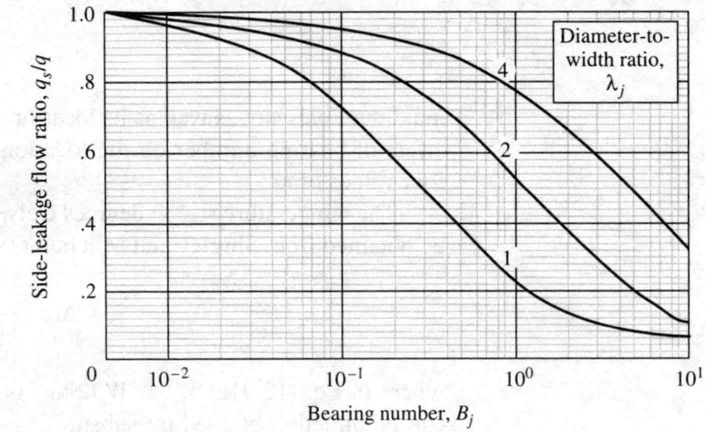

Figure 12.32 Effect of bearing number on side-leakage flow ratio for four diameter-to-width ratios. [A. A. Raimondi and J. Boyd, "A Solution for the Finite Journal Bearing and Its Application to Analysis and Design," *ASME Transactions,* 1958, vol. I, no. 1, pp. 159–74.: II, pp. 175–193: III, 194–209. Reprinted with permission of the American Association of Mechanical Engineers International.]

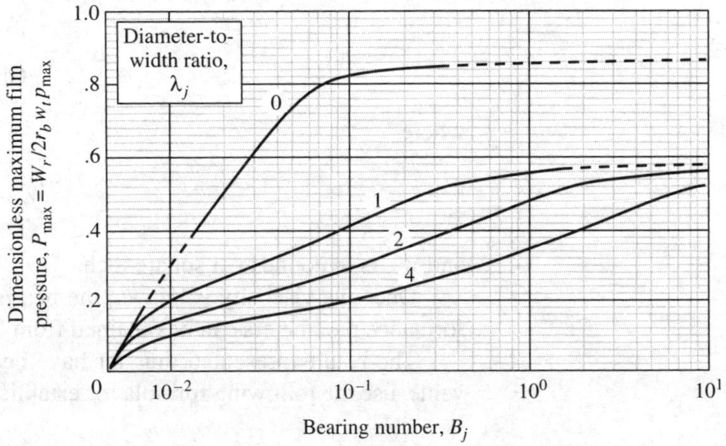

Figure 12.33 Effect of bearing number on dimensionless maximum film pressure for four diameter-to-width ratios. [A. A. Raimondi and J. Boyd, "A Solution for the Finite Journal Bearing and Its Application to Analysis and Design," *ASME Transactions,* 1958, vol. I, no. 1, pp. 159–74.: II, pp. 175–193: III, 194–209. Reprinted with permission of the American Association of Mechanical Engineers International.]

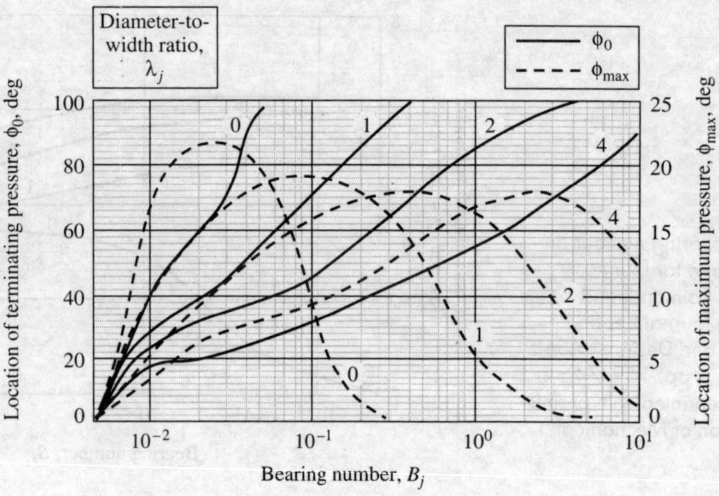

Figure 12.34 Effect of bearing number on location of terminating and maximum pressures for four diameter-to-width ratios. [A. A. Raimondi and J. Boyd, "A Solution for the Finite Journal Bearing and Its Application to Analysis and Design," *ASME Transactions*, 1958, vol. I, no. 1, pp. 159–74.: II, pp. 175–193: III, 194–209. Reprinted with permission of the American Association of Mechanical Engineers International.]

maximum pressure as well as its location are shown in Fig. 12.25. Figure 12.34 shows the effect of bearing number on the location of the terminating and maximum pressures for four values of λ_j.

The temperature rise in degrees Celsius of the lubricant from the inlet to the outlet can be obtained from Shigley and Mitchell (1983) as

$$\Delta t_m = \frac{8.3W_r^*(r_b/c)\mu}{Q(1-0.5\,q_s/q)} \tag{12.91a}$$

where in Eq. (12.91a) $W_r^* = W_r/2r_b w_t$ is in megapascals. Therefore, the temperature rise can be directly obtained by substituting the values of $r_b\mu/c$ obtained from Fig. 12.30, Q from Fig. 12.31, and q_s/q from Fig. 12.32 into Eq. (12.91a). The temperature rise in degrees Fahrenheit is given by

$$\Delta t_m = \frac{0.103W_r^*(r_b/c)\mu}{Q(1-0.5\,q_s/q)} \tag{12.91b}$$

where

$$W_r^* = \frac{W_r}{2r_b w_t} \tag{12.91c}$$

and W_r^* is in pounds per square inch.

Once the viscosity is known, the bearing number can be calculated and then the performance parameters can be obtained from Figs. 12.28 to 12.34 and Eqs. (12.91).

The results presented thus far have been for λ_j of 0, 1, 2, and 4. If λ_j is some other value, use the following formula for establishing the performance parameters:

$$y=\frac{1}{(w_t/2r_b)^3}\left[-\frac{1}{8}\left(1-\frac{w_t}{2r_b}\right)\left(1-\frac{w_t}{r_b}\right)\left(1-2\frac{w_t}{r_b}\right)y_0+\frac{1}{3}\left(1-\frac{w_t}{r_b}\right)\left(1-2\frac{w_t}{r_b}\right)y_1\right.$$
$$\left.-\frac{1}{4}\left(1-\frac{w_t}{2r_b}\right)\left(1-2\frac{w_t}{r_b}\right)y_2+\frac{1}{24}\left(1-\frac{w_t}{2r_b}\right)\left(1-\frac{w_t}{r_b}\right)y_4\right] \quad \textbf{(12.92)}$$

where y is any one of the performance parameters (H_{min}, Φ, $r_b\mu/c$, Q, q_s/q, p_{max}, ϕ_0, or ϕ_{max}) and where the subscript on y is the λ_j value; for example, y_1 is equivalent to y evaluated at $\lambda_j = 1$. All the results presented are valid for a full journal bearing.

Example 12.5

GIVEN A full journal bearing has the specifications SAE 60 oil with an inlet temperature of 40°C, $N_a = 30$ rps, $W_r = 2200$ N, $r_b = 2$ cm, and $w_t = 4$ cm.

FIND From the figures given in this section establish the operating and performance parameters for this bearing while designing for maximum load.

Solution

The angular speed can be expressed as

$$\omega_b = 2\pi N_a = 2\pi(30) = 60\pi \text{ rad/s}$$

The diameter-to-width ratio is

$$\lambda_j = \frac{2r_b}{w_t} = \frac{2(2)}{4} = 1$$

From Fig. 12.28 for $\lambda_j = 1$ and designing for maximum load

$$B_j = 0.2 \qquad \frac{h_{min}}{c} = 0.53 \qquad \text{and} \qquad \varepsilon = 0.47 \quad \textbf{(a)}$$

For $B_j = 0.2$ and $\lambda_j = 1$ from Figs. 12.30 to 12.32

$$\frac{r_b\mu}{c} = 4.9 \qquad Q = 4.3 \qquad \text{and} \qquad \frac{q_s}{q} = 0.6 \quad \textbf{(b)}$$

From Eq. (12.91c) the radial load per area is

$$W_r^* = \frac{W_r}{2r_b w_t} = \frac{2200}{2(2)(4)(10^{-4})} \text{ Pa} = 1.375 \text{ MPa} \quad \textbf{(c)}$$

The lubricant temperature rise in degrees Celsius obtained by using Eq. (12.91a) and the results from Eqs. (b) and (c) is

$$\Delta t_m = \frac{8.3W_r^*(r_b/c)\mu}{Q(1-0.5q_s/q)} = \frac{8.3(1.375)(4.9)}{4.3[1-(0.5)(0.6)]} = 18.58°\text{C}$$

From Eq. (12.87) the mean temperature in the lubricant conjunction is

$$\tilde{t}_m = t_{mi} + \frac{\Delta t_m}{2} = 40 + \frac{18.58}{2} = 49.29°\text{C}$$

From Fig. 8.13 for SAE 60 oil at 49.3°C the absolute viscosity is

$$5\times10^{-5} \text{ reyn} = 3.45\times10^2 \text{ centipose} = 0.345 \text{ N-s/m}^2$$

From Eq. (12.85) the radial clearance can be expressed as

$$c = r_b\sqrt{\frac{\eta_0\omega_b r_b w_t}{\pi W_r B_j}} = (2)(10^{-2})\sqrt{\frac{(0.345)60\pi(2)(10^{-2})(4)(10^{-2})}{\pi(2.2)(10^3)(0.2)}} = 0.1227\times10^{-3}\ \text{m}$$

The coefficient of friction from Eq. (b) is

$$\mu = \frac{4.9c}{r_b} = \frac{(4.9)(0.1227)(10^{-3})}{(2)(10^{-2})} = 0.0301$$

The circumferential volumetric flow rate is

$$q = \frac{Q r_b c w_t \omega_b}{2\pi} = \frac{(4.3)(2)(10^{-2})(0.1227)(10^{-3})(4)(10^{-2})60\pi}{2\pi} = 12.67\times10^{-6}\ \text{m}^3/\text{s}$$

$$q_s = 7.288\times10^{-6}\ \text{m}^3/\text{s}$$

From Fig. 12.29 for $B_j = 0.2$ and $\lambda_j = 1$ the attitude angle is 61°. From Fig. 12.33 for $B_j = 0.2$ and $\lambda_j =$ 1 the dimensionless maximum pressure is $P_{\max} = 0.46$. The maximum pressure is

$$p_{\max} = \frac{W_r}{2 r_b w_t P_{\max}} = \frac{2200}{2(2)(10^{-2})(4)(10^{-2})(0.46)}\ \text{Pa} = 2.989\ \text{MPa}$$

From Fig. 12.34 for $B_j = 0.2$ and $\lambda_j = 1$ the location of the maximum pressure from the applied load is 21.5° and the location of the terminating pressure from the applied load is 86°.

12.4.5 OPTIMIZATION TECHNIQUES

The radial clearance c is the most difficult of the performance parameters to control accurately during manufacturing, and it may increase because of wear. Figure 12.35 shows the performance of a particular bearing for a range of radial clearances. If the clearance is too tight, the temperature will be too high and the minimum film thickness too low. High temperature may cause the bearing to fail by fatigue. If the oil film is too thin, dirt particles may not pass through without scoring or may embed themselves in the bearing. In either event there will be excessive wear and friction, resulting in high temperatures and possible seizing. A large clearance will permit dirt particles to pass through and also permit a large flow of oil, lowering the temperature and lengthening bearing life. However, if the clearance becomes too large, the bearing becomes noisy and the minimum film thickness begins to decrease again.

Figure 12.35 shows the best compromise, when both the production tolerance and the future wear on the bearing are considered, to be a clearance range slightly to the left top of the minimum-film-thickness curve. Future wear will move the operating point to the right, increasing the film thickness and decreasing the operating temperature.

Tutorial 7

12.4.6 NONPLAIN CONFIGURATIONS

Thus far, this chapter has focused on plain full journal bearings. As applications have demanded higher speeds, vibration problems due to critical speeds, imbalance, and instability have created a need for journal bearing geometries other than plain journal bearings.

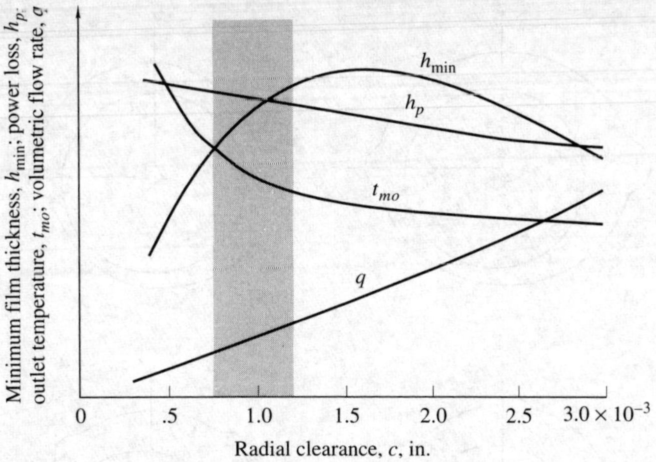

Figure 12.35 Effect of radial clearance on some performance parameters for a particular case.

These geometries have various patterns of variable clearance so as to create pad film thicknesses that have more strongly converging and diverging regions. Figure 12.36 shows elliptical, offset-half, three-lobe, and four-lobe bearings—bearings different from the plain journal bearing. Allaire and Flack (1980) presents an excellent discussion of the performance of these bearings, and some of their conclusions are presented here. In Fig. 12.36 each pad is moved toward the bearing center some fraction of the pad clearance to make the fluid film thickness more converging and diverging than that occurring in a plain journal bearing. The pad center of curvature is indicated by a cross. Generally, these bearings suppress instabilities in the system well but can be subject to subsynchronous vibration at high speeds. They are not always manufactured accurately.

A key parameter used in describing these bearings is the fraction of length in which the film thickness is converging to the full pad length, called the *offset factor* and defined as

$$\alpha_a = \frac{\text{Length of pad with converging film thickness}}{\text{Full pad length}}$$

In an elliptical bore bearing [Fig. 12.36(a)] the two pad centers of curvature are moved along the vertical axis. This movement creates a pad with one-half the film shape converging and the other half diverging (if the shaft were centered), corresponding to an offset factor α_a of 0.5. The offset-half bearing [Fig. 12.36(b)] is a two-axial-groove bearing that is split by moving the top half horizontally for $\alpha_a = 1.125$ resulting in low vertical stiffness. Generally, the vibration characteristics of this bearing are such as to avoid oil whirl, which can drive a machine unstable. The offset-half bearing has a purely converging film thickness with a converged pad arc length of 160° and the point opposite the center of curvature at 180°. Both the three-lobe and four-lobe bearings [Fig. 12.36(c) and (d)] have $\alpha_a = 0.5$.

The fractional reduction of the film clearance when the pads are brought in is called the **preload factor** m_p. Let the bearing clearance at the pad minimum film thickness (with

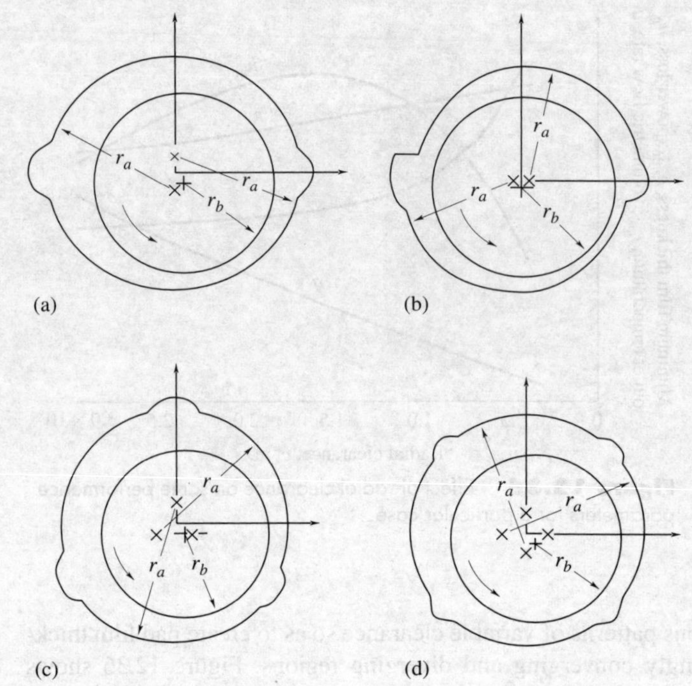

Figure 12.36 Types of fixed-incline preloaded journal bearing and their offset factors α_a. Preload factor m_p = 0.4. (a) Elliptical bore bearing (α_a = 0.5); (b) offset-half bearing (α_a = 1.125); (c) three-lobe bearing (α_a = 0.5); (d) four-lobe bearing (α_a = 0.5). [P. E. Allaire and R. D. Flack, "Journal Bearing Design for High Speed Turbomachinery," *Bearing Design—Historical Aspects, Present Technology and Future Problems*, W. J. Anderson (ed.), American Society of Mechanical Engineers, 1980, pp. 111–60. Reprinted with permission of the American Society of Mechanical Engineers International.]

the shaft center) be denoted by c_b. Figure 12.37(a) shows that the largest shaft which can be placed in the bearing has a radius $r_b + c_b$, thereby establishing the definition of c_b. The preload factor m_p is given by

$$m_p = \frac{c - c_b}{c}$$

A preload factor of zero corresponds to all the pad centers of curvature coinciding at the bearing center; a preload factor of 1.0 corresponds to all the pads touching the shaft. Figure 12.37(b) and (c) illustrates these extreme situations.

12.5 SQUEEZE FILM BEARINGS

In Secs. 12.3 and 12.4 slider bearings were considered, where the positive bearing pressure is due to the physical wedge mechanism covered in Sec. 12.2.2.3. Here the focus is on the development of positive pressure in a fluid contained between two surfaces moving toward each other. A finite time is required to squeeze the fluid out of the gap, and this action provides a useful cushioning effect in bearings. The reverse effect, which occurs when the surfaces are moving apart, can lead to cavitation in the fluid film.

The concept of **squeeze film bearings** was first introduced in this text in Fig. 8.4(b), and the pressure-generating mechanism was covered earlier in this chapter (Sec. 12.2.2.4).

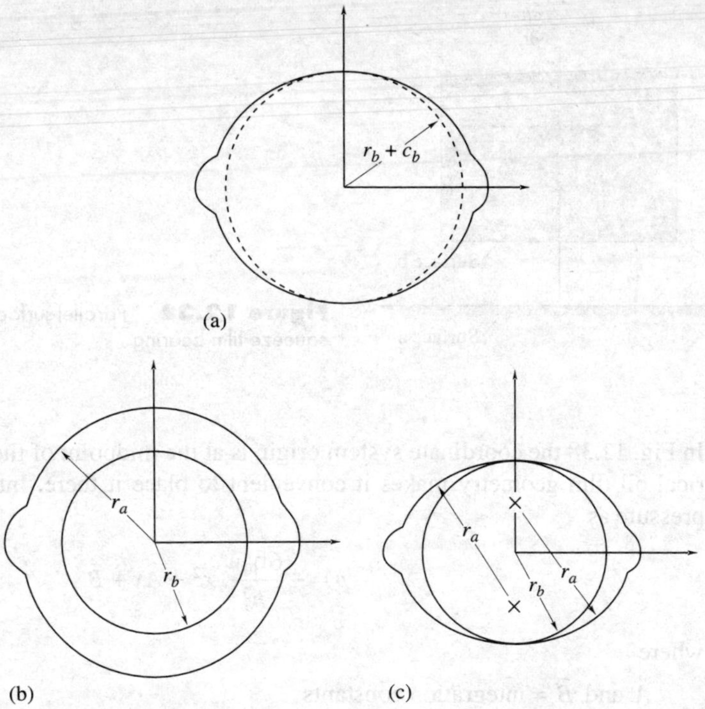

Figure 12.37 Effect of preload factor m_p on two-lobe bearings. (a) Largest shaft that fits in bearing; (b) $m_p = 0$, largest shaft r_a, bearing clearance $c_b = c$; (c) $m_p = 1.0$, largest shaft r_a, bearing clearance $c_b = 0$. [P. E. Allaire and R. D. Flack, "Journal Bearing Design for High Speed Turbomachinery," *Bearing Design—Historical Aspects, Present Technology and Future Problems*, W. J. Anderson (ed.), American Society of Mechanical Engineers, 1980, pp. 111–60. Reprinted with permission of the American Society of Mechanical Engineers International.]

For squeeze film bearings a relationship needs to be developed between load and normal velocity at any instant. The time required for the separation of the surfaces to change by a specified amount can be determined by a single integration with respect to time.

The appropriate Reynolds equation for squeeze film bearings is

$$\frac{\partial}{\partial x}\left(h^3 \frac{\partial p}{\partial x}\right) + \frac{\partial}{\partial y}\left(h^3 \frac{\partial p}{\partial y}\right) = 12\eta_0 \frac{\partial h}{\partial t} = -12\eta_0 w \tag{12.93}$$

where

w = squeeze velocity, m/s

12.5.1 PARALLEL-SURFACE SQUEEZE FILM THRUST BEARING

Figure 12.38 shows a simple parallel-surface squeeze film bearing. When side leakage is neglected [flow is in the y direction (the direction out from the paper)], the second term in Eq. (12.93) is neglected. This assumption is valid for bearings whose width w_t in the y direction is much larger than their length l, or $w_t/l > 4$. Thus, if side leakage is neglected and a parallel film is assumed, Eq. (12.93) reduces to

$$\frac{\partial^2 p}{\partial x^2} = -\frac{12\eta_0 w}{h_o^3} \tag{12.94}$$

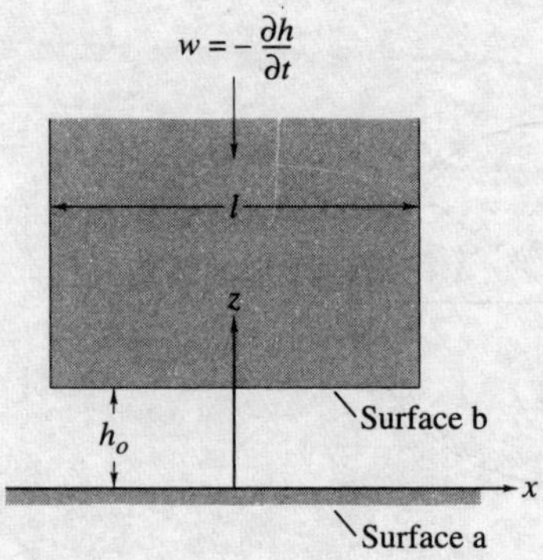

Figure 12.38 Parallel-surface squeeze film bearing.

In Fig. 12.38 the coordinate system origin is at the midpoint of the bearing, since symmetrical oil film geometry makes it convenient to place it there. Integrating twice gives the pressure as

$$p = -\frac{6\eta_0 w}{h_o^3}x^2 + \tilde{A}x + \tilde{B} \tag{12.95}$$

where

$\tilde{A}$ and $\tilde{B}$ = integration constants

The boundary conditions are

$$p = 0 \qquad \text{when } x = \pm l/2$$

Making use of these boundary conditions gives

$$p = \frac{3\eta_0 w}{2h_o^3}\left(l^2 - 4x^2\right) \tag{12.96}$$

The normal load-carrying capacity per unit width is

$$W_z' = \int_{-l/2}^{l/2} p\,dx = \frac{3\eta_0 w}{2h_o^3}\int_{-l/2}^{l/2}\left(l^2 - 4x^2\right)dx = \frac{\eta_0 l^3 w}{h_o^3} \tag{12.97}$$

For time-independent loads (W_z' not a function of t) Eq. (12.97) can be used to determine the time for the gap between the parallel surfaces to be reduced by a given amount. The squeeze velocity w can be expressed as $-\partial h/\partial t$. Since the film thickness does not vary with x but does vary with time, Eq. (12.97) becomes

$$W_z' = -\frac{\eta_0 l^3}{h_o^3}\frac{\partial h_o}{\partial t} \tag{12.98}$$

Rearranging terms and integrating give

$$-\frac{W_z'}{\eta_0 l^3}\int_{t_1}^{t_2} dt = \int_{h_{o,1}}^{h_{o,2}} \frac{dh_o}{h_o^3}$$

$$\therefore \Delta t = t_2 - t_1 = \frac{\eta_0 l^3}{2W_z'}\left(\frac{1}{h_{o,2}^2} - \frac{1}{h_{o,1}^2}\right) \qquad \textbf{(12.99)}$$

The final outlet film thickness $h_{o,2}$ can be expressed in terms of the initial film thickness $h_{o,1}$ and the time interval Δt as

$$h_{o,2} = \frac{h_{o,1}}{\left[1+\left(2W_z'\Delta t\, h_{o,1}^2/\eta_0 l^3\right)\right]^{1/2}} \qquad \textbf{(12.100)}$$

Example 12.6

GIVEN Two parallel plates are 0.025 m long and infinitely wide. The oil separating the plates is initially 25 μm thick and has a viscosity of 0.5 Pa-s. The load per unit width is 20 000 N/m.

FIND Calculate the time required to reduce the film thickness to (a) 2.5 μm, (b) 0.25 μm, and (c) zero.

Solution

Making use of Eq. (12.99) gives

$$\Delta t = \frac{\eta_0 l^3}{2W_z'}\left(\frac{1}{h_{o,2}^2} - \frac{1}{h_{o,1}^2}\right) = \frac{(0.5)(0.025)^3}{2(20\ 000)}\left[\frac{1}{(2.5\times10^{-6})^2} - \frac{1}{(25\times10^{-6})^2}\right]$$

a) For $h_o = 2.5$ μm

$$\Delta t = 1.95\times10^{-10}(0.16-0.0016)(10^{12}) = 30.9 \text{ s} \qquad \textbf{(a)}$$

b) For $h_o = 0.25$ μm

$$\Delta t = 1.95\times10^{-10}\left[\frac{1}{(0.25\times10^{-6})^2} - \frac{1}{(25\times10^{-6})^2}\right] = 3120 \text{ s} \qquad \textbf{(b)}$$

c) For $h_o = 0$

$$\Delta t = 1.95\times10^{-10}\left[\frac{1}{0} - \frac{1}{(25\times10^{-6})^2}\right] = \infty \qquad \textbf{(c)}$$

Equation (c) implies that theoretically the oil would never be squeezed out of the space between the parallel plates.

Example 12.7

GIVEN A compact disk is, by mistake, dropped into a container of punch during a party and falls slowly to the bottom. The disk remains horizontal as it approaches the horizontal flat bottom of the container. Neglect flow through the central hole. The disk is 120 mm in diameter and 1 mm thick, and the density of the disk material is 1100 kg/m^3. The punch has a density of 930 kg/m^3 (40% alcohol) and a viscosity of 0.0015 N-s/m^2.

FIND The time it takes for the compact disk to fall from 2 mm above the bottom to 1 mm above the bottom.

Solution

As the disk falls to the bottom, the volume inside the radius r is squeezed out with the flow rate $\pi r^2 w = -\pi r^2\,\partial h/\partial t$. That flow must equal the pressure flow at radius r from the center.

$$\therefore -2\pi r\left(\frac{h^3}{12\eta}\frac{\partial p}{\partial r}\right) = -\pi r^2\frac{\partial h}{\partial t} \tag{a}$$

The pressure gradient $\partial p/\partial r$ is in the direction of the radius. It is negative as the pressure falls from the center toward the outside of the disk.

$$\therefore \frac{\partial p}{\partial r} = \frac{6\eta r}{h^3}\frac{\partial h}{\partial t} \tag{b}$$

Integrating gives

$$p = \frac{3\eta r^2}{h^3}\frac{\partial h}{\partial t} + C_1 \tag{c}$$

The boundary condition is $r = r_o$ when $p = 0$

$$\therefore C_1 = -\frac{3\eta r_o^2}{h^3}\frac{\partial h}{\partial t} \tag{d}$$

Substituting Eq. (d) into Eq. (c) gives

$$p = -\frac{3\eta}{h^3}\frac{\partial h}{\partial t}\left(r_o^2 - r^2\right) \tag{e}$$

The normal applied load is

$$W_z = \int_0^{r_o} p2\pi r dr = -\frac{3\pi\eta r_o^4}{2h^3}\frac{\partial h}{\partial t} \tag{f}$$

The buoyancy force of the disk in the punch gives

$$W_z = \pi r_o^2 t_w\left(\rho_d - \rho_p\right)g \tag{g}$$

Equating Eqs. (f) and (g) gives

$$-\frac{3\pi\eta r_o^4}{2h^3}\frac{\partial h}{\partial t} = \pi r_o^2 t_w\left(\rho_w - \rho_p\right)g$$

$$\therefore -\frac{dh}{h^3} = \frac{2t_w g}{3\eta r_o^2}\left(\rho_d - \rho_p\right)dt$$

Integrating gives

$$\frac{1}{2h_2^2} - \frac{1}{2h_1^2} = \frac{2t_w g\left(\rho_d - \rho_p\right)(t_2 - t_1)}{3\eta r_o^2}$$

$$t_2 - t_1 = \frac{3\eta r_o^2}{4t_w g\left(\rho_d - \rho_p\right)}\left(\frac{1}{h_2^2} - \frac{1}{h_1^2}\right)$$

$$= \frac{(3)(0.0015)(0.060)^2}{4(0.001)9.807(1100 - 930)}\left[\frac{1}{(0.001)^2} - \frac{1}{(0.002)^2}\right]$$

$$= 1.822 \text{ s}$$

It takes a little less than 2 s for the disk to fall from 2-mm height to 1-mm height in the punch.

12.5.2 GENERAL COMMENTS ABOUT SQUEEZE FILM BEARINGS

For a thrust bearing subjected to a squeeze velocity a parallel film shape produces the largest load-carrying capacity of all possible film shapes. In contrast, for a thrust slider bearing a parallel film should produce no positive pressure and no load-carrying capacity. Recall that a slider bearing uses a physical wedge mechanism (see Sec. 12.2.2.3) to produce a positive pressure. Squeeze film bearings can be applied to a number of bearing configurations other than the parallel-surface thrust bearing considered earlier. Hamrock (1993) applies squeeze films to a number of conformal and nonconformal surfaces, such as journal bearings, parallel circular plates, and an infinitely long cylinder near a plate.

The second major observation about normal squeeze film bearings is that as the bearing surfaces move toward each other, the viscous fluid shows great reluctance to be squeezed out the sides of the bearing. The tenacity of the squeeze film is remarkable, and the survival of many modern bearings depends on this phenomenon.

The third remarkable feature of squeeze film bearings is that a small approach velocity will provide an extremely large load-carrying capacity. The reason is mainly that in Eq. (12.97) the normal applied load per unit width is inversely proportional to the film thickness raised to the third power and directly proportional to the squeeze velocity.

12.6 HYDROSTATIC BEARINGS

Slider bearings, considered in Secs. 12.3 and 12.4, use a physical wedge pressure-generating mechanism to develop pressure within the bearing. Such bearings, in addition to having low frictional drag and hence low power loss, have the great advantage that they are basically simple and therefore are reliable and cheap and require little attention. Slider bearings have, however, certain important disadvantages:

1. If the design speed is low, it may not be possible to generate sufficient hydrodynamic pressure.
2. Fluid film lubrication may break down during starting, direction changing, and stopping.
3. In a journal slider bearing, as considered in Sec. 12.4, the shaft runs eccentrically and the bearing location varies with load, thus implying low stiffness.

In **hydrostatic** (also called **externally pressurized**) **lubricated bearings** the bearing surfaces are separated by a fluid film maintained by a pressure source outside the bearing. Hydrostatic bearings avoid disadvantages 1 and 2 and reduce the variation of bearing location with load mentioned in disadvantage 3. The characteristics of hydrostatically lubricated bearings are

1. Extremely low friction
2. Extremely high load-carrying capacity at low speeds
3. High positional accuracy in high-speed, light-load applications
4. A lubrication system more complicated than that for slider bearings (considered in Secs. 12.3 and 12.4)

Therefore, hydrostatically lubricated bearings are used when the requirements are extreme, as in large telescopes and radar tracking units, where extremely heavy loads and extremely low speeds are used; or in machine tools and gyroscopes, where extremely high speeds, light loads, and gas lubricants are used.

Figure 12.39 shows how a fluid film forms in a hydrostatically lubricated bearing system. In a simple bearing system under no pressure [Fig. 12.39(a)] the runner, under the influence of load W_z, is seated on the pad. As the source pressure builds up [Fig. 12.39(b)], the pressure in the pad recess also increases. The recess pressure is built up to a point [Fig. 12.39(c)] where the pressure on the runner over an area equal to the pad recess area is just sufficient to lift the load. This is commonly called the *lift pressure* p_l. Just after the runner separates from the bearing pad [Fig. 12.39(d)] the recess pressure is less than that required to lift the bearing runner ($p_r < p_l$). After lift, flow commences through the system. Therefore, a pressure drop exists between the pressure source and the bearing (across the restrictor) and from the recess to the bearing outlet. If more load is added to the bearing [Fig. 12.39(e)], the film thickness will decrease and the recess pressure will rise until the integrated pressure across the land equals the load. If the load is then reduced to less than the original [Fig. 12.39(f)], the film thickness will increase to some higher value and the recess pressure will decrease accordingly. The maximum load that can be supported by the pad will be reached theoretically when the recess pressure is equal to the source pressure. If a load greater than this is applied, the bearing will seat and remain seated until the load is reduced and can again be supported by the supply pressure.

The normal load, flow, and power loss can be obtained for a number of different shapes, such as a circular step bearing pad, an annular thrust bearing, or rectangular sectors. Consult Hamrock (1993) for details about the design of these specific types of hydrostatic bearing.

Example 12.8

GIVEN A flat, circular hydrostatic thrust bearing is used to carry a load as high as 25 tons in a large milling machine. The purpose of the hydrostatic bearing is to position the workpiece accurately without any friction. The vertical positioning of the workpiece is influenced by the oil film thickness and its variation with load. To minimize the variation for different loads, a step is made in the bearing as shown in Fig. 12.40. The pressure in the step is constant and $S_h/h_o > 10$. Thus, $p = p_r$ through the step or for $0 < r < r_i$. The step height $s_h = 5$ mm and $r_o = 200$ mm. The oil viscosity is 0.25 N-s/m^2 and the smallest value of h_o allowed is 50 μm. The oil flow through the bearing is constant at 0.1×10^{-3} m^3/s.

FIND Determine r_i such that the oil outlet (minimum) film thickness h_o varies as little as possible when the load varies between 5 and 25 tons.

Solution

Assuming that the film thickness is the same in any radial or angular position and that the pressure does not vary in the θ direction, the appropriate Reynolds equation is

$$\frac{\partial}{\partial r}\left(r\frac{\partial p}{\partial r}\right) = 0$$

Integrating once gives

$$r\frac{dp}{dr} = \tilde{A}$$

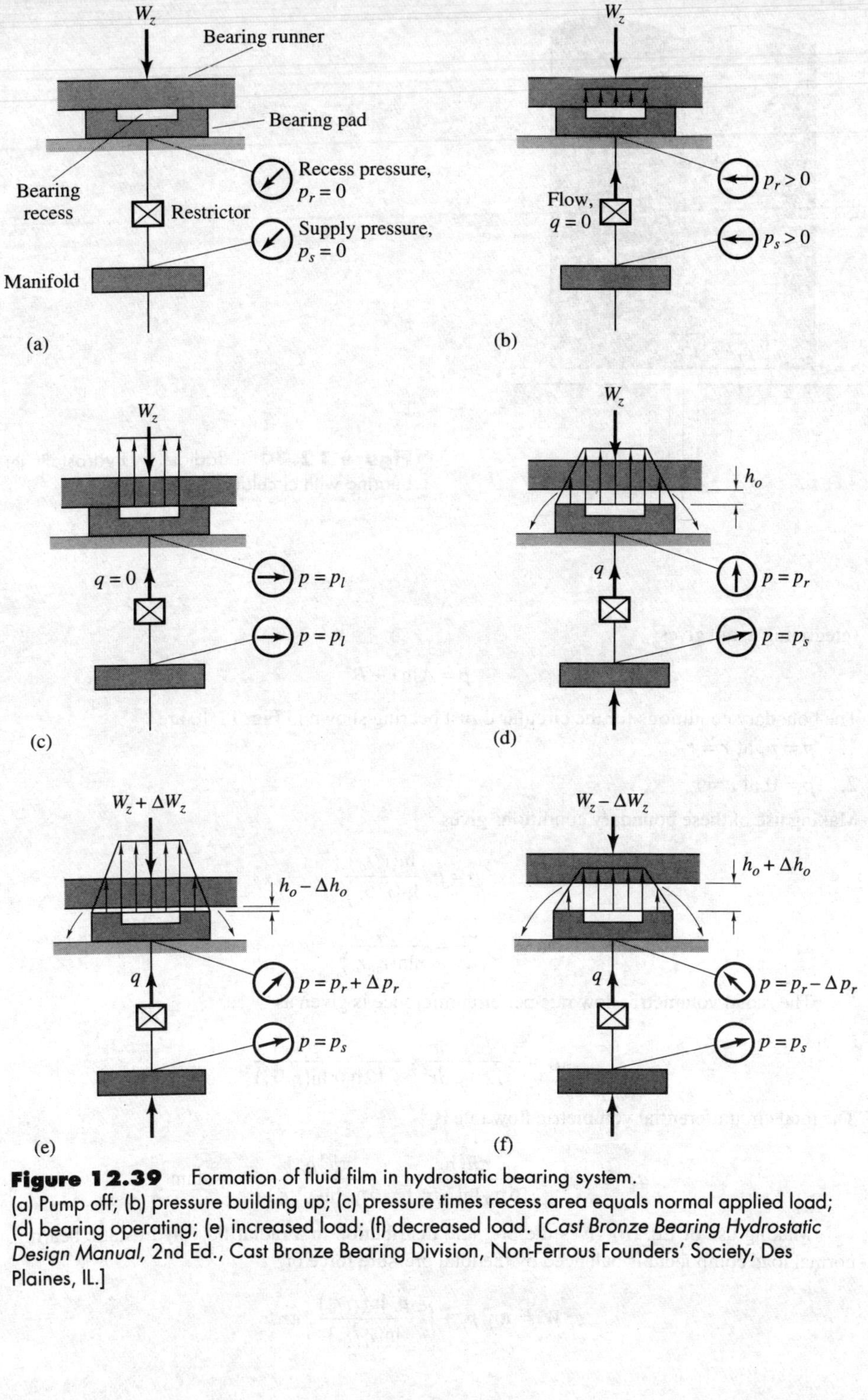

Figure 12.39 Formation of fluid film in hydrostatic bearing system.
(a) Pump off; (b) pressure building up; (c) pressure times recess area equals normal applied load; (d) bearing operating; (e) increased load; (f) decreased load. [*Cast Bronze Bearing Hydrostatic Design Manual,* 2nd Ed., Cast Bronze Bearing Division, Non-Ferrous Founders' Society, Des Plaines, IL.]

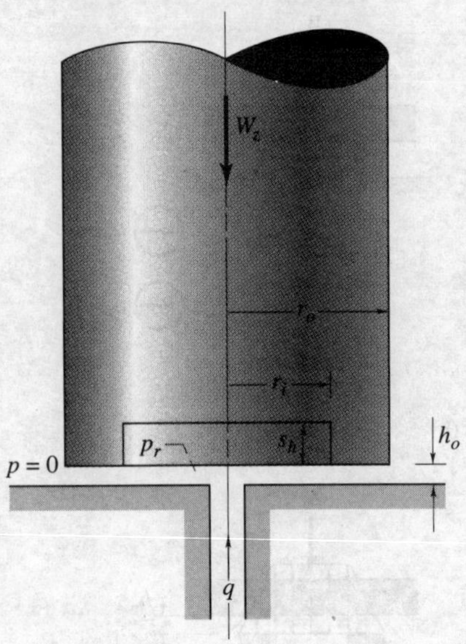

Figure 12.40 Radial-flow hydrostatic thrust bearing with circular step pad.

Integrating again gives

$$p = \tilde{A} \ln r + \tilde{B} \tag{a}$$

The boundary conditions for the circular thrust bearing shown in Fig. 12.40 are

1. $p = p_r$ at $r = r_i$
2. $p = 0$ at $r = r_o$

Making use of these boundary conditions gives

$$p = p_r \frac{\ln(r/r_o)}{\ln(r_i/r_o)} \tag{b}$$

$$\frac{dp}{dr} = \frac{p_r}{r \ln(r_i/r_o)} \tag{c}$$

The radial volumetric flow rate per circumference is given as

$$q_r' = -\frac{h_o^3}{12\eta_0}\frac{dp}{dr} = -\frac{h_o^3 p_r}{12\eta_0 r \ln(r_i/r_o)} \tag{d}$$

The total circumferential volumetric flow rate is

$$q = 2\pi r q_r' = -\frac{\pi h_o^3 p_r}{6\eta_0 \ln(r_i/r_o)} = \frac{\pi h_o^3 p_r}{6\eta_0 \ln(r_o/r_i)} = \text{constant} \tag{e}$$

Making use of Eq. (b) gives the pressure distribution in a radial-flow hydrostatic bearing. The normal load component is balanced by the total pressure force or

$$W_z = \pi r_i^2 p_r + \int_{r_i}^{r_o} \frac{p_r \ln(r/r_o)}{\ln(r_i/r_o)} 2\pi r dr$$

This equation reduces to

$$W_z = \frac{\pi p_r \left(r_o^2 - r_i^2\right)}{2\ln(r_o/r_i)} \quad \textbf{(f)}$$

From Eqs. (e) and (f)

$$W_z = \frac{\left(r_o^2 - r_i^2\right)6q\eta_0}{2h_o^3} \quad \textbf{(g)}$$

From Eq. (g), $(W_z)_{max}$ occurs when $(h_o)_{min}$ and $(r_i)_{min}$ whereas $(W_z)_{min}$ occurs when $(h_o)_{max}$ and $(r_i)_{max}$.

$$\therefore (h_o)_{min}^3 = \frac{6\eta q}{(W_z)_{max}} \frac{\left(r_o^2 - r_{i,max}^2\right)}{2}$$

$$(50)^3\left(10^{-6}\right)^3 = \frac{(6)(0.25)(0.1)\left(10^{-3}\right)}{(25\ 000)(9.807)} \frac{\left(0.200^2 - r_{i,max}^2\right)}{2}$$

$$\therefore r_{i,max} = 0.1990 \text{ m}$$

Also,

$$(h_o)_{max}^3 = \frac{6\eta q}{(W_z)_{min}} \frac{\left(r_o^2 - r_{i,min}^2\right)}{2} = \frac{(6)(0.25)\left(10^{-4}\right)}{(5000)(9.807)} \frac{\left(0.200^2 - 0.1990^2\right)}{2}$$

$$(h_o)_{max} = 84.82 \times 10^{-6} \text{ m} = 84.82\ \mu\text{m}$$

The inner radius should be 199 mm, which gives a minimum film thickness variation of 84.82 – 50 = 34.82 μm.

12.7 Gas-Lubricated Bearings

Gas-lubricated film bearings have commanded considerable attention in recent years because they have characteristics that make their use advantageous in many bearing applications. They are analogous to the hydrodynamic oil-lubricated bearings considered thus far except that the fluid is compressible. Furthermore, since air is 1000 times less viscous than even the thinnest mineral oils, the viscous resistance is very much less. However, the distance of nearest approach between the bearing surfaces is also correspondingly smaller, so that special precautions must be taken in manufacturing gas-lubricated bearings.

Some *strengths* of gas-lubricated bearings are as follows:

1. Their friction or viscous resistance is extremely low.
2. The lubricant is ample and clean.
3. The lubricant does not contaminate surfaces.
4. The lubricant operates well from extremely low to extremely high temperatures.
5. The film does not break down from cavitation or ventilation.

Some *weaknesses* of gas-lubricated bearings are as follows:

1. For the same-size bearing the load-carrying capacity of a gas-lubricated bearing is many times less than that of an oil-lubricated bearing.
2. The surfaces must have an extremely fine finish.
3. The alignment must be extremely good.
4. Dimensions and clearances must be extremely accurate.
5. Speed must be high.
6. Loading must be low.
7. Their stability characteristics are poor.

Gas bearings are finding increasing usage in gas-cycle machinery, where the cycle gas is used in the bearings, thus eliminating the need for a conventional lubrication system; in gyros, where precision and constancy of torque are critical; in food- and textile-processing machinery, where cleanliness and absence of contaminants are critical; and in high-speed dental drills. Obviously, gas bearings will find applications only where their strengths outweigh their weaknesses. Although their use will always be limited in scope because of inherent weaknesses, it is broad enough to merit an examination of their behavior as a machine element.

Design details of gas-lubricated bearings are beyond the scope of this text. Consult Hamrock (1993) for more details about these bearings.

Case Study | DESIGN OF A CRANKSHAFT MAIN BEARING

An automotive crankshaft is shown inside an internal combustion engine in Fig. 12.41. The crankshaft is located in one of the most difficult-to-access portions of the engine. Therefore, it is imperative that the crankshaft and its bearings be designed so that maintenance will never be needed in the lifetime of the automobile. This requires that the bearings be designed to operate in the full-film lubrication regime.

The load acting on one of the bearings is shown in Fig. 12.42, where both magnitude and direction are given as functions of time. The loading is quite complex and reflects the large dynamic loads associated with ignition within each cylinder. The peak loads are actually softened by the grudgeon pin bearing in the connecting rod, an example of a squeeze film bearing.

It is desired to specify a clearance between the crankshaft and the bearing and to estimate the temperature rise in the lubricant so that a proper cooling system can be designed. Obviously, the lubricant is not the only temperature source in the engine. However, it is an important component, since oil temperatures should not become too high inside a bearing or the lubricant will chemically degrade (break down).

GIVEN A ground and lapped crankshaft with the loading shown. The shaft diameter is 3.0 in. The lubricant is SAE 40 at a maximum inlet temperature of 200°F. Shaft speeds vary from 800 rpm at idle to 4500 rpm maximum. The greatest challenge is to design a bearing that operates at high speed and high temperature and is suddenly brought to idle. Thus, only the idle speed is considered in the problem, recognizing that the inlet temperature specified reflects high operating temperatures. The bearing width dimension must lie between the extremes of 1 and 2 in. for the design to remain compact.

FIND A bearing clearance if the surfaces are ground and lapped, and a bearing width subject to the constraint that the maximum lubricant temperature in the bearing does not exceed 250°F.

(continued)

Case Study (CONTINUED)

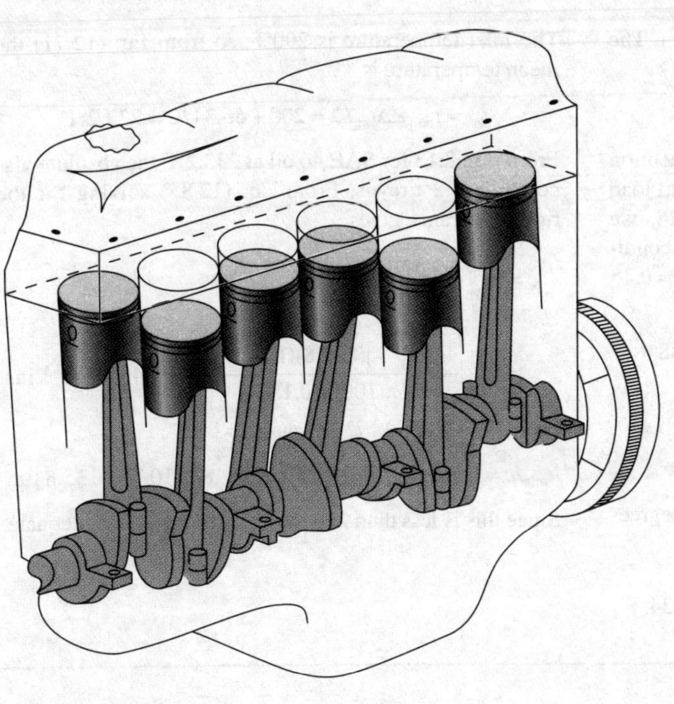

Figure 12.41 Cross section of internal combustion engine, showing main bearing.

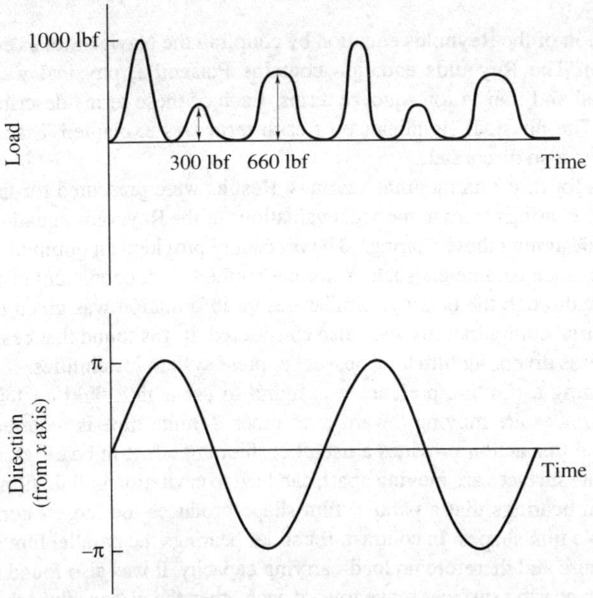

Solution

A full solution of the Reynolds equation for this problem would be extremely difficult because the time-dependent load would require a time-dependent solution of a highly nonlinear differential equation. Clearly, the approach of a worst-case scenario will be useful in making this problem tractable.

It is recognized that the larger the load, the smaller the film thickness if a steady state is assumed. Thus, if the bearing is analyzed by using the constant peak load, the calculated film thickness will be less than the film thickness actually encountered in application. If this film thickness leads to acceptable frictional and thermal properties in the bearing, the bearing will be at least as robust in practice. Therefore, the maximum load is taken as 1000 lbf.

The bearing dimensions in the designer's direct control are the width and the clearance ratio. Obviously, increasing the width will also increase the film thickness for a given set of operating conditions. These increases would allow an increase in the radial clearance, which is beneficial for manufacturing and assembly reasons. Also, to maintain a full lubricant film, the minimum thickness must be ten times the surface roughness in the bearing. Since it was specified that the surface is ground and lapped, the achievable surface roughness from Table 12.1 is approximately 10 μin. Thus, the design film thickness in the bearing must be larger than the allowable film thickness.

Since we are unsure which value of either the width or clearance to use, we will need to make an assumption. Only one example of the analysis will be explained in depth, as this can be repeated by the student.

Figure 12.42 Load and direction of load as a function of time at main bearing location under consideration.

(continued)

Case Study (CONCLUDED)

As a first try, we use a bearing width of 1.0 in. The diameter-to-width ratio is

$$\lambda_j = 2r/w_t = 3/1 = 3$$

Since the statement of the problem does not mention whether the design is to be optimized for maximum load or minimum coefficient of friction in Fig. 12.28, we choose a value midway between these optimum conditions. Thus, we find that $B_j = 0.15$. For $\lambda_j = 3$ and $B_j = 0.15$ from Figs. 12.30 to 12.32

$$r_b\mu/c = 5.5 \qquad Q = 5.4 \qquad \text{and} \qquad q_s/q = 0.88$$

The radial load per area is

$$W_r^* = \frac{W_r}{2r_b w_t} = \frac{1000}{2(1.5)(1)} = 333 \text{ psi}$$

From Eq. (12.91b) the temperature rise in degrees Fahrenheit is

$$\Delta t_m = \frac{0.103W_r^*(r_b\mu/c)}{Q(1-0.5q_s/q)} = \frac{0.103(333)(5.5)}{5.4[1-0.5(0.88)]} = 66.34°\text{F}$$

The inlet temperature is 200°F, so from Eq. (12.71) the mean temperature is

$$\tilde{t}_m = t_{mi} + \Delta t_m/2 = 200 + 66.34/2 = 233.2°\text{F}$$

From Fig. 8.13 for SAE 40 oil at 233.2°F the absolute viscosity is 1.2 μreyns. From Eq. (12.85) solving for the radial clearance gives

$$c = r_b\sqrt{\frac{\eta_0\omega_b r_b w_t}{\pi W_r B_j}}$$

$$= 1.5\sqrt{\frac{1.2(10^{-6})(83.78)(1.5)(1.0)}{\pi(1000)(0.15)}} = 0.8485\times10^{-3} \text{ in.}$$

From Fig. 12.28 for $\lambda_j = 3$ and $B_j = 0.15$

$$h_{\min}/c = 0.16 \quad \text{or} \quad h_{\min} = 0.16(0.8485)(10^{-3}) = 135.8\ \mu\text{m}$$

Since this is less than 250 μin., the design is not adequate.

12.8 SUMMARY

The chapter began with the derivation of the Reynolds equation by coupling the Navier-Stokes equations with the continuity equation. The Reynolds equation contains Poiseuille, physical wedge, stretch, local expansion, and normal and translation squeeze terms. Each of these terms describes a specific type of physical motion. The physical significance of each term was explained. Standard forms of the Reynolds equation were also discussed.

Design information was given for thrust and journal bearings. Results were presented for fixed-incline and pivoted-pad thrust slider bearings from numerical evaluations of the Reynolds equation. A procedure was outlined to assist in designing these bearings. The procedure provided an optimum pad configuration and described performance parameters such as normal applied load, coefficient of friction, power loss, and lubricant flow through the bearing. Similar design information was given for a plain journal bearing. Nonplain journal configurations were also considered. It was found that bearing designs with more converging and less diverging film thicknesses suppress system instabilities.

In a normal squeeze film bearing a positive pressure was found to occur in a fluid contained between two surfaces when the surfaces are moving toward each other. A finite time is required to squeeze the fluid out of the gap, and this action provides a useful cushioning effect in bearings. The reverse effect, which occurs when the surfaces are moving apart, can lead to cavitation in fluid films. It was found for normal squeeze film bearings that a parallel film shape produces the largest normal load-carrying capacity of all possible film shapes. In contrast, for slider bearings the parallel film was shown to produce no positive pressure and therefore no load-carrying capacity. It was also found that for the normal squeeze action, as the bearing surfaces move toward each other, the viscous fluid shows great reluctance to be squeezed out thc sides of the bearing. The tenacity of a squeeze film is remarkable and the survival of many modern bearings depends on this phenomenon. It was found that a relatively small approach velocity will provide an extremely large load-carrying capacity.

Hydrostatic bearings were briefly considered and were found to offer certain operating advantages over other types of bearing. Probably the most useful characteristics of hydrostatic bearings are their high load-carrying capacity and inherently low friction at any speed, even zero. The principles and basic concepts of film formation were discussed.

The last section briefly discussed gas-lubricated bearings. Gas bearings are analogous to hydrodynamic oil-lubricated bearings except that the fluid is compressible and the perfect-gas equation of state must be applied. Furthermore, since air is 1000 times less viscous than even the thinnest mineral oils, the viscous resistance is much less.

KEY WORDS

adiabatic temperature rise temperature rise in lubricant if all heat is carried away through convection

bearing number dimensionless operating parameter, specific to each type of bearing, that is important for determining bearing performance

Couette flow velocity-driven fluid flow

density wedge term dealing with density changes in bearing

eccentricity displacement of shaft center relative to bearing center

externally pressurized bearings same as hydrostatic bearings

fluid film lubrication lubrication condition where two surfaces transmitting a load are separated by a pressurized fluid layer or film

gas-lubricated film bearings bearings where lubricating fluid is a gas

hydrodynamic lubrication lubrication activated by motion of bearing surfaces

hydrostatic lubricated bearings slider bearings where lubricant is provided at elevated pressure

journal bearings bearings whose surfaces are parallel to axis of rotation

line of centers line containing bearing and shaft centers

local expansion term involving time rate of density change

normal squeeze term dealing with approach of two bearing surfaces toward one another

Petrov's equation equation for frictional torque of concentric journal bearings, given by

$$T = Fr = \frac{4\pi^2 \eta_0 r^3 w_t N_a}{c} = \frac{2\pi \eta_0 r^3 w_t \omega}{c}$$

physical wedge term dealing with rate of channel convergence

Poiseuille flow pressure-driven fluid flow

preload factor fractional reduction of film clearance when pads are brought in

Reynolds equation differential equation governing pressure distribution and film thickness in bearings

Sommerfeld number bearing characteristic number for a journal bearing

squeeze film bearings bearing where one dominant film-generating mechanism is squeeze effect (see terms in Reynolds equation)

stretch term term dealing with rate of velocity change in sliding direction

thrust bearings bearings whose surfaces are perpendicular to axis of rotation

translation squeeze term dealing with translation of two surfaces

RECOMMENDED READINGS

Halling, J. (ed.) (1978) *Principles of Tribology,* Macmillan, New York.
Hamrock, B.J. (1994) *Fundamentals of Fluid Film Lubrication,* McGraw-Hill, New York.
Juvinall, R.C. (1991) *Fundamentals of Machine Component Design,* Wiley, New York.
Mott, R. L. (1985) *Machine Elements in Mechanical Design,* Merrill Publishing Co., Columbus, OH.
Norton, R.L. (1996) *Machine Design,* Prentice-Hall, Englewood Cliffs, NJ.
Radzimovski, E.I. (1959) *Lubrication of Bearings,* Ronald Press, New York.
Shigley, J.E., and Mischke, C.R. (1989) *Mechanical Engineering Design,* McGraw-Hill, New York.
Szeri, A.Z. (ed.) (1980) *Tribology: Friction, Lubrication and Wear,* Hemisphere Publishing Co., Washington, DC.

REFERENCES

Allaire, P.E., and Flack, R.D. (1980) Journal Bearing Design for High Speed Turbomachinery, *Bearing Design—Historical Aspects, Present Technology and Future Problems,* W.J. Anderson (ed.), American Society of Mechanical Engineers, New York, pp. 111–160.
Currie, I.G. (1993) *Fundamental Mechanics of Fluids,* 2nd ed., McGraw-Hill, New York.
Dowson, D. (1998) *History of Tribology,* 2nd ed., Longman, London.
Engineering Sciences Data Unit (ESDU) (1967) *General Guide to the Choice of Thrust Bearing Type,* Item 67033, Institution of Mechanical Engineers, London.
Hamrock, B.J. (1991) *Fundamentals of Fluid Film Lubrication,* NASA Reference Publication 1255.
Hamrock, B.J. (1993) *Fundamentals of Fluid Film Lubrication,* McGraw-Hill, New York.
Petrov, N.P (1883) Friction in Machines and the Effect of the Lubricant, *Inzh. Zh. St. Petersburg,* vol. 1, pp. 71–140; vol. 2, pp. 227–279; vol. 3, pp. 377–463; vol. 4, pp. 535–564.
Raimondi, A.A., and Boyd, J. (1955) Applying Bearing Theory to the Analysis and Design of Pad-Type Bearings, *ASME Trans.,* vol. 77, no. 3, pp. 287–309.
Raimondi, A.A., and Boyd, J. (1958) A Solution for the Finite Journal Bearing and Its Application to Analysis and Design-I, -II, -III, *ASLE Trans.,* vol. 1, no. 1, I—pp. 159–174; II—pp. 175–193; III—pp. 194–209.
Reynolds, O. (1886) On the Theory of Lubrication and Its Application to Mr. Beauchamp Tower's Experiments, Including an Experimental Determination of the Viscosity of Olive Oil, *Philos. Trans. R. Soc.,* vol. 177, pp. 157–234.

Rippel, H.C. (1963) *Cast Bronze Hydrostatic Bearing Design Manual,* 2nd ed., Cast Bronze Bearing Institute, Inc., Cleveland, OH.
Shigley J.E., and Mitchell, L.D. (1983) *Mechanical Engineering Design,* 4th ed., McGraw-Hill, New York.
Tower, B. (1883) First Report on Friction Experiments (Friction of Lubricated Bearings), *Proc. Inst. Mech. Eng. (London),* pp. 632–659.

PROBLEMS

Section 12.2

12.1 To decrease the leakage past a seal of the type discussed in Example 12.1, the sealing gap is tapered and the film shape is

$$h(x) = h_o + x\frac{dh}{dx} = h_o + kx$$

Because of the roughness of the surfaces the outlet (minimum) film thickness h_o has to be at least 10 μm. Find the optimum oil film shape to minimize the total leakage past the 10-mm-wide seal during a full work cycle (back and forth) of the piston when the fluid viscosity is 0.075 Pa-s, the piston speed is 0.8 m/s in both directions, the stroke length is 85 mm, the sealing pressure is 1 MPa, and the piston diameter is 50 mm.

12.2 The oil viscosity is 0.01 Pa-s in a very wide slider bearing with an exponential oil film shape

$$h = h_o e^{-x/l}$$

Find the pressure distribution and calculate the load-carrying capacity per unit width when the sliding speed is 10 m/s if $h_o = 60$ μm and the bearing is 0.10 m long in the direction of motion.

Section 12.3

12.3 A plain journal bearing has a diameter of 2 in. and a length of 1 in. The full journal bearing is to operate at a speed of 2000 rpm and carries a load of 750 lbf. If SAE 20 oil at an inlet temperature of 110°F is to be used, determine the following:

a) The radial clearance for optimum load-carrying capacity, the temperature rise, and the mean temperature

b) The performance parameters: coefficient of friction, flow rate, side flow, and attitude angle

c) The kinematic viscosity of the oil at the mean temperature if the oil density is 0.89 g/cm^3

12.4 The torque transmission in a speedometer is through two circular plates, with radius *r,* placed in an oil bath at a distance *h* from each other (see sketch *a*). When shaft 1 suddenly starts to rotate, shaft 2 will be moved by the shear forces in the oil. Find the angular speed ω_2 as a function of time if $\omega_2(t = 0) = 0$ and $\omega_1(t) = 0$ for $t \le 0$. Shaft 2 has moment of inertia J_2 and the oil has viscosity η.

12.5 A fixed-incline slider bearing and a pivoted-pad slider bearing are optimized to give minimum relative power loss. The pad length l equals the width w_t. The bearing carries the load $W_z = 10$ kN. The viscosity has been chosen to avoid surface contact through the lubricant film. The sliding speed $u_b = 15$ m/s. Find the minimum film thickness for the bearing if $\eta =$ 0.08 N-s/m^2, $\lambda = w_t/l = 1$, and $l = 0.070$ m.

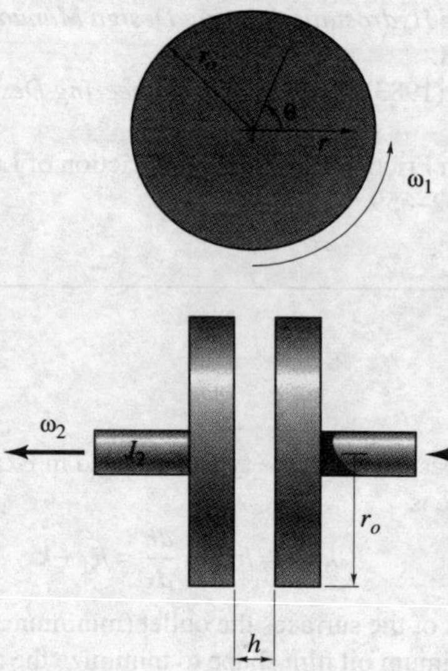

Sketch *a*, for Problem 12.4

12.6 The flat slider bearing in Problem 12.5 is split into two equal halves, each with width w_t = 0.035 m. Find the viscosity needed at the running temperature to get the same minimum film thickness as in Problem 12.5. The load is equally split between the two halves, with 5 kN per half. The bearing halves should both be optimized for minimum relative power loss.

12.7 The flat slider bearing in Problem 12.5 has load W_z = 10 kN and sliding speed u_b = 15 m/s. The oil viscosity η = 0.08 N-s/m^2, the bearing width w_t = 0.070 m, and the bearing length l = 0.070 m. Find the spring constant for the bearing if the force variation is small. The spring constant is defined as

$$c_s = |\Delta W_z / \Delta h_p|$$

where

ΔW_z = load variation

Δh_p = oil film thickness variation at load application point

12.8 A very wide slider bearing lubricated with an oil having a viscosity of 0.03 N-s/m^2 consists of two parts, each with a constant film thickness (see sketch *b*). Find the pressure distribution in the oil and the position and size of the bearing load per unit width of the bearing. All dimensions are in millimeters.

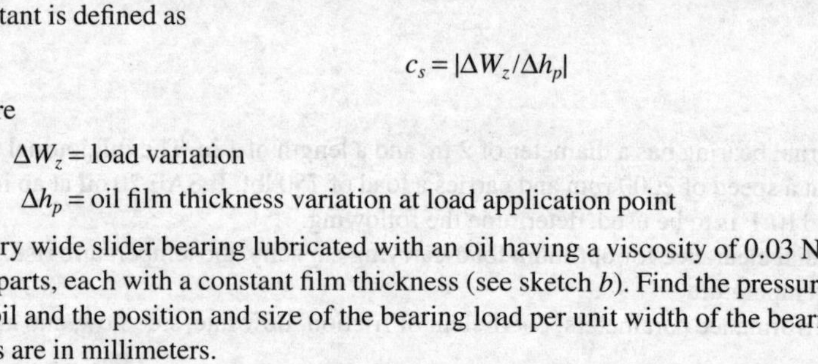

Sketch *b*, for Problem 12.8

12.9 An infinitely wide slider bearing, shown in sketch *c,* has three sections with film thicknesses $2h_1$, αh_1, and h_1, respectively. Find which value of α gives the maximum load-carrying capacity for given values of h_1, l, u_b, and η. Use the Reynolds equation and the oil flow equation.

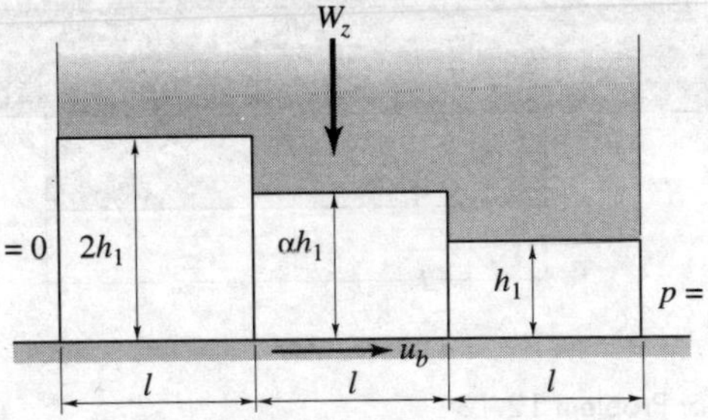

Sketch *c,* for Problem 12.9

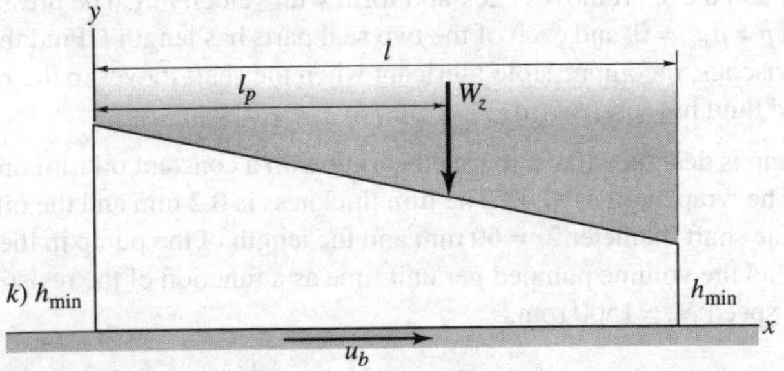

Sketch *d,* for Problem 12.10

12.10 Given the infinitely wide slider bearing shown in sketch *d,* find the pressure distribution by using the Reynolds equation.

12.11 A pivoted-pad thrust bearing carrying the weight of a water turbine for a hydroelectric powerplant is lubricated by water. The design should be optimized to give minimum power loss when the roughness of the bearing surfaces is $R_a = 7$ μm, indicating that the bearing minimum film thickness should be 100 μm. The load on the bearing from the weight of the turbine and the water is 1 MN. The viscosity of water is 0.001 Pa-s and the bearing length-to-width ratio is 1, which gives minimum power loss for five pads circumferentially around the bearing with a mean radius equal to the bearing pad width. Determine the bearing dimensions and calculate the coefficient of friction at a rotating speed of 100 rpm.

12.12 A propeller shaft thrust bearing for an oil tanker is a pivoted-pad slider bearing with 12 pads. Each pad is 300 mm wide and 300 mm long and has a mean radius of 800 mm. The shaft speed is 220 rpm. Find the bearing pivot position and the optimum lubricant viscosity to get minimum power loss. Calculate the coefficient of friction and the power loss. The lubricant inlet temperature is 63°C, the heat capacity of the oil is $C_s = 1.72 \times 10^6$ J/(m^3-°C), and the propeller thrust load is 1.3 MN.

Section 12.4

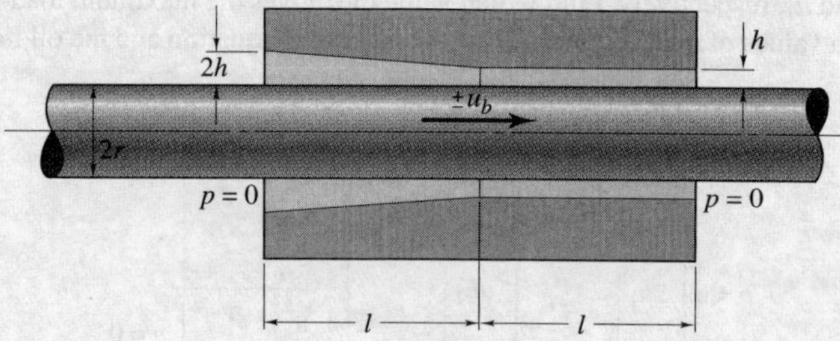

(c)

Sketch *e*, for Problem 12.13

12.13 The sealing for an axially moving circular shaft is shown in sketch *e*. The seal has rotational symmetry and the shaft moves back and forth with velocity u_b. The pressure on each side of the seal is $p = p_{atm} = 0$, and each of the two seal parts has length l. Find the maximum pressure in the isoviscous, incompressible lubricant when the shaft moves to the right in the sketch. The sealed fluid has viscosity η.

12.14 An oil pump is designed like a journal bearing with a constant oil film thickness according to sketch *f*. The wrap angle is 320°. The film thickness is 0.2 mm and the oil viscosity is 0.022 N-s/m^2. The shaft diameter $2r = 60$ mm and the length of the pump in the axial direction is 50 mm. Find the volume pumped per unit time as a function of the resisting pressure at the rotational speed $N_a = 1500$ rpm.

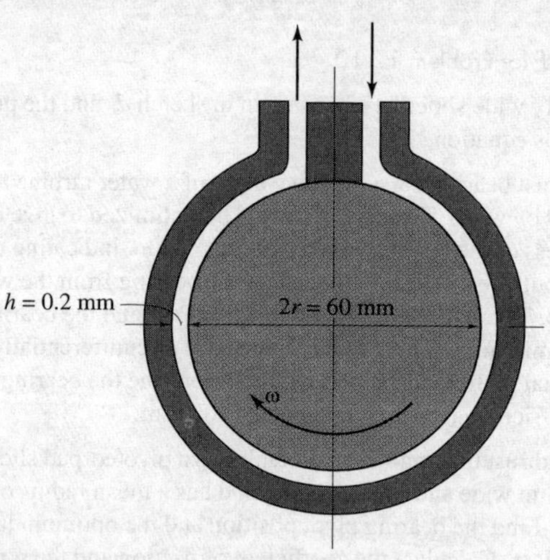

Sketch *f*, for Problem 12.14

12.15 An oil pump, shown in sketch *g*, consists of a shaft rotating with surface speed u_b in two bearing halves each with constant film thicknesses h_1 and h_2, respectively. The bearing halves have the same length l. A lightly spring-loaded release valve connects the part with oil film thickness h_1 to the pump outlet. Find and draw the pressure p at the outlet as a function of the oil flow rate q per unit width of the pump.

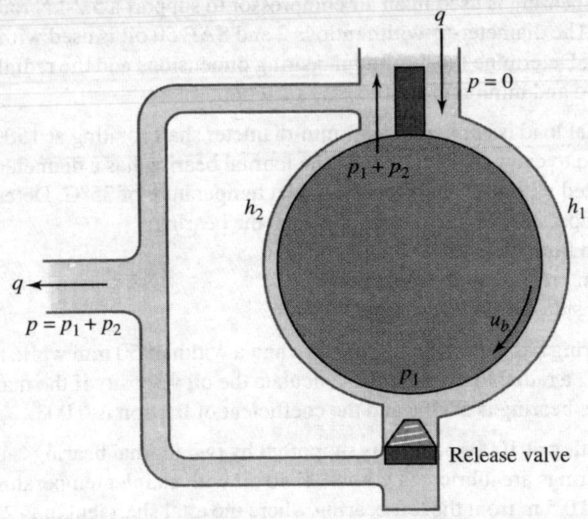

Sketch g, for Problem 12.15

12.16 A full journal bearing is used in a powerplant generator in the United States. The generator has four poles and delivers electricity with 60-Hz frequency. The rotor weighs 30 000 kg and is equally carried by two journal slider bearings. The shaft diameter is 300 mm and each bearing is also 300 mm wide. The bearing and shaft surfaces are ground to a surface finish $R_a = 1$ μm. Dimension the bearings for minimum power loss and calculate the coefficient of friction, the side-leakage ratio, and the location of the terminating pressure to find where the oil feed groove can be placed. To make sure that the bearings are dynamically stable, choose an eccentricity ratio of 0.82. Calculate the power loss in the bearing.

12.17 The powerplant generator considered in Problem 12.16 is rebuilt and moved to Europe. The rebuilt generator has two poles and the frequency of the electricity is 50 Hz. Redimension the bearings for minimum power loss and calculate the coefficient of friction, the side-leakage ratio, and the location of the terminating pressure. Let the eccentricity ratio be 0.82 to make sure that the bearings are dynamically stable. Calculate the power loss.

12.18 A motor speed of 110 rpm is used in a gearbox to provide a propeller speed of 275 rpm. The propeller shaft diameter is 780 mm, and the shaft is directly coupled to the outgoing shaft of the gearbox. The combined influence of the propeller shaft weight and the gear forces gives a 1.1-MN load on the journal bearing. Dimension the journal bearing for a diameter-to-width ratio of 2, and find the viscosity and radial clearance needed to give a minimum power loss. Also, calculate the power loss in the bearing. The journal bearing diameter is 780 mm and the surface roughness $R_a = 3$ μm.

12.19 A steam turbine rotor for a small powerplant in a paper factory has two journal bearings each carrying half the weight of a 20 000-N rotor. The rotor speed is 3000 rpm, the bearing diameters are both 120 mm, and the bearings are 120 mm wide. Design the bearing with surface roughness $R_a = 0.6$ μm and minimum power loss. Calculate the coefficient of friction, the attitude angle, the oil flow rate, and the side-leakage flow rate. Choose an eccentricity ratio of 0.82 to avoid dynamic instability.

12.20 A full journal bearing is used in an air compressor to support a 5.2-kN radial load at a speed of 2300 rpm. The diameter-to-width ratio is 2 and SAE 50 oil is used with an inlet temperature of 100°C. Determine the minimum bearing dimensions and the radial clearance for both maximum load and minimum coefficient of friction.

12.21 A 3.2-kN radial load is applied to a 50-mm-diameter shaft rotating at 1500 rpm. A journal bearing is used to carry the radial load. The journal bearing has a diameter-to-width ratio of 1 and is lubricated with SAE 20 oil with an inlet temperature of 35°C. Determine the following:

a) Mean temperature and temperature rise in the bearing
b) Minimum film thickness and its location
c) Maximum pressure and its location
d) Total and side-leakage flow rates

12.22 A journal bearing has a diameter of 200 mm and a width of 50 mm while rotating at 300 rpm and supporting a radial load of 10 kN. Calculate the oil viscosity if the maximum pressure obtained in the bearing is 5 MPa and the coefficient of friction is 0.003.

12.23 A shaft is rotating at 1000 rpm and is supported by two journal bearings at the two ends of the shaft. The bearings are lubricated with SAE 40 oil with an inlet temperature of 25°C. A 5-kN load is applied 0.5 m from the left bearing where the total shaft length is 2.5 m. The bearing width is 25 mm, the diameter is 50 mm, and the radial clearance is 0.0315 mm. Determine

a) Temperature rise
b) Minimum film thickness
c) Maximum pressure
d) Side-leakage flow rate
e) Bearing power loss

Section 12.5

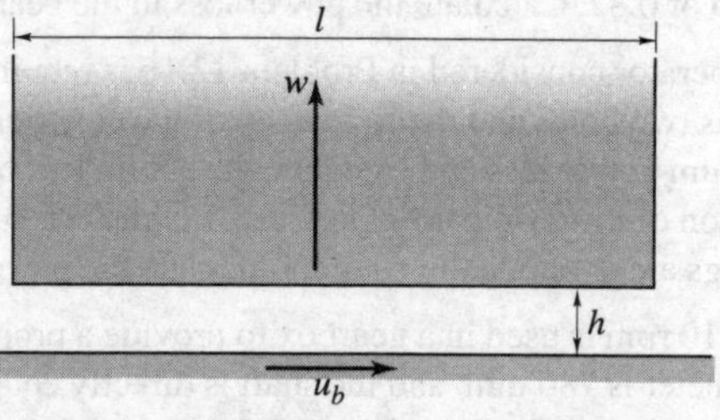

Sketch *h*, for Problem 12.24

12.24 The infinitely wide slider bearing shown in sketch *h* has parallel surfaces, sliding speed u_b, velocity across the film *w*, oil film thickness *h*, and bearing length *l*. Find the bearing damping constant $B = -\partial P/\partial w$, where *P* is the bearing load per unit width. Use the Reynolds equation for squeeze motion.

12.25 A squeeze film damper is used to stop the motion and damp the kinetic energy of an overhead crane. The damper consists of two flat, parallel steel plates 2 m in the side-leakage direction and 0.1 m long separated by a 1-mm-thick layer of oil with viscosity 0.1 Pa-s. When the overhead crane hits the damper, the oil is pressed out from between the plates and the oil layer becomes thinner. Calculate the damping force at the first moment of impact and the instantaneous retardation if the speed at impact is 0.2 m/s and the crane mass is 2000 kg. Also calculate the damping force at the first moment of impact if the crane carries a 10-ton load in a wire.

Section 12.6

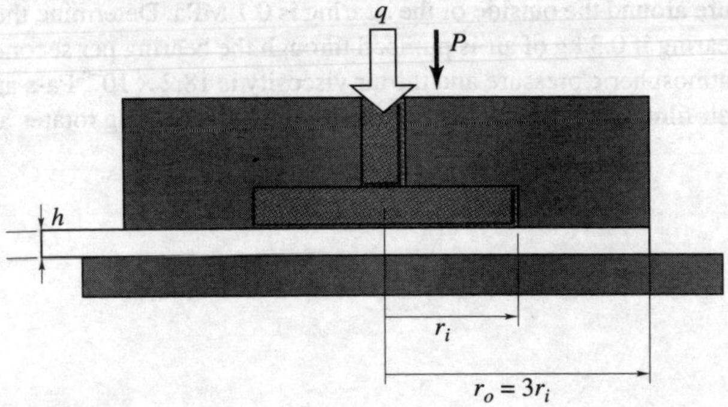

Sketch *i*, for Problem 12.26

12.26 The circular hydrostatic thrust bearing shown in sketch *i* has inner radius r_i and outer radius r_o, a constant oil film thickness *h*, and a constant oil flow rate *q*. The bearing load is *P* and the oil viscosity is η. Find the bearing stiffness coefficient $c_s = -\partial P/\partial h$.

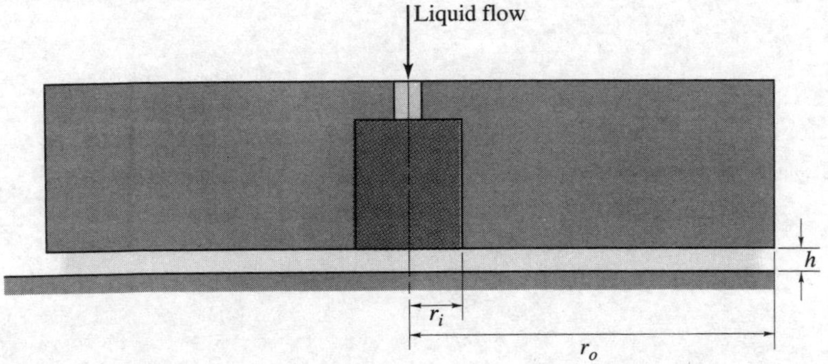

Sketch *j*, for Problem 12.27

12.27 A circular hydrostatic thrust bearing, shown in sketch *j*, has a constant oil film thickness *h*. A central lubricant reservoir, whose height is much larger than the oil film thickness, is fed by a liquid with pressure p_r. Find the pump power needed if the bearing carries load *P*. Given that $r_i = 27$ mm, $\eta = 0.00452$ N-s/m^2, $h = 0.120$ mm, $r_0 = 200$ mm, $P = 3.08 \times 10^4$ N.

Section 12.7

12.28 When some early satellites were launched, the bearings used in gyroscopes were high-frequency squeeze film gas bearings consisting of two flat, circular disks vibrating against each other. The lubricating gas film is described by

$$h = h_m + \frac{a \sin \omega t}{|\sin \omega t|}$$

where the vibration frequency ω is extremely high and the amplitude can be expressed as $a = 0.8\, h_m$. Calculate the bearing load-carrying capacity as a function of the air pressure in the satellite and the bearing diameter when ω is extremely high and the gas is adiabatic with the pressure-volume relationship

$$pV^{1.4} = p_m V_m^{1.4} = \text{constant}$$

12.29 A flat, circular hydrostatic air bearing has an outer diameter of 160 mm and a 5-mm-deep recess from the 50-mm diameter to the bearing center. The air pressure in the recess is 2 MPa, and the pressure around the outside of the bearing is 0.1 MPa. Determine the pressure distribution in the bearing if 0.3 kg of air is pumped through the bearing per second. The air density is 1 kg/m^3 at atmospheric pressure and the air viscosity is 18.2×10^{-6} Pa-s at 22°C. Also, calculate the air film thickness and the friction torque if the bearing rotates at 8000 rpm.

CHAPTER

13

ROLLING-ELEMENT BEARINGS

An assortment of rolling-element bearings. (Courtesy of SKF)

Since there is no model in nature for guiding wheels on axles or axle journals, man faced a great task in designing bearings—a task which has not lost its importance and attraction to this day.

[From *Rolling Bearings and Their Contribution to the Progress of Technology* (1986).]

SYMBOLS

A	area, m^2
$\tilde{A}$	constant
B	total conformity ratio
b_w	bearing width, m
b^*	contact semiwidth, m
$\tilde{b}$	semiwidth in y direction of uniform pressure, m
C	dynamic load rating, N
$\overline{C}$	specific dynamic capacity or load rating of bearing, N
C_0	static load rating, N
c	radial clearance, m
c_d	diametral clearance, m
c_e	free endplay, m
c_r	distance between race curvature centers, m
D_x	diameter of contact ellipse along x axis, m
D_y	diameter of contact ellipse along y axis, m
$\overline{D}$	material factor
d	diameter of rolling element, m
d_e	pitch diameter, m
d_i, d_o	inner- and outer-race diameters, m
E'	effective elastic modulus, $2\left(\frac{1-\nu_a^2}{E_a}+\frac{1-\nu_b^2}{E_b}\right)^{-1}$, Pa
$\overline{E}$	metallurgical processing factor
$\mathcal{E}$	complete elliptic integral of second kind
$\tilde{e}$	plane strain component in solid
e_1	constant
$\overline{F}_l$	lubrication factor
$\mathcal{F}$	complete elliptic integral of first kind
f_0	dimensionless factor
G	dimensionless materials parameter, $\xi E'$
$\overline{G}$	speed effect factor
H	dimensionless film thickness
H_{max}	dimensionless maximum film thickness
H_{min}	dimensionless minimum film thickness
$\overline{H}_m$	misalignment factor
$\tilde{H}_{e,min}$	dimensionless minimum film thickness for elliptical conjunctions obtained from curve-fitting results, $\tilde{h}_{min}/R_x$
h	film thickness, m
h_{min}	minimum film thickness, m
$\tilde{h}_{min}$	minimum film thickness obtained from curve-fitting results, m
h_0	central film thickness, m
K_1	constant defined in Eq. (13.36)
$K_{1.5}$	constant defined in Eq. (13.35)
k_e	ellipticity parameter, D_y/D_x
k_o	flow rate constant of orifice, $m^4/(s\text{-}N^{1/2})$
$\overline{k}_e$	ellipticity parameter from approximate formula, $\alpha_r^{2/\pi}$
L_A	adjusted life, millions of cycles
$\tilde{L}$	fatigue life, millions of cycles
$\tilde{L}_{10}$	fatigue life for probability of survival of 0.90, millions of cycles
l	length of rolling element, m
l_e	length of roller land, m
l_l	roller effective length, m
l_t	roller length, m
l_v	length dimension in stressed volume, m
$\tilde{M}$	probability of failure
m_k	load-life exponent
$\overline{m}$	slope
N_a	rotational speed, rps
N_b	rotational speed, rpm
N_c	number of cycles to failure
n	number of rolling elements per row
n_s	static safety factor
P	normal applied load, N
P_a	dimensionless pressure at inlet, $p_a h_o^2/\eta_0 l u_b$; axial component of load, N
P_e	dimensionless pressure, p/E'
P_{z_1}	normal applied load, N
P_z	normal applied load, N
P_z'	normal applied load per unit width, N/m
P_r	radial component of load, N
P_t	system thrust load, N
P_0	equivalent static load, N
$\overline{P}$	bearing equivalent load, N
p	pressure, Pa
p_a	ambient pressure, Pa
p_H	maximum Hertzian pressure, Pa
p_{max}	maximum pressure, Pa
q_a	constant, $(\pi/2) - 1$
R	curvature sum, m
R_e	radius ratio
R_q	root-mean-square (rms) surface roughness, m
R_r	race conformity, $r/2r_r$
R_x, R_y	effective radii in x and y directions, respectively, m
r	radius, m
r_c	corner radius, m
r_i	inner radius, m
r_o	outer radius, m
r_r	crown radius, m
$\tilde{S}$	probability of survival
s_h	shoulder height, m

s_1	r/h_0
t_0	characteristic time, s
U	dimensionless speed parameter, $\eta_0 \tilde{u}/E'R_x$
u	velocity in x direction, m/s
u_a	surface a velocity, m/s
u_b	surface b velocity, m/s
u_0	characteristic velocity in x direction, m/s
$\bar{u}$	dimensionless velocity in x direction, u/u_0
$\tilde{u}$	mean surface velocity in x direction, $(u_a + u_b)/2$, m/s
v_i, v_o	linear velocities of inner and outer contacts, m/s
W	dimensionless load parameter, $P'_z/E'R_x$
W_t	total load, N
W_ψ	load at ψ, N
w	bearing load, N
w_e	equivalent bearing load, N
w_0	equivalent static bearing load, N
w_t	total thrust load of bearing, N
X	radial factor
X_0	radial factor for static loading
Y	thrust factor
Y_0	thrust factor for static loading
$\tilde{x}, \tilde{y}$	factors for calculating equivalent load
x	Cartesian coordinate in direction of sliding, m
y	Cartesian coordinate in side-leakage direction, m
Z_w	constant defined in Eq. (13.48)
z	Cartesian coordinate in direction of film, m
z_o	depth of maximum shear stress, m
α_r	load angle, deg
β	contact angle, deg
β_f	free contact angle, deg
γ_d	angle defined in Fig. 13.9
δ	elastic deformation, m
δ_a	axial deflection, m
δ_m	elastic deformation, m
δ_{max}	interference, or total elastic compression on load line, m
δ_t	deflection due to thrust load, m
δ_φ	elastic compression of ball, m
η_0	absolute viscosity at $p = 0$ and constant temperature, Pa-s
θ_s	angle used to define shoulder height, deg
Λ	dimensionless film parameter, $h_{min}/(R_{qa}^2 + R_{qb}^2)^{1/2}$
ν	Poisson's ratio
ξ	pressure-viscosity coefficient, m²/N
ρ	density of lubricant, kg/m³
ρ_0	density at $p = 0$, kg/m³
τ_0	shear stress at zero pressure, Pa
ϕ_s	angle locating ball-spin vector, deg
ψ	angle to load line, deg
ψ_t	angular extent of bearing loading, deg
ω	angular velocity, rad/s
ω_b	angular velocity of surface b or of rolling element about its own axis, rad/s
ω_c	angular velocity of separator or of ball set, rad/s
ω_i, ω_o	angular velocities of inner and outer races, rad/s
ω_l	rotational speed of load vector, rad/s
ω_r	angular velocity of race, rad/s
ω_s	angular velocity due to spinning, rad/s

Subscripts

a	solid a
all	allowable
b	solid b
i	inner; inlet
max	maximum
o	outer; outlet
x,y,z	coordinates
0	static loading

13.1 INTRODUCTION

Chapter 13 considered hydrodynamic bearings, which assume that the surfaces are conformal and that the motion is sliding. This chapter considers **rolling-element bearings**, precise, yet simple, machine elements of great utility where the surfaces are nonconformal and the motion is primarily rolling. First, the history of rolling-element bearings is briefly reviewed. Subsequent sections describe the types of rolling-element bearing and their geometry and kinematics. The organization of this chapter is such that unloaded and unlubricated rolling-element bearings are considered in the next five sections; loaded but unlubricated rolling-element bearings, in Sec. 13.7; and loaded and lubricated rolling-element bearings, in Secs. 13.8 and 13.9. Case studies following Sec. 13.9 give applications of a cylindrical roller bearing and a radial ball bearing.

Animations of rolling-element bearing motion

13.2 HISTORICAL OVERVIEW

The purpose of a bearing is to provide relative positioning and rotational freedom while transmitting a load between two structures, usually a shaft and a housing. The basic form and concept of the rolling-element bearing are simple. If loads are to be transmitted between surfaces in relative motion in a machine, the action can be facilitated most effectively if rolling elements are interposed between the sliding members. The frictional resistance encountered in sliding is then largely replaced by the much smaller resistance associated with rolling, although the arrangement is inevitably afflicted with high stresses in the restricted regions of effective load transmission.

The precision rolling-element bearing of the twentieth century, a product of exacting technology and sophisticated science, is extremely effective in reducing friction and wear in a wide range of machinery. The spectacular development of numerous forms of rolling-element bearing in the twentieth century is well known and documented. However, the origins of these vital machine elements can be traced to long before there was a large industrial demand for such devices and certainly long before adequate machine tools for their effective manufacture existed in large quantities. A complete history of rolling-element bearings is given in Hamrock and Dowson (1981), and therefore only a brief overview is presented here.

The influence of general technological progress on the development of rolling-element bearings, particularly those used in moving heavy stone building blocks and carvings and in road vehicles, precision instruments, water-raising equipment, and windmills is discussed in Hamrock and Dowson (1981). The concept of rolling-element bearings emerged in Roman times, faded from memory during the Middle Ages, revived during the Renaissance, developed steadily in the seventeenth and eighteenth centuries for various applications, and was firmly established for individual road carriage bearings during the Industrial Revolution. Toward the end of the nineteenth century the great merit of ball bearings for bicycles promoted interest in the manufacture of accurate steel balls. Initially, the balls were turned from bars on special lathes, with individual general machine manufacturing companies making their own bearings. Growing demand for both ball and roller bearings encouraged the formation of specialist bearing manufacturing companies at the turn of the century and thus laid the foundations of a great industry. The advent of precision grinding techniques and the availability of improved materials did much to confirm the future of the new industry.

The essential features of most forms of the modern rolling-element bearing were therefore established by the second half of the nineteenth century, but it was the formation of specialist precision-manufacturing companies in the early years of the twentieth century that finally established the rolling-element bearing as a most valuable, high-quality, readily available machine component. The availability of ball and roller bearings in standard sizes has had a tremendous impact on machine design throughout the twentieth century. Such bearings still provide a challenging field for further research and development, and many engineers and scientists are currently engaged in exciting and demanding research projects in this area. In many cases new materials or enlightened design concepts have extended the life and range of application of rolling-element bearings. In other respects much remains to be done in explaining the extraordinary operating characteristics of these bearings, which have served our technology so well for almost a century.

13.3 BEARING TYPES

Ball and roller bearings are available to the designer in a great variety of designs and size ranges. The intent of this chapter is not to duplicate the complete descriptions given in manufacturers' catalogs but rather to present a guide to representative bearing types along with the approximate range of sizes available. Tables 13.1 to 13.5 illustrate some of the more widely used bearing types. In addition, numerous types of specialty bearings are available; space does not permit a complete cataloging of all available bearings. Size ranges are generally given in metric units. Traditionally, most rolling-element bearings have been manufactured to metric dimensions, predating the efforts toward a metric standard.

In addition to bearing types and approximate size ranges available, Tables 13.1 to 13.5 also list approximate relative load-carrying capacities, both radial and thrust, and, where relevant, approximate tolerances to misalignment.

Rolling-element bearings are an assembly of several parts: an inner race, an outer race, a set of balls or rollers, and a cage or separator. The cage or separator maintains even spacing of the rolling elements. A cageless bearing, in which the annulus is packed with the maximum rolling-element complement, is called a full-complement bearing. Full-complement bearings have high load-carrying capacity but lower speed limits than bearings equipped with cages. Tapered-roller bearings are an assembly of a cup, a cone, a set of tapered rollers, and a cage.

Although rolling-element bearings will work well without a lubricant and are sometimes operated that way, it is often advantageous to provide a lubricant film to extend life. Since lubricant supply is difficult with rolling-element bearing geometry, bearings are often packed with thick greases and then sealed so that the bearing operates under lubrication. The seals are a critical component, both to keep the lubricant inside the bearing pack and also to prevent the lubricant from being washed away (e.g., as in an automobile axle when the car is driven through a puddle of water). Although lubrication is often a critical issue in bearing performance, this subject is not investigated in depth in this chapter. It should be recognized, however, that rolling-element bearings usually operate under elastohydrodynamic lubrication, often with thick greases as the fluid, and that proper lubrication is essential for long service life.

13.3.1 BALL BEARINGS

Ball bearings are used in greater quantity than any other type of rolling-element bearing. For an application where the load is primarily radial with some thrust load present, one of the types in Table 13.1 can be chosen. A **Conrad**, or **deep-groove, bearing** has a ball complement limited by the number of balls that can be packed into the annulus between the inner and outer **races** with the inner race resting against the inside diameter of the outer race. A stamped and riveted two-piece **cage**, piloted on the ball set, or a machined two-piece cage, ball piloted or race piloted, is almost always used in a Conrad bearing. The only exception is a one-piece cage with open-sided pockets that is snapped into place. A filled-notch bearing most often has both inner and outer races notched so that a ball complement limited only by the annular space between the races can be used. It has low thrust capacity because of the filling notch.

The **self-aligning internal bearing** (Table 13.1) has an outer-race ball path ground in a spherical shape so that it can accept high levels of misalignment. The **self-aligning external bearing** (Table 13.1) has a multipiece outer race with a spherical inner race. It too can accept high misalignment and has higher capacity than the self-aligning internal bearing. However, the self-aligning external bearing is somewhat less self-aligning than its internal counterpart because of friction in the multipiece outer race.

Representative **angular-contact ball bearings** are illustrated in Table 13.2. An angular-contact ball bearing has a two-shouldered ball groove in one race and a single-shouldered ball groove in the other race. Thus, it can support only a unidirectional thrust load. The cutaway shoulder allows bearing assembly by snapping the inner race over the ball set after it is positioned in the cage and outer race. It also permits use of a one-piece, machined, race-piloted cage that can be balanced for high-speed operation. Typical contact angles vary from 15 to 40°.

Table 13.1 Characteristics of representative radial ball bearings. [B. J. Hamrock and W. J. Anderson, *Rolling-Element Bearings,* NASA Reference Publication 1105, 1983.]

Type	Approximate range of bore sizes, mm		Relative capacity		Limiting speed factor	Tolerance to misalignment
	Minimum	Maximum	Radial	Thrust		
Conrad or deep groove 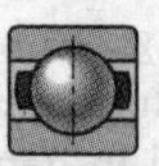	3	1060	1.00	[a]0.7	1.0	± 0°15′
Maximum capacity or filled notch 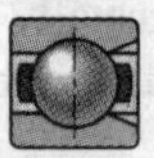	10	130	1.2–1.4	[a]0.2	1.0	± 0°3′
Self-aligning internal 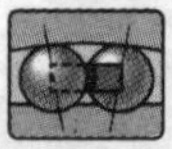	5	120	0.7	[b]0.2	1.0	± 2°30′
Self-aligning external 			1.0	[a]0.7	1.0	High
Double row, maximum 	6	110	1.5	[a]0.2	1.0	± 0°3′
Double row, deep groove 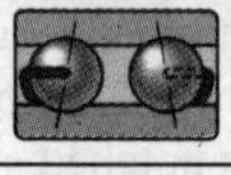	6	110	1.5	[a]1.4	1.0	0°

[a]Two directions.
[b]One direction.

Table 13.2 Characteristics of representative angular-contact ball bearings. [Minimum bore size; 10 mm. B. J. Hamrock and W. J. Anderson, *Rolling-Element Bearings*, NASA Reference Publication 1105, 1983.]

Type		Approximate maximum bore size, mm	Relative capacity: Radial	Relative capacity: Thrust	Limiting speed factor	Tolerance to mis-alignment
One-directional thrust		320	[b]1.00–1.15	[a,b]1.5–2.3	[b]1.1–3.0	± 0°2′
Duplex, back to back	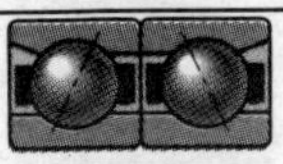	320	1.85	[c]1.5	3.0	0°
Duplex, face to face	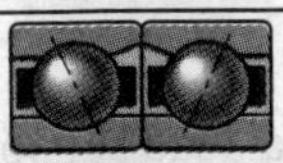	320	1.85	[c]1.5	3.0	0°
Duplex, tandem	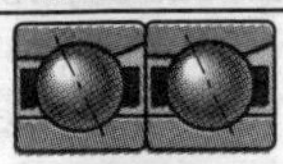	320	1.85	[a]2.4	3.0	0°
Two directional or split ring		110	1.15	[c]1.5	3.0	± 0°2′
Double row	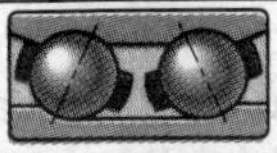	140	1.5	[c]1.85	0.8	0°

[a]One direction.
[b]Depends on contact angle.
[c]Two directions.

Angular-contact ball bearings are used in **duplex pairs** mounted either back to back or face to face, as shown in Table 13.2. Duplex bearing pairs are manufactured so that they "preload" each other when clamped together in the housing and on the shaft. Preloading provides stiffer shaft support and helps prevent bearing skidding at light loads. Proper levels of preload can be obtained from the manufacturer. A duplex pair can support bidirectional thrust load. The back-to-back arrangement offers more resistance to moment or overturning loads than does the face-to-face arrangement.

Where thrust loads exceed the load-carrying capacity of a simple bearing, two bearings can be used in tandem, with each bearing supporting part of the thrust load. Three or more bearings are occasionally used in tandem, but this is discouraged because of the difficulty in achieving good load sharing. Even slight differences in operating temperature will cause an uneven distribution of load sharing.

The **split-ring bearing** (Table 13.2) offers several advantages. The split ring (usually the inner race) has its ball groove ground as a circular arc with a shim between the ring halves. The shim is then removed when the bearing is assembled so that the split-ring ball groove has the shape of a gothic arch. This shape reduces the axial play for a given radial play and results in more accurate axial positioning of the shaft. The bearing can support bidirectional thrust loads but must not be operated for prolonged times at predominantly radial loads. This restriction results in three-point, ball-race contact and relatively high frictional losses. As with the conventional angular-contact bearing, a one-piece, precision-machined cage is used.

Thrust ball bearings (90° contact angle), shown in Table 13.3, are used almost exclusively for machinery with vertically oriented shafts. The flat-race bearing allows eccentricity of the fixed and rotating members. An additional bearing must be used for radial positioning. It has low thrust load-carrying capacity because of its small ball-race contact and consequent high Hertzian stress. Grooved-race bearings have higher thrust load-carrying capacity and can support low-magnitude radial loads. All the pure-thrust ball bearings have modest speed capability because of the 90° contact angle and the consequent high level of ball spinning and frictional losses.

13.3.2 ROLLER BEARINGS

Cylindrical roller bearings (Table 13.4) provide purely radial load support in most applications. An N or U type of bearing will allow free axial movement of the shaft relative to the housing to accommodate differences in thermal growth. An F or J type of bearing will support a light thrust load in one direction; and a T type of bearing, a light bidirectional thrust load. Cylindrical roller bearings have moderately high radial load-carrying capacity as well as high speed capability. Their speed capability exceeds that of either spherical or tapered-roller bearings. A commonly used bearing combination for support of a high-speed rotor is an angular-contact ball bearing (or a duplex pair) and a cylindrical roller bearing. As explained in the following section on bearing geometry, the rollers in cylindrical roller

Table 13.3 Characteristics of representative thrust ball bearings. [Relative radial capacity, O. B. J. Hamrock and W. J. Anderson, *Rolling-Element Bearings,* NASA Reference Publication 1105, 1983.]

Type		Approximate range of bore sizes, mm		Relative thrust capacity	Limiting speed factor	Tolerance to mis-alignment
		Minimum	Maximum			
One directional, flat race		6.45	88.9	[a]0.7	0.10	[b]0°
One directional, grooved race	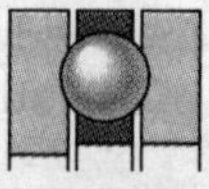	6.45	1180	[a]1.5	0.30	0°
Two directional, grooved race	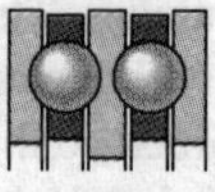	15	220	[c]1.5	0.30	0°

[a]One direction.
[b]Accepts eccentricity.
[c]Two directions.

Table 13.4 Characteristics of representative cylindrical roller bearings. [B. J. Hamrock and W. J. Anderson, *Rolling-Element Bearings,* NASA Reference Publication 1105, 1983.]

Type	Approximate range of bore sizes, mm: Minimum	Approximate range of bore sizes, mm: Maximum	Relative capacity: Radial	Relative capacity: Thrust	Limiting speed factor	Tolerance to misalignment
Separable outer ring, nonlocating (N) 	10	320	1.55	0	1.20	± 0°5′
Separable inner ring, nonlocating (NU) 	12	500	1.55	0	1.20	± 0°5′
Separable inner ring, one-direction locating (NJ) 	12	320	1.55	[a]Locating	1.15	± 0°5′
Separable inner ring, two-direction locating (NUP) 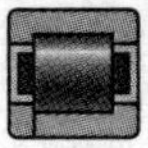	20	320	1.55	[b]Locating	1.15	± 0°5′

[a]One direction.
[b]Two directions.

bearings are seldom pure cylinders. They are usually crowned, or made slightly barrel shaped, to relieve stress concentrations on the roller ends when any misalignment of the shaft and housing is present. Cylindrical roller bearings may be equipped with one- or two-piece cages, usually race piloted. Full-complement bearings can be used for greater load-carrying capacity but at a significant sacrifice in speed capability.

Spherical roller bearings (Table 13.5) are made as either single- or double-row bearings. The more popular bearing design (convex) uses barrel-shaped rollers. An alternative design (concave) employs hourglass-shaped rollers. Spherical roller bearings combine extremely high radial load-carrying capacity with modest thrust load-carrying capacity (with the exception of the thrust type) and excellent tolerance to misalignment. They find widespread use in heavy-duty rolling mill and industrial gear drives, where all these bearing characteristics are requisite.

Table 13.5 Characteristics of representative spherical roller bearings. [B. J. Hamrock and W. J. Anderson, *Rolling-Element Bearings,* NASA Reference Publication 1105, 1983.]

Type		Approximate range of bore sizes, mm		Relative capacity		Limiting speed factor	Tolerance to mis-alignment
		Minimum	Maximum	Radial	Thrust		
Single row, barrel or convex	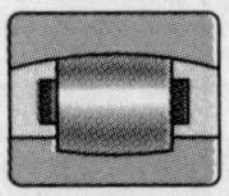	20	320	2.10	0.20	0.50	± 2°
Double row, barrel or convex	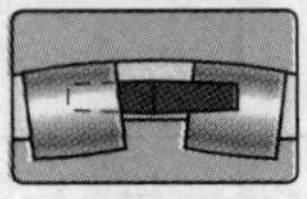	25	1250	2.40	0.70	0.50	± 1°30′
Thrust		85	360	[a]0.10 [b]0.10	[a]1.80 [b]2.40	0.35–0.50	± 3°
Double row, concave	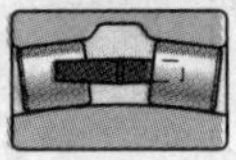	50	130	2.40	0.70	0.50	± 1°30′

[a]Symmetric rollers.
[b]Asymmetric rollers.

Other types of roller bearing, such as tapered-roller and needle roller bearings, are not considered here but are discussed in Hamrock (1993) and Hamrock and Anderson (1983).

Example 13.1

GIVEN The geometry of the rolling surfaces in rolling-element bearings is determined from the load-carrying capacity of the elements and the type of motion. The load in ball bearings is carried by the balls. Depending on the geometry of the races, either radial, axial, or moment loads can be transmitted through a ball bearing.

FIND Describe which ball bearing type should be used for

a) Pure radial load
b) Pure axial load
c) Bending moment load
d) High radial stiffness
e) High axial stiffness

Solution

a) For pure radial loading a radial ball bearing is recommended. Conrad (deep-groove) ball bearings are the least recommended. A filled-notch bearing offers additional radial capacity. If there are no space limitations and the bearing width is not restricted, a double-row bearing should be considered. Table 13.1 indicates bearing type preferences for radially loaded bearings.

b) For pure axial loading the force transmitted through the balls should be as close as possible to the bearing axis. This location gives the smallest possible internal force acting through the balls. Table 13.3 shows thrust bearings that are good for pure axial loading.
c) For bending moment loading a bearing must take loads in both the axial and radial directions. Deep-groove ball bearings and double deep-groove ball bearings can be used. If two bearings are mounted so far apart that the center distance between the balls is about the same as the bearing outer diameter, any type of bearing shown in Tables 13.1 and 13.2 can be used.
d) Equation (8.27) shows that the elastic deformation (δ_{max}) in a ball bearing is proportional to the load W raised to the 2/3 power.

$$\therefore \delta_{max} \propto W^{2/3}$$

The contact stiffness is

$$\frac{\partial W}{\partial \delta_{max}} = \frac{1}{\dfrac{\partial \delta_{max}}{\partial W}} = \frac{1}{\frac{2}{3}W^{-\frac{1}{3}}} = \frac{3}{2}W^{\frac{1}{3}}$$

When W approaches zero, the contact stiffness also approaches zero. Thus, the only way to get high radial stiffness in a ball bearing is to preload the balls. To get high radial and high axial stiffness and high running accuracy, pairs of angular-contact ball bearings are used with axial preloading against each other. The same effect can be achieved by using a single bearing with a split ring, as shown in Table 13.2.
e) High axial stiffness can be achieved by preloading thrust ball bearings, such as deep-groove or angular-contact ball bearings, against each other. The higher the preload, the higher the stiffness becomes, but at the same time the shorter the life becomes and the higher the power loss (temperature) becomes.

13.4 GEOMETRY

The operating characteristics of a rolling-element bearing depend greatly on its diametral clearance. This clearance varies for the different types of bearing discussed in the preceding section. In this section the principal geometrical relationships governing the operation of unloaded rolling-element bearings are developed. This information will be of vital interest when stress, deflection, load-carrying capacity, and life are considered in subsequent sections. Although bearings rarely operate in the unloaded state, an understanding of this section is essential to appreciation of the remaining sections.

13.4.1 BALL BEARINGS

The geometry of ball bearings involves pitch diameter and clearance, race conformity, contact angle, endplay, shoulder height, and curvature sum and difference.

13.4.1.1 PITCH DIAMETER AND CLEARANCE The cross section through a radial single-row ball bearing shown in Fig. 13.1 depicts the radial clearance and various diameters. The **pitch diameter** d_e is the mean of the inner- and outer-race contact diameters and is given by

$$d_e = d_i + \frac{d_o - d_i}{2}$$

or

$$d_e = \frac{d_o + d_i}{2} \tag{13.1}$$

Also, from Fig. 13.1 the diametral clearance c_d can be written as

$$c_d = d_o - d_i - 2d \tag{13.2}$$

Diametral clearance may therefore be thought of as the maximum distance that one race can move diametrally with respect to the other when nonmeasurable force is applied and both races lie in the same plane. Although diametral clearance is generally used in connection with single-row radial bearings, Eq. (13.2) is also applicable to angular-contact bearings.

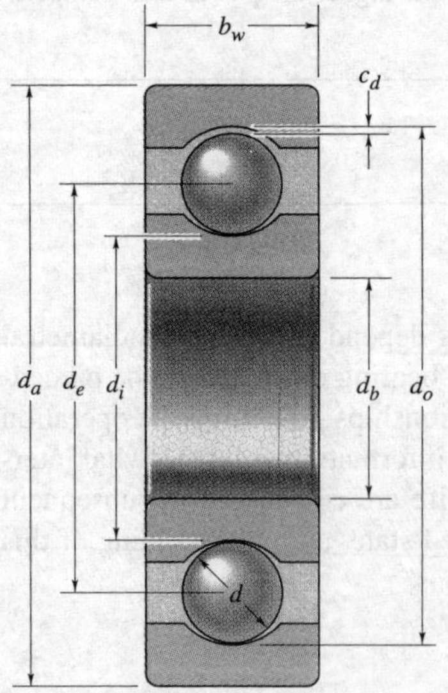

Figure 13.1 Cross section through radial single-row ball bearings. [B. J. Hamrock and W. J. Anderson, *Rolling-Element Bearings*, NASA Reference Publication 1105, 1983.]

13.4.1.2 RACE CONFORMITY Race conformity is a measure of the geometrical conformity of the race and the ball in a plane passing through the bearing axis, which is a line passing through the center of the bearing perpendicular to its plane and transverse to the race. Figure 13.2 is a cross section of a ball bearing showing race conformity, expressed as

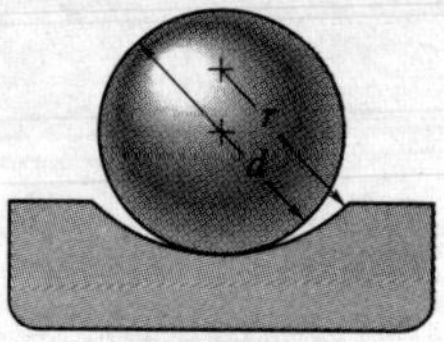

Figure 13.2 Cross section of ball and outer race, showing race conformity. [B. J. Hamrock and W. J. Anderson, *Rolling-Element Bearings,* NASA Reference Publication 1105, 1983.]

$$R_r = \frac{r}{d} \tag{13.3}$$

For perfect conformity, where the race radius is equal to the ball radius, R_r is equal to 1/2. The closer the race conforms to the ball, the greater the frictional heat within the contact. On the other hand, open-race curvature and reduced geometrical conformity, which reduce friction, also increase the maximum contact stresses and consequently reduce the bearing fatigue life. For this reason most ball bearings made today have race conformity ratios in the range $0.51 \le R_r \le 0.54$, with $R_r = 0.52$ being the most common value. The race conformity ratio for the outer race is usually made slightly larger than that for the inner race to compensate for the closer conformity in the plane of the bearing between the outer race and the ball than between the inner race and the ball. This larger ratio tends to equalize the contact stresses at the inner- and outer-race contacts. The difference in race conformities does not normally exceed 0.02.

13.4.1.3 CONTACT ANGLE Radial bearings have some axial play, since they are generally designed to have a diametral clearance, as shown in Fig. 13.3. Axial play implies a free contact angle different from zero. **Angular-contact bearings** are specifically designed to operate under thrust loads. The clearance built into the unloaded bearing, along with the race conformity ratio, determines the bearing free contact angle. Figure 13.3 shows a radial bearing with contact due to the axial shift of the inner and outer races when no measurable force is applied.

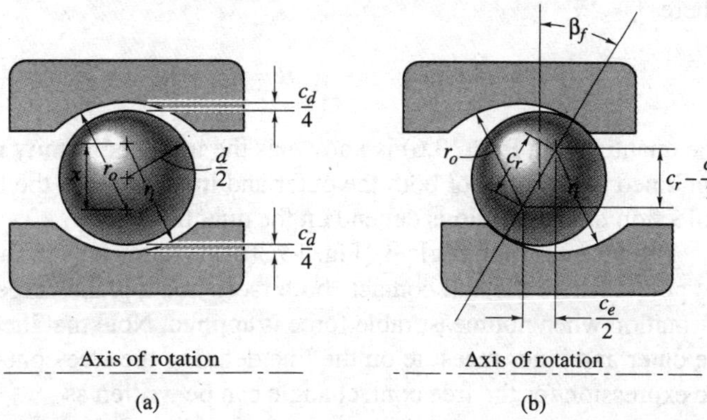

Figure 13.3 Cross section of radial bearing, showing ball-race contact due to axial shift of inner and outer races.
(a) Initial position; (b) shifted position. [B. J. Hamrock and W. J. Anderson, *Rolling-Element Bearings,* NASA Reference Publication 1105, 1983.]

Before the free contact angle is discussed, it is important to define the distance between the centers of curvature of the two races in line with the center of the ball in Fig. 13.3(a) and (b). This distance—denoted by x in Fig. 13.3(a) and by c_r in Fig. 13.3(b)—depends on race radius and ball diameter. When quantities referred to the inner and outer races are denoted by the subscripts i and o, respectively, Fig. 13.3(a) and (b) shows that

$$\frac{c_d}{4} + d + \frac{c_d}{4} = r_o - x + r_i$$

$$x = r_o + r_i - d - \frac{c_d}{2}$$

$$d = r_o - c_r + r_i$$

or

$$c_r = r_o + r_i - d \qquad \textbf{(13.4)}$$

From these equations

$$x = c_r - \frac{c_d}{2}$$

This distance, shown in Fig. 13.3(b), will be useful in defining the contact angle. Equation (13.3) can be used to write Eq. (13.4) as

$$c_r = Bd \qquad \textbf{(13.5)}$$

where

$$B = R_{ro} + R_{ri} - 1 \qquad \textbf{(13.6)}$$

The quantity B in Eq. (13.6) is known as the total conformity ratio and is a measure of the combined conformity of both the outer and inner races to the ball. Calculations of bearing deflection in later sections depend on the quantity B.

The **free contact angle** β_f [Fig. 13.3(b)] is defined as the angle made by a line through the points where the ball contacts both races and a plane perpendicular to the bearing axis of rotation when nonmeasurable force is applied. Note that the centers of curvature of both the outer and inner races lie on the line defining the free contact angle. From Fig. 13.3(b) the expression for the free contact angle can be written as

$$\cos\beta_f = 1 - \frac{c_d}{2c_r} \qquad \textbf{(13.7)}$$

Equations (13.2) and (13.4) can be used to write Eq. (13.7) as

$$\beta_f = \cos^{-1} \frac{r_o + r_i - \dfrac{d_o - d_i}{2}}{r_o + r_i - d} \tag{13.8}$$

Equation (13.8) shows that if the size of the balls is increased and everything else remains constant, the free contact angle is decreased. Similarly, if the ball size is decreased, the free contact angle is increased.

From Eq. (13.7) the diametral clearance c_d can be written as

$$c_d = 2c_r\left(1 - \cos\beta_f\right) \tag{13.9}$$

This is an alternative definition of the diametral clearance given in Eq. (13.2).

13.4.1.4 ENDPLAY Free **endplay** c_e is the maximum axial movement of the inner race with respect to the outer when both races are coaxially centered and no measurable force is applied. Free endplay depends on total curvature and contact angle, as shown in Fig. 13.3, and can be written as

$$c_e = 2c_r \sin\beta_f \tag{13.10}$$

Figure 13.4 shows the variation of free contact angle and endplay with the diametral clearance ratio $c_d/2d$ for four values of the total conformity ratio normally found in single-row ball bearings. Eliminating β_f in Eqs. (13.9) and (13.10) enables the following relationships between free endplay and diametral clearance to be established:

$$c_d = 2c_r - \left[(2c_r)^2 - c_e^2\right]^{1/2}$$

$$c_e = \sqrt{4c_r c_d - c_d^2}$$

13.4.1.5 SHOULDER HEIGHT Figure 13.5 illustrates the **shoulder height** of ball bearings. Shoulder height, or **race depth**, is the depth of the race groove measured from the shoulder to the bottom of the groove and is denoted by s_h in Fig. 13.5. From this figure the equation defining the shoulder height can be written as

$$s_h = r\left(1 - \cos\theta_s\right) \tag{13.11}$$

The maximum possible diametral clearance for complete retention of the ball-race contact within the race under zero thrust load is given by the condition $(\beta_f)_{max} = \theta_s$. Making use of Eqs. (13.9) and (13.11) gives

$$\left(c_d\right)_{max} = \frac{2c_r s_h}{r} \tag{13.12}$$

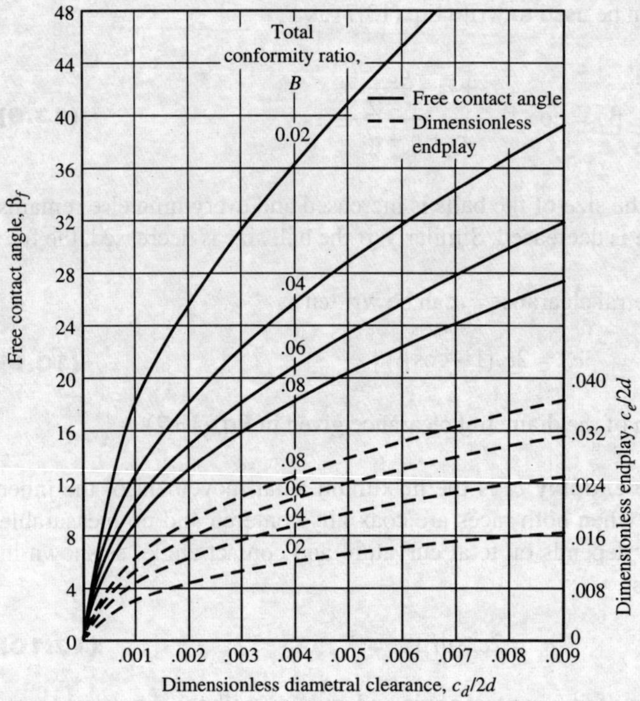

Figure 13.4 Free contact angle and endplay as function of $c_d/2d$ for four values of total conformity radio. [B. J. Hamrock and W. J. Anderson, *Rolling-Element Bearings,* NASA Reference Publication 1105, 1983.]

13.4.1.6 CURVATURE SUM AND DIFFERENCE The undeformed geometry of contacting solids in a ball bearing can be represented by two ellipsoids as discussed in Sec. 8.7.1. This section will apply the information provided there to a ball bearing.

Figure 13.6 shows a cross section of a ball bearing operating at a contact angle. Equivalent radii of curvature for both inner- and outer-race contacts in, and normal to, the direc-

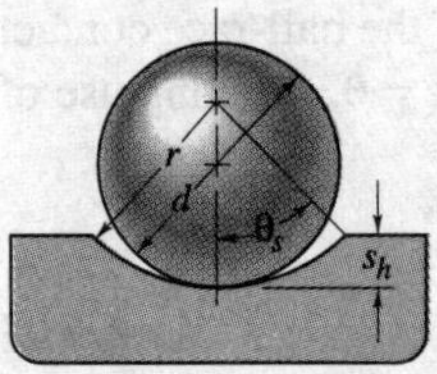

Figure 13.5 Shoulder height in ball bearing. [B. J. Hamrock and W. J. Anderson, *Rolling-Element Bearings,* NASA Reference Publication 1105, 1983.]

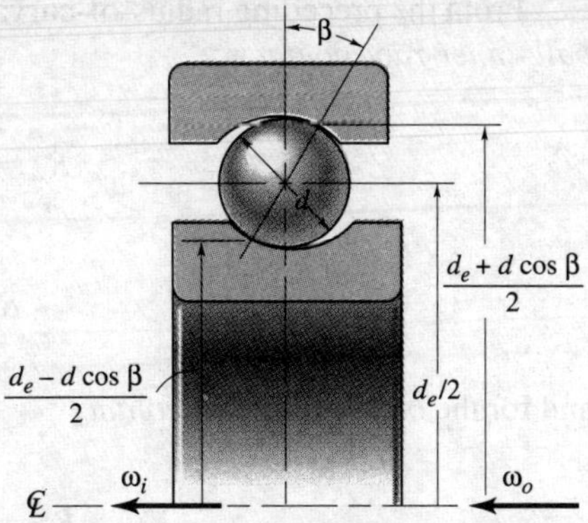

Figure 13.6 Cross section of ball bearing. [B. J. Hamrock and W. J. Anderson, *Rolling-Element Bearings*, NASA Reference Publication 1105, 1983.]

tion of rolling can be calculated from this figure. The radii of curvature for the *ball–inner-race contact* are

$$r_{ax} = r_{ay} = \frac{d}{2} \tag{13.13}$$

$$r_{bx} = \frac{d_e - d\cos\beta}{2\cos\beta} \tag{13.14}$$

$$r_{by} = -R_{ri}d = -r_i \tag{13.15}$$

The radii of curvature for the *ball–outer-race contact* are

$$r_{ax} = r_{ay} = \frac{d}{2} \tag{13.16}$$

$$r_{bx} = \frac{-(d_e + d\cos\beta)}{2\cos\beta} \tag{13.17}$$

$$r_{by} = -R_{ro}d = -r_o \tag{13.18}$$

In Eqs. (13.14) and (13.17) β is used instead of β_f because these equations are also valid when a load is applied to the contact. By setting $\beta = 0$, Eqs. (13.13) to (13.18) are equally valid for radial ball bearings. For thrust ball bearings $r_{bx} = \infty$ and the radii are defined as given in the preceding equations.

From the preceding radius-of-curvature expressions and Eqs. (8.17) and (8.18), for the *ball–inner-race contact*

$$R_{xi} = \frac{d(d_e - d\cos\beta)}{2d_e} \tag{13.19}$$

$$R_{yi} = \frac{R_{ri}d}{2R_{ri} - 1} \tag{13.20}$$

and for the *ball–outer-race contact*

$$R_{xo} = \frac{d(d_e + d\cos\beta)}{2d_e} \tag{13.21}$$

$$R_{yo} = \frac{R_{ro}d}{2R_{ro} - 1} \tag{13.22}$$

Substituting these equations into Eqs. (8.15) and (8.16) enables the curvature sum and difference to be obtained.

Example 13.2

GIVEN A single-row, deep-groove ball bearing, shown in Fig. 13.1, has a pitch diameter of 100 mm, an inner-race diameter of 80 mm, and a ball diameter of 19.9 mm. The outer-race radius is 10.5 mm, and the inner-race radius is 10.2 mm. The shoulder height angle is 30°.

FIND

a) Diametral clearance
b) Inner- and outer-race conformities as well as total conformity
c) Free contact angle and endplay
d) Shoulder heights for the inner and outer races
e) Ellipticity parameter and the curvature sum and difference of the contacts

Solution

a) Using Eqs. (13.1) and (13.2) gives the diametral clearance as

$$c_d = d_o - d_i - 2d = 2d_e - 2d_i - 2d = 200 - 160 - 2(19.9) = 0.2\,\text{mm}$$

b) Using Eq. (13.3) gives the conformities of the outer and inner races as

$$R_{ro} = \frac{r_o}{d} = \frac{10.5}{19.9} = 0.528$$

$$R_{ri} = \frac{r_i}{d} = \frac{10.2}{19.9} = 0.513$$

From Eq. (13.6) the total conformity is

$$B = R_{ro} + R_{ri} - 1 = 0.528 + 0.513 - 1 = 0.041$$

From Eq. (13.5) the distance between the race curvature centers is

$$c_r = Bd = (0.041)(19.9) = 0.816 \text{ mm}$$

c) From Eq. (13.7) the free contact angle is

$$\cos\beta_f = 1 - \frac{c_d}{2c_r} = 1 - \frac{0.2}{2(0.816)} = 0.877$$

$$\therefore \beta_f = 28.7°$$

From Eq. (13.10) the endplay is

$$c_e = 2c_r \sin\beta_f = 2(0.816)\sin(28.7°) = 0.783 \text{ mm}$$

d) From Eq. (13.11) the shoulder heights are

$$s_{hi} = r_i(1 - \cos\theta_s) = 10.2(1 - \cos 30°) = 1.367 \text{ mm}$$

$$s_{ho} = r_o(1 - \cos\theta_s) = 10.5(1 - \cos 30°) = 1.407 \text{ mm}$$

From Eq. (13.12) the minimum shoulder height for complete retention of the ball-race contact is

$$(s_{hi})_{\min} = \frac{c_d r_i}{2c_r} = \frac{(0.2)(10.2)}{2(0.816)} = 1.25 \text{ mm}$$

$$(s_{ho})_{\min} = \frac{c_d r_o}{2c_r} = \frac{(0.2)(10.5)}{2(0.816)} = 1.29 \text{ mm}$$

e) Ball–inner-race contact
From Eqs. (13.3), (13.4), and (13.15)

$$r_{ax} = r_{ay} = \frac{d}{2} = \frac{19.9}{2} = 9.95 \text{ mm}$$

$$r_{bx} = \frac{d_e - d\cos\beta}{2\cos\beta} = \frac{100 - (19.9)\cos 28.7°}{2\cos 28.7°} = 47.1 \text{ mm}$$

$$r_{by} = -R_{ri}d = -r_i = -10.2 \text{ mm}$$

The curvature sums in the x and y directions are

$$\frac{1}{R_x} = \frac{1}{r_{ax}} + \frac{1}{r_{bx}} = \frac{1}{9.95} + \frac{1}{47.1} = 0.122/\text{mm}$$

$$\therefore R_x = 8.215 \text{ mm}$$

$$\frac{1}{R_y} = \frac{1}{r_{ay}} + \frac{1}{r_{by}} = \frac{1}{9.95} - \frac{1}{10.2} = (2.46)(10^{-3})/\text{mm}$$

$$\therefore R_y = 406.0 \text{ mm}$$

From Eqs. (8.15) and (8.16) the curvature sum and difference are

$$\frac{1}{R} = \frac{1}{R_x} + \frac{1}{R_y} = 0.122 + 0.00246 = 0.0124/\text{mm}$$

$$\therefore R = 8.052 \text{ mm}$$

$$R_d = R\left(\frac{1}{R_x} - \frac{1}{R_y}\right) = 8.052(0.122 - 0.00246) = 0.9625$$

From Eqs. (8.19) and (8.24) the radius ratio and the ellipticity parameter are

$$\alpha_{ri} = \frac{R_y}{R_x} = \frac{406.0}{8.215} = 49.42$$

$$k_{ei} = \alpha_{ri}^{2/\pi} = (49.42)^{2/\pi} = 12.0$$

Ball–outer-race contact

$$r_{ax} = r_{ay} = \frac{d}{2} = 9.95 \text{ mm}$$

$$r_{bx} = -\frac{d_e + d\cos\beta}{2\cos\beta} = -\frac{100 + 19.9\cos(28.7°)}{2\cos(28.7°)}$$

$$r_{bx} = -66.95 \text{ mm}$$

$$r_{by} = -r_o = -10.5 \text{ mm}$$

The curvature sums in the x and y directions are

$$\frac{1}{R_x} = \frac{1}{r_{ax}} + \frac{1}{r_{bx}} = \frac{1}{9.95} - \frac{1}{66.95} = 0.0856/\text{mm}$$

$$\therefore R_x = 11.69 \text{ mm}$$

$$\frac{1}{R_y} = \frac{1}{r_{ay}} + \frac{1}{r_{by}} = \frac{1}{9.95} - \frac{1}{10.5} = 0.0053/\text{mm}$$

$$\therefore R_y = 190.0 \text{ mm}$$

The curvature sum and difference are

$$\frac{1}{R}=\frac{1}{R_x}+\frac{1}{R_y}=0.0856+0.0053=0.0909/\text{mm}$$

$$\therefore R=11.00 \text{ mm}$$

$$R_d=R\left(\frac{1}{R_x}-\frac{1}{R_y}\right)=11.00(0.0856-0.0053)=0.8833$$

The radius ratio and the ellipticity parameter are

$$\alpha_{ro}=\frac{R_y}{R_x}=\frac{190.0}{11.69}=16.25$$

$$k_{eo}=\alpha_r^{2/\pi}=(16.25)^{2/\pi}=5.90$$

13.4.2 ROLLER BEARINGS

The equations developed for the pitch diameter d_e and diametral clearance c_d for ball bearings [Eqs. (13.1) and (13.2), respectively] are directly applicable for roller bearings.

13.4.2.1 CROWNING High stresses at the edges of the rollers in cylindrical roller bearings are usually prevented by machining a **crown** into the rollers, either fully or partially as shown in Fig. 13.7. The crown curvature is greatly exaggerated for clarity. Crowning the rollers also protects the bearing against the effects of slight misalignment. For

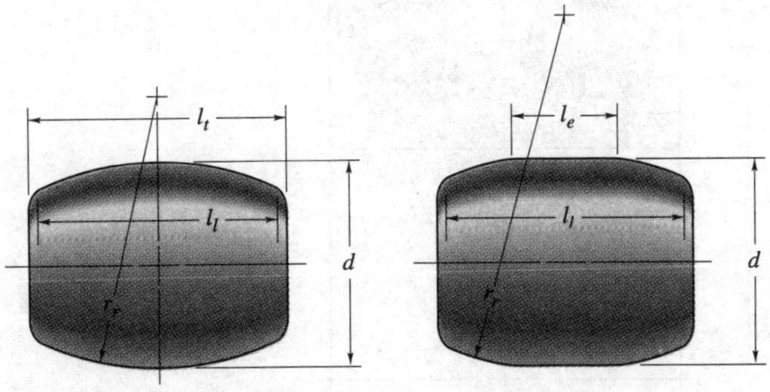

Figure 13.7
(a) Spherical roller (fully crowned) and (b) cylindrical roller (partially crowned). [B. J. Hamrock and W. J. Anderson, *Rolling-Element Bearings,* NASA Reference Publication 1105, 1983.]

cylindrical rollers $r_r/d \approx 10^2$. In contrast, for spherical rollers in spherical roller bearings, as shown in Fig. 13.7(a), $r_r/d \approx 4$. Observe in Fig. 13.7 that the roller effective length l_l is the length presumed to be in contact with the races under loading. Generally, the roller effective length can be written as

$$l_l = l_t - 2r_c$$

where r_c is the roller corner radius or the grinding undercut, whichever is larger.

13.4.2.2 RACE CONFORMITY Race conformity applies to roller bearings much as it applies to ball bearings. It is a measure of the geometrical conformity of the race and the roller. Figure 13.8 shows a cross section of a spherical roller bearing. From this figure the race conformity can be written as

$$R_r = \frac{r}{2r_r}$$

In this equation if R_r and r are subscripted with i or o, the race conformity values are for the inner- or outer-race contacts, respectively.

13.4.2.3 FREE ENDPLAY AND CONTACT ANGLE Cylindrical roller bearings have a contact angle of zero and may take thrust load only by virtue of axial flanges. Tapered-roller bearings must be subjected to a thrust load or the inner and outer races (the cone and the

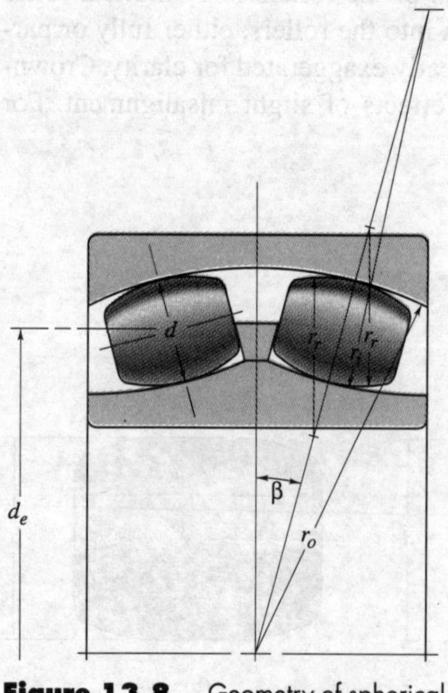

Figure 13.8 Geometry of spherical roller bearing. [B. J. Hamrock and W. J. Anderson, *Rolling-Element Bearings*, NASA Reference Publication 1105, 1983.]

cup) will not remain assembled; therefore, tapered-roller bearings do not exhibit free diametral play. Radial spherical roller bearings are, however, normally assembled with free diametral play and hence exhibit free endplay. The diametral play c_d for a spherical roller bearing is the same as that obtained for ball bearings as expressed in Eq. (13.2). Figure 13.9 shows diametral play, as well as endplay, for a spherical roller bearing. From this figure

$$r_o \cos\beta = \left(r_o - \frac{c_d}{2}\right)\cos\gamma_d$$

or

$$\beta = \cos^{-1}\left[\left(1 - \frac{c_d}{2r_o}\right)\cos\gamma_d\right]$$

Also from Fig. 13.9 the free endplay can be written as

$$c_e = 2r_o(\sin\beta - \sin\gamma_d) + c_d \sin\gamma_d$$

13.4.2.4 Curvature Sum and Difference The same procedure used for ball bearings will be used for defining the curvature sum and difference for roller bearings. For

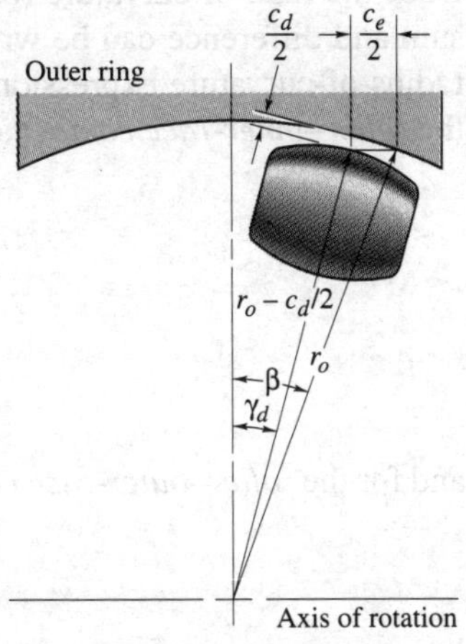

Figure 13.9 Schematic diagram of spherical roller bearing, showing diametral play and endplay. [B. J. Hamrock and W. J. Anderson, *Rolling-Element Bearings,* NASA Reference Publication 1105, 1983.]

spherical roller bearings (Fig. 13.8) the radii of curvature for the *roller–inner-race contact* can be written as

$$r_{ax} = \frac{d}{2}$$

$$r_{ay} = \frac{r_i}{2R_{ri}} = r_r$$

$$r_{bx} = \frac{d_e - d\cos\beta}{2\cos\beta}$$

$$r_{by} = -2R_{ri}r_r = -r_i$$

And the radii of curvature for the *roller–outer-race contact* can be written as

$$r_{ax} = \frac{d}{2}$$

$$r_{ay} = \frac{r_o}{2R_{ro}} = r_r$$

$$r_{bx} = -\frac{d_e + d\cos\beta}{2\cos\beta}$$

$$r_{by} = -2R_{ro}r_r = -r_o$$

Once the radii of curvature for the respective contact conditions are known, the curvature sum and difference can be written directly from Eqs. (8.15) and (8.16). Furthermore, the radius-of-curvature expressions R_x and R_y for spherical roller bearings can be written for the *roller–inner-race contact* as

$$R_{xi} = \frac{d(d_e - d\cos\beta)}{2d_e} \quad \textbf{(13.23)}$$

$$R_{yi} = \frac{r_i r_r}{r_i - r_r} \quad \textbf{(13.24)}$$

and for the *roller–outer-race contact* as

$$R_{xo} = \frac{d(d_e + d\cos\beta)}{2d_e} \quad \textbf{(13.25)}$$

$$R_{yo} = \frac{r_o r_r}{r_o - r_r} \quad \textbf{(13.26)}$$

Substituting these equations into Eqs. (8.15) and (8.16) enables the curvature sum and difference to be obtained.

13.5 KINEMATICS

The relative motions of the separator, the balls or rollers, and the races of rolling-element bearings are important to understanding their performance. The relative velocities in a ball bearing are somewhat more complex than those in roller bearings, the latter being analogous to the specialized case of a zero- or fixed-contact-angle ball bearing. For that reason the ball bearing is used as an example here to develop approximate expressions for relative velocities. These expressions are useful for rapid but reasonably accurate calculation of elastohydrodynamic film thickness, which can be used with surface roughnesses to calculate the lubrication life factor.

The precise calculation of relative velocities in a ball bearing in which speed or centrifugal force effects, contact deformations, and elastohydrodynamic traction effects are considered requires a large computer to numerically solve the relevant equations. Such a treatment is beyond the scope of this section.

When a ball bearing operates at high speeds, the centrifugal force acting on the ball creates a divergency of the inner- and outer-race contact angles, as shown in Fig. 13.10, in order to maintain force equilibrium on the ball. For the most general case of rolling and spinning at both inner- and outer-race contacts, the rolling and spinning velocities of the ball are as shown in Fig. 13.11.

Jones (1964) developed the equations for ball and separator angular velocity for all combinations of inner- and outer-race rotation. Without introducing additional relationships to describe the elastohydrodynamic conditions at both ball-race contacts, however, the ball-spin axis orientation angle ϕ_s cannot be obtained. As mentioned, this requires a lengthy numerical solution except for the two extreme cases of outer- or inner-race control. Figure 13.12 illustrates these cases.

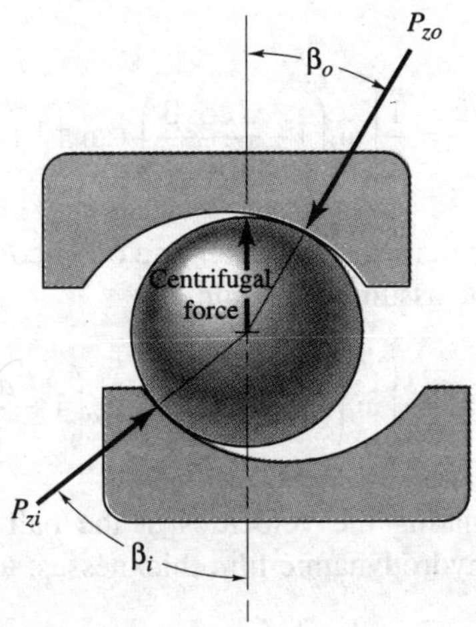

Figure 13.10 Contact angles in ball bearing at appreciable speeds. [B. J. Hamrock and W. J. Anderson, *Rolling-Element Bearings,* NASA Reference Publication 1105, 1983.]

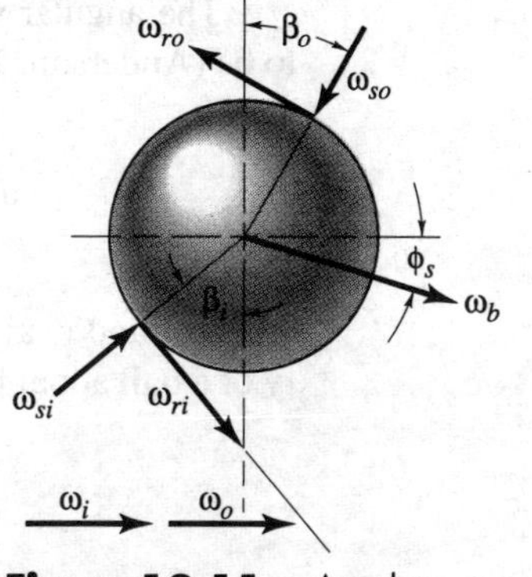

Figure 13.11 Angular velocities of ball. [B. J. Hamrock and W. J. Anderson, *Rolling-Element Bearings,* NASA Reference Publication 1105, 1983.]

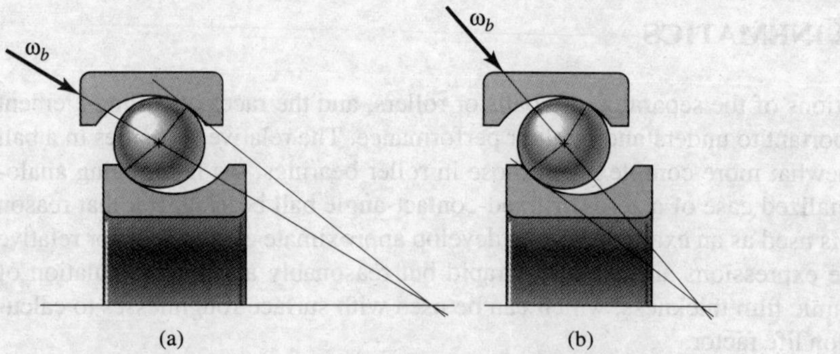

Figure 13.12 Ball-spin axis orientations for (a) outer-race control and (b) inner-race control. [B. J. Hamrock and W. J. Anderson, *Rolling-Element Bearings,* NASA Reference Publication 1105, 1983.]

Race control assumes that pure rolling occurs at the controlling race, with all the ball spin occurring at the other race contact. The orientation of the ball rotation axis can then be easily determined from bearing geometry. Race control probably occurs only in dry bearings or dry-film-lubricated bearings, where Coulomb friction conditions exist in the ball-race contact ellipses. The spin-resisting moment will always be greater at one of the race contacts. Pure rolling will occur at the race contact with the higher magnitude spin-resisting moment, usually the inner race at low speeds and the outer race at high speeds.

In oil-lubricated bearings in which elastohydrodynamic films exist in both ball-race contacts, rolling with spin occurs at both contacts. Therefore, precise ball motions can only be determined through a computer analysis. The situation can be approximated with a reasonable degree of accuracy, however, by assuming that the ball rolling axis is normal to the line drawn through the centers of the two ball-race contacts [Fig. 13.12(b)].

The angular velocity of the separator or ball set ω_c about the shaft axis can be shown to be (Anderson, 1970)

$$\omega_c = \frac{(v_i + v_o)/2}{d_e/2} = \frac{1}{2}\left[\omega_i\left(1 - \frac{d\cos\beta}{d_e}\right) + \omega_o\left(1 + \frac{d\cos\beta}{d_e}\right)\right] \tag{13.27}$$

where v_i and v_o are the linear velocities at the inner and outer contacts. The angular velocity of a ball about its own axis ω_b, assuming no spin, is

$$\omega_b = \frac{v_i - v_o}{d_e/2} = \frac{d_e}{2d}\left[\omega_i\left(1 - \frac{d\cos\beta}{d_e}\right) - \omega_o\left(1 + \frac{d\cos\beta}{d_e}\right)\right] \tag{13.28}$$

It is convenient for calculating the velocities at the ball-race contacts, which are required for calculating elastohydrodynamic film thicknesses, to use a coordinate system that rotates at ω_c. This system fixes the ball-race contacts relative to the observer. In the rotating coordinate system the angular velocities of the inner and outer races become

$$\omega_{ri} = \omega_i - \omega_c = \frac{\omega_i - \omega_o}{2}\left(1 + \frac{d\cos\beta}{d_e}\right)$$

$$\omega_{ro} = \omega_o - \omega_c = \frac{\omega_o - \omega_i}{2}\left(1 - \frac{d\cos\beta}{d_e}\right)$$

The surface velocities entering the *ball–inner-race contact* for pure rolling are

$$u_{ai} = u_{bi} = \frac{\omega_{ri}(d_e - d\cos\beta)}{2} \tag{13.29}$$

$$u_{ai} = u_{bi} = \frac{d_e(\omega_i - \omega_o)}{4}\left(1 - \frac{d^2\cos^2\beta}{d_e^2}\right) \tag{13.30}$$

and those at the *ball–outer-race* contact are

$$u_{ao} = u_{bo} = \omega_{ro}\frac{d_e + d\cos\beta}{2} \tag{13.31}$$

$$u_{ao} = u_{bo} = \frac{d_e(\omega_o - \omega_i)}{4}\left(1 - \frac{d^2\cos^2\beta}{d_e^2}\right) \tag{13.32}$$

Thus,

$$|u_{ai}| = |u_{ao}|$$

For a cylindrical roller bearing $\beta = 0$ and Eqs. (13.27) to (13.32) become, if d is roller diameter,

$$\left.\begin{aligned}
\omega_c &= \frac{1}{2}\left[\omega_i\left(1 - \frac{d}{d_e}\right) + \omega_o\left(1 + \frac{d}{d_e}\right)\right] \\
\omega_b &= \frac{d_e}{2d}\left[\omega_i\left(1 - \frac{d}{d_e}\right) + \omega_o\left(1 + \frac{d}{d_e}\right)\right] \\
u_{ai} = u_{bi} &= \frac{d_e(\omega_i - \omega_o)}{4}\left(1 - \frac{d^2}{d_e^2}\right) \\
u_{ao} = u_{bo} &= \frac{d_e(\omega_o - \omega_i)}{4}\left(1 - \frac{d^2}{d_e^2}\right)
\end{aligned}\right\} \tag{13.33}$$

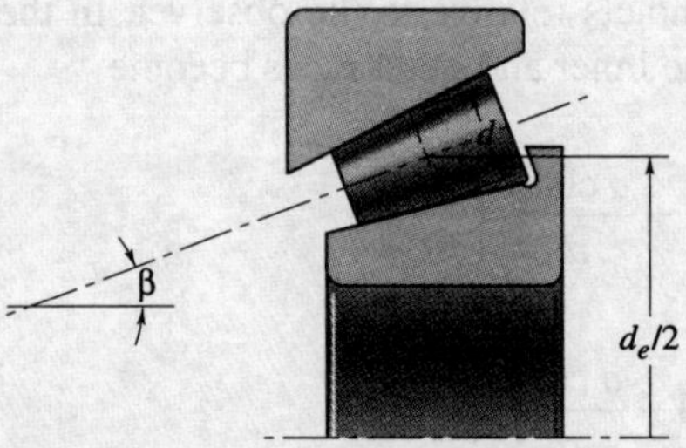

Figure 13.13 Simplified geometry for tapered-roller bearing. [B. J. Hamrock and W. J. Anderson, *Rolling-Element Bearings*, NASA Reference Publication 1105, 1983.]

Equations directly analogous to those for a ball bearing can be used for a tapered-roller bearing if d is the average diameter of the tapered roller, d_e is the diameter at which the geometric center of the rollers is located, and β is the contact angle, as shown in Fig. 13.13.

Example 13.3

GIVEN As shown in Fig. 13.12 the motion of the balls can be guided by the inner-race contact or by the outer-race contact or by both. This event is dependent on both the surface roughness and the surface geometry of angular-contact bearings. At extremely high speeds the centrifugal forces acting on the balls change both the contact angle of the outer race and the friction forces between the ball and the outer race. The outer-race contact angle is 20° and the inner-race contact angle is 31°. The ball diameter is 14 mm and the pitch diameter is 134 mm.

FIND The angular speed of the balls and the separator in an angular-contact ball bearing at an inner-race speed of 10 000 rpm

a) When the balls are outer-race guided

b) When they are inner-race guided

Solution

a) From Eq. (13.27) the separator angular speed when the balls are outer-race guided is

$$\omega_{co} = \frac{1}{2}\left[\omega_i\left(1-\frac{d\cos\beta_o}{d_e}\right)+\omega_o\left(1+\frac{d\cos\beta_o}{d_e}\right)\right]$$

$$= \frac{(10\ 000)\left(\frac{2\pi}{60}\right)}{2}\left[1-\frac{(14)\cos 20^\circ}{134}\right] = 472.2 \text{ rad/s} = 4509 \text{ rpm}$$

From Eq. (13.28) the ball speed when the balls are outer-race guided is

$$\omega_{bo} = \frac{d_e}{2d}\left[\omega_i\left(1 - \frac{d\cos\beta_o}{d_e}\right)\right]$$

$$= \frac{(134.0)(10\ 000)\left(\frac{2\pi}{60}\right)}{2(14.0)}\left(1 - \frac{14.0\cos 20^\circ}{134.0}\right) = 4520 \text{ rad/s} = 43\ 160 \text{ rpm}$$

b) From Eq. (13.27)

$$\omega_{ci} = \frac{1}{2}\left[\omega_i\left(1 - \frac{d\cos\beta_i}{d_e}\right)\right] = \frac{(10\ 000)\left(\frac{2\pi}{60}\right)}{2}\left(1 - \frac{14\cos 31^\circ}{134}\right) = 476.7 \text{ rad/s} = 4552 \text{ rpm}$$

From Eq (13.28)

$$\omega_{bi} = \frac{134.0}{2(14)}\left[(10\ 000)\frac{2\pi}{60}\left(1 - \frac{14\cos 31^\circ}{134}\right)\right] = 4563 \text{ rad/s} = 43\ 570 \text{ rpm}$$

13.6 SEPARATORS

Ball and roller bearing **separators** (sometimes called cages or **retainers**) are bearing components that, although they never carry load, can exert a vital influence on bearing efficiency. In a bearing without a separator the rolling elements contact each other during operation and experience severe sliding and friction. The primary functions of a separator are to maintain the proper distance between the rolling elements and to ensure proper load distribution and balance within the bearing. Another function of the separator is to maintain control of the rolling elements so as to produce the least possible friction through sliding contact. Furthermore, a separator is necessary for several types of bearing to prevent the rolling elements from falling out of the bearing during handling. Most separator troubles occur from improper mounting, misaligned bearings, or improper (inadequate or excessive) clearance in the rolling-element pocket.

The materials used for separators vary according to the type of bearing and the application. In ball bearings and some sizes of roller bearing the most common type of separator is made from two strips of carbon steel that are pressed and riveted together. Called ribbon separators, they are the least expensive separators to manufacture and are entirely suitable for many applications. They are also lightweight and usually require little space.

The design and construction of angular-contact ball bearings allow the use of a one-piece separator. The simplicity and inherent strength of one-piece separators permit their fabrication from many desirable materials. Reinforced phenolic and bronze are the two most commonly used materials. Bronze separators offer strength and low-friction characteristics and can be operated at temperatures to 230°C (450°F). Machined, silver-plated ferrous alloy separators are used in many demanding applications. Because reinforced cotton-based phenolic separators combine the advantages of low weight, strength, and nongalling properties, they are used for such high-speed applications as gyroscope bearings. Lightness and strength are particularly desirable in high-speed bearings, since the stresses increase with speed but may be greatly minimized by reducing separator weight. A limitation of phenolic separators, however, is that they have an allowable maximum temperature of about 135°C (275°F).

Example 13.4

GIVEN A lightly loaded lubricated roller on a flat surface. The minimum oil film thickness is directly proportional to the roller radius. The cage pocket for a cylindrical roller bearing can be made with straight sides or with sides having a radius 5% larger than the roller radius.

FIND How much thicker will the oil film between the roller and the pockets be with a radius than with the straight pockets?

Solution

The geometry for a cage pocket with 5% larger radius than the roller is equivalent to a much larger roller on a flat. The equivalent radius R giving the same oil film is

$$\frac{1}{R} = \frac{1}{r} - \frac{1}{1.05r} = \frac{0.05}{1.05r}$$

$$\therefore R = 21r$$

Since the oil film thickness is directly proportional to the equivalent radius of the contact, the oil film thickness will be 21 times as large as that for a straight cage pocket.

13.7 STATIC LOAD DISTRIBUTION

Since a simple analytical expression for the deformation in terms of load was defined in Sec. 8.7.1.4, it is possible to consider how the bearing load is distributed among the elements. Most rolling-element bearing applications involve steady-state rotation of either the inner or outer race or both. However, the rotational speeds are usually not so great as to cause ball or roller centrifugal forces or gyroscopic moments of significant magnitudes. In analyzing the loading distribution on the rolling elements, it is usually satisfactory to ignore these effects in most applications. In this section the load deflection relationships for ball and roller bearings are given, along with radial and thrust load distributions of statically loaded rolling elements.

13.7.1 LOAD DEFLECTION RELATIONSHIPS

For an elliptical conjunction the load deflection relationship given in Eq. (8.27) can be written as

$$w_z = K_{1.5}\delta^{3/2} \qquad \textbf{(13.34)}$$

where

$$K_{1.5} = \pi k_e E' \left(\frac{2\mathscr{E}R}{9\mathscr{F}^3} \right)^{1/2} \tag{13.35}$$

Similarly, for a rectangular conjunction from Eq. (8.31)

$$w_z = K_1 \delta_m$$

where

$$K_1 = \frac{\frac{\pi}{2} l E'}{2 \ln\left(\frac{4R_x}{b^*} \right) - 1} \tag{13.36}$$

and l = length of rolling element

In general then,

$$w_z = K_j \delta_m^j \tag{13.37}$$

in which j is 1.5 for ball bearings and 1.0 for roller bearings. The total normal approach between two races separated by a rolling element is the sum of the deformations under load between the rolling element and both races. Therefore,

$$\delta_m = \delta_{mo} + \delta_{mi} \tag{13.38}$$

where

$$\delta_{mo} = \left[\frac{w_z}{(K_j)_o} \right]^{\frac{1}{j}} \tag{13.39}$$

$$\delta_{mi} = \left[\frac{w_z}{(K_j)_i} \right]^{\frac{1}{j}} \tag{13.40}$$

Substituting Eqs. (13.38), (13.39), and (13.40) into Eq. (13.37) gives

$$K_j = \frac{1}{\left\{ \left[1/(K_j)_o \right]^{1/j} + \left[1/(K_j)_i \right]^{1/j} \right\}^j} \tag{13.41}$$

Recall that $(K_j)_o$ and $(K_j)_i$ are defined by Eqs. (13.35) and (13.36) for an elliptical and a rectangular conjunction, respectively. These equations show that $(K_j)_o$ and $(K_j)_i$ are functions only of the contact geometry and the material properties. The radial and thrust load analyses are presented in the following two sections and are directly applicable for radially loaded ball and roller bearings and thrust-loaded ball bearings.

13.7.2 RADIALLY LOADED BALL AND ROLLER BEARINGS

Figure 13.14 shows a radially loaded rolling-element bearing with radial clearance $c_d/2$. In the concentric arrangement [Fig. 13.14(a)] a uniform radial clearance between the rolling element and the races of $c_d/2$ is evident. The application of an arbitrarily small radial load to the shaft causes the inner race to move a distance $c_d/2$ before contact is made between a rolling element located on the load line and the inner and outer races. At any angle there will still be a radial clearance c that, if c_d is small relative to the radius of the groove, can be expressed with adequate accuracy by

$$c = (1 - \cos\psi)\frac{c_d}{2}$$

On the load line when $\psi = 0°$, the clearance is zero; but when $\psi = 90°$, the clearance retains its initial value of $c_d/2$.

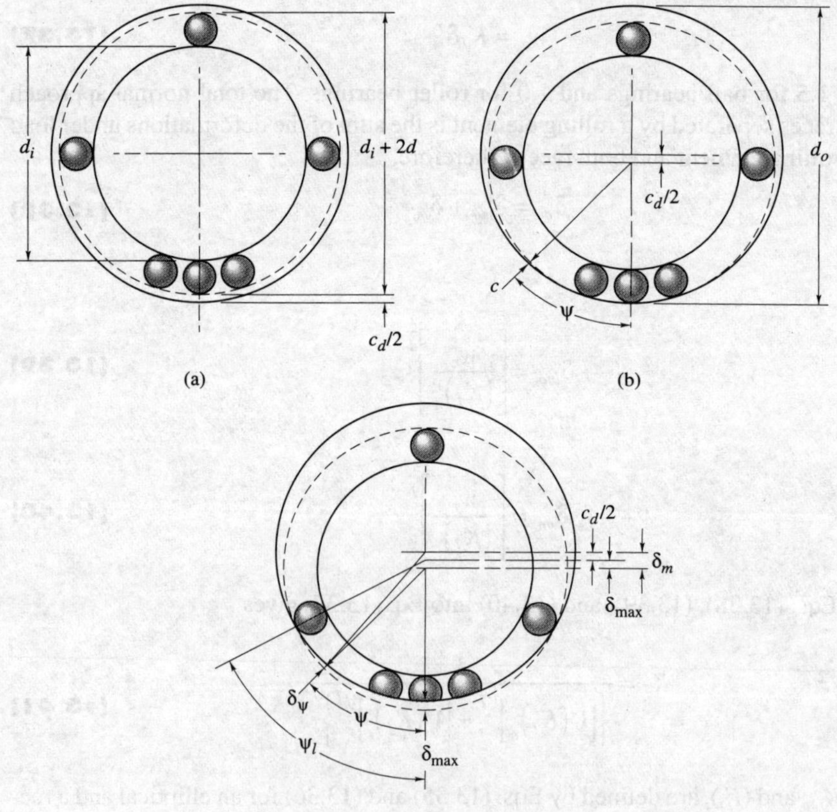

Figure 13.14 Radially loaded rolling-element bearing.
(a) Concentric arrangement; (b) initial contact; (c) interference. [B. J. Hamrock and W. J. Anderson, *Rolling-Element Bearings*, NASA Reference Publication 1105, 1983.]

The application of further load will elastically deform the balls and eliminate clearance around an arc $2\psi_l$. If the interference or total elastic compression on the load line is δ_{max}, the corresponding elastic compression of the ball δ_ψ along a radius at angle ψ to the load line will be given by

$$\delta_\psi = (\delta_{max}\cos\psi - c) = \left(\delta_{max} + \frac{c_d}{2}\right)\cos\psi - \frac{c_d}{2}$$

This assumes that the races are rigid. Now, it is clear from Fig. 13.14(c) that $\delta_{max} + c_d/2$ represents the total relative radial displacement of the inner and outer races. Hence,

$$\delta_\psi = \delta_m \cos\psi - \frac{c_d}{2} \quad \textbf{(13.42)}$$

The relationship between load and elastic compression along the radius at angle ψ to the load vector is given by Eq. (13.37) as

$$w_\psi = K_j \delta_\psi^j$$

Substituting Eq. (13.42) into this equation gives

$$w_\psi = K_j\left(\delta_m \cos\psi - \frac{c_d}{2}\right)^j$$

For static equilibrium the applied load must equal the sum of the components of the rolling-element loads parallel to the direction of the applied load.

$$w_t = \sum w_\psi \cos\psi$$

Therefore,

$$w_t = K_j \sum\left(\delta_m \cos\psi - \frac{c_d}{2}\right)^j \cos\psi \quad \textbf{(13.43)}$$

The angular extent of the bearing arc $2\psi_l$ in which the rolling elements are loaded is obtained by setting the root expression in Eq. (13.43) equal to zero and solving for ψ.

$$\psi_l = \cos^{-1}\left(\frac{c_d}{2\delta_m}\right)$$

The summation in Eq. (13.43) applies only to the angular extent of the loaded region. This equation can be written for a roller bearing as

$$w_t = \left(\psi_l - \frac{c_d}{2\delta_m}\sin\psi_l\right)\frac{nK_1\delta_m}{2\pi} \quad \textbf{(13.44)}$$

and similarly in integral form for a ball bearing as

$$w_t = \frac{n}{(w_z)_{max}} K_{1.5}\delta_m^{3/2} \int_0^{\psi_l} \left(\cos\psi - \frac{c_d}{2\delta}\right)^{\frac{3}{2}} \cos\psi \, d\psi$$

The integral in the equation can be reduced to a standard elliptic integral by the hypergeometric series and the beta function. If the integral is numerically evaluated directly, the following approximate expression is derived:

$$\int_0^{\psi_l} \left(\cos\psi - \frac{c_d}{2\delta}\right)^{\frac{3}{2}} \cos\psi \, d\psi = 2.491\left\{\left[1+\left(\frac{c_d/2\delta_m - 1}{1.23}\right)^2\right]^{\frac{1}{2}} - 1\right\}$$

which fits the exact numerical solution to within 2% for a complete range of $c_d/2\delta_m$.

The load carried by the most heavily loaded rolling element is obtained by substituting $\psi = 0°$ in Eq. (13.43) and dropping the summation sign.

$$(w_z)_{max} = K_j \delta_m^j \left(1 - \frac{c_d}{2\delta_m}\right)^j \tag{13.45}$$

Dividing this maximum load by the total radial load for a roller bearing [Eq. (13.44)] gives

$$w_z = \frac{\left(\psi_l - \dfrac{c_d}{2\delta_m}\sin\psi_l\right)\dfrac{n(w_z)_{max}}{2\pi}}{1 - c_d/2\delta_m} \tag{13.46}$$

and similarly for a ball bearing

$$w_z = \frac{n(w_z)_{max}}{Z_w} \tag{13.47}$$

where

$$Z_w = \frac{\pi(1 - c_d/2\delta_m)^{3/2}}{2.491\left\{\left[1+\left(\dfrac{1 - c_d/2\delta_m}{1.23}\right)^2\right]^{1/2} - 1\right\}} \tag{13.48}$$

For roller bearings when the diametral clearance c_d is zero, Eq. (13.46) gives

$$w_z = \frac{n(w_z)_{max}}{4} \tag{13.49}$$

For ball bearings when the diametral clearance c_d is zero, the value of Z_w in Eq. (13.47) becomes 4.37. This is the value derived by Stribeck (1901) for ball bearings of zero diametral clearance. The approach used by Stribeck was to evaluate the finite summation for various numbers of balls. He then derived the celebrated Stribeck equation for static load-carrying capacity by writing the more conservative value of 5 for the theoretical value of 4.37:

$$w_z = \frac{n(w_z)_{\max}}{5} \tag{13.50}$$

In using Eq. (13.50) remember that Z_w was considered to be a constant and that the effects of clearance and applied load on load distribution were not taken into account. These effects were, however, considered in obtaining Eq. (13.47).

13.7.3 THRUST-LOADED BALL BEARINGS

The **static thrust load-carrying capacity** of a ball bearing may be defined as the maximum thrust load that the bearing can endure before the contact ellipse approaches a race shoulder, as shown in Fig. 13.15, or as the load at which the allowable mean compressive stress is reached, whichever is smaller. Both the limiting shoulder height and the mean compressive stress must be calculated to find the static thrust load-carrying capacity.

Each ball is subjected to an identical thrust component W_t/n, where W_t is the total thrust load. The initial free contact angle before the application of a thrust load is denoted by β_f. Under load the normal ball thrust load W_t acts at the contact angle β and is written as

$$W = W_t/n \sin\beta \tag{13.51}$$

Figure 13.16 shows a cross section through an angular-contact bearing under a thrust load W_t. From this figure the contact angle after the thrust load has been applied can be written as

$$\beta = \cos^{-1}\left(\frac{c_r - c_d/2}{c_r + \delta_m}\right) \tag{13.52}$$

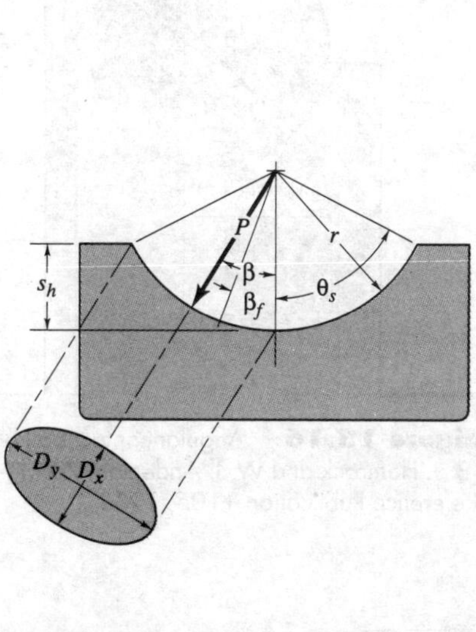

Figure 13.15 Contact ellipse in bearing race under load. [B. J. Hamrock and W. J. Anderson, *Rolling-Element Bearings*, NASA Reference Publication 1105, 1983.]

The initial free contact angle was given in Eq. (13.7). Using that equation and rearranging terms in Eq. (13.52) give, solely from geometry (Fig. 13.16),

$$\delta_m = c_r\left(\frac{\cos\beta_f}{\cos\beta} - 1\right) \quad \textbf{(13.53)}$$

$$\delta_m = \delta_{mo} + \delta_{mi}$$

$$\delta_m = \left[\frac{W}{(K_j)_o}\right]^{\frac{1}{j}} + \left[\frac{W}{(K_j)_i}\right]^{\frac{1}{j}}$$

$$K_{1.5} = \frac{1}{\left\{\left[\dfrac{(4.5\mathcal{F}_o^3)^{1/2}}{\pi k_{eo}E_o'(R_o\mathcal{E}_o)^{1/2}}\right]^{2/3} + \left[\dfrac{(4.5\mathcal{F}_i^3)^{1/2}}{\pi k_{ei}E_i'(R_i\mathcal{E}_i)^{1/2}}\right]^{2/3}\right\}^{3/2}} \quad \textbf{(13.54)}$$

$$W = K_{1.5}c_r^{3/2}\left(\frac{\cos\beta_f}{\cos\beta} - 1\right)^{3/2} \quad \textbf{(13.55)}$$

where Eq. (13.35) for $K_{1.5}$ is replaced by Eq. (13.53) and k, $\mathcal{E}$, and $\mathcal{F}$ are given in Table 8.6.

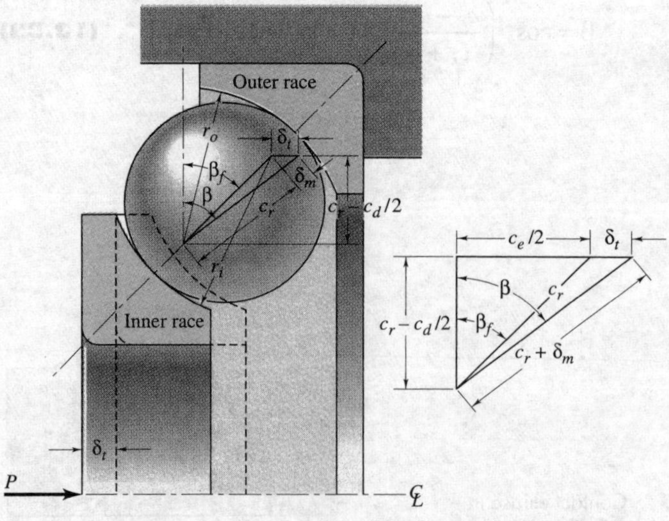

Figure 13.16 Angular-contact ball bearing under thrust load. [B. J. Hamrock and W. J. Anderson, *Rolling-Element Bearings*, NASA Reference Publication 1105, 1983.]

From Eqs. (13.51) and (13.55)

$$\frac{W_t}{n\sin\beta} = W \tag{13.56}$$

$$\frac{W_t}{nK_{1.5}c_r^{3/2}} = \sin\beta\left(\frac{\cos\beta_f}{\cos\beta} - 1\right)^{3/2} \tag{13.57}$$

This equation can be numerically solved by the Newton-Raphson method. The iterative equation to be satisfied is

$$\beta' - \beta = \frac{\dfrac{W_t}{nK_{1.5}c_r^{3/2}} - \sin\beta\left(\dfrac{\cos\beta_f}{\cos\beta} - 1\right)^{3/2}}{\cos\beta\left(\dfrac{\cos\beta_f}{\cos\beta} - 1\right)^{3/2} + \dfrac{3}{2}\cos\beta_f\tan^2\beta\left(\dfrac{\cos\beta_f}{\cos\beta} - 1\right)^{1/2}} \tag{13.58}$$

In this equation convergence is satisfied when $\beta' - \beta$ becomes essentially zero.

When a thrust load is applied, the shoulder height limits the axial deformation, which can occur before the pressure-contact ellipse reaches the shoulder. As long as the following inequality is satisfied, the contact ellipse will not exceed this limit:

$$\theta_s > \beta + \sin^{-1}\left(\frac{D_y}{R_r d}\right)$$

From Fig. 13.5 and Eq. (13.11) the angle θ_s used to define the shoulder height can be written as

$$\theta_s = \cos^{-1}\left(1 - \frac{s_h}{R_r d}\right)$$

From Fig. 13.16 the axial deflection δ_t corresponding to a thrust load can be written as

$$\delta_t = (c_r + \delta_m)\sin\beta - c_r \sin\beta_f \tag{13.59}$$

Substituting Eq. (13.53) into Eq. (13.59) gives

$$\delta_t = \frac{c_r \sin(\beta - \beta_f)}{\cos\beta} \tag{13.60}$$

Once β has been determined from Eq. (13.57) and β_f from Eq. (13.7), the relationship for δ_t can be easily evaluated.

13.7.4 PRELOADING

The use of angular-contact bearings as duplex pairs preloaded against each other is discussed in Sec. 13.3. As shown in Table 13.2 duplex bearing pairs are used in either back-to-back or face-to-face arrangements. Such bearings are usually preloaded against each other by providing what is called "stickout" in the manufacture of the bearing. Figure 13.17 illustrates this for a bearing pair used in a back-to-back arrangement. The magnitude of the stickout and the bearing design determine the level of preload on each bearing when the bearings are clamped together as in Fig. 13.17. The magnitude of preload and the load deflection characteristics for a given bearing pair can be calculated by using Eqs. (13.7), (13.34), (13.51), (13.54), and (13.57).

Figure 13.18 shows the relationship of initial preload, system load, and final load for bearings a and b. The load deflection curve follows the relationship $\delta_m = KP^{2/3}$. When a system thrust load P_t is imposed on the bearing pairs, the magnitude of load on bearing b increases while that on bearing a decreases until the difference equals the system load. The physical situation demands that the change in each bearing deflection be the same ($\Delta a = \Delta b$ in Fig. 13.18). The increments in bearing load, however, are not the same. This is important because it always requires a system thrust load far greater than twice the preload before one bearing becomes unloaded. Prevention of bearing unloading, which can result in skidding and early failure, is an objective of preloading.

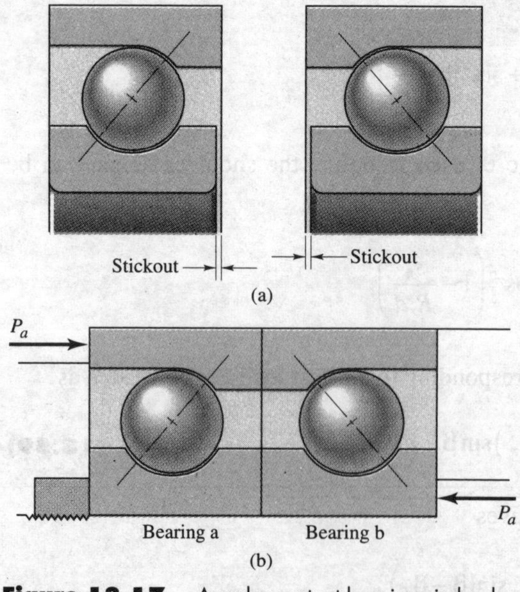

Figure 13.17 Angular-contact bearings in back-to-back arrangement, shown (a) individually as manufactured and (b) as mounted with preload. [B. J. Hamrock and W. J. Anderson, *Rolling-Element Bearings*, NASA Reference Publication 1105, 1983.]

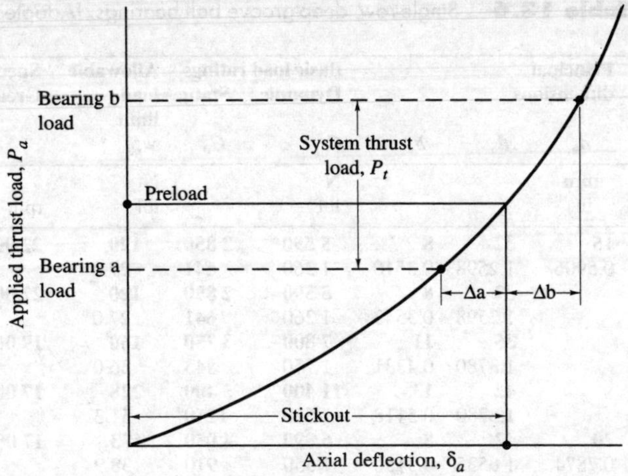

Figure 13.18 Thrust load–axial deflection curve for typical ball bearing. [B. J. Hamrock and W. J. Anderson, *Rolling-Element Bearings,* NASA Reference Publication 1105, 1983.]

13.7.5 STATIC LOAD RATING

The **static load rating** for rolling-element bearings represents the loads at which, under defined operating conditions, the load carried by the most heavily loaded rolling element w_{max} equals the allowable rolling-element load w_{all}, which is the maximum load before plastic (rather than elastic) deformation occurs within the bearing.

The static load rating C_0 can be expressed as

$$C_0 = n_{s0} w_0 \tag{13.61}$$

where

n_{s0} = static safety factor

w_0 = equivalent static bearing load

Table 13.6, which was obtained from the SKF catalog, gives the static load rating for a single-row, deep-groove ball bearing. This table gives the overall dimensions of the bearing d_b, d_a, and b_w (see Fig. 13.1 for details), both static and dynamic load ratings, the allowable load limit, speed ratings for greases and oil, and the bearing designation. **Static thrust load-carrying capacity** is the load that gives small plastic deformation of the races.

The allowable load limit may be viewed as the maximum bearing load where fatigue failure will not occur. Thus, if the design load were equivalent to the allowable load limit, the safety factor would be 1 for infinite life. Also, in this table the speed ratings correspond to the maximum speed the bearing should experience before the bearing design must be altered to handle the high temperature developed in the bearing.

Table 13.6 Single-row, deep-groove ball bearings. [Adapted from SKF Catalog (1991).]

Additional data is contained on the CD

Principal dimensions			Basic load ratings Dynamic	Static	Allowable load limit	Speed ratings Grease	Oil	Mass	Designation
d_b	d_a	b_w	C	C_0	w_{all}				
mm in.			N lbf		N lbf	rpm		kg lbm	—
15	32	8	5 590	2 850	120	22 000	28 000	0.025	16002
0.5906	1.2598	0.3510	1 260	641	27.0			0.055	
	32	8	5 590	2 850	120	22 000	28 000	0.030	6002
	1.2598	0.3543	1 260	641	27.0			0.066	
	35	11	7 800	3 750	160	19 000	24 000	0.045	6202
	1.3780	0.4331	1 750	843	36.0			0.099	
	42	13	11 400	5 400	228	17 000	20 000	0.082	6302
	1.3780	0.5118	2 560	1210	51.3			0.18	
20	42	8	6 890	4 050	173	17 000	20 000	0.050	16004
0.7874	1.6535	0.3150	1 550	910	38.9			0.11	
	42	12	9 360	5 000	212	17 000	20 000	0.090	6004
	1.6535	0.4724	2 100	1 120	47.7			0.15	
	47	14	12 700	6 550	280	15 000	18 000	0.11	6204
	1.8504	0.5512	2 860	1 470	62.9			0.24	
	52	15	15 900	7 800	335	13 000	16 000	0.14	6304
	2.0472	0.5906	3 570	1 750	75.3			0.31	
	72	19	30 700	15 000	640	10 000	13 000	0.40	6404
	2.8346	0.7480	6 900	3 370	144			0.88	
25	47	12	11 200	6 550	275	15 000	18 000	0.080	6005
0.9843	1.8504	0.4724	5 520	1 470	61.8			0.18	
	52	15	14 000	7 800	335	12 000	15 000	0.13	6205
	2.0472	0.5906	3 150	1 750	75.3			0.29	
	62	17	22 500	11 600	490	11 000	14 000	0.23	6305
	2.4409	0.6693	5 060	2 610	110			0.51	
	80	21	35 800	19 300	815	9 000	11 000	0.53	6405
	3.1496	0.8268	8 050	4 340	183			1.15	
30	55	15	13 300	8 300	355	12 000	15 000	0.12	6006
1.1811	2.1654	0.5118	2 990	1 870	79.8			0.26	
	62	16	19 500	11 200	475	10 000	13 000	0.20	6206
	2.4409	0.6299	4 380	2 520	107			0.44	
	72	19	28 100	16 000	670	9 000	11 000	0.35	6306
	2.8346	0.7480	6 320	3 600	151			0.77	
	90	23	43 600	23 600	1000	8 500	10 000	0.74	6406
	3.5433	0.9055	9 800	5 310	225			1.653	
35	62	14	15 900	10 200	440	10 000	13 000	0.16	6007
1.3780	2.4409	0.5512	3 570	2 290	98.9			0.35	
	72	17	25 500	15 300	655	9 000	11 000	0.29	6207
	2.8346	0.6693	5 370	3 440	147			0.64	
	80	21	33 200	19 000	815	8 500	10 000	0.46	6307
	3.1496	0.8268	7 460	4 270	183			1.00	
	100	25	55 300	31 000	1290	7 000	8 500	0.95	6407
	3.9370	0.9843	12 400	6 970	290			2.10	
40	68	15	16 800	11 600	490	9 500	12 000	0.19	6008
1.5748	2.6672	0.5906	3 780	2 610	110			0.42	
	80	18	30 700	19 000	800	8 500	10 000	0.37	6208
	3.1496	0.7087	6 900	4 270	147			0.82	
	90	23	41 000	24 000	1020	7 500	9 000	0.63	6308
	3.5433	0.9055	9 220	5 400	229			1.40	
	110	27	63 700	36 500	1530	6 700	8 000	1.25	6408
	4.3307	1.0630	14 300	8 210	344			2.75	

(continued)

Table 13.6 *(continued)*

Principal dimensions d_b	d_a	b_w	Basic load ratings Dynamic C	Static C_0	Allowable load limit w_{all}	Speed ratings Grease	Oil	Mass	Designation
mm in.			N lbf		N lbf	rpm		kg lbm	—
45	**75**	**16**	**20 800**	**14 600**	**640**	**9 000**	**11 000**	**0.25**	**6009**
1.7717	2.9528	0.6299	4 680	3 280	144			0.55	
	85	**19**	**33 200**	**21 600**	**915**	**7 500**	**9 000**	**0.41**	**6209**
	3.3465	0.7480	7 400	4 860	206			0.90	
	100	**25**	**52 700**	**31 500**	**1340**	**6 700**	**8 000**	**0.83**	**6309**
	3.9370	0.9843	11 900	7 080	301			1.85	
	120	**29**	**76 100**	**45 000**	**1900**	**6 000**	**7 000**	**0.55**	**6409**
	4.7244	1.1417	17 100	10 100	427			3.40	
50	**80**	**16**	**21 600**	**16 000**	**710**	**8 500**	**10 000**	**0.26**	**6010**
1.9685	3.1496	0.6299	4 860	3 600	160			0.57	
	90	**20**	**35 100**	**23 200**	**1600**	**7 000**	**8 500**	**0.46**	**6210**
	3.5433	0.7874	7 890	5 220	220			1.00	
	110	**27**	**61 800**	**38 000**	**980**	**6 300**	**7 500**	**1.05**	**6310**
	4.3307	1.0630	13 900	8 540	360			2.30	
	130	**31**	**87 100**	**52 000**	**2200**	**5 300**	**6 300**	**1.90**	**6410**
	5.1181	1.2205	19 600	11 700	495			4.20	

A sampling of bearing data is contained on the CD

Table 13.7 gives similar information as Table 13.6 but for a single-row, cylindrical roller bearing. In this text consideration is limited to these two types of bearing. Bearing catalogs are readily available to handle other types of bearing, of which the number available is endless.

Tutorial 8

13.7.6 EQUIVALENT STATIC LOAD

A load P acting on a rolling-element bearing at an angle α_p (Fig. 13.19) is a combined load, since it contains both a radial and an axial component. The radial component is $P_r = P \cos \alpha_p$ and the axial component is $P_a = P \sin \alpha_p$. The two components P_r and P_a combine to form the **equivalent static load** P_0, which can be expressed as

$$P_0 = X_0 P_r + Y_0 P_a \tag{13.62}$$

where

X_0 = radial factor obtained from Table 13.8

Y_0 = thrust factor obtained from Table 13.8

In Table 13.8 β is the contact angle.

Table 13.7 Single-row, cylindrical roller bearing. [Adapted from SKF Catalog (1991).]

Principal dimensions			Basic load ratings Dynamic	Static	Allowable load limit	Speed ratings Grease	Oil	Mass	Designation
d_b	d_a	b_w	C	C_0	w_{all}				
mm in.			N lbf		N lbf	rpm		kg lbm	—
15	35	11	12 500	10 200	1 200	18 000	22 000	0.047	NU 202 EC
0.5906	1.3780	0.4331	2 810	2 290	274			0.10	
	42	13	19 400	15 300	1 860	16 000	19 000	0.086	NU 302 EC
	1.6535	0.5118	4 360	3 440	418			0.19	
20	47	14	25 100	22 000	2 750	13 000	16 000	0.11	NU 204 EC
0.7874	1.8504	0.5512	5 640	4 950	618			0.24	
	52	15	30 800	26 000	3 250	12 000	15 000	0.15	NU 304 EC
	2.0472	0.5906	6 920	5 850	731			0.33	
25	52	15	28 600	27 000	3 350	11 000	14 000	0.13	NU 205 EC
0.9843	2.0472	0.5906	6 430	6 070	753			0.29	
	62	17	40 200	36 500	4 550	9 500	12 000	0.24	NU 305 EC
	2.4409	0.6693	9 040	8 210	1 020			0.53	
30	62	16	38 000	36 500	4 450	9 500	12 000	0.20	NU 206 EC
1.811	2.4409	0.6299	8 540	8 210	1 020			0.44	
	72	19	51 200	48 000	6 200	9 000	11 000	0.36	NU 306 EC
	2.8346	0.7480	11 500	10 800	1 390			0.79	
35	72	17	48 400	48 000	6 100	8 500	10 000	0.30	NU 207 EC
1.3780	2.8346	0.6693	10 900	10 800	1 370			0.66	
	80	21	64 400	63 000	8 150	8 000	9 500	0.48	NU 307 EC
	3.1496	0.8268	14 500	14 200	1 830			1.05	
40	80	18	53 900	53 000	6 700	7 500	9 000	0.37	NU 208 EC
1.5748	3.1496	0.7087	12 100	11 900	1 510			0.82	
	90	23	80 900	78 000	10 200	6 700	8 000	0.65	NU 308 EC
	3.5433	0.9055	18 200	17 500	2 290			1.05	
45	85	19	60 500	64 000	8 150	6 700	8 000	0.43	NU 209 EC
1.7717	3.3465	0.7480	13 600	14 400	1 830			0.95	
	100	25	99 000	100 000	12 900	6 300	7 500	0.90	NU 309 EC
	3.9370	0.9843	22 300	22 500	2 900			2.00	
50	90	20	64 400	69 500	8 800	6 300	7 500	0.48	NU 210 EC
1.9685	3.5433	0.7874	14 500	15 600	1 980			1.05	
	110	27	110 000	112 000	15 000	5 000	6 000	1.15	NU 310 EC
	4.3307	1.0630	24 700	25 200	3 370			2.55	

Additional data is contained on the CD

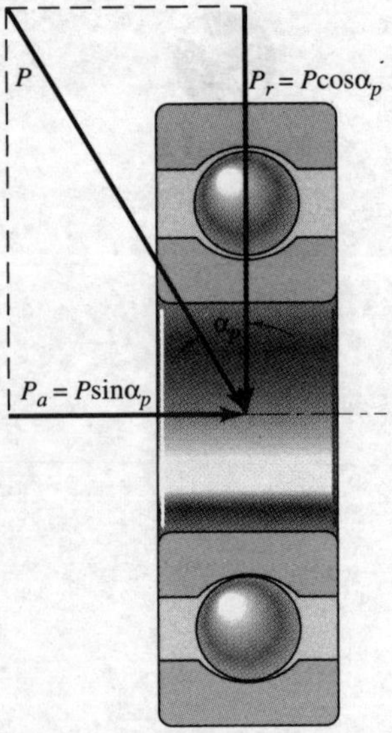

Figure 13.19 Combined load acting on a radial deep-groove ball bearing.

Table 13.8 Radial factor X_0 and thrust factor Y_0 for statically stressed radial bearings

Bearing type		Single row X_0	Y_0	Double row X_0	Y_0
Radial deep-groove ball		0.6	0.5	0.6	0.5
Radial angular-contact ball	$\beta = 20°$	0.5	0.42	1	0.84
	$\beta = 25°$	0.5	0.38	1	0.76
	$\beta = 30°$	0.5	0.33	1	0.66
	$\beta = 35°$	0.5	0.29	1	0.58
	$\beta = 40°$	0.5	0.26	1	0.52
Radial self-aligning ball		0.5	$0.22 \cot\beta$	1	$0.44 \cot\beta$
Radial spherical roller		0.5	$0.22 \cot\beta$	1	$0.44 \cot\beta$
Radial tapered roller		0.5	$0.22 \cot\beta$	1	$0.44 \cot\beta$

Example 13.5

GIVEN A single-row, angular-contact ball bearing loaded with a static load having an axial component of 1000 N and a radial component of 500 N. The static basic load rating is 1500 N.
FIND The contact angle between 20 and 40° that produces the largest static safety factor.

Solution

From Eq. (13.62) for a radial angular-contact bearing (see Table 13.8)

$$P_0 = X_0 P_r + Y_0 P_a = (0.5)(500) + Y_0(1000)$$

$$\therefore P_0 = 250 + 1000 Y_0$$

The safety factor is the largest when P_0 is the smallest, which means when Y_0 is the smallest. This implies that the contact angle should be 40°, thus producing $Y_0 = 0.26$.

$$\therefore P_0 = 250 + (0.26)(1000) = 510 \text{ N}$$

13.8 ELASTOHYDRODYNAMIC LUBRICATION

Nonconformal surfaces, such as those in rolling-element bearings, are lubricated elastohydrodynamically as first presented in Sec. 8.3.2. There it was shown that elastohydrodynamic lubrication is a form of fluid film lubrication where elastic deformation of the bearing surfaces becomes significant. It is usually associated with highly stressed machine components, such as rolling-element bearings. Historically, elastohydrodynamic lubrication may be viewed as one of the major developments in the field of lubrication in the twentieth century. It not only revealed the existence of a previously unsuspected regime of lubrication in highly stressed nonconforming machine elements, but it also brought order to the complete spectrum of lubrication regimes, ranging from boundary to hydrodynamic.

13.8.1 RELEVANT EQUATIONS

The relevant equations used in elastohydrodynamic lubrication are given here.
Lubrication equation (Reynolds equation) first developed in Chapter 12:

$$\frac{\partial}{\partial x}\left(\frac{\rho h^3}{\eta}\frac{\partial p}{\partial x}\right) + \frac{\partial}{\partial y}\left(\frac{\rho h^3}{\eta}\frac{\partial p}{\partial y}\right) = 12\tilde{u}\frac{\partial}{\partial x}(\rho h) \tag{12.23}$$

where

$$\tilde{u} = \frac{u_a + u_b}{2}$$

Viscosity variation:

$$\eta = \eta_0 e^{\xi p} \tag{13.63}$$

where η_0 is the coefficient of absolute or dynamic viscosity at atmospheric pressure and ξ is the pressure-viscosity coefficient of the fluid.

Density variation (for mineral oils):

$$\rho = \rho_0\left(1+\frac{0.6p}{1+1.17p}\right) \tag{13.64}$$

where ρ_0 is the density at atmospheric conditions and p is in gigapascals.

Elasticity equation:

$$\delta = \frac{2}{E'}\iint_A \frac{p(x,y)dxdy}{\sqrt{(x-x_1)^2+(y-y_1)^2}} \tag{13.65}$$

where

$$E' = \frac{2}{\dfrac{1-\nu_a^2}{E_a}+\dfrac{1-\nu_b^2}{E_b}} \tag{13.66}$$

Film thickness equation:

$$h = h_0 + \frac{x^2}{2R_x}+\frac{y^2}{2R_y}+\delta(x,y) \tag{13.67}$$

where

$$\frac{1}{R_x} = \frac{1}{r_{ax}}+\frac{1}{r_{bx}} \tag{8.17}$$

$$\frac{1}{R_y} = \frac{1}{r_{ay}}+\frac{1}{r_{by}} \tag{8.18}$$

The different radii are expressed in Eqs. (13.13) to (13.18).

The elastohydrodynamic lubrication solution therefore requires calculating the pressure distribution within the conjunction, at the same time allowing for the effects that this pressure will have on the properties of the fluid and on the geometry of the elastic solids. The solution will also provide the shape of the lubricant film, particularly the minimum clearance between the solids. Hamrock (1993) gives the complete elastohydrodynamic lubrication theory.

13.8.2 DIMENSIONLESS GROUPING

The variables resulting from the elastohydrodynamic lubrication theory are

E' = effective elastic modulus, Pa

P = normal applied load, N

h = film thickness, m

R_x = effective radius in x (motion) direction, m

R_y = effective radius in y (transverse) direction, m

$\tilde{u}$ = mean surface velocity in x direction, m/s

ξ = pressure-viscosity coefficient of fluid, m^2/N

η_0 = atmospheric viscosity, $N\text{-}s/m^2$

From these variables the following five dimensionless groupings can be established:

Dimensionless film thickness:

$$H = \frac{h}{R_x} \tag{13.68}$$

Ellipticity parameter:

$$k_e = D_y/D_x = \left(R_y/R_x\right)^{2/\pi} \tag{8.24}$$

Dimensionless load parameter:

$$W = \frac{P}{E'R_x^2} \tag{13.69}$$

Dimensionless speed parameter:

$$U = \frac{\eta_0 \tilde{u}}{E'R_x} \tag{13.70}$$

Dimensionless materials parameter:

$$G = \xi E' \tag{13.71}$$

The dimensionless minimum film thickness can thus be written as a function of the other four parameters

$$H = f(k_e, U, W, G)$$

The most important practical aspect of elastohydrodynamic lubrication theory is the determination of the minimum film thickness within a conjunction. That is, maintaining a fluid film thickness of adequate magnitude is extremely important to the operation of nonconformal machine elements such as rolling-element bearings. Specifically, elastohydrodynamic film thickness influences fatigue life, as discussed in Sec. 13.9.

13.8.3 MINIMUM-FILM-THICKNESS FORMULA

By using the numerical procedures outlined in Hamrock (1993), the influence of the ellipticity parameter and the dimensionless speed, load, and materials parameters on minimum film thickness has been investigated. The ellipticity parameter k_e was varied from 1 (a ball-on-plate configuration) to 8 (a configuration approaching a rectangular contact). The dimensionless speed parameter U was varied over a range of nearly two orders of magnitude; and the dimensionless load parameter W, over a range of one order of magnitude. Situations equivalent to using solid materials of bronze, steel, and silicon nitride and lubricants of paraffinic and naphthenic oils were considered in the investigation of the role of the dimensionless materials parameter G. Thirty-four cases were used in generating the minimum-film-thickness formula given here.

$$H_{\min} = 3.63U^{0.68}G^{0.49}W^{-0.073}\left(1 - e^{-0.68k_e}\right) \quad \textbf{(13.72)}$$

In this equation the most dominant exponent occurs on the speed parameter, and the exponent on the load parameter is very small and negative. The materials parameter also carries a significant exponent, although the range of this variable in engineering situations is limited.

Similarly, from Hamrock (1993) for rectangular contacts the dimensionless minimum-film-thickness formula is

$$H_{\min,r} = \frac{h_{\min}}{R_x} = 1.714(W')^{-0.128}U^{0.694}G^{0.568} \quad \textbf{(13.73)}$$

where

$$W' = \frac{P_z'}{E'R_x} \quad \textbf{(13.74)}$$

where

P_z' = normal applied load per unit width, N/m

Example 13.6

GIVEN A deep-groove ball bearing has 12 balls radially loaded with 12 000 N. The groove and ball radii are matched to given an ellipticity ratio of 10. The lubricant viscosity is 0.075 N-s/m^2, and its pressure-viscosity coefficient is 1.6×10^{-8} m^2/N. The rotational speed is 1500 rpm, the mean surface velocity is 3 m/s, and the effective radius in the x direction is 8 mm. The modulus of elasticity for the ball and races is 207 GPa and Poisson's ratio is 0.3.

FIND The minimum oil film thickness in the bearing.

Solution

The Stribeck equation [Eq. (13.50)] gives the load acting on the most heavily loaded ball as

$$(w_z)_{\max} = \frac{5W_z}{n} = \frac{5(12\ 000)}{12} = 5000 \text{ N}$$

The effective elasticity modulus is

$$E' = \frac{E}{1-\nu^2} = \frac{(207)(10^9)}{1-(0.3)^2} = 227 \text{ GPa}$$

The dimensionless load, speed, and materials parameters from Eqs. (13.69), (13.70), and (13.71), respectively, can be expressed as

$$U = \frac{\eta_0 \tilde{u}}{E'R_x} = \frac{(0.075)(3)}{(227)(10^9)(8)(10^{-3})} = 1.236\times 10^{-10}$$

$$W = \frac{P}{E'R_x^2} = \frac{5000}{(227)(10^9)(8)^2(10^{-6})} = 3.442\times 10^{-4}$$

$$G = \xi E' = (1.6)(10^{-8})(227)(10^9) = 3632$$

The dimensionless minimum-film-thickness formula for elliptical contacts [Eq. (13.72)] produces

$$H_{\min} = \frac{h_{\min}}{R_x} = 3.63U^{0.68}G^{0.49}W^{-0.073}\left(1-e^{-0.68k_e}\right)$$

$$= 3.63\left(1.236\times 10^{-10}\right)^{0.68}(3632)^{0.49}\left(3.442\times 10^{-4}\right)^{-0.073}\left(1-e^{-0.68(10)}\right)$$

$$= 6.596\times 10^{-5}$$

The dimensional minimum film thickness is

$$h_{\min} = H_{\min}R_x = (6.596)(10^{-5})(8)(10^{-3}) = 0.53\times 10^{-6}\text{ m} = 0.53\ \mu\text{m}$$

Thus, the minimum oil film thickness in the deep-groove ball bearing considered in this example is 0.53 μm.

Example 13.7

GIVEN A ball bearing is having problems with an oil film that is too thin. The geometries are such that the ellipticity parameter is 2. By decreasing the race radius of curvature in the axial direction, the balls fit more tightly with the races and thus the ellipticity parameter increases from 2 to 10.

FIND How much thicker the oil film becomes in changing the ellipticity parameter from 2 to 10. Assume that all the other operating parameters remain constant.

Solution

From Eq. (13.72)

$$\frac{\left(H_{\min}\right)_{k_e=10}}{\left(H_{\min}\right)_{k_e=2}} = \frac{1-e^{-0.68(10)}}{1-e^{-0.68(2)}} = 1.344$$

The film thickness will increase by 34.4% when the ellipticity parameter is increased from 2 to 10.

13.9 FATIGUE LIFE

13.9.1 CONTACT FATIGUE THEORY

Rolling fatigue is a material failure caused by the application of repeated stresses to a small volume of material. It is a unique failure type—essentially a process of seeking out the weakest point at which the first failure will occur. A typical spall is shown in Fig. 13.20. On a microscopic scale there will probably be a wide dispersion in material strength or resistance to fatigue because of inhomogeneities in the material. Because bearing materials are complex alloys, they are not homogeneous or equally resistant to failure at all points. Therefore, the fatigue process can be expected to be one in which a group of supposedly identical specimens exhibit wide variations in failure time when stressed in the same way. For this reason it is necessary to treat the fatigue process statistically.

Predicting how long a particular bearing will run under a specific load requires the following essential pieces of information:

1. An accurate, quantitative estimate of life dispersion or scatter
2. The life at a given survival rate or reliability level

This translates into an expression for the "load-carrying capacity," or the ability of the bearing to endure a given load for a stipulated number of stress cycles or revolutions. If a group of supposedly identical bearings is tested at a specific load and speed, the distribution in bearing life shown in Fig. 13.21 will occur.

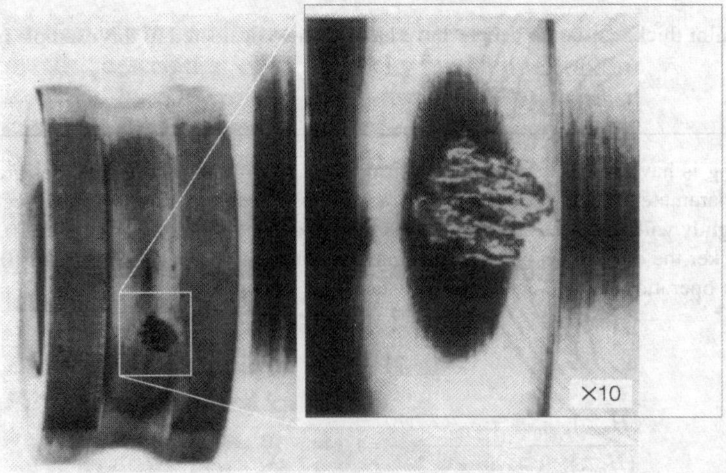

Figure 13.20 Typical fatigue spall. [B. J. Hamrock and W. J. Anderson, *Rolling-Element Bearings*, NASA Reference Publication 1105, 1983.]

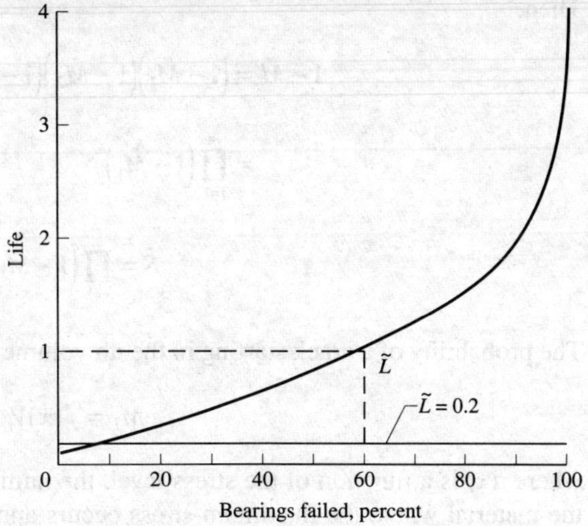

Figure 13.21 Distribution of bearing fatigue failures. [B. J. Hamrock and W. J. Anderson, *Rolling-Element Bearings,* NASA Reference Publication 1105, 1983.]

13.9.2 WEIBULL DISTRIBUTION

Weibull (1949) postulated that the fatigue lives of a homogeneous group of rolling-element bearings are dispersed according to the following relation:

$$\ln \ln \frac{1}{\tilde{S}} = e_1 \ln \frac{\tilde{L}}{\tilde{A}} \tag{13.75}$$

where $\tilde{S}$ is the probability of survival, $\tilde{L}$ is the fatigue life, and e_1 and $\tilde{A}$ are constants. The Weibull distribution results from a statistical theory of strength based on probability theory, where the dependence of strength on volume is explained by the dispersion in material strength. This is the "weakest link" theory.

Figure 13.22 considers a volume being stressed that is broken up into m similar volumes. The $\tilde{M}$'s represent the probability of failure, and the $\tilde{S}$'s the probability of survival. For the entire volume

$$\tilde{S} = \tilde{S}_1 \tilde{S}_2 \tilde{S}_3 \ldots \tilde{S}_m$$

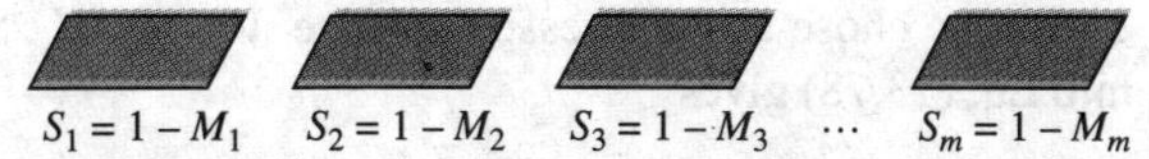

Figure 13.22 Representation of m similar stressed volumes.

Then,

$$1-\tilde{M}=\left(1-\tilde{M}_1\right)\left(1-\tilde{M}_2\right)\left(1-\tilde{M}_3\right)\ldots\left(1-\tilde{M}_m\right)$$
$$=\prod_{i=1}^{m}\left(1-\tilde{M}_i\right)$$

$$\tilde{S}=\prod_{i=1}^{m}\left(1-\tilde{M}_i\right) \quad \textbf{(13.76)}$$

The probability of a crack starting in the *i*th volume is

$$\tilde{M}_i = f(x)\tilde{V}_i \quad \textbf{(13.77)}$$

where $f(x)$ is a function of the stress level, the number of stress cycles, and the depth into the material where the maximum stress occurs and $\tilde{V}_i$ is the elementary volume. Therefore, substituting Eq. (13.77) into Eq. (13.76) gives

$$\tilde{S}=\prod_{i=1}^{m}\left[1-f(x)\tilde{V}_i\right]$$

$$\ln\tilde{S}=\sum_{i=1}^{m}\ln\left[1-f(x)\tilde{V}_i\right] \quad \textbf{(13.78)}$$

Now, if $f(x)\tilde{V}_i << 1$, then $\ln[1-f(x)\tilde{V}_i]=-f(x)\tilde{V}_i$ and $\ln\tilde{S}=-\sum_{i=1}^{m}f(x)\tilde{V}_i$. Let $\tilde{V}_i\to 0$; then,

$$\sum_{i=1}^{\infty}f(x)\tilde{V}_i=\int f(x)d\tilde{V}=\tilde{f}(x)\tilde{V} \quad \textbf{(13.79)}$$

where

$\tilde{f}(x)$ = volume-average value of $f(x)$

Lundberg and Palmgren (1947) assumed that $f(x)$ could be expressed as a power function of shear stress τ_0, number of stress cycles $\tilde{J}$, and depth of the maximum shear stress z_0:

$$f(x)=\frac{\tau_0^{c_1}\tilde{J}^{c_2}}{z_0^{c_3}} \quad \textbf{(13.80)}$$

They also chose as the stressed volume $\tilde{V}=D_y z_0 l_v$. Substituting Eqs. (13.79) and (13.80) into Eq. (13.78) gives

$$\ln\tilde{S}=-\frac{\tau_0^{c_1}\tilde{J}^{c_2}D_y l_v}{z_0^{c_3-1}} \quad \textbf{(13.81)}$$

$$\ln \frac{1}{\tilde{S}} = \frac{\tau_0^{c_1} \tilde{J}^{c_2} D_y l_v}{z_0^{c_3 - 1}} \tag{13.82}$$

For a specific bearing and load (e.g., stress) τ_0, D_y, l_v, and z_0 are all constant so that

$$\ln \frac{1}{\tilde{S}} \approx \tilde{J}^{c_2}$$

Designating $\tilde{J}$ as fatigue life $\tilde{L}$ in stress cycles gives

$$\ln \frac{1}{\tilde{S}} = \left(\frac{\tilde{L}}{\tilde{A}} \right)^{c_2}$$

$$\ln \ln \frac{1}{\tilde{S}} = c_2 \ln \frac{\tilde{L}}{\tilde{A}} \tag{13.83}$$

This is the **Weibull distribution**, which relates probability of survival and life. It has two principal functions. First, bearing fatigue lives plot as a straight line on Weibull coordinates (log-log versus log), so that the life at any reliability level can be determined. Of most interest are the $\tilde{L}_{10}$ life ($\tilde{S} = 0.9$) and the $\tilde{L}_{50}$ life ($\tilde{S} = 0.5$). Bearing load ratings are based on the $\tilde{L}_{10}$ life. Second, Eq. (13.83) can be used to determine what the $\tilde{L}_{10}$ life must be or to obtain a required life at any reliability level. The $\tilde{L}_{10}$ life is calculated, from the load on the bearing and the bearing dynamic capacity or the load rating given in manufacturers' catalogs and engineering journals, by using the equation

$$\tilde{L} = \left(\frac{\bar{C}}{P} \right)^{m_k} \tag{13.84}$$

where

$\bar{C}$ = specific dynamic capacity or basic dynamic load rating, N

P = equivalent bearing load, N

m_k = load-life exponent; 3 for elliptical contacts and 10/3 for rectangular contacts

Note that $\tilde{L}$ in Eq. (13.84) is bearing life expressed in million of revolutions. The bearing life in hours is

$$\left(\tilde{L}\right)_{\text{hour}} = 10^6 \tilde{L}/60 N_b \tag{13.85}$$

where

N_b = rotational speed, rpm

A typical Weibull plot is shown in Fig. 13.23.

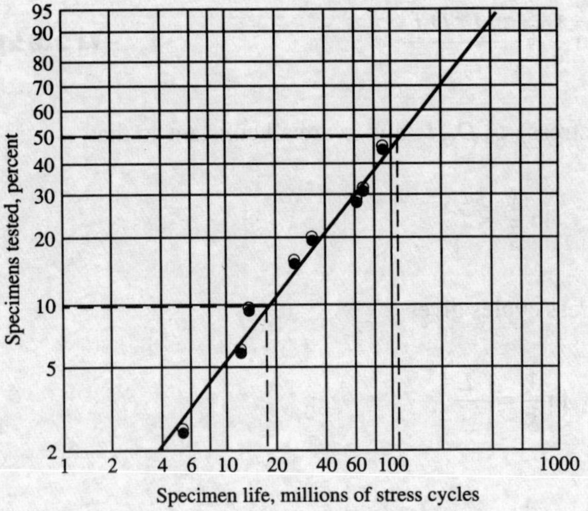

Figure 13.23 Typical Weibull plot of bearing fatigue failures. [B. J. Hamrock and W. J. Anderson, *Rolling-Element Bearings,* NASA Reference Publication 1105, 1983.]

13.9.3 DYNAMIC LOAD RATING

The Lundberg-Palmgren theory, on which bearing ratings are based, is expressed by Eq. (13.82). The exponents in this equation are determined experimentally from the dispersion of bearing lives and the dependence of life on load, geometry, and bearing size. As a standard of reference, all bearing load ratings are expressed in terms of the dynamic load rating $\overline{C}$, which, by definition, is the load that a bearing can carry for 1 million inner-race revolutions with a 90% chance of survival.

Factors on which specific dynamic capacity and bearing life depend are

1. Size of rolling element
2. Number of rolling elements per row
3. Number of rows of rolling elements
4. Conformity between rolling elements and races
5. Contact angle under load
6. Material properties
7. Lubricant properties
8. Operating temperature
9. Operating speed

Only factors 1 to 5 are incorporated in bearing dynamic load ratings developed from the Lundberg-Palmgren theory (1947, 1952).

Table 13.6 gives the dynamic load ratings for single-row, deep-groove ball bearings; and Table 13.7, those for single-row, cylindrical roller bearings. The dynamic load ratings are based on the material and manufacturing techniques used for producing standard production bearings. They apply to loads that are constant in both magnitude and direction; for radial bearings these are radial loads, and for thrust bearings they are axial loads that act centrally.

13.9.4 EQUIVALENT DYNAMIC LOAD

Equation (13.82), used for calculating the fatigue life, assumes a purely radial load in radial bearings and a purely axial load in thrust bearings, both loads of constant direction and magnitude. In many rolling-element bearing applications the load does not satisfy these conditions, since the force acts on the bearing obliquely or changes in magnitude. In such cases a constant radial or axial force must be determined for the rating life calculation, representing, with respect to the rating life, an equivalent stress.

The **equivalent load** is expressed as

$$P = XP_r + YP_a \tag{13.86}$$

where

X = radial factor
Y = thrust factor
P_r = radial load component, N
P_a = axial load component, N

Table 13.9 shows values of X and Y for various radial bearing types at respective contact angles. Also, C_0 is the static load rating given in Sec. 13.7.5. Table 13.10 shows values of X and Y for various thrust bearing types.

13.9.5 LIFE ADJUSTMENT FACTORS

This section presents the factors affecting the fatigue life of bearings that were not taken into account in the Lundberg-Palmgren theory. It is assumed that the various environmental or bearing design factors are multiplicative in their effects on bearing life. The following equation results:

$$L_A = \overline{D}\overline{E}\overline{F}_l\overline{G}\overline{H}_m\overline{L}_{10} \tag{13.87}$$

$$L_A = \overline{D}\overline{E}\overline{F}_l\overline{G}\overline{H}_m\left(\frac{\overline{C}}{\overline{P}}\right)^{m_k} \tag{13.88}$$

where

$\overline{D}$ = material factor
$\overline{E}$ = metallurgical processing factor
$\overline{F}_l$ = lubrication factor
$\overline{G}$ = speed effect factor

Table 13.9 Capacity formulas for rectangular and elliptical conjunctions for radial and angular bearings.

Bearing type		e	Single-row bearings				Double-row bearings			
			$\frac{P_a}{P_r} \le e$		$\frac{P_a}{P_r} > e$		$\frac{P_a}{P_r} \le e$		$\frac{P_a}{P_r} > e$	
			X	Y	X	Y	X	Y	X	Y
Deep-groove ball bearings	$P_a/C_0 = 0.025$	0.22	1	0	0.56	2.0				
	$P_a/C_0 = 0.04$	0.24	1	0	0.56	1.8				
	$P_a/C_0 = 0.07$	0.27	1	0	0.56	1.6				
	$P_a/C_0 = 0.13$	0.31	1	0	0.56	1.4				
	$P_a/C_0 = 0.25$	0.37	1	0	0.56	1.2				
	$P_a/C_0 = 0.50$	0.44	1	0	0.56	1				
Angular-contact ball bearings	$\beta = 20°$	0.57	1	0	0.43	1	1	1.09	0.70	1.63
	$\beta = 25°$	0.68	1	0	0.41	0.87	1	0.92	0.67	1.41
	$\beta = 30°$	0.80	1	0	0.39	0.76	1	0.78	0.63	1.24
	$\beta = 35°$	0.95	1	0	0.37	0.66	1	0.66	0.60	1.07
	$\beta = 40°$	1.14	1	0	0.35	0.57	1	0.55	0.57	0.93
	$\beta = 45°$	1.33	1	0	0.33	0.50	1	0.47	0.54	0.81
Self-aligning ball bearings		$1.5 \cdot \tan\beta$					1	$0.42 \cdot \cot\beta$	0.65	$0.65 \cdot \cot\beta$
Spherical roller bearings		$1.5 \cdot \tan\beta$					1	$0.45 \cdot \cot\beta$	0.67	$0.67 \cdot \cot\beta$
Tapered-roller bearings		$1.5 \cdot \tan\beta$	1	0	0.40	$0.4 \cdot \cot\beta$	1	$0.42 \cdot \cot\beta$	0.67	$0.67 \cdot \cot\beta$

Table 13.10 Radial factor X and thrust factor Y for thrust bearings.

Bearing type		e	Single acting		Double acting			
			$\frac{P_a}{P_r} > e$		$\frac{P_a}{P_r} \le e$		$\frac{P_a}{P_r} > e$	
			X	Y	X	Y	X	Y
Thrust ball	$\beta = 45°$	1.25	0.66	1	1.18	0.59	0.66	1
	$\beta = 60°$	2.17	0.92	1	1.90	0.55	0.92	1
	$\beta = 75°$	4.67	1.66	1	3.89	0.52	1.66	1
Spherical roller thrust		$1.5 \tan\beta$	$\tan\beta$	1	$1.5 \tan\beta$	0.67	$\tan\beta$	1
Tapered roller		$1.5 \tan\beta$	$\tan\beta$	1	$1.5 \tan\beta$	0.67	$\tan\beta$	1

$\overline{H}_m$ = misalignment factor
$\overline{P}$ = bearing equivalent load, N
m_k = load-life exponent; 3 for ball bearings or 10/3 for roller bearings
$\overline{C}$ = specific dynamic capacity, N

Factors $\overline{D}$, $\overline{E}$, and $\overline{F}_l$ are reviewed briefly here. Refer to Bamberger (1971) for a complete discussion of all five life adjustment factors.

13.9.5.1 Material Factors

For over a century AISI 52100 steel has been the predominant material for rolling-element bearings. In fact, the specific dynamic capacity as defined by the Anti-Friction Bearing Manufacturers Association (AFBMA) in 1949 is based on an air-melted 52100 steel, hardened to at least Rockwell C58. Since that time better control of air-melting processes and the introduction of vacuum-remelting processes have resulted in more homogeneous steels with fewer impurities. Such steels have extended rolling-element bearing fatigue lives to several times the catalog life. Life extensions of three to eight times are not uncommon. Other steel compositions, such as AISI M–1 and AISI M–50, chosen for their higher temperature capabilities and resistance to corrosion, also have shown greater resistance to fatigue pitting when vacuum-melting techniques are employed. Case-hardened materials, such as AISI 4620, AISI 4118, and AISI 8620, used primarily for rolling-element bearings, have the advantage of a tough, ductile steel core with a hard, fatigue-resistant surface.

The recommended material factors $\overline{D}$ for various through-hardened alloys processed by air melting are shown in Table 13.11. Insufficient definitive life data were found for case-hardened materials to recommend values of $\overline{D}$ for them. Refer to the bearing manufacturer for the choice of a specific case-hardened material.

The metallurgical processing variables considered in the development of the metallurgical processing factor $\overline{E}$ included melting practice (air and vacuum melting) and metalworking (thermomechanical working). Thermomechanical working of M–50 has also been shown to lengthen life, but in a practical sense it is costly and still not fully developed as a processing technique. Bamberger (1971) recommends an $\overline{E}$ of 3 for consumable-electrode-vacuum-melted materials.

Table 13.11 Material factors for through-hardened bearing materials. [E. N. Bamberger, *Adjustment Factors for Ball and Roller Bearings—An Engineering Design Guide*, ASME, 1971. Reprinted by permission of the American Society of Mechanical Engineers International.]

Material	Material factor, $\overline{D}$
52100	2.0
M–1	.6
M–2	.6
M–10	2.0
M–50	2.0
T–1	.6
Halmo	2.0
M–42	.2
WB 49	.6
440C	0.6–0.8

13.9.5.2 Lubrication Factor

Until approximately 1960 the role of the lubricant between surfaces in rolling contact was not fully appreciated. Metal-to-metal contact was presumed to occur in all applications, with attendant required boundary lubrication. The development of elastohydrodynamic lubrication theory showed that lubricant films with thicknesses on the order of microinches and tenths of microinches occur in rolling contact. Since surface finishes are of the same order of magnitude as the lubricant film thicknesses, the significance of rolling-element bearing surface roughnesses to bearing performance became apparent. Tallian (1967) was the first to report on the importance to bearing life of the ratio of elastohydrodynamic lubrication film thickness to surface roughness. Figure 13.24 shows calculated $\tilde{L}_{10}$ life as a function of the dimensionless film parameter Λ, which was introduced in Chapter 8, where

$$\Lambda = \frac{h_{\min}}{\left(R_{qa}^2 + R_{qb}^2\right)^{1/2}} \qquad \textbf{(8.8)}$$

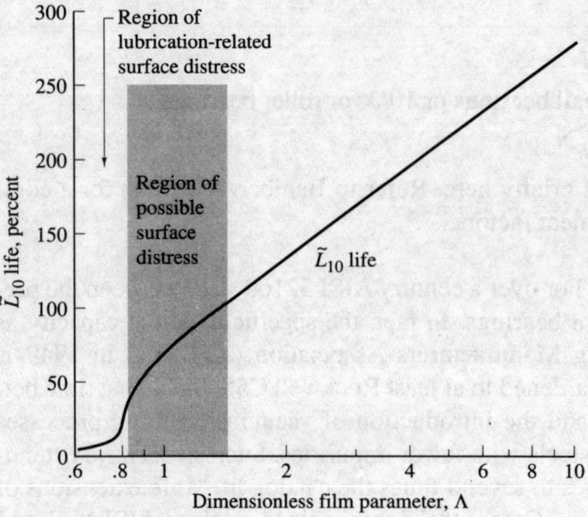

Figure 13.24 Group fatigue life $\tilde{L}_{10}$ as function of dimensionless film parameter. [T. E. Tallian "On Competing Failure Modes in Rolling Contact," *ASME Transactions,* vol. 10, 1967, pp. 418–39. Reprinted with permission of the American Society of Mechanical Engineers, International.]

Figure 13.25, from Bamberger (1971), presents a curve of the recommended $\overline{F}_l$ as a function of Λ. A mean of the curves presented in Tallian (1967) for ball bearings and in Skurka (1970) for roller bearings is recommended for use. A formula for calculating the

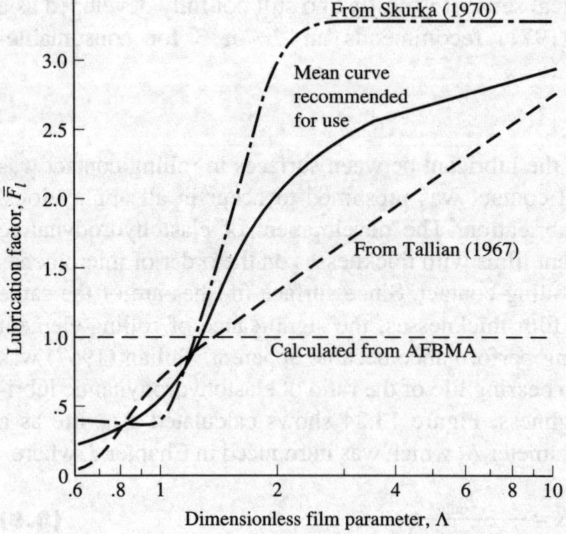

Figure 13.25 Lubrication factor as function of dimensionless film parameter. [E. N. Bamberger, *Adjustment Factors for Ball and Roller Bearings—An Engineering Guide*, ASME, 1971. Reprinted by permission of the American Society of Mechanical Engineers International.]

minimum film thickness H_{min} is given in Eq. (13.72). Procedures for calculating $\overline{D}$ and $\overline{F}_l$ are at present less than definitive, reflecting the need for additional research, life data, and operating experience.

Example 13.8

GIVEN A deep-groove 6304 ball bearing has a load rating of 15 900 N. It is radially loaded with 4000 N and axially loaded with 2000 N. The rotational speed is 3000 rpm.
FIND How many hours will the bearing survive with 90% probability?

Solution

From Table 13.6 for a 6304 deep-groove ball bearing the static load rating is 7800 N.

$$\therefore \frac{P_a}{C_0} = \frac{2000}{7800} = 0.256$$

From Table 13.9 for $P_a/C_0 = 0.256$ for a deep-groove ball bearing the constant e is 0.37. Since $P_a/P_r = 0.5 > 0.37$, the radial and thrust factors are $X = 0.56$ and $Y = 1.2$. From Eq. (13.86) the equivalent bearing load is

$$P = XP_r + YP_a = 0.56(4000) + 1.2(2000) = 4640 \text{ N}$$

The rotational speed in revolutions per minute is

$$N_b = 3000 \text{ rpm}$$

From Eqs. (13.84) and (13.85) the bearing life in hours is

$$\left(\tilde{L}_{10}\right)_{\text{hour}} = \frac{10^6}{60\,N_b}\left(\frac{\overline{C}}{P}\right)^3 = \frac{10^6}{(60)(3000)}\left(\frac{15\,900}{4640}\right)^3 = 223.5 \text{ hr}$$

The bearing has a 90% probability of surviving for 223.5 hr.

Example 13.9

GIVEN A machine manufacturer has dimensioned the roller bearings in a paper-making machine containing 400 bearings such that the $\tilde{L}_{10}$ for each bearing is 40 years.
FIND Use Fig. 13.23 to calculate the $\tilde{L}_{50}$ life in years. Also, determine the mean time between failures in the steady state for an old paper-making machine where bearings are replaced when they fail.

Solution

From Fig. 13.23 the $\tilde{L}_{50}$ life is about 110 million stress cycles while the $\tilde{L}_{10}$ life is 17 million stress cycles. The ratio is

$$\frac{\tilde{L}_{50}}{\tilde{L}_{10}} = \frac{110}{17} = 6.47$$

The $\tilde{L}_{50}$ life is thus 6.47 times the $\tilde{L}_{10}$ life, which in this example is 40 years.

$$\therefore \tilde{L}_{50} = (6.47)(40) = 258.8 \text{ years}$$

As there are 400 independent bearings in the machine, the mean time between failures is

$$\frac{\tilde{L}_{50}}{400} = \frac{258.8}{400} = 0.647 \text{ year} = 236 \text{ days}$$

Case Study 13.1 ROLLER BEARING DESIGN FOR LOCOMOTIVE WHEEL AXLE BOX

GIVEN The applications for roller bearings are truly everywhere. They can be found in washing machines, jet engines, process equipment of all kinds, lawnmowers, roller blades, etc. One of the most historically successful applications of rolling-element bearings is in vehicle axles, such as the railroad boxcar axle bearing shown in Fig. 13.26. Such bearings are used for prolonged periods of maintenance-free operation and are universally packed with a suitable bearing grease. This case study involves determining film thickness in the bearing so that wear can be assessed. The bearings are highly polished; a typical surface roughness of 0.05 μm is assumed.

The equations for elastohydrodynamic film thickness developed in Sec. 13.8.3 relate primarily to elliptical and rectangular conjunctions. The rectangular conjunctions correspond to line-contact problems, as would be found in such a cylindrical roller bearing. Figures 13.1 and 13.7 show a roller bearing describing the various dimensions used in this case study. Therefore, the minimum elastohydrodynamic film thicknesses are calculated on the inner and outer races of a cylindrical roller bearing with the following dimensions—for both the elliptical conjunction and the rectangular conjunction:

Inner-race diameter d_i, mm	64
Outer-race diameter d_o, mm	96
Diameter of cylindrical rollers d, mm	16
Effective axial length of cylindrical rollers l_t, mm	16
Number of rollers in complete bearing n	9

A railroad bearing of this kind might well experience the following static operating conditions for extended periods:

Radial load P, N	10 800
Inner-race angular velocity ω_i, rad/s	52.4
Outer-race angular velocity ω_o, rad/s	0
Absolute viscosity at $p = 0$ and bearing effective operating temperature η_0, Pa-s	0.1
Viscosity-pressure coefficient ξ, m²/N	2.2×10^{-8}
Modulus of elasticity for both rollers and races E, Pa	2.075×10^{11}
Poisson's ratio for both rollers and races ν	0.3

FIND The minimum film thickness for such a cylindrical roller bearing.

Solution

Since the diametral clearance c_d is zero, from Eq. (13.49) the most heavily loaded roller can be expressed as

$$P_{z,\max} = \frac{4P_z}{n} = \frac{4(10\ 800 \text{ N})}{9} = 4800 \text{ N}$$

Therefore, the radial load per unit length on the most heavily loaded roller can be expressed as

$$(P_z')_{\max} = \frac{4800 \text{ N}}{0.016 \text{ m}} = 0.3 \text{ MN/m}$$

(continued)

Case Study 13.1 (CONTINUED)

(a)

(b)

Figure 13.26
(a) Railroad boxcar wheel assembly and (b) disassembled axle, showing bearing components. (Courtesy of FAG Kugelfischer Georg Schafer Kga. Schweinfurt, W. Germany)

From Fig. 8.14 the radii of curvature are

$$r_{ax} = 0.008 \text{ m} \qquad r_{ay} = \infty$$

$$r_{bxi} = 0.032 \text{ m} \qquad r_{byi} = \infty$$

$$r_{bxo} = -0.048 \text{ m} \qquad r_{byo} = \infty$$

Then,

$$\frac{1}{R_{xi}} = \frac{1}{0.008} + \frac{1}{0.032} = \frac{5}{0.032}$$

giving $R_{xi} = 0.0064$ m,

$$\frac{1}{R_{xo}} = \frac{1}{0.008} - \frac{1}{0.048} = \frac{5}{0.048}$$

giving $R_{xo} = 0.0096$ m, and

$$\frac{1}{R_{yi}} = \frac{1}{R_{yo}} = \frac{1}{\infty} + \frac{1}{\infty} = 0$$

giving $R_{yi} = R_{yo} = \infty$.

(continued)

Case Study 13.1 (CONTINUED)

From the input information the effective modulus of elasticity can be written as

$$E' = \frac{2}{\dfrac{1-\nu_a^2}{E_a} + \dfrac{1-\nu_b^2}{E_b}} = 2.28\times10^{11}\,\text{Pa}$$

For pure rolling the mean surface velocity $\tilde{u}$ relative to the lubricated conjunctions for a cylindrical roller is

$$\tilde{u} = \frac{|\omega_i - \omega_o|(d_e^2 - d^2)}{4d_e}$$

where

d_e = pitch diameter
d = roller diameter

$$d_e = \frac{d_o + d_i}{2} = \frac{0.096 + 0.064}{2} = 0.08\text{ m}$$

Hence,

$$\tilde{u} = \frac{0.08^2 - 0.016^2}{4(0.08)}|52.4 - 0| = 1.006\text{ m/s}$$

The dimensionless speed, materials, and load parameters expressed in Eqs. (13.68) to (13.70) for the inner- and outer-race conjunctions thus become

$$U_i = \frac{\eta_0\tilde{u}}{E'R_{xi}} = \frac{(0.10)(1.006)}{(2.28)(10^{11})(0.0064)} = 6.895\times10^{-11}$$

$$G_o = G_i = \xi E' = 5016$$

$$W_i = \frac{(P_z)_{\max}}{E'R_{xi}^2} = \frac{4800}{(2.28)(10^{11})(0.0064)^2} = 5.140\times10^{-4}$$

$$U_o = \frac{\eta_0\tilde{u}}{E'R_{xo}} = \frac{(0.10)(1.006)}{(2.28)(10^{11})(0.0096)} = 4.597\times10^{-11}$$

$$W_o = \frac{(P_z)_{\max}}{E'(R_{xo})^2} = \frac{4800}{(2.28)(10^{11})(0.0096)^2} = 2.284\times10^{-4}$$

From Eq. (13.71) the appropriate elastohydrodynamic film thickness equation for a fully flooded elliptical conjunction is

$$H_{\min} = \frac{\tilde{h}_{\min}}{R_x} = 3.63U^{0.68}G^{0.49}W^{-0.073}\left(1 - e^{-0.68k_e}\right)$$

For a roller bearing $k_e = \infty$ and this equation reduces to

$$H_{\min} = 3.63U^{0.68}G^{0.49}W^{-0.073}$$

The dimensionless film thickness for the *roller–inner-race conjunction* is

$$\tilde{H}_{\min} = \frac{h_{\min}}{R_{xi}}$$

$$= (3.63)(1.231)(10^{-7})(65.04)(1.738) = 50.5\times10^{-6}$$

and hence,

$$h_{\min} = (0.0064)(50.5)(10^{-6}) = 0.32\times10^{-6}\text{ m} = 0.32\ \mu\text{m}$$

The dimensionless film thickness for the *roller–outer-race conjunction* is

$$H_{\min} = \frac{h_{\min}}{R_{xo}}$$

$$= (3.63)(9.343)(10^{-8})(65.04)(1.844) = 40.7\times10^{-6}$$

and hence,

$$h_{\min} = (0.0096)(40.7)(10^{-6}) = 0.3905\times10^{-6} = 0.39\ \mu\text{m}$$

It is clear from these calculations that the smaller minimum film thickness in the bearing occurs at the roller–inner-race conjunction, where the geometrical conformity is less favorable. It was found that if the ratio of minimum film thickness to composite surface roughness

(continued)

Case Study 13.1 (CONCLUDED)

is greater than 3, an adequate elastohydrodynamic film is maintained. This result implies that a composite surface roughness of less than 0.1 μm is needed to ensure that an elastohydrodynamic film is maintained.

Now using, instead of the elliptical conjunction results, the rectangular conjunction results from Eq. (13.73) gives

$$H_{min} = \frac{h_{min}}{R_x} = 1.714(W')^{-0.128} U^{0.694} G^{0.568}$$

where

$$W' = \frac{(P_z')_{max}}{E' R_x}$$

For the *roller–inner-race conjunction*

$$W' = \frac{(P_z')_{max}}{E' R_x} = \frac{(0.3)(10^6)}{(2.28)(10^{11})(0.0064)} = 2.056 \times 10^{-4}$$

$$H_{min} = 1.714(2.056 \times 10^{-4})^{-0.128}(6.895 \times 10^{-11})^{0.694} \times (5016)^{0.568} = 56.98 \times 10^{-6}$$

$$h_{min} = 0.0064(56.98)(10^{-6}) = 0.3647 \times 10^{-6} \text{ m} = 0.3647 \text{ μm}$$

For the *roller–outer-race conjunction*

$$W' = \frac{(P_z')_{max}}{E' R_x} = \frac{(0.3)(10^6)}{(2.28)(10^{11})(0.0096)} = 1.371 \times 10^{-4}$$

$$H_{min} = 1.714(1.371 \times 10^{-4})^{-0.128}(4.597 \times 10^{-11})^{0.694} \times (5016)^{0.568} = 45.29 \times 10^{-6}$$

so that

$$h_{min} = (0.0096)(45.29 \times 10^{-6}) = 0.4348 \times 10^{-6} \text{ m} = 0.4348 \text{ μm}$$

The film thickness obtained from the rectangular conjunction results produces approximately a 12% increase from the elliptical conjunction results.

Since the original surface roughness is 0.05 μm, the ratio of film thickness to surface roughness is approximately 6.0. This ratio still ensures that the bearing is in a full film lubrication state and that adhesive wear is far less likely than fatigue wear.

Case Study 13.2 SUPPORT BEARING FOR FLYWHEEL SHAFT BEARING

GIVEN Figure 12.41 shows a cutaway view of an internal combustion engine. On the right side of the figure is the flywheel, which is mounted on the crankshaft. A rolling-element bearing supports the shaft next to the flywheel to maximize rigidity. To the right of the figure is a bearing pack that supports the flywheel. Just as in the case study for Chapter 12, it is imperative that the bearing in this application be designed for long life. Consider a single-row, radial, deep-groove ball bearing with the following dimensions:

Inner-race diameter d_i, m	0.052291
Outer-race diameter d_o, m	0.077706
Ball diameter d, m	0.012700
Number of balls in complete bearing n	9
Inner-groove radius r_i, m	0.006604
Outer-groove radius r_o, m	0.006604
Contact angle β, deg	0
rms surface roughness of balls R_{qb}, μm	0.0625
rms surface roughness of races R_{qa}, μm	0.175

(continued)

Case Study 13.2 (CONTINUED)

A bearing of this kind might well experience the following operating conditions:

Radial load P_z, N	8900
Inner-race angular velocity ω_i, rad/s	400
Outer-race angular velocity ω_o, rad/s	0
Absolute viscosity at $p = 0$ and bearing effective operating temperature η_0, Pa-s	0.04
Viscosity-pressure coefficient ξ, m²/N	2.3×10^{-8}
Modulus of elasticity for both ball and races, Pa	2.0×10^{11}
Poisson's ratio for both balls and races ν	0.3

FIND

Whether or not the design is suitable based on a lubrication rationale.

Solution

The essential geometric features of the inner and outer conjunctions (Figs. 13.1 and 13.6) can be ascertained as follows:

Pitch diameter:

$$d_e = \frac{d_o + d_i}{2} = 0.065 \text{ m}$$

Diametral clearance:

$$c_d = d_o - d_i - 2d = 1.5 \times 10^{-5} \text{ m}$$

Race conformity:

$$R_{ri} = R_{ro} = \frac{r}{d} = 0.52$$

Equivalent radii:

$$R_{xi} = \frac{d(d_e - d\cos\beta)}{2d_e} = 0.00511 \text{ m}$$

$$R_{xo} = \frac{d(d_e + d\cos\beta)}{2d_e} = 0.00759 \text{ m}$$

$$R_{yi} = \frac{R_{ri}d}{2R_{ri} - 1} = 0.165 \text{ m}$$

$$R_{yo} = \frac{R_{ro}d}{2R_{ro} - 1} = 0.165 \text{ m}$$

The curvature sum of the *ball–inner-race contact*

$$\frac{1}{R_i} = \frac{1}{R_{xi}} + \frac{1}{R_{yi}} = 201.8$$

gives R_i = 4.956 mm, and the curvature sum of the *ball–outer-race contact*

$$\frac{1}{R_o} = \frac{1}{R_{xo}} + \frac{1}{R_{yo}} = 137.8$$

gives R_o = 7.256 mm. Also, $\alpha_{ri} = \frac{R_{yi}}{R_{xi}} = 32.29$ and $\alpha_{ro} = \frac{R_{yo}}{R_{xo}} = 21.74$.

The nature of the Hertzian contact conditions can now be assessed.

Ellipticity parameters:

$$k_{ei} = \alpha_{ri}^{2/\pi} = 9.135$$

$$k_{eo} = \alpha_{ro}^{2/\pi} = 7.10$$

Elliptic integrals:

$$\mathscr{E}_i = 1 + \frac{q_a}{\alpha_{ri}} = 1.018$$

$$\mathscr{E}_o = 1 + \frac{q_a}{\alpha_{ro}} = 1.026$$

$$\mathscr{F}_i = \frac{\pi}{2} + q_a \ln \alpha_{ri} = 3.554$$

$$\mathscr{F}_o = \frac{\pi}{2} + q_a \ln \alpha_{ro} = 3.328$$

(continued)

Case Study 13.2 (CONTINUED)

The effective elastic modulus E' is given by

$$E' = \frac{2}{\dfrac{1-\nu_a^2}{E_a} + \dfrac{1-\nu_b^2}{E_b}} = 2.198 \times 10^{11}\,\text{Pa}$$

To determine the load carried by the most heavily loaded ball in the bearing, it is necessary to adopt an iterative procedure based on the calculation of local static compression. From Eq. (13.47)

$$(P_z)_{\max} = Z_w \frac{P_z}{n}$$

where

$(P_z)_{\max}$ = load of most heavily loaded ball
P_z = radial load on bearing
n = number of balls

However, it is customary to adopt a value of $Z_w = 5$ in simple calculations to produce a conservative design, and this value will be used to begin the iterative procedure.

Stage 1: Assume that $Z_w = 5$, Then,

$$(P_z)_{\max} = \frac{5P_z}{9} = 5\frac{(8900)}{9} = 4944\ \text{N}$$

The maximum local elastic compression is

$$\delta_i = \mathcal{F}_i \left[\frac{9}{2\,\mathcal{E}_i R_i} \left(\frac{(P_z)_{\max}}{\pi k_{ei} E'} \right)^2 \right]^{1/3} = 2.906 \times 10^{-5}\,\text{m}$$

$$\delta_o = \mathcal{F}_o \left[\frac{9}{2\,\mathcal{E}_o R_o} \left(\frac{(P_z)_{\max}}{\pi k_{eo} E'} \right)^2 \right]^{1/3} = 2.832 \times 10^{-5}\,\text{m}$$

The sum of the local compressions on the inner and outer races is

$$\delta = \delta_i + \delta_o = 5.738 \times 10^{-5}\,\text{m}$$

A better value for Z_w can now be obtained from

$$Z_w = \frac{\pi(1 - c_d/2\delta)^{3/2}}{2.491\left\{\left[1 + \left(\dfrac{1 - c_d/2\delta}{1.23}\right)^2\right]^{1/2} - 1\right\}}$$

since $c_d/2\delta = 1.5(10^{-5})/11.48(10^{-5}) = 0.1307$. Thus, $Z_w = 4.552$.

Stage 2:

$$(P_z)_{\max} = \frac{(4.552)(8900)}{9} = 4502\ \text{N}$$

$$\delta_i = 2.730 \times 10^{-5}\,\text{m}$$

$$\delta_o = 2.661 \times 10^{-5}\,\text{m}$$

$$\delta = 5.391 \times 10^{-5}\,\text{m}$$

$$c_d/2\delta = 0.1391$$

Thus, $Z_w = 4.567$.

Stage 3:

$$(P_z)_{\max} = \frac{(4.567)(8900)}{9} = 4516\ \text{N}$$

$$\delta_i = 2.736 \times 10^{-5}\,\text{m}$$

$$\delta_o = 2.666 \times 10^{-5}\,\text{m}$$

$$\delta = 5.402 \times 10^{-5}\,\text{m}$$

$$c_d/2\delta = 0.1388$$

and hence, $Z_w = 4.566$. This value is almost exactly the same as the previous value from stage 2 of 4.567 and hence, $Z_w = 4.567$. The load on the most heavily loaded ball is thus 4515 N.

(continued)

Case Study 13.2 (CONCLUDED)

The elastohydrodynamic minimum film thickness is calculated as follows: For pure rolling

$$\tilde{u} = \frac{|\omega_o - \omega_i|(d_e^2 - d^2)}{4d_e} = 6.252\,\text{m/s}$$

The dimensionless speed, materials, and load parameters for the inner- and outer-race conjunctions thus become, when the velocity is only in the rolling direction ($\tilde{v} = 0$),

$$U_i = \frac{\eta_0 \tilde{u}}{E' R_{xi}} = \frac{(0.04)(6.252)}{(2.198)(10^{11})(5.11)(10^{-3})} = 2.227 \times 10^{-10}$$

$$G_i = G_o = \xi E' = (2.3)(10^{-8})(2.198)(10^{11}) = 5055$$

$$W_i = \frac{P_z}{E'(R_{xi})^2}$$

$$= \frac{4515}{(2.198)(10^{11})(5.11)^2(10^{-6})} = 7.866 \times 10^{-4}$$

$$U_o = \frac{\eta_0 \tilde{u}}{E'(R_{xo})}$$

$$= \frac{(0.04)(6.252)}{(2.198)(10^{11})(7.59)(10^{-3})} = 1.499 \times 10^{-10}$$

$$W_o = \frac{P_z}{E'(R_{xo})^2}$$

$$= \frac{4515}{(2.198)(10^{11})(7.59)^2(10^{-6})} = 3.566 \times 10^{-4}$$

The dimensionless minimum elastohydrodynamic film thickness in a fully flooded elliptical conjunction is again obtained from Eq. (13.72) as

$$H_{\min} = \frac{h_{\min}}{R_x} = 3.63 U^{0.68} G^{0.49} W^{-0.073}\left(1 - e^{-0.68 k_e}\right)$$

For the *ball–inner-race contact* it is

$$(H_{\min})_i = 3.63(2.732)(10^{-7})$$

$$\times (65.29)(1.685)(0.9980) = 1.09 \times 10^{-4}$$

Thus,

$$(h_{\min})_i = 1.09(10^{-4}) R_{xi} = 0.556\ \mu\text{m}$$

The dimensionless film parameter Λ [Eq. (8.8)] was found to play a significant role in determining the fatigue life of rolling-element bearings. In this case

$$\Lambda_i = \frac{(h_{\min})}{\left(R_{qa}^2 + R_{qb}^2\right)^{1/2}}$$

$$= \frac{(0.556)(10^{-6})}{\left[(0.175)^2 + (0.0625)^2\right]^{1/2}(10^{-6})} = 3.0$$

For the *ball–outer-race contact* it is

$$(H_{\min})_o = \frac{(h_{\min})_o}{R_{xo}}$$

$$= 3.63(2.087)(10^{-7})(65.29)(1.785)(0.9920)$$

Thus,

$$(H_{\min})_o = 0.876 \times 10^{-4}$$

$$(h_{\min})_o = (0.876)(10^{-4}) R_{xo} = 0.665\ \mu\text{m}$$

In this case the dimensionless film parameter is given by

$$\Lambda_o = \frac{0.665 \times 10^{-6}}{\left[(0.175)^2 + (0.0625)^2\right]^{1/2}(10^{-6})} = 3.58$$

It is evident that the smaller minimum film thickness occurs between the most heavily loaded ball and the inner race. However, in this case the minimum elastohydrodynamic film thickness is about three times the composite surface roughness, and the bearing lubrication can be deemed to be entirely satisfactory. Indeed, it is clear from Fig. 13.25 that little improvement in the lubrication factor $\overline{F}_l$ and thus in the fatigue life of the bearing can be achieved by further increasing the minimum film thickness and hence the dimensionless film parameter.

13.10 SUMMARY

Rolling-element bearings are precise, yet simple, machine elements of great utility. This chapter drew together the current understanding of rolling-element bearings and attempted to present it in a concise manner. The detailed precise calculations of bearing performance were only summarized and appropriate references were given. The history of rolling-element bearings was briefly reviewed, and subsequent sections described the types of rolling-element bearing, their geometry, and their kinematics. Having defined ball bearing operation under unloaded and unlubricated conditions, the chapter then focused on static loading of rolling-element bearings. Most rolling-element bearing applications involve steady-state rotation of either the inner or outer race or both; however, the rotational speeds are usually not so great as to cause ball or roller centrifugal forces or gyroscopic moments of significant magnitudes. Thus, these were neglected. Radial, thrust, and preloaded bearings that are statically loaded were considered in the second major thrust of the chapter, which was on loaded but unlubricated rolling-element bearings.

The last major thrust of the chapter dealt with loaded and lubricated rolling-element bearings. Topics covered were fatigue life and dynamic analyses. The chapter concluded by applying the knowledge of this and previous chapters to roller and ball bearing applications. The use of the elastohydrodynamic lubrication film thickness was integrated with the rolling-element bearing ideas developed in this chapter. It was found that the most critical conjunctions of both ball and roller bearings occurred between the rolling elements and the inner races.

KEY WORDS

angular-contact ball bearings bearings with two-shouldered groove in one race and single-shouldered groove in other

angular-contact bearings bearings that have clearance built into unloaded bearing, which allows operation under high thrust loads

ball bearings rolling-element bearings using spheres as rolling elements

cage fitting or spacer to keep proper distance between balls in bearing track

Conrad bearing same as deep-groove bearing

crown curvature machined into rollers to eliminate high edge stresses

cylindrical roller bearings bearings using cylinders as rolling elements

deep-groove bearing ball bearing with race containing pronounced groove for rolling elements

duplex pairs sets of two angular-contact ball bearings that preload each other upon assembly to shaft

endplay maximum axial movement of inner race with respect to outer race under small forces

equivalent static load load equivalent to resultant load when considering thrust and radial components

free contact angle angle made by line through points where ball contacts both races and plane perpendicular to bearing axis rotation under low loads

pitch diameter mean of inner- and outer-race contact diameters

races grooves within bearing rings for rolling elements to roll in

race control condition where pure rolling occurs at controlling race

race depth same as shoulder height

retainers same as cage

rolling-element bearings machinery elements where surfaces are nonconformal and motion is primarily rolling

self-aligning bearings bearings with one race having spherical shape to allow for large misalignment

separators same as cage

shoulder height depth of race groove

spherical roller bearings same as self-aligning bearings

split-ring bearing bearing that has one race made from two halves allowing for accurate axial positioning of shafts

static load rating load at which, under defined operating conditions, load carried by most heavily loaded rolling element W_{max} equals allowable rolling-element load w_{all}, or maximum load before plastic (rather than elastic) deformation occurs within bearing

static thrust load-carrying capacity maximum thrust load that bearing can endure before contact ellipse approaches race shoulder

thrust ball bearings bearings with race grooves arranged to support large axial or thrust forces

Weibull distribution relationship between survival and life, applied in this chapter to bearings

RECOMMENDED READINGS

Halling, J. (ed.) (1978) *Principles of Tribology*, Macmillan, New York.

Hamrock, B.J. (1993) *Fundamentals of Fluid Film Lubrication*, McGraw-Hill, New York.

Hamrock, B.J., and Dowson, D. (1981) *Ball Bearing Lubrication—The Elastohydrodynamics of Elliptical Contacts*, Wiley, New York.

Harris, T. A. (1991) *Rolling Bearing Analysis*, 3rd ed., Wiley, New York.

Johnson, K.L. (1985) *Contact Mechanics*, Cambridge University Press, Cambridge, England.

Juvinall, R.C., and Marshek, K.M. (1991) *Fundamentals of Machine Component Design*, Wiley, New York.

Mott, R. L. (1985) *Machine Elements in Mechanical Design*, Merrill Publishing Co., Columbus, OH.

Norton, R.L. (1996) *Machine Design*, Prentice-Hall, Englewood Cliffs, NJ.

Shigley, J.E., and Mischke, C.R. (1989) *Mechanical Engineering Design*, McGraw-Hill, New York.

REFERENCES

Anderson, W.J. (1970) Elastohydrodynamic Lubrication Theory as a Design Parameter for Rolling Element Bearings, *ASME Paper* 70–DE–19.

Bamberger, E.N. (1971) *Life Adjustment Factors for Ball and Roller Bearings—An Engineering Design Guide*, American Society for Mechanical Engineers, New York.

Hamrock, B.J. (1993) *Fundamentals of Fluid Film Lubrication*, McGraw-Hill, New York.

Hamrock, B.J., and Anderson, W.J. (1983) *Rolling-Element Bearings*, NASA Reference Publication 1105.

Hamrock, B.J., and Dowson, D. (1981) *Ball Bearing Lubrication—The Elastohydrodynamics of Elliptical Contacts*, Wiley-Interscience, New York.

Jones, A.B. (1964) The Mathematical Theory of Rolling Element Bearings, *Mechanical Design and Systems Handbook*, H.A. Rothbart (ed.), McGraw-Hill, New York, pp. 13–1 to 13–76.

Lundberg, G., and Palmgren, A. (1947) Dynamic Capacity of Rolling Bearings, *Acta Polytech. Mech. Eng. Sci.*, vol. 1, no. 3. (Also, Ingeniörs Vetenskaps Akadamien Handlingar, no. 196.)

Lundberg, G., and Palmgren, A. (1952) Dynamic Capacity of Roller Bearing, *Acta Polytech. Mech. Eng. Sci.*, vol. 2, no. 4. (Also, Ingeniörs Vetenskaps Akadamien Handlingar, no. 210.)

SKF Catalog (1991), SKF USA, Inc.

Skurka, J.C. (1970) Elastohydrodynamic Lubrication of Roller Bearings, *J. Lubr. Technol.*, vol. 92, no. 2, pp. 281–291.

Stribeck, R. (1901) Kugellager für beliebige Belastungen, *Z. Ver. Deutsch. Ing.*, vol. 45, pp. 73–125.

Tallian, T.E. (1967) On Competing Failure Modes in Rolling Contact, *Trans. ASLE*, vol. 10, pp. 418–439.

Weibull, W. (1949) A Statistical Representation of Fatigue Failures in Solids, *Transactions of the Royal Institute of Technology, Stockholm*, no. 27.

PROBLEMS

Section 13.3

13.1 In roller bearings the load is carried by rollers and the contacts between rollers and races are rectangular. The bearing macrogeometry determines which types of load can be carried by the bearing: radial, axial, or moment loads. Find which type of load can be carried by the different types of bearing shown in Tables 13.4 and 13.5.

13.2 Too high a power loss can be a problem in extremely high-speed applications of angular-contact ball bearings. A large part of that power loss comes from the churning of the oil when the rolling balls displace the oil and splash it around the bearing. It is therefore important not to collect too much oil in the rolling tracks. For a duplex arrangement with the lubricating oil passing through both bearings for cooling, find if a back-to-back or face-to-face arrangement is preferred from a low-power-loss point of view.

Section 13.4

13.3 A 214 deep-groove ball bearing with its dimensions is shown in sketch *a*. All dimensions are in millimeters. Also, $n = 11$, $d = 21/32$ in. = 16.67 mm. Find the maximum ball force and stress as well as the relative displacement of the race.

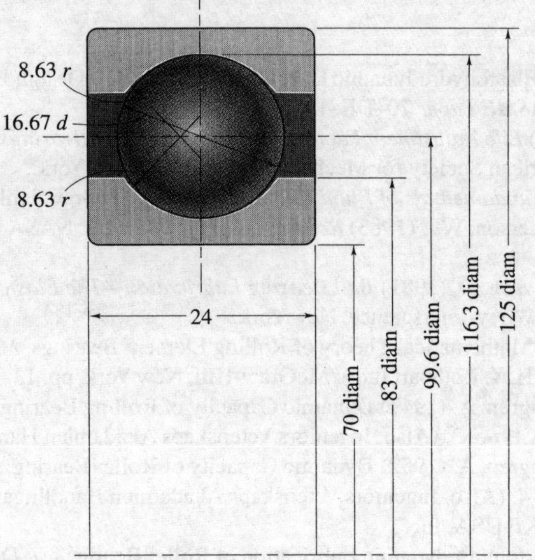

Sketch *a*, for Problem 13.3

13.4 In determining how large the contact areas are in a rolling-element bearing at a certain load, the geometry of each bearing part must be known. For a deep-groove ball bearing the width-to-length ratio of the contact area is given by the race conformity as expressed in Eq. (13.3). Calculate the race conformity for both the inner- and outer-race contacts when the ball diameter is 17 mm and the radii of curvature in the axial direction are 8.840 mm for the inner race and 9.180 mm for the outer race. Also calculate the free contact angle and the endplay when the diametral clearance is 136 μm.

13.5 In a spherical roller bearing the inner radius of the outer race (the radius to the spherical surface) is 175 mm. The roller diameter is 41 mm and the crowning radius is 168 mm. The diametral play is 0.33 mm for the outer race. Calculate the race conformity, the free endplay, and the free contact angle when the angle between each roller center and the bearing center plane is 12°.

Section 13.5

13.6 For a cylindrical roller bearing the separator has the form of a cylinder with outer diameter of 80 mm, inner diameter of 74 mm, and width of 16 mm. Fourteen 13- by 13-mm rectangular pockets are evenly distributed around the circumference of the separator. In each pocket is a 12.5- by 12.5-mm cylindrical roller. The inner-race outer diameter is 65 mm and the outer-race inner diameter is 90 mm. How fast can the bearing inner race rotate before the polyester separator fails? Use a safety factor of 10 and the lowest values in Table A.4 for polyester.

13.7 For the cylindrical roller bearing in Problem 13.6 calculate the sliding speed at the contact between the rollers and the separator if the inner race rotates at 10 000 rpm.

Section 13.6

13.8 A separator for a cylindrical roller bearing is made of nylon 66 with an ultimate strength of 80 MPa at 20°C. At the maximum service temperature of 260°C the ultimate strength falls to 1 MPa. The cage material can withstand a bearing speed of 16 000 rpm at 50°C. What speed can it withstand at 150°C if the ultimate strength varies linearly with the temperature?

13.9 A deep-groove ball bearing has a nylon 66 snap cage that is snapped in between the balls after they have been mounted. If the arms separating the balls have a square 3- by 3-mm cross section and are 12 mm long, calculate how fast a 100-mm-diameter cage can rotate without overstressing the ball-separating arms. The maximum allowable bending stress is 20 MPa.

Section 13.8

13.10 Using the results from Problem 13.3, determine the minimum film thickness and the film parameter for the ball–inner-race contact and the ball–outer-race contact. The inner race rotates at 20 000 rpm and the outer race is stationary. The absolute viscosity at $p = 0$ and bearing effective temperature is 0.04 Pa-s. Also, the viscosity-pressure coefficient is 2.3×10^{-8} m^2/N. The rms surface roughnesses are 0.0625 μm for the balls and 0.175 μm for the races.

13.11 A 0.4-in.-diameter ball is loaded against a plane surface with 25 lbf. The other parameters are

rms surface finish of ball	2.5 μin.
rms surface finish of plane	7.5 μin.
Absolute viscosity at $p = 0$ and bearing effective temperature	6×10^{-6} lbf-s/in.^{2}
Viscosity-pressure coefficient	1.6×10^{-4} in.^{2}/lbf
Modulus of elasticity for ball and plane	3×10^7 psi
Poisson's ratio for ball and plane	0.3
Film parameter	3.0

Determine the following:

a) Minimum film thickness
b) Mean velocity necessary to achieve the above film thickness
c) Contact dimensions
d) Maximum pressure
e) Maximum deformation

13.12 A deep-groove ball bearing has steel races and silicon nitride balls. It is lubricated with a mineral oil that has a viscosity of 0.026 N-s/m^2 at the application temperature. The pressure-viscosity coefficient of the oil is 2×10^{-8} m^2/N. The ball diameter is 17 mm, and the radius of curvature in the axial direction is 8.84 mm for both races. The radii in the rolling direction are 30 mm for the inner race and 47 mm for the outer race. Calculate the minimum oil film thickness when the ball lòad is 20 000 N and the rolling speed is 3 m/s at the inner-race contact.

What should the radius of curvature in the axial direction be for the outer race to get the same film thickness as that of the inner race?

13.13 For the bearing considered in Problem 13.12 the silicon nitride balls are changed to steel balls. Calculate how large the load can be and still maintain the same oil film thickness as obtained in Problem 13.12.

13.14 A steel roller, shown in sketch *b*, is used for rolling steel sheets that have an ultimate strength of 400 MPa. The 1-m-long roller has a diameter of 20 cm. What load will cause plastic deformation of the sheets?

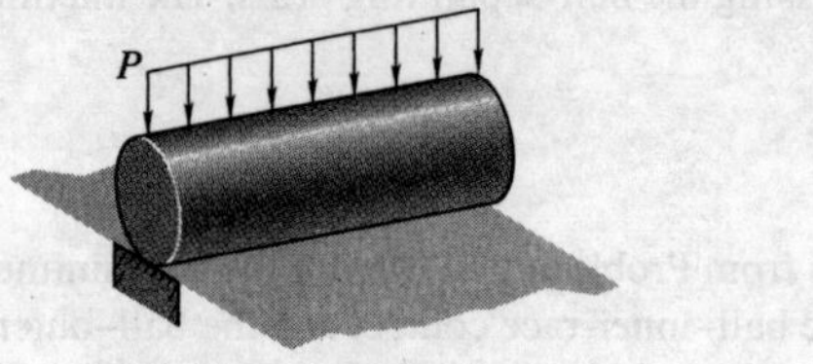

Sketch *b*, for Problem 13.14

Section 13.9

13.15 A shaft with rolling bearings and loads is shown in sketch *c*. The bearings are mounted such that both the axial and radial forces caused by the external loads are absorbed. For bearing NU205EC the geometry is cylindrical with no flange on the inner race, so that only radial load can be absorbed. The full axial thrust load P_a is thus taken up by the deep-groove ball bearing 6205. Find the $\tilde{L}_{10}$ lives of each of the two bearings for the following conditions:

a) For bearing NU205EC the dimensions are $d = 25$ mm, $D = 52$ mm, $B = 15$ mm; and the basic dynamic load rating $C = 28\ 600$ N.

b) For bearing 6205 the inner diameter d, the outer diameter D, and the bearing width B are the same as for bearing NU205EC; but the basic static load rating $C_0 = 7800$ N, the basic dynamic load rating $C = 14\ 000$ N, $P_r = 1000$ N, and $P_a = 500$ N.

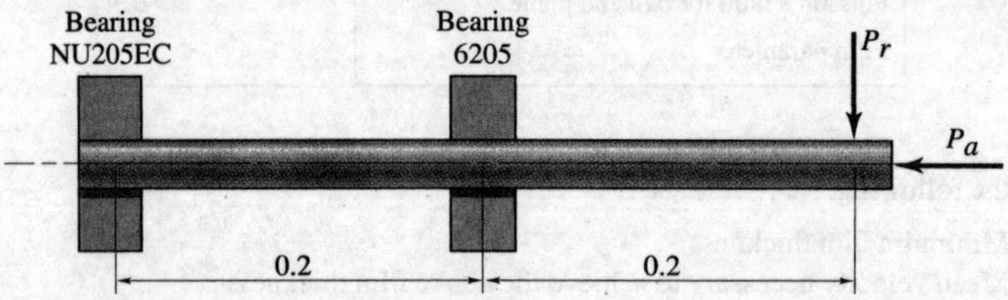

Sketch c, for Problem 13.15

13.16 Given the shaft and bearings in Problem 13.15 but with the bearing positions shifted so the 6205 deep-groove ball bearing is in the left end of the shaft and bearing NU205EC in the middle, find the $\tilde{L}_{10}$ lives for the two bearings in their new positions.

13.17 A bearing arrangement for a shaft in a gearbox is shown in sketch *d*. Bearing *A* is a cylindrical roller bearing (NU202EC) with $d = 15$ mm, $D = 35$ mm, $B = 11$ mm, and $C = 12\ 500$ N. Bearing *B* is a deep-groove ball bearing (6309) with $d = 45$ mm, $D = 100$ mm, $B = 25$ mm, $C = 52\ 700$ N, and $C_0 = 31\ 500$ N. Find which bearing is likely to fail first and what the $\tilde{L}_{10}$ is for that bearing. Length dimensions are in millimeters.

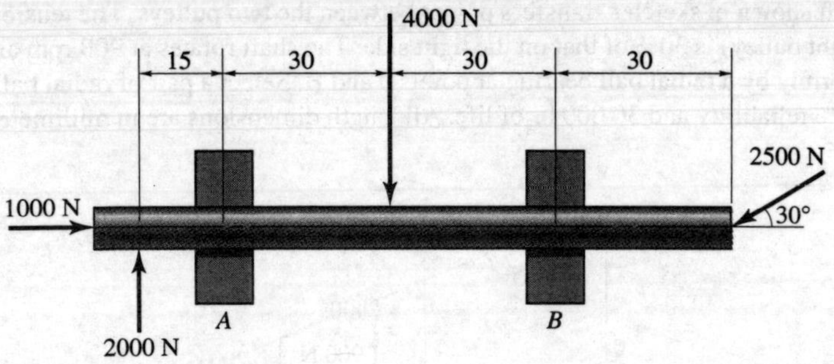

Sketch *d*, for Problem 13.17

13.18 The two bearings in Problem 13.17 have the same composite surface roughness of the inner-race, rolling-element contact, $(R_{qa}^2 + R_{qb}^2)^{0.5} = 0.10$ μm. For bearing NU202EC the outer radius of the inner race is 9.65 mm, the roller diameter is 5.7 mm, and inner radius of the outer race is 15.35 mm. For bearing 6309 the ball diameter is 17 mm, the outer radius of the inner race is 27.4 mm, and the inner radius of the outer race is 44.4 mm. The lubricant used has the absolute viscosity $\eta_0 = 0.1$ N-s/m^2 and the pressure-viscosity coefficient $\xi = 2 \times 10^{-8}$ m^2/N. The bearing rotational speed is 1500 rpm. The cylindrical roller bearing has 13 rollers and the ball bearing has 10 balls. The ball bearing is made of AISI 440C ($E = 200$ GPa), and the roller bearing is made of AISI 52100 ($E = 207$ GPa). Both materials have $\nu = 0.3$. Find the adjusted bearing lives $\tilde{L}_{10,\text{adj}}$ for the two bearings when the ellipticity ratios are 10 and ∞, respectively.

13.19 Given the shaft arrangement in Fig. 11.3 with the applied forces indicated, bearing B is a cylindrical roller bearing of type NU304EC with $C = 30\,800$ N, and bearing A is a deep-groove ball bearing of type 6404 ($C = 30\,700$ N, $C_0 = 15\,000$ N). Find the L_{10} life for each bearing and judge if they are strong enough to be used at 1500 rpm for 20 years.

13.20 Given the shaft arrangement in Problem 13.19 but with the belt drive at D changed to a worm gear giving the same radial force but also an additional axial force of 800 N directed from D to A, find the $\tilde{L}_{10}$ lives of the bearings and judge if they are strong enough to last 20 years at 1500 rpm.

13.21 A flywheel has deep-groove ball bearings located close to the rotating disk. The disk mass is 50 kg, and it is mounted out of balance, so that the center of mass is situated at a radius 1 mm from the bearing centerline. The shaft mass is neglected. The bearings are two 6305 deep-groove ball bearings with $C = 22\,500$ N. Find the nominal bearing $\tilde{L}_{10}$ in hours for each bearing at

a) 500 rpm
b) 2500 rpm
c) 12 500 rpm

13.22 A ball bearing manufacturer wants to make sure that its bearings get long service lives on different components. Extra polishing operation can be applied to

a) the ring surfaces
b) the ball surfaces
c) all contact surfaces

The polishing decreases the roughness by one-half the original roughness, which for the races is 0.08 μm and for the balls is 0.04 μm. Find how much the Λ value changes when either *a*, *b*, or *c* is polished.

13.23 The shaft shown in sketch *e* transfers power between the two pulleys. The tension on the slack side (right pulley) is 30% of that on the tight side. The shaft rotates at 900 rpm and is supported uniformly by a radial ball bearing at points 0 and B. Select a pair of radial ball bearings with 99% reliability and 30 000 hr of life. All length dimensions are in millimeters.

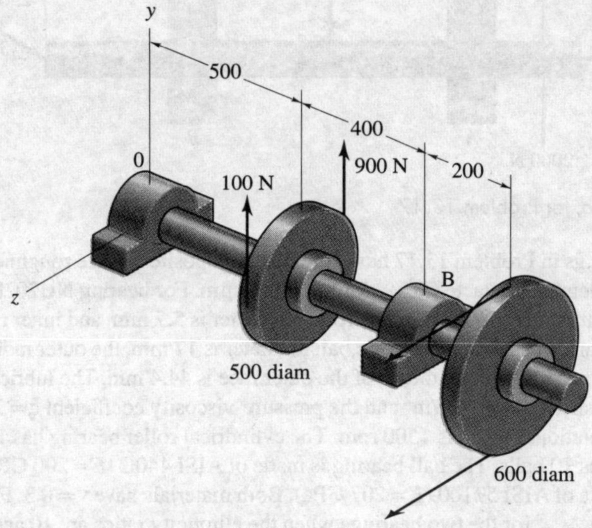

Sketch *e*, for Problem 13.23

13.24 In sketch *f* the tension on the slack side of the left pulley is 20% of that on the tight side. The shaft rotates at 720 rpm. Select a pair of roller bearings to support the shaft for 99% reliability and a life of 24 000 hr. All length dimensions are in millimeters.

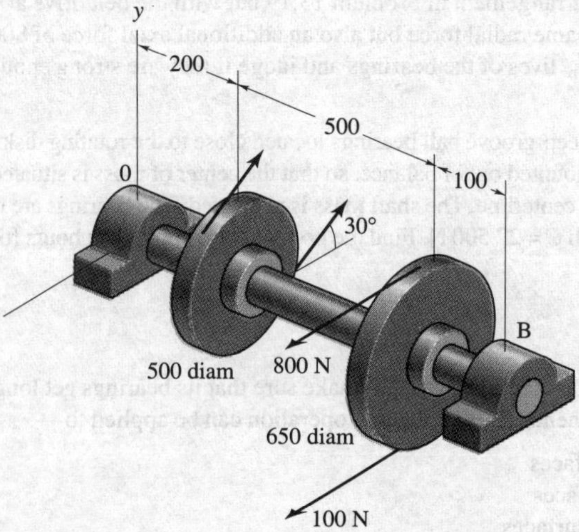

Sketch *f*, for Problem 13.24

13.25 In sketch *g* the shaft rotates at 1100 rpm and transfers power with light shock from a 600-mm-diameter pulley to a 300-mm-diameter sprocket. Select a radial ball bearing for support at A and a roller bearing for support at 0. The bearings should have a life of 24 000 hr at 95% reliability. All length dimensions are in millimeters.

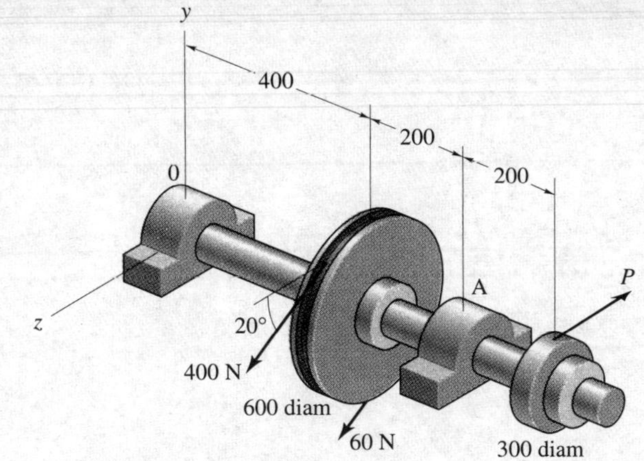

Sketch *g*, for Problem 13.25

13.26 The power-transmitting system shown in sketch *h* consists of a helical gear, a bevel gear, and a shaft that rotates at 600 rpm and is supported by two roller bearings. The load on the bevel gear is $P = -0.5P\bar{i} - 0.41P\bar{j} + 0.44P\bar{k}$. The left bearing supports the axial load on the shaft. Select bearings for 36 000-hr life at 98% reliability. All length dimensions are in millimeters.

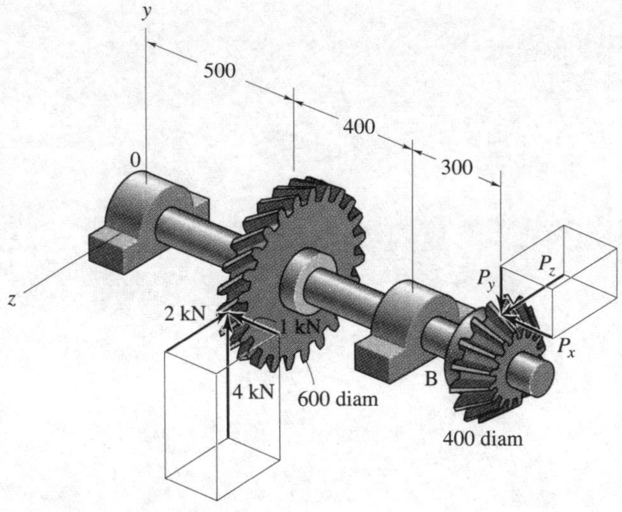

Sketch *h*, for Problem 13.26

13.27 A cylindrical roller bearing is used to support an 8-kN load for 1200 hr at 600 rpm. Calculate the load-carrying capacity of this bearing for 95% reliability. What size bearing would you recommend?

13.28 A deep-groove ball bearing is used to support a 6-kN load for 1000 hr at 1200 rpm. For 99% reliability what should be the load-carrying capacity of the bearing? What size bearing should be chosen?

Sketch e for Problem 13-25

13-26 The power-transmitting system shown in sketch f consists of a helical gear with a bevel [illegible] and a shaft that rotates at 600 rpm and is supported by two [illegible] bearings. The load on the [illegible] K_r = 0.5 [illegible] 0.4 [illegible]. The left bearing supports the axial load on the shaft. Select bearings for 36 000 h with a 95 percent reliability. All length dimensions are in millimeters.

Sketch f for Problem 13-26

13-27 Ninety percent of the bearings in a group [illegible] load for [illegible] 000 [illegible]. Calculate the load-carrying capacity of this bearing for [illegible] h of life. What size bearing [illegible] this application?

13-28 A deep-groove ball bearing is used to support a 5-kN load for 10 000 h at 1200 rpm. For 95 percent reliability, what should be the catalog rating capacity of the bearing? What size bearing should be chosen?

CHAPTER

14

GEARS

An assortment of gears (Courtesy of Quality Transmission Components).

Just stare at the machine. There is nothing wrong with that. Just live with it for a while. Watch it the way you watch a line when fishing and before long, as sure as you live, you'll get a little nibble, a little fact asking in a timid, humble way if you're interested in it. That's the way the world keeps on happening. Be interested in it.

Robert Pirsig, Zen and the Art of Motorcycle Maintenance

SYMBOLS

a	addendum, m
b	dedendum, m
b^*	Hertzian half-width, m
b_w	face width, m
b_l	backlash, m
C_r	contact ratio
c	clearance, m
c_d	center distance, m
d	pitch diameter, m
d_b	base diameter, m
d_{bg}	base diameter of gear, m
d_{bp}	base diameter of pinion, m
d_g	pitch diameter of gear, m
d_{og}	outside diameter of gear, m
d_{op}	outside diameter of pinion, m
d_p	pitch diameter of pinion, m
d_r	root or fillet diameter, m
E	modulus of elasticity, Pa
E'	effective modulus of elasticity, Pa
g_r	gear ratio
HB	Brinell hardness
h	height, m
h_k	working depth, m
h_p	transmitted horsepower, hp
h_t	total depth, m
I	area moment of inertia, m^4
K_a	application factor
K_g	geometry factor for contact stress evaluation
K_m	load distribution factor
K_s	size factor
K_v	dynamic factor
L_{ab}	length of line of action, m
$L_{a^*b^*}$	length of line from a^* to b^*, m
m	module, d/N, mm
N	number of teeth
N_a	gear rotational speed, rpm
N_g	number of teeth in gear
N_p	number of teeth in pinion
N_v	equivalent number of teeth
n_s	safety factor
P	power, W
P_z	normal load, N
P_z'	normal load per face width, N/m
p_a	axial pitch for helical gears, m
p_b	base pitch, m
p_c	circular pitch, $\pi d/N$, m
p_{cn}	normal circular pitch, m
p_d	diametral pitch, N/d, in.$^{-1}$
p_{dn}	normal diametral pitch, in.$^{-1}$
p_H	maximum Hertzian contact pressure, Pa
p_p	pitch point
Q_v	transmission accuracy level number
R_q	rms surface finish of gear, m
R_x	curvature sum in x direction, m
r	pitch radius, m
r_b	base radius, m
r_{bg}	base radius of gear, m
r_{bp}	base radius of pinion, m
r_g	pitch radius of gear, m
r_o	outside radius, m
r_{og}	outside radius of gear, m
r_{op}	outside radius of pinion, m
r_p	pitch radius of pinion, m
T	torque, N-m
t	tooth thickness, m
t_h	circular tooth thickness measured on pitch circle, m
t_{ha}	actual circular tooth thickness measured on pitch circle, m
$\tilde{u}$	mean velocity, m/s
v_t	pitch-line velocity, ft/min
W	load, N
W_r	radial load, N
W_t	tangential load, N
w'	load per unit width, N/m
x	unknown distance, m
Y	Lewis form factor
Y_j	geometry factor
Z	angular velocity ratio
α	coefficient of thermal expansion, $(°C)^{-1}$
α_{ag}	angle of approach of gear, deg
α_{ap}	angle of approach of pinion, deg
α_{rg}	angle of recess of gear, deg
α_{rp}	angle of recess of pinion, deg
η_0	atmospheric viscosity, Pa-s
Λ	film parameter
ξ	pressure-viscosity coefficient, m^2/N
σ	stress, Pa
σ_d	design stress, Pa
σ_g	bending stress of gear, Pa
σ_{max}	maximum normal stress, Pa
σ_r	contact stress of gear, Pa
ϕ	pressure angle, deg
ϕ_n	pressure angle in normal direction for helical gears, deg
ψ	helix angle, deg
Ω	angular velocity of shaft, rad/s
ω	angular speed of gear, rad/s

Subscripts

a	actual; axial
all	allowable
b	bending
c	contact
g	gear
max	maximum
o	outside
p	pinion
t	total

14.1 INTRODUCTION

A **gear** may be thought of as a toothed wheel that when meshed with another smaller-in-diameter toothed wheel (the **pinion**) will transmit rotation from one shaft to another. The primary function of a gear is to transfer power from one shaft to another while maintaining a definite ratio between the velocities of the shaft rotations. The teeth of a driving gear mesh push on the driven gear teeth, exerting a force component perpendicular to the gear radius. Thus, a torque is transmitted, and because the gear is rotating, power is transferred. Gears are the most rugged and durable torque transmitters listed in Chapter 1 (Sec. 1.2). Their power transmission efficiency is as high as 98%. On the other hand, gears are usually more costly than other torque transmitters, such as chain drives and belt drives. Gears are highly standardized as to tooth shape and size. The American Gear Manufacturers Association (AGMA) publishes standards for gear design (AGMA, 1990), manufacture (AGMA, 1988), and assembly (AGMA, 1989). This chapter attempts to follow the methods and recommendations defined by the AGMA. English units are the primary units and SI units are secondary, opposite from what is done in the rest of the text.

14.2 TYPES OF GEAR

Gears can be divided into three major classes: parallel-axis gears, nonparallel but coplanar gears, and nonparallel and noncoplanar gears. This section describes gears in each of these classes.

14.2.1 PARALLEL-AXIS GEARS

Parallel-axis gears are the simplest and most popular type of gear. They connect parallel shafts and can transfer large amounts of power with high efficiency. **Spur** and **helical gears** are two of the primary gears in this classification.

14.2.1.1 SPUR GEARS Figure 14.1 shows a spur gear drive with teeth on the outside of a cylinder and parallel to the cylinder axis. Spur gears are the simplest and the most common type of gear.

14.2.1.2 HELICAL GEARS Figure 14.2 shows a helical gear drive, with gear teeth cut on a spiral that wraps around a cylinder. Helical teeth enter the meshing zone progressively and, therefore, have a smoother action than spur gear teeth. Helical gears also tend to be quieter. Another positive feature of helical gears (relative to spur gears) is that the

Animation

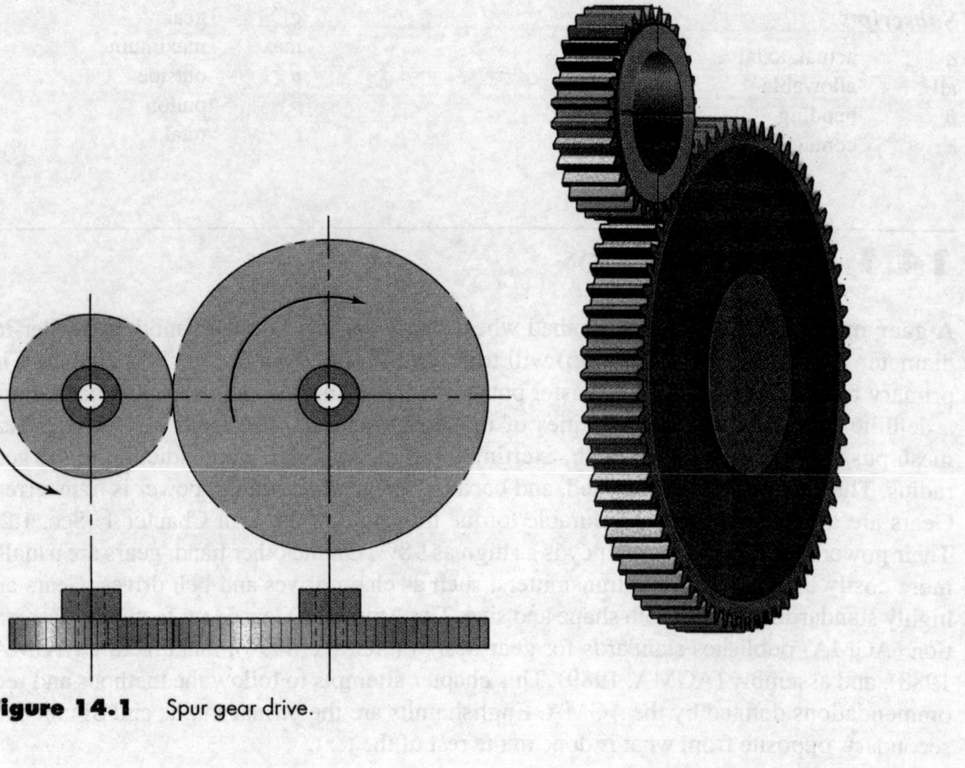

Figure 14.1 Spur gear drive.

Animation

Figure 14.2 Helical gear drive.

transmitted load is somewhat larger, thus implying that helical gear life will be longer for the same load. A smaller helical gear can transmit the same load as a larger spur gear.

A disadvantage of helical gears (relative to spur gears) is that they produce an additional end thrust along the shaft axis that is not present with spur gears. This end thrust may require an additional component, such as a thrust collar, ball bearings, or tapered-roller bearings. Another disadvantage is that helical gears have slightly lower efficiency than equivalent spur gears. Efficiency depends on total *normal* tooth load, which is higher for spur gears. Although the total load-carrying capacity is larger for helical gears, the load is distributed normally and axially, whereas for a spur gear all load is distributed normally.

14.2.2 NONPARALLEL, COPLANAR GEARS

This section discusses gears that have nonparallel axes (unlike spur and helical gears, which have parallel axes) and that are coplanar (like spur and helical gears). **Bevel,** straight, Zerol, and **spiral gears** are all in the nonparallel, coplanar class. The common feature of this class is the redirection of power around a corner, as might be required, for example, when connecting a horizontally mounted engine to the vertically mounted rotor shaft on a helicopter. Figure 14.3 shows a bevel gear drive with straight teeth and illustrates the gears in this classification. Observe that the axes of the gear drive are nonparallel but coplanar.

14.2.3 NONPARALLEL, NONCOPLANAR GEARS

Nonparallel, noncoplanar gears are more complex in both geometry and manufacturing than the two previous gear classifications. As a result, these gears are more expensive than the other gears discussed.

Figure 14.4, a **worm gear** drive with cylindrical teeth, illustrates this class of gear. Note that the axes are nonparallel and noncoplanar. These gears can provide considerably higher reduction ratios than coplanar or simple crossed-axis gear sets, but their load-carry-

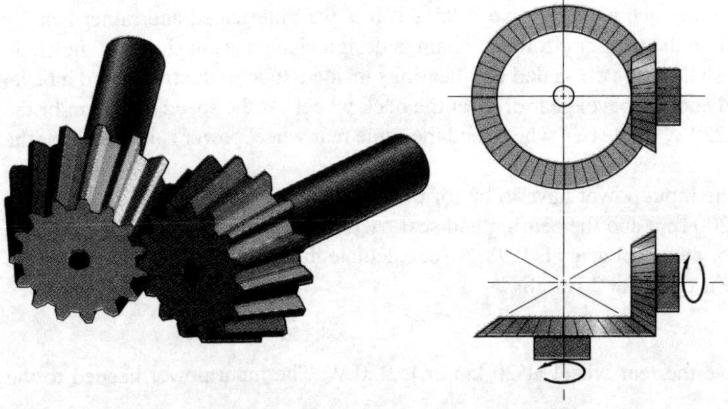

Figure 14.3 Bevel gear drive with straight teeth.

Animation

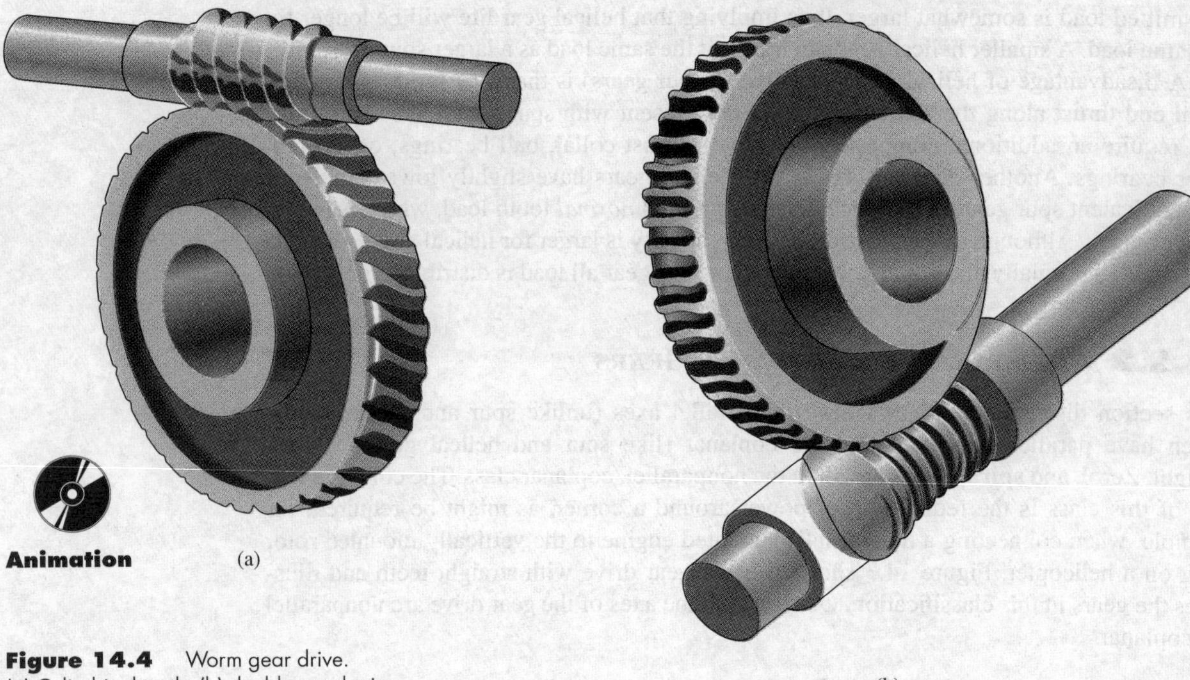

Animation

(a) (b)

Figure 14.4 Worm gear drive.
(a) Cylindrical teeth; (b) double enveloping.

ing capacity is low, their contact pressure is extremely high, and their wear rate is high. Thus, worm gear drives are used only for light-load applications.

This brief discussion explains the various gear classes and is not meant to be all-inclusive. More detailed and inclusive discussion can be found in Drago (1992). The rest of this chapter focuses primarily on spur gears with a section on helical gears at the end.

Example 14.1

GIVEN A bicycle is normally driven by the pedals through a roller chain transmission to the back wheel. The power efficiency of such a chain drive is 95% if it is well lubricated and rather heavily loaded. To avoid getting oil on the trousers from the chain, a design change is considered. The chain drive is to be changed to a shaft with two sealed ball bearings inside a tube in the frame and a bevel gear drive at the pedals and another bevel gear drive at the back wheel. At the speed of 20 km/hr the chain-driven bike requires 220 W to the real wheel, and the same rear wheel power is required for the new design.

FIND How large does the input power have to be for the new drive to run 20 km/hr if the newly designed shaft rotates at 1200 rpm and the bearing and seal friction torque in each bearing is 0.1 N-m? The power efficiency for the bevel gears is 0.98. Also, calculate the total power efficiency for the new drive when the bicycle is driven at 20 km/hr.

Solution

The power required to drive the rear wheel at 20 km/hr is 220 W. The input power needed to the bevel gear is

$$\frac{P}{0.98} = \frac{220}{0.98} = 224.5\,\text{W}$$

The power needed to overcome the friction in the bearings and seals is

$$T\omega = 2(0.1)\left(\frac{1200}{60}\right)(2\pi) = 25.13\ \text{W}$$

The output power from the bevel gear is 224.5 + 25.13 = 249.6 W, and the input power needed from the pedals is

$$\frac{249.6}{0.98} = 254.7\,\text{W}$$

The power efficiency is

$$\frac{P_{\text{out}}}{P_{\text{in}}} = \frac{220}{254.7} = 0.864 \text{ or } 86.4\%$$

This efficiency is lower than that for the chain drive.

14.3 GEAR GEOMETRY

Figure 14.5 shows the basic spur gear geometry, and Fig. 14.6 shows gear teeth with the nomenclature to be used. The geometry of gear teeth enables a normal to the tooth profiles at their point of contact to pass through a fixed point on the line of centers called the **pitch point.** From Figs. 14.5 and 14.6 the addendum a is the distance from the top land to the pitch circle, and the dedendum b is the radial distance from the bottom land to the pitch circle. The clearance shown in Fig. 14.6 is the amount by which the dedendum exceeds the addendum. A clearance is needed to prevent the end of one gear tooth from riding the bottom of the mating gear.

14.3.1 CENTER DISTANCE, CIRCULAR PITCH, AND DIAMETRAL PITCH

From Fig. 14.5 the pitch circles of the gear and pinion make contact at the pitch point. The center distance between the axes of rotation of the rotating gears can be expressed as

$$c_d = \left(d_p + d_g\right)/2 \qquad \textbf{(14.1)}$$

where

d_p = pitch diameter of pinion, m

d_g = pitch diameter of gear, m

The **circular pitch** p_c is the distance measured on the pitch circle from one point on one tooth to a corresponding point on the adjacent tooth. Mathematically, it is

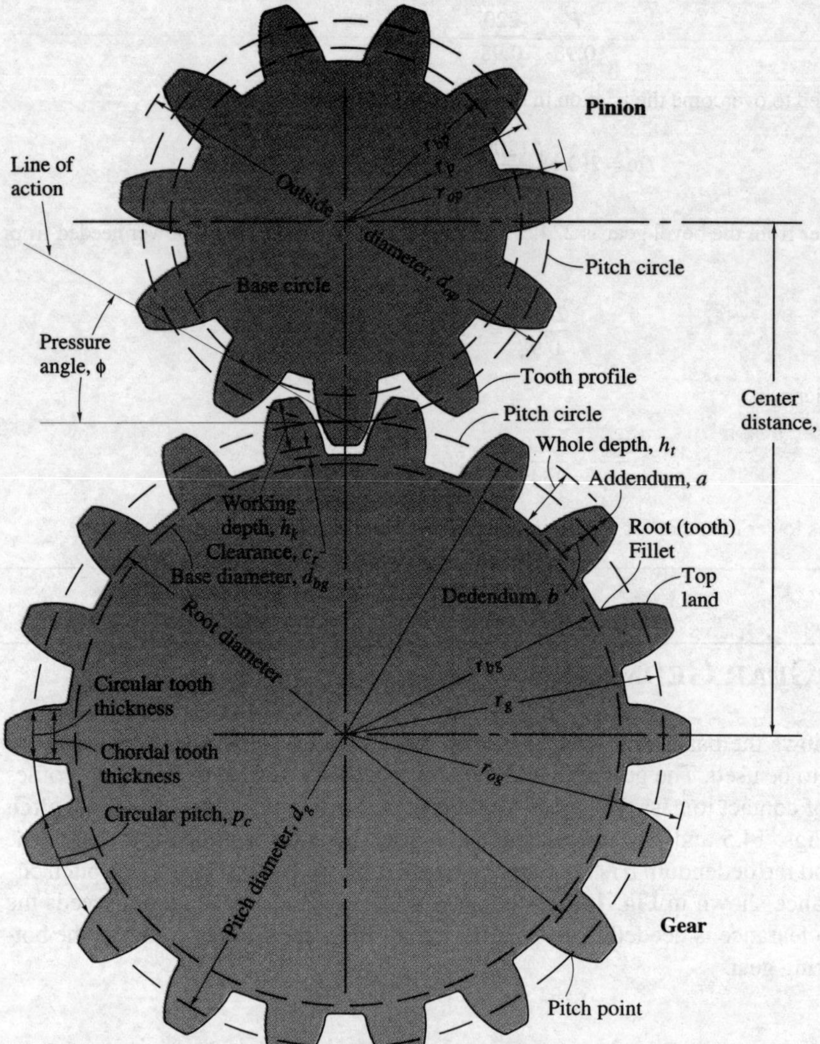

Figure 14.5 Basic spur gear geometry.

$$p_c = \text{Total width of gear} + \text{Width of spacing} = \frac{\pi d}{N} = \frac{\pi d_p}{N_p} = \frac{\pi d_g}{N_g} \tag{14.2}$$

where

N_p = number of pinion teeth

N_g = number of gear teeth

From Eq. (14.2) the **gear ratio** is

$$g_r = \frac{d_g}{d_p} = \left(\frac{p_c N_g}{\pi}\right)\left(\frac{\pi}{p_c N_p}\right) = \frac{N_g}{N_p} \tag{14.3}$$

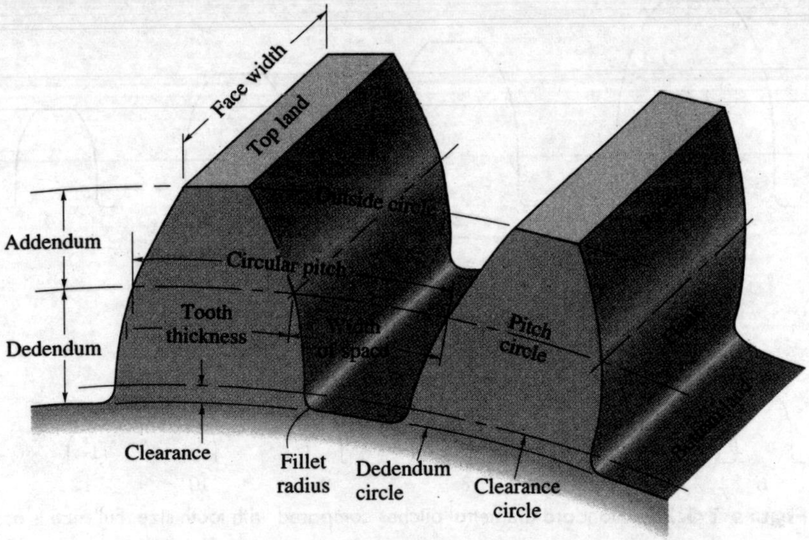

Figure 14.6 Nomenclature of gear teeth.

Substituting Eqs. (14.2) and (14.3) into Eq. (14.1) gives

$$c_d = \frac{p_c}{2\pi}(N_p + N_g) = \frac{p_c N_p}{2\pi}\left(1 + \frac{N_g}{N_p}\right) = \frac{p_c N_p}{2\pi}(1 + g_r) \quad \textbf{(14.4)}$$

The **diametral pitch** p_d is the number of teeth in the gear per pitch diameter. In the English system the unit of diametral pitch is reciprocal inches. The expression for diametral pitch is

$$p_d = \frac{N_p}{d_p} = \frac{N_g}{d_g} \quad \textbf{(14.5)}$$

Figure 14.7 shows standard diametral pitches and compares them with tooth size for full-size gears. The smaller the tooth, the larger the diametral pitch. Table 14.1 shows

Table 14.1 Preferred diametral pitches for four tooth classes.

Class	Diametral pitch, p_d, in.$^{-1}$
Coarse	1/2, 1, 2, 4, 6, 8, 10
Medium coarse	12, 14, 16, 18
Fine	20, 24, 32, 48, 64 72, 80, 96, 120, 128
Ultrafine	150, 180, 200

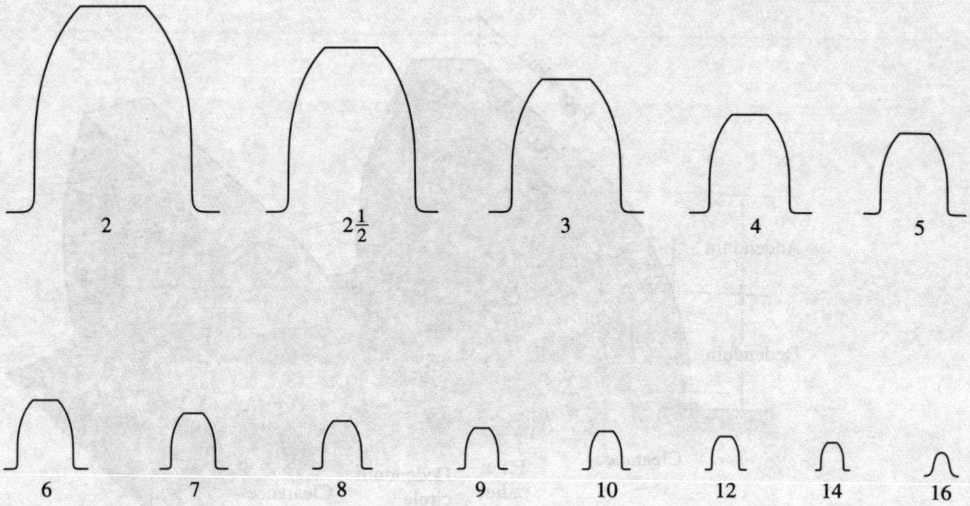

Figure 14.7 Standard diametral pitches compared with tooth size. Full size is assumed.

preferred diametral pitches for four tooth classes. Coarse and medium-coarse gears are used for power transmission applications; fine and ultrafine gears are used for instrument and control systems.

Comparing Eqs. (14.2) and (14.5) yields the following:

$$p_c = \pi / p_d \tag{14.6}$$

Thus, the circular pitch is inversely proportional to the diametral pitch.

The **module** m is the ratio of the pitch diameter to the number of teeth and is only expressed in the SI unit of millimeters.

$$m = \frac{d_g}{N_g} = \frac{d_p}{N_p} \tag{14.7}$$

From Eqs. (14.2) and (14.7)

$$p_c = \pi m \tag{14.8}$$

Also, from Eqs. (14.6) and 14.8)

$$m = 1/p_d \tag{14.9}$$

Example 14.2 **GIVEN** A 20-tooth pinion with diametral pitch of 5 per inch meshes with a 63-tooth gear.

FIND The center distance, the circular pitch, and the gear ratio.

Solution

From Eq. (14.5)

$$d_p = \frac{N_p}{p_d} = \frac{20}{5} = 4 \text{ in.}$$

$$d_g = \frac{N_g}{p_d} = \frac{63}{5} = 12.6 \text{ in.}$$

From Eq. (14.1) the center distance is

$$c_d = \left(d_p + d_g\right)/2 = (4 + 12.6)/2 = 8.3 \text{ in.}$$

From Eq. (14.2) the circular pitch is

$$p_c = \frac{\pi d_p}{N_p} = \frac{\pi 4}{20} = 0.6283 \text{ in.}$$

$$p_c = \frac{\pi d_g}{N_g} = \frac{\pi(12.6)}{63} = 0.6283 \text{ in.}$$

From Eq. (14.3) the gear ratio is

$$g_r = \frac{d_g}{d_p} = \frac{N_g}{N_p}$$

$$\therefore g_r = \frac{12.6}{4} = \frac{63}{20} = 3.150$$

14.3.2 ADDENDUM, DEDENDUM, AND CLEARANCE

Table 14.2 shows values of addendum, dedendum, and clearance for a spur gear with a 20° pressure angle and a full-depth involute. Both coarse-pitch and fine-pitch formulations are given. Note from Table 14.2 that the following is true:

$$c = b - a \qquad \textbf{(14.10)}$$

Once the addendum, dedendum, and clearance are known, a number of other parameters can be obtained. Some of these are

Outside diameter: $$d_o = d + 2a \qquad \textbf{(14.11)}$$

Root diameter: $$d_r = d - 2b \qquad \textbf{(14.12)}$$

Total depth: $$h_t = a + b \qquad \textbf{(14.13)}$$

Working depth: $$h_k = a + a = 2a \qquad \textbf{(14.14)}$$

Table 14.2 Formulas for addendum, dedendum, and clearance (pressure angle, 20°; full-depth involute).

Parameter	Symbol	Coarse pitch ($p_d < 20$ in.$^{-1}$)	Fine pitch ($p_d \geq 20$ in.$^{-1}$)	Metric module system
Addendum	a	$1/p_d$	$1/p_d$	1.00 m
Dedendum	b	$1.25/p_d$	$1.200/p_d + 0.002$	1.25 m
Clearance	c	$0.25/p_d$	$0.200/p_d + 0.002$	0.25 m

These parameters can also be expressed in terms of diametral pitch and number of teeth rather than pitch diameter by using Eq. (14.5). For example, the outside diameter given in Eq. (14.11) can be expressed, by making use of the formulas given in Table 14.2 and Eq. (14.5), as

$$d_o = \frac{N}{p_d} + 2\left(\frac{1}{p_d}\right) = \frac{N+2}{p_d} \tag{14.15}$$

All formulas given in this section are equally applicable for the gear and the pinion.

14.3.3 LINE OF ACTION, PRESSURE ANGLE, AND GEAR INVOLUTE

The purpose of meshing gear teeth is to provide constant, instantaneous relative motion between the engaging gears. To achieve this tooth action, the common normal of the curves of the two meshing gear teeth must pass through a common point, called the pitch point, on the pitch circles.

Figure 14.8 shows the pitch and base circles for both the pinion and the gear. Also shown in this figure is the line of action, the straight line from *a* to *b*, and the pressure angle ϕ. The pressure angle is the angle between the line of centers and a line perpendicular to

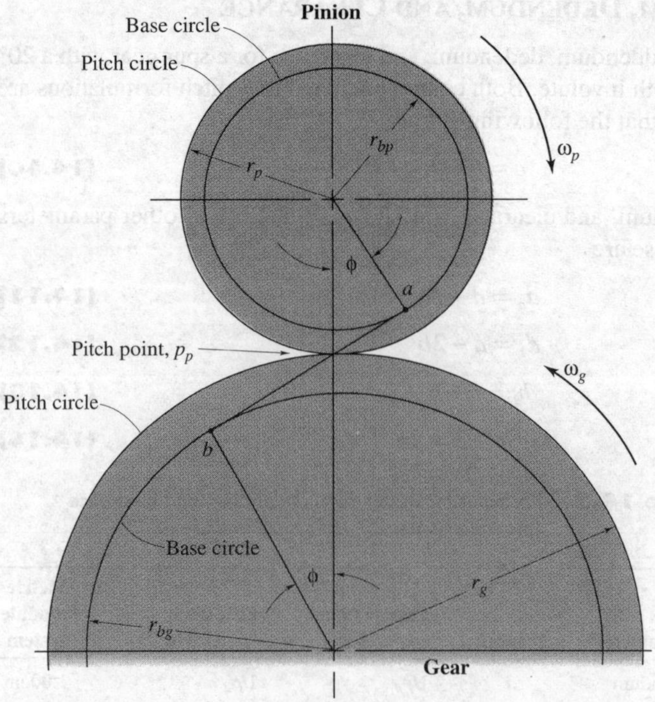

Figure 14.8 Pitch and base circles for pinion and gear as well as line of action and pressure angle.

the line of action and going through the center of the gear or pinion. From Fig. 14.8 observe that the radius of the base circle is

$$r_b = r\cos\phi$$

$$r_{bp} = r_p \cos\phi \quad \text{and} \quad r_{bg} = r_g \cos\phi \tag{14.16}$$

The base pitch p_b is the distance from a point on one tooth to the corresponding point on the adjacent tooth measured around the base circle. From Eq. (14.2) the base pitch is

$$p_b = p_c \cos\phi = \frac{\pi d}{N}\cos\phi = \frac{\pi d_p}{N_p}\cos\phi = \frac{\pi d_g}{N_g}\cos\phi \tag{14.17}$$

The shape of the gear tooth is obtained from the involute curve shown in Fig. 14.9. The following procedure is used for constructing the involute curve:

1. Divide the base circle into a number of equal distances, thus constructing A_0, A_1, A_2,...
2. Beginning at A_1, construct the straight line A_1B_1, perpendicular with $0A_1$, and likewise beginning at A_2 and A_3.
3. Along A_1B_1, lay off the distance A_1A_0, thus establishing C_1. Along A_2B_2, lay off twice A_1A_0, thus establishing C_2, etc.
4. Establish the involute curve by using points A_0, C_1, C_2, C_3,... Gears made from the involute curve have at least one pair of teeth in contact with each other.

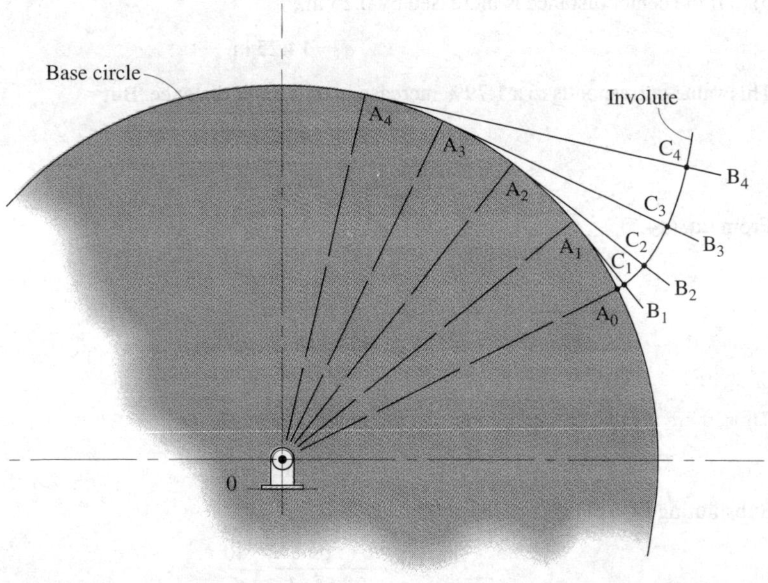

Figure 14.9 Construction of involute curve.

Example 14.3

GIVEN A gear with 40 teeth, a pinion with 16 teeth, a diametral pitch of 2 per inch, and a pressure angle of 20°.

FIND

a) Determine the circular pitch, the center distance, and the base diameter for the pinion and the gear.

b) If the center distance is increased by 0.25 in., find the pitch diameters for the gear and pinion and the pressure angle.

Solution

a) From Eq. (14.6) the circular pitch is $p_c = \pi/p_d = \pi/2$ in. = 1.571 in. From Eq. (14.5) the pitch diameters of the gear and pinion are

$$d_g = \frac{N_g}{p_d} = \frac{40}{2} = 20 \text{ in.}$$

$$d_p = \frac{N_p}{p_d} = \frac{16}{2} = 8 \text{ in.}$$

From Eq. (14.1) the center distance is

$$c_d = \left(d_p + d_g\right)/2 = (8+20)/2 = 14 \text{ in.}$$

From Eq. (14.16) the radii of the base circles for the pinion and the gear are

$$r_{bp} = \frac{d_p}{2}\cos\phi = \frac{8}{2}\cos 20° = 3.759 \text{ in.}$$

$$r_{bg} = \frac{d_g}{2}\cos\phi = \frac{20}{2}\cos 20° = 9.397 \text{ in.}$$

b) If the center distance is increased by 0.25 in.,

$$c_d = 14.25 \text{ in.}$$

This value corresponds to a 1.79% increase in the center distance. But

$$c_d = \left(d_p + d_g\right)/2$$

$$\therefore d_p + d_g = 2(14.25 \text{ in.}) = 28.5 \text{ in.} \qquad \textbf{(a)}$$

From Eq. (14.5)

$$\frac{N_p}{d_p} = \frac{N_g}{d_g}$$

$$\therefore \frac{16}{d_p} = \frac{40}{d_g} \qquad \textbf{(b)}$$

Thus, there are two equations and two unknowns. From Eq. (a)

$$d_p = 28.5 - d_g \qquad \textbf{(c)}$$

Substituting Eq. (c) into Eq. (b) gives

$$\frac{16}{28.5 - d_g} = \frac{40}{d_g}$$

$$16 d_g = 40(28.5) - 40 d_g$$

$$56d_g = 1140$$

$$d_g = 20.36 \text{ in.} \quad \textbf{(d)}$$

Substituting Eq. (d) into Eq. (a) gives

$$d_p = 28.50 - 20.36 = 8.143 \text{ in.} \quad \textbf{(e)}$$

From Eq. (14.5) the diametral pitch is

$$p_d = \frac{N_p}{d_p} = \frac{16}{8.143} = 1.965/\text{in.}$$

Also,

$$p_d = \frac{N_g}{d_g} = \frac{40}{20.36} = 1.965/\text{in.}$$

From Eq. (14.6) the circular pitch is

$$p_c = \frac{\pi}{p_d} = \frac{\pi}{1.965} = 1.599 \text{ in.}$$

Changing the center distance does not affect the base circle of the pinion or the gear. Thus, from Eq. (14.16)

$$\phi = \cos^{-1}\left(\frac{r_{bp}}{r_p}\right) = \cos^{-1}\left(\frac{3.759}{\frac{8.143}{2}}\right) = \cos^{-1}(0.9232) = 22.60°$$

Also,

$$\phi = \cos^{-1}\left(\frac{r_{bp}}{r_p}\right) = \cos^{-1}\left(\frac{9.397}{\frac{20.36}{2}}\right) = \cos^{-1}(0.9232) = 22.60°$$

Thus, a 1.79% increase in center distance results in a 2.60° increase in pressure angle.

14.4 KINEMATICS

For meshing gears the rotational motion and the power must be transmitted from the driving gear to the driven gear with a smooth and uniform positive motion and with minor loss of power due to friction. The **fundamental law of gearing** states that the common normal to the tooth profile at the point of contact must always pass through a fixed point, called the pitch point, in order to maintain a constant velocity ratio of the two meshing gear teeth. The pitch point p_p is shown in Fig. 14.8. Since the velocity at this point must be the same for both the gear and the pinion,

$$r_p \omega_p = r_g \omega_g$$

Expressing this equation in diameter rather than in radius and making use of Eq. (14.3) give the gear ratio as

$$g_r = \frac{d_g}{d_p} = \frac{\omega_p}{\omega_g} = \frac{N_g}{N_p} \tag{14.18}$$

This equation assumes no slip between the meshing pinion and the gear.

Example 14.4

GIVEN When two involute gears work properly, the contact point between the gear flanks moves along the straight line of action. Let $N_p = 20$, $N_g = 38$, and $\phi = 20°$. The diametral pitch is 8 per inch and the pinion speed is 3600 rpm.

FIND The gear speed.

Solution

From Eq. (14.5) the diametral pitch is

$$p_d = \frac{N_p}{d_p} \qquad \text{or} \qquad d_p = \frac{N_p}{p_d} = \frac{20}{8} = 2.5 \text{ in.}$$

The speed at the contact point along the straight line of action is

$$\omega_p r_{bp} = \omega_p \left(\frac{d_p}{2}\right)\cos 20° = \left(\frac{3600}{60}\right) 2\pi \left(\frac{2.5}{2}\right)\cos 20°$$

$$= 442.8 \text{ in./s} = 11.25 \text{ m/s}$$

14.5 CONTACT RATIO

To ensure smooth, continuous tooth action, as one pair of teeth ceases contact, a succeeding pair of teeth must already have come into contact. As much overlap as possible is desired. Figure 14.10 illustrates the important parameters in defining the contact ratio, a measure of this overlapping action. The contact ratio is a ratio of the length of the line of action to the base pitch.

Figure 14.11 shows more details of the length of the line of action shown in Fig 14.10. From this figure

$$r_{og}^2 = r_{bg}^2 + L_{ab^*}^2 = r_{bg}^2 + \left(L_{ac} + L_{cb^*}\right)^2$$

But

$$\sin\phi = L_{cb^*}/r_g \qquad \text{or} \qquad L_{cb^*} = r_g \sin\phi$$

$$\therefore r_{og}^2 = r_{bg}^2 + \left(L_{ac} + r_g \sin\phi\right)^2$$

$$L_{ac} = \sqrt{r_{og}^2 - r_{bg}^2} - r_g \sin\phi$$

Also from Fig 14.11

$$r_{op}^2 = r_{bp}^2 + L_{a^*b}^2 = r_{bp}^2 + \left(L_{a^*c} + L_{cb}\right)^2$$

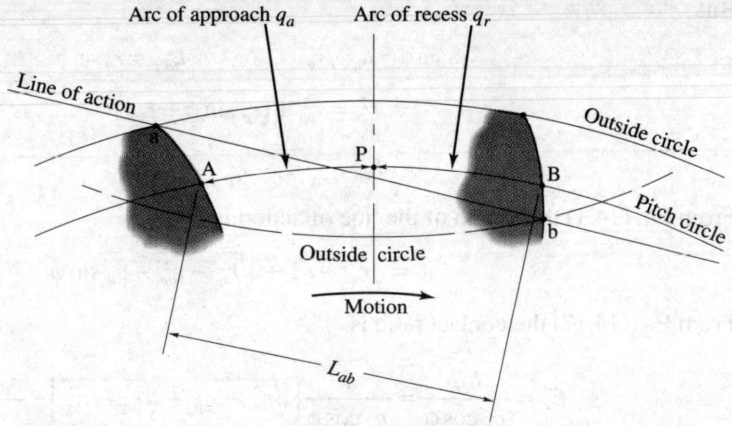

Figure 14.10 Illustration of parameters important in defining contact ratio.

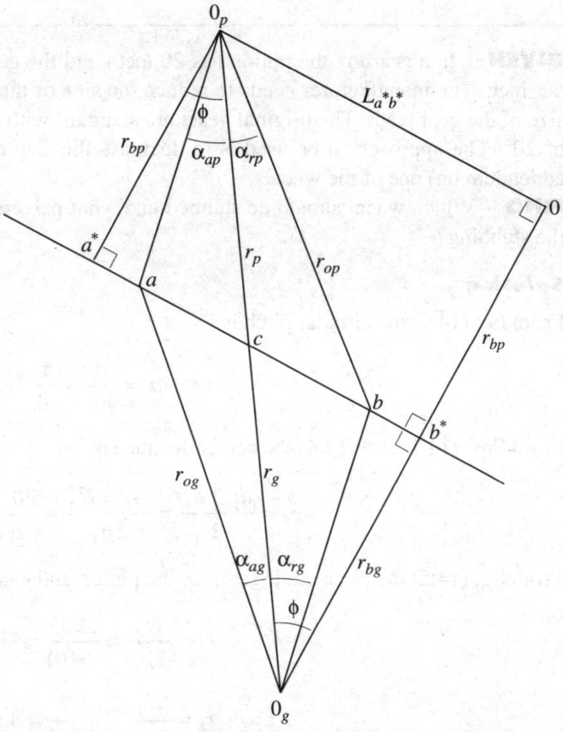

Figure 14.11 Details of line of action, showing angles of approach and recess for both pinion and gear.

But

$$\sin\phi = L_{a^*c}/r_p \qquad \text{or} \qquad L_{a^*c} = r_p \sin\phi$$

$$\therefore r_{op}^2 = r_{bp}^2 + \left(r_p \sin\phi + L_{cb}\right)^2$$

$$L_{cb} = \sqrt{r_{op}^2 - r_{bp}^2} - r_p \sin\phi$$

From Eq. (14.1) the length of the line of action is

$$L_{ab} = \sqrt{r_{op}^2 - r_{bp}^2} + \sqrt{r_{og}^2 - r_{bg}^2} - c_d \sin\phi \tag{14.19}$$

From Eq. (14.17) the contact ratio is

$$C_r = \frac{L_{ab}}{p_c \cos\phi} = \frac{1}{p_c \cos\phi}\left[\sqrt{r_{op}^2 - r_{bp}^2} + \sqrt{r_{og}^2 - r_{bg}^2}\right] - \frac{c_d \tan\phi}{p_c} \tag{14.20}$$

It is good practice to maintain a contact ratio of 1.2 or greater. Under no circumstances should the contact ratio be below 1.1.

A contact ratio between 1 and 2 means that part of the time two pairs of teeth are in contact and during the remaining time one pair is in contact. A contact ratio between 2 and 3 means that part of the time two or three pairs of teeth are always in contact.

Example 14.5

GIVEN In a gearbox the pinion has 20 teeth and the gear has 67 teeth. The diametral pitch is 10 per inch. The manufacturer needs to reduce the size of the gearbox but does not want to change the size of the gear teeth. The original gears are standard with an addendum of $1/p_d$ and a pressure angle of 20°. The approach to be used is to decrease the size of the gearbox by stubbing (reducing the addendum on) one of the wheels.

FIND Which wheel should be stubbed and what percentage smaller would the gearbox be due to the stubbing?

Solution

From Eq. (14.6) the circular pitch is

$$p_c = \frac{\pi}{p_d} = \frac{\pi}{10} \text{ in.}$$

From Eqs. (14.1) and (14.4) the center distance is

$$c_d = \frac{d_p + d_g}{2} = \frac{N_p + N_g}{2p_d} = \frac{20 + 67}{2(10)} = 4.35 \text{ in.}$$

From Eq. (14.5) the pitch circle radii for the pinion and gear are

$$r_p = \frac{N_p}{2p_d} = \frac{20}{2(10)} = 1 \text{ in.}$$

$$r_g = \frac{N_g}{2p_d} = \frac{67}{2(10)} = 3.35 \text{ in.}$$

From Eq. (14.16) the base circle radii for the pinion and gear are

$$r_{bp} = r_p \cos\phi = (1)\cos 20° = 0.9397 \text{ in.}$$

$$r_{bg} = r_g \cos\phi = (3.35)\cos 20° = 3.148 \text{ in.}$$

The outside radii for the pinion and gear are

$$r_{op} = r_p + a$$

$$r_{og} = r_g + a$$

a) *Standard gears:* For standard gears the addendum is

$$a = \frac{1}{p_d} = \frac{1}{10} = 0.1 \text{ in.}$$

$$r_{op} = r_p + a = 1 + 0.1 = 1.1 \text{ in.} \quad \textbf{(a)}$$

$$r_{og} = r_g + a = 3.35 + 0.1 = 3.45 \text{ in.} \quad \textbf{(b)}$$

$$\therefore r_{op} + r_{og} = 4.55 \text{ in.}$$

b) *Pinion stubbed:* If the pinion is stubbed, the addendum radius of the pinion is

$$r_{op} = r_p + xa \quad \textbf{(c)}$$

where

x = constant between 0 and 1

Note that if $x = 1$ a standard pinion exists and if $x = 0$ the pinion is stubbed to the pitch circle. Either the pinion or gear is stubbed but not both. The smallest contact ratio should be 1.1. Thus, using Eqs. (14.20), (b), and (c) gives

$$1.1 = \frac{10}{\pi\cos 20°}\left[\sqrt{(1+0.1x)^2 - (0.9397)^2} + \sqrt{(3.45)^2 - (3.148)^2}\right] - \frac{4.35\tan 20°}{\pi/10}$$

Solving for x gives $x = 0.2164$. The outside radius for the pinion is

$$r_{op} = 1 + 0.1(0.2164) = 1.0216 \text{ in.}$$

The sum of the outside radii is

$$r_{op} + r_{og} = 1.0216 + 3.45 = 4.472 \text{ in.}$$

Thus, the percent reduction of the gearbox if the pinion is stubbed is

$$\frac{(4.55 - 4.472)100}{4.55} = 1.714\%$$

c) *Gear stubbed:* If the gear is stubbed, the addendum radius of the gear is

$$r_{og} = r_g + ya \quad \textbf{(d)}$$

where

y = constant between 0 and 1

Using Eqs. (14.20), (d), and (a) while setting the contact ratio to 1.1 gives

$$1.1 = \frac{10}{\pi\cos 20°}\left[\sqrt{(1.1)^2 - (0.9397)^2} + \sqrt{(3.35+0.1y)^2 - (3.148)^2}\right] - \frac{4.35\tan 20°}{\pi/10}$$

Solving for y gives $y = 0.3370$. The outside radius for the gear is

$$r_{og} = 3.35 + 0.1(0.3370) = 3.384 \text{ in.}$$

The sum of the outside radii is

$$r_{op} + r_{og} = 1.1 + 3.384 = 4.484 \text{ in.}$$

Thus, the percent reduction of the gearbox if the gear is stubbed is

$$\frac{(4.55 - 4.484)(100)}{4.55} = 1.451\%$$

Therefore, the pinion should be stubbed because it will result in a larger (1.714%) reduction of the gearbox.

14.6 TOOTH THICKNESS, BACKLASH, AND INTERFERENCE

The circular tooth thickness t_h is the arc length, measured on the pitch circle, from one side of the tooth to the other. Theoretically, the circular tooth thickness is one-half of the circular pitch.

$$t_h = \frac{p_c}{2} = \frac{\pi}{2p_d} \qquad \textbf{(14.21)}$$

For perfect meshing of pinion and gear the tooth thickness measured on the pitch circle must be exactly one-half of the circular pitch. Because of unavoidable inaccuracies it is necessary to cut the teeth slightly thinner to provide some clearance so that the gears will not bind but will mesh together smoothly. This clearance measured on the pitch circle is called **backlash.** Figure 14.12 illustrates backlash in meshing gears.

To obtain the amount of backlash desired, it is necessary to decrease the tooth thickness given in Eq. (14.21). This decrease must be greater than the desired backlash owing to

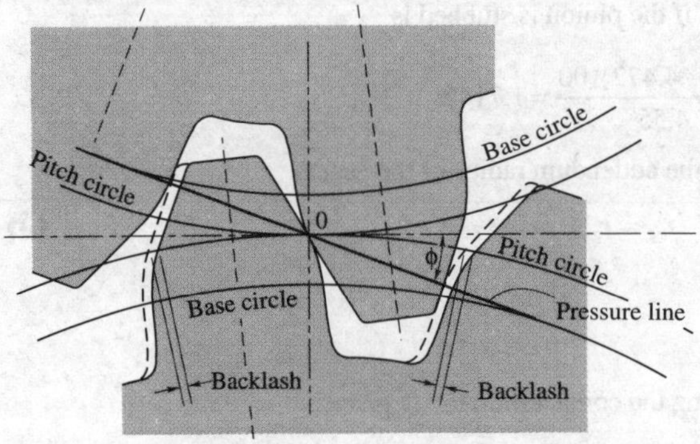

Figure 14.12 Illustration of backlash in gears.

Table 14.3 Recommended minimum backlash for coarse-pitch gears.

Diametral pitch P_d, in.$^{-1}$	Center distance, c_d, in. 2	4	8	16	32
	Backlash, b_l, in.				
18	0.005	0.006	—	—	—
12	0.006	0.007	0.009	—	—
8	0.007	0.008	0.010	0.014	—
5	—	0.010	0.012	0.016	—
3	—	0.014	0.016	0.020	0.028
2	—	—	0.021	0.025	0.033
1.25	—	—	—	0.034	0.042

errors in manufacturing and assembly. Since the amount of the decrease in tooth thickness depends on machining accuracy, the allowance for a specified backlash varies according to manufacturing conditions.

In the following example fundamental equations are given for determining backlash in a single meshing gear and pinion. To determine backlash in gear trains (covered later in this chapter), it is necessary to sum the backlash of each mated gear pair. However, to obtain the total backlash for a series of meshes, it is necessary to take into account the gear ratio of each mesh relative to a chosen reference shaft in the gear train. This is beyond the scope of this text, but more details can be found in Mechalec (1962) or Townsend (1992).

Backlash might occur in two ways: (1) if the tooth thickness is less than that given in Eq. (14.21), and (2) if the operating center distance is greater than that obtained when Eqs. (14.21) and (14.1) are valid. Table 14.3 gives recommended minimum backlash for coarse-pitch gears. All dimensions are given in inches. Thus, when the backlash b_l from Table 14.3 and the theoretical circular tooth thickness t_h from Eq. (14.21) are known, the actual circular tooth thickness is

$$t_{ha} = t_h - b_l \qquad \textbf{(14.22)}$$

Example 14.6

GIVEN The two situations given in Example 14.3.

FIND Determine the tooth thickness, the recommended backlash, and the contact ratio.

Solution

a) For the original center distance, from Eq. (14.21) the tooth thickness is

$$t_h = p_c/2 = 1.571/2 = 0.7855 \text{ in.}$$

From Table 14.3 the interpolated recommended backlash for a diametral pitch of 2 per inch and a center distance of 14 in. is

$$b_l = 0.024 \text{ in.}$$

Thus, the actual tooth thickness, from Eq. (14.22), is

$$t_{ha} = t_h - b_l = 0.7855 - 0.024 = 0.7615 \text{ in.}$$

From Table 14.2 the addendum for both the pinion and gear is

$$a = 1/p_d = 1/2 = 0.5 \text{ in.}$$

The outside radii for the pinion and gear are

$$r_{op} = r_p + a = 4 + 0.5 = 4.5 \text{ in.}$$

$$r_{og} = r_g + a = 10 + 0.5 = 10.5 \text{ in.}$$

By making use of the above and Eq. (14.20), the contact ratio can be expressed as

$$C_r = \frac{1}{1.571\cos 20°}\left[\sqrt{4.5^2 - 3.759^2} + \sqrt{10.5^2 - 9.397^2}\right] - \frac{14\tan 20°}{1.571} = 1.606$$

This implies that part of the time two pairs of teeth are in contact and during the remaining time one pair is in contact.

b) When the center distance is increased by 0.25 in., from Eq. (14.21) the tooth thickness is

$$t_h = p_c/2 = 1.599/2 = 0.800$$

The actual tooth thickness is

$$t_{ha} = t_h - b_l = 0.8000 - 0.024 = 0.776 \text{ in.}$$

The outside radii for the pinion and gear are

$$r_{op} = r_p + a = 4.072 + 0.5 = 4.572 \text{ in.}$$

$$r_{og} = r_g + a = 10.18 + 0.5 = 10.68 \text{ in.}$$

$$C_r = \frac{1}{1.599\cos 22.6°}\left[\sqrt{4.572^2 - 3.759^2} + \sqrt{10.68^2 - 9.397^2}\right] - \frac{14.25\tan 22.6°}{1.599} = 1.491$$

Thus, increasing the center distance increases the pressure angle and decreases the contact ratio.

14.7 GEAR TRAINS

Gear trains are machine elements consisting of multiple gears. They are used to obtain a desired angular velocity of an output shaft while the input shaft rotates at a different angular velocity. The angular velocity ratio between input and output gears is constant.

14.7.1 SINGLE GEAR MESH

Figure 14.13 shows *externally* meshing spur gears with fixed centers. Using Eq. (14.18) and the sign convention that positive means the same direction and negative means the opposite direction gives the angular velocity ratio for the situation in Fig. 14.13 as

$$Z_{21} = \frac{\omega_2}{\omega_1} = -\frac{N_1}{N_2} \tag{14.23}$$

Externally meshing gear teeth are the type of meshing gear discussed throughout this chapter. The gears rotate in opposite directions. The center distance for externally meshing spur gears is

$$c_d = r_1 + r_2 \tag{14.24}$$

Figure 14.14 shows *internally* meshing spur gears. Note that both gears are moving in the same direction and thus are positive. The angular velocity ratio is

$$Z_{21} = \frac{\omega_2}{\omega_1} = \frac{N_1}{N_2} \tag{14.25}$$

The difference in sign in Eqs. (14.23) and (14.25) accounts for the fact that externally meshing gears rotate in opposite directions whereas internally meshing gears rotate in the same direction.

The center distance for internally meshing spur gears is

$$c_d = r_2 - r_1 \tag{14.26}$$

14.7.2 Simple Spur Gear Trains

Animation

A simple gear train, as shown in Fig. 14.15, has only one gear mounted on each shaft. If the ith gear ($i = 1,2,3,\ldots,n$) has N_i teeth and rotates with angular velocity ω_i measured positive counterclockwise, the angular velocity ratio Z_{ji} of the jth gear relative to the ith gear is

$$Z_{ji} = \frac{\omega_j}{\omega_i} = \pm\frac{N_i}{N_j} \tag{14.27}$$

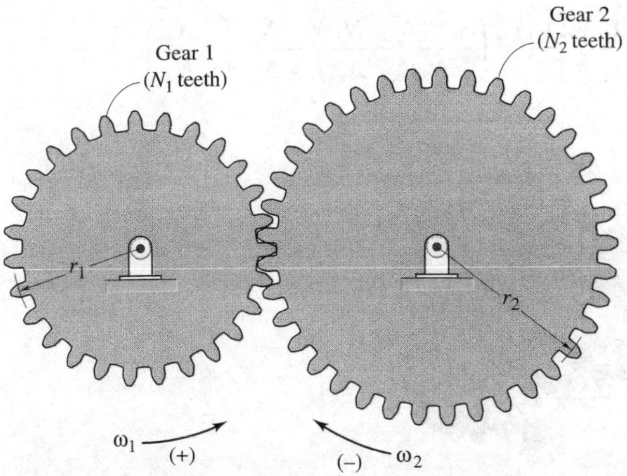

Figure 14.13 Externally meshing spur gears.

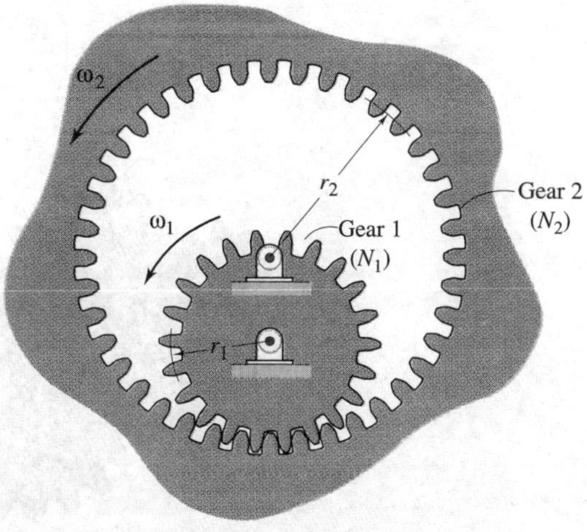

Figure 14.14 Internally meshing spur gears.

Figure 14.15 Simple gear train.

A positive sign is used in the above equation when j is odd, and a negative sign is used when j is even. The reason for this is that in Fig. 14.15 all the gears are external, implying that the directions of rotation of adjacent shafts are opposite.

14.7.3 Compound Spur Gear Trains

In a compound gear train (Fig. 14.16) at least one shaft carries two or more gears. In this type of train all gears are keyed to their respective shafts, so that the angular velocities of all gears are equal to that of the shaft on which they are mounted. The angular velocities of adjacent shafts are governed by the gear ratio of the associated mesh.

Let Ω_i represent the angular velocity of the ith shaft. Using Fig. 14.16 while going from left to right and assuming that positive is counterclockwise and negative is clockwise give

$$\frac{\Omega_2}{\Omega_1} = -\frac{N_1}{N_2} \qquad \frac{\Omega_3}{\Omega_2} = \frac{N_3}{N_4} \qquad \frac{\Omega_4}{\Omega_3} = -\frac{N_5}{N_6} \qquad \text{and} \qquad \frac{\Omega_5}{\Omega_4} = -\frac{N_7}{N_8}$$

Hence, the angular velocity ratio Z_{51} of shaft 5 relative to shaft 1 is given by

$$Z_{51} = \Omega_5/\Omega_1 \tag{14.28}$$

where

$$\frac{\Omega_5}{\Omega_1} = \left(\frac{\Omega_5}{\Omega_4}\right)\left(\frac{\Omega_4}{\Omega_3}\right)\left(\frac{\Omega_3}{\Omega_2}\right)\left(\frac{\Omega_2}{\Omega_1}\right) = \left(-\frac{N_7}{N_8}\right)\left(-\frac{N_5}{N_6}\right)\left(-\frac{N_3}{N_4}\right)\left(-\frac{N_1}{N_2}\right)$$

Figure 14.16 Compound gear train.

For the compound gear train shown in Fig. 14.16

$$Z_{51} = \frac{N_1 N_3 N_5 N_7}{N_2 N_4 N_6 N_8} \quad \textbf{(14.29)}$$

Note that

$$Z_{51} = \omega_8/\omega_1 \quad \textbf{(14.30)}$$

Example 14.7

GIVEN The gear train shown in Fig. 14.17.

FIND Determine the angular velocity ratio for the gear train. If the shaft carrying gear A rotates at 1750 rpm clockwise, determine the speed and direction of the shaft carrying gear E.

Solution

From Fig. 14.17 observe that

$$Z_{14} = \left(-\frac{N_B}{N_A}\right)\left(-\frac{N_D}{N_C}\right)\left(-\frac{N_E}{N_D}\right) = -\frac{N_B N_E}{N_A N_C} = -\frac{(70)(54)}{(20)(18)} = -10.5$$

Gear D is an *idler,* since it has no effect on the magnitude of the angular velocity ratio, but it does cause a direction reversal. Therefore, if gear D were not present, the directionality of gear E would be in the opposite direction, but the angular velocity ratio would have the same numerical value but opposite sign.

The output speed is

$$Z_{14} = \frac{\omega_A}{\omega_E} = -\frac{N_B N_E}{N_A N_C} = -10.5$$

$$\therefore \omega_E = \frac{\omega_A}{Z_{14}} = \frac{1750}{-10.5} = -166.7 \text{ rpm}$$

The negative sign implies that the direction is counterclockwise.

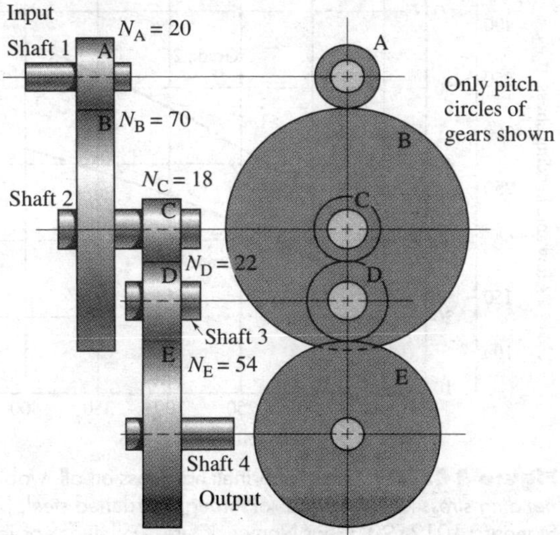

Figure 14.17 Gear train used in Example 14.7.

14.8 GEAR MATERIALS AND ALLOWABLE STRESSES

The variety of materials used in gears is extensive and depends on the application of the gear. Recall that Chapter 3 focused on materials. This section assumes the use of a through-hardened steel gear material, a somewhat restrictive assumption, and focuses on other factors besides material. More information about different materials, different manufacturing processes, and different coatings applied to gear teeth can be found in AGMA (1988, 1989, 1990).

Section 14.10 explains that a gear tooth acts like a cantilevered beam in resisting the force exerted on it by the mating tooth. The point of highest tensile bending stress is at the root of the tooth, where the involute curve bends with the fillet. AGMA (1990) presents a set of allowable bending stresses for two grades of through-hardened steel for various Brinell hardnesses. These allowable bending stresses are shown in Fig. 14.18. The grades differ in degree of control of the microstructure, alloy composition, cleanness, prior heat treatment, nondestructive testing performed, core hardness values, and other factors. Grade 2 materials are more tightly controlled than grade 1 materials.

A second, independent form of gear failure is caused by pitting and spalling of the tooth surfaces, usually near the pitch line, where the contact stresses are extremely high. The nonconformal contact of the meshing gear teeth produces high contact stresses and eventual failure due to fatigue caused by the cyclic variation of stresses. AGMA (1990) also presents allowable contact stresses for various Brinell hardnesses for two grades of steel. These allowable contact stresses are shown in Fig. 14.19. Both Figs. 14.18 and 14.19 will be used to predict the failure of gears.

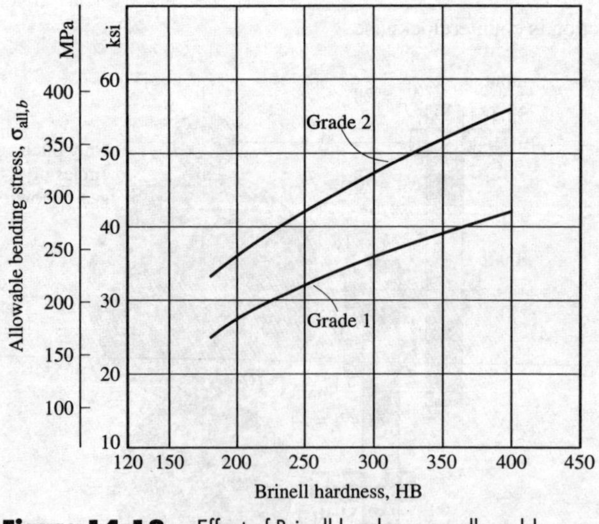

Figure 14.18 Effect of Brinell hardness on allowable *bending stress* for two grades of through-hardened steel. [AGMA Standard 1012-F90, Gear Nomenclature, Definitions of Terms with Symbols, 1998.]

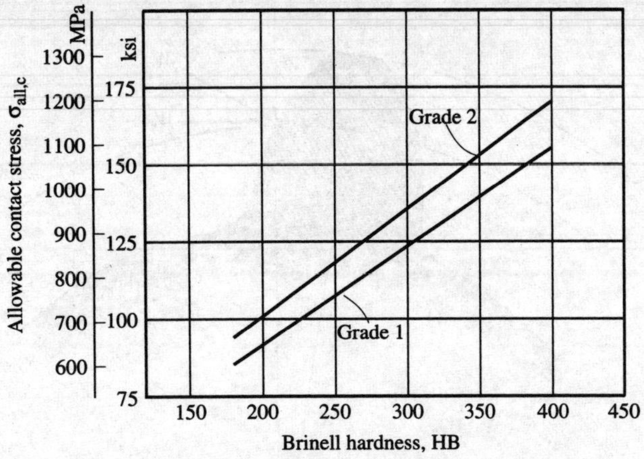

Figure 14.19 Effect of Brinell hardness on allowable *contact stress* for two grades of through-hardened steel. [ANSI/AGMA Standard 1012-F90, Gear Nomenclature, Definitions of Terms with Symbols, 1998.]

Recall from Chapter 1, Eq. (1.1), that the safety factor is

$$n_s = \frac{\sigma_{all}}{\sigma_d} = \frac{\sigma_{all}}{\sigma_{max}} \tag{14.31}$$

Once the allowable stress is known, the stress of a particular design can be determined. Before that, however, the forces acting on a gear tooth must be considered.

14.9 Loads Acting on a Gear Tooth

Gears transfer power by the driving teeth exerting a load on the driven teeth while the reaction load acts back on the teeth of the driving gear. Figure 14.20 shows the loads acting on an individual gear tooth. The tangential W_t, radial W_r, and normal W loads are shown along with the pressure angle and the pitch circle. The tangential load W_t, or transmitted force, in pounds force, can be obtained from Eq. (4.40) as

$$W_t = \frac{126\ 050\ h_p}{dN_a} \tag{14.32}$$

where

h_p = transmitted horsepower, hp
d = pitch diameter, in.
N_a = gear rotational speed, rpm

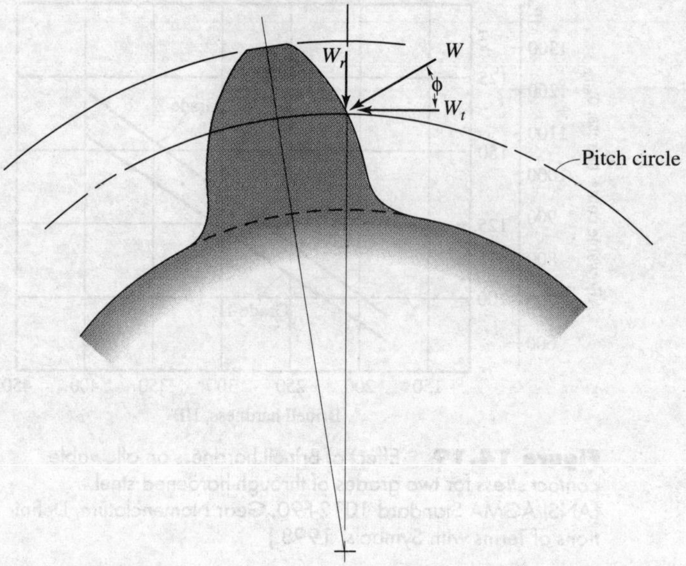

Figure 14.20 Loads acting on individual gear tooth.

Equation (14.32) is valid only for English units. The comparable equation to Eq. (14.32) in SI units can be obtained directly from Eq. (4.41). The normal load W and the radial load W_r can be computed from the following:

$$W = W_t/\cos\phi \quad \textbf{(14.33)}$$

$$W_r = W_t \tan\phi \quad \textbf{(14.34)}$$

The normal load is used in determining contact stress, contact dimensions, and elastohydrodynamic lubrication film thickness.

14.10 BENDING STRESSES IN GEAR TEETH

Gear failure can occur from tooth breakage, which results when the design stress due to bending is greater than the allowable stress. Figure 14.21(a) shows the actual tooth with the forces and dimensions used in determining bending tooth strength; Fig. 14.21(b) shows a cantilevered beam simulating the forces and dimensions acting on the gear tooth. Note that the load components shown in Fig. 14.20 have been moved to the tooth tip in Fig. 14.21. Using the bending of a cantilevered beam to simulate the stresses acting on a gear tooth is only approximate at best, since the situations present in a long, thin uniform beam loaded by moments at the ends are quite different from those found in a gear tooth.

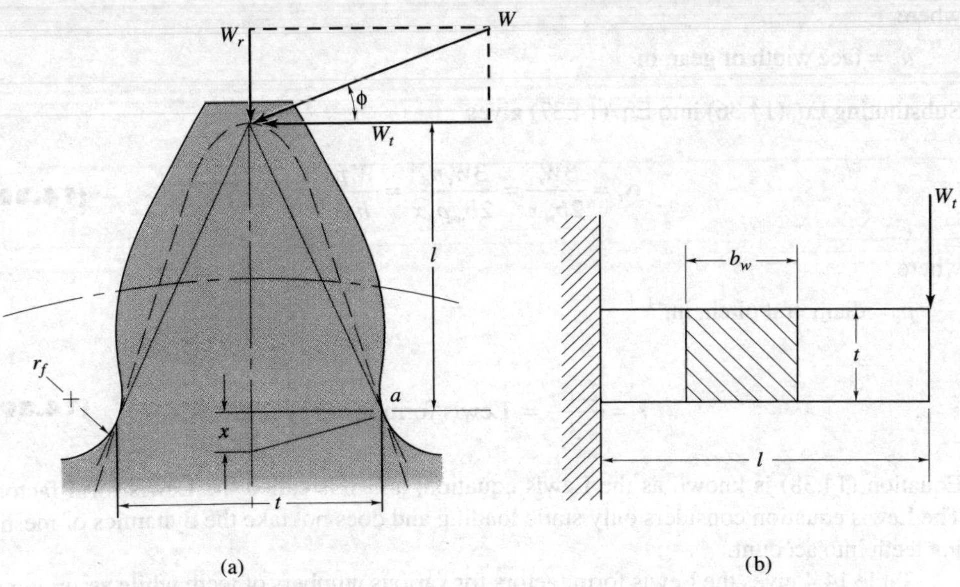

Figure 14.21 Loads and length dimensions used in determining bending tooth stress. (a) Tooth; (b) cantilevered beam.

Wilfred Lewis was the first to use this approach (Lewis, 1892), and thus the equation that results is often called the **Lewis equation.** This approach has had universal acceptance and will therefore be used.

From Eq. (4.48) the bending stress can be written as

$$\sigma_t = Mc/I \tag{14.35}$$

From similar triangles

$$\tan\alpha = \frac{x}{t/2} = \frac{t/2}{l}$$

$$l = t^2/4x \tag{14.36}$$

For a rectangular section from Table 4.1 with base b_w and height t, the area moment of inertia is

$$I = \frac{bh^3}{12} = \frac{b_w t^3}{12}$$

If $M = W_t l$ and $c = t/2$, by making use of Eq. (14.36), Eq. (4.48) becomes

$$\sigma_t = \frac{W_t l(t/2)}{b_w t^3/12} = \frac{6W_t l}{b_w t^2} \tag{14.37}$$

Table 14.4 Lewis form factor for various numbers of teeth (pressure angle, 20°; full-depth involute).

Number of teeth	Lewis form factor
10	0.176
11	0.192
12	0.210
13	0.223
14	0.236
15	0.245
16	0.256
17	0.264
18	0.270
19	0.277
20	0.283
22	0.292
24	0.302
26	0.308
28	0.314
30	0.318
32	0.322
34	0.325
36	0.329
38	0.332
40	0.336
45	0.340
50	0.346
55	0.352
60	0.355
65	0.358
70	0.360
75	0.361
80	0.363
90	0.366
100	0.368
150	0.375
200	0.378
300	0.382

where

b_w = face width of gear, m

Substituting Eq. (14.36) into Eq. (14.37) gives

$$\sigma_t = \frac{3W_t}{2b_w x} = \frac{3W_t p_d}{2b_w p_d x} = \frac{W_t p_d}{b_w Y} \tag{14.38}$$

where

p_d = diametral pitch, in.$^{-1}$

$$Y = \frac{2xp_d}{3} = \text{Lewis form factor} \tag{14.39}$$

Equation (14.38) is known as the Lewis equation, and Y is called the Lewis form factor. The Lewis equation considers only static loading and does not take the dynamics of meshing teeth into account.

Table 14.4 gives the Lewis form factors for various numbers of teeth while assuming a pressure angle of 20° and a full-depth involute. The Lewis form factor given in Eq. (14.39) is dimensionless. It is also independent of tooth size and only a function of shape.

The Lewis equation [Eq. (14.38)] does not take the stress concentration K_c that exists in the tooth fillet into account. Introducing a geometry factor Y_j, where $Y_j = Y/K_c$, changes the Lewis equation to

$$\sigma_t = \frac{W_t p_d}{b_w Y_j} \tag{14.40}$$

Figure 14.22 gives values of geometry factors Y_j for a spur gear with a pressure angle of 20° and a full-depth involute. The discontinuity at fewer than 18 teeth is due to the undercutting of pinions with fewer than 18 teeth. Equation (14.40) is known as the **modified Lewis equation.**

Other modifications to Eq. (14.38) are recommended in AGMA (1989) for practical design to account for the variety of conditions that can be encountered in service. The following design equation developed by Mott (1992) is used here:

$$\sigma_t = \frac{W_t p_d K_a K_s K_m}{b_w Y_j K_v} \tag{14.41}$$

where

K_a = application factor

K_s = size factor

K_m = load distribution factor

K_v = dynamic factor

Each of these factors is discussed here.

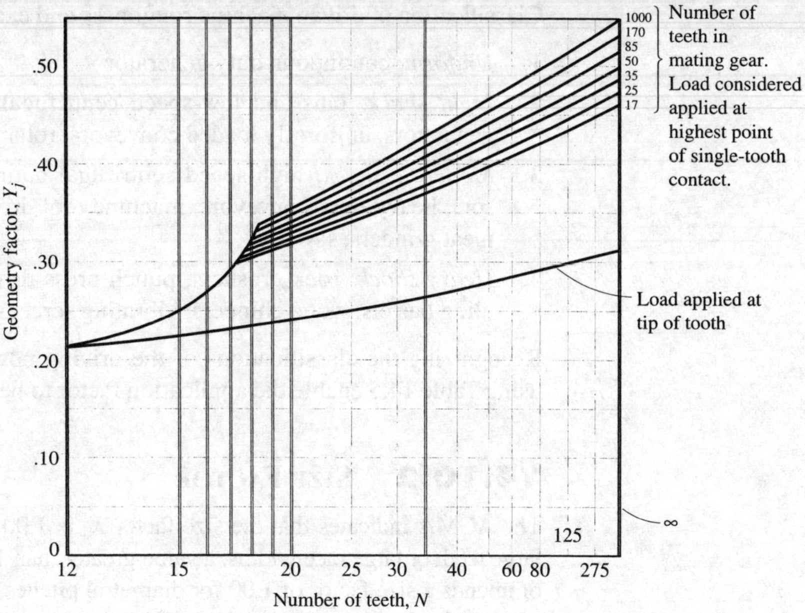

Figure 14.22 Spur gear geometry factors for pressure angle of 20° and full-depth involute [AGMA Standard 1012-F90, Gear Nomenclature, Definitions of Terms with Symbols, 1998.]

14.10.1 APPLICATION FACTOR

The application factor K_a considers the probability that load variations, vibrations, shock, speed changes, and other application-specific conditions may result in peak loads greater than W_t being applied to the gear teeth during operation. Table 14.5 will be used, where the application factor is a function of both the driving power source and the driven machine. An application factor of 1.00 would be applied for a perfectly smooth-operating electric motor driving a perfectly smooth-operating generator through a gear speed reducer. Rougher conditions produce a value of K_a greater than 1.00. Power source classifications with typical examples are

1. *Uniform:* electric motor with constant-speed turbine
2. *Light shock:* water turbine with variable-speed drive
3. *Moderate shock:* multicylinder engine

Table 14.5 Application factor as function of driving power source and driven machine.

	Driven machines			
Power source	**Uniform**	**Light shock**	**Moderate shock**	**Heavy shock**
		Application factor, K_a		
Uniform	1.00	1.25	1.50	1.75
Light shock	1.20	1.40	1.75	2.25
Moderate shock	1.30	1.70	2.00	2.75

Classification of driven machine roughness and examples are

1. *Uniform:* continuous-duty generator
2. *Light shock:* fans and low-speed centrifugal pumps, liquid agitators, variable-duty generators, uniformly loaded conveyors, rotary positive-displacement pumps
3. *Moderate shock:* high-speed centrifugal pumps, reciprocating pumps, and compressors, heavy-duty conveyors, machine tool drives, concrete mixers, textile machinery, meat grinders, saws
4. *Heavy shock:* rock crushers, punch press drives, pulverizers, processing mills, tumbling barrels, wood chippers, vibrating screens, railroad car dumpers

Recognizing the classifications of the driving power source and the driven machine and using Table 14.5 enable the application factor to be obtained.

14.10.2 SIZE FACTOR

The AGMA indicates that the size factor K_s is 1.00 for most gears. However, for gears with large teeth or large face widths, a value greater than 1.00 is recommended. Dudley (1984) recommends a size factor of 1.00 for diametral pitches of 5 or greater or metric modules of 5 or smaller. For larger teeth the values of the size factors shown in Table 14.6 are recommended.

14.10.3 LOAD DISTRIBUTION FACTOR

The value of the load distribution factor K_m is based on many variables in the design of gears, making it the most difficult factor to determine. Figure 14.23 gives values of K_m as a function of face width b_w and ratio of face width to pitch diameter b_w/d. It is assumed that the gears are of general commercial quality.

14.10.4 DYNAMIC FACTOR

The dynamic factor K_v accounts for the fact that the load is higher than the transmitted load, since there is some impact loading that is not accounted for in Eq. (14.40). The value of K_v depends on the accuracy of the tooth profile, the elastic properties of the tooth, and the speed

Table 14.6 Size factor as function of diametral pitch or module.

Diametral pitch p_d, in.$^{-1}$	Module, m, mm	Size factor, K_s
≥5	≤5	1.00
4	6	1.05
3	8	1.15
3	12	1.25
1.25	20	1.40

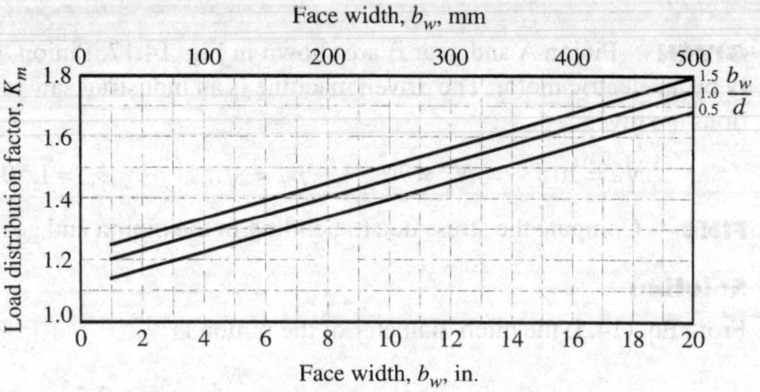

Figure 14.23 Load distribution factor as function of face width and ratio of face width to pitch diameters. Commercial quality gears assumed. [*Machine Elements in Mechanical Design* by Mott, ©1992. Reprinted by permission of Prentice-Hall, Inc., Upper Saddle River, NJ.]

with which the teeth come into contact. Figure 14.24 gives the dynamic factors for pitch-line velocity and transmission accuracy level number. This figure is intended to account for the effect of tooth inaccuracies. Gears having a transmission accuracy level number Q_v of 5, 6, or 7 are gears in typical machine design applications that are made by hobbing or shaping with average to good tooling. If the teeth are finish ground or shaved to improve the accuracy of the tooth profile and spacing, a transmission accuracy level number of 8, 9, 10, or 11 should be used. If the teeth are cut by form milling, gears having factors lower than those found from curve $Q_v \leq 5$ should not be used at a pitch-line velocity above 2500 ft/min.

The pitch-line velocity v_t used in Fig. 14.24 is the linear velocity of a point on the pitch circle of the gear. The expression for the pitch-line velocity, while making use of Eq. (4.39), is

$$v_t = \frac{\pi d_p N_{ap}}{12} = \frac{\pi d_g N_{ag}}{12} \tag{14.42}$$

where

v_t = pitch-line velocity, ft/min
N_{ap} = rotational speed of pinion, rpm
N_{ag} = rotational speed of gear, rpm
d_p = pitch diameter of pinion, in.
d_g = pitch diameter of gear, in.

Note the English units.

Figure 14.24 Dynamic factor as function of pitch-line velocity and transmission accuracy level number.

Example 14.8 **GIVEN** Pinion A and gear B are shown in Fig. 14.17. Pinion A rotates at 1750 rpm, driven directly by an electric motor. The driven machine is an industrial saw requiring 25 hp. The following conditions are given:

$$N_p = 20 \qquad N_g = 70 \qquad p_d = 8/\text{in.} \qquad b_w = 1.50 \text{ in.} \qquad \text{and} \qquad Q_v = 6$$

FIND Compute the stress due to bending in the pinion and gear.

Solution

From Eq. (14.5) the pitch diameter of the pinion is

$$d_p = \frac{N_p}{p_d} = \frac{20}{8} = 2.5 \text{ in.}$$

From Eq. (14.42) the pitch-line velocity in feet per minute is

$$v_t = \frac{\pi d_p N_{ap}}{12} = \frac{\pi(2.5)(1750)}{12} = 1145 \text{ ft/min}$$

Making use of Eq. (14.32) gives the transmitted load as

$$W_t = \frac{126\,050 \text{ hp}}{d_p N_{ap}} = \frac{(126\,050)(25)}{(2.5)(1750)} = 720.3 \text{ lbf}$$

From Fig. 14.22 the geometry factors for the pinion and gear are

$$Y_{jp} = 0.34 \qquad \text{and} \qquad Y_{jg} = 0.42$$

The application factor K_a from Table 14.5 for a smooth-operating, uniform electric motor driving an industrial saw generating moderate shock is 1.5. The size factor K_s from Table 14.6 for a diametral pitch of 8 is 1.0. From Fig. 14.23 for $b_w = 1.5$ in. and $b_w/d_p = 0.6$, the load distribution factor $K_m = 1.18$. From Fig. 14.24 for $v_t = 1145$ ft/min and $Q_v = 6$, the dynamic factor $K_v = 0.68$.

Equation (14.41) gives the bending stress for the pinion as

$$\sigma_{tp} = \frac{W_t p_d K_a K_s K_m}{b_w Y_{jp} K_v} = \frac{(720.3)(8)(1.5)(1.0)(1.18)}{(1.5)(0.34)(0.68)} = 29\,410 \text{ psi}$$

The bending stress for the gear is

$$\sigma_{tg} = \sigma_{tp} Y_{jp} / Y_{ig} = 29\,410(0.34)/0.42 = 23\,810 \text{ psi}$$

The stress will always be higher in the pinion than in the gear.

From Fig. 14.18 and Eq. (14.31) observe that the allowable stress is larger than the design bending stress (thus implying a safety factor greater than 1) if a grade 2 steel is used or if the Brinell hardness of a grade 1 steel is greater than HB = 210.

14.11 CONTACT STRESSES IN GEARS

In addition to considering the critical bending stress in a gear design, we need to also check for failure caused by too large contact stresses. These stresses can cause pitting, scoring, and scuffing of the surfaces. Pitting is the removal of small bits of metal from the gear surface due to fatigue, leaving small holes or pits. Pitting is caused by excessive surface stress due to high normal loads, a high local temperature due to high rubbing speeds, or inadequate lubricant.

Scoring is heavy scratch patterns extending from the tooth root to the tip. It appears as if a heavily loaded tooth pair has dragged particles between sliding teeth. It can be caused by lubricant failure, incompatible materials, and overload.

Scuffing is a surface destruction comprising plastic material flow plus superimposed gouges and scratches caused by loose metallic particles acting as an abrasive between teeth. Both scoring and scuffing are associated with welding or seizing and plastic deformation. The resulting contact stress in meshing gear teeth is high because of the nonconformal surfaces. This high stress results in pitting and spalling of the surfaces.

The contact stress is calculated by using a Hertzian contact stress analysis. The Hertzian contact stress of gear teeth is based on the analysis of two cylinders under a radial load. The radii of the cylinders are the radii of curvature of the involute tooth forms of the mating teeth at the point of contact. The load on the teeth is the normal load (see Figs. 14.20 and 14.21) and is given in Eq. (14.33).

The maximum Hertzian pressure in the contact can be written from Eq. (8.32) as

$$p_H = E'\left(\frac{W'}{2\pi}\right)^{\frac{1}{2}} \tag{14.43}$$

where

$$E' = \text{effective modulus of elasticity} = \frac{2}{\dfrac{1-\nu_a^2}{E_a} + \dfrac{1-\nu_b^2}{E_b}}$$

$$W' = \text{dimensionless load} = \frac{w'}{E'R_x}$$

$$w' = \text{load per unit width} = P_z/b_w$$

$$\frac{1}{R_x} = \left(\frac{1}{r_p} + \frac{1}{r_g}\right)\frac{1}{\sin\phi} = \left(\frac{1}{d_p} + \frac{1}{d_g}\right)\frac{2}{\sin\phi} \tag{14.44}$$

The parameter R_x is the effective radius and for gears is a function of the pitch radius of the pinion and the gear as well as the pressure angle. Figure 14.8 should be used for a visualization of these parameters. Equation (14.44) thus states that 1 over the effective radius of the meshing spur gears is equal to 1 over the radius from b to p_p plus 1 over the radius from a to p_p in Fig. 14.8.

The contact made between meshing spur gears is rectangular, having a Hertzian half-width b^*, and the length of the contact is the face width b_w. Equation (8.29) can be used to calculate the Hertzian half-width. Note that the maximum deformation in the center of the contact is given by Eq. (8.31).

Just as modification factors [see Eq. (14.41)] were used to describe the bending stress acting on the meshing gears, there are also modification factors when dealing with contact stresses. Thus, the contact stress given in Eq. (14.43) is modified to

$$\sigma_c = E'\left(\frac{W'K_aK_sK_m}{2\pi K_v}\right)^{\frac{1}{2}} = p_H\left(\frac{K_aK_sK_m}{K_v}\right)^{\frac{1}{2}} \tag{14.45}$$

The values for the application factor K_a, the size factor K_s, the load distribution factor K_m, and the dynamic factor K_v are obtained the same way as those for the bending stress in Eq. (14.41).

14.12 ELASTOHYDRODYNAMIC FILM THICKNESS

The minimum-film-thickness formula for elastohydrodynamic lubrication (EHL) developed in Chapter 13 (Sec. 13.8) for rolling-element bearings is equally applicable for gears or any other nonconformal machine element as long as the appropriate representations of the geometry, load, and speed are made. The best way to observe how EHL is applied is through the following example.

Example 14.9

GIVEN A meshing spur gear has pitch radii of 50 and 75 mm and a pressure angle of 20°. The angular velocity of the larger gear is 210 rad/s, the face width of the gear is 30 mm, and the transmitted load is 2250 N. The module is 5.0 mm, the transmission accuracy level number is 9, and the power source and the driven machine are assumed to be uniform. The properties of the lubricant and the gear materials are $\eta_0 = 0.075$ Pa-s, $\xi = 2.2 \times 10^{-8}\,\text{m}^2/\text{N}$, $E = 207$ GPa, and $\nu = 0.3$. The surface roughness of the gear is 0.3 μm.

FIND Calculate the minimum film thickness and the dimensionless film parameter of the meshing gear teeth. Also, determine the maximum Hertzian stress, the safety factor for contact stress analysis, the dimensions of the contact, and the maximum deformation in the contact.

Solution

From Eq. (14.44) the effective radius of the meshing gears is

$$\frac{1}{R_x} = \left(\frac{1}{r_p} + \frac{1}{r_g}\right)\frac{1}{\sin\phi} = \left(\frac{1}{50} + \frac{1}{75}\right)\frac{1}{\sin 20°} = 0.0975/\text{mm}$$

$$\therefore R_x = 10.26 \text{ mm} = 0.01026 \text{ m}$$

From Eq. (14.3) the gear ratio is

$$g_r = \frac{r_g}{r_p} = \frac{75}{50} = 1.5$$

From Eq. (14.18)

$$\omega_p = g_r \omega_g = 1.5(210) = 315 \text{ rad/s}$$

The mean velocity is

$$\tilde{u} = \frac{\omega_p r_p \cos\phi + \omega_g r_g \cos\phi}{2} = \frac{(315)(50)\sin 20° + (210)(75)\sin 20°}{2}$$

$$= 5387 \text{ mm/s} = 5.387 \text{ m/s}$$

Since no slip occurs between meshing gear teeth at the pitch point, the numerators of the two terms are equal.

The effective modulus of elasticity is

$$E' = \frac{2}{\dfrac{1-\nu_p^2}{E_p} + \dfrac{1-\nu_g^2}{E_g}} = \frac{E}{1-\nu^2} = \frac{(2.07)(10^{11})}{1-0.3^2} = 2.275\times 10^{11}\ \text{Pa}$$

The transmitted load W_t was given as 2250 N. The normal load, from Eq. (14.33), is

$$W = W_t/\cos\phi = 2250/\cos 20° = 2394\ \text{N}$$

The normal load per face width is

$$P_z' = \frac{W}{b_w} = \frac{2394}{0.03} = 79\ 800\ \text{N/m}$$

The dimensionless parameters for load W', speed U, and materials G can be obtained from Eqs. (13.69), (13.70), and (13.71) for a rectangular contact.

$$W' = \frac{P_z'}{E'R_x} = \frac{(7.980)(10^4)}{(2.275)(10^{11})(1.026)(10^{-2})} = 3.419\times 10^{-5}$$

$$U = \frac{\eta_0 \tilde{u}}{E'R_x} = \frac{(0.075)(5.387)}{(2.275)(10^{11})(1.026)(10^{-2})} = 1.731\times 10^{-10}$$

$$G = \xi E' = (2.2)(10^{-8})(2.275)(10^{11}) = 5005$$

Making use of Eq. (13.73) gives the dimensionless minimum film thickness for a rectangular contact as

$$H_{\min} = \frac{h_{\min}}{R_x} = 1.714(W')^{-0.128}U^{0.694}G^{0.568}$$
$$= 135.3\times 10^{-6}$$

The dimensional minimum film thickness is

$$h_{\min} = R_x H_{\min} = (1.026)(10^{-2})(1.353)(10^{-4})\text{m} = 1.388\ \mu\text{m}$$

From Eq. (8.8) the dimensionless film parameter is

$$\Lambda = \frac{h_{\min}}{\left(R_{qa}^2 + R_{qb}^2\right)^{\frac{1}{2}}} = \frac{(1.388)(10^{-6})}{(10^{-6})(0.3^2 + 0.3^2)^{\frac{1}{2}}} = 3.272$$

Therefore, the meshing spur gears will enjoy the significant benefits of elastohydrodynamic lubrication. Having $\Lambda \approx 3$ is an optimum design as discussed in Chapter 13 for rolling-element bearings.

From Eq. (8.32) the maximum Hertzian stress in the contact of meshing gear teeth is

$$\sigma_c = p_H = E'\left(\frac{W'}{2\pi}\right)^{\frac{1}{2}} = 2.275(10^{11})\left(\frac{3.419\times 10^{-5}}{2\pi}\right)^{\frac{1}{2}}$$
$$= 5.307\times 10^8\ \text{Pa} = 530.7\ \text{MPa}$$

Incorporating the modification factors from Eq. (14.45) gives

$$\sigma_c = (530.7)(10^6)\left(\frac{K_a K_s K_m}{K_v}\right)^{\frac{1}{2}}$$

Both the power source and the driven machine have uniform operation, so that the application factor from Table 14.5 is 1.0. The size factor from Table 14.6 for $m = 5$ mm is $K_s = 1.0$. From Fig. 14.23 for $b_w = 30$ mm and $b_w/d_p = 0.30$, the load distribution factor K_m is 1.12.

The module was given as 5.0 mm. From Eq. (14.7)

$$N_g = \frac{d_g}{m} = \frac{150}{5} = 30 \text{ teeth}$$

$$N_p = \frac{d_p}{m} = \frac{100}{5} = 20 \text{ teeth}$$

The speed of the pinion is

$$N_{ap} = 60\left(\frac{\omega}{2\pi}\right) = \frac{30(315)}{\pi} = 3008 \text{ rpm}$$

The pitch diameter of the pinion is

$$d_p = 100 \text{ mm} = 0.1 \text{ m} = 3.937 \text{ in.}$$

From Eq. (14.42) the pitch-line velocity is

$$v_t = \frac{\pi d_p N_{ap}}{12} = \frac{\pi(3.937)(3008)}{12} = 3100 \text{ ft/min}$$

From Eq. (14.32) the transmitted horsepower is

$$h_p = \frac{d_p N_{ap} W_t}{126\ 050} = \frac{(3.937)(3008)(505.4)}{126\ 050} = 47.52 \text{ hp}$$

From Fig. 14.24 for $v_t = 3100$ ft/min and $Q_v = 9$ the dynamic factor $K_v = 0.75$.

Thus, the modified contact stress is

$$\sigma_c = (530.7)\left(10^6\right)\left[\frac{(1.0)(1.0)(1.12)}{0.75}\right]^{\frac{1}{2}} \text{Pa} = 0.649 \text{ GPa}$$

From Fig. 14.19 for a Brinell hardness of 400 and grade 2 steel the allowable contact stress is

$$\sigma_{\text{all}} = 1.2 \text{ GPa}$$

The safety factor is

$$n_s = \frac{\sigma_{\text{all}}}{\sigma_c} = \frac{1.2}{0.649} = 1.85$$

Thus, fatigue failure due to contact stress is not anticipated.

14.13 HELICAL GEARS

The material covered previously in this chapter related primarily to spur gears. This section describes helical gears. Although helical gears were introduced in Sec. 14.2.1.2, this section describes them in more detail. However, instead of providing as much detail as was given for spur gears, this section builds upon the information obtained from spur gears and describes only the differences that exist between spur gears and helical gears.

Tooth strength is improved with helical gears (relative to spur gears) because of the elongated helical wrap-around tooth base support, as shown in Fig. 14.2. For helical gears the contact ratio is higher owing to the axial tooth overlap. Helical gears, therefore, tend to have greater load-carrying capacity than spur gears of the same size. Spur gears, on the other hand, have a somewhat higher efficiency.

14.13.1 HELICAL GEAR RELATIONSHIPS

All the relationships governing spur gears apply to helical gears with some slight modifications to account for the axial twist of the teeth caused by the helix angle. The helix angle varies from the base of the tooth to the outside radius. The helix angle ψ is defined as the angle between an element of the pitch cylinder and the tangent to the helicoidal tooth at the intersection of the pitch cylinder and the tooth profile. Figure 14.25 defines the helix angle. Front and side views are shown.

14.13.2 PITCHES OF HELICAL GEARS

Helical gears have two related pitches: one in the plane of rotation and the other in a plane normal to the tooth. For spur gears the pitches were described only in terms of the plane of rotation, but there is an additional axial pitch for helical gears. Figure 14.26 shows the circular and axial pitches of helical gears, which are related by the normal circular pitch

$$p_{cn} = p_c \cos \psi \tag{14.46}$$

where

ψ = helix angle, deg

The normal diametral pitch is

$$p_{dn} = \frac{p_d}{\cos \psi} \tag{14.47}$$

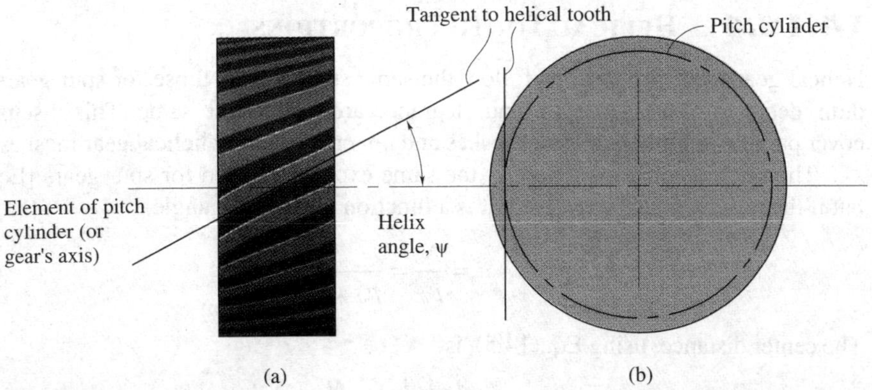

Figure 14.25 Helical gear.
(a) Front view; (b) side view.

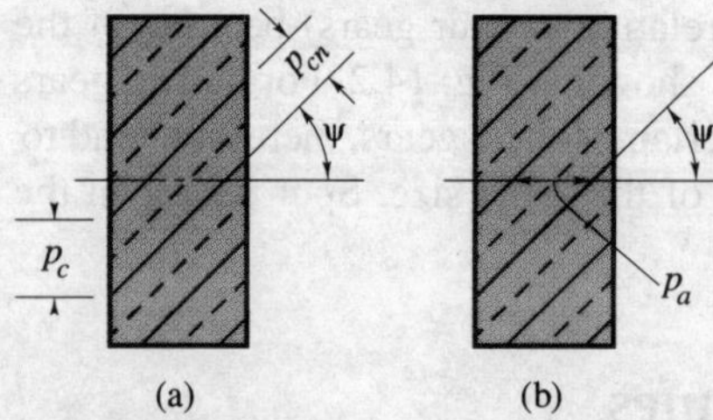

Figure 14.26 Pitches of helical gears.
(a) Circular; (b) axial.

The axial pitch of a helical gear is the distance between corresponding points on adjacent teeth measured parallel to the gear's axis [see Fig. 14.26(b)]. The axial pitch is related to the circular pitch by the following expression:

$$p_a = p_c \cot\psi = \frac{p_{cn}}{\sin\psi} \tag{14.48}$$

14.13.3 Equivalent Number of Teeth and Pressure Angle

The number of teeth for helical gears equivalent to that found for spur gears is given as

$$N_n = \frac{N}{\cos^3\psi} \tag{14.49}$$

For helical gears there is a normal pressure angle as well as the usual pressure angle in the plane of rotation, and they are related to the helix angle, or

$$\tan\phi = \frac{\tan\phi_n}{\cos\psi} \tag{14.50}$$

14.13.4 Helical Tooth Proportions

Helical gear tooth proportions follow the same standards as those for spur gears. Addendum, dedendum, whole depth, and clearance are exactly the same. This discussion will cover parallel-shaft helical gear meshes and not crossed-shaft helical gear meshes.

The pitch diameter is given by the same expression used for spur gears [Eq. (14.5)], but if the normal pitch is involved, it is a function of the helix angle.

$$d = \frac{N}{p_d} = \frac{N}{p_{dn}\cos\psi} \tag{14.51}$$

The center distance, using Eq. (14.5), is

$$c_d = \frac{d_p + d_g}{2\cos\psi} = \frac{N_p + N_g}{2p_{dn}\cos\psi} \tag{14.52}$$

Note that changing the value of the helix angle in Eq. (14.52) causes the center distance to change by a large range of values.

The contact ratio of helical gears is enhanced by the axial overlap of the teeth. The total contact ratio is the sum of the circumferential contact ratios when calculated in the same manner as done for spur gears [Eq. (14.20)] and added to the axial contact ratio C_{ra} or

$$C_{rt} = C_r + C_{ra} \qquad \textbf{(14.53)}$$

where

$$C_{ra} = \frac{b_w}{p_a} = \frac{b_w \tan\psi}{p_c} = \frac{b_w \sin\psi}{p_{cn}} \qquad \textbf{(14.54)}$$

Tutorial 9

C_r = circumferential contact ratio obtained from Eq. (14.20)

b_w = face width, m

14.13.5 LOADS AND STRESSES

The thrust load W_t is the same for spur or helical gears. Recall that the thrust load is the force that transmits power from the driver to the driven gear. It acts perpendicular to the axis of the shaft carrying the gear. Thus, Eqs. (14.32) and (14.40) are equally valid for spur gears or helical gears.

The *axial,* or thrust, load in a helical gear is

$$W_a = W_t \tan\psi \qquad \textbf{(14.55)}$$

Note that the axial load increases as the helix angle increases. Helix angles typically range from 15 to 45°.

The *radial* load is the force that acts toward the center of the gear (i.e., radially). The direction of the force is to push the two gears apart. Making use of Eq. (14.34), the radial load is

$$W_r = W_t \tan\phi \qquad \textbf{(14.56)}$$

The *normal* load used in the elastohydrodynamic lubrication evaluation is

$$W = \frac{W_t}{\cos\phi_n \cos\psi} \qquad \textbf{(14.57)}$$

The bending and contact stresses for helical gears are the same as those given for spur gears in Eqs. (14.41) and (14.43) with the coefficients in those equations having the same meaning but different numerical values. This chapter explained the methodology for spur gears and gave a brief discussion of the difference in going from a spur gear to a helical gear. The same methodology can be applied to helical gears with the use of figures and tables available in AGMA (1990).

Example 14.10

GIVEN An involute gear drives a high-speed centrifuge. The speed of the centrifuge is 18 000 rpm. It is driven by a 3000-rpm electric motor through a 6:1 speedup gearbox. The pinion has 21 teeth and the gear has 126 teeth with a diametral pitch of 14 per inch. The width of the gears is 1.8 in. and the pressure angle is 20°.

FIND The gear contact ratio for

a) A spur gear

b) A helical gear with helix angle of 30°

Also find the helix angle if the contact ratio is 3.

Solution

a) *Spur gear:* From Eq. (14.6) the circular pitch is

$$p_c = \frac{\pi}{p_d} = \frac{\pi}{14} = 0.224 \text{ in.} = 5.700 \text{ mm}$$

From Eqs. (14.1) and (14.4) the center distance is

$$c_d = \frac{d_p + d_g}{2} = \frac{N_p + N_g}{2p_d} = \frac{21 + 126}{2(14)} = 5.25 \text{ in.} = 133.4 \text{ mm}$$

From Eq. (14.5) the pitch circle radii for the pinion and gear are

$$r_p = \frac{N_p}{2p_d} = \frac{21}{2(14)} = 0.75 \text{ in.}$$

$$r_g = \frac{N_g}{2p_d} = \frac{126}{2(14)} = 4.5 \text{ in.}$$

The radii of the base circles for the pinion and gear while using Eq. (14.16) are

$$r_{bp} = r_p \cos\phi = 0.75 \cos 20^\circ = 0.7048 \text{ in.} = 17.90 \text{ mm}$$

$$r_{bg} = r_g \cos\phi = 4.5 \cos 20^\circ = 4.2286 \text{ in.} = 107.4 \text{ mm}$$

The outside radii for the pinion and gear are

$$r_{op} = r_p + a = r_p + \frac{1}{p_d} = 0.75 + \frac{1}{14} = 0.8214 \text{ in.} = 20.86 \text{ mm}$$

$$r_{og} = r_g + a = r_g + \frac{1}{p_d} = 4.5 + \frac{1}{14} = 4.5714 \text{ in.} = 116.1 \text{ mm}$$

From Eq. (14.20) the contact ratio for the spur gear is

$$C_r = \frac{1}{(0.2244)\cos 20^\circ}\left[\sqrt{(0.8214)^2 - (0.7048)^2} + \sqrt{(4.571)^2 - (4.229)^2}\right] - \frac{5.250 \tan 20^\circ}{0.2244}$$

$$= 1.712$$

b) *Helical gear:* The total contact ratio for a helical gear is given by Eqs. (14.53) and (14.54) as

$$C_{rt} = C_r + C_{ra} = 1.712 + \frac{(1.8)\tan 30^\circ}{0.2244} = 6.343$$

To get a contact ratio of 3, the helix angle has to be such that

$$C_{ra} = \frac{b_w \tan\psi}{p_c} = 3.00 - 1.712 = 1.288$$

$$\therefore \tan\psi = (1.288)(0.2244)/1.8$$

$$\psi = 9.122^\circ$$

Therefore, the helix angle should be 9.122° to get a contact ratio of 3.

Case Study | PORTION OF GEAR TRAIN FOR INDUSTRIAL FOOD MIXER

Figure 14.27 depicts a view of the gearing present in a common food mixer of the type found in restaurants and small bakeries. Different speeds can be achieved by engaging different gears with the drive, each gear with a prescribed number of teeth. The different speeds are required to perform various operations, such as kneading, mixing, and whipping. This problem involves the specification of a bevel gear to effect mixing or kneading of dough.

GIVEN The motor is attached to a speed reducer whose output speed is 200 rpm and whose torque and speed curves are as shown in Fig. 14.28. Use an allowable bending stress of 40 ksi and an allowable contact stress of 125 ksi.

FIND A helical gear mating pair, as shown in the figure, so that the output shaft speed is near 90 rpm. (This speed is then the input to a planet of a planetary gear train.)

Solution

The torque and speed characteristics are extremely important. If the mixer has been overloaded by using improper ingredients or by overfilling, a torque high enough to stall the motor may be developed. The highest torque that can be developed is that at the maximum current provided by the power supply, or 20 ft-lbf. (The power supply circuitry is designed to limit the stall current through placement of resistive elements.) It is important that a stalled motor does not lead to gear tooth breakage, since gears are a relatively expensive component and difficult to replace. Thus, the gears will be designed to survive the stall torque, even though they should ideally never be subjected to such high loading.

Given the required speed ratios, the number of teeth in the driven gears must be approximately 2.22 times the number of teeth in the driving gear. At this point an infinite number of gear tooth combinations will serve to fulfill the design requirements, since a face width has not been specified. However, because it is difficult to manufacture gears with few teeth, the driving gear will be arbitrarily assigned 24 teeth; and the driven gear, 52 teeth. Note that this is a particular solution to a problem with many solutions.

This application is not so demanding as to require fine or ultrafine tooth classes; a coarse spacing may even be sufficient, but a medium-coarse spacing will be chosen as a balance between performance and cost. Using a diametral pitch of 12 per inch gives the gear pitch diameters as 2 and 4.32 in. for the driving and driven gears,

Figure 14.27 Gears inside industrial mixer. [Courtesy of Hobart]

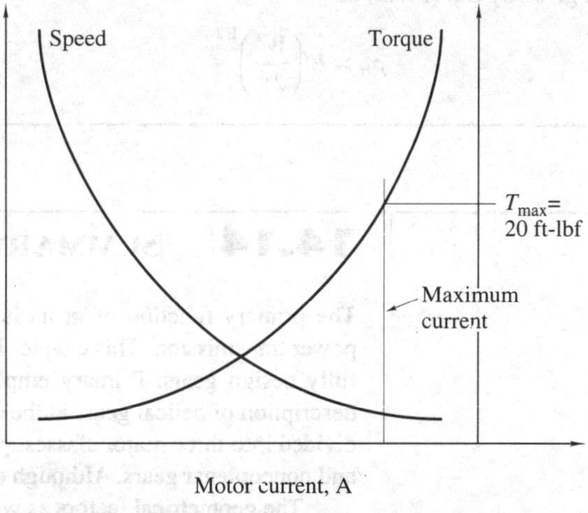

Figure 14.28 Torque and speed of motor as function of current for industrial mixer used in case study.

(continued)

Case Study (CONCLUDED)

respectively. The torque exerted onto the drive shaft causes a load of $W_t = T/r = T/(1) = 240$ lbf.

The application factor for this machine can be taken from Table 14.5 as 1.00, since the power source is uniform as is the machine operation. The size factor, from Table 14.6, is clearly 1.00, since the diametral pitch is so much larger than 5. The dynamic factor is taken as 1.00 as well, since at motor stall the pitch-line velocity will be extremely low, if not zero. The geometry factor Y_j is taken from Fig. 14.22 as 0.35. The only unknown factor from Eq. (14.41) is the load distribution factor, since this value depends on face width, which has yet to be determined.

The face width can be iteratively obtained from Eq. (14.41) as

$$\sigma_t = \frac{W_t p_d K_a K_s K_m}{b_w Y_j K_v}$$

$$= \frac{(240 \text{ lbf})(12 \text{ in.}^{-1})(1)(1)(K_m)}{b_w (0.35)(1.0)} = 40 \text{ ksi}$$

$$\therefore \frac{b_w}{K_m} = 0.21$$

Because K_m is fairly insensitive to low values of face width and is approximately equal to 1.1, $b_w = 0.25$ in. is chosen.

The maximum Hertzian pressure in the contact zone is given by Eq. (14.43) as

$$p_H = E'\left(\frac{W'}{2\pi}\right)^{1/2}$$

where $E' = 33.0 \times 10^6$ psi for two steel gears.

$$\frac{1}{R_x} = \left(\frac{1}{r_p} + \frac{1}{r_g}\right)\frac{1}{\sin\phi} \qquad R_x = 0.467 \text{ in.}$$

and therefore

$$W' = \frac{w'}{E'R_x}$$

$$= \frac{(240 \text{ lbf})/(0.25 \text{ in.})}{(33.0\times 10^6 \text{ psi}^2)(0.467 \text{ in.})} = 3.89\times 10^{-6}$$

Thus, the maximum Hertzian pressure is

$$p_H = (33.0\times 10^6 \text{ psi})\left(\frac{3.89\times 10^{-6}}{2\pi}\right)^{1/2} = 25.9\times 10^3 \text{ psi}$$

From Eq. (14.45) the resulting AGMA contact stress is

$$\sigma_c = p_H\left(\frac{K_a K_s K_m}{K_v}\right)^{1/2} = 27\times 10^3 \text{ psi}$$

Because this value is well below the allowable contact stress, a face width of 0.25 in. is selected to complete the design.

This problem is typical of most design problems in that an infinite number of solutions exist. This case study merely demonstrates the design procedure used to obtain a reasonable answer, not *the* right answer.

14.14 SUMMARY

The primary function of gears is to provide constancy of angular velocities or proportionality of power transmission. This chapter logically developed the important methodology needed to successfully design gears. Primary emphasis was on the methodology of spur gear design with a brief description of helical gears at the end of the chapter. Various types of gear were discussed. Gears are divided into three major classes: parallel-axis gears, nonparallel and coplanar gears, and nonparallel and noncoplanar gears. Although spur gears were the major focus, other gear types were explained.

The geometrical factors as well as the kinematics, loads, and stresses acting in a spur gear were discussed. Gears transfer power by the driving teeth exerting a load on the driven teeth while the reacting loads act back on the teeth of the driving gear. The thrust load is the force that transmits power from the driving to the driven gear. It acts perpendicular to the axis of the shaft carrying the gear. This thrust load is used in establishing the bending stress in the tooth. The highest tensile bending stress occurs at the root of the tooth, where the involute curve blends with the fillet. The design bending stress is then compared with an allowable stress to establish whether failure will occur.

A second, independent, form of gear failure results from surface pitting and spalling, usually near the pitch line, where the contact stresses are high. Thus, failure due to fatigue occurs. The Hertzian contact pressure with modification factors was used to establish the design stress, which was then compared with an allowable stress to determine if failure due to fatigue would occur. It was found that if an adequate protective elastohydrodynamic lubrication film exists, gear life is greatly extended.

KEY WORDS

backlash clearance measured on pitch circle

bevel gears nonparallel, coplanar gears with straight teeth

circular pitch distance from one tooth to corresponding point on adjacent tooth

diametral pitch number of teeth per pitch diameter

fundamental law of gearing law stating that common normal to tooth profile must pass through fixed point, called pitch point

gear toothed wheel that, when mated with another toothed wheel, transmits power between shafts

gear ratio ratio of number of teeth between two meshing gears

gear trains machine elements of multiple gears used to obtain desired velocity ratio

helical gears parallel-axis gears with teeth cut on spiral that wraps around cylinder

Lewis equation equation for bending stresses on gear teeth based on cantilever beam theory

modified Lewis equation Lewis equation with correction factors for gear geometry

module ratio of pitch diameter to number of teeth

pinion smaller of two meshed gears

pitch point point defined by normal to tooth profiles of meshing gears at point of contact on line of centers; see also fundamental law of gearing

spiral gears nonparallel coplanar gears with teeth cut on spiral

spur gears parallel-axis gears with straight teeth

worm gear nonparallel, noncoplanar gear with typically high reduction ratio

RECOMMENDED READINGS

Howes, M.A., ed. (1980) *Source Book on Gear Design Technology and Performance,* American Society for Metals, Metals Park, OH.

Juvinall, R.C., and Marshek, K.M. (1991) *Fundamentals of Machine Component Design,* Wiley, New York.

Litvin, F.L. (1994) *Gear Geometry and Applied Theory,* Prentice-Hall, Englewood Cliffs, NJ.

Mott, R.L. (1985) *Machine Elements in Mechanical Design,* Merrill Publishing Co., Columbus, OH.
Norton, R.L. (1996) *Machine Design,* Prentice-Hall, Englewood Cliffs, NJ.
Shigley, J.E., and Mischke, C.R. (1989) *Mechanical Engineering Design,* McGraw-Hill, New York.
Stokes, A. (1992) *Gear Handbook,* Butterworth Heinemann, Oxford, England.
Townsend, D.P. (1992) *Dudley's Gear Handbook,* 2nd ed., McGraw-Hill, New York.

REFERENCES

AGMA (1988) *Fundamental Rating Factors and Calculation Methods for Involute Spur and Helical Gear Teeth,* ANSI/AGMA Standard 2001–B88, American Gear Manufacturers Association, Alexandria, VA.
AGMA (1989) *Geometry Factors for Determining the Pitting Resistance and Bending Strength of Spur, Helical, and Herringbone Gear Teeth,* ANSI/AGMA Standard 908–B89, American Gear Manufacturers Association, Alexandria, VA.
AGMA (1990) *Gear Nomenclature, Definitions of Terms with Symbols,* ANSI/AGMA Standard 1012–F90, American Gear Manufacturers Association, Alexandria, VA.
Drago, R.J. (1992) Gear Types and Nomenclature, Chapter 2 in *Dudley's Gear Handbook,* D.P. Townsend, ed., 2nd ed., McGraw-Hill, New York.
Dudley, D.W. (1984) *Handbook of Practical Gear Design,* McGraw-Hill, New York.
Lewis, W. (1892), "Investigation of the Strength of Gear Teeth," an Address to the Engineer's Club of Philadelphia, Oct. 15, 1892.
Mechalec, G.W. (1962) *Gear Handbook,* McGraw-Hill, New York, Chapter 9.
Mott, R.L. (1992) *Machine Elements in Mechanical Design,* 2nd ed., Merrill, New York.
Townsend, D.P. (1992) *Dudley's Gear Handbook,* 2nd ed., McGraw-Hill, New York.

PROBLEMS

Section 14.2

14.1 For gear transmissions there is a direct proportionality between the local sliding speeds experienced by the contacting surfaces and the power loss. For a normal spur or helical gear the mean sliding speed is typically 20% of the load transmission speed. The coefficient of friction in lubricated gears is typically between 0.05 and 0.1. This leads to a power loss of 1 to 2% and a power efficiency of 99 to 98%. For a hypoid gear in the rear axle of a car the sliding speed can be 60% of the load transmission speed, and for a worm gear it can be 1200%. Estimate the power efficiencies for a hypoid gear and a worm gear.

Section 14.3

14.2 A pinion has a pressure angle of 20°, a module of 3 mm, and 22 teeth. It is meshed with a gear having 39 teeth. The center distance between the shafts is 81 mm. Determine the gear ratio and the diametral pitch.

14.3 A gearbox has permanent shaft positions due to the bearing positions, but the gear ratio can be changed by changing the gear wheels. To achieve similar power transmission ability for different gear ratios, a manufacturer chooses to have the same module for two different gearboxes. One of the gearboxes has a pinion with 22 teeth, a gear with 68 teeth, and a center distance of 225 mm. How large is the gear module and which gear ratios are possible for a pinion with 22 or more teeth for the same module?

14.4 The blank for a normal gear with 20 teeth was made too large during manufacturing. Therefore, the addendum for the gear teeth was moved outward, and the tooth thickness at the tops of the teeth was decreased. The module for the rack tool was 2 mm. Find how large the maximum gear outside radius can be when the tool flank angle is 20° to avoid zero thickness at the gear teeth tops.

14.5 The inlet and outlet speeds for a reduction gear train are 2000 and 320 rpm, respectively. The pressure angle is designed to be 25°, but due to shaft deflections becomes 30°. The gears have a 10-mm module and their center distance is 425 mm.

a) Determine the number of teeth for both gears, the angular pitch, and the radii of the base, addendum, and dedendum circles.

b) Find the maximum allowable change in center distance.

14.6 The base circle radii of a gear pair are 100 and 260 mm, respectively. Their module is 8 mm and the pressure angle is 20°. Mounting inaccuracy has caused a 5-mm change in center distance. Calculate the change in the pitch radii, the center distance, and the pressure angle.

14.7 A gear train has a 50.3-mm circular pitch and a 25° pressure angle. Design this gear train for the lowest volume occupied and the fewest teeth. Obtain the pitch diameters, the number of teeth, the speed ratio, the center distance, and the module of this gear train. The gearbox housing has a diameter of 620 mm.

Section 14.4

14.8 For the gears in Example 14.4 calculate the maximum sliding speed between the contacting gear teeth, and give the position along the line of action where the maximum sliding speed occurs.

14.9 The pinion of the gear train for an endurance strength test machine (sketch *a*) has 5 kW of power and rotates clockwise at 1200 rpm. The gears have a 12.57-mm circular pitch and a 20° pressure angle.

a) Find the magnitude and direction of the output speed.

b) Calculate the center distance between the input and output shafts.

c) Draw free-body diagrams of forces acting on gears 3 and 4.

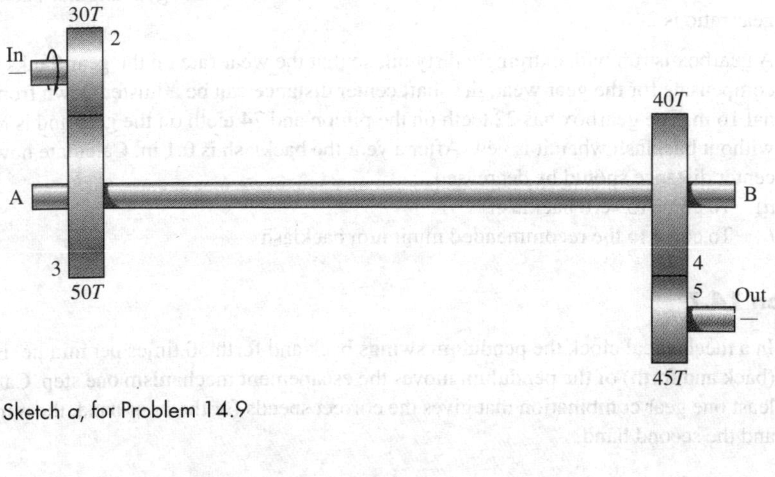

Sketch *a*, for Problem 14.9

Section 14.5

14.10 A standard straight gear system consists of two gears with tooth numbers $N_g = 51$ and $N_p = 20$. Pressure angle $\phi = 20°$, diametral pitch $p_d = 5.08$ per inch, and addendum $a = 1/p_d = 0.1969$ in. Find the center distance c_d and the contact ratio C_r.

14.11 Given a straight gear with diametral pitch $p_d = 6.35$ per inch, pressure angle $\phi = 20°$, $N_g = 28$, and $N_p = 18$, find the contact ratio C_r. Assume the base circle radii r_{bp} and r_{bg} do not change. Also $\phi = 20°$ and $\Delta c_d/c_d = 0.1$. Determine $\Delta\phi$.

14.12 Given an internal straight gear with diametral pitch $p_d = 5.08$ per inch and $N_p = 22$, find the contact ratio C_r if center distance $c_d = 4.33$ in. and pressure angle $\phi = 20°$.

14.13 A straight gear with diametral pitch $p_d = 6.35$ per inch and face width $b_w = 1.575$ in. has center distance $c_d = 4.724$ in. The pinion has 20 teeth, the speed is 1000 rpm, and pressure angle $\phi = 20°$. Find the contact stress at the pitch point when the gear transmits 20 kW. Neglect friction forces. The gear steel has modulus of elasticity $E = 207$ GPa and a Poisson's ratio of 0.3.

14.14 A spur gear has $N_g = 40$, $N_p = 20$, and $\phi = 20°$. The diametral pitch $p_d = 5.08$ per inch. Find the center distances between which the gear can work (i.e., from the backlash-free state to $C_r = 1$).

14.15 Given a gear with the following data: $N_g = 45$; $N_p = 18$; $p_d = 6.35$ per inch; and $\phi = 20°$, find

a) Center distance c_d

b) Contact ratio C_r

c) Tooth thickness on the pitch circle for both pinion and gear

14.16 Given a gear with the center distance c_d, find how much the pressure angle ϕ increases if the center distance increases by a small amount Δc_d.

Section 14.6

14.17 A straight gear is free of backlash at theoretical center distance $c_d = a = 0.264$ m $= 10.394$ in. Pressure angle $\phi = 20°$, $N_g = 47$, $N_p = 19$, and $\alpha = 11 \times 10^{-6}$ (°C)$^{-1}$. Find the center distance needed to give a backlash of 0.008378 in. (0.2128 mm). What gear temperature increase is then needed relative to the housing temperature to make the gear free from backlash?

14.18 A gear with a center distance of 8 in. is going to be manufactured. Different diametral pitches can be chosen depending on the transmitted torque and speed and on the required running accuracy. Which diametral pitch should be chosen to minimize gear angular backlash if the gear ratio is 2?

14.19 A gearbox is run with extremely dirty oil, so that the wear rate on the gear flanks is high. To compensate for the gear wear, the shaft center distance can be adjusted down from the original 16 in. The gearbox has 22 teeth on the pinion and 74 teeth on the gear and is mounted without backlash when it is new. After a year the backlash is 0.1 in. Calculate how much the center distance should be decreased

a) To come to zero backlash

b) To come to the recommended minimum backlash

Section 14.7

14.20 In a mechanical clock the pendulum swings back and forth 30 times per minute. Each swing (back and forth) of the pendulum moves the escapement mechanism one step. Calculate at least one gear combination that gives the correct speeds for the hour hand, the minute hand, and the second hand.

14.21 In a mangle for pressing household linen the 1500-rpm speed of the drive motor should be reduced to 15 rpm for the gear. Choose a gear that gives the correct gear ratio.

14.22 A drive line needs to be developed for a rear-wheel-driven car. The gear ratio in the hypoid gear from the car transmission to the rear wheels is 4.5:1. The rear wheel diameter is 550 mm. The motor can be used in the speed interval 1500 to 6000 rpm. At 6000 rpm the speed of the car should be 200 km/hr in fifth gear. In first gear it should be possible to drive as slow as 10 km/hr without slipping the clutch. Choose the gear ratios for the five gears in the gearbox.

14.23 In the train drive shown in sketch *b*, gear 2 is the inlet of the gear train and transfers 2.1 kW of power at 2000 rpm. Each gear contact reduces the speed ratio by 5:3.

a) Find the number of teeth for gears 3 and 4.

b) Determine the rotational speed of each gear.

c) Calculate the torque of each shaft, when because of frictional losses there is a 5% power loss in each matching pair.

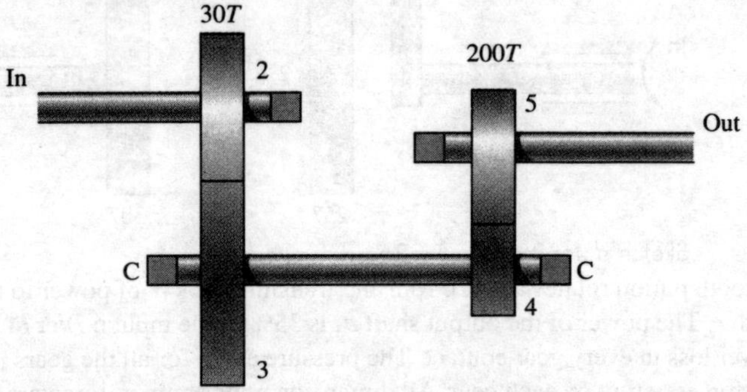

Sketch *b*, for Problem 14.23

14.24 In Problem 14.23 the module of all gears in the train drive is 8 mm and the pressure angle is 25°. Draw a free-body diagram of the forces acting on each gear in the train.

14.25 In the gear train shown in sketch *c* the pinion rotates at 420 rpm and transfers 5.5 kW to the train. The circular pitch is 31.4 mm and the pressure angle is 25°. Find the output speed and draw a free-body diagram of the forces acting on each gear.

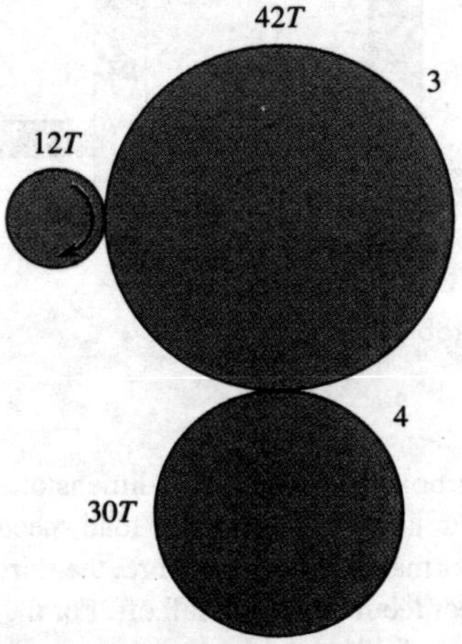

Sketch *c*, for Problem 14.25

14.26 The input shaft of the planetary gear shown in sketch *d* has 10 kW of power and is rotating clockwise at 500 rpm. The modules are 3 mm for gear 4 and 4 mm for gear 5. The pressure angle is 20°. The inlet and outlet shafts are coaxial. Find the following:

a) The output speed and its direction

b) The minimum inside diameter of the gearbox housing

c) The contact loads acting on each gear

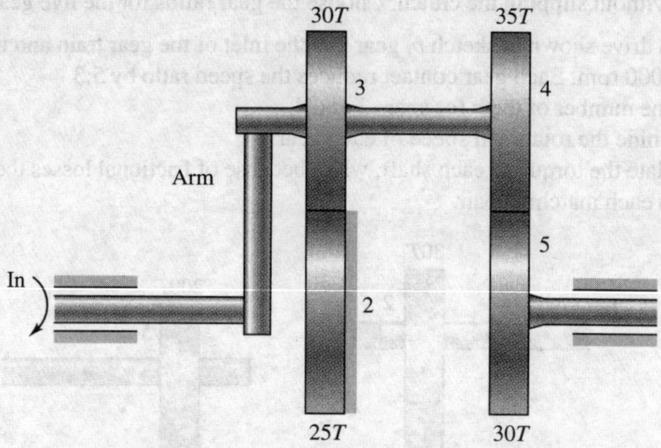

Sketch *d*, for Problem 14.26

14.27 An 18-tooth pinion rotates at 1420 rpm and transmits 52 kW of power to the gear train shown in sketch *e*. The power of the output shaft o_1 is 35% of the input power of the train. There is 3% power loss in every gear contact. The pressure angle for all the gears is 25°. Calculate the contact forces acting on each gear. All dimension units are in millimeters.

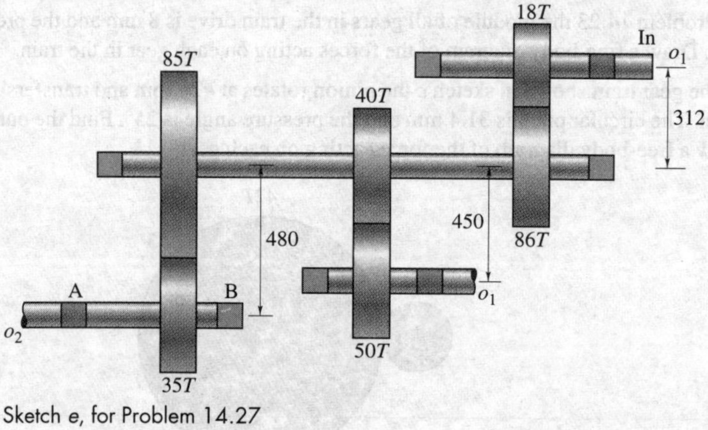

Sketch *e*, for Problem 14.27

Section 14.8

14.28 The third gear in the gearbox of a sports car is dimensioned to have a finite life in order to save weight. When the car is used in a race, the load spectrum is totally different from the spectrum produced by normal driving. Therefore, the third gear fails by bending stresses at the gear roots and the gear teeth crack and fall off. For the load spectrum used to dimension the gearbox, bending stress failure at the gear tooth root has the same stress safety factor as surface pitting at the contact areas. The gears are made of grade 2 steel with Brinell hardness

HB = 250. By changing the material to 400 Brinell hardness, the bending stress failures disappear. Is there then any risk of surface pitting?

14.29 A speed of 520 rpm is used in a 2:1 gear reduction pair to transfer 3 kW of power for a useful life of 50 million turns with 90% reliability. A pressure angle of 20° is to be used along with a safety factor of 2. The loads are uniform with a regular mounting and hub-machined teeth. On the basis of fatigue strength determine the proper material, module, and number of teeth for this pair, and calculate the pitch diameters and the face width.

Section 14.9

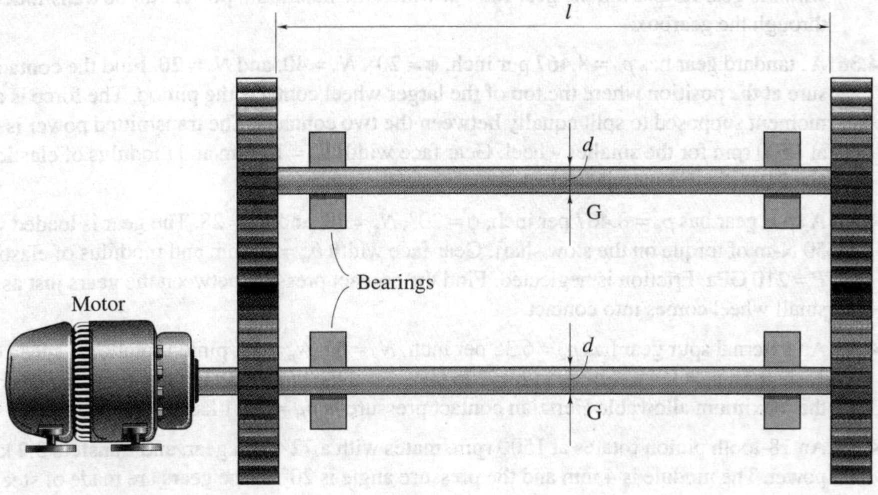

Sketch *f*, for Problem 14.30

14.30 An endurance strength test machine (see sketch *f*) consists of a motor, a gearbox, two elastic shafts, bearings, and the test gears. The gear forces are created by twisting the shafts when the gears are mounted. At one location the shafts were twisted at an angle representing one gear tooth. Find the gear force in the test gears as a function of the parameters given in the sketch.

14.31 For a 2:1 gear reduction the cast iron pinion is rotating at 320 rpm and transfers 5.2 kW of power. The gears are standard full depth with a 25° pressure angle. Design this gear pair for a safety factor of 2.

Section 14.10

14.32 To increase the power density of a machine, the rotational speeds are increased. Typical examples of this practice are high-speed electric motors and vacuum cleaners, where the speeds recently have been increased more than tenfold by using modern frequency converters. If the output shaft from a high-speed motor must have a low speed, a reduction gear is used. The quality of the reduction gear determines how large the dynamic impact forces will be in the output shaft gear. Hobbed gears of quality level 5 are used for a gear rotating at 500 rpm at a pitch diameter of 50 mm. To increase the throughput power, the speed is increased to 5000 rpm. How much higher power can be transferred through the gear at that speed? How high would the power be with extremely accurate gearing?

14.33 A 25-tooth pinion running at 1200 rpm and transferring 13 kW gives a 4:1 reduction gear pair. The gears are made of steel. The circular pitch is 15.7 mm, the face width is 49 mm, and the pressure angle is 20°. Determine the bending stress of the gears.

Section 14.11

14.34 A standard gear with pressure angle $\phi = 20°$ and diametral pitch $p_d = 8.47$ per inch transfers 24 kW at a pinion rotational speed of 1500 rpm. Face width $b_w = 15$ mm, $N_p = 24$, $N_g = 42$, and $E = 207$ GPa for the gear steel. Find the contact pressure at the pitch point when friction and power loss are neglected.

14.35 A gearbox with spur gears has a fixed shaft center distance c_d. The contact pressure between the gears at the pitch point limits the power being transmitted through the gearbox. The rotational speed of the incoming shaft is fixed. The rotational speed of the outgoing shaft varies with the gear ratio. Find the gear ratio at which the maximum power can be transmitted through the gearbox.

14.36 A standard gear has $p_d = 8.467$ per inch, $\phi = 20°$, $N_g = 40$, and $N_p = 20$. Find the contact pressure at the position where the top of the larger wheel contacts the pinion. The force is at that moment supposed to split equally between the two contacts. The transmitted power is 16 kW at 1500 rpm for the smaller wheel. Gear face width $b_w = 18$ mm and modulus of elasticity $E = 210$ GPa. Friction is neglected.

14.37 A spur gear has $p_d = 8.467$ per inch, $\phi = 20°$, $N_g = 58$, and $N_p = 28$. The gear is loaded with 50 N-m of torque on the slow shaft. Gear face width $b_w = 20$ mm and modulus of elasticity $E = 210$ GPa. Friction is neglected. Find the contact pressure between the gears just as the small wheel comes into contact.

14.38 An external spur gear has $p_d = 6.35$ per inch, $N_g = 52$, $N_p = 28$, pinion rotational speed of 1500 rpm, $b_w = 0.7874$ in., and $E = 206$ GPa. Find the largest power the gear can transmit if the maximum allowable Hertzian contact pressure is $p_H = 142.1$ ksi.

14.39 An 18-tooth pinion rotates at 1500 rpm, mates with a 72-tooth gear, and transfers 3.2 kW of power. The module is 4 mm and the pressure angle is 20°. If the gears are made of steel with a yield strength of 720 MPa, calculate the face width for a safety factor of 5. Also calculate the contact stress between the mating teeth.

14.40 A 16-tooth pinion rotates at 720 rpm and transfers 5 kW to a 3:1 speed reduction gear pair. The circular pitch is 9.4 mm, the face width is 38 mm, and the pressure angle is 20°. Calculate the contact stresses for this gear pair.

Section 14.13

14.41 A 16-tooth, right-handed helical pinion turns at 2000 rpm and transfers 8 kW to a 38-tooth gear. The helix angle is 30°, the normal pressure angle is 25°, and the normal module is 20 mm. For a safety factor of 3 calculate the pitch diameters; the normal, axial, and tangential pitch; the tangential pressure angle; and the face width. The material is AISI 3140 hot-rolled steel.

CHAPTER

15

FASTENERS AND POWER SCREWS

A collection of threaded fasteners. (Courtesy of Clark Craft Fasteners)

Engineers need to be continually reminded that nearly all engineering failures result from faulty judgments rather than faulty calculations.

Eugene S. Ferguson, *Engineering and the Mind's Eye*

SYMBOLS

A	cross-sectional area, m^2
A_g	gasket area per bolt, m^2
A_i	constant
A_s	shear stress area, m^2
A_t	tensile stress area, m^2
a	distance from wall to applied load, m
B_i	constant
b	member width, m
C_k	dimensionless stiffness parameter
c	distance from neutral axis to outer fiber, m

d	major diameter, largest possible diameter of thread, m
d_c	crest diameter, largest actual diameter of thread, m
d_i	smallest diameter of frustum cone, m
d_p	pitch diameter, m
d_r	root diameter, smallest diameter of thread, m
d_s	distance traveled in axial direction, m
d_w	washer diameter, m
E	modulus of elasticity, Pa
e	efficiency of raising load, percent
e_k	elongation, m
h	maximum cantilever thickness, m
h_e	weld leg length, m
h_p	power, W
h_t	largest possible thread height, m
I	area moment of inertia, m^4
I_u	unit area moment of inertia, m^3
J	polar area moment of inertia, m^4
J_u	unit polar area moment of inertia, m^3
K_f	fatigue stress concentration factor
k	stiffness, N/m
k_f	surface finish factor
k_m	miscellaneous factor
k_r	reliability factor
k_s	size factor
L	length, m
L_g	grip length, m
L_s	shank length, m
L_{se}	shank effective length, m
L_t	threaded length, m
L_{te}	effective threaded length, m
L_w	weld length, m
l	lead, m
M	bending moment, N-m
m	1, 2, 3,… for single-, double-, and triple-threaded screws, respectively
N_a	rotational speed, rpm
N_r	number of rivets in width of member
n	number of threads per inch, in.$^{-1}$
n_o	number of revolutions
n_s	safety factor
P	load; friction force; deflection force; stretching force; press force; shear force, N
P_i	preload, N
P_n	load acting on diagonal of parallelepiped, N
P_0	load neglecting friction, N
p	pitch, in.
p_0	minimum gasket seal pressure, Pa
r	distance from centroid of weld group to farthest point in weld, m
S_e	modified endurance limit, Pa
S_e'	endurance limit, Pa
S_p	proof strength, Pa
S_y	yield strength, Pa
S_u	ultimate tensile strength, Pa
T	torque, N-m
T_l	torque needed to lower load, N-m
T_r	torque needed to raise load, N-m
t_e	weld throat length, m
t_m	thickness of thinnest member, m
V	shear force, N
W	load; mating force; tensile force, N
w'	force per unit length of weld, N/m
x, y	coordinates, m
$\bar{x}, \bar{y}$	centroid, m
y	maximum cantilever deflection, m
Z_m	section modulus, m^3
α	lead angle; wedge angle of cantilever tip; angle of inclination, deg
α_f	angle of frustum cone stress representation, deg
β	thread angle, deg
γ	cone angle, deg
ε	maximum allowable strain
θ_n	angle created by force P_n for diagonal of parallelepiped shown in Fig. 15.7(a), deg
μ	coefficient of friction between threads
μ_c	coefficient of friction of collar
σ	normal stress, Pa
τ	shear stress, Pa
ω	rotational speed, rad/s

Subscripts

a	alternating
all	allowable
b	bolt
c	collar
cr	critical
d	direct
g	gasket
i	preload; inner
j	joint
m	mean
max	maximum
min	minimum
n	normal
o	outer
p	proof
s	shear; secant
t	thread; torsional
x, y	coordinates

15.1 INTRODUCTION

The primary focus of this chapter is on fasteners. Since the complications of manufacturing intricate parts often require assembly of components, engineers are then confronted with the task of fastening various members together. A number of fastener types can be used for this task, including threaded fasteners, riveted joints, welded joints, and adhesive joints. Each of these types is considered in this chapter. Various fastening methods are presented, and information is given about selecting a suitable type for a specific task. Interesting information about fastener materials indicates that the bolt, rivet, or weld material should be strong and tough whereas the members being fastened should ideally be soft and ductile. This chapter also considers power screws.

15.2 THREAD TERMINOLOGY, CLASSIFICATION, AND DESIGNATION

Each machine element has its unique terminology. Figure 15.1 describes the terminology and dimensions of threaded parts. Note in particular the differences between the various diameters (major, crest, pitch, and root), the largest possible thread height h_t, and the thread angle β. The **pitch** p shown in Fig. 15.1 is the distance from a point on one thread to the same point on an adjacent thread, and its unit is meters or inches. A parameter that can be used instead of pitch is **threads per inch** n. An important relationship between pitch and threads per inch is

$$p = 1/n \tag{15.1}$$

Another thread parameter is **lead**, which is the distance that the screw would advance relative to the nut in one revolution. For a single-threaded screw $l = p$ and for a double-threaded screw $l = 2p$, etc. Figure 15.2 shows the differences between single-, double-, and triple-threaded screws.

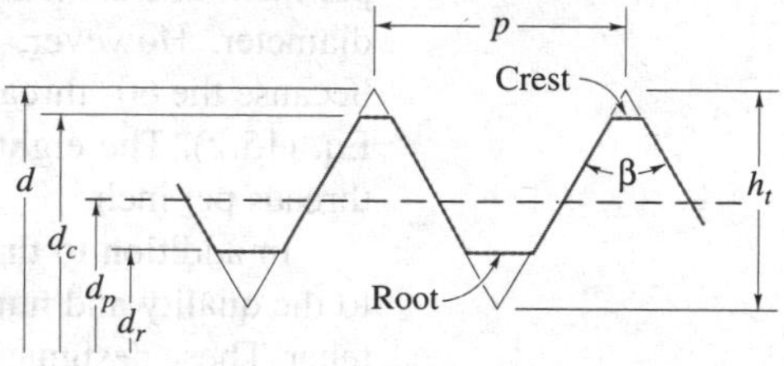

Figure 15.1 Parameters used in defining terminology of thread profile.

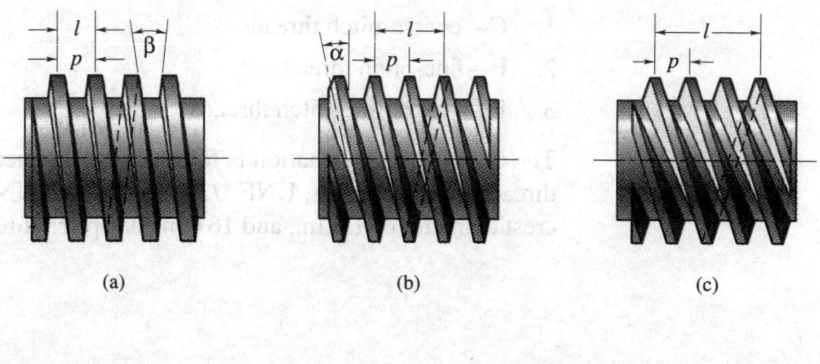

Figure 15.2 (a) Single-, (b) double-, and (c) triple-threaded screws.

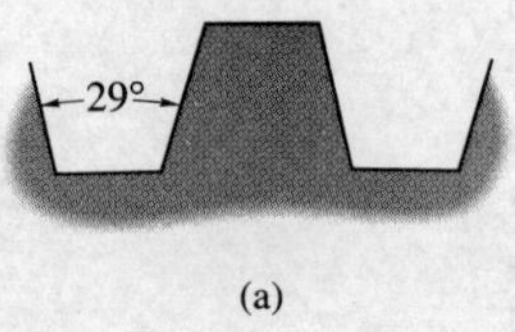

Figure 15.3 Thread profiles. (a) Acme; (b) UN.

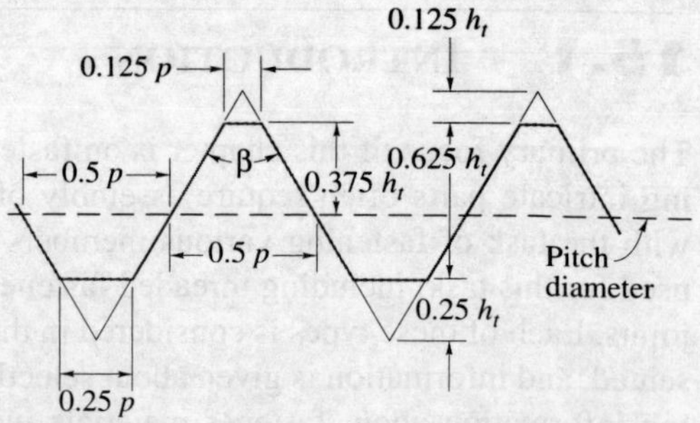

Figure 15.4 Details of M and UN thread profiles.

A number of different thread profiles can be used for a wide variety of applications. Figure 15.3 shows two types. The Acme thread profile is used for power screws and machine tool threads. The second profile, the unified (UN), is used extensively today. The Acme profile has a thread angle of 29°, whereas the UN profile has a thread angle of 60°. The metric profile (M) is popular and is quite similar to the UN profile.

Figure 15.4 shows more details of the M and UN thread profiles than are shown in Fig. 15.3. From this figure

$$h_t = \frac{0.5p}{\tan 30°} = 0.8660\,p \tag{15.2}$$

Once pitch p and the largest possible thread height h_t are known, the various dimensions of the UN and M thread profiles can be obtained. This text focuses on the UN thread profile while using English units and the M profile for metric units. Information obtained from these thread profiles can be directly applied to other thread profiles.

The term **thread series** can be applied to threads of any size. The UN thread profile has eight thread series of constant pitch. Each thread series has the same number of threads per inch. The helix angle that the thread makes around the fastener varies with the fastener diameter. However, the thread depth is constant, regardless of the fastener diameter, because the 60° thread angle remains constant. This situation can be clearly observed from Eq. (15.2). The eight constant-pitch UN thread series are 4, 6, 8, 12, 16, 20, 28, and 32 threads per inch.

In addition to thread series, thread profiles are classified by **coarseness**, which refers to the quality and number of threads per inch produced on a common diameter of the fastener. These designations after UN have the following meanings:

1. C—coarse-pitch threads
2. F—fine-pitch threads
3. EF—extra-fine-pitch threads

The coarseness designation is followed by the mean diameter in inches and the number of threads per inch. Thus, UNF 1/2 × 16 means a UN thread profile with fine-pitch threads, a crest diameter of 1/2 in., and 16 constant-pitch threads per inch.

Table 15.1 Inch and metric equivalent thread classifications.

Inch series		Metric series	
Bolts	**Nuts**	**Bolts**	**Nuts**
1A	1B	8g	7H
2A	2B	6g	6H
3A	3B	8h	5H

For metric thread profiles the coarseness designation is usually considered only as coarse or fine, omitting the extra-fine designation. Instead of using threads per inch, the metric thread series simply uses the pitch distance between two threads in millimeters. For example, MF 6×1 means an M thread profile with fine-pitch threads, a crest diameter of 6 mm, and a pitch distance of 1 mm.

The classifications given above are only applicable for individual threads and do not consider how the male and female parts of the fastener fit together. That is, is the fit loose and sloppy or is it tight? In English units the letter A designates external threads and B designates internal threads. There are three fits: 1(the loosest fit), 2 (a normal fit), and 3 (a tight fit). Thus, for a very dirty environment a 1 fit would be specified; a 3 fit requires an extra degree of precision and a very clean environment. A 2 fit is usually designated for normal applications, somewhere between the extremes of 1 and 3. Thus, the designation UNC 2×8–1B means a UN thread profile with coarse-pitch threads, a crest diameter of 2 in., eight constant-pitch threads per inch, and a loose fit, with the internal part of the fastener being specified.

The fit designation has more options in metric units than in English units. There G and H denote internal threads, and e, f, g, and h denote external threads. For example, G and e define the loosest fit and the greatest clearance, and H and h define zero allowance, no deviation from the basic profile. Thus, a very dirty environment would require a G and e fit and an extra degree of precision, and a very clean environment would require an H and h fit.

The tolerance grade for metric threads is denoted by a number symbol. Seven tolerance grades have been established and are identified by numbers 3 to 9. Nine defines the loosest fit (the most generous tolerance) and 3 defines the tightest. Thus, MF 8×2–G6 means an M thread profile with fine-pitch threads, a crest diameter of 8 mm, a pitch of 2 mm, and a normal fit of the internal threads.

Table 15.1 gives approximate equivalent fits of English and metric unit threads. This table does not show all possible equivalencies but does give an idea of how the two systems work.

Example 15.1

GIVEN A screw thread is going to be chosen for a nuclear power plant pressure vessel lid. The outside diameter of the screws is limited to 160 mm, and the length of the threaded part screwed into the pressure vessel wall is 220 mm. For metric threads M160 three different pitches are preferred: 6, 4, and 3 mm. The manufacturing tolerance for the screw thread outer diameter is 0.6 mm, and the difference in thermal expansion of the pressure vessel wall relative to the screw is 0.2 mm because the screw heats more slowly than the wall during start-up. The manufacturing tolerance for the thread in the pressure vessel wall is 0.9 mm.

FIND Which pitch should be chosen?

Solution

The highest mean shear stress in the thread is equal to the force divided by π times the square root diameter of the screw (if there is no play between the screw and nut) times the length of the thread. If there is play between the screw and nut, a smaller part consisting of only the tops of the threads needs to be sheared off; and if a weaker material is used in the pressure vessel than in the screws, that material will break before the screw threads.

If the pitch is 3 mm, the height of the thread profile is

$$(0.625)\left(\frac{\sqrt{3}}{2}\right)(3) = 1.624 \text{ mm}$$

which is less than the combined thermal expansion and manufacturing error. For maximum safety the largest pitch of 6 mm should be chosen.

15.3 POWER SCREWS

Power screws are a device used to change angular motion into linear motion and usually to transmit power. More specifically, power screws are used

1. To obtain greater mechanical advantage in order to lift weight, as in a screw type of jack for cars
2. To exert large forces, as in a home compactor or a press
3. To obtain precise positioning of an axial movement, as in a micrometer screw or the lead screw of a lathe

In each of these applications a torque is applied to the ends of the screws through a set of gears, thus creating a load on the device.

Power screws use the Acme thread profile, first shown in Fig. 15.3(a). Figure 15.5 shows more details of Acme threads. The thread angle is 29°, and the thread dimensions can be easily determined once the pitch is known.

Table 15.2 gives the crest diameter, the number of threads per inch, and the tensile and shear stress areas for Acme power screw threads. The tensile stress area is

$$A_t = \frac{\pi}{4}\left(\frac{d_r + d_p}{2}\right)^2 \tag{15.3}$$

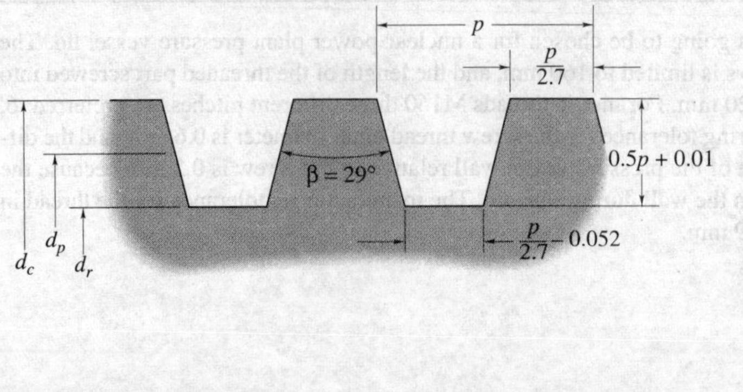

Figure 15.5 Details of Acme thread profile. (All dimensions are in inches.)

Table 15.2 Crest diameters, threads per inch, and stresses for Acme thread

Crest diameter, d_c, in.,	Number of threads per inch, n	Tensile stress area, A_t, in.^{2}	Shear streess areas, A_s, in.^{2}
1/4	16	0.02632	0.3355
5/16	14	.04438	.4344
3/8	12	.06589	.5276
7/16	12	.09720	.6396
1/2	10	.1225	.7278
5/8	8	.1955	.9180
3/4	6	.2732	1.084
7/8	6	.4003	1.313
1	5	.5175	1.493
1 1/8	5	.6881	1.722
1 1/4	5	.8831	1.952
1 3/8	4	1.030	2.110
1 1/2	4	1.266	2.341
1 3/4	4	1.811	2.803
2	4	2.454	3.262
2 1/4	3	2.982	3.610
2 1/2	3	3.802	4.075
2 3/4	3	4.711	4.538
3	2	5.181	4.757
3 1/2	2	7.338	5.700
4	2	9.985	6.640
4 1/2	2	12.972	7.577
5	2	16.351	8.511

The shear stress area A_s is also given in Table 15.2, but a simple formulation of it is not readily available. Note that A_s represents the area in shear approximately at the pitch line of the threads for a 1.0-in. length of engagement. The length of engagement is the axial length over which the threads are engaged with each other. Other lengths than 1.0 in. would require that the shear stress area be modified by the ratio of the actual length to 1.0 in. Because of this, the modified shear stress area differs considerably from the tensile stress area. Table 15.2 and Fig. 15.5 should be used jointly in evaluating Acme power screws. The pitch diameter of an Acme power screw thread is

$$d_p = d_c - 0.5p - 0.01 \tag{15.4}$$

This equation is valid only when inches are used for d_c and p.

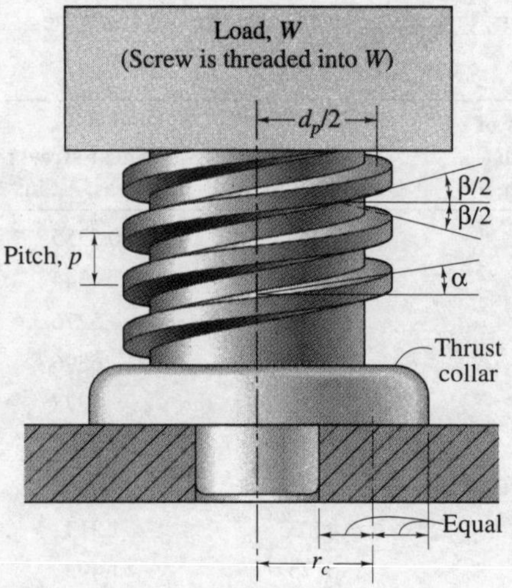

Figure 15.6 Dimensions and angles of power screw with collar.

15.3.1 FORCES AND TORQUE

Figure 15.6 shows a load W into which the supporting screw is threaded and which can be raised or lowered by rotating the screw. It is assumed that the load W is prevented from turning when the screw rotates. Also shown in Fig. 15.6 is the thread angle of an Acme power screw, which is 29°, and the lead angle. The lead angle α relates the lead to the pitch circumference through the following equation:

$$\alpha = \tan^{-1}\left(\frac{l}{\pi d_p}\right) \tag{15.5}$$

where

$$l = \text{lead} = mp \tag{15.6}$$

$m = 1$, single-threaded screw

$m = 2$, double-threaded screw

$m = 3$, triple-threaded screw

d_p = pitch diameter, m

Figure 15.6 also shows a thrust collar, which will be covered later in this chapter.

The distance traveled in the axial direction can be expressed as

$$d_s = n_o l = n_o mp \tag{15.7}$$

where

n_o = number of revolutions

Thus, the power screw uses rotary motion to obtain a uniform axial linear motion.

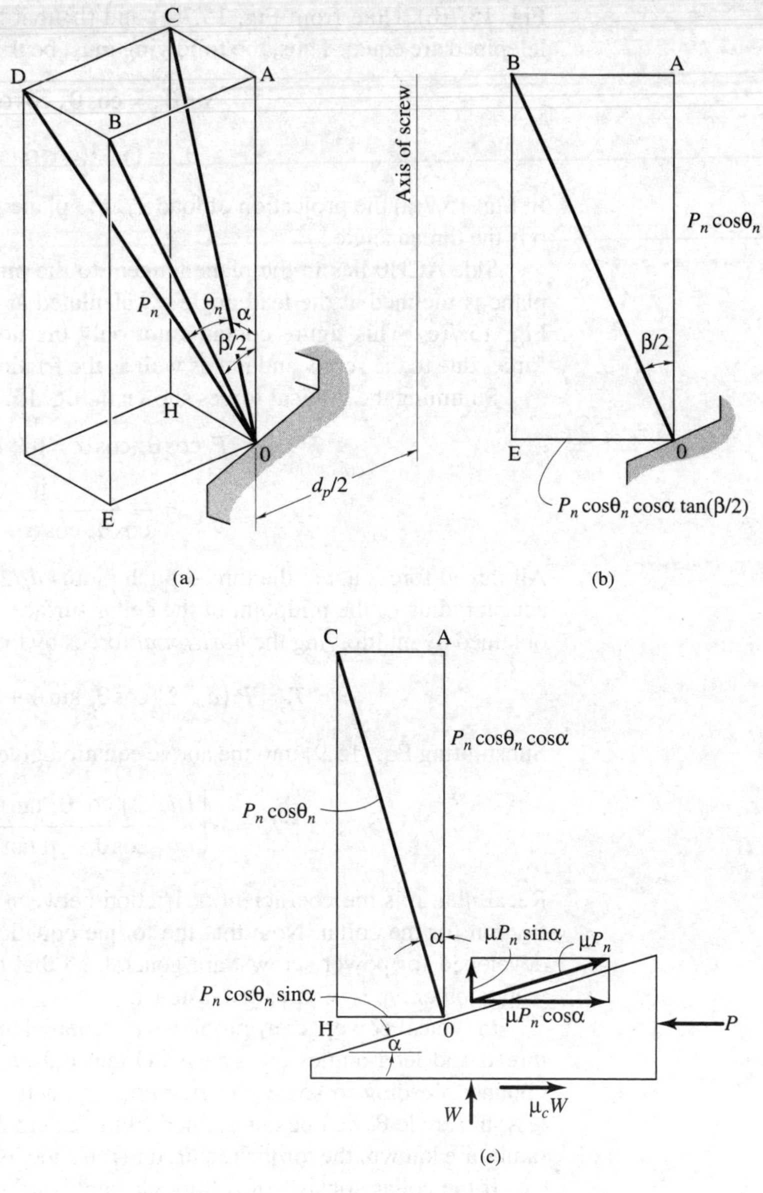

Figure 15.7 Forces acting in raising load of power screw.
(a) Forces acting on parallelepiped;
(b) forces acting on axial section;
(c) forces acting on tangential plane.

15.3.1.1 RAISING THE LOAD To determine the force P required to overcome a certain load W, it is necessary to observe the relationship of the load to the distance traveled. The total force on the threads is represented by a single source P_n, shown in Fig. 15.7(a), that is normal to the thread surface. From Fig. 15.7 recall the representation of the lead angle α and the thread angle β. The angle θ_n is created by the force P_n, which is the diagonal of the parallelepiped. Side ABE0 is an axial section through the bolt and is shown in

Fig. 15.7(b). Note from Fig. 15.7(a) and (b) that the sides DC, AB, and 0E of the parallelepiped are equal. Thus, the following must be true:

$$\sin\theta_n = \cos\theta_n \cos\alpha \tan\beta/2$$
$$\theta_n = \tan^{-1}(\cos\alpha \tan\beta/2) \qquad \textbf{(15.8)}$$

In Fig. 15.7(a) the projection of load P_n into plane ABE0 is inclined at an angle $\beta/2$, where β is the thread angle.

Side ACH0 lies in the plane tangent to the pitch point. The projection of P_n into this plane is inclined at the lead angle α calculated at the screw pitch radius and is shown in Fig. 15.7(c). This figure contains not only the normal components but also the friction forces due to the screw and nut as well as the friction forces due to the collar.

Summing the vertical forces shown in Fig. 15.7(c) gives

$$P_n \cos\theta_n \cos\alpha = \mu P_n \sin\alpha + W$$
$$P_n = \frac{W}{\cos\theta_n \cos\alpha - \mu\sin\alpha} \qquad \textbf{(15.9)}$$

All thread forces act at the thread pitch radius $d_p/2$. From Fig. 15.6 the force of the collar acts at radius r_c, the midpoint of the collar surface. The torque required to raise the load is obtained by multiplying the *horizontal* forces by the appropriate radii. Thus,

$$T_r = P_n(d_p/2)(\cos\theta_n \sin\alpha + \mu\cos\alpha) + r_c\mu_c W$$

Substituting Eq. (15.9) into the above equation gives

$$T_r = W\left[\frac{(d_p/2)(\cos\theta_n \tan\alpha + \mu)}{\cos\theta_n - \mu\tan\alpha} + r_c\mu_c\right] \qquad \textbf{(15.10)}$$

Recall that μ is the coefficient of friction between the threads and μ_c is the coefficient of friction for the collar. Note that the torque equations [Eqs. (15.10) and (15.8)], although developed for power screws, are general, so that these equations can be applied equally well to other types of threaded fastener.

In typical power screw problems the coefficients of friction μ and μ_c are given and the thread and lead angles are known. Often, only approximate values of μ and μ_c can be obtained, leading to some uncertainty in the actual forces and torques required. Regardless, the angle θ_n can be determined from Eq. (15.8). Once the pitch radius and the collar radius are known, the torque required to raise the load can be calculated.

If the collar consists of rolling-element bearings, the collar friction is extremely low and can be neglected ($\mu_c = 0$).

15.3.1.2 LOWERING THE LOAD Lowering the load differs from raising the load, given in Fig. 15.7, only in that the friction force components [Fig. 15.7(c)] become negative. These changes result in the summation of the vertical forces being

$$P_n = \frac{W}{\cos\theta_n \cos\alpha + \mu\sin\alpha} \qquad \textbf{(15.11)}$$

The torque required to lower the load is obtained by multiplying the horizontal forces by the appropriate radii. Thus,

$$T_l = P_n(d_p/2)(\cos\theta_n \sin\alpha - \mu_n \cos\alpha) - r_c\mu_c W$$

Substituting Eq. (15.11) into the above equation gives

$$T_l = -W\left[\frac{(d_p/2)(\mu - \cos\theta_n \tan\alpha)}{\cos\theta_n + \mu \tan\alpha} + r_c\mu_c\right] \tag{15.12}$$

Using Eqs. (15.8) and (15.12) gives the torque required to lower the load.

15.3.2 POWER AND EFFICIENCY

Once the torque is known, the power transferred by the power screw can be obtained from Eqs. (4.40) and (4.42). From Eq. (4.40) the power h_p in horsepower can be expressed as

$$h_p = TN_a/63\,025$$

where

T = torque, N-m

N_a = rotational speed, rpm

For metric units from Eq. (4.42) the power h_p in watts is

$$h_p = T\omega$$

where

ω = rotational speed, rad/s

The efficiency of a screw mechanism, expressed in percent, is the ratio of work output to work input, or

$$e = \frac{100Wl}{2\pi T} \tag{15.13}$$

where

l = lead, m

W = load, N

Once the torque required to raise the load is known, the efficiency of the power screw mechanism can be calculated. Note from Eqs. (15.10) and (15.12) that if the coefficients of friction of the threads and the collar are equal to zero,

$$e = \frac{100Wl}{2\pi W(d_p/2)\tan\alpha} = \frac{(100)l}{2\pi(d_p/2)\tan\alpha}$$

Therefore, making use of Eq. (15.5) shows that if μ and μ_c can be reduced to zero, the efficiency will be 100%.

Example 15.2 **GIVEN** A double-threaded Acme power screw has a load of 1000 lbf, a crest diameter of 1 in., a collar diameter of 1.5 in., a thread coefficient of friction of 0.16, and a collar coefficient of friction of 0.12.

FIND Determine the torque required to raise and lower the load. Also determine the efficiency for raising the load.

Solution

From Table 15.2 for $d_c = 1$ in. the number of threads per inch $n = 5$, or $p = 0.2$ in. The lead $l = 2p = 0.4$ in. Also, Eq. (15.4) gives

$$d_p = d_c - 0.5p - 0.01 = 1.0 - 0.5/5 - 0.01 = 0.89 \text{ in.}$$

From Eq. (15.5) the lead angle is

$$\alpha = \tan^{-1}\left(\frac{l}{\pi d_p}\right) = \tan^{-1}\left(\frac{0.4}{\pi 0.89}\right) = 8.142°$$

Recall that for an Acme screw the thread angle $\beta = 29°$. Making use of Eq. (15.8) gives

$$\theta_n = \tan^{-1}\left[\cos\alpha \tan(\beta/2)\right] = \tan^{-1}\left[\cos 8.142° \tan 14.5°\right] = 14.36°$$

Note that θ_n and $\beta/2$ are nearly equal.

Using Eq. (15.10) gives the torque required to *raise* the load as

$$T_r = W\left[\frac{(d_p/2)(\cos\theta_n \tan\alpha + \mu)}{\cos\theta_n - \mu\tan\alpha} + r_c\mu_c\right]$$

$$= 1000\left[\frac{(0.445)(\cos 14.36° \tan 8.142° + 0.16)}{\cos 14.36° - 0.16\tan 8.142°} + 0.75(0.12)\right]$$

$$= 1000[0.1405 + 0.0900] = 230.5 \text{ in.-lbf}$$

Using Eq. (15.12) gives the torque required to *lower* the load as

$$T_l = -W\left[\frac{(d_p/2)(\mu - \cos\theta_n \tan\alpha)}{\cos\theta_n + \mu\tan\alpha} + r_c\mu_c\right]$$

$$= -1000\left[\frac{(0.445)(0.16 - \cos 14.36° \tan 8.142°)}{\cos 14.36° + 0.16\tan 8.142°} + 0.75(0.12)\right]$$

$$= -1000[0.0096 + 0.0900] = -99.60 \text{ in.-lbf.}$$

Note that 90.4% of the torque is due to collar friction. As would be expected, less torque is required to lower the load than to raise the load. The minus sign indicates the opposite direction in lowering the load relative to raising the load.

Using Eq. (15.13) gives the efficiency in raising the load as

$$e = \frac{100Wl}{2\pi T_r} = \frac{100(1000)(0.4)}{2\pi(230.5)} = 27.62\%$$

Example 15.3

GIVEN The exact situation in Example 15.2 except that a single-threaded power screw is used instead of a double-threaded one.

FIND Determine the torque required to raise and lower the load. Also determine the efficiency for raising the load.

Solution

The lead equals the pitch ($l = p = 0.2$ in.). The lead angle is

$$\alpha = \tan^{-1}\left(\frac{l}{\pi d_p}\right) = \tan^{-1}\left(\frac{0.2}{0.89\pi}\right) = 4.091°$$

From Eq. (15.8)

$$\theta_n = \tan^{-1}\left[\cos\alpha\tan(\beta/2)\right] = \tan^{-1}(\cos 4.091° \tan 14.5°) = 14.46°$$

Using Eq. (15.10) gives the torque required to *raise* the load as

$$T_r = W\left[\frac{(d_p/2)(\cos\theta_n \tan\alpha + \mu)}{\cos\theta_n - \mu\tan\alpha} + r_c\mu_c\right]$$

$$= 1000\left[\frac{(0.445)(\cos 14.46° \tan 4.091° + 0.16)}{\cos 14.46° - 0.16\tan 4.091°} + 0.75(0.12)\right]$$

$$= 1000[0.1066 + 0.0900] = 196.6 \text{ in.-lbf}$$

Note that to raise the load, the single-threaded screw requires lower torque than the double-threaded screw.

Using Eq. (15.13) gives the efficiency in raising the load as

$$e = \frac{100Wl}{2\pi T_r} = \frac{100(1000)(0.2)}{2\pi(196.6)} = 16.19\%$$

The single-threaded screw is thus less efficient (by 11 percentage points) than the double-threaded screw given in Example 15.2.

Using Eq. (15.12) gives the torque required to *lower* the load as

$$T_l = -W\left[\frac{(d_p/2)(\mu - \cos\theta_n \tan\alpha)}{\cos\theta_n + \mu\tan\alpha} + r_c\mu_c\right]$$

$$= -1000\left[\frac{(0.445)(0.16 - \cos 14.46° \tan 4.091°)}{\cos 14.46° + 0.16\tan 4.091°} + 0.75(0.12)\right]$$

$$= -1000[0.0412 + 0.0900] = -131.2 \text{ in.-lbf}$$

15.3.3 Self-Locking Screws

If the screw thread is very steep (i.e., has a large lead angle), the friction force may not be able to stop the tendency of the load to slide down the plane, and gravity will cause the load to fall. Usually, for power screws the lead angle is small, and the friction force of the thread interaction is large enough to oppose the load and keep it from sliding down. Such a screw is called **self-locking**, a desirable characteristic to have in power screws.

If the collar uses a rolling-element bearing, the collar friction can be assumed to be zero. From Eq. (15.12) the torque required to lower the load is negative if

$$\mu - \cos\theta_n \tan\alpha \geq 0$$

Thus, under static conditions self-locking occurs if

$$\mu > \cos\theta_n \tan\alpha = \frac{l\cos\theta_n}{\pi d_p} \qquad \textbf{(15.14)}$$

Vibrations may cause the power screw to move.

Example 15.4

GIVEN The double-threaded screw of Example 15.2 and the single-threaded screw of Example 15.3.

FIND Determine the thread coefficient of friction required to ensure that self-locking occurs. Assume that rolling-element bearings have been installed at the collar.

Solution

For the double-threaded screw self-locking occurs if

$$\mu > \cos\theta_n \tan\alpha = \cos 14.36° \tan 8.142° = 0.1386$$

For the single-threaded screw self-locking occurs if

$$\mu > \cos\theta_n \tan\alpha = \cos 14.36° \tan 4.091° = 0.0693$$

Thus, the double-threaded screw requires twice the friction coefficient of the single-threaded screw.

Example 15.5

GIVEN A triple-threaded Acme power screw has a crest diameter of 2 in. and four threads per inch. A rolling-element bearing is used at the collar, so that the collar coefficient of friction is zero. The thread coefficient of friction is 0.15.

FIND Determine the load that can be raised by a torque of 400 in.-lbf. Also, check to see if self-locking occurs.

Solution

The pitch and the lead are

$$p = 1/n = 1/4 \text{ in.} = 0.25 \text{ in.}$$

$$l = mp = 3(0.25) = 0.75 \text{ in.}$$

The pitch diameter is

$$d_p = d_c - 0.5p - 0.01 = 2 - 0.5(0.25) - 0.01 = 1.865 \text{ in.}$$

From Eq. (15.5) the lead angle is

$$\alpha = \tan^{-1}\left(\frac{l}{\pi d_p}\right) = \tan^{-1}\left(\frac{0.75}{\pi(1.865)}\right) = 7.295°$$

From Eq. (15.8)

$$\theta_n = \tan^{-1}(\cos\alpha \tan\beta/2) = \tan^{-1}(\cos 7.295° \tan 14.5°) = 14.39°$$

Using Eq. (15.10) gives the torque required to *raise* the load as

$$T_r = W\left[\frac{(d_p/2)(\cos\theta_n \tan\alpha + \mu)}{\cos\theta_n - \mu\tan\alpha}\right]$$

$$400 = W\left[\frac{(1.865/2)(\cos 14.39° \tan 7.295° + 0.15)}{\cos 14.39° - 0.15 \tan 7.295°}\right]$$

$$W = \frac{400}{0.2691} = 1486 \text{ lbf}$$

From Eq. (15.14) self-locking occurs if

$$\mu > \cos\theta_n \tan\alpha$$

$$0.15 > \cos 14.39° \tan 7.295° = 0.1240$$

Therefore, self-locking occurs.

15.4 THREADED FASTENERS

A fastener is a device used to connect or join two or more members. Many fastener types and variations are available for use by the engineer. This chapter considers threaded, riveted, welded, and adhesive fasteners. The most common are threaded fasteners.

15.4.1 TYPES OF THREADED FASTENER

Figure 15.8 shows three types of threaded fastener: the bolt and nut, the cap screw, and the stud. Most threaded fasteners consist of a **bolt** passing through a hole in the members being joined and mating with a **nut**, as shown in Fig. 15.8(a). Occasionally, the bolt mates with threads in one of the members rather than with a nut, as shown in Fig. 15.8(b). This type is called a **cap screw**. A **stud** [Fig. 15.8(c)] is threaded on both ends and screwed, more or less permanently, into the threaded hole in one of the members being joined.

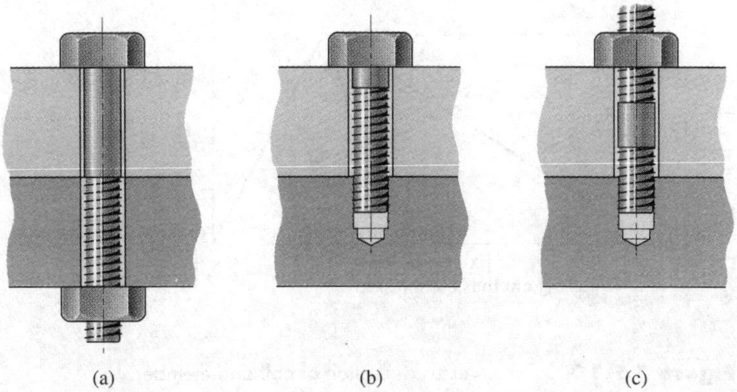

Figure 15.8
Three types of threaded fastener.
(a) Bolt and nut;
(b) cap screw; (c) stud.

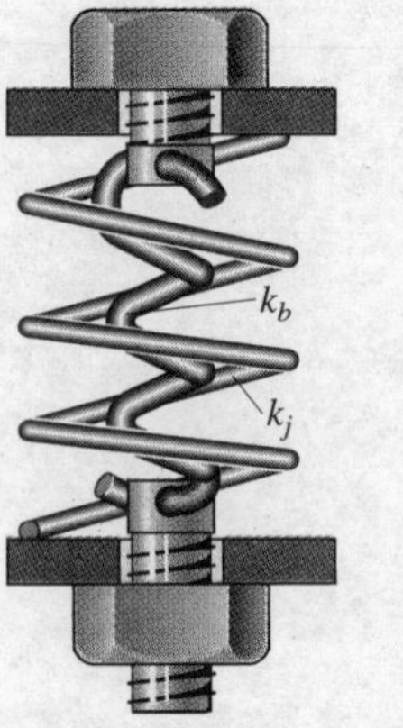

Figure 15.9 Bolt-and-nut assembly simulated as bolt-and-joint spring.

Most bolt heads (and nuts) are hexagonal, and this text considers only this type. A more detailed coverage of bolted joints can be found in Bickford (1995).

15.4.2 Load Analysis of Bolts and Nuts

The bolt and nut shown in Fig. 15.8(a) can be thought of as a spring system, as shown in Fig. 15.9. The bolt is viewed as a spring in tension with stiffness k_b. The joint, with a number of members being joined, is viewed as a compressive spring with stiffness k_j.

Figure 15.10 shows the forces and deflections acting on the bolt and the joint. In Fig. 15.10(a) the bolt and the joint are considered to be unassembled, and the load and the deflection are shown for each. For the bolt the force is tensile and the deflection is an extension of the bolt, whereas for the joint the deflection is a contraction. Figure 15.10(b) considers the assembled bolt, nut, and joint. The slopes of the load-deflection lines are the same as in Fig. 15.10(a), but the slopes for the bolt and joint are not only opposite in sign but have quite different values. The bolt and joint load-deflection lines intersect at a preload P_i.

Figure 15.11 shows the forces versus deflection for a bolt and a joint when an external load has been applied. The bolt is elongated by e_k and the contraction of the joint is reduced, as shown in Fig. 15.11. The force on the bolt becomes $P_i + k_b e_k$, and the force on the joint becomes $P_i - k_j e_k$.

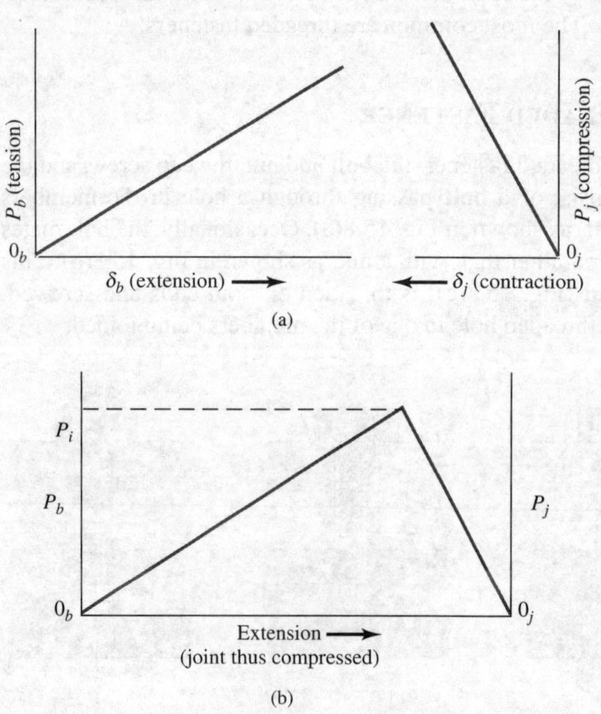

Figure 15.10 Force versus deflection of bolt and member. (a) Separated bolt and joint; (b) assembled bolt and joint.

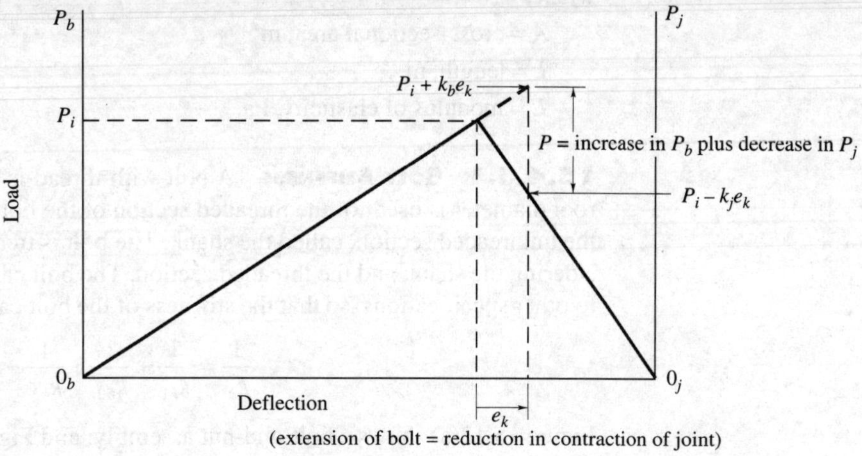

Figure 15.11 Forces versus deflection of bolt and joint when external load is applied.

From the equilibrium of the bolt

$$P + P_i - k_j e_k - P_i - k_b e_k = 0$$

$$\therefore e_k = \frac{P}{k_b + k_j} \tag{15.15}$$

Making use of Eq. (15.15) gives the load on the bolt as

$$P_b = P_i + k_b e_k = P_i + \frac{P k_b}{k_b + k_j} = P_i + C_k P \tag{15.16}$$

where the dimensionless stiffness parameter

$$C_k = k_b/(k_b + k_j) \tag{15.17}$$

From Eq. (15.17) C_k is less than 1.

The load on the joint is

$$P_j = P_i - k_j e_k = P_i - \frac{P k_j}{k_j + k_b} = P_i - (1 - C_k)P \tag{15.18}$$

The preceding analysis assumes that the external force is applied under the head of the bolt and under the head of the nut.

15.4.3 STIFFNESS PARAMETERS

Recall from Chapter 4 that the spring rate is the normal load divided by the elastic deflection for axial loading [Eq. (4.26)] and that the elastic deflection for axial loading is given in Eq. (4.25). Making use of these equations gives

$$k = \frac{P}{\dfrac{PL}{AE}} = \frac{AE}{L} \tag{15.19}$$

where

A = cross-sectional area, m^2
L = length, m
E = modulus of elasticity, Pa

15.4.3.1 BOLT STIFFNESS A bolt with thread is considered as a stepped shaft. The root diameter is used for the threaded section of the bolt, and the crest diameter is used for the unthreaded section, called the shank. The bolt is treated as a spring in series when considering the shank and the threaded section. The bolt can also have different diameters due to other specifications, so that the stiffness of the bolt can be expressed as

$$\frac{1}{k_b} = \frac{1}{k_{b1}} + \frac{1}{k_{b2}} + \frac{1}{k_{b3}} + \dots \tag{15.20}$$

Figure 15.12(a) shows a bolt-and-nut assembly; and Fig. 15.12(b), the stepped-shaft representation of the shank and the threaded section. In Fig. 15.12(b) the effective length of the shaft and the threaded section includes additional lengths extending into the bolt head and the nut. Making use of Eq. (15.19) and Fig. 15.12(b) gives

$$\frac{1}{k_b} = \frac{4}{\pi E}\left[\frac{L_{se}}{d_c^2} + \frac{L_{te}}{d_r^2}\right] = \frac{4}{\pi E}\left[\frac{L_s + 0.4d_c}{d_c^2} + \frac{L_t + 0.4d_r}{d_r^2}\right] \tag{15.21}$$

where

d_c = crest diameter, m
d_r = root diameter, m

Equation (15.21) is valid for a shank having a constant crest diameter. In some designs the shank has different diameters. This situation can be easily analyzed by using Eq. (15.20) and the appropriate lengths and diameters. Note that $0.4d_c$ is added to the length

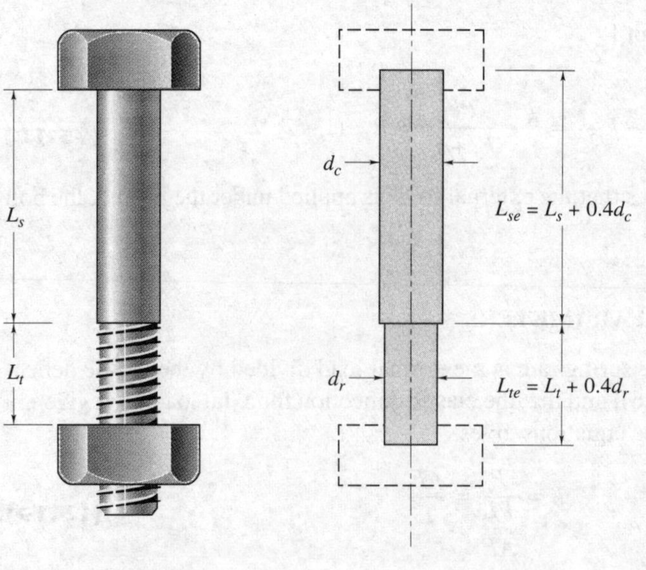

Figure 15.12 Bolt and nut. (a) Assembled; (b) stepped-shaft representation of shank and threaded section.

only for the section closest to the bolt head. The other lengths in the shank should be the actual lengths.

Often, a threaded fastener dedicated to a particular application will be especially fabricated, and any combination of diameter and thread length is possible. However, it is usually much more economical to use mass-produced threaded fasteners; these threads are generally of higher quality (as they may be rolled instead of cut) and they have a more repeatable strength. Usually, a bolt or cap screw will have as little of its length threaded as practicable to maximize bolt stiffness. However, for standardized threads the threaded length is given by

$$L_t = \begin{cases} 2d_c + 6 & L \le 125 \\ 2d_c + 12 & 125 < L \le 200 \\ 2d_c + 25 & L > 200 \end{cases} \quad \text{(metric threads)} \tag{15.22}$$

$$L_t = \begin{cases} 2d_c + 0.25 \text{ in.} & L \le 6 \text{ in.} \\ 2d_c + 0.50 \text{ in.} & L > 6 \text{ in.} \end{cases} \quad \text{(inch series)} \tag{15.23}$$

where

L_t = threaded length

L = total bolt length

d_c = crest diameter

15.4.3.2 JOINT STIFFNESS Determining joint stiffness is much more complicated than determining bolt stiffness (which was covered in the preceding section). Thus, approximations must be employed. One of the most frequent of these approximations is that the stress induced in the joint is uniform throughout a region surrounding the bolt hole with zero stress outside this region. Two conical frustums symmetric about the joint midplane and each having a vertex angle of $2\alpha_f$ are often used to represent the stresses in the joint. Figure 15.13 shows the conical frustum stress representation of the joint in a bolt-and-nut assembly. Note that d_w is the diameter of the washer through which the load

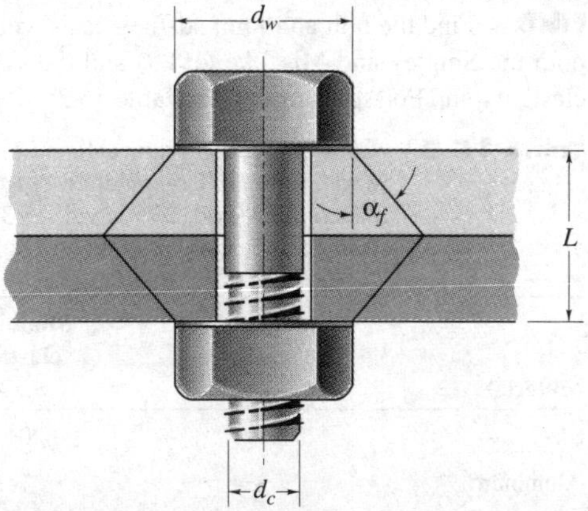

Figure 15.13 Bolt-and-nut assembly with conical frustum stress representation of joint.

is transferred. Shigley and Mischke (1989) arrived at the following expression for joint stiffness:

$$k_{ji} = \frac{\pi E_i d_c \tan\alpha_f}{2\ln\left[\frac{(L_i \tan\alpha_f + d_i - d_c)(d_i + d_c)}{(L_i \tan\alpha_f + d_i + d_c)(d_i - d_c)}\right]} \quad \textbf{(15.24)}$$

where

L_i = axial length of frustum cone, m

d_i = frustum cone diameter, m

Always use the smallest of the frustum cone diameters. For the member closest to the bolt head or to the nut $d_i = d_w = 1.5d_c$.

The resulting joint stiffness is

$$\frac{1}{k_j} = \frac{1}{k_{j1}} + \frac{1}{k_{j2}} + \frac{1}{k_{j3}} + \dots \quad \textbf{(15.25)}$$

Wileman et al. (1991) obtained an exponential expression for the joint stiffness that curve fits results from a finite-element analysis which uses a range of materials and geometry. The expression for joint stiffness is

$$k_{ji} = \sum_{i=1,\dots} E_i d_i A_i e^{B_i d_c / L_i} \quad \textbf{(15.26)}$$

where A_i, B_i = numerical constants given in Table 15.3.

Note that A_i and B_i are dimensionless and that Eq. (15.26) is equally applicable in English units.

Once the joint and bolt stiffnesses are known, the dimensionless stiffness parameter can be calculated from Eq. (15.17).

Example 15.6

GIVEN A hexagonal bolt-and-nut assembly, shown in Fig. 15.14(a), is used to join two members. The bolt and the nut are made of steel and the frustum cone angle is 30°. The thread crest diameter is 14 mm and the root diameter is 12 mm.

FIND Find the bolt and joint stiffnesses as well as the dimensionless stiffness parameter by using both the Shigley and Mischke (1989) and the Wileman et al. (1991) methods. Use the modulus of elasticity and Poisson's ratio from Table 15.3.

Table 15.3 Constants used in joint stiffness formula [Eq. (15.26)]. [J. Wileman, M. Choudhury, and I. Green, "Computation of Member Stiffness in Bolted Connections," *Journal of Machine Design*, 1991, vol. 193, pp. 432–37. Reprinted with permission of the American Society of Mechanical Engineers International.]

Material	Poisson's ratio, ν	Modulus of elasticity, *E*, GPa	Numerical constants, A_i	B_i
Steel	0.291	206.8	0.78715	0.62873
Aluminum	.334	71.0	.79670	.63816
Copper	.326	118.6	.79568	.63553
Gray cast iron	.211	100.0	.77871	.61616

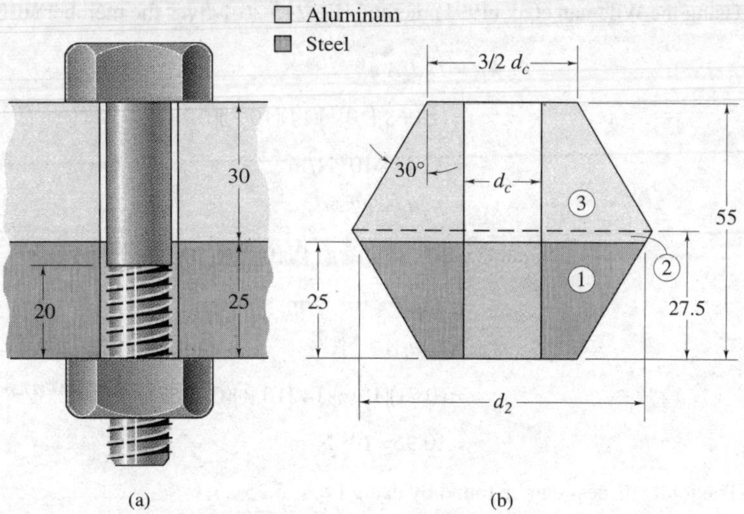

Figure 15.14 Hexagonal bolt-and-nut assembly used in Example 15.6. (a) Assembly and dimensions; (b) dimension of frustum cone. (All dimensions are in millimeters.)

Solution

From Eq. (15.21) the bolt stiffness is

$$\frac{1}{k_b} = \frac{4}{\pi(206.8)(10^9)}\left[\frac{(35)(10^{-3}) + 0.4(14)(10^{-3})}{(14)^2(10^{-6})} + \frac{20(10^{-3}) + (0.4)(12)(10^{-3})}{(12)^2(10^{-6})}\right]$$

$$k_b = 0.4281 \times 10^9 \text{ N/m} = 4.281 \times 10^8 \text{ N/m}$$

Using the Shigley and Mischke (1989) method [Eq. (15.24)] and Fig. 15.14(b) gives the joint stiffnesses as

$$k_{j1} = \frac{\pi \tan 30°(2.068)(10^{11})(14)(10^{-3})}{2\ln\left[\frac{(25\tan 30° + 21 - 14)(21+14)}{(25\tan 30° + 21 + 14)(21-14)}\right]} = 33.93 \times 10^8 \text{ N/m}$$

$$k_{j2} = \frac{\pi \tan 30°(0.71)(10^{11})(14)(10^{-3})}{2\ln\left[\frac{(2.5\tan 30° + 49.87 - 14)(49.87+14)}{(2.5\tan 30° + 49.87 + 14)(49.87-14)}\right]} = 527.1 \times 10^8 \text{ N/m}$$

$$k_{j3} = \frac{\pi \tan 30°(0.71)(10^{11})(14)(10^{-3})}{2\ln\left[\frac{(27.5\tan 30° + 21 - 14)(21+14)}{(27.5\tan 30° + 21 + 14)(21-14)}\right]} = 11.13 \times 10^8 \text{ N/m}$$

Substituting the above equations into Eq. (15.25) gives

$$\frac{1}{k_j} = \frac{1}{(33.93)(10^8)} + \frac{1}{(527)(10^8)} + \frac{1}{(11.13)(10^8)}$$

$$\therefore k_j = 8.248 \times 10^8 \text{ N/m}$$

Using the Wileman et al. (1991) method [Eq. (15.26)] gives the member stiffnesses as

$$k_{j1} = E_1 d_c A_1 e^{B_1 d_c / L_1}$$
$$= (2.068)(10^{11})(14)(10^{-3})(0.7872)e^{(0.6287)(14)/(25)}$$
$$= 32.41 \times 10^8 \text{ N/m}$$
$$k_{j2} = E_2 d_c A_2 e^{B_2 d_c / L_2}$$
$$= (0.71)(10^{11})(14)(10^{-3})(0.7967)e^{(0.6382)(14)/(2.5)}$$
$$= 282.4 \times 10^8 \text{ N/m}$$
$$k_{j3} = E_3 d_c A_3 e^{B_3 d_c / L_3}$$
$$= (0.71)(10^{11})(14)(10^{-3})(0.7967)e^{(0.6382)(14)/(27.5)}$$
$$= 10.96 \times 10^8 \text{ N/m}$$

The joint stiffness can be found by using Eq. (15.25) as

$$\frac{1}{k_j} = \frac{1}{k_{j1}} + \frac{1}{k_{j2}} + \frac{1}{k_{j3}} = \frac{1}{(32.41)(10^8)} + \frac{1}{(282.4)(10^8)} + \frac{1}{(10.96)(10^8)}$$
$$= (0.03086 + 0.00354 + 0.09124)(10^{-8})$$
$$k_j = 7.959 \times 10^8 \text{ N/m}$$

Note that the Wileman et al. (1991) calculations are simple and quite close to the Shigley and Mischke (1989) calculations. The Wileman et al. (1991) method is therefore recommended for calculating joint stiffness.

The dimensionless stiffness parameter [Eq. (15.17)] gives

$$C_k = \frac{k_b}{k_b + k_j} = \frac{4.281}{4.281 + 7.959} = 0.3498$$

15.4.4 STRENGTH

The **proof load** of a bolt is the maximum load that a bolt can withstand without acquiring a permanent set. The **proof strength** is the limiting value of the stress determined by using the proof load and the tensile stress area. Although proof strength and yield strength have something in common, the yield strength is usually higher because it is based on a 0.2% permanent deformation.

The proof strength S_p, as defined by the Society of Automotive Engineers (SAE), the American Society for Testing and Materials (ASTM), and International Organization for Standardization (ISO) specifications, defines bolt grades or classes that specify material, heat treatment, and minimum proof strength for the bolt or screw. Table 15.4 gives the strength information for several SAE grades of bolt, and Table 15.5 gives similar information for metric bolts. SAE grade numbers range from 1 to 8 and metric grade numbers from 4.6 to 12.9, with higher numbers indicating greater strength.

Proof stress is comparable to the allowable stress, which is used throughout the text. Thus, the safety factor is equivalent to the proof stress divided by the design stress.

Table 15.4 Strength of steel bolts for various sizes in inches.

SAE grade	Range of crest diameters, in.	Ultimate tensile strength, S_u, ksi	Yield strength, S_y, ksi	Proof strength, S_p, ksi
1	$^1/_4$–$1^1/_2$	60	36	33
2	$^1/_4$–$^3/_4$	74	57	55
	>$^3/_4$–$1^1/_2$	60	36	33
4	$^1/_4$–$1^1/_2$	115	100	65
5	$^1/_4$–1	120	92	85
	>1–$1^1/_2$	105	81	74
7	$^1/_4$–$1^1/_2$	133	115	105
8	$^1/_4$–$1^1/_2$	150	130	120

Tables 15.6 and 15.7 give the dimensions and the tensile stress areas for UN and M coarse and fine threads. The equation for the tensile stress areas for UN thread profiles given in Table 15.6 is

$$A_t = (0.7854)\left(d_c - \frac{0.9743}{n}\right)^2 \quad \textbf{(15.27)}$$

In Eq. (15.27) the crest diameter is in inches and n is the number of threads per inch. The equation for the tensile stress areas for M thread profiles given in Table 15.7 is

$$A_t = (0.7854)(d_c - 0.9382p)^2 \quad \textbf{(15.28)}$$

In Eq. (15.28) both the crest diameter d_c and the pitch p are in millimeters.

Table 15.5 Strength of steel bolts for various sizes in millimeters.

Metric grade	Crest diameter, d_c, mm	Ultimate tensile strength, S_u, MPa	Yield strength, S_y, MPa	Proof strength, S_p, MPa
4.6	M5–M36	400	240	225
4.8	M1.6–M16	420	[a]340	310
5.8	M5–M24	520	[a]415	380
8.8	M17–M36	830	660	600
9.8	M1.6–M16	900	[a]720	650
10.9	M6–M36	1040	940	830
12.9	M1.6–M36	1220	1100	970

[a]Yield strengths approximate and not included in standard.

Table 15.6 Dimensions and tensile stress areas for UN coarse and fine threads.

Crest diameter, d_c, in.	Coarse threads (UNC) Number of threads per inch, n	Coarse threads (UNC) Tensile stress area, A_t, in.^{2}	Fine threads (UNF) Number of threads per inch, n	Fine threads (UNF) Tensile stress area, A_t, in.^{2}
0.0600	—	—	80	0.00180
0.0730	64	0.00263	72	0.00278
0.0860	56	0.00370	64	0.00394
0.0990	48	0.00487	56	0.00523
0.1120	40	0.00604	48	0.00661
0.1250	40	0.00796	44	0.00830
0.1380	32	0.00909	40	0.01015
0.1640	32	0.0140	36	0.01474
0.1900	24	0.0175	32	0.0200
0.2160	24	0.0242	28	0.0258
0.2500	20	0.0318	28	0.0364
0.3125	18	0.0524	24	0.0580
0.3750	16	0.0775	24	0.0878
0.4735	14	0.1063	20	0.1187
0.5000	13	0.1419	20	0.1599
0.5625	12	0.182	18	0.203
0.6250	11	0.226	18	0.256
0.7500	10	0.334	16	0.373
0.8750	9	0.462	14	0.509
1.000	8	0.606	12	0.663
1.125	7	0.763	12	0.856
1.250	7	0.969	12	1.073
1.375	6	1.155	12	1.315
1.500	6	1.405	12	1.581
1.750	5	1.90	—	—
2.000	$4^1/_2$	2.50	—	—

15.4.5 BOLT PRELOAD—STATIC LOADING

This section makes use of Figs. 15.10 and 15.11 and the material covered in Sec. 15.4.2. Equation (15.16) can be written in terms of bolt stress as

$$\sigma_b = \frac{P_b}{A_t} = \frac{P_i}{A_t} + C_k \frac{P}{A_t} \tag{15.29}$$

The limiting value of the bolt stress is the proof strength, which is given in Tables 15.4 and 15.5. Also, introducing the safety factor for bolt stress gives the proof strength as

Table 15.7 Dimensions and tensile stress areas for metric coarse and fine threads.

	Coarse threads (MC)		Fine threads (MF)	
Crest diameter, d_c, mm	Pitch, p, mm	Tensile stress area, A_t, mm^2	Pitch, p, mm	Tensile stress area, A_t, mm^2
1	0.25	0.460	—	—
1.6	.35	1.27	0.20	1.57
2	.4	2.07	.25	2.45
2.5	.45	3.39	.35	3.70
3	.5	5.03	.35	5.61
4	.7	8.78	.5	9.79
5	.8	14.2	.5	16.1
6	1	20.1	.75	22
8	1.25	36.6	1	39.2
10	1.5	58.0	1.25	61.2
12	1.75	84.3	1.25	92.1
16	2	157	1.5	167
20	2.5	245	1.5	272
24	3	353	2	384
30	3.5	561	2	621
36	4	817	3	865
42	4.5	1121	—	—
48	5	1473	—	—

$$S_p = \frac{P_i}{A_t} + \frac{P_{\max} n_s C_k}{A_t} \tag{15.30}$$

where

A_t = tensile stress area given by Eq. (15.27) or (15.28) and Table 15.6 or 15.7

P_i = preload, N

The safety factor is not applied to the preload stress. Equation (15.30) can be rewritten to give the bolt failure safety factor as

$$n_{sb} = \frac{A_t S_p - P_i}{P_{\max,b} C_k} \tag{15.31}$$

where

$P_{\max,b}$ = maximum load applied to bolt, N

Equation (15.31) suggests that the safety factor is maximized by having zero preload on the bolt. For statically loaded bolts where separation is not a concern, this is certainly true. However, a preload is often applied to bolted connections to make certain that

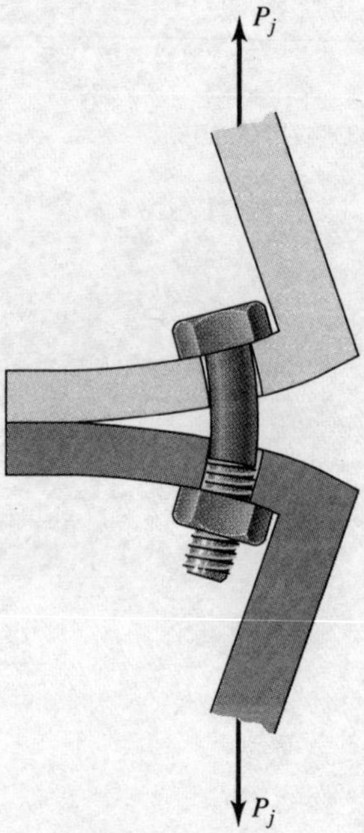

Figure 15.15
Separation of joint.

members are tightly joined in order to minimize tolerances and provide tight fits. Further, many applications require members to avoid separation, as shown in Figure 15.15. Separation occurs when in Eq. (15.18) $P_j = 0$. Thus, the safety factor guarding against separation is

$$n_{sj} = \frac{P_i}{P_{\max,j}(1 - C_k)} \tag{15.32}$$

where

$P_{\max,j}$ = maximum load applied to joint, N

The amount of preload that is in practice applied to bolts under static conditions is therefore a compromise between bolt overloading (where zero preload is most beneficial) and separation (where a large preload is desirable). The preload is given for reused and permanent connections as

$$P_i = \begin{cases} 0.75P_p & \text{for reused connections} \\ 0.90P_p & \text{for permanent connections} \end{cases} \tag{15.33}$$

where

P_p = proof load, $S_p A_t$

The above equation is applicable for static as well as dynamic conditions. Recall that the tensile stress area is obtained from either Table 15.6 or 15.7 depending on whether UN or metric threads are used. Also, the proof strength S_p is obtained from Table 15.4 or 15.5 depending on whether thread size is expressed in inches or millimeters.

In design practice, preloads are rarely specified because preloads are extremely difficult to measure during assembly of bolted connections. Two alternatives are used:

1. Specify a torque to be applied during tightening, which is then controlled by using a torque wrench.
2. Define a number of rotations from a "snug state," such as a half-turn.

Both alternatives have difficulties associated with them: specifying a torque does not accommodate changes in thread friction, which clearly affect the resulting preload, and a snug state is inherently subjective. Thus, although Eq. (15.33) can be used in design problems and for determining a pretorque or deformation, it should be recognized that bolts in applications have a wide variance in preloads.

Example 15.7

GIVEN The results of Example 15.6.

FIND Determine the maximum load for bolt-and-joint failure while assuming reused connections and a static safety factor of 2.5. Assume a 5.8 grade and coarse threads.

Solution

From Table 15.5 for 5.8 grade, $S_p = 380$ MPa, and from Table 15.7 for a crest diameter of 14 mm and coarse threads, $A_t = (84.3 + 157)/2 = 120.7\ \text{mm}^2$. For reused connections from Eq. (15.33)

$$P_i = 0.75P_p = 0.75A_tS_p = (0.75)(120.7)(10^{-6})(380)(10^6) = 34\ 400\ \text{N}$$

From Eq. (15.31) the maximum load that the bolt can carry while assuming a safety factor of 2.5 is

$$P_{\max,b} = \frac{A_t S_p - P_i}{n_{sb} C_k} = \frac{(120.7)(380) - 34\ 400}{(2.5)(0.3498)} = 13\ 111\ \text{N}$$

From Eq. (15.32) the maximum load before separation occurs is

$$P_{\max,j} = \frac{P_i}{n_{sj}(1 - C_k)} = \frac{34\ 400}{(2.5)(1 - 0.3498)} = 21\ 163\ \text{N}$$

Thus, failure due to separation will not occur before bolt failure.

15.4.6 Bolt Preload—Dynamic Loading

The preload effect is even greater for dynamically loaded joints than for statically loaded joints. Figure 15.16 attempts to describe how the forces versus deflection of the bolt and joint will vary as a function of time. While considering Fig. 15.16 recall Figs. 15.10 and 15.11, which consider static loading. From Fig. 15.16 observe that the bolt load is in tension that varies with time from the preload P_i to some maximum bolt load $P_{b,\max}$. Simultaneously, the joint is in compression that varies with time from the preload P_i to the minimum joint load $P_{j,\min}$. When bolt load is a maximum, the joint load has a minimum value and vice versa. Also, because of the differences in slope of the load and deflection curves for the bolt and the joint, the amount of load variation for one cycle is much larger for the joint than for the bolt.

Figure 15.16 also shows the deflection variation with time for both the bolt and the joint. The quantitative amount is the same, but for the bolt the deflection is an increased elongation whereas for the joint the contraction is decreased.

Since failure due to cyclic loading is more apt to occur to the bolt, we only consider the bolt here. Equation (15.16) gives the alternating and mean loads acting on the bolt as

$$P_{ba} = \frac{P_{b,\max} - P_{b,\min}}{2} = \frac{C_k(P_{\max} - P_{\min})}{2} = C_k P_a \qquad \textbf{(15.34)}$$

$$P_{bm} = \frac{P_{b,\max} + P_{b,\min}}{2} = P_i + \frac{(P_{\max} + P_{\min})C_k}{2} = P_i + C_k P_m \qquad \textbf{(15.35)}$$

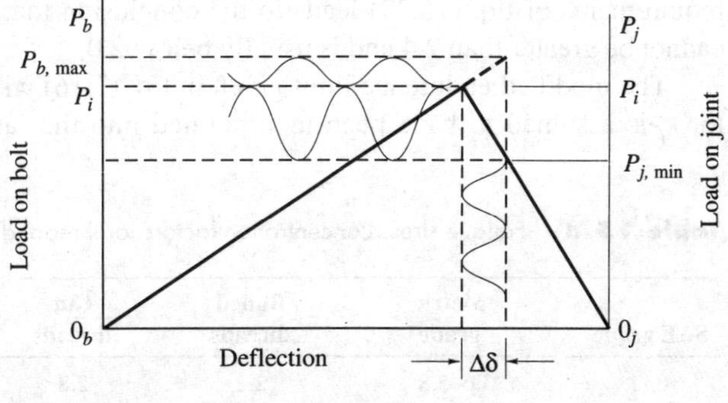

Figure 15.16 Forces versus deflection of bolt and joint as function of time.

The alternating and mean stresses can be expressed as

$$\sigma_a = \frac{C_k P_a n_s}{A_t} \tag{15.36}$$

$$\sigma_m = \frac{P_i + C_k P_m n_s}{A_t} \tag{15.37}$$

The safety factor is not applied to the preload.

The Goodman fatigue failure criterion given in Sec. 7.9.1.2 is used. Since the safety factor has been incorporated into the alternating and mean stresses [Eqs. (15.36) and (15.37)], the Goodman equation [Eq. (7.28) when $n_s = 1$] is

$$\frac{K_f \sigma_a}{S_e} + \frac{\sigma_m}{S_{ut}} = 1 \tag{15.38}$$

Substituting Eqs. (15.36) and (15.37) into Eq. (15.38) gives

$$\frac{K_f C_k P_a n_s}{S_e A_t} + \frac{P_i + C_k P_m n_s}{A_t S_{ut}} = 1 \tag{15.39}$$

Eq. (15.39) can also be expressed in terms of stresses as

$$\frac{n_s K_f C_k \sigma_a}{S_e} + \frac{\sigma_i + C_k \sigma_m n_s}{S_{ut}} = 1$$

Solving for the safety factor gives

$$n_s = \frac{S_{ut} - \sigma_i}{C_k \left[K_f \sigma_a \left(\frac{S_{ut}}{S_e} \right) + \sigma_m \right]} \tag{15.40}$$

The fatigue stress concentration factor K_f for threaded elements can be obtained from Table 15.8. Cutting is the simplest method of producing the threads. However, rolling the threads instead of cutting them provides a smoother thread finish (fewer initial cracks). Rolling provides an unbroken flow of material grain in the thread region. A generous fillet between the head and the shank reduces the fatigue stress concentration, as shown in Table 15.8. It is usually safe to assume that the threads have been rolled, unless specific information is available.

The smallest safety factor for fatigue loading is the ratio of ultimate stress to prestress (if the mean and alternating stresses are zero). Examining the values in Table 15.4 and the requirements of Eq. (15.33) leads to the conclusion that the safety factor for bolt fatigue cannot be greater than 2.4 and is usually below 2.0.

The modified endurance limit given in Eq. (7.16) will be used. The surface finish factor k_f is assumed to have been incorporated into the fatigue stress concentration factor.

Table 15.8 Fatigue stress concentration factors for threaded elements.

SAE grade	Metric grade	Rolled threads	Cut threads	Fillet
0–2	3.6–5.8	2.2	2.8	2.1
4–8	6.6–10.9	3.0	3.8	2.3

Also, the size factor is 1 for axial loading. If no specific information is given about the reliability factor, it is recommended to assume a survival probability of 90%. The endurance limit is obtained from Eq. (7.7) for various types of loading.

The previous section discussed the importance of preload in statically loaded connections. With dynamic or cyclic loads a preload is especially important. The reasons can be seen in Eqs. (15.36) and (15.37) and by recalling some subtleties of fatigue analysis discussed in Chapter 7. As long as separation does not occur, the alternating stress experienced by the bolt is reduced by the dimensionless stiffness parameter C_k. The mean stress is increased by the preload. However, the failure theories in Sec. 7.9 (e.g., Gerber, Goodman, Soderberg, etc.) show that the alternating stress has a much greater effect on service life than the mean stress. That is, the endurance limit is usually a small fraction of the yield or ultimate strength. Thus, a large enough preload to prevent separation is critical. For the purposes of this text this preload can be obtained by using the Russell relations for preload given by Eq. (15.33).

Example 15.8

GIVEN The hexagonal bolt-and-nut assembly shown in Fig. 15.14(a) and considered in Examples 15.6 and 15.7 has an external load of 11 000 N applied cyclically in a released-tension manner.

FIND Find the safety factor guarding against fatigue failure of the bolt. Assume a survival probability of 90%, rolled threads, and a 5.8 metric grade.

Solution

For released tension with a maximum load of 11 000 N

$$P_a = P_m = 5500 \text{ N}$$

The endurance limit for axial loading from Eq. (7.7) and Table 15.5 for 5.8-metric-grade thread is

$$S'_e = 0.45 S_u = 0.45(520)(10^6) \text{ Pa} = 234 \text{ MPa}$$

From Table 15.8 for 5.8-metric-grade thread and rolled thread the fatigue stress concentration factor $K_f = 2.2$. From Table 7.4 the reliability factor k_r for 90% survival probability is 0.90. Thus, the modified endurance limit from Eq. (7.16) is

$$S_e = k_f k_s k_r k_m S'_e = (1)(1)(0.9)(1)(234)(10^6) \text{ Pa} = 210.6 \text{ MPa}$$

From Table 15.7 for a crest diameter of 14 mm and coarse threads the tensile stress area is 120.7 mm^2. From Eq. (15.40)

$$n_s = \frac{S_{ut} - \sigma_i}{\sigma_a C_k \left[\dfrac{K_f S_{ut}}{S_e} + 1\right]} = \frac{520(10^6) - 285(10^6)}{45.57(10^6)(0.3498)\left[2.2\left(\dfrac{520}{210.6}\right) + 1\right]} = 2.292$$

Bolt fatigue failure is not expected.

15.4.7 GASKETED JOINTS

Figure 15.17 shows a threaded fastener with an unconfined **gasket** joining two members. An unconfined gasket extends over the entire diameter (as shown in Fig. 15.17) of a joint, whereas a confined gasket extends over only a part of the joint diameter. A confined gasket may be considered as if the gasket is not present.

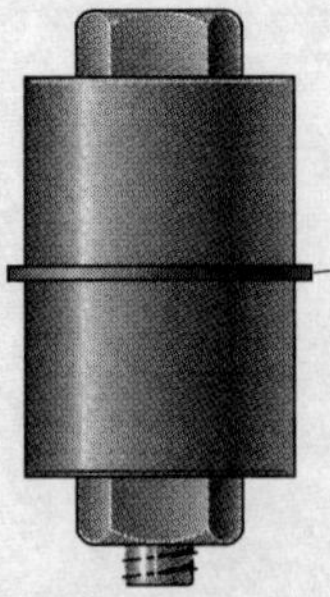

Figure 15.17 Threaded fastener with unconfined gasket and two other members.

An unconfined gasket is made of a soft elastomeric material. Recall that the various members of the joint can be considered as individual springs with different spring rates depending on their materials. Gaskets are made of material that is soft relative to the other joint members. The modulus of elasticity is much less for the gasket than for the other joint members ($E_g << E_1, E_2, \ldots$). Since the modulus of elasticity is directly proportional to the stiffness,

$$k_g << k_1, k_2, \ldots \tag{15.41}$$

Furthermore,

$$\frac{1}{k_g} >> \frac{1}{k_1}, \frac{1}{k_2}, \ldots$$

Making use of the above and Eq. (15.25) gives for an unconfined gasket

$$\frac{1}{k_j} = \frac{1}{k_1} + \frac{1}{k_2} + \ldots + \frac{1}{k_g} \approx \frac{1}{k_g}$$
$$\therefore k_j = k_g \tag{15.42}$$

Thus, for unconfined gaskets the joint stiffness is equivalent to the gasket stiffness.

A gasketed joint must satisfy Eqs. (15.17) and (15.18) for preload. In addition, the preload must be large enough to achieve the minimum sealing pressure required for the gasket material. Thus,

$$P_i \geq A_g p_0 \tag{15.43}$$

where

A_g = gasket area per bolt, m^2

p_0 = minimum gasket seal pressure, Pa

Thus, for gasketed joints Eq. (15.43) must be checked.

The material covered thus far is for axial loading. Threaded fasteners can also experience shear. Shear is considered in the next section, since riveted joints in shear are treated the same as threaded fasteners in shear.

15.5 RIVETED FASTENERS

Rivets are in wide use as fasteners of members in buildings, bridges, boilers, tanks, ships, and (in general) for any framework. In design the rivet diameter is used rather than the hole diameter, even through the rivet diameter may expand on assembly and nearly fill the hole.

The reasons for choosing rivets over threaded fasteners are the following:

1. They will not shake loose.
2. They are cheap, especially in assembly time.
3. They are lightweight.
4. They can be assembled from the blind side.

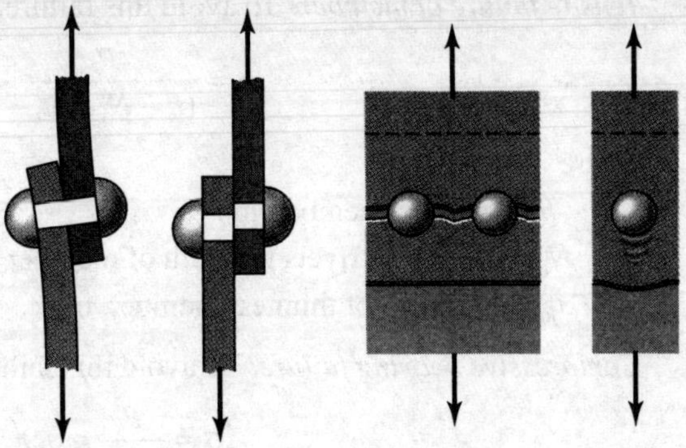

Figure 15.18 Failure modes due to shear loading of riveted fasteners. (a) Bending of member; (b) shear of rivet; (c) tensile failure of member; (d) bearing of rivet on member or bearing of member on rivet.

However, rivets cannot provide as strong an attachment as a threaded fastener of the same diameter. Also, rivets do not allow any disassembly of the joint.

Rivets must be spaced neither too close nor too far apart. The minimum rivet spacing, center to center, for structural steel work is usually taken as three rivet diameters. A somewhat closer spacing is used in boilers. Rivets should not be spaced too far apart or plate buckling will occur. The maximum spacing is usually taken as 16 times the thickness of the outside plate.

Riveted and threaded fasteners in shear are treated exactly alike in design and failure analysis. Failure due to a shear force is primarily caused by the four different modes of failure shown in Fig. 15.18: bending of a member, shearing of the rivet, tensile failure of a member, and the bearing of the rivet on a member or the bearing of a member on the rivet. Each of these four modes is presented here:

1. *Bending of a member:* To avoid this failure, the following should be valid:

$$\sigma = \frac{PL_g c}{2I} < 0.6\left(S_y\right)_j \tag{15.44}$$

where

L_g = length of grip, $L_s + L_t$ (see Fig. 15.12), m
Z_m = section modulus of weakest member, I/c, m^3
$(S_y)_j$ = yield strength of weakest member, Pa

2. *Shear of rivets:* To avoid this failure, the following should be valid:

$$\tau = \frac{4P}{\pi d_c^2} < S_{sy} = 0.4S_y \tag{15.45}$$

where

d_c = crest diameter, m
S_{sy} = rivet yield strength in shear, Pa

In Eq. (15.45) the diameter of the rivet rather than the diameter of the hole is used. Shear of rivets is a major failure mode considered in the design of threaded or riveted fasteners. This mode is also called the direct shear mode of failure.

3. *Tensile failure of member:* To avoid this failure, the following should be valid:

$$\sigma = \frac{P}{(b - N_r d_c)t_m} < (S_y)_j \quad \textbf{(15.46)}$$

where

b = width of member, m

N_r = number of rivets in width of member

t_m = thickness of thinnest member, m

4. *Compressive bearing failure:* To avoid this failure, the following should be valid:

$$\sigma = \frac{P}{d_c t_m} < 0.9(S_y)_j \quad \textbf{(15.47)}$$

These four failure modes relate to failure of an individual rivet or an individual member. Threaded and riveted fasteners are used in groups, and the torsional shear of the group must be considered as a failure mode. The result of the shear stress acting on the rivet is the vectorial sum of the direct and torsional shear stresses

$$\tau = \tau_d + \tau_t \quad \textbf{(15.48)}$$

where

τ_d = shear stress of rivet, or direct shear, Pa

τ_t = shear stress of rivet due to torsional loading, Pa

When dealing with a group of threaded or riveted fasteners, the centroid of the group must be evaluated. Equilibrium of vertical and horizontal loads must be observed.

Example 15.9

GIVEN The riveted member shown in Fig. 15.19(a) is fastened to a beam, and an eccentric load is applied. Rivets A and C are 5/8 in. in diameter and rivets B and D are 7/8 in. in diameter.

FIND Determine the following:

a) Centroid of the rivet assembly

b) Direct and resultant shear stresses acting on each rivet

c) Safety factor guarding against shear of the rivet

d) Safety factor guarding against bending of the member

Assume that the yield stress of the rivet is 85 ksi and that of the member is 54 ksi.

Solution

a) The total area of the four rivets is

$$\sum A_i = \frac{\pi}{2}\left[\left(\frac{5}{8}\right)^2 + \left(\frac{7}{8}\right)^2\right] = 1.816 \text{ in.}^2$$

In Fig. 15.19(a) the *x-y* coordinate is at the center of the C rivet, the rivet in the lower left corner. The centroid of the rivet group is

$$\bar{y} = \frac{\sum A_i \bar{y}_i}{\sum A_i} = \frac{0.3068(0+6) + 0.6013(0+6)}{1.816} = 3.000 \text{ in.}$$

$$\bar{x} = \frac{\sum A_i \bar{x}_i}{\sum A_i} = \frac{0.3068(0+0) + 0.6013(7+7)}{1.816} = 4.635 \text{ in.}$$

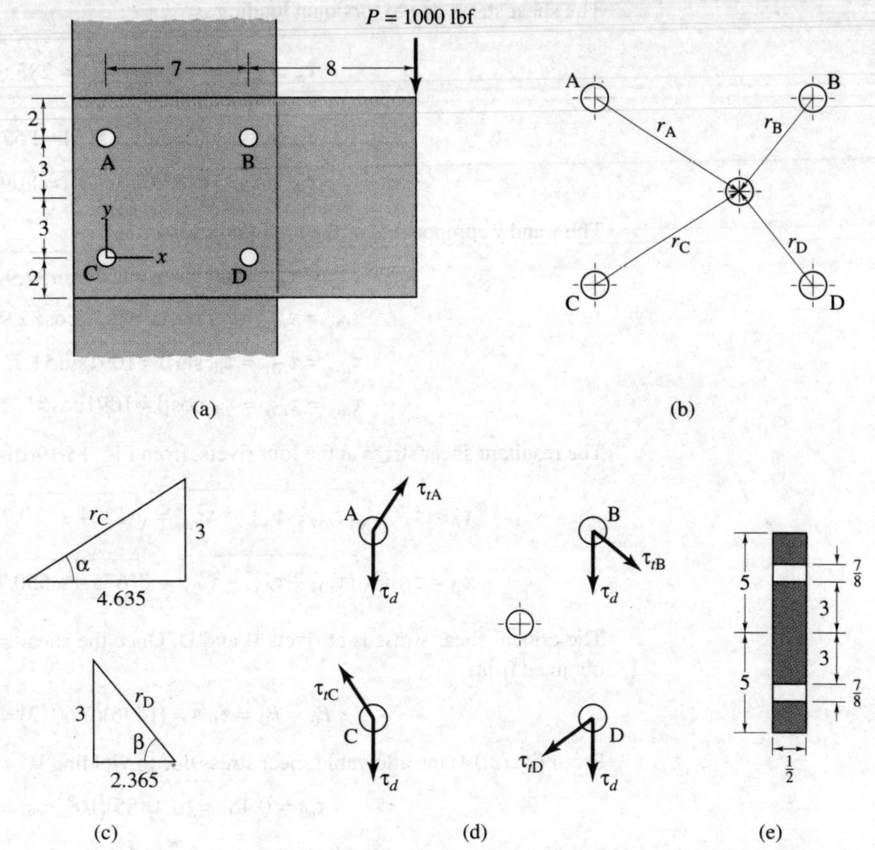

Figure 15.19 Group of riveted fasteners used in Example 15.9. (a) Assembly of rivet group; (b) radii from centroid to center of rivets; (c) resulting triangles; (d) direct and torsional shear acting on each rivet; (e) side view of member. (All dimensions are in inches.)

The eccentricity of the loading is

$$e_c = 8 + 7 - 4.635 = 10.36 \text{ in.}$$

Figure 15.19(b) shows the distances from the center of the rivets to the centroid, and Fig. 15.19(c) shows the resulting triangles. From these figures

$$r_A = r_C = \sqrt{3^2 + 4.635^2} = 5.521 \text{ in.}$$

$$r_B = r_D = \sqrt{3^2 + 2.365^2} = 3.820 \text{ in.}$$

b) The direct shear stress acting on each of the rivets is

$$\tau_d = \frac{P}{\sum A} = \frac{1000 \text{ lbf}}{1.816 \text{ in.}^2} = 550.7 \text{ psi}$$

Figure 15.19(d) illustrates the direct and torsional shear stresses acting on each rivet. The torque due to eccentric loading is

$$T = Pe_c = (1000 \text{ lbf})(10.36 \text{ in.}) = 10\,360 \text{ lbf-in.}$$

$$\sum r_j^2 A_j = 2\left(r_C^2 A_C + r_D^2 A_D\right)$$

$$\sum r_j^2 A_j = 2\left[(5.525)^2(0.3068) + (3.820)^2(0.6013)\right] = 36.28 \text{ in.}^4$$

The shear stress due to torsional loading is

$$\tau_{ti} = \frac{Tr_i}{\sum r_j^2 A_j} = \frac{10\,360 r_i}{36.28} = 285.6 r_i$$

$$\tau_{tA} = \tau_{tC} = (285.6)(5.521) = 1\,577\ \text{lbf/in.}^2$$

$$\tau_{tB} = \tau_{tD} = (285.6)(3.820) = 1\,091\ \text{lbf/in.}^2$$

The x and y components of the torsional stresses are

$$\tau_{tAx} = \tau_{tCx} = \tau_{tA} \sin\alpha = 1577 \sin 32.91^\circ = 856.9\ \text{psi}$$

$$\tau_{tAy} = \tau_{tCy} = \tau_{tA} \cos\alpha = 1577 \cos 32.91^\circ = 1324\ \text{psi}$$

$$\tau_{tBx} = \tau_{tDx} = \tau_{tB} \sin\beta = 1091 \sin 51.75^\circ = 856.8\ \text{psi}$$

$$\tau_{tBy} = \tau_{tDy} = \tau_{tB} \cos\beta = 1091 \cos 51.75^\circ = 675.4\ \text{psi}$$

The resultant shear stress at the four rivets, from Fig. 15.19(d), is

$$\tau_A = \tau_C = \sqrt{(\tau_{tAy} - \tau_d)^2 + \tau_{tAx}^2} = \sqrt{(1324 - 550.7)^2 + (856.8)^2} = 1154\ \text{psi}$$

$$\tau_B = \tau_D = \sqrt{(\tau_{tDy} + \tau_d)^2 + \tau_{tDx}^2} = \sqrt{(675.4 + 550.7)^2 + (856.8)^2} = 1496\ \text{psi}$$

The critical shear stress is at rivets B and D. Once the shear stress is known, the shear force can be obtained from

$$P_B = P_D = \tau_B A = (1496)(0.6013) = 899.5\ \text{lbf}$$

From Eq. (3.14) the allowable shear stress due to yielding is

$$\tau_{\text{all}} = 0.4 S_y = (0.4)(85)(10^3)\ \text{psi} = 34.0\ \text{ksi}$$

c) The safety factor guarding against shear of the rivets is

$$n_s = \frac{\tau_{\text{all}}}{\tau_D} = \frac{34.0}{1.496} = 22.73$$

d) Figure 15.19(e) shows a side view of the member at a critical section where there are two rivets across the width. The distance from the neutral axis to the outer fiber is 5 in. The bending of the member, from Fig. 15.19(a), is

$$M_B = (1000)(8) = 8000\ \text{lbf-in.}$$

The area moment of inertia for the critical section shown in Fig. 15.19(e) is

$$I = \frac{1}{12}\left(\frac{1}{2}\right)(10^3) - 2\left[\left(\frac{1}{12}\right)\left(\frac{1}{2}\right)\left(\frac{7}{8}\right)^3 + \left(\frac{1}{2}\right)\left(\frac{7}{8}\right)(3)^2\right]$$

$$= 41.67 - 2(0.0279 + 3.938)$$

$$= 33.74\ \text{in.}^4$$

The bending stress at rivet B is

$$\sigma_B = \frac{M_B c}{I} = \frac{(8000)(5)}{33.74}\ \text{psi} = 1.186 \times 10^3\ \text{psi}$$

From Eq. (3.15) the allowable stress due to bending is

$$\sigma_{\text{all}} = 0.6S_y = (0.6)(54)(10^3)\text{psi} = 32.4\times 10^3 \text{ psi}$$

The safety factor guarding against bending of the member is

$$n_s = \frac{\sigma_{\text{all}}}{\sigma_{\text{B}}} = \frac{32.4}{1.186} = 27.3$$

Therefore, if failure should occur, it would occur first from shear of the rivets. However, because the safety factor guarding against shear of the rivets is 22.73, it is highly unlikely that failure will occur.

Example 15.10

GIVEN A pedestrian walkway is riveted to a steel bridge as shown in Fig. 15.20. The maximum load on the walkway is equivalent to a 3000-N load located 2 m from the steel bridge side for each pair of rivets. A safety factor of 5 is assumed.

FIND The rivet diameter needed if the rivets are made of AISI 1040 steel.

Solution

The load on each rivet is

$$P = \frac{3000}{2} = 1500 \text{ N}$$

The shear stress is

$$\tau = \frac{P}{A} = \frac{1500}{A} \tag{a}$$

The tensile forces acting on the two rivets are proportional to the distance from the lower end of the bracket. Elastic deformation occurs in the rivets but not in the bracket or the bridge.

$$(P_{\text{A}})(0.75) + (P_{\text{B}})(0.25) - (3000)(2) = 0$$

$$\therefore \frac{0.75}{0.25} P_{\text{B}}(0.75) + (P_{\text{B}})(0.25) = 6000$$

$$\therefore P_{\text{B}} = 2400 \text{ N} \tag{b}$$

$$P_{\text{A}} = 3(2400) = 7200 \text{ N}$$

$$\therefore \sigma = \frac{P_{\text{A}}}{A} = \frac{7200}{A}$$

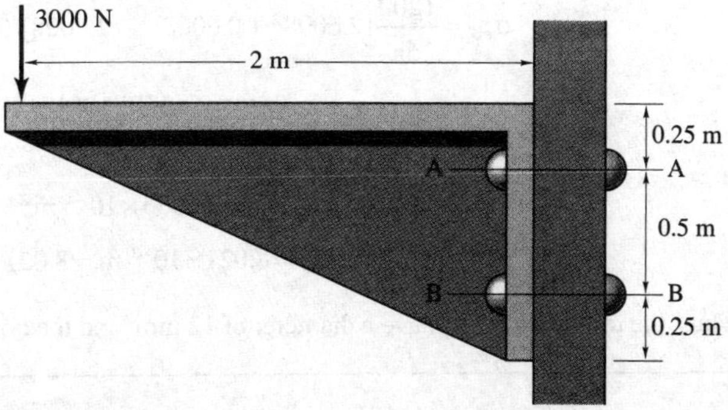

Figure 15.20 Pedestrian walkway riveted to steel bridge. Used in Example 15.10.

a) *Top rivet or rivet* A: From Eq. (2.16) the principal stresses for a biaxial stress state are

$$\sigma_1,\sigma_2 = \frac{\sigma_x+\sigma_y}{2} \pm \sqrt{\tau_{xy}^2 + \left(\frac{\sigma_x-\sigma_y}{2}\right)^2}$$

$$= \frac{7200}{2A_A} \pm \sqrt{\left(\frac{1500}{A_A}\right)^2 + \left(\frac{7200}{2A_A}\right)^2}$$

$$= \frac{3600}{A_A}(1 \pm 1.083)$$

From Eq. (6.10) the von Mises stress for a biaxial stress state for the top or A rivet is

$$\sigma_{eA} = \left(\sigma_1^2+\sigma_2^2-\sigma_1\sigma_2\right)^{1/2} = \frac{3600}{A_A}\left[2.083^2+0.083^2+(2.083)(0.083)\right]^{1/2} = \frac{7653}{A_A}$$

From Table A.1 for 1040 steel the yield strength is 350 MPa. From Eq. (6.11) the distortion-energy theory (DET) predicts failure if

$$\sigma_{eA} \geq \frac{S_y}{n_s} = \frac{(350)(10^6)}{5}\ \text{Pa} = 70\ \text{MPa}$$

$$\frac{7653}{A_A} = (70)(10^6)$$

$$A_A = \frac{7653}{(70)(10^6)} = 1.09\times10^{-4}\ \text{m}^2$$

$$\frac{\pi d_A^2}{4} = A_A = 1.09\times10^{-4}$$

$$\therefore d_A = 11.80\times10^{-3}\ \text{m} = 11.80\ \text{mm}$$

b) *Bottom rivet or rivet* B

$$\sigma_1,\sigma_2 = \frac{2400}{2A_B} \pm \sqrt{\left(\frac{1500}{A_B}\right)^2 + \left(\frac{2400}{2A_B}\right)^2} = \frac{1200}{A_B}(1 \pm 1.6008)$$

$$\sigma_{eB} = \frac{1200}{A_B}\left[2.6008^2+0.6008^2+(2.6008)(0.6008)\right]^{1/2} = \frac{3537}{A_B}$$

$$\frac{3537}{A_B} = \frac{(350)(10^6)}{5}$$

$$A_B = 0.5053\times10^{-4}\ \text{m}^2$$

$$\therefore d_B = 8.021\times10^{-3}\ \text{m} = 8.021\ \text{mm}$$

Thus, the top rivet should have a diameter of 12 mm; and the bottom rivet, a diameter of 9 mm.

15.6 WELDED JOINTS

A **weld** is accomplished by joining two members together and then agitating the molecules at their junctions (usually by heat) until they fuse together. When the source of agitation is removed, the members are permanently joined. There are a number of welding methods, but this text focuses on one, shielded metal arc welding (generally called arc welding). In this method the two members are placed in close proximity, both of them in contact with an electrical conductor. A low-voltage, high-current arc is struck with an electrode to complete the electric circuit at the joint. The distance between the electrode and the work is controlled such that the electric arc is sustained, creating temperatures high enough to melt the members at the joint. The electrode is gradually consumed in the process, supplying additional metal to the joint as it melts. A flux covering the electrode provides a shielding gas (hence, the term "shielded metal arc welding") and a chalky covering (slag), both of which serve to protect the metal from oxidation. The slag is usually chipped off once the weld has cooled sufficiently.

Some advantages of welded joints over threaded fasteners are that they are inexpensive and there is no danger of the joint loosening. Some disadvantages of welded joints over threaded fasteners are that they produce residual stresses, they distort the shape of the member, metallurgical changes occur, and disassembly is a problem.

There are many configurations of welds, but only the fillet weld is considered here. This section examines one configuration with the expectation that, once it is understood, application to other configurations will be relatively easy.

A fillet weld, shown in Fig. 15.21(a), is made with equal legs. The thinnest section is then at the **throat** of the weld, at 45° from the legs. The governing stress in fillet welds is shear on the throat of the weld as shown in Fig. 15.21(b). Observe from this figure that in the fillet weld aligned parallel to the load, the shear stress occurs along the throat of the fillet parallel to the load. In a fillet weld aligned transverse to the load, the shear stress occurs at 45° to the load, acting transverse to the axis of the fillet.

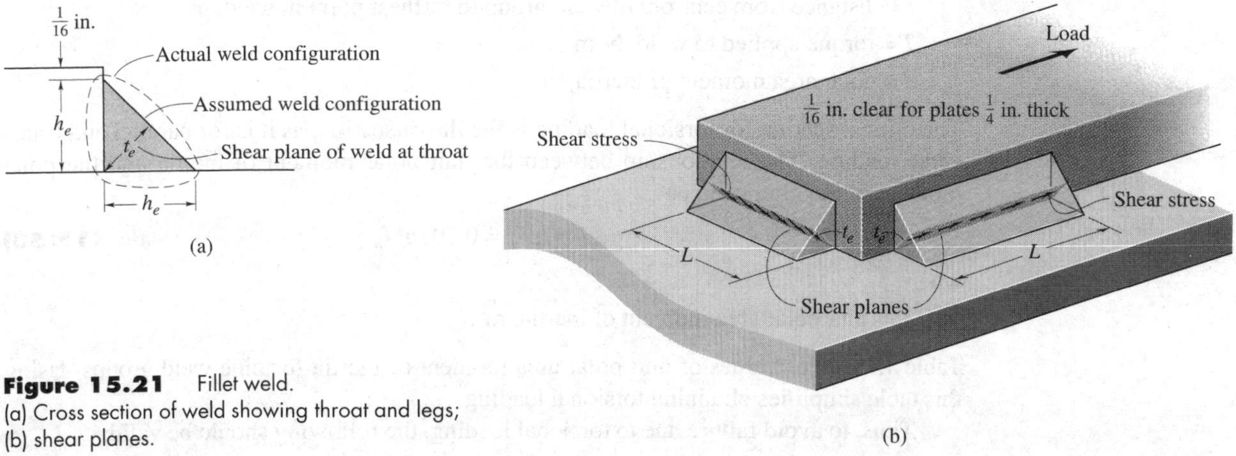

Figure 15.21 Fillet weld.
(a) Cross section of weld showing throat and legs;
(b) shear planes.

15.6.1 PARALLEL AND TRANSVERSE LOADING

Fillet welds fail by shearing at the minimum section, which is at the throat of the weld, shown in Fig. 15.21(a). This is true whether the weld has parallel (on the side) or transverse [at the end, as shown in Fig. 15.21(b)] loading. The shear stress from these types of loading is

$$\tau = \frac{P}{t_e L_w} = \frac{P}{0.707 h_e L_w} = \frac{1.414P}{h_e L_w} \quad \textbf{(15.49)}$$

where

t_e = weld throat length, $h_e \sin 45° = 0.707h_e$, m

h_e = weld leg length, m

L_w = weld length, m

Thus, to avoid failure, the following should be valid:

$$\tau = \frac{P}{t_e L_w} < \left(S_{sy}\right)_{\text{weld}} \quad \textbf{(15.50)}$$

15.6.2 TORSIONAL LOADING

For torsional loading of a weld group the resultant shear stress acting on the weld group is the vectorial sum of the direct and torsional shear stresses, as shown in Eq. (15.48) for rivets. The direct (or transverse) shear stress in the weld is

$$\tau_d = \frac{V}{A} = \frac{\text{Shear force}}{\text{Total throat area}} \quad \textbf{(15.51)}$$

The torsional shear stress is

$$\tau_t = Tr/J \quad \textbf{(15.52)}$$

where

r = distance from centroid of weld group to farthest point in weld, m

T = torque applied to weld, N-m

J = polar area moment of inertia, m^4

The critical section for torsional loading is the throat section, as it is for parallel and transverse loading. The relationship between the unit polar moment of inertia and the polar moment of the fillet weld is

$$J = t_e J_u = 0.707 h_e J_u \quad \textbf{(15.53)}$$

where

J_u = unit polar area moment of inertia, m^3

Table 15.9 gives values of unit polar area moment of inertia for nine weld groups. Using this table simplifies obtaining torsional loading.

Thus, to avoid failure due to torsional loading, the following should be valid:

$$\tau = \tau_d + \tau_t < \left(S_{sy}\right)_{\text{weld}} \quad \textbf{(15.54)}$$

Table 15.9 Geometry of welds and parameters used when considering various types of loading. [*Machine Elements in Mechanical Design* by Mott, © 1992. Reprinted by permission of Prentice-Hall, Inc., Upper Saddle River, NJ.]

Dimensions of weld	Bending	Torsion
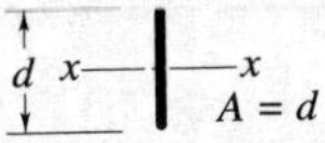 $A = d$	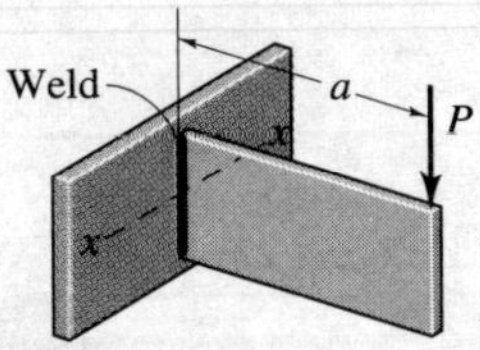 $I_u = d^2/6$, $M = Pa$	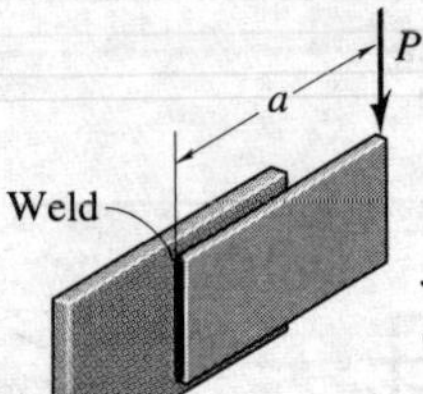 $J_u = d^3/12$, $T = Pa$, $c = d/2$
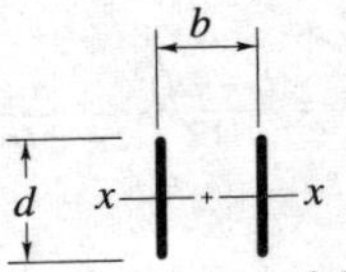 $A = 2d$	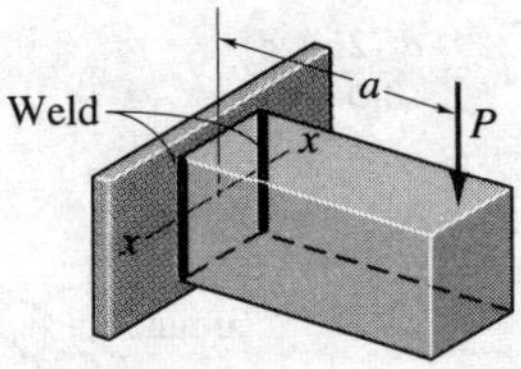 $I_u = d^2/3$	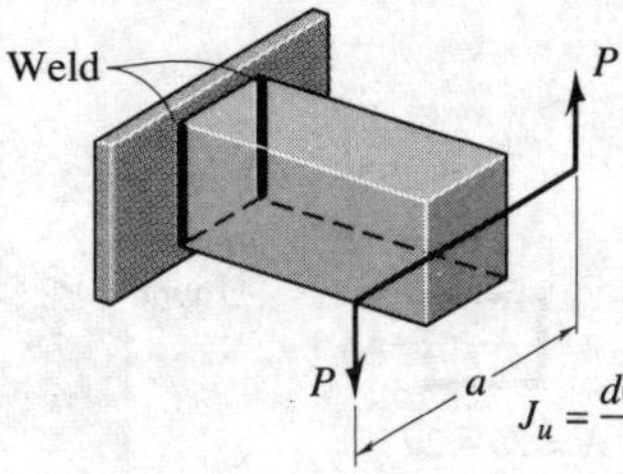 $J_u = \dfrac{d(3b^2 + d^2)}{6}$
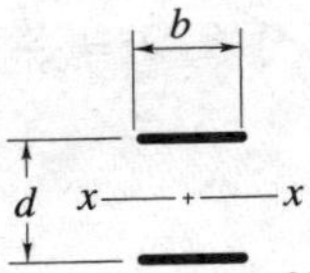 $A = 2b$	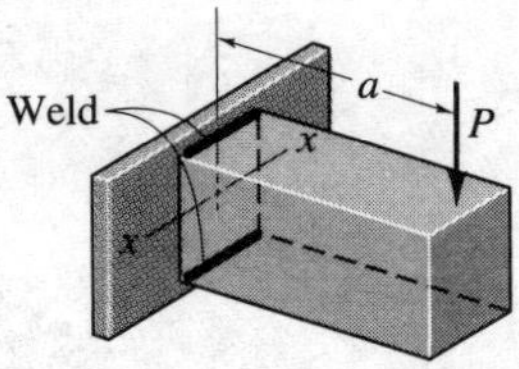 $I_u = bd$	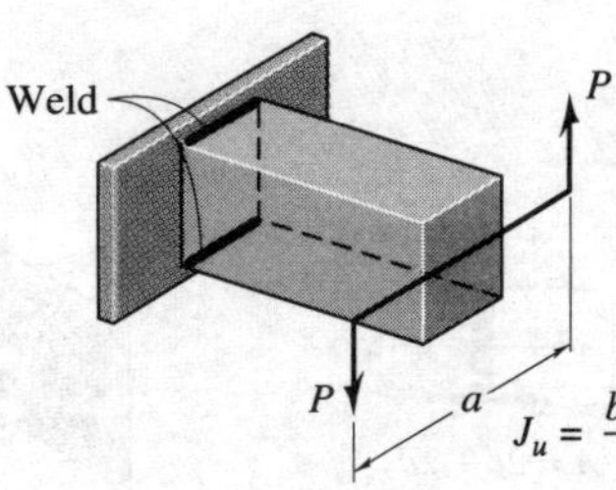 $J_u = \dfrac{b^3 + 3bd^2}{6}$
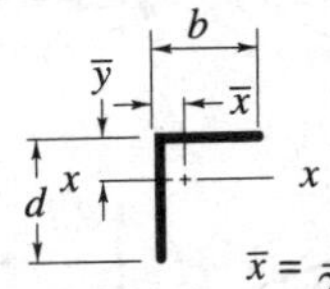 $A = b + d$, $\bar{x} = \dfrac{b^2}{2(b+d)}$, $\bar{y} = \dfrac{d^2}{2(b+d)}$	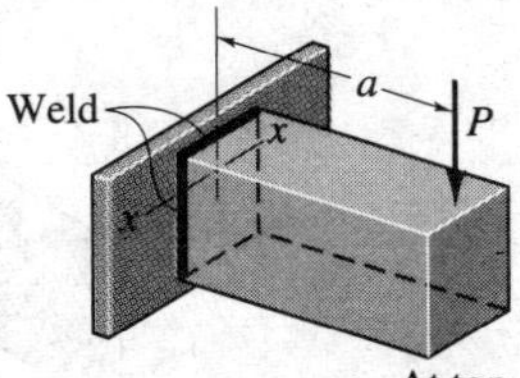 At top $I_u = \dfrac{4bd + d^2}{6}$; At bottom $I_u = \dfrac{d^2(4b+d)}{6(2b+d)}$	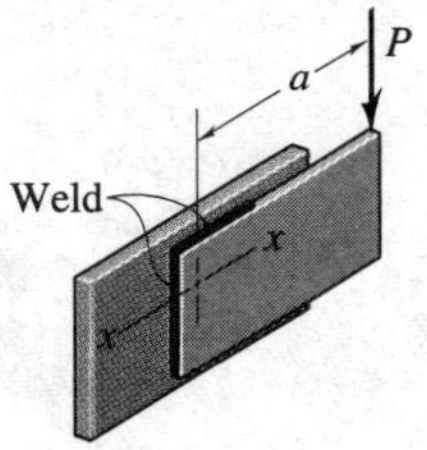 $J_u = \dfrac{(b+d)^4 - 6b^2d^2}{12(b+d)}$
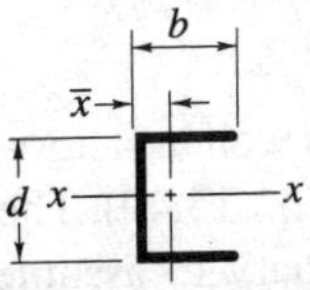 $A = d + 2b$, $\bar{x} = \dfrac{b^2}{2(b+d)}$	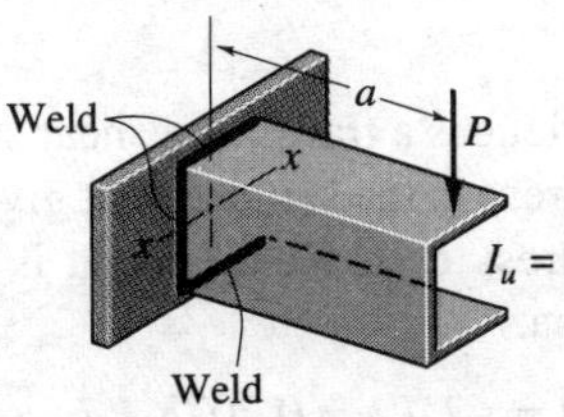 $I_u = bd + d^2/6$	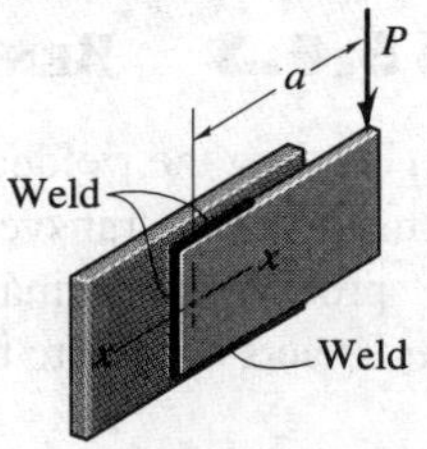 $J_u = \dfrac{(2b+d)^3}{12} - \dfrac{b^2(b+d)^2}{(2b+d)}$

Table 15.9 *(concluded)*

Dimensions of weld	Bending	Torsion
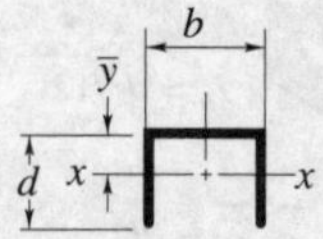 $A = b + 2d \quad \bar{y} = \frac{d^2}{(b+2d)}$	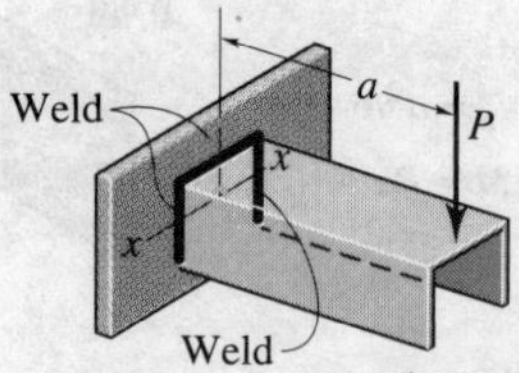 At top $I_u = \frac{2bd + d^2}{3}$ At bottom $I_u = \frac{d^2(2b+d)}{3(b+d)}$	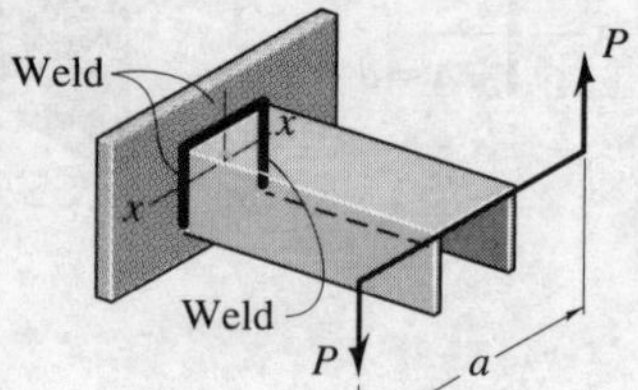 $J_u = \frac{(b+2d)^3}{12} - \frac{d^2(b+d)^2}{(b+2d)}$
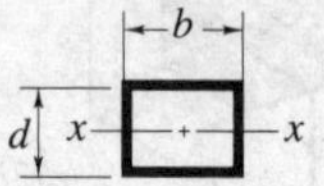 $A = 2b + 2d$	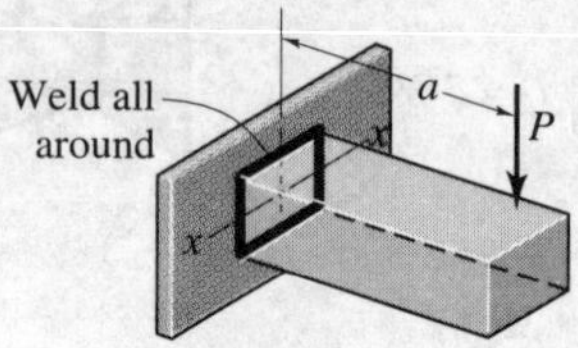 $I_u = bd + d^2/3$	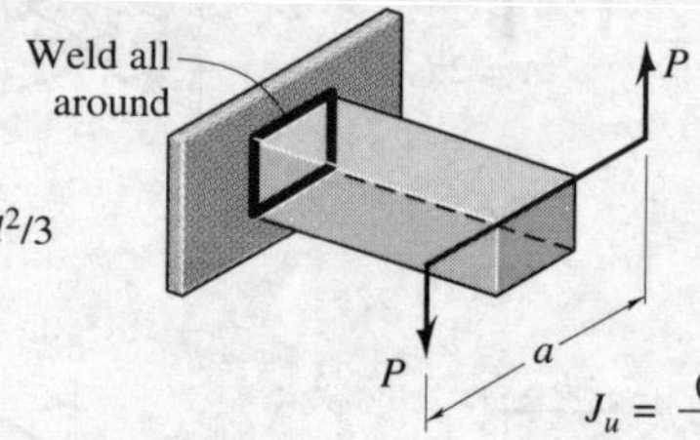 $J_u = \frac{(b+d)^3}{6}$
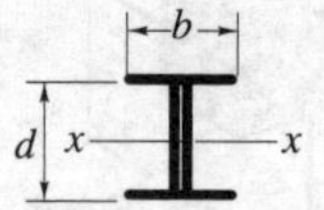 $A = 2b + 2d$	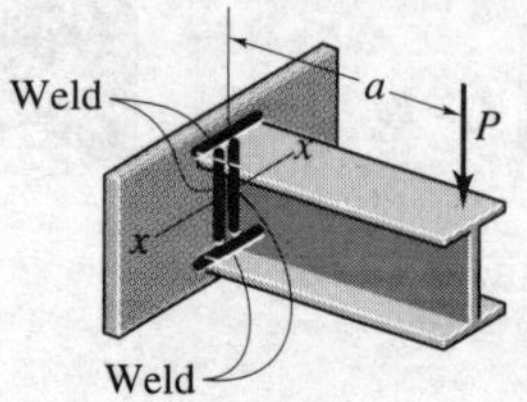 $I_u = bd + d^2/3$	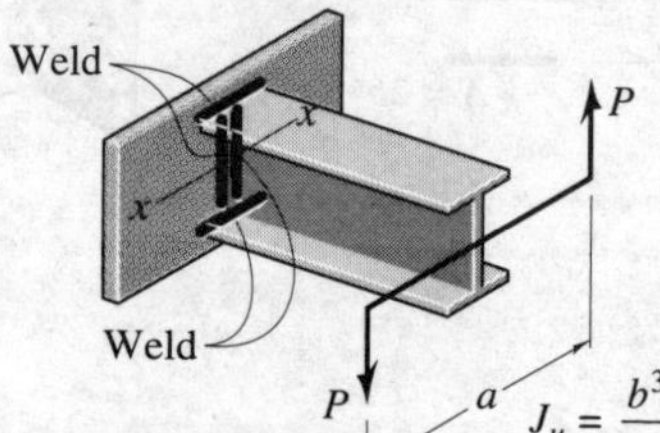 $J_u = \frac{b^3 + 3bd^2 + d^3}{6}$
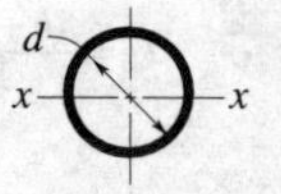 $A = \pi b$	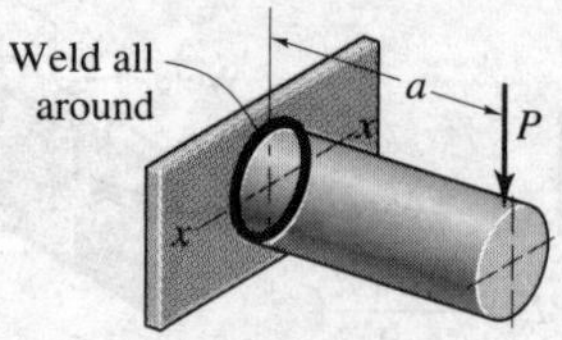 $I_u = \pi(d^2/4)$	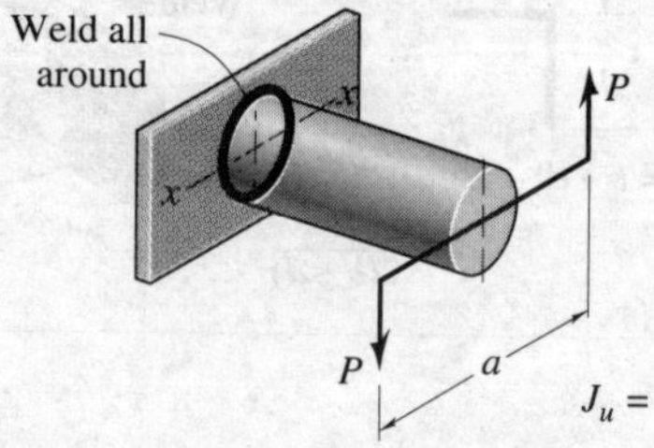 $J_u = \pi(d^3/4)$

15.6.3 BENDING

In bending the welded joint experiences a transverse shear stress as well as a normal stress. The direct (or transverse) shear stress is the same as that given in Eq. (15.49). The moment M produces a normal bending stress σ in the welds. It is customary to assume that the stress acts normal to the throat area.

$$I = t_e I_u L_w = 0.707 h_e I_u L_w \tag{15.55}$$

where

I_u = unit area moment of inertia, m^2

L_w = length of weld, m

Table 15.9 gives values of unit area moment of inertia for nine weld groups. The force per unit length of the weld is

$$w' = Pa/I_u \tag{15.56}$$

where

a = distance from wall to applied load, m

The normal stress due to bending is

$$\sigma = Mc/I \tag{15.57}$$

where

c = distance from neutral axis to outer fiber, m

Once the shear stress and the normal stress are known, the principal shear stresses [from Eq. (2.18)] or the principal normal stresses [from Eq. (2.16)] can be determined. Once these principal stresses are obtained, the maximum-shear-stress theory (MSST) can be determined from Eq. (6.6), or the distortion-energy theory (DET) can be determined from Eq. (6.7) to establish if the weld group will fail.

15.6.4 Weld Strength

The electrodes used in arc welding are identified by the letter E followed by a four-digit number, such as E6018. The first two numbers indicate the strength of the deposited material in thousands of pounds per square inch. The last digit indicates variables in the welding technique, such as current supply. The next-to-last digit indicates the welding position as, for example, flat, vertical, or overhead. This text considers the first two digits.

Table 15.10 lists the minimum strength properties for some electrode classes. Once the yield strength of the weld and the type of loading are known from Eqs. (3.13) to (3.16), the allowable stress can be determined. Having the allowable and design stresses, the safety factor can be determined from Eq. (1.1).

Table 15.10 Minimum strength properties of electrode classes.

Electrode number	Ultimate tensile strength, S_u, ksi	Yield strength, S_y, ksi	Elongation, e_k, percent
E60XX	62	50	17–25
E70XX	70	57	22
E80XX	80	67	19
E90XX	90	77	14–17
E100XX	100	87	13–16
E120XX	120	107	14

Example 15.11 **GIVEN** A bracket is welded to a beam as shown in Fig. 15.22(a). Assume a steady loading of 20 kN and weld lengths $l_1 = 150$ mm and $l_2 = 100$ mm. Assume an electrode number of E60XX and a fillet weld.

FIND Determine the weld leg length for the eccentric loading shown in Fig. 15.22(a) while considering torsion but no bending for a safety factor of 2.5.

Solution

The sum of the weld areas is

$$\sum A_i = t_e(l_1 + l_2) = 0.7071h_e(150+100) = 176.8h_e = 250t_e$$

The centroid of the weld group from Eqs. (4.5) and (4.6) is

$$\bar{x} = \frac{\sum A_i x_i}{\sum A_i} = \frac{0(150)t_e + 50(100)t_e}{250t_e} = 20 \text{ mm}$$

$$\bar{y} = \frac{\sum A_i y_i}{\sum A_i} = \frac{75(150)t_e + 150(100)t_e}{250t_e} = 105 \text{ mm}$$

The torque being applied is

$$T = Pr = (20\ 000)(0.3 - 0.02) \text{ N-m} = 5.6 \text{ kN-m}$$

From Table 15.9 for torsional consideration and the weld shown in Fig. 15.22(a), the unit polar area moment of inertia is

$$J_u = \frac{(b+d)^4 - 6b^2d^2}{12(b+d)} = \frac{(100+150)^4 - 6(100)^2(150)^2}{12(100+150)} = 852\ 083 \text{ mm}^3$$

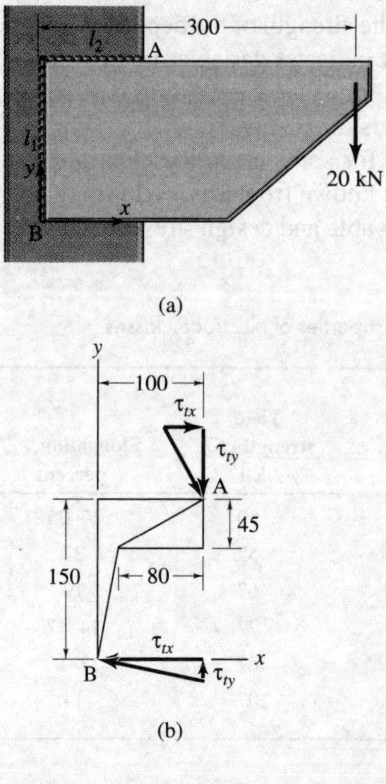

Figure 15.22 Welded bracket used in Example 15.11. (a) Dimensions, load, and coordinates; (b) torsional shear stress components at points A and B. (All dimensions are in millimeters.)

The formulas given in Table 15.9 can be derived by using the parallel-axis theorem given in Eqs. (4.11) and (4.12). From Eq. (4.11) for *weld 1*

$$I_{x'} = \frac{l_1^3}{12} + l_1\left(\bar{y} - \frac{l_1}{2}\right)^2 = \frac{(150)^3}{12} + 150(105-75)^2 = 416\ 250\ \text{mm}^3$$

From Eq. (4.12) for *weld 1*

$$I_{y'} = l_1(\bar{x})^2 = 150(20)^2 = 60\ 000\ \text{mm}^3$$

The unit polar area moment for *weld 1* is

$$(J_u)_1 = I_{x'} + I_{y'} = 476\ 250\ \text{mm}^3$$

For *weld 2*

$$I_{x'} = l_2(l_2 - \bar{y})^2 = 202\ 500\ \text{mm}^3$$

$$I_{y'} = \frac{l_2^3}{12} + l_2\left(\bar{x} - \frac{l_2}{2}\right)^2 = \frac{(100)^3}{12} + 100(50-20)^2 = 173\ 333\ \text{mm}^3$$

The unit polar area moment of inertia for *weld 2* is

$$(J_u)_2 = I_{x'} + I_{y'} = 375\ 833\ \text{mm}^3$$

Thus, the unit polar area moment of inertia for welds 1 and 2 is

$$J_u = (J_u)_1 + (J_u)_2 = 852\ 083\ \text{mm}^3$$

This is the same result as that obtained by using the formula from Table 15.9.

The polar area moment of inertia is

$$J = t_e J_u = 0.7071 h_e J_u = 602\ 514 h_e$$

The J is expressed in millimeters raised to the fourth power.

From Eq. (15.51) the direct (or transverse) shear at locations A and B shown in Fig. 15.22 is

$$\tau_{dA} = \tau_{dB} = \frac{P}{A} = \frac{20\ 000}{176.8 h_e} = \frac{113.1}{h_e}$$

The weld leg length is in millimeters and the shear stress is in megapascals.

From Fig. 15.22(b) the torsional shear stress components at *point A* are

$$\tau_{tAx} = \frac{T(45)}{J} = \frac{(5.6)(10^6)(45)}{(0.6024)h_e}\ \text{Pa} = \frac{418.3}{h_e}\ \text{MPa}$$

$$\tau_{tAy} = \frac{T(80)}{J} = \left(\frac{80}{45}\right)\left(\frac{418.3\times10^6}{h_e}\right)\ \text{Pa} = \frac{743.7}{h_e}\ \text{MPa}$$

$$\tau_{Ax} = \tau_{tAx} = \frac{418.3}{h_e}\ \text{MPa}$$

$$\tau_{Ay} = \tau_{dA} + \tau_{tAy} = \left(\frac{113.1}{h_e} + \frac{743.7}{h_e}\right)(10^6)\ \text{Pa} = \frac{856.8}{h_e}\ \text{MPa}$$

$$\therefore \tau_A = \sqrt{\tau_{Ax}^2 + \tau_{Ay}^2} = \frac{1}{h_e}\sqrt{(418.3)^2 + (856.8)^2}\ (10^6)\text{Pa} = \frac{953.5}{h_e}\ \text{MPa}$$

Similarly, at *point B*

$$\tau_{tBx} = \frac{T(105)}{J} = \frac{(5.6)(10^6)(105)}{(0.6024)h_e}\ \text{Pa} = \frac{976.1}{h_e}\ \text{MPa}$$

$$\tau_{tBy} = \frac{T(20)}{J} = \left(\frac{20}{105}\right)\left(\frac{976.1}{h_e}\right)\ \text{MPa} = \frac{185.9}{h_e}\ \text{MPa}$$

$$\tau_{Bx} = \tau_{tBx} = \frac{976.1}{h_e}\ \text{MPa}$$

$$\tau_{By} = \tau_{dB} - \tau_{tBy} = \left(\frac{185.9}{h_e} - \frac{113.1}{h_e}\right)(10^6)\text{Pa} = \frac{72.82}{h_e}\ \text{MPa}$$

$$\tau_B = \sqrt{\tau_{Bx}^2 + \tau_{By}^2} = \sqrt{\left(\frac{976.1}{h_e}\right)^2 + \left(\frac{72.82}{h_e}\right)^2}\ \text{MPa} = \frac{978.8}{h_e}\ \text{MPa}$$

Because the shear stress is larger at point B, this point is selected for design.

From Table 15.10 for electrode E60XX the yield strength is 50 ksi. From Eq. (3.14) for shear the allowable shear stress is

$$\tau_{all} = 0.40S_y = 0.40(50)(10^3) = 20\ \text{ksi} = 138.0\ \text{MPa}$$

From Eq. (1.1) the safety factor is

$$n_s = \frac{\tau_{all}}{\tau_B} = \frac{138.0}{\frac{978.8}{h_e}}$$

Given that the safety factor is 2.5,

$$\therefore h_e = \frac{(978.8)(2.5)}{138.0} = 17.73\ \text{mm}$$

15.6.5 FATIGUE STRENGTH OF WELDS

When welded members are placed into an environment where cyclic loadings are experienced, the welds fail long before the welded members. Since the electrode material contains a large amount of alloying elements, it is relatively strong, and it is not clear why the weld strength is suspect. However, because a stress concentration is associated with every weld, the stresses are highest in the immediate vicinity of the weld. Further, cracks rarely propagate through the weld material, but rather fatigue failure arises by crack propagation through the heat-affected zone (HAZ) of the welded material. The HAZ is the area around the weld that melts and recrystallizes as a wrought structure. The HAZ material is weaker than the cold-worked substrate and has micropores typical of castings.

Numerous machine design textbooks suggest that welded members should not be used at all in fatigue applications because of these shortcomings. However, since welded connections are routinely placed in situations where fatigue is possible, such admonitions provide no assistance to machine designers.

Recognizing the difficulties associated with determining the actual magnitude of the stress concentration associated with welds, Shigley and Mischke (1989) recommend the

Table 15.11 Fatigue strength reduction factors for welds [J.E. Shigley and L.D. Mitchell, *Mechanical Engineering Design*, 4E, © 1983 by McGraw-Hill. Reproduced with permission of The McGraw-Hill Companies.]

Type of weld	Fatigue stress concentration factor, K_f
Reinforced butt weld	1.2
Toe of transverse fillet weld	1.5
End of parallel fillet weld	2.7
T-butt joint with sharp corners	2.0

fatigue strength reduction factors shown in Table 15.11. These factors should be applied to the wrought material strength (regardless of the manufacturing history prior to welding) as well as to the weld material strength.

15.7 ADHESIVE BONDING

Adhesive bonding is the process of joining materials chemically through the formation of interatomic or intermolecular bonds. Adhesive bonding can be used to join a wide variety of materials (metal to metal, metal to ceramic, metal to polymer, etc.) for both structural and nonstructural uses. The application of adhesive technology in the aerospace industry is responsible for significant recent progress in the understanding of adhesive bonding and its use as a material joining method.

Some of the advantages of adhesive bonding over other fastening techniques covered earlier in this chapter are

1. Uniform stress distribution with resultant increased life
2. Reduced weight
3. Fatigue resistance
4. Ability to join thick or thin materials
5. Ability to join dissimilar materials
6. No stress concentration, since no perforation required
7. Leakproof joints
8. Vibration-damping and insulation properties
9. Load spreading over large area

Some of the limitations of adhesive bonding are

1. Possible need for surface preparation
2. Temperature limitation of 500°F (260°C)
3. Tendency to creep under sustained load
4. Questionable long-term durability

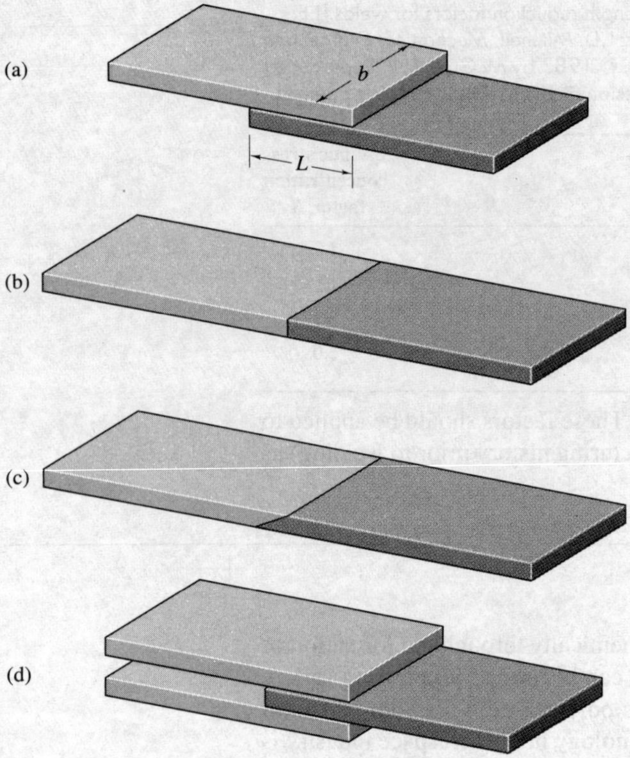

Figure 15.23 Four methods of applying adhesive bonding. (a) Lap; (b) butt; (c) scarf; (d) double lap.

Upon seeing these advantages and disadvantages it is understandable why adhesive bonding is so popular.

Four methods of applying adhesive bonding are shown in Fig. 15.23. Figure 15.23(a) shows a lap joint with a large bond area, whereas Fig. 15.23(b) shows a butt joint with a small bond area. The scarf joint shown in Fig. 15.23(c) has a larger bond area than the butt joint, but scarf joints are impractical for thin plates. The double lap joint shown in Fig. 15.23(d) has the largest bonding area.

If the adhesion for a lap joint is uniform over the lapped surfaces, as shown in Fig. 15.23(a), the average shear stress is

$$\tau_{\text{avg}} = \frac{P}{A} = \frac{P}{bL} \qquad \textbf{(15.58)}$$

The shear stress would not be constant over the area and would be highest at the edges and lowest in the center. For a typical configuration we might expect that

$$\tau_{\max} = 2\tau_{\text{avg}} = \frac{2P}{bL} \qquad \textbf{(15.59)}$$

The stress distribution factor varies with aspect ratio b/L, but for an aspect ratio of 1 it is 2.

For a scarf joint adhesively bonded under *axial* load, as shown in Fig. 15.24(a), while assuming a uniform stress distribution

$$\sigma_x = \frac{P}{b\,t_m/\sin\theta} = \frac{P\sin\theta}{bt_m} \qquad \textbf{(15.60)}$$

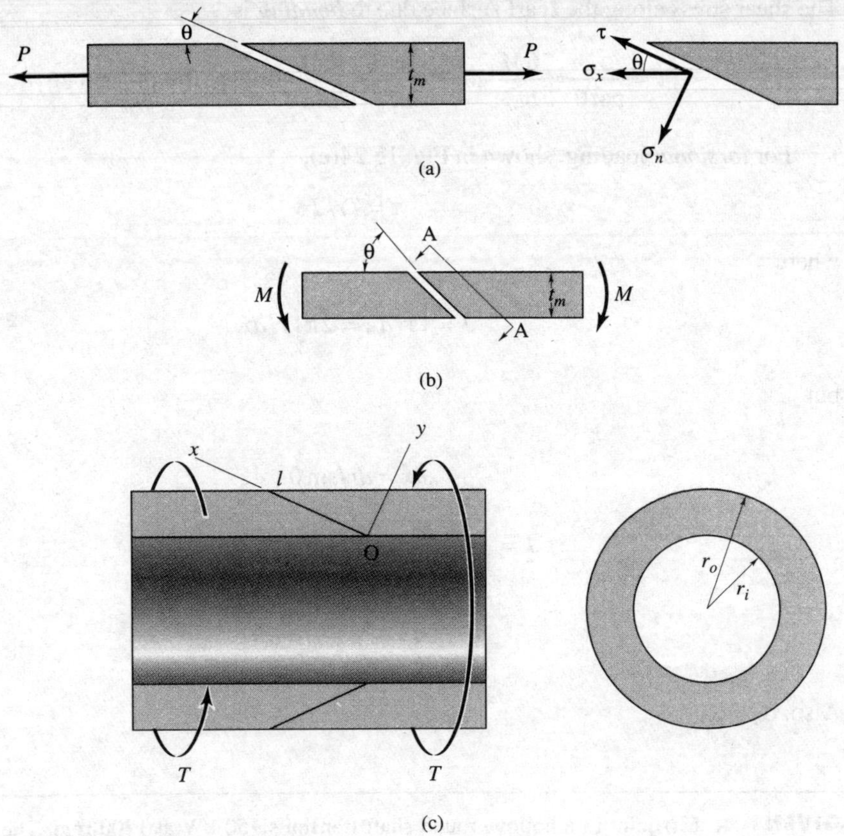

Figure 15.24 Scarf joint.
(a) Axial loading; (b) bending; (c) torsion.

where

t_m = thickness of thinnest member, m

The normal and shear stress components can be written as

$$\sigma_n = \sigma_x \sin\theta = \frac{P}{bt_m}\sin^2\theta \tag{15.61}$$

$$\tau = \sigma_x \cos\theta = \frac{P}{bt_m}\sin\theta\cos\theta = \frac{P}{2bt_m}\sin 2\theta \tag{15.62}$$

For *bending*, shown in Fig. 15.24(b), the stress distribution along the scarf surface (A–A) is

$$\sigma_n = \frac{Mc}{I}$$

$$\therefore \sigma_n = \frac{M(t_m/2\sin\theta)}{\dfrac{b}{12}\left(\dfrac{t_m}{\sin\theta}\right)^3} = \frac{6M\sin^2\theta}{bt_m^2} \tag{15.63}$$

The shear stress along the scarf surface due to *bending* is

$$\tau = \frac{\sigma_n}{\tan\theta} = \frac{6M}{bt_m^2}\sin^2\theta\frac{\cos\theta}{\sin\theta} = \frac{6M}{bt_m^2}\sin\theta\cos\theta = \frac{3M}{bt_m^2}\sin 2\theta \tag{15.64}$$

For *torsional* loading, shown in Fig. 15.24(c),

$$\tau = Tr/J$$

where

$$J = \int r^2 dA = 2\pi\int_0^l r^3 dx$$

but

$$dx = dr/\sin\theta$$

$$J = \frac{2\pi}{\sin\theta}\int_{r_i}^{r_o} r^3 dr = \frac{\pi}{2\sin\theta}\left(r_o^4 - r_i^4\right)$$

$$\therefore \tau = \frac{2Tr\sin\theta}{\pi\left(r_o^4 - r_i^4\right)} \tag{15.65}$$

Also, $\sigma_n = 0$.

Example 15.12 **GIVEN** A scarf joint in a hollow, round shaft transmits 150 kW at 1200 rpm. The shaft has an inner- to outer-radius ratio of 0.8, and the allowable shear stress is 30 MPa. The scarf joint has a 30° orientation with respect to the shaft axis. Use a safety factor of 2.0.

FIND The contact area.

Solution

From Eq. (4.42) the torque in newton-meters is

$$T = h_p/\omega$$

where

h_p = power, W

ω = rotational speed, rad/s

Converting from revolutions per minute to radians per second requires that

$$\omega = 2\pi N_a/60$$

$$\therefore \omega = 2\pi(1200)/60 = 125.7 \text{ rad/s}$$

Equation (4.42) becomes

$$T = \frac{(150)(10^3)}{125.7} = 1194 \text{ N-m}$$

From Eq. (15.65) when $r = r_o$

$$\tau_{\max} = \frac{2Tr_o \sin\theta}{\pi\left(r_o^4 - r_i^4\right)} = \frac{2T\sin\theta}{\pi r_o^3\left[1-\left(\frac{r_i}{r_o}\right)^4\right]} \le \frac{\tau_{\text{all}}}{n_s}$$

$$\left(r_o\right)_{\text{cr}} = \left[\frac{2Tn_s \sin\theta}{\pi\tau_{\text{all}}\left[1-\left(\frac{r_i}{r_o}\right)^4\right]}\right]^{\frac{1}{3}} = \left[\frac{2(1194)(2)\sin 30^\circ}{\pi(30)(10^6)(1-0.8^4)}\right]^{\frac{1}{3}} = 0.035 \text{ m} = 3.5 \text{ cm}$$

The normal area is

$$A_n = \pi\left(r_o^2 - r_i^2\right) = \pi r_o^2\left(1-0.8^2\right) = \pi 3.5^2\left(1-0.8^2\right) = 13.85 \text{ cm}^2$$

The scarf area is

$$A = \frac{A_n}{\sin\theta} = \frac{13.85}{\sin 30^\circ} = 27.71 \text{ cm}^2$$

15.8 INTEGRATED SNAP FASTENERS

Although not a separate machinery element, **snap fasteners** are extremely popular when integrated into machine components because they greatly simplify assembly. Figure 15.25 shows common integrated snap fasteners. Integrated snap fasteners are commonly used with plastic housings or components, since they can be molded directly into the elements. This is one of the most powerful mechanisms for simplifying assemblies and reducing component and manufacturing costs.

Analyzing snap fasteners is complicated by the highly nonlinear elastic behavior of the polymers used. Although Renaud and Karr (1996) discuss in detail some of the difficulties and pitfalls of snap fastener design, here attention is restricted to the analysis of a cantilever snap joint as depicted in Fig. 15.26. For cantilever snap joints it is recommended that the following design rules be followed:

1. $\frac{l}{h} \ge 5$. [LNP (1992) recommends a ratio of 10.] If this is not the case, deformations will not be restricted to the cantilever, and the beam stiffness can be much larger than expected because of nonnegligible shear in the cantilever.

2. A secant modulus must be used because the polymers are usually highly nonlinear. The secant modulus is the slope of the straight line drawn from zero strain to ε on the stress-strain curve.

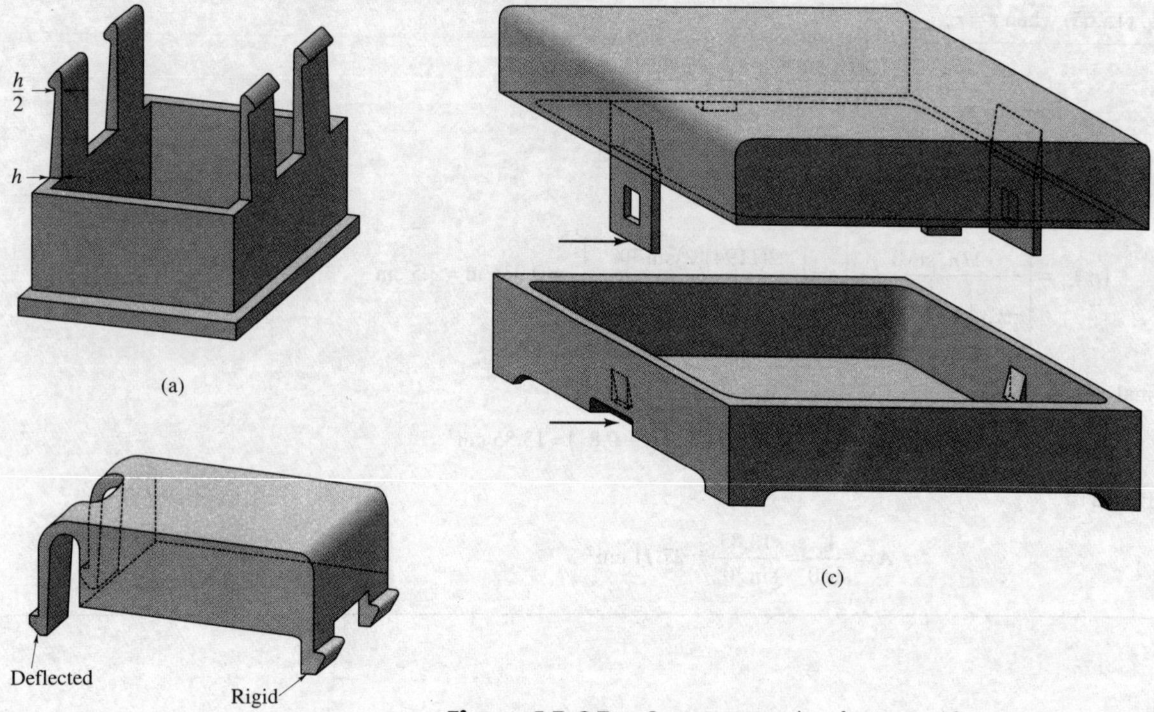

Figure 15.25 Common examples of integrated fasteners. (a) Module with four cantilever lugs; (b) cover with two cantilever and two rigid lugs; (c) separable snap joints for chassis cover.

3. $\dfrac{l}{b} \geq 5$. If the beam is wider, it approaches a plate configuration and will be much stiffer due to the Poisson effect.
4. Snaps should be located away from stress risers such as sharp corners and manufacturing complications such as mold gates or weld lines.

The maximum deflection for the cantilever shown is given by [Bayer (1996); LNP (1992)]

$$y = A\frac{\varepsilon l^2}{h} \tag{15.66}$$

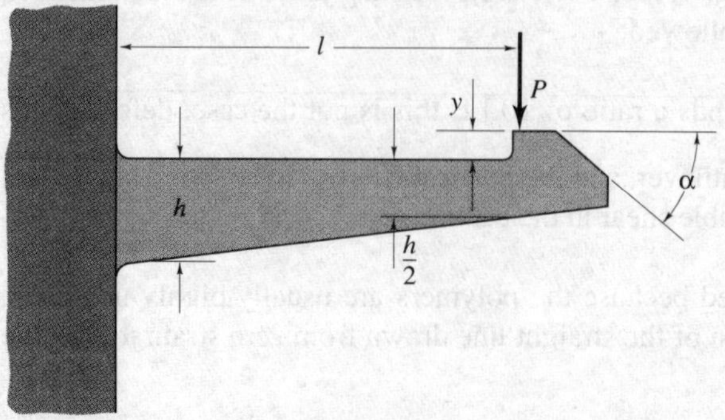

Figure 15.26 Cantilever snap joint.

(Permissible) deflection	Type of design	Shape of cross section: A	Shape of cross section: B
		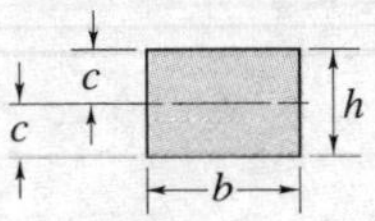 Rectangle	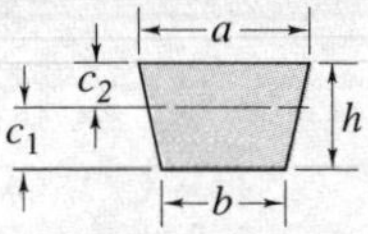 Trapezoid
1	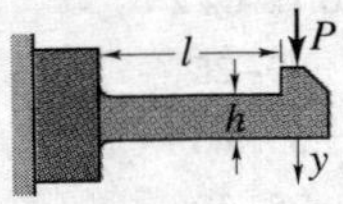 Cross section constant over length	$y = 0.67\dfrac{\varepsilon l^2}{h}$	$y = \dfrac{a+b_{(1)}}{2a+b}\dfrac{\varepsilon l^2}{h}$
2	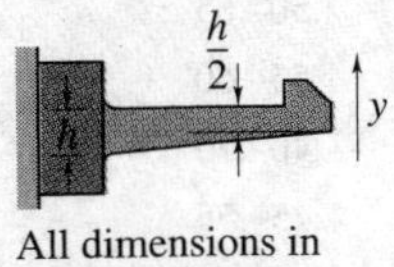 All dimensions in direction y (e.g., h) decrease to one-half	$y = 1.09\dfrac{\varepsilon l^2}{h}$	$y = 1.64\dfrac{a+b_{(1)}}{2a+b}\dfrac{\varepsilon l^2}{h}$
3	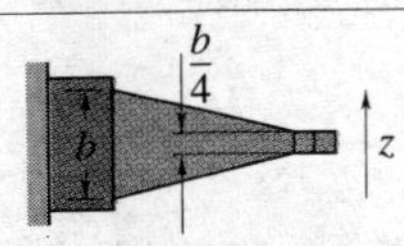 All dimensions in direction z (e.g., b and a) decrease to one-quarter	$y = 0.86\dfrac{\varepsilon l^2}{h}$	$y = 1.28\dfrac{a+b_{(1)}}{2a+b}\dfrac{\varepsilon l^2}{h}$

Figure 15.27 Permissible deflection of different snap fastener cantilever shapes.

where

A = constant for given cantilever shape, as given in Fig. 15.27
ε = allowable strain for polymer
l = cantilever length
h = maximum cantilever thickness

For a rectangular cross section the force required to obtain this deformation is

$$P = \frac{bh^2}{6}\frac{E_s\varepsilon}{l} \qquad \textbf{(15.67)}$$

where

b = cantilever width
E_s = secant modulus of elasticity
l = cantilever length

During assembly, however, the mating force that must be overcome is directed along the cantilever axis and also includes frictional effects. The mating force is given by

$$W = P\frac{\mu + \tan\alpha}{1 - \mu\tan\alpha} \qquad \textbf{(15.68)}$$

Table 15.12 Coefficients of friction for common snap fastener polymers. [*Bayer Snap-Fit Joints for Plastics, A Design Guide*, Bayer Corporation Polymers division, 1996.]

Material	Coefficient of friction: On steel	Coefficient of friction: On self-mated polymer
Polytetrafluoroethylene PTFE (Teflon)	0.12–0.22	—
Polyethylene (rigid)	.20–.25	0.40–0.50
Polyethylene (flexible)	.55–.60	.66–.72
Polypropylene	.25–.30	.38–.45
Polymethylmethacrylate (PMMA)	.50–.60	.60–.72
Acrylonitrile-butadiene-styrene (ABS)	.50–.65	.60–.78
Polyvinylchloride (PVC)	.55–.60	.55–.60
Polystyrene	.40–.50	.48–.60
Polycarbonate	.45–.55	.54–.66

where

W = mating force, N
μ = coefficient of friction
α = wedge angle of cantilever tip, deg

Table 15.12 gives some typical friction coefficients for polymers used in integrated snap fasteners.

Example 15.13

GIVEN A snap-fitting hook must be designed for a rectangular-cross-section cantilever with a constant decrease in thickness from h at the root to $h/2$ at the end of the hook. The undercut dimension serves as the maximum deformation needed for assembly and is decided based on disassembly force requirements. The material is polycarbonate on polycarbonate, length l = 19 mm, width b = 9.5 mm, angle of inclination $\alpha = 30°$, undercut y = 2.4 mm, secant modulus of elasticity $E_s = 1815 \text{ N/mm}^2$.

FIND

a) The thickness h at which full deflection causes a strain of 2%
b) Deflection force P
c) Mating force W

Solution

a) From Eq. (15.66) and using A =1.09 as shown in Fig. 15.27

$$y = 1.09\frac{\varepsilon l^2}{h} \qquad h = 1.09\frac{\varepsilon l^2}{y} = 1.09\frac{0.02(19\text{ mm})^2}{2.4\text{ mm}} = 3.28\text{ mm}$$

b) From Eq. (15.67)

$$P = \frac{bh^2}{6}\frac{E_s\varepsilon}{l} = \frac{(9.5\text{ mm})(3.28\text{ mm})^2}{6}\frac{(1815\text{ N/mm}^2)(0.02)}{19\text{ mm}} = 32.5\text{ N}$$

c) For polycarbonate from Table 15.12 an average coefficient of friction is 0.600. Using this value yields

$$\frac{\mu + \tan\alpha}{1 - \mu\tan\alpha} = 1.801$$

Thus, the mating force is given by

$$W = P\frac{\mu + \tan\alpha}{1 - \mu\tan\alpha} = (32.5\ \text{N})(1.801) = 58.54\ \text{N}$$

Case Study | DESIGN OF BOLTS FOR A HYDRAULIC BALER

Balers are thought of mainly for their use in agriculture to tie hay into manageable shapes. However, balers are often used in industrial applications as well, usually to compress and bind solid materials or wastes into manageable bales before recycling or disposal. A baler uses a hydraulic ram to compress material fed through a hopper and extrude it through an opening. By placing a binding material, usually wire or cable, over the opening, the material is tied into the baled shape.

The key to a baler is the hydraulic cylinder used to compress the material. This case study examines the design of a component of the cylinder, the bolts used to fasten the end plate onto the cylinder.

GIVEN One of the more strenuous applications for industrial balers is the baling of scrap paper from a manufacturer of cardboard boxes. Such a machine will typically produce two completed bales per hour, with 10 to 12 compressions of scrap per bale, and the baler will be operated continuously. To achieve the required baling force, a cylinder with a 150-mm internal bore will be pressurized to a maximum of 7 MPa during operation, although usually much lower than this value. The baler should have a design life of 30 years. The end cap and cylinder flange is shown in Fig. 15.28, and each is 25 mm thick.

FIND Design the bolts needed to affix the end caps to the cylinder of the hydraulic system for the baler. Specify the bolts needed so that they will not fail during the service life of the baler. Use a static safety factor of 3 and make sure that fatigue failure will not occur. Because of the importance of bolt tightness, use a fine thread. Restrict yourself to the standard (rolled) threads in Table 15.7.

Solution

Obviously, the loading is cyclic and the number of total loadings and unloadings is given by

$$\begin{aligned} N &= (12\ \text{cycles/bale})(2\ \text{bales/hr})(24\ \text{hr/day}) \\ &\quad \times (365\ \text{days/yr})(30\ \text{yr}) \\ &= 6.31 \times 10^6\ \text{cycles} \end{aligned}$$

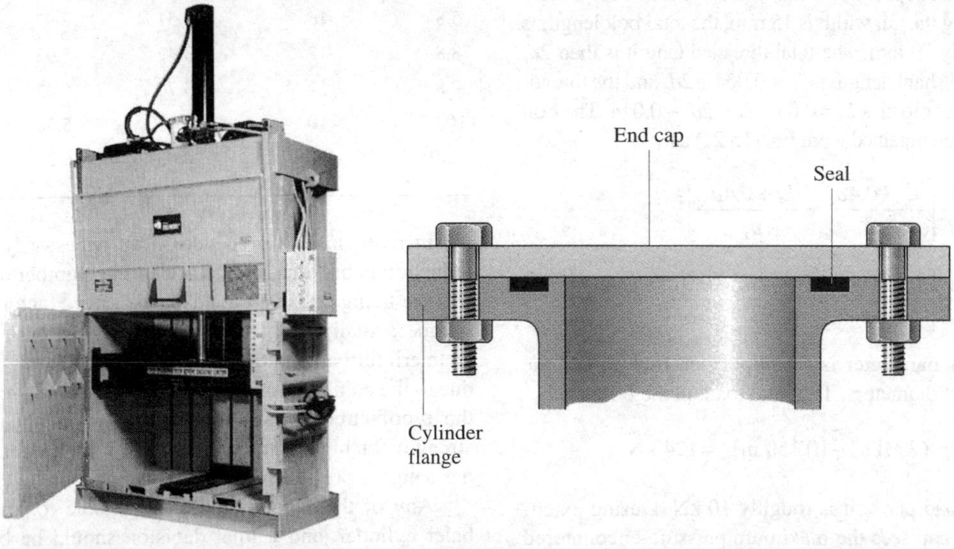

Figure 15.28 End cap and cylinder flange. (Photo courtesy of Pacific Compactor Corporation)

(continued)

Case Study (CONCLUDED)

Thus, these bolts should be designed for high-cycle fatigue. Also, the O-ring seal is placed so that it does not affect the joint stiffness, although the effective pressurized diameter will be greater than the cylinder bore. As a worst-case analysis, the pressurized area will be set equal to the O-ring diameter, or 300 mm.

It is important to use a sufficient number of bolts so that warping in the end plate or cylinder flange will not lead to unloading of the O-ring gasket and leakage of the hydraulic fluid. At this point 12 bolts will be chosen, giving a bolt spacing of just over 85 mm. A denser bolt spacing can be used if an excessively large bolt diameter is required. Since the washer diameter is 1.5 times the crest diameter, each of the joint members has the following stiffness from the Wileman method [Eq. (15.26]:

$$k_{1,2} = E(1.5d_c)(0.78715)e^{(0.62873)d_c/0.025}$$

From Eq. (15.25) the combined joint stiffness is

$$\frac{1}{k_j} = \frac{1}{k_1} + \frac{1}{k_2}$$

$$k_j = \frac{E(1.5d_c)(0.78715)e^{(0.62873)d_c/0.025}}{2}$$

The bolt stiffness cannot be determined until the threaded length is known. The necessary bolt length is the joint length plus the nut width plus a few millimeters to allow for assembly. Assuming the nut width is 15 mm, the total bolt length is approximately 70 mm. The total threaded length is then $2d_c + 6$. Thus, the shank length is $L_s = 0.064 - 2d_c$ and the threaded section under load is $L_t = 0.05 - L_s = 2d_c - 0.014$. The bolt stiffness is then obtained from Eq. (15.21) as

$$\frac{1}{k_b} = \frac{4}{\pi E}\left[\frac{L_s + 0.4d_c}{d_c^2} + \frac{L_t + 0.4d_r}{d_r^2}\right]$$

$$= \frac{4}{\pi E}\left[\frac{0.064 - 1.6d_c}{d_c^2} + \frac{2d_c - 0.014 + 0.4d_r}{d_r^2}\right]$$

The stiffness parameter is a complicated function of the root and crest diameters. The total load on the bolts is

$$P = (7\text{ MPa})\frac{\pi}{4}(0.150\text{ m})^2 = 124\text{ kN}$$

so that the load per bolt is roughly 10 kN. During extension the end cap sees the maximum pressure encountered by the cylinder. During retraction the pressure falls to a great extent. Assuming that the pressure during retraction falls to zero, P_a and P_m are both 5 kN. The alternating and mean stresses acting on the bolt are given by Eqs. (15.36) and (15.37), respectively, and are functions of the root and crest diameters.

Since the safety factor against fatigue failure must be 3, Eq. (15.40) will allow solution. The approach is as follows:

a) Select a grade for the bolts. This defines the ultimate yield and proof strengths.

b) Select a potential crest diameter and associated root diameter. Evaluate the joint and bolt stiffnesses, dimensionless stiffness parameters, alternating and mean stresses, and the safety factor.

c) If the safety factor is greater than 3 [per Eq. (15.31)], select a smaller bolt crest diameter. If the safety factor is less than 3, choose a larger crest diameter.

The results from this approach are as follows:

Metric grade	Minimum required crest diameter, mm	Safety factors		
		Bolt overload	Joint separation	Bolt fatigue
4.6	20	3.17	6.8	1.64
4.8	20	4.37	9.4	1.25
5.8	16	3.51	6.97	1.27
8.8	12	3.42	5.93	1.30
9.8	12	3.70	6.43	1.31
10.9	10	3.45	5.36	1.20
12.9	10	4.03	6.26	1.21

An increase in bolt grade does not necessarily mean a reduction in bolt diameter, since a finite number of diameters are being considered. Also, the safety factor for bolt fatigue actually drops with an increase in bolt grade, a counterintuitive situation. Recall that the safety factor is due to the combination of applied loading and preloading; the proof strength is a higher fraction of the ultimate strength for higher grade bolts; and a high-strength bolt has a higher preload [per Eq. (15.33)].

Any of these bolts would be suitable for use in the baler cylinder, and a final decision should be based on availability and cost.

15.9 SUMMARY

Members can be joined by using a number of different techniques. This chapter investigated threaded, riveted, welded, and adhesive joining of members. Power screws were also covered. A power screw is a device used to change angular motion into linear motion and to transmit power. A general analysis of the forces and torques required to raise and lower a load was presented. It was assumed that an Acme thread is specified for power screws. Collar friction was also considered. Power and efficiency of power screws, as well as when self-locking occurs, were discussed.

The types and classes of thread were given for threaded fasteners. It was found that fatigue strength can be increased by applying initial tension in the bolt. The lead and stiffness of the bolt and joint were analyzed. The most difficult of these analyses is determining the joint stiffness. A needed assumption is that the stress induced in the joint is uniform throughout a region surrounding the bolt hole, with zero stress outside this region. Once the load is known, the design stresses imposed by the threaded fastener can be determined. The allowable stress for threaded fasteners was obtained by considering the proof stress and proof load. Having the design and allowable stresses, the safety factor can be determined. Both static and dynamic considerations were made and gasketed joints were covered.

Riveted and threaded fasteners in shear are treated alike in design and failure analysis. Four modes of failure were presented: bending of a member, shear of a rivet, tensile failure of a member, and compressive bearing failure.

The design of welded fasteners was considered. Fillet welds were focused on, since they are the most frequently used type of weld. For axial loading when the fillet weld is aligned parallel to the load, the shear stress occurs along the throat of the fillet parallel to the load. The load-and-stress analysis occurs along the critical section, which is the plane created by the throat dimension extending the length of the weld. Failure could be predicted for various types of loading while also considering the strength of the weld.

The last two fastener methods to be considered were adhesive bonding and snap fasteners. Adhesively bonded lap and scarf joints were briefly analyzed. In addition, integrated snap fasteners were examined.

KEY WORDS

adhesive bonding process of joining materials chemically through formation of interatomic or intermolecular bonds

bolt externally threaded fastener intended to be used with nut

cap screw externally threaded fastener intended to be used with internally threaded hole

coarseness quality and threads per inch of thread profiles on bolt, nut, or screw

gasket compliant member intended to provide seal

lead distance that screw would advance in one revolution

nut internally threaded mating member for bolt

pitch distance from point on one thread to same point on adjacent thread

power screws power transmission device using mechanical advantage of threads to apply large loads

proof load maximum load that bolt can withstand without acquiring permanent set

proof strength limiting value of stress determined by proof load and tensile stress area

rivets fasteners that function through mechanical interference, usually by upsetting one end of rivet extending outside of free hole

self-locking power screw where thread friction is high enough to prevent loads from lowering in absence of externally applied torque

snap fasteners integrated fasteners that operate through elastic deformation and recovery of part of structure after insertion into proper retainer

stud externally threaded member with threads on both ends in lieu of cap

threads per inch number of threads in one inch, related to the pitch by $p = 1/n$

thread series standardized thread profile, either UN (unified) or M (metric)

throat thinnest section of weld

weld junction formed through diffusion of two materials to be joined, combined with optional filler material

Recommended Readings

ASM Handbook Volume 6: *Welding, Brazing, and Soldering* (1993) American Society for Metals, Metals Park, OH.

Bickford, J.H. (1995) *An Introduction to the Design and Behavior of Bolted Joints*, 3rd ed., Marcel Dekker, New York.

Blodgett, O.W. (1976) *Design of Weldments*, James F. Lincoln Arc Welding Foundation, Cleveland, OH.

Blodgett, O.W. (1976) *Design of Welded Structures*, James F. Lincoln Arc Welding Foundation, Cleveland, OH.

Fastener Standards, 6th ed. (1988) Industrial Fastener Institute, Cleveland, OH.

Fisher, J.W., and Struik, J.H.A. (1974) *Guide to Design Criteria for Bolted and Riveted Joints*, Wiley, New York.

Juvinall, R.C., and Marshek, K.M. (1991) *Fundamentals of Machine Component Design*, Wiley, New York.

Krutz, G.W., Schuelle, J.K., and Claar, P.W. (1994) *Machine Design for Mobile and Industrial Applications*, Society of Automotive Engineers, Warrendale, MI.

Mott, R.L. (1985) *Machine Elements in Mechanical Design*, Merrill Publishing Co., Columbus, OH.
Norton, R.L. (1996) *Machine Design*, Prentice-Hall, Englewood Cliffs, NJ.
Parmly, R.O. (1989) *Standard Handbook of Fastening and Joining*, 2nd ed., McGraw-Hill, New York.
Shigley, J.E., and Mischke, C.R. (1989) *Mechanical Engineering Design*, 5th ed., McGraw-Hill, New York.

REFERENCES

Bayer (1996) *Bayer Snap-fit Joints for Plastics, A Design Guide*, Bayer Polymers Division, Pittsburgh, PA.
Bickford, J.H. (1995) *An Introduction to the Design and Behavior of Bolted Joints*, 3rd ed., Marcel Dekker, New York.
LNP (1992) *Design and Engineering Guide*, LNP Engineering Plastics, Inc.
Mott, R.L. (1992) *Machine Elements in Mechanical Design*, 2nd ed., Macmillan, New York.
Renaud, J.E., and Karr, P. (1996) "Development of an Integrated Plastic Snap Fastener Design Advisor," *Engineering Design and Automation*, vol. 2, pp. 291–304.
Shigley, J.E., and Mischke, C.R. (1989) *Mechanical Engineering Design*, 5th ed., McGraw-Hill, New York.
Wileman, J., Choudhury, M., and Green, I. (1991) "Computation of Member Stiffness in Bolted Connections," *J. Machine Design*, ASME, vol. 113, pp. 432–437.

PROBLEMS

Section 15.3

15.1 An Acme-threaded power screw with a pitch diameter of 1 in. and single thread is used to raise a load of 25 000 lbf. The collar mean diameter is 1.5 in. The coefficient of friction is 0.1 for both the thread and the collar. Determine the following:

a) Crest diameter of the screw
b) Screw torque required to raise the load
c) Maximum thread coefficient of friction needed to prevent the screw from self-locking if collar friction is eliminated

15.2 A car jack consists of a screw and a nut. The car is lifted by turning the screw. Calculate the torque needed to lift a load of 1 ton. The lead of the thread $l = 9$ mm, its pitch diameter is 22 mm, and its thread angle is 30°. The coefficient of friction is 0.10 in the threads and zero elsewhere.

15.3 A power screw gives the axial tool motions in a numerically controlled lathe. To get high accuracy in the motions, the heating and power loss in the screw have to be low. Determine the power efficiency of the screw if the coefficient of friction is 0.10, pitch diameter is 30 mm, lead is 6 mm, and thread angle is 25°.

15.4 Sketch *a* shows a stretching device for steel wires used to stabilize the mast of a sailing boat. Both front and side views are shown and all dimensions are in millimeters. A screw with square threads ($\beta = 0$), a lead and pitch of 4 mm, and an outer diameter of 20 mm is used. The screw can move axially but is prevented from rotating by flat guiding pins (side view in sketch *a*). Derive an expression for the tightening torque as a function of the stretching force *P* when coefficient of friction of all surface contacts is 0.20. Also, calculate the torque needed when the tightening force is 1000 N.

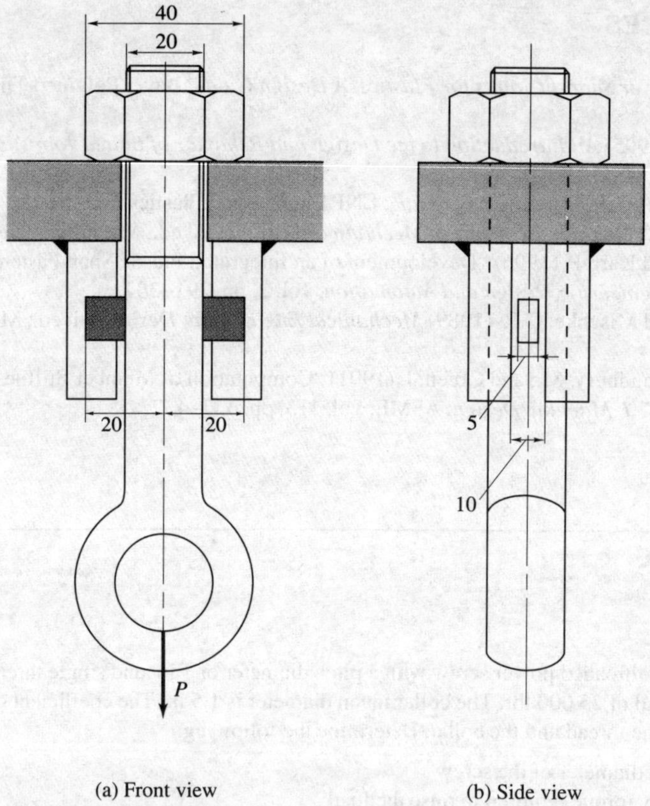

Sketch *a*, for Problem 15.4

15.5 A flywheel of a motorbike is fastened by a thread manufactured directly in the center of the flywheel as shown in sketch *b*. The flywheel is mounted by applying a torque *T*. The cone angle is γ. Calculate the tensile force *W* in the shaft between the contact line at *N* and the thread as a function of D_1, D_2, γ, and *T*. The lead angle is α at the mean diameter D_1. The shaft is assumed not to deform.

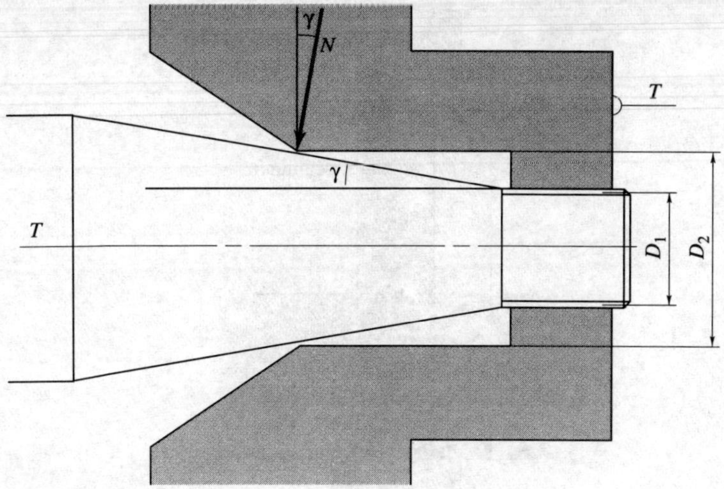

Sketch *b*, for Problem 15.5

15.6 To change its oil, a 10-ton truck is lifted a height of 1.7 m by a screw jack. The power screw has Acme threads and a crest diameter of 5 in. with two threads per inch, and the lead equals the pitch. Calculate how much energy has been used to lift and lower the truck if the only friction is in the threads, where the coefficient of friction is 0.08.

15.7 A single-threaded M32×3.5 power screw is used to raise a 12-kN load at a speed of 25 mm/s. The coefficients of friction are 0.08 for the thread and 0.12 for the collar. The collar mean diameter is 55 mm. Determine the power required. Also determine how much power is needed for lowering the load at 40 mm/s.

15.8 A double-threaded Acme power screw is used to raise a 1350-lbf load. The outer diameter of the screw is 1.25 in. and the mean collar diameter is 2.0 in. The coefficients of friction are 0.13 for the thread and 0.16 for the collar. Determine the following:

a) Required torque for raising and lowering the load
b) Geometrical dimensions of the screw
c) Efficiency in raising the load
d) Load corresponding to the efficiency if the efficiency in raising the load is 18%

15.9 A 25-kN load is raised by two Acme-threaded power screws with a minimum speed of 35 mm/s and a maximum power of 1750 W. Because of space limitations the screw diameter should not be larger than 45 mm. The coefficient of friction for both the thread and the collar is 0.09. The collar mean diameter is 65 mm. Assuming that the loads are distributed evenly on both sides, select the size of the screw to be used and calculate its efficiency.

Section 15.4

15.10 A screw with Acme thread can have more than one entrance to the thread per screw revolution. A single thread means that the pitch and the lead are equal, but for double and triple threads the lead is larger than the pitch. Determine the relationship between the number of threads per inch n, the pitch p, and the lead l.

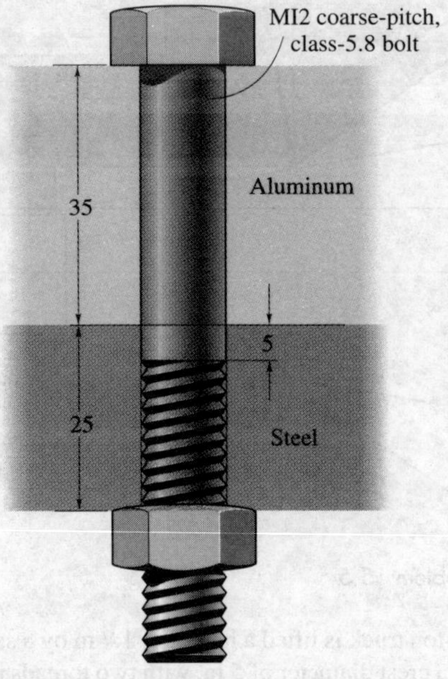

Sketch *c*, for Problem 15.11

15.11 An M12, coarse-pitch, class-5.8 bolt with a hexagonal nut assembly is used to keep two machine parts together as shown in sketch *c*. Determine the following:

a) Bolt stiffness and clamped member by using Wileman's method
b) Maximum external load that the assembly can support for a load safety factor of 2.5
c) Safety factor guarding against separation of the members
d) Safety factor guarding against fatigue if a repeated external load of 10 kN is applied to the assembly

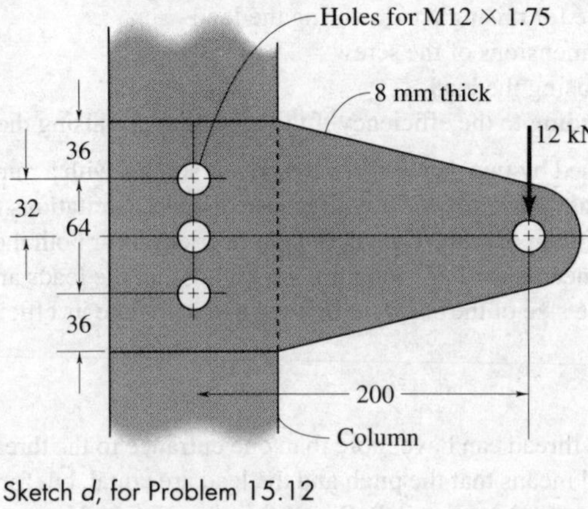

Sketch *d*, for Problem 15.12

15.12 Find the total shear load on each of the three bolts for the connection shown in sketch *d*. Also, compute the shear stress and the bearing stress. Find the area moment of inertia for the 8-mm-thick plate on a section through the three bolt holes.

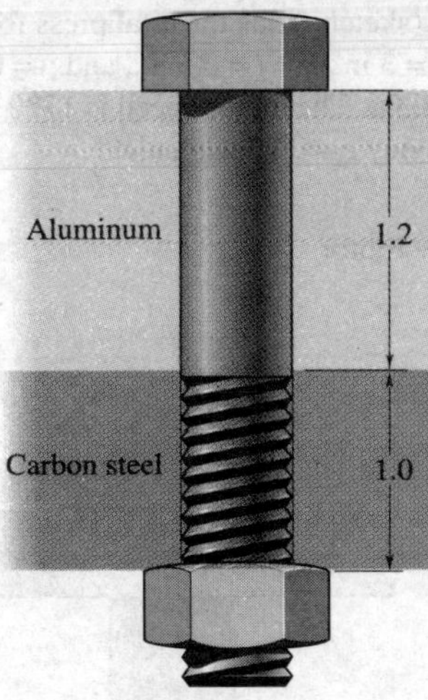

Sketch *e*, for Problem 15.13

15.13 A coarse-pitch, SAE grade-5 bolt with a hexagonal nut assembly is used to keep two machine parts together as shown in sketch *e*. The major diameter of the bolt is 0.5 in. The bolt and the bottom member are made of carbon steel. Assume that the connection is to be reused. Length dimension is in inches. Determine the following:

a) Length of the bolt
b) Stiffnesses of the bolt and the member
c) Safety factor guarding against separation of the members when the maximum external load is 5000 lbf
d) Safety factor guarding against fatigue if the repeated maximum external load is 2500 lbf in a released-tension loading cycle

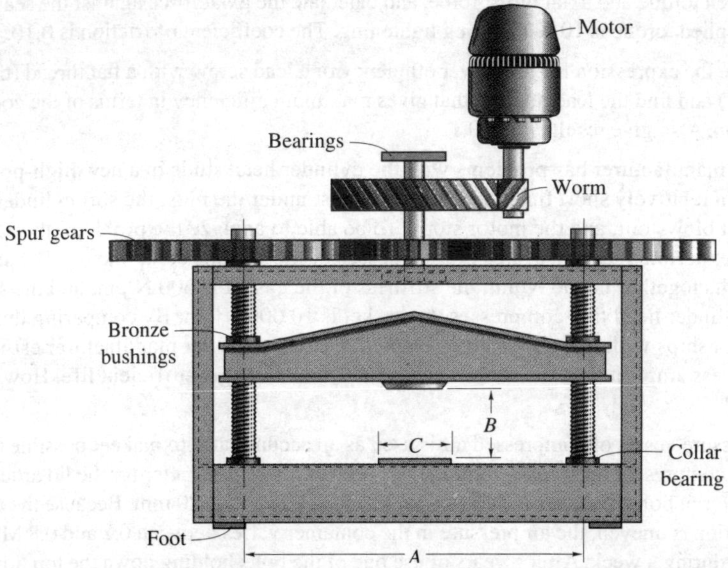

Sketch *f*, for Problem 15.14

15.14 An electric-motor-driven press (sketch *f*) has the total press force $P = 5000$ lbf. The screws are Acme type with $\beta = 29°$, $d_p = 3$ in., $p = l = 0.5$ in., and $\mu = 0.05$. The thrust bearings for the screws have $d_c = 5$ in. and $\mu_c = 0.06$. The motor speed is 1720 rpm, the total speed ratio is 75:1, and the mechanical efficiency $e = 0.95$. Calculate

a) Press head speed
b) Power rating needed for the motor

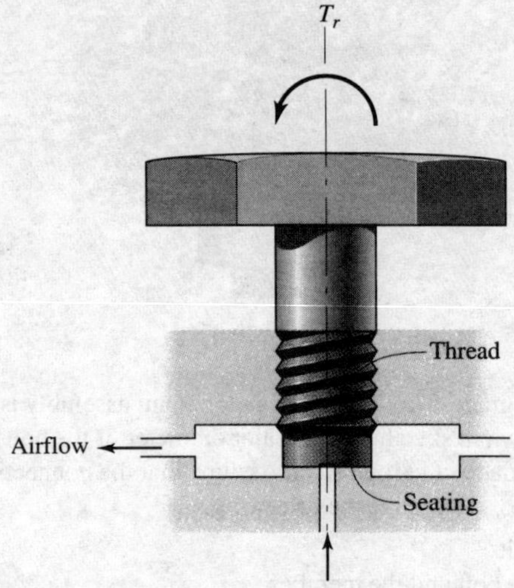

Sketch *g*, for Problem 15.15

15.15 A valve for high-pressure air is shown in sketch *g*. The spindle has thread M12 with a pitch diameter of 10.9 mm, lead $l = 1.75$ mm, and a thread angle of 60°. Derive the relationship between torque and axial thrust force, and calculate the axial force against the seating when the applied torque is 10 N-m during tightening. The coefficient of friction is 0.10.

15.16 Derive the expression for the power efficiency of a lead screw with a flat thread (thread angle $\beta = 0°$) and find the lead angle α that gives maximum efficiency in terms of the coefficient of friction. Also give results if $\mu = 0.1$.

15.17 A car manufacturer has problems with the cylinder head studs in a new high-power motor. After a relatively short time the studs crack just under the nuts, the soft cylinder head gasket blows out, and the motor stops. To be able to analyze the problem, the car manufacturer experimentally measures the stiffnesses of the various components. The stiffness for all bolts together is 400 N/μm, the stiffness of the gasket is 600 N/μm, and the stiffness of the cylinder head that compresses the gasket is 10 000 N/μm. By comparing the life-stress relationships with those for rolling-element bearings, the car manufacturer estimates that the stress amplitude in the screws needs to be halved to get sufficient life. How can that be done?

15.18 A pressure vessel of compressed air is used as an accumulator to make it possible to use a small compressor that works continuously. The stiffness parameter for the lid around each of the 10-mm bolt diameters is 900 MN/m. The shank length is 20 mm. Because the air consumption is uneven, the air pressure in the container varies between 0.2 and 0.8 MPa many times during a week. After 5 years of use one of the bolts holding down the top lid of the pressure vessel cracks off. A redesign is then made, decreasing the stress variation amplitude by 25%, to increase the life of the bolts to at least 50 years. The stress variation amplitude is decreased by lengthening the bolts and using circular tubes with the same cross-sectional area as the solid circular cross section of the bolt to transfer the compressive force from the bolt head to the lid. Calculate how long the tubes should be.

15.19 A loading hook of a crane is fastened to a block hanging in six steel wires. The hook and block are bolted together with four 10-mm-diameter screws prestressed to 20 000 N each. The shank length is 80 mm and the thread length is 5 mm. The stiffness of the material around each screw is 1 GN/m. One of the screws of the crane cracks due to fatigue after a couple of years of use. Will it help to change the screws to 12-mm diameter while other dimensions are unchanged if the stress variation needs to be decreased by at least 20%?

15.20 Depending on the roughness of the contacting surfaces of a bolted joint, some plastic deformation takes place on the tops of the roughness peaks when the joint is loaded. The rougher the surfaces are, the more pressure in the bolted joint is lost by plastic deformation. For a roughness profile depth of 20 μm on each of the surfaces a plastic deformation of 6.5 μm can be expected for the two surfaces in contact. For a bolt-and-nut assembly as shown in Fig. 15.13 three sets of two surfaces are in contact. The stiffness of the two steel plates together is 700 MN/m when each is 40 mm thick. The bolt diameter is 16 mm with metric thread. The shank length is 700 mm. The bolt is prestressed to 25 kN before plastic deformation sets in. Calculate how much of the prestress is left after the asperities have deformed.

15.21 An ISO M12 × 1.75 class = 12.9 bolt is used to fasten three members as shown in sketch *h*. The first member is made of cast iron, the second is low-carbon steel, and the third is aluminum. The static loading safety factor is 2.5. Dimensions are in millimeters. Determine

a) Total length, threaded length, and threaded length in the joint
b) Bolt-and-joint stiffness using a 30° cone
c) Preload for permanent connections
d) Maximum static load that the bolt can support

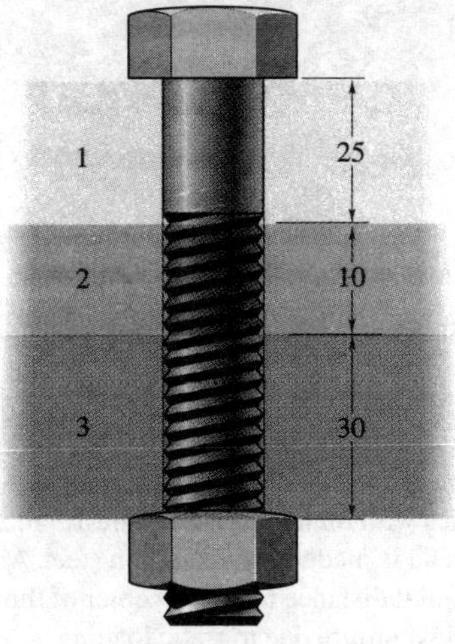

Sketch *h*, for Problem 15.21

15.22 A pressurized cast iron cylinder shown in sketch *i* is used to hold pressurized gas at a static pressure of 8 MPa. The cylinder is joined to a low-carbon-steel cylinder head by bolted joints. The bolt to be used is metric grade 12.9 with a safety factor of 3. Dimensions are in millimeters. Determine the bolt dimensions and the required number of bolts.

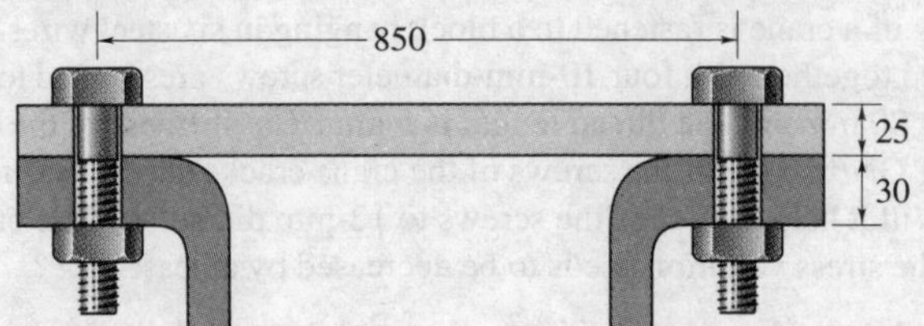

Sketch *i*, for Problem 15.22

15.23 In the bolted joint shown in sketch *j* the first member is made of low-carbon steel, the second member is aluminum, and the third member is cast iron. Assuming that the members can be rearranged and the frustum cone angle is 45°, find the arrangement that can support the maximum load. Dimensions are in millimeters.

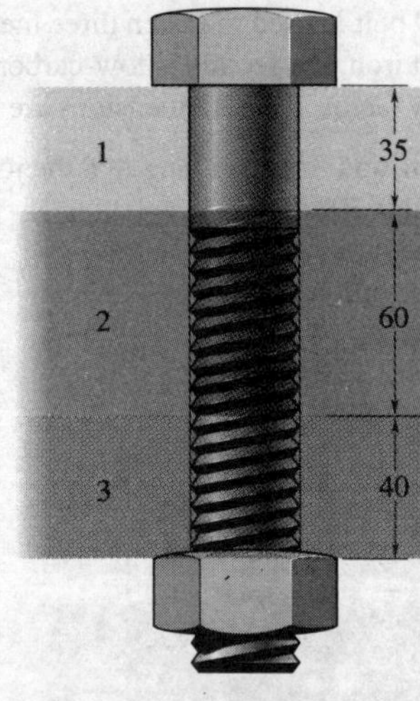

Sketch *j*, for Problem 15.23

15.24 The cylinder shown in sketch *k* is pressurized up to 7 MPa and is connected to the cylinder head by sixteen M24 × 3 metric-grade-8.8 bolts. The bolts are evenly spaced around the perimeters of the two circles with diameters of 1.2 and 1.5 m, respectively. The cylinder is made of cast iron and its head is made of high-carbon steel. Assume that the force in each bolt is inversely related to its radial distance from the center of the cylinder head. Calculate the safety factor guarding against failure due to static loading.

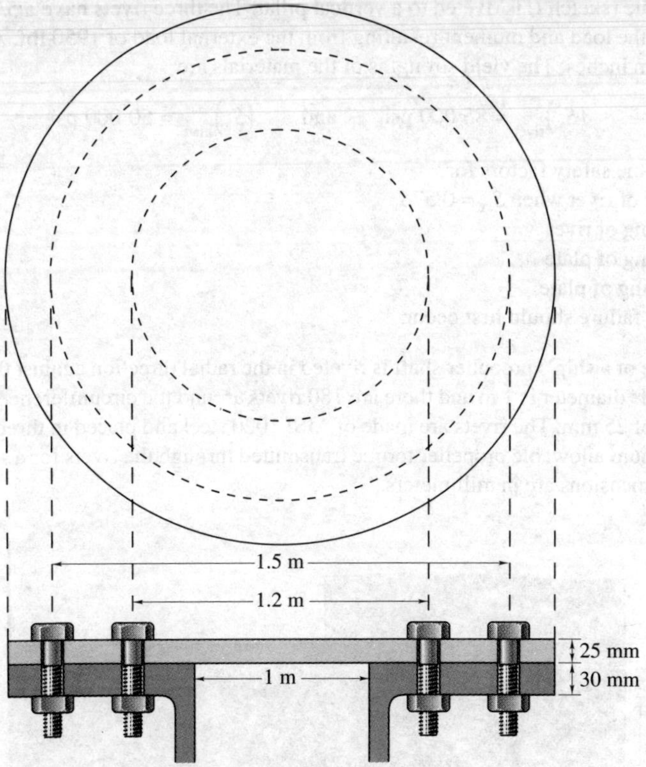

Sketch *k*, for Problem 15.24

Section 15.5

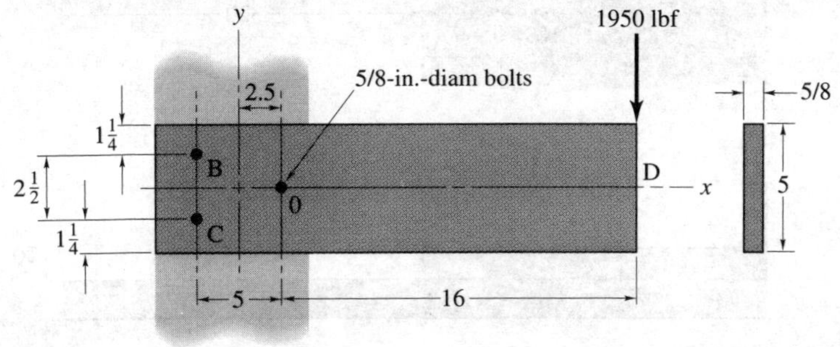

Sketch *l*, for Problem 15.25

15.25 A steel plate (sketch *l*) is riveted to a vertical pillar. The three rivets have a 5/8-in. diameter and carry the load and moment resulting from the external load of 1950 lbf. All length dimensions are in inches. The yield strengths of the materials are

$$(S_y)_{\text{rivet}} = 85\ 000 \text{ psi} \qquad \text{and} \qquad (S_y)_{\text{plate}} = 50\ 000 \text{ psi}$$

Calculate the safety factors for

a) Shear of rivet when $S_{sy} = 0.57S_y$

b) Bearing of rivet

c) Bearing of plate

d) Bending of plate

State how failure should first occur.

15.26 The flange of a ship's propeller shaft is riveted in the radial direction against the hollow shaft. The outside diameter is 1 m and there are 180 rivets around the circumference, each with a diameter of 25 mm. The rivets are made of AISI 1020 steel and placed in three rows. Calculate the maximum allowable propeller torque transmitted through the rivets for a safety factor of 3. Length dimensions are in millimeters.

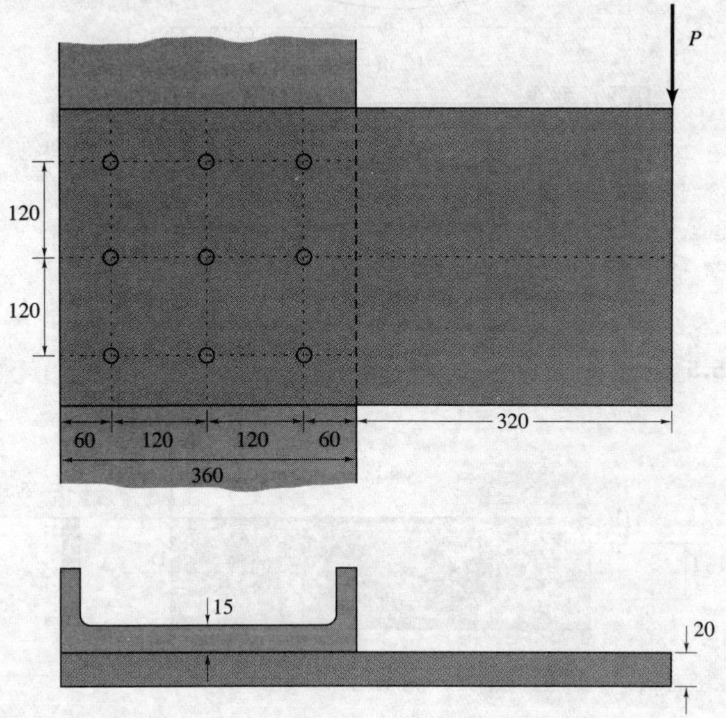

Sketch *m*, for Problem 15.27

15.27 A rectangular steel plate is connected with rivets to a steel beam as shown in sketch *m*. Assume the steel to be low-carbon steel. The rivets have a yield strength of 600 MPa. A load of 24 kN is applied. For a safety factor of 3 calculate the diameter of the rivets.

15.28 Repeat Problem 15.27 but with the plate and beam shown in sketch *n*.

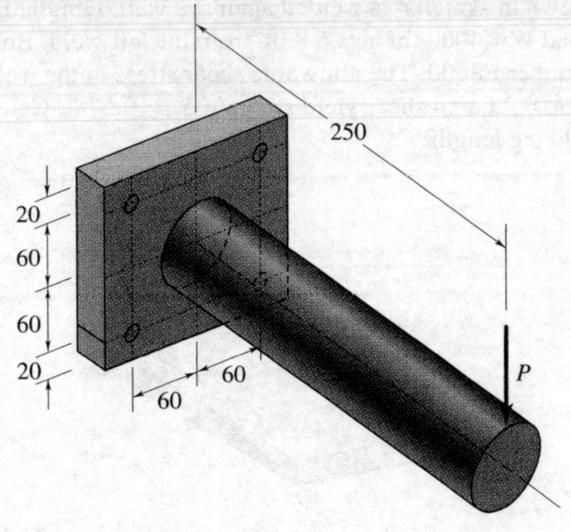

Sketch *n*, for Problem 15.28

Section 15.6

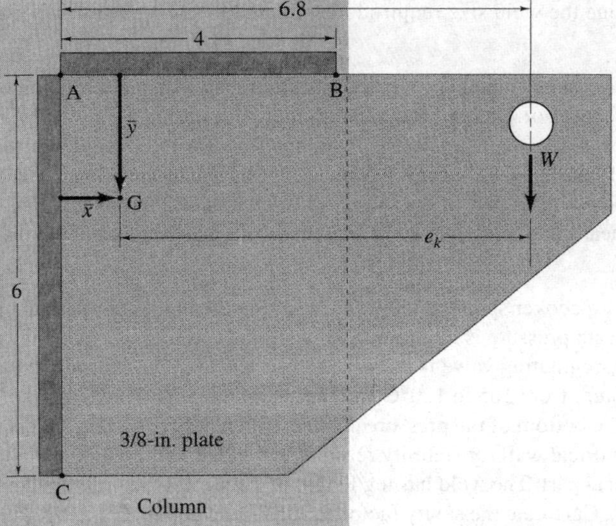

Sketch *o*, for Problem 15.29

15.29 The steel plate shown in sketch o is welded against a wall. Length dimensions are in inches. The vertical load $W = 4000$ lbf acts 6.8 in. from the left weld. Both welds are made by AWS electrode number E8000. The allowable shear stress in the weld should be calculated with safety factor $n_s = 3.0$ and shear yield strength $S_{sy} = 0.57S_u$. Calculate the weld throat length and the weld leg length.

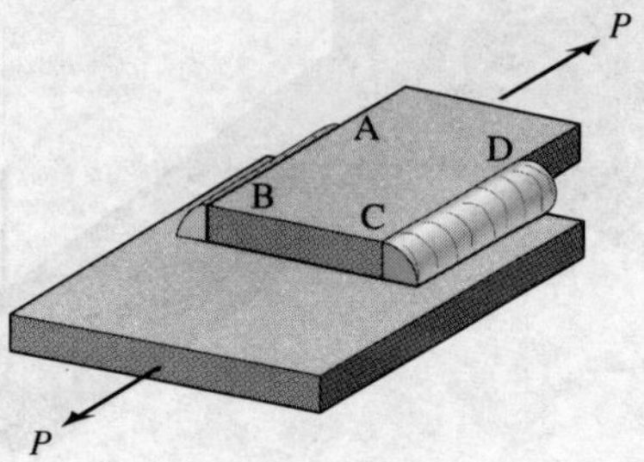

Sketch p, for Problem 15.30

15.30 Two medium-carbon steel (AISI 1040) plates are attached by parallel-loaded fillet welds as shown in sketch p. E60 series welding rods are used. Each of the welds is 3 in. long. What minimum leg length must be used if a load of 15.0 kN is to be applied with a safety factor of 3.5?

15.31 Determine the weld size required if only the top (AB) portion is welded in Problem 15.30.

15.32 A bicycle frame is made of welded circular tubes with an outer diameter of 30 mm and a wall thickness of 2.5 mm. A cyclist going uphill during a Tour de France race pulls vertically upward on the right handle to compensate for the large downward thrust on the right pedal. By doing so he applies a torque of 500 N-m to the bicycle frame. Neglecting direct shear, how large a part of the tube wall thickness of 2.5 mm has to be covered by the weld if the shear yield stress is 180 MPa in the weld material and 250 N-m is applied to each weld?

15.33 A pressure cooker speeds cooking by utilizing the higher temperature that can be reached when steam pressure is increased. To speed the cooking of a beef stew, the weight of the pressure-regulating valve is increased from 100 to 500 g. This action increases the cooking temperature from 105 to 120°C and the steam pressure from 0.12 to 0.20 MPa (1.2 to 2.0 bars). The bottom of the pressure cooker, which is slightly larger than the outer diameter of the cylindrical wall for stability reasons, is welded with a fillet weld of class E80 to the cylindrical part. The weld has leg length of 2 mm and the pressure cooker outer diameter is 250 mm. Calculate the safety factor at 105°C and at 120°C.

15.34 The universal joint on a car axle is welded to the 60-mm-outside-diameter tube and should be able to transfer 1500 N-m of torque from the gearbox to the rear axle. Calculate how large the weld leg should be to give a safety factor of 10 if the weld metal is of class E70.

15.35 The steel bar shown in sketch q is welded by an E60XX electrode to the wall. A 3.5-kN load is applied in the y direction at the end of the bar. Calculate the safety factor against yielding. Also, would the safety factor change if the direction of load P is changed to the z direction? Dimensions are in millimeters.

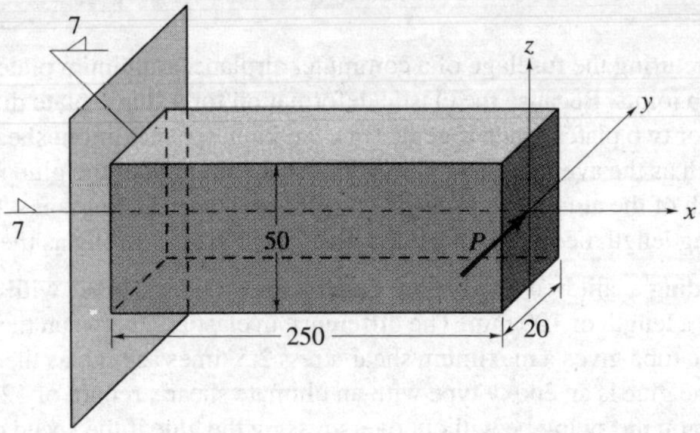

Sketch *q*, for Problem 15.35

15.36 The bar shown in sketch *r* is welded to the wall by AWS electrodes. A 5-kN load is applied at the top of the bar. Dimensions are in millimeters. For a safety factor of 2.5 against yielding, determine the electrode number that must be used and the weld throat length, which should not exceed 10 mm.

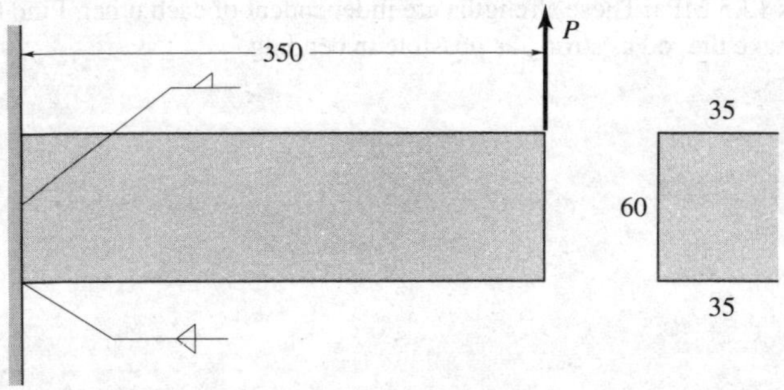

Sketch *r*, for Problem 15.36

15.37 AWS electrode number E100XX is used to weld a bar, shown in sketch *s*, to the wall. For a safety factor of 3 against yielding, find the maximum load that can be supported. What is the maximum stress in the weld and where does it occur? Dimensions are in millimeters.

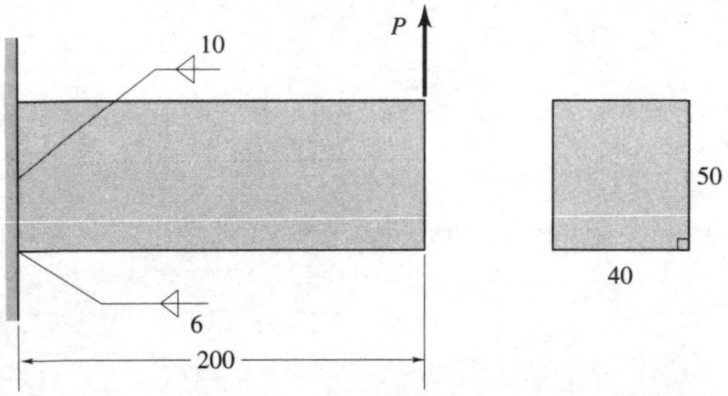

Sketch *s*, for Problem 15.37

Section 15.7

15.38 When manufacturing the fuselage of a commuter airplane, aluminum plates are glued together with lap joints. Because the elastic deformation for a single plate differs from the deformation for two plates glued together in a lap joint, the maximum shear stress in the glue is twice as high as the average shear stress. The shear strength of the glue is 20 MPa, the tensile strength of the aluminum plates is 95 MPa, and their thickness is 4.0 mm. Calculate the overlapping length needed to make the glue joint twice as strong as the aluminum plate.

15.39 The ropes holding a children's swing are glued into two plastic tubes with an inner diameter of 10 mm and a length of 100 mm. The difference in elasticity between the rope and the rope plus the plastic tube gives a maximum shear stress 2.5 times as high as the mean shear stress in the glue. The glue is an epoxy type with an ultimate shear strength of 12 MPa. How heavy can the person on the swing be without overstressing the glue if the speed of the swing at its lowest point is 6.5 m/s and the distance from the center of gravity of the person to the fastening points of the ropes is 2 m? The safety factor is 10.

15.40 A fishing rod is made of carbon-fiber-reinforced plastic tube. To get optimum elastic properties along the length of the rod, and to therefore be able to make long and accurate casts, the concentrations of the fibers in the various parts of the rod have to be different. It is necessary to scarf joint the rod parts. The tensile strength of the epoxy glue joint is 10 MPa and its shear strength is 12.5 MPa. These strengths are independent of each other. Find the optimum scarf angle to make the rod as strong as possible in bending.

CHAPTER

16

SPRINGS

A collection of helical compression springs. (Courtesy of Danly Die)

Entia non multiplicantor sunt prater necessitatum.
(Do not complicate matters more than necessary.)

Galileo Galilei

SYMBOLS

A	cross-sectional area, m^2
A_p	intercept, Pa
a	length dimension, m
b	width of leaf spring, m
b_s	slope
C	spring index, D/d
$\overline{C}$	intercept
c	distance from neutral axis to outer fiber, m
D	mean coil diameter, m
D_i	coil inside diameter, m
D_i'	coil inside diameter after loading, m
D_o	coil outside diameter, m
d	wire diameter, m
E	modulus of elasticity, Pa
f	bump height, m
f_n	lowest natural frequency, Hz
G	shear modulus of elasticity, Pa
g	gravitational acceleration, 9.807 m/s^2 (32.174 ft/s^2)
g_a	gap, m
h	height, m
I	area moment of inertia, m^4

J	polar area moment of inertia, m^4
K_d	transverse shear factor, $(C+0.5)/C$
K_i	spring stress concentration factor, $(4C^2-C-1)/4C(C-1)$
K_w	Wahl curvature correction factor, $\frac{4C-1}{4C-4}+\frac{0.615}{C}$
K_1	defined in Eq. (16.54)
k	spring rate, N/m
k_r	reliability factor
k_t	spring rate considering torsional loading, N/m
k_θ	angular spring rate, N/m
l	length, m
l_w	length of wire in spring, m
M	moment, N-m
m	slope
N	number of coils
N_a	number of active coils
N'_a	number of active coils after loading
N_b	number of coils in body
N_e	number of end coils
N_t	total number of coils
n	number of leaves
n_s	safety factor
P	force, N
P_i	preload, N
p	pitch, m; gas pressure, Pa
R	radius used in applying torque; linear arm length for bracket, m
R_d	diameter ratio, D_o/D_i
r	radius, m
r_i	hook inner radius, m
$r_1, \ldots, r_4$	hook radii, m
S	strength, Pa
S_{se}	modified endurance limit, Pa
S'_{se}	endurance limit, Pa
S_{sf}	modified shear fatigue strength, Pa
S_{su}	shear ultimate strength, Pa
S_{sy}	shear yield strength, Pa
S_{ty}	tensile yield stress, Pa
S_y	yield stress, Pa
S_{ut}	tensile ultimate stress, Pa
T	torque, N-m
t	thickness, m
U	stored elastic energy, N-m
ΔU	change in energy, N-m
v	volume, m^3
x	Cartesian coordinate, m
y	deflection from unloaded state, m
$\gamma_{\theta z}$	shear strain due to torsional loading
Δ_v	loss coefficient
δ	deflection, m
δ_t	deflection due to torsional loading, m
ς	cone angle
θ_{rad}	angular deflection in radians
θ_{rev}	angular deflection in revolutions
ν	Poisson's ratio
ρ	density, kg/m^3
σ	stress, Pa
θ	angle of twist, rad
τ	shear stress, Pa

Subscripts

a	alternating
all	allowable
b	body
c	conical
d	transverse (or direct)
f	free (without load)
h	hook
i	installed; inside; preload
l	loop
m	mean
max	maximum
min	minimum
o	operating; outside
s	solid
t	torsional; total
u	ultimate
0	initial

16.1 INTRODUCTION

A **spring** is a flexible machine element used to exert a force or a torque and, at the same time, to store energy. Energy is stored in the solid that is bent, twisted, stretched, or compressed to form the spring. The energy is recoverable by the elastic return of the distorted material. Springs must have the ability to withstand large deflections elastically. The force can be a linear push or pull, or it can be radial. The torque can be used to cause rotation or to provide a counterbalance force for a machine element pivoting on a hinge. Springs frequently operate with high working stresses and with loads that are continuously varying.

Some more specific applications of springs are

1. To store and return energy, as in a gun recoil mechanism
2. To apply and maintain a definite force, as in relief valves and governors
3. To isolate vibrations, as in an automobile
4. To indicate and/or control load, as in a scale
5. To return or displace a component, as in a brake pedal or engine valve

16.2 SPRING MATERIALS

Strength is one of the most important characteristics to consider when selecting a spring material. Figure 3.19 shows modulus of elasticity E against strength S. Strength means yield strength for metals and polymers, compressive crushing strength for ceramics, tear strength for elastomers, and tensile strength for composites and woods. As Ashby (1992) pointed out, the normalized strength S/E is the parameter to be used in evaluating strength. If just evaluating strength, from Fig. 3.19 we note that engineering polymers have values of S/E in the range 0.01 to 0.1. The values for metals are 10 times lower. Even ceramics in compression are not as strong as engineering polymers, and in tension they are far weaker. Composites and woods lie on the 0.01 contour, as good as the best metals. Because of their exceptionally low elastic modulus elastomers have higher S/E, between 0.1 and 1.0, than does any other class of material.

The *loss coefficient* Δ_v is the second parameter that is important in deciding what material a spring should be made of. The loss coefficient measures the fractional energy dissipated in a stress-strain cycle. Figure 16.1 shows the stress-strain variation for a complete cycle. The loss coefficient is

$$\Delta_v = \Delta U / 2U \tag{16.1}$$

where

ΔU = change of energy over one cycle, N-m

U = stored elastic energy, N-m

The loss coefficient is a dimensionless parameter. A material used for springs should have a *low* loss coefficient. Elastomers, as indicated by Ashby (1992), have the highest loss

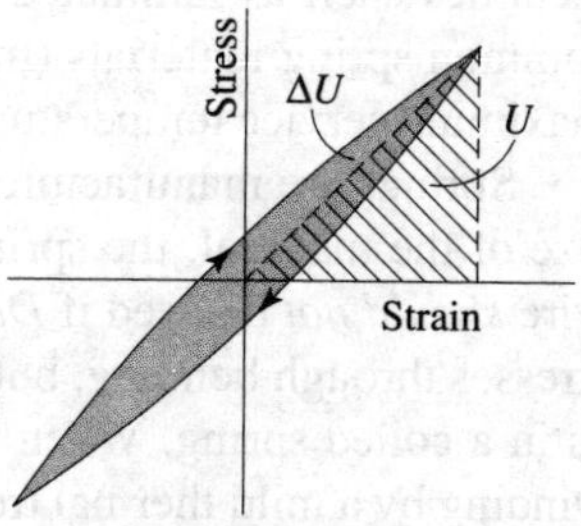

Figure 16.1 Stress-strain curve for one complete cycle.

Table 16.1 Typical properties of common spring materials. [Adapted from Relvas (1996).]

Common name	Specification	Modulus of elasticity, E, psi	Shear modulus of elasticity, G, psi	Density, ρ, lbf/in.^{3}	Maximum service temperature, °F	Principal characteristics
High-carbon steels						
Music wire	ASTM A228	30×10^6	11.5×10^6	0.283	250	High strength; excellent fatigue life
Hard drawn	ASTM A227	30×10^6	11.5×10^6	.283	250	General purpose use; poor fatigue life
Stainless steels						
Martensitic	AISI 410, 420	29×10^6	11×10^6	.280	500	Unsatisfactory for subzero applications
Austenitic	AISI 301, 302	28×10^6	10×10^6	.282	600	Good strength at moderate temperatures; low stress relaxation
Copper-based alloys						
Spring brass	ASTM B134	16×10^6	6×10^6	.308	200	Low cost; high conductivity; poor mechanical properties
Phosphor bronze	ASTM B159	15×10^6	6.3×10^6	.320	200	Ability to withstand repeated flexures; popular alloy
Beryllium copper	ASTM B197	19×10^6	6.5×10^6	.297	400	High elastic and fatigue strength; hardenable
Nickel-based alloys						
Inconel 600	—	31×10^6	11×10^6	.307	600	Good strength; high corrosion resistance
Inconel X–750	—	31×10^6	11×10^6	.298	1100	Precipitation hardening; for high temperatures
Ni-Span C	—	27×10^6	9.6×10^6	.294	200	Constant modulus over a wide temperature range

coefficients, and ceramics have the lowest with a four-order-of-magnitude range between them. Ceramics are not a suitable spring material, since they are brittle in tension. High-carbon steels have just slightly higher loss coefficients than ceramics and are a more suitable spring material.

For the reasons stated above most commercial spring materials come from the group of high-strength, low-loss-coefficient materials that includes high-carbon steel; cold-rolled, precipitation-hardened stainless steel; nonferrous alloys; and a few specialized nonmetallics such as laminated fiberglass. Table 16.1 gives values of typical properties for common spring materials (modulus of elasticity, shear modulus of elasticity, density, and maximum service temperature) and gives additional design information.

Springs are manufactured by either hot- or cold-working processes, depending on the size of the material, the spring index, and the properties desired. In general, prehardened wire *should not* be used if $D/d < 4$ in. or if $d > 1/4$ in. Winding the spring induces residual stresses through bending, but these are normal to the direction of torsional working stresses in a coiled spring. When a spring is manufactured, it is quite frequently relieved after winding by a mild thermal treatment.

The ultimate strength of a spring material varies significantly with wire size, so that the ultimate strength cannot be specified unless the wire size is known. The material and its processing also have an effect on tensile strength. Results of extensive testing by Associated Spring Corp., Barnes Group Inc., show that a semilogarithmic plot of wire strength versus wire diameter is almost a straight line for some materials. The data for five materials can be fitted closely to the following exponential form:

$$S_{ut} = A_p / d^m \tag{16.2}$$

where

A_p = intercept of straight line
m = slope

Table 16.2 gives values of A_p and m for five materials. Equation (16.2) is valid only for the limited range given in Table 16.2. Also, note that A_p in thousand pounds per square inch requires d to be in inches and that A_p in megapascals requires d to be in millimeters.

In the design of springs the allowable stress is the torsional yield strength rather than the ultimate strength. Once the ultimate strength is known from Eq. (16.2), the shear yield stress, which is the allowable shear stress, can be obtained from Eq. (3.14) as

$$S_{sy} = \tau_{\text{all}} = 0.40 S_y \tag{16.3}$$

Example 16.1

GIVEN A spring arrangement is being considered for opening the ashtray in a car. There can be either one spring behind the ashtray or one weaker spring on each side of the ashtray pushing it out after a snap mechanism is released. Except for the spring wire diameter the spring force and the spring geometry are the same for the two options. The material chosen is music wire. For the one-spring option the wire diameter needed is 1 mm.

FIND

a) Which option is cheaper to make if the cost of the springs is proportional to the weight of the spring material?

Table 16.2 Coefficients used in Eq. (16.2) for five spring materials. [*Engineering Guide to Spring Design*, Barnes Group, Inc., 1987.]

	Size range		Exponent,	Constant, A_p	
Material	in.	mm	m	ksi	MPa
Music wire[a]	0.004–0.250	0.10–6.5	0.146	196	2170
Oil-tempered wire[b]	0.020–0.500	0.50–12	0.186	149	1880
Hard-drawn wire[c]	0.028–0.500	0.70–12	0.192	136	1750
Chromium vanadium[d]	0.032–0.437	0.80–12	0.167	169	2000
Chromium silicone[e]	0.063–0.375	1.6–10	0.112	202	2000

[a]Surface is smooth and free from defects and has a bright, lustrous finish.
[b]Surface has a slight heat-treating scale that must be removed before plating.
[c]Surface is smooth and bright with no visible marks.
[d]Aircraft-quality tempered wire; can also be obtained annealed.
[e]Tempered to Rockwell C49 but may also be obtained untempererd.

b) Is there any risk that the two-spring option will suffer from fatigue if the one-spring option does not fatigue?

Solution

a) From Table 16.2 for music wire

$$m = 0.146 \quad \text{and} \quad A_p = 2170 \text{ MPa}$$

Substituting the above into Eq. (16.2) gives

$$S_{ut} = 2170/1^{0.146} = 2170 \text{ MPa}$$

This value is the allowable stress for option 1. From Table 5.1(a) for a concentrated load at the end of the beam ($a = x = l$) the deflection at the free end of the beam is

$$y = \frac{Pl^3}{3EI} \tag{a}$$

The spring has a circular cross section, so that the area moment of inertia is

$$I = \frac{\pi r^4}{4} \tag{b}$$

Substituting Eq. (b) into Eq. (a) gives

$$y = \frac{Pl^3 4}{3E\pi r^4} = \left(\frac{4l^3}{3\pi E}\right)\left(\frac{P}{r^4}\right) \tag{c}$$

The material used and the spring length are the same for options 1 and 2. Thus, if subscript 1 refers to option 1 and subscript 2 refers to option 2, from Eq. (c)

$$\frac{P_1}{r_1^4} = \frac{P_1/2}{r_2^4}$$

$$\therefore r_2 = \frac{r_1}{(2)^{0.25}} = 0.8409 r_1 \tag{d}$$

The cost will change proportionally to the weight or

$$\frac{\pi r_1^2 l\rho}{2\pi r_2^2 l\rho} = \frac{r_1^2}{2r_2^2} = \frac{r_1^2}{2(0.8409 r_1)^2} = 0.7071 \tag{e}$$

Thus, the one-spring option costs only 0.7071 times the two-spring option and is therefore 29.3% cheaper.

b) From Eq. (4.48) the bending (design) stress for one spring (option 1) is

$$\sigma_1 = \frac{Mc}{I} = \frac{Plr_1}{\frac{\pi r_1^4}{4}} = \frac{4Pl}{\pi r_1^3} \tag{f}$$

For two springs (option 2) the design stress is

$$\sigma_2 = \frac{4(P/2)l}{\pi r_2^3} = \frac{2Pl}{\pi(0.8409)^3 r_1^3} = \frac{\sigma_1}{2(0.8409)^3} = 0.8409\sigma_1$$

Using Eq. (16.2) gives the allowable stress for option 2 as

$$S_{ut} = 2170/(0.8409)^{0.146} = 2226 \text{ MPa}$$

Thus, the allowable stress is higher for option 2 than for option 1 and the design stress is lower for option 2 than for option 1; so the safety factor for option 2 is much larger than that for option 1. Recall from Eq. (1.1) that the safety factor is the allowable stress divided by the design stress. Therefore, if there is no risk of fatigue failure for option 1, there is no risk of fatigue failure for option 2.

16.3 HELICAL COMPRESSION SPRINGS

The helix is the spiral form of spring wound with constant coil diameter and uniform pitch. **Pitch** is the distance, measured parallel to the coil axis, from the center of one coil to the center of the adjacent coil. In the most common form of the **helical compression spring** a round wire is wrapped into a cylindrical form with a constant pitch between adjacent coils. The best way to visualize this is to first begin with a straight wire of length l and wire diameter d, to fit a bracket of linear arm length R to each end, and then to subject the device to force P, as shown in Fig. 16.2(a). The force P applied at each end at a distance R from the center of the wire produces a torsional moment

$$T = PR \tag{16.4}$$

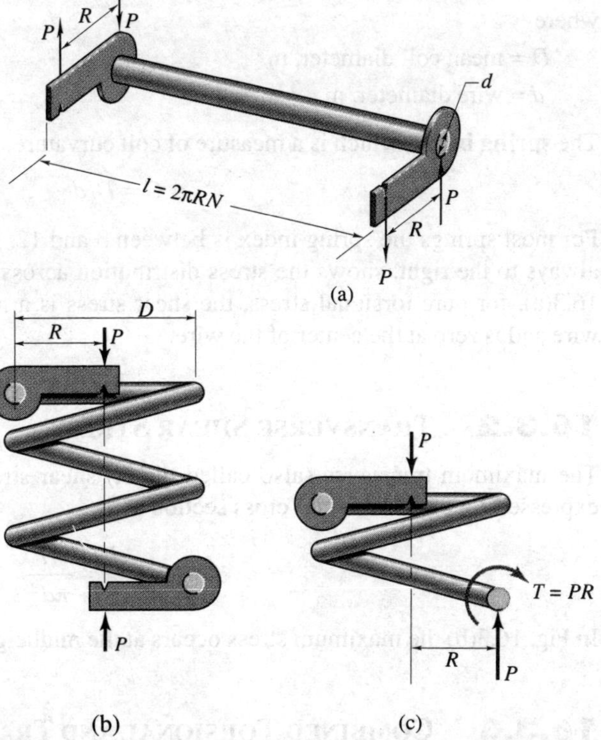

Figure 16.2 Helical coil.
(a) Straight wire before coiling; (b) coiled wire showing transverse (or direct) shear; (c) coiled wire showing torsional shear.

In Fig. 16.2(a) the wire is in equilibrium under the action of the forces P, and the length is described in the number of coils this straight wire is to form.

Figure 16.2(b) shows that the wire has been formed into a helix of N coils each with mean coil radius R. The coiled wire is in equilibrium under the action of two equal and opposite forces P. The important difference between the parts of Fig. 16.2 is that the straight wire in Fig. 16.2(a) experiences only torsional shear whereas the helical coil spring in Fig. 16.2(b) and (c) experiences transverse and torsional shear. Figure 16.2(b) illustrates transverse (sometimes called direct) shear; Fig. 16.2(c) illustrates torsional shear. Torsional shear in a helical coil spring is the primary stress, but transverse (or direct) shear is important enough to consider.

16.3.1 Torsional Shear Stress

The stress in the straight wire shown in Fig. 16.2(a) is shear caused by the torque given in Eq. (16.4). The major stress in the helically coiled spring shown in Fig. 16.2(c) is also torsional shear stress. From Chapter 4, Eq. (4.34), the maximum torsional shear stress while assuming that the cross section of the wire is circular and using Table 4.1 is

$$\tau_{t,\max} = \frac{Tc}{J} = \frac{Td(32)}{(2)\pi d^4} = \frac{16PR}{\pi d^3} = \frac{8PD}{\pi d^3} \tag{16.5}$$

where

D = mean coil diameter, m

d = wire diameter, m

The **spring index,** which is a measure of coil curvature, is

$$C = D/d \tag{16.6}$$

For most springs the spring index is between 3 and 12. Figure 16.3, where the coil axis is always to the right, shows the stress distribution across the wire cross section. From Fig. 16.3(a), for pure torsional stress, the shear stress is a maximum at the outer fiber of the wire and is zero at the center of the wire.

16.3.2 Transverse Shear Stress

The maximum transverse (also called direct) shear stress shown in Fig. 16.3(b) can be expressed for a solid circular cross section as

$$\tau_{d,\max} = \frac{P}{A} = \frac{4P}{\pi d^2} \tag{16.7}$$

In Fig. 16.3(b) the maximum stress occurs at the midheight of the wire.

16.3.3 Combined Torsional and Transverse Shear Stresses

The maximum shear stress resulting from summing the torsional and transverse shear stresses, using Eqs. (16.5) and (16.7), is

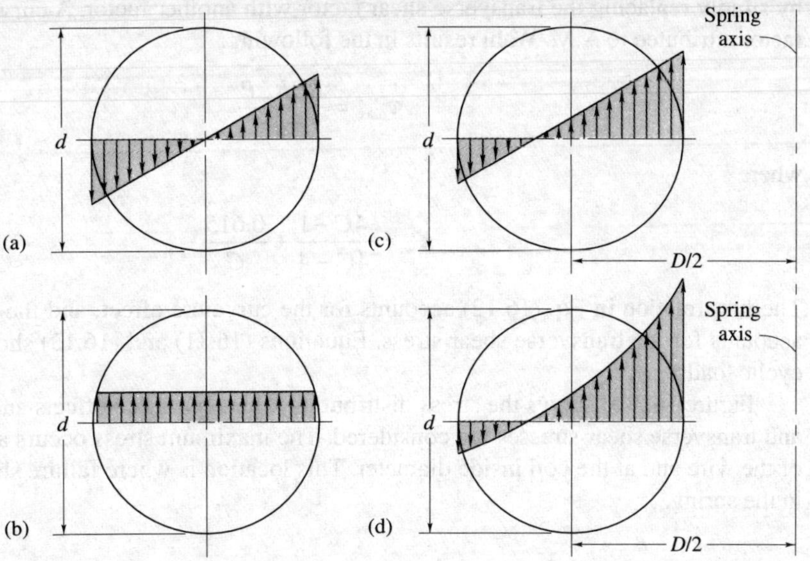

Figure 16.3 Shear stresses acting on wire and coil.
(a) Pure torsional loading; (b) transverse loading; (c) torsional and transverse loading with no curvature effects; (d) torsional and transverse loading with curvature effects.

$$\tau_{\max} = \tau_{t,\max} + \tau_{d,\max} = \frac{8PD}{\pi d^3} + \frac{4P}{\pi d^2} = \frac{8DP}{\pi d^3}\left(1 + \frac{d}{2D}\right) \quad \textbf{(16.8)}$$

Figure 16.3(c) shows that the maximum shear stress occurs at the midheight of the wire and at the coil inside diameter. Curvature effects *are not* considered in Eq. (16.8) and are not shown in Fig. 16.3(c).

Equation (16.8) can be rewritten as

$$\tau_{\max} = \frac{8DK_d P}{\pi d^3} \quad \textbf{(16.9)}$$

where

K_d = transverse shear factor = $(C + 0.5)/C$ **(16.10)**

Note from Eq. (16.8) that if the transverse shear were small relative to the torsional shear, then as it relates to Eq. (16.10) K_d would be equal to 1. Any contribution from the transverse shear term would make the transverse shear factor greater than 1.

Also, recall that the spring index C is usually between 3 and 12. If that range is used, the range of transverse shear factor is 1.0417 to 1.1667. Thus, the contribution due to transverse shear is indeed small relative to that due to torsional shear. Equations (16.9) and (16.10) are used for static loading conditions and also to check if buckling is a problem.

From Sec. 4.5.2 we found that for a curved member the stresses can be considerably higher at the inside surface than at the outside surface. Thus, incorporating curvature can play a significant role in the spring design. Considering curvature effects alters Eq. (16.9) by simply replacing the transverse shear factor with another factor. A curvature correction factor attributed to A.M. Wahl results in the following:

$$\tau_{max} = \frac{8DK_wP}{\pi d^3} \tag{16.11}$$

where

$$K_w = \frac{4C-1}{4C-4} + \frac{0.615}{C} \tag{16.12}$$

The first fraction in Eq. (16.12) accounts for the curvature effect, and the second fraction accounts for the transverse shear stress. Equations (16.11) and (16.12) should be used for cyclic loading.

Figure 16.3(d) shows the stress distribution when curvature effects and both torsional and transverse shear stresses are considered. The maximum stress occurs at the midheight of the wire and at the coil inside diameter. This location is where failure should first occur in the spring.

16.3.4 DEFLECTION

From Eq. (4.27) the shear strain due to torsional loading is

$$\gamma_{\theta z} = \frac{r\theta}{l} = \frac{\text{Deflection}}{\text{Length}} \tag{4.27}$$

The deflection due to torsional loading is

$$\delta_t = r\theta = \frac{D}{2}\theta \tag{16.13}$$

Making use of Eq. (4.31) gives

$$\delta_t = \left(\frac{D}{2}\right)\left(\frac{TL}{GJ}\right) \tag{16.14}$$

Applying this equation to a helical coil spring while making use of Fig. 16.2 and assuming that the wire has a circular cross section gives

$$\delta_t = \left(\frac{D}{2}\right)\left[\frac{(D/2)P(2\pi)(D/2)N_a}{G(\pi d^4/32)}\right] = \frac{8PD^3N_a}{Gd^4} = \frac{8PC^3N_a}{Gd} \tag{16.15}$$

where

C = spring index, D/d
N_a = number of active coils
G = shear modulus of elasticity, Pa

The circumferential deflection due to torsional and transverse shear loading can be derived by using Castigliano's theorem (covered in Chapter 5, Sec. 5.6). The total strain energy from Table 5.2 is

$$U = \frac{T^2 l}{2GJ} + \frac{P^2 l}{2AG} \tag{16.16}$$

The first term on the right of the equal sign corresponds to torsional loading and the second term corresponds to transverse shear for a circular cross section.

Applying Eq. (16.16) to a helical coil spring of circular cross section gives the total strain energy as

$$U = \frac{(PD/2)^2(\pi D N_a)}{2G(\pi d^4/32)} + \frac{P^2(\pi D N_a)}{2G(\pi d^2/4)} = \frac{4P^2 D^3 N_a}{Gd^4} + \frac{2P^2 D N_a}{Gd^2}$$

Using Castigliano's theorem [Eq. (5.30)] gives

$$\delta = \frac{\partial U}{\partial P} = \frac{8PD^3 N_a}{Gd^4} + \frac{4PDN_a}{Gd^2} = \frac{8PD^3 N_a}{Gd^4}\left(1 + \frac{d^2}{2D^2}\right)$$

$$= \frac{8PC^3 N_a}{Gd}\left(1 + \frac{0.5}{C^2}\right) \tag{16.17}$$

Comparing Eq. (16.15) with (16.17) shows that the second term in Eq. (16.17) is the transverse shear term and that for a spring index in the normal range $3 \le C \le 12$ the deflection due to transverse shear is extremely small.

From Eqs. (16.15) and (16.17) the spring rate is

$$k_t = \frac{P}{\delta_t} = \frac{Gd}{8C^3 N_a} \tag{16.18}$$

$$k = \frac{P}{\delta} = \frac{Gd}{8C^3 N_a\left(1 + \frac{0.5}{C^2}\right)} \tag{16.19}$$

Equation (16.18) is applicable only for torsional shear; Eq. (16.19) is applicable for torsional as well as transverse shear loading.

The difference between spring index C and **spring rate** k is important. Spring index is dimensionless whereas spring rate is in newtons per meter. Also, a stiff spring has a small spring index and a large spring rate. (Recall that the spring index for most conventional springs varies between 3 and 12.) A spring with excessive deflection has a large spring index and a small spring rate.

16.3.5 End Conditions and Spring Length

Figure 16.4 shows four types of end commonly used in compression springs. The end coils produce an eccentric application of the load, increasing the stress on one side of the spring. Different end conditions have different amounts of eccentric loading, and compensation must be made in designing a spring. This design alteration is made by designating the number of active coils in a spring. A deduction is made from the total number of coils to account for the turns at the ends, which do not affect the deflection. It is difficult to identify just how much the deduction should be; however, an average number based on experimental results is used.

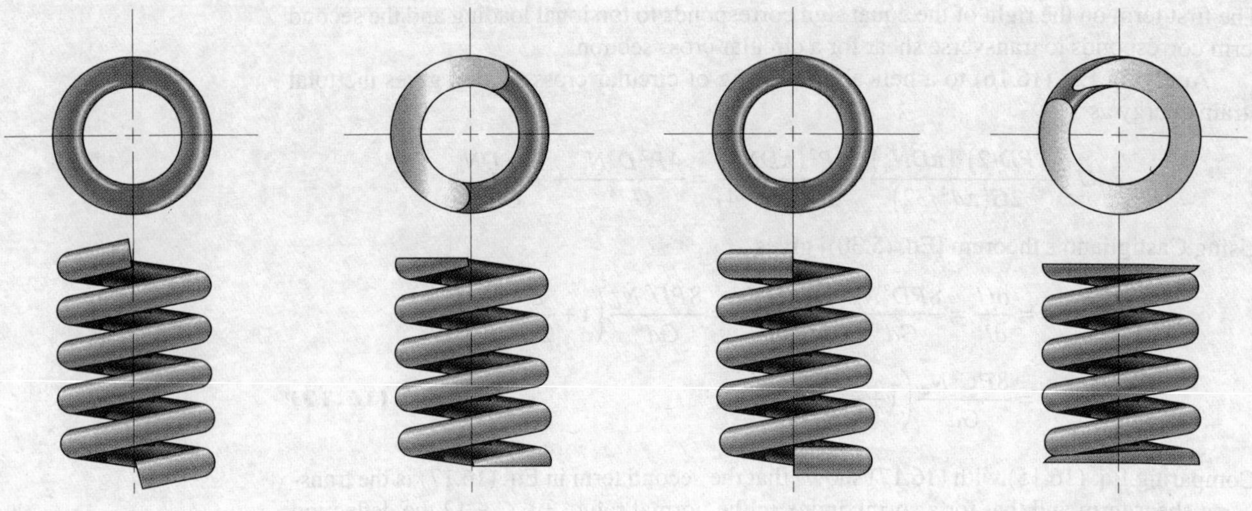

Figure 16.4 Four end types commonly used in compression springs. (a) Plain; (b) plain and ground; (c) squared; (d) squared and ground.

Figure 16.4(a) shows plain ends that have a noninterrupted helicoid; the spring rates for the ends are same as if they were not cut from a longer coil. Figure 16.4(b) shows a plain end that has been ground. In Fig. 16.4(c) a spring with plain ends that are squared (or closed) is obtained by deforming the ends to a 0° helix angle. Figure 16.4(d) shows squared and ground ends. A better load transfer into the spring is obtained by using ground ends.

Table 16.3 shows useful formulas for the pitch, length, and number of coils for compression springs for the four end conditions shown in Fig. 16.4. Recall that the *pitch* is the distance measured parallel to the spring axis from the center of one coil to the center of an adjacent coil. In Table 16.3 the formulas differ for the various end conditions. The **solid length** is the length of the spring when all adjacent coils are in metal-to-metal contact. The

Table 16.3 Useful formulas for compression springs with four end conditions.

	Type of spring end			
Term	**Plain**	**Plain and ground**	**Squared or closed**	**Squared and ground**
Number of end coils, N_e	0	1	2	2
Total number of coils, N_t	N_a	$N_a + 1$	$N_a + 2$	$N_a + 2$
Free length, l_f	$pN_a + d$	$p(N_a + 1)$	$pN_a + 3d$	$pN_a + 2d$
Solid length, l_s	$d(N_t + 1)$	dN_t	$d(N_t + 1)$	dN_t
Pitch, p	$(l_f - d)/N_a$	$l_f/(N_a + 1)$	$(l_f - 3d)/N_a$	$(l_f - 2d)/N_a$

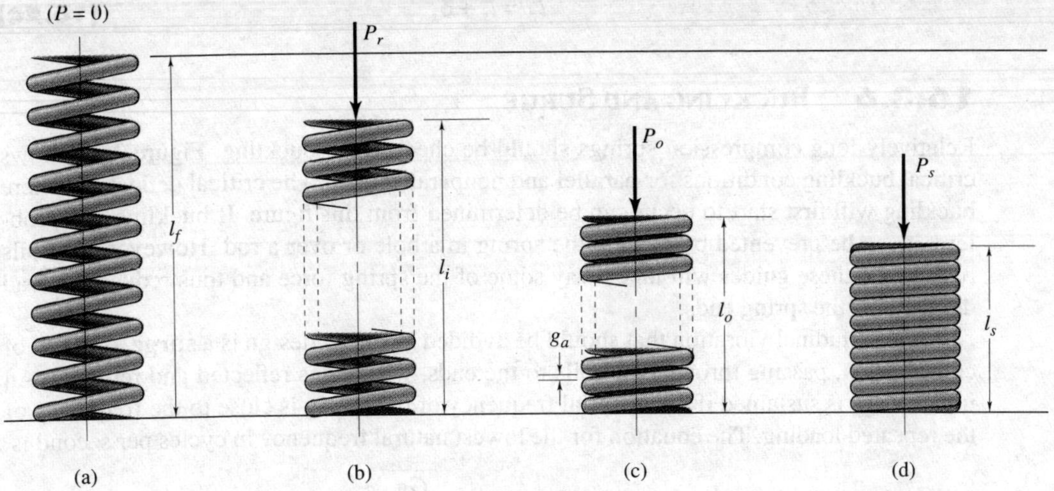

Figure 16.5 Various lengths and forces applicable to helical compression springs. (a) Unloaded; (b) under initial load; (c) under operating load; (d) under solid load.

free length is the length of the spring when no external forces are applied to it. Figure 16.5 shows the various lengths and forces applicable to helical compression springs.

Figure 16.6 shows the interdependent relationships between force, deflection, and spring length for four distinctively different positions: free, initial, operating, and solid. The length is zero at the top right corner and moves to the left for positive lengths. Also, the free length l_f is equal to the solid deflection plus the solid length.

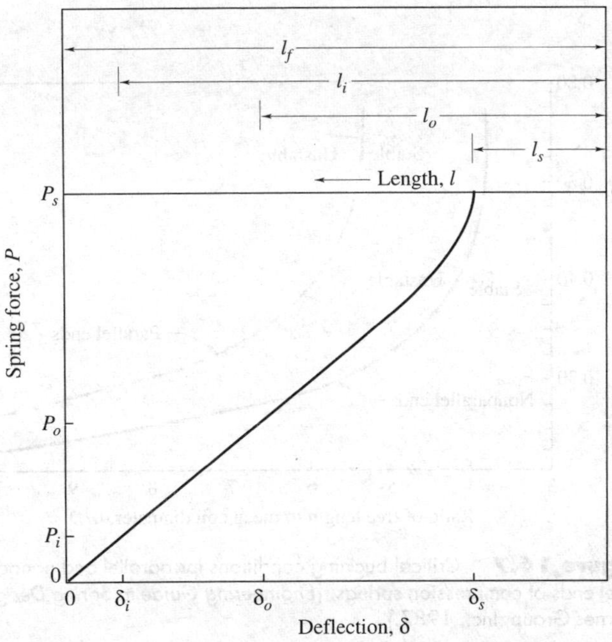

Figure 16.6 Graphical representation of deflection, force, and length for four spring positions.

$$l_f = l_s + \delta_s \tag{16.20}$$

16.3.6 BUCKLING AND SURGE

Relatively long compression springs should be checked for buckling. Figure 16.7 shows critical buckling conditions for parallel and nonparallel ends. The critical deflection where buckling will first start to occur can be determined from this figure. If buckling is a problem, it can be prevented by placing the spring in a hole or over a rod. However, the coils rubbing on these guides will take away some of the spring force and thus reduce the load delivered at the spring ends.

A longitudinal vibration that should be avoided in spring design is a **surge,** or pulse of compression, passing through the coils to the ends, where it is reflected and returned. An initial surge is sustained if the material frequency of the spring is close to the frequency of the repeated loading. The equation for the lowest natural frequency in cycles per second is

$$f_n = \frac{2}{\pi N_a} \frac{d}{D^2} \sqrt{\frac{Gg}{32\rho}} \tag{16.21}$$

where

G = shear modulus of elasticity, Pa
g = gravitational acceleration, m/s^2
ρ = density, kg/m^3

Vibrations may also occur at whole multiples, such as two, three, and four times the lowest frequency. The spring design should avoid these frequencies.

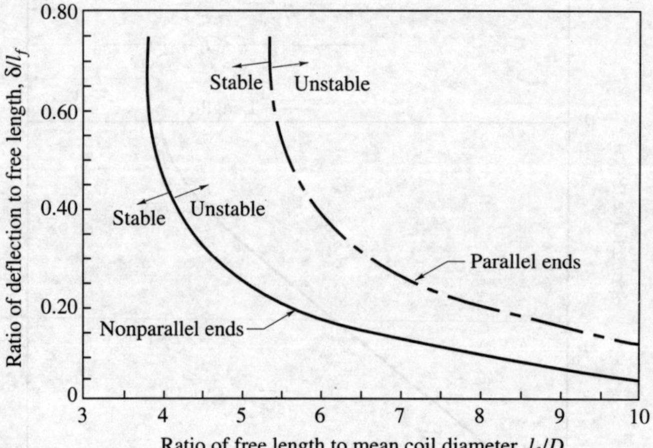

Figure 16.7 Critical buckling conditions for parallel and nonparallel ends of compression springs. [*Engineering Guide to Spring Design,* Barnes Group, Inc., 1987.]

Example 16.2

GIVEN A helical compression spring with plain ends is made to have a spring rate of 100 000 N/m, a wire diameter of 10 mm, and a spring index of 5. The maximum allowable shear stress is 480 N/mm^2 and the shear modulus of elasticity is 80 GPa.

FIND The number of active coils, the maximum allowable static load, and the manufactured pitch so that the maximum load just compresses the spring to its solid length.

Solution

From Eq. (16.19) the number of active coils is

$$N_a = \frac{Gd}{8C^3k\left(1+\dfrac{0.5}{C^2}\right)} = \frac{(80)(10^9)(10)(10^{-3})}{8(5)^3(10^5)\left(1+\dfrac{0.5}{25}\right)} = 7.843 \text{ coils}$$

Therefore, let the number of active coils be 8. From Eq. (16.10) the transverse shear factor is

$$K_d = \frac{C+0.5}{C} = \frac{5+0.5}{5} = 1.10$$

If $\tau_{max} = \tau_{all} = 480$ N/mm^2, from Eq. (16.9) the maximum force is

$$P_{max} = \frac{\pi d^3 \tau_{max}}{8DK_d} = \frac{\pi(10^{-2})^3(480)}{(10)^{-6}\,8(50)(10^{-3})(1.10)} = (3.427)(10^3)\text{ N} = 3.427 \text{ kN}$$

The maximum deflection just to bring the spring to solid length is

$$\delta_s = \delta_{max} = \frac{P_{max}}{k} = \frac{3427}{10^5} = (34.27)(10^{-3})\text{ m} = 34.27 \text{ mm}$$

From Table 16.3 the solid length for plain ends is

$$l_s = d(N_a+1) = (10)(10^{-3})(9) = (90)(10^{-3})\text{ m} = 90 \text{ mm}$$

From Eq. (16.20) the free length is

$$l_f = l_s + \delta_s = 90 + 34.27 = 124.27 \text{ mm}$$

From Table 16.3 the pitch is

$$p = (l_f - d)/N_a = (124.27-10)/8 = 14.28 \text{ mm}$$

Example 16.3

GIVEN A compression coil spring is made of music wire with squared and ground ends. The spring is to have a spring rate of 1250 N/m. The force corresponding to a solid length is 60 N. The spring index is fixed at 10. Static loading conditions are assumed.

FIND Find the wire diameter and the mean coil diameter for the limit when the spring is compressed solid. Give the free and solid lengths and indicate whether buckling is a problem. Also give a design recommendation.

Solution

The tensile ultimate strength can be determined from Eq. (16.2) and Table 16.2 as

$$S_{ut} = \frac{A_p}{d^m} = \frac{(2.170)(10^9)}{d^{0.146}}$$

where d is in millimeters. If d were expressed in meters rather than in millimeters, the above equation would become

$$S_{ut} = \frac{(2.170)(10^9)}{(1000)^{0.146} d^{0.146}} = \frac{(791.5)(10^6)}{d^{0.146}}$$

where d is in meters. From Eq. (16.3) the shear yield stress is

$$S_{sy} = 0.40 S_{ut} = \frac{(316.6)(10^6)}{d^{0.146}} \quad \textbf{(a)}$$

For static loading and given that the spring index is 10, the transverse shear factor from Eq. (16.10) is

$$K_d = 1 + \frac{0.5}{C} = 1.05$$

From Eq. (16.9) the maximum design shear stress is

$$\tau_{max} = \frac{8CK_d P}{\pi d^2} = \frac{8(10)(1.05)(60)}{\pi d^2} = \frac{1604}{d^2} \quad \textbf{(b)}$$

Equating Eqs. (a) and (b) gives

$$\frac{(316.6)(10^6)}{d^{0.146}} = \frac{1604}{d^2} \quad \textbf{(c)}$$

The mean coil diameter is

$$D = Cd = 10(1.393) = 13.93 \text{ mm}$$

Since the spring rate is 1250 N/m and the force corresponding to a solid length is 60 N,

$$\delta_s = \frac{P_s}{k} = \frac{60}{1250} = 0.0480 \text{ m} = 48 \text{ mm}$$

From Table 16.1 the shear modulus of elasticity for music wire is

$$G = 11.5 \times 10^6 \text{ psi} = 0.7935 \times 10^{11} \text{ Pa}$$

From Eq. (16.18) the number of active coils is

$$N_a = \frac{Gd}{8k_t C^3} = \frac{(0.7935)(10^{11})(1.393)(10^{-3})}{8(1250)(10^3)} = 11.05 \text{ coils}$$

From Table 16.3 for squared and ground ends the total number of coils is

$$N_t = N_a + 2 = 13.05 \text{ coils}$$

The solid length is

$$l_s = dN_t = (1.393)(13.05) = 18.18 \text{ mm}$$

The free length from Eq. (16.20) is

$$l_f = l_s + \delta_s = 18.18 + 48 = 66.18 \text{ mm}$$

The pitch is

$$p = (l_f - 2d)/N_a = (66.18 - 2.786)/11.05 = 5.737 \text{ mm}$$

Making use of Fig. 16.7 when

$$\frac{l_f}{D} = \frac{66.18}{13.93} = 4.751 \quad \text{and} \quad \frac{\delta_s}{l_f} = \frac{48.00}{66.18} = 0.7253$$

and assuming parallel ends shows that stability *should not* be a problem.

Recognize that in Eq. (c) the allowable and maximum designs are assumed to have a safety factor of 1. If instead the safety factor is 2, the wire diameter would be

$$\frac{(316.6)(10^6)}{d^{0.146}} = \frac{3208}{d^2}$$

$$\therefore d = 2.017 \text{ mm} \quad \text{and} \quad D = Cd = 20.17 \text{ mm}$$

From Eq. (16.18) the number of active coils is

$$N_a = \frac{Gd}{8kC^3} = 16.00 \text{ coils}$$

$$N_t = N_a + 2 = 18.00 \text{ coils}$$

The solid and torsional lengths are

$$l_s = dN_t = (2.017)(18.00) = 36.31 \text{ mm}$$

$$l_f = l_s + \delta_s = 36.31 + 48 = 84.31 \text{ mm}$$

The pitch is

$$p = \left(l_f - 2d\right)/N_a = (84.31 - 4.034)/16 = 5.017 \text{ mm}$$

Also,

$$\frac{l_f}{D} = \frac{84.31}{20.17} = 4.180$$

$$\frac{\delta_s}{l_f} = \frac{48.00}{84.31} = 0.5693$$

There is thus no problem with stability for parallel ends.

16.3.7 Cyclic Loading

Because spring loading is most often continuously fluctuating, allowance must be made in designing for fatigue and stress concentration. The material developed in Chapter 7 is applied in this section. Helical springs *are never* used as both compression and extension springs. Furthermore, springs are assembled with a preload in addition to the working stress, thus preventing the stress from being zero. The cyclic variation is that on the nonzero mean considered in Sec. 7.3. The worst condition would occur if there were no preload (i.e., when $\tau_{min} = 0$).

For cyclic loading the Wahl curvature correction factor given in Eq. (16.12) should be used instead of the transverse shear factor defined in Eq. (16.10), which is used for static conditions. The Wahl curvature correction factor can be viewed as a fatigue stress concentration factor. However, a major difference for springs in contrast to other machine elements is that for springs the Wahl curvature correction factor is applied to both the mean stress and the stress amplitude. The reason is that in its truest sense the Wahl curvature correction factor is not really a fatigue stress concentration factor (as considered in Sec. 7.7.1), but a way of calculating the shear stress inside the coil.

When Eqs. (7.1) and (7.2) are applied to force rather than to stress, the alternating and mean forces can be expressed as

$$P_a = \frac{P_{max} - P_{min}}{2} \tag{16.22}$$

$$P_m = \frac{P_{max} + P_{min}}{2} \tag{16.23}$$

The alternating and mean stresses, from Eq. (16.11), are

$$\tau_a = \frac{8DK_w P_a}{\pi d^3} \tag{16.24}$$

$$\tau_m = \frac{8DK_w P_m}{\pi d^3} \tag{16.25}$$

For springs the safety factor guarding against torsional endurance limit fatigue is

$$n_s = S_{se} / \tau_a \tag{16.26}$$

Against torsional yielding it is

$$n_s = S_{sy} / (\tau_a + \tau_m) \tag{16.27}$$

And against torsional fatigue strength it is

$$n_s = S_{sf} / \tau_a \tag{16.28}$$

where

S_{sf} = modified shear fatigue strength, Pa

Equations (7.12) to (7.14) should be used to determine the torsional fatigue strength.

The best data for the torsional endurance limits of spring steels are those of Zimmerli (1957). The surprising fact about these data is that size, material, and tensile strength have no effect on the endurance limits of spring steels with wire diameters under 3/8 in. (10 mm). Zimmerli's (1957) results are

$$S'_{se} = 45.0 \text{ ksi (310 MPa)} \quad \text{for unpeened springs}$$

$$S'_{se} = 67.5 \text{ ksi (465 MPa)} \quad \text{for peened springs} \tag{16.29}$$

These results are valid for all materials in Table 16.2. Shot peening is working the surface material to cause compressive residual stresses that toughen the surface. The endurance limit given in Eq. (16.29) is corrected for all factors given in Eq. (7.16) except the reliability factor.

The modulus of rupture, or shear ultimate strength S_{su}, for spring steels can be expressed as

$$S_{su} = 0.60S_u \tag{16.30}$$

Example 16.4

GIVEN A helical compression spring has 14 active coils, a free length of 1.25 in., and an outside diameter of 7/16 in. The ends of the spring are squared and ground and the end plates are constrained to be parallel. The material is music wire with presetting. The wire diameter is fixed at 0.042 in.

FIND

a) For *static* conditions compute the spring rate, the solid length, and the stress when the spring is compressed to the solid length. Will static yielding occur while bringing the spring to its solid length?

b) For *dynamic* conditions with $P_{min} = 0.9$ lbf and $P_{max} = 2.9$ lbf, will the spring experience torsional endurance limit fatigue, torsional yielding, or torsional fatigue failure? Assume a survival probability of 90%, unpeened coils, and fatigue failure based on 50×10^3 cycles.

Solution

a) The spring index from Eq. (16.6) is

$$C = \frac{D}{d} = \frac{D_o - d}{d} = \frac{\frac{7}{16} - 0.042}{0.042} = 9.417$$

From Eq. (16.10) the transverse shear factor is

$$K_d = 1 + \frac{0.5}{C} = 1 + \frac{0.5}{9.417} = 1.053$$

From Eq. (16.12) the Wahl curvature correction factor is

$$K_w = \frac{4C-1}{4C-4} + \frac{0.615}{C} = \frac{4(9.417)-1}{4(9.417)-4} + \frac{0.615}{9.417} = 1.154$$

From Table 16.3 for squared and ground ends

$$N_t = N_a + 2 = 14 + 2 = 16 \text{ coils}$$

The solid length is

$$l_s = dN_t = (0.042)(16) = 0.6720 \text{ in.}$$

The pitch is

$$p = \left(l_f - 2d\right)/N_a = (1.25 - 0.084)/14 = 0.0833 \text{ in.}$$

Also, from Eq. (16.20) the deflection to solid length is

$$\delta_s = l_f - l_s = 1.25 - 0.6720 = 0.5780 \text{ in.}$$

From Eq. (16.2) the ultimate strength in tension is

$$S_{ut} = A_p / d^m$$

From Table 16.2 for music wire

$$S_{ut} = \frac{(196)(10^3)}{(0.042)^{0.146}} \text{ psi} = 311.4 \times 10^3 \text{ psi}$$

From Eq. (16.3) the torsional yield stress, or the allowable shear stress for static loading, is

$$\tau_{all} = S_{sy} = 0.40 S_{ut} = 0.40(311.4)(10^3) = 124.6 \times 10^3 \text{ psi}$$

From Eq. (16.15) the force required to compress the coils to a solid length while using the shear modulus of elasticity for music wire in Table 16.1 of 11.5×10^6 psi is

$$P_s = \frac{G d \delta_s}{8 C^3 N_a} = \frac{(11.5)(10^6)(0.042)(0.5780)}{8(9.417)^3(14)} = 2.985 \text{ lbf}$$

From Eq. (16.9) the maximum design stress is

$$\tau_{max} = \frac{8 D K_d P_s}{\pi d^3} = \frac{8 C K_d P_s}{\pi d^2} = \frac{8(9.417)(1.053)(2.985)}{\pi (0.042)^2} = 42\ 729 \text{ psi}$$

The safety factor guarding against static yielding is

$$n_s = \frac{\tau_{all}}{\tau_{max}} = \frac{124\ 600}{42\ 729} = 2.916$$

Thus, failure should not occur due to static yielding.

Checking for buckling gives

$$\frac{\delta_s}{l_f} = \frac{0.5780}{1.25} = 0.4624$$

$$\frac{l_f}{D} = \frac{1.25}{7/16 - 0.042} = 3.161$$

From Fig. 16.7 buckling is not a problem.

b) The alternating and mean forces are

$$P_a = \frac{P_{max} - P_{min}}{2} = \frac{2.9 - 0.9}{2} = 1.0 \text{ lbf}$$

$$P_m = \frac{P_{max} + P_{min}}{2} = \frac{2.9 + 0.9}{2} = 1.9 \text{ lbf}$$

From Eq. (16.11) the alternating and mean stresses are

$$\tau_a = \frac{8 C K_w P_a}{\pi d^2} = \frac{8(9.417)(1.154)(1.0)}{\pi (0.042)^2} = 15.69 \times 10^3 \text{ psi}$$

$$\tau_m = \frac{8 C K_w P_m}{\pi d^2} = \tau_a \left(\frac{P_m}{P_a} \right) = 29.81 \times 10^3 \text{ psi}$$

The safety factor guarding against torsional yielding is

$$n_s = \frac{S_{sy}}{\tau_a + \tau_m} = \frac{124.6}{15.69 + 29.81} = 2.738$$

Therefore, failure should not occur from torsional yielding.

From Table 7.4 and Eq. (16.29) the modified endurance limit is

$$S_{se} = k_r S'_{se} = (0.9)(45)(10^3)\text{psi} = 40.5 \times 10^3 \text{ psi}$$

The safety factor guarding against torsional endurance limit fatigue is

$$n_s = \frac{S_{se}}{\tau_a} = \frac{(40.5)(10^3)}{(15.69)(10^3)} = 2.581$$

Therefore, failure should not occur from torsional endurance limit fatigue.

From Eq. (16.30) the modulus of rupture for spring steel is

$$S_{su} = 0.60 S_{ut} = 0.60(311.4)(10^3)\text{psi} = 186.8 \times 10^3 \text{ psi}$$

From Eq. (7.12) the slope used to calculate the fatigue strength is

$$b_s = -\frac{1}{3}\log\left(\frac{0.72 S_{su}}{S_{se}}\right) = -\frac{1}{3}\log\left[\frac{(0.72)(186.8)(10^3)}{(40.5)(10^3)}\right] = -0.1738$$

From Eq. (7.13) the intercept used to calculate the fatigue strength is

$$\overline{C} = \log\left[\frac{(0.72 S_{su})^2}{S_{se}}\right] = \log\left[\frac{(0.72)^2(186.8)^2(10^3)^2}{(40.5)(10^3)}\right] = 5.650$$

From Eq. (7.14) the fatigue strength is

$$\begin{aligned} S_{sf} &= 10^{\overline{C}}(N_t)^{b_s} = 10^{5.650}\left[(50)(10^3)\right]^{-0.1738} \\ &= 68.12 \times 10^3 \text{ psi} \end{aligned}$$

From Eq. (16.28) the safety factor guarding against torsional fatigue strength failure is

$$n_s = \frac{S_{sf}}{\tau_a} = \frac{68.12}{15.69} = 4.342$$

Tutorial 11

Therefore, failure should not occur under either static or dynamic conditions.

16.4 HELICAL EXTENSION SPRINGS

In applying the load in **helical extension springs** the shape of the hooks, or end turns, must be designed so that the stress concentration effects caused by the presence of sharp bends are decreased as much as possible. In Fig. 16.8(a) and (b) the end of the extension spring is formed by merely bending up a half-loop. If the radius of the bend is small, the stress concentration at the cross section, at B in Fig. 16.8(b), will be large.

The most obvious method for avoiding these severe stress concentrations is to make the mean radius of the hook r_2 larger. Figure 16.8(c) shows another method of achieving this goal. Here, the hook radius is smaller, which increases the stress concentration. The stress is greatly reduced, however, because the coil diameter is reduced. There is a lower stress because of the shorter moment arm. The largest stress occurs at cross section B shown in Fig. 16.8(d), a side view of Fig. 16.8(c).

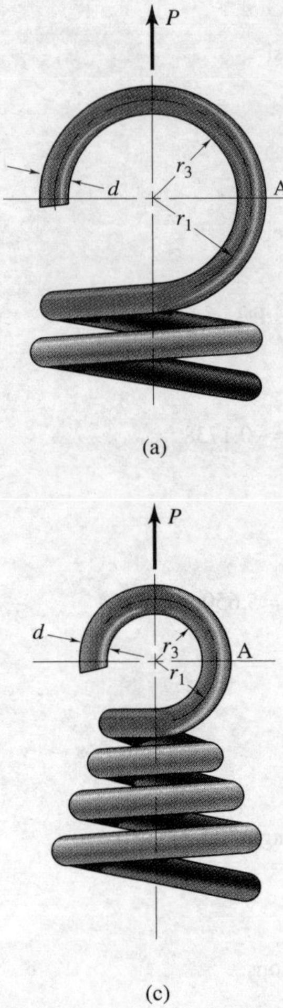

Figure 16.8 Ends for extension springs.
(a) Conventional design; (b) side view of Fig. 16.8(a);
(c) improved design over Fig. 16.8(a); (d) side view of Fig. 16.8(c).

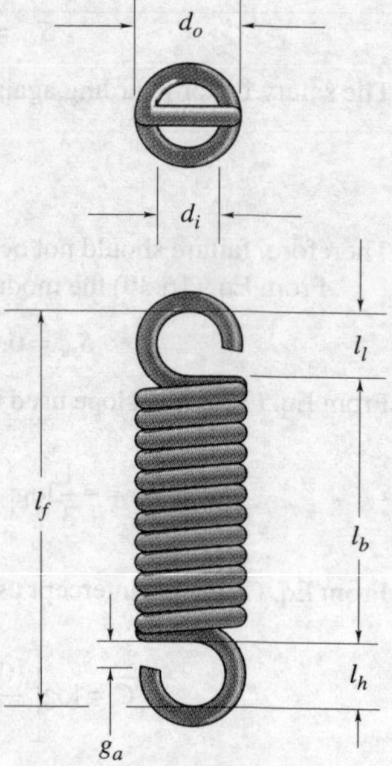

Figure 16.9 Dimensions of helical extension spring.

Figure 16.9 shows some of the important dimensions of a helical extension spring. All the coils in the body are assumed to be active. The total number of coils is

$$N_t = N_a + 1 \tag{16.31}$$

The length of the body is

$$l_b = dN_t \tag{16.32}$$

The free length is measured from the inside of the end loops, or hooks, and is

$$l_f = l_b + l_h + l_l \tag{16.33}$$

For close-wound extension springs the force-deflection curve is such that some initial force is required before any deflection occurs, and after the initial force the force-deflection curve is linear. Thus, the force is

$$P = P_i + \frac{\delta G d^4}{8 N_a D^3} \tag{16.34}$$

where

P_i = preload, N

The spring rate is

$$k = \frac{P - P_i}{\delta} = \frac{d^4 G}{8 N_a D^3} = \frac{dG}{8 N_a C^3} \tag{16.35}$$

The shear stresses for static and dynamic conditions are given by Eqs. (16.9) and (16.11), respectively. The preload stress τ_i is obtained by using Eq. (16.5).

The spring index given in Eq. (16.6) and the preferred range of 3 to 12 for the spring index are equally valid for extension springs and for compression springs. The initial tension wound into the spring, giving a preload, is described by

$$P_i = \frac{\pi \tau_i d^3}{8D} = \frac{\pi \tau_i d^2}{8C} \tag{16.36}$$

Recommended values of τ_i depend on the spring index and are given in Fig. 16.10. Springs should be designed for midway in the preferred spring index range in Fig. 16.10.

The critical stresses in the hook occur at sections A and B, as shown in Fig. 16.8. At section A the stress is due to bending and transverse shear, and at section B the stress is due to torsion. The bending moment and transverse shear stresses acting at section A can be expressed as

$$\sigma_A = \left(\frac{Mc}{I}\right)\left(\frac{r_1}{r_3}\right) + \frac{P_A}{A} = \left(\frac{32 P_A r_1}{\pi d^3}\right)\left(\frac{r_1}{r_3}\right) + \frac{4 P_A}{\pi d^2} \tag{16.37}$$

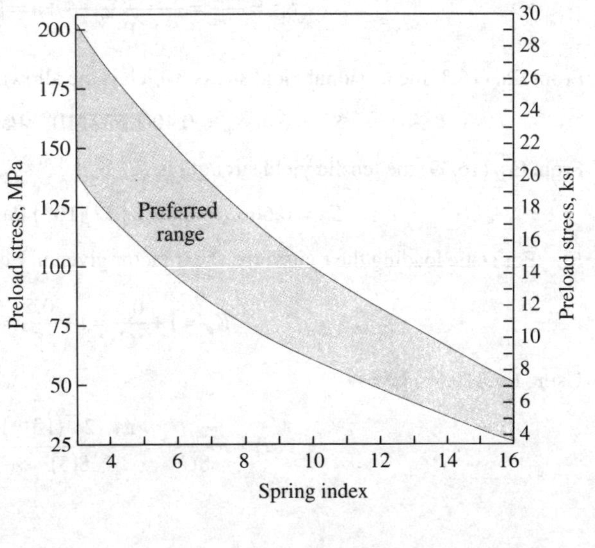

Figure 16.10 Preferred range of preload stress for various spring indexes. [Adapted from Almen and Laszlo (1936).]

The shear stress acting at section B is

$$\tau_B = \frac{8P_B C}{\pi d^2}\left(\frac{r_2}{r_4}\right) \tag{16.38}$$

Radii r_1, r_2, r_3, and r_4 are given in Fig. 16.8. Recommended design practice is that $r_4 > 2d$. Values of σ_A and τ_B given in Eqs. (16.37) and (16.38) are the design stresses. These stresses are compared with the allowable stresses, taking into account the type of loading to determine if failure will occur. The allowable shear stress to be used with Eq. (16.38) is the shear yield stress given in Eq. (16.3). The allowable stress to be used with Eq. (16.37) is the yield strength in tension, given as

$$S_{ty} = 0.60 S_{ut} \tag{16.39}$$

Example 16.5

GIVEN A helical extension spring similar to that shown in Figs. 16.8(a) and (b) is made of hard-drawn wire with a mean coil diameter of 10 mm, a wire diameter of 2 mm, and 120 active coils. The hook radius is 6 mm ($r_1 = 6$ mm) and the bend radius is 3 mm ($r_2 = 3$ mm). The preload is 30 N and the free length is 264 mm.

FIND

Determine the following:

a) Tensile and torsional yield strength of the wire
b) Initial torsional shear stress of the wire
c) Spring rate
d) Force required to cause the torsional stress in the hook to reach the yield stress
e) Force required to cause the normal stress in the hook to reach the tensile yield strength
f) Distance between hook ends if the smaller of the two forces found in parts *d* and *e* is applied

Solution

a) From Table 16.2 for hard-drawn wire $m = 0.192$ and $A_p = 1750$ MPa. From Eq. (16.2) the ultimate strength in tension is

$$S_{ut} = \frac{A_p}{d^m} = \frac{(1750)(10^6)}{(2)^{0.192}} \text{ Pa} = 1.532 \text{ GPa}$$

From Eq. (16.3) the torsional yield stress, which is the allowable shear stress, is

$$S_{sy} = 0.40 S_{ut} = 0.40(1.532)(10^9) \text{ Pa} = 0.6128 \text{ GPa}$$

From Eq. (16.39) the tensile yield strength is

$$S_{ty} = 0.60 S_{ut} = 0.60(1.532)(10^9) \text{ Pa} = 0.9192 \text{ GPa}$$

b) For static loading the transverse shear factor given in Eq. (16.10) is

$$K_d = 1 + \frac{0.5}{C} = 1 + \frac{0.5}{5} = 1.1$$

Using Eq. (16.36) gives

$$P_i = \frac{\pi \tau_i d^2}{8C} = \frac{\pi \tau_i (2)^2 (10^{-6})}{8(5)} = 30 \text{ N}$$

$$\tau_i = 95.50 \text{ MPa}$$

c) From Table 16.1 for hard-drawn wire $G = 11.5 \times 10^6$ psi $= 79.35 \times 10^9$ Pa. From Eq. (16.35) the spring rate is

$$k = \frac{dG}{8N_a C^3} = \frac{(2)(10^{-3})(79.35)(10^9)}{8(120)(5)^3} = 1323 \text{ N/m}$$

d) The critical bending stress in the hook, from Eq. (16.37) and the yield stress in tension given above, is

$$\sigma_A = \left(\frac{32P_A r_1}{\pi d^3}\right)\left(\frac{r_1}{r_3}\right) + \frac{4P_A}{\pi d^2} = S_{ty}$$

$$\frac{32P_A(10^{-3})(6)^2}{\pi(2)^3(10^{-9})(6-1)} + \frac{4P_A}{\pi(2)^2(10^{-6})} = (0.9192)(10^9)$$

$$P_A(9.167 + 0.3183)(10^6) = (0.9192)(10^9)$$

The transverse shear stress is only 3.5% of the stress due to both bending and transverse shear.

Solving for P_A gives the critical load where failure will first start to occur by normal stress in the hook.

$$P_A = 96.91 \text{ N}$$

e) The critical torsional shear stress in the hook can be determined from Eq. (16.38), or

$$\tau_B = \frac{8P_B C}{\pi d^2}\left(\frac{r_2}{r_4}\right) = S_{sy}$$

$$\therefore \frac{8P_B(5)}{\pi(2)^2(10^{-6})}\left(\frac{3}{2}\right) = (0.6128)(10^9)$$

$$P_B = 128.3 \text{ N}$$

The smaller load is $P_A = 96.91$ N, which indicates that failure will first occur by normal stress in the hook.

f) Using P_A in Eq. (16.35) gives the deflection as

$$\delta = \frac{P_A - P_i}{k} = \frac{96.91 - 30}{1323} = 0.05057 \text{ m} = 50.57 \text{ mm}$$

The distance between hook ends is

$$l_f + \delta = 264 + 50.57 = 314.6 \text{ mm}$$

Example 16.6

GIVEN The ends of a helical extension spring are manufactured according to the improved design shown in Fig. 16.8(c). The conical parts have five coils at each end with a winding diameter changing according to

$$D = D_o\left(1 - \frac{\varsigma}{15\pi}\right)$$

The cylindrical part of the spring has an outside diameter D_o and 20 coils.

FIND How much softer does the spring become with the conical parts at the ends?

Solution

From Eq. (16.4) the torque in the spring is

$$T = PR = PD/2$$

From Eq. (4.31) the change of angular deflection for a small length is

$$d\theta = \frac{T}{GJ}dl$$

The axial compression of the conical part of the spring is

$$d\delta_c = R\,d\theta = \frac{RT}{GJ}dl = \frac{PR^2}{GJ}R\,d\varsigma = \frac{PR^3}{GJ}d\varsigma$$

The total deflection of the two conical ends is

$$\delta_c = 2\int_0^{10\pi}\frac{PR^3}{GJ}d\varsigma = \frac{2P}{GJ}\int_0^{10\pi}\left(\frac{D_o}{2}\right)^3\left(1-\frac{\varsigma}{15\pi}\right)^3 d\varsigma$$

$$= \frac{PD_o^3(-15\pi)}{4GJ(4)}\left\{\left[1-\frac{\varsigma}{15\pi}\right]^4\right\}_0^{10\pi} = \frac{75\pi}{81}\frac{PD_o^3}{GJ}$$

From Eq. (16.14) the deflection in the main part of the spring is

$$\delta = \left(\frac{D_o}{2}\right)\frac{Tl}{GJ} = \left(\frac{D_o}{2}\right)\frac{PD_o\pi D_o(20)}{2GJ} = \frac{5\pi PD_o^3}{GJ}$$

$$\frac{\delta_t}{\delta} = \frac{\delta+\delta_c}{\delta} = \frac{\left(5\pi+\frac{75\pi}{81}\right)\frac{PD_o^3}{GJ}}{5\pi\frac{PD_o^3}{GJ}} = 1.185$$

∴ The spring becomes 18.52% softer by including the conical ends.

16.5 HELICAL TORSION SPRINGS

Tutorial 12

The previous section shows that the helical coil spring can be loaded either in compression or in tension. This section considers a torsional load being applied. Figure 16.11 shows a typical example of a **helical torsion spring.** The coil ends can have a great variety of shapes to suit the various applications. The coils are usually close wound like an extension spring, as shown in Fig. 16.11, but differ from extension springs in that torsion springs do not have any initial tension. Also, note from Fig. 16.11 that the ends are shaped to transmit torque rather than force as is the case for compression and tension springs. The torque is applied about the axis of the helix. The torque acts as a bending moment on each section of the wire. The primary stress in a torsional spring is bending. The bending moment $M = Pa$ produces a normal stress in the wire. (In contrast, in a compression or extension spring the

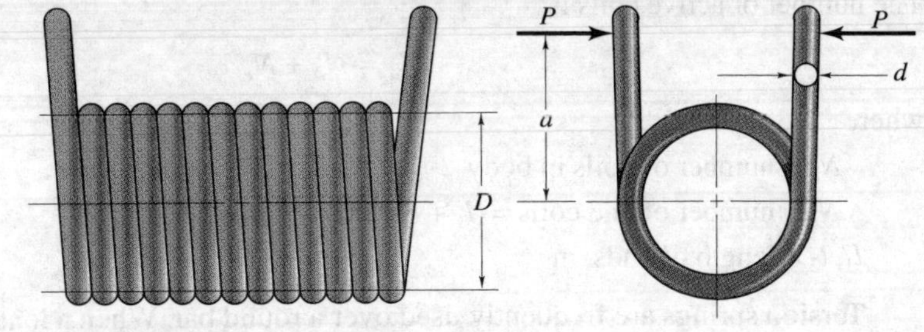

Figure 16.11 Helical torsion spring.

load produces a torsional stress in the wire.) This equality means that the residual stresses built in during winding are in the same direction, but of opposite sign, as the working stresses that occur during use. These residual stresses are useful in making the spring stronger by opposing the working stress, provided that the load is always applied to cause the spring to wind up. Because the residual stress opposes the working stress, torsional springs are designed to operate at stress levels that equal or even exceed the yield strength of the wire.

The maximum bending stress occurs at the inner fiber of the coil and is given as

$$\sigma = \frac{K_i Mc}{I} = \frac{32 K_i M}{\pi d^3} \quad \textbf{(16.40)}$$

where

$$K_i = \frac{4C^2 - C - 1}{4C(C-1)} \quad \textbf{(16.41)}$$

and a circular cross section of the wire is assumed. The angular deflection in radians is

$$\theta_{\text{rad}} = \frac{Ml_w}{EI} \quad \textbf{(16.42)}$$

where

$l_w = \pi D N_a$ = length of wire in spring, m

The angular deflection in revolutions is

$$\theta_{\text{rev}} = \frac{M\pi D N_a}{E\left(\frac{\pi d^4}{64}\right)}\left(\frac{1\ \text{rev}}{2\pi\ \text{rad}}\right) = \frac{32 MDN_a}{\pi E d^4} = \frac{10.18 MDN_a}{Ed^4} \quad \textbf{(16.43)}$$

The angular spring rate is

$$k_\theta = \frac{M}{\theta_{\text{rev}}} = \frac{Ed^4}{10.18 DN_a} \quad \textbf{(16.44)}$$

The number of active coils is

$$N_a = N_b + N_e \tag{16.45}$$

where

N_b = number of coils in body
N_e = number of end coils = $(l_1 + l_2)/3\pi D$
l_1, l_2 = length of ends, m

Torsion springs are frequently used over a round bar. When a load is applied to a torsion spring, the spring winds up, causing a decrease in the inside diameter. For design purposes the inside diameter of the spring must never become equal to the diameter of the bar; otherwise, a spring failure will occur. The inside diameter of a loaded torsion spring is

$$D_i' = N_a D_i / N_a' \tag{16.46}$$

where

N_a = number of active coils at no load
D_i = coil inside diameter at no load, m
N_a' = number of active coils when loaded
D_i' = coil inside diameter when loaded, m

Example 16.7

GIVEN A torsion spring similar to that shown in Fig. 16.11 is made of 0.055-in.-diameter music wire and has six coils in the body of the spring and straight ends. The distance a in Fig. 16.11 is 2 in. The outside diameter of the coil is 0.654 in.

FIND

a) If the maximum stress is set equal to the yield strength of the wire, what is the corresponding moment?
b) With load applied, what is the angular deflection in revolutions?
c) When the spring is loaded, what is the resulting inside diameter?
d) What size bar should be placed inside the coils?

Solution

a) The mean coil diameter is

$$D = D_o - d = 0.654 - 0.055 = 0.599 \text{ in.}$$

The coil inside diameter without a load is

$$D_i = D - d = 0.599 - 0.055 = 0.544 \text{ in.}$$

The spring index is

$$C = \frac{D}{d} = \frac{0.599}{0.055} = 10.89$$

From Eq. (16.41)

$$K_i = \frac{4C^2 - C - 1}{4C(C-1)} = \frac{4(10.89)^2 - 10.89 - 1}{4(10.89)(10.89-1)} = 1.074$$

From Eq. (16.2) and Table 16.2 for music wire

$$S_{ut} = \frac{A_p}{d^m} = \frac{196(10^3)}{(0.055)^{0.146}} \text{ psi} = 299.3\times10^3 \text{ psi}$$

$$S_y = 0.6S_{ut} = 179.6\times10^3 \text{ psi}$$

After equating the bending stress to the yield strength, Eq. (16.40) gives

$$M = \frac{\pi d^3 S_y}{32K_i} = \frac{\pi(0.055)^3(179.6)(10^3)}{32(1.074)} = 2.731 \text{ lbf-in.}$$

b) The number of active coils is

$$N_a = N_b + N_e = 6 + \frac{4}{3\pi(0.599)} = 6.709$$

From Table 16.1 the modulus of elasticity for music wire is 30×10^6 psi. From Eq. (16.44) the angular spring rate is

$$k_\theta = \frac{Ed^4}{10.18DN_a} = \frac{(30)(10^6)(0.055)^4}{10.18(0.599)(6.709)} = 6.710 \text{ lbf-in./rev}$$

The angular deflection in revolutions is

$$\theta_{\text{rev}} = \frac{M}{k_\theta} = \frac{2.731}{6.710} = 0.4070 \text{ rev}$$

c) The number of coils when loaded is

$$N_a' = N_a + \theta_{\text{rev}} = 6.709 + 0.4070 = 7.116$$

From Eq. (16.46) the inside diameter after loading is

$$D_i' = \frac{N_a}{N_a'} D_i = \frac{6.709}{7.116}(0.544) = 0.5129 \text{ in.}$$

d) Thus, if the bar inside the coils is 1/2 in. in diameter, the spring should not fail.

Example 16.8

GIVEN A hand-squeeze grip training device with two 100-mm-long handles is connected to a helical torsion spring with 4.375 coils. The handles and coils are made of music wire. The spring rate at the end of the spring should be 2000 N/m.

FIND The spring dimensions for a safety factor of 2 against breakage when the handles are moved together so that the helical torsion spring has 4.5 coils.

Solution

From Eq. (16.40) the maximum bending stress is

$$\sigma = \frac{K_i Mc}{I} = \frac{K_i Mr}{I} \tag{a}$$

Using Eq. (16.42) gives the moment needed to deform the spring to 4.5 coils as

$$M = \frac{EI\theta_{\text{rad}}}{l_w} = \frac{EI(4.5-4.375)2\pi}{l_w} = \frac{0.25\pi EI}{l_w} \tag{b}$$

Substituting Eq. (b) into Eq. (a) gives

$$\sigma = \frac{\pi}{8}\frac{K_i dE}{l_w}$$

But $l_w = \pi D N_a$

$$\sigma = \frac{\pi K_i dE}{8\pi D N_a} = \frac{K_i E}{8 C N_a}$$

For music wire $E = (2.07)(10^{11})$ Pa and $N_a = 4.375$

$$\therefore \sigma = \frac{K_i}{C}\left[\frac{(2.07)(10^{11})}{8(4.375)}\right] \text{Pa} = 5.914\frac{K_i}{C}\ \text{GPa} \qquad \textbf{(c)}$$

To get the spring rate of 2000 N/m at the ends of the handles, the angular spring rate is

$$k_\theta = \left(2000\ \frac{\text{N}}{\text{m}}\right)(0.1\ \text{m})(0.1\ \text{m})/\text{rad} = 20\ \text{N-m/rad}$$

From Eq. (16.44)

$$d^3 = \frac{(10.18)CN_a k_\theta}{E} = C\frac{(10.18)(4.375)(20)}{(2.07)(10^{11})} = 4.303\times10^{-9}C \qquad \textbf{(d)}$$

Guess $C = 6$. From Eq. (16.41)

$$K_i = \frac{4C^2 - C - 1}{4C(C-1)} = \frac{4(6)^2 - 6 - 1}{4(6)(5)} = 1.142$$

From Eq. (d)

$$d^3 = 25.82\times10^{-9}$$

$$d = 2.956\times10^{-3}\ \text{m} = 2.956\ \text{mm}$$

$$\therefore D = 6(2.956)(10^{-3}) = 17.73\times10^{-3}\ \text{m} = 17.73\ \text{mm}$$

From Eq. (c) the maximum design stress is

$$\sigma = (5.914)\frac{1.142}{6} = 1.126\ \text{GPa} \qquad \textbf{(e)}$$

From Eq. (16.2) the allowable stress is

$$\sigma_{\text{all}} = S_{ut} = A_p/d^m = \frac{(2170)(10^6)}{(2.956)^{0.146}}\ \text{Pa} = 1.852\ \text{GPa}$$

The safety factor is

$$n_s = \frac{\sigma_{\text{all}}}{\sigma} = \frac{1.852}{1.126} = 1.645$$

Guess $C = 7.5$. From Eq. (16.41)

$$K_i = \frac{4C^2 - C - 1}{4C(C-1)} = \frac{4(7.5)^2 - 7.5 - 1}{4(7.5)(6.5)} = 1.110$$

From Eq. (d)

$$d^3 = 4.303\times10^{-9}(7.5) = 32.27\times10^{-9}$$

$$d = 3.184\times10^{-3}\ \text{m} = 3.184\ \text{mm}$$

$$\therefore D = 7.5(3.184\times10^{-3}) = 23.88\times10^{-3}\ \text{m} = 23.88\ \text{mm}$$

From Eq. (c) the maximum design stress is

$$\sigma = (5.914)\frac{1.110}{7.5} = 0.8753\ \text{GPa}$$

From Eq. (16.2) the allowable stress is

$$\sigma_{\text{all}} = S_{ut} = A_p/d^m = \frac{(2170)(10^6)}{(3.184)^{0.146}}\ \text{Pa} = 1.832\ \text{GPa}$$

$$\therefore n_s = \frac{\sigma_{\text{all}}}{\sigma} = \frac{1.832}{0.8753} = 2.09$$

The design is adequate.

16.6 LEAF SPRINGS

Multiple-**leaf springs** are in wide use, especially in the automobile and railway industries. An exact analysis of this type of spring is complicated. A multiple-leaf spring can be considered as a simple cantilever type, as shown in Fig. 16.12(b). It can also be considered as a triangular plate, as shown in Fig. 16.12(a). The triangular plate is cut into n strips of width b and stacked in the graduated manner shown in Fig. 16.12(b).

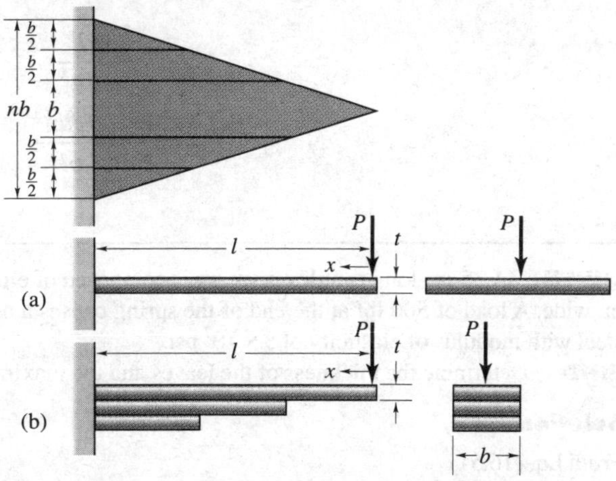

Figure 16.12 Leaf spring.
(a) Triangular-plate, cantilever spring; (b) equivalent multiple-leaf spring.

Before analyzing a multiple-leaf spring first consider a single-leaf, cantilevered spring with constant rectangular cross section. For bending of a straight beam from Eq. (4.48)

$$\sigma = \frac{Mc}{I} \tag{16.47}$$

For a rectangular cross section of base b and height t the above equation becomes

$$\sigma = \frac{6M}{bt^2} = \frac{6Px}{bt^2} \tag{16.48}$$

The moment is just Px. The maximum moment occurs at $x = l$ and at the outer fiber of the cross section, or

$$\sigma_{max} = \frac{6Pl}{bt^2} \tag{16.49}$$

From Eq. (16.48) the stress is a function of x along the beam. The design engineer can insist that the stress be constant all along the beam. To accomplish this, either t is held constant and b is allowed to vary, or conversely, so that the stress is constant for any x.

$$\frac{b(x)}{x} = \frac{6P}{t^2\sigma} = \text{constant} \tag{16.50}$$

Equation (16.50) is linear, giving the triangular shape shown in Fig. 16.12(a) and a constant stress for any x.

The triangular-plate spring and the equivalent multiple-leaf spring have identical stresses and deflection characteristics with two exceptions:

1. Interleaf friction provides damping in the multiple-leaf spring.
2. The multiple-leaf spring can carry a full load in only one direction.

The deflection and spring rate for the ideal leaf spring are

$$\delta = \frac{6Pl^3}{Enbt^3} \tag{16.51}$$

$$k = \frac{P}{\delta} = \frac{Enbt^3}{6l^3} \tag{16.52}$$

Example 16.9

GIVEN A 35-in.-long cantilever spring is composed of eight graduated leaves. The leaves are 7/4 in. wide. A load of 500 lbf at the end of the spring causes a deflection of 3 in. The spring is made of steel with modulus of elasticity of 3×10^7 psi.

FIND Determine the thickness of the leaves and the maximum bending stress.

Solution

From Eq. (16.51)

$$\delta = \frac{6Pl^3}{Enbt^3} \quad \text{or} \quad t^3 = \frac{6Pl^3}{\delta Enb}$$

$$t^3 = \frac{6(500)(35)^3}{3(3)(10^7)(8)(1.75)} = 0.1021 \text{ in.}^3$$

$$t = 0.4674 \text{ in.}$$

From Eq. (16.49) the maximum bending stress is

$$\sigma_{max} = \frac{6Pl}{nbt^2} = \frac{6(500)(35)}{(8)(1.75)(0.9347)^2} = 73\ 469 \text{ psi}$$

Example 16.10

GIVEN A leaf spring for a locomotive wheel set is made of spring steel with a thickness of 20 mm and an allowable bending stress of 1050 MPa. The spring has a modulus of elasticity of 207 GPa, is 1.6 m long from tip to tip, and carries a weight of 12.5 tons at the middle of each leaf spring.

FIND

a) The width of the spring for a safety factor of 3

b) How high the locomotive has to be lifted during overhauls to unload the springs

Solution

a) From Eq. (1.1) the design stress is

$$\sigma_{max} = \frac{\sigma_{all}}{3} = \frac{(1.05)(10^9)}{3} = 0.350 \text{ GPa} \qquad \textbf{(a)}$$

From Eq. (16.49) the maximum bending stress is

$$\sigma_{max} = \frac{6Pl}{bt^2}$$

Because P is force applied at the tip, the load is 6.25 tons and the length is 0.8 m in the locomotive spring.

$$\therefore \sigma_{max} = \frac{6Pl}{bt^2} = \frac{6(6.25)(10^3)(9.807)(0.8)}{b(0.020)^2} = \frac{(0.7355)(10^9)}{b}$$

Making use of Eq. (a) gives

$$b = 2.101 \text{ m}$$

Splitting into 10 leaves gives the width of the leaf spring as

$$\frac{b}{10} = 0.2101 \text{ m} = 210.1 \text{ mm}$$

b) Using Eq. (16.52) gives the amount the locomotive has to be lifted to unload the spring (deflection) as

$$\delta = \frac{P}{k} = \frac{6Pl^3}{Ebt^3} = \frac{6(6250)(9.807)(0.8)^3}{(2.07)(10^{11})(2.101)(0.020)^3}$$

$$\delta = 0.05409 \text{ m} = 54.1 \text{ mm}$$

16.7 BELLEVILLE SPRINGS

Belleville springs are named after their inventor, J.F. Belleville, who patented their design in 1867. Shaped like a coned disk, these springs are especially useful where large forces are desired for small spring deflections. In fact, because many lock washers used with bolts follow this principle to obtain a bolt preload, these springs are often referred to as Belleville washers. Typical applications include clutch plate supports, gun recoil mechanisms, and a wide variety of bolted connections.

Figure 16.13 shows the cross section of a Belleville spring. Two of the critical parameters affecting a Belleville spring are the diameter ratio $R_d = D_o/D_i$ and the height-to-thickness ratio h/t. From Fig. 16.14 the behavior of a Belleville spring is highly nonlinear and varies considerably with a change in h/t. For low h/t values the spring acts almost linearly, whereas large h/t values lead to highly nonlinear behavior. A phenomenon occurring at large h/t is "snapthrough buckling." In snapthrough the spring deflection is unstable once the maximum force is applied; the spring quickly deforms, or snaps, to the next stable position.

The force-deflection curves for Belleville springs are given by

$$P = \frac{4E\delta}{K_1 D_o^2 (1-\nu^2)}\left[(h-\delta)\left(h-\frac{\delta}{2}\right)t + t^3\right] \tag{16.53}$$

where E is the elastic modulus, δ is the deflection from the unloaded state, D_o is the coil outside diameter, ν is Poisson's ratio for the material, h is the spring height, and t is the spring thickness. The factor K_1 is given by

$$K_1 = \frac{6}{\pi \ln R_d}\left[\frac{(R_d-1)^2}{R_d^2}\right] \tag{16.54}$$

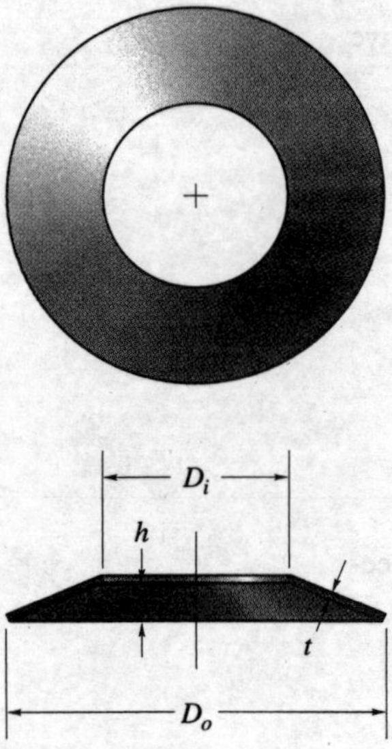

Fig. 16.13 Typical Belleville spring.

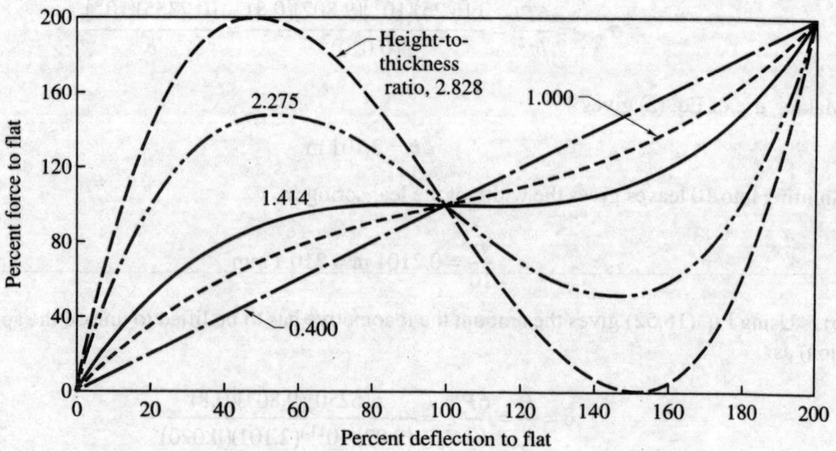

Figure 16.14 Force-deflection response of Belleville spring. [Norton, R. L., *Machine Design*, ©1996. Reprinted by permission of Prentice-Hall, Inc., Upper Saddle River, NJ.]

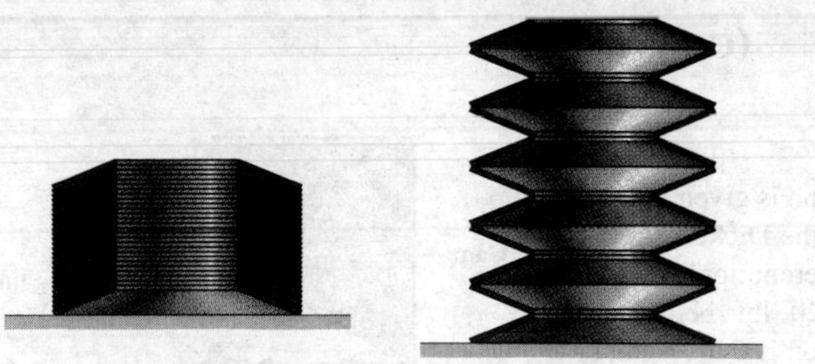

(a) (b)

Figure 16.15 Stacking of Belleville spring. (a) In parallel; (b) in series.

where R_d is the diameter ratio, given by

$$R_d = \frac{D_o}{D_i} \tag{16.55}$$

The force required to totally flatten a Belleville spring is given by

$$P_{\text{flat}} = \frac{4Eht^3}{K_1 D_o^2 \left(1-\nu^2\right)} \tag{16.56}$$

The forces associated with a Belleville spring can be multiplied by stacking them in parallel, as shown in Fig. 16.15(a). The deflection associated with a given force can be increased by stacking the springs in series, as shown in Fig. 16.15(b). Because such a configuration may be susceptible to buckling, a central support is necessary.

Case Study | Design of Compression Spring for Dickerman Feed Unit

GIVEN Figure 16.16 shows a Dickerman feed unit, used to automatically feed material into mechanical presses for progressive die work. During a stroke of a mechanical press, a cam attached to the press ram or punch holder lowers, pushing the follower shown to the right and causing one or more springs to compress. When the ram rises, the compressed spring forces the gripping unit to the left, and since the gripping unit mechanically interferes with notched strip or sheet stock, it forces new material into the press.

Dickerman feeds are among the oldest and least expensive feeding devices in wide use. They can be mounted to feed material in any direction, on both mechanical and hydraulic presses, and the feed length can be adjusted by changing the low-cost, flat cams. Grippers can be blades (as shown) or cylinder assemblies.

The case study involves designing a helical compression spring for a Dickerman feed unit. It is desired to use a spring that will provide at least 20 lbf and have a 6-in. stroke. The pin diameter is 0.5 in. Also, the Dickerman unit should be compactly designed; the chamber for the spring will be not more than 8 in. long.

FIND All necessary dimensions of the helical compression spring.

(continued)

Case Study (CONTINUED)

Solution

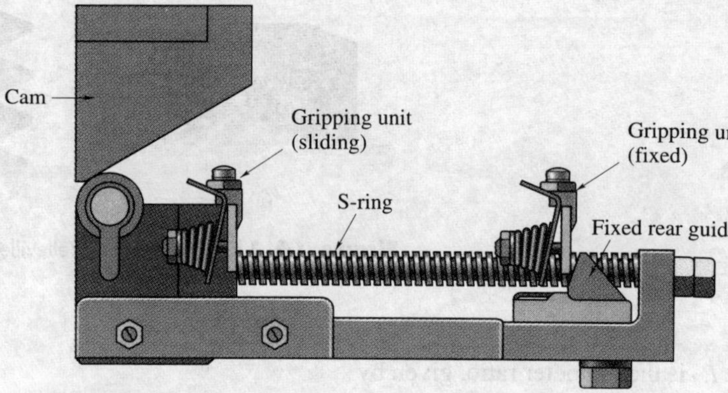

Figure 16.16 Dickerman feed unit. [Reprinted with permission of the Society of Manufacturing Engineers, *Tool and Manufacturing Engineering Handbook*, Fourth Edition, Volume 2, Forming, Copyright 1984.]

Not much information is given about the spring. Noting that the Dickerman drive can be used for different applications, it is necessary that 20 lbf be provided whether the deflection is 6 in., 4 in., or a very small distance. Thus, the spring will be chosen so that at a length of 8 in., it is loaded with 20 lbf of compressive force. Another requirement on the spring is that it must have a solid length not greater than 4 in. or it will bottom out with every stroke of the press. Fortunately, buckling of the spring does not need to be considered because of the guide rod. However, the spring should not be very much longer than the guide rod, since this complicates the Dickerman unit motion.

Economics is not a critical issue here. As long as the spring is selected with standard dimensions and made from spring steel, the cost difference between candidate springs will be low, especially when spread over the total number of parts that will be produced with the Dickerman feed. It is much more essential that a highly reliable spring be chosen because an hour of downtime (required to replace a spring and set up the Dickerman drive) costs far more than a spring. Thus, the ends will be squared and ground for, as discussed in Sec. 16.3.5, better load transfer into the spring. The material used will be music wire, an outstanding spring material.

No maximum force requirement is given; indeed it appears that the maximum force is inconsequential. However, if the compressed spring contains too large a force, it could buckle the press stock, or else dynamic effects could cause too much material to feed with each stroke. An inconsequential effect is the reduction of press capacity; loads on the order of tens of pounds are not high enough to affect press capacity.

Since mechanical presses used for progressive die work are operated unattended for long periods and typically will operate at 100 strokes per minute, the Dickerman feed unit must be designed for fatigue. The important equations are (16.22) to (16.28). The solution method involves a variation of the unknown independent parameters (i.e., d, D, and N_a). All other quantities can be calculated from these parameters. However, the problem is actually much simpler, for the following reasons:

1. The more active coils in the spring, the lower the spring stiffness and hence the lower the maximum force. In addition to reducing the stress levels encountered by the spring, a large number of coils makes a more smoothly operating drive for the reasons discussed above. In terms of this case study the number of active coils is given by

$$N_a = \frac{4}{d} - 2$$

because the ends are squared and ground.

2. The spring diameter is not specified. However, the guide rod in the center of the spring is 0.5 in. in diameter, so that the minimum spring diameter is $0.5 + d$. Normally, the spring diameter would be treated as another parameter to be varied, but in this problem it has been set at 0.75 in. The reasons for this are twofold: First, this dimension ensures adequate clearance and ease of assembly onto the guide rod. Second, this standard size will be more readily available, so that suppliers are more likely to provide high-quality springs quickly.

3. A Goodman failure criterion is used, so that the safety factor is given by

$$\frac{\tau_a}{S_{se}} + \frac{\tau_m}{S_{su}} = \frac{1}{n_s} \quad \text{or} \quad n_s = \frac{S_{se} S_{su}}{\tau_a S_{su} + \tau_m S_{se}}$$

(continued)

Case Study (CONCLUDED)

The solution approach is as follows:

a) Given D choose a value for the wire diameter d. For this case the wire diameter varies over the useful range 0.02 to 0.2 in.

b) Calculate $C = D/d$.

c) Calculate the Wahl and tranverse shear stress factors, given by

$$K_w = \frac{4C-1}{4C-4} + \frac{0.615}{C}$$

$$K_d = (C+0.5)/C$$

d) Obtain the number of active coils N_a from

$$N_a = \frac{4}{d} - 2$$

e) The wire properties are

$$S_{ut} = A_p / d^m$$

$$S'_{se} = 45.0 \text{ ksi}$$

$$S_{su} = 0.60 S_u$$

f) By using the results from steps a to e, calculate the following:

$$k_t = \frac{P}{\delta_t} = \frac{Gd}{8C^3 N_a}$$

$$P_a = \frac{P_{max} - P_{min}}{2}$$

$$P_m = \frac{P_{max} + P_{min}}{2}$$

$$\tau_a = \frac{8DK_w P_a}{\pi d^3} \qquad \textbf{(16.24)}$$

$$\tau_m = \frac{8DK_w P_m}{\pi d^3}$$

$$n_s = \frac{S_{se} S_{su}}{\tau_a S_{su} + \tau_m S_{se}}$$

g) Output the maximum force, the safety factor, and all other desired quantities. Figure 16.17 shows maximum force and the safety factor as a function of wire diameter.

From the charts the safety factor is highest for a spring using a wire diameter of approximately 0.08 in. Thus, a wire diameter of 0.081 in. (2 mm) is chosen. The resulting spring will have 61 active coils and result in a maximum force (when compressed to a length of 4 in.) of 34 lbf, with a free length of approximately 16 in. The safety factor for this case is 1.23.

The fact that the safety factor is small for extremely thin wires should not be surprising, since the stresses are inversely proportional to the cube of wire diameter. However, it is not immediately obvious that the safety factor would eventually decrease for increasing wire diameters. The ultimate strength of the wire decreases rapidly with increasing wire diameter, so that there is in fact an optimum wire diameter.

This analysis should be repeated for additional spring diameters to obtain a true optimum solution but is not done here because of space limitations. In this circumstance it was quite fortunate that the forces were reasonable at the maximum safety factor location. If this were not the case we would have to consider another spring diameter, another spring material, or an entirely different class of spring.

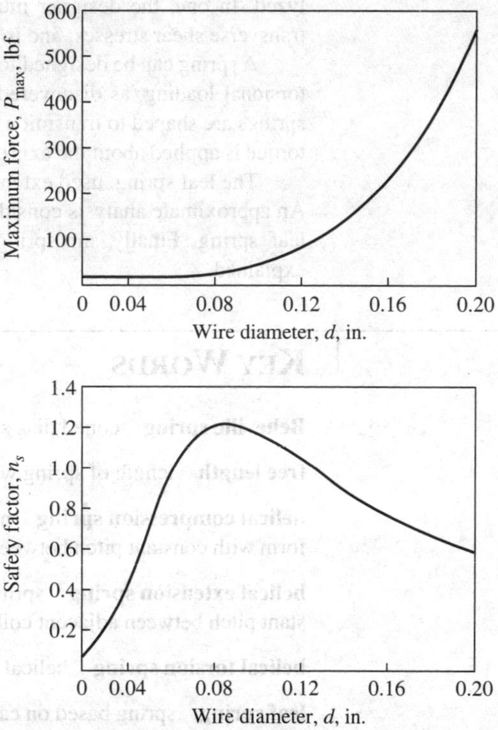

Figure 16.17 Figure used in case study.

16.8 SUMMARY

When a machine element needs flexibility or deflection, some form of spring is usually selected. Springs are used to exert forces or torques in a mechanism or to absorb the energy of suddenly applied forces. Springs frequently operate with high stresses and continuously varying forces.

This chapter provided information about spring design without considering the endless types of spring available to engineers. Different spring materials were presented. Spring strength is a material parameter that is obviously important in spring design. The loss coefficient, which measures the fractional energy dissipated in a stress-strain cycle, is an equally important parameter.

This chapter emphasized helical compression springs. Both torsional shear stress and transverse shear stress were considered. Transverse shear is small relative to torsional shear. The maximum stress occurs at the midheight of the wire at the coil inside diameter.

The spring index is dimensionless and is the mean coil diameter divided by the wire diameter. The spring rate is the force divided by the deflection, thus having the unit of newtons per meter. The difference between these terms is important. Different conditions cause different amounts of eccentric loading that must be compensated for in the design of a compression spring. Relatively long compression springs must be checked for buckling. Also, a surge, or longitudinal vibration, should be avoided in spring design. Avoiding the natural frequency is thus recommended. Spring loading is most often continuously fluctuating, so this chapter considered the design allowance that must be made for fatigue and stress concentrations.

In helical extension springs it was found that the hooks need to be shaped so that the stress concentration effects are decreased as much as possible. Two critical locations within a hook were analyzed. In one, the designer must consider normal stress caused by the bending moment and the transverse shear stresses; and in the other, shear stress must be considered. Both should be checked.

A spring can be designed for axial loading, either compressive or tensile, or it can be designed for torsional loading, as discovered when helical torsion springs were considered. The ends of torsion springs are shaped to transmit a torque rather than a force as for compression and tensile springs. The torque is applied about the axis of the helix and acts as a bending moment on each section of the wire.

The leaf spring, used extensively in the automobile and railway industries, was also considered. An approximate analysis considered a triangular-plate, cantilever spring and an equivalent multiple-leaf spring. Finally, air springs and Belleville springs were considered, and their advantages explained.

KEY WORDS

Belleville spring coned disk spring

free length length of spring when no forces are applied to it

helical compression spring most common spring, wherein round wire is wrapped into cylindrical form with constant pitch between adjacent coils and is loaded in compression

helical extension spring spring wherein round wire is wrapped into cylindrical form with constant pitch between adjacent coils and is loaded in tension

helical torsion spring helical coil spring loaded in torsion

leaf spring spring based on cantilever action

pitch distance, measured parallel to coil axis, from center of one coil to center of adjacent coil

solid length length of spring when all adjacent coils are in metal-to-metal contact

spring flexible machine element used to exert force or torque and, at same time, to store energy

spring index ratio of coil to wire diameter, measure of coil curvature

spring rate ratio of applied force to spring deflection

surge stress pulse that propagates along spring

RECOMMENDED READINGS

Carlson, H. (1978) *Spring Designers Handbook,* Marcel Dekker, New York.

Juvinall, R.C., and Marshek, K.M. (1994) *Fundamentals of Machine Component Design,* Wiley, New York.

Krutz, G.W., Schuelle, J.K., and Claar, P.W. (1994) *Machine Design for Mobile and Industrial Applications,* Society of Automotive Engineers, Warrendale, PA.

Mott, R.L. (1985) *Machine Elements in Mechanical Design,* Merrill Publishing Co., Columbus, OH.

Rothbart, H.A., ed. (1996) *Mechanical Design Handbook,* McGraw-Hill, New York.

SAE (1982) *Manual on Design and Application of Leaf Springs,* Society of Automotive Engineers, Warrendale, PA.

SAE (1996) *Spring Design Manual,* 2nd ed., Society of Automotive Engineers, Warrendale, PA.

Shigley, J.E., and Mischke, C.R. (1989) *Mechanical Engineering Design,* 5th ed., McGraw-Hill, New York.

REFERENCES

Almen, J.O., and Laszlo, A. (1936) "The Uniform Section Disc Springs," Trans. ASME, vol 58, no. 4, pp. 305–314.

Ashby (1992) *Materials Selection in Mechanical Design,* Pergamon Press, Oxford.

Design Handbook (1987) *Engineering Guide to Spring Design.* Associated Spring Corp., Barnes Group Inc., Bristol, CT.

Norton, R.L. (1996) *Machine Design,* Prentice-Hall, Englewood Cliffs, NJ.

Relvas, A.A. (1996) Springs, Chapter 28 in *Mechanical Design Handbook,* H.A. Rothbart, ed., McGraw-Hill, New York.

SME (1984) *Tool and Manufacturing Engineers Handbook,* vol. 2, Society of Manufacturing Engineers, Dearborn, MI, p. 58.

Zimmerli, F.P. (1957) "Human Failures in Spring Applications," *The Mainspring,* no. 17, Associated Spring Corp., Barnes Group, Inc., Bristol, CT.

PROBLEMS

Section 16.2

16.1 A guitar string is made of ASTM A228 steel music wire. To get the correct tone from the string, a tension of 200 N has to be applied. When the string breaks, it is replaced by a string made of ASTM B159 phosphor bronze. What tension will the new string need to get the same tone if the string diameter is the same?

16.2 A tuning fork is made of ASTM A227 steel and vibrates at an eigenfrequency of 400 Hz. To avoid corrosion, the material is changed to phosphor bronze. At what frequency will the tuning fork vibrate?

Section 16.3

16.3 An overflow valve, shown in sketch *a*, has a piston diameter of 15 mm and a slit length of 5 mm. The spring has mean coil diameter $D = 10$ mm and wire diameter $d = 2$ mm. The valve should open at 1 bar pressure and be totally open at 3 bar pressure when the spring is fully compressed. Calculate the number of active coils, the free length, and the pitch of the spring. The shear modulus for the spring material $G = 80$ GPa. The spring ends are squared and ground. Determine the maximum shear stress for this geometry.

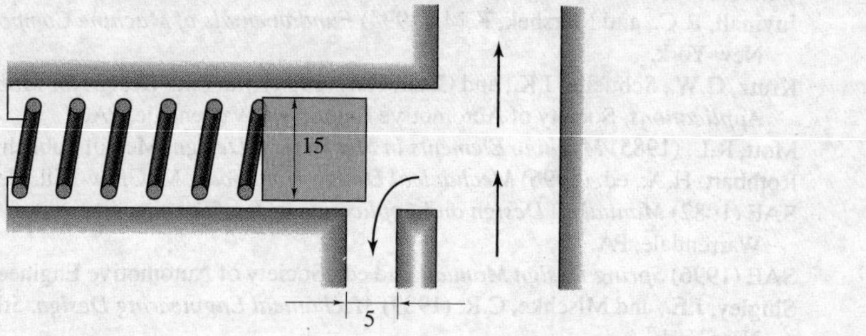

Sketch *a*, for Problem 16.3

16.4 A helical compression spring has mean coil diameter $D = 90$ mm and wire diameter $d = 15$ mm. The spring has 12 active coils. Derive the equation for the spring deflection and calculate it for $P = 4000$ N. Also, find the shear stress in the wire if the stress concentration factor $K_i = 1.26$. The shear modulus of elasticity $G = 80$ GPa.

16.5 A vehicle has individual wheel suspension in the form of helical springs. The free length of the spring $l_f = 360$ mm and the solid length $l_s = 160$ mm at a compressive force of 5000 N. The shear modulus $G = 80$ GPa. Use $D/d = 9$ and calculate the shear stress for pure torsion of the spring wire. The spring ends are squared and ground. Find N_a, p, d, D, and τ_{max}.

16.6 Two equally long cylindrical helical compression springs are placed one inside the other (see sketch *b*) and loaded in compression. How should the springs be dimensioned to get the same shear stresses in both springs?

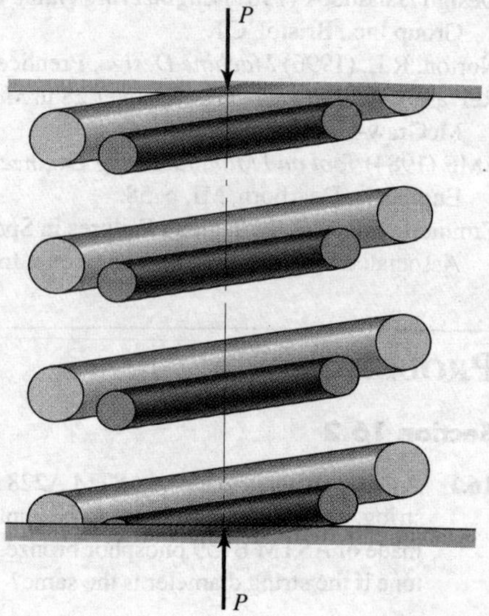

Sketch *b*, for Problem 16.6

16.7 A mechanism is used to press as hard as possible against the moving horizontal surface shown in sketch *c*. The mechanism consists of a stiff central beam and two flexible bending springs made of circular rods with length l, diameter d, modulus of elasticity E, and allowable stress σ_{all}. Wheels are mounted on these rods and can roll over a bump with height f. Calculate the diameter of the springs so that the prestress of the wheels against the moving surface is as high as possible without plastically deforming the springs when the bump is rolled over. The deflection of a spring is shown in sketch *d*.

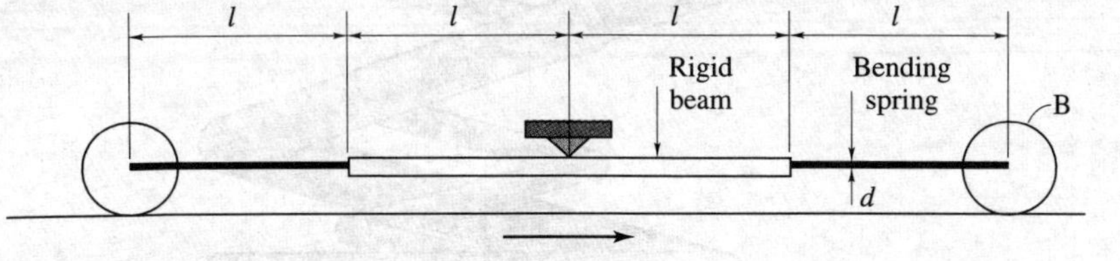

Sketch c, for Problem 16.7

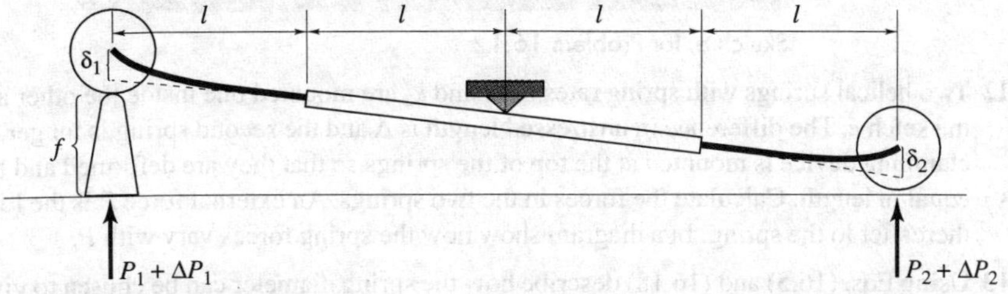

Sketch d, for Problem 16.7

16.8 A helical compression spring is used as a catapult. Calculate the maximum speed of a body weighing 10 kg being thrown by the catapult, given that τ_{max} = 500 MPa, D = 50 mm, d = 8 mm, N_a = 20, and G = 80 GPa.

16.9 A spring is preloaded with force P_i and is then exposed to a force increase whereby the shear stress increases to a certain value τ. Choose the mean coil diameter D to maximize the energy absorption caused by the force increase.

16.10 A compression spring made of music wire is used for static loading. Wire diameter d = 1.4 mm, coil outside diameter D_o = 12.1 mm, and there are eight active coils. Also, assume that the spring ends are squared and parallel. Find the following:

a) Spring rate and solid length

b) Greatest load that can be applied without causing a permanent set in excess of 2%

c) Spring free length with load determined in part b that causes the spring to be solid

d) Whether buckling is a problem. If it is, recommend what you would change in the redesign.

16.11 Design a helical compression spring made of music wire with squared and parallel ends. Assume a spring index C = 12, a spring rate k = 300 N/m, and a force to solid length of 60 N. Find the wire diameter and the mean coil diameter while assuming a safety factor n_s = 1.5. Also, check whether buckling is a problem. Assume steady loading.

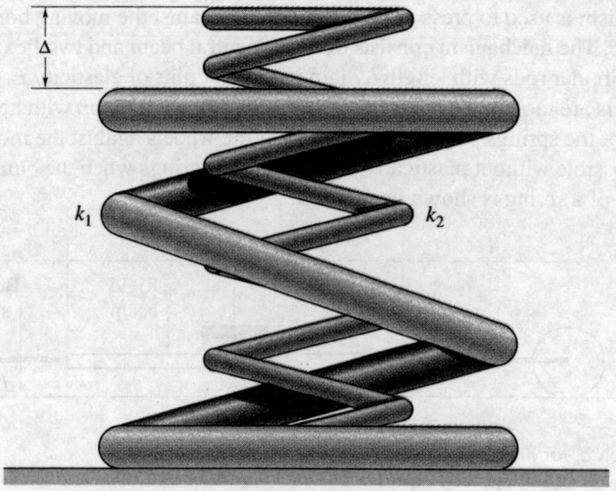

Sketch *e*, for Problem 16.12

16.12 Two helical springs with spring rates of k_1 and k_2 are mounted one inside the other as shown in sketch *e*. The difference in unstressed length is Δ and the second spring is longer. A loose clamping device is mounted at the top of the springs so that they are deformed and become equal in length. Calculate the forces in the two springs. An external force P is the load applied thereafter to the spring. In a diagram show how the spring forces vary with P.

16.13 Using Eqs. (16.5) and (16.18) describe how the spring diameter can be chosen to give as low a maximum shear stress as possible.

16.14 Consider a helical compression spring made of hard-drawn wire with a spring index of 12 and a stiffness of 0.3 kN/m. The applied load on the spring is 60 N. For a safety factor of 1.5 guarding against yielding find the spring diameter, the wire diameter, the number of coils, and the free and solid lengths of the spring. Does this spring have buckling and/or dynamic instability problems?

16.15 An 18-mm-diameter helical compression spring has 22 coils, has 2-mm wire diameter, and is made of chromium vanadium. Determine the following:

a) The maximum load-carrying capacity for a safety factor of 1.5 guarding against yielding
b) The maximum deflection of the spring
c) The free length for squared and ground ends

16.16 A 60-mm-diameter helical compression spring is made of hard-drawn steel and has a wire diameter of 2.5 mm. The shear strength of the wire material is 750 MPa and the spring has 20 coils. Find the following:

a) The required load needed to compress the spring to its solid length
b) By applying the load found in part a) and then unloading it, whether the spring will return to its free length

Section 16.4

16.17 A desk lamp has four helical extension springs to make it possible to position it over different parts of the desk. Each spring is preloaded to 15 N, so that no deflection takes place for forces below 15 N. Above 15-N force the spring rate is 100 N/m. The mean coil diameter is 10 mm and the wire diameter is 1 mm. Calculate the tension needed on the wire during manufacturing to get the correct preload, and calculate how many coils are needed to get the correct spring rate.

16.18 A spring balance for weighing fish needs to be dimensioned. The weighing mechanism is a sharp hook hanging in a helical extension spring. To make it easy to read the weight of the fish in the range from zero to 10 kg, the length of the scale should be 100 mm. The spring material is music wire.

16.19 A muscle-training device consists of two handles with three parallel 500-mm-long springs in between. The springs are tightly wound but without prestress. When the springs are fully extended to 1600 mm, the force in each spring is 100 N. The springs are made of music wire. Dimension the springs.

16.20 A 45-mm-diameter extension spring [similar to that shown in Fig. 16.8(a) and (b)] has 102 coils and is made of 4-mm-diameter music wire. The stress due to preload is equivalent to 10% of the yield shear strength. The hook radius is 5 mm and the bend radius is 2.5 mm. Determine the following:

a) The solid length of the spring and the spring stiffness
b) The preload and the load that causes failure

16.21 The extension spring shown in sketch *f* is used in a cyclic motion in turning on and off a power switch. The spring has a 15-mm outer diameter and is made of a 1.5-mm-diameter wire of hard-drawn steel. The spring has no preload. In a full stroke of the spring the force varies between 25 and 33 N. Determine the following:

a) The number of coils, the free length, the maximum and minimum lengths during cyclic loading, and the spring rate
b) For infinite life with 99% reliability the safety factors guarding against static and fatigue failure

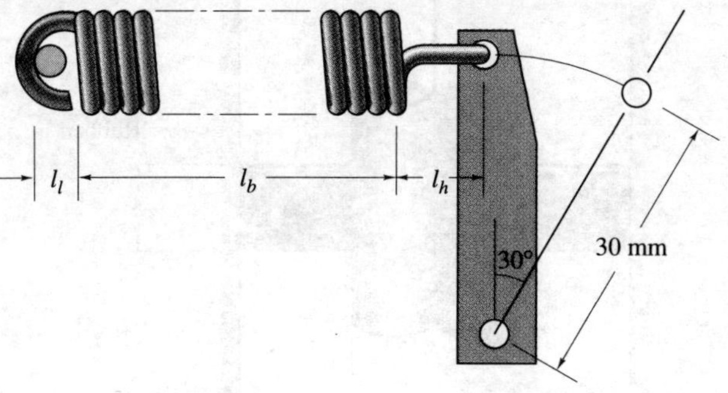

Sketch *f*, for Problem 16.21

16.22 Calculate the safety factor guarding against fatigue failure if the spring given in Problem 16.21 is designed for 50 000 strokes of motion with 50% reliability.

16.23 The extension spring shown in sketch *g* is used in a car braking system. The spring is made of 2-mm-diameter wire of BS5216 steel, has a mean coil diameter of 10 mm, and has a free length of 106 mm. The spring has 18 active coils and a deflection under braking force of 11 mm. Determine the following:

a) The safety factor of the spring
b) The torsional and normal stresses at the hook

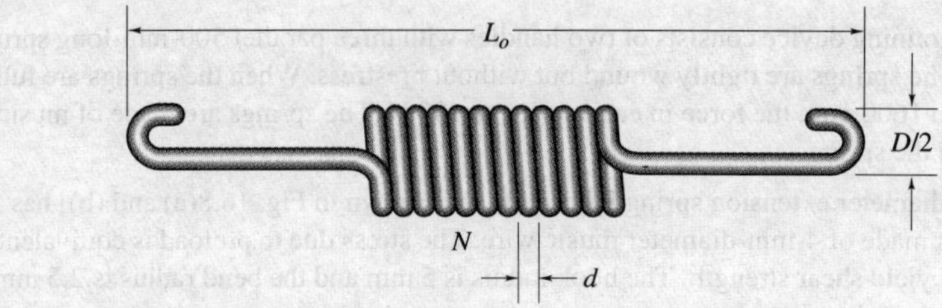

Sketch *g*, for Problem 16.23

Section 16.5

16.24 Derive a simple expression for the spring rate of a spring in pure shear. The force *P* is evenly distributed over the surface. All deflections are small.

16.25 A torsion rod is used as a vehicle spring. The torque on the rod is created by a force $P = 1500$ N acting on radius $R = 200$ mm. Maximum torsion angle is 45°. Calculate the diameter and length of the rod if the maximum shear stress is 500 MPa.

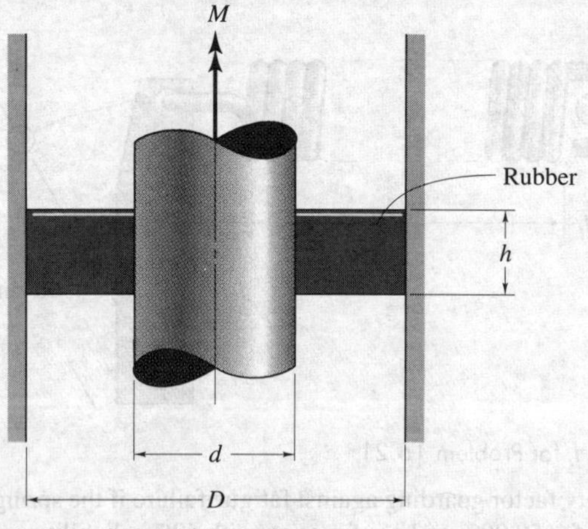

Sketch *h*, for Problem 16.26

16.26 A torsional spring, shown in sketch *h*, consists of a steel cylinder with a rubber ring glued to it. The dimensions of the ring are $D = 45$ mm, $d = 15$ mm, and $h = 20$ mm. Calculate the torque as a function of the angular deflection. The shear modulus of elasticity for rubber is 150 N/cm^2.

16.27 To keep a sauna door shut, helical torsion springs are mounted at each of the two door hinges. The friction torque in each hinge is 0.2 N-m. It should be possible to open the door 180° without plastically deforming the spring. Dimension the springs so that a 10-kg door 700 mm wide will shut itself in 2 s. The length of the wire ends can be neglected. The air drag on the door is also neglected. The springs are manufactured from music wire and have a diameter of 2.5 mm.

16.28 A helical torsion spring used in a door handle is made of hard-drawn steel wire with a diameter of 3 mm. The spring has 10 active coils and a coil diameter of 21 mm. After 5 years of use the spring cracks due to fatigue. To get longer fatigue life, the stresses should be decreased by 5%, but the only available cylindrical space is used by the present spring, which has a outside diameter of 24 mm, an inside diameter of 18 mm, and a length of 30 mm. The spring rate for the new spring should be the same as for the old one. Find a solution.

16.29 The oven door on a kitchen stove is kept shut by a helical torsion spring. The spring torque is 1 N-m when the door is shut. When the oven door is fully open, it must stay open by gravitational force. The door height is 450 mm and it weighs 4 kg. Dimension the helical torsion spring using music wire as the spring material and a wire diameter of 4.5 mm. Is it also possible to use 3-mm-diameter wire?

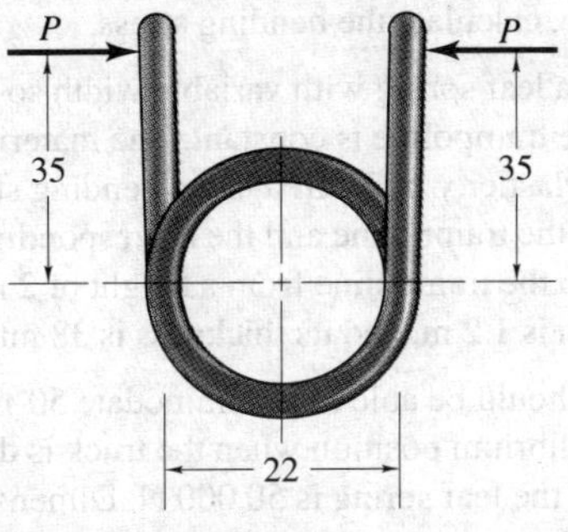

Sketch *i*, for Problem 16.30

16.30 A helical torsion spring, shown in sketch *i*, is made from hard-drawn steel with a wire diameter of 2.2 mm and 8.5 turns. Dimensions are in millimeters.

a) Using a safety factor of 2, find the maximum force and the corresponding angular displacement.

b) What would the coil inside diameter be when the maximum load is applied?

c) For 100 000 loading cycles calculate the maximum moment and the corresponding angular displacement for a safety factor of 2.5 guarding against fatigue failure.

16.31 The helical torsion spring shown in sketch *j* has a coil outside diameter of 22 mm, 8.25 turns, and a wire diameter of 2 mm. The spring material is hard-drawn steel. What would the applied moment be if the maximum stress equaled the yield limit? Calculate the inside diameter of the loaded spring and the corresponding angular displacement. Dimensions are in millimeters.

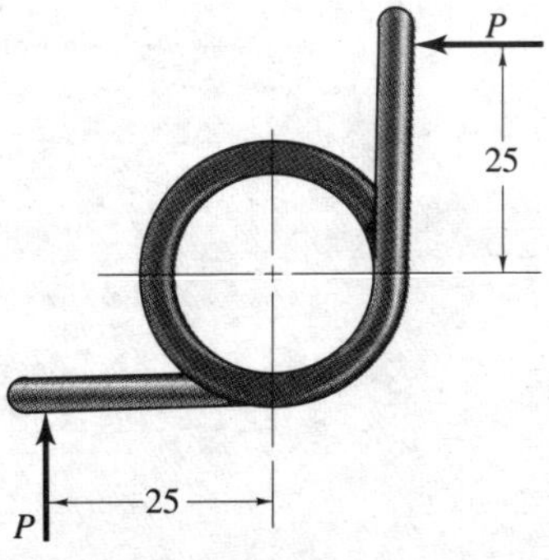

Sketch *j*, for Problem 16.31

16.32 A helical torsion spring made of music wire has a coil diameter of 17.5 mm and a wire diameter of 1.5 mm while supporting a 1.5-N-m moment with a 20% fluctuation. The maximum number of turns is 12, and the load is 22 mm from the center of the spring. For infinite life with 99% reliability find the safety factors guarding against static and fatigue failure. Also, determine the inside diameter of the spring when load has been applied.

Section 16.6

16.33 A fishing rod is made like an ideal leaf spring with rectangular cross sections. It is made of carbon-fiber-reinforced plastic with a 150-GPa modulus of elasticity. The thickness is constant at 8 mm and the length is 2 m. Find how large the cross section must be at the handle to carry a 0.3-kg fish by the hook without bending the top of the rod more than 200 mm. Neglect the weight of the lure. Also, calculate the bending stress.

16.34 A trampoline is made like a leaf spring with variable width so that the maximum bending stress in each section of the trampoline is constant. The material is glass-fiber-reinforced plastic with a modulus of elasticity of 28 GPa and a bending strength of 300 MPa. Calculate the spring rate at the tip of the trampoline and the corresponding safety factor if a swimmer weighing 80 kg jumps onto the trampoline from a height of 2 m. The active length of the trampoline is 3 m, its width is 1.2 m, and its thickness is 38 mm.

16.35 The leaf spring of a truck should be able to accommodate 50-mm deflections (up and down) of the wheels from an equilibrium position when the truck is driven on a rough road. The static load in the middle of the leaf spring is 50 000 N. Dimension the spring so that the static stress is 400 MPa and the dynamic stress amplitude is 200 MPa. The modulus of elasticity for the spring is 207 GPa.

CHAPTER

17

BRAKES AND CLUTCHES

A long-shoe, expanding, internal rim clutch. (Courtesy of Bucyrus Erie)

Nothing has such power to broaden the mind as the ability to investigate systematically and truly all that comes under thy observation in life.

Marcus Aurelius, Roman Emperor

SYMBOLS

A	area, m^2; constant
a	distance, m
b	cone or face width, m
C	cost
c	constant
D	largest diameter of cone, m
d	smallest diameter of cone, m
d_1, d_2, d_3, d_4	distances used for short-shoe brakes, m
d_5, d_6, d_7	distances used for long-shoe rim brakes, m
d_8, d_9, d_{10}	distances used for band brakes, m
F	friction force, N

F_1	pin reaction force for band brakes, N
F_2	actuating force for band brakes, N
h_p	work or energy conversion rate, W
M	moment, N-m
N	number of sets of disks
n_s	safety factor
P	normal force, N
p	contact pressure, Pa
p_0	uniform pressure, Pa
R	reaction force, N
r	radius, m
T	torque, N-m
$\bar{T}$	dimensionless torque, $T/(2\mu P r_o)$
t_m	temperature, °C
u	sliding velocity, m/s
W	actuating force, N
α	half-cone angle of cone clutch, deg
β	radius ratio, r_i/r_o
β_o	optimum radius ratio
δ	wear depth, m
θ	circumferential coordinate, deg
θ_a	angle where $p = p_{max}$, deg
θ_1	location where shoe begins, deg
θ_2	location where shoe ends, deg
μ	coefficient of friction
ϕ	wrap angle, deg
ω	angular velocity, rad/s

Subscripts

B	braking
d	deenergizing
F	friction force
H	horizontal
i	inner
m	mean
max	maximum
o	outer
P	normal force
p	uniform pressure
s	self-energizing
V	vertical
w	uniform wear
x, y	coordinates

17.1 INTRODUCTION

Brakes and clutches are examples of machine elements that use friction in a useful way. **Clutches** are required when shafts must be frequently connected and disconnected. The function of a clutch is twofold: first, to provide a gradual increase in the angular velocity of the driven shaft, so that its speed can be brought up to the speed of the driving shaft without shock; second, when the two shafts are rotating at the same angular velocity, to act as a coupling without slip or loss of speed in the driving shaft. A **brake** is a device used to bring a moving system to rest, to slow its speed, or to control its speed to a certain value under varying conditions. The function of the brake is to turn mechanical energy into heat. The design of brakes and clutches is subjected to uncertainties in the value of the coefficient of friction that must necessarily be used. Materials covered in Secs. 3.5 and 3.7 should be recalled for use in this chapter.

Figure 17.1 shows five types of brake and clutch that are covered in this chapter. These include a **rim type** that has internal expanding and external contracting shoes, a **band brake**, a **thrust disk**, and a **cone disk**. This figure also shows the actuating forces being applied to each brake or clutch.

Brakes and clutches are all similar, but different from other machinery elements, in that they are tribological systems where friction is intended to be high. Therefore, much effort has been directed toward identifying and developing materials that result in high coefficients of friction and low wear so that a reasonable service life can occur. In previous years brake and clutch materials were asbestos-fiber-containing composites, but the wear particles associated with these materials resulted in excessive health hazards

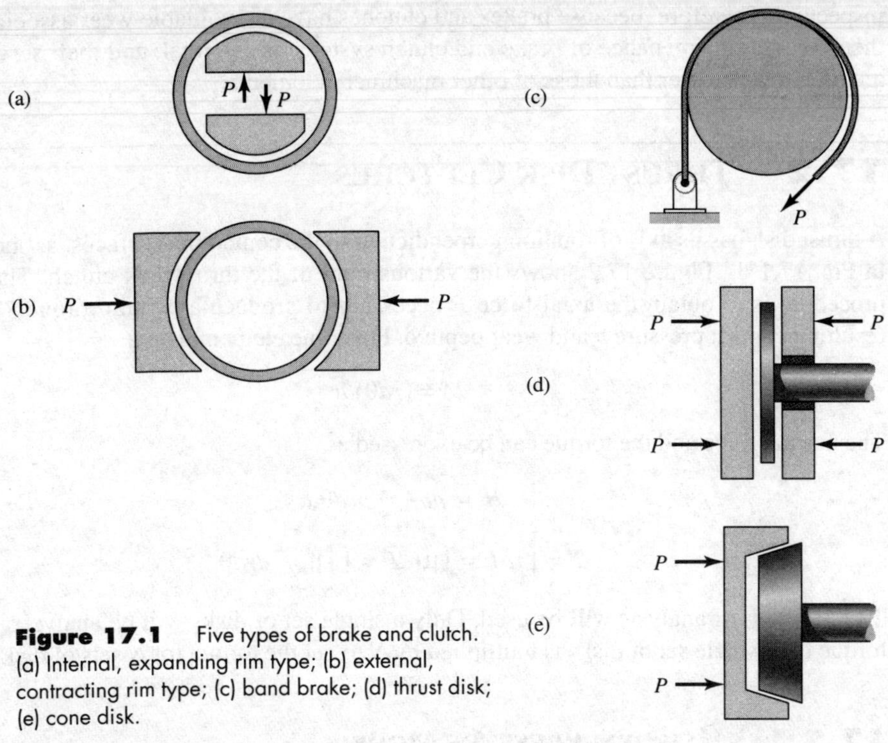

Figure 17.1 Five types of brake and clutch. (a) Internal, expanding rim type; (b) external, contracting rim type; (c) band brake; (d) thrust disk; (e) cone disk.

to maintenance personnel. Modern brakes and clutches use "semimetallic" materials (i.e., metals produced using powder metallurgy techniques) in the tribological interface, even though longer life could be obtained by using the older asbestos-based linings. This substitution is a good example of multidisciplinary design, in that a consideration totally outside of mechanical engineering has eliminated a class of materials from consideration.

Typical brake and clutch design also involves selecting components of sufficient size and capacity to attain reasonable service life. Many of the problems are solid mechanics oriented; the associated theory is covered in Chapters 4 to 7 and not repeated here. Because this chapter is mainly concerned with the performance of brake and clutch systems, the focus here is on the actuating forces and resultant torques. Component size will not be emphasized.

A critical consideration in the design of brake and clutch components is temperature. Whenever brakes or clutches are activated, one high-friction material slides over another with a large normal force. The large amounts of energy created must be dissipated into heat. If the components become too hot, their performance and life can be severely reduced. For example, when brakes are overused, the surfaces can become glazed or "glassy" and will have a marked reduction in friction. The glazed surface must be removed either by replacing the component or by sanding the damaged surface. Further, overheating brake drums and rotors can cause heat checking (small cracks caused by excessively high local temperatures), warping, and even cracking of the component. Obviously, such circumstances should be avoided, but they can only be discovered through periodic inspection. Therefore, because brakes and clutches have unavoidable wear associated with them, regular maintenance of brake and clutch systems is essential, and their service lives are often much lower than those of other machinery elements.

17.2 THRUST DISK CLUTCHES

A thrust disk has its axis of rotation perpendicular to the contacting surfaces, as shown in Fig. 17.1(d). Figure 17.2 shows the various radii of the thrust disk clutch. The design procedure is to obtain the axial force P necessary to produce a certain torque T and the resulting contact pressure p and wear depth δ. For some elemental area

$$dA = (rd\theta)dr \tag{17.1}$$

The normal force and the torque can be expressed as

$$dP = pdA = prd\theta dr \tag{17.2}$$

$$T = \int rdF = \int \mu rdP = \int\int \mu pr^2 drd\theta \tag{17.3}$$

Two methods of analysis will be used. Only a single set of disks will be analyzed, but the torque for a single set of disks is multiplied by N to get the torque for N sets of disks.

17.2.1 UNIFORM PRESSURE MODEL

For new, accurately flat, and aligned disks the pressure will be uniform, or $p = p_0$. Substituting this into Eqs. (17.2) and (17.3) gives

$$P_p = \pi p_0 \left(r_o^2 - r_i^2\right) \tag{17.4}$$

$$T_p = \frac{2\pi\mu p_0}{3}\left(r_o^3 - r_i^3\right) = \frac{2\mu P_p\left(r_o^3 - r_i^3\right)}{3\left(r_o^2 - r_i^2\right)} \tag{17.5}$$

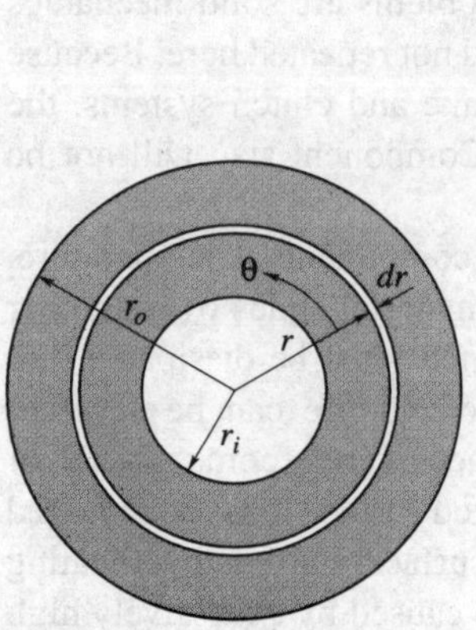

Figure 17.2 Thrust disk clutch surface with various radii.

Thus, expressions for the normal load and torque can be determined from the uniform pressure, the geometry (r_o and r_i), and the coefficient of friction μ.

17.2.2 UNIFORM WEAR MODEL

If the mating surfaces of the clutch are sufficiently rigid, it can be assumed that uniform wear will occur. This assumption generally holds true after some initial running in. After initial wear has taken place, the disks have worn down to the point where uniform wear becomes possible. Thus, for the disk shown in Fig. 17.2 the wear is constant over the surface area $r_i \le r \le r_o$ and around the circumference of the disk.

If the rate of wear is assumed to be proportional to the work or energy, the conversion rate is

$$h_p = Fu = \mu Pu = \mu pAu \tag{17.6}$$

where

F = friction force, N
μ = coefficient of friction
P = normal force, N
u = velocity, m/s
A = area, m^2

A new clutch wears most rapidly at the outer radius, where the velocity u is the greatest. Where the surfaces wear the most, the pressure decreases the most. Thus, multiplying pressure times velocity will produce a constant work or energy conversion, implying that the wear should be uniform at any radius. Then, μpAu remains constant; and if μA is constant, p is inversely proportional to u and hence for any radius r

$$p = c/r \tag{17.7}$$

Substituting Eq. (17.7) into Eq. (17.2) gives

$$P_w = 2\pi c(r_o - r_i) \tag{17.8}$$

Since $p = p_{\max}$ at $r = r_i$, it follows from Eq. (17.7) that

$$c = p_{\max} r_i \tag{17.9}$$

Substituting Eq. (17.9) into Eq. (17.8) gives

$$P_w = 2\pi p_{\max} r_i (r_o - r_i) \tag{17.10}$$

Substituting Eq. (17.7) into Eq. (17.3) gives

$$T_w = \mu c \iint r\,dr\,d\theta = \frac{2\pi\mu c}{2}\left(r_o^2 - r_i^2\right) \tag{17.11}$$

Substituting Eq. (17.9) into Eq. (17.11) gives

$$T_w = \pi\mu r_i p_{\max}\left(r_o^2 - r_i^2\right) \tag{17.12}$$

Substituting Eq. (17.10) into Eq. (17.12) gives

$$T_w = F_w r_m = \frac{\mu P_w (r_o + r_i)}{2} \quad \textbf{(17.13)}$$

By coincidence, Eq. (17.13) gives the same result as if the torque was obtained by multiplying the mean radius $r_m = (r_o + r_i)/2$ by the friction force F.

Equations (17.5) and (17.13) can be expressed as the dimensionless torque for uniform pressure $\overline{T}_p$ and for uniform wear $\overline{T}_w$ by the following equations:

$$\overline{T}_w = T_w / 2\mu P_w r_o = (1+\beta)/4 \quad \textbf{(17.14)}$$

$$\overline{T}_p = T_p / 2\mu P_p r_o = (1-\beta^3)/3(1-\beta^2) \quad \textbf{(17.15)}$$

where

β = radius ratio, r_i/r_o **(17.16)**

Figure 17.3 shows the effect of radius ratio on dimensionless torque for the uniform pressure and uniform wear models. The largest difference between these models occurs at a radius of zero, and the smallest difference occurs at a radius of 1. Also, for the same dimensionless torque the uniform wear model requires a larger radius ratio than does the uniform pressure model. This larger radius ratio implies that a larger area is needed for the uniform wear model. Thus, the uniform wear model may be viewed as the safer approach. This safety, along with its simplicity, makes it often the preferred equation to be used.

Table 17.1 gives the coefficient of friction for several materials rubbing against smooth cast iron or steel under dry conditions. It also gives the maximum contact pressure and the maximum bulk temperature for these materials. Table 17.2 gives the coefficient of friction for several materials, including those in Table 17.1, rubbing against smooth cast iron or steel in oil. As would be expected, the coefficients of friction are much smaller in oil than under dry conditions.

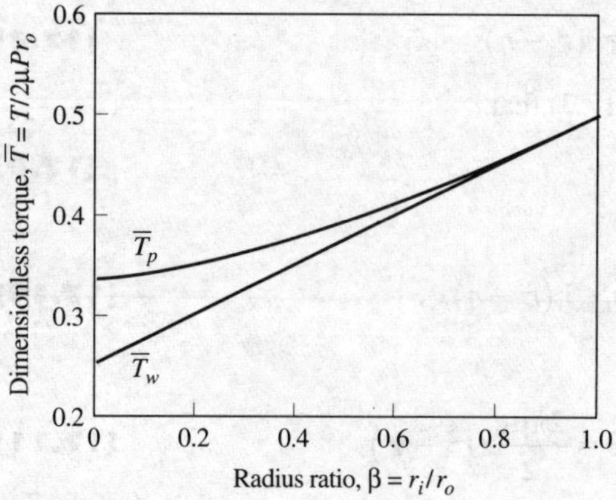

Figure 17.3 Effect of radius ratio on dimensionless torque for uniform pressure and uniform wear models.

Table 17.1 Representative properties of contacting materials operating dry.

Friction material[a]	Coefficient of friction, μ	Maximum contact pressure,[b] p_{max} psi	kPa	Maximum bulk temperature, $t_{m,max}$ °F	°C
Molded	0.25–0.45	150–300	1030–2070	400–500	204–260
Woven	0.25–0.45	50–100	345–690	400–500	204–260
Sintered metal	0.15–0.45	150–300	1030–2070	400–1250	232–677
Cork	0.30–0.50	8–14	55–95	180	82
Wood	0.20–0.30	50–90	345–620	200	93
Cast iron; hard steel	0.15–0.25	100–250	690–1720	500	260

[a]When rubbing against smooth cast iron or steel.
[b]Use of lower values will give longer life.

Table 17.2 Coefficient of friction for contacting materials operating in oil.

Friction material[a]	Coefficient of friction, μ
Molded	0.06–0.09
Woven	0.08–0.10
Sintered metal	0.05–0.08
Paper	0.10–0.14
Graphitic	0.12 (avg.)
Polymeric	0.11 (avg.)
Cork	0.15–0.25
Wood	0.12–0.16
Cast iron; hard steel	0.03–0.16

[a]When rubbing against smooth cast iron or steel.

Example 17.1

GIVEN A single set of thrust disk clutches is to be designed for use in an engine with a maximum torque of 150 N-m. A woven material will contact steel in a dry environment. A safety factor of 1.5 is assumed to account for slippage at full engine torque. The outside diameter should be as small as possible.

FIND Determine the appropriate values for r_o, r_i, and P.

Solution

For woven material in contact with steel in a dry environment Table 17.1 gives the coefficient of friction as $\mu = 0.35$ and the maximum contact pressure as $p_{max} = 345$ kPa $= 0.345$ MPa. The average coefficient of friction has been used but the smallest pressure.

Making use of the above and Eq. (17.12) gives

$$r_i\left(r_o^2 - r_i^2\right) = \frac{n_s T_w}{\pi \mu p_{max}} = \frac{(1.5)(150)}{\pi(0.35)(0.345)(10^6)} = 0.5931\times10^{-3}\ \text{m}^3$$

Solving for the outside radius r_o gives

$$r_o = \sqrt{\frac{(5.931)(10^{-4})}{r_i} + r_i^2} \qquad \textbf{(a)}$$

The minimum outside radius is obtained by taking the derivative of the outside radius with respect to the inside radius and setting it equal to zero to give

$$\frac{dr_o}{dr_i} = \frac{0.5}{\sqrt{\dfrac{(5.931)(10^{-4})}{r_i} + r_i^2}}\left(-\frac{5.931\times10^{-4}}{r_i^2} + 2r_i\right) = 0$$

$$r_i^3 = 2.966\times10^{-4} = 0.2966\times10^{-3}\ \text{m}^3$$

$$r_i = 0.06669\ \text{m} = 66.69\ \text{mm} \qquad \textbf{(b)}$$

Substituting Eq. (b) into Eq. (a) gives

$$r_o = \sqrt{\frac{(5.931)(10^{-4})}{(666.9)(10^{-4})} + (6.669)^2(10^{-4})} = 0.1155\ \text{m} = 115.5\ \text{mm}$$

The radius ratio is

$$\beta = \frac{r_i}{r_o} = \frac{66.69}{115.5} = 0.5774$$

The radius ratio required to maximize the torque capacity is the same as the radius ratio required to minimize the outside radius for a given torque capacity. Thus, the radius ratio for maximizing the torque capacity or for minimizing the outside radius is

$$\beta_o = \sqrt{\frac{1}{3}} = 0.5774$$

From Eq. (17.13) the maximum normal force that can be applied to the clutch is

$$P = \frac{2n_s T_w}{\mu(r_o + r_i)} = \frac{2(1.5)(150)}{(0.35)(0.1155 + 0.06669)} = 7057\ \text{N}$$

Tutorial 13

Equations (17.4) and (17.5) for the uniform pressure model and Eqs. (17.10) and (17.13) for the uniform wear model, which are applicable for thrust disk clutches, are also applicable for thrust disk brakes provided that the disk shape is similar to that shown in Fig. 17.2. A detailed analysis of disk brakes gives equations that result in slightly larger torques than those resulting from the clutch equations. This text assumes that the brake and clutch equations are identical.

17.3 CONE CLUTCHES

The cone clutch uses wedging action to increase the normal force on the lining, thus increasing the friction force and the torque. Figure 17.4 shows a conical surface with forces acting on an element. The area of the element and the normal force on the element are

$$dA = (rd\theta)(dr/\sin\alpha) \qquad \textbf{(17.17)}$$

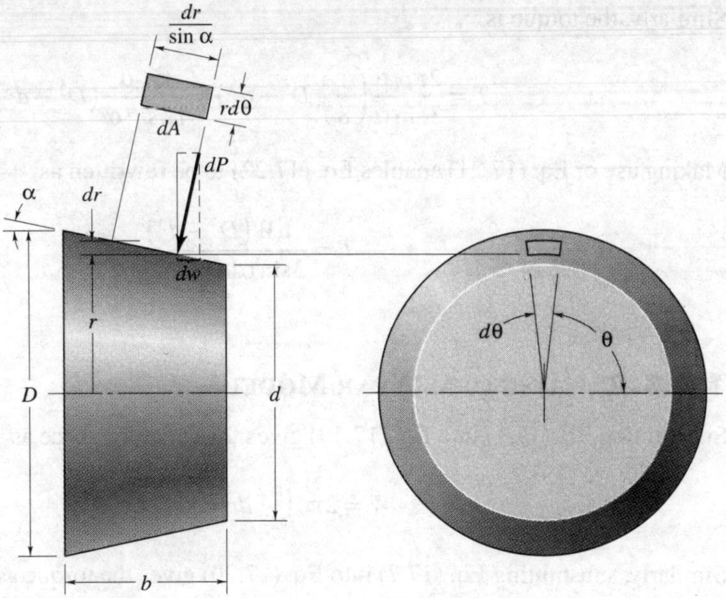

Figure 17.4 Forces acting on elements of cone clutch.

$$dP = pdA \tag{17.18}$$

where

α = half-cone angle, deg

The actuating force is the thrust component dW of the normal force dP, or

$$dW = dP\sin\alpha = pdA\sin\alpha = prdrd\theta$$

Using Eq. (17.2) gives the actuating force as

$$W = \iint prdrd\theta = 2\pi\int_{d/2}^{D/2} prdr \tag{17.19}$$

Similarly, Eq. (17.3) gives the torque as

$$T = \int \mu rdP = \frac{2\pi}{\sin\alpha}\int_{d/2}^{D/2} \mu pr^2 dr \tag{17.20}$$

17.3.1 UNIFORM PRESSURE MODEL

As was discovered in Sec 17.2.1 the pressure for a thrust disk clutch is assumed to be uniform over the surfaces, or $p = p_0$. Substituting this into Eq. (17.19) gives the actuating force as

$$W = \frac{\pi p_0}{4}\left(D^2 - d^2\right) \tag{17.21}$$

Similarly, the torque is

$$T = \frac{2\pi p_0 \mu}{3\sin\alpha}\left(\frac{1}{8}\right)\left(D^3 - d^3\right) = \frac{\pi p_0 \mu}{12\sin\alpha}\left(D^3 - d^3\right) \tag{17.22}$$

Making use of Eq. (17.21) enables Eq. (17.22) to be rewritten as

$$T = \frac{\mu W\left(D^3 - d^3\right)}{3\sin\alpha\left(D^2 - d^2\right)} \tag{17.23}$$

17.3.2 UNIFORM WEAR MODEL

Substituting Eq. (17.7) into Eq. (17.19) gives the actuating force as

$$W = 2\pi c \int_{d/2}^{D/2} dr = \pi c(D - d) \tag{17.24}$$

Similarly, substituting Eq. (17.7) into Eq. (17.20) gives the torque as

$$T = \frac{2\pi\mu c}{\sin\alpha}\int_{d/2}^{D/2} r\,dr = \frac{\pi\mu c}{4\sin\alpha}\left(D^2 - d^2\right) \tag{17.25}$$

Making use of Eq. (17.24) enables Eq. (17.25) to be rewritten as

$$T = \frac{\mu W}{4\sin\alpha}(D + d) \tag{17.26}$$

Example 17.2 **GIVEN** A cone clutch similar to that shown in Fig. 17.4 has the following dimensions: $D = 330$ mm, $d = 306$ mm, and $b = 60$ mm. The coefficient of friction is assumed to be 0.26, and the torque transmitted is 200 N-m.

FIND Determine the actuating force and the contact pressure by using the uniform pressure and uniform wear models.

Solution

Uniform wear: From Fig. 17.4 the half-cone angle of the cone clutch is

$$\tan\alpha = \frac{D - d}{2b} = \frac{165 - 153}{60} = \frac{12}{60} = 0.2000$$

$$\alpha = 11.31^\circ$$

The pressure is a maximum when $r = d/2$. Thus, making use of Eqs. (17.25) and (17.7) gives

$$T = \frac{\pi\mu d p_{max}}{8\sin\alpha}\left(D^2 - d^2\right)$$

$$p_{max} = \frac{8T\sin\alpha}{\pi\mu d\left(D^2 - d^2\right)} = \frac{8(200)\sin 11.31^\circ}{\pi(0.26)(0.306)\left(0.330^2 - 0.306^2\right)} \text{ Pa} = 82.25 \text{ kPa}$$

From Eq. (17.26) the actuating force can be written as

$$W = \frac{4T\sin\alpha}{\mu(D+d)} = \frac{4(200)\sin 11.31°}{(0.26)(0.330+0.306)} = 948.8\ \text{N}$$

Uniform pressure: From Eq. (17.23) the actuating force can be expressed as

$$W = \frac{3T\sin\alpha\left(D^2-d^2\right)}{\mu\left(D^3-d^3\right)} = \frac{3(200)\sin 11.31°\left(0.330^2-0.306^2\right)}{(0.26)\left(0.330^3-0.306^3\right)} = 948.4\ \text{N}$$

From Eq. (17.21) the uniform pressure, which is also the maximum pressure, is

$$p_{\max} = p_0 = \frac{4P_a}{\pi\left(D^2-d^2\right)} = \frac{4(948.4)}{\pi\left(0.330^2-0.306^2\right)}\ \text{Pa} = 79.11\ \text{kPa}$$

17.4 Block, or Short-Shoe, Brakes

A block, or short-shoe, brake can be guided to move radially against a cylindrical drum, as shown in Fig. 17.5. A normal force P develops a friction force $F = \mu P$ on the drum, where μ is the coefficient of friction. The actuating force W is also shown in Fig. 17.5 along with critical hinge pin dimensions d_1, d_2, d_3, and d_4. The normal force P and the friction force μP (Fig. 17.5) are the forces acting on the brake. For block, or short-shoe, brakes a constant pressure is assumed over the pad surface. As long as the pad is short relative to the circumference of the drum, this assumption is relatively accurate.

A brake is considered to be **self-energizing** if the friction moment assists the actuating moment in applying the brake. This implies that the signs of the friction and actuating moments are the same. **Deenergizing** effects occur if the friction moment counteracts the actuating moment in applying the brake. Figure 17.5 can be used to illustrate self-energizing and deenergizing effects.

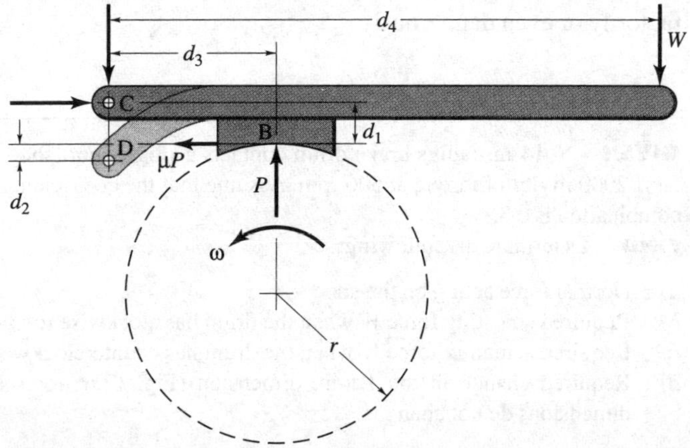

Figure 17.5 Block, or short-shoe brake, with two configurations.

Summing the moments about the hinge at C (see Fig. 17.5) and setting the sum equal to zero give

$$d_4W + \mu P d_1 - d_3 P = 0$$

Since the signs of the friction and actuating moments are the same, the brake hinged at C is self-energizing. Solving for the normal force gives

$$P = \frac{d_4 W}{d_3 - \mu d_1} \quad \textbf{(17.27)}$$

The braking torque at C (see Fig. 17.5) is

$$T = Fr = \mu r P = \frac{\mu r d_4 W}{d_3 - \mu d_1} \quad \textbf{(17.28)}$$

where

r = radius of drum, m

Summing the moments about the hinge at D (see Fig. 17.5) and setting the sum equal to zero give

$$-Wd_4 + \mu P d_2 + d_3 P = 0$$

Since the signs of the friction and actuating moments are opposite, the brake hinged at D is deenergizing. Solving for the normal force gives

$$P = Wd_4 / (d_3 + \mu d_2) \quad \textbf{(17.29)}$$

The torque of the brake hinged at D in Fig. 17.5 is

$$T = (\mu d_4 r W) / (d_3 + \mu d_2) \quad \textbf{(17.30)}$$

A brake is considered to be *self-locking* if the actuating force (W in Fig. 17.5) equals zero. Self-locking brakes are not desired, since they seize or grab, thus operating unsatisfactorily or even dangerously.

Example 17.3

GIVEN A 14-in.-radius brake drum contacts a single short shoe, as shown in Fig. 17.6, and sustains 2000 in.-lbf of torque at 500 rpm. Assume that the coefficient of friction for the drum and shoe combination is 0.3.

FIND Determine the following:

a) Normal force acting on the shoe
b) Required actuating force W when the drum has clockwise rotation
c) Required actuating force W when the drum has counterclockwise rotation
d) Required change in the 1.5-in. dimension (Fig. 17.6) for self-locking to occur if the other dimensions do not change

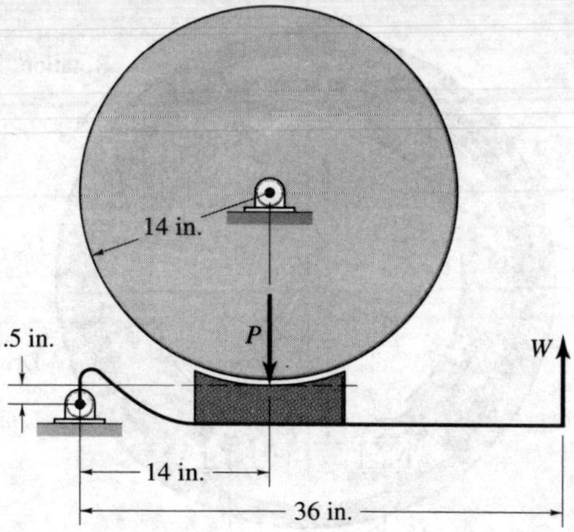

Figure 17.6 Short-shoe brake used in Example 17.3.

Solution

a) The torque of the brake is

$$T = rF = r\mu P$$

$$P = \frac{T}{\mu r} = \frac{2000}{(0.3)(14)} = 476.2 \text{ lbf}$$

$$\mu P = (0.3)(476.2) = 142.9 \text{ lbf}$$

b) For clockwise rotation, summing the moments about the hinge pin and setting the sum equal to zero give

$$(1.5)(142.9) + 36W - 14(476.2) = 0 \qquad \textbf{(a)}$$

$$\therefore W = 179.2 \text{ lbf}$$

Since the signs of the friction and actuating moments are the same, the brake is self-energizing.

c) For counterclockwise rotation, summing the moments about the hinge pin and setting the sum equal to zero give

$$(1.5)(142.9) - 36W + 14(476.2) = 0$$

$$\therefore W = 191.1 \text{ lbf}$$

Since the signs of the friction and actuating moments are not the same, the brake is deenergizing.

d) If in Eq. (a) $W = 0$ and 1.5 is set to x,

$$x = \frac{(14)(476.2)}{142.9} = 46.65 \text{ in}$$

Therefore, self-locking will occur if the distance of 1.5 in. in Fig. 17.6 is changed to 46.65 in. Since self-locking is *not* a desirable effect in a brake and 1.5 in. is quite a distance from 46.65 in., we would not expect the brake to have a self-locking effect.

Internal Brake Animation

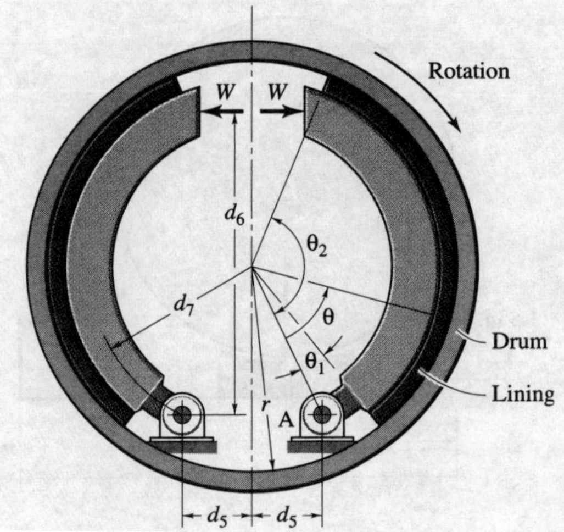

Figure 17.7 Long-shoe, internal, expanding rim brake with two shoes.

17.5 LONG-SHOE, INTERNAL, EXPANDING RIM BRAKES

Figure 17.7 shows a long-shoe, internal, expanding rim brake with two pads. The hinge pin for the right shoe is at A in Fig. 17.7. The *heel* of the pad is the region closest to the hinge pin, and the *toe* is the region closest to the actuating force W. A major difference between a short shoe (Fig. 17.5) and a long shoe (Fig. 17.7) is that the pressure can be considered constant for a short shoe but *not* for a long shoe. In a long shoe no or little pressure is applied at the heel, and the pressure increases as the toe is approached. This sort of pressure variation suggests that the pressure may vary sinusoidally. Thus, a relationship of the contact pressure p in terms of the maximum pressure $p_{\max}$ may be written as

$$p = p_{\max}\left(\frac{\sin\theta}{\sin\theta_a}\right) \tag{17.31}$$

where

θ_a = angle where $p = p_{\max}$

Observe from Eq. (17.31) that $p = p_{\max}$ when $\theta = 90°$ and that for any angular extent of the shoe less than 90° (in Fig. 17.7, $\theta_2 < 90°$) $p = p_{\max}$ when, for example, $\theta_a = \theta_2$.

Observe also in Fig. 17.7 that the distance d_6 is perpendicular to the actuating force W. Figure 17.8 shows the forces and critical dimensions of a long-shoe, internal, expanding rim brake. In Fig. 17.8 the θ coordinate begins with a line drawn from the center of the drum and the center of the hinge pin. Also, the shoe lining does not begin at $\theta = 0°$ but at some θ_1 and extends until θ_2. At any angle θ of the lining the differential normal force dP is

$$dP = pbrd\theta \tag{17.32}$$

where

b = face width, m

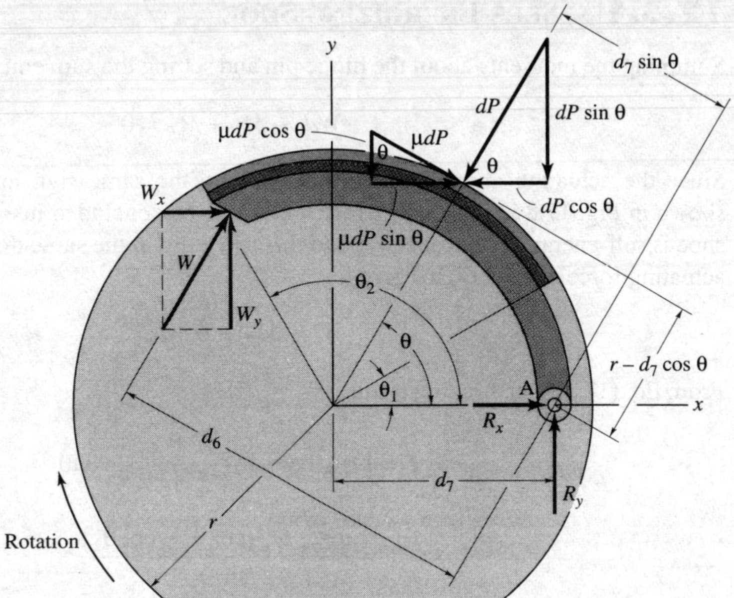

Figure 17.8 Forces and dimensions of long-shoe, internal, expanding rim brake.

The face width is the distance perpendicular to the paper. Substituting Eq. (17.31) into Eq. (17.32) gives

$$dP = \frac{p_{\max} br \sin\theta d\theta}{\sin\theta_a} \tag{17.33}$$

From Eq. (17.33) the normal force moment with moment arm $d_7 \sin\theta$ is

$$M_P = \int d_7 \sin\theta dP = \frac{d_7 b r p_{\max}}{\sin\theta_a} \int_{\theta_1}^{\theta_2} \sin^2\theta d\theta$$

$$= \frac{b r d_7 p_{\max}}{4 \sin\theta_a} \left[2(\theta_2 - \theta_1) \frac{\pi}{180^\circ} - \sin 2\theta_2 + \sin 2\theta_1 \right] \tag{17.34}$$

where θ_1 and θ_2 are in degrees. From Eq. (17.33) the friction force moment with moment arm $r - d_7 \cos\theta$ is

$$M_F = \int (r - d_7 \cos\theta) \mu dP = \frac{\mu p_{\max} br}{\sin\theta_a} \int_{\theta_1}^{\theta_2} (r - d_7 \cos\theta) \sin\theta d\theta$$

$$= \frac{\mu p_{\max} br}{\sin\theta_a} \left[-r(\cos\theta_2 - \cos\theta_1) - \frac{d_7}{2} (\sin^2\theta_2 - \sin^2\theta_1) \right] \tag{17.35}$$

17.5.1 SELF-ENERGIZING SHOE

Summing the moments about the hinge pin and setting the sum equal to zero give

$$-Wd_6 - M_F + M_P = 0 \tag{17.36}$$

Since the actuating and friction moments have the same sign in Eq. (17.36), the shoe shown in Fig. 17.8 is self-energizing. It can also be concluded just from Fig. 17.8 that the shoe is self-energizing because W_x and μdP sin θ are in the same direction. Solving for the actuating force in Eq. (17.36) gives

$$W = (M_P - M_F)/d_6 \tag{17.37}$$

From Eq. (17.33) the braking torque is

$$T = \int r\mu dP = \frac{\mu p_{\max} br^2}{\sin\theta_a}\int_{\theta_1}^{\theta_2}\sin\theta d\theta$$

$$T = \frac{\mu p_{\max} br^2(\cos\theta_1 - \cos\theta_2)}{\sin\theta_a} \tag{17.38}$$

Figure 17.8 shows the reaction forces as well as the friction force and the normal force. Summing the forces in the x direction and setting the sum equal to zero give

$$R_{xs} + W_x - \int\cos\theta dP + \int\mu\sin\theta dP = 0 \tag{17.39}$$

Substituting Eq. (17.33) into Eq. (17.39) gives the reaction force in the x direction for a self-energizing shoe as

$$R_{xs} = \frac{p_{\max,s} br}{\sin\theta_a}\int_{\theta_1}^{\theta_2}\sin\theta\cos\theta d\theta - \frac{\mu p_{\max,s} br}{\sin\theta_a}\int_{\theta_1}^{\theta_2}\sin^2\theta d\theta - W_x$$

$$= -W_x + \frac{p_{\max,s} br}{4\sin\theta_a} \tag{17.40}$$

$$\times\left\{2(\sin^2\theta_2 - \sin^2\theta_1) - \mu\left[2(\theta_2 - \theta_1)\frac{\pi}{180} - \sin 2\theta_2 + \sin 2\theta_1\right]\right\}$$

where θ_1 and θ_2 are in degrees. From Fig. 17.8 summing the forces in the y direction and setting the sum equal to zero give

$$R_{ys} + W_y - \int\mu dP\cos\theta - \int dP\sin\theta = 0 \tag{17.41}$$

$$R_{ys} = -W_y + \frac{p_{\max,s} br}{4\sin\theta_a} \tag{17.42}$$

$$\times\left[2(\theta_2 - \theta_1)\frac{\pi}{180} - \sin 2\theta_2 + \sin 2\theta_1 + 2\mu(\sin^2\theta_2 - \sin^2\theta_1)\right]$$

In Eqs. (17.39) to (17.42) the reference system has its origin at the center of the drum. The positive x axis is taken to be through the hinge pin. The positive y axis is in the direction of the shoe.

17.5.2 Deenergizing Shoe

If in Fig. 17.8 the direction of rotation is changed from clockwise to counterclockwise, the friction forces change direction. Thus, summing the moments about the hinge pin and setting the sum equal to zero give

$$-Wd_6 + M_F + M_P = 0 \tag{17.43}$$

The only difference between Eq. (17.36) and Eq. (17.43) is the sign of the friction moment. Solving for the actuating force in Eq. (17.43) gives

$$W = (M_P + M_F)/d_6 \tag{17.44}$$

The evaluations of the normal moment in Eq. (17.34) and the friction moment in Eq. (17.35) are the same whether the shoe is self-energizing or deenergizing with the exception of the maximum pressure.

For a deenergizing shoe the only changes from the equations derived for the self-energizing shoe are that in Eqs. (17.40) and (17.42) a sign change occurs for terms containing the coefficient of friction μ, resulting in the following:

$$R_{xd} = -W_x + \frac{p_{\max,d}br}{4\sin\theta_a} \times \left\{2(\sin^2\theta_2 - \sin^2\theta_1) + \mu\left[2(\theta_2-\theta_1)\frac{\pi}{180} - \sin 2\theta_2 + \sin 2\theta_1\right]\right\} \tag{17.45}$$

$$R_{yd} = -W_y + \frac{p_{\max,d}br}{4\sin\theta_a} \times \left[2(\theta_2-\theta_1)\frac{\pi}{180} - \sin 2\theta_2 + \sin 2\theta_1 - 2\mu(\sin^2\theta_2 - \sin^2\theta_1)\right] \tag{17.46}$$

The maximum contact pressure used in evaluating a self-energizing shoe is taken from Table 17.1. The maximum contact pressure used in evaluating a deenergizing shoe is less than that for the self-energizing shoe, since the actuating force is the same for both types of shoe.

Example 17.4

GIVEN Figure 17.9 shows four long shoes in an internal, expanding rim brake. The brake drum has a 400-mm diameter. Each hinge pin (A and B) supports a pair of shoes. The actuating mechanism is to be arranged to produce the same actuating force W on each shoe. The shoe face width is 75 mm. The material of the shoe and drum produces a coefficient of friction of 0.24 and a maximum contact pressure of 1 MPa. Dimensions are the following: $d = 50$ mm, $b = 165$ mm, and $a = 150$ mm.

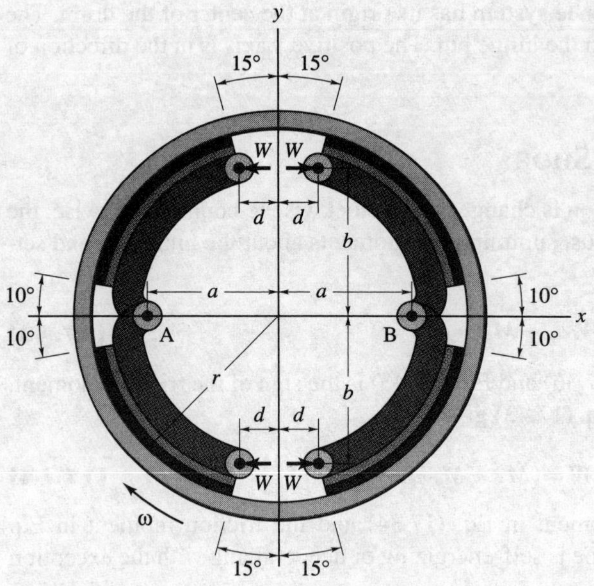

Figure 17.9 Four-long-shoe, internal, expanding rim brake used in Example 17.4.

FIND

a) Which shoes are self-energizing and which are deenergizing?
b) What are the actuating forces and total torques for the four shoes?
c) What are the hinge pin reactions as well as the resultant reaction?

Solution

a) With the drum rotating in the clockwise direction (Fig. 17.9) the top-right and bottom-left shoes have their actuating and friction moments acting in the same direction. Thus, they are self-energizing shoes. The top-left and bottom-right shoes have their actuating and friction moments acting in opposite directions. Thus, they are deenergizing shoes.

b) The dimensions given in Fig. 17.9 correspond to the dimensions given in Figs. 17.7 and 17.8 as $d_5 = d = 50$ mm, $d_6 = b = 165$ mm, and $d_7 = a = 150$ mm. Also, since $\theta_2 < 90°$, then $\theta_2 = \theta_a$. Because the hinge pins in Fig. 17.9 are at A and B, $\theta_1 = 10°$ and $\theta_2 = \theta_a = 75°$.

Self-energizing shoes: Making use of the above and Eq. (17.34) gives the normal force moment as

$$M_{Ps} = \frac{brd_7 p_{\max,s}}{4\sin\theta_a}\left[2(\theta_2-\theta_1)\frac{\pi}{180°} - \sin 2\theta_2 + \sin 2\theta_1\right]$$

$$= \frac{(0.075)(0.2)(0.15)(10^6)}{4\sin 75°}\left[2(75-10)\frac{\pi}{180} - \sin 150° + \sin 20°\right]$$

$$= 1229 \text{ N-m}$$

From Eq. (17.35) the friction moment is

$$M_{Fs} = \frac{\mu p_{\max,s} br}{\sin\theta_a}\left[-r(\cos\theta_2 - \cos\theta_1) - \frac{d_7}{2}(\sin^2\theta_2 - \sin^2\theta_1)\right]$$

$$M_{Fs} = \frac{(0.24)(10^6)(0.075)(0.20)}{\sin 75°} \times\left[-(0.20)(\cos 75° - \cos 10°) - 0.075\left(\sin^2 75° - \sin^2 10°\right)\right]$$

$$= 288.8 \text{ N-m}$$

From Eq. (17.37) the actuating force for both the self-energizing and deenergizing shoes is

$$W_s = W_d = \frac{M_P - M_F}{d_6} = \frac{1229 - 288.8}{0.165} = 5698 \text{ N}$$

From Eq. (17.38) the braking torque for each self-energizing shoe is

$$T_s = \mu p_{\max} b r^2 (\cos\theta_1 - \cos\theta_2)/\sin\theta_a$$

$$= (0.24)(10^6)(0.075)(0.2)^2(\cos 10° - \cos 75°)/\sin 75° = 541.2 \text{ N-m}$$

Deenergizing shoes: The only change in the calculation of the normal and friction moments for the deenergizing shoes is the maximum pressure.

$$M_{Pd} = M_{Ps} p_{\max,d} / p_{\max,s} = (1229) p_{\max,d} / (10^6)$$

$$M_{Fd} = M_{Fs} p_{\max,d} / p_{\max,s} = (288.8) p_{\max,d} / (10^6)$$

From Eq. (17.44) the actuating load for the deenergizing shoes is

$$W_d = \frac{M_{Pd} + M_{Fd}}{d_6} = \left(\frac{1229}{10^6} + \frac{288.8}{10^6}\right)\frac{p_{\max,d}}{0.165} = W_s = 5698 \text{ N}$$

$$\therefore p_{\max,d} = \frac{(5698)(0.165)(10^6)}{1229 + 288.8} \text{ Pa} = 0.6194 \text{ MPa}$$

The braking torque for the deenergizing shoes is

$$T_d = T_s\left(\frac{p_{\max,d}}{p_{\max,s}}\right) = 541.2\left(\frac{0.6194}{1.000}\right) = 335.2 \text{ N-m}$$

The total braking torque of the four shoes, two of which are self-energizing and two of which are deenergizing, is

$$T = 2(T_s + T_d) = 2(541.2 + 335.2) = 1753 \text{ N-m}$$

c) Self-energizing shoes: From Eq. (17.40)

$$R_{xs} = -W_x + \frac{p_{\max,s} b r}{4\sin\theta_a} \times\left\{2\left(\sin^2\theta_2 - \sin^2\theta_1\right) - \mu\left[2(\theta_2 - \theta_1)\frac{\pi}{180} - \sin 2\theta_2 + \sin 2\theta_1\right]\right\}$$

$$R_{xs} = -5698 + \frac{(10^6)(0.075)(0.2)}{4\sin 75°}$$

$$\times\left\{2(\sin^2 75° - \sin^2 10°) - 0.24\left[2(75-10)\frac{\pi}{180} - \sin 150° + \sin 20°\right]\right\}$$

$$= -654.6 \text{ N}$$

From Eq. (17.42)

$$R_{ys} = -W_y + \frac{p_{\max,s} br}{4\sin\theta_a}$$

$$\times\left[2(\theta_2 - \theta_1)\frac{\pi}{180} - \sin 2\theta_2 + \sin 2\theta_1 + 2\mu(\sin^2\theta_2 - \sin^2\theta_1)\right]$$

$$= -0 + \frac{(10^6)(0.075)(0.2)}{4\sin 75°}$$

$$\times\left[2(75-10)\frac{\pi}{180} - \sin 150° + \sin 20° + 2(0.24)(\sin^2 75° - \sin^2 10°)\right]$$

$$= 9878 \text{ N}$$

Deenergizing shoes: From Eq. (17.45)

$$R_{xd} = -5698 + \frac{(0.6194)(10^6)(0.075)(0.2)}{4\sin 75°}$$

$$\times\left\{2(\sin^2 75° - \sin^2 10°) + 0.24\left[2(75-10)\frac{\pi}{180} - \sin 150° + \sin 20°\right]\right\}$$

$$= 5560 - 5698 = -137.5 \text{ N}$$

From Eq. (17.46)

$$R_{yd} = -0 + \frac{(0.6194)(10^6)(0.075)(0.2)}{4\sin 75°}$$

$$\times\left[2(75-10)\frac{\pi}{180} - \sin 150° + \sin 20° - 2(0.24)(\sin^2 75° - \sin^2 10°)\right]$$

$$= 4034 \text{ N}$$

The resultant forces of the reactions in the hinge pin in the horizontal and vertical directions are

$$R_H = -654.6 - 137.5 = -792.1 \text{ N}$$

$$R_V = 9878 - 4034 = 5844 \text{ N}$$

The resultant force at the hinge pin is

$$R = \sqrt{R_H^2 + R_V^2} = \sqrt{(-792.1)^2 + (5844)^2} = 5897 \text{ N}$$

17.6 LONG-SHOE, EXTERNAL, CONTRACTING RIM BRAKES

Figure 17.10 shows the forces and dimensions of a long-shoe, external, contracting rim brake. In Fig. 17.8 the brake is internal to the drum, whereas in Fig. 17.10 the brake is external to the drum. The symbols used in these figures are similar. The equations developed in Sec. 17.5 for internal shoe brakes are exactly the same as those for external shoe brakes as long as one properly identifies whether the brake is self-energizing or deenergizing.

The *internal* brake shoe in Fig. 17.8 was found to be *self-energizing* for clockwise rotation, since in the moment summation [(Eq. (17.36)] the actuating and friction moments have the same sign. The *external* brake shoe in Fig. 17.10 is *deenergizing* for clockwise rotation. Summing the moments and equating the sum to zero give

$$Wd_6 - M_F - M_P = 0 \tag{17.47}$$

The actuating and friction moments have opposite signs and thus the external brake shoe shown in Fig. 17.10 is deenergizing.

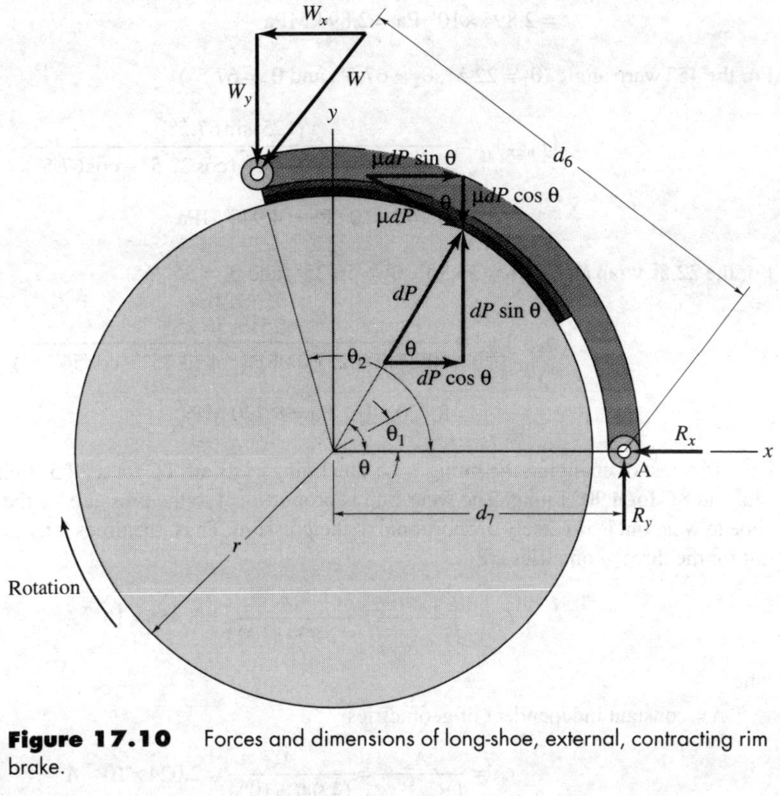

Figure 17.10 Forces and dimensions of long-shoe, external, contracting rim brake.

If in Figs. 17.8 and 17.10 the direction of rotation were changed from clockwise to counterclockwise, the friction moments in Eqs. (17.36) and (17.47) would have opposite signs. Therefore, the internal brake shoe would be deenergizing and the external brake shoe would be self-energizing.

Example 17.5 **GIVEN** An external, long-shoe rim brake is to be cost optimized. Three lining geometries are being considered, covering the entire 90° of the shoe, covering only 45° of the central portion of the shoe, or covering only 22.5° of the central portion of the shoe. The braking torque must be the same for all three geometries, and the cost of changing any of the brake linings is half of the cost of a 22.5° lining. The cost of the lining material is proportional to the wrap angle. The wear rate is proportional to the pressure. The input parameters for the 90° lining are $d_7 = 100$ mm, $r = 80$ mm, $b = 25$ mm, $\theta_1 = 0°$, $\theta_2 = 90°$, $\mu = 0.27$, and $T = 125$ N-m.

FIND Which of the wrap angles (90°, 45°, or 22.5°) would be the most economical?

Solution

The braking torque is given by Eq. (17.38) and is the same for all three geometries. For the 90° wrap angle ($\theta_1 = 0°$, $\theta_2 = 90°$, and $\theta_a = 90°$)

$$(p_{\max})_{90} = \frac{T\sin\theta_a}{\mu b r^2(\cos\theta_1 - \cos\theta_2)} = \frac{(125)\sin 90°}{(0.27)(0.025)(0.08)^2(\cos 0° - \cos 90°)}$$

$$= 2.894\times10^6 \text{ Pa} = 2.894 \text{ MPa}$$

For the 45° warp angle ($\theta_1 = 22.5°$, $\theta_2 = 67.5°$, and $\theta_a = 67.5°$)

$$(p_{\max})_{45} = \frac{(125)\sin 67.5°}{(0.27)(0.025)(0.080)^2(\cos 22.5° - \cos 67.5°)}$$

$$= 4.940\times10^6 \text{ Pa} = 4.940 \text{ MPa}$$

For the 22.5° wrap angle ($\theta_1 = 33.75°$, $\theta_2 = 56.25°$, and $\theta_a = 56.25°$)

$$(p_{\max})_{22.5} = \frac{125\sin 56.25°}{(0.27)(0.025)(0.08)^2(\cos 33.75° - \cos 56.25°)}$$

$$= 8.720\times10^6 \text{ Pa} = 8.720 \text{ MPa}$$

The cost of changing the lining is C. The lining costs are $2C$ for a 22.5° lining, $4C$ for a 45° lining, and $8C$ for a 90° lining. The wear rate is proportional to the pressure, or the time it takes for the shoe to wear out is inversely proportional to the pressure. Thus, the times it takes for the shoe to wear out for the three geometries are

$$t_{90} = \frac{A}{(p_{\max})_{90}} = \frac{A}{(2.894)(10^6)} = 3.455\times10^{-7} A$$

where

A = constant independent of geometries

$$t_{45} = \frac{A}{(p_{\max})_{45}} = \frac{A}{(4.940)(10^6)} = 2.024\times10^{-7} A$$

$$t_{22.5} = \frac{A}{(p_{max})_{22.5}} = \frac{A}{(8.720)(10^6)} = 1.147 \times 10^{-7} A$$

The costs per unit time for the three geometries are

90° wrap angle:

$$\frac{8C + C}{(3.455)(10^{-7})A} = 26.05 \times 10^6 \, C/A$$

45° wrap angle:

$$\frac{4C + C}{(2.024)(10^{-7})A} = 24.70 \times 10^6 \, C/A$$

22.5° wrap angle:

$$\frac{2C + C}{(1.147)(10^{-7})A} = 26.16 \times 10^6 \, C/A$$

The 45°-wrap-angle shoe gives the lowest cost, 5.6% lower than the 22.5°-wrap-angle shoe and 5.2% lower than the 90°-wrap-angle shoe.

17.7 SYMMETRICALLY LOADED PIVOT-SHOE BRAKES

Figure 17.11 shows a symmetrically loaded pivot-shoe brake. The major difference between the internal and external rim brakes shown in Figs. 17.8 and 17.10, respectively, and the symmetrically loaded pivot-shoe brake shown in Fig. 17.11 is the pressure distribution

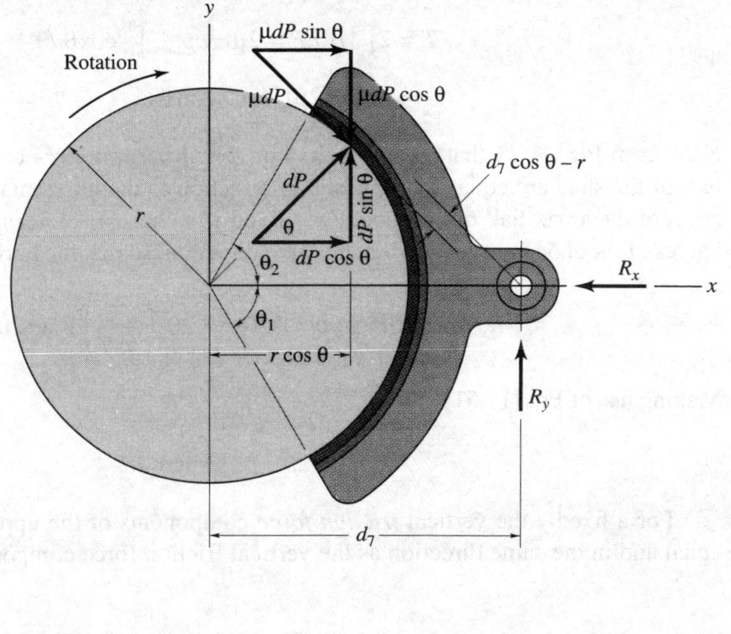

Figure 17.11
Symmetrically loaded pivot-shoe brake.

around the shoe. Recall from Eq. (17.31) for the internal rim brake that the maximum pressure was at $\theta = 90°$ and the pressure distribution from the heel to the top of the brake was sinusoidal. For the symmetrically loaded pivot-shoe brake (Fig. 17.11) the maximum pressure is at $\theta = 0°$, which suggests the pressure variation to be

$$p = p_{\max} \cos\theta / \cos\theta_a = p_{\max} \cos\theta \tag{17.48}$$

For any θ from the pivot a differential normal force dP acts with a magnitude of

$$dP = pbrd\theta = p_{\max} br \cos\theta d\theta \tag{17.49}$$

The design of a symmetrically loaded pivot-shoe brake is such that the distance d_7, which is measured from the center of the drum to the pivot, is chosen such that the resulting friction moment acting on the brake shoe is zero. From Fig. 17.11 the friction moment, when set equal to zero, is

$$M_F = 2\int_0^{\theta_2} \mu dP(d_7 \cos\theta - r) = 0 \tag{17.50}$$

Substituting Eq. (17.49) into Eq. (17.50) gives

$$2\mu p_{\max} br \int_0^{\theta_2} \left(d_7 \cos^2\theta - r\cos\theta\right) = 0$$

This reduces to

$$d_7 = \frac{4r\sin\theta_2}{2\theta_2\left(\dfrac{\pi}{180}\right) + \sin 2\theta_2} \tag{17.51}$$

This value of d_7 produces a friction moment equal to zero ($M_F = 0$).

The braking torque is

$$\begin{aligned} T &= 2\int_0^{\theta_2} r\mu dP = 2\mu r^2 bp_{\max} \int_0^{\theta_2} \cos\theta d\theta \\ &= \mu r P = 2\mu r^2 bp_{\max} \sin\theta_2 \end{aligned} \tag{17.52}$$

Note from Fig. 17.11 that for any x the horizontal *friction force* components of the upper half of the shoe are equal and opposite in direction to the horizontal friction force components of the lower half of the shoe. For a fixed x the horizontal *normal* components of both halves of the shoe are equal and in the same direction, so that the horizontal reaction force is

$$R_x = 2\int_0^{\theta_2} dP\cos\theta = \frac{p_{\max} br}{2}\left[2\theta_2\left(\frac{\pi}{180}\right) + \sin 2\theta_2\right] \tag{17.53}$$

Making use of Eq. (17.51) gives

$$R_x = 2br^2 p_{\max} \sin\theta_2 / d_7 \tag{17.54}$$

For a fixed y the vertical *friction force* components of the upper half of the shoe are equal and in the same direction as the vertical friction force components of the lower half of the shoe. For a fixed y the vertical *normal* components of both halves of the shoe are equal and opposite in direction, so that the vertical reaction force is

$$R_y = 2\int_0^{\theta_2} \mu dP\cos\theta = \frac{\mu p_{max} br}{2}\left[2\theta_2\left(\frac{\pi}{180}\right) + \sin 2\theta_2\right] \quad \textbf{(17.55)}$$

Making use of Eq. (17.53) gives

$$R_y = 2\mu br^2 p_{max}\sin\theta_2 / d_7 = \mu R_x$$

Example 17.6

GIVEN A symmetrically loaded, pivot-shoe brake has the distance d_7 shown in Fig 17.11 optimized for a 180° wrap angle. When the brake lining is worn out, it is replaced with a 90°-wrap-angle lining symmetrically positioned in the shoe. The actuating force is 11 000 N, the coefficient of friction is 0.31, the brake drum radius is 100 mm, and the brake width is 45 mm.

FIND Calculate the pressure distribution in the brake shoe and the braking torque.

Solution

The distance d_7 can be expressed from Eq. (17.51) for the 180° wrap angle as

$$(d_7)_{180} = \frac{4r\sin\theta_2}{2\theta_2\left(\frac{\pi}{180}\right) + \sin 2\theta_2} = \frac{4(0.1)\sin 90°}{2(90)\left(\frac{\pi}{180}\right) + \sin 180°} = 0.1273\text{ m}$$

For the 90°-wrap-angle, symmetrically loaded, pivot-shoe brake the pressure distribution will be unsymmetrical. The maximum pressure, which will occur at θ_2, needs to be determined from the shoe equilibrium.

$$\therefore \int_{-\pi/4}^{\pi/4} prd\theta(b)(d_7\sin\theta) - \int_{-\pi/4}^{\pi/4} \mu prd\theta(b)(d_7\cos\theta - r) = 0 \quad \textbf{(a)}$$

If the wear rate is proportional to the pressure and the pressure maximum is at $\theta = \theta_0$, the pressure distribution is

$$p = p_{max}\cos(\theta - \theta_0) \quad \textbf{(b)}$$

Substituting Eq. (b) into Eq. (a) gives

$$\int_{-\pi/4}^{\pi/4} d_7\cos(\theta-\theta_0)\sin\theta d\theta - \int_{-\pi/4}^{\pi/4} \mu\cos(\theta-\theta_0)(d_7\cos\theta - r)d\theta = 0$$

But

$$\cos(\theta-\theta_0) = \cos\theta\cos\theta_0 + \sin\theta\sin\theta_0$$

$$\therefore d_7\int_{-\pi/4}^{\pi/4}\left[\cos\theta_0\cos\theta\sin\theta + \sin\theta_0\sin^2\theta\right]d\theta = \mu\int_{-\pi/4}^{\pi/4}(\cos\theta_0\cos\theta + \sin\theta_0\sin\theta)(d_7\cos\theta - r)d\theta$$

Integrating gives

$$\frac{d_7\cos\theta_0}{2}\left[\sin^2\frac{\pi}{4} - \sin^2\left(-\frac{\pi}{4}\right)\right] + d_7\sin\theta_0\left(\frac{\pi}{4} - \frac{1}{2}\right) = \mu d_7\cos\theta_0\left(\frac{\pi}{4} + \frac{1}{2}\right) +$$

$$+\mu d_7 \sin\theta_0 \left(\frac{\sin^2 \dfrac{\pi}{4}}{2} - \frac{\sin^2\left(-\dfrac{\pi}{4}\right)}{2} \right) - \mu r \cos\theta_0 \sqrt{2}\mu r \sin\theta_0 \left(\frac{1}{\sqrt{2}} - \frac{1}{\sqrt{2}} \right)$$

This reduces to

$$d_7 \sin\theta_0 \left(\frac{\pi}{4} - \frac{1}{2} \right) = \mu \cos\theta_0 \left[d_7 \left(\frac{1}{\pi} + \frac{1}{2} \right) - r\sqrt{2} \right]$$

$$\tan\theta_0 = \frac{\mu\left[d_7 \left(\dfrac{\pi}{4} + \dfrac{1}{2} \right) - \sqrt{2} r \right]}{d_7 \left(\dfrac{\pi}{4} - \dfrac{1}{2} \right)} = \frac{0.31\left[0.1273 \left(\dfrac{\pi}{4} + \dfrac{1}{2} \right) - \sqrt{2}(0.100) \right]}{0.1273 \left(\dfrac{\pi}{4} - \dfrac{1}{2} \right)}$$

$$\therefore \theta_0 = 10.75°$$

The actuating force is

$$W = \int_{-\pi/4}^{\pi/4} prd\theta b \cos\theta + \int_{-\pi/4}^{\pi/4} \mu prd\theta b \sin\theta \qquad \textbf{(c)}$$

Substituting Eq. (b) into Eq. (c) gives

$$W = rbp_{max} \left[\int_{-\pi/4}^{\pi/4} \cos(\theta - \theta_0) \cos\theta d\theta + \int_{-\pi/4}^{\pi/4} \cos(\theta - \theta_0) \sin\theta d\theta \right] \qquad \textbf{(d)}$$

But

$$\int_{-\pi/4}^{\pi/4} \cos(\theta - \theta_0)\cos\theta d\theta = \int_{-\pi/4}^{\pi/4} (\cos\theta\cos\theta_0 + \sin\theta\sin\theta_0)\cos\theta d\theta = \left(\frac{\pi}{4} + \frac{1}{2} \right)\cos\theta_0 \qquad \textbf{(e)}$$

and

$$\int_{-\pi/4}^{\pi/4} \cos(\theta - \theta_0)\sin\theta d\theta = \left(\frac{\pi}{4} - \frac{1}{2} \right)\sin\theta_0 \qquad \textbf{(f)}$$

Substituting Eqs. (e) and (f) into Eq. (d) while solving for the maximum pressure gives

$$p_{max} = \frac{W}{rb\left[\left(\dfrac{\pi}{4} + \dfrac{1}{2} \right)\cos\theta_0 + \mu\left(\dfrac{\pi}{4} - \dfrac{1}{2} \right)\sin\theta_0 \right]}$$

$$= \frac{11\,000}{(0.100)(0.045)\left[\left(\dfrac{\pi}{4} + \dfrac{1}{2} \right)\cos 10.75° + 0.31\left(\dfrac{\pi}{4} - \dfrac{1}{2} \right)\sin 10.75° \right]}$$

$$= 1.911 \times 10^6 \text{ Pa} = 1.911 \text{ MPa}$$

From Eq. (b) the pressure distribution can be expressed as

$$p = (1.911)(10^6)\cos(\theta - 10.75°)$$

The braking torque is

$$T = \mu p_{max} b r^2 \int_{-\pi/4}^{\pi/4} \cos(\theta - \theta_0) d\theta$$

$$= \mu p_{max} b r^2 \left[\sin\left(\frac{\pi}{4} - \theta_0\right) + \sin\left(\frac{\pi}{4} + \theta_0\right) \right]$$

$$= (0.31)(1.911)(10^6)(0.045)(0.1)^2 [\sin 34.25° + \sin 55.75°]$$

$$= 370.4 \text{ N-m}$$

17.8 BAND BRAKES

Figure 17.12 shows a band brake, which consists of a band wrapped partly around a drum. The brake is actuated by pulling the band tighter against the drum, as shown in Fig. 17.12(a). The band is assumed to be in contact with the drum over the entire wrap angle, ϕ in Fig. 17.12(a). The pin reaction force is given as F_1 and the actuating force as F_2. In Fig. 17.12(a) the heel of the brake is near F_1 and the toe is near F_2. Since some friction will exist between the band and the drum, the actuating force will be less than the pin reaction force, or $F_2 < F_1$.

Figure 17.12(b) shows the forces acting on an element of the band. The forces are the normal force P and the friction force F. Summing the forces in the *vertical* (radial) direction while using Fig. 17.12(b) gives

$$(F + dF) \sin\left(\frac{d\theta}{2}\right) + F \sin\left(\frac{d\theta}{2}\right) - dP = 0$$

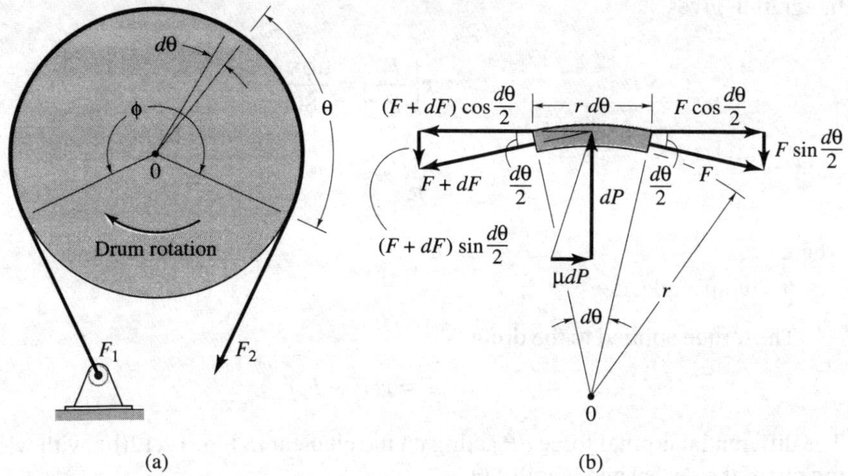

Figure 17.12 Band brake.
(a) Forces acting on band; (b) forces acting on element.

$$dP = 2F\sin\left(\frac{d\theta}{2}\right) + dF\sin\left(\frac{d\theta}{2}\right)$$

Since $dF \ll F$,

$$dP = 2F\sin\left(\frac{d\theta}{2}\right)$$

Since $d\theta/2$ is small, then $\sin d\theta/2 \approx d\theta/2$. Therefore,

$$dP = F d\theta \tag{17.56}$$

Summing the forces in the *horizontal* (tangential) direction while using Fig. 17.12(b) gives

$$(F + dF)\cos\left(\frac{d\theta}{2}\right) - F\cos\left(\frac{d\theta}{2}\right) - \mu dP = 0$$

$$dF\cos\left(\frac{d\theta}{2}\right) - \mu dP = 0$$

Since $d\theta/2$ is small, then $\cos(d\theta/2) = 1$. Therefore,

$$dF - \mu dP = 0 \tag{17.57}$$

Substituting Eq. (17.56) into Eq. (17.57) gives

$$dF - \mu F d\theta = 0$$

$$\int_{F_2}^{F_1} \frac{dF}{F} = \mu \int_0^{\phi} d\theta = 0$$

Integrating gives

$$\ln\left(\frac{F_1}{F_2}\right) = \frac{\mu\phi\pi}{180}$$

$$\frac{F_1}{F_2} = e^{\mu\phi\pi/180^\circ} \tag{17.58}$$

where

ϕ = wrap angle, deg

The torque applied to the drum is

$$T = r(F_1 - F_2) \tag{17.59}$$

The differential normal force dP acting on the element in Fig. 17.12(b), with width b (coming out of the paper) and length $r d\theta$, is

$$dP = pbr d\theta \tag{17.60}$$

where

p = contact pressure, Pa

Substituting Eq. (17.60) into Eq. (17.56) gives

$$p = \frac{F}{br} \tag{17.61}$$

The pressure is proportional to the tension in the band. The maximum pressure occurs at the heel, or near the pin reaction force, and has the value

$$p_{max} = \frac{F_1}{br} \tag{17.62}$$

Example 17.7

GIVEN The band brake shown in Fig. 17.13 has $r = 4$ in., $b = 1$ in., $d_9 = 9$ in., $d_8 = 2$ in., $d_{10} =$ 0.5 in., $\phi = 270°$, $\mu = 0.2$, and $p_{max} = 75$ psi.

FIND Determine the following:

a) Brake torque
b) Actuating force
c) Value of d_{10} when the brake force locks

Solution

From Eq. (17.62) the pin reaction force is

$$F_1 = p_{max}br = (75)(1)(4) = 300 \text{ lbf}$$

From Eq. (17.58) the actuating force is

$$F_2 = F_1 e^{-\mu\phi\pi/180} = 300\, e^{-0.2(270)\pi/180} = 116.9 \text{ lbf}$$

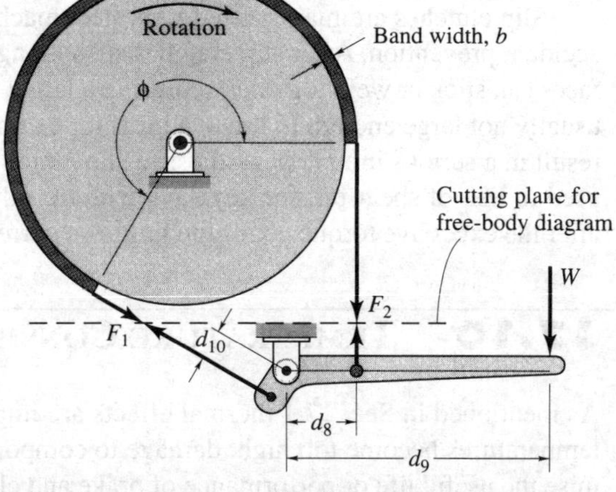

Figure 17.13 Band brake used in Example 17.7.

From Eq. (17.59) the braking torque is

$$T = r(F_1 - F_2) = 4(300 - 116.9) = 732.4 \text{ lbf-in.}$$

Summing the moments about the hinge pin and setting the sum equal to zero give

$$-d_9 W + d_8 F_2 - d_{10} F_1 = 0$$

Solving for the actuating force W gives

$$W = \frac{d_8 F_2 - d_{10} F_1}{d_9} = \frac{2(116.9) - (0.5)(300)}{9} = 9.311 \text{ lbf}$$

If $W = 0$, the brake will self-lock.

$$d_8 F_2 - d_{10} F_1 = 0$$

$$d_{10} = \frac{d_8 F_2}{F_1} = \frac{2(116.9)}{300} = 0.7793 \text{ in.}$$

The brake will self-lock if $d_{10} \geq 0.7793$ in.

17.9 SLIP CLUTCHES

A clutch will often be used as a torque-limiting device, usually to prevent machinery damage in a malfunction or undesired event. A **slip clutch** consists of two surfaces held together by a constant force so that they slip when a preset level of torque is applied to them. Slip clutches come in a wide variety of sizes but are very compact. They are designed to be actuated only rarely and thus the friction elements do not need to be sized for wear. Also, slip clutches are almost always contacting disks, mainly because it is imperative to prevent the possibility of a self-energizing shoe (which would compromise the torque-limiting control).

Slip clutches are mainly used to protect machinery elements and are not relied on for accident prevention. After slip clutches spend long time periods in contact, the friction surfaces can stick or weld together, requiring a larger torque to initiate slippage. This torque is usually not large enough to break a gear, for example, but can be enough of an increase to result in a serious injury. Regardless, a slip clutch is an effective torque limiter and can be used in lieu of shear pins or keyways with the advantage that no maintenance is required after the excessive torque condition has been corrected.

17.10 TEMPERATURE CONSIDERATIONS

As mentioned in Sec. 17.1 thermal effects are important in braking and clutch systems. If temperatures become too high, damage to components could result, which could compromise the useful life or performance of brake and clutch systems.

Predicting temperatures of brake and clutch systems is extremely difficult in practice because they are operated under widely varying conditions. The first law of thermodynamics requires that

$$Q_{\text{friction}} = Q_{\text{conduction}} + Q_{\text{convection}} + Q_{\text{storage}}$$

where

Q_{friction} = energy input into brake or clutch system from friction between sliding elements

$Q_{\text{conduction}}$ = heat lost from conduction through machinery elements

$Q_{\text{convection}}$ = heat lost from convection to surrounding environment

Q_{storage} = energy stored in brake and clutch components, resulting in temperature increase

If conduction and convection are negligible, the temperature rise in the brake or clutch material is given by

$$\Delta t_m = \frac{Q_{\text{friction}}}{C_p m_a}$$

where

C_p = specific heat of material, J/kg-°C

m_a = mass, kg

This equation is useful for determining the instantaneous temperature rise in a brake or clutch pad, since the frictional energy is dissipated directly on the contacting surfaces and does not have time to be conducted or conveyed. Brake pads and clutches usually have an area in contact that then moves out of contact and can cool. The maximum operating temperature is a complicated function of heat input and cooling rates.

The main difficulty in predicting brake system temperatures is that the heat conducted and the heat convected depend on the machine ambient temperature and the brake or clutch geometry and vary widely. In previous circumstances in this text a worst-case analysis would be performed, which in this case quickly reduces to circumstances where brakes and clutches become obviously overheated. This result is not incorrect: most brake and clutch systems are overheated when abused and can sustain serious damage as a result. The alternative is to use such massive brake and clutch systems as to make the economic burden unbearable to responsible users. It is far more reasonable to use brake systems that require periodic maintenance and can be damaged through abuse than to incur the economic costs of surviving worst-case analyses. This differs from previous circumstances, where a worst-case analysis still resulted in a reasonable product.

Some clutches are intended to be used with a fluid (wet clutches) to aid in cooling the clutch. Similarly, some pads or shoes will include grooves so that air or fluid can be better entrained, resulting in increased heat convection from the clutch or brake. Predicting brake temperatures is a complex problem and requires numerical, usually finite element, methods.

Obviously, the proper size of brake components is extremely difficult to determine with certainty. For the purposes of this text the values given by Juvinall and Marshek

Table 17.3 Product of contact pressure and sliding velocity for brakes and clutches. [*Fundamentals of Machine Component Design* by Juvinall and Marshek. Copyright ©1991 by Juvinall and Marshek. Reprinted by permission of John Wiley & Sons.]

	pu	
Operating condition	**(kPa)(m/s)**	**(psi)(ft/min)**
Continuous; poor heat dissipation	1050	30 000
Occasional; poor heat dissipation	2100	60 000
Continuous; good heat dissipation as in oil bath	3000	85 000

(1991) for the product of brake shoe or pad contact pressure and sliding velocity *pu* can be used to estimate component sizes (Table 17.3). Most manufacturers rely heavily on experimental verification of designs; the application of these numbers in the absence of experimental verification requires extreme caution.

Case Study | SELECTION OF CLUTCH FOR MOBILE HYDRAULIC CRANE

GIVEN Mobile hydraulic cranes are extremely common and valuable construction and material-handling vehicles. This case study illustrates the selection of a brake for one of the crane's functions. A crane can perform a number of different functions, such as hoisting (by reeling in the hoist line), elevating or lowering the boom (by reeling in the gantry line), and rotating. Each of these functions requires its own control system (i.e., brakes and clutches). This case study is restricted to a tractable problem, choosing a brake for the hoist line.

Crane capacities are given in terms of the maximum load they can lift while hoisting, which occurs when the boom is at maximum elevation. At lower elevations the crane will tip over at much lower loads, so the crane capacity is the true maximum load which will be experienced by the crane. This problem will deal with clutch selection for a 30-ton crane, which has a 12-in.-diameter drum for the cable. The drum diameter has been chosen to give reasonable wire rope service life, a topic further explored in Chapter 18.

FIND Select a brake for the hoist line.

Solution

A large crane achieves its full capacity only if multiple-part hoist lines are used. That is, the hoist line is run through multiple sheaves so that each part supports a fraction of the total load. A 30-ton crane may or may not use multiple-part hoist lines. A worst-case analysis suggests that the brake must withstand the entire hoist line force.

It is imperative that the crane be able to hold a load that has been hoisted. Also, dynamic loads are possible. For example, in dam construction wooden forms are placed, concrete is poured, and the forms are then removed by attaching the crane's hoist line and prying the forms off the now-cured concrete. If the hoist line has insufficient tension, the form will drop a certain distance until the wire rope is taut, thereby causing a dynamic load to be applied to the hoist line. For this reason an application factor of 1.5 will be used. In most cases this allows for large safety factors, since the hoist line is rarely supporting the crane's capacity because of tipping as discussed above.

The type of brake to be used is a long-shoe, internal, expanding rim type. A brake used for a swing mechanism on a similar crane is shown in the chapter's opening illustration. The brake used for this application is similar and is sketched in Fig. 17.14. The reasons for using such a brake are multiple and include the following:

1. Space is really not an issue; the crane superstructure has sufficient room for any kind of brake discussed in this chapter.

(continued)

Case Study (CONTINUED)

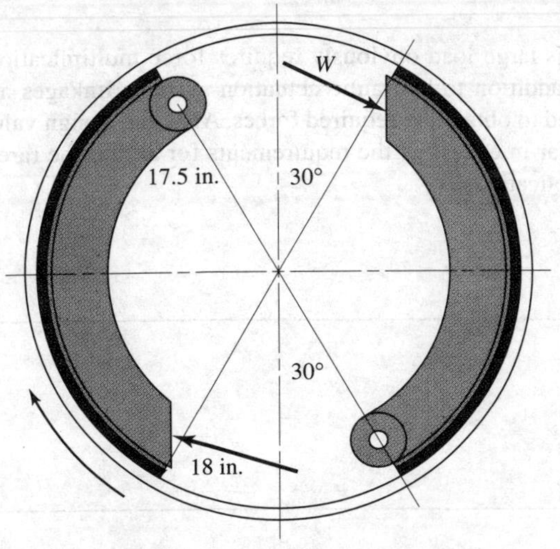

Figure 17.14 Hoist line brake for mobile hydraulic crane. The cross section of brake with relevant dimensions is given.

2. To maximize maintenance-free service life, drum or disk brakes are the most obvious candidates. This statement is not meant to imply that long lives are not possible from band brakes or the like, merely that drum or disk brakes have long lives with easily replaceable shoes and pads.
3. Self-actuating behavior is desirable in this case, so a drum brake with an internal shoe is chosen.
4. A long-shoe brake efficiently uses the braking material, since little pressure occurs near the hinge pin.

Although a crane manufacturer will occasionally manufacture its own brake components, usually design work is performed in collaboration with a component supplier. The sizes of components are selected based on previous design experience for crane applications and on stress analyses using relations like the Archard wear law. Such a design synthesis problem has been discussed in previous case studies.

In this design a mechanical spring normally would apply the brakes, and the actuating cylinder would force the shoes away from the drum. Thus, in actuating the brakes, an operator really is disengaging the hydraulic cylinder. The reason for this approach is that it is imperative that loads not drop in the event of a power failure or if all the crane's fuel is consumed. The brake is a fail-safe system in this respect. Further, through proper linkage design both shoes can be self-energizing, limiting the required force multiplication in the hydraulic actuation system.

The design problem becomes one of specifying the actuating force required to obtain the desired braking torque on the cable drum. This actuating force will then be achieved through proper hydraulic multiplication of the operator's exertions on the hoist control. The required torque is given by

$$T = 1.5F\left(\frac{D}{2}\right) = 1.5(60\ 000)(6)$$
$$= 540\ 000 \text{ lbf-in.}$$

Because both shoes are self-energizing, each shoe must develop one-half the required torque. Using Eq. (17.38) gives

$$270\ 000 = \frac{\mu p_{\max} b r^2 (\cos\theta_1 - \cos\theta_2)}{\sin\theta_a}$$

From the defined geometry the maximum pressure occurs at $\theta_a = 90°$ where $\theta_1 = 30°$ and $\theta_2 = 150°$. Similar cranes have brake shoes approximately 8 in. wide, so $b = 8$ in. The drum radius has been specified as 18 in., although in a general design problem this dimension could be altered to ensure proper performance. Table 17.1 gives the coefficients of friction and the maximum temperatures and pressures for a wide variety of friction materials. For applications such as this mobile crane sintered powder metal brake components are usually used because they allow the largest pressures and are most forgiving in terms of local heating. Therefore, recognizing that the brake lining is a sintered metal, $\mu = 0.25$ will be used. The coefficient of friction can be smaller than 0.25, but for the brake components used in craning this is a reasonable lower bound. Equation (17.38) then yields, upon rearranging,

$$p_{\max} = \frac{270\ 000}{(0.25)(8)(18)^2(\cos 30° - \cos 150°)}$$
$$= 240.6 \text{ psi}$$

(continued)

Case Study (CONCLUDED)

The maximum pressure is also a reasonable value, as can be seen from Table 17.1. The actuating force is given by Eq. (17.37). Thus,

$$M_P = \frac{brd_7 p_{\max}}{4}\left[2(\theta_2 - \theta_1)\frac{\pi}{180} - \sin 2\theta_2 + \sin 2\theta_1\right]$$
$$= 895\ 500 \text{ lbf-in.}$$

$$M_F = \mu p_{\max} br\left[-r(\cos\theta_2 - \cos\theta_1) - \frac{d_7}{2}(\sin^2\theta_2 - \sin^2\theta_1)\right]$$
$$= 270\ 000 \text{ lbf-in.}$$

$$W = \frac{M_P - M_F}{d_6} = 20\ 700 \text{ lbf}$$

This large load obviously requires force multiplication; in addition to hydraulic actuation systems linkages are used to obtain the required forces. Also, the design value is far in excess of the requirements for all but the rarest applications.

17.11 SUMMARY

This chapter focused on two machine elements, clutches and brakes, that are associated with rotation and have the common function of storing and/or transferring rotating energy. In analyzing the performance of clutches and brakes the actuating force, the torque transmitted, and the reaction force at the hinge pin were the major focuses of this chapter. The torque transmitted is related to the actuating force, the coefficient of friction, and the geometry of the clutch or brake. This is a problem in statics, where different geometries were studied separately.

For clutches two theories were studied: the uniform pressure model and the uniform wear model. It was found that for the same dimensionless torque the uniform wear model requires a larger radius ratio than does the uniform pressure model for the same maximum pressure. This larger radius ratio implies that a larger area is needed for the uniform wear model. Thus, the uniform wear model was viewed as a safer approach.

KEY WORDS

band brake brake that uses contact pressure of flexible band against outer surface of drum

brake device used to bring moving system to rest through dissipation of energy to heat by friction

clutches power transfer devices that allow coupling and decoupling of shafts

cone disk brake or clutch that uses shoes pressed against convergent surface of cone

deenergizing brake or clutch shoe where frictional moment hinders engagement

rim type brake or clutch that uses internal shoes which expand onto inner surface of drum

self-energizing brake or clutch shoe where frictional moment assists engagement

slip clutch clutch where maximum transferred torque is limited

thrust disk brake or clutch that uses flat shoes pushed against rotating disk

Recommended Readings

Army (1976) *Analysis and Design of Automotive Brake Systems*, U.S. Department of the Army Manual DARCOM–P706–358, Alexandria, VA.

Baker, A.K. (1986) *Vehicle Braking*, Pentech Press, London.

Crouse, W.H., and Anglin, D.L. (1983) *Automotive Brakes, Suspension and Steering*, 6th ed., McGraw-Hill, New York.

Krutz, G.W., Schuelle, J.K, and Claar, P.W. (1994) *Machine Design for Mobile and Industrial Applications*, Society of Automotive Engineers, Warrendale, PA.

Monroe, T. (1977) *Clutch and Flywheel Handbook*, H.P. Books, Tucson, AZ.

Mott, R.L. (1985) *Machine Elements in Mechanical Design*, Merrill Publishing Co., Columbus, OH.

Norton, R.L. (1996) *Machine Design*, Prentice-Hall, Englewood Cliffs, NJ.

Orthwein, W.C. (1986) *Clutches and Brakes: Design and Selection*. Marcel Dekker, New York.

Shigley, J.E., and Mischke, C.R. (1989) *Mechanical Engineering Design*, McGraw-Hill, New York.

Reference

Juvinall, R.C., and Marshek, K.M. (1991) *Fundamentals of Machine Component Design*, Wiley, New York.

Problems

Section 17.2

17.1 The disk brake shown in sketch *a* has brake pads in the form of circular sections with inner radius r, outer radius $2r$, and section angle $\pi/4$. Calculate the brake torque when the pads are applied with normal force P. The brake is worn in so that pu is constant, where p is the contact pressure and u is the sliding velocity. The coefficient of friction is μ.

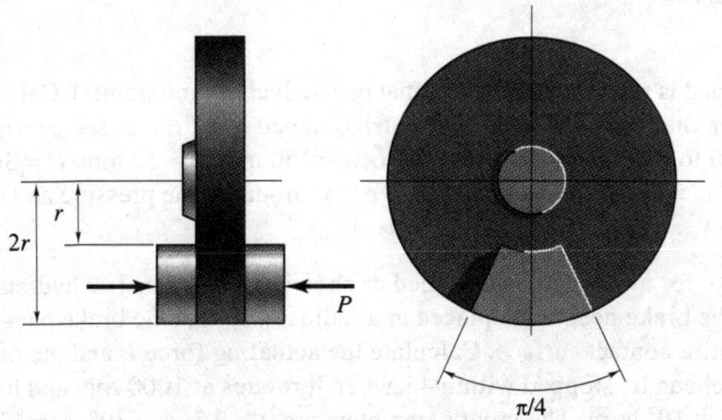

Sketch *a*, for Problem 17.1

17.2 An automotive clutch with a single friction surface is to be designed with a maximum torque of 140 N-m. The materials are chosen such that $\mu = 0.35$ and $p_{max} = 0.35$ MPa. Use safety factor $n_s = 1.3$ with respect to slippage at full engine torque and as small an outside diameter as possible. Determine appropriate values of r_o, r_i, and P by using both the uniform pressure and uniform wear models.

17.3 The brakes used to stop and turn a tank are built like a multiple-disk clutch with three loose disks connected through splines to the drive shaft and four flat rings connected to the frame of the tank. The brake has an outer contact diameter of 600 mm, an inner contact diameter of 300 mm, and six surface contacts. The wear of the disks is proportional to the contact pressure multiplied by the sliding distance. The coefficient of friction of the brake is 0.12, and the friction between the caterpillar and the ground is 0.16, which gives a braking torque of 12 800 N-m needed to block one caterpillar track so that it slides along the ground. Calculate the force needed to press the brake disks together to block one caterpillar track. Also calculate the force when the brake is new.

17.4 A disk brake used in a printing machine is designed as shown in sketch *b*.

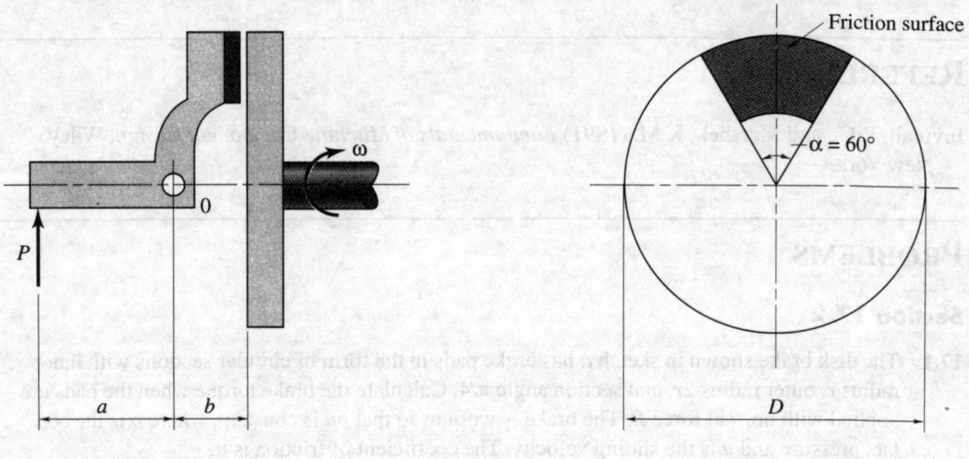

Sketch *b*, for Problem 17.4

The brake pad is mounted on an arm that can swivel around point 0. Calculate the braking torque when the force $P = 5000$ N. The friction pad is a circular sector with the inner radius equal to half the outer radius. Also, $a = 150$ mm, $b = 50$ mm, $D = 300$ mm, and $\mu = 0.25$. The wear of the brake lining is proportional to the pressure and the sliding distance.

17.5 A disk brake for a flywheel is designed as shown in sketch *c*. The hydraulic pistons actuating the brake need to be placed at a radius r_p so that the brake pads wear evenly over the entire contact surface. Calculate the actuating force P and the radius so that the flywheel can be stopped within 4 s when it rotates at 1000 rpm and has a kinetic energy of 5×10^5 N-m. The input parameters are $\mu = 0.3$, $\alpha = 30°$, $r_o = 120$ mm, $r_i = 60$ mm.

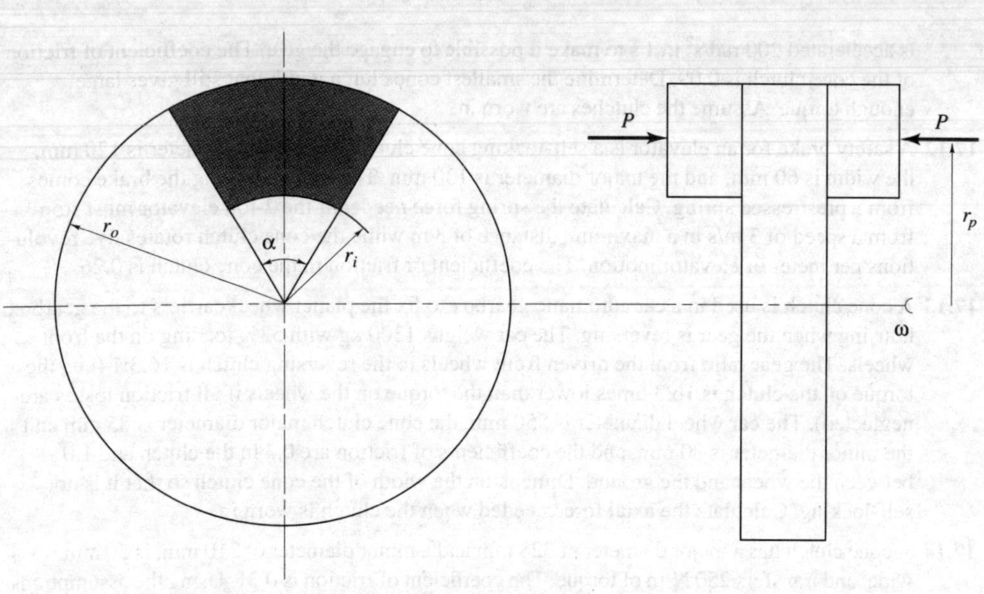

Sketch c, for Problem 17.5

17.6 Three pairs of thrust disk clutches are mounted on a shaft. Each has a pair of frictional surfaces. The hardened-steel clutches are identical, with an inside diameter of 100 mm and an outside diameter of 245 mm. What is the torque capacity of these clutches based on a) uniform wear and b) uniform pressure?

17.7 A pair of disk clutches has an inside diameter of 250 mm and an outside diameter of 420 mm. A normal force of 18.5 kN is applied and the coefficient of friction of the contacting surfaces is 0.215. Using the uniform wear and uniform pressure assumptions determine the maximum pressure acting on the clutches. Which of these assumptions would produce results closer to reality?

17.8 A disk clutch is made of cast iron and has a maximum torque of 210 N-m. Because of space limitations the outside diameter must be minimized. Using the uniform wear assumption and a safety factor of 1.3, determine

a) The inner and outer radii of the clutch
b) The maximum actuating force needed

17.9 A disk clutch produces a torque of 125 N-m and a maximum pressure of 315 kPa. The coefficient of friction of the contacting surfaces is 0.28. Assume a safety factor of 1.8 and design the smallest disk clutch for the above constraints. What should the actuating force be?

Section 17.3

17.10 A leather-face cone clutch is to transmit 1200 lbf-in. of torque. The half-cone angle $\alpha = 10°$, the mean diameter of the friction surface is 12 in., and face width $b = 2$ in. For coefficient of friction $\mu = 0.25$ find the normal force P and the maximum contact pressure p by using both the uniform pressure and uniform wear models.

17.11 The synchronization clutch for the second gear of a car has a major cone diameter of 50 mm and a minor diameter of 40 mm. When the stick shift is moved to second gear, the synchronized clutch is engaged with an axial force of 100 N, and the moment of inertia of 0.005 kg-m^2

is accelerated 200 rad/s^2 in 1 s to make it possible to engage the gear. The coefficient of friction of the cone clutch is 0.09. Determine the smallest cone clutch width that still gives large enough torque. Assume the clutches are worn in.

17.12 A safety brake for an elevator is a self-locking cone clutch. The minor diameter is 120 mm, the width is 60 mm, and the major diameter is 130 mm. The force applying the brake comes from a prestressed spring. Calculate the spring force needed if the 2-ton elevator must stop from a speed of 3 m/s in a maximum distance of 3 m while the cone clutch rotates five revolutions per meter of elevator motion. The coefficient of friction in the cone clutch is 0.26.

17.13 A cone clutch is used in a car automatic gearbox to fix the planet wheel carriers to the gearbox housing when the gear is reversing. The car weighs 1300 kg with 53% loading on the front wheels. The gear ratio from the driven front wheels to the reversing clutch is 16.3:1 (i.e., the torque on the clutch is 16.3 times lower than the torque on the wheels if all friction losses are neglected). The car wheel diameter is 550 mm, the cone clutch major diameter is 85 mm and the minor diameter is 80 mm, and the coefficients of friction are 0.3 in the clutch and 1.0 between the wheel and the ground. Dimension the width of the cone clutch so that it is not self-locking. Calculate the axial force needed when the clutch is worn in.

17.14 A cone clutch has a major diameter of 328 mm and a minor diameter of 310 mm, is 50 mm wide, and transfers 250 N-m of torque. The coefficient of friction is 0.31. Using the assumptions of uniform pressure and uniform wear, determine the actuating force and the contact pressure.

17.15 The coefficient of friction of a cone clutch is 0.25. It can support a maximum pressure of 410 kPa while transferring a maximum torque of 280 N-m. The width of the clutch is 65 mm. Minimize the major diameter of the clutch. Determine the clutch dimension and the actuating force.

Section 17.4

17.16 A block brake is used to stop and hold a rope used to transport skiers from a valley to the top of a mountain. The distance between cars used to transport the skiers is 100 m, the length of the rope from the valley to the top of the mountain is 4 km, and the altitude

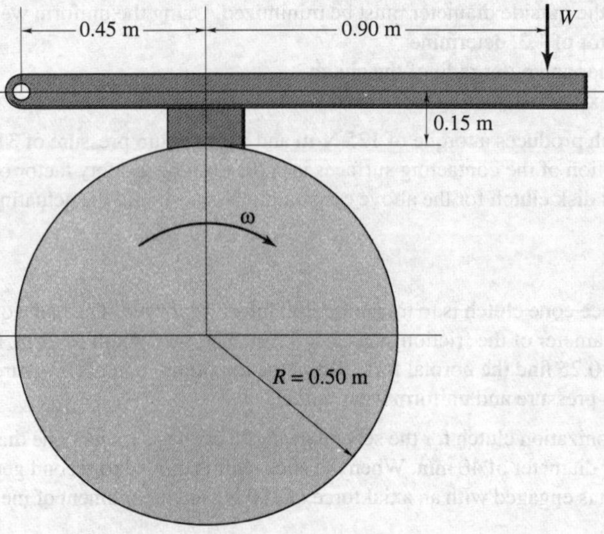

Sketch *d*, for Problem 17.16

difference is 1.4 km. The rope is driven by a V-groove wheel with a diameter of 2 m. The rope is stopped and held with a block brake mounted on the shaft of the V-groove wheel, shown in sketch *d*. Neglect all friction in the different parts of the ropeway except the friction in the driving sheave, and assume that the slope of the mountain is constant. Dimension the brake for 20 passengers with each passenger's equipment weighing 100 kg, and assume that all ropeway cars descending from the top are empty of people. The direction of rotation for the drive motor is shown in the figure. Calculate the braking force W needed to hold the ropeway still if all passengers are on their way up. Do the same calculation if all passengers continue on down to the valley with the ropeway. The coefficient of friction in the brake is 0.23.

17.17 The motion of an elevator is controlled by an electric motor and a block brake. On one side the rotating shaft of the electric motor is connected to the gearbox driving the elevator, and on the other side it is connected to the block brake. The motor has two magnetic poles and can be run on either 60- or 50-Hz electricity (3600 or 3000 rpm). When the elevator motor is driven by 50-Hz electricity, the braking distance needed to stop is 52 cm when going down and 31 cm when going up with maximum load in the elevator. To use it with 60-Hz electricity and still be able to stop exactly at the different floor levels without changing the electric switch positions, the brake force at the motor should be changed. How should it be changed for going up and for going down? The brake geometry is like that shown in Fig. 17.5 with $d_1 =$ 0.030 m, $d_3 = 0.100$ m, $d_4 = 0.400$ m, $r = 0.120$ m, and $\mu = 0.20$. Only the inertia of the elevator must be considered, not the rotating parts. Make the brake self-energizing when the elevator is going down.

17.18 An anchor winding is driven by an oil hydraulic motor with a short-shoe brake to stop the anchor machinery from rotating and letting out too much anchor chain when the wind moves the ship. The maximum force transmitted from the anchor through the chain is 1.1 MN at a radius of 2 m. Figure 17.5 describes the type of block brake used, which is self-energizing. Calculate the brake force W needed when the the brake dimensions are $d_1 = 0.9$ m, $d_3 = 1.0$ m, $d_4 = 6$ m, $r = 3$ m, and $\mu = 0.31$. Also calculate the contact force between the brake shoe and the drum.

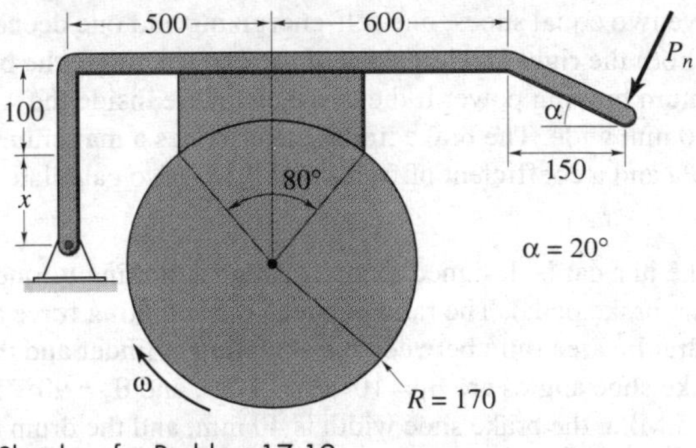

Sketch e, for Problem 17.19

17.19 The hand brake shown in sketch *e* has an average pressure of 600 kPa across the shoe and is 50 mm wide. The wheel runs at 150 rpm and the coefficient of friction is 0.25. Dimensions are in millimeters. Determine the following:

a) If $x = 150$ mm, what should the actuating force be?
b) What value of x causes self-locking?
c) What torque is transferred?
d) If the direction of rotation is reversed, how would the answers to parts a to c change?

17.20 The short-shoe brake shown in sketch *f* has an average pressure of 1 MPa and a coefficient of friction of 0.32. The shoe is 250 mm long and 45 mm wide. The drum rotates at 310 rpm and has a diameter of 550 mm. Dimensions are in millimeters.

a) Obtain the value of x for the self-locking condition
b) Calculate the actuating force if $x = 275$ mm
c) Calculate the braking torque
d) Calculate the reaction at point A

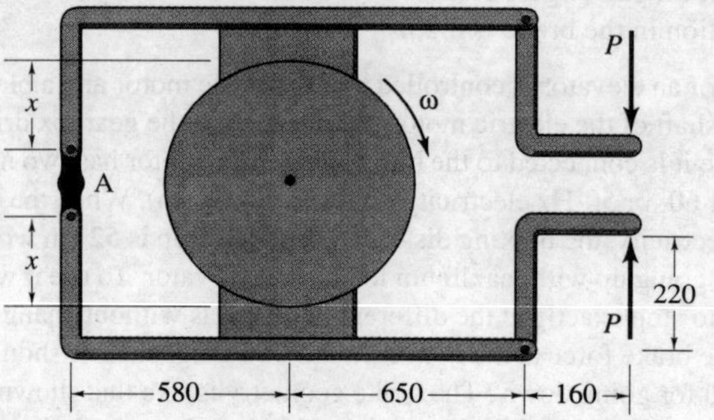

Sketch *f*, for Problem 17.20

Section 17.5

17.21 The brake on the rear wheel of a car is the long-shoe internal type. The brake dimensions according to Fig. 17.7 are $\theta_1 = 10°$, $\theta_2 = 120°$, $r = 95$ mm, $d_7 = 73$ mm, $d_6 = 120$ mm, and $d_5 =$ 30 mm. The brake shoe lining is 38 mm wide, and the maximum allowable contact pressure is 5 MPa. Calculate the braking torque and the fraction of the torque produced from each brake shoe when the brake force is 5000 N. Also calculate the safety factor against contact pressure that is too high.

17.22 The maximum volume of the long-shoe internal brake on a car is given as 10^{-3} m^3. The brake should have two equal shoes, one self-energizing and one deenergizing, so that the brake can fit on both the right and left sides of the car. Calculate the brake width and radius for maximum braking power if the space available inside the wheel is 400 mm in diameter and 100 mm wide. The brake lining material has a maximum allowed contact pressure of 4 MPa and a coefficient of friction of 0.18. Also calculate the maximum braking torque.

17.23 A long-shoe brake in a car is designed to give as high a braking torque as possible for a given force on the brake pedal. The ratio between the actuating force and the pedal force is given by the hydraulic area ratio between the actuating cylinder and the cylinder under the pedal. These brake shoe angles are $\theta_1 = 10°$, $\theta_2 = 170°$, and $\theta_a = 90°$. The maximum brake shoe pressure is 5 MPa, the brake shoe width is 40 mm, and the drum radius is 100 mm. Find the distance d_7 that gives the maximum braking power for a coefficient of friction of 0.2 at any pedal force. What braking torque would result if the coefficient of friction were 0.25?

17.24 A long-shoe internal brake similar to that shown in Fig. 17.7 should be optimized for maximum wear life. The brake lining wear is proportional to the sliding distance times the contact pressure times the coefficient of friction. The actuating force, applied by a hydraulic cylinder, is equally large for the two shoes. The contact pressure is limited to 4 MPa. The two brake shoes are made geometrically equal to be able to use the brake equally well for both directions of rotation, but the coefficient of friction might be different. The brake is 120 mm in radius and 30 mm wide. The brake torque needed is 900 N-m. Assume that $\theta_1 = 20°$, $\theta_2 = 160°$, and $d_7 = 85$ mm.

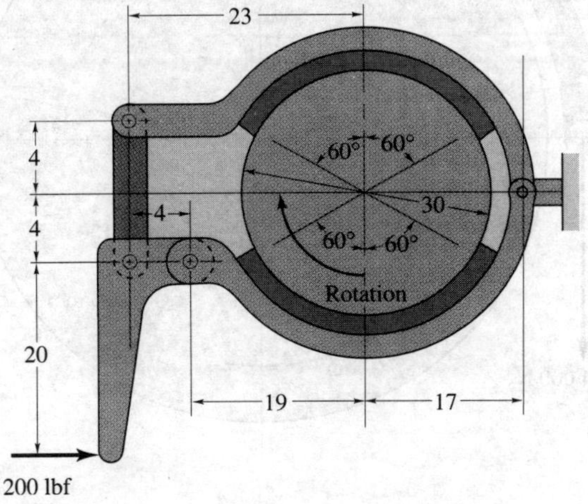

Sketch *g*, for Problem 17.25

Section 17.6

17.25 An external drum brake assembly (see sketch *g*) has a normal force $P = 200$ lbf acting on the lever. Dimensions are in inches. Assume that coefficient of friction $\mu = 0.25$ and maximum contact pressure $p = 100$ psi. Determine the following from long-shoe calculations:

a) Free-body diagram with the directionality of the forces acting on each component
b) Which shoe is self-energizing and which is deenergizing
c) Total braking torque
d) Pad width as obtained from the self-energizing shoe (deenergizing shoe width equals self-energizing shoe width)
e) Pressure acting on the deenergizing shoe

17.26 A long-shoe external brake as shown in Fig. 17.10 has a pivot point such that $d_7 = 4r$, $d_6 = 2r$, $\theta_1 = 5°$, and $\theta_2 = 45°$. Find the coefficient of friction needed to make the brake self-lock if the rotation is in the direction shown in Fig. 17.10. If the shaft rotates in the opposite direction, calculate the drum radius needed to get a braking torque of 180 N-m for the actuating force of 10 000 N.

17.27 An external, long brake shoe is mounted on an elastic arm. When a load is applied, the arm and the brake lining bend and redistribute the pressure. Instead of the normal sine pressure distribution the pressure becomes constant along the length of the lining. For a given actuating force calculate how the brake torque changes when the pressure distribution changes from sinusoidal to a constant pressure. Also assume that $d_7 = 110$ mm, $r = 90$ mm, $b = 40$ mm, $\theta_1 = 20°$, $\theta_2 = 160°$, $d_6 = 220$ mm, $\mu = 0.25$, and $W = 12$ kN.

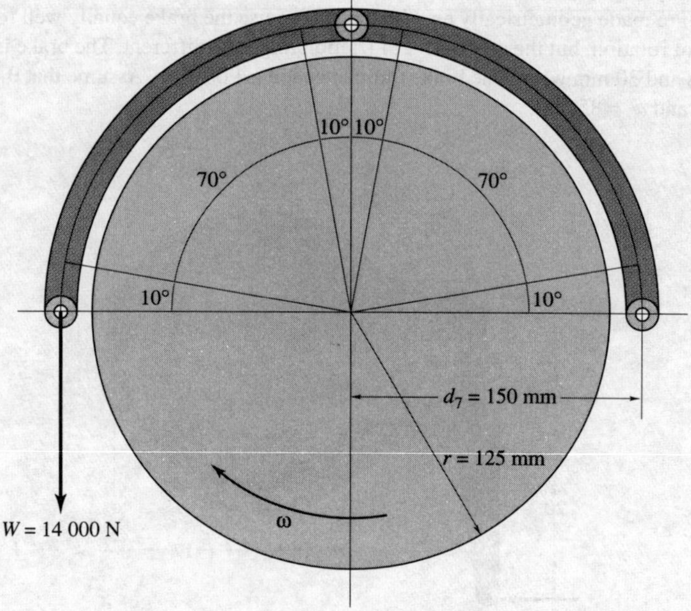

Sketch *h*, for Problem 17.28

17.28 A long-shoe external brake has two identical shoes coupled in series so that the peripheral force from the first shoe is directly transferred to the second shoe. No radial force is transmitted between the shoes. Each of the two shoes covers 90° of the circumference, and the brake linings cover the central 70° of each shoe, leaving 10° at each end without lining as shown in sketch *h*. The actuating force is applied tangentially to the brake drum at the end of the loose shoe, 180° from the fixed hinge point of the other shoe. Calculate the braking torques for both rotational directions when $d_7 = 150$ mm, $r = 125$ mm, $b = 50$ mm, $W = 14\ 000$ N, and $\mu = 0.2$. Also show a free-body diagram of these forces acting on the two shoes.

17.29 A special type of brake is used in a car factory to hold the steel panels during drilling operations so that the forces from the drill bits cannot move the panels. The brake is shown in sketch *i*. Calculate the braking force P_B on the steel panel when it moves to the right with the speed u_b between drilling operations. The actuating force is P_M. The brake lining is thin relative to the other dimensions.

17.30 Redo Problem 17.19 for long-shoe assumptions. The average contact pressure occurs at 40°. Determine the maximum contact pressure and its location. Assume that the distance *x* is

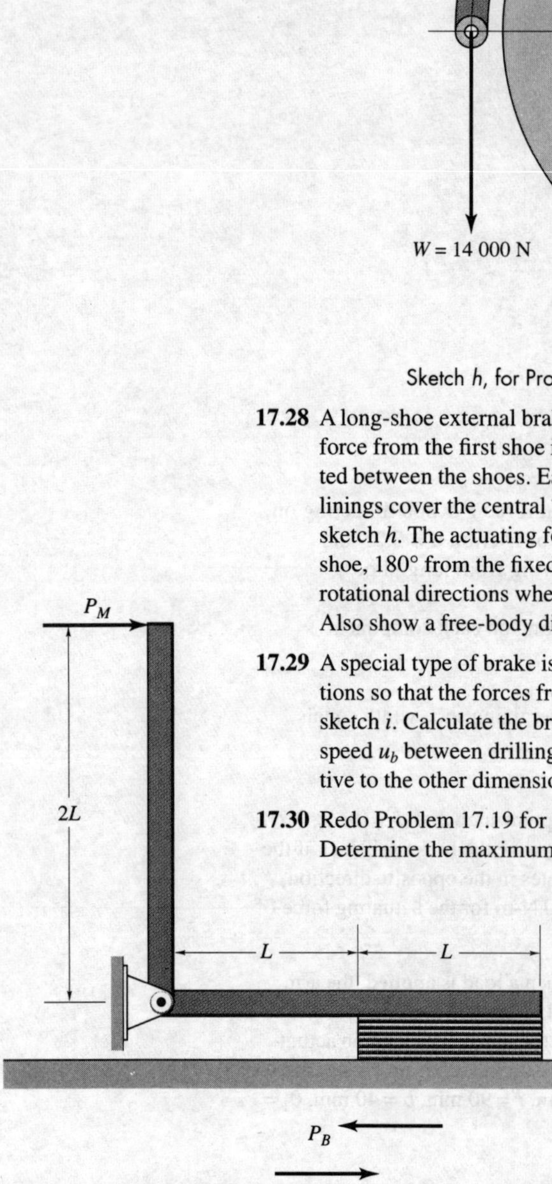

Sketch *i*, for Problem 17.29

150 mm. What is the braking torque? Also, repeat this problem while reversing the direction of rotation. Discuss the changes in the results.

Section 17.7

17.31 A symmetrically loaded, pivot-shoe brake has a wrap angle of 180° and the optimum distance d_7, giving a symmetrical pressure distribution. The coefficient of friction of the brake lining is 0.30. A redesign is considered that will increase the breaking torque without increasing the actuating force. The wrap angle is decreased to 80° (+40° to –40°) and the d_7 distance is decreased to still give a symmetrical pressure distribution. How much does the brake torque change?

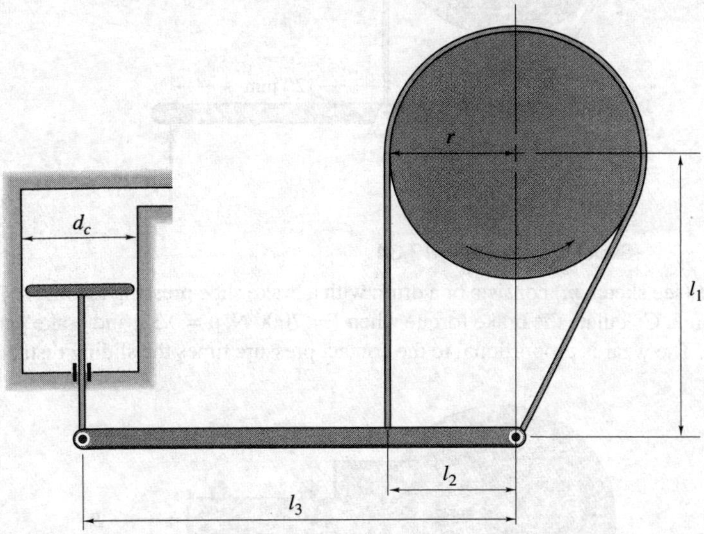

Sketch *j*, for Problem 17.32

Section 17.8

17.32 The band brake shown in sketch *j* is activated by a compressed-air cylinder with diameter d_c. The brake cylinder is driven by air pressure $p = 0.7$ MPa. Calculate the maximum possible brake moment if the coefficient of friction between the band and the drum is 0.25. The mass force on the brake arm is neglected, $d_c = 50$ mm, $r = 200$ mm, $l_1 = 500$ mm, $l_2 = 200$ mm, and $l_3 = 500$ mm.

17.33 The band brake shown in sketch *k* has wrap angle $\phi = 225°$ and cylinder radius $r = 80$ mm. Calculate the brake torque when the lever is loaded by 100 N and coefficient of friction $\mu = 0.3$. How long is the braking time from 1200 rpm if the rotor mass moment of inertia is 2.5 kg-m^2?

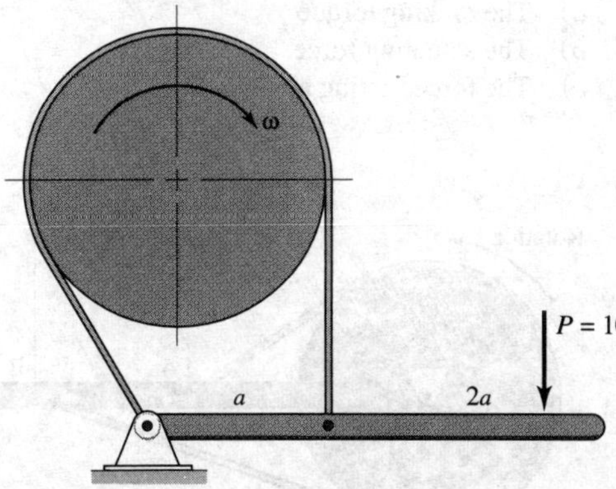

Sketch *k*, for Problem 17.33

17.34 The band brake shown in sketch *l* has wrap angle $\phi = 215°$ and cylinder radius $r = 60$ mm. Calculate the brake torque when coefficient of friction $\mu = 0.25$. How long is the braking time from 1500 rpm if the rotor moment of inertia is $J = 2$ kg-m^2?

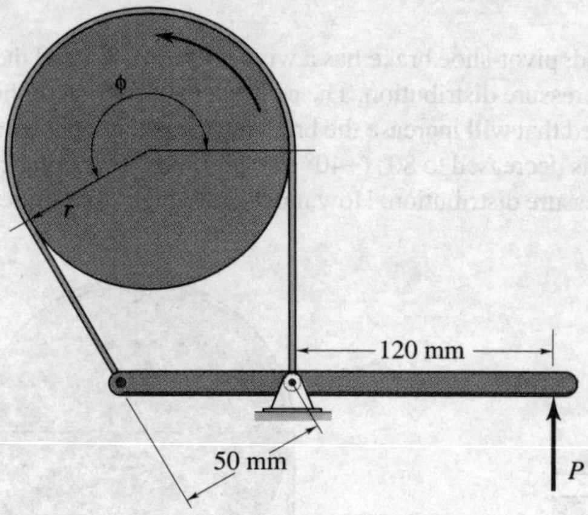

Sketch *l*, for Problem 17.34

17.35 A brake (see sketch *m*) consists of a drum with a brake shoe pressing against it. Drum radius $r = 80$ mm. Calculate the brake torque when $P = 7000$ N, $\mu = 0.35$, and brake pad width $b =$ 40 mm. The wear is proportional to the contact pressure times the sliding distance.

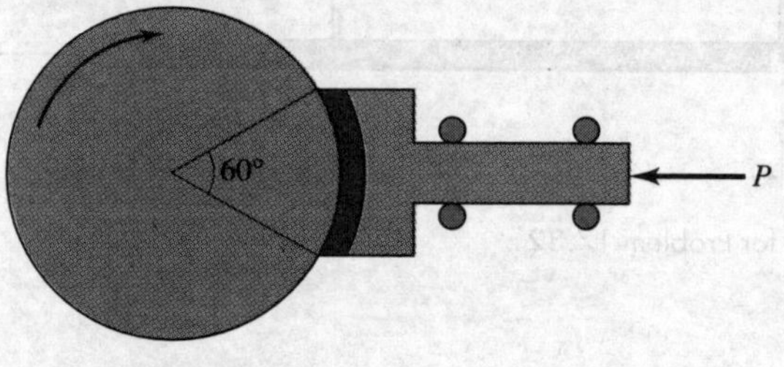

Sketch *m*, for Problem 17.35

17.36 For the band brake shown in sketch *n* the following conditions are given: $d = 350$ mm, $p_{max} =$ 1.2 MPa, $\mu = 0.25$, and $b = 50$ mm. All dimensions are in millimeters. Determine the following:

a) The braking torque
b) The actuating force
c) The forces acting at hinge 0

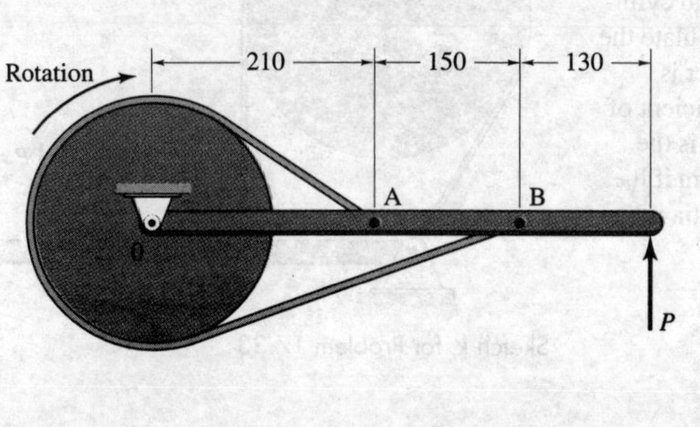

Sketch *n*, for Problem 17.36

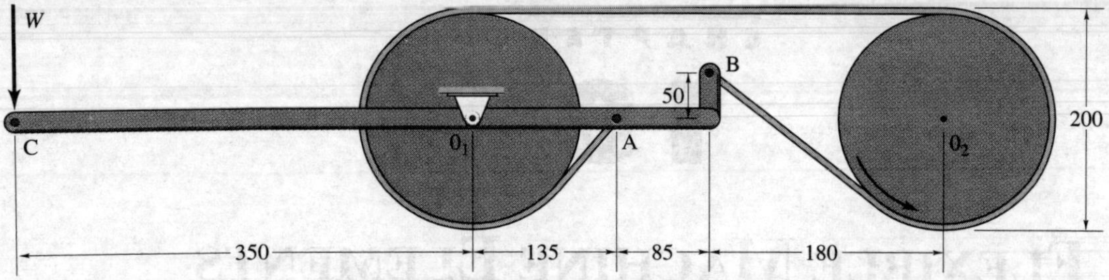

Sketch o, for Problem 17.37

17.37 The band brake shown in sketch *o* is 40 mm wide and can take a maximum pressure of 1.1 MPa. All dimensions are in millimeters. The coefficient of friction is 0.3. Determine the following:

a) The maximum allowable actuating force

b) The braking torque

c) The reaction supports at 0_1 and 0_2

d) Whether it is possible to change the distance 0_1A in order to have self-locking. Assume point A can be anywhere on line $C0_1$A.

CHAPTER

18

FLEXIBLE MACHINE ELEMENTS

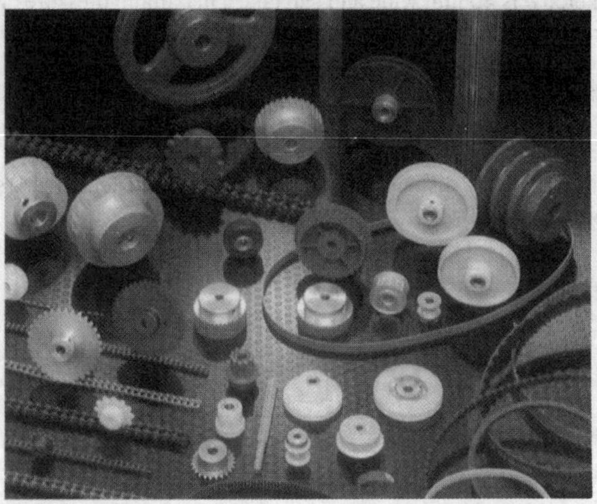

An assortment of chains and belts and their associated sprockets and pulleys. (Courtesy of Sterling Instrument Co.)

Scientists study the world as it is, engineers create the world that never has been.

Theodore von Karman

SYMBOLS

A	$\dfrac{L}{p_t} - \dfrac{N_1 + N_2}{2}$
A_m	cross-sectional area of metal strand in rope, m^2
a	link plate thickness, m
a_1	service factor
a_2	multiple-strand factor in rolling chain
B	$(N_2 - N_1)/2\pi$
c	distance from neutral axis to outer fiber, m
c_d	center distance, m
D	sheave or pulley diameter, m
d	diameter, m
d_w	wire diameter, m
E	modulus of elasticity, Pa
F	friction force, N
F_a	force due to acceleration, N
F_c	centrifugal force, N
F_f	fatigue allowable force, N
F_i	initial tensile force, N
F_r	rope weight, N
F_t	total friction force, N
F_w	deadweight, N
F_1	driver friction force, N
F_2	driven friction force, N
f_1	overload service factor

f_2	extra overload factor
G	gantry line tension, N
g	gravitational acceleration, 32.2 ft/s^2 = 9.807m/s^2
g_r	velocity ratio
H_B	Brinell hardness
h_p	power transmitted, W
h_{pr}	rated power transmitted, W
h_t	V-belt height, m
I	area moment of inertia, m^4
L	belt length, m
l_p	hoist rope length, m
M	bending moment, N-m
m	mass, kg
m'	mass per unit length, kg/m
N	number of teeth in sprocket
N_a	speed of revolution, rpm
n_s	safety factor
p	bearing pressure, Pa
p_{all}	allowable bearing pressure, Pa
p_t	pitch, m
r	sheave or pulley radius, m
Δr	chordal rise, m
r_c	chordal radius, m
S_u	ultimate strength, Pa
T	torque, N-m
t	thickness, m
u	belt velocity, m/s
w	belt weight, N
w_t	belt width, m
w_z	normal load, N
w'_z	normal load per unit length, N/m
α	angle used to describe loss in arc of contact, deg
β	sheave or pulley angle, deg
ζ	mast angle, deg
θ	boom angle, deg
θ_r	angle of rotation to give chordal rise, deg
μ	coefficient of friction
σ	normal stress, Pa
σ_{all}	allowable normal stress, Pa
σ_b	bending stress, Pa
σ_t	tensile normal stress, Pa
ϕ	wrap angle; gantry line angle, deg
ψ	angle from horizontal; gantry line system angle, deg
ω	angular velocity, s^{-1}

Subscripts

1	driver pulley or sheave
2	driven pulley or sheave

18.1 INTRODUCTION

The machine elements considered in this chapter, like the machine elements considered in Chapter 17, use friction as a useful agent to produce a high and uniform force and to transmit power. A belt, rope, or chain provides a convenient means for transferring power from one shaft to another. Belts are frequently necessary to reduce the higher rotative speeds of electric motors to the lower values required by mechanical equipment. Chains are effective in transferring power between parallel shafts. The design of belts and ropes is subject to uncertainties in the value of the coefficient of friction that must be used.

18.2 FLAT BELTS

Flat belts find considerable use in applications requiring small pulley diameters, high belt surface speeds, low noise levels, low weight, or low inertia. They cannot be used where absolute synchronization between pulleys must be maintained because they rely on friction for their proper functioning. All flat belts are subject to creep because relative motion occurs between the pulley surface and the adjacent belt surface that is under load deformation from the combined tension and flexural stresses. Flat belts must be kept under tension to function, and they therefore require tensioning devices.

18.2.1 BELT LENGTH

Figure 18.1 shows the dimensions, angles of contact, and center distance of an open flat belt. The term "open" is used to distinguish it from the geometry of a crossed belt, where the belt forms a figure eight. Note that distance 0_2D is equal to r_1, distance BD is equal to $r_2 - r_1$, and distance AD is the center distance c_d. Also, angle ABD is a right triangle so that

$$AB^2 + BD^2 = AD^2$$

$$\therefore AB^2 + (r_2 - r_1)^2 = c_d^2$$

$$AB = \sqrt{c_d^2 - (r_2 - r_1)^2} \tag{18.1}$$

The length of the open flat belt can be expressed as

$$L = 2AB + r_1\phi_1 \frac{\pi}{180} + r_2\phi_2 \frac{\pi}{180} \tag{18.2}$$

where

ϕ = wrap angle, deg

The wrap angles can be expressed as

$$\phi_1 = 180° - 2\alpha \qquad \text{and} \qquad \phi_2 = 180° + 2\alpha \tag{18.3}$$

Also from right triangle ABD in Fig. 18.1 the angle used to describe the loss in arc of contact is

$$\sin\alpha = \frac{D_2 - D_1}{2c_d} \quad \text{or} \quad \alpha = \sin^{-1}\left(\frac{D_2 - D_1}{2c_d}\right) \tag{18.4}$$

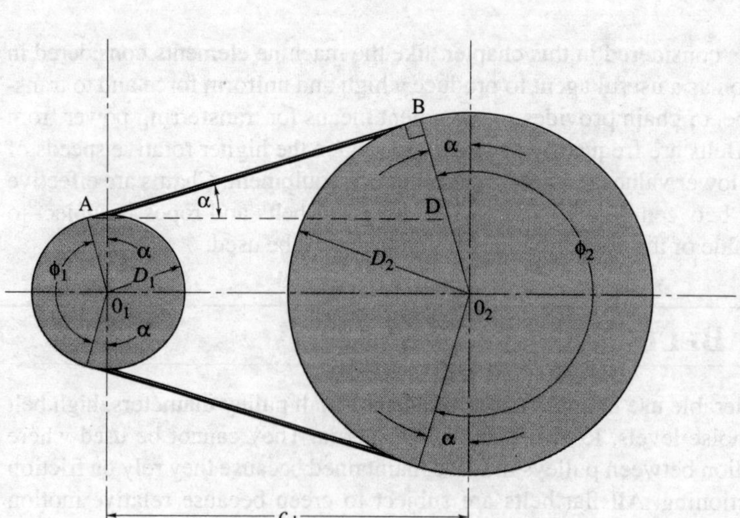

Figure 18.1 Dimensions, angles of contact, and center distance of open flat belt.

The angle α is in degrees and is created because the pulleys do not have a 1-to-1 ratio. By making use of Eqs. (18.3) and (18.4) the belt length in Eq. (18.2) can be expressed as

$$L = \sqrt{(2c_d)^2 - (D_2 - D_1)^2} + \frac{D_1\pi}{360}\left[180° - 2\sin^{-1}\left(\frac{D_2 - D_1}{2c_d}\right)\right] + \frac{D_2\pi}{360}\left[180° + 2\sin^{-1}\left(\frac{D_2 - D_1}{2c_d}\right)\right]$$

$$= \sqrt{(2c_d)^2 - (D_2 - D_1)^2} + \frac{\pi}{2}(D_1 + D_2) + \frac{\pi(D_2 - D_1)}{180}\sin^{-1}\left(\frac{D_2 - D_1}{2c_d}\right) \quad \textbf{(18.5)}$$

18.2.2 BELT FORCES

The basic equations developed for band brakes are also applicable here. From Sec. 17.8 the following torque and friction equations can be written:

$$T = (F_1 - F_2)D_1/2 \quad \textbf{(17.59)}$$

$$\frac{F_1}{F_2} = e^{\mu\phi\,\pi/180°} \quad \textbf{(17.58)}$$

where

ϕ = wrap angle, deg

μ = coefficient of friction

F_1 = tight-side or driver friction force, N

F_2 = slack-side or driven friction force, N

In obtaining the preceding equations it is assumed that the coefficient of friction on the belt is uniform throughout the entire wrap angle and that centrifugal forces on the belt can be neglected.

The required initial belt tension (or tensile force) F_i depends on the elastic characteristics of the belt but can be approximated by

$$F_i = (F_1 + F_2)/2 \quad \textbf{(18.6)}$$

In transmitting power from one shaft to another by means of a flat belt and pulleys, the belt must have the initial tensile force F_i given in Eq. (18.6). Also, from Eq. (17.59) note that when power is being transmitted, the tensile force F_1 in the tight side exceeds the tension in the slack side F_2.

The power transmitted, in SI units of watts, is

$$h_p = (F_1 - F_2)u \quad \textbf{(18.7)}$$

where

u = belt velocity, m/s

The power transmitted, in English units, is

$$h_p = \frac{(F_1 - F_2)u}{33\,000} \quad \textbf{(18.8)}$$

where F_1 and F_2 are in pounds force and u is in feet per minute.

The centrifugal force can be expressed as

$$F_c = m'u^2 = \frac{w'_z}{g}u^2 \tag{18.9}$$

where

m' = mass per unit length, kg/m
u = belt velocity, m/s
w'_z = weight per unit length, N/m

When the centrifugal force is considered Eq. (17.58) becomes

$$\frac{F_1 - F_c}{F_2 - F_c} = e^{\mu\phi\pi/180°} \tag{18.10}$$

Equation (17.59) is applicable whether or not centrifugal forces are considered.

Example 18.1

GIVEN A flat belt is 6 in. wide and 1/3 in. thick and transmits 15 hp. The center distance is 8 ft. The driving pulley is 6 in. in diameter and rotates at 2000 rpm such that the loose side of the belt is on top. The driven pulley is 18 in. in diameter. The belt material weighs 0.035 lbf/in.^{3}.

FIND Determine the following:

a) If $\mu = 0.30$, determine F_1 and F_2.

b) If μ is reduced to 0.20 because of oil getting on part of the pulley, what are F_1 and F_2? Would the belt slip?

c) What is the belt length?

Solution

a) The belt velocity is

$$u = \frac{\pi D_1 N_1}{12} = \frac{\pi(6)(2000)}{12} = 3142 \text{ ft/min}$$

From Eq. (18.8)

$$F_1 - F_2 = \frac{(33\,000)h_p}{u} = \frac{(33\,000)(15)}{3142} = 157.6 \text{ lbf} \tag{a}$$

The weight per volume was given as

$$\frac{w_z}{Lw_t t} = 0.035 \text{ lbf/in.}^3$$

$$\therefore w_z/L = 0.035 w_t t = (0.035)(6)(1/3) = 0.070 \text{ lbf/in.} = 0.84 \text{ lbf/ft}$$

The centrifugal force acting on the belt is

$$F_c = \frac{w_z}{L}\frac{u^2}{g} = \frac{(0.84)(3142)^2}{(32.2)(60)^2} = 71.54 \text{ lbf}$$

From Eq. (18.4)

$$\alpha = \sin^{-1}\left(\frac{D_2 - D_1}{2c_d}\right) = \sin^{-1}\left(\frac{18-6}{(2)(8)(12)}\right) = 3.583° \tag{b}$$

The wrap angle is

$$\phi = 180° - 2(\alpha) = 180° - 2(3.583°) = 172.8° \quad \textbf{(c)}$$

Making use of Eq. (18.10) gives

$$\frac{F_1 - 71.54}{F_2 - 71.54} = e^{(0.3)(172.8)\pi/180} = 2.472$$

$$\therefore F_1 - 71.54 = 2.472F_2 - 176.8$$

$$\therefore F_1 = 2.472F_2 - 105.3 \quad \textbf{(d)}$$

Substituting Eq. (d) into Eq. (a) gives

$$2.472F_2 - F_2 = 105.3 + 157.6$$

$$F_2 = \frac{262.9}{1.472} = 178.6 \text{ lbf}$$

$$\therefore F_1 = 157.6 + 178.6 = 336.2 \text{ lbf}$$

From Eq. (18.6) the initial belt tension is

$$F_i = (F_1 + F_2)/2 = (336.2 + 178.6)/2 = 257.4 \text{ lbf}$$

b) If $\mu = 0.20$ instead of 0.30,

$$\frac{F_1 - 71.54}{F_2 - 71.54} = e^{(0.2)(172.8)\pi/180} = 1.828$$

$$F_1 = 1.828F_2 - (71.54)(1.828) + 71.54$$

$$= 1.828F_2 - 59.23 \quad \textbf{(e)}$$

Substituting Eq. (e) into Eq. (18.6) gives

$$\frac{1.828F_2 - 59.23 + F_2}{2} = 257.4$$

$$\therefore F_2 = 203.0 \text{ lbf}$$

Substituting this into Eq. (e) gives

$$F_1 = 1.828(203.0) - 59.23 = 311.8 \text{ lbf}$$

From Eq. (18.8) the power transmitted, in English units, is

$$h_p = \frac{(F_1 - F_2)u}{33\,000} = \frac{(311.8 - 203.0)(3142)}{33\,000} = 10.36 \text{ hp}$$

Since this is less than 15 hp, the belt will slip.

c) From Eq. (18.5) while making use of Eqs. (b) and (c)

$$L = 2\sqrt{8^2(12)^2 - (9-3)^2} + \frac{3\pi}{180°}(172.8°) + \frac{9\pi}{180°}(187.2°)$$

$$= 191.6 + 9.048 + 29.41 = 230.1 \text{ in.}$$

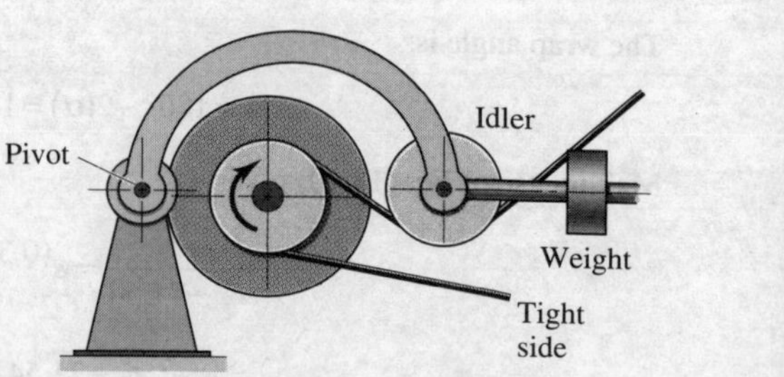

Figure 18.2 Weighted idler used to maintain desired belt tension.

18.2.3 SLIP

To eliminate slip, the initial belt tension needs to be retained. But as the belt stretches over time, some of the initial tension will be lost. One solution might be to have excessive initial tension, but this would put large loads on the pulley and shaft and also shorten the belt life. Some better approaches are the following:

1. Develop ways of adjusting tension during operation.
2. Increase the wrap angle.
3. Change the belt material to increase the coefficient of friction.
4. Use a larger belt section.

Figure 18.2 illustrates a way of maintaining belt tension. The slack side of the belt is on the top, so that the sag of the belt acts to increase its wrap angle.

18.3 SYNCHRONOUS BELTS

Synchronous belts, or **timing belts**, are basically flat belts with a series of evenly spaced teeth on the inside circumference, thereby combining the advantages of flat belts with the excellent traction of gears and chains (covered later in this chapter, Sec. 18.6). A synchronous, or timing, belt is shown in Fig. 18.3.

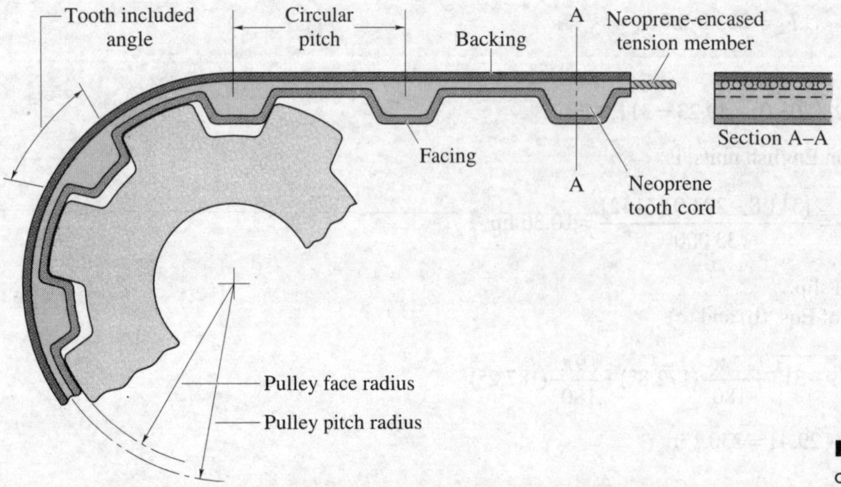

Figure 18.3 Synchronous, or timing, belt.

Because synchronous belts do not slip or creep, unlike flat belts, the required belt tension is low, producing very small bearing loads. Synchronous belts will not stretch and require no lubrication. Speed is transmitted uniformly because there is no chordal rise and fall of the pitch line as in rolling chains (Sec. 18.6). The equations developed for flat-belt length and forces in Sec. 18.2 are equally valid for synchronous belts.

Example 18.2

GIVEN A timing belt is used to transfer power from a high-speed turbine to a grinding wheel. The timing belt is 750 mm long and weighs 180 g. The maximum allowable force in the belt is 2000 N, and the speed of the turbine is the same as the speed of the grinding wheel, 5000 rpm.

FIND Calculate the optimum pulley pitch diameter for maximum power transfer.

Solution

If the total maximum force in the belt is F_1 and the centrifugal force is F_c, the maximum power transmitted is

$$h_p = u(F_1 - F_c)$$

Letting F_1 be equal to the maximum allowable force and making use of Eq. (18.9) for the centrifugal force gives

$$h_p = u\left(2000 - \frac{0.180}{0.750}u^2\right) = u(2000 - 0.24u^2)$$

The optimum power transmitted occurs when

$$\frac{\partial h_p}{\partial u} = 2000 - (0.24)(3)u^2 = 0$$

$$\therefore u = 52.70 \text{ m/s}$$

The pulley diameter that produces the maximum power transfer is

$$\omega D/2 = u$$

$$D = \frac{2u}{\omega} = \frac{2(52.70)}{(5000)\left(\frac{2\pi}{60}\right)} = 0.2013 \text{ m} = 201.3 \text{ mm}$$

Thus, the optimum pulley diameter for maximum power transfer is 201.3 mm.

18.4 V-BELTS

V-belts are used with electric motors to drive a great number of components, such as blowers, compressors, or machine tools. One or more V-belts are used to drive accessories on automotive and most other internal-combustion engines. V-belts are made to standard lengths and with standard cross-sectional sizes, the details of which can be found in catalogs. The grooved pulleys that V-belts run in are called **sheaves**. They are usually made of cast iron, pressed steel, or die-cast metal. Figure 18.4 shows a V-belt in a sheave groove.

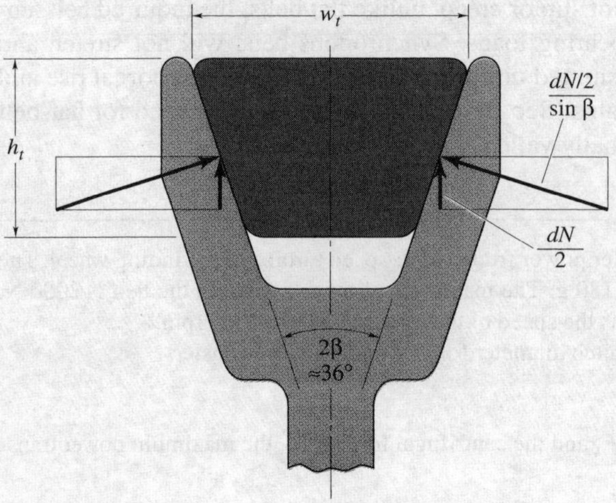

Figure 18.4 V-belt in sheave groove.

A number of different types of V-belt are available. The various types will not be discussed here. Instead, information about the light-duty type with three different sections 2L, 3L, and 4L will be considered. Once the design of the light-duty type of V-belt is known, other types of V-belt can be easily designed by using company catalogs.

V-belts find frequent application where synchronization between shafts is not important. V-belts are easily installed and removed, quiet in operation (but not quite as quiet as flat belts), and low in maintenance and provide shock absorption between the driver and driven shafts. V-belt drives normally operate best at belt velocities between 1500 and 6500 ft/min. Optimum velocity (peak capacity) is 4500 ft/min.

V-belts can operate satisfactorily at velocity ratios

$$g_r = \frac{N_1}{N_2} = \frac{r_2}{r_1} \tag{18.11}$$

to approximately 7 to 1. V-belts typically operate at 90 to 98% efficiency, lower than that found for flat belts.

V-belts have a Fiberglas-reinforced neoprene core and a fabric-impregnated neoprene jacket that protects the interior and provides a wear-resistant surface for the belt. Because their interior tension cords are stretch resistant, V-belts (unlike flat belts) *do not* require frequent adjustment of initial tension.

A major advantage of V-belts over flat belts is their wedging action, which increases the normal force from dN for flat belts to $(dN/2)/\sin\beta$ (as shown in Fig. 18.4) for V-belts, where β is the sheave angle. Because the V-belt has a trapezoidal cross section, the belt rides on the side of the groove and a wedging action increases the traction. Pressure and friction forces act on the side of the belt. The force equations developed in Sec. 18.2 for flat belts are equally applicable for V-belts if the coefficient of friction μ is replaced with $\mu/\sin\beta$. Also, the belt length equation for flat belts [Eq. (18.5)] is equally valid for V-belts. The only difference when using Eq. (18.5) for V-belts is that pitch radius or pitch diameter is used, whereas for flat belts the outside radius or outside diameter of the pulley is used.

18.4.1 INPUT NORMAL POWER RATING

It is essential that the possible maximum load conditions be considered in designing a V-belt. The normal power rating of the driver is

$$\left(h_{pr}\right)_{\text{input}} = h_p\left(f_1 + f_2\right) \quad \textbf{(18.12)}$$

where

f_1 = overload service factors for various types of driven unit given in Table 18.1
f_2 = extra overload factor of 0.4 if any of the following conditions prevail:

Table 18.1 Overload service factors(f_1) for various types of driven unit.

Driven unit	Overload factor
Agitators	
Liquid	1.2
Semiliquid	1.4
Compressor	
Centrifugal	1.2
Reciprocating	1.4
Conveyors and elevators	
Package and oven	1.2
Belt	1.4
Fans and blowers	
Centrifugal, calculating	1.2
Exhausters	1.4
Food machinery	
Slicers	1.2
Grinders and mixers	1.4
Generators	
Farm lighting and exciters	1.2
Heating and ventilating	
Fans and oil burners	1.2
Stokers	1.4
Laundry machinery	
Dryers and ironers	1.2
Washers	1.4
Machine tools	
Home workshop and woodworking	1.4
Pumps	
Centrifugal	1.2
Reciprocating	1.4
Refrigeration	
Centrifugal	1.2
Reciprocating	1.4
Worm gear speed reducers, input side	1.0

1) Frequent starting and stopping
2) High starting motor torque
3) Gas engine

Thus, a properly designed belt drive should use the power rating from Eq. (18.12).

18.4.2 DRIVE SIZE

The design of a V-belt drive should use the largest possible pulleys. Small pulleys are less efficient and greatly reduce belt life because of slip and extreme flexing of the belt. Where small pulleys must be used and speed is high, select the lightest belt ($2L$). Table 18.2 shows minimum pitch diameters in inches, and Table 18.3 shows recommended minimum pulley diameters in inches for electric motors. The criteria used in arriving at these critical sizes are bearing and shaft loading and not belt flexing life. Tables 18.2 and 18.3 give belt and pulley sizes for a specific belt drive.

18.4.3 ARC CORRECTION FACTOR

The rated power tables to be considered in the next section (Sec. 18.4.4) are based on a 1-to-1 ratio with 180° of belt wrap around each pulley. When pulleys of different diameters are used, there is a loss in traction effort on the smaller pulley. The angle α used to describe the loss in arc of contact is expressed in Eq. (18.4) and shown in Fig. 18.1. Recall that α is

Table 18.2 Recommended minimum pitch diameters of pulley for three belt sizes.

Belt type	Size of belt, in.	Minimum pitch diameter, in. Recommended	Absolute
$2L$	$^1/_4 \times {}^1/_8$	1.0	1.0
$3L$	$^3/_8 \times {}^7/_{32}$	1.5	1.5
$4L$	$^1/_2 \times {}^5/_{16}$	2.5	1.8

Table 18.3 Recommended pulley diameters in inches for three electric motor sizes.

	Motor speed, rpm				
Motor horsepower, hp	575	695	870	1160	1750
	Recommended pulley diameter, in.				
0.50	2.50	2.50	2.50	—	—
0.75	3.00	2.50	2.50	2.50	—
1.00	3.00	3.00	2.50	2.50	2.25

Table 18.4 Arc correction factors for various angles of loss in arc of contact.

Loss in arc of contact, deg.	Correction factor
0	1.00
5	.99
10	.98
15	.96
20	.95
25	.93
30	.92
35	.89
40	.89
45	.87
50	.86
55	.84
60	.83
65	.81
70	.79
75	.76
80	.74
85	.71
90	.69

expressed in degrees rather than radians. Table 18.4 shows the arc correction factors for various values of α.

18.4.4 Design Power Rating and Center Distance

Table 18.5 gives the rated power for three light-duty belt types. The three types considered are $2L$ with $w_t = 1/4$ in. and $h_t = 1/8$ in., $3L$ with $w_t = 3/8$ in. and $h_t = 1/4$ in., and $4L$ with $w_t = 1/2$ in. and $h_t = 9/32$ in. (Figure 18.4 clarifies the width w_t and height h_t of the belt.) From these tables for a designated speed of the faster shaft, given in the first column, and the driver pulley's effective outside diameter, given in the top row of the table, the power rating can be obtained.

Table 18.6 shows the center distances of the $3L$ and $4L$ V-belt types. All dimensions are in inches. These tables give the center distance for various sizes of the driven and driver pulleys. The belt velocity expressed in feet per minute is

$$u = \pi D_1 N_{a1}/12 \tag{18.13}$$

Table 18.5 Power ratings for light duty V-belt. (a) $2L$ section with $w_t = 1/4$ in. and $h_t = 1/8$ in.; (b) $3L$ section with $w_t = 3/8$ in. and $h_t = 1/4$ in.; (c) $4L$ section with $w_t = 1/2$ in. and $h_t = 9/32$ in.

Speed of faster shaft, rpm	Pulley effective outside diameter, in.					
	1.00	2.00	3.00	4.00	5.00	6.00
	Rated horsepower, hp					
500	0.04	0.05	0.06	0.08	0.10	0.12
1000	0.05	0.08	0.12	0.16	0.19	0.21
1500	0.06	0.12	0.17	0.22	0.25	0.30
2000	0.08	0.15	0.23	0.28	0.32	0.39
2500	0.10	0.18	0.27	0.34	0.39	0.44
3000	0.12	0.21	0.31	0.40	0.41	0.47
3500	0.14	0.24	0.35	0.44	0.44	
4000	0.16	0.28	0.38	0.46		
4500	0.18	0.31	0.42			
5000	0.20	0.35				

(a)

Speed of faster shaft, rpm	Pulley effective outside diameter, in.						
	1.50	1.75	2.00	2.25	2.50	2.75	3.00
	Rated horsepower, hp						
500	0.04	0.07	0.09	0.12	0.14	0.17	0.19
1000	0.07	0.12	0.16	0.21	0.25	0.30	0.34
1500	0.09	0.15	0.22	0.29	0.35	0.41	0.47
2000	0.10	0.19	0.27	0.35	0.43	0.51	0.59
2500	0.11	0.21	0.31	0.41	0.51	0.60	0.69
3000	0.11	0.23	0.35	0.45	0.57	0.68	0.78
3500	0.11	0.25	0.38	0.50	0.62	0.74	0.84
4000	0.11	0.26	0.40	0.54	0.66	0.78	0.88
4500	0.10	0.25	0.42	0.56	0.68	0.80	0.90
5000	0.09	0.26	0.42	0.57	0.69	0.80	0.89

(b)

(continued)

Table 18.5 *(concluded)*

Speed of faster shaft, rpm	Pulley effective outside diameter, in.								
	2.00	2.25	2.50	2.75	3.00	3.25	3.50	3.75	4.00
	Rated horsepower, hp								
500	0.08	0.14	0.19	0.24	0.29	0.34	0.39	0.44	0.49
1000	0.11	0.21	0.31	0.41	0.50	0.60	0.69	0.78	0.87
1500	0.12	0.26	0.40	0.54	0.67	0.81	0.94	1.07	1.20
2000	0.11	0.30	0.47	0.65	0.82	0.99	1.15	1.31	1.47
2500	0.09	0.31	0.53	0.73	0.94	1.13	1.32	1.51	1.69
3000	0.06	0.31	0.56	0.79	1.02	1.24	1.45	1.65	1.84
3500	0.02	0.30	0.57	0.83	1.07	1.31	1.53	1.73	1.92
4000	—	0.27	0.56	0.83	1.09	1.33	1.55	1.75	1.92
4500	—	0.22	0.54	0.81	1.07	1.30	1.51	1.69	1.84
5000	—	0.15	0.47	0.75	1.01	1.23	1.41	1.66	1.65

(c)

Table 18.6 Center distances for various pitch diameters of driver and driven pulleys. (a) 3*L* type of V-belt; (b) 4*L* type of V-belt.

Pulley combination		Nominal center distance			
		Short center		Medium center	
Driver pitch diameter, in.	Driven pitch diameter, in.	Belt number	Center distance, in.	Belt number	Center distance, in.
2.0	2.0	3L200	6.4	3L250	9.4
3.0	3.0	3L250	7.4	3L310	10.4
2.0	2.5	3L210	6.6	3L270	9.6
2.0	3.0	3L220	6.7	3L280	9.7
3.0	4.5	3L290	8.2	3L350	11.2
2.0	3.5	3L240	7.3	3L300	10.3
2.0	4.0	3L250	7.2	3L310	10.3
2.25	4.5	3L270	7.7	3L330	10.7
2.5	5.0	3L290	8.1	3L350	11.1
3.0	6.0	3L330	8.9	3L390	11.9
2.0	5.0	3L250	8.0	3L340	11.0
2.0	6.0	3L310	8.6	3L370	11.6
3.0	9.0	3L410	10.3	3L470	13.4
2.0	7.0	3L340	9.2	3L400	12.2
2.0	9.0	3L390	9.9	3L450	13.0
2.0	10.0	3L420	10.4	3L480	13.6
1.5	9.0	3L390	10.1	3L450	13.3

(a)

(continued)

Table 18.6 *(concluded)*

Pulley combination		Nominal center distance					
		Minimum center		Short center		Medium center	
Driver pitch diameter, in.	Driven pitch diameter, in.	Belt number	Center distance, in.	Belt number	Center distance, in.	Belt number	Center distance, in.
2.5	2.5	4L170	4.0	4L150	8.0	4L330	12.0
3.0	3.0	4L200	4.8	4L280	8.8	4L360	12.8
3.0	4.5	4L240	5.5	4L320	9.6	4L400	13.6
4.0	6.0	4L300	6.5	4L380	10.6	4L460	14.5
3.0	6.0	4L280	6.2	4L360	10.3	4L440	14.3
3.5	7.0	4L320	7.0	4L400	11.1	4L480	15.1
3.0	7.5	4L320	6.8	4L400	11.0	4L480	15.0
4.0	10.0	4L410	8.5	4L490	12.5	4L570	16.7
3.0	9.0	4L360	7.5	4L440	11.7	4L520	15.8
3.5	10.5	4L420	8.8	4L500	13.0	4L580	17.1
4.0	12.0	4L470	9.6	4L550	13.8	4L630	18.0
3.0	10.5	4L410	8.6	4L490	12.9	4L570	17.0
4.0	14.0	4L530	10.5	4L610	15.0	4L690	19.1
3.0	12.0	4L450	9.1	4L530	13.4	4L610	17.6
3.5	14.0	4L520	10.4	4L600	14.8	4L680	19.0
2.0	9.0	4L350	7.5	4L430	11.8	4L510	15.9
4.0	18.0	4L650	12.8	4L730	17.3	4L810	21.6
2.4	12.0	4L440	8.9	4L520	13.3	4L600	17.5
2.5	18.0	4L480	9.3	4L560	14.3	4L640	18.5
2.8	14.0	4L510	10.3	4L590	14.7	4L670	19.0
2.0	11.0	4L400	8.0	4L480	12.5	4L560	16.7
2.0	12.0	4L430	8.5	4L510	13.0	4L590	17.3
2.5	15.0	4L530	10.3	4L610	14.9	4L690	19.2
3.0	13.0	4L630	12.2	4L710	16.8	4L790	21.2
2.0	14.0	4L490	9.5	4L570	14.1	4L650	18.4

(b)

Example 18.3

GIVEN A 1/3-hp, capacitor, alternating-current motor is driving a reciprocating air compressor at 720 rpm. The motor speed is 1750 rpm on approximately 12-in. centers. The maximum pitch diameter of the compressor pulley is to be 8 in. Design the belt drive with a light-duty V-belt.

FIND Give the size of the pulleys, the belt dimensions, and the rated horsepower.

Solution

From Table 18.1 for a reciprocating compressor $f_1 = 1.4$. Since the motor has a high starting torque, $f_2 = 0.4$. The given power was 1/3 hp. The normal power rating of the drive, from Eq. (18.12), is

$$\left(h_{pr}\right)_{\text{input}} = h_p(f_1 + f_2) = \frac{1}{3}(1.4 + 0.4) = 0.600 \text{ hp}$$

The velocity ratio input is

$$g_r = \frac{\text{Speed of motor}}{\text{Speed of compressor}} = \frac{1750}{720} = 2.43$$

Given that the maximum allowable pitch diameter of the driven pulley is 8 in., assume that the driven pitch diameter at the limit is $D_2 = 8$ in. Then the driver diameter is

$$D_1 = \frac{D_2}{g_r} = \frac{8.0}{2.43} = 3.292 \text{ in.}$$

A pitch diameter of 3.292 in. corresponds to an outside diameter of 3.5 in. From Table 18.5(c) the best belt type to handle the requirements is 4*L*. Specifically, for a pulley outside diameter of 3.5 and speed of 1750 rpm, the design rated power is 1.05 hp.

Table 18.6(b) does not contain the identical situation in this example, a speed ratio of 2.4 and pulley pitch diameters of 8 and 3.292 in. However, from this table one could estimate that a center distance of 12 in. should work. With the above values the angle used to describe the loss in arc contact and the belt length can be expressed from Eqs. (18.4) and (18.5) as

$$\alpha = \sin^{-1}\left(\frac{D_2 - D_1}{2c_d}\right) = \sin^{-1}\left(\frac{8 - 3.292}{24}\right) = 11.31°$$

$$L = \sqrt{(2c_d)^2 - (D_2 - D_1)^2} + \frac{\pi}{2}(D_1 + D_2) + \frac{\pi(D_2 - D_1)\alpha}{180}$$

$$= \sqrt{(2)^2(12)^2 - (8 - 3.292)^2} + \frac{\pi}{2}(8 + 3.292) + \frac{\pi(8 - 3.292)(11.31°)}{180}$$

$$= 23.53 + 17.74 + 0.814 = 42.08 \text{ in.}$$

Recall that the design power of 1.05 hp is for a 1-to-1 speed rating. From Table 18.4 for $2\alpha = 22.62°$ the arc correction factor is 0.94. The corrected design power rating is 0.94 (1.05) = 0.99 hp.

The belt velocity can be determined from Eq. (18.13) or

$$u = \frac{\pi}{12} D_1 N_{a1} = \frac{\pi}{12}(3.292)(1750) = 1508 \text{ ft/min}$$

Summarizing our 4*L* V-belt design,

$$D_1 = 3.292 \text{ in.} \quad D_2 = 8 \text{ in.} \quad c_d = 12 \text{ in.} \quad L = 42 \text{ in.} \quad g_r = 2.4 \quad N_{a1} = 1750 \quad u = 1508 \text{ ft/min}$$

$$(h_p)_{\text{input}} = 0.6 \text{ hp} \quad (h_p)_{\text{design}} = 1 \text{ hp}$$

18.5 WIRE ROPES

Wire ropes are used instead of flat belts or V-belts when power must be transmitted over long center distances as in hoists, elevators, and ski lifts. Figure 18.5 shows a cross section of a wire rope. The center portion (dark section) is the **core** of the rope and is often made of hemp (a tall Asiatic herb). The purposes of the core are to lubricate and thus prevent excessive wire wear and to elastically support the strands. Some other core materials are polypropylene wire and steel wire.

Figure 18.5 Cross section of wire rope.

The **strands** are groupings of wires placed around the core. In Fig. 18.5 there are six strands and each strand consists of 19 wires. Wire ropes are typically designated as, for example, "1 ⅛, 6 × 19 hauling rope." The 1 ⅛ gives the wire rope diameter in inches, designated by the symbol *d*. The 6 designates the number of strands and 19

Figure 18.6 Two lays of wire rope. (a) Lang; (b) regular.

designates the number of wires in a strand. The wire diameter is designated by the symbol d_w. The term "hauling rope" designates the application in which the wire rope is to be used.

Figure 18.6 shows two lays of wire rope. The regular **lay** [Fig. 18.6(b)] has the wire twisted in one direction to form strands and the strands twisted in the opposite direction to form the rope. Visible wires are approximately parallel to the rope axis. The major advantage of the regular lay is that the rope *does not kink or untwist* and is easy to handle. The Lang lay [Fig. 18.6(a)] has the wires in the strands and the strands in the rope twisted in the same direction. This type of lay has *more resistance to abrasive wear and bending fatigue* than the regular lay. Lang-lay ropes are, however, more susceptible to handling abuses, pinching in undersized grooves, and crushing when improperly wound on drums. Also, the twisting moment acting in the strand tends to unwind the strand, causing excessive rope rotation. Lang-lay ropes should therefore always be secured at the ends to prevent the rope from unlaying.

Although steel is most popular, wire rope is made of many kinds of metal, such as copper, bronze, stainless steel, and wrought iron. Table 18.7 lists some of the various ropes that are available, together with their characteristics and properties. The cross-sectional area of the metal strand in standard hoisting and haulage ropes is $A_m \approx 0.38d^2$.

18.5.1 TENSILE STRESS

The total force acting on the rope is

$$F_t = F_w + F_r + F_a \tag{18.14}$$

where

F_w = deadweight being supported, N
F_r = rope weight, N
F_a = force due to acceleration, N

Table 18.7 Wire rope data. [J.E. Shigley and L.D. Mitchell, *Mechanical Engineering Design*, 4e, © 1983 by McGraw-Hill. Reproduced with permission from The McGraw-Hill Companies.]

Rope	Weight per height, lb/ft	Minimum sheave diameter, in.	Rope diameter, d, in.	Material	Size of outer wires	Modulus of elasticity,[a] psi	Strength,[b] psi
6 × 7 Haulage	$1.50d^2$	$42d$	$\frac{1}{4}-1\frac{1}{2}$	Monitor steel	$d/9$	14×10^6	100×10^3
				Plow steel	$d/9$	14	88
				Mild plow steel	$d/9$	14	76
6 × 19 Standard hoisting	$1.60d^2$	$26d - 34d$	$\frac{1}{4}-2\frac{3}{4}$	Monitor steel	$d/13 - d/16$	12	106
				Plow steel	$d/13 - d/16$	12	93
				Mild plow steel	$d/13 - d/16$	12	80
6 × 37 Special flexible	$1.55d^2$	$18d$	$\frac{1}{4}-3\frac{1}{2}$	Monitor steel	$d/22$	11	100
				Plow steel	$d/22$	11	88
8 × 19 Extra flexible	$1.45d^2$	$21d - 26d$	$\frac{1}{4}-1\frac{1}{2}$	Monitor steel	$d/15 - d/19$	10	92
				Plow steel	$d/15 - d/19$	10	80
7 × 7 Aircraft	$1.70d^2$	—	$\frac{1}{16}-\frac{3}{8}$	Corrosion-resistant steel	—	—	124
				Carbon steel	—	—	124
7 × 9 Aircraft	$1.75d^2$	—	$\frac{1}{8}-1\frac{3}{8}$	Corrosion-resistant steel	—	—	135
				Carbon steel	—	—	143
19-Wire aircraft	$2.15d^2$	—	$\frac{1}{32}-\frac{5}{16}$	Corrosion-resistant steel	—	—	165
				Carbon steel	—	—	165

[a]The modulus of elasticity is only approximate; it is affected by the loads on the rope and, in general, increases with the life of the rope.

[b]The strength is based on the nominal area of the rope. The figures given are only approximate and are based on 1-in. rope sizes and 1/4-in. aircraft cable sizes.

The tensile stress is

$$\sigma_t = F_t / A_m \tag{18.15}$$

where

A_m = cross-sectional area of metal strand in standard hoisting and haulage ropes

The allowable stress is obtained from Table 18.7.

From Eq. (1.1) the safety factor is

$$n_s = \sigma_{\text{all}} / \sigma_t \tag{18.16}$$

Table 18.8 gives minimum safety factors for a variety of wire rope applications. The safety factor obtained from Eq. (18.16) should be larger than the safety factor obtained from Table 18.8.

Table 18.8 Minimum safety factors for variety of wire rope applications. [J.E. Shigley and L.D. Mitchell, *Mechanical Engineering Design*, 4e, © 1983 by McGraw-Hill. Reproduced with permission from The McGraw-Hill Companies.]

Application	Safety factor,[a] n_s
Track cables	3.2
Guys	3.5
Mine shafts, ft	
Up to 500	8.0
1000–2000	7.0
2000–3000	6.0
Over 3000	5.0
Hoisting	5.0
Haulage	6.0
Cranes and derricks	6.0
Electric hoists	7.0
Hand elevators	5.0
Private elevators	7.5
Hand dumbwaiters	4.5
Grain elevators	7.5
Passenger elevators, ft/min	
50	7.60
300	9.20
800	11.25
1200	11.80
1500	11.90
Freight elevators, ft/min	
50	6.65
300	8.20
800	10.00
1200	10.50
1500	10.55
Powered dumbwaiters, ft/min	
50	4.8
300	6.6
800	8.0

[a]Use of these factors does not preclude a fatigue failure.

18.5.2 BENDING STRESS

From Sec. 4.5 [Eqs. (4.48) and (4.50)] the bending moment of the wires in a rope passing over a pulley is

$$M = \frac{EI}{r} \quad \text{and} \quad M = \frac{\sigma I}{c} \tag{18.17}$$

where

E = modulus of elasticity of wire, Pa

The symbols are the same as those used in Chapter 4. Equating the two equations and solving for the bending stress gives

$$\sigma_b = \frac{Ec}{r}$$

where

r = radius of curvature that rope will experience, m

c = distance from neutral axis to outer fiber of wire, m

The radius of curvature that the rope experiences is similar to the pulley radius $D/2$, and c is similar to $d_w/2$.

$$\therefore \sigma_b = \frac{Ed_w}{D} \tag{18.18}$$

where

d_w = wire diameter, m

D = pulley diameter, m

Note from Eq. (18.18) that the bigger D/d_w, the smaller the bending stress. Suggested minimum pulley diameters in Table 18.7 are based on a D/d_w ratio of 400. If possible, the pulleys should be designed for a larger ratio. If the ratio D/d_w is less than 200, heavy loads will often cause a permanent set of the rope. Thus, for a safe design assume that $D/d_w \geq 400$.

These design rules are hardly ever followed, for a number of reasons. First, consider a 2-in.-diameter wire rope, such as would typically be used in the crane of the case study in Chapter 17. According to the rules just stated, for safety the pulleys on the crane would have to be 80 in., or over 6 ft, in diameter. Further, since the wire rope is wound on a drum and failure can occur in the rope adjacent to the drum, such a large diameter would also be needed for the drum. Because the required motor torque is the product of the hoist rope tension and the drum radius, a huge motor would be required for lifting relatively light loads.

The design rules just stated are recommendations for applications where the wire rope should attain infinite life. The economic consequences of infinite-life wire rope are usually too large to bear, and smaller pulleys are usually prescribed. To prevent failures that result in property damage or personal injury, the wire ropes are periodically examined for damage. Since a broken wire will generally be easily detected with a cotton cloth dragged over the rope's surface, rope life requirements are often expressed in terms of the number of broken wires allowed per length of wire rope. For example, the American Society of Mechanical Engineers [ASME (1994)] requires inspections of crane wire ropes every 6 months, and if any section has more than six broken wires within one lay of the rope, or three in any strand within a lay, the entire wire rope must be replaced. The same standard calls for pulley-to-wire diameter ratios of 12:1, which obviously results in wire rope with finite service life. Figures 18.7 and 18.8 relate the decrease in strength and service life associated with smaller pulley diameters.

18.5.3 BEARING PRESSURE

The rope stretches and rubs against the pulley, causing wear of both the rope and pulley. The amount of wear depends on the pressure on the rope in the pulley groove, or

$$p = \frac{2F_t}{dD} \tag{18.19}$$

where

d = rope diameter, m

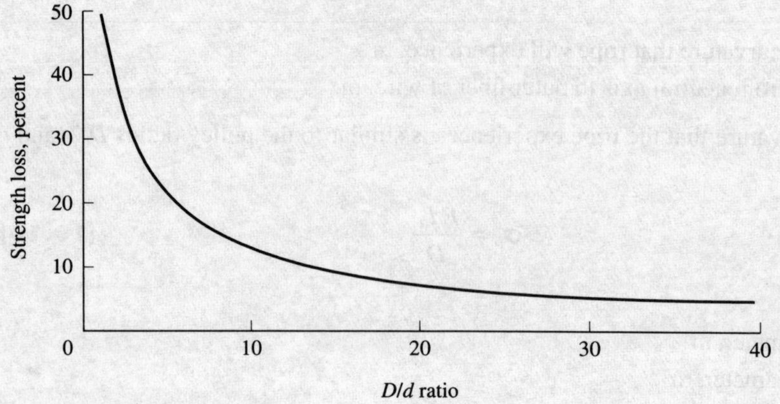

Figure 18.7 Percent strength loss in wire rope for different D/d ratios.

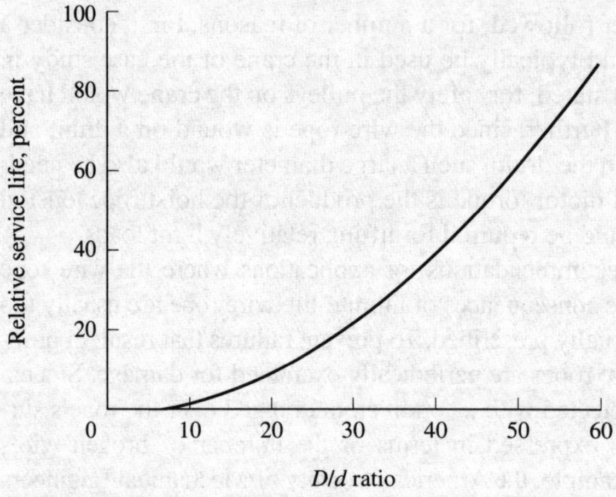

Figure 18.8 Service life for different D/d ratios.

The pressure obtained from Eq. (18.19) should be less than the maximum pressure obtained from Table 18.9 for various pulley materials and types of rope.

18.5.4 FATIGUE

For the rope to have a long life, the total friction force F_t must be less than the fatigue allowable force ($F_t \leq F_f$), where

$$F_f = \frac{S_u dD}{2000} \tag{18.20}$$

Table 18.9 Maximum allowable bearing pressures for various sheave materials and types of rope. [J.E. Shigley and L.D. Mitchell, *Mechanical Engineering Design*, 4e, © 1983 by McGraw-Hill. Reproduced with permission from The McGraw-Hill Companies.]

	Material				
	Wood[a]	Cast iron[b]	Cast steel[c]	Chilled cast iron[d]	Manganese steel[e]
Rope	Allowable bearing pressure, p_{all}, psi				
Regular lay					
6×7	150	300	550	650	1470
6×19	250	480	900	1100	2400
6×37	300	585	1075	1325	3000
8×19	350	680	1260	1550	3500
Lang lay					
6×7	165	350	600	715	1650
6×19	275	550	1000	1210	2750
6×37	330	660	1180	1450	3300

[a]On end grain of beech, hickory, or gum.
[b]For H_B (min.) = 125.
[c]30–40 carbon; H_B (min.) = 160.
[d]Use only with uniform surface hardness.
[e]For high speeds with balanced sheaves having ground surfaces.

The ultimate strength given by Eq. (18.20) for three materials is

$$\begin{aligned} &\text{Improved plow steel (monitor)} && 240 \le S_u \le 280\times10^3 \text{ psi} \\ &\text{Plow steel} && 210 \le S_u \le 240\times10^3 \text{ psi} \\ &\text{Mild plow steel} && 180 \le S_u \le 210\times10^3 \text{ psi} \end{aligned} \qquad \textbf{(18.21)}$$

Example 18.4

GIVEN A hand elevator is to travel a height of 90 ft. The maximum load to be hoisted is 3000 lbf at a velocity not exceeding 2 ft/s and an acceleration of 4 ft/s². Use 1-in. plow steel, 6×19 standard hoisting ropes.

FIND Determine the safety factor while considering

a) Tensile stress
b) Bending stress
c) Bearing pressure
d) Fatigue

Solution

From Table 18.7 for 6×19 standard hoisting wire rope, assuming the use of only one rope,

$$F_r = 1.60d^2h_t = 1.60(1)^2(90) = 144.0 \text{ lbf/rope}$$

$$F_w = \text{Deadweight} = W_{max} = 3000 \text{ lbf}$$

The force due to acceleration is

$$F_a = ma = \frac{W}{g}a = \frac{(3000+144)(4)}{32.2} = 390.6 \text{ lbf}$$

The total force on the rope is

$$F_t = F_w + F_r + F_a = 3000 + 144 + 390.6 = 3535 \text{ lbf}$$

a) *Tensile stress*:

$$\sigma_t = \frac{F_t}{A_m} = \frac{3535}{0.38} = 9301 \text{ psi}$$

From Table 18.7 for 6×19 standard hoisting wire rope made of plow steel

$$\sigma_{\text{all}} = 93\,000 \text{ psi}$$

The safety factor is

$$n_s = \frac{\sigma_{\text{all}}}{\sigma_t} = \frac{93\,000}{9301} = 10.0$$

From Table 18.8 the recommended safety factor for hand elevators is 5.0. Thus, one rope is sufficient as far as the tensile stress is concerned.

b) *Bending stress*: From Table 18.7 the minimum pulley diameter for 6×19 standard hoisting wire rope is 26*d* to 34*d*. Choose $D = 34d = 34$ in. Also from the same table the wire diameter should be between $d/13$ and $d/16$. Choose $d_w = d/16 = 1/16$ in.

$$\therefore \frac{D}{d_w} = \frac{34}{\frac{1}{16}} = 544$$

Permanent set should be avoided, since $D/d_w \geq 400$. From Table 18.7 the modulus of elasticity is 12×10^6 psi. The bending stress is

$$\sigma_b = E\frac{d_w}{D} = \frac{12(10^6)}{544} = 22.06 \times 10^3 \text{ psi}$$

The safety factor due to bending is

$$n_s = \frac{\sigma_{\text{all}}}{\sigma_b} = \frac{93\,000}{22\,060} = 4.22$$

This safety factor is less than the 5.0 recommended, and increasing the number of ropes will *not* alter the results. Changing the material from plow steel to monitor steel would produce a safety factor of 4.81, closer to 5. However, to use a lower than recommended safety factor, a larger pulley is needed.

c) *Bearing pressure*: From Eq. (18.19)

$$p = \frac{2F_t}{dD} = \frac{2(3535)}{(1)(34)} = 207.9 \text{ psi}$$

From Table 18.9 for a 6×19 Lang lay for a cast steel pulley, $p_{\text{all}} = 1000$ psi. The safety factor is

$$n_s = \frac{p_{\text{all}}}{p} = \frac{1000}{207.9} = 4.81$$

d) *Fatigue*: For monitor steel the ultimate strength is 280×10^3 psi. The allowable fatigue force from Eq. (18.20) is

$$F_f = \frac{S_u dD}{2000} = \frac{(280)(10^3)(1)(34)}{2000} = 4760 \text{ lbf}$$

The safety factor is

$$n_s = \frac{F_f}{F_t} = \frac{4760}{3535} = 1.35$$

Thus, fatigue failure is the most likely failure to occur, since it produced the smallest safety factor. Using four ropes instead of one would produce a safety factor greater than 5.

18.6 ROLLING CHAINS

Rolling chains are used to transmit power through sprockets rotating in the same plane. The machine element that it most resembles is a timing belt (shown in Fig. 18.3). The major advantage of using a rolling chain over a belt is that rolling chains do not slip. Large center distances can be dealt with more easily with rolling chains with fewer elements and in less space than with gears. Rolling chains also have high efficiency. No initial tension is necessary and shaft loads are therefore smaller than with belt drives. The only maintenance required after careful alignment of the elements is lubrication, and with proper lubrication long life can be ensured.

18.6.1 OPERATION OF ROLLING CHAINS

Figure 18.9 shows the various parts of a rolling chain with pins, bushings, rollers, and link plates. The rollers turn on bushings that are press fitted to the inner link plates. The pins are prevented from rotating in the outer plates by the press-fit assembly. Table 18.10 gives dimensions for standard sizes. The size range given in Table 18.10 is large, so that chains can be used for both large and small amounts of power transmission. A large reduction in

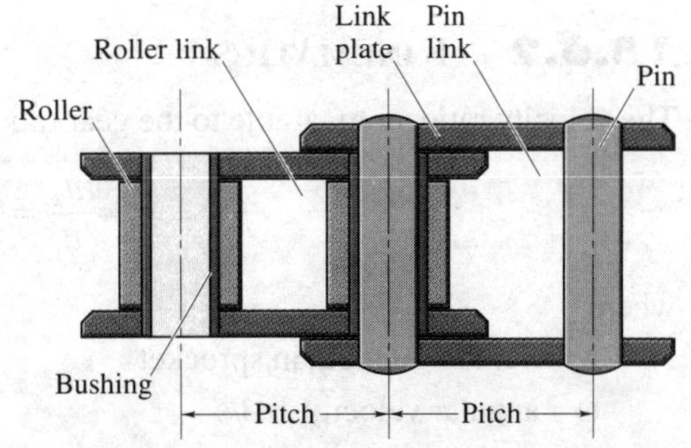

Figure 18.9 Various parts of rolling chain.

Table 18.10 Standard sizes and strengths of rolling chains.

Chain number	Pitch, p_t, in.	Roller Diameter, in.	Roller Width, in.	Pin diameter, d, in.	Link plate thickness, a, in.	Average ultimate strength, S_u, lbf	Weight per foot, lbf
[a]25	1/4	0.130	1/8	0.0905	0.030	875	0.084
[a]35	3/8	[a]0.200	3/16	0.141	0.050	2100	0.21
[a]41	1/2	0.306	1/4	0.141	0.050	2000	0.28
40	1/2	5/16	5/16	0.156	0.060	3700	0.41
50	5/8	2/5	3/8	0.200	0.080	6100	0.68
60	3/4	15/32	1/2	0.234	0.094	8500	1.00
80	1	5/8	5/8	0.312	0.125	14 500	1.69
100	$1\frac{1}{4}$	3/4	3/4	0.375	0.156	24 000	2.49
120	$1\frac{1}{2}$	7/8	1	0.437	0.187	34 000	3.67
140	$1\frac{3}{4}$	1	1	0.500	0.219	46 000	4.93
160	2	$1\frac{1}{8}$	$1\frac{1}{4}$	0.562	0.250	58 000	6.43
180	$2\frac{1}{4}$	$1\frac{13}{32}$	$1\frac{13}{32}$	0.687	0.281	76 000	8.70
200	$2\frac{1}{2}$	$1\frac{9}{16}$	$1\frac{1}{2}$	0.781	0.312	95 000	10.51
240	3	$1\frac{7}{8}$	$1\frac{7}{8}$	0.937	0.375	130 000	16.90

[a]Without rollers.
[b]Lightweight chain.

speed can be obtained with rolling chains if desired. The tolerances for a chain drive are larger than for gears, and the installation of a chain is relatively easy.

The minimum wrap angle of the chain on the smaller sprocket is 120°. A smaller wrap angle can be used on idler sprockets, which are used to adjust the chain slack where the center distance is not adjustable. Horizontal drives (the line connecting the axes of sprockets is parallel to the ground) are recommended; vertical drives (the line connecting the axes of sprockets is perpendicular to the ground) are less desirable. Vertical drives, if used, should be used with idlers to prevent the chain from sagging and to avoid disengagement from the lower sprocket.

18.6.2 KINEMATICS

The velocity ratio, comparable to the gear ratio given in Eq. (14.18), is

$$g_r = \frac{d_2}{d_1} = \frac{\omega_1}{\omega_2} = \frac{N_2}{N_1} \tag{18.22}$$

where

N = number of teeth in sprocket

ω = angular velocity, rad/s

d = diameter, m

For a one-step transmission it is recommended that $g_r < 7$. Values of g_r between 7 and 10 can be used at low speed (<650 ft/min).

18.6.3 CHORDAL RISE

An important factor affecting the smoothness of rolling chain drive operation, especially at high speeds, is chordal rise, shown in Fig. 18.10. From right triangle 0CA

$$r_c = r\cos\theta_r \tag{18.23}$$

The chordal rise while using Eq. (18.23) gives

$$\Delta r = r - r_c = r(1-\cos\theta_r) = r\left[1-\cos\left(\frac{180}{N}\right)\right] \tag{18.24}$$

where

N = number of teeth in sprocket

Note also from triangle 0CA

$$\sin\theta_r = \frac{p_t/2}{r} \qquad \text{or} \qquad p_t = 2r\sin\theta_r = D\sin\theta_r \tag{18.25}$$

18.6.4 CHAIN LENGTH

The number of links is

$$\frac{L}{p_t} = \frac{2c_d}{p_t} + \frac{N_1+N_2}{2} + \frac{(N_2-N_1)^2}{4\pi^2\left(\dfrac{c_d}{p_t}\right)} \tag{18.26}$$

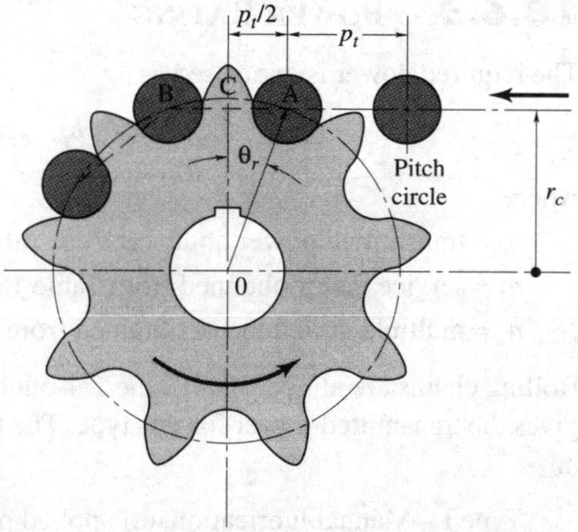

Figure 18.10 Chordal rise in rolling chains.

where

c_d = center distance between sprockets, m

It is normally recommended that c_d/p_t lie between 30 and 50 pitches. If the center distance per pitch is not given, the designer is free to fix c_d/p_t and to calculate L/p_t from Eq. (18.26). The nearest larger (preferably even) integer L/p_t should be chosen. With L/p_t as an integer the center distance per pitch becomes

$$\frac{c_d}{p_t} = A + \sqrt{A^2 - \frac{B^2}{2}} \tag{18.27}$$

where

$$A = \frac{L}{p_t} - \frac{N_1 + N_2}{2} \tag{18.28}$$

$$B = \frac{N_2 - N_1}{2\pi} \tag{18.29}$$

The value of c_d/p_t obtained from Eq. (18.27) should be decreased by about 1% to provide slack in the nondriving chain strand.

The chain velocity in feet per minute is

$$u_1 = \frac{\pi N_{a1} D_1}{12} \tag{18.30}$$

$$u_1 = \frac{N_{a1} p_t N}{12} \tag{18.31}$$

where

N_{a1} = speed of member 1, rpm

18.6.5 Power Rating

The required power is expressed as

$$h_{pr} = h_p a_1 a_2 \tag{18.32}$$

where

h_p = transmitted power obtained from Table 18.11

a_1 = service factor obtained from Table 18.12

a_2 = multiple-strand factor obtained from Table 18.13

Rolling chains are almost always the 1/4-pitch, no. 25, single-strand type, and Table 18.11 gives the transmitted power for this type. The four types of lubrication given in Table 18.11 are

Type I—Manual lubrication, oil applied periodically with brush or spout can

Table 18.11 Transmitted power of single-strand, no. 25 rolling chain.

Number of teeth in small sprocket	Small sprocket speed, rpm: 100	500	900	1200	1800	2500	3000	3500	4000	4500	5000	5500	6000	6500	7000	7500	8000	8500	9000	10 000
	Type 1				Type II lubrication															
11	0.054	0.23	0.39	0.50	0.73	0.98	1.15	1.32	1.42	1.19	1.01	0.88	0.77	0.68	0.61	0.55	0.50	0.46	0.42	0.36
12	0.059	0.25	0.43	0.55	0.80	1.07	1.26	1.45	1.62	1.36	1.16	1.00	0.88	0.78	0.70	0.63	0.57	0.52	0.48	0.41
13	0.064	0.27	0.47	0.60	0.87	1.17	1.38	1.58	1.78	1.53	1.30	1.13	0.99	0.88	0.79	0.71	0.64	0.59	0.54	0.46
14	0.070	0.30	0.50	0.65	0.94	1.27	1.49	1.71	1.93	1.71	1.46	1.26	1.11	0.94	0.88	0.79	0.72	0.66	0.60	0.51
15	0.075	0.32	0.54	0.70	1.01	1.36	1.61	1.85	2.06	1.89	1.62	1.40	1.23	1.09	0.96	0.88	0.80	0.73	0.67	0.57
16	0.081	0.34	0.58	0.75	1.09	1.46	1.72	1.88	2.23	2.08	1.78	1.54	1.35	1.20	1.07	0.97	0.88	0.80	0.74	0.63
17	0.086	0.37	0.62	0.81	1.16	1.56	1.84	2.11	2.38	2.28	1.95	1.69	1.48	1.31	1.18	1.06	0.96	0.88	0.81	0.69
18	0.097	0.39	0.66	0.86	1.23	1.66	1.85	2.25	2.53	2.49	2.12	1.84	1.52	1.43	1.28	1.16	1.05	0.96	0.88	0.75
19	0.097	0.41	0.70	0.91	1.31	1.76	2.07	2.38	2.69	2.70	2.30	2.00	1.75	1.55	1.39	1.25	1.14	1.04	0.96	0.81
20	0.103	0.44	0.74	0.96	1.38	1.86	2.19	2.52	2.84	2.91	2.49	2.16	1.89	1.68	1.50	1.35	1.23	1.12	1.03	0.88
21	0.108	0.46	0.78	1.01	1.46	1.96	2.31	2.65	2.99	3.13	2.68	2.32	2.04	1.80	1.61	1.46	1.32	1.21	1.11	0.95
22	0.114	0.48	0.82	1.06	1.53	2.06	2.43	2.79	3.15	3.36	2.87	2.49	2.18	1.93	1.73	1.56	1.42	1.29	1.19	1.01
23	0.119	0.51	0.86	1.12	1.61	2.16	2.55	2.93	3.30	3.59	3.07	2.66	2.33	2.07	1.85	1.67	1.51	1.38	1.27	1.06
24	0.125	0.53	0.90	1.17	1.69	2.26	2.67	3.07	3.46	3.83	3.27	2.83	2.48	2.20	1.97	1.78	1.61	1.47	1.35	1.16
25	0.131	0.56	0.94	1.22	1.76	2.37	2.79	3.20	3.61	4.07	3.48	3.01	2.64	2.34	2.10	1.89	1.72	1.52	1.44	1.23
28	0.148	0.63	1.07	1.38	1.99	2.67	3.15	3.62	4.28	4.54	4.12	3.57	3.13	2.78	2.49	2.24	2.04	1.86	1.71	1.46
30	0.159	0.68	1.15	1.49	2.14	2.88	3.39	3.90	4.40	4.89	4.57	3.96	3.47	3.08	2.76	2.49	2.26	2.06	1.89	1.62
32	0.170	0.73	1.23	1.60	2.30	3.09	3.64	4.18	4.71	5.24	5.03	4.36	3.83	3.39	3.04	2.74	2.49	2.27	2.06	1.78
35	0.188	0.80	1.36	1.76	2.53	3.40	4.31	4.62	5.19	5.78	5.76	4.99	4.38	3.88	3.48	3.13	2.85	2.60	2.38	2.04
40	0.217	0.92	1.57	2.03	2.93	3.93	4.63	5.32	6.00	6.67	7.04	6.10	5.35	4.75	4.25	3.83	3.48	3.17	2.91	2.49
45	0.246	1.05	1.78	2.31	3.32	4.46	5.26	6.04	6.81	7.58	8.33	7.28	6.39	5.66	5.07	4.57	4.15	3.79	3.48	2.97
50	0.276	1.18	1.99	2.58	3.72	5.00	5.89	6.77	7.64	8.49	9.33	8.52	7.48	6.63	5.93	5.35	4.96	4.44	4.07	3.48
55	0.306	1.30	2.21	2.96	4.12	5.54	6.53	7.51	8.46	9.41	10.3	9.83	8.63	7.65	6.85	6.17	5.60	5.12	4.76	4.01
60	0.336	1.43	2.43	3.15	4.53	6.09	7.18	8.25	9.30	10.3	11.3	11.2	9.83	8.72	7.80	7.03	6.38	5.83	5.35	4.57
	Type II				III	Type IV lubrication														

Table 18.12 Service factors for rolling chains.

Type of driven load	Type of input power: Internal combustion engine with hydraulic drive	Electric motor or turbine	Internal combustion engine with mechanical drive
Smooth	1.0	1.0	1.2
Moderate shock	1.2	1.3	1.4
Heavy shock	1.4	1.5	1.7

Table 18.13 Multiple-strand factors for rolling chains.

Number of strands	Multiple-strand factor, a_2
2	1.7
3	2.5
4	3.3

Type II—Drip lubrication, oil applied between link plate edges from a drop lubricator

Type III—Oil bath or oil slinger, oil level maintained in casing at predetermined height

Type IV—Oil stream, oil supplied by circulating pump inside chain loop or lower span

18.6.6 SELECTION OF SPROCKET SIZE AND CENTER DISTANCE

To ensure smooth performance and long life, the sprockets should consist of at least 17 teeth and no more than 67 teeth. For special situations where the speed is low or where space limitation is a factor, sprockets with fewer than 17 teeth can be used. Experience has shown that the velocity ratio given in Eq. (18.22) should not exceed 7.

The small sprocket should have a chain wrap angle of 120° to have satisfactory operation and performance. The chain wrap angle of the smaller sprocket of a two-sprocket drive with speed ratio less than 3.5 will always be 120° or more. Chain wrap angle increases as the center distance is increased. For normal applications the center distance is between 30 and 50 pitches (30 and 50 p_t) of the chain.

Example 18.5

GIVEN A four-strand, no. 25 rolling chain transmits power from a 21-tooth driving sprocket that turns at 1200 rpm. The speed ratio is 4:1.

FIND Determine the following:

a) Rated power of this drive

b) Tension in the chain

c) Safety factor of the chain based on minimum tensile strength
d) Chain length if center distance is about 10 in.

Solution

a) From Table 18.11 for a smaller sprocket of 21 teeth and a speed of 1200 rpm, the transmitted power is 1.01 hp and type II lubrication is required. From Table 18.13 for four strands $a_2 = 3.3$. Nothing is mentioned about the type of input power or drive load, so assume that $a_1 = 1$. From Eq. (18.32) the required power is

$$h_{pr} = h_p a_1 a_2 = (1.01)(1)(3.3) = 3.333 h_p$$

From Table 18.10 for no. 25 chain the pitch is 0.25 in.

b) The velocity in feet per minute can be obtained from Eq. (18.31) as

$$u_1 = \frac{N_{a1} p_t N_1}{12} = \frac{(1200)(0.25)(21)}{12} = 525.0 \text{ ft/min}$$

Making use of Eq. (4.37) gives

$$P_t = \frac{(33\,000) h_{pr}}{u_1} = \frac{(33\,000)(3.333)}{525} = 209.5 \text{ lbf}$$

c) From Table 18.10 the ultimate strength for a single strand is 875 lbf. For four strands the ultimate strength is 4(875) = 3500 lbf. The safety factor is

$$n_s = \frac{P_{\text{all}}}{P_t} = \frac{3500}{209.5} = 16.71$$

d) The number of teeth on the larger sprocket given a speed ratio of 4 is 4(21) = 84 teeth. However,

$$\frac{c_d}{p_t} = \frac{10}{\frac{1}{4}} = 40$$

which is between the 30 and 50 pitches recommended. From Eq. (18.26)

$$\frac{L}{p_t} = 2\left(\frac{c_d}{p_t}\right) + \frac{N_1 + N_2}{2} + \frac{(N_2 - N_1)^2}{4\pi^2\left(\frac{c_d}{p_t}\right)}$$

$$= 80 + \frac{21 + 84}{2} + \frac{(84 - 21)^2}{4\pi^2(40)}$$

$$= 80 + 52.5 + 2.513 = 135.0$$

$$\therefore L = 135 \text{ pitches}$$

Case Study | DESIGN OF GANTRY FOR DRAGLINE

GIVEN A dragline, often used for mining or dredging operations, uses a large bucket to remove material and transport it elsewhere, usually into a trailer or a train boxcar. The gantry line system is the portion of the dragline that supports the boom and fixes the boom angle, as shown in Fig. 18.11. The boom weighs 10 200 lbf, is 100 ft long, and can be considered to have its center of gravity at its geometric center. The dragline tips if the moment from the hoist rope and the boom exceeds 960 kip-ft. Figure 18.11 shows the dimensions of the dragline. The gantry line lengths are selected so that the 20-ft mast is vertical at a boom angle of 30°.

FIND Design a gantry line system for the dragline.

Solution

Since draglines have booms, many people confuse them with cranes. There are significant differences, however. Dragline load is limited by the size of the bucket, whereas a crane can attempt to lift extremely large loads. Further, a dragline will operate at a set boom angle for extended periods of time and will rarely, if ever, operate at very high or very low boom angles. Therefore, the range of boom angles that must be achieved is somewhat limited.

Just as with cranes, however, the gantry line system serves to eliminate bending forces on the boom. Thus, the gantry line must attach at the tip of the boom, known as the boom point. Attaching the gantry line on the crane's superstructure near the boom pivot will translate into extremely large forces because of static equilibrium. A mast is used to offset the gantry line system and obtain reasonably low stresses in the gantry line.

A number of clever design features have been incorporated into draglines in the past (Fig. 18.11). The lines closest to the crane have multiple pulleys (and hence are called a multiple-part line) to share the load and reduce the stress on the cable. Above the multiple-part line is a single line attached by proper couplings to the boom. Because this single line is never wound over a pulley, its size is determined by the rated strength of the cable.

The gantry line length can be obtained from geometry by considering the case when the boom is at a 30° elevation. The boom point is then 86.6 ft from the boom pivot and elevated 50 ft above the boom pivot. Since the mast is vertical and 20 ft long, the hoist rope length is given by

$$l_p = \sqrt{(86.6)^2 + (50-20)^2} = 91.6 \text{ ft}$$

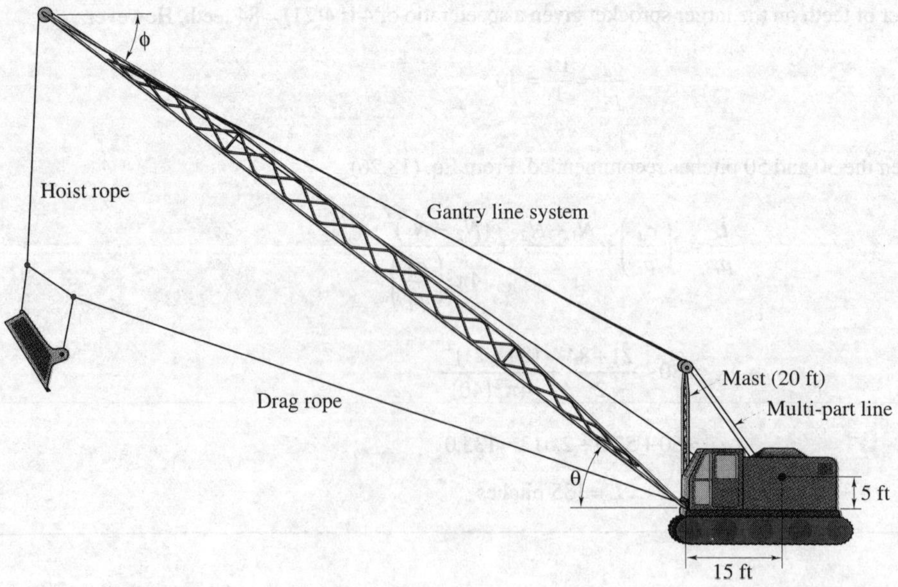

Figure 18.11 Typical dragline.

(continued)

Case Study (CONCLUDED)

The angle of the gantry line system from horizontal ψ is then 19.1°. At any other boom angle θ the gantry line angle is given by

$$\phi = 19.1° + \theta - 45° = \theta - 25.9°$$

The gantry line load can be expressed as a function of θ by taking the moment equilibrium of the boom about the boom pivot:

$$L(100)\cos\theta + W(50)\cos\theta - G\cos\phi(100\sin\theta) + G\sin\phi(100\cos\theta) = 0$$

But the dragline tips over when the load and boom moments combined exceed 960 kip-ft. Thus, the gantry line tension is given by

$$G = \frac{960}{100(\sin\theta\cos\phi - \cos\theta\sin\phi)} = \frac{9.6}{\sin(\theta-\phi)} = 22 \text{ kips}$$

The gantry line system must support 22 kips. Two lines, each to support 11 kips, are selected.

Because the mast rotates as the boom angle changes, the load supported by the gantry lines will change with the boom angle. From moment equilibrium of the mast the total load on the gantry lines is

$$T_g = \frac{T_p \cos 19.1°}{\sin(\zeta - \psi)}$$

where ζ is the angle of the mast from horizontal and ψ is the angle of the gantry line system from horizontal as shown above. These angles can be shown to be functions of θ:

$$\zeta = \theta + 60°$$

$$\psi = \tan^{-1}\left(\frac{20\sin\zeta - 5}{20\cos\zeta + 15}\right)$$

The gantry line tension is highest at a boom angle of 0°. At this angle the gantry line system must support a tension of 37.4 kips. Although draglines are not operated at such low boom angles because the lifting capacity is severely limited, this result places an upper bound on the load in the gantry line system.

In Fig. 18.11 a number of equalizer pulleys (a multiple-part line) are used so that the load in the gantry line is reduced. The load in the gantry line depends on the number of pulleys or parts, each of which supports an equal share of the load. Thus, for n parts to the line the gantry line system tension is

$$T_g = 37.4 \text{ kips}/n$$

The earlier discussion (Sec. 18.5.2) regarding pulley sizing is relevant. If the pulley and drum are too large, the motor capacity will be excessive. Per ASME requirements, pulley and drum diameters of 12d are to be chosen. These diameters reduce gantry line strength by approximately 12% (Fig. 18.7). Table 18.14 summarizes the results. From this summary, a number of reasonable alternatives exist. A five-part gantry line is selected as a good compromise between cost and safety.

Table 18.14 Summary of results of case study.

Number of parts in gantry line	Minimum diameter of 6 × 37 line with 100-ksi breaking strength, in.	Diameter selected (next largest 0.25-in. increment), in.	Actual safety factor
4	0.90	1.0	7.4
5	0.80	1.0	9.2
6	0.73	0.75	6.24
7	0.68	0.75	7.27
8	0.63	0.75	10.7

18.7 SUMMARY

Belts and ropes are machine elements (like brakes and clutches) that use friction as a useful agent, in contrast to other machine elements in which friction is to be kept as low as possible. Belts, ropes, and chains provide a convenient means for transferring power from one shaft to another. This chapter discussed both flat and V-belts. All flat belts are subject to creep due to the relative motion between the pulley surface and the adjacent belt surface that is under load deformation from the combined tension and flexural stresses. Also, flat belts must be kept under tension to function and therefore require tensioning devices. One major advantage of V-belts over flat belts is that the wedging action of V-belts increases the normal force from dN to $dN/\sin\beta$, where β is the sheave angle.

Wire ropes are used instead of flat or V-belts when power must be transmitted over very long center distances. The major advantage of using rolling chains over flat or V-belts is that rolling chains do not slip. Another advantage of rolling chains over belt drives is that no initial tension is necessary and thus the shaft loads are smaller. The required length, power rating, and modes of failure were considered for belts, ropes, and rolling chains.

KEY WORDS

core center of wire rope, mainly intended to support outer strands

flat belts power transmission device that consists of loop of rectangular cross section placed under tension between pulleys

lay type of twist in wire rope (regular or Lang); distance for strand to revolve around wire rope

rolling chains power transmission device using rollers and links to form continuous loop, used with sprockets

sheaves grooved pulleys that V-belts run in

strands grouping of wire used to construct wire rope

synchronous belt flat belt with series of evenly spaced teeth on inside circumference, intended to eliminate slip and creep

timing belt same as synchronous belt

V-belt power transmission device with trapezoidal cross section placed under tension between grooved sheaves

wire rope wound collection of strands

RECOMMENDED READINGS

American Chain Association, *Chains for Power Transmission and Material Handling*, Marcel Dekker, New York.

Dickie, D.E. (1985) *Rigging Manual*, Construction Safety Association of Ontario, Toronto.

Juvinall, R.C., and Marshek, K.M. (1991) *Fundamentals of Machine Component Design*, Wiley, New York.

Krutz, G.W., Schuelle, J.K., and Claar, P.W. (1994) *Machine Design for Mobile and Industrial Applications*, Society of Automotive Engineers, Warrendale, PA.

Mott, R.L. (1985) *Machine Elements in Mechanical Design*, Merrill Publishing Co., Columbus, OH.
Rossnagel, W.E., Higgins, L.R., and MacDonald, J.A. (1988) *Handbook of Rigging*, 4th ed., McGraw-Hill, New York.
Shapiro, H.I., et al. (1991) *Cranes and Derricks*, 2nd ed., McGraw-Hill, New York.
Shigley, J.E., and Mischke, C.R. (1989) *Mechanical Engineering Design*, 5th ed., McGraw-Hill, New York.
Wire Rope Users Manual (1972) Armco Steel Corp.

REFERENCES

ASME (1994) B30.5 *Mobile and Locomotive Cranes*, American Society of Mechanical Engineers, New York.
Shigley and Mitchell (1983) *Mechanical Engineering Design*, 4th ed., McGraw-Hill, New York.

PROBLEMS

Section 18.2

18.1 An open flat belt 10 in. wide and 0.13 in. thick connects a 16-in.-diameter pulley with a 36-in.-diameter pulley over a center distance of 15 ft. The belt speed is 3600 ft/min. The allowable preload per unit width of the belt is 100 lbf/in., and the weight per volume is 0.042 lbf/in.^{3}. The coefficient of friction between the belt material and the pulley is 0.8. Find the length of the belt, the maximum forces acting on the belt before failure is experienced, and the maximum power that can be transmitted.

18.2 The driving and the driven pulley in an open-flat-belt transmission each have a diameter of 160 mm. Calculate the preload needed in the belt if a power of 7 kW should be transmitted at 1000 rpm by using only half the wrap angle on each pulley. The coefficient of friction is 0.25, the density of the belt material is 1500 kg/m^3, and the allowable belt preload stress is 5 MPa. Also calculate the cross-sectional area of the belt.

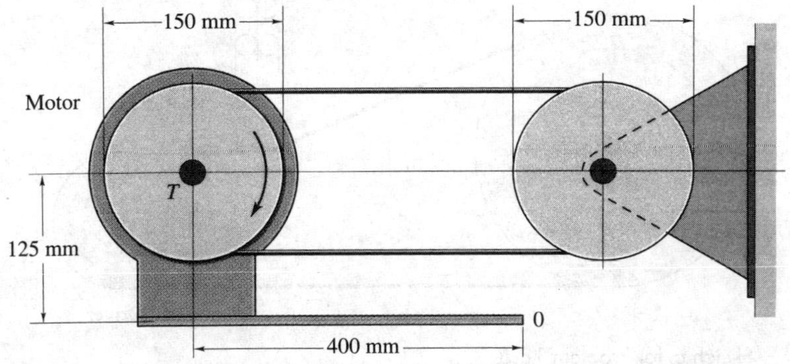

Sketch *a*, for Problem 18.3

18.3 A belt transmission is driven by a motor hinged at point 0 in sketch *a*. The motor mass is 50 kg and its speed is 1400 rpm. Calculate the maximum transferable power for the transmission. Check, also, that the largest belt stress is lower than σ_{all} when the bending stresses are included. Belt dimensions are 100 by 4 mm^2, the coefficient of friction is 0.32, the modulus of elasticity is 100 MPa, the belt density is 1200 kg/m^3, and the allowable stress is 20 MPa.

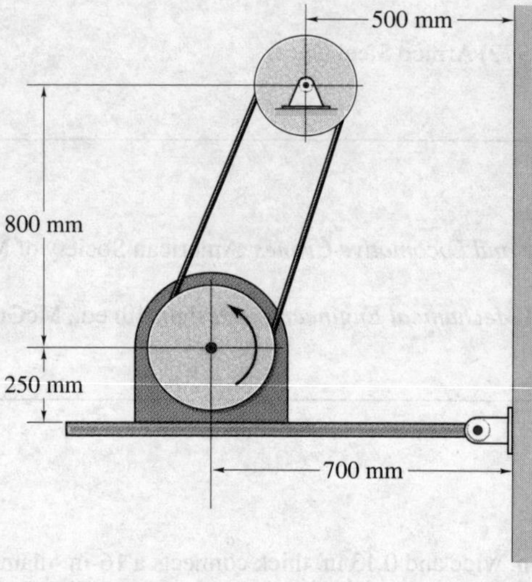

Sketch *b*, for Problem 18.4

18.4 The tension in a flat belt is given by the motor weight, as shown in sketch *b*. The mass is 80 kg and is assumed to be concentrated at the motor shaft position. The motor speed is 1405 rpm and the pulley diameter is 400 mm. Calculate the belt width when the allowable belt stress is 6.00 MPa, the coefficient of friction is 0.5, the belt thickness is 5 mm, the modulus of elasticity is 150 MPa, and the density is 1200 kg/m^3.

18.5 Calculate the maximum possible power transmitted when a full wrap angle is used as shown in sketch c. The belt speed is 5 m/s and the coefficient of friction is 0.19.

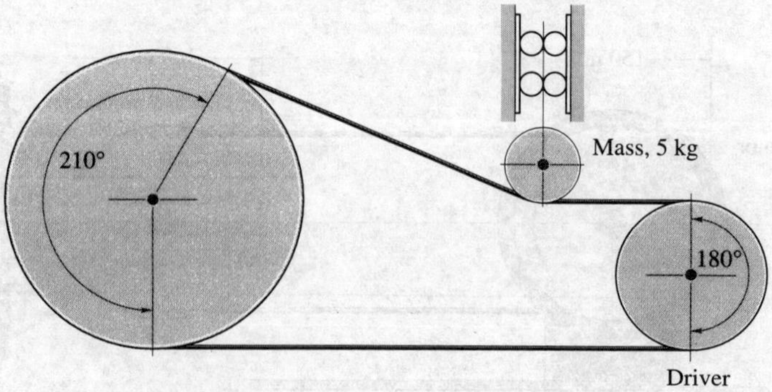

Sketch *c*, for Problem 18.5

18.6 Equation (18.10) gives the belt forces, where F_1 is the largest belt force. What is the maximum power the belt can transmit for a given value of F_1?

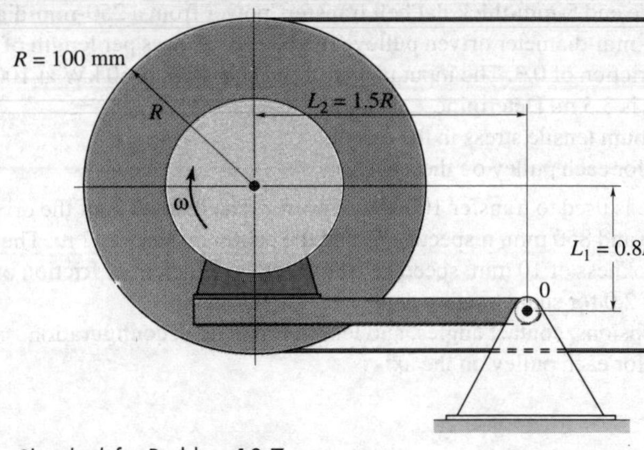

Sketch *d*, for Problem 18.7

18.7 In a flat belt drive with parallel belts the motor and pulley are pivoted at point 0 as shown in sketch *d*. Calculate the maximum possible power transmitted at 1200 rpm. The motor and pulley together have a 50-kg mass at the center of the shaft. The centrifugal forces on the belt have to be included in the analysis. Coefficient of friction is 0.3, belt area is 300 mm^2, and density is 1500 kg/m^3.

18.8 A flat belt drive, shown in sketch *e*, has the top left pulley driven by a motor and driving the other two pulleys. Calculate the wrap angles for the two driven pulleys when the full wrap angle is used on the driving pulley and the power from pulley 2 divided by the power from pulley 3 is equal to 2. The coefficient of friction is 0.3 and $l = 4r$.

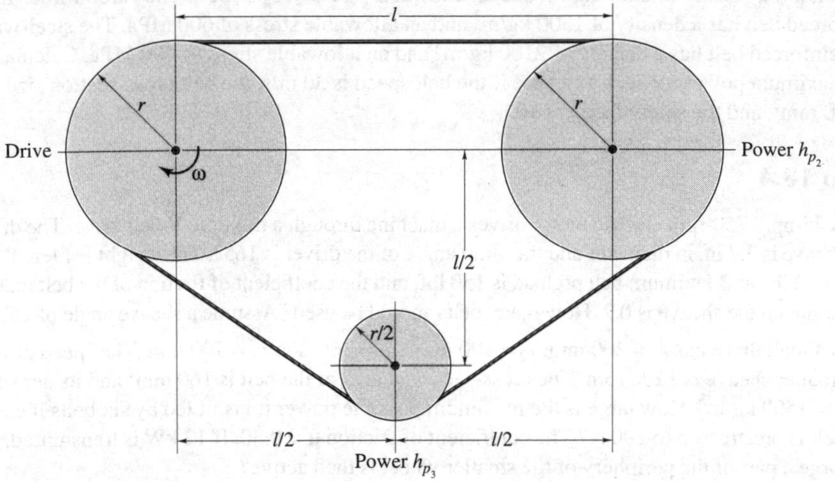

Sketch *e*, for Problem 18.8

18.9 A flat belt 6 mm thick and 60 mm wide is used in a belt transmission with a speed ratio of 1. The radius of the pulleys is 100 mm, their angular speed is 50 rad/s, the center distance is 1000 mm, the belt modulus of elasticity is 200 MPa, and the density is 1 kg/dm^3. The belt is pretensioned by increasing the center distance. Calculate this increase in center distance if the transmission should use only half the wrap angle to transmit 1 kW. Centrifugal effects should be considered. The coefficient of friction is 0.3.

18.10 A 120-mm-wide and 5-mm-thick flat belt transfers power from a 250-mm-diameter driving pulley to a 700-mm-diameter driven pulley. The belt has a mass per length of 2.1 kg/m and a coefficient of friction of 0.3. The input power of the belt drive is 60 kW at 1000 rpm. The center distance is 3.5 m. Determine

a) The maximum tensile stress in the belt

b) The loads for each pulley on the axis

18.11 A flat belt drive is used to transfer 100 kW of power. The diameters of the driving and driven pulleys are 300 and 850 mm, respectively, and the center distance is 2 m. The belt has a width of 200 mm, thickness of 10 mm, speed of 20 m/s, and coefficient of friction of 0.35. For a safety factor of 2.0 for static loading determine the following:

a) The belt tensions, contact angles, and length for an open configuration

b) The loads for each pulley on the axis

Section 18.3

18.12 A synchronous belt transmission is used as a timing belt for an overhead camshaft on a car motor. The belt is elastically prestressed at standstill to make sure it does not slip at high speeds due to centrifugal forces. The belt is 1100 mm long and weighs 200 g. The elastic spring constant for 1 m of belt material is 10^5 N/m. The belt prestress elongates it 2 mm. It needs to elongate 4 mm more to start slipping. Calculate the maximum allowable motor speed if the pulley on the motor shaft has a diameter of 100 mm.

18.13 A timing belt for power transfer should be used at a velocity ratio of one-third, so that the outgoing speed should be three times as high as the incoming speed. The material for the belt reinforcement can be chosen from glass fiber, carbon fiber, and steel wire. These materials give different belt densities as well as different tensile strengths. The glass-fiber-reinforced belt has a density of 1400 kg/m^3 and an allowable stress of 300 MPa. The carbon-fiber-reinforced belt has a density of 1300 kg/m^3 and an allowable stress of 600 MPa. The steel-wire-reinforced belt has a density of 2100 kg/m^3 and an allowable stress of 400 MPa. Calculate the maximum power for each belt type if the belt speed is 30 m/s, the belt cross-section size is 50 mm^2, and the safety factor is 10.

Section 18.4

18.14 A 25-hp, 1750-rpm electric motor drives a machine through a multiple V-belt drive. The driver sheave is 3.7 in. in diameter and the wrap angle of the driver is 165°. The weight per length is 0.012 lbf/in. Maximum belt preload is 150 lbf, and the coefficient of friction of the belt material acting on the sheave is 0.2. How many belts should be used? Assume a sheave angle of 18°.

18.15 A V-belt drive has $r_1 = 200$ mm, $r_2 = 100$ mm, $2\beta = 36°$, and $c_d = 700$ mm. The speed of the smaller sheave is 1200 rpm. The cross-sectional area of the belt is 160 mm^2 and its density $\rho = 1500$ kg/m^3. How large is the maximum possible power transmitted by six belts if each belt is prestressed to 200 N? The coefficient of friction $\mu = 0.30$. If 12 kW is transmitted, how large a part of the periphery of the smaller wheel is then active?

18.16 A combined V-belt and flat belt drive, shown in sketch *f*, has a speed ratio of 4. Three V-belts drive the sheave in three grooves but connect to the larger sheave like flat belts on the cylindrical surface. The center distance is six times the smaller sheave radius, which is 80 mm with a speed of 1500 rpm. Determine which sheave can transmit the largest power without slip. Determine the ratio between the powers transmitted through the two sheaves. The angle 2β is 36° and the coefficient of friction is 0.30.

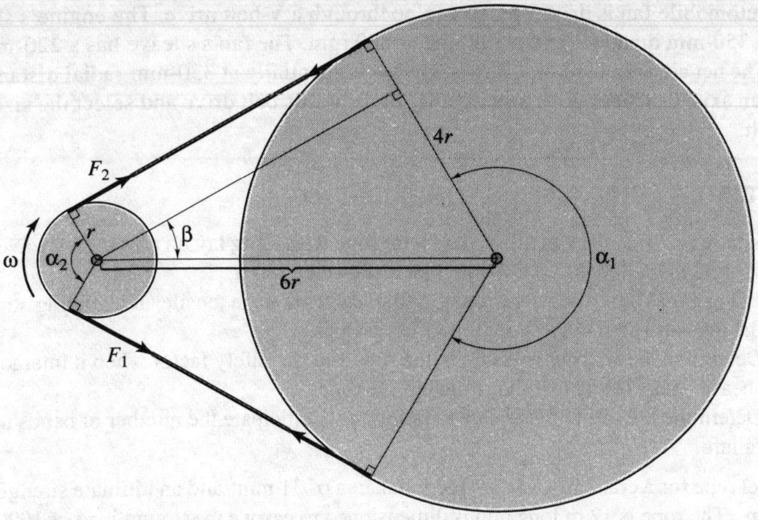

Sketch *f*, for Problem 18.16

18.17 A V-belt drive for an industrial fan transmits 4 kW at a motor speed of 3000 rpm through four parallel V-belts of the 4*L* type. The sheave effective pitch diameter is 4 in. and the speed ratio is 1. The fan is running continuously. Calculate what the manufacturers presumed the coefficient of friction to be to get the rated power for the belt drive. Neglect power variations and assume that all friction forces between the belt and the sheaves are tangent to the sheaves. The density of the V-belt is assumed to be 1000 kg/m^3 and the prestress gives $F_1 + F_2$ = constant independent of speed. Use the power ratings at 3000 and 5000 rpm for the calculations.

18.18 The input power to shaft A, shown in sketch *g*, is transferred to shaft B through a pair of mating spur gears and then to shaft C through a 2*L*-section V-belt drive. The sheaves on shafts B and C have 76- and 200-mm diameters. The center distance is approximately 200 mm. For the maximum power the belt can transfer, determine the following:

a) The input and output power torques of the system

b) The belt length for an approximate center distance of 550 mm

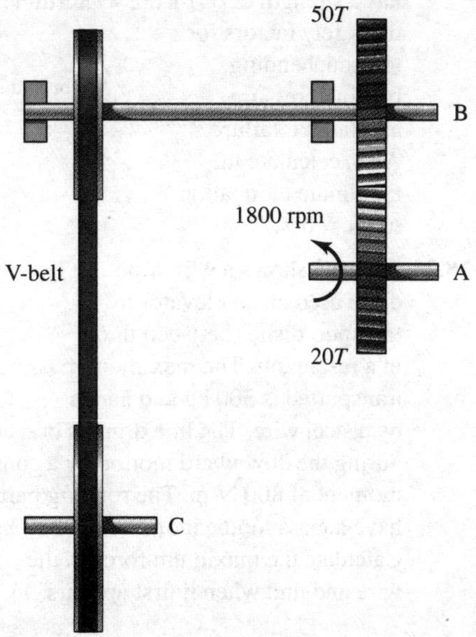

Sketch *g*, for Problem 18.18

18.19 An automobile fan is driven by an engine through a V-belt drive. The engine's sheave has a 350-mm diameter and is running at 880 rpm. The fan's sheave has a 220-mm diameter. The net air drag load of 2.3 N is applied tangentially at 330-mm radial distance from the fan axis. Consider 20% loss in other parts of this belt drive and select the size of the V-belt.

Section 18.5

18.20 A hoist uses a 2 in., 6×19 monitor steel wire rope with Lang lay on cast steel sheaves. The rope is used to haul loads up to 8000 lbf a distance of 480 ft.

a) Using a maximum acceleration of 2 ft/s^2, determine the tensile and bending stresses in the rope and their corresponding safety factors.

b) Determine the bearing pressure in the rope and the safety factor when it interacts with the sheave. Also, determine the stretch of the rope.

c) Determine the safety factor due to fatigue and anticipate the number of bends until failure.

18.21 A steel rope for a crane has a cross-sectional area of 31 mm^2 and an ultimate strength of 1500 N/mm^2. The rope is 12 m long and is dimensioned to carry a maximum load of 1000 kg. If that load is allowed to free-fall 1 m before the rope is tightened, how large will the stress be in the rope? The modulus of elasticity is 68 GPa.

18.22 Two 2-cm, 6×19 plow-steel wire ropes are used to haul mining material to a depth of 150 m at a speed of 8 m/s and an acceleration of 1 m/s^2. The pulley diameter is 80 cm. Using a proper safety factor, determine the maximum hauling load for these ropes. Are two ropes enough for infinite life?

18.23 Using six 16-mm, 6×19 wire ropes, an elevator is to lift a 2500-kg weight to a height of 81 m at a speed of 4 m/s and an acceleration of 1 m/s^2. The 726-mm-diameter pulley has a strength of 641 MPa. Determine the safety factors for tension, bending, bearing pressure, and fatigue failure. Also, calculate the maximum elongation of the ropes.

18.24 Sketch *h* shows a wire rope drive used on an elevator to transport dishes between floors in a restaurant. The maximum mass being transported is 300 kg and hangs by a steel wire. The line drum is braked during the downward motion by a constant moment of 800 N-m. The rotating parts have a mass moment of inertia of 13 kg-m^2. Calculate the maximum force in the wire and find when it first appears.

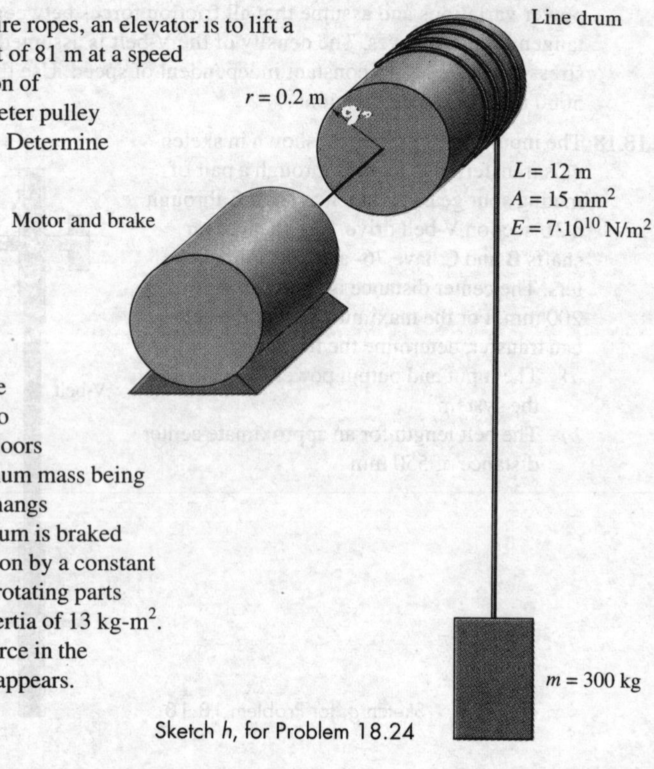

Sketch *h*, for Problem 18.24

18.25 End and front views of a hoisting machine are shown in sketch *i*. Calculate the lifting height per drum revolution. Calculate the maximum force in the wire when the drum is instantly started with angular speed ω to lift a mass m. The free length of the wire is L and its cross-sectional area is A. The modulus of elasticity of the wire is E.

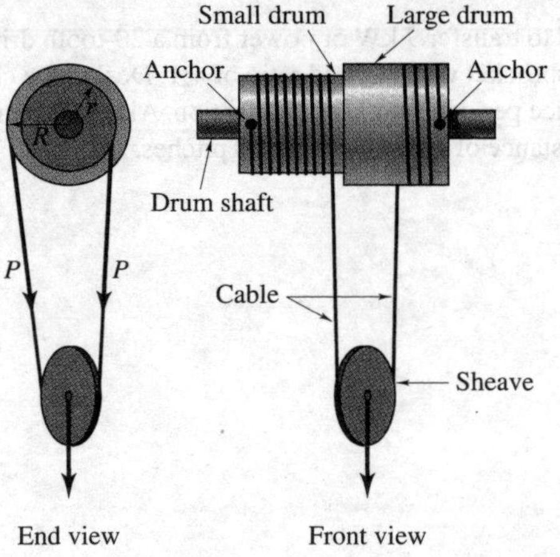

Sketch *i*, for Problem 18.25

18.26 A hauling unit with 25-mm, 8×19 wire rope made of fiber-core plow steel is used to raise a 45-kN load to a height of 150 m at a speed of 5 m/s and an acceleration of 2 m/s^2. Design for a minimum pulley diameter while determining the safety factors for tensile strength, bending stress, bearing pressure, and fatigue endurance.

18.27 A building elevator operates at a speed of 5 m/s and 1.2-m/s^2 acceleration and is designed for 10-kN deadweight, five 70-kg passengers, and 150-kg overload. The building has twelve 4-m stories. Using the proper safety factor, design the wire rope required for the elevator. A maximum of four ropes may be used.

Section 18.6

18.28 A three-strand, ANSI 50 roller chain is used to transfer power from a 20-tooth driving sprocket that rotates at 500 rpm to a 65-tooth driven sprocket. The input power is from an internal combustion engine, and the chain experiences medium shock. Determine the following:

a) The normal power
b) The length of the chain for an approximate center distance of 550 mm
c) The chordal speed variation

18.29 To transfer 100 kW of power at 400 rpm, two strands of roller chain are needed. The load characteristics are heavy shock, poor lubrication, average temperature, and 16 hr of service per day. The driving sprocket has 13 teeth and the driven sprocket has 42 teeth. Determine the kind, length, and size of the chain for a center distance close to 50 pitches.

18.30 A two-strand AISI 80 roller chain system is used to transmit the power of an electric motor rotating at 500 rpm. The driver sprocket has 12 teeth and the driven sprocket has 60 teeth. The operating condition is specified by precision mounting and light shock. Calculate the power rating, the chordal speed variation, and the chain length for the maximum center distance.

18.31 A roller chain is used to transfer 9 kW of power from a 20-tooth driving sprocket running at 600 rpm to a driven sprocket with a speed ratio of 4:1. Design the chain drive for heavy shocks, 20 hr of service per day, and poor lubrication. Also, determine the required chain length for a center distance of approximately 70 pitches.

CHAPTER

19

DESIGN PROJECTS

A roller coaster. The brake system for such a roller coaster is considered in Section 19.3. (AP/Wide World Photos)

In your further operations you must of course be guided by the circumstances by which you are surrounded, and the information you are able to collect, and must not consider yourself committed to any particular line of conduct, but be governed by your good judgment.

Robert E. Lee

SYMBOLS

A	area, m^2
a_d	deceleration, equivalent to $\ddot{x}$, m/s^2
b	brake width, m
b_w	face width, m
C_r	contact ratio
C_1, C_2	integration constants
c	distance from neutral axis to outer fiber, m
D	mean thread diameter, m
D_1, D_2	mean spring diameters, m
d	diameter, m
d_c	crest diameter, m
d_h	hydraulic cylinder diameter, m
d_w	wheel diameter, m
d_6, d_7	distance from neutral axis to outer fiber, m
F	friction force, N
G	shear modulus of elasticity, Pa
g	gravitational acceleration, 9.807 m/s^2
g_r	gear ratio
h_p	power, hp
h_t	height mass acts above ground, m
I	area moment of inertia, m^4
J_1, J_2	flywheel and shaft mass moments of inertia, respectively
L_g	grip length, m
l	wheel base length, m
m	module, d/N, mm
m_a	mass, kg
N	number of teeth; number of coils
N_a	rotational speed, rpm
n_l	portion of base length from front wheel to center of gravity divided by base length
n_s	safety factor
n_μ	used portion of maximum friction
P	force, N
p	pressure, Pa; pitch
$p_{\max}$	maximum pressure, Pa
q_n	notch sensitivity factor
r	drum or roller radius, m
S_r	speed reduction
S_{sy}	shear yield strength, Pa
S_{ut}	ultimate tensile strength, Pa
S_y	yield strength, Pa
T	torque, N-m
t	time, s
t_w	thickness of thinnest member, m
V	volume, m^3
v	velocity, m/s
W	actuating force, N
w_z	normal load, N
w_{zf}	normal load applied to the front wheel, N
w_{zr}	normal load applied to the rear wheel, N
x	stopping or braking distance, m
$\dot{x}$	velocity, m/s
$\ddot{x}$	deceleration, m/s^2
Z_m	section modulus, m^3
α	lead angle; backplate angle, deg
β	thread angle; sheave angle, deg
θ	circumferential coordinate, deg
θ_a	angle where $p = p_{\max}$, deg
θ_1, θ_2	locations where shoe begins and ends, respectively
μ	maximum coefficient of friction between road and tire
μ_b	coefficient of friction of brake and drum
μ_n	used portion of maximum coefficient of friction
σ	normal stress, Pa
σ_r	contact pressure, Pa
τ	shear stress, Pa
ϕ	pressure angle, deg
ψ	helix angle, deg
ω	angular speed, rad/s

Subscripts

f	front
g	gear
i	ingoing
max	maximum
n	normal
r	rear
s	spring
t	tangential
I	shoe I
II	shoe II

19.1 INTRODUCTION

This textbook has summarized the most common machine elements used to form mechanical systems. However, the approach to system design has not been formally expounded upon aside from occasional notes in case studies. It is difficult to address both machine element design and mechanical system design in the same book, let alone in a university course. Whereas machine element design requires familiarity with analysis and the development of analytical skills, mechanical system design also requires analysis as well as synthesis, or the ability to creatively reach new solutions.

A notion popular among many educators and practicing engineers is that synthesis of good designs requires the ability to analyze components. Once a component has been analyzed several times (e.g., to size a gear from given torques and speeds), an engineer can give a "ballpark" estimate of gear size. Then, when laying out a power transmission system the engineer can evaluate whether a gear train is viable (e.g., for space reasons) or whether shafting, rubber belts, chains, or wire ropes are better alternatives. Often, some design alternatives are eliminated without extensive calculations. The surviving candidates are then considered, and detailed design work in the form of robust analysis takes place on superior alternatives. Thus, the analysis portion of system design follows the synthesis, but the synthesis cannot be performed without analysis of components.

This chapter presents three system design projects that are completely considered. When possible, some of the design engineer's reasons for pursuing a specific alternative are discussed in detail. At the end of the chapter 26 additional projects are suggested.

19.2 DESIGN OF VERTICAL MIXER

Vertical mixers are common equipment in the food preparation industry. Figure 19.1 shows a typical vertical mixer. The following points should be made regarding its operation:

1. Vertical mixers, with the proper agitators, can be used to mix doughs, icings, frostings, etc. Each product may require a different agitator and a different agitator speed. Thus, the speed must be adjustable, and the agitator must be easily removable without using tools.
2. Sanitation standards and codes must be followed. A complete description of these codes is beyond the scope of this section, but all surfaces that contact food must be readily accessible for cleaning. All surfaces that do not contact food should be isolated behind sealed housings when possible. Further, if lubricated gears are used, they must be effectively sealed so that lubricant cannot drip into food contact zones.
3. Vertical mixers have their capacities given in terms of the mixer bowl volume, typically 20, 30, 40, 60, or 100 quarts. After a batter or dough has been mixed, it must be transferred to a work surface or conveyor. Therefore, access to the bowl must be large enough for material-handling equipment (although the smaller bowls are usually lifted manually by one or two persons). However, if the bowl must be lifted or lowered with respect to the agitator, a mechanical force multiplication system will be needed.

The design of a vertical mixer, indeed of any industrial equipment, requires multiple disciplines and in-depth knowledge of the end use. Certainly, in this case the complete

Figure 19.1 Typical 30-quart vertical mixer with common agitator attachments. [Courtesy of Hobart.]

design problem would involve electrical engineering and materials science contributions that will not be discussed here. However, in an industrial setting the mechanical engineer would have to communicate effectively with these groups to ensure a smooth product development cycle. This example merely focuses on certain mechanical elements of a small, commercial vertical mixer.

The bowl volume is 30 quarts, and the intended customer is a commercial restaurant that will use the mixer for doughs, sauces, batters, and icings. Desired attributes of such a machine are difficult to obtain except from knowledge of previous applications and direct communication with end users. Often, such information is labeled as "market information" and is collected with more or less diligence depending on the firm. Regardless, it is assumed that the information contained in Table 19.1 was obtained from a source other than an engineering department, either from end users or from marketing department personnel.

Table 19.1 Requirements for vertical mixer.

Machine bowl capacity, quarts	30
Maximum height, ft	6
Maximum floor footprint, ft	2.5 by 3
Agitator path	Circular, clockwise as viewed from top
Agitator rotation	Counterclockwise as viewed from top, approx. 3 revolutions per agitator path revolution, but not critical (1 to 5 is acceptable)
Number of speeds	2
High-speed target range (rotation), rpm	100
Low-speed target range (rotation), rpm	50
Bowl infeed	Rack-and-pinion lift to agitator position
Motor operating speed, rpm	1000

Because of space limitations Table 19.1 lists only a small subset of the usual starting points for the design synthesis portion of product development. These basic design attributes can be thought of as the minimum requirements of the desired machine. They are used by engineering designers in selecting components and systems, by marketing personnel in drafting sales and product literature, and by procurement departments in purchasing parts or prototypes from vendors.

19.2.1 KINEMATICS OF AGITATOR GEARS

Agitator movement within the bowl of a vertical mixer can be simply achieved through a number of mechanisms, but the added requirement of agitator rotation while following a circular path eliminates some possibilities and makes others extremely difficult. For example, it is certainly possible, but extremely complicated, to mount an agitator assembly on a pulley (to achieve the circular path) and incorporate a motor in the agitator assembly (to achieve rotation). A far more elegant and straightforward solution is to use a planetary gear system. If the agitator is directly mounted on a properly designed planet gear, it will naturally follow a circular path and simultaneously rotate as desired.

Using existing agitators is a marketing advantage because mass-produced agitators are less expensive. The potential layout of such a system reveals that the radius of the circular agitator path must be approximately 5 in. Figure 19.2 illustrates a planetary gear system. Although only one planet gear is used in the analysis, the system will have three planet gears. The sun gear will be powered, and the ring gears will be stationary and fixed to the machine frame. The agitator will be mounted onto a shaft attached to a planet gear.

The following equation relates the relative speeds of the ring and sun gears:

$$\frac{\omega_{ring} - \omega_{arm}}{\omega_{sun} - \omega_{arm}} = -\frac{N_{sun}}{N_{ring}} \tag{19.1}$$

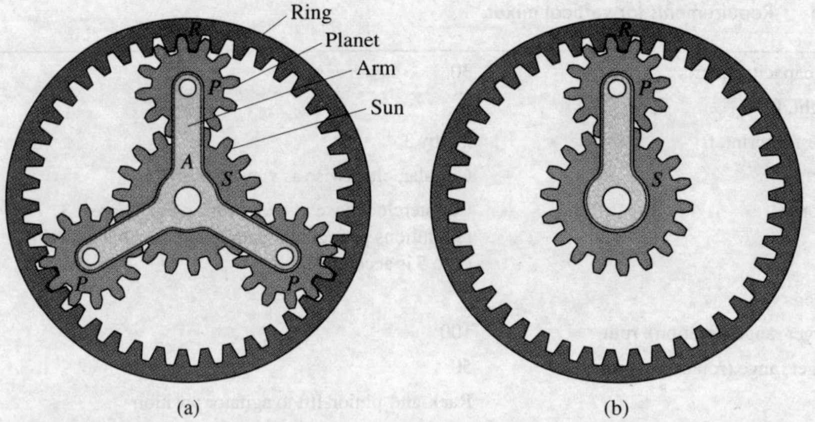

Figure 19.2 Illustration of planetary gear.
(a) With three planets (typical); (b) with one planet (for analysis only). (The agitator is attached to a planet.)

Since the arm must rotate at or near 100 rpm for the high-speed setting and 50 rpm for the low-speed setting, with the ring gear stationary ($\omega_{ring} = 0$), this becomes a relationship between ω_{sun}, N_{sun}, and N_{ring}. The following equation holds for the planet gear:

$$\frac{\omega_{planet} - \omega_{arm}}{\omega_{sun} - \omega_{arm}} = -\frac{N_{sun}}{N_{planet}} \tag{19.2}$$

The teeth in the sun, planet, and ring gears must have the same diametral pitch, and the number of teeth are related by

$$N_{ring} = N_{sun} + 2N_{planet} \tag{19.3}$$

Within the housing of the vertical mixer, there is room for a ring gear if it is less than 20 in. in diameter. The data in Table 19.2 can then be produced for a gear with a diametral pitch of 5 using an arm angular velocity of 100 rpm as an input with an 18-in.-diameter ring gear. These data show that the required input velocity on the sun gear is higher than the motor output for small sun gear diameters. This situation could in theory be achieved with speed-increasing gearing between the sun gear and the motor, but this is poor design practice because of torque limitations on the motor. Usually, electric motors provide low torque at high speed; using speed reducers increases the torque. A small sun gear would have the opposite effect.

Therefore, an 8-in.-diameter sun gear is selected, with 5-in.-diameter planet gears, operating inside an 18-in.-diameter ring. For the low-speed setting the sun gear must be driven by an input velocity of 325/2 = 162 rpm. The arm radius is approximately 6.5 in. in this setup, and the planet gear rotation is sufficient to allow for proper agitator spin while it follows a circular path. Figure 19.3 is a detailed schematic of the planetary gear head. The incoming speed to the sun gear is a requirement that must still be met. The approach to this is discussed in Sec. 19.2.2.

Table 19.2 Data for vertical mixer.

Number of sun gear teeth	Sun gear diameter, in.	Number of planet gear teeth	Planet gear speed, ω_{planet}, rpm	Sun gear speed, ω_{sun}, rpm
10	2	40	325	1000
20	4	35	357	550
30	6	30	400	400
40	8	25	460	325
50	10	20	550	280
60	12	15	700	250

So far the kinematic nature of the agitator gears has been specified to meet the design criteria, but no mention has been made of the proper gear size. Indeed, no information exists regarding what torques are to be transmitted by the gear. To define the proper gear thickness, a study would have to be made of the torques encountered in service. A more direct approach is to use the motor manufacturer's data on the motor stall torque as a worst-case scenario and design the gears to survive this torque.

19.2.2 POWER TRANSMISSION SYSTEM

The mixer as designed thus far requires two speeds to be input to the sun gear, either 325 or 162 rpm. Electric motors commonly use either a gear-based speed reduction system or a belt drive. Gear speed reducers usually reduce speed much more than needed here, so a belt drive is the obvious choice.

The power transmission approach has a gear speed reducer transmitting power to a shaft onto which are mounted several bevel gears. Additional gears needed for grinding attachments are at the front of the mixer. The bevel gears mesh with a second set of bevel gears mounted on an axially translatable shaft. Thus, either set of bevel gears can contact the pulley shaft, but not both. This driven shaft then meshes with additional gears mounted to the sun gear of the planetary gear train discussed in the earlier section. All that is necessary to obtain the proper agitator speeds are the gear ratios needed for high- and low-speed operation.

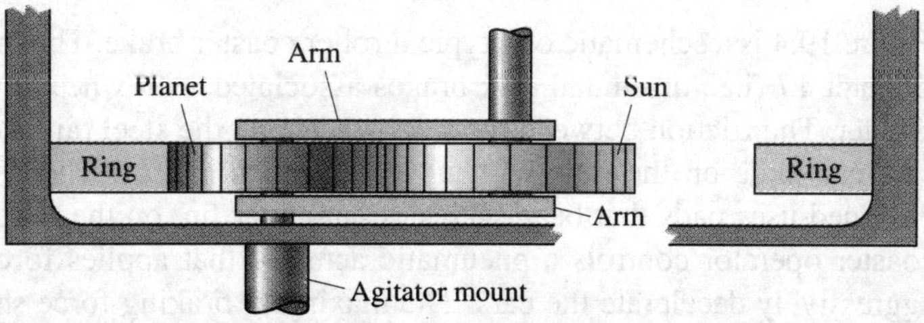

Figure 19.3 Schematic of planetary gear head.

The sun gear must be driven at 325/1000 = 0.32 times the shaft speed for high-speed operation and at 0.16 times the shaft speed for low-speed operation. Speed is reduced by one-third through the gear reducer attached to the motor. Therefore, the driven gears must have 1.04 times the number of teeth needed prior to speed reduction for high-speed operation and 2.08 times for low-speed operation. This requirement can be easily accomplished for a 12-diametral-pitch gear system by increasing the teeth of the pulley-driven shaft gears from 24 to 25 and those of the sun gear shaft gears from 36 to 50.

19.2.3 FURTHER NOTES

Selecting the proper gears for a power transmission system would normally require calculating tooth contact stresses and shear stresses as discussed in Chapter 14. However, this approach has not been followed here because of space limitations.

The final mixer design does not require a clutch. The operator merely engages the motor while the mixer is at the proper speed setting. Although this procedure could cause significant dynamic loads in most applications, the agitator is either free to move when starting or is constrained by an extremely compliant material (batter or dough). Thus, the major dynamic forces are inertial and can be safely overcome without a clutch.

Although the power transmission system naturally calls for a planetary gear system, a number of gears transmit power elsewhere in the mixer where belts or chains would initially appear to serve the purpose. Gears are preferred for their reliability; gears rarely fail. Indeed, some commercial mixers use only gearing for power transmission. Then why use a belt drive at all? Belts can be a safety device to prevent motor damage. If the agitator jams, the belt slips but does not stall the motor. Since jams are so rare in food mixers, it is reasonable to pursue either a gear or a belt design.

The shafting used for the gears can be based on deflection criteria as in Chapter 11, but in the vertical mixer the gears are so close together that the extremely small torsional deflection requires only a very small diameter shaft. But, as is typical of many industrial applications, similar equipment in use has much larger shafts than necessary. The reason is that the bevel gears used are mass produced and therefore of standard size. Part of the size designation is a hub size. If the hub is much larger than the shaft, a sleeve must be fabricated around the shaft and the gear placed over the sleeve to assemble the system. An alternative is to use a shaft the same size as the gear hub. This approach results in an overdesigned shaft but is still economically superior because of the resulting ease in assembly.

19.3 ROLLER COASTER BRAKING SYSTEM

Figure 19.4 is a schematic of a typical roller coaster brake. The brake systems discussed in Chapter 17 (i.e., drum and disk brakes associated with wheels) will not work for a roller coaster. The friction between the steel wheel and the steel rail is extremely low so that the cars can slide on the rails without losing much energy. Therefore, the brake system as designed uses pads that bear against a flange, or fin, on the roller coaster cars. The roller coaster operator controls a pneumatic actuator that applies force through the linkage to aggressively decelerate the cars. The maximum braking force should not expose the passengers to greater than 1/2 g of deceleration, preferably not more than 8 ft/s^2 or so. What brake components are needed to bring the roller coaster to a stop at the desired location?

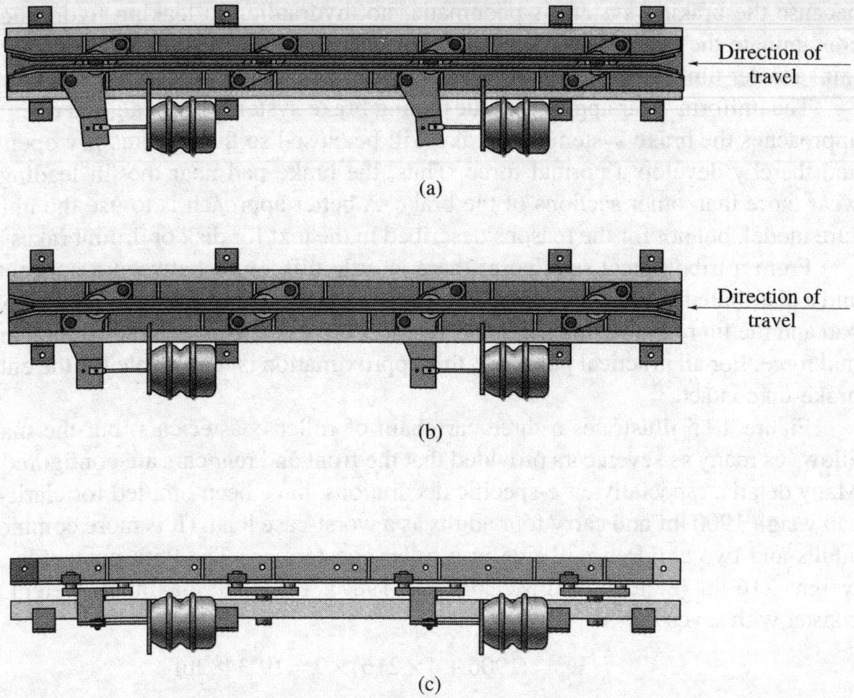

Figure 19.4 Schematic of roller coaster brake.
(a) Open, allowing cars to pass through; (b) closed, so that braking is achieved with flange on cars; (c) side elevation. (Courtesy of Custom Coasters International.)

Typically, several brakes are used at the point of access to a roller coaster. The system under consideration uses four brakes, which is not unusual. Roller coasters do not have a power source associated with the cars. The maximum energy to the cars is input by hoisting them on an inclined section of track. Most roller coasters rely on this initial hill for energy. This reliance simplifies the brake design considerably, since they can be designed for a particular circumstance. Also, because roller coaster brakes are actuated only intermittently, they avoid the thermal checking and temperature problems typical of automotive disk or drum brakes.

Roller coasters are usually placed onto wooden lattice structures. Therefore, the bolts attaching the brake to the wooden lattice need not be designed for fatigue according to the applicable standards. Such rides are routinely inspected, and the wooden lattice is known to fatigue far more rapidly than steel bolts with the required safety factor. This situation is obviously peculiar to wooden structures.

19.3.1 BRAKE PAD–FIN INTERACTION

The first part of the design problem is to correlate brake system variables (actuating force, pad area, etc.) with performance (braking force developed). The coefficient of friction can be reasonably taken as at least 0.25 for sintered brass on steel (it is usually higher for these materials). Contaminants on the roller coaster are not a concern for two reasons. First, because the braking system is pneumatic, not hydraulic, no leaking hydraulic fluids can contaminate the brake pads. Second, because such roller coasters are not operated in the rain, a water film cannot develop between friction elements.

The uniform wear approach to designing brake systems is not a good one. When a car approaches the brake system, the brake will be closed so that the fins pry open the brakes and thereby develop a normal force. Thus, the brake pad near the fin leading edge will wear more than other sections of the brake. A better approach is to use the uniform pressure model, but not for the reasons described in the text for disk or drum brakes.

From a tribological standpoint there is little difference between a concentrated load and a distributed load in generating friction. As long as the real area of contact between the pad and the fin remains small, a linear relationship exists between the friction and the normal force. For all practical purposes, this approximation is reasonable for the entire time of brake-fin contact.

Figure 19.5 illustrates a three-car chain of roller coaster cars, but the manufacturer allows as many as seven cars provided that the front and rear cars are configured as shown. Many details, especially case-specific decorations, have been omitted for clarity. Each car can weigh 1900 lbf and carry four adults as a worst-case load. (It is more common that two adults and two children will ride in a roller coaster car.) The 95th percentile adult male weighs 216 lbf (Sanders and McCormick, 1987). Thus, the maximum weight of a roller coaster with seven cars is

$$w_z = (1900 + 4 \times 216) \times 7 = 19\ 348 \text{ lbf}$$

The maximum incoming velocity is approximately 30 mph (44 ft/s) on the basis of the energy attainable from the first roller coaster hill. For the target deceleration of 8 ft/s^2 the braking force must equal

$$F = \frac{w_z a_d}{g} = \frac{(19.348 \times 10^3)(8)}{32.2} = 4807 \text{ lbf} \qquad \textbf{(19.4)}$$

The normal force required to achieve this braking force is

$$P_n = \frac{F/2}{\mu} = \frac{4807/2}{0.25} = 9614 \text{ lbf} \qquad \textbf{(19.5)}$$

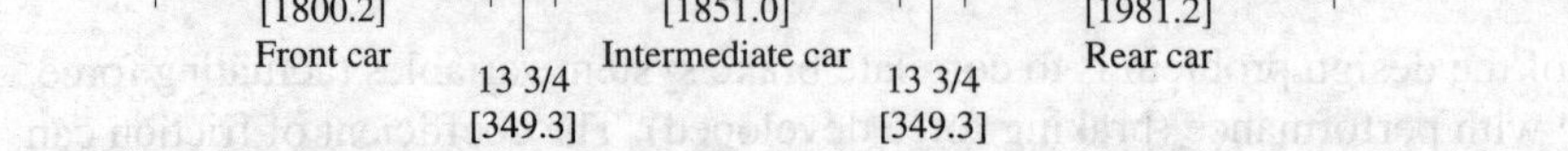

Figure 19.5 Schematic of typical roller coaster cars. Dimensions are in inches. (Courtesy of Custom Coasters International.)

The brake pad arrangement shown in Fig. 19.4 develops frictional force on both sides of the fin. Hence, only one-half the braking force in Eq. (19.4) need be considered.

This normal force represents the worst-case design load in the cars. No increase is necessary from a safety standpoint for the following reasons:

1. For most cases this friction force is more than adequate.
2. Typical installations will have a series of at least three brakes spaced in alternate bays. If deceleration can be maintained throughout the entire contact length between the cars and the brakes, extra braking capacity is inherently placed in the system.
3. Even in the event of insufficient braking (perhaps due to excessive wear of the braking elements), emergency brakes in the operators station can stop the cars. The alternative is to generate larger braking forces with the subject braking system, leading to excessive deceleration and possible injuries to riders from jarring and jerking. The target deceleration has been chosen to provide good braking potential while minimizing the risk of such injuries. A stronger brake would actually lead to a less safe system.

19.3.2 Brake Actuation System

Figure 19.6 shows a detail of the brake-actuating cylinder. Recall from Fig. 19.4 that two cylinders are used on each brake and two unpowered stabilizing elements are used to distribute the contact forces over the whole shoe. For the linkage shown, and moment equilibrium about the pivot, the normal force is

$$\frac{P_n}{3}(3.5) - P_t(9) = 0 \tag{19.6}$$

$$\therefore\ P_t = 1246 \text{ lbf} \tag{19.7}$$

This force already recognizes that two cylinders will be used per brake and three brakes will be used per roller coaster. The brake is based on a pneumatic cylinder/spring system

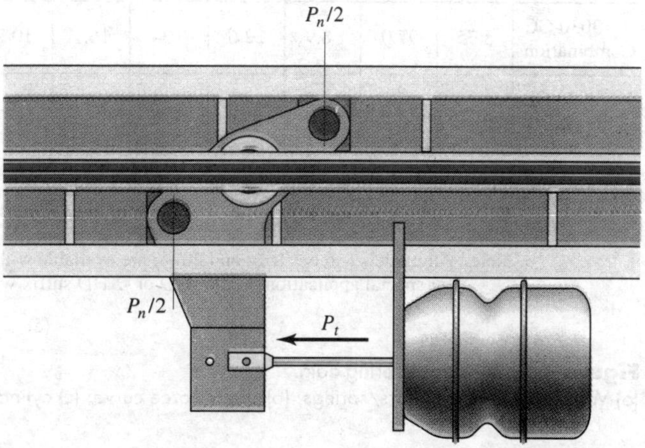

Figure 19.6 Detail of brake-actuating cylinder with forces shown. Scale is 1/8 in. = 1 in. (Courtesy of Custom Coasters International.)

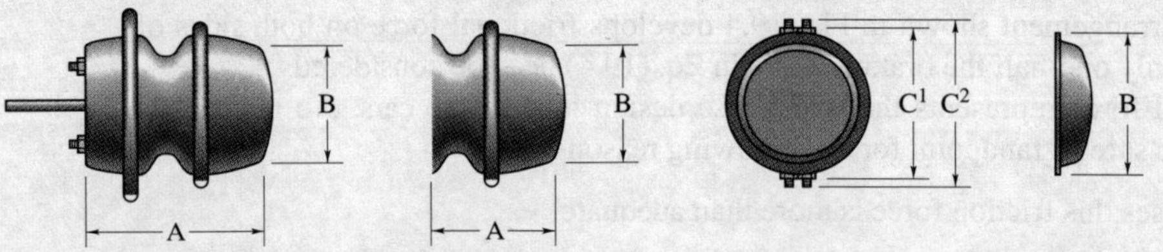

	Max. chamber volume at 100 psig		A		B		*C^1		*C^2		Service diaphragm D		Mass	
Gold Seal®	Service cu. in.	Spring cu. in.	in.	cm.	in.	cm.	in.	cm.	in.	cm.	in.	cm.	lb.	kg.
1624 GP Piggyback	—	72.5	6.3	16.0	5.9	15.0	—	—	7.5	19.0	—	—	10.0	4.5
2024 GC Combination	53.0	72.5	8.5	21.6	5.9	15.0	8.9	22.6	7.5	19.0	6.5	16.5	17.3	7.8
2024 GP Piggyback	—	72.5	6.3	16.0	5.9	15.0	—	—	7.5	19.0	—	—	10.3	4.7
2424 GC Combination	63.0	72.5	8.5	21.6	5.9	15.0	9.25	23.5	7.5	19.0	7.1	18.0	17.5	7.9
2424 GP Piggyback	—	72.5	6.3	16.0	5.9	15.0	—	—	7.5	19.0	—	—	10.5	4.7
2430 GCLS Combination	69.5	103.5	8.9	22.6	6.34	16.1	9.25	23.5	8.35	21.3	† 7.1	18.0	19.5	8.8
2430 GPLS Piggyback	—	103.5	6.7	16.9	6.34	16.1	—	—	8.35	21.3	—	—	12.1	5.5
2430 GC Combination	63.0	93.2	8.5	21.6	6.34	16.1	9.25	23.5	8.35	21.3	† 7.1	18.1	18.2	8.2
2430 GP Piggyback	—	93.2	6.4	16.3	6.34	16.1	—	—	8.35	21.3	—	—	12	5.4
3030 GC Combination	8.75	97.0	8.9	22.6	6.34	16.1	10.2	26.0	8.35	21.3	7.8	19.9	19.6	8.9
3030 GP Piggyback	—	97.0	6.8	17.3	6.34	16.1	—	—	8.35	21.3	—	—	11.8	5.3

* C^1 DIM. at service clamp. C^2 DIM. at rolled seal.

† Use long-stroke diaphragm only. The use of a standard diaphragm may impair brake performance.

Note: All models (except 1624 and 2024) are available with heavy-duty and extra heavy-duty housings for special applications. Add -HD or -XHD suffix when ordering.

(a)

Figure 19.7 Brake spring data.
(a) Various sizes of cylinders/springs; (b) spring force curve; (c) cylinder assembly.

(continued)

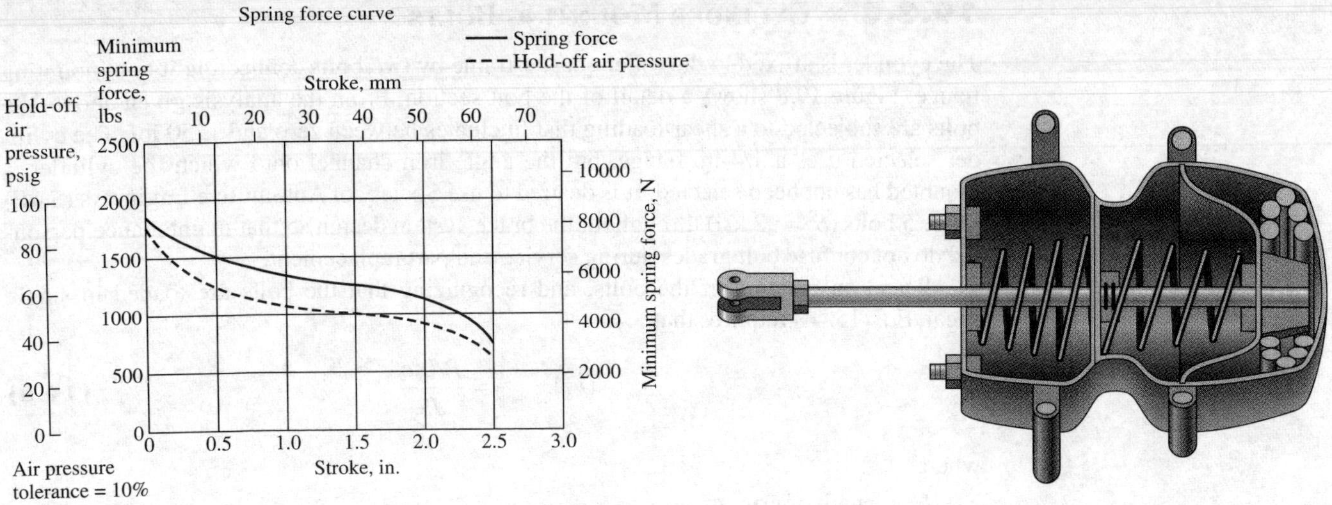

(b) (c)

Figure 19.7 (concluded)

so that the normal cylinder position is with the piston totally extended, corresponding to the brake generating maximum brake force. The reason for such a system is that many difficulties can arise with pneumatic systems: air hoses can leak, pneumatic couplings can decouple, power can fail, etc. The system is fail safe if the brakes are engaged in such a reasonably foreseeable event.

Pneumatic brake cylinders are available in a range of capacities from a brake manufacturer. Figure 19.7(a) shows the various sizes. Each brake system has a characteristic spring force curve, one example of which is shown in Fig. 19.7(b). The nonlinear nature of this curve is typical of pneumatic springs, but this cylinder uses a combination of air pressure and helical springs for actuation as shown in the breakaway view in Fig. 19.7(c).

A subtlety in sizing such brake cylinders must be explained. During assembly the piston is totally extended and may be difficult to affix to the hinge mechanism. Therefore, a piston extension is threaded for a distance on its length. The cylinder can be assembled and then the collar on the piston extension adjusted for brake actuation. However, there is some variability in the degree to which the brakes can be adjusted, shifting the force curve in Fig. 19.7(b) horizontally.

The highly nonlinear nature of the spring force curve consists of a rapid drop in actuating force, after which the actuating force is relatively insensitive to stroke. The brake is intended to operate where a relatively constant force is required. Indeed, this is also a safety requirement, in that the brake force is then fairly insensitive to both wear and air pressure fluctuations. Therefore, the brake cylinder with the force curve shown in Fig. 19.7(b) is a logical choice because the spring force is comparable to the required force of 1250 lbf.

19.3.3 CYLINDER MOUNTING BOLTS

The cylinder is affixed to the roller coaster frame by two bolts connecting it to a mounting flange. Figure 19.8 shows a detail of the bolt section. From the analysis given above, the bolts are subjected to a shear loading that fluctuates between zero and 1250 lbf. The cylinder selected uses a 1/4-in. flange, but the angle iron channel onto which the cylinder is mounted has not been selected. It is desired to use Society of Automotive Engineers (SAE) grade 5 bolts ($S_y = 92$ ksi) throughout the brake system design so that maintenance personnel do not confuse bolt grades during service and part replacement.

Based on bending of the bolts, and recognizing that the bolts are loaded in single shear, Eq. (15.44) requires that

$$(n_s)_{\text{bending}} = \frac{P_n L_g c}{2IS_y} \tag{19.8}$$

where

L_g = length of grip, m

$\dfrac{I}{c} = Z_m =$ section modulus of weaker member, m^3

From Eq. (15.45) the safety factor based on shear of the bolts is

$$(n_s)_{\text{shear}} = \frac{4P_t}{\pi d_c^2 S_{sy}} \tag{19.9}$$

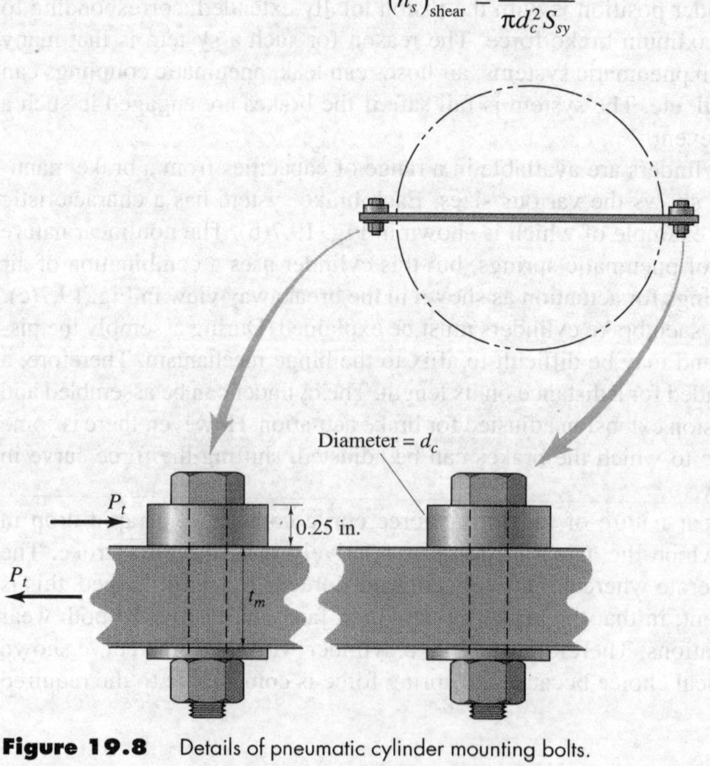

Figure 19.8 Details of pneumatic cylinder mounting bolts.

where

d_c = crest diameter, m

From Eq. (15.47) the safety factor based on compressive bearing failure is

$$(n_s)_{\text{bearing}} = \frac{P_n}{d_c t_w S_y} \tag{19.10}$$

where

t_w = thickness of thinnest member, m

It is the experience of the manufacturer that a safety factor of 6 is reasonable for such applications.

Applying these equations makes clear that, for reasonable angle iron flange thicknesses, the critical failure mode is bending of the bolts. The data of Table 19.3 result from analysis where the safety factor for bending is set equal to 6. Given the flange width, Eq. (15.44) is solved iteratively for the bolt diameter, rounded to the next largest 1/8 in. Once the bolt diameter is known, obtaining the remaining safety factors is straightforward.

From the results shown in Table 19.3, it is reasonable to use a 4- by 4- by 1/4-in. angle iron as the support bracket for the pneumatic cylinder. Because this is short, bending is not a serious issue. Using a 1/2-in. bolt results in generous safety factors, and thus bolt failure is not expected.

19.4 CAR BRAKE DIMENSIONING

A car with a mass m_a has a wheel base length l as shown in Fig. 19.9. The center of gravity at which the concentrated mass acts is a height h_t above the ground. The base length from the front wheel to the center of gravity divided by the base length is n_l, a dimensionless parameter between zero and one. The wheel diameter including the tires is d_w, and the coefficient of friction between the tires and the road is μ. The car is to use long-shoe brakes on all four wheels.

The calculation procedure used in dimensioning car brakes is given here. The brake design requires that the allowable range for each input parameter is

$$800 \le m_a \le 1800 \text{ kg}$$

$$2.0 \le l \le 3.5 \text{ m}$$

Table 19.3 Results from analysis of brake cylinder mounting bolts.

Thickness of thinnest member, t_w, in.	Crest diameter, d_c, in.	Bolt diameter (rounded), d, in.	Safety factor, n_s		
			Bending	Shear	Bearing
0.125	0.43	0.5	9.6	16	9.2
0.25	0.46	0.5	7.2	17	18
0.375	0.50	0.50	6.0	17	27
0.5	0.54	0.625	9.4	26	46

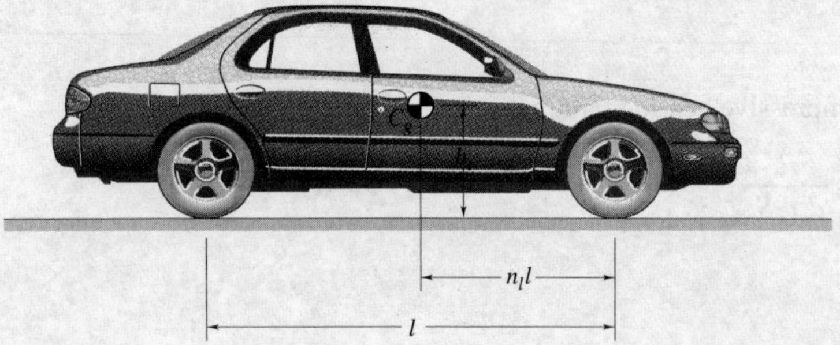

Figure 19.9 Car with critical parameters used in dimensioning brakes.

$$\begin{aligned} 0.4 \le h_t \le 0.7\,\text{m} \\ 0.5 \le d_w \le 0.75\,\text{m} \\ 0.3 \le n_l \le 0.6 \\ 0.60 \le \mu \le 1.00 \end{aligned} \tag{19.11}$$

19.4.1 DECELERATION AND BRAKING FORCE

The design requires a knowledge of how the deceleration depends on the used portion of the maximum coefficient of friction or

$$\mu_n = n_\mu \mu \tag{19.12}$$

where

n_μ = used portion of maximum friction (between zero and 1)

μ = maximum coefficient of friction between road and tire

Assume that the speed of the car is between zero and 150 km/hr, the latter being the speed at which applying the brakes almost locks the wheels. Important points in brake design relating to deceleration are the following:

1. Braking distance as a function of car velocity.
2. Braking distance as a function of car velocity changes in a snowstorm, when μ is decreased by 75%.
3. Different braking torques can be applied to the front and rear wheels.

If it is assumed that all wheels use the same μ_n, the equation of motion gives

$$m_a \ddot{x} = -n_\mu \mu m_a g$$

$$\therefore a_d = \ddot{x} = -n_\mu \mu g \tag{19.13}$$

Figure 19.10 shows a plot of deceleration as a function of the used portion of the maximum coefficient of friction.

The torque applied to the front and rear wheels is

$$T_f = \mu \left(\frac{w_{zf}}{2}\right)\left(\frac{d_w}{2}\right) \tag{19.14}$$

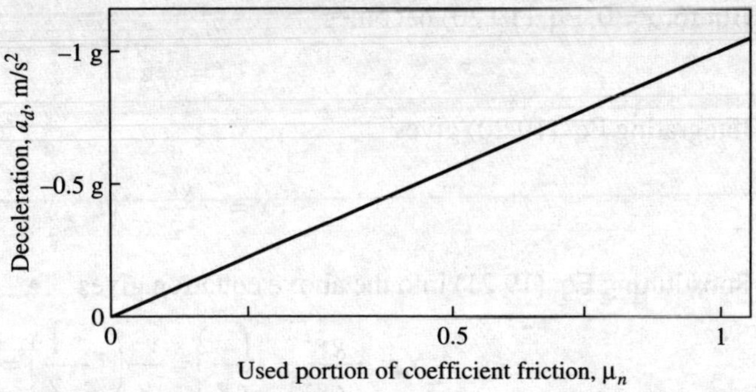

Figure 19.10 Deceleration as function of used portion of maximum coefficient of friction.

$$T_r = \mu\left(\frac{w_{zr}}{2}\right)\left(\frac{d_w}{2}\right) \quad \textbf{(19.15)}$$

If the maximum friction is applied to all four wheels, $n_\mu = 1$ and $\mu = 1$ and thus $\mu_n = 1$. Thus, from Eq. (19.13)

$$a_d = \ddot{x} = -g$$

Integrating once gives

$$\dot{x} = -gt + C_1 \quad \textbf{(19.16)}$$

Integrating again gives

$$x = -\frac{gt^2}{2} + C_1 t + C_2 \quad \textbf{(19.17)}$$

where

C_1 and C_2 = integration constants

At $t = 0$, $\dot{x} = 150$ km/hr = 150 000/3600 m/s = 41.67 m/s and

$$\therefore C_1 = 41.67 \text{ m/s}$$

At $t = 0$, $x = 0$ and

$$\therefore C_2 = 0$$

Then the stopping distance for $t < 4.29$ s is

$$x = -\frac{gt^2}{2} + 41.67t \quad \textbf{(19.18)}$$

For speeds lower than 150 km/hr, or 41.67 m/s, Eq. (19.17) becomes

$$x = -\frac{gt^2}{2} + vt \quad \textbf{(19.19)}$$

$$\dot{x} = -gt + v \quad \textbf{(19.20)}$$

But for $\dot{x} = 0$, Eq. (19.20) becomes

$$t = v/g \tag{19.21}$$

Integrating Eq. (19.20) gives

$$x = -\frac{gt^2}{2} + vt$$

Substituting Eq. (19.21) into the above equation gives

$$x = -\frac{gv^2}{2g^2} + v\left(\frac{v}{g}\right) = \frac{v^2}{g}\left(1 - \frac{1}{2}\right) = \frac{v^2}{2g} \tag{19.22}$$

Figure 19.11 shows a plot of the velocity as a function of the braking distance from Eq. (19.22).

Figure 19.12 shows the normal forces acting on the car. From force equilibrium the following equations are obtained:

$$w_{zf} - m_a g + w_{zr} = 0 \tag{19.23}$$

$$-\mu w_{zf} + m_a a_d - \mu w_{zr} = 0 \tag{19.24}$$

Multiplying Eq. (19.23) by μ and then adding Eqs. (19.23) and (19.24) gives

$$a_d = \mu g \tag{19.25}$$

From moment equilibrium

$$1.5 w_{zr} - 1.0 w_{zf} + 0.4 m_a a_d = 0$$

By making use of Eqs. (19.23) and (19.24) the above equations become

$$1.5 w_{zr} - 1.0(m_a g - w_{zr}) + 0.4 m_a \mu g = 0$$

$$\therefore 2.5 w_{zr} - m_a g(1 - 0.4\mu) = 0$$

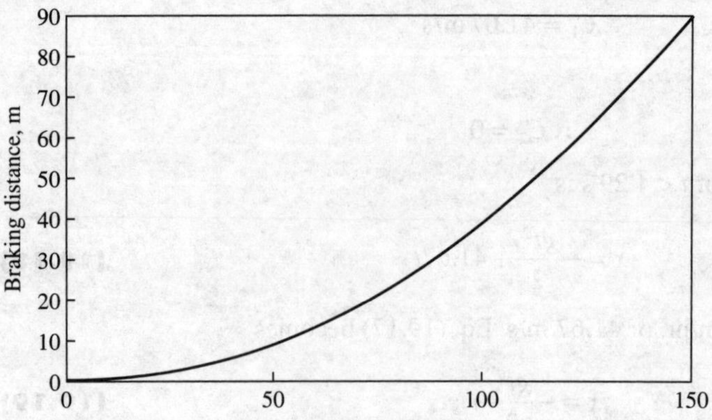

Figure 19.11 Velocity as function of braking distance.

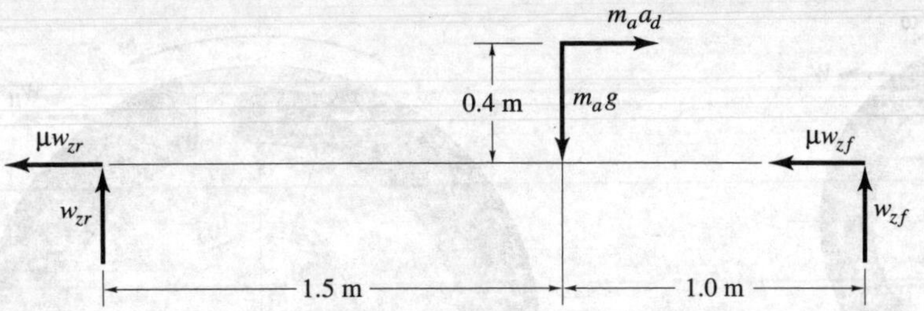

Figure 19.12 Forces acting on car.

$$w_{zr} = \frac{m_a g(1-0.4\mu)}{2.5} = 0.4 m_a g(1-0.4\mu) \tag{19.26}$$

Substituting this into Eq. (19.23) gives

$$w_{zf} - m_a g + 0.4 m_a g(1-0.4\mu) = 0$$

$$\therefore w_{zf} = m_a g(0.6 + 1.6\mu) \tag{19.27}$$

For $\mu = 1$, $n = 0.4$, $l = 2.5$ m, $h_t = 0.4$ m, $d_w = 0.6$ m, and $m_a = 1000$ kg.

$$w_{zr} = 0.24 m_a g \tag{19.28}$$

$$w_{zf} = 0.76 m_a g \tag{19.29}$$

Substituting Eq. (19.28) and (19.29) into Eqs. (19.14) and (19.15) gives the torque applied to the front and rear wheels as

$$T_f = \frac{\mu w_{zf} d_w}{4} = \frac{(1)(0.76)(1000)(9.807)(0.6)}{4} = 1118 \text{ N-m}$$

$$T_r = \left(\frac{w_{zr}}{w_{zf}}\right) T_f = \frac{0.24}{0.76}(1118) = 353.1 \text{ N-m}$$

In a snowstorm the coefficient of friction is decreased by 75% ($\mu \rightarrow \mu/4$), implying that the braking distance increases by a factor of 4. Thus, if snowstorm conditions exist, the distance for a specific velocity in Fig. 19.11 must be multiplied by 4.

19.4.2 BRAKE DIMENSIONS

Figure 19.13 shows an internal long-shoe brake with two coupled shoes. The left shoe is hinged to the backplate at A, and the right shoe is hinged to the left shoe at B. The brake should be able to handle the largest possible moment from the wheel. The hydraulic oil pressure in the brake system experiences a maximum of 5 MPa. It is assumed that the brake linings wear in proportion to the pressure times the sliding distance. The maximum allowable brake shoe pressure is 2 MPa. The coefficient of friction for contact between the brake shoe lining and the drum is 0.3 ($\mu_b = 0.3$). The design consideration for this section is to determine the braking torque for the two shoes from the information given.

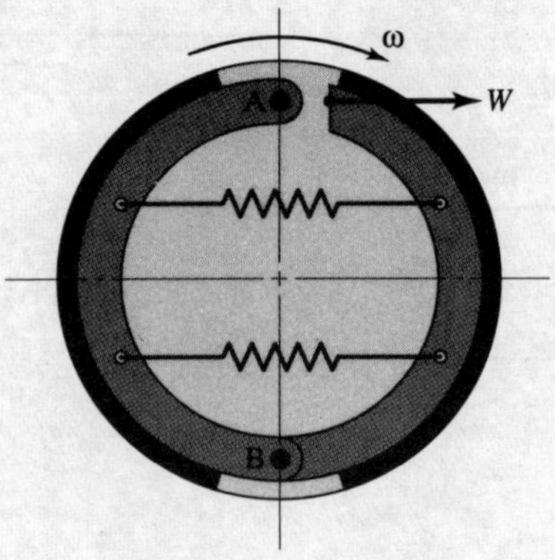

Figure 19.13 Schematic of car brake shoes.

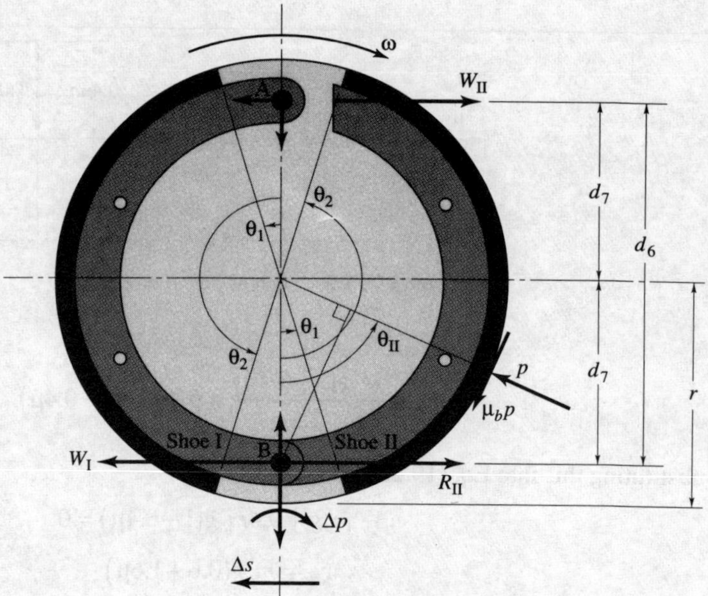

Figure 19.14 Forces acting on brake shoes, as well as critical angles.

Figure 19.14 shows the forces acting on the brake shoes as well as the critical angles. Brake shoe I is similar to the long-shoe, internal, expanding rim brakes considered in Sec. 17.5. Since the angle where $p = p_{max}$ is greater than 90°, then $\theta_a = 90°$. Thus, Eq. (17.33) gives the pressure distribution around the shoe as

$$p = p_{max} \sin\theta \tag{19.30}$$

Shoe I is self-energizing. Using Eqs. (17.39), (17.36), and (17.37) gives the actuating force on shoe I as

$$W_I = \frac{br^2 p_{max}}{2d_6}\left\{\frac{d_7}{r}\left[\frac{(\theta_2-\theta_1)\pi}{180} + \frac{\sin 2\theta_1 - \sin 2\theta_2}{2} + \mu_b(\sin^2\theta_2 - \sin^2\theta_1)\right] + 2\mu_b(\cos\theta_2 - \cos\theta_1)\right\} \tag{19.31}$$

Also, from Eq. (17.38) the braking torque of shoe I is

$$T_I = \mu_b p_{max} br^2(\cos\theta_1 - \cos\theta_2) \tag{19.32}$$

Brake shoe II is not standard, and thus the material covered in Chapter 17 cannot be directly applied here. The main reason is that the hinge, around which the shoe moves, can also move. Thus, the hinge moves a distance Δs to the left as the shoe turns an angle $\Delta\varphi$. The contact pressure on shoe II is

$$p_{II} = \frac{2p_{max}}{\Delta s}(-\Delta s \sin\theta + \Delta\varphi d_7 \sin\theta)$$

$$\therefore\ p_{\text{II}} = 2p_{\max} \sin\theta\left(\frac{d_7 \Delta\varphi}{\Delta s} - 1\right) \qquad \textbf{(19.33)}$$

Moment equilibrium applied about point B in Fig. 19.14 gives the actuating force for shoe II as

$$W_{\text{II}} = 2p_{\max}\left(\frac{d_7\Delta\varphi}{\Delta s} - 1\right)\frac{br^2}{2d_6} \times \left\{\frac{d_7}{r}\left[\frac{(\theta_2 - \theta_1)\pi}{180} + \frac{\sin 2\theta_1 - \sin 2\theta_2}{2} + \mu_b\left(\sin^2\theta_2 - \sin^2\theta_1\right)\right] + 2\mu_b\left(\cos\theta_2 - \cos\theta_1\right)\right\} \qquad \textbf{(19.34)}$$

From horizontal force equilibrium while noting that $R_{\text{II}} = W_{\text{I}}$

$$0 = W_{\text{I}} + W_{\text{II}} - \int_{\theta_1}^{\theta_2} p_{\text{II}} \sin\theta br d\theta - \int_{\theta_1}^{\theta_2} \mu_b p_{\text{II}} \cos\theta br d\theta$$

Making use of Eq. (19.33) while recognizing that $\theta_2 = 180° - \theta_1$ and making use of Eqs. (19.31) and (19.34) after some manipulation give

$$\frac{p_{\text{II}_{\max}}}{p_{\text{I}_{\max}}} = \frac{180° - 2\theta_1 + \sin 2\theta_1 - 4\mu_b\left(\dfrac{r}{d_7}\right)\cos\theta_1}{180° - 2\theta_1 + \sin 2\theta_1 + 4\mu_b\left(\dfrac{r}{d_7}\right)\cos\theta_1}$$

Choosing $\theta_1 = 30°$, $\dfrac{r}{d_7} = 1.2$, and $\mu_b = 0.3$ gives

$$\frac{p_{\text{II}_{\max}}}{p_{\text{I}_{\max}}} = 0.4072$$

Thus, shoe I experiences the highest pressure. If the maximum pressure at shoe I is set at the maximum allowable pressure,

$$p_{\text{I}_{\max}} = 2 \text{ MPa}$$

$$p_{\text{II}_{\max}} = 2(0.4072) = 0.8144 \text{ MPa}$$

From Eq. (17.38) the total braking torque of the two shoes is

$$T = \mu_b br^2(\cos\theta_1 - \cos\theta_2)\left(p_{\text{I}_{\max}} + p_{\text{II}_{\max}}\right)$$

If $b = 50$ mm, $r = 125$ mm, $\theta_1 = 30°$, $\theta_2 = 150°$, and $\mu_b = 0.3$, the total braking torque is

$$T = 0.3(0.050)(0.125)^2(\cos 30° - \cos 150°)(2\times10^6 + 0.8144\times10^6)$$

$$\therefore T = 1142 \text{ N-m}$$

The hydraulically applied brake force is equivalent to the actuating force in shoe II and is found to be 2180 N while using Eq. (19.34). The pressure in the hydraulic cylinder is 5 MPa, and the cylinder area needed is

$$A = \frac{W_{\text{II}}}{p} = \frac{2180}{5\times10^6} = 4.360\times10^{-4}\,\text{m}^2$$

Thus, the diameter of the hydraulic cylinder needed is

$$d_h = \sqrt{\frac{4(4.360\times10^{-4})}{\pi}} = 0.02356\,\text{m} = 23.56\text{ mm}$$

19.4.3 SELF-LOCKING

With the brake dimensions obtained from the previous sections the design then considers how large the coefficient of friction of the brake lining μ_b can be without experiencing brake self-locking. Thus, if the actuating force equals zero, self-locking will occur. From Eqs. (19.31) and (19.34) observe that this occurs when

$$0 = \frac{d_7}{r}\left[\frac{(\theta_2-\theta_1)\pi}{180}+\frac{\sin 2\theta_1 - 2\theta_2}{2}+\mu_b(\sin^2\theta_2-\sin^2\theta_1)\right]+2\mu_b(\cos\theta_2-\cos\theta_1)$$

But, $r/d_7 = 1.2$, $\theta_1 = 30°$, and $\theta_2 = 150°$. Substituting this into the above equation gives

$$\mu_b = 0.7122$$

The brake will self-lock if the coefficient of friction between the shoe and the drum is above 0.7122.

19.5 SUMMARY

The projects presented in this chapter were intended to demonstrate the relationship between design analysis and design synthesis. Much of design synthesis follows from an understanding and familiarity with analysis, and in the generation of workable designs such understanding is extremely useful. Ultimately, design analysis is needed after synthesis to complete the total design process. Thus, design analysis and design synthesis are irrevocably intertwined, and to do either well, one must understand both.

Many design decisions are not wholly based on analysis. For example, in the vertical mixer the shafts were not analyzed for torsional deflection with the rationale that the hub diameters on standard components are much larger than needed for deflection and the shafts in the mixer were very short. An attempt to justify a shaft diameter with a safety factor argument or a minimum-deflection requirement would be inherently wrong, since these factors played no role in the design process. The final shaft was selected for ease of manufacture and has the economic rationale of minimizing assembly costs by using low-cost, mass-produced components. This important illustration holds for many if not most products; form does not necessarily follow function or analysis. However, although different automobile models may look very different, the elements that make up the machines are identical. This textbook has attempted to place a number of tools in the design engineer's toolbox, and this chapter has briefly illustrated how some of the tools are applied.

Even so, the examples presented were merely small portions of the total design burden associated with the total products. The reason is partly space constraints, but also that some design challenges were from a different discipline entirely, such as in the design of the control circuitry for the roller coaster braking system. Many times input information came from outside sources, either marketing personnel or practical experiences. Design is an exciting, complicated endeavor requiring multidisciplinary efforts and is also an activity where experience helps considerably.

REFERENCES

Sanders, M.S., and McCormick, E.J. (1987) *Human Factors in Engineering Design*, McGraw-Hill, New York.

PROJECTS

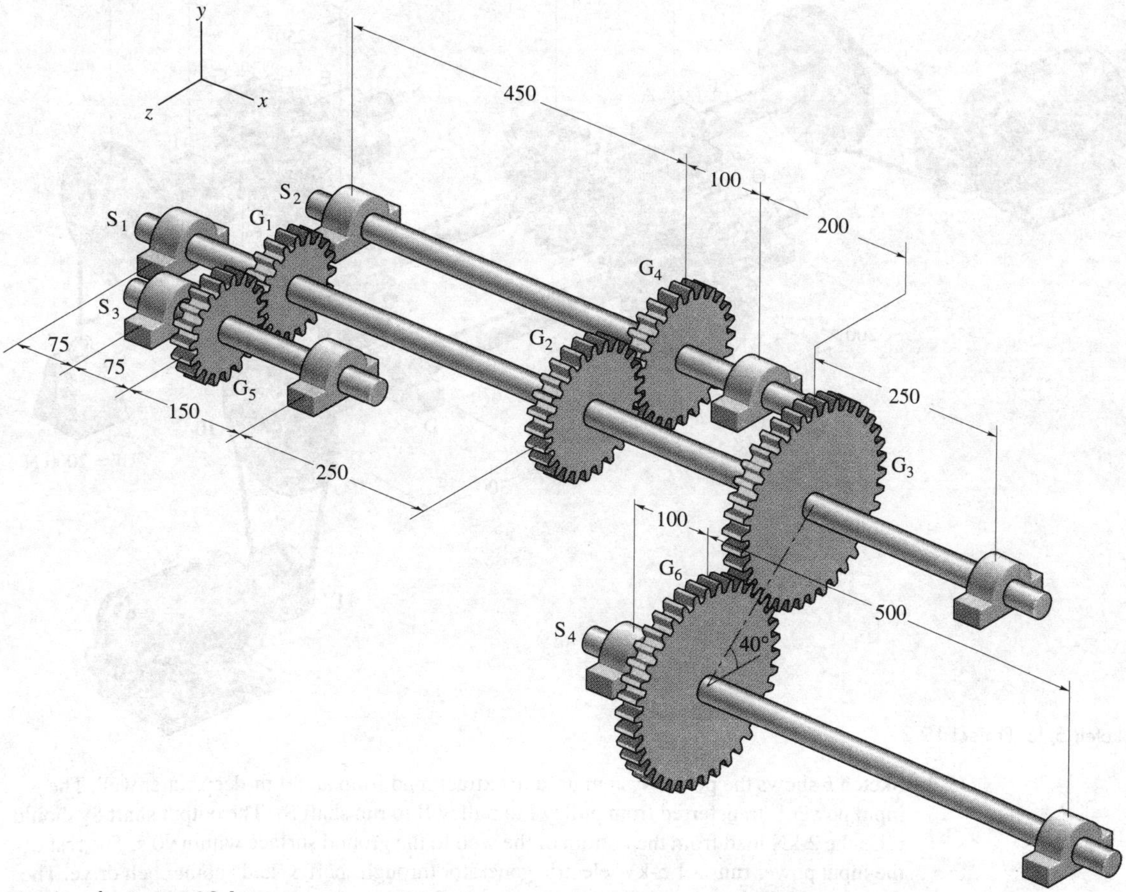

Sketch *a*, for Project 19.1

19.1 Sketch *a* shows part of a textile machine. Dimensions are in millimeters. The input power to the system is 5 kW at shaft S_1 and has a speed of N_a. Twenty percent of this power is transferred to shaft S_3 through gear G_1, 30% to shaft S_2 through gear G_2, and the rest to shaft S_4 through gear G_3. The output of shafts S_2 and S_3 is to the left and that of shaft S_4 is to the right. The following information is given for all gears: $\phi = 20°$, $N_a = 1200$ rpm, $N_{g1} = 22$, $N_{g2} = 24$, $N_{g3} = 60$, $N_{g4} = 26$, $N_{g5} = 22$, $N_{g6} = 40$, $m_1 = 3$ mm, $m_2 = m_3 = 4$ mm. All shafts are made of hot-rolled steel ($S_y = 650$ MPa and $S_{ut} = 800$ MPa), and the gears are made of SAE annealed steel.

a) Design shafts S_1, S_2, S_3, and S_4 completely for a safety factor of 2.5 guarding against fatigue failure.

b) Design all the bearings (preferably ball bearings) for a 5000-hr life.
c) Design all the gears (only the pinions) for a safety factor of 3.5 guarding against fatigue failure.

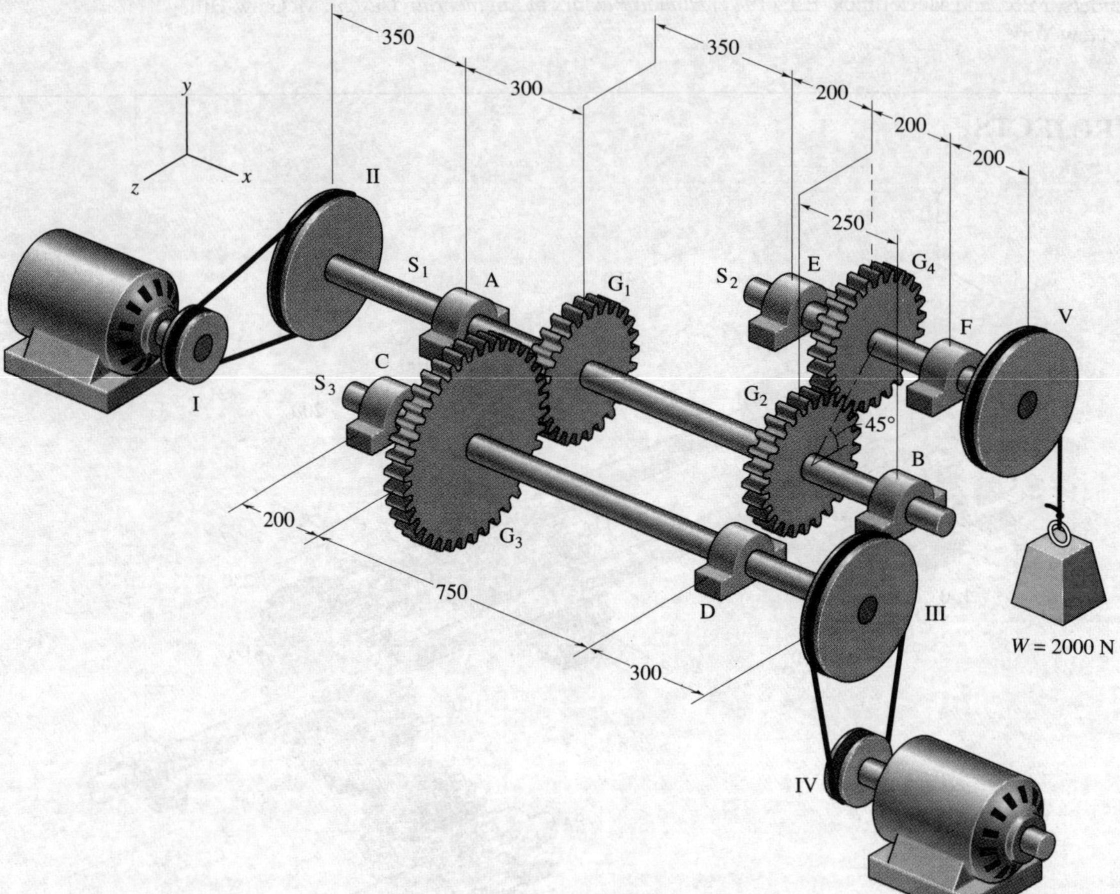

Sketch *b*, for Project 19.2

19.2 Sketch *b* shows the power system used to extract mud from an 80-m-deep water well. The input power is transferred from pulley I to pulley II to run shaft S_1. The output shaft S_2 should raise the 2-kN load from the bottom of the well to the ground surface within 90 s. The rest of the input power runs a 1.5-kW electric generator through shaft S_3 and another belt drive. The dimensions are given in millimeters. Design shafts S_1, S_2, S_3, and S_4 for a safety factor of 3, and select the appropriate ball bearings for each shaft. Also design the belt system and the wind rope. The following information can be used: $d_I = 100$ mm, $d_{II} = 200$ mm, $d_{III} = 300$ mm, $d_{IV} = 100$ mm, $d_V = 500$ mm, $\phi = 20°$, $N_{g1} = 32$, $N_{g2} = 32$, $N_{g3} = 80$, $N_{g4} = 64$, $d_1 = 100$ mm, $d_2 = 100$ mm, $d_3 = 250$ mm, $d_4 = 200$ mm.

19.3 In the power train shown in sketch *c* the 7.5-kW input power to shaft S_1 is transferred to shaft S_2 by a pair of mating spur gears and from shaft S_2 to shaft S_3 through three V-belts. Shaft S_2 rotates at 900 rpm and is made of hot-rolled steel with $S_y = 500$ MPa and $S_{ut} = 800$ MPa. Gear C has 16 teeth and a pressure angle of 20°. The larger pulley, mounted on shaft S_2, is 200 mm

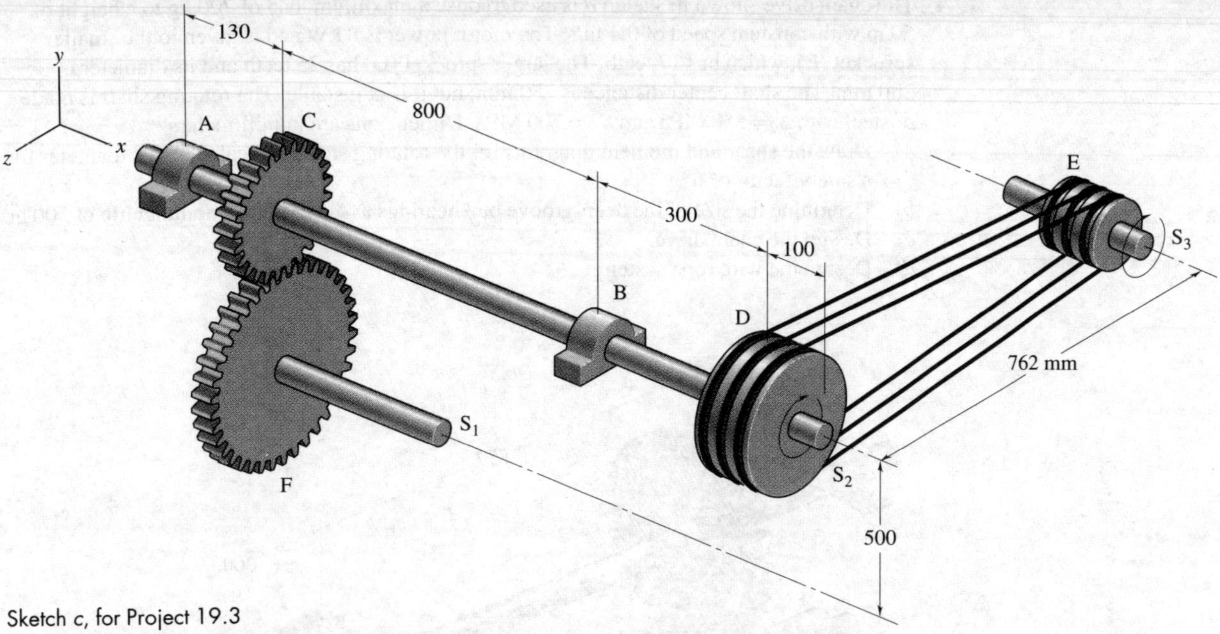

Sketch c, for Project 19.3

in diameter and the smaller pulley is 127 mm in diameter. All dimensions are in millimeters. Assume that the tension on the slack side of the V-belts is 20% of that on the tight side.

a) Draw the moment diagrams of shaft S_2, and calculate its diameter for a safety factor of 4.

b) Determine the appropriate deep-groove ball bearings for supports A and B for a life of 3000 hr.

c) Design the pair of spur gears C and F for a safety factor of 3.

d) Design the V-belt drive for a safety factor of 2.5.

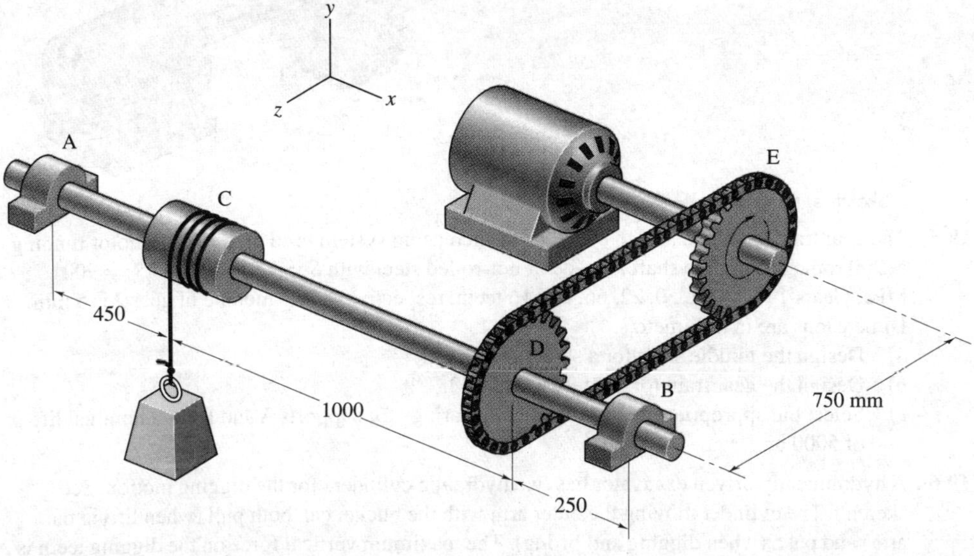

Sketch d, for Project 19.4

19.4 The chain drive shown in sketch *d* is used to raise a maximum load of 700 kg to a height of 50 m with constant speed of 0.4 m/s. The motor power is 8 kW and is given to the smaller sprocket (E), which has 17 teeth. The larger sprocket (D) has 25 teeth and its diameter is 200 mm. The shaft center distance is 750 mm, but it is adjustable. The rotating shaft is made of steel with $S_y = 550$ MPa and $S_{ut} = 700$ MPa. Dimensions are in millimeters.

a) Draw the shear and moment diagrams for the rotating shaft, and calculate its diameter for a safety factor of 6.

b) Determine the size of the deep-groove ball bearings at A and B for a nominal life of 500 hr.

c) Design the chain drive.

d) Design the wire rope system.

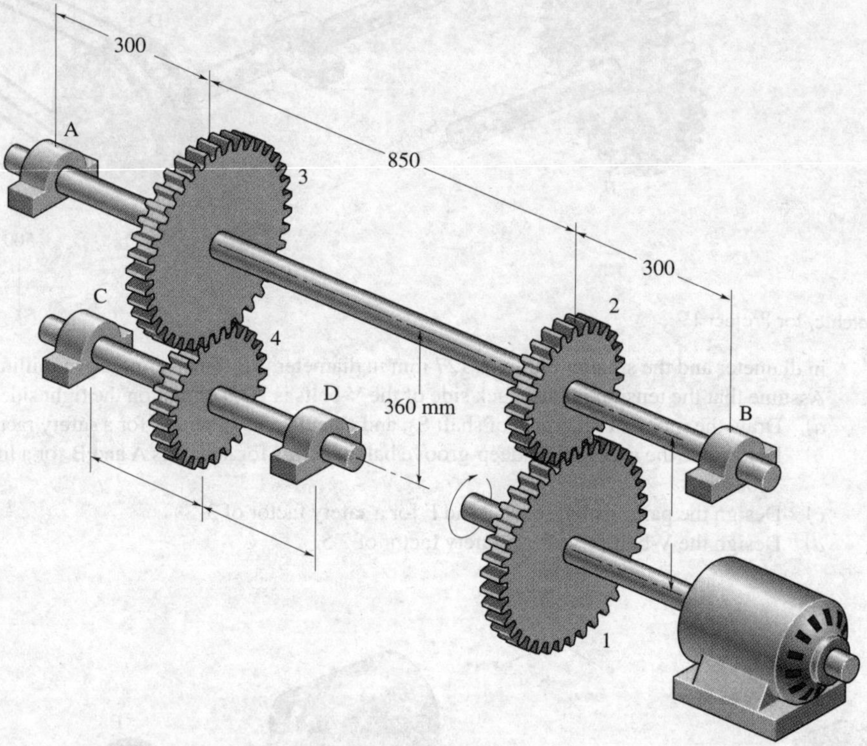

Sketch *e*, for Project 19.5

19.5 The gear train shown in sketch *e* is a speed-increasing system used on a 10-kW motor running at 900 rpm. The middle shaft is made of hot-rolled steel with $S_y = 750$ MPa and $S_{ut} = 900$ MPa. Gears 1 to 4 have 50, 22, 66, and 16 teeth, respectively. The module of gear 2 is 5 mm. Dimensions are in millimeters.

a) Design the middle shaft for a safety factor of 4.

b) Design the gear train for a safety factor of 3.

c) Select the appropriate deep-groove ball bearings for supports A and B for a nominal life of 5000 hr.

19.6 A hydraulically driven excavator has two hydraulic cylinders for the digging motion. See sketch *f*. The cylinder moving the outer arm with the bucket can both pull (when lifting the arm) and push (when digging and lifting). The maximum vertical force on the digging teeth is

25 kN, and the maximum horizontal force is 18 kN. Dimension the hydraulic cylinders so that they do not buckle. Dimensions are in millimeters.

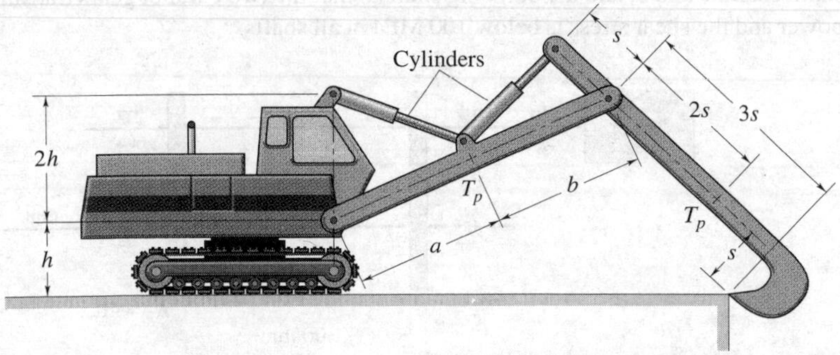

Sketch *f*, for Project 19.6

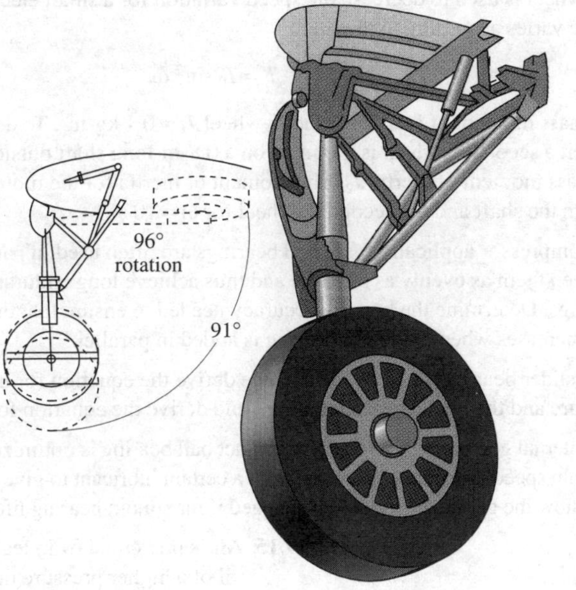

Sketch *g*, for Project 19.7

19.7 The linkage to an aircraft landing gear is shown in sketch *g*. The vertical force component is directly transmitted to the aircraft frame, but the horizontal friction force of 15 kN is transmitted through the hydraulic cylinders and the struts. Design them so that they do not buckle.

19.8 A simple damping device consisting of a cylindrical ring press-fitted onto a 50-mm-diameter vertical shaft is used as a safety feature for an elevator. If the elevator starts to fall and comes below the lowest normal level, it will be braked by the press-fitted ring when that slides down along the shaft. Dimension the ring so that a 200-kg elevator falling at a speed of 5 m/s can be stopped in a distance of 1 m.

19.9 Two hollow shafts are to be joined by a shrink-fitted tube of variable thickness along its length. The shaft-tube joints should be able to transfer 100 N-m in both directions without any sliding or microslipping between the shafts and the tube. Design the shafts and the tube and dimension the shrink fit.

19.10 A motor drives two gears on the same shaft. See sketch h. The total transmitted power is 20 kW at a motor speed of 900 rpm. Each gear has 98% power efficiency. Design the ingoing shaft, shafts 1 and 2, and the outgoing shaft so that the two parallel gears transmit equal power and the shear stress is below 100 MPa in all shafts.

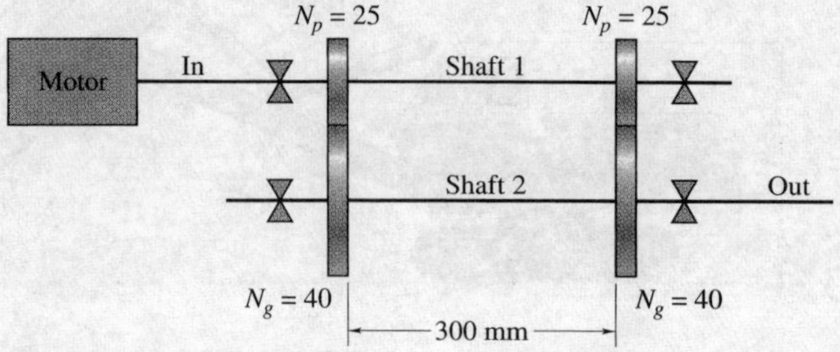

Sketch h, for Project 19.10

19.11 A flywheel is used to decrease the speed variation for a small electric motor when the loading torque varies according to the rule

$$T = T_0 \sin^2 \omega t$$

The mass moment of inertia for the flywheel $J_1 = 0.1$ kg-m^2. To decrease the speed variation further, a second flywheel is mounted on a 0.8-m-long shaft outside the first flywheel having the mass moment of inertia J_2. The moment of inertia for the motor and the shaft is small. Design the shaft and the second flywheel for $\omega = 10\ \text{s}^{-1}$.

19.12 For compressor applications several bearings are often used in parallel to split the total load between them as evenly as possible and thus achieve long endurance life for the group of bearings. Determine the bearing accuracy needed to ensure that the life of the group of bearings increases when one more bearing is added in parallel.

19.13 For a slider bearing without side leakage derive the equation for the position of maximum pressure and the load application point, and derive the equation for bearing friction.

19.14 The internal geometry of an angular-contact ball bearing is optimized for a certain thrust load at a certain speed and temperature and with a certain lubricant to give minimum power loss. Determine how the geometry should be changed if maximum bearing life were required instead.

19.15 Air is prevented from leaking past a seal by pumping oil of a higher pressure than the air through the sealing gap. See sketch i. Optimize the seal geometry when the force in the spring is P_s, the pressure drop across the seal is Δp, and the angular speed is ω so that the total power loss is minimized.

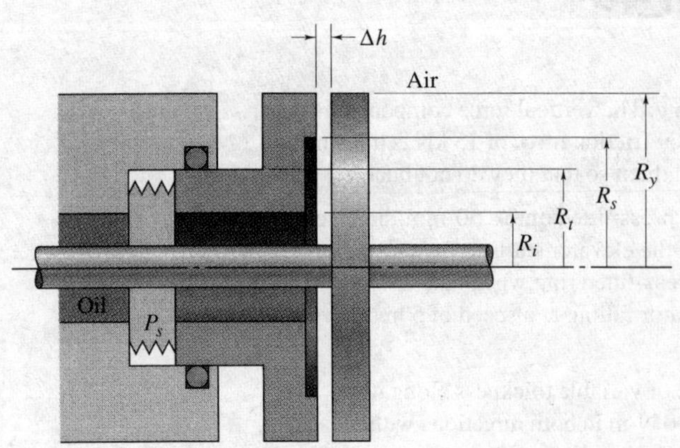

Sketch i, for Project 19.15

19.16 An external gear wheel and an *internal* gear ring together comprise a gear with the ratio g_r. Design the gear for power h_p at pinion speed N_a. The gear material is American Iron and Steel Institute (AISI) 1080 steel hardened so that the notch sensitivity factor $q_n =$ 1 and the maximum allowable contact pressure is σ_r. The gear face width $b_w = 10/p_d$. Avoid undercutting of the gear teeth, and dimension the gears for both contact pressures and bending stresses. Calculate the contact ratio C_r and the center distance.

19.17 To check the manufacturing accuracy of an internal spur gearbox, precision rollers are placed in the clearings between the teeth as close as possible to the diametral positions, and the distance over the rollers is measured. The standard spur gear wheel has a module m and a pressure angle $\phi = 20°$. The precision rollers have a diameter of $2r$. Calculate the control measure for an even number of gear teeth N and for an odd number of pinion teeth $N + 1$.

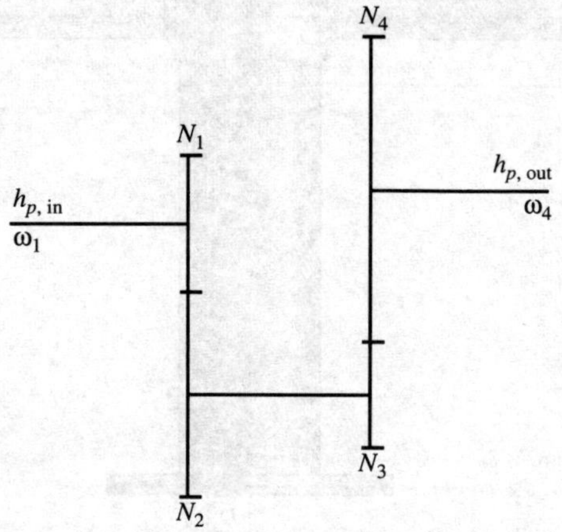

Sketch *j*, for Project 19.18

19.18 Design a gearbox with four helical gears and total speed reduction S_r in the two steps. The gear wheels are manufactured with helix angle ψ and pressure angle ϕ. Sketch *j* shows the gear setup and the input output power. Design the gearbox for power h_p and ingoing shaft speed N_{ai} by using the maximum allowable contact pressure σ_r. The transmitted force is assumed to be evenly distributed along the whole contact length, which varies during the meshing. Neglect friction. Calculate the bending stress according to Lewis, calculate the bearing reaction forces, and design bearings for the shafts. It is given that

$$10 \le h_p \le 80 \text{ kW}$$
$$800 \le N_{ai} \le 1600 \text{ rpm}$$
$$4 \le S_r \le 10$$
$$1 \le \sigma_r \le 1.5 \text{ GPa}$$
$$\psi = 20°,\ 25°,\ 30°,\ 35°,\ \text{or } 40°$$

19.19 Design bag-sealing equipment for plastic bags used by a bakery to wrap bread. The sealing action should be done by using a standard, 10-mm-wide, single-sided adhesive tape. The power and force necessary to seal the bags is provided by manually moving the bag through the sealing equipment. The design should be simple and adapted for easy manufacture in large numbers. The equipment should be able to handle different-size plastic bags. The cross-sectional diameter of a twisted bag should be allowed to vary between 3 and 5 mm.

19.20 Design the thread to the mechanical jack shown in sketch *k*. The mean thread diameter is D and the thread angle is β. The coefficient of friction at all contact areas is μ. Determine at which lead angle α the jack is self-locking. Calculate the power efficiency for the jack if the lead angle is chosen as one-half the above value. By replacing the screw with a low-friction ball-screw, find how much higher the power efficiency can become if the screw lead angle is one-half of the maximum possible for self-locking.

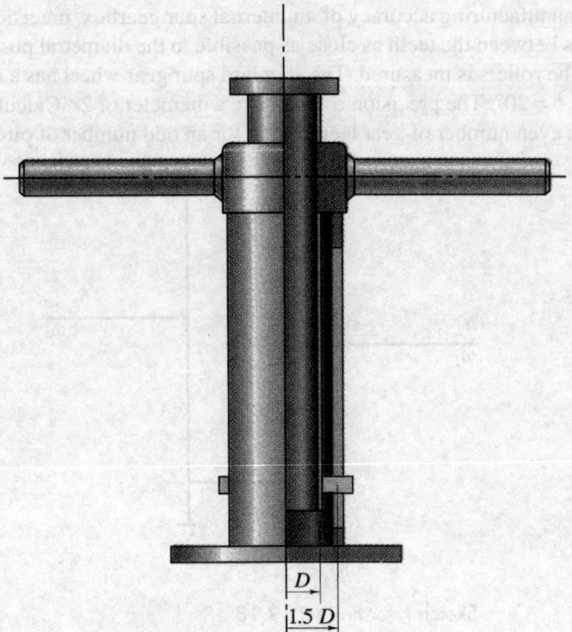

Sketch *k*, for Project 19.20

19.21 The conical compression spring shown in sketch *l* has *N* free coils and constant pitch *p*. Determine the spring rate when the smallest mean diameter is D_1 and the largest is D_2 and the wire diameter is *d*. The shear modulus of elasticity for the spring steel is *G*. Calculate and show in a diagram the spring force as a function of the deformation from zero load to a totally compressed spring.

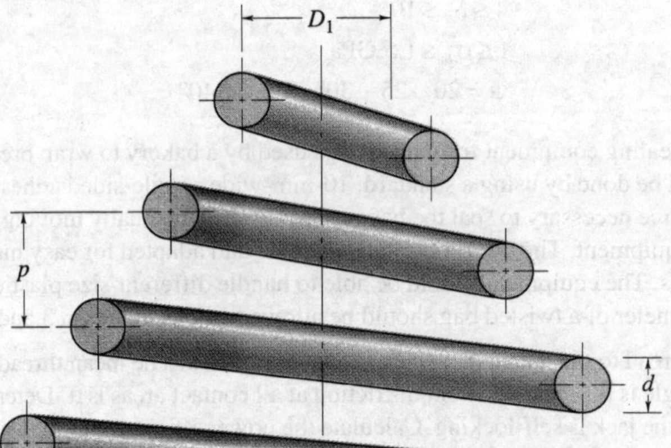

Sketch *l*, for Project 19.21

19.22 A gas-filled series of bellows is used as a combined spring and damper. The spring force comes from the damping caused by the heat generated in the gas when it is compressed and the subsequent heat loss from the bellows. The compression of the gas is adiabatic, so $pV^{1.4}$ = constant, where p is pressure and V is volume. During the expansion after the gas has cooled to room temperature, the temperature is constant and so pV = constant. Determine the spring rate at high- and low-frequency oscillations, and calculate the damping if the compression is fast and the expansion is slow.

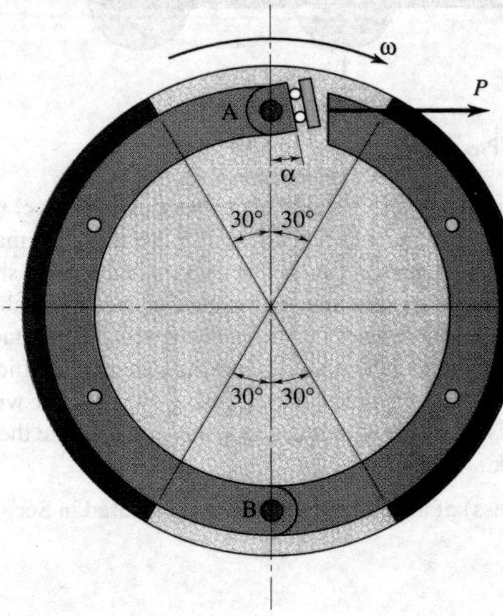

Sketch *m*, for Project 19.23

19.23 An internal long-shoe brake has the two shoes coupled together at point B as shown in sketch *m*. The left shoe is hinged at A and can slide without friction against a stop in the backplate having the angle α. The right-hand shoe is hinged to the left-hand shoe at B. The wear of the brake lining is proportional to the contact pressure times the sliding distance. The maximum allowable contact pressure is 2 MPa. The coefficient of friction between the brake shoes and the brake drum is μ. Design the brake so it can transmit the torque T. Check that the brake is not self-locking. Calculate and draw the pressure distributions for the two shoes. Calculate the actuating force P needed.

19.24 A V-belt drive is used at its limit for force transmission, so a more accurate calculation of its power transmission ability is needed. The simplified derivation given in Chapter 18 builds on the assumption that all friction forces between the sheave and the V-belt and all sliding speeds are in the circumferential direction. In a V-belt drive, when the belt goes into the pulley, the friction force on the belt has a radial component outward, causing the friction to be directed inward. The tension in the V-belt increases its length and at the same time decreases its cross-sectional dimensions. The belt thus moves deeper into the sheave when the tight part enters the sheave (the driving sheave) than when the slack part enters the sheave (the driven sheave). Determine the coefficient of friction needed at the sheave angle β = 18° to ensure that a small driving sheave, which has a small wrap angle, can transmit as much power as a large driven sheave.

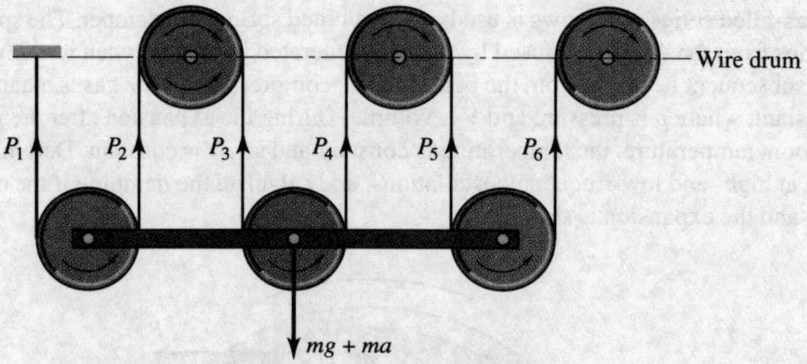

Sketch *n*, for Project 19.25

19.25 A transverse crane in a switchyard has a six-part sheaving for a steel wire of regular type. The principle of wire motion over the wheels is shown in sketch *n*. The mass to be lifted is 20 000 kg, and the speed of the line drum is 1 m/s. The friction in the cable sheave makes the cable force 4% larger on the outgoing side than on the ingoing side for each sheave. Design the sheaves and the wires so that the safety factor is 8 when the lifting starts from the ground and the total free length of the wire is 60 m. Calculate the total power efficiency for the crane and also how the force in the wire fluctuates during lifting owing to the wire elasticity. Calculate for the given speed how short the wire must be at the start to make the lifted mass jump and the wires become slack.

19.26 Select the size (thickness) of all gears in the mixer considered in Sec. 19.2 if a 2-hp motor is used.

APPENDIX

A

MATERIAL PROPERTIES

Table A.1 Properties of ferrous metals.

Material	Density, kg/m^3	Modulus of elasticity, psi $\times 10^6$ (GPa)	Yield strength, ksi (MPa)	Ultimate strength, ksi (MPa)	Ductility, %EL in 2 in.	Poisson's ratio	Thermal conductivity, W/m-°C	Coefficient of thermal expansion, (°C)$^{-1} \times 10^{-6}$
Iron	7 870	30 (207)	19 (130)	38 (260)	45	0.29	80	11.8
Gray cast iron	7 150	Variable	—	18 (125)	—	Variable	46	10.8
Nodular cast iron	7 120	24 (165)	40 (275)	60 (415)	18	0.28	33	11.8
Malleable cast iron	7 200–7 450	25 (172)	32 (220)	50 (345)	10	0.26	51	11.9
Low-carbon steel (AISI 1020)	7 860	30 (207)	43 (295)	57 (395)	37	0.30	52	11.7
Medium-carbon steel (1040)	7 850	30 (207)	51 (350)	75 (520)	30	0.30	52	11.3
High-carbon steel (AISI 1080)	7 840	30 (207)	55 (380)	89 (615)	25	0.30	48	11.0
Stainless steels								
Ferritic, type 446	7 500	29 (200)	50 (345)	80 (552)	20	0.30	21	10.4
Austenitic, type 316	8 000	28 (193)	30 (207)	80 (552)	60	0.30	16	16.0
Martensitic, type 410	7 800	29 (200)	40 (275)	70 (483)	30	0.30	25	9.9

Table A.2 Properties of nonferrous metals.

Material	Density, kg/m³	Modulus of elasticity, psi ×10⁶ (GPa)	Yield strength, ksi (MPa)	Ultimate strength, ksi (MPa)	Ductility, %EL in 2 in.	Poisson's ratio	Thermal conductivity, W/m-°C	Coefficient of thermal expansion, (°C)⁻¹ × 10⁻⁶
Aluminum (>99.5%)	2 710	10 (69)	2.5 (17)	8 (55)	25	0.33	231	23.6
Aluminum alloy 2014	2 800	10.5 (72)	14 (97)	27 (186)	18	0.33	192	22.5
Copper (99.95%)	8 940	16 (110)	10 (69)	32 (220)	45	0.35	398	16.5
Brass (70Cu–30Zn)	8 530	16 (110)	11 (75)	44 (303)	68	0.35	120	20.0
Bronze (92Cu–8Sn)	8 800	16 (110)	22 (152)	55 (380)	70	0.35	62	18.2
Magnesium (>99%)	1 740	6.5 (45)	6 (41)	24 (165)	14	0.29	122	27.0
Molybdenum (>99%)	10 220	47 (324)	82 (565)	95 (655)	35	—	142	4.9
Nickel (>99%)	8 900	30 (207)	20 (138)	70 (483)	40	0.31	80	13.3
Silver (>99%)	10 490	11 (76)	8 (55)	18 (125)	48	0.37	418	19.0
Titanium (>99%)	4 510	15.5 (107)	35 (240)	48 (330)	30	0.34	17	9.0

Table A.3 Properties of ceramics.

Material	Density kg/m^3	Modulus of elasticity, psi × 10^6 (GPa)	Poisson's ratio	Approximate hardness (Knoop)	Fracture strength, ksi (MPa)	Thermal conductivity, W/m-°C	Coefficient of thermal expansion (°C)$^{-1}$ × 10^{-6}
Alumina (Al_2O_3)	3 970	57 (393)	0.27	2100	40–80 (275–550)	30	8.8[a]
Magnesia (MgO)	3 580	30 (207)	0.36	370	15[b] (105)	48	13.5[a]
Spinel ($MgAl_2O_4$)	3 550	36 (284)	—	1600	12–32[b] (83–220)	15.0[a]	7.6[a]
Zirconia[c] (ZrO_2)	5 560	22 (152)	0.32	1200	20–35[b] (138–240)	2.0	10.0[a]
Fused silica (SiO_2)	2 200	11 (75)	0.16	500	16 (110)	1.3	0.5[a]
Soda–lime glass	2 500	10 (69)	0.23	550	10 (69)	1.7	9.0[d]
Borosilicate glass	2 230	9 (62)	0.20	—	10 (69)	1.4	3.3[d]
Silicon carbide (SiC)	3 220	60 (414)	0.19	2500	65–75[b] (450–520)	90	4.7
Silicon nitride (Si_3N_4)	3 440	44 (304)	0.24	2200	60–80[b] (414–580)	16–33[a]	3.6[a]
Titanium carbide (TiC)	4 920	67 (462)	—	2600	40–65[b] (275–450)	17.2	7.4

[a]Mean value taken over the temperature range 0–1000°C.
[b]Sintered and containing approximately 5% porosity.
[c]Stabilized with calcium oxide.
[d]Mean value taken over the temperature range 0–300°C.

Table A.4 Properties of polymers.

Polymers	State	Density, kg/m^3	Modulus of elasticity, ksi (GPa)	Ultimate strength, ksi (MPa)	Elongation at break, percent	Glass transition temperature, °C	Maximum service temperature, °C	Thermal conductivity, W/m-°C	Coefficient of thermal expansion, $(°C)^{-1} \times 10^{-6}$
Thermoplastics									
Polyethylene	High density, 70–80% crystalline	952.1–965.0	155–158 (1.07–1.09)	3.2–4.5 (22–31)	10–1200	–90	130–137	0.48	60–110
	Low density, 40–50% crystalline	917.1–932.1	25–41 (0.17–0.28)	1.2–4.5 (8.3–31.0)	100–650	–110	98–115	0.33	100–220
Polytetrafluoroethylene	50–70% crystalline	2 140–2 200	58–80 (0.40–0.55)	2.0–5.0 (14–34)	200–400	–90	327	0.25	70–120
Polyvinyl chloride	Highly amorphous	1 300–1 580	350–600 (2.4–4.1)	6.0–7.5 (41–52)	40–80	75–105	212	0.18	50–100
Polypropylene	50–60% crystalline	900–910	165–225 (1.14–1.55)	4.5–6.0 (31–41)	100–600	–20	168–175	0.12	80–100
Polystyrene	Amorphous	1 040–1 050	330–475 (2.28–3.28)	5.2–7.5 (36–52)	1.2–2.5	74–105	—	0.13	50–83
Polymethyl methacrylate	Amorphous	1 170–1 200	325–470 (2.24–3.24)	7–11 (48–76)	2–10	85–105	—	0.21	50–90
Nylon 6,6 Poly(hexamethylene adipamide)	30–40% crystalline	1 130–1 150	230–550 (1.58–3.79)	11–13.7 (76–94)	15–300	57	255–265	0.24	80
Polyethylene terephthalate (PET, a polyester)	0–30% crystalline	1 290–1 400	400–600 (2.76–4.14)	7.0–10.5 (48–72)	30–300	73–80	245–265	0.14	65
Polycarbonate (Polybisphenol, –A carbonate)	Amorphous	1 200	345 (2.38)	9.5 (65.5)	110	150	—	0.20	68
Thermosets									
Epoxy	Complex network, amorphous	1110–1140	350 (2.41)	4.0–13.0 (28–90)	3–6	—	—	0.19	45–65
Phenolic	Complex network, amorphous	1240–1320	400–700 (2.76–4.83)	5–9 (34–62)	1.5–2.0	—	—	0.15	68
Polyester	Complex network, amorphous	1040–1460	300–640 (2.07–4.41)	6–13 (41–90)	<2	—	—	0.19	55–100

Table A.5 Properties of natural rubbers. [Adapted from *Materials Engineering,* a Penton Publication.]

Name/Repeat unit	Density, kg/m³	Ultimate strength, ksi (MPa)	Maximum elongation, percent	Modulus, at 100% elongation, psi (MPa)	Minimum service temperature, °C (°F)	Maximum service temperature, °C (°F)	Abrasion resistance	Tear resistance	Oxidation resistance
Natural polyisoprene (natural rubber, NR)	920–1 037	3.5–4.6 (24–32)	500–760	480-850 (3.3–5.9)	–60 (–75)	120 (250)	Excellent	Excellent	Good
Styrene-butadiene (SBR, GRS)	940	1.8–3.0 (12–21)	450–500	300–1 500 (2.1–10.3)	–60 (–75)	120 (250)	Excellent	Fair	Good
Acrylonitrile-butadiene (nitrile, Buna A, NBR)	980	1.0–3.5 (7–24)	400–600	490 (3.4)	–50 (–60)	150 (300)	Excellent	Good	Fair-Good
Chloroprene (neoprene, CR)	1 230–1 250	0.5–3.5 (3.5–24)	100–800	100–3 000 (0.7–20)	–50 (–60)	105 (225)	Excellent	Good	Very good
Polybutadiene (BR)	910	2.0–2.5 (14–17)	450	300–1 500 (2.1–10.3)	–100 (–150)	90 (200)	Excellent	Good	Good
Polyurethane	1 020–1 250	0.8–8.0 (5.5–55)	250–800	25–5 000 (0.17–34.5)	–55 (–65)	120 (250)	Excellent	Outstanding	Excellent
Polydimethylsiloxane (silicone)	1 100–1 600	1.5 (10)	100–800	—	–115 (–175)	315 (600)	Poor	Fair	Excellent

APPENDIX B

STRESS-STRAIN RELATIONSHIPS

SYMBOLS

C	elastic material coefficients, Pa
E	modulus of elasticity, Pa
G	shear modulus, Pa
K	bulk modulus, Pa
x, y, z	Cartesian coordinate system, m
x', y', z'	rotated Cartesian coordinate system, m
γ	shear strain
δ	deformation, m
ε	normal strain
λ	Lamé's constant, Pa
ν	Poisson's ratio
σ	normal stress, Pa
τ	shear stress, Pa

Subscripts

x, y, z	Cartesian coordinates
x', y', z'	rotated Cartesian coordinates
1, 2, 3	principal axes

B.1 LAWS OF STRESS TRANSFORMATION

Let a new orthogonal coordinate system $0'x'y'z'$ be placed having origin P, having the z' axis coincident with the normal n to the plane, and having the x' and y' axes (which must parallel the plane with normal n) coincident with the desired shear stresses directions. Such a coordinate system is shown in Fig. B.1, where the plane ABC is imagined to pass through point P, since the tetrahedron is very small. Figure B.1 also shows the three mutually perpendicular stress components $\sigma_{z'}$, $\tau_{z'x'}$, and $\tau_{z'y'}$ acting on the inclined plane. The laws of stress transformation give these stresses in terms of σ_x, σ_y, σ_z, τ_{xy}, τ_{yz}, and τ_{zx}. The previously established rules for shear stress subscripts are equally applicable here. For example, $\tau_{z'x'}$ is the shear stress directed in the x' direction and acting in a plane through P whose normal is directed in the z' direction.

Because the normal direction coincides with the z' direction,

$$\begin{aligned} p_{nz} &= \sigma_z \cos(z',z) + \tau_{xz}\cos(z',x) + \tau_{yz}\cos(z',y) \\ p_{ny} &= \tau_{zx}\cos(z',z) + \sigma_x \cos(z',x) + \tau_{yx}\cos(z',y) \\ p_{nx} &= \tau_{zy}\cos(z',z) + \tau_{xy}\cos(z',x) + \sigma_y \cos(z',y) \end{aligned} \tag{B.1}$$

where

$\cos(z', x)$ = cosine of angle between z' and x

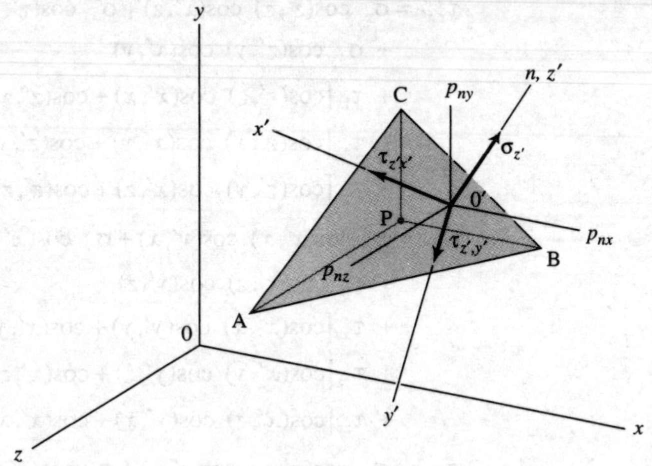

Figure B.1 Small tetrahedronal element cut from body at point P. The three perpendicular stress components $\sigma_{z'}$, $\tau_{z'x'}$, and $\tau_{z'y'}$ acting on the inclined plane are shown.

The stress $\sigma_{z'}$ must equal the sum of the projections of p_{nz}, p_{ny}, and p_{nx} onto the z' axis,

$$\sigma_{z'} = p_{nz}\cos(z',z) + p_{nx}\cos(z',x) + p_{ny}\cos(z',y) \quad \textbf{(B.2)}$$

Similarly,

$$\tau_{z'x'} = p_{nz}\cos(x',z) + p_{nx}\cos(x',x) + p_{ny}\cos(x',y) \quad \textbf{(B.3)}$$

By substituting Eqs. (B.1) into Eqs. (B.2) and (B.3) while also making use of the fact developed earlier that the shear stresses are symmetrical (e.g., $\tau_{x'y'} = \tau_{y'x'}$), Eqs. (B.4) and (B.5) can be developed. Equations (B.6) to (B.9) can be derived in an entirely similar way by using two more tetrahedrons with inclined planes having outer normals parallel to the x' and y' axes. The result of all of the above is

$$\begin{aligned}\sigma_{z'} &= \sigma_z\cos^2(z',z) + \sigma_x\cos^2(z',x) + \sigma_y\cos^2(z',y)\\ &\quad + 2\tau_{zx}\cos(z',z)\cos(z',x) + 2\tau_{xy}\cos(z',x)\cos(z',y)\\ &\quad + 2\tau_{yz}\cos(z',y)\cos(z',z)\end{aligned} \quad \textbf{(B.4)}$$

$$\begin{aligned}\sigma_{x'} &= \sigma_x\cos^2(x',x) + \sigma_y\cos^2(x',y) + \sigma_z\cos^2(x',z)\\ &\quad + 2\tau_{xy}\cos(x',x)\cos(x',y) + 2\tau_{yz}\cos(x',y)\cos(x',z)\\ &\quad + 2\tau_{zx}\cos(x',z)\cos(x',x)\end{aligned} \quad \textbf{(B.5)}$$

$$\begin{aligned}\sigma_{y'} &= \sigma_y\cos^2(y',y) + \sigma_z\cos^2(y',z) + \sigma_x\cos^2(y',x)\\ &\quad + 2\tau_{yz}\cos(y',y)\cos(y',z) + 2\tau_{zx}\cos(y',z)\cos(y',x)\\ &\quad + 2\tau_{xy}\cos(y',x)\cos(y',y)\end{aligned} \quad \textbf{(B.6)}$$

$$
\begin{aligned}
\tau_{z'x'} = {} & \sigma_z \cos(z',z)\cos(x',z) + \sigma_x \cos(z',x)\cos(x',x) \\
& + \sigma_y \cos(z',y)\cos(x',y) \\
& + \tau_{zx}\left[\cos(z',z)\cos(x',x) + \cos(z',x)\cos(x',z)\right] \\
& + \tau_{xy}\left[\cos(z',x)\cos(x',y) + \cos(z',y)\cos(x',x)\right] \\
& + \tau_{yz}\left[\cos(z',y)\cos(x',z) + \cos(z',z)\cos(x',y)\right]
\end{aligned}
\quad \textbf{(B.7)}
$$

$$
\begin{aligned}
\tau_{x'y'} = {} & \sigma_x \cos(x',x)\cos(y',x) + \sigma_y \cos(x',y)\cos(y',y) \\
& + \sigma_z \cos(x',z)\cos(y',z) \\
& + \tau_{xy}\left[\cos(x',x)\cos(y',y) + \cos(x',y)\cos(y',x)\right] \\
& + \tau_{yz}\left[\cos(x',y)\cos(y',z) + \cos(x',z)\cos(y',y)\right] \\
& + \tau_{zx}\left[\cos(x',z)\cos(y',x) + \cos(x',x)\cos(y',z)\right]
\end{aligned}
\quad \textbf{(B.8)}
$$

$$
\begin{aligned}
\tau_{y'z'} = {} & \sigma_y \cos(y',y)\cos(z',y) + \sigma_z \cos(y',z)\cos(z',z) \\
& + \sigma_x \cos(y',x)\cos(z',x) \\
& + \tau_{yz}\left[\cos(y',y)\cos(z',z) + \cos(y',z)\cos(z',y)\right] \\
& + \tau_{zx}\left[\cos(y',z)\cos(z',x) + \cos(y',x)\cos(z',z)\right] \\
& + \tau_{xy}\left[\cos(y',x)\cos(z',y) + \cos(y',y)\cos(z',x)\right]
\end{aligned}
\quad \textbf{(B.9)}
$$

It is important to recognize the meaning of these equations. For example, Eq. (B.7) can be used to find the shear stress acting in the x' direction on a surface having an outer normal directed in the z' direction if the six Cartesian stress components σ_x, σ_y, σ_z, τ_{xy}, τ_{yz}, and τ_{zx} at the point and the orientation of the coordinate system $0'x'y'z'$ are known.

B.2 LAWS OF STRAIN TRANSFORMATION

In the study of stresses we found that three normal stresses σ_x, σ_y, and σ_z and three shear stresses τ_{xy}, τ_{yz}, and τ_{zx} act on planes parallel to the Cartesian planes. Similarly, three normal strains ε_x, ε_y, and ε_z and three shear strains γ_{xy}, γ_{yz}, and γ_{zx} characterize the behavior of line segments originally parallel to the Cartesian axes. From Durelli et al. (1958) the laws of strain transformation can be written directly from the laws of stress transformation given in Eqs. (B.4) to (B.9) if the following replacements are made to these equations:

$$
\sigma_x \leftarrow \varepsilon_x, \quad \sigma_y \leftarrow \varepsilon_y, \quad \sigma_z \leftarrow \varepsilon_z, \quad 2\tau_{xy} \leftarrow \gamma_{xy}, \quad 2\tau_{yz} \leftarrow \gamma_{yz}, \quad 2\tau_{zx} \leftarrow \gamma_{zx} \quad \textbf{(B.10)}
$$

B.3 HOOKE'S LAW GENERALIZED

What is the relationship between stresses and strains when the stress system is not in simple tension or compression as in Eq. (3.22)? The stress system can be defined by the six components σ_x, σ_y, σ_z, τ_{xy}, τ_{yz}, and τ_{zx}; and the strain system, by the six components ε_x, ε_y,

ε_z, γ_{xy}, γ_{yz}, and γ_{zx}. Thus, a generalization of Hooke's law is to make each stress component a linear function of the strain components or

$$\begin{aligned}
\sigma_x &= C_{11}\varepsilon_x + C_{12}\varepsilon_y + C_{13}\varepsilon_z + C_{14}\gamma_{xy} + C_{15}\gamma_{yz} + C_{16}\gamma_{zx} \\
\sigma_y &= C_{21}\varepsilon_x + C_{22}\varepsilon_y + C_{23}\varepsilon_z + C_{24}\gamma_{xy} + C_{25}\gamma_{yz} + C_{26}\gamma_{zx} \\
&\quad\vdots \\
\tau_{zx} &= C_{61}\varepsilon_x + C_{62}\varepsilon_y + C_{63}\varepsilon_z + C_{64}\gamma_{xy} + C_{65}\gamma_{yz} + C_{66}\gamma_{zx}
\end{aligned} \qquad \textbf{(B.11)}$$

where $C_{11}, C_{12}, \ldots, C_{66}$ = elastic material coefficients independent of stress or strain. Equation (B.11) is valid for many hard materials over a strain range of practical interest in designing machine elements.

The assumption of an isotropic material implies that the stress and strain components referred to a coordinate system $0x'y'z'$ of any arbitrary orientation must be related by the same elastic material coefficients $C_{11}, C_{12}, \ldots, C_{66}$. Thus, the six Eqs. (B.11) imply that the subscripts change, $x \rightarrow x'$, $y \rightarrow y'$, and $z \rightarrow z'$, while the constants $C_{11}, C_{12}, \ldots, C_{66}$ remain the same. Thus,

$$\begin{aligned}
\sigma_{x'} &= C_{11}\varepsilon_{x'} + C_{12}\varepsilon_{y'} + C_{13}\varepsilon_{z'} + C_{14}\gamma_{x'y'} + C_{15}\gamma_{y'z'} + C_{16}\gamma_{z'x'} \\
\sigma_{y'} &= C_{21}\varepsilon_{x'} + C_{22}\varepsilon_{y'} + C_{23}\varepsilon_{z'} + C_{24}\gamma_{x'y'} + C_{25}\gamma_{y'z'} + C_{26}\gamma_{z'x'} \\
&\quad\vdots \\
\tau_{z'x'} &= C_{61}\varepsilon_{x'} + C_{62}\varepsilon_{y'} + C_{63}\varepsilon_{z'} + C_{64}\gamma_{x'y'} + C_{65}\gamma_{y'z'} + C_{66}\gamma_{z'x'}
\end{aligned} \qquad \textbf{(B.12)}$$

Recall that $0x'y'z'$ is an arbitrary coordinate system.

First, obtaining a new coordinate system $0x'y'z'$ by rotating $0z$ through a 180° angle results in Fig. B.2. Recall that

$$(x,y) = \text{Angle between } x \text{ and } y$$

From Fig. B.2 and the above notation

$$\begin{aligned}
(x',x) &= 180^\circ, & (x',y) &= 90^\circ, & (x',z) &= 90^\circ \\
(y',y) &= 180^\circ, & (y',z) &= 90^\circ, & (y',x) &= 90^\circ \\
(z',z) &= 0^\circ, & (z',x) &= 90^\circ, & (z',y) &= 90^\circ
\end{aligned} \qquad \textbf{(B.13)}$$

Making use of Eq. (B.13) reduces Eqs. (B.4) to (B.9) to

$$\sigma_{z'} = \sigma_z, \quad \sigma_{x'} = \sigma_x, \quad \sigma_{y'} = \sigma_y, \quad \tau_{x'y'} = \tau_{xy}, \quad \tau_{y'z'} = -\tau_{yz}, \quad \tau_{z'x'} = -\tau_{zx} \qquad \textbf{(B.14)}$$

Similarly, for the strains

$$\varepsilon_{z'} = \varepsilon_z, \quad \varepsilon_{x'} = \varepsilon_x, \quad \varepsilon_{y'} = \varepsilon_y, \quad \gamma_{x'y'} = \gamma_{xy}, \quad \gamma_{y'z'} = -\gamma_{yz}, \quad \gamma_{z'x'} = -\gamma_{zx} \qquad \textbf{(B.15)}$$

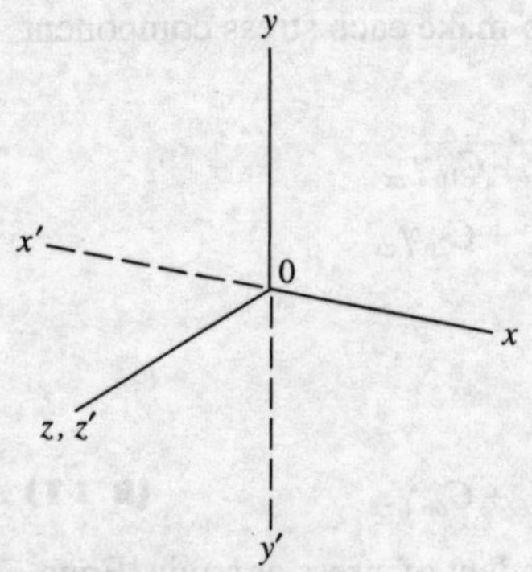

Figure B.2 Coordinate system $0x'y'z'$ obtained by rotating $0z$ 180°.

Substituting Eqs. (B.14) and (B.15) into Eq. (B.12) gives

$$\begin{aligned}
\sigma_x &= C_{11}\varepsilon_x + C_{12}\varepsilon_y + C_{13}\varepsilon_z + C_{14}\gamma_{xy} - C_{15}\gamma_{yz} - C_{16}\gamma_{zx} \\
\sigma_y &= C_{21}\varepsilon_x + C_{22}\varepsilon_y + C_{23}\varepsilon_z + C_{24}\gamma_{xy} - C_{25}\gamma_{yz} - C_{26}\gamma_{zx} \\
&\cdot \quad \cdot \quad \cdot \quad \cdot \quad \cdot \quad \cdot \\
&\cdot \quad \cdot \quad \cdot \quad \cdot \quad \cdot \quad \cdot \\
&\cdot \quad \cdot \quad \cdot \quad \cdot \quad \cdot \quad \cdot \\
\tau_{zx} &= -C_{61}\varepsilon_x - C_{62}\varepsilon_y - C_{63}\varepsilon_z - C_{64}\gamma_{xy} + C_{65}\gamma_{yz} + C_{66}\gamma_{zx}
\end{aligned} \tag{B.16}$$

Comparing Eqs. (B.11) and (B.16) shows that

$$C_{15} = -C_{15} \qquad C_{16} = -C_{16}$$

implying that

$$\begin{aligned}
C_{15} = C_{16} = C_{25} = C_{26} = C_{35} = C_{36} = C_{45} = C_{46} = 0 \\
C_{51} = C_{52} = C_{53} = C_{54} = C_{61} = C_{62} = C_{63} = C_{64} = 0
\end{aligned} \tag{B.17}$$

Substituting Eqs. (B.17) into Eqs. (B.11) gives

$$\begin{aligned}
\sigma_x &= C_{11}\varepsilon_x + C_{12}\varepsilon_y + C_{13}\varepsilon_z + C_{14}\gamma_{xy} \\
\sigma_y &= C_{21}\varepsilon_x + C_{22}\varepsilon_y + C_{23}\varepsilon_z + C_{24}\gamma_{xy} \\
\sigma_z &= C_{31}\varepsilon_x + C_{32}\varepsilon_y + C_{33}\varepsilon_z + C_{34}\gamma_{xy} \\
\tau_{xy} &= C_{41}\varepsilon_x + C_{42}\varepsilon_y + C_{43}\varepsilon_z + C_{44}\gamma_{xy} \\
\tau_{yz} &= C_{55}\gamma_{yz} + C_{56}\gamma_{zx} \\
\tau_{zx} &= C_{65}\gamma_{yz} + C_{66}\gamma_{zx}
\end{aligned} \tag{B.18}$$

Thus, using the coordinate system of Fig. B.2 with the isotropic assumption reduces the 36 elastic material constants of Eqs. (B.11) to the 20 expressed in Eqs. (B.18).

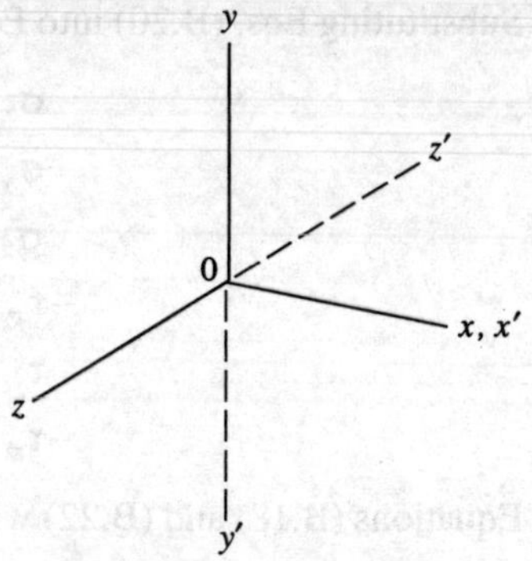

Figure B.3 Coordinate system $0x'y'z'$ obtained by rotating $0x$ 180°.

Obtaining a new coordinate system by rotating $0x$ through a 180° angle results in Fig. B.3, giving

$$\begin{aligned}
&(x',x)=0^\circ, && (x',y)=90^\circ, && (x',z)=90^\circ \\
&(y',y)=180^\circ, && (y',x)=90^\circ, && (y',z)=90^\circ \\
&(z',z)=180^\circ, && (z',x)=90^\circ, && (z',y)=90^\circ
\end{aligned} \tag{B.19}$$

Making use of Eqs. (B.19) reduces Eqs. (B.4) to (B.9) to

$$\begin{aligned}
&\sigma_{z'}=\sigma_z, && \sigma_{x'}=\sigma_x, && \sigma_{y'}=\sigma_y \\
&\tau_{z'x'}=-\tau_{zx}, && \tau_{x'y'}=-\tau_{xy}, && \tau_{y'z'}=\tau_{yz} \\
&\varepsilon_{x'}=\varepsilon_x, && \varepsilon_{y'}=\varepsilon_y, && \varepsilon_{z'}=\varepsilon_z \\
&\gamma_{x'y'}=-\gamma_{xy}, && \gamma_{y'z'}=\gamma_{yz}, && \gamma_{z'x'}=-\gamma_{zx}
\end{aligned} \tag{B.20}$$

The assumption of an isotropic material requires that the new coordinate system $0x'y'z'$ shown in Fig. B.3 must be related to the same elastic material constants C_{11}, C_{12}, . . ., C_{66} expressed in Eqs. (B.18), implying that

$$\begin{aligned}
\sigma_{x'} &= C_{11}\varepsilon_{x'} + C_{12}\varepsilon_{y'} + C_{13}\varepsilon_{z'} + C_{14}\gamma_{x'y'} \\
\sigma_{y'} &= C_{21}\varepsilon_{x'} + C_{22}\varepsilon_{y'} + C_{23}\varepsilon_{z'} + C_{24}\gamma_{x'y'} \\
\sigma_{z'} &= C_{31}\varepsilon_{x'} + C_{32}\varepsilon_{y'} + C_{33}\varepsilon_{z'} + C_{34}\gamma_{x'y'} \\
\tau_{x'y'} &= C_{41}\varepsilon_{x'} + C_{42}\varepsilon_{y'} + C_{43}\varepsilon_{z'} + C_{44}\gamma_{x'y'} \\
\tau_{y'z'} &= C_{55}\gamma_{y'z'} + C_{56}\gamma_{z'x'} \\
\tau_{z'x'} &= C_{65}\gamma_{y'z'} + C_{66}\gamma_{z'x'}
\end{aligned} \tag{B.21}$$

Substituting Eqs. (B.20) into Eqs. (B.21) gives

$$\begin{aligned}
\sigma_x &= C_{11}\varepsilon_x + C_{12}\varepsilon_y + C_{13}\varepsilon_z - C_{14}\gamma_{xy} \\
\sigma_y &= C_{21}\varepsilon_x + C_{22}\varepsilon_y + C_{23}\varepsilon_z - C_{24}\gamma_{xy} \\
\sigma_z &= C_{31}\varepsilon_x + C_{32}\varepsilon_y + C_{33}\varepsilon_z - C_{34}\gamma_{xy} \\
-\tau_{xy} &= C_{41}\varepsilon_x + C_{42}\varepsilon_y + C_{43}\varepsilon_z - C_{44}\gamma_{xy} \\
\tau_{yz} &= C_{55}\gamma_{yz} - C_{56}\gamma_{zx} \\
-\tau_{zx} &= C_{65}\gamma_{yz} - C_{66}\gamma_{zx}
\end{aligned} \tag{B.22}$$

Equations (B.18) and (B.22) will agree only if

$$C_{14} = C_{24} = C_{34} = C_{41} = C_{42} = C_{43} = C_{56} = C_{65} = 0 \tag{B.23}$$

Substituting Eqs. (B.13) into Eqs. (B.8) gives

$$\begin{aligned}
\sigma_x &= C_{11}\varepsilon_x + C_{12}\varepsilon_y + C_{13}\varepsilon_z \\
\sigma_y &= C_{21}\varepsilon_x + C_{22}\varepsilon_y + C_{23}\varepsilon_z \\
\sigma_z &= C_{31}\varepsilon_x + C_{32}\varepsilon_y + C_{33}\varepsilon_z \\
\tau_{xy} &= C_{44}\gamma_{xy} \\
\tau_{yz} &= C_{55}\gamma_{yz} \\
\tau_{zx} &= C_{66}\gamma_{zx}
\end{aligned} \tag{B.24}$$

Thus, using the coordinate system of Fig. B.3 with the isotropic assumption reduces the 20 elastic material constants of Eqs. (B.18) to the 12 expressed in Eqs. (B.24).

Third, rotating $0x$ through a 90° angle results in Fig. B.4, implying that

$$\begin{aligned}
&(x', x) = 0^\circ, \quad (x', y) = 90^\circ, \quad (x', z) = 90^\circ \\
&(y', y) = 90^\circ, \quad (y', x) = 90^\circ, \quad (y', z) = 0^\circ \\
&(z', z) = 90^\circ, \quad (z', x) = 90^\circ, \quad (z', y) = 180^\circ
\end{aligned} \tag{B.25}$$

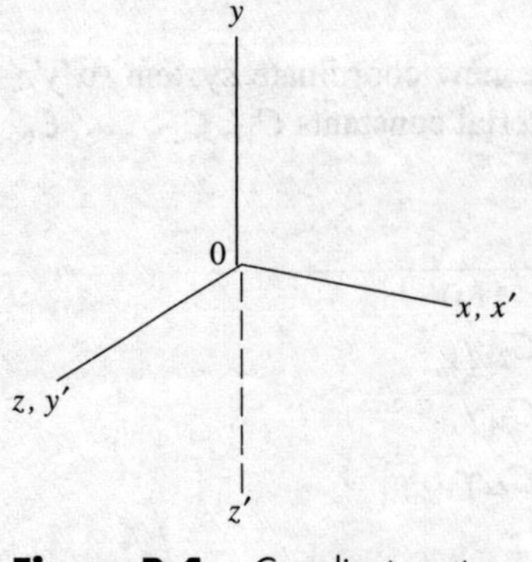

Figure B.4 Coordinate system $0x'y'z'$ obtained by rotating $0x$ 90°.

Making use of Eqs. (B.25) reduces Eqs. (B.4) to (B.9) and Eq. (2.37) to

$$\begin{array}{lll} \sigma_{z'} = \sigma_y, & \sigma_{x'} = \sigma_x, & \sigma_{y'} = \sigma_z \\ \tau_{z'x'} = -\tau_{xy}, & \tau_{x'y'} = \tau_{zx}, & \tau_{y'z'} = -\tau_{yz} \\ \varepsilon_{z'} = \varepsilon_y, & \varepsilon_{x'} = \varepsilon_x, & \varepsilon_{y'} = \varepsilon_z \\ \gamma_{z'x'} = -\gamma_{xy}, & \gamma_{x'y'} = \gamma_{zx}, & \gamma_{y'z'} = -\gamma_{yz} \end{array} \tag{B.26}$$

The assumption of an isotropic material requires that the new coordinate system $0x'y'z'$ shown in Fig. B.4 must be related to the same elastic material constants expressed in Eqs. (B.24), implying that

$$\begin{aligned} \sigma_{x'} &= C_{11}\varepsilon_{x'} + C_{12}\varepsilon_{y'} + C_{13}\varepsilon_{z'} \\ \sigma_{y'} &= C_{21}\varepsilon_{x'} + C_{22}\varepsilon_{y'} + C_{23}\varepsilon_{z'} \\ \sigma_{z'} &= C_{31}\varepsilon_{x'} + C_{32}\varepsilon_{y'} + C_{33}\varepsilon_{z'} \\ \tau_{x'y'} &= C_{44}\gamma_{x'y'} \\ \tau_{y'z'} &= C_{55}\gamma_{y'z'} \\ \tau_{z'x'} &= C_{66}\gamma_{z'x'} \end{aligned} \tag{B.27}$$

Substituting Eqs. (B.26) into Eqs. (B.27) gives

$$\begin{aligned} \sigma_x &= C_{11}\varepsilon_x + C_{12}\varepsilon_z + C_{13}\varepsilon_y \\ \sigma_z &= C_{21}\varepsilon_x + C_{22}\varepsilon_z + C_{23}\varepsilon_y \\ \sigma_y &= C_{31}\varepsilon_x + C_{32}\varepsilon_z + C_{33}\varepsilon_y \\ \tau_{xy} &= C_{66}\gamma_{xy} \\ \tau_{yz} &= C_{55}\gamma_{yz} \\ \tau_{zx} &= C_{44}\gamma_{zx} \end{aligned} \tag{B.28}$$

Equations (B.24) and (B.28) will agree only if

$$C_{12} = C_{13} \quad C_{21} = C_{31} \quad C_{23} = C_{32} \quad C_{22} = C_{33} \quad C_{44} = C_{66} \tag{B.29}$$

Substituting Eqs. (B.29) into Eqs. (B.28) gives

$$\begin{aligned} \sigma_x &= C_{11}\varepsilon_x + C_{12}(\varepsilon_y + \varepsilon_z) \\ \sigma_y &= C_{21}\varepsilon_x + C_{22}\varepsilon_y + C_{23}\varepsilon_z \\ \sigma_z &= C_{21}\varepsilon_x + C_{22}\varepsilon_z + C_{23}\varepsilon_y \\ \tau_{xy} &= C_{44}\gamma_{xy} \\ \tau_{yz} &= C_{55}\gamma_{yz} \\ \tau_{zx} &= C_{44}\gamma_{zx} \end{aligned} \tag{B.30}$$

Thus, using the coordinate system of Fig. B.4 with the isotropic assumption reduces the 12 elastic material constants of Eqs. (B.24) to the 7 expressed in Eqs. (B.30).

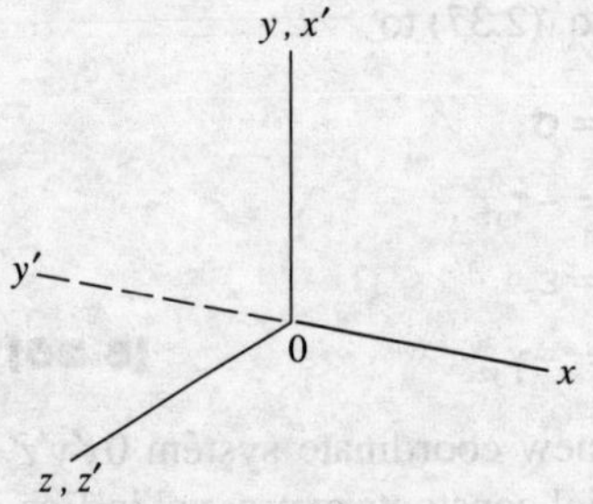

Figure B.5 Coordinate system $0x'y'z'$ obtained by rotating $0z$ 90°.

Fourth, rotating $0z$ through a 90° angle results in Fig. B.5, giving

$$
\begin{aligned}
(x',x) &= 90^\circ, \quad (x',y) = 0^\circ, \quad (x',z) = 90^\circ \\
(y',y) &= 90^\circ, \quad (y',x) = 180^\circ, \quad (y',z) = 90^\circ \\
(z',z) &= 0^\circ, \quad (z',x) = 90^\circ, \quad (z',y) = 90^\circ
\end{aligned} \tag{B.31}
$$

Making use of Eqs. (B.31) reduces Eqs. (B.4) to (B.9) and Eq. (2.37) to

$$
\begin{aligned}
\sigma_{z'} &= \sigma_z, \quad \sigma_{x'} = \sigma_y, \quad \sigma_{y'} = \sigma_x \\
\tau_{z'x'} &= \tau_{yz}, \quad \tau_{x'y'} = -\tau_{xy}, \quad \tau_{y'z'} = -\tau_{zx} \\
\varepsilon_{z'} &= \varepsilon_z, \quad \varepsilon_{x'} = \varepsilon_y, \quad \varepsilon_{y'} = \varepsilon_x \\
\gamma_{z'x'} &= \gamma_{yz}, \quad \gamma_{x'y'} = -\gamma_{xy}, \quad \gamma_{y'z'} = -\gamma_{zx}
\end{aligned} \tag{B.32}
$$

The assumption of an isotropic material requires that Eqs. (B.30) be written as

$$
\begin{aligned}
\sigma_{x'} &= C_{11}\varepsilon_{x'} + C_{12}(\varepsilon_{y'} + \varepsilon_{z'}) \\
\sigma_{y'} &= C_{21}\varepsilon_{x'} + C_{22}\varepsilon_{y'} + C_{23}\varepsilon_{z'} \\
\sigma_{z'} &= C_{21}\varepsilon_{x'} + C_{22}\varepsilon_{z'} + C_{23}\varepsilon_{y'} \\
\tau_{x'y'} &= C_{44}\gamma_{x'y'} \\
\tau_{y'z'} &= C_{55}\gamma_{y'z'} \\
\tau_{z'x'} &= C_{44}\gamma_{z'x'}
\end{aligned} \tag{B.33}
$$

Substituting Eqs. (B.32) into Eqs. (B.33) gives

$$
\begin{aligned}
\sigma_y &= C_{11}\varepsilon_y + C_{12}(\varepsilon_x + \varepsilon_z) \\
\sigma_x &= C_{21}\varepsilon_y + C_{22}\varepsilon_x + C_{23}\varepsilon_z \\
\sigma_z &= C_{21}\varepsilon_y + C_{22}\varepsilon_z + C_{23}\varepsilon_x \\
\tau_{xy} &= C_{44}\gamma_{xy} \\
\tau_{yz} &= C_{44}\gamma_{yz} \\
\tau_{zx} &= C_{55}\gamma_{zx}
\end{aligned} \tag{B.34}
$$

Equations (B.30) and (B.34) will agree only if

$$C_{22} = C_{11} \qquad C_{21} = C_{12} \qquad C_{23} = C_{12} \qquad C_{55} = C_{44} \tag{B.35}$$

Substituting Eqs. (B.37) into Eqs. (B.34) gives

$$\begin{aligned}
\sigma_x &= C_{11}\varepsilon_x + C_{12}(\varepsilon_y + \varepsilon_z)\\
\sigma_y &= C_{11}\varepsilon_y + C_{12}(\varepsilon_x + \varepsilon_z)\\
\sigma_z &= C_{11}\varepsilon_z + C_{12}(\varepsilon_x + \varepsilon_y)\\
\tau_{xy} &= C_{44}\gamma_{xy}\\
\tau_{yz} &= C_{44}\gamma_{yz}\\
\tau_{zx} &= C_{44}\gamma_{zx}
\end{aligned} \tag{B.36}$$

Thus, using the coordinate system of Fig. B.5 with the isotropic assumption reduces the seven elastic material constants of Eqs. (B.30) to the three expressed in Eqs. (B.36).

Fifth and finally, rotating $0z$ through a 45° angle results in Fig. B.6, implying that

$$\begin{aligned}
&(x',x) = 45^\circ, \quad (x',y) = 45^\circ, \quad (x',z) = 90^\circ\\
&(y',y) = 45^\circ, \quad (y',x) = 135^\circ, \quad (y',z) = 90^\circ\\
&(z',z) = 0^\circ, \quad (z',x) = 90^\circ, \quad (z',y) = 90^\circ
\end{aligned} \tag{B.37}$$

Making use of Eqs. (B.37) reduces Eqs. (B.4) to (B.9) and Eq. (2.37) to

$$\begin{aligned}
&\sigma_{z'} = \sigma_z, && \sigma_{x'} = \tfrac{1}{2}(\sigma_x + \sigma_y) + \tau_{xy}, && \sigma_{y'} = \tfrac{1}{2}(\sigma_x + \sigma_y) - \tau_{xy}\\
&\tau_{z'x'} = \tfrac{1}{\sqrt{2}}(\tau_{zx} + \tau_{yz}), && \tau_{x'y'} = \tfrac{1}{2}(\sigma_y - \sigma_x), && \tau_{y'z'} = \tfrac{1}{\sqrt{2}}(\tau_{yz} - \tau_{zx})\\
&\varepsilon_{z'} = \varepsilon_z, && \varepsilon_{x'} = \tfrac{1}{2}(\varepsilon_x + \varepsilon_y + \gamma_{xy}), && \varepsilon_{y'} = \tfrac{1}{2}(\varepsilon_x + \varepsilon_y - \gamma_{xy})\\
&\gamma_{z'x'} = \tfrac{1}{\sqrt{2}}(\gamma_{zx} + \gamma_{yz}), && \gamma_{x'y'} = \tfrac{1}{2}(\varepsilon_y - \varepsilon_x), && \gamma_{y'z'} = \tfrac{1}{\sqrt{2}}(\gamma_{yz} - \gamma_{zx})
\end{aligned} \tag{B.38}$$

The assumption of an isotropic material requires that Eqs. (B.37) be written as

$$\begin{aligned}
\sigma_{x'} &= C_{11}\varepsilon_{x'} + C_{12}(\varepsilon_{y'} + \varepsilon_{z'})\\
\sigma_{y'} &= C_{11}\varepsilon_{y'} + C_{12}(\varepsilon_{x'} + \varepsilon_{z'})\\
\sigma_{z'} &= C_{11}\varepsilon_{z'} + C_{12}(\varepsilon_{x'} + \varepsilon_{y'})\\
\tau_{x'y'} &= C_{44}\gamma_{x'y'}\\
\tau_{y'z'} &= C_{44}\gamma_{y'z'}\\
\tau_{z'x'} &= C_{44}\gamma_{z'x'}
\end{aligned} \tag{B.39}$$

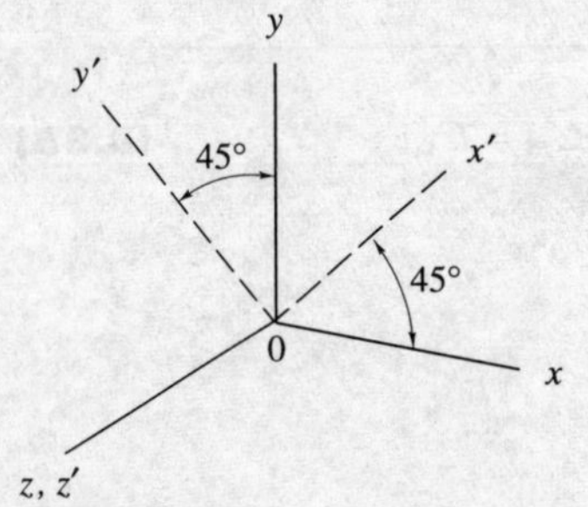

Figure B.6 Coordinate system $0x'y'z'$ obtained by rotating $0z$ 45°.

Substituting Eqs. (B.38) into the first of Eqs. (B.39) gives

$$\tfrac{1}{2}(\sigma_x + \sigma_y) + \tau_{xy} = C_{11}\left(\frac{\gamma_{xy}}{2} + \frac{\varepsilon_x + \varepsilon_y}{2}\right) + C_{12}\left(\frac{\varepsilon_x + \varepsilon_y}{2} - \frac{\gamma_{xy}}{2} + \varepsilon_z\right)$$

and substituting the expressions for σ_x and σ_y in Eqs. (B.36) into the above equation gives

$$\tau_{xy} = (C_{11} - C_{12})\frac{\gamma_{xy}}{2} \tag{B.40}$$

Comparing Eq. (B.40) with the τ_{xy} expression in Eqs. (B.36) gives

$$C_{11} = C_{12} + 2C_{44} \tag{B.41}$$

Letting $C_{12} = \lambda$ and $C_{44} = G$ and making use of Eq. (B.41) gives Eqs. (B.36) as

$$\begin{aligned}
\sigma_x &= (\lambda + 2G)\varepsilon_x + \lambda(\varepsilon_y + \varepsilon_z) \\
\sigma_y &= (\lambda + 2G)\varepsilon_y + \lambda(\varepsilon_x + \varepsilon_z) \\
\sigma_z &= (\lambda + 2G)\varepsilon_z + \lambda(\varepsilon_x + \varepsilon_y) \\
\tau_{xy} &= G\gamma_{xy} \\
\tau_{yz} &= G\gamma_{yz} \\
\tau_{zx} &= G\gamma_{zx}
\end{aligned} \tag{B.42}$$

Thus, using Fig. B.6 with the isotropic assumption reduces the number of elastic material constants from the 36 expressed in Eqs. (B.11) to the 2 expressed in Eqs. (B.42).

Equation (B.42) can be solved for strains to give

$$\varepsilon_x = \frac{(\lambda + G)\sigma_x}{G(3\lambda + 2G)} - \frac{\lambda(\sigma_y + \sigma_z)}{2G(3\lambda + 2G)}$$

$$\varepsilon_y = \frac{(\lambda + G)\sigma_y}{G(3\lambda + 2G)} - \frac{\lambda(\sigma_x + \sigma_z)}{2G(3\lambda + 2G)}$$

$$\varepsilon_z = \frac{(\lambda + G)\sigma_z}{G(3\lambda + 2G)} - \frac{\lambda(\sigma_x + \sigma_y)}{2G(3\lambda + 2G)}$$

$$\gamma_{xy} = \frac{\tau_{xy}}{G}$$

$$\gamma_{yz} = \frac{\tau_{yz}}{G}$$

$$\gamma_{zx} = \frac{\tau_{zx}}{G} \qquad \textbf{(B.43)}$$

For isotropic but *nonhomogeneous* materials the constants λ and G are functions of the space coordinates x, y, and z and vary from point to point. For isotropic and *homogeneous* materials these constants are not functions of the space coordinates and do not vary from point to point. They depend only on the particular material.

If the $0x$, $0y$, and $0z$ axes are chosen along the principal axes of stress, $\tau_{xy} = \tau_{yz} = \tau_{zx} = 0$. From Eqs. (B.43) it follows that $\gamma_{xy} = \gamma_{yz} = \gamma_{zx} = 0$. Thus, for isotropic materials the principal axes of stress and strain coincide.

B.4 PHYSICAL SIGNIFICANCE OF ELASTIC MATERIAL CONSTANTS

The first elastic material constant is shear modulus, or modulus of rigidity, G in pascals. The second elastic material constant λ, known as Lamé's constant, is of no particular physical significance.

In a uniaxial stress state where $\sigma_y = \sigma_z = \tau_{xy} = \tau_{yz} = \tau_{zx} = 0$ and σ_x is the applied uniaxial stress, Eqs. (B.43) give

$$\varepsilon_x = \frac{(\lambda + G)\sigma_x}{G(3\lambda + 2G)} \qquad \textbf{(B.44)}$$

$$\varepsilon_y = \varepsilon_z = \frac{-\lambda\sigma_x}{2G(3\lambda + 2G)} \qquad \textbf{(B.45)}$$

Comparing Eq. (B.44) with Eq. (3.23) gives

$$\lambda = \frac{G(E - 2G)}{3G - E} \qquad \textbf{(B.46)}$$

where E = modulus of elasticity, the third elastic material constant, covered in Sec. 3.5.2.

Relating the transverse strain in Eq. (B.45) to Poisson's ratio in Eq. (3.4) gives

$$\nu = \frac{\lambda}{2(\lambda + G)} \tag{B.47}$$

or solving for λ

$$\lambda = \frac{2G\nu}{1 - 2\nu} \tag{B.48}$$

Thus, Eqs. (B.46) and (B.48) express λ in terms of two known elastic material constants. Equating Eqs. (B.46) and (B.48) gives

$$\nu = \frac{E - 2G}{2G} \tag{B.49}$$

where Poisson's ratio ν is the fourth elastic material constant. Equation (B.49) is equivalent to Eq. (3.6). The range of λ is between zero and 0.5. At $\nu = 0$ no transverse deformation, but rather longitudinal deformation, occurs. At $\nu = 0.5$ the material exhibits constant volume. The volume increase longitudinally is the same as the shrinkage in the transverse direction.

Besides the four elastic material constants G, λ, E, and ν a fifth is provided by bulk modulus K, the ratio of applied hydrostatic pressure to observed volume shrinkage per unit volume, which is in pascals (pounds per square inch). The bulk modulus is obtained from hydrostatic compression or when

$$\sigma_x = \sigma_y = \sigma_z = -p \qquad \text{for } p > 0$$
$$\tau_{xy} = \tau_{yz} = \tau_{zx} = 0$$

Substituting the above into Eqs. (B.43) gives

$$\varepsilon_x = \varepsilon_y = \varepsilon_z = \frac{-p}{3\lambda + 2G} \tag{B.50}$$

But the total strain is

$$\varepsilon = \varepsilon_x + \varepsilon_y + \varepsilon_z \rightarrow p = \frac{-(3\lambda + 2G)\varepsilon}{3} = -K\varepsilon \tag{B.51}$$

where

$$K = \frac{3\lambda + 2G}{3} \tag{B.52}$$

It is possible to express any of the elastic material constants (G, λ, E, ν, and K) if two of these five constants are given. These relationships are expressed in Table B.1.

Table B.1 Relationships between elastic material constants for isotropic materials.

Constants involved	Lamé's constant, λ	Shear modulus, G	Modulus of elasticity, E	Poisson's ratio, ν	Bulk modulus, K
λ G	—	—	$E = \frac{G(3\lambda + 2G)}{\lambda + G}$	$\nu = \frac{\lambda}{2(\lambda + G)}$	$K = \frac{3\lambda + 2G}{3}$
λ E	—	$G = \frac{A^{\dagger} + (E - 3\lambda)}{4}$	—	$\nu = \frac{A - (E + \lambda)}{4\lambda}$	$K = \frac{A + (3\lambda + E)}{6}$
λ ν	—	$G = \frac{\lambda(1 - 2\nu)}{2\nu}$	$E = \frac{\lambda(1 + \nu)(1 - 2\nu)}{\nu}$	—	$K = \frac{\lambda(1 + \nu)}{3\nu}$
λ K	—	$G = \frac{3(K - \lambda)}{2}$	$E = \frac{9K(K - \lambda)}{3K - \lambda}$	$\nu = \frac{\lambda}{3K - \lambda}$	—
G E	$\lambda = \frac{G(2G - E)}{E - 3G}$	—	—	$\nu = \frac{E - 2G}{2G}$	$K = \frac{GE}{3(3G - E)}$
G ν	$\lambda = \frac{2G\nu}{1 - 2\nu}$	—	$E = 2G(1 + \nu)$	—	$K = \frac{2G(1 + \nu)}{3(1 - 2\nu)}$
G K	$\lambda = \frac{3K - 2G}{3}$	—	$E = \frac{9KG}{3K + G}$	$\nu = \frac{3K - 2G}{2(3K + G)}$	—
E ν	$\lambda = \frac{\nu E}{(1 + \nu)(1 - 2\nu)}$	$G = \frac{E}{2(1 + \nu)}$	—	—	$K = \frac{E}{3(1 - 2\nu)}$
E K	$\lambda = \frac{3K(3K - E)}{9K - E}$	$G = \frac{3EK}{9K - E}$	—	$\nu = \frac{3K - E}{6K}$	—
ν K	$\lambda = \frac{3K\nu}{1 + \nu}$	$G = \frac{3K(1 - 2\nu)}{2(1 + \nu)}$	$E = 3K(1 - 2\nu)$	—	—

$^{\dagger}A = \sqrt{(E + \lambda)^2 + 8\lambda^2}$

B.5 STRESS-STRAIN EQUATIONS IN TERMS OF MODULUS OF ELASTICITY AND POISSON'S RATIO

From Table B.1, λ and G can be expressed in terms of E and ν as

$$\lambda = \frac{\nu E}{(1+\nu)(1-2\nu)} \qquad \text{and} \qquad G = \frac{E}{2(1+\nu)} \tag{B.53}$$

Substituting Eqs. (B.53) into Eqs. (B.43) gives

$$\begin{aligned}
\varepsilon_x &= \frac{1}{E}\left[\sigma_x - \nu\left(\sigma_y + \sigma_z\right)\right] \\
\varepsilon_y &= \frac{1}{E}\left[\sigma_y - \nu\left(\sigma_z + \sigma_x\right)\right] \\
\varepsilon_z &= \frac{1}{E}\left[\sigma_z - \nu\left(\sigma_x + \sigma_y\right)\right] \\
\gamma_{xy} &= \frac{2(1+\nu)}{E}\tau_{xy} \\
\gamma_{yz} &= \frac{2(1+\nu)}{E}\tau_{yz} \\
\gamma_{zx} &= \frac{2(1+\nu)}{E}\tau_{zx}
\end{aligned} \tag{B.54}$$

Similarly, the stress components can be expressed in terms of strains as

$$\begin{aligned}
\sigma_x &= \frac{E}{(1+\nu)(1-2\nu)}\left[(1-\nu)\varepsilon_x + \nu\left(\varepsilon_y + \varepsilon_z\right)\right] \\
\sigma_y &= \frac{E}{(1+\nu)(1-2\nu)}\left[(1-\nu)\varepsilon_y + \nu\left(\varepsilon_z + \varepsilon_x\right)\right] \\
\sigma_z &= \frac{E}{(1+\nu)(1-2\nu)}\left[(1-\nu)\varepsilon_z + \nu\left(\varepsilon_x + \varepsilon_y\right)\right] \\
\tau_{xy} &= \frac{E}{2(1+\nu)}\gamma_{xy} \\
\tau_{yz} &= \frac{E}{2(1+\nu)}\gamma_{yz} \\
\tau_{zx} &= \frac{E}{2(1+\nu)}\gamma_{zx}
\end{aligned} \tag{B.55}$$

Although the stress and strain equations given in Eqs. (B.54) and (B.55), respectively, are expressed in terms of E and ν, by using Table B.1 these equations can be rewritten in terms of any two of the five elastic material constants (G, λ, E, ν, and K). The reason for using E and ν is that data for a particular material can be readily obtained for these constants as demonstrated in Tables 3.2 and 3.3. For rubber ($\nu \to 0.5$) it is more accurate to use the bulk modulus K together with the shear modulus G, since $1 - 2\nu$ appears in the denominator.

For the special case in which the x, y, and z axes are coincidental with the principal axes 1, 2, and 3 (the *triaxial stress state*), Eqs. (B.54) and (B.55) are simplified by virtue of all shear stresses and shear strains being equal to zero:

$$\varepsilon_1 = \frac{1}{E}[\sigma_1 - \nu(\sigma_2 + \sigma_3)]$$
$$\varepsilon_2 = \frac{1}{E}[\sigma_2 - \nu(\sigma_3 + \sigma_1)]$$
$$\varepsilon_3 = \frac{1}{E}[\sigma_3 - \nu(\sigma_1 + \sigma_2)] \qquad \textbf{(B.56)}$$

$$\sigma_1 = \frac{E}{(1+\nu)(1-2\nu)}[(1-\nu)\varepsilon_1 + \nu(\varepsilon_2 + \varepsilon_3)]$$
$$\sigma_2 = \frac{E}{(1+\nu)(1-2\nu)}[(1-\nu)\varepsilon_2 + \nu(\varepsilon_3 + \varepsilon_1)]$$
$$\sigma_3 = \frac{E}{(1+\nu)(1-2\nu)}[(1-\nu)\varepsilon_3 + \nu(\varepsilon_1 + \varepsilon_2)] \qquad \textbf{(B.57)}$$

For the commonly encountered *biaxial stress state*, one of the principal stresses (say, σ_3) is zero and Eqs. (B.56) become

$$\varepsilon_1 = \frac{\sigma_1 - \nu\sigma_2}{E}$$
$$\varepsilon_2 = \frac{\sigma_2 - \nu\sigma_1}{E}$$
$$\varepsilon_3 = \frac{-\nu(\sigma_1 + \sigma_2)}{E} \qquad \textbf{(B.58)}$$

For $\sigma_3 = 0$ the third of Eqs. (B.57) gives

$$\varepsilon_3 = -\frac{\nu(\varepsilon_1 + \varepsilon_2)}{1-\nu} \qquad \textbf{(B.59)}$$

Substituting Eq. (B.59) into Eqs. (B.57) gives

$$\sigma_1 = \frac{E(\varepsilon_1 + \nu\varepsilon_2)}{1-\nu^2}$$
$$\sigma_2 = \frac{E(\varepsilon_2 + \nu\varepsilon_1)}{1-\nu^2}$$
$$\sigma_3 = 0 \qquad \textbf{(B.60)}$$

For the *uniaxial stress state* Eqs. (B.58) and (B.60) must reduce to

$$\varepsilon_1 = \frac{\sigma_1}{E} \qquad \varepsilon_2 = \varepsilon_3 = \frac{-\nu\sigma_1}{E} \qquad \textbf{(B.61)}$$

$$\sigma_1 = E\varepsilon_1 \qquad \sigma_2 = \sigma_3 = 0 \qquad \textbf{(B.62)}$$

Table B.2 Principal stresses and strains in terms of modulus of elasticity and Poisson's ratio for uniaxial, biaxial, and triaxial stress states. (It is assumed that the *x*, *y*, and *z* axes are coincident with the principal axes 1, 2, and 3, thus implying that the shear stresses and strains are equal to zero.)

Type of stress	Principal strains	Principal stresses
Uniaxial	$\varepsilon_1 = \dfrac{\sigma_1}{E}$	$\sigma_1 = E\varepsilon_1$
	$\varepsilon_2 = -\nu\varepsilon_1$	$\sigma_2 = 0$
	$\varepsilon_3 = -\nu\varepsilon_1$	$\sigma_3 = 0$
Biaxial	$\varepsilon_1 = \dfrac{\sigma_1}{E} - \dfrac{\nu\sigma_2}{E}$	$\sigma_1 = \dfrac{E(\varepsilon_1 + \nu\varepsilon_2)}{1-\nu^2}$
	$\varepsilon_2 = \dfrac{\sigma_2}{E} - \dfrac{\nu\sigma_1}{E}$	$\sigma_2 = \dfrac{E(\varepsilon_2 + \nu\varepsilon_1)}{1-\nu^2}$
	$\varepsilon_3 = -\dfrac{\nu\sigma_1}{E} - \dfrac{\nu\sigma_2}{E}$	$\sigma_3 = 0$
Triaxial	$\varepsilon_1 = \dfrac{\sigma_1}{E} - \dfrac{\nu\sigma_2}{E} - \dfrac{\nu\sigma_3}{E}$	$\sigma_1 = \dfrac{E\varepsilon_1(1-\nu) + \nu E(\varepsilon_2 + \varepsilon_3)}{1-\nu-2\nu^2}$
	$\varepsilon_2 = \dfrac{\sigma_2}{E} - \dfrac{\nu\sigma_1}{E} - \dfrac{\nu\sigma_3}{E}$	$\sigma_2 = \dfrac{E\varepsilon_2(1-\nu) + \nu E(\varepsilon_1 + \varepsilon_3)}{1-\nu-2\nu^2}$
	$\varepsilon_3 = \dfrac{\sigma_3}{E} - \dfrac{\nu\sigma_1}{E} - \dfrac{\nu\sigma_2}{E}$	$\sigma_3 = \dfrac{E\varepsilon_3(1-\nu) + \nu E(\varepsilon_1 + \varepsilon_2)}{1-\nu-2\nu^2}$

These expressions for the uniaxial, biaxial, and triaxial stress states can be expressed in tabular form as shown in Table B.2. Recall that we chose to express the stress and strain in terms of the modulus of elasticity and Poisson's ratio. Furthermore, we are considering the special case in which the *x*, *y*, and *z* axes are coincidental with the principal axes 1, 2, and 3, thus implying that the shear stresses and strains are equal to zero. For this case the shear stress and strain are equal to zero.

Example B.1

GIVEN Equations (B.42) and (B.43).

FIND Determine the modulus of elasticity E as a function of G and ν, where ν is given by Hooke's law for uniaxial tension.

$$\varepsilon_x = \frac{\sigma_x}{E}$$

$$\varepsilon_y = -\frac{\nu\sigma_x}{E}$$

$$\varepsilon_z = -\frac{\nu\sigma_x}{E}$$

Solution

Equation (B.43) gives

$$\varepsilon_x = \frac{(\lambda+G)\sigma_x}{G(3\lambda+2G)} - \frac{\lambda(\sigma_y+\sigma_z)}{2G(3\lambda+2G)} = \frac{\sigma_x}{E} - \frac{\nu(\sigma_y+\sigma_z)}{E}$$

$$\varepsilon_y = \frac{(\lambda+G)\sigma_y}{G(3\lambda+2G)} - \frac{\lambda(\sigma_x+\sigma_z)}{2G(3\lambda+2G)} = \frac{\sigma_y}{E} - \frac{\nu(\sigma_x+\sigma_z)}{E}$$

$$\varepsilon_z = \frac{(\lambda+G)\sigma_z}{G(3\lambda+2G)} - \frac{\lambda(\sigma_x+\sigma_y)}{2G(3\lambda+2G)} = \frac{\sigma_z}{E} - \frac{\nu(\sigma_x+\sigma_y)}{E}$$

$$\left.\begin{aligned} \frac{\lambda+G}{G(3\lambda+2G)} &= \frac{1}{E} \\ \frac{\lambda}{2G(3\lambda+2G)} &= \frac{\nu}{E} \end{aligned}\right\} \frac{\lambda}{2(\lambda+G)} = \nu$$

$$(\lambda+G)E = 3\lambda G + 2G^2$$

$$G(E-2G) = \lambda(3G-E) \qquad \lambda = \frac{G(E-2G)}{3G-E}$$

$$\nu = \frac{G(E-2G)}{2(3G-E)\left[\dfrac{G(E-2G)}{3G-E}+G\right]}$$

$$\nu = \frac{E-2G}{2G} = \frac{E}{2G} - 1$$

$$E = (\nu+1)2G \qquad \text{and} \qquad G = \frac{E}{2(1+\nu)}$$

The modulus of elasticity $E = 2G(\nu + 1)$, the shear modulus $G = E/2(1 + \nu)$, and Poisson's ratio $\nu = E/2G - 1$.

B.7 SUMMARY

Hooke's law was generalized in this appendix. That is, the linear relationship between stress and strain in the elastic range was generalized for the six components of stress: three normal stresses and three shear stresses. By using the laws of stress and strain transformation, the 36 elastic material constants were reduced to 2 through five different coordinate orientations. It was also found that, for the isotropic and homogeneous materials assumed throughout this chapter, these elastic material constants are not functions of the space coordinates and do not vary from point to point. They depend only on the particular material.

Relationships were presented between the four elastic material constants (the shear modulus or modulus of rigidity G, the modulus of elasticity E, the bulk modulus K, and Poisson's ratio ν) and

Lamé's constant λ, which has no particular physical significance. It was shown how any two of the five constants can be expressed in terms of the other constants. For the special case in which the *x*, *y*, and *z* axes are coincidental with the principal axes, simplified equations were expressed where all shear stresses and shear strains were zero. For these situations uniaxial, biaxial, and triaxial principal stresses could be expressed. These equations serve as the foundation for the main text of this book.

REFERENCE

Durelli, A.J., Phillips, E.A., and Tsao, C.H. (1958) *Introduction to the Theoretical and Experimental Analysis of Stress and Strain*, McGraw-Hill, New York.

INDEX

NAME AND SUBJECT

Centroid, area moment of inertia, and area for seven cross sections.

Cross section	Centroid	Area moment of inertia	Area
Circular area 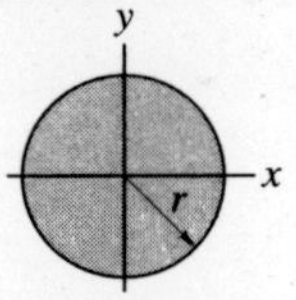	$\bar{x} = 0$ $\bar{y} = 0$	$I_x = I_{\bar{x}} = \frac{\pi}{4} r^4$ $I_y = I_{\bar{y}} = \frac{\pi}{4} r^4$ $J = \frac{\pi}{2} r^4$	$A = \pi r^2$
Hollow circular area 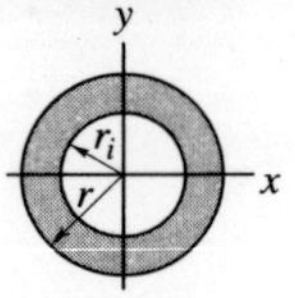	$\bar{x} = 0$ $\bar{y} = 0$	$I_x = I_{\bar{x}} = \frac{\pi}{4}\left(r^4 - r_i^4\right)$ $I_y = I_{\bar{y}} = \frac{\pi}{4}\left(r^4 - r_i^4\right)$ $J = \frac{\pi}{2}\left(r^4 - r_i^4\right)$	$A = \pi\left(r^2 - r_i^2\right)$
Triangular area 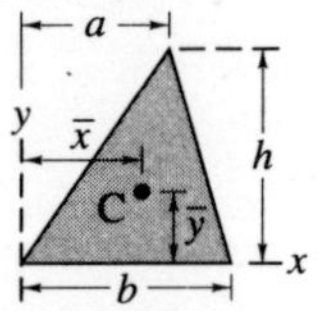	$\bar{x} = \frac{a+b}{3}$ $\bar{y} = \frac{h}{3}$	$I_x = \frac{bh^3}{12},\ I_{\bar{x}} = \frac{bh^3}{36}$ $I_y = \frac{bh\left(b^2 + ab + a^2\right)}{36}$ $I_{\bar{y}} = \frac{bh\left(b^2 - ab + a^2\right)}{36}$ $\bar{J} = \frac{bh}{36}\left(b^2 + h^2\right)$	$A = \frac{bh}{2}$
Rectangular area 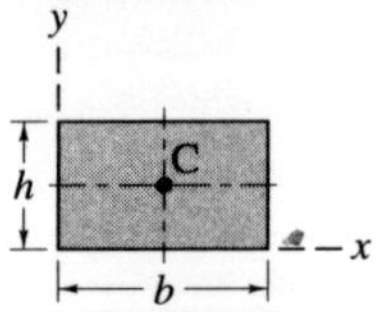	$\bar{x} = \frac{b}{2}$ $\bar{y} = \frac{h}{2}$	$I_x = \frac{bh^3}{3},\ I_{\bar{x}} = \frac{bh^3}{12}$ $I_y = \frac{hb^3}{3},\ I_{\bar{y}} = \frac{hb^3}{12}$ $\bar{J} = \frac{bh}{12}\left(b^2 + h^2\right)$	$A = bh$
Area of circular sector 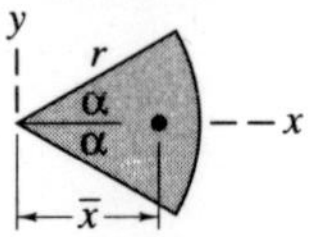	$\bar{x} = \frac{2}{3}\,\frac{r \sin\alpha}{\alpha}$	$I_x = \frac{r^4}{4}\left(\alpha - \frac{1}{2}\sin 2\alpha\right)$ $I_y = \frac{r^4}{4}\left(\alpha + \frac{1}{2}\sin 2\alpha\right)$ $J = \frac{1}{2} r^4 \alpha$	$A = r^2\alpha$
Quarter-circular area 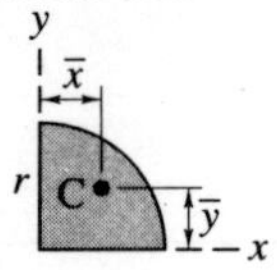	$\bar{x} = \bar{y} = \frac{4r}{3\pi}$	$I_x = I_y = \frac{\pi r^4}{16}$ $I_{\bar{x}} = I_{\bar{y}} = \left(\frac{\pi}{16} - \frac{4}{9\pi}\right) r^4$ $J = \frac{\pi r^4}{8}$	$A = \frac{\pi r^2}{4}$
Area of elliptical quadrant 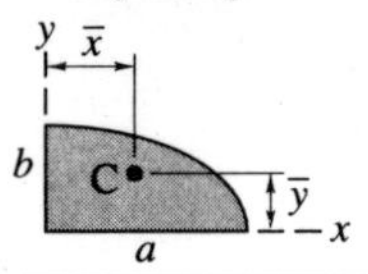	$\bar{x} = \frac{4a}{3\pi}$ $\bar{y} = \frac{4b}{3\pi}$	$I_x = \frac{\pi ab^3}{16},\ I_{\bar{x}} = \left(\frac{\pi}{16} - \frac{4}{9\pi}\right) ab^3$ $I_y = \frac{\pi a^3 b}{16},\ I_{\bar{y}} = \left(\frac{\pi}{16} - \frac{4}{9\pi}\right) a^3 b$ $J = \frac{\pi ab}{16}\left(a^2 + b^2\right)$	$A = \frac{\pi ab}{4}$